3rd International Conference on Metamaterials and Nanophotonics (METANANO 2018)

Journal of Physics: Conference Series
Volume 1092

Sochi, Russia
17 – 21 September 2018

ISBN: 978-1-5108-7339-1
ISSN: 1742-6588

Printed from e-media with permission by:

Curran Associates, Inc.
57 Morehouse Lane
Red Hook, NY 12571

proceedings.com

Some format issues inherent in the e-media version may also appear in this print version.

Printed by Curran Associates, Inc. (2018)

For permission requests, please contact the Institute of Physics
at the address below.

Institute of Physics
Dirac House, Temple Back
Bristol BS1 6BE UK

Phone: 44 1 17 929 7481
Fax: 44 1 17 920 0979

techtracking@iop.org

Additional copies of this publication are available from:

Curran Associates, Inc.
57 Morehouse Lane
Red Hook, NY 12571 USA
Phone: 845-758-0400
Fax: 845-758-2633
Email: curran@proceedings.com
Web: www.proceedings.com

Table of contents

Volume 1092

METANANO 2018

17–21 September 2018, Sochi, Russian Federation

Accepted papers received: 24 August 2018
Published online: 15 October 2018

Preface

Papers

This issue of *Journal of Physics: Conference Series* contains extended contributions presented at the 3rd International Conference on Metamaterials and Nanophotonics **METANANO 2018** held in Sochi, Russia, 17-21 September 2018.

METANANO is an annual international conference where the latest research in the fields of metamaterials, nanophotonics, plasmonics, terahertz technologies and optical engineering is presented. METANANO 2018 is a hub for researchers working in these fields and studying nanoscience, photonics, non-linear optics and related topics. The aim of the conference is to encourage discussion and sharing of knowledge and experience between the scientific community, engineers, industry and academic members around the world, with the prospect of collaboration between these groups in the future. METANANO covers a broad range of fascinating topics including metamaterials and metasurfaces for microwave and optical spectrum, topological, graphene and plasmonic structures, THz metamaterials and their application in communication, medicine, biology, security, and energy.

The first METANANO in Russia was held in Anapa in 2016 and accumulated nearly 100 speakers. Next METANANO 2017 in Vladivostok gathered over 230 participants from Europe, Asia, and Russia. METANANO 2018 includes already over 380 submitted talks of which almost 200 have been included into this issue. The impetuous growth of the number of conference participants indicates enlargement of the interest in the conference, recognition of its quality and at the same time the importance and relevance of the scientific fields covered by the conference.

To ensure high publication standard mandated by IOP every paper was peer-reviewed by one or two reviewers with expertise before it was accepted for publication. As a general chair of the conference, I would like to express my sincere appreciation to all the reviewers involved in the evaluation of the papers for their invaluable contribution. Special thanks go also to the organizers of the special sessions who added hot topics to the conference program and provided the participation of top-notch experts.

I acknowledge all the speakers of the conference, members of the Technical Program Committee and Advisory Board who have conducted enormous work reviewing the conference papers and forming its program. I am also thankful to the conference sponsors and co-organizers who have made an invaluable input to the growing quality of this conference.

I hope you will enjoy reading these Proceedings and discovering new and original research results. I also welcome you to attend the forthcoming METANANO conferences.

Prof. Pavel Belov

General Chair, METANANO conference

COMMITTEES

General Chair

Pavel Belov, Russia

Technical Program Committee Chair

Polina Kapitanova, Russia

Technical Program Committee

Carsten Rockstuhl, Germany
Isabelle Staude, Germany
Mario Silveirinha, Portugal
Valentyn Volkov, Denmark
Alexey Kavokin, Russia
Nicolas Bonod, France
Andrei Lavrinenko, Denmark
Andrey Miroshnichenko, Australia
Alexey Akimov, USA
Pavel Ginzburg, Israel
Maxim Shcherbakov, Russia
Sergey Makarov, Russia
Aleksandr Kuchmizhak, Russia
Ivan Iorsh, Russia
Stanislav Glybovski, Russia
Alexey Yulin, Russia

Advisory Board

Yuri Kivshar, Australia
Mikhail Lapine, Australia
Dragomir Neshev, Australia

Conference Secretaries

Anastasia Kaptsova, Russia
Mikhail Odit, Russia

PR Manager

Julia Kuznetsova, Russia

IOP Conf. Series: Journal of Physics: Conf. Series **1092** (2018) 011002 doi:10.1088/1742-6596/1092/1/011002

Program Manager

Evgeniia Cherotchenko, Russia

Reviewers

A. Poddubny, A. Akimov, C. Rockstuhl, A. Lavrinenko, A. Evlyukhin, O. Shramkova, M. Silveirinha, A. Miroshnichenko, I. Iorsh, I. Nefedov, A. Yulin, S. Makarov, V. Milichko, I. Shadrivov, I. Staude, N. Bonod, M. Shcherbakov, M. Zyuzin, M. Zubkov, I. Mukhin, A. Kuchmizhak, Yu. Stebunov, A. Bogdanov, M. Petrov, M. Gorlach, A. Samusev, V. Volkov, V. Leiman, A. Vyshnevyy, A. Sayanskiy, G.Solomakha, A. Slobozhanyuk, K. Baryshnikova, A. Hurshkainen, D. Yudin, M. Rybin, M. Limonov, V. Shahnazaryan, K. Koshelev, A. Markvart, A. Schelokova, A. Andreychenko, S. Glybovski, S. Kosulnikov, D. Zuev, A. Shalin, P. Ginzburg

IOP Conf. Series: Journal of Physics: Conf. Series **1092** (2018) 011004 doi:10.1088/1742-6596/1092/1/011004

ACKNOWLEDGMENTS

The organizers of the conference are thankful to all the supporters who have made valuable impact to METANANO 2018 conference, in particular ITMO University (www.ifmo.ru), Far Eastern Federal University (www.dvfu.ru), Moscow Institute of Physics and Technology (mipt.ru), neaspec GmbH (www.neaspec.com) and Inscience (in-science.ru).

ITMO University is thankful to the Russian Foundation for Basic Research (www.rfbr.ru), project № 18-02-20114.

Informational support was provided by the Metamorphose VI, The Virtual Institute for Artificial Electromagnetic Materials and Metamaterials (metamorphose-vi.org).

ITMO University is one of the Russia's leading higher education and research institutions, specializing in Information Technology, Optical Design and Engineering. Founded in 1900, one hundred years later tireless work of several generations of professors, lecturers and staff have brought our university a well-deserved success.

In 2011 we were named St. Petersburg National Research University of Information Technologies, Mechanics and Optics and today we are home to 11 125 students blending the culture of innovation and discovery with world-class education. Our focus is on preparing elite scientists, engineers and programmers, notably in the areas of IT, Photonics, Optics, Mechatronics, Robotics, Smart Technologies, Art & Science, and Urban Studies.

ITMO University's mission is to provide opportunities for the holistic development of individuals and to inspire them to tackle global challenges, as the university's philosophy is to be "More than a University".

IOP Conf. Series: Journal of Physics: Conf. Series **1092** (2018) 011004 doi:10.1088/1742-6596/1092/1/011004

Far Eastern Federal University is one of the leading federal universities of Russia. FEFU is the bearer of traditions of the Russian classical culture, science and education. Far Eastern Federal University has its origins in 1899 at the time when the Oriental Institute was founded in Vladivostok.

Today FEFU is the only classical university in Eastern Russia. In incorporates Schools with generations-old traditions and areas of research that form the core of our intellectual base. One of the FEFU's advantages is its geographic location in the center of the Asia Pacific region, which is now one of the most dynamically developing parts of the world. This place is the second pole of a world civilization, which includes the leading counties of the globe, such as China, the United States, Japan, India and many world-famed metropolitan cities, such as Singapore, Hong Kong, Sydney and Seoul.

Moscow Institute of Physics and Technology (MIPT), known informally as PhysTech is one of the leading Russian universities in the areas of physics and technology. The university holds a reputable position in the country and abroad in qualified graduate training.

Students and graduates of MIPT are representatives of an elite circle who, thanks to their interdisciplinary scientific surroundings, are able to fully realize their potential. The strong alumni community is spread all over and includes top range scientists, businessmen, politicians, people of art, many of who are awardees of prestigious international prizes. PhysTech (the university) and phystechs (its alumni) are strongly proud and supportive of each other.

MIPT has a very rich history. Its founders included academicians Pyotr Kapitsa, Nikolay Semenov and Sergey Khristianovich. Its first professors were Nobel Prize winners Kapitsa, Semenov and Lev Landau, and its first rector was Ivan Petrov. There are Nobel Prize winners among MIPT's graduates as well.

Many MIPT professors are leading Russian scientists, including over 80 members of the Russian Academy of Sciences. From the outset, MIPT has used a unique system for training specialists, known as the PhysTech System, which combines fundamental science, engineering disciplines and student research.

IOP Conf. Series: Journal of Physics: Conf. Series **1092** (2018) 011004 doi:10.1088/1742-6596/1092/1/011004

Platinum Sponsors

neaspec was founded to revolutionize nanoscale analytics, with the ultimate goal to enable technological & scientific progress in every lab around the world. Neaspec has gained worldwide attention in the scientific community for inventing a series of breakthrough technologies in nanoscale analytics. Founded in 2007 by the world's leading experts of near-field microscopy as a spin-off from the Max-Planck-Institute of Biochemistry (Germany), Neaspec's engineers have developed the neaSNOM microscope, an easy-to-use optical near-field microscope platform using ground-breaking and patented optical background-filtering techniques. NeaSNOM combines the best of two worlds – the nanoscale resolution of atomic force microscopy (AFM) with the analytical power of visible, infrared and even THz imaging & spectroscopy. At a spatial resolution of only 10nm, the method only requires standard AFM sample preparation and is non-destructive. This opens a new era for modern nano-analytical applications such as chemical nano-identification (IR), nano-plasmonic field mapping (VIS & IR) or free charge carrier nano-mapping (THz). Nowadays, Neaspec is rapidly expanding globally and serves leading researchers and scientists in highly-regarded research labs around the globe. The unique expertise of Neaspec's engineers guarantees not only the top performance of neaSNOM microscope-systems, but also a professional service including on-site training and personalized customer support.

IOP Conf. Series: Journal of Physics: Conf. Series **1092** (2018) 011004 doi:10.1088/1742-6596/1092/1/011004

INSCIENCE
SCIENTIFIC•SOLUTIONS

INSCIENCE company is reliable supplier of various equipment for customers in Russia - state universities, research laboratories and industrial manufacturers, wich activity is focused on photonics, optics and optoelectronics. INSCIENCE company is mostly targeted on lasers and accessories, optical tables and optomechanics, spectroscopy equipment and signal analysis hardware.

We offer not just equipment supply, but we always try to get deeply involved in customer`s goal, we support technically and legally on each phase of our work. Well-established logistics facilitate supply of any volumes from all over the world and help to solve any kind of challenge safely and fine.

IOP Conf. Series: Journal of Physics: Conf. Series **1092** (2018) 011006 doi:10.1088/1742-6596/1092/1/011006

Peer review statement

All papers published in this volume of *Journal of Physics: Conference Series* have been peer reviewed through processes administered by the proceedings Editors. Reviews were conducted by expert referees to the professional and scientific standards expected of a proceedings journal published by IOP Publishing.

IOP Conf. Series: Journal of Physics: Conf. Series **1092** (2018) 012001 doi:10.1088/1742-6596/1092/1/012001

Modeling of C_{60} pseudopotential with the different basis sets

K. B. Agapev[1], I. I. Vrubel[1], R. G. Polozkov[1], V. K. Ivanov[2]
[1] ITMO University, 49 Kronverksky Pr., St. Petersburg, 197101, Russia
[2] Peter the Great Saint-Petersburg Polytechnic University, 29 Politekhnicheskaya, St. Petersburg, 195251, Russia

cyr96@mail.ru

Abstract. As a result of *ab-initio*-based calculations, we can purpose the description of the fullerene C_{60} in term of pseudopotential. In the paper, the results of calculations based on the basis set composed of atomic orbitals and plane waves basis set have been compared. Also, the pseudopotential was approximated by analytical function, which can be applied in future calculation

1. Introduction

Fullerene C_{60} is being extensively researched all over the world from the first prediction [1] due to its unique properties and wide availability of such compounds. Because of the high affinity to electron, C_{60} can play the role of electron acceptor in the complex systems. The fullerenes can act as a semiconductor, conductor and superconductor under specific conditions. Fullerenes can display the photochromic effect, which is a change in light transmission based on intensity

Modeling of extended nanostructured systems that contain C_{60} is very difficult, and for some applied problems a simple analytical description of fullerene potential is highly desirable. Modeling of C_{60} potential starts from introducing the jelly model [2, 3] and still stays actual. Due to the fact that the fullerene symmetry is close to spherical, the natural choice of the effective potential is the angular averaged total electrostatic potential that obtained from *ab-initio* calculations.

2. Details of calculation and results

Ab-initio calculations were performed in 2 ways: with basis set of atomic orbitals and basis set of plane wave that can be seen from figure 1.

First of all, the role of electronic correlation effects to the total electrostatic potential of C_{60} should be investigated. For this purpose we carried out calculation with atomic orbitals within the hybrid Becke-Lee-Yang-Parr three-parameter [4] (B3LYP) density functional treatment. That includes HF-type exchange and exchange-correlation functionals formulated within the Generalized Gradient Approximation (GGA). Moreover, the attractive dispersion corrections to the DFT treatment via the

B3LYP-D3 [5] approach was taken into account since dispersion-corrected DFT is the current standard. Total electrostatic potential was averaged over a solid angle to clarify the radial dependence.

Ab-initio computations within atomic orbitals were performed with the FireFly quantum-chemistry package [6]. The initial optimization of the structure was realized through the semi-empirical PM3 approach [7]. The resulting geometry was used as a zero-order approximation for calculation in B3LYP/6-31G(d) and B3LYP-D3/6-31G(d) treatments (Popple notation).

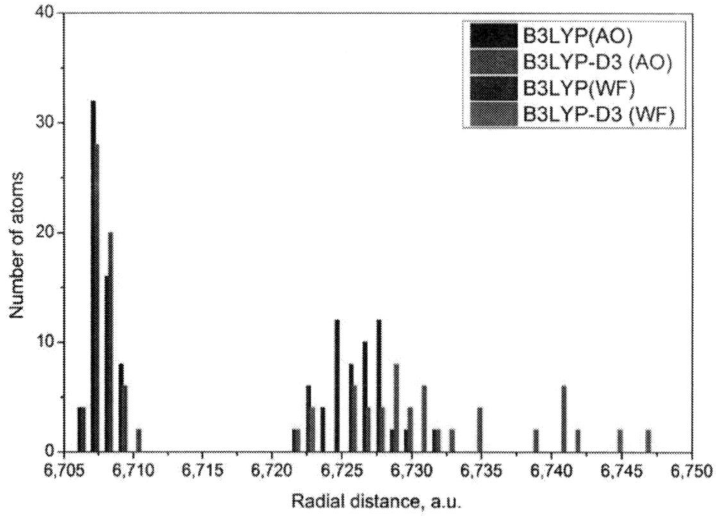

Figure 1. The histograms of distances between the center of C_{60} and positions of carbon nuclei obtained with atomic orbital (AO) basis and wave function (WF) basis.

As it was mentioned above the pseudopotentials of C_{60} and were constructed on the basis of the corresponding total electrostatic potentials. The latter was presented as a sum of two terms: the potential of nuclei $U_n(r)$, which depends on the positions of sixty carbon atoms, and the potential $U_{el}(r)$ created by electron density $\rho(r)$:

$$U_{\mathrm{tot}}(\mathbf{r}) = U_{\mathrm{n}}(\mathbf{r}) + U_{\mathrm{el}}(\mathbf{r}) = -\sum_{i=1}^{60} \frac{6}{|r-R_i|} + \int \frac{\rho(r')}{|r-r'|} dr' \qquad (1)$$

Positions of the carbon atoms within the optimized geometry and corresponding charge density were extracted from results of ab-initio FireFly QC package calculations by using a Multifunctional Wavefunction Analyzer (Multiwfn) [25]. The same software package was then used for computations of the corresponding electrostatic potentials on a specified grid of the r positions. Finally we have averaged the electrostatic potential over the r directions to make the radial dependence of the pseudopotentials:

$$U_{\mathrm{pseudo}}(r) = U_{\mathrm{tot}}(r) = U_{\mathrm{n}}(r) + U_{\mathrm{el}}(r) \qquad (2)$$

$$U_i(r) = \frac{1}{4\pi} \int U_i(r) d\Omega \qquad \text{(i=tot, n, el)}$$

Calculations within plane wave basis set were performed with Quantum Espresso package [8-9] also with DFT and similar approximations.

DFT-based approximations lead to an increase of the effective radius C_{60} that associated with the accounting of the electronic correlations and the B3LYP-D3 approach within atomic orbital basis set

IOP Conf. Series: Journal of Physics: Conf. Series **1092** (2018) 012001 doi:10.1088/1742-6596/1092/1/012001

produces the lowest total energy. The account of electron correlations makes potential deeper keeping its width. Also the position of pseudopotential minimum is shifted to the larger radial distances when correlations are taken into consideration. Since the sharp shape of potential leads to very high sensitivity of the electronic level position, even this small discrepancy should be treated carefully. Thus, the result of B3LYP-D3 calculations should be used for construction of analytical approximation of pseudopotential of C_{60}.

For the future purposes, numerical solution was approximated by the analytical Chesler-Cram peak function [8, 9] that generally used for fitting of chromatographic peaks.

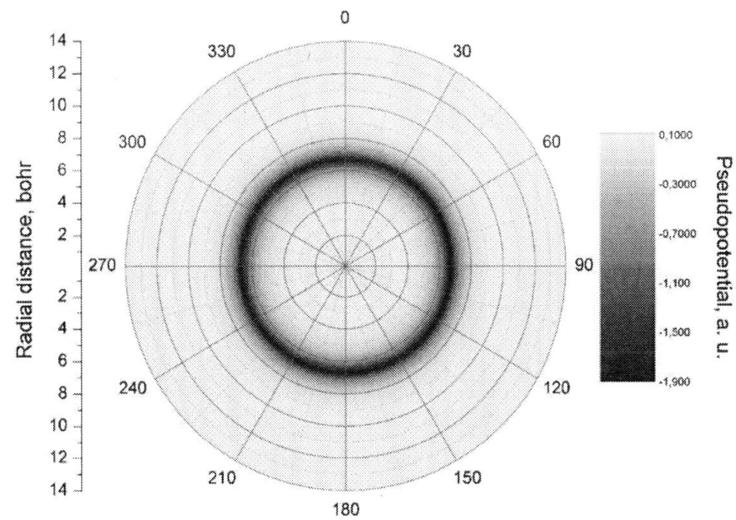

Figure 2. Color map of the pseudopotentials C_{60} obtained within the approximation of B3LYP-D3. The considered fullerene exhibit the strong attraction near the fullerene radius and have weak attraction force far from fullerene cage.

For the future purposes, numerical solution was approximated by the analytical Chesler-Cram peak function [10]. Chesler-Cram peak function is eight parameter function that originally used to fit the chromatographic peaks. We used this function, since it is able to reproduce the discontinuity point and consists of elementary functions.

$$y = y_0 + A\{e^{-\frac{(x-x_{c1})^2}{2w}} + B[1 - 0.5(1 - \tanh(k2(x - x_{c2})))]e^{-0.5k3(|x-x_{c3}|+(x-x_{c3}))}\} \quad (3),$$

where x is the radial distance. The sets of constants allowing to achieve accurate approximation are, y_0=-0,01239, x_{c1}=6,66442, A=-0,15527, w=0,04559, B=27,94053, k_2=0,89397, x_{c2}=6,85974, k_3=2,25645, x_{c3}=6,71233 are the constants of approximation.

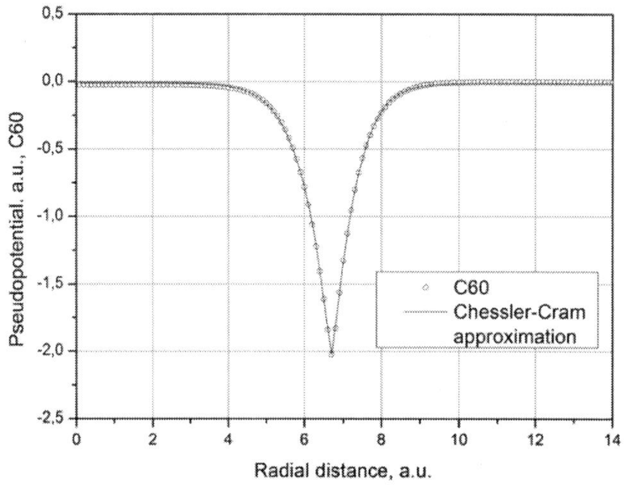

Figure 3. Chesler-Cram approximation for pseudopotentials obtained from ab-initio B3LYP-D3 calculations. Chesler-Cram peak function fits well within C_{60} pseudopotential at any distance.

Conclusion

We have calculated the electronic and spatial structure of the fullerene C_{60}. By averaging the total electrostatic potential, the radial dependent pseudopotential were constructed. The calculated pseudopotential have been fitted by the analytical Chesler-Cram peak function and can be applied for further tasks.

Acknowledgments

This work was supported by the Horizon2020 RISE project CoExAN(Grant No. 644076), mega-grant No 14.Y26.31.0015, Project 3.2614.2017/4.6 of the Ministry of Education and Science of Russian Federation and by Goverment of Russian Federation (Grant 08-08).

References

[1] Schultz HP 1965 *J. Org. Chem.* **30**(5) 1361-4.
[2] Yabana K and Bertsch GF 1993 *Phys. Scr.* **48**(5) 633.
[3] Yannouleas C and Landman U 1994 *Chem. Phys. Lett.* **217**(3) 175–85.
[4] Becke AD 1993 *J. Chem. Phys* **98** 5648–52.
[5] Grimme S, Antony J, Ehrlich S and Krieg H 2010 *J. Chem. Phys.* **132**(15) 154104.
[6] Granovsky AA. Firefly version 8. http://classic.chem.msu.su/gran/firefly/index.html
[7] Stewart JJ 1989 *J. Comput. Chem.* **10**(2) 221–64.
[8] Giannozzi P, Baroni S, Bonini N, Calandra M, Car R, Cavazzoni C, Ceresoli D, Chiarotti GL, Cococcioni M, Dabo I and Dal Corso A 2009 *J. Phys. Condens. Matter* **21**(39) 395502.
[9] Giannozzi P, Andreussi O, Brumme T, Bunau O, Nardelli MB, Calandra M, Car R, Cavazzoni C, Ceresoli D, Cococcioni M and Colonna N 2017 *J. Phys. Condens. Matter* **29**(46) 465901.
[10] Chesler SN and Cram SP 1973 *Anal. Chem.* **45**(8) 1354–9.

IOP Conf. Series: Journal of Physics: Conf. Series **1092** (2018) 012002 doi:10.1088/1742-6596/1092/1/012002

Optical and electrical properties of the $MoSe_2$/graphene heterostructures

P A Alekseev[1], B R Borodin[1], F A Benimetskiy[2], A N Smirnov[1], V Yu Davydov[1], S P Lebedev[1], A A Lebedev[1] and M S Dunaevskiy[1,2]

[1] Ioffe Institute, Saint-Petersburg, 194021, Russia
[2] ITMO University, Saint-Petersburg, 197101, Russia

npoxep@gmail.com

Abstract. Optical and electrical properties of $MoSe_2$ monolayer transferred to graphene/SiC substrate are studied by scanning probe microscopy and photoluminescence spectroscopy. By Kelvin probe microscopy it was measured a work function of the $MoSe_2$ monolayer (4.3 eV) and it was revealed the n-type of doping. With the increasing of $MoSe_2$ layers quantity the work function also increases which leads to the reducing of the Schottky barrier height between the SPM probe and the layers. Monolayer graphene quenches photoluminescence of the $MoSe_2$ monolayer while the bilayer graphene does not perturb the photoluminescence.

1. Introduction

Van-der-Waals (vdW) materials are promising for designing electronic and optoelectronic devices with improved characteristics[1]. From monolayers of these materials a layered heterostructures can be formed. The most studied vdW material is a graphene[2]. However narrow bandgap of the graphene impedes its using in optoelectronic devices in a visible range. Transition metal dichalcogenides (TMD) are another vdW materials which have a wider band gap suitable for visible optoelectronics and polaritonics[3]. Concept of the vdW heterostructures is in formation of multilayer sandwiches from monolayers of conducting (graphene), semiconducting (TMD) and dielectric (h-BN) materials. Due to the 2D nature of the vdW materials their electronic properties are extremely sensitive to the interface and surface configuration of the heterostructure. The aim of our work was to study optical and electrical properties of the vDW heterostructure formed by $MoSe_2$ monolayer and a graphene/SiC substrate.

2. Samples and methods

2.1. Samples

$MoSe_2$ layers were exfoliated from bulk sample using a scotch tape and then transferred to the graphene/SiC substrate. This substrate was covered by high-quality monolayer graphene (MLG) with randomly distributed bilayer (BLG) islands (10% of the area). Graphene film was obtained using a method of the thermal decomposition of a 6*H*-SiC surface[4]. Note that this method allows to form wafer-scale graphene with the quality exceeding the quality of the CVD grown graphene[5]. Moreover, SiC substrate is a semi-insulating and a transparent in a visible range.

2.2. Methods

Study of the electrical properties of the $MoSe_2$/graphene heterostructure was based on scanning probe microscopy (SPM) methods. These methods include Kelvin probe microscopy (KPM) and conductive atomic force microscopy (C-AFM)[6]. By the KPM method[7] it was measured a surface potential distribution of the heterostructure providing the values of work function of $MoSe_2$ and graphene layers. The C-AFM method allows to measure I-V curves in regions with different number of $MoSe_2$ layers on graphene. The study was performed using a Ntegra Aura (NT-MDT) scanning probe microscope in ambient conditions. During the measurements a surface of the graphene was grounded.

Optical properties were studied by measuring of microphotoluminescence (PL) spectra. The experiments were carried out at room temperature (300 K) using Horiba Jobin Yvon T64000 and LabRAM HR spectrometers equipped with a Linkam THMS600 temperature-controlled microscope stage. The measurements were performed with continuous-wave (cw) excitation using the 532 nm laser line of a Nd: YAG laser. We used a Mitutoyo 100× NIR (NA=0.90) long working-distance objective lens to focus the incident beam into a spot of ~0.7 μm diameter.

3. Results and discussion

Figure 1 (a) shows a surface potential distribution measured in the area with the transferred $MoSe_2$ layers. With the measured surface potential, a surface topography map was simultaneously obtained (not shown here). From the topography it was defined a quantity of the $MoSe_2$ layers in the area. A local number of layers is labeled in figure 1 (a), in assuming that thickness of monolayer is of 0.7 nm.

Surface potential distribution in figure 1 (a) is non-uniform. The value of the surface potential correlates with the number of $MoSe_2$ layers. To estimate a work function of the $MoSe_2$ it was analyzed a surface potential of the graphene substrate. It is known that the work functions of the graphene monolayer and bilayer on SiC are of 4.55 and 4.44 eV, respectively[8]. Fortunately, during the transfer the $MoSe_2$ layer covered graphene with monolayer (MLG) and bilayer (BLG) regions. The BLG regions can be find in figure 1(a) as diagonal stripes with higher surface potential. Since the surface potential it is a difference between work function of the probe and the studied surface, the work function of the 1ML $MoSe_2$ was defined as 4.30±0.05 eV. With increasing of layers number, the work function also increases (4.35 eV for 2 ML and 4.37 eV for 3 ML). These values were not presented in the literature before. Comparison with the work function and electron affinity of the bulk $MoSe_2$ (not shown here) reveal the n-type doping of the layers.

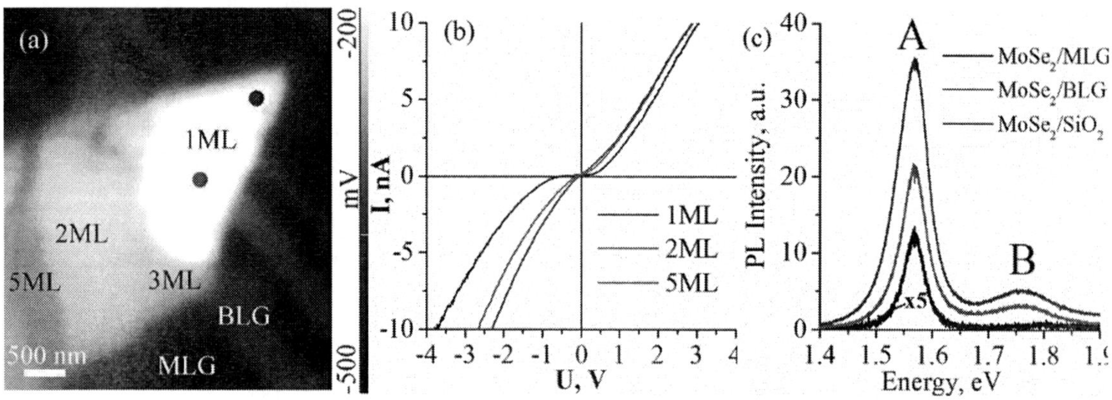

Figure 1. (a) Surface potential distribution of the $MoSe_2$ layers on the graphene. MLG and BLG are the regions with monolayer and bilayer graphene, respectively. 1-5ML indicate the number of the $MoSe_2$ layers. (b) I-V curves measured by C-AFM by contacting a probe to the regions with different number of the $MoSe_2$ layers. (c) PL spectra measured in the area pointed by black and red dots in figure 1(a), respectively.

To further study the electrical properties, the I-V curves were measured. For the measurement, the first contact was created to the graphene and the second was a SPM probe connected to the regions with the different number of the $MoSe_2$ layers. Measured I-V curves are presented in figure 1 (b). From the figure it follows that the Schottky barrier between the probe and a surface is reduces with increasing of the layers quantity. For the explanation of this effect it was developed a simple model based on the work function of the probe and the data obtained by KPM (figure 1 (a)). From the model it follows that increasing of the work function leads to a decreasing of the Schottky barrier height. Such an approach paves the way to tune the metal work function for creating an ohmic contact to $MoSe_2$ monolayer.

Figure 1 (c) shows PL spectra measured on the $MoSe_2$ monolayer. Black curve was measured in the area pointed by the black dot (figure 1(a)), where the $MoSe_2$ contacted with the MLG. Red curve (red dot) was measured in the area covering the BLG. For comparison a $MoSe_2$ monolayer was also transferred to a SiO_2 (300nm)/Si substrate (blue curve). The spectra exhibit two broad peaks which can be attributed to A (1.58 eV) and B (1.76 eV) excitons. From the figure it follows that the MLG significantly quench the PL from $MoSe_2$, while the BLG do not affect the PL. Additionally, the ratio between the A and B peaks intensity is the same for the $MoSe_2$ on BLG and SiO_2, while for MLG it is changed. Such behavior is consistent with the recent results obtained on $MoSe_2$ on CVD monolayer graphene[9] and the WS_2 on MLG and BLG/SiC[10]. In the work[9] it was shown a PL quenching by monolayer CVD graphene. KPM and PL study of the WS_2 on graphene/SiC substrates reveal the quenching of the WS_2 PL only by MLG.

4. Conclusions

Optical and electrical properties of the $MoSe_2$/graphene heterostructures were studied by means of the scanning probe microscopy and the photoluminescence spectroscopy methods. The work function of the $MoSe_2$ monolayer was measured as 4.30 ± 0.05 eV. With the increasing of $MoSe_2$ layers quantity the work function also increases which leads to the reducing of the Schottky barrier height between the SPM probe and the layers. It was shown that a monolayer graphene quenches a PL Intensity of the $MoSe_2$ monolayer, while the bilayer graphene does not perturb the PL. Obtained results open the way to improve parameters of $MoSe_2$/graphene heterostructures for optoelectronic devices.

Acknowledgments

This work is supported by Presidential Grant MK-5852.2018.2.

References

[1] Geim A K and Grigorieva I V 2013 *Nature* **499** 419
[2] Novoselov K S, Geim A K, Morozov S V, Jiang D, Zhang Y, Dubonos S V, Grigorieva I V and Firsov A A 2004 *Science* **306** 666–669
[3] Lundt N, Maryński A, Cherotchenko E, Pant A, Fan X, Tongay S, Sęk G, Kavokin A, Höfling S and Schneider C 2016 *2D Materials* **4** 015006
[4] Davydov V Y, Usachov D Y, Lebedev S, Smirnov A, Levitskii V, Eliseyev I, Alekseev P, Dunaevskiy M, Vilkov O Y, Rybkin A and others 2017 *Semiconductors* **51** 1072–1080
[5] Emtsev K V, Bostwick A, Horn K, Jobst J, Kellogg G L, Ley L, McChesney J L, Ohta T, Reshanov S A, Röhrl J and others 2009 *Nature materials* **8** 203
[6] Alekseev P A, Dunaevskiy M S, Ulin V P, Lvova T V, Filatov D O, Nezhdanov A V, Mashin A I and Berkovits V L 2015 *Nano letters* **15** 63–68
[7] Dunaevskiy M, Alekseev P, Girard P, Lashkul A, Lahderanta E and Titkov A 2012 *Journal of Applied Physics* **112** 064112
[8] Panchal V, Pearce R, Yakimova R and Kazakova O 2013 *Scientific reports* **3** 2597
[9] Shim G W, Yoo K, Seo S-B, Shin J, Jung D Y, Kang I-S, Ahn C W, Cho B J and Choi S-Y 2014 *ACS nano* **8** 6655–6662
[10] Giusca C E, Rungger I, Panchal V, Melios C, Lin Z, Lin Y-C, Kahn E, Elías A L, Robinson J A, Terrones M and others 2016 *ACS nano* **10** 7840–7846

Highly Efficient Visible Hologram through Dielectric Metasurface

Muhammad Afnan Ansari[1,2], Muhammad Hamza Waseem[1,6], Inki Kim[4], Nasir Mahmood[1,3], Heongyeong Jeong[4], Farooq Ahmad Tahir[3], Tauseef Tauqeer[1*], Muhammad Qasim Mehmood[1*], Junsuk Rho[4,5*]

[1]Information Technology University of the Punjab, Lahore, Pakistan.
[2]Department of Electrical and Electronics Engineering, Middle East Technical University, 06800 Çankaya/Ankara, Turkey.
[3]National University of Sciences and Technology (NUST), Islamabad, Pakistan.
[4]Department of Mechanical Engineering, Pohang University of Science and Technology (POSTECH), Pohang 37673, Republic of Korea.
[5]Department of Chemical Engineering, Pohang University of Science and Technology (POSTECH), Pohang 37673, Republic of Korea.
[6]Department of Electrical Engineering, University of Engineering and Technology, Lahore 54000, Pakistan.

*Corresponding authors: tauseef.tauqeer@itu.edu.pk, qasim.mehmood@itu.edu.pk, jsrho@postech.ac.kr

Abstract. To achieve applied aspect of metasurfaces in the visible regime, dielectric materials with low absorption are indispensable. This work presents highly efficient generation of hologram via processed amorphous silicon, which exhibits significantly low absorption in the region of interest. The phase and the polarization of transmitted light are tailored by varying the orientation of dielectric nanorods whereas their conversion efficiency is optimized by adjusting their structural parameters. Better image fidelity and higher conversion efficiency (up-to 75%) are achieved as compared to previously reported work. The proposed design methodology paves a way toward on-chip realization of various novel phenomena with substantially enhanced performance.

1. Introduction

Metasurfaces are two dimensional (2D) metamaterials with the ability to control the wavefront of light through spatial modulation of phase, amplitude and/or polarization at sub-wavelength scale [1]–[4]. This unique property of wavefront engineering facilitates in demonstrating diverse phenomena in the visible domain, like flat lensing [5]–[8] , vortex beam generation [9]–[12] and holography [13]–[17]. Meta-holograms are used to reconstruct holographic images, leading to different applications e.g. multi optical switching and image processing. Initially metallic nanoantennas were employed to realize visible meta-holograms, both in transmission [16], [18] and reflection modes [13], [14]. However, these plasmonic meta-holograms were afflicted by unavoidable Ohmic losses causing deteriorated efficiency [15]. Although reflection-type metasurfaces exhibited enhanced efficiency [14], they were impractical for transmission based applications [19].

Hence, lossless dielectric materials with high refractive index proved to be an ideal candidate to demonstrate highly efficient transmission-type metasurfaces in the visible regime. In this regard, recently Devlin et al. implemented highly efficient meta-holograms using dielectric nanoresonators made up of TiO_2 [20]. However, expensive and complex fabrication methodology of TiO_2 limit its mass-scale realization and compatibility with existing semiconductor fabrication processes. Among reported dielectric materials, silicon (Si) provides significant advantage in terms of its cost-effectiveness and CMOS compatibility. Owing to this unique advantage, Huang et al. [15] and Yoon [21] demonstrated transmission-type, Si-based, visible meta-hologram. However, significant dielectric losses of amorphous Si, in the visible spectrum, deteriorated image fidelity and efficiency (3%), hence limiting their desired applicability [15]. Therefore, low-cost, highly-efficient and CMOS compatible dielectric metasurfaces are required for the visible regime. The ellipsometry data of our processed amorphous silicon (a-Si) reveals comparatively higher refractive index and considerably lower absorption coefficient ($3.25+0.047i$) at a wavelength of 633 *nm*, hence making it a better candidate for low cost and highly efficient metasurface holograms. In this paper, we have proposed a highly efficient transmission-type hologram, using low-loss processed a-Si, for the red light (633 *nm*).

2. Methodology

To achieve 2π phase coverage for full wavefront manipulation, Pancharatnam-Berry Phase Modulation (PBPM) is implemented. Amplitude modulation of our hologram is two level. As suggested by PBPM, the orientation of the amorphous silicon nanorods with respect to x-axis (φ) is varied to achieve the desired phase. For instance, if left circularly polarized light (LCP) is impinged on nanorod (with an orientation angle φ), the transmitted wave will have one component of scattered wave with the same helicity and the other component of scattered wave with the opposite helicity, carrying an extra phase of 2φ (known as Pancharatnam-Berry (PB) phase) as shown in the Equation. 1 [22].

$$E_L^t = \frac{t_p+t_s}{2}\widehat{e_L} + \frac{t_p-t_s}{2}e^{i2\varphi}\widehat{e_R} \qquad (1)$$

To maintain highest conversion efficiency of the cross-polarized component, the optimization of unit cell (Figure. 1(a)) is performed using Finite Difference Time Domain (FDTD) analysis with periodic boundaries along x and y directions. However, perfect match layer (PML) boundary is used along the propagation direction.

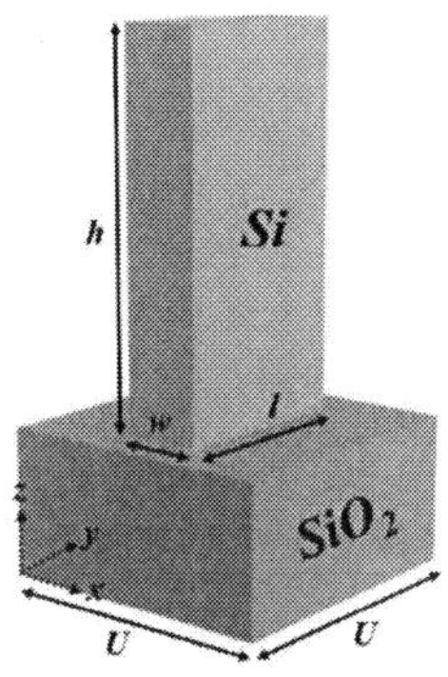

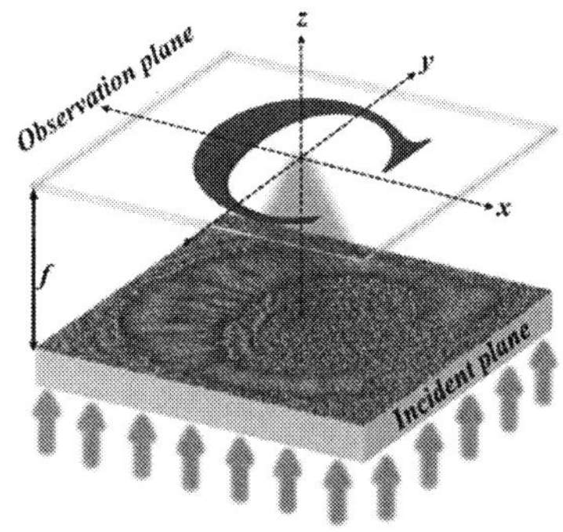

Figure 1. (a) Optimised unit cell of a-Si nanorod with glass substrate. *l =200 nm*, *w=80 nm*, and *h =380 nm* denote length, width and height of a-Si rod. Periodicity of unit cell is denoted by *U = 290 nm*. (b) A metasurface producing hologram at an observation plane (height $f = 200\ \mu m$). Letter 'C' is formed when the incident light is RCP.

Length (l), width (w) and periodicity (U) of unit cell are optimized for maximum transmission (T) and maximum conversion efficiency (ξ) of the cross-polarized light at 633 *nm* wavelength. Cross-polarized ξ is defined as the ratio of power of cross-polarized light and power of incident light. An iterative method known as particle swarm algorithm is employed to optimize the structural parameters of corresponding unit cell in FDTD Lumerical. The optimization of unit cell aligns the electric and magnetic dipole resonances, i.e. their identical phase and amplitude, of nanorod as described by Mie theory [23]. This optimization also ensures maximum forward scattering for 2π phase coverage. FDTD simulations showed that the highest value of T and ξ at 633 *nm* are 76% and 75% respectively. The average value of ξ for the proposed design is 73%. Schematic diagram of our proposed hologram is shown in Figure. 1(b).

3. Results and Discussion

The phase only hologram is used to generate a metasurface using FDTD Lumerical (Figure 1(b)). Due to limited computational power, only 35 x 35 μm^2 hologram is simulated in FDTD Lumerical to prove the concept. PML boundary conditions are used along each dimension for hologram simulation. Image ('C') on the projection plane (at a distance f) can be seen if the incident light is right circularly polarized (RCP). Figures 2(a) and 2(b) depict the simulation results of the designed hologram at 633 *nm* wavelength. Letter 'C' is observed with a high image fidelity at an on-axis focal plane.

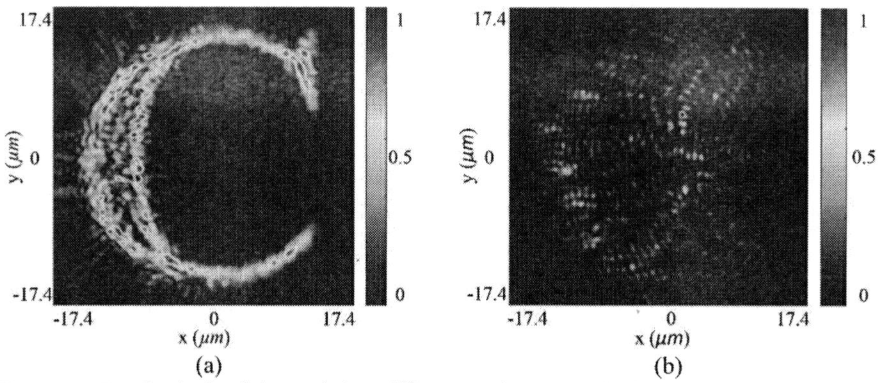

Figure 2. Result obtained from Finite Difference Time Domain (FDTD) simulations when incident light is (a) RCP and (b) LCP.

Moreover, it is clearly observed from the simulation results, that significant portion of the incident light has contributed to the target image. Only a minor portion is dissipated/absorbed due to absorption coefficient of the material. Unlike previously reported silicon based metasurfaces, this hologram possesses high quality image formation with the highest conversion efficiency at a visible wavelength (633 *nm*). Due to polarization dependency, no information is observed for incident light with left circular polarization, as shown in the Figure 2(b). FDTD Lumerical results can be further improved by increasing the size of the hologram.

4. Summary

In summary, processed amorphous silicon meta-hologram at a wavelength of 633 *nm* is demonstrated. The transmission type hologram instructs Pancharatnam-Berry (PB) phase to the transmitted wave by spatially controlling the orientation of nanorods. We have successfully presented high optical efficiency and good image quality by using a-Si nanorods, which are easy to fabricate and are compatible with CMOS fabrication. Hence, all silicon-based metasurface devices can be integrated into photonic circuits without degrading the performance, which paves a path toward applications of highly efficient integrated photonic circuits based on metasurfaces/metamaterials.

Acknowledgements

The authors would like to acknowledge TÜBİTAK for 2216 - Research Grant (Number: B.14.2.TBT.0.06.01.02-216-10539) and acknowledge POSTECH for their support.

References

[1] N. Yu *et al.*, "Light propagation with phase discontinuities: Generalized laws of reflection and

refraction," *Science (80-.).*, vol. 334, no. 6054, pp. 333–337, 2011.

[2] A. V. Kildishev, A. Boltasseva, and V. M. Shalaev, "Planar photonics with metasurfaces," *Science*, vol. 339, no. 6125. pp. 12320091–12320096, 2013.

[3] C. Della Giovampaola and N. Engheta, "Digital metamaterials," *Nat. Mater.*, vol. 13, no. 12, pp. 1115–1121, 2014.

[4] N. Yu and F. Capasso, "Flat optics: Controlling wavefronts with optical antenna metasurfaces," in *IEEE Antennas and Propagation Society, AP-S International Symposium (Digest)*, 2013, pp. 2341–2342.

[5] F. Aieta *et al.*, "Aberration-Free Ultrathin Flat Lenses and Axicons at Telecom Wavelengths Based on Plasmonic Metasurfaces," *Nano Lett.*, vol. 12, no. 9, pp. 4932–4936, Sep. 2012.

[6] X. Chen *et al.*, "Dual-polarity plasmonic metalens for visible light," *Nat. Commun.*, vol. 3, 2012.

[7] X. Ni, S. Ishii, A. V. Kildishev, and V. M. Shalaev, "Ultra-thin, planar, Babinet-inverted plasmonic metalenses," *Light Sci. Appl.*, vol. 2, no. APRIL, 2013.

[8] S. Mei, M. Q. Mehmood, K. Huang, and C.-W. Qiu, "Multi-foci metalens for spin and orbital angular momentum interaction," 2015, vol. 9544, p. 95441J.

[9] M. Q. Mehmood *et al.*, "Visible-Frequency Metasurface for Structuring and Spatially Multiplexing Optical Vortices," *Adv. Mater.*, vol. 28, no. 13, pp. 2533–2539, 2016.

[10] S. Mei *et al.*, "Flat Helical Nanosieves," *Adv. Funct. Mater.*, vol. 26, no. 29, pp. 5255–5262, 2016.

[11] J. T. Kun Huang, Hong Liu, Sara Restuccia, M. Q. Mehmood, Shengtao mei, Daniel Giovannini, Aaron Danner, Miles Padgett and C.-W. Qiu, "Spiniform-Phase-Encoded Metagratings Entangling Arbitrarily Rational-order Orbital Angular Momentum," *Light Sci. Appl.*, 2017.

[12] M. Q. MEHMOOD, C.-W. QIU, A. DANNER, and J. TENG, "Generation of Optical Vortex Beams by Compact Structures," *J. Mol. Eng. Mater.*, vol. 2, no. 2, p. 1440013, 2014.

[13] D. Wen *et al.*, "Helicity multiplexed broadband metasurface holograms," *Nat. Commun.*, vol. 6, no. May, pp. 1–7, 2015.

[14] G. Zheng, H. Mühlenbernd, M. Kenney, G. Li, T. Zentgraf, and S. Zhang, "Metasurface holograms reaching 80% efficiency," *Nat. Nanotechnol.*, vol. 10, no. 4, pp. 308–312, 2015.

[15] K. Huang *et al.*, "Silicon multi-meta-holograms for the broadband visible light," *Laser Photonics Rev.*, vol. 10, no. 3, pp. 500–509, 2016.

[16] X. Ni, A. V. Kildishev, and V. M. Shalaev, "Metasurface holograms for visible light," *Nat. Commun.*, vol. 4, 2013.

[17] Z. Li *et al.*, "Dielectric Meta-Holograms Enabled with Dual Magnetic Resonances in Visible Light," *ACS Nano*, vol. 11, no. 9, pp. 9382–9389, 2017.

[18] K. Huang *et al.*, "Ultrahigh-capacity non-periodic photon sieves operating in visible light," *Nat. Commun.*, vol. 6, 2015.

[19] D. B. Williams and C. B. Carter, "The Transmission Electron Microscope," in *Transmission Electron Microscopy*, 1996, pp. 3–17.

[20] R. C. Devlin, M. Khorasaninejad, W. T. Chen, J. Oh, and F. Capasso, "Broadband high-efficiency dielectric metasurfaces for the visible spectrum," *Proc. Natl. Acad. Sci.*, vol. 113, no. 38, pp. 10473–10478, 2016.

[21] G. Yoon, D. Lee, K. T. Nam, and J. Rho, "Pragmatic metasurface hologram at visible wavelength: the balance between diffraction efficiency and fabrication compatibility," *ACS Photonics*, p. acsphotonics.7b01044, 2017.

[22] H.-H. Hsiao, C. H. Chu, and D. P. Tsai, "Fundamentals and Applications of Metasurfaces," *Small Methods*, vol. 1, no. 4, p. 1600064, 2017.

[23] J. A. Schuller, R. Zia, T. Taubner, and M. L. Brongersma, "Dielectric metamaterials based on electric and magnetic resonances of silicon carbide particles," *Phys. Rev. Lett.*, vol. 99, no. 10, 2007.

IOP Conf. Series: Journal of Physics: Conf. Series **1092** (2018) 012004 doi:10.1088/1742-6596/1092/1/012004

Core-shell biochar-bearing iron ore powder model for calculation of effective electrodynamic parameters

Anton P Anzulevich[1], Svetlana N Anzulevich[1], Igor V Bychkov[1], Vasiliy D Buchelnikov[1], Leonid N Butko[1], Dmitry A Kalganov[1], Dmitry A Pavlov[1], Sergey G Moiseev[2,3] and Zhiwei Peng[4]

[1] Chelyabinsk State University, 129 Br. Kashirinykh Str., Chelyabinsk 454021, Russia
[2] Ulyanovsk State University, 42 L.Tolstoy Str., Ulyanovsk 432017, Russia
[3] Kotelnikov Institute of Radio Engineering and Electronics of Russian Academy of Sciences, Ulyanovsk Branch, 48/2 Goncharov Str., Ulyanovsk 432011, Russia
[4] Central South University, Changsha, Hunan, China

E-mail: anzul@list.ru

Abstract. The simplified model of biochar-bearing iron ore with binder was studied. It was considered to be a mixture of two types of core-shell particles, iron ore - binder particles and biochar - binder particles. The expressions to calculate complex effective permittivity and permeability was derived by effective medium approximation (EMA). The corresponding dependencies on volume fraction of iron ore in the mixture are provided.

1. Introduction

In this work, we consider biochar as a carbon-containing filler of iron ore. By using biochar, which is obtained via low-temperature pyrolysis of waste, abundant and sustainable biomass, to replace coal and coke, faster reduction and low reduction temperature can be expected with much lower emission of CO2 and hazardous gas (such as SO2). If the biochar-bearing pellets are heated by microwave energy which is characterized by volumetric heating, selective heating and non-thermal effect, reduction temperature and time can be further reduced with better quality of reduced pellets. The metallized pellet can then be used as a good raw material for steelmaking in electric arc furnace.

Given that biochar is a relatively new material, a huge amount of publications [1,2] studying the possibilities of its application in various fields of science and technology can be observed. In work [2], the absorbing capacity of biochar with respect to heavy metal oxyanions (Cr(VI)) was improved by impregnating the negatively charged surface of the biochar with Fe3+ particles to neutralize the surface. In addition, a composite consisting of carboxymethylcellulose, iron sulfide and biochar (CMC-FeS@biochar) as an absorber of heavy metal particles such as Cr(VI) was proposed in [3]. Using a composite of polyvinyl alcohol and biochar as a part of pressure sensors is proposed in [4]. In [5], the influence of the pyrolysis temperature of biochar and the distribution of metal oxyhydrochlorides on the surface of biochar particles pretreated with Al and Fe trichlorides on the ability to exchange ions in a neutral medium is investigated. Finally, a review [6] considers all possible aspects of the use of biochar as a catalyst.

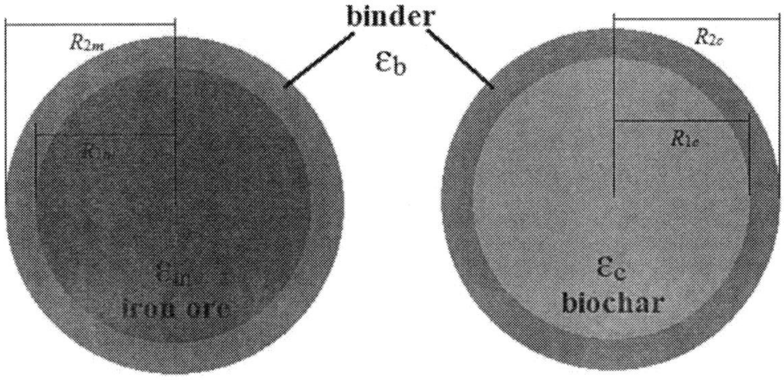

Figure 1. Simplified model of biochar-bearing iron ore powder compacted by binder.

In connection with the foregoing, the theoretical calculations of electrodynamic parameters of biochar-bearing materials are of particular interest and modeling the propagation of electromagnetic radiation of a wide frequency range in such composite structures is topical.

2. Theoretical model

Let us consider the following simplified theoretical model of biochar-bearing iron ore compacted with addition of binder - see figure 1.

Here ε_b is permittivity of binder, ε_c - permittivity of biochar, ε_m - permittivity of iron ore; R_{1m} - core radius of iron ore particle, R_{2m} - radius of entire core-shell iron ore - binder particle, R_{1c} - core radius of biochar particle, R_{2c} - radius of entire core-shell biochar - binder particle. Both types of particles are identical except of values of permittivity, permeability, and radius. Hence, the solutions of Maxwell equations with standard boundary conditions for core-shell particles look like following [7]:

$$\vec{E}_{m,c} = \frac{9\varepsilon_b \varepsilon_{eff} \zeta_{m,c}}{2\alpha_{m,c}\varepsilon_{eff} + \beta_{m,c}\varepsilon_b} \vec{E}_0, r < R_{1m,1c}, \tag{1}$$

$$\begin{gathered} \vec{E}_{bm,bc} = \frac{3\zeta_{m,c}\varepsilon_{eff}}{2\alpha_{m,c}\varepsilon_{eff}+\beta_{m,c}\varepsilon_b} \times \\ \times \left[\left(\varepsilon_{m,c} + 2\varepsilon_b - (\varepsilon_{m,c} - \varepsilon_b)\frac{R_{1m,1c}^3}{r^3} \right) \vec{E}_0 + \right. \\ \left. +3(\varepsilon_{m,c} - \varepsilon_b)\frac{R_{1m,1c}^3}{r^5}\vec{r}\left(\vec{E}_0 \cdot \vec{r}\right) \right], R_{1m,1c} < r < R_{2m,2c}, \end{gathered} \tag{2}$$

Here bm and bc indexes correspond to the field inside the shell of iron ore particle and biochar particle correspondingly;

$$\begin{gathered} \zeta_{m,c} = (R_{2m,2c}/R_{1m,1c})^3 = (1 + l_{m,c})^3, \\ l_{m,c} = (R_{2m,2c} - R_{1m,1c})/R_{1m,1c} \\ \alpha_{m,c} = (\zeta_{m,c} - 1)\varepsilon_{m,c} + 2(\zeta_{m,c} + 1)\varepsilon_b, \\ \beta_{m,c} = (2 + \zeta_{m,c})\varepsilon_{m,c} + 2(\zeta_{m,c} - 1)\varepsilon_b, \end{gathered} \tag{3}$$

IOP Conf. Series: Journal of Physics: Conf. Series **1092** (2018) 012004 doi:10.1088/1742-6596/1092/1/012004

In EMA we have deal with a mixture of two types of spherical particles, which are randomly distributed in the effective medium. In this paper, we deal with three types of particles. Two of them are identical and considered above. As third type of particles we will consider spherical inclusions of gas (vacuum). It is considered that the permittivity of such a composite is equal to the permittivity of the effective medium. These gas particles can be any radius so that to fill all the remaining space. They are spherical and electric field inside of them is determined as:

$$\vec{E}_g = \frac{3\varepsilon_{eff}}{\varepsilon_g + 2\varepsilon_{eff}} \vec{E}_0, \tag{4}$$

Here ε_g is the permittivity of gas or vacuum. According to the EMA an average value of electric displacement of effective medium connects with an average value of electric field strength as:

$$\langle \vec{D} \rangle = \varepsilon_{eff} \langle \vec{E} \rangle = \varepsilon_{eff} \vec{E}_0, \tag{5}$$

After substitution of electrical fields (1), (2) and (4) into Eq. (5) and their integration we can find the final equation for calculation of the effective permittivity:

$$\begin{gathered}
(1 - p_m\zeta_m - p_c\zeta_c)\frac{\varepsilon_g - \varepsilon_{eff}}{\varepsilon_g + 2\varepsilon_{eff}} + \\
+ p_m\zeta_m \frac{\varepsilon_b[3\varepsilon_m + (\zeta_m - 1)(\varepsilon_m + 2\varepsilon_b)] - \varepsilon_{eff}[3\varepsilon_b + (\zeta_m - 1)(\varepsilon_m + 2\varepsilon_b)]}{2\alpha_m\varepsilon_{eff} + \beta_m\varepsilon_b} + \\
+ p_c\zeta_c \frac{\varepsilon_b[3\varepsilon_c + (\zeta_c - 1)(\varepsilon_c + 2\varepsilon_b)] - \varepsilon_{eff}[3\varepsilon_b + (\zeta_c - 1)(\varepsilon_c + 2\varepsilon_b)]}{2\alpha_c\varepsilon_{eff} + \beta_c\varepsilon_b} - \\
- p_m\zeta_m \frac{\frac{9}{2}\varepsilon_b(\varepsilon_m - \varepsilon_b)ln(1 + l_m)}{2\alpha_m\varepsilon_{eff} + \beta_m\varepsilon_b} - \\
- p_c\zeta_c \frac{\frac{9}{2}\varepsilon_b(\varepsilon_c - \varepsilon_b)ln(1 + l_c)}{2\alpha_c\varepsilon_{eff} + \beta_c\varepsilon_b} = 0,
\end{gathered} \tag{6}$$

where p_m and p_c are volume fractions of iron ore and biochar in effective medium mixture.

3. Results and discussion

This equation (6) can be used to determine the complex effective permittivity of biochar-bearing iron ore if permittivities and concentrations of initial components are known. Effective permeability of this mixture can be determined by the same equation (6) replacing ε with μ.

On the basis of the following papers [8–10], let's consider that monolithic biochar (containing 85% of C) and iron ore (containing 65% of Fe in oxide form) have following complex values of permittivity and permeability:

$$\begin{gathered}
\varepsilon_m = 14.2 + i0.2, \mu_m = 1.8 + i0.9, \varepsilon_c = 3 + i0.12245, \mu_c = 1, \\
\varepsilon_b = 1.2, \mu_b = 1, \varepsilon_g = 1, \mu_g = 1, \\
R_{1m} = 25\mu m, R_{1c} = 30\mu m
\end{gathered} \tag{7}$$

We considered that $\varepsilon = \varepsilon' + i\frac{4\pi\sigma}{\omega}$ and $\omega = 2\pi\nu$ where ν is frequency of microwave radiation that equals to $2.45GHz$. Taking into account that volume fraction of binder is about 1-2% it is easy to calculate that $l_m \approx l_c \approx 0.0066227$. Volume fraction of biochar component is assumed to be $p_c = 0.9 - p_m$.

Eventually, on figure 2 we can see that wave impedance $Z = \sqrt{\frac{\mu'_{eff}}{\varepsilon'_{eff}}}$ decreases from 0.6 to 0.38 when volume fraction of iron ore in the mixture increases from 0 to 1.

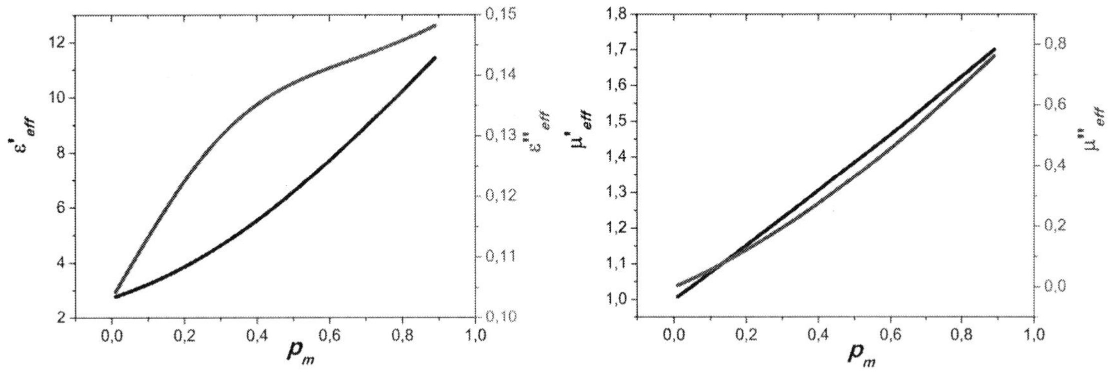

Figure 2. Real (black) and imaginary (red) parts of effective permittivity (left) and permeability (right) of biochar-bearing iron ore powder on volume fraction of iron ore p_m.

4. Conclusion

Using the obtained on figure 2 dependencies it is possible to optimize matching of impedances on the surface of studied biochar-bearing iron ore powder by changing volume fraction of iron ore depending on depth. That will allow to improve penetration of microwave radiation into the volume of powdered sample and enhance heating and, hence, reduction of iron ore.

Acknowledgments

This work was supported by the Russian Foundation for Basic Research (Project No. 18-58-53055, 16-29-14045, 17-02-01382) and by the Ministry of Education and Science of the Russian Federation (State Contract No. 3.5698.2017/9.10).

References

[1] Zhiwei Peng, Jiann-Yang Hwang, and Matthew Andriese, Absorber Impedance Matching in Microwave Heating, Applied Physics Express 5 (2012) 077301, pp. 1-3

[2] Wang H., Tian Z., Jiang L., Luo W., Wei Z., Li S., Cui J., Wei W., Highly efficient adsorption of Cr(VI) from aqueous solution by Fe3+ impregnated biochar, Journal of Dispersion Science and Technology, Volume 38, Issue 6, 3 June 2017, pp. 815-825

[3] Lyu H., Tang J., Huang Y., Gai L., Zeng E.Y., Liber, K., Gong Y., Removal of hexavalent chromium from aqueous solutions by a novel biochar supported nanoscale iron sulfide composite, Chemical Engineering Journal, Volume 322, 15 August 2017, pp. 516-524

[4] Nan N., DeVallance D.B., Development of poly(vinyl alcohol)/wood-derived biochar composites for use in pressure sensor applications, Journal of Materials Science, Volume 52, Issue 13, 1 July 2017, pp. 8247-8257

[5] Lawrinenko M., Jing D., Banik C., Laird D.A., Aluminum and iron biomass pretreatment impacts on biochar anion exchange capacity, Carbon, Volume 118, 1 July 2017, pp. 422-430

[6] Lee J., Kim K.-H., Kwon E.E., Biochar as a Catalyst, Renewable and Sustainable Energy Reviews, Volume 77, 1 September 2017, pp. 70-79

[7] V.D. Buchelnikov, D.V. Louzguine-Luzgin, A.P. Anzulevich, G. Xie, S. Li, N. Yoshikawa, M. Sato, I.V. Bychkov, A. Inoue. Heating of metallic powders b microwaves: Experiment and theory, Journal of Applied Physics, College Park, MD: AIP, 2008, 1, 104, pp 1-10. ISSN 1089-7550 / ISSN 0021-8979

[8] V.A. Sotskov, A.N. Zabavin, Experimental study of concentration dependence of the permittivity in three-component disordered systems, Journal of Technical Physics, 2012, Volume 82, Issue 7, pp. 137-139

[9] M.P., Dielectric constant and magnetic permeability of various ferrites at ultrahigh frequencies, Physics-Uspekhi (UFN), 50, pp. 152-155, doi: 10.3367/UFNr.0050.195305j.0152

[10] Randeep S. Gabhi, Donald W. Kirk, Charles Q. Jia, Preliminary Investigation of Electrical Conductivity of Monolithic Biochar, Carbon, 11678, 21 January 2017, doi: 10.1016/j.carbon.2017.01.069

Theoretical evaluation of the thermostimulated intrinsic radiation in a dipole phase transition

Svetlana N Anzulevich, Anton P Anzulevich, Igor V Bychkov
Chelyabinsk State University, 129 Br. Kashirinykh Str, Chelyabinsk 454021, Russia
E-mail: anzulic@gmail.com, anzul@list.ru, bigv@mail.ru

Abstract. Our research is devoted to a more detailed study of the generation of radiation by two-level quantum medium. An electric dipole transition is considered. From the equations of Maxwell-Bloch, the amplitude of the radiation as a function of the phase transition frequency was obtained.

1. Introduction

Currently, many scientists are studying the problem of thermally stimulated self-radiation of matter at phase transitions (PT) and using this phenomenon in a variety of practical devices. Such devices include generators of electromagnetic waves on the basis of solids, paramagnetic amplifiers with low intrinsic noise of the lasers. For the first time, the effect of strong over-temperature radiation was observed for a layer of boiling water in the near and infrared (IR) region of the spectrum [1]. A similar effect takes place when boiling metals [2]. Flash radiation is recorded as well during the crystallization of the alkali halide solids and sapphire [3, 4]. In addition, an internal radiation in the temperature interval close to magnetostructural PT 1-St and 2-nd type were observed for alloys Ni-Mn-Ga-Fe and Ti-Ni [5].

Independently, the physics and technology of vacuum tunnel structures develop, in which the temperature difference in the vacuum interval causes the appearance of a tunnel current, which, in turn, can generate a high specific power of electric energy with a strong temperature gradient and serve as a source of energy with a high efficiency parameter due to the fact that the heat transfer in vacuum is strongly suppressed [6, 7].

Thus, there is an experimental material confirming the existence of such radiation. However, the phenomenon of phase-transition radiation does not follow from the existence of the PT concept and is not taken into account when considering their kinetics. Such effects can be considered on the model of a two-level quantum medium. Especially that the interaction of radiation with such a model explains many phenomena related to the properties of substances such as absorption and dispersion of light, saturation and enlightenment transition, generation of electromagnetic radiation in lasers. Moreover, this model allows us to consider both electric and magnetic dipole transitions [8].

For the implementation of PT radiation it is necessary that the optical transition time of particles from the ground state of one phase to the ground state of another phase is less than or the order of the temperature of non-radiative relaxation [9]. This can be achieved in a large group of particles. For the first time, the possibility of radiative PT for a group of particles was considered in the frame of quantum electrodynamics even before the experimental detection of the phenomenon of collective spontaneous radiation or over emission [10–13].

2. Theoretical model

To model the self radiation in the PT, consider the two-level model of a quantum dipole transition. The system of Maxwell-Bloch equations in this case has the following form [?]:

$$\ddot{\vec{P}} + \frac{2}{T_2}\dot{\vec{P}} + \Omega^2\vec{P} = \frac{2\Omega}{\hbar}\frac{L|\mu_{12}|^2}{3}\Delta N\vec{E}, \tag{1}$$

$$\frac{\partial \Delta N}{\partial t} + \frac{\Delta N - \Delta N^e}{T_1} = -\frac{2}{\hbar\Omega}\left(\dot{\vec{P}}\cdot\vec{E}\right), \tag{2}$$

$$\nabla \times \left(\nabla \times \vec{E}\right) + \eta\frac{\mathcal{A}}{c}\frac{\partial \vec{E}}{\partial t} + \frac{\eta^2}{c^2}\frac{\partial^2\vec{E}}{\partial t^2} = -\mu_0\frac{\partial^2\vec{P}}{\partial t^2}, \tag{3}$$

For plane wave distributing along z axis solution of these equations for the amplitude of self-radiated wave can be found as:

$$E_x = \sqrt{\left[\frac{3i\hbar^2\Omega\left[(\Omega-\omega)-\frac{i}{T_2}\right]}{2\omega L T_1|\mu_{12}|^2} - \frac{i\mu_0\omega\Delta N^e\hbar\Omega\left[k^2+\frac{i\omega\eta\mathcal{A}}{c}-\frac{\eta^2\omega^2}{c^2}\right]}{2T_1}\right]}e^{-i(\omega t-kz)}, \tag{4}$$

where T_1 is a longitudinal relaxation time, T_2 - a transverse relaxation time, Ω - transition frequency, L - Lorentz correction coefficient, η - refractive index of matter, $\mathcal{A}$ - power attenuation factor, ΔN - population difference per unit volume, ΔN^e - population difference per unit volume under equilibrium conditions, $|\mu_{12}|$ - matrix element of the dipole moment operator connecting the states $|1\rangle$ and $|2\rangle$.

3. Conclusion

The obtained equation (4) can be used to estimate thermally stimulated self-radiation of matter at PT that is determined by quantum dipole transition at the boundary of different phases.

Acknowledgments

This work was supported by the Russian Foundation for Basic Research (Project No. 18-58-53055, 16-29-14045, 17-02-01382), by the Ministry of Education and Science of the Russian Federation (State Contract No. 3.5698.2017/9.10), and by the Foundation for Advanced Scientific Research of the CSU (FPNI).

References

[1] Potter W.R., Hoffman J.G., Infr. Phys. 1968. Vol. 18. N 6. P. 265269

[2] Burcev V.A., Kalinin N.V., Luchinskii A.V., Electrical explosion of conductors and its application in electrophysical installations., M.: Energoatomizdat, 1990. 288 c.

[3] Tatarchenko V.A., Crystallography. 1979. T. 24. 2. C. 408409.

[4] Tatarchenko V.A., Umarov L.M., Crystallography. 1980. T. 25. 6. C. 13111313.

[5] I. V. Bychkov, V. A. Golunov, D. S. Kalenov, A. P. Kamantsev, D. S. Kuchin, V. V. Koledov, D. A. Kuzmin, V. V. Meriakri, S. V. von Gratowski, M. P. Parkhomenko, A. V. Mashirov, V. G. Shavrov, The intrinsic radiation and reflection coefficient of the EMW in the range of 8 mm Ni2,14Mn0,81GaFe0,05 and TiNi alloys in the temperature range near the first and second-order phase transitions, Journal Of Radioelectronics, 12, (2014).

[6] Belbachir, R.Y., An, Z., Ono, T. Thermal investigation of a micro-gap thermionic power generator, Journal of Micromechanics and Microengineering. 2014. V. 24, 085009.

[7] Giazotto, F., Heikkil, T.T., Luukanen, et al. Opportunities for mesoscopics in thermometry and refrigeration, Reviews of Modern Physics. 2006. V. 78 (1), P. 217-274.

[8] R. H. Pantell, H. E. Puthoff, Fundamentals of Quantum Electronics, John Wiley Sons, Inc., New York, London, Sydney, Toronto (1969).
[9] S.A. Sall, A.P. Smirnov, Phase-transition radiation and the growth of a new phase, Journal of Technical Physics, 2000
[10] Perelman M.E., DAN SSSR. 1972. T. 203. 5. C. 1030 1034.
[11] Scribanowitz N., Hermann I.P., MacGillivray I.C., Feld M.S., Phys. Rev. Lett. 1973. Vol. 30. N 3. P. 309316
[12] Andreev A.V., Emelyanov V.I., Ilinsky Yu.A., Cooperative phenomena in optics. .: Science, 1988. 288 C.
[13] Zheleznyakov V.V., Kocharovsky V.V., Kocharovsky Vl.V., UFN., 1989. T. 159. 2. C. 194260.

Multi-field asymptotic homogenization approach for Bloch wave propagation in periodic thermodiffusive elastic materials

Andrea Bacigalupo and Giorgio Gnecco

MUSAM and AXES Research Units
IMT School for Advanced Studies - Piazza S. Francesco, 19 - 55100 Lucca, Italy

{andrea.bacigalupo,giorgio.gnecco}@imtlucca.it

Abstract. Multi-field asymptotic homogenization methods are proposed to describe the behaviour of periodic Cauchy materials subject to several physical phenomena, focusing on thermodiffusion. The resulting homogenized models provide the overall constitutive tensors and overall inertial terms. Moreover, they allow one to investigate the complex band structures associated with damped Bloch waves travelling in periodic materials, avoiding the challenging computations needed by the adoption of micromechanical approaches.

1. Introduction

Periodic composite materials subject to thermodiffusive phenomena [1,2] are of great interest for many important engineering and technological applications. For example, several renewable energy devices, energy harvesters, battery devices, like lithium-ion batteries and solid oxide fuel cells (see Figure 1), are characterized, among the others, by periodic configurations possessing many phases made of different materials. Furthermore, since high operational temperatures can be reached in such devices, and intense particle fluxes are needed to sustain their electrical current, their components can be subject to severe thermomechanical stresses, as well as to stresses induced by the intense particle diffusion and to thermal-shock waves, which can compromise an adequate performance of these devices. For such reason, a characterization of their overall constitutive equations can be extremely useful to predict these phenomena, to ensure a successful manufacturing of such systems and their reliability, and for the optimal design of innovative materials used to build these devices.

Multi-field asymptotic homogenization methods are rigorous and consistent tools for determining the overall constitutive tensors characterizing the behaviour of the periodic material, also in the presence of several physical phenomena, in a synthetic and accurate way [3-9]. The homogenized model provides a useful tool for the damped wave propagation analysis in both bounded and unbounded domains, avoiding the challenging computations required by micromechanical approaches. Moreover, the nonlocal homogenized models provide an accurate description of the acoustic behaviour of periodic Cauchy materials. By increasing the order of the nonlocal continuum in which proper nonlocal constitutive and inertial terms are involved, they allow one to approximate in a consistent way the complex frequency spectrum (complex band structure) of materials with periodic microstructures in the high frequency and short wavelengths domain.

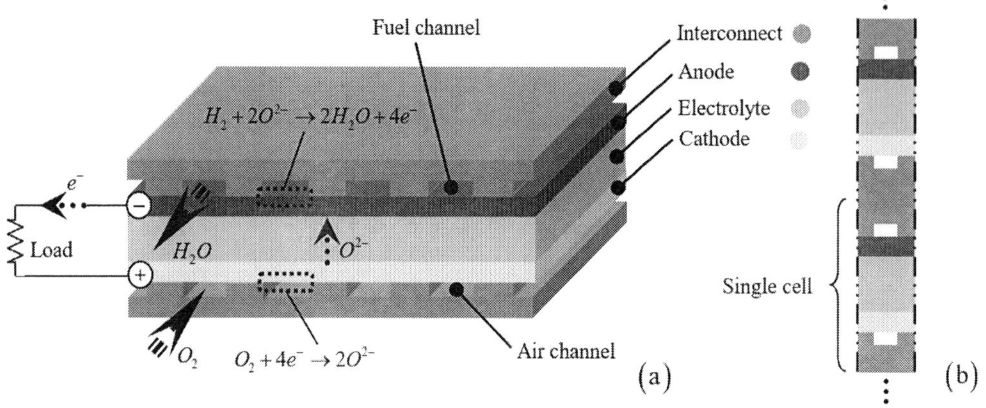

Figure 1: (a) Solid oxide fuel cell (SOFC); (b) Schematic multi-layer SOFC device (adapted from [4]).

2. Multiscale modelling of periodic Cauchy materials vs. local or nonlocal average governing equations

Dynamic microscopic field equations, expressed in terms of the physical and mechanical properties of the periodic microstructure, are asymptotically expanded, obtaining *recursive differential problems* (formulated in terms of microscopic variables, i.e. microscopic displacement, temperature and chemical potential fields), defined on the periodic cell A of characteristic size ε, which is representative of the material. Imposing the A-periodicity solvability conditions for recursive differential problems, suitable down-scaling relations are determined. They relate the microscopic variables to macroscopic ones and to their gradients, by means of A-periodic perturbation functions. Such perturbation functions depend exclusively on geometrical and physical-mechanical properties of the microstructure, and are determined by solving suitable *hierarchical cell problems*. These are obtained by inserting, in a suitable way, the different orders of the down-scaling relations (expressed as asymptotic expansions in ε) into the recursive differential problems expressed in terms of the microscopic variables. The solvability conditions of the hierarchical cell problems in the class of A-periodic solutions allow one to obtain the macroscopic governing equations of a homogeneous equivalent continuum, or rather to obtain higher order approximations of microscopic field equations, by considering nonlocal terms of the averaged field equations of infinite order. Appropriate time-dependent down-scaling and up-scaling relations correlating the microscopic fields to the macroscopic displacement, temperature and chemical potential fields are consistently determined, and the effects of the material inhomogeneities are described by perturbation functions. In particular, a closed form of the overall constitutive tensors and of the overall inertial terms is obtained, involving the perturbation functions and the microscopic constitutive tensors.

Finally, the complex band structure associated with damped Bloch waves travelling in periodic thermodiffusive elastic materials is analytically determined, applying the Fourier-Bloch transform method in accordance to Floquet-Bloch theory (see Figure 2 for some curves obtained using this procedure). The actual complex frequency spectrum in the low frequency and long/medium wavelengths domain is compared with the one obtained using the nonlocal continuum derived from a second order homogenization scheme. Moreover, its asymptotic approximation, also in the high frequency and short wavelengths domain, is expressed in closed form through perturbation techniques.

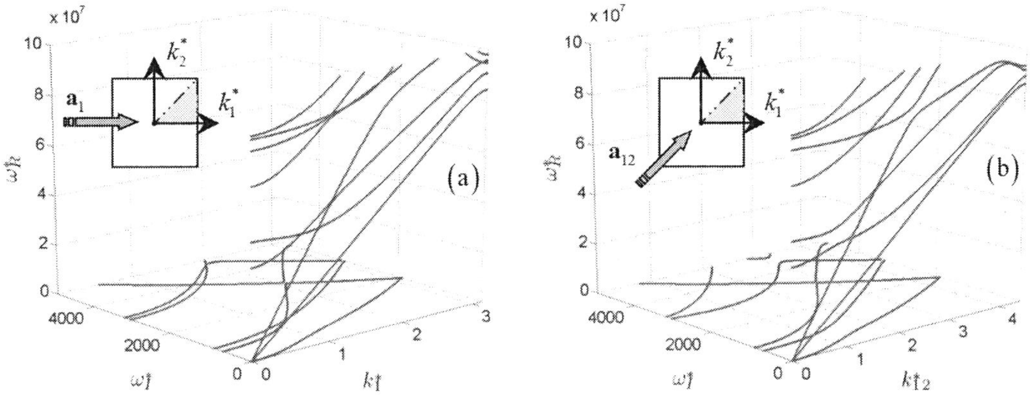

Figure 2: Complex band structure of the thermodiffusive material obtained via the Fourier-Bloch transform method. Wave propagation along: (a) $\mathbf{a}_1$ and (b) $\mathbf{a}_{12}$ directions.

References

[1] W. Nowacki, Dynamical problems of thermodiffusion in solids, *Bulletin of Polish Academy of Sciences Series, Science and Technology*, **22**, 55-64, 1974.

[2] W. Nowacki, Thermoelasticity, second edition, Pergamon Press, Oxford, 1986.

[3] A. Salvadori, E. Bosco, D. Grazioli, A computational homogenization approach for Li-ion battery cells: Part 1 - formulation, *Journal of the Mechanics and Physics of Solids*, **65**, 114-137, 2014. doi: 10.1016/j.jmps.2013.08.010.

[4] A. Bacigalupo, L. Morini, A. Piccolroaz, Effective elastic properties of planar SOFCs: A non-local dynamic homogenization approach, *International Journal of Hydrogen Energy*, **39** (Issue 27), 15017-15030, 2014. doi: 10.1016/j.ijhydene.2014.06.104.

[5] A. Bacigalupo, L. Morini, A. Piccolroaz, Multiscale asymptotic homogenization analysis of thermo-diffusive composite materials, *International Journal of Solids and Structures*, **85-86**, 15-33, 2016. doi: 10.1016/j.ijsolstr.2016.01.016.

[6] A. Bacigalupo, L. Morini, A. Piccolroaz, Overall thermomechanical properties of layered materials for energy devices applications, *Composite Structures*, **257**, 366-385, 2016. doi: 10.1016/j.compstruct.2016.07.048.

[7] M. L. De Bellis, A. Bacigalupo, Auxetic behavior and acoustic properties of microstructured piezoelectric strain sensors, *Smart Materials and Structures*, **26** (Issue 8), article no. 085037, 2017. doi: 10.1088/1361-665X-aa7772.

[8] F. Fantoni, A. Bacigalupo, M. Paggi, Multi-field asymptotic homogenization of thermo-piezoelectric materials with periodic microstructure, *International Journal of Solids and Structures*, **120**, 31-56, 2017. doi: 10.1016/j.ijsolstr.2017.04.009.

[9] F. Fantoni, A. Bacigalupo, M. Paggi, Design of thermo-piezoelectric microstructured bending actuators via multi-field asymptotic homogenization, *International Journal of Mechanical Sciences*, 2018, submitted. https://arxiv.org/abs/1803.05346.

IOP Conf. Series: Journal of Physics: Conf. Series **1092** (2018) 012007 doi:10.1088/1742-6596/1092/1/012007

Asymmetrical optical effects in anapole metamaterials

Maxim K. Myagkih, Kseniia V. Baryshnikova

ITMO University, 49 Kronverksky Pr., St. Petersburg, 197101, Russia

k.baryshnikova@metalab.ifmo.ru

Abstract. Anapole metamaterials attract great scientific interest due to their possibility to concentrate light without scattering. They are applied for sensing, nanolasing, light manipulation and so on. Asymmetrical effects can increase sensibility of sensors based of anapole states or give additional route to manage light propagation in metamaterial.

1. Introduction

All-dielectric nanophotonics attracts great interest due to the ability of nanoparticles made of materials with high refractive index and low absorption to resonantly enhance both electric and magnetic components of light. When in all-dielectric system electric and magnetic responses are induced simultaneously, optical effects can be observed like a directive scattering or hot spots enhancement (in case of standalone nanoparticles), perfect transmission or absorption, phase shifting (in case of metasurface), bianistropy and so on [1–4]. One of the wonderful examples of optical effects in all-dielectric structures is an anapole state in which scattering is suppressed down to zero and distribution of fields inside the nanoparticle is non-trivial [5].

Previously it was shown numerically that scattering pattern of asymmetrical dielectric nanoparticles can depend on the incident wave propagation direction [6]. Thus, one can observe Kerker-type scattering when silicon pyramid is irradiated from the top and different scattering pattern when is irradiated from base side.

Here we consider another case of excitation inside all-dielectric structures, associated with anapole-type states. We show that it is possible to construct metasurface based on the anapole states with strongly asymmetrical optical response.

2. Standalone nanoparticles asymmetrical scattering

If the size of nanoparticles is comparing to the wavelength of incident light, scattering cross section can exceed geometrical cross-section many times depending on the shape, size, and material properties of the system and scattering features which are associated with the multipoles enhanced in the nanoparticle. The nature of this effect in case of dielectric nanoparticles under light illumination is in enhancement of non-homogeneous displacement currents [7]. Radiation of these currents to the far-field zone can be expand into multipoles contribution [7,8].

In general, optical properties of asymmetrical particles system depend on the incidence direction [6]. However, scattering of standalone nanoparticles has symmetrical properties correlated with symmetry of particle system. Obviously, scattering pattern is preserved when particles system rotate by the angle of system's symmetry [9]. However, there is a general low about scattering symmetry, which valid for the nanoparticles of the arbitrary shape. It concludes following: if we interchange the direction of the incident wave and the direction of observation, the component of the scattered wave in the direction of polarization is unchanged; and this strictly follows from reciprocity principle [10]. As a direct consequence of this low, forward scattering does not change after inversion of incidence direction. Herewith backward scattering can be changed dramatically.

This scattering property, which looks rather obvious, should be stressed due to considering asymmetrical dielectric nanoparticles and clusters as bianisotropical systems [3,11]. How it is common-known, reciprocity can be ruined for bianisotropical systems, and it can be used for designing of non-reciprocal devices including metasurfaces based on bianisotropical structures.

3. Asymmetric metasurfaces based on anapole states

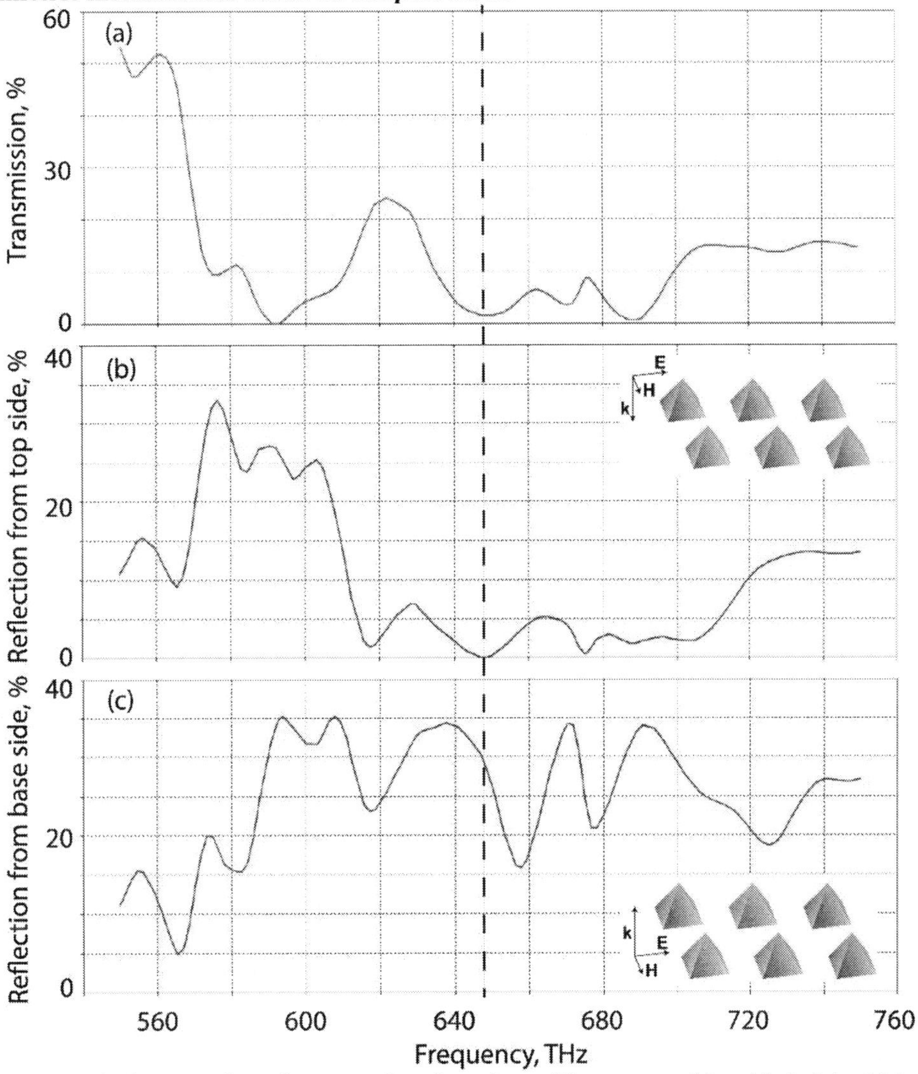

Figure 1. Optical properties of metasurface based on silicon pyramids with height 315 nm, base side 376 nm, period of metasurface 450 nm: (a) transmission, irradiation from both normal sides; (b) reflection, transmission from top side; (c) reflection, irradiation from base side. Vertical dashed line shows anapole state (minimum reflection and transmission).

Constructing of meta-atoms is usually most important step in designing of dielectric metasurfaces because diffraction plays minor role and interaction between dielectric nanoparticles rather insignificant, especially comparing with plasmonic ones. Optical properties of nanoparticles, in general, define properties of metasurface based on these nanoparticles. Let's consider a case, when metasurface based on the silicon pyramids with following parameters: height 315 nm, base side 376 nm, period of metasurface 450 nm.

It can be shown analytically that transmission is proportional to the forward scattering of each nanoparticle. So, the transmission stays the same for both normal directions of incidence due to reciprocity. But if absorption inside the nanoparticle is not equal to zero, reflection can be changed.

In case of anapole state realization in metasurface, electric dipole radiation of each metaatom is suppressed down to zero, and other multipoles radiation also suppressed. So, metasurface based on anapole states should work as a perfect absorber or as device form transformation of light energy in third harmonics energy [12]. For considered metasurface, an anapole-type state is observed at wavelength 463 nm in case of irradiation from the top side (corresponds to frequency 648 THz, see Fig. 1). Herewith in case of irradiation from base side reflection is enhanced, so here there is no anapole state. Let us consider multipole decomposition in both cases (see Fig. 2). Actually, in case of irradiation from top size electric dipole radiation is suppressed due to interference with toroidal one. In contradistinction, in case of base side irradiation, there is a maximum of electric dipole irradiation

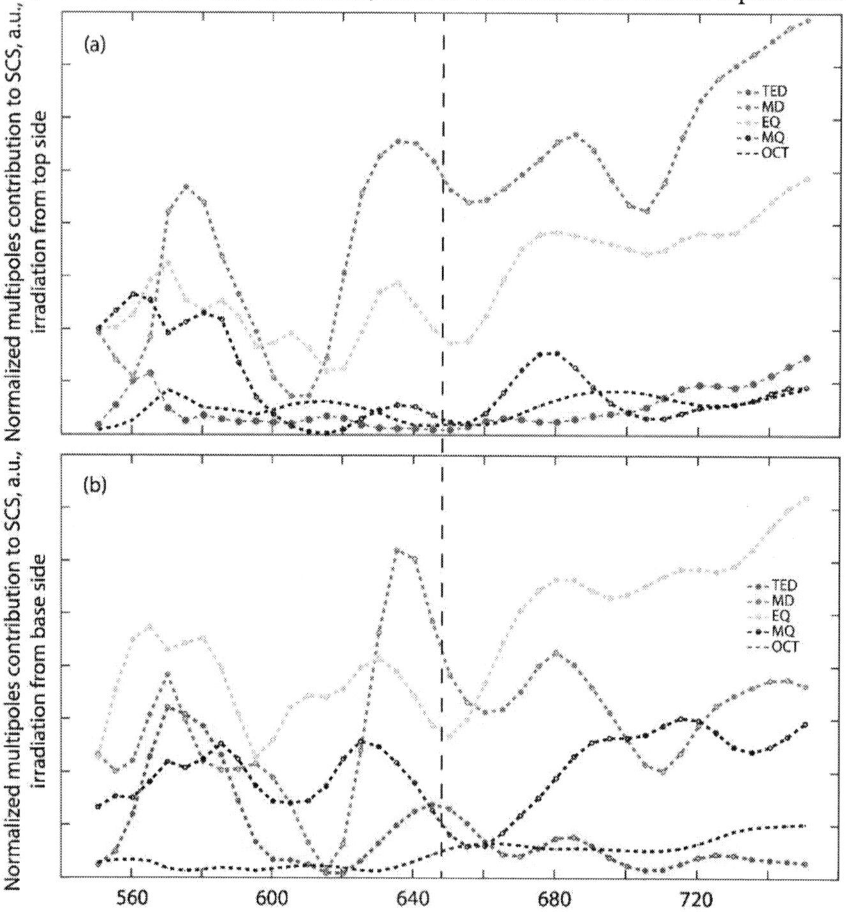

Figure 2. Multipole decomposition of scattering cross section of pyramidal metaatom. It is clearly shown that while in the case of irradiation from top side anapole state correspond to the minimum of electric dipole moment (a), in case of irradiation from the base side there is a maximum at the same frequency (b). Vertical dashed line shows anapole state frequency for both graphs.

4. Conclusion

In conclusion, highly absorptive metasurfaces working on anapole states can have asymmetrical optical response while the shape of nanoparticles is asymmetrical. However, this asymmetry is not associated with non-reciprocal effects, but with different multipoles excitation and different absorption inside each metaatom. It can be used for the design of asymmetrical perfect absorbers, beam splitters and so on.

Acknowledgements
Work has been supported by the Russian Science Foundation Grant No. 17-72-10230.

References

[1] Baryshnikova K, Evlyukhin A B, Ginzburg P and Shalin A S Magnetic hot- spots in hollow silicon cylinders *Saint Petersbg. OPEN, B. Abstr. – St. Petersburg, 2016.* 358

[2] Cheng J, Ansari-Oghol-Beig D and Mosallaei H 2014 Wave manipulation with designer dielectric metasurfaces *Opt. Lett.* **39** 6285–8

[3] Markovich D, Baryshnikova K, Shalin A, Samusev A, Krasnok A, Belov P and Ginzburg P 2016 Enhancement of artificial magnetism via resonant bianisotropy *Sci. Rep.* **6** 22546

[4] Kivshar Y, Miroshnichenko A, Kuznetsov A and Lukyanchuk B 2014 Magnetic Light and All-Dielectric Nanophotonics **35** 2014

[5] Miroshnichenko A E, Evlyukhin A B, Yu Y F, Bakker R M, Chipouline A, Kuznetsov A I, Luk'yanchuk B, Chichkov B N and Kivshar Y S 2015 Nonradiating anapole modes in dielectric nanoparticles *Nat. Commun.* **6** 9069

[6] Terekhov P D, Baryshnikova K V., Artemyev Y A, Karabchevsky A, Shalin A S and Evlyukhin A B 2017 Multipolar response of nonspherical silicon nanoparticles in the visible and near-infrared spectral ranges *Phys. Rev. B* **96** 035443

[7] Evlyukhin A B, Reinhardt C and Chichkov B N 2011 Multipole light scattering by nonspherical nanoparticles in the discrete dipole approximation *Phys. Rev. B - Condens. Matter Mater. Phys.* **84** 1–8

[8] Alaee R, Rockstuhl C and Fernandez-Corbaton I 2018 An electromagnetic multipole expansion beyond the long-wavelength approximation *Opt. Commun.* **407** 17–21

[9] Hu C, Kattawar G W, Parkin M E and Herb P 1987 Symmetry theorems on the forward and backward scattering Mueller matrices for light scattering from a nonspherical dielectric scatterer *Appl. Opt.* **26** 4159–73

[10] Saxon D S 1955 Tensor scattering matrix for the electromagnetic field *Phys. Rev.* **100** 1771–5

[11] Alaee R, Albooyeh M, Rahimzadegan A, Mirmoosa M S, Kivshar Y S and Rockstuhl C 2015 All-dielectric reciprocal bianisotropic nanoparticles *Phys. Rev. B* **92** 245130

[12] Grinblat G, Li Y, Nielsen M P, Oulton R F and Maier S a. 2016 Enhanced third harmonic generation in single germanium nanodisks excited at the anapole mode *Nano Lett.* **16** 4635–40

Analytical solutions for multi-wave transfer matrices in layered structures

Yu N Belyayev

Department of Mathematical Modelling and Cybernetics, Syktyvkar State University, 55 Oktyabrskii pr., Syktyvkar, 167001, Russia

E-mail: ybelyayev@mail.ru

Abstract. The matrix approach to the calculation of n-beam diffraction in a periodic structure is discussed. The possibility of representing N-layered transfer matrix T^N through the principal minors of the matrix T is investigated. The method of computing integer powers of a matrix by means of symmetric polynomials of the latter is developed. New analytic formulas for integer powers of matrices are represented. These solutions are demonstrated using examples of matrices of the sixth order.

1. Introduction

The physical properties of many natural and artificial objects are well described within the framework of one-dimensional periodic models [1, 2]. Examples of such structures that have found wide application are optical thin films [3], liquid crystals [4], superlattices [5], photonic crystals [6]. Matrix approaches have become one of the main ones in the calculations of wave propagations in such media. The problem of calculating the wave transmission through N-layered structure is reduced to finding so-called transfer matrix T (or characteristic matrix [7], or scattering matrix [8], or propagator matrix [9]) of one separate layer and matrix T^N for the whole structure. In this regard, of great interest are analytical solutions for integer powers of matrices. The Baker-Vandermonde formula [10] represents the solution of this problem in terms of matrix eigenvalues. A remarkable result for unimodular second-order matrices was obtained by Abelés [11] in the study of interference optical films. In his paper the elements of the characteristic matrix M^N have been expressed through the trace of the matrix M using the Chebyshev polynomials of the second kind. Similar formulas for the integer powers of second order matrices of general form and certain matrices of the third and fourth orders were found in [12]. In this paper, we present new analytical solutions for integer powers of nth order matrices. Their proofs are based on the properties of symmetric polynomials of nth order [13].

2. Polynomial approach to the calculation of integer powers of matrices

According to one of the corollaries of the Cayley-Hamilton theorem [14], integer power j of any matrix A can be expressed as a linear combination of the first n powers $A^0 \equiv I, A, \ldots, A^{n-1}$:

$$A^j = \sum_{h=0}^{n-1} c_{jh} A^h, \quad j \geq 0. \tag{1}$$

We use the following representation of the coordinates $\mathcal{C}_{jh}$ [13]:

$$\mathcal{C}_{j\,h} = \sum_{g=0}^{h} p_{n-h+g}\mathscr{B}_{j-1-g}(n)\,,\ h = 0, 1, \ldots, n-1,\ j = 0, 1, 2, \ldots, \tag{2}$$

where p_m, $m = 1, \ldots, n$, are coefficients of characteristic equation $\lambda^n = \sum_{i=1}^{n} p_i\lambda^{n-i}$ of matrix $A = \|a_{gh}\|$ and functions $\mathscr{B}_g(n)$ are defined by recurrence relations

$$\mathscr{B}_g(n) = 0,\ \ 0 \le g \le n-2;\ \ \ \mathscr{B}_{n-1}(n) = 1;\ \ \ \mathscr{B}_g(n) = \sum_{j=1}^{n} p_j\mathscr{B}_{g-j}(n),\ \ g \ge n. \tag{3}$$

Two representations of the coefficients p_m, namely $p_1 = \sum_{g=1}^{n}\lambda_g$, $p_2 = -\sum_{l\neq j}\lambda_l\lambda_j, \ldots,$ $p_n = (-1)^{n-1}\prod_{j=1}^{n}\lambda_j$ and $p_1 = \sum_{g=1}^{n} a_{gg}$, $p_2 = -\sum_{j>i}\begin{vmatrix} a_{ii} & a_{ij} \\ a_{ji} & a_{jj}\end{vmatrix}, \ldots, p_n = (-1)^{n-1}\det A$ give respectively two interpretation of the functions $\mathscr{B}_g(n)$: *symmetric polynomials of nth order* [13] (symmetric with respect to the eigenvalues λ_j, $j = 1, \ldots, n$) and *polynomials of principal minors* of the matrix A.

2.1. Expression of A^N in terms of symmetric polynomials

Theorem 1. *Symmetric polynomials of nth order are expressed in terms of the eigenvalues λ_i, $i = 1, 2, \ldots, n$ of the matrix by the formulas*

$$\mathscr{B}_g(n) = \frac{\Delta_g}{\Delta_V},\quad \Delta_V = \begin{vmatrix} 1 & \lambda_1 & \ldots & \lambda_1^{n-2} & \lambda_1^{n-1} \\ 1 & \lambda_2 & \ldots & \lambda_2^{n-2} & \lambda_2^{n-1} \\ \vdots & \vdots & \vdots & \vdots & \vdots \\ 1 & \lambda_n & \ldots & \lambda_n^{n-2} & \lambda_n^{n-1} \end{vmatrix},\quad \Delta_g = \begin{vmatrix} 1 & \lambda_1 & \ldots & \lambda_1^{n-2} & \lambda_1^{g} \\ 1 & \lambda_2 & \ldots & \lambda_2^{n-2} & \lambda_2^{g} \\ \vdots & \vdots & \vdots & \vdots & \vdots \\ 1 & \lambda_n & \ldots & \lambda_n^{n-2} & \lambda_n^{g} \end{vmatrix}, \tag{4}$$

where g is any integer, if all values $\lambda_i \neq 0$, and $g \ge 0$, if some eigenvalue $\lambda_i = 0$.

Proof. The properties of the determinants make it obvious that the formulas (4) correspond to the definition (3) of polynomials $\mathscr{B}_g(n)$ for $g = 0, 1, \ldots, n-1$.

Suppose that the equality (4) holds for $g = j-n+1, \ldots, j-1, j$, where $j \ge n-1$. We show that then it will also true (i.e. functions (4) satisfy the definition (3)) for $g = j+1$. Indeed, according to characteristic equation, $\lambda^{j+1} = p_1\lambda^j + p_2\lambda^{j-1} + \ldots + p_n\lambda^{j-n+1}$. Therefore

$$\mathscr{B}_{j+1}(n) = \frac{1}{\Delta_V}\begin{vmatrix} 1 & \lambda_1 & \ldots & \lambda_1^{n-3} & \lambda_1^{n-2} & p_1\lambda_1^j + p_2\lambda_1^{j-1} + \ldots + p_n\lambda_1^{j-n+1} \\ 1 & \lambda_2 & \ldots & \lambda_2^{n-3} & \lambda_2^{n-2} & p_1\lambda_2^j + p_2\lambda_2^{j-1} + \ldots + p_n\lambda_2^{j-n+1} \\ \vdots & \vdots & \vdots & \vdots & \vdots & \vdots \\ 1 & \lambda_n & \ldots & \lambda_n^{n-3} & \lambda_n^{n-2} & p_1\lambda_n^j + p_2\lambda_n^{j-1} + \ldots + p_n\lambda_n^{j-n+1} \end{vmatrix}$$
$$= \frac{1}{\Delta_V}\sum_{h=1}^{n} p_h\Delta_{j+1-h} = \sum_{h=1}^{n} p_h\mathscr{B}_{j+1-h}(n),$$

which agrees with the recurrence formulas (3).

Note that substitutions of the expressions (4) into the formulas (2) and (1) lead to one more proof of the Baker-Vandermonde formula.

A more explicit form of symmetric polynomials $\mathscr{B}_g(n)$ follows from (4) if we use the value of the Vandermonde determinant $\Delta_V = \prod_{j>i}(\lambda_j - \lambda_i)$. For example, suppose that the characteristic equation of a sixth-order matrix has the form: $\lambda^6 - p_2\lambda^4 - p_4\lambda^2 - p_6 = 0$. In this case, the eigenvalues are pairwise equal to each other with a precision up to the sign, say

$\lambda_1 = -\lambda_4$, $\lambda_2 = -\lambda_5$, $\lambda_3 = -\lambda_6$. These eigenvalues are expressed in the radicals in terms of the coefficients of the characteristic equation. In turn, the coefficients of the characteristic equation p_2, p_4, p_6 are expressed in terms of the eigenvalues λ_1, λ_2, λ_3 by the formulas: $p_2 = \lambda_1^2+\lambda_2^2+\lambda_3^2$, $p_4 = \lambda_1^2\lambda_2^2 + \lambda_1^2\lambda_3^2 + \lambda_2^2\lambda_3^2$, $p_6 = \lambda_1^2\lambda_2^2\lambda_3^2$. In this case, the symmetric polynomials of sixth order are expressed through the matrix eigenvalues as:

$$\mathscr{B}_j(6) = \frac{1+(-1)^{j-1}}{2}\frac{\Lambda_{j-1}}{\Lambda_4}, \quad \Lambda_i = \lambda_1^i(\lambda_3^2-\lambda_2^2) - \lambda_2^i(\lambda_3^2-\lambda_1^2) + \lambda_3^i(\lambda_2^2-\lambda_1^2). \tag{5}$$

In accordance with (1)

$$A^j = \begin{cases} p_6\mathscr{B}_{j-2}(6)A + [p_4\mathscr{B}_{j-2}(6) + p_6\mathscr{B}_{j-4}(6)]A^3 + \mathscr{B}_j(6)A^5, & \text{if } j = 7, 9, 11, \ldots, \\ p_6\mathscr{B}_{j-1}(6)I + [p_4\mathscr{B}_{j-1}(6) + p_6\mathscr{B}_{j-3}(6)]A^2 + \mathscr{B}_{j+1}(6)A^4, & \text{if } j = 6, 8, 10, \ldots, \end{cases} \tag{6}$$

where the polynomials $\mathscr{B}_g(6)$ are defined by formulas (5) and I is identity matrix.

2.2. *Representation of integer powers of a matrix by means of the coefficients of its characteristic equation*

Theorem 2. *If only two of the coefficients of the characteristic equation p_h and p_l, $l = 2h$, are nonzero, then the polynomials $\mathscr{B}_g(n)$ are expressed by the formulas:*

$$\mathscr{B}_g(n) = \begin{cases} \sqrt{(-p_l)^{j-1}}U_{j-1}(b), & \text{if } g = h(j-1)+n-1, \\ 0, & \text{if } g \neq h(j-1)+n-1, \end{cases} \quad j = 1, 2, \ldots, \tag{7}$$

where

$$U_j(b) = \frac{\sin[(j+1)\arccos b]}{\sqrt{1-b^2}}, \quad b = \frac{p_h}{2\sqrt{-p_l}}. \tag{8}$$

Proof. It's obvious that $h(j-1)+n-1 \geq n-1$. Therefore, it follows from (7) that $\mathscr{B}_g(n) = 0$, if $g \leq n-2$, which agrees with (3). If $g = n-1$, then, by formulas (7) and (8), we find $\mathscr{B}_{n-1}(n) = U_0(b) = 1$. This also satisfies definition (3). Finally, if $g \geq n$, then, according to the definition (3) and the condition of the theorem, the function (7) must be a solution of equation

$$\mathscr{B}_g(n) = p_h\mathscr{B}_{g-h}(n) + p_l\mathscr{B}_{g-l}(n), \quad g \geq n. \tag{9}$$

It's easy to check. Let $g \neq h(j-1)+n-1$. Then $g-h$ and $g-2h$ not equal to $h(j-1)+n-1$. In this case formulas (7) give values $\mathscr{B}_g(n) = \mathscr{B}_{g-h}(n) = \mathscr{B}_{g-l}(n) = 0$, which satisfy equation (9). At last, if $g = h(j-1)+n-1$, then substitution of (7) into (9) gives equality

$$U_{j-1}(b) = 2bU_{j-2}(b) - U_{j-3}(b), \tag{10}$$

whose identity is established by elementary transformations using the functions (8). Thus, the functions (7) are solutions of equations (9), which was to be proved.

If $-1 \leq b \leq 1$, the functions $U_j(b)$ with indexes $j \geq 0$ are orthogonal Chebyshev polynomials of the second kind.

For instance for sixth order matrix A with $p_1 = p_2 = 0$, $p_3 \neq 0$, $p_4 = p_5 = 0$, $p_6 \neq 0$, we find from (1), (2) and (7)

$$A^{6+i+3k} = p_6\mathscr{B}_{5+3k}(6)A^i + \mathscr{B}_{8+3k}(6)A^{i+3}, \quad i = 0, 1, 2, \quad k = 0, 1, \ldots \tag{11}$$

where $\mathscr{B}_g(6) = (\det A)^{\frac{g-5}{6}}\sin[((g-2)\arccos b)/3]/\sqrt{1-b^2}$, $b = p_3/(2\sqrt{\det A})$ for $g = 5, 8, 11, \ldots$ and $\mathscr{B}_g(6) = 0$ for $g \neq 5, 8, 11, \ldots$

Example. Transfer matrix T of elastic waves in a crystalline layer of thickness d is expressed by the formula $T = \exp(Wd)$, where sixth-order matrix $W = \|w_{ij}\|$ is determined [15] from the system of equations of motion and Hooke's law. For many geometries of elastic waves scattering by a cubic crystal, the symmetric polynomials $\mathscr{B}_g(6)$ of the matrix W and matrices W^j satisfy relations (5) and (6) respectively. For these cases, the analytical solution for the transfer matrix has the form

$$T = \exp(Wd) = \sum_{j=1}^{\infty} \frac{(Wd)^j}{j!} = \sum_{l=0}^{5} W^l \sum_{g=1}^{3} \frac{\tau_{gl}}{\Lambda_4} \times \begin{cases} \cosh(\lambda_g d), \text{ if } l = 0, 2, 4, \\ \sinh(\lambda_g d), \text{ if } l = 1, 3, 5, \end{cases} \tag{12}$$

where $\tau_{10} = \lambda_2^2\lambda_3^2(\lambda_3^2-\lambda_2^2)$, $\tau_{20} = \lambda_1^2\lambda_3^2(\lambda_1^2-\lambda_3^2)$, $\tau_{30} = \lambda_1^2\lambda_2^2(\lambda_2^2-\lambda_1^2)$, $\tau_{g1} = \tau_{g0}/\lambda_g$, $\tau_{12} = \lambda_2^4-\lambda_3^4$, $\tau_{22} = \lambda_3^4 - \lambda_1^4$, $\tau_{32} = \lambda_1^4 - \lambda_2^4$, $\tau_{g3} = \tau_{g2}/\lambda_g$, $\tau_{14} = \lambda_3^2 - \lambda_2^2$, $\tau_{24} = \lambda_1^2 - \lambda_3^2$, $\tau_{34} = \lambda_2^2 - \lambda_1^2$, $\tau_{g5} = \tau_{g4}/\lambda_g$. In the case when the matrix W has the form (11), an analytic solution for the elastic wave transfer matrix can be found in a similar way.

For sufficiently small thickness d the series $\sum_{i=1}^{\infty}(Wd)^j/(j!)$ is well approximated by the sum of the first two or three terms, i.e. $T = I + Wd + W^2d^2/2$. Such definition of the matrix T, if $\max|w_{ij}d| \ll 1$, provides calculation of matrix T elements with a relative error of less than $6(\max|w_{ij}d|)^3$. This allows us to consider a homogeneous layer of thickness Nd as N-layered periodic structure and calculate its transfer matrix T^N in accordance with the formulas (1), (2), (3). It is easy to show that in this case the coefficients p_m and the polynomials of principal minors up to terms of the second order of smallness in $\max|w_{ij}d|$ are expressed by the formulas $p_m = (-1)^{m-1}6!(1+mx)/[m!(6-m)!]$, $m = 1, \ldots, 6$; $\mathscr{B}_g(6) = c_g[1 + (g-5)x]$, $g \geq 5$, where $x = \sum_{i=1}^{6} w_{ii}d/6$ and the coefficients c_g are determined from the recurrence relations: $c_g = 0$, $g \leq 4$; $c_5 = 1$; $c_g = \sum_{j=1}^{6} c_{g-j}6!/[j!(6-j)!]$, $g \geq 5$.

3. Conclusion

As the examples of two- and four-beam scattering of electromagnetic waves in layered media show, analytical formulas for transfer matrices facilitate a theoretical study of the features of the corresponding diffraction spectra. Such formulas as (6), (11), (12) play no less important role in the verification of numerical algorithms for calculating matrices T^N with large N, since round-off errors during matrix calculations depend strongly on the number of matrix multiplications.

References

[1] Elachi C 1976 Waves in active and passive periodic structures *Proc. IEEE* **64** 1666–1698
[2] Karpov S Yu, Stolyarov S N 1993 Propagation and transformation of electromagnetic waves in one-dimensional periodic structures *Physics-Uspekhi* **36** 1–22
[3] Knittl Z 1975 *Optics of thin films* (London: J. Wiley)
[4] Belyakov V A 1992 *Diffraction Optics of Complex-Structured Periodic Media* (New York: Springer)
[5] Herman M 1986 *Semiconductor Superlattices* (Berlin: Akademie)
[6] Sakoda K 2001 *Optical Properties of Photonic Crystals* (Berlin: Springer)
[7] Born M, Wolf E 1964 *Principles of optics* (Oxford: Pergamon press)
[8] Cowley J M 1975 *Diffraction Physics* (New York: N.-H.Publ. Comp.)
[9] Brekhovskikh L M, Godin O A 1990 *Acoustics of layered media* (Berlin: Springer-Verlag)
[10] Baker H F 1903 On the integration of linear differential equation *Proc. London Math. Soc.* **35** 333–378
[11] Abelés F 1950 Recherehes sur la propagation des ondes electromagnetiques sinusoidals dans les milieux stratifiés. Application aux conches minces *Ann. Phys.* **5** 596–640 706–782
[12] Belyayev Yu N 2015 Characteristic matrices of layered periodic structures *Proc. Int. Conf. Days on Diffraction 2015 May 25–29 St. Petersburg Russia* 42–48
[13] Belyayev Yu N 2013 On the calculation of functions of matrices *Mathematical Notes* **94** 177–184
[14] Gantmacher F R 1959 *The Theory of Matrices* vol 1 (New York: Chelsea)
[15] Belyayev Yu N 2018 Calculation of the six-beam diffraction in layered media using polynomials of principal minors *Journal of theoretical and computational acoustics* **26** 1850017-1—1850017-21

Optical properties of spatially dispersive Mie-resonant halide perovskite nanoparticles

A S Berestennikov[1], I V Iorsh[1] and S V Makarov[1]

[1] Department of Nanophotonics and Metamaterials, ITMO University, St. Petersburg 199034, Russia

E-mail: a.berestennikov@metalab.ifmo.ru

Abstract. We have proposed a mechanism explaining the unusual blue shift of photoluminescence peak in relatively large perovskite nanoparticles, where the quantum confinement effect is insignificant. We performed theoretical analysis of the absorption and scattering cross-sections of incident light in such nanoparticles in classical (local) and nonlocal cases. The blue shift for the scattering and absorption features in spectra around the exciton peak was obtained from the model, which can be related to that the previously observed shift of the luminescence peak in perovskite nanoparticles with a diameter less than 100 nm owing to the exciton nonlocality.

1. Introduction

Organic-inorganic perovskites, such as $CH_3NH_3PbX_3$ family, where X is I, Br or Cl, are very promising class of material for effective nanoscale light sources, solar cells, light emitters and photodetectors, which have excitonic states at room temperature, high enough refractive index, low losses at the exciton wavelength [1]. These excellent properties along with low cost of their fabrication by wet-chemistry allowed for creation of hybrid perovskite nanoantennas [2] and metasurfaces [3].

The materials based on perovskite nanoparticles have a favourable combination of quantum-size effects (if the radius of the particle is less than or of the order of the Bohr radius of exciton), improving their optical properties in comparison with bulk analogues, Mie resonances and the ability to disperse in a variety of solvents and matrices, which allows them to be introduced into various devices. In order to create effective and tunable light-emitting nanoantennas, it is necessary to be able to change the optical properties of nanoparticles, such as the wavelength of the luminescence, the absorption/scattering cross-sections and the radiation direction [4].

In recent works [5, 6], it was found that in nanocrystals of perovskites with a radius less than 100 nm, the photoluminescence peak is a blue-shifted. This effect is known for nanoparticles of several nanometers and associated with quantum-size effects. However, this effect has not yet been explained for the case of large perovskite nanoparticles.

In this work, we show theoretically that exciton-induced nonlocality can affect resonant properties of $CH_3NH_3PbI_3$ perovskite nanoparticles with relatively big sizes and shift the photoluminescence, absorption and scattering cross-section peaks to shorter wavelengths.

2. Model of spatially dispersive dielectric spheres

The Mie theory explains the optical properties of sphere of arbitrary diameters, consisting of an isotropic material, which is characterized by a local dielectric constant. It depends only on the frequency of the incident wave:

$$\varepsilon(\omega, \mathbf{k}) = \varepsilon_0 + \frac{\omega_p^2}{\omega_{exc}^2 - \omega^2 - i\gamma\omega}, \tag{1}$$

where ε_0 is background dielectric constant, ω_p is the dipole oscillator strength, γ is the damping term and ω_{exc} is the excitonic transition frequency.

However, in the general case the dielectric function of the medium ε depends on the wave vector, or in other words has a nonlocal response. Therefore, metallic and dielectric spheres can have interesting optical effects, which cannot be explained by the classical local model. In this case, the observed effects can be explained by using a modified Mie theory, in which spatial dispersion is included by adding a nonlocal term to the equation of the classical local dielectric function [7]:

$$\varepsilon(\omega, \mathbf{k}) = \varepsilon_0 + \frac{\omega_p^2}{\omega_{exc}^2 - \omega^2 + Dk^2 - i\gamma\omega}, \tag{2}$$

where $Dk^2 = \hbar^2 k^2 \omega_{exc}/2m_{exc}$ is the spatial dispersion term.

By using classic Maxwell boundary conditions, which require the tangential components of E and H be continuous at r = R and Pekar additional boundary conditions, which impose a condition of zero polarization at the surface of the particle P = 0, we can determine the expansion coefficients of the scattered wave and obtain the scattering, extinction and absorption cross-sections of spatially dispersive dielectric nanoparticles

$$Q_{scat} = \frac{2}{k_0^2 r^2} \sum_{n=1}^{\infty} (2n+1)(|a_n|^2 + |b_n|^2), \tag{3}$$

$$Q_{abs} = \frac{2}{k_0^2 r^2} \sum_{n=1}^{\infty} (2n+1)\mathrm{Re}(a_n + b_n) - Q_{scat}, \tag{4}$$

where a_n and b_n are expansion coefficients, r is the radius of the particle, $k_0 = \omega/c\sqrt{\varepsilon_M}$ and ε_M is the dielectric permittivity of surrounding medium.

3. Results and discussions

To explain the blue shift of the photoluminescence we applied the generalized Mie theory of spatially dispersive spheres to the calculation of the optical properties of the$CH_3NH_3PbI_3$ (or $MAPbI_3$) nanoparticles. In figure 1 the plots of local and nonlocal absorption and scattering cross-sections of the $MAPbI_3$ nanoparticles with different radii are calculated. In the classical (local) case with dielectric function from equation (1), the blue shift of the absorption (figure 1a) and scattering (figure 1(c,d) black line) peaks are not observed. However, we found that in nonlocal case with $\varepsilon(\omega, k)$ from equation (2), the main scattering (figure 1c, red line) and absorption (figure 1b) peaks shifts to lower wavelengths by 45 nm with decreasing of the nanoparticle radius. This shift is due to the fact that the exciton is not stationary and has the kinetic energy, so it emits photons at a higher frequency.

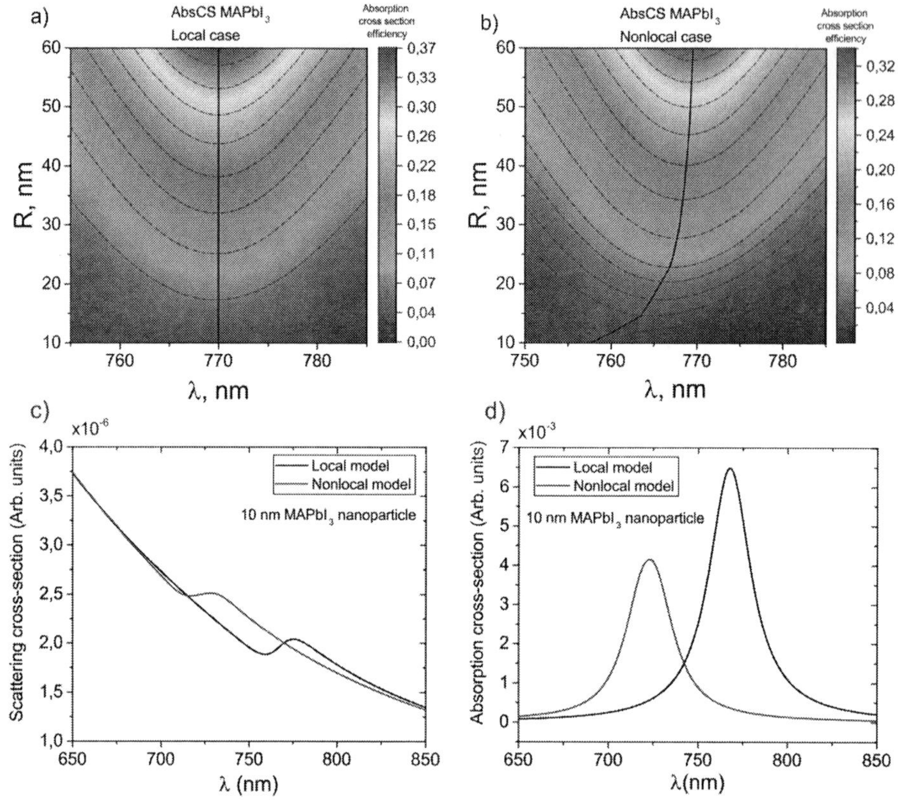

Figure 1. Calculated maps of absorption cross sections of $MAPbI_3$: (a) local and (b) nonlocal models; scattering (c) and absorption (d) cross-sections of 10 nm $MAPbI_3$ nanoparticle .

4. Conclusions

We proposed that a mechanism where the blueshift of the luminescence peak in perovskite nanoparticles is associated with nonlocality of material around an exciton. We performed analytical calculations of the absorption and scattering cross-sections of incident light in such nanoparticles in the local and nonlocal cases. Using this results, we plotted the spectral and radii dependences of the absorption and scattering cross-sections in both cases. Our investigations showed that the possible mechanism of the photoluminescence peak blue shift in perovskite nanoparticles is associated with the nonlocality of excitons.

5. Acknowledgements

This work was supported by Russian Science Foundation (grant 17-73-20336).

References

[1] Huang H, Polavarapu L, Sichert J A, Susha A S, Urban A S and Rogach A L 2016 *NPG Asia Materials* **8** e328
[2] Tiguntseva E Y, Zograf G P, Komissarenko F E, Zuev D A, Zakhidov A A, Makarov S V and Kivshar Y S 2018 *Nano Lett.* **18** 1185–90
[3] Makarov S V, Milichko V, Ushakova E V, Omelyanovich M, Cerdan P A, Haroldson R, Balachandran B, Wang H, Hu W, Kivshar Y S and Zakhidov A A 2017 *ACS Photonics* **4** 728–735
[4] Kovalenko M V, Protesescu L and Bodnarchuk M I 2017 *Science* **358** 745–750
[5] Di D, Musselman K P, Li G, Sadhanala A, Ievskaya Y, Song Q, Tan Zhi-Kuang, Lai M L, MacManus-Driscoll J L, Greenham N C and Friend R H 2015 *J. Phys. Chem. Lett.* **6** 446–450

[6] DInnocenzo V, Srimath Kandada A R, De Bastiani M, Gandini M and Petrozza A 2014 *J. Am. Chem. Soc.* **136** 17730–33
[7] Ruppin R 1981 *J. Opt. Soc. Am.* **71** 755–758

IOP Conf. Series: Journal of Physics: Conf. Series **1092** (2018) 012010 doi:10.1088/1742-6596/1092/1/012010

Integrated W-type structure for spectral and spatial filtering of optical radiation propagating in a slab waveguide

E A Bezus[1,2], L L Doskolovich[1,2] and D A Bykov[1,2]

[1] Image Processing Systems Institute of RAS — Branch of the FSRC "Crystallography and Photonics" RAS, 151 Molodogvardeyskaya st., Samara 443001, Russia

[2] Samara National Research University, 34 Moskovskoe shosse, Samara 443086, Russia

evgeni.bezus@gmail.com

Abstract. We present a simple and compact integrated narrowband filter that enables spectral or spatial filtering of optical radiation propagating in a slab waveguide. The filter consists of two grooves on the waveguide surface and constitutes an integrated analogue of the so-called W-type structure. The filtering is performed in transmission at oblique incidence due to the excitation of a leaky eigenmode of the structure localized at a ridge separating the grooves. For the considered parameters, neither "parasitic" out-of-plane scattering nor polarization conversion occurs, and strictly zero reflectance and unity transmittance are achieved at resonant conditions. The quality factor and location of the resonance can be controlled by changing the widths of the grooves and of the ridge between them, respectively. The proposed filter may find applications in novel devices of integrated nanophotonics.

1. Introduction

Spectral and spatial filters that selectively transmit or reflect incident radiation are indispensable in various optical devices including, among others, spectrum analyzers, wavelength multiplexers and demultiplexers, and remote sensing systems. In many integrated optoelectronic systems, the filtering is performed in a slab waveguide [1–4]. Such geometry is suitable for the creation of fully integrated optical devices. Waveguide-integrated Bragg gratings (BGs) and phase-shifted Bragg gratings (PSBGs) are widely used for spectral filtering of optical radiation propagating in slab waveguides [1]. Recently, novel structures that can be applied in integrated spectrometers, the so-called digital planar holograms (DPHs), were proposed [3]. DPHs comprise a sophisticated superposition of narrowband low-contrast curvilinear BG-based reflectors fabricated on the surface of the guiding layer.

In this work, we propose a planar spectral filter operating in transmission and consisting of two grooves on the surface of a slab waveguide. In contrast to the BG- or PSBG-based filters and DPHs, our filter has a much simpler geometry and is substantially more compact.

2. Geometry of the structure

The proposed filter shown in figure 1(a) is a planar version of dielectric three-layer W-type resonant structure, which was studied in [5] for achieving large optical field enhancement over a large volume.

The effective refractive index profile of the structure is shown in figure 1(b). The structure consists of two grooves (length h_{cl}) on the surface of a single-mode dielectric ridge waveguide (dielectric permittivity ε_f, thickness $w_{inc} > w_g$) and a ridge (length h_{wg}) separating them. Theoretical analysis performed by the present authors [6, 7] shows that, similarly to the conventional W-type structure [5], the investigated planar structure supports a leaky eigenmode localized at the central ridge, which in this case acts as a resonant cavity. The excitation of this mode leads to the appearance of resonant transmittance peaks and reflectance dips, which suggests that this effect can be used for spectral or spatial filtering of the incident radiation propagating in the waveguide.

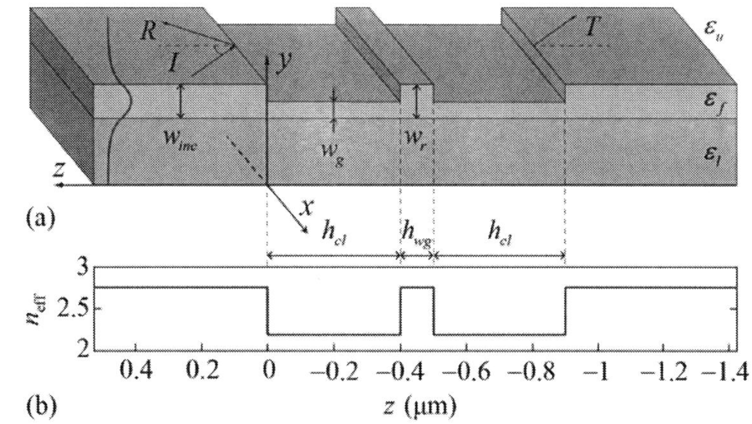

Figure 1. Geometry of the proposed two-groove planar filter (a) and the effective refractive index profile of the structure (b). The dimensions in (a) and the values in (b) correspond to one of the examples considered below.

3. Results and discussion

Let us now numerically investigate the performance of the proposed integrated W-type filter. In the simulations, the following parameters were used: free-space wavelength $\lambda_0 = 630$ nm, TE-polarization, $\varepsilon_f = 3.32^2$ (GaP), $\varepsilon_l = 1.45^2$ (fused silica), $\varepsilon_u = 1$, $w_{inc} = 100$ nm. At these parameters, the considered slab waveguide is single-mode. The filter consists of two grooves with depth $w_{inc}/2$ (so that $w_g = w_{inc}/2 = 50$ nm) and a ridge between them. In the case of TE-polarization, the effective refractive index of the incident guided mode and the mode in the ridge region is $n_{eff}(w_{inc}) = n_{eff}(w_r) = 2.7563$, while the effective refractive index in the groove region amounts to $n_{eff}(w_g) = 2.19$.

As mentioned above, the resonances occurring in the proposed filter are associated with the excitation of a leaky eigenmode localized at the central ridge of the structure and propagating along the x axis. This eigenmode is close to a mode of a photonic rib waveguide. The quality factor of the resonance (the width of the resonant peak) is determined by the width of the grooves h_{cl}. To illustrate this, we calculated the angular and wavelength transmission spectra of the investigated structure at two groove widths $h_{cl} = 400$ nm and $h_{cl} = 600$ nm (figure 2). At these groove widths, the angular and spectral widths of the resonant peaks differ by approximately 10 times. The width of the ridge is fixed and equals $h_{wg} = 100$ nm. The spectra were calculated using an in-house implementation of the aperiodic rigorous coupled-wave analysis (aRCWA) technique [8]. RCWA, also called the Fourier modal method, is an established numerical technique for solving Maxwell's equations. The angular spectra in figure 2(a) were calculated at a fixed wavelength $\lambda_0 = 630$ nm. The transmittance reaches unity at the angles of incidence $\theta_1 = 55.201°$ (at $h_{cl} = 400$ nm) and $\theta_2 = 55.226°$ (at $h_{cl} = 600$ nm). The full width of the resonant transmittance peaks at half maximum (FWHM) amounts to $\Delta_{\theta,1} = 0.091°$ and $\Delta_{\theta,2} = 0.0092°$ at $h_{cl} = 400$ nm and $h_{cl} = 600$ nm, respectively. The dependence of the FWHM of the resonant peak on the width h_{cl} is shown inset. Let us note that the angles of incidence $\theta_{1,2}$ depend on the thickness of the waveguide in the groove regions w_g. With a decrease in w_g (decrease in $n_{eff}(w_g)$), the angles $\theta_{1,2}$ also decrease, and at $w_g = 0$ ($n_{eff}(0) = 1$ in the groove regions) they approach 33°.

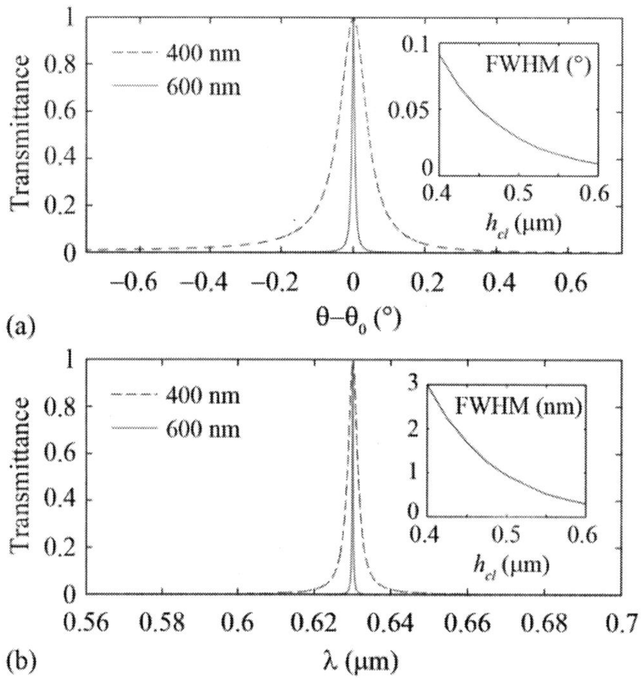

Figure 2. Angular (a) and wavelength (b) transmission spectra of the studied filter at $h_{cl} = 400$ nm (dashed blue curves) and $h_{cl} = 600$ nm (solid red curves). The insets show the FWHM of the resonant peaks vs. h_{cl}.

It is worth noting that in the case of incidence of a TE-polarized mode at angles of incidence greater than $\theta > 46.34°$, there is neither out-of-plane scattering nor polarization conversion in the structure. Indeed, one can show that at $\theta > 46.34°$ the slab waveguide does not support any propagating TM-polarized modes with the propagation constants $k_{TM} > k_{x,TE}$, where $k_{x,TE} = k_0 n_{\text{eff}}(w_{\text{inc}})\sin\theta$, $k_0 = 2\pi/\lambda_0$ is the tangential component of the wave vector of the incident TE-polarized mode. Moreover, the plane waves under and over the structure (in the substrate and superstrate regions with dielectric permittivities ε_l and ε_u) with the x wave vector component $k_{x,TE}$ are evanescent. Thus, the energy of the incident TE-polarized mode will be divided only between two modes (reflected and transmitted) having the same polarization. Due to this, strictly zero reflectance and unity transmittance are achieved in the structure at resonant conditions.

The wavelength spectra in figure 2(b) were calculated at the angles of incidence θ_1 and θ_2 taking into account the dielectric permittivity dispersion of the core layer. FWHM values of the resonant peaks are $\Delta_{\lambda,1} = 3.0$ nm (at $h_{cl} = 400$ nm) and $\Delta_{\lambda,2} = 0.3$ nm (at $h_{cl} = 600$ nm). Similarly to figure 2(a), the inset in figure 2(b) shows the FWHM of the peak vs. the width h_{cl}.

Thus, the presented results confirm that the proposed integrated W-type structure can be utilized as a narrowband spectral or spatial filter. It is worth noting that the proposed structure is substantially more compact than the integrated filters based on BGs or PSBGs. In particular, in comparison with the integrated PSBGs designed for the waveguide configuration described above and providing transmittance peaks with the same spectral FWHM values, the total width of the proposed W-type structures ($2h_{cl} + h_{wg}$) is 2.5–5 times smaller.

4. Conclusion

In this work, we presented a simple and compact transmission filter consisting of two grooves on the surface of a slab waveguide and intended for spectral filtering of optical radiation propagating in the waveguide. The spectral filtering is performed due to resonant excitation of an eigenmode of the structure localized at the cavity corresponding to the ridge between the grooves. The possibility to control the spectral width and location of the resonant peak by changing the widths of the grooves and the ridge between them was demonstrated. Moreover, for the considered parameters, there is no unwanted parasitic scattering and polarization conversion in the filter. The proposed structure can be used for the creation of planar analogues of linear variable filters for waveguide-integrated spectrometers.

Acknowledgments

This work was funded by Russian Science Foundation (RSF) under grant no. 14-19-00796.

References

[1] Haus H A 1984 *Waves and Fields in Optoelectronics* (Hemel Hempstead: Prentice-Hall)
[2] Mossberg T 2001 *Opt. Lett.* **26** 414–6
[3] Calafiore G, Koshelev A, Dhuey S, Goltsov A, Sasorov P, Babin S, Yankov V, Cabrini S and Peroz C 2014 *Light: Sci. & Appl.* **3** e203
[4] Ma X, Li M and He J J 2013 *IEEE Photon. J.* **5** 6600807
[5] Sainidou R, Renger J, Teperik T V, González M U, Quidant R and García de Abajo F J 2010 *Nano Lett.* **10** 4450–5
[6] Doskolovich L L, Bezus E A and Bykov D A 2018 *Photon. Res.* **6** 61–5
[7] Doskolovich L L, Bezus E A, Golovastikov N V, Bykov D A and Soifer V A 2017 *Opt. Express* **25** 22328–40
[8] Silberstein E, Lalanne P, Hugonin J-P and Cao Q 2001 *J. Opt. Soc. Am. A* **18** 2865–75

IOP Conf. Series: Journal of Physics: Conf. Series **1092** (2018) 012011 doi:10.1088/1742-6596/1092/1/012011

Chiral comma-shaped gold nanoparticles: fabrication and refractive index sensitivity

V E Bochenkov

Department of Chemistry, Lomonosov Moscow State University, Leninskie gory 1/3 Moscow 119991, Russia

Email: boch@kinet.chem.msu.ru

Abstract. Chiral gold nanoparticles with asymmetric crescent shape have been fabricated using the In-situ resist colloidal lithography approach. Optical properties of these plasmonic particles have been studied experimentally and using the finite-difference time-domain simulations. Nanocommas exhibit a considerable circular dichroism with high anisotropic g-factor of about 0.2. Numerically studied refractive index sensitivity of the material has been demonstrated.

1. Introduction

Chiral metamaterials have been attracting a lot of attention in the past decade, due to the interesting and promising optical properties, such as the extraordinary high or broadband optical activity, and switchable circular dichroism [1]. A large number of materials with different 2D and 3D particles have been studied so far, indicating the importance of using the accurate fabrication methods to exploit the tunability of their optical properties. Besides, practical application of chiral metamaterials relies upon robustness and cost-efficiency of the fabrication process. In this work we demonstrate the fabrication of comma-shaped 2D chiral gold nanoparticles using the self-assembly based colloidal lithography approach, followed by the results of experimental and numerical investigation of their optical properties.

2. Fabrication

Chiral plasmonic crescent nanostructures have been fabricated using the in-situ resist colloidal lithography (IRCL) approach [2]. In this method, sulfate latex microspheres are electrostatically adsorbed onto a substrate covered by triple polyelectrolyte layer; then a series of glancing-angle deposition (GLAD) steps is needed to produce nanostructures with desired shape. Each microsphere serves as a lithographic mask in further vacuum deposition process. A distinctive feature of IRCL is the deposition of a dielectric resist layer, which patterns the substrate's surface prior to the metal deposition. Particularly, silica has been successfully used as a such resist to define the areas for further deposition of gold.

For the fabrication of comma-shaped particles, 300 nm latex sulfate particles have been used. Then, after 5 min UV/ozone oxidation, 25 nm of silica resist layer has been deposited at a glancing angle of 40°. After that, the substrate has been turned by 90° around the normal axis and 20 nm of gold has been deposited at 35°.

Latex microspheres and the top layer of gold have been removed after taking samples to air by multiple tape-stripping. The remaining gold nanoparticles were located inside the holes in silica film.

3. Optical properties

Optical properties of comma-shaped nanoparticles have been studied using the UV-vis-NIR spectroscopy, circular dichroism spectroscopy and finite-difference time-domain (FDTD) simulations. A considerable circular dichroism has been detected in experiment and simulations at the wavelength of the low-energy resonance mode. The CD spectrum is presented in figure 1. The corresponding value of anisotropic g-factor was about 0.2.

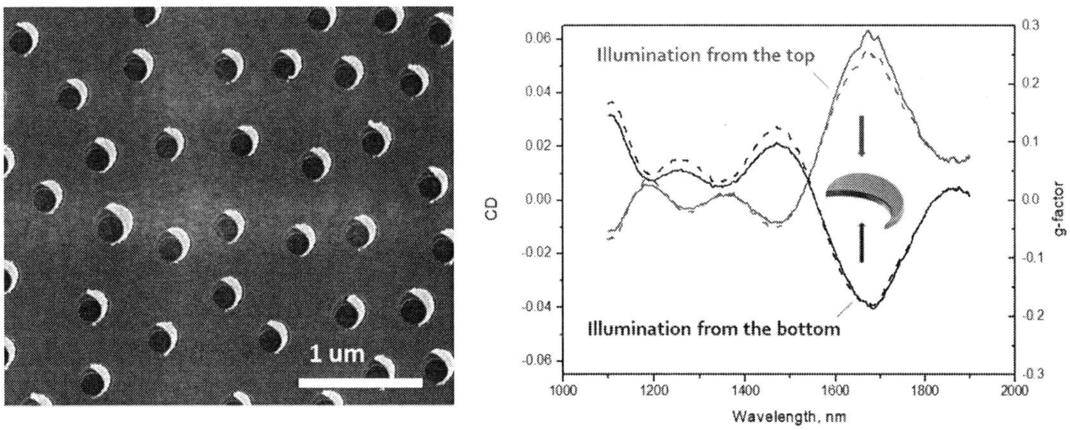

Figure 1. SEM image of comma-shaped gold nanoparticles (left) and their circular dichroism spectra (right) upon illumination from the top (red) and the bottom (black).

Since the calculated local electric field enhancement reaches two orders of magnitude near the sharp tip, these structures can be used for refractive index sensing. Extinction spectra of comma nanoparticle have been calculated at different refractive index of the media (see figure 2). The largest values of the peak shift have been observed for the low-energy mode peak near 1800 nm. The calculated refractive index sensitivity of this peak was about 600 nm/RIU.

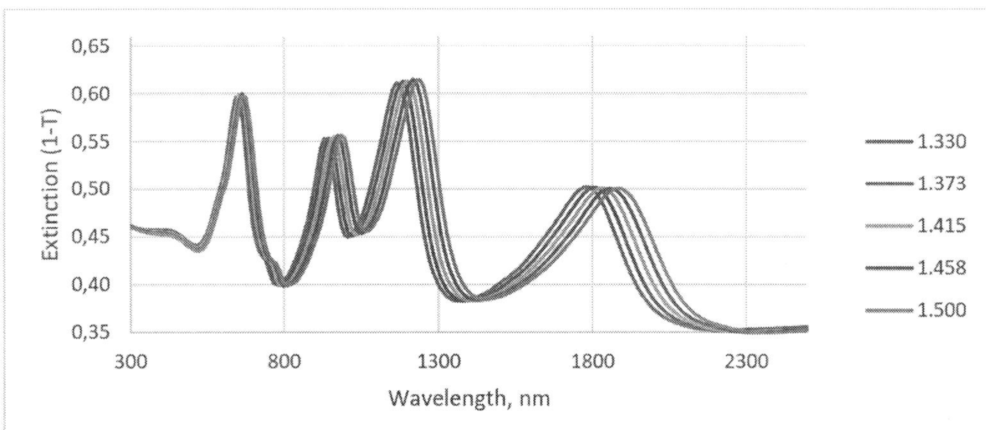

Figure 2. Calculated extinction spectra of gold comma-shaped particle at different refractive index of the media

To prove the capability of such structures to detect small concentrations of biomolecules, the peak shifts induced by the presence of a small 5 nm sphere with refractive index n=1.5 near the comma surface, has been studied. The observed peak shift was 5 nm compared to the peak position without the sphere.

4. Conclusion.

Arrays of comma-shaped chiral nanoparticles have been fabricated using the IRCL technique. The structures exhibit circular dichroism and possess high anisotropy factor g=0.2. The particles are evenly distributed on the surface with short-range order, resembling an amorphous metasurface. Besides, these plasmonic nanostructures can be used for refractive index sensing, exhibiting sensitivity of 600 nm/RIU at ~1800 nm. The capability to detect biomolecules has been demonstrated.

Acknowledges
The work has been supported by the Russian Science Foundation grant 17-13-01276.

References

[1] Valev V K, Baumberg J J, Sibilia C and Verbiest T 2013 *Adv Mater* **25** 18 2517
[2] Bochenkov V E 2017 *AIP Conf Proc* **1874** 030004

Bound state in the continuum in 1D chain of dielectric disks: theory and experiment

M. Balyzin[1], Z. Sadrieva[1], M. Belyakov[1], P. Kapitanova[1], A. Sadreev[2], A. Bogdanov[1]

[1]Department of Nanophotonics and Metamaterials, ITMO University, Saint-Petersburg, Russia
[2]Kirensky Institute of Physics, Federal Research Center KSC SB RAS, Krasnoyarsk, Russia

E-mail: a.bogdanov@metalab.ifmo.ru

Abstract. In this work we experimentally observe a symmetry protected optical bound state in the continuum (BIC) with zero angular momentum in 1D array of ceramic disks at GHz frequencies. We analyze the dependence of Q factor of BIC on the number of the disks and the level of the material losses. We confirmed theoretical prediction about quadratic growth of the Q factor with the number of the disks and its following saturation due to material losses.

1. Introduction

It is well known that the spectrum of a dielectric waveguide consists of the guided and leaky modes. Usually, the leaky modes are unwanted since they result in considerable losses and, therefore, hinder the operation of optical devices. However, the radiation losses can be totally canceled. In this case, we have a bound state embedded in the radiation continuum. The simplest way to cancel the radiation losses and get *bound state in the continuum* in a slab or cylindrical waveguide is to periodically modulate their permittivity along the direction of the translation symmetry. The periodicity discretises the number of open (diffraction) channels and makes them finite at each frequency. By choice of the proper design of the unit cell it is possible to make all amplitudes of (diffraction channels) outgoing waves equal to zero. This mechanism is analyzed in details in [1, 2, 3].

First, the concept of BIC was proposed in quantum mechanics in 1929 by E. Wigner and J. von Neuman [4]. In optics, it has been actively studied over the last 10 years. The significant results were achieved for this short period. It was shown that BICs are very prospective for many applications from lasing [5, 6], and filtering [7, 8] and biosensing [9, 10]. The latest achievements in physics of BIC are reviewed in [11].

To date, BICs in 1D periodic arrays were considered only theoretically in spite of these system are very prospective for generation of light with orbital angular momentum. In this work, we demonstrate first experimental observation and provide detailed analysis of BICs in 1D axially symmetric finite array of dielectric scatterers.

2. Sample and Setup

As an experimental sample we use an array of disks fabricated from BaO-TiO_2 microwave ceramic with the permittivity of ε=40 at GHz frequencies. The radius and the height of the disks are D=10.2 mm and h=10.1 mm, respectively. The period of the array is L=15.1 mm.

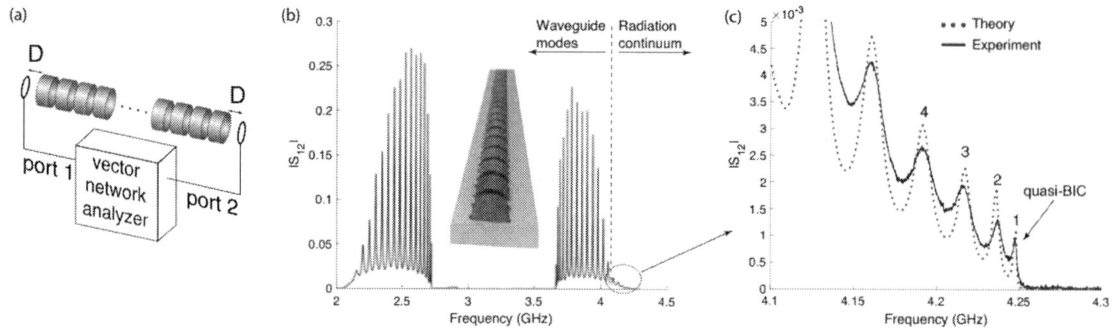

Figure 1. (a) Experimental setup for measurements of transmission spectra through the 1D array of dielectric disks. (b) Spectrum of S_{12} amplitude. The insets shows the experimental sample. (c) Zoomed part of panel (b) showing the resonances in the transmission spectra corresponding to leaky modes.

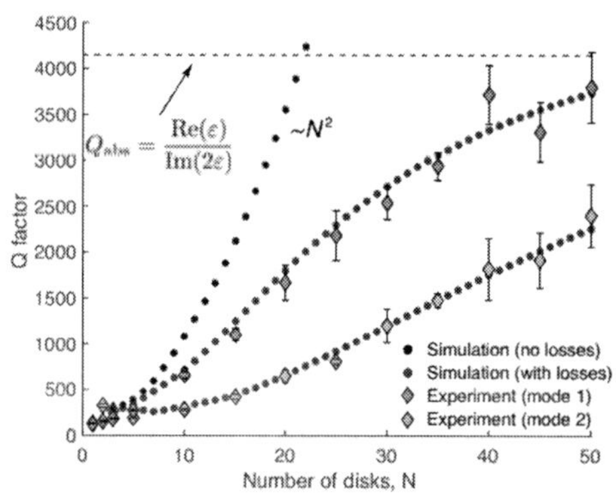

Figure 2. Experimental measured dependence of Q factors of the symmetry protected quisi-BIC (mode 1) and the neighbour resonance (mode 2) on the number of the period N in the chain of ceramics disks. Numeration of the resonances is shown in Fig. 1(c). Dotted line shows the results of numerical simulation in Comsol Multiphysics.

To fix the disks a custom holder with groves was fabricated from a styrofoam material with a permittivity close to 1.1 at the microwave frequencies. The refractive indexes of the holder is close to the refractive index of air that justifies a neglecting of effect of substrate onto the BIC. The fabricated prototype is shown in the inset of Fig. 1(b).

To analyse the resonances of the system, we measure its transmission spectra S_{12}. Here we focus on BIC with zero angular momentum. Thus, to avoid excitation of the states with nonzero angular momentum we use two identical shielded-loop antennas placed concentric with the axis of the array [see Fig. 1(a)]. Both antennas were connected to the ports of the vector network analyser (VNA). The shielded-loop antennas with the outer diameter of 10 mm have been fabricated from 086 Semi-rigid Coax Cable. The antennas have been placed at the distance 5 mm away from the edges. This distance is enough to provide a weak coupling with the array and not distort Q factors of the resonances.

3. Results

The transmission spectra $|S_{12}|$ of the array consisting of 20 disks placed between two loop antenna is shown in Fig. 1(b). The vertical dashed line separates regions of true leaky resonances and quasi-guided modes, which turn into guided modes as number of disks N tends to infinity. One can see that amplitude of the leaky resonances is much less that one of the quasi-guided modes. Figure 1(c) shows zoomed part of Fig. 1(b). The red dotted line shows the results of the simulation carried out in Comsol Multiphysics. The last peak in the series (with the highest frequency) corresponds to quasi-BIC, which turns into symmetry protected BIC as $N \to \infty$.

The simulated dependence of its Q factor on the N in absence of material losses is shown in Fig. 2 by black dotted lines. We found that $Q \sim N^2$ at small number of disks and then Q saturates due to material losses. The maximal Q factor can be estimated as $Re(\varepsilon)/Im(2\varepsilon)$. The dependance of Q factor on N for two last resonances extracted form the experimental spectra is shown in Fig. 2 by green and blue squares. The experimental results are in good agreement with the results of the simulations.

4. Conclusions

We report the first experimental observation of the symmetry protected BIC in 1D periodic array of dielectric scatterers. We analyze the dependence of Q factor of quasi-BIC on the number of the disks in the array. We observe quadratic growth of Q with N at small number of the disks and following saturation at the level $Q_{max} = Re(\varepsilon)/Im(2\varepsilon)$ at N of several tens.

Acknowledgments

This work is supported by RSF (17-12-01581).

References

[1] Bulgakov E N and Sadreev A F 2015 *Phys. Rev. A* **92** 023816
[2] Bulgakov E N and Sadreev A F 2017 *Phys. Rev. A* **96** 013841
[3] Sadrieva Z F, Sinev I S, Koshelev K L, Samusev A, Iorsh I V, Takayama O, Malureanu R, Bogdanov A A and Lavrinenko A V 2017 *Acs Photonics* **4** 723–727
[4] von Neumann J and Wigner E P 1993 Über merkwürdige diskrete eigenwerte *The Collected Works of Eugene Paul Wigner* (Springer) pp 291–293
[5] Kodigala A, Lepetit T, Gu Q, Bahari B, Fainman Y and Kanté B 2017 *Nature* **541** 196
[6] Bahari B, Vallini F, Lepetit T, Tellez-Limon R, Park J, Kodigala A, Fainman Y and Kante B 2018 *B. Am. Phys. Soc.*
[7] Foley J M, Young S M and Phillips J D 2014 *Phys. Rev. B* **89** 165111
[8] Cui X, Tian H, Du Y, Shi G and Zhou Z 2016 *Sci. Rep.* **6** 36066
[9] Romano S, Torino S, Coppola G, Cabrini S and Mocella V 2017 Optical sensors based on photonic crystal: a new route *Optical Sensors 2017* vol 10231 (International Society for Optics and Photonics) p 102312J
[10] Liu Y, Zhou W and Sun Y 2017 *Sensors* **17** 1861
[11] Hsu C W, Zhen B, Stone A D, Joannopoulos J D and Soljačić M 2016 *Nat. Rev. Mater.* **1** 16048

GaN nanowires on Si (111) substrates via molecular beam epitaxy: growth, electronic and optical properties

A.D. Bolshakov[1], V.V. Fedorov[1], G.A. Sapunov[1], A.M. Mozharov[1], , L.N. Dvoreckaia[1], K. Shugurov[1], V. Shkoldin[1], I.V. Shtrom[3,4], M.S. Mukhin[2], G.E. Cirlin[1-3], I.S. Mukhin[1,2]

1. St. Petersburg Academic University, Khlopina 8/3, 194021, St. Petersburg, Russia

2. ITMO University, Kronverkskij 49, 197101, St. Petersburg, Russia

3. Ioffe Institute, Politekhnicheskaya 29, 194021, Saint Petersburg, Russia

4. Institute for Analytical Instrumentation RAS, Rizhsky Pr., 26, 190103, St. Petersburg, Russia

bolshakov@live.com

Abstract. In this report, we study influence of the Si (111) substrate surface preparation on the growth, electronic and optical properties of the GaN nanowires (NWs) obtained via plasma-assisted molecular beam epitaxy. The substrate preparation varied from bare Si (111) surface and its deliberately nitridated counterpart to growth on AlN and Ga_2O_3 buffer layers and Ga droplets seeding layers. Statistical data on the morphology of the synthesized arrays was obtained and analyzed. The most homogeneous NW array in terms of length distribution was obtained on AlN buffer layer. It was demonstrated that the NWs surface density drastically depends on the surface preparation method. Electrical properties of the arrays were studied via analysis of volt-ampere characteristics and optical properties were investigated with photoluminescence. The highest conductivity and optical response were obtained with AlN buffer layer.

1. Introduction

Attention to the nitride semiconductor nanostructures is evoked by the possibility of their use in photovoltaic and optoelectronic devices operating in the ultraviolet range [1]. As it was shown in our previous work, one of the promising practical applications of the GaN NWs/Si heterostructure is a solar cell, where NWs array can act as an effective light harvesting and antireflection coating [2]. An ability to control the NW aspect ratio and tapering allows the fabrication of the functional structures with strong light confinement and given distribution of electric field, promising for future laser devices [3], or for use as nanoscale terahertz emitters and detectors [4], [5] based on Gunn effect.

The high cost of GaN substrates promotes researchers to develop the nitride compounds growth techniques on more accessible wafers and find new technological solutions in order to reduce the effect of lattice mismatch [6]. Geometric confinement of the nanostructures can significantly affect physical properties and high surface area to volume ratio leads to the effective strain relaxation compare to epitaxial films even on highly mismatched substrates. As an example of the nanostructure advantages in the field of epitaxial growth - it was demonstrated that GaN nanowires (NWs) grown on Si substrate can have nearly defect-free crystal structure [7].

Instead of the most III-Vs the growth of GaN NWs by plasma-assisted molecular beam epitaxy (PA-MBE) can be performed without use of a catalyst which role is usually played by golden droplets. This technique eliminates the inclusion of the catalyst material in the NW structure which can negatively affect the device characteristics.

Typically, Si (111) and (100) substrates are used for GaN NWs growth. In both cases, GaN NWs represent high quality wurtzite (hexagonal) crystal structure with the growth direction parallel to the [0 0 0 1] crystallographic direction and perpendicular to the substrate surface [8]. It was demonstrated that growth on the Si (100) leads to a less regular morphology of the NWs array compare to the Si (111), while the optoelectronic properties of the nanostructures are similar according to the photoluminescence spectra analysis [6]. One of the main drawbacks of GaN NWs synthesis on Si substrate in terms of further device application is virtually inevitable formation of the 1-2.5 nm thick amorphous silicon nitride layer at the heterointerface due to a large difference between the bond energies of Si-N (4.5 eV) and Ga-N (2.2 eV) [9].

Despite crystalline structure of the NWs it was demonstrated experimentally that their growth can be obtained on non-crystalline surfaces. To this end in [9] Si (100) substrate was thermally oxidized to form an amorphous ~ 320 nm SiO2 layer prior to deposition of GaN. The resulting NWs had an insufficient diameter dispersion, good crystal structure and a growth direction along the c-axis perpendicular to the substrate surface. Furthermore, 2 nm of silicon nitride was formed on the oxide.

Growth of GaN NWs can also be obtained via deposition of the Ga wetting layer prior to the NWs synthesis [2, 10]. As was demonstrated the presence of Ga puddles with a diameter up to 340 nm lead to the NWs formation at the surface free of Ga and the puddles serve as reservoirs of Ga adatoms [2].

In order to prevent reaction between metallic Ga and Si, a thin AlN buffer layer (few nm) can be used to increase the homogeneity of the nucleation and orientation of the NWs [6, 11]. The growth of GaN NWs on 15 nm thick amorphous AlxO layer on Si (111) substrate was also demonstrated [12].

Despite vast experimental data on synthesis and investigation of the GaN NWs properties discrepancy of the results takes place due to difference in experimental setups used for the nanostructures synthesis. In this report we study mutual influence of the substrate preparation of Si(111) substrate on geometrical, electronic and optical properties of GaN NWs.

2. Experimental

GaN nanostructures were grown by plasma-assisted molecular beam epitaxy (PAMBE). The experimental setup can be found elsewhere [13]. Silicon (111) p-type wafers with a 4° miscut oriented towards <1 -1 0>, cleaned with Shiraki method were used as substrates. After thermal annealing (20 min @ 950 °C) atomically clean Si surface was obtained as confirmed by observation of a 7x7 reflection high energy electron diffraction (RHEED) reconstruction pattern and smooth surface morphology on AFM images.

Typical nitrogen molecular beam flux equivalent pressure (BEP) was in the range of $2\cdot10^{7}$ to $3\cdot10^{7}$ Torr. The Ga source BEP was kept sufficiently small $1\cdot10^{-8}$ Torr in all of the experiments.

In our experiments after the oxide removal we carried out different substrate surface treatment procedures. In the first growth experiment (sample 1) we deposited GaN immediately after the oxide removal. The second sample was subjected under activated nitrogen plasma and annealed for 20 minutes with a temperature rising from 630°C to 810°C for formation of a thin SiN_x amorphous layer which is commonly used to obtain GaN NWs growth. The third substrate was covered with a thin AlN buffer layer. In order to do that the substrate was heated to 650°C and deposition of Al for 1 min was carried out. After 1 minute growth interruption the nitridation of the metal layer was carried out. All of the GaN deposition experiments were carried out at 810°C. Prior to placement into the MBE growth chamber the fourth sample was covered with Ga_2O_3 layer with atomic-layer deposition (ALD).

The rest of the samples (5th-7th) were covered with Ga droplet seeding layer performed prior to the ignition of the RF-plasma. Ga droplets were formed after Ga deposition on the reconstructed Si (111)-(7x7) substrate surface at 630°C. All of the samples were doped with Si for the last few hours for further investigation of the NWs and heterointerface electrical properties.

Morphology of the synthesized nanoheterostructures was studied with scanning electron microscopy (SEM) (Zeiss SUPRA 25-30-63). The back electrical contact was made via Al thermal vacuum deposition with the following thermal annealing. At the next stage a dielectric layer was deposited on the top surface of the substrate. The latter was then etched with oxygen plasma to unsheathe the NW top parts. Later the samples were covered through metal mask with ITO transparent conducting layer in the shape of 2.5mm disk mesa to provide top contact to the NWs.

To study device application potential of the synthesized nanostructures volt-ampere characteristics (VACs) were measured. The VACs were measured on the thermo-stabilized table at 25°C with the use of Keithley 2400 SourceMeter multimeter.

3. Results

All of the used growth techniques provided good verticality of the synthesized NWs. The best vertical orientation homogenity is demonstrated with 0.3ML and 0.6ML thick Ga seeding layers. Slight inclination of the grown nanostructures is observed on both clean and nitridated Si (111) substrates and on Ga_2O_3 buffer layer. Further increase of the inclination is registered with AlN buffer, while for the last sample with a relatively thick Ga seeding layer presence of randomly inclined NWs is observed. The third sample grown with the use of a thin AlN buffer layer is the only one that possess a 3D GaN layer. NWs demonstrate hexagonal cross-section which is an indication of their high crystal quality. All of the synthesized NWs exhibit vertical sidewalls – no sufficient anisotropic lateral extension along the NW length was observed.

The most homogeneous is the array of GaN nanowires synthesized on AlN buffer layer with a mean length distribution of only 5%. At the same time these NWs are the less sparsely distributed with the corresponding surface density of 69 mkm^{-2}. Such a phenomenon demonstrates crucial influence of the substrate preparation on the NWs nucleation and adatoms diffusion leading to the highest NWs elongation rate (40.3 nmh^{-1}) in case of AlN buffer layer. All of the mentioned geometrical properties of the latter array together with its good verticality demonstrate its device application potential.

After post-processing of the synthesized arrays explained in details in the experimental section the VACs were measured. Analysis of the transport properties of the synthesized arrays necessitates consideration of different current channels. In the case of n-GaN/p-Si heterostructure there is no sufficient contribution from the hot holes due to high value of the potential barrier for these carriers. At the same time, in this case exists a contribution of electrons and holes recombination at the heterointerface due to presence of the surface states. This current in the first approximation is proportional to concentration of electrons and holes at the interface. Thus properties of the interface affect both quantity of the surface states and material properties in the region close to it, e.g. doping and presence of deep levels in the band gap.

Analysis of the experimental data demonstrates strong influence of the buffer and seeding layers on transport properties of the heterointerface. AlN buffer layer provides the best electrical properties in our experimental series. Aluminum is a p-type dopant for Si having close value of covalent radius with the latter providing good incorporation of this material in Si. That is why in sample 3 Si surface layer has high doping level. At the same time Al induces formation of the deep levels in Si enhancing the recombination current through the diode. According to the VAC analysis recombination current plays the main role in this particular case. The latter fact most probably corresponds to high electron concentration at the heterointerface due to doping effect of Si during initial stages of the NWs growth.

Compare to AlN buffer 2 ML thick Ga seeding layer provides worse conductivity than in the first case due to lower capabilities of Ga compare to Al in terms of Si doping. Thus recombination contribution is lower in case of Ga presence but it still plays the main role. Decrease of Ga layer equivalent thickness leads to drop of the recombination and total currents. When the seeding layer thickness is only 0.3ML Si-Ga chemical bond adopts sp3 hybridization equivalent to native Si leading to decrease of Ga bulk diffusion into Si with the corresponding drop of the doping level.

In case of G_a2O_3 buffer layer formation of n-GaN/Ga_2O_3/p-Si heterostructure takes place with a negligible barrier for electrons in the conduction band and high barrier for holes in the valence band. The aim of this layer deposition was to spatially separate free electrons in GaN and holes in Si to decrease the interface recombination. Nevertheless, current through this structure at high bias is lower compare to the other samples due to high resistance of Ga_2O_3 layer. Improvement of the layer conductivity can be realized via doping during the PECVD deposition.

Analysis of the photoluminescence spectra taken at 10K demonstrates strong influence of the surface preparation on optical properties of the synthesized nanostructures. Spectra shows that no yellow luminescence occurs in our samples, indicating high crystal quality of the synthesized GaN nanostructure arrays. Three PL lines corresponding to free excitons of light X_A, heavy X_B holes are observed as well as two PL lines corresponding to D_0X_A and D_0X_B excitons bound to neutral donors. The highest intensity excitation was obtained again with AlN buffer layer.

Conclusions

In this letter we study effects of the Si (111) substrate preparation on the growth and physical properties of GaN NWs. It is demonstrated that geometry and surface density of the nanostructures can be tailored via the surface preparation. AlN buffer layer is considered to be the most appropriate option in terms of potential semiconductor device application due to several reasons. First is the length homogeneity of the array and its sparse surface density together with a good verticality of the nanostructures. Second is the good electronic properties with the highest achieved conductivity of the heterostructure. The optical response from the discussed heterostructure was the highest among all of the synthesized nanoheterostructures most likely due to highest crystalline quality of the array.

Acknowledgements

This work was carried out with the support of the Russian Federation President grants (MK-6492.2018.2 and MK-3632.2017.2), the Russian Foundation for Basic Research (16-32-60094 mol_a_dk and 18-32-00899 mol_a), the leading universities of the Russian Federation (grant 074-U01), grant of government of the Russian Federation (3.9796.2017/8.9, 16.2593.2017/4.6 and 16.2483.2017/4.6).

References

[1] Nakamura, S. (1998). Science, 281(5379), 956-961.
[2] Ristić, J., Calleja, E., Fernández-Garrido, S., Cerutti, L., Trampert, A., Jahn, U., & Ploog, K. H. (2008). Journal of crystal growth, 310(18), 4035-4045.
[3] Fernández-Garrido, S., Kong, X., Gotschke, T., Calarco, R., Geelhaar, L., Trampert, A., & Brandt, O. (2012). Nano letters, 12(12), 6119-6125.
[4] Reznik, R., Kotlyar, K., Ilkiv, I., Soshnikov, I., Kukushkin, S., Osipov, A., ... & Cirlin, G. (2016, June). In AIP Conference Proceedings (Vol. 1748, No. 1, p. 040003). AIP Publishing.
[5] Calabrese, G. et al. (2016). Applied Physics Letters, 108(20), 202101.
[6] Bertness, K. A. et al. (2011). IEEE Journal of selected topics in quantum electronics, 17(4), 847-858.
[7] Bertness, K. et al. (2011). (No. SPIE Newsroom).
[8] Reznik, R. et al. (2016, June). In AIP Conference Proceedings (Vol. 1748, No. 1, p. 040003). AIP Publishing.
[9] Stoica, T., et al. (2008). Small, 4(6), 751-754.
[10] Guo, W. et al. (2010). Nano letters, 10(9), 3355-3359.
[11] Songmuang, R. et al. Applied Physics Letters, 91(25), 251902.
[12] Sobanska, M. et al. (2016). Nanotechnology, 27(32), 325601.
[13] Bolshakov, A. D. et al. (2018). Beilstein Journal of Nanotechnology, 9(1), 146-154.

IOP Conf. Series: Journal of Physics: Conf. Series **1092** (2018) 012014 doi:10.1088/1742-6596/1092/1/012014

HIGH FREQUENCY ELASTIC PROPERTIES OF NITRIDE NANOWIRES-BASED STRUCTURES

O. Boyko, E. Charron, B. Bonello,

Sorbonne Université, Institut des NanoSciences de Paris, 4, place Jussieu 75252 Paris, France

L. Lu, M. Tchernycheva, N. Gogneau

University Paris Sud, Université Paris Saclay, Centre for Nanoscience and Nanotechnology 15, rue Georges Clémenceau 91405 Orsay Cedex, France

E. Glushkov, N. Glushkova

Kuban State University, Krasnodar, Russia

Olga.Boyko@insp.jussieu.fr

Abstract. The present research aims at studying the elastic properties of III-Nitride nanowires (NWs) embedded in HSQ matrix, capable to efficiently convert the mechanical energy into electricity. We report on elastic properties of such novel nano-engineered functional materials studied by using "transient grating" acoustic experiment.

1. Introduction

Flexible piezo-electric nanowires are today a topic of intense research, motivated by their numerous economically relevant applications (e.g., rollable displays, wearable intelligent electronics, piezo-generators, and so forth). Nitride nanowires show remarkable mechanical and optoelectronic properties stemming from their high aspect-ratio. They are mechanically flexible and can stand high deformations without plastic relaxation. Thanks to this promising performance, polymer-embedded nanowires offer an elegant solution to create a piezo-generator material, which converses the environment energy to electric one.

A large number of works are devoted to study optoelectronic properties, there have been few research works on elastic properties of nanowires. A few researches are mainly focused on the study of mechanic-electric energy conversion at kHz domain. The most of environment vibrations covers a kHz scale frequency spectrum, and the transformation of ambient kHz noise into the electric power is a promising way to create any self-power micro-devices. Nevertheless, related on the nanosize, the mechanical resonances are expected at frequencies within a range stretching from a few hundreds of MHz up to several GHz.

The current work proposes to explore ultrasonic elastic properties of nitride nanowires at MHz and GHz frequency range. The interest of this research is fundamental: the mechanical behavior of materials at the nanoscale is different from that at the macroscopic scale due to the increasing ratio of the surface to the volume. The systematic study of elastic properties leads to an accurate understanding of the deformation of nanowires allowing thereby for the prediction of the piezoelectric behavior, the enhancement of the energy conversion and in return to an improved design of the nanowires-bases structures.

2. Growth of GaN nanowires.

Self-assembled GaN NWs were grown by Plasma-Assisted Molecular Beam Epitaxy (PA-MBE) on a thin n-doped AlN buffer layer [1] deposited on n-type doped Si (111) substrate (the NW morphology is illustrated in Fig. 1.

The growth was performed at 760°C, under N-rich conditions. NWs have a cylindrical shape with a hexagonal cross section delimited by {10-10} planes [2]. They are characterized by height and diameter of the order of 1 μm ± 120 nm and 45 nm ± 20 nm, respectively and by a density of the order of 10· NWs/cm^2.

Figure 1. Carpet of Nitride Nanowires (SEM photo).

3. Fabrication and characterization of elastic properties of piezogenerators.

3.1. Fabrication.

The sample consists in piezoelectric GaN nanowires grown on a thick silicon substrate. The NW arrays are encapsulated into spin-on glass (HSQ), which is annealed at 400°C. Then the NW tops are uncovered by doing reactive ion etching and the Schottky contact is deposited, though a shadow-mask, on the NW top side. Finally, an ohmic contact is directly deposited on the substrate (Figure 3).

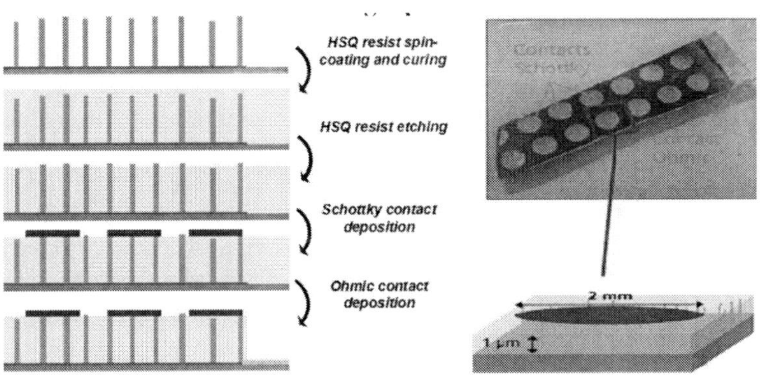

Figure 2. Process steps to realize the piezogenerator (left) and the picture of sample (right).

3.2. Experimental set-up.

Experimental techniques well suited to measure the sound velocity in composite systems, and in turn the dispersion, are not that many. When the device includes piezoelectric elements, electromechanical transducers and network analyzers can be used for characterization but with the disadvantage of a poor tuning of the frequency. The non-contact technique that we describe here allows for characterizing the elastic properties of almost any kind of materials, transparent or opaque, ordered or disordered, piezoelectric or not… in a frequency range that stretches from a few tens of MHz up to several GHz.

Basically, the so-called Transient Grating Method [3-5] is a four waves mixing technique which consists in exciting the sample into vibration by illuminating its surface with two IR light pulses (λ_{exc}=1064nm; 30ps in duration). These IR light pulses interfere to produce interference fringes with a period of Λ. As a consequence of photoelastic processes, standing elastic waves with wavelength of Λ and angular frequency of Ω are thus created within the illuminated area. The wavelength Λ can be easily tuned by setting the angle between the two incident beams.

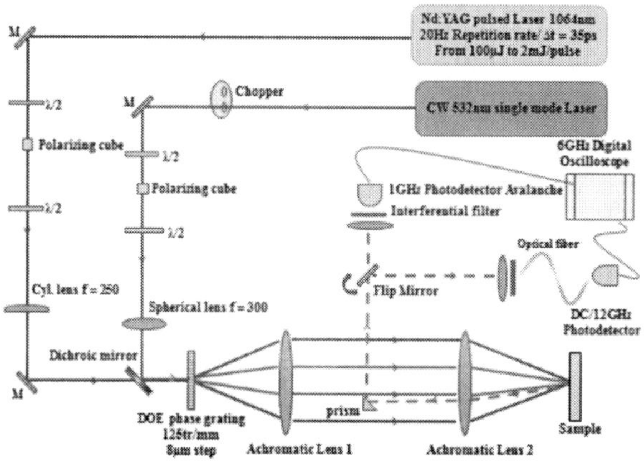

Figure 3. Scheme of the experimental setup. M and DOE stand for Mirror and Diffractive Optical Element respectively, λ/2 is for half wave plate.

The detection of the elastic waves is made using a heterodyne scheme. This involves two continuous laser beams (respectively the probe and the reference beam) both incident on the sample with an optical angular frequency ω. The probe beam gets diffracted upon reflection on the sample, giving rise to diffraction orders ±1 at optical angular frequencies ω±Ω. The beam at ω+Ω is further mixed with the specularly reflected reference beam, whereas the beam at ω−Ω is ignored. It is then straightforward to show that a signal proportional to cos(Ωt) results from the mixing of both first order diffracted probe beam and specularly reflected reference beam. This signal is recorded with a fast photodiode connected to a broadband oscilloscope and further processed using conventional FFT techniques.

3.3. Characterization of elastic properties

We show in Fig. 4 (left panel) a typical signal recorded on the sample. The spectrum of the signal obtained by Fourier transform is shown in inset.

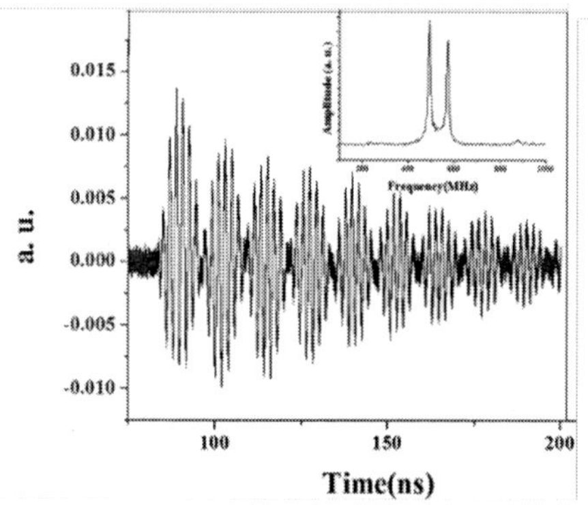

Figure 4. typical signal recorded in the sample shown in the Figure 1. The wave vector is set to $k=2\pi/\Lambda=1.07\mu m^{-1}$ ($L=5.85\mu m$). Inset: Fourier transform of the data.

The frequency spectrum features two peaks at 495 and 575MHz respectively. They correspond to the first two Rayleigh modes (the higher order Rayleigh modes are sometimes called Sezawa modes; they require a minimum layer thickness to propagate). The effective phase velocities V of these modes can be very simply deduced from the measured frequencies ν through $V = \Lambda\nu$. Regarding the dispersion, it may be recorded by tuning Λ owing to a proper choice of the angle between the two incident optical pulses. As an example of the achievements of this experimental technique, we show in Fig. 5 the dispersion for three carpets of nanowires similar to the one shown in Fig. 2 but with different curing temperatures).

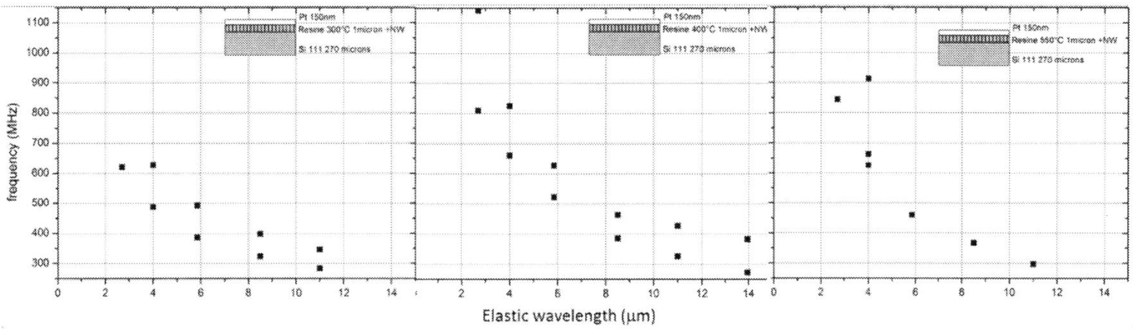

Figure 5. Dispersion of the first two Rayleigh modes propagating in layers of nanowires deposited on a silicon substrate. The samples differ each other by the annealed temperature (300, 400 and 550°C).

The stress of HSQ matrix increases with increasing curing temperature. Furthermore, HSQ may have become SiO_2 based on the change in the chemical structure and larger increase in modulus and hardness when cured above 650°C. Here, the interest is in keeping the nanowires flexible within the HSQ matrix, thus, temperatures below 650° are chosen.

Analysing these dispersion curves, one sees that with increasing temperature, the velocity of Rayleigh waves also increase, and the rapid change of frequencies from wavelength in MHz diapason allows the elastic constants to be determined with accuracy.

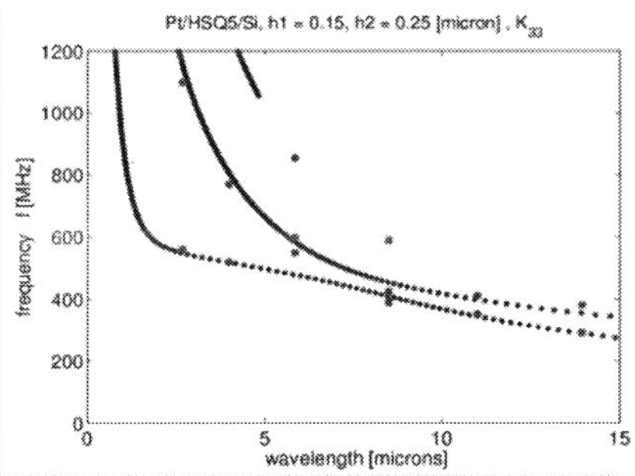

Figure 6. Dispersion curves of Pt/HSQ/Si structure: theoretically predicted (blue line); experimental results (red points).

In order to identify all modes and to determine elastic constants, theoretical investigations are carried out for a three-layers elastic half-space. Figure 6 represents analytical behaviour of dispersion curves (blue lines) and experimental data for one sample cured at 400°C (red points). The theoretical curves obtained from the cross-section of the $|K33(\lambda,f)|$ surface along vertical lines λ=constant in the range from 0 to 15 µm [6]. The elastic constant C_{11}, C_{44} as well as velocities in all layers can be deduced directly from comparison between the theory and experiment.

4. Conclusion.

Beside the dispersion properties of the nanostructures, it is worth noting that this experimental technique allows for a precise determination of the effective elastic parameters of such a layer on a substrate. It should also be noted that this non-destructive contactless technique is efficient to investigate the effective elastic properties at very high frequency (several GHz), along any directions of an anisotropic sample as the phononic crystals and the elastic metamaterials are.

The theoretical part (SAW simulation) is supported by the Russian Science Foundation (Project No. 17-11-01191).

References

[1] L. Largeau, E. Galopin, N. Gogneau, L. Travers, F. Glas, J.-C. Harmand, "N-Polar GaN Nanowires Seeded by Al Droplets on Si(111)", Cryst. Growth Des. 12, 2724 (2012)

[2] O. Schneegans, P. Chrétien, F. Houzé, French Patent FR 10 01940, 05/05/2010, international PCT WO 2011/138738, 10/11/20A

[3] A. Taschin, P. Bartolini, M. Ricci, and R. Torre, Philosophical Magazine **84**, 1471 (2003).

[4] L. Dhar and J.A. Rogers. Appl. Phys. Lett. **77**, 1402 (2000).

[5] A.A. Maznev. Phys. Rev. B **78**, 155323 (2008).

[6] E. Glushkov, N. Glushkova, Ch. Zhang, Journal of Applied Physics 112, 064911 (2012).

IOP Conf. Series: Journal of Physics: Conf. Series **1092** (2018) 012015 doi:10.1088/1742-6596/1092/1/012015

Feasibility of metasurface assisted magnetic resonance spectroscopy at 1.5 T

E. Brui[1], A. Shchelokova[1], A. Slobozhanyuk[1], I. Melchakova[1], A. Andreychenko[1]

[1]Department of Nanophotonics and Metamaterials, ITMO University, 197101 Saint-Petersburg, Russia

E-mail: e.brui@metalab.ifmo.ru

Abstract.
A hybrid metasurface realized by brass wires coupled to two high-permittivity dielectric slabs at both sides was used together with a conventional birdcage body coil for transmission and reception in 1.5 T magnetic resonance spectroscopy. A feasibility of improvement of spectral quality has been shown. Up to 7.5 times increase of signal-to-noise ratio for the spectra obtained for a spherical MSR phantom containing a composition of prostate metabolites was achieved. The most significant increase has been shown for the area close to the metasurface.

1. Introduction

Magnetic resonance spectroscopy (MRS) is a noninvasive method that allows studying of composition and distribution of metabolites in human organs and tissues in vivo. Since the in vivo concentration of metabolites is usually low (more than thousand times lower than the water concentration in human tissue), the proton MR-signal received from the metabolites after a single excitation has a low signal-to-noise (SNR) ratio, when compared to the water signal. This makes reliable metabolite detection challenging. For this reason, a pulse sequence has to be repeated and the signal should be averaged in a series of multiple acquisitions (typically, 128 for 1.5T spectroscopy). However, even with multiple averaging 1.5 T MR-spectroscopy remains an additional and only supportive method in MR scanning due to the long acquisition. The most radical and effective way to increase the signal-to-noise ratio is the use of high-field MR scanners (above 3T), which is difficult to perform due to the low prevalence of such devices. However, in addition to the transition to higher fields, there are other ways to improve the characteristics of MR-spectrum, such as the signal-to-noise ratio. One of the possible ways is to increase the efficiency of the receiving coils. The best reception characteristics among currently used in clinical practice are multichannel coils based on phased arrays [1]. However, despite the improvement in reception characteristics, such coils have a number of significant drawbacks: an increase in the number of channels leads to a serious weighting of the structure, which affects the convenience of their use by medical personnel and the comfort of patients; with the complication of the construction increases the possibility of breakage. Thus, the task of facilitating the construction is also important. In addition, the cost of such RF coils can reach hundreds of thousands of dollars, and repairs, even in the case of minor breakdowns, dozens. The solution to these problems probably lies in the transition to wireless technologies based on

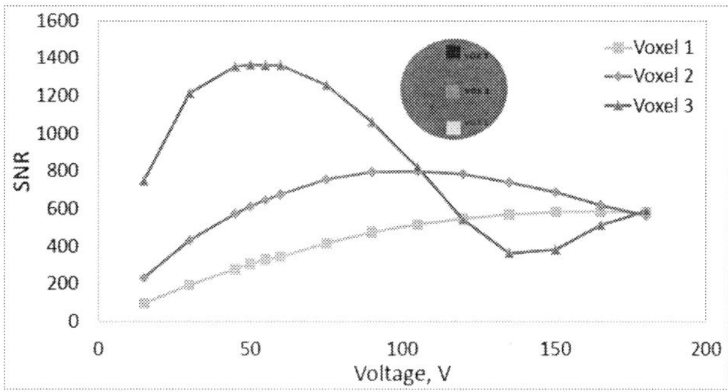

Figure 1. Manual power calibration procedure for three voxels located along y - axis. SNR dependence on reference voltage.

new materials and structures. An improvement in the distribution of the radio-frequency field can be achieved by placing, in the immediate vicinity of the area of investigation, materials with high dielectric permittivity [2, 3]. For 7 T in vivo MRS spectral quality has been improved by using high-dielectric materials [4]. Also, in recent years, a lot of works have been devoted to the use of metamaterials [5, 6, 7] - materials that have a dielectric permittivity and / or magnetic susceptibility that do not exist in nature, - and devices based on this principle in magnetic resonance. Artificially created materials that allow to control the distribution of the electromagnetic field, as a rule, have a high Q, which greatly narrows the field of their application, but, however, is a very successful property for use in studies using the phenomenon of nuclear magnetic resonance. Recently, several types of metasurfaces have been proposed for solving the problems of redistribution of the RF magnetic field and improving the quality of the experiment in MRI [8, 9]. Possibility of SNR enhancement in MRS by means of metasurfaces had logically followed from the obtained results and was shown for ultra-high field [10]. The aim of this study was to show the feasibility of metasurface assisted MRS at 1.5 T and to compare the quality of the acquired metabolite spectra of a prostate MRS-phantom obtained with and without metasurface.

2. Methods and Materials

A hybrid metasurface realized by brass wires coupled to two high-permittivity dielectric slabs at both sides [11] was used for MR signal enhancement. Distilled water was used as a dielectric media with a permittivity of 78 and conductivity of $\sigma = 5.55{\times}10^{-6}$ S/m. The hybrid metasurface was placed inside a 1.5 T Siemens Symphony MR-scanner, so that the wires were parallel to the direction of the main magnetic field B0 (z-direction), and its eigenmode was tuned to the Larmor frequency (63.66 MHz) by adapting the amount of water. A spherical MSR phantom containing a composition of prostate metabolites was placed 2 cm bellow the center of wires in the isocenter of the magnet. The body coil birdcage was used both for transmission and reception. Due to the inhomogeneity of transmitted radiofrequency field in case of metasurface presence, a careful manual power calibration was performed for three voxels (inset on Fig 1) in order to obtain true 90 degree excitation pulses in these areas. A series of GRE MR-images were acquired with different voltage of excitation pulse (the reference voltage was manually changed in Adjustments menu). For every of three voxels the SNR dependence on reference voltage was plotted and the position of maximum was associated with 90 degree excitation pulse (Fig 1).

MR-spectra for three voxels positioned along the y-axis of the phantom (close to the top of the phantom, in the center, and close to the bottom of the phantom) were acquired using the single voxel spectroscopy method. Then the metasurface was removed and the same procedure

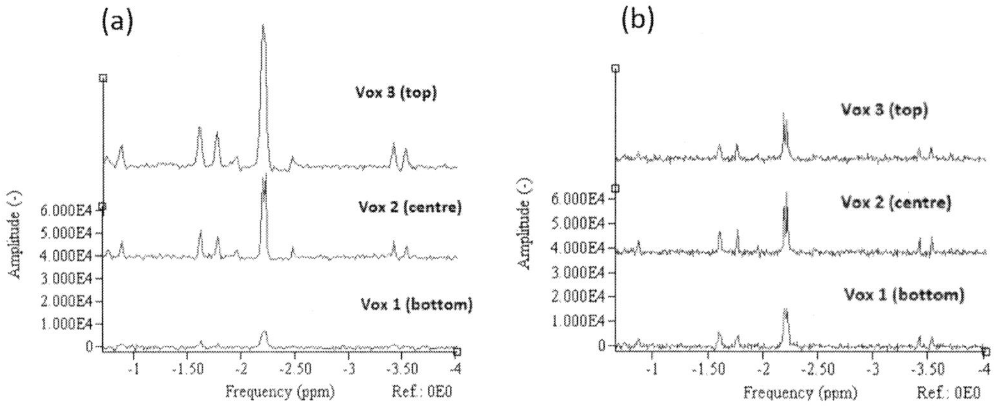

Figure 2. MR-spectra of metabolites of a spherical prostate MRS-phantom in case of (a) both birdcage body coil and metasurface working on transmission and reception, (b) only birdcage body coil working on transmission and reception.

was repeated. Much attention was paid to the fact that the phantom should remain in the fixed position. Parameters of a single voxel pulse sequence (STEAM – stimulated echo) were the following: TR = 2310 ms, TE = 20 ms, TM = 10 ms, voxel size - 20 x 20 x 20 mm, number of averages – 128, bandwidth – 2048Hz. Metabolite spectra were subsequently exported to jMRUI software package for MRS (http://www.jmrui.eu) and processed using Amares [12] quantitation algorithm. The individual peaks were fitted and their integrated amplitudes and linewidth were calculated. The standard deviation of noise was measured at signal free areas of spectra and the corresponding signal-to-noise ratios were calculated.

3. Results and discussion

Three metabolite spectra of voxels located in a prostate phantom in the presence of metasurface were acquired (Fig 3a) and compared to the spectra obtained without the metasurface (i.e. only by birdcage body coil) (Fig 3b). The SNR measurement showed an almost 7.5-fold enhancement in a voxel located close to the wires which is in a good correspondence with previously reported MRI data [11]. For the central voxel the enhancement was around 2.4 times, and for the bottom voxel – 0.77. Figure 4 illustrates the signal-to noise ratios for choline (Fig 4a) and creatine (Fig 4b) peaks and the errors of their estimation.

The linewidth for the case of metasurfase presence was up to 2 times higher for the creatine and choline peaks in comparison with the case without metasurface. This may be explained by the poorer shimming of the main magnetic field in the presence of a huge bulk of water which was used as a dielectric media in the current design of a metasurface.

4. Conclusion

In this work we have shown the feasibility of significant enhancement of signal-to-noise ratio of metabolite signals by means of a metasurface assisted. The most significant increase has been shown for the area close to the metasurface, that is peculiar for all surface receive RF-coils. However, due to the huge amount of distilled water, the linewidth of individual peaks has increased. The authors consider that the use of a different type of dielectric media (e.g. ceramics) may allow to narrow down the linewidth. The procedure of power calibration used here was complicated and time consuming, that is why the passive detuning of the used structure

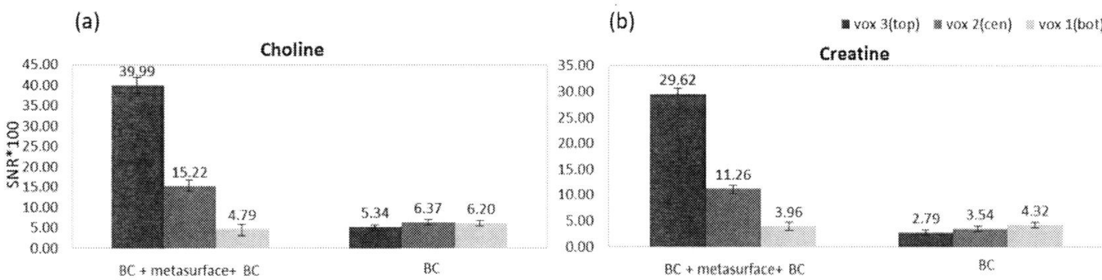

Figure 3. SNR data for (a) choline and (b) creatine peaks with errors of their estimation.

(e.g. by means of crossed diodes) is a subject for the further research.

References

[1] Fujita H, et al., 2007 *Magn. Reson. Med. Sci.* **6** 29–42
[2] Webb A.G, et al., 2011 *Concepts in Magnetic Resonance Part A* **38A** 4
[3] Webb A.G, et al., 2013 *NMR Biomed.* **26,11** 1555–1561
[4] Snaara J.E.M., et al., 2011 *NMR Biomed.* **24(7)** 873–879
[5] Smith D.R., et al., 2004 *Science* **305** 788–792
[6] Wiltshire M. C. K., et. al., 2001 *Science* **291** 849
[7] Radu X., et. al., 2009 *Metamaterials* **3** 90
[8] Slobozhanyuk A.P., et. al., 2016 *Adv. Mater.* **28** 9
[9] Glybovski S.B., et. al., 2015 *(Conference paper). V.A. Radio and Antenna Days of the Indian Ocean (RADIO).* **1-2** 21–24
[10] Schmidt R., et. al., 2017 *Sci. Rep.* **7** 1678
[11] Shchelokova A.V., et. al., 2018 *Physical Review Applied* **9(1)** 014020
[12] Vanhamme, L. et. al., 1997 *Journal of Magnetic Resonance* **129** 35–43

Negative group velocity waves in rectilinear thin wires array

Leonid N. Butko,[1] Anton P. Anzulevich,[1] Igor V. Bychkov,[1] Dmitriy A. Pavlov,[1] Sergey G. Moiseev,[2,3]

[1] Faculty of Physics, Chelyabinsk State University, Chelyabinsk, 454001, Russia
[2] Ulyanovsk State University, 42 L.Tolstoy Str., 432017 Ulyanovsk, Russia
[3] Kotelnikov Institute of Radio Engineering and Electronics of Russian Academy of Sciences, Ulyanovsk Branch, 48/2 Goncharov Str., 432011 Ulyanovsk, Russia

E-mail: `lnbutko@yandex.ru`

Abstract. A well-known metamaterial consisting of rectilinear thin wires array forming a 2D-square lattice within the non-conductive matrix was theoretically investigated in this paper. The frequency dependencies of wavelength of smoothed electromagnetic field distributions inside the wired structure, show that this material has a wide band in which there are negative group velocity waves. This effect is clearly observed in the case when the wavelength of the smoothed distribution of the electromagnetic field is much longer than the distance between the wires. For this case, also figures of field distribution from point source were modeled. They show that negative refraction and focusing of electromagnetic radiation observed, ie. the studied material is left-handed. Unlike most of the left-handed metamaterials that are composed of two separate inter-elements (typically, a split ring and a rod), in this paper it is proposed to use only one type of inclusion, namely the conductive rod.

1. Introduction

When electrical permittivity $\epsilon(\omega)$ and magnetic permeability $\mu(\omega)$ are both simultaneously negative, the direction of the phase velocity is opposite to the direction of energy current. Thus, the directions of the field vectors **E** and **H**, and the direction of the propagation wave vector **k** form a left-handed coordinate system. Therefore, these media are left-handed media (LHM) [1]. Furthermore, the index of refraction $n(\omega)$ used in the interpretation of Maxwell's equations should be taken as the negative square root as following $n = -\sqrt{\epsilon(\omega)\mu(\omega)}$. In [2, 3, 4] to create a LHM it was proposed to use the structures consisting of miniature split ring resonators, acting as magnetic dipoles, and straight segments of wire. Structures proposed in [5] were made of only one S-shaped element which yields an overlapping negative permittivity and negative permeability response over a frequency band of about 2.6 GHz. In this paper, we assume that a metamaterial consisting of an array of only conductive wires forming a square lattice within a non-conducting matrix can be the LHM. Similar structures were considered in [6, 7, 8]. In [6], effective permittivity at low frequencies was analytically determined, and, in [7], bandgaps for wire-medium were analytically obtained. However, currently, there are no papers containing such data as frequency dependencies of wavelength of smoothed electromagnetic field distributions (SEFD)- λ_{sefd} inside the wired structure at $0.65 < a/\lambda_m < 1$ (λ_m - a wavelength of an electromagnetic field in the non-conducting matrix, a - the distance between the wires)

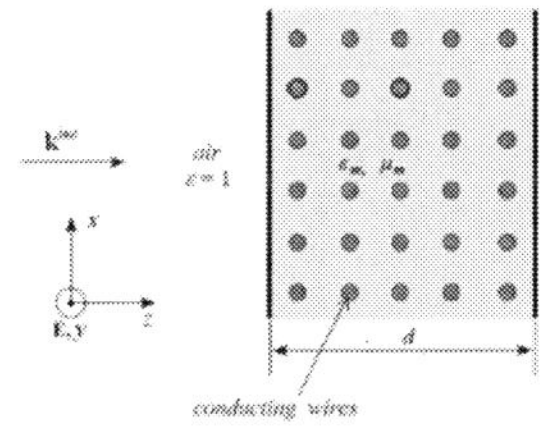

Figure 1. The geometry of a model and electric field distribution in two 2D unit cells. The color change corresponds to a change in the electric field intensity along the OY axis for an arbitrary frequency (E_y).

and a statement that array of wires can be a left-handed material on its own, where wavelength of SEFD much more distance between the wires.

2. Model of composite material based on a wire-structure

Let us consider a composite structure consisting of an array of rectilinear conductive wires (Fig. 1) forming a square lattice in a non-conducting matrix. The distance between the wires (a) and their length is much greater than the radius (r). Such a structure was previously studied in [6, 7, 8]. A response of the structure under investigation (Fig. 1) to external electromagnetic radiation depends on the direction and phase of this radiation. In the present paper the case when electric field $\mathbf{E}$ is parallel to the axis of a wire (Fig. 1) is considered. Let an incident electromagnetic wave propagates along axis z, and the vector of electric field be parallel to axis y. Thus, the model shown in Fig. 1 corresponds to a composite layer which is infinite in xy plane and which is formed by periodically arranged rectilinear conducting wires.

3. Wavelength of smoothed electromagnetic field distributions in a slab of wire metamaterial. Focusing of electromagnetic waves through the wired structure

In the case when reflection at the slab boundary is small for frequencies of 2nd photonic transparency range at $0.65 < a/\lambda_m < 1$ [9], the electric field distribution is obtained in the form of a "comb" (Fig. 2). In the figure there are long-wave oscillations of the effective field, on the background of which small interference oscillation can be seen.

For the comprehension of nature of field oscillations, the interference oscillations can be smoothed with various filters, for example, the Fourier filter. Fig. 2 shows that the wavelength obtained after filtering (λ_{sefd}) coincides with the wavelength outside the structure and with the values obtained in [8, 9] .

From Fig. 2, 3 it is seen that where wavelength of SEFD much more distance between the wires the wave vector and the phase velocity of the oscillations of the effective field are opposite to the direction of energy current. That is, the smoothed electromagnetic field distributions is the same as the field distribution in the left-hand metamaterial. The ability of such kind of structures to be left-handed metamaterial was also shown in the research of the EMW negative refraction of EMW [10, 11].For additional verification of this result, figures of field distribution for the case of point source were modeled. They show that in this frequency band negative refraction and focusing of electromagnetic radiation from a point source are observed. That cannot be done with the help of Bragg scattering mechanism.

4. Physical explanation for the magnetic response

For are left-handed media (LHM) [1] the index of refraction $n(\omega)$ used in the interpretation of Maxwells equations should be taken as the negative square root as following $n = -\sqrt{\epsilon(\omega)\mu(\omega)}$.

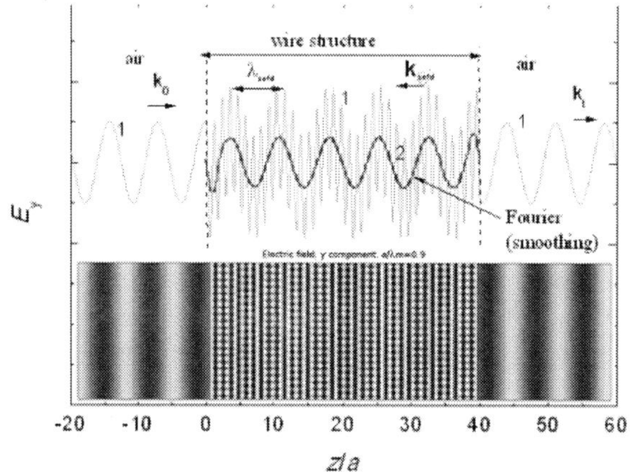

Figure 2. Curve 1 - the dependence of an alternating electric field on coordinate z along the line passing in between the wires, in sizes of period a. $a/\lambda_m = 0.9$. Curve 2: - a signal inside the wired structure, smoothed with Fourier filter (λ_{sefd}).

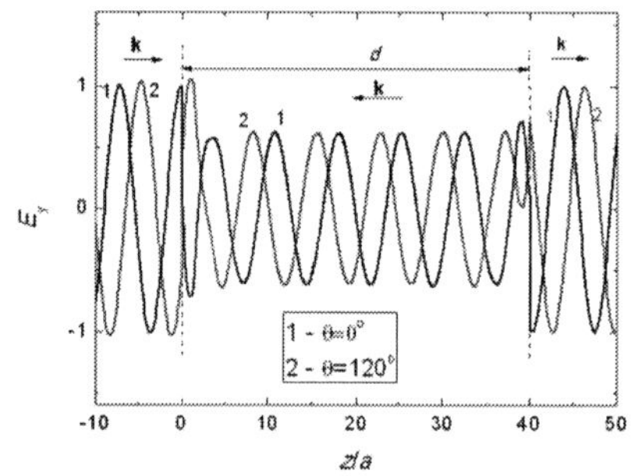

Figure 3. A signal inside the wired structure, smoothed with Fourier filter at different phases.

In [2, 3, 4] to create a LHM it was proposed to use the structures consisting of miniature split ring resonators, acting as magnetic dipoles, and straight segments of wire,acting as electric dipoles. But difficulty with the definition of physical meaning of permeability at higher frequencies, which is important for theory as well as for the interpretation of experiments, is related in [12] to the fact that it may be impossible to 'measure' the permeability by measuring the total induced magnetic moment of a macroscopic body. Indeed, the induced macroscopic current density **J** in time-dependent fields arises not only from the magnetization but also from the dielectric polarization **P** [13]: $\mathbf{J} = c \bigtriangledown \times \mathbf{M} + \partial P/\partial t$. $\mathbf{M}^{tot} = \mathbf{M}_1^{tot} + \mathbf{M}_2^{tot}$, where $\mathbf{M}_1^{tot} = \int \mathbf{M} dV$ and $\mathbf{M}_2^{tot} = (1/2c) \int (\mathbf{r} \times \partial \mathbf{P}/\partial t) dV$.

More often $\mathbf{M}_2^{tot}$ can be neglected, and the quantity $\mu(\omega)$ with reasonable accuracy defines the magnetic moment of a unit volume in the field of the plane electromagnetic wave propagating in the medium. However, for metamaterials created from elements which sizes are much greater than the sizes of molecules, frequencies where effect of $\mathbf{M}_2^{tot}$ becomes sufficient are much lower than in natural materials. Using Maxwell equation and the definitions of **M** and of **P**, we can immediately evaluate relative contributions to the induced current for a monochromatic electromagnetic wave. For the contribution from the time-dependent dielectric polarization to dominate, i.e., to have $\partial P/\partial t >> c \bigtriangledown \times \mathbf{M}$, which allows neglecting the term $\mathbf{M}_1^{tot}$, the following inequality must be satisfied: $|\epsilon_{eff} - 1| >> |\mu_{eff} - 1| |\epsilon_{eff}|$. The values from [9] , this inequality is satisfied. Also, it is satisfied for the values of effective permittivity of wire medium from works [3, 6, 7] if the permeability is equal to one. Hence, in this case the effective permeability is determined mainly by the polarization current which contribution is probably significant. Effective permeability no longer corresponds to magnetic moment of unit volume and can be unequal to one. According to [12, 13, 14] such permeability is a natural reaction of inhomogeneous fields on magnetic field, i.e. a reaction to the spatial dispersion, which is observed in figures 1, 2. In conclusion, it can be noted that , in general, the permeability determined by us cannot be associated with the total magnetic moment of a macroscopic body. Nevertheless, from a applied point of view (which means that we can estimate the wavelength in the medium, the phase velocity of the wave in the medium and its direction, refraction angles, refractive index, reflection and transmission coefficients), we decided to use this term. In paper [13] it is shown that the appearance of negative group velocity waves (fig. 2) with negative refraction (fig. 3), and, hence, the availability of negative effective permeability is associated with spatial dispersion.

5. Conclusions

Show that well-known metamaterial consisting of rectilinear thin wires array forming a 2D-square lattice within the non-conductive matrix has a wide band with negative group velocity waves, where wavelength of smoothed electromagnetic field distributions much more distance between the wires. Also negativ refraction and focusing of electromagnetic radiation from a point source are observed. Unlike most of the left-handed metamaterials that are composed of two separate inter-elements (typically, a split ring and a rod), in this paper it is proposed to use only one type of inclusion, namely the conductive rod.

6. Acknowledgments

This work was supported by the Russian Foundation for Basic Research (Project No. 118-58-53055 GFEN_a, 16-29-14045 ofi_m,17-02-01382), the Foundation for advanced scientific research of the "CSU" (FPNI) and by the Ministry of Education and Science of the Russian Federation (State Contract No. 3.5698.2017/9.10).

References

[1] Veselago V G 1968 *SOV. PHYS. USPEKHI.* **10** 509
[2] Pendry J B, Holden A J, Robbins D J and Stewart W J 1999 *IEEE Transactions on Microwave Theory and Techniques* **47** 2075–2084
[3] Smith D R, Padilla W J, Vier D C, Nemat-Nasser S C and Schultz S 2000 *Phys. Rev. Lett.* **84** 4184
[4] Dong Z G, Zhu S N, Liu H, Zhu J and Cao W 2005 *Phys. Rev. E.* **72** 016607
[5] Chen H, Ran L, Huangfu J, Zhang X, Chen K and Grzegorczyk T M 2004 *Phys. Rev. E.* **70** 057605
[6] Pendry J B, Holden A J, Stewart W J and Youngs I 1996 *Phys. Rev. Lett.* **76** 4773
[7] Belov P A, Tretyakov S A and Viitanen A J 2002 *Journal of Electromagnetic Waves and Applications* **16** 1153–1170
[8] Davidovich M V, Stephuk J V and Shilovskii P A 2012 *Tech. Phys.* **57** 320
[9] Butko L N 2017 *AIP Conference Proceedings* **1874** 040002-1–040002-4
[10] Pavlov D A, Butko L N and Fedy A A 2015 *Journal of radio electronics* **11** 7
[11] Parimi P V, Lu W T, Vodo P, Sokoloff J, Derov J S and Sridha S 2004 *Phys. Rev. Lett.* **93** 073902
[12] Landau L D and Lifshitz E M 1960 *Electrodynamics of Continuous Media* (Pergamon Press)
[13] Agranovich V M and Gartstein Y N 2006 *Physics-Uspekhi* **49** 1029
[14] Pamyatnykh E A and Turov E A 2000 *Foundations of the Electrodynamics of Material Media in Variable and Inhomogeneous Fields* (Nauka)

IOP Conf. Series: Journal of Physics: Conf. Series **1092** (2018) 012017 doi:10.1088/1742-6596/1092/1/012017

Experimental study of graphene-based plasmonic crystals in a strong-coupling regime

Andrey Bylinkin[1], Michail A. Kashchenko[1], Vitaly V. Mikheev[2], Dmitry Svintsov[1]

[1] Laboratory of 2D Materials' Optoelectronics,
Moscow Institute of Physics and Technology, Dolgoprudny 141700, Russia
[2] Moscow Institute of Physics and Technology, Dolgoprudny 141700, Russia

E-mail: `Bylinkin@phystech.edu`

Abstract. We have fabricated a large-area graphene-based transistor with top metal grating, which represents plasmonic crystal containing two-dimensional electron system gated by periodic metal grating. Using Fourier-transform infrared spectroscopy, we have studied the graphene plasmon modes in transmission spectra, which were efficiently excited by the incident light with the perpendicular polarization to metal grating. We have researched the dependence of resonance frequency of graphene plasmons on the bottom gate voltage. Here we demonstrate that excited graphene plasmons are strongly coupled to top metal grating and have the linear dispersion. These results not only prove the previous theoretical works but also demonstrate the possibility of significant reducing photon wavelength which can be controlled by modulating the carrier density of graphene via electrical gating. This result is important for design of highly effective tunable terahertz and far-infrared detectors.

1. Introduction

Most previous structures for excitation of graphene plasmons dealt with patterned graphene, which hampers its electronic properties and disables the possibility of controlling optical properties with gates. The most natural structure free of these drawbacks represents a continuous 2d layer covered by metal or dielectric grating [1]. Previously, such structures were studied in a weak couplig regime, where the properties of graphene plasmon are unaffected by metal gate [2]. By fabricating metal gratings in extreme proximity to graphene ($\sim$15 nm hBN), we study experimentally the opposite regime of graphene plasmons strongly coupled to metal gates.

2. Methods

The sample is used in this study consists of a chemical vapour deposition (CVD) graphene, multilayer hBN, periodic metal grating is made of Ti and Au electrodes. CVD graphene was transferred to p-doped Si/SiO_2 substrates and covered by multilayer hBN using standart wet transfer technique [3]. Thickness of hBN d is around 15 nm with effective dielectric screening $\epsilon = 4.3$. Au electrodes of 30 nm thickness were deposited onto graphene sheets in vacuum using ion-beam sputtering. Periodic metal gratings were fabricated in a 5×5 mm^2 area at the centre of the device using e-beam litography. Period of the metal grating is equal to $T = 500$ nm, width of metal strip is $W = 250$ nm.

The carrier concentration in graphene are controlled by silicon bottom gate. The gate-dependent measurement of the electrical resistance of graphene shows that the charge neutrality point (CNP) $V_{CNP} = 14$V and device have mobility μ around 1,000 $cm^2\ V^{-1}\ s^{-1}$. Gate voltage-dependent transmission spectra in the spectral range from 50 to 450 cm^{-1} were measured using Fourier spectrometer Bruker Vertex 80v. Infrared absorption is directly related to $\Delta T/T_0$, where T_0 is the transmission spectra at CNP and $\Delta T = T_0 - T$. In order to extract only plasmon contribution to attenuation of transmission spectra we studied the ratio of two infrared absorptions with the perpendicular polarization $(\Delta T/T_0)_\perp$ to parallel polarization $(\Delta T/T_0)_\parallel$ of the incident radiation to metal grating.

3. Result and Discussion

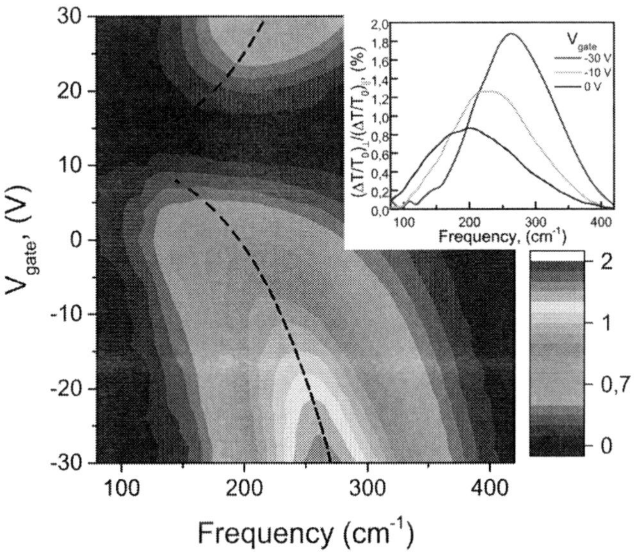

Figure 1. Intensity map of the attenuation spectra corresponding to graphene plasmon resonance as a function of gate voltage and frequency. The black dashed lines shows the calculated graphene plasmon dispersion law using equation (1). The inset shows the cuts of the intensity map along certain gate voltage.

Our measurements shows that the excited graphene plasmon mode can be tunable by gate voltage Figure 1. The experimental plasmon resonance frequency had good agreement with theoretical prediction [4], that plasmons is strongly coupled to metal grating. We have used the dispersion law of gated graphene plasmons, assuming $qd \ll 1$:

$$\omega_p = q_p\sqrt{\frac{4e^2 d v_0 k_F}{\hbar\epsilon}}, \tag{1}$$

where $v_0 = 10^6$ m/s is Fermi velocity in graphene, $k_F = \sqrt{\pi|n|}$ is the Fermi wavenumber, n is the charge carrier density in graphene, $\hbar$ is Plancks constant, and e is the elementary charge. Charge carrier density calculated $n = \alpha \times (V_{gate} - V_{CNP})$, where $\alpha = 3.5 \times 10^{-4}$ for our device.

IOP Conf. Series: Journal of Physics: Conf. Series **1092** (2018) 012017 doi:10.1088/1742-6596/1092/1/012017

4. Conclusion

We have experimentally studied the plasmon mode in a two-dimensional plasmonic crystal. The graphene plasmons in this device have the linear dispersion law and strongly coupled to metal grating. Moreover, we have studied the plasmon resonance frequency dependence on charge carrier concentration ($\omega_p \propto n^{1/4}$), which is signature of massless Dirac electrons. The tunable plasmon resonances observed here in graphene with the symmetric metal grating can be generalized for designing plasmonic crystals with an asymmetric grating for plasmon-enhanced terahertz and far-infrared photodetection.

5. Acknowledgments

This work was performed using equipment of MIPT Shared Facilities Center and with financial support from the Ministry of Education and Science of the Russian Federation (Grant No. RFMEFI59417X0014). AB, MK, and DS acknowledge Russian Foundation for Basic Research, grants # 16-37-60110 and 16-29-03402.

References

[1] Gao W, Shi G, Jin Z, Shu J, Zhang Q, Vajtai R, Ajayan P M, Kono J and Xu Q 2013 *Nano letters* **13** 3698–3702
[2] Gao W, Shu J, Qiu C and Xu Q 2012 *ACS nano* **6** 7806–7813
[3] Suk J W, Kitt A, Magnuson C W, Hao Y, Ahmed S, An J, Swan A K, Goldberg B B and Ruoff R S 2011 *ACS nano* **5** 6916–6924
[4] Popov V, Fateev D, Ivchenko E and Ganichev S 2015 *Physical Review B* **91** 235436

Bilayer graphenes with antidots: structures, properties and applications

L A Chernozatonskii[1], V A Demin[1], S V Erohin[2], D G Kvashnin[1,2], A V Krasheninnikov[3], P.B. Sorokin[1,2]

[1]Emanuel Institute of Biochemical Physics RAS, Moscow 199339, Russian Federation
[2]National University of Science and Technology MISIS, Moscow 119049, Russian Federation
[3]Helmholtz-Zentrum Dresden-Rossendorf, Institute of Ion Beam Physics and Materials Research, 01328, Dresden, Germany

cherno@sky.chph.ras.ru

Abstract. We consider structures, properties and possible applications of next bilayer graphene (BLG) materials with nanoholes. Formation of folded nanoholes in bilayer graphene: modelling and comparison with experiments will discussed. The most advantageous shape of the BLG holes was determined in a manner analogous to the method used by Woolf to find the optimal shape of the crystals, unlike the method, the edge energy depends not only on the direction, but also on the cut along the direction. This suggests that the optimal shape of the holes depends on their size. The comparisons with the experiments of M.Takeguchi group (Japan) was carried out. The holes in single and few-layer graphene films can be the impacts of heavy high-energy ions. To carry out the comparison with the experiments, a collaboration with N. Nebogatikova group (Russia), was carried out. We have studied the atomic structures of nanoholes in both bilayer and multilayer graphene using two-temperature molecular dynamics simulations requires proper treating of dynamic energy exchange between electronic and ionic subsystems in the irradiated system. Different stable forms of nanoholes with closed edges are considered in BLG structures of AA, AB and Moiré stacking's. We have found stable BLG nanomeshs with both semimetal, metal and semiconductor properties. Some applications of similar BLG structures are proposed.

1. Introduction

Successful synthesis and intensive investigation of graphene [1-3] yields many outstanding properties of this materials which can be used in the various fields of science and technology. Like graphene monolayer bilayered graphene (BLG) also attracts a lot of attention to the worldwide scientific community. In BLG the Dirac cone transforms to parabolic bands in which small band gap (~10 meV) can be opened by molecular doping or applying an external electric field [3-6]. Very recently, for the first time for a purely carbon structure on a graphene with Muar structure, a superconductivity was discovered [7].

One of the promising way of the opening of band gap in graphene monolayer by formation of periodically arranged antidots (nanoholes) [8] significantly increases the speed of development of BLG structures [9]. Unlike graphene monolayer where holes have open edges which are unstable and passivated another atoms, holes in bilayered graphene tend to closing edges [10–12]. Moreover, ab initio calculations predicted that the connection of the edges even does not require overcoming of any barrier

[11] and therefore hollow sp2-hybridized graphene structure forms spontaneously. The possibility of formation of such structures was confirmed in the experiments [13, 14]. Also, it is worth noting that, bilayer graphene with periodically located holes is a very interesting and promising object for use in the field of semiconductor optoelectronics. Presence of the holes in the structure of hollow sp2 carbon structures drastically changed the electronic properties of BLG by appearance of additional mini-bands in the vicinity of Fermi energy level. Such changes in the band structure could be explained in the terms of the redistribution of the electron density directly on the newly formed topological defects. In the report we consider structures, properties and possible applications of next bilayer graphene materials with nanoholes.

2. Formation of folded nanoholes in BLG

Other paragraphs Different stable forms of nanoholes with closed edges are studied in BLG structures of AA, AB and Moiré stacking's. The modeling examples and their comparisons with experiments are shown in papers [10-12, 14, 16].

The most advantageous shape of the BLG holes was determined in a manner analogous to the method used by G. Wulff to find the optimal shape of the crystals, which can also be used to determine the shape of the pores for two-dimensional crystals [17,18]. However, in our case, unlike the Wulff method, the edge energy depends not only on the direction, but also on the cut along the direction. This suggests that the optimal shape of the holes depends on their size. A certain size leaves only one definite optimal cut for any direction. Some examples of such constructions are shown in figure1: There are AA (a) and twisted 5°tBG (b), and predicted optimal shape comparison with the observed nanoholes by electron beam preparation in 5°tBG (Japanise M.Takeguchi group [14]).

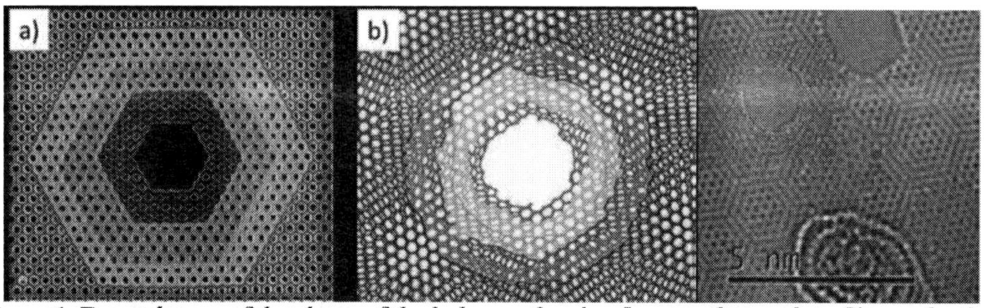

Figure1. Dependences of the shape of the holes on the size for a random point of its nucleation in a BLG and predicted optimal shape comparison with the observed nanoholes. [14]

3. The holes in single and few-layer graphene films

Some The holes in single and few-layer graphene films can be the impacts of heavy high-energy ions. To carry out the comparison with the experiments of N. Nebogatikova group (Novosibirsk, RF) [15], we have studied the atomic structures of nanoholes in both bilayer and multilayer graphene films. To interpretation of the experimental results through atomistic simulations requires proper treating of dynamic energy exchange between electronic and ionic subsystems in the irradiated system. We used two-temperature molecular dynamics model in which atomic subsystem is described by classical molecular dynamics simulations, whereas electronic subsystem is characterized by local electronic temperature and treated as a continuum on a regular grid. The model account for energy transfer between these subsystems. For the calculations we used the LAMMPS code.

In the Fig. 2 are shown: a) hole size as a function of the number of Xe ions irradiated considered region; the range of experimental values of the pore size is denoted by orange; upper and down series of the insets (relatively to the data) represent atomic structures of holes formed by ions with energy 167 and 26 MeV, respectively; b) temperature and structural changes of three-layered graphene in MD simulations after the irradiation at Xe^{+} energy 100 MeV; the evolution of the graphene structure at the chosen steps is shown. In the inset the zoomed side view connected edges is shown; c) band gap of the infinite ribbon with connected edges as a function of the distance between connected regions.

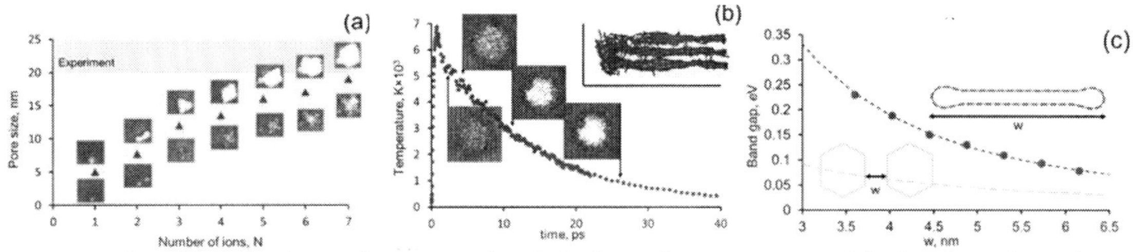

Figure 2. The schematic views of corresponding hole formation structures and the band gap dependence via the region W between folded pores in bilayer graphene.

4. Examples of 30°tBG nanomeshes

We have modeled different stable types of BGNMs and found stable BLG nanomesh structures with both semi metal, metal and semiconductor properties [10-12,14,16].

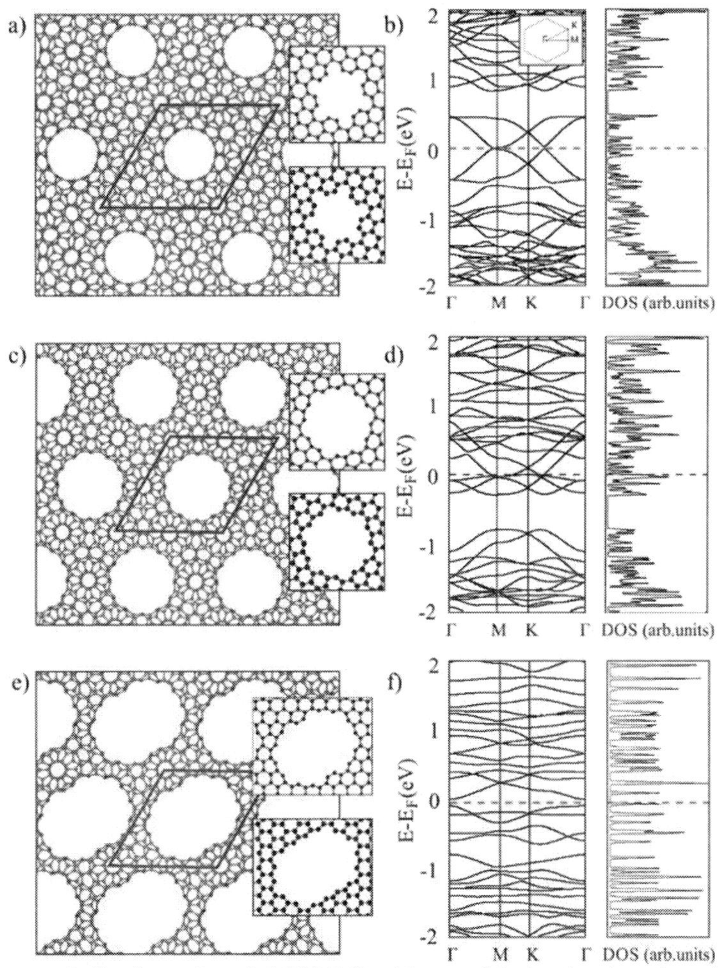

Fugure 3. Perforated moiré 30°tBG with round holes (a, c) and oval holes (e), their electronic spectra and the density of electronic states (DOS) (b, d, f).

Some their examples of 30°BG nanomeshes [16] are shown in figure 3: a) a structure with a period of l = 17.7Å and a smallest hole radius r_1 = 0.4 nm, c) a structure with l = 17.6Å and a radius of r_2 = 0.6 nm., e) periodic "oval 1" 30°tBG structure with l = 16.91Å, (b, d, f) electronic spectrum and the density of electronic states (DOS) of each nanomeshes. A parallelogram designates a unit cell; a dashed line is the Fermi level. On the inset - fragments of the upper and lower layers of the structure

5. Conclusion

Thus, we considered examples of a new class of bi-graphite nanostructures with different shapes and sizes of holes with closed boundaries. These structures are multiply connected systems of completely sp2-hybridized carbon atoms with inherent features in electronic spectra. So the band structures of BGNM nano-networks differ sharply from the spectra of monographene meshes, whose pairs, bound along the edges of the holes, constitute these bi-graphene meshes.

We note that applying a transverse voltage to semimetallic nanoscreens 30°tGNMr1 and 30°tGNMr2, will result in a shift of the Fermi level, analogous to the shift in graphene [1]. Thus, it is possible to obtain a conduction-change effect on the semiconductor, and, consequently, to use these nanoscale as switching elements in nanoelectronics. The result of the presence in all tGNM meshes of quasi-local modes in the electronic spectrum, the number of doubly larger modes localized near similar holes in single-layer graphene nanomeshes. This leads to the appearance in a wide range of peak energies of electronic states. In this connection, there appears the prospect of effective use of the considered 30°tGNMs in photovoltaics and optoelectronics, as well as in thermoelectric elements like the use of moiré twisted BG structures [19, 20]. Obtained structure with close edges holes can display "p- or n-doped" metallic or semiconductor conductivity. Note that, because of the absence of a center of symmetry, the similar BLG nanomeshes should display significant piezoelectric effect, which increases the interest in the use of these structures in electro- or opto- mechanical nanodevices.

Acknowledgments
The work was supported by RFBR RFBR 17-02-01095 and grant of president of Russian Federation for government support of young PhD scientists (D.G.K.) - MK- 3326.2017.2.

References

[1] K S Novoselov, A K Geim, S V Morozov *et al.* 2004 *Science* **306** 666
[2] K S Novoselov, D Jiang, F Schedin, *et al.* 2005 *Proc. Natl. Acad. Sci.* **102** 10451
[3] K S Novoselov 2011 *Rev. Mod. Phys.* **83** 837
[4] T Ohta et al. 2006 *Science* **313** 951
[5] W Zhang et al. 2011 *ACS Nano* **5** 7517
[6] Y Zhang et al. 2009 *Nature* **459** 820
[7] Y Cao et al. 2018 *Nature* doi:10.1038/nature26160
[8] W Oswald, Z Wu 2012 *Phys. Rev.* B **85** 115431
[9] R Zan, Q M Ramasse, U Bangert, K S Novoselov 2012 *Nano Lett.* **12** 3936
[10] L A Chernozatonskii, V A Demin, A A Artyukh 2014 *JETP Lett.* **99** 3094
[11] D G Kvashnin, et al. 2015 *Nano Res.* **8** 1250
[12] L A Chernozatonskii, V A Demin, Ph Lambin 2016 Phys. Chem. Chem. Phys.18 27432.
[13] K He, et al. 2015 *Nanoscale* **7** 11602
[14] Kano E., et al.. 2018 (submitted in *Small*)
[15] N A Nebogatikova., et al. 2018 (accepted to *Nanoscale*)
[16] L A Chernozatonskii, V A Demin 2018 *JETP Lett.* **107** 333
[17] V Artyukhov, Y Liua, B Yakobson 2012 *Proc. Natl. Acad. Sci.* **109** 15136
[18] M Teng, et al. 2013 *Proc. Natl. Acad. Sci.* **110** 20386
[19] Z Tan, et al. 2016 *ACS Nano* **10** 6725
[20] P S Mahapatra, et al. 2017 *Nano Lett.* **17** 6822

IOP Conf. Series: Journal of Physics: Conf. Series **1092** (2018) 012019 doi:10.1088/1742-6596/1092/1/012019

On laser assisted stabilization of 1D and 2D quantum inverted harmonic oscillators

V Chistyakov

Department of Mechatronics, University ITMO, Kronverksky Avenue, 49, Saint-Petersburg, 197101, Russia

v.chistyakov@corp.ifmo.ru

Abstract. Studied is the behaviour of a particle in negative 1D and 2D quadratic potentials driven by sinusoidal electrical field of laser irradiation in order to investigate the quantum analogue of dynamical stabilization. It is shown that for certain phases of the laser field depending on its frequency and initial coordinate uncertainty the mass centre of the evolving packet is stabilized near the potential maximum for a time of about two natural periods. The phases serve as bifurcation points switching the direction of final escape to the infinity.

1. Introduction. Dynamical stabilization in classical case

As a classical example may serve a famous Kapitza pendulum however commonly any straight standing or walking human may be treated as an IHO from biomechanical point of view.

When been harmonically driven on 'resonant' frequency an inverted oscillator is described by the Hamiltonian function $H(p,q)=\frac{p^2}{2}-\frac{q^2}{2}-f_0\sin(\tau\ \ \varphi)q$; and the matrix dynamical equation is as follows

$$\frac{d}{d\tau}\begin{pmatrix}q\\p\end{pmatrix}=\begin{pmatrix}0&1\\1&0\end{pmatrix}\begin{pmatrix}q\\p\end{pmatrix}+f_0\sin(\tau+\varphi)\begin{pmatrix}0\\1\end{pmatrix} \quad (1).$$

Its solution for general φ is as follows

$$q(\tau)=0.5\left(q_0+p_0+0.5(\sin\phi+\cos\phi)\right)e^{\tau}-0.5f_0\sin(\tau+\phi)+0.5\left(q_0-p_0+0.5(\sin\phi\ \ \cos\phi)\right)e^{-\tau} \quad (2).$$

It is spatially limited if and only if

$$q_0+p_0+0.5(\sin\phi+\cos\phi)=0 \quad (3).$$

Example. (*University ITMO Olympiad on theoretical mechanics, 2015, Chistyakov V.V.*) A thin straight pipe of length l rotates evenly at angular speed of ω in vertical plane around its middle where a tiny piece of matter is attached. What inclination angle ϕ of the pipe to horizon should it be released at in order not to leave out the pipe?

The solution of appropriate dynamical equation with initial conditions (ICs) $x(0)=x_0, \dot{x}(0)=u_0$ gives the space limited dependence

IOP Conf. Series: Journal of Physics: Conf. Series **1092** (2018) 012019 doi:10.1088/1742-6596/1092/1/012019

$x(t) = 0.5g\omega^{-2}\sin(\omega t+\phi)+\left(x_0 - 0.5g\omega^{-2}\sin(\phi)\right)e^{-\omega t}, t \quad 0$ provided the condition $u_0 + \omega x_0 = \dfrac{g\sin\left(\phi+\frac{\pi}{4}\right)}{\sqrt{2}\omega}, x_0 \in \left(-\frac{l}{2},\frac{l}{2}\right)$ is fulfilled. In particular for zero ICs $x(0)=0, \dot{x}(0)=0$ the stabilization arises at angles (phase shifts of gravity force) of $\phi = -0.25\pi; 0.75\pi$.

From the fundamental viewpoint of correspondence of classical and quantum mechanics it would be of great interest to explore whether this dynamical stabilization is inherited in a quantum IHO. And if it takes place then what its features will be conserved and what new in comparison with classical mechanics?

2. Quantum driven IHO

The significance of quantum inverted oscillator is extremely great in modern physics as an universal model of unstable quantum state of any particle. This model is used when modelling the tunnel effect or the «short time behavior in *superfluorescence*» and so on (see *e.g.* [1, 2]). Commonly, any metastable quantum state may be considered in first order as an IHO.

One dimensional "resonantly" driven inverted oscillator is described in the dimensionless variables of $\xi = x\sqrt{\frac{m\omega}{\hbar}}$, $f_0 = \frac{F_0}{\omega\sqrt{\hbar m\omega}}$, $\tau=\omega t$ by the following non-stationary Schroedinger equation (NSE):

$$i\frac{\partial\Psi(\xi,\tau)}{\partial\tau} = -\frac{1}{2}\frac{\partial^2\Psi(\xi,\tau)}{\partial\xi^2} - \frac{\xi^2}{2}\Psi(\xi,\tau) - f_0\xi\sin(\tau+\varphi)\Psi(\xi,\tau), \qquad (4).$$

The chain of substitutions such as $\rho = \xi + f_0\sin(\tau+\varphi), \Psi(\rho,\tau) = \psi(\rho)g(\rho,\tau)$ and $g(\rho,\tau) = e^{-i\left(E(\rho)+\frac{f_0^2}{4}\right)\tau + i\frac{f_0^2\sin(2\tau+2\varphi)}{8}} z(\rho,\tau)$ with the $E(\rho) = -\frac{1}{2\psi_0(\rho)}\frac{d^2\psi_0(\rho)}{d\rho^2} - \frac{\rho^2}{2}$ (see below) transforms this NSE into following exactly integrable partial differential equation (PDE):

$$i\frac{\partial z(\rho,\tau)}{\partial\tau}\cdot\frac{1}{z} + \left(\frac{i}{z}\frac{\partial z(\rho,t)}{\partial\rho} + \frac{dE}{d\rho}\tau\right)f_0\cos(\tau+\varphi) + i\frac{d\psi(\rho)}{d\rho}\frac{1}{\psi}f_0\cos(\tau+\varphi) + \frac{1}{z}\frac{\partial z(\rho,\tau)}{\partial\rho}\cdot\left(-i\frac{dE(\rho)}{d\rho}\tau\right)$$

$$+\frac{1}{2z}\frac{\partial^2 z(\rho,\tau)}{\partial\rho^2} - \frac{1}{2}\left(\frac{dE(\rho)}{d\rho}\right)^2\tau^2 - \frac{i}{2}\frac{d^2E(\rho)}{d\rho^2}\tau + \frac{1}{z}\frac{\partial z(\rho,t)}{\partial\rho}\cdot\frac{d\psi}{d\rho}\cdot\frac{1}{\psi} + \left(-i\frac{dE(\rho)}{d\rho}\tau\right)\frac{d\psi}{d\rho}\cdot\frac{1}{\psi} \quad 0 \qquad (5).$$

As for an initial wave function (w.f.) in ideal case it should be an eigen function of dimensionless ICs' operator: $\hat{c} = \hat{p} + \hat{\xi} = -i\frac{\partial}{\partial\xi} + \xi$. But such w.f. $\psi_{eig} = C\exp\left(-0.5i\xi^2\right)$ doesn't correspond to whatever physical state so the quantum ICs above may only be fitted by choosing a converging to zero point 1D Gaussian "flux"

$$\Psi(\xi,0) = \psi_0(\xi) = \left(\frac{1}{\sqrt{2\pi}\sigma_0}\right)^{\frac{1}{2}}\exp\left(-\frac{i\xi^2}{2} - \frac{\xi^2}{4\sigma_0^2}\right), \sigma_0 = \sqrt{\overline{\xi^2(0)}} >> 1.$$

The PDE (5) may be solved by substitution $z(\rho,\tau) = \exp\left(\alpha(\tau)+\beta(\tau)\rho+\gamma(\tau)\rho^2\right)$, and for the major coefficient γ which defines the current half width $\sigma(\tau)$ (Fig. 2) of w.f. packet (Re $\gamma(\tau)$) and re-distributing 1D flow $\vec{j}(\xi)$ of probability density inside it (Im $\gamma(\tau)$) it is received

$$\gamma(\tau)=\frac{1}{4\sigma_0^{\,2}}+\frac{i}{2}+\left(\frac{1}{2\sigma_0^{\,2}}-\frac{i}{8\sigma_0^{\,4}}\right)\tau+\frac{1}{2}tg\left(2i\tau-arctg\left(+i\ \frac{1}{2\sigma_0^{\,2}}\right)\right) \quad (6).$$

This formula describes a strong collapse of the packet at time interval $\tau \le \tau_0$. And the collapse is the stronger the larger is an initial half width of σ_0 serving the quantum analogue of $l/2$ of the pipe (Fig. 1, 2). Just while collapsing the "beam" converts to diverging one so quick flux switching occurs.

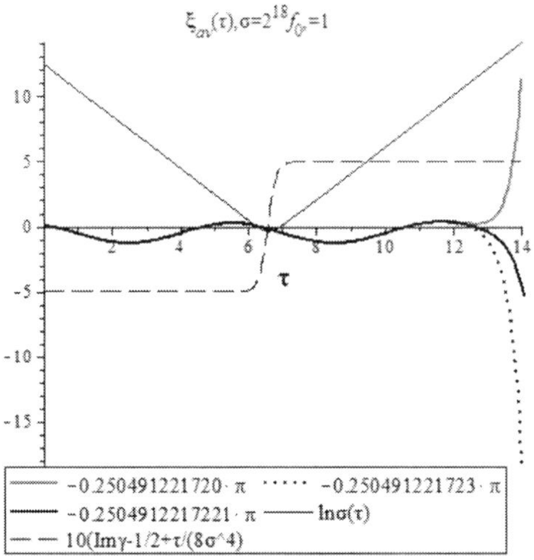

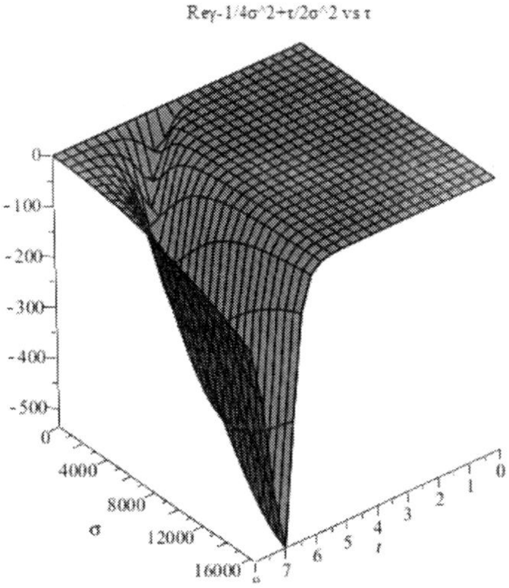

Figure 1, Short time stabilization and bifurcation of $\xi_{av}(\tau)$ at "resonant" frequency $\Omega=\omega$ for $\sigma_0=2^{18}$ and $f_0=1$; the curves for $\ln\sigma(\tau)$ and $\mathrm{Im}\,\gamma(\tau)-\frac{1}{2}+\frac{8\tau}{\sigma_0^4}$.

Figure 2. 3D plot for the value of $\mathrm{Re}\,\gamma(\tau)-1/4\sigma_0^2+\tau/2\sigma_0^2\equiv-1/4\sigma^2(\tau)$

The numerical solution in wide range of $n=\log_2\sigma_0\in[-8,21]$ gives for the stabilizing phases the inverted Fermi – Dirac dependence $\phi^-(\sigma_0)=-\frac{\pi}{4}-\frac{\pi}{4\cdot(1+a2^n)}$ (Fig. 3) and $\phi^+(\sigma_0)=\phi^-(\sigma_0)+\pi$, ($a$ = 0.076962 (0.001744), $adj-R^2$ =0.99965). At these phases: a) the w.f. packet center $\xi_{av}(\tau)=\overline{\xi(\tau)}$ oscillates near the current equilibrium point for about of two "natural time periods", b) takes place a clearly expressed bifurcation of final outlet of the center to infinity, and c) it takes also place the peculiarity of $\xi_{av}(\tau)$ at the time of the packet maximal constriction (Fig. 1,3).

The observed pattern gives a hint on possibility of a long quantum separatrix though not revealed yet because of precision limitation of used program Maple. As for observed quasi-separatrixes they are the longer the larger (not shorter!) was an initial wave packet width σ_0 and consequently the closer was the initial w.f. $\psi_0(\xi)$ to ideal nonphysical one $\psi_{eig}(\xi)$. However the paradox here seems only for the simple reason that it is the case of $\sigma_0\to\infty$ which corresponds to classical limit and not of $\sigma_0\to 0$ what in common agrees completely with the behavior of $\phi^{\pm}(\sigma_0\to\infty)$-values coinciding with the classical ones $\phi^{\pm}=-0.25\pi;0.75\pi$.

It's worth also noting that on similar effect of wave packet delay near top point of repelling quadratic potential had indicated the author of research [3] devoted to exactly solvable discrete Schroedinger models.

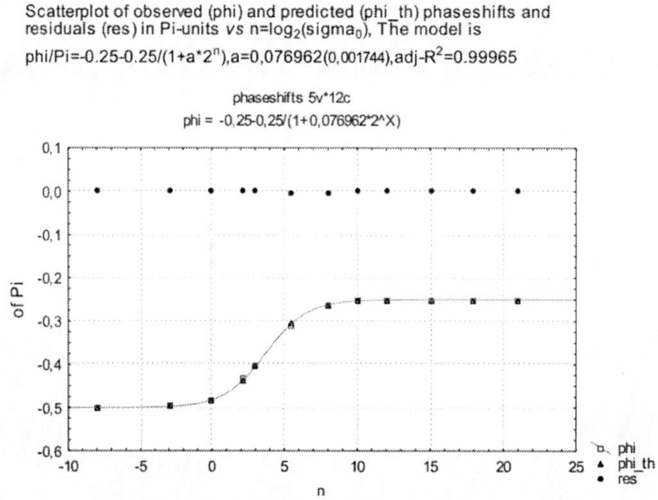

Figure 3. An inverted *Fermi–Dirac*-like dependence for the stabilizing phase ϕ^{-} *vs* $n = \log_2 \sigma_0$.

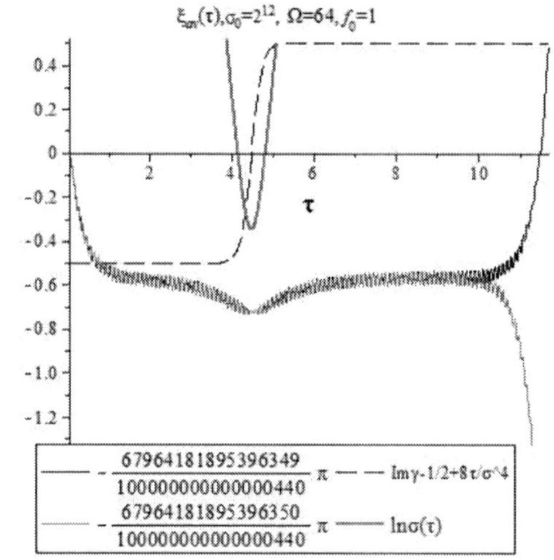

Figure 4. Quasi-stabilization and bifurcation at high frequency $\Omega = 64\omega$, $\sigma_0 = 2^{12}$.

References

[1] Serban I and Wilhelm F 2007 *Phys. Rev. Lett.*, **10**, 101 (see at https://www.researchgate.net/publication/5913852)
[2] Nogami Y and Toyama F 1994 *Phys. Rev.* **E 49**(5) 4497
[3] Zakhar'ev B 1992 *Sov. J. Part. Nucl.* **23**(5), 1387 (Russian)

Acknowledgement
The research is supported by RBRF grant 16-08-00997 «Investigation of nonlinear multiple controlled mechanical systems by means of mathematical and computer modelling».

Effectivity of nanovaccine against tick-borne encephalitis

N Chopenko[1], A Mazeika[1], L Davydova[1], A Stenkova[2], G Leonova[3], E Kostetsky[1], N Sanina[1]

[1] Far Eastern Federal University, Sukhanova St., 8, Vladivostok, 690950, Russia
[2] G.B. Elyakov Pacific Institute of Bioorganic Chemistry, FEB RAS, Prospect stoletiya Vladivostoka 159, Vladivostok, 690022, Russia
[3] G.P. Somov Institute of Epidemiology and Microbiology, Selskaya str., 1,Vladivostok 690087, Russia

Email: natali_1389@mail.ru

Abstract. The tubular immunostimulating complex (TI-complex) is a new nanoparticulate antigen delivery system, which was developed to enhance immunogenicity of different subunit antigens within anti-infectious vaccines and increase their economic efficacy and safety. TI-complexes are self-organized from mixture of triterpene glycosides from *Cucumaria japonica*, cholesterol and monogalactosyldiacylglycerol (MGDG) from marine macrophytes. MGDG plays role of lipid matrix for subunit protein antigen interrupted in TI-complexes. Microviscosity of MGDG was shown to influence the conformation and immunogenicity of protein antigen. Present work was aimed to study adjuvant effect of TI-complexes on immunogenicity of chimeric protein antigen based on E protein domain III of tick-borne encephalitis (TBE) virus and OmpF porin of *Yersinia pseudotuberculosis* depending on MGDGs isolated from different marine algae and seagrass. It was shown that TI-complex including MGDG from *Ulva lactuca* was most effective. Immunization by chimeric antigen incorporated in TI-complexes provided a high level of animal protection in experimental infection with TBE. Thus, the proposed construction is promising for the development of vaccines against TBE.

1. Introduction

The search for new approaches to the prevention of infectious diseases is a priority problem of modern fundamental medicine due to the constant growth of economic damage from infectious diseases in the world. In particular, new approaches are needed to prevent such widespread and deadly viral illness as tick-borne encephalitis (TBE), since classical inactivated vaccines have various side effects and, moreover, do not provide reliable protection against a neuroviral infection. The use of safe isolated antigens of microbial pathogens in subunit vaccines allows to avoid this problem. However, most subunit antigens have insufficient immunogenicity. The obtained results showed the prospect of using nanoparticulate tubular immunostimulating complexes (TI-complexes) to deliver subunit antigens and to enhance immune response against them and microbial pathogen as a whole [1, 2]. TI-complexes are self-organized from the mixture of cholesterol with biologically active components - triterpene glycoside from marine invertebrate *Cucumaria japonica* and glycolipid monogalactosyldiacylglycerol (MGDG) from marine macrophytes (algae and seagrasses). These tubular nanoparticles are characterized by outer and inner diameters of about 16 nm and 6 nm, respectively, and the length of about 500 nm [3]. The glycolipid matrix is necessary to incorporate antigen in TI-complex. Different physicochemical properties of MGDGs isolated from different marine macrophytes allow to influence the conformation of the incorporated protein antigen and to obtain the most immunogenic anti-infectious vaccine construction [1]. Present work was aimed to elaborate effective and safe anti-TBE vaccine construction based on TI-complex and recombinant chimeric protein, consisting of the sequence part of protein E (III domain) of TBE virus (TBEV) connected with porin OmpF of *Yersinia pseudotuberculosis* by linker sequence (OmpF-EIII) [4]. The resulting chimeric protein has a number of advantages. Thus, E protein domain III contains the main antigenic determinants of the protective antigen of TBEV. Membrane protein OmpF plays role of anchor, which provides better incorporation of the chimeric antigen in the glycolipid matrix of TI-complexes. In addition, the bacterial protein in the content of the chimeric antigen facilitates its synthesis in the expression system of *Escherihia coli*.

2. Material and methods

2.1. Animals and immunization

The mice were divided into six experimental groups (ten animals in each): (1) mice, which were injected with phosphate-buffered saline (PBS) containing 0.125% n-octylglucoside (control); (2) mice immunized with individual OmpF-EIII; (3–6) mice immunized with OmpF-EIII incorporated into TI-complexes based on MGDGs from *Ulva lactuca*, *Sargassum pallidum*, *Laminaria japonica* or *Zostera marina*, respectively. The adult BALB/c (females) mice with the body weight of 18–20 g were immunized subcutaneously twice, applying a dose of 20 μg of OmpF-EIII *per* mouse, at an interval of 14 days in all groups. Experiment was terminated 28 days after the first immunization.

2.2. ELISA

Blood sera of the experimental mice were obtained to determine the content of the anti-EIII antibodies. The blood of the mice was collected immediately after decapitation and incubated at 37 °C for 2 h. After clot retraction, the samples were centrifuged at 1500 rpm for 10 min. The supernatant was taken into plastic tubes and stored at −20 °C. The content of anti-EIII antibodies in mice blood serum was determined by ELISA as described in [5].

2.3. Protective activity against TBE

Three groups each by ten mice were immunized subcutaneously according to the procedure described above (section *2.1.*): (1) PBS containing 0.125% *n*-octylglucoside (control); mice immunized with individual OmpF-EIII; (3) mice immunized with OmpF-EIII incorporated into TI-complex. Two weeks after the second immunization, the mice were subcutaneously infected with TBEV strain Dal'negorsk [6]. Then mice were monitored daily for 21 days to survival rate.

3. Results and discussion

The adjuvant effect of TI-complexes on immunogenicity of chimeric protein OmpF-EIII [4] were evaluated by production of anti-EIII antibodies in the blood serum of mice immunized with individual chimeric antigen OmpF-EIII and OmpF-EIII incorporated in TI-complexes comprising MGDG from *S. pallidum*, *U. lactuca*, *L. japonica* or *Z. marina*. Earlier, it was shown that TI-complexes exhibit a different adjuvant effect depending on the physicochemical properties of MGDG [1]. Results of present study confirm this conclusion. So, TI-complexes containing the most viscous MGDG from *Z. marina* contributed to a decrease in the immune response in comparison with the effect of the individual chimeric antigen, while the TI-complex based on MGDG from *U. lactuca* with lower microviscosity showed the highest adjuvant activity (Figure 1).

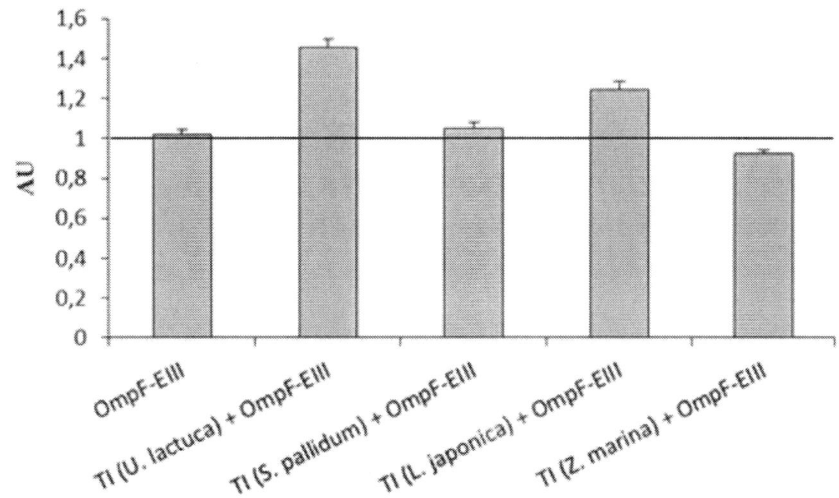

Figure 1. The content of anti-EIII antibodies in mice blood serum depending on MGDG in the composition of TI-complexes. X-axis: experimental groups of mice immunized with individual EIII-OmpF (OmpF-EIII) and OmpF-EIII incorporated in TI-complexes based on MGDG from *Ulva lactuca* (TI(U. lactuca + OmpF-EIII)), *Sargassum pallidum* (TI(S. pallidum + OmpF-EIII)), *Laminaria japonica* (TI(L. japonica + OmpF-EIII)) or *Zostera marina* (TI(Z. marina + OmpF-EIII)). Y-axis: the ratio between the content of anti-EIII antibodies in experimental groups and the control (mice injected with phosphate-buffered saline). Data are expressed in arbitrary units (AU) relative to the control value equal to 1 (the horizontal line).

A study of the protective activity of vaccine constructs based on the chimeric protein OmpF-EIII incorporated in the TI-complex showed that the mortality of mice immunized with both individual OmpF-EIII and OmpF-EIII incorporated in TI-complex was significantly reduced till 30% and 20%, respectively, against 70% in the non-immunized mice.

Hence, the chimeric protein OmpF-EIII has a high protective activity which significantly increases as a result of incorporation of the antigen into TI-complexes.

Acknowledgments
This work was supported by Russian Science Foundation (grant No. 15-15-00035-P).

References

[1] Sanina NM, Kostetsky EY, Shnyrov VL, Tsybulsky AV, Novikova OD, Portniagina OY, Vorobieva NS, Mazeika AN and Bogdanov MV 2012 *Biochimie* **94** 1048–1056
[2] Vorobieva N, Sanina N, Vorontsov V, Kostetsky E, Mazeika A, Tsybulsky A, Kim N and Shnyrov V 2014 *J Mol Microbiol Biotechnol* **24** 202–209
[3] Kostetsky EY, Sanina NM, Mazeika AN, Tsybulsky AV, Vorobyeva NS and Shnyrov VL 2011 *J. Nanobiotechnology* **9**:35
[4] Stenkova AM, Chopenko NS, Davydova LA, Mazeika AN, Bystritskaya EP, Portnyagina OY, Anastyuk SD, Kulbatskii DS, Lyukmanova EN, Dolgikh DA, Kostetsky EY and Sanina NM 2017 *Protein Pept Lett* **24** 974-981
[5] Vorobyeva NS, Mazeika AN, Davydova LA, Velansky PV, Tsybulsky AV, Kostetsky EY and Sanina NM 2015 *Russian Journal of Marine Biology* **41** 69–77
[6] Belikov SI, Kondratov IG, Potapova UV and Leonova GN 2014 *PLoS One* **9**:e94946

IOP Conf. Series: Journal of Physics: Conf. Series **1092** (2018) 012021 doi:10.1088/1742-6596/1092/1/012021

2D and 2.5D plasmonic metastructures for surface-enhanced infrared absorption applications: fabrication and properties

D A Chubich[1,*], D A Kolymagin[1], R D Zvagelsky[1], A V Pisarenko[1], E S Zhukova[1], E V Korostylev[1], D V Negrov[1], A G Vitukhnovsky[1,2]

[1]Moscow Institute of Physics and Technology (State University), Institutskii per. 9, Moscow Region, Dolgoprudny, 141700, Russian Federation
[2]Lebedev Physical Institute, Russian Academy of Sciences, Leninskii Prospect 53, Moscow 119991, Russian Federation

*chubich.conference@gmail.com

Abstract. Metamaterials possesses novel electromagnetic properties and have great potential for applications. Different type of nanosized plasmonic antennas considered to be effective functional materials in infrared spectrum. Still, strict requirements on dimensions and periodicity of nanostructures limits their practical implementation. Here, we have fabricated 2D and 2.5D different topology arrays of nanosized plasmonic antennas using e-beam and direct laser writing (DLW) lithography techniques. Appling the DLW method we created periodic polymer structures that act as a template for creating plasmonic nanoantennas with metallization and lift-off techniques. Improvement of the morphological parameters of such structures is possible due to STED-DLW (STimulated Emission Depletion assisted Direct Laser Writing) compatible photoresist. With e-beam lithography and lift-off techniques we obtained arrays of gold plasmonic nanoantennas. Using standard Fourier-transform microscopy we measure the spectral properties of fabricated nanostructures. We show that observed position, intensity and Q-factor of the collective surface plasmon resonance are directly conditioned by fabrication process. Thus, the studied structures can be applied as substrates for surface-enhanced infrared absorption (SEIRA). SEIRA antenna arrays in its turn possess considerable potential in real-world applications such as chemical and biological vibrational sensing.

1. Introduction

Special designs of metastructures, known as plasmonic antennas, are employed to overcome limitations of IR-spectroscopy. Such nanoantennas provide huge electromagnetic near fields, which can be used to enhance vibrational signals of molecules located in such hot spots [1]. Following this approach of surface-enhanced infrared absorption (SEIRA), one can increased the sensitivity by up to 5 orders of magnitude in comparison to conventional infrared transmittance spectroscopy. The development of new approach for creating SEIRA-active structures is a relevant scientific task.

2. STED-inspired direct laser writing

It was shown that optical lithography known as direct laser writing (DLW) competes (both in lithography speed and manufacturing cost) with e-beam lithography in usability efficiency for creation of the array of nanoantennas [2]. The DLW method is based on the phenomenon of photopolymerization as a result of two-photon absorption of light when the photoresist is exposed to focused femtosecond laser radiation. The subsequent raster scanning allows to create arbitrary 2D and 2.5D-structures. The resolution of a direct laser letter is limited by diffraction and is estimated by half the wavelength of the radiation used for writing. Thus, the DLW lithography allows to create large-sized antennas for a limited time, and hence increases the effective area for collective plasmonic excitations.

2.1. STED idea

The resolution of direct laser writing directly affects the roughness and other morphological defects of the obtained antennas. Accordingly, increasing the resolution of DLW makes it possible to improve the Q-factor of the plasmon resonance and, as a consequence, the selectivity and sensitivity of the IR-absorption.

The DLW resolution at a constant excitation wavelength can be increased applying the idea of STED (STimulated Emission Depletion) in order to reduce the axial and lateral dimensions of an effectively excited region. The essence of this method is to add a second depletion laser with a certain configuration of the electromagnetic field to stimulate radiative transitions along the edge of the focal spot, that is, to reduce the size of the excited region by using depletion field.

It should be noted that the use of the STED idea in luminescent microscopy was distinguished in 2014 by the Nobel Prize in chemistry [3]. A record resolution of 2.4 nm has been demonstrated in STED-nanoscopy with excitation wavelength of 775 nm. Thus, it is possible to create structures whose dimensions are less than the excitation wavelength by 2 orders of magnitude.

2.2. Experimental setup

We have created a setup for STED-DLW lithography based on a femtosecond Ti:Sapphire laser with a tunable excitation wavelength of 720-860 nm and a pulse duration of 50 fs. To realize the STED-idea we have used depletion CW-laser with a wavelength of 532 nm. The composition of the photoresist and the parameters were selected for reasons of suitable mode of laser writing [4].

With the help of improved technology of direct laser writing we fabricated 2D- (figure 1) and 2.5D-periodic structures with different spatial periods. The morphology of obtained structures was studied by scanning electron microscopy and atomic force microscopy.

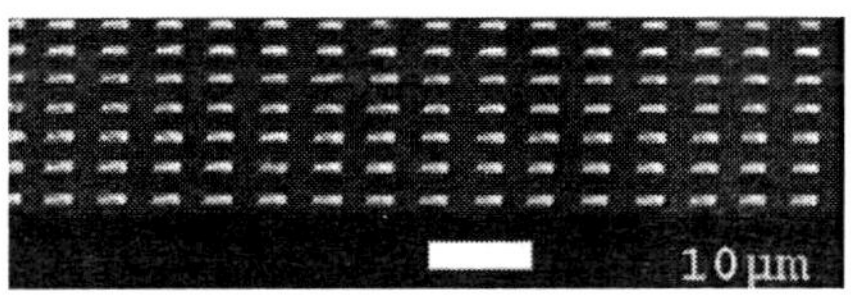

Figure 1. SEM-image of the periodic 2D "strokes" structure.

3. E-beam lithography

One of the most common technologies for the production of metastructures is an electron-beam lithography. Further, the sample is metallized by the lift-off method to form a gold/silver plasmonic structure of the required thickness.

3.1. Results and discussion

We have chosen the optimal dimensional nanoantenna parameters and have fabricated arrays of periodic golden strokes with lengths of 2 µm and 2.3 µm with a height of approximately 100 nm. We constructed a physical model of such antennas with an absorbing medium to further perform computer simulations of the interaction of a single antenna with an incident plane wave. In addition, we have fabricated c-shaped structures with a characteristic radius of 0.5 µm that are split-ring resonators (figure 2). Morphological properties of the obtained nanoantennas were studied by SEM, AFM, and dark field optical microscopy. Spectral properties of the fabricated structures was checked in mid-infrared spectral range. With standard Fourie-transform spectroscopic technique we measured the transmission spectra, as well as the dependence on the direction of polarization of the incident light.

Figure 3 shows typical transmission spectra of C-shape antenna array measured for different polarization of light. Reference spectra were taken on unstructured part of CaF_2 substrate. Minima in the spectra corresponds to collective plasmonic resonance of nanoantennas. Its position shows strong dependence on the relation between direction of polarization of electrical field and antenna gap. It shifts from 3.7 μm for E perpendicular to the gap of individual antenna to 7.7 μm for parallel orientation.

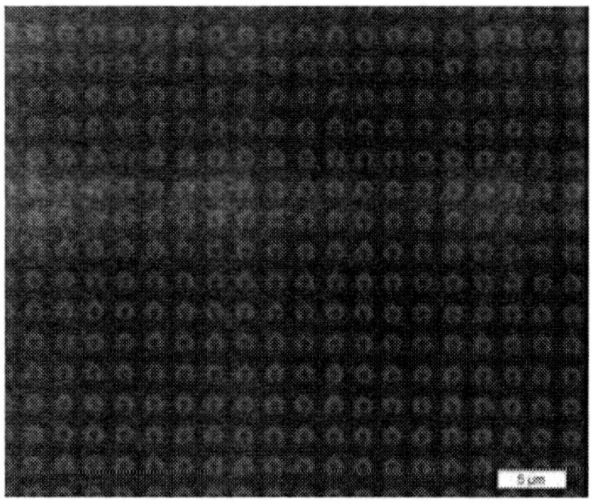

Figure 2. Dark field image of c-shaped antennas array from optical microscope.

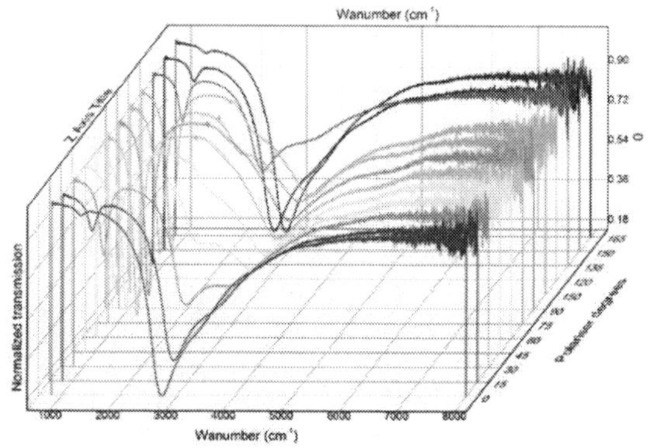

Figure 3. Spectra of normalized transmission of c-shaped antennas array for different polarizations of light.

Thus, the fine-tuning of incident light polarization allows one to adjust the plasmon resonance of nanoantenna to the required vibrational frequency of chemical.

4. Conclusion

We considered two methods of lithography for the fabricating of plasmonic metastructures for SEIRA. The STED-DLW setup assembled and writing parameters for polymer structures selected. Both methods used to fabricate polymer templates and gold plasmonic nanoantennas for SEIRA applications. A physical model developed, the idea of which is the coincidence of the plasmonic resonance frequency of the antenna and the absorption line of the chemical to study. The PETTA (pentaerythritol tetraacrylate) oligomer, which used in the photoresist for DLW process, studied as a test chemical. The interaction of plasmons in nanoantennas and the absorbing medium can be discussed in the terms of the Fano effect. The interaction of nanoantennas with linearly polarized radiation studied experimentally and computer simulation of the interaction with IR radiation has carried out to determine the enhancement factor for such a system.

Acknowledgments

The research work was granted by the Russian Science Foundation under project no. 17-79-20418. E-beam lithography was performed using the equipment of the MIPT Shared Facilities Centre and with financial support from the Ministry of Education and Science of the Russian Federation (Grant No. RFMEFI59417X0014). The authors are grateful to S S Vergeles for useful consultations concerning numerical calculations of metastructures.

References

[1] Thomas G. Mayerhöfer and Jürgen Popp 2018 Periodic array-based substrates for surface-enhanced infrared spectroscopy *Nanophotonics* Vol 7(1) pp 39–79

[2] Bagheri S, Weber K, Gissibl T, Weiss T, Neubrech F, Giessen H 2015 Fabrication of Square-Centimeter Plasmonic Nanoantenna Arrays by Femtosecond Direct Laser Writing Lithography: Effects of Collective Excitations on SEIRA Enhancement *ACS Photonics* Vol 2(6), pp 779–786
[3] Hell S W 2015 Nanoscopy with Focused Light (Nobel Lecture) *Angew. Chem. Int. Ed.* Vol 54 p 8054
[4] Vitukhnovsky A G, Chubich D A, Eliseev S P, Sychev V V, Kolymagin D A, Selyukov A S 2017 Advantages of STED-Inspired 3D Direct Laser Writing for Fabrication of Hybrid Nanostructures *Journal of Russian Laser Research* Vol 38 (4), pp 375-382

Isotropic Huygens sources made of clusters of nanoparticles for metasurfaces applications

R Dezert [1], P Richetti [1] and A Baron [1]

[1] University of Bordeaux, CNRS, CRPP-UPR8941,115 Avenue Schweitzer, F-33600 Pessac, France

E-mail: dezert@crpp-bordeaux.cnrs.fr, richetti@crpp-bordeaux.cnrs.fr, baron@crpp-bordeaux.cnrs.fr

Abstract. We present spherical clusters, composed by spherical dielectric or metallic inclusions, as a new kind of efficient and isotropic Huygens sources. We demonstrate that this design can allow a large and multimode overlapping of the electric and magnetic resonances excited at visible frequencies. They may also serve as building blocks for metasurface applications. We investigate their possible uses in high transmittance devices requiring a local phase control, and in thin absorber metalattices. They are particularly suited to bottom-up fabrication and self-assembly, offering an alternative to the classical lithography compatible Huygens meta-atom.

1. Introduction

Huygens sources have, during the last years, experienced a renewed interest as they open new possibilities for light manipulation at the nanoscale. Acting as nanoantennas, those objects are characterized by a highly asymmetrical scattering resulting from interferences between their optically induced modes. Dielectric nanoparticles of high refractive index were first proposed as promising candidates for Huygens sources realizations as they can exhibit strong electric and magnetic Mie resonances in the visible or infrared domain. Experimental investigation on Si, GaAs... dielectric spheres [1,2] has indeed demonstrate that unidirectional forward scattering was obtain at the Kerker condition, when the particle possess electric and magnetic dipolar response equal in both magnitude and phase.

However, for those spherical objects, as the involved modes do not spectrally overlap, Kerker condition is satisfied far from the resonances, where light is only weakly scattered. One solution to overcome this low efficiency is to consider anisotropic particles such as ellipsoids, nanodisks, nanobars, cones,... [3-5]. Tuning the aspect ratio of those structures can bring the resonances at the same wavelength, but, because of the break of symmetry, their properties then become dependent on the polarization and incident angle of the light. On the other hand, metallo-dielectric core-shell particles [6] or multi-layered spheres [7] were shown to work as isotropic resonant directional sources.

Our work [8] presents an alternative structure, acting as an efficient isotropic Huygens source, consisting of compact spherical nanoclusters of particles. In section 2, We demonstrate that this design allows spectral overlap of the electric and magnetic resonances of same order in the visible domain or IR. We provide different examples to illustrate that dipolar or extremely broadband sources can be achieved with either metallic or dielectric inclusions. Furthermore, by considering arrays of clusters, we show in section 3, that those elements might constitute interesting building blocks for metasurfaces.

Indeed, silicon clusters in the IR can be used to locally control the phase shift, in a range close to the full interval of 2π, while conserving high transmission, opening the path to the design of wavefront shaping devices. On the other hand, dissipative losses in silicon doubly resonant clusters in visible can be exploited to obtain perfect absorption in this spectral range.

2. Clusters of particles as resonant or broadband Huygens sources with high efficiency

As previously mentioned, we investigate the optical properties of spherical clusters of nanospheres, generated by considering the closest packing of small inclusions interacting repulsively into a large sphere. The scattering of those object is calculated using the T-matrix solver provided by Mackowski [9], based on generalized Mie theory.

Both dielectric and plasmonic clusters can exhibit optically induced magnetic response as well as an electric answer in the visible. In fact, an effective medium theory approach of the problem reveals that such clusters behave as a bulk dielectric homogenous medium which refractive index is mostly determined by the volume fraction of particles. Although it is known that high-refracted index spherical structures possess spectrally isolated resonances, recent work have shown that those resonance can spectrally merge for spherical particles with low index [10,11]. As the effective refractive index can deliberately be engineered in a cluster structure, it is indeed possible to target an optimum value of this parameter to ensure overlapped resonances while keeping a high value of the scattering efficiency. Such a cluster structure also provide more flexibility in term of choice of dispersion compare to the limited choice of natural material of low index for which dispersion are fixed.

Following this idea, we target a peak resonance wavelength of 500 nm, and explored the size and density parameters for the cases of silver and silicon inclusions with the aim of getting overlapped electric and magnetic dipolar resonances. Two examples of designs are presented in figure 1. The first, is a 60-cluster made of silver inclusions in water of radius $r = 15$ nm, with a filling ratio $f = 20\%$. The cluster radius is $R \approx 100$ nm. The second is a 13-cluster with radius $R \approx 123$ nm made of silicon inclusions of radius $r = 41$ nm, with $f = 47\%$ in air. For both designs, Maxwell Garnett effective medium theory indicates an effective refractive index for the structure close to $n+ik = 2+0.1i$.

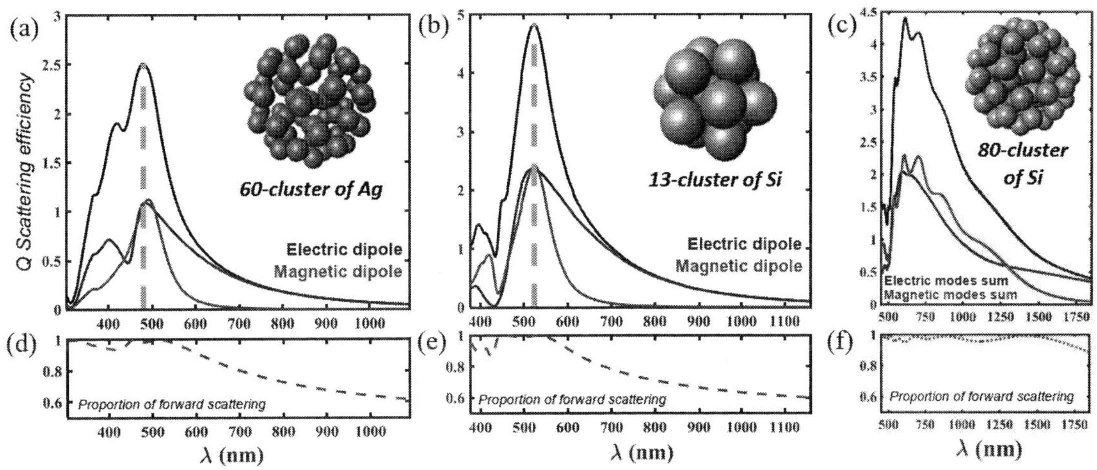

Figure 1. (a), (b), and (c) are the scattering efficiencies for 3 different clusters. The black curve is the total efficiency, the blue (red) curve is the efficiency of the electric (magnetic) modes. (d), (e) and (f) are the corresponding fractions of total energy scattered in the forward direction

Figures 1(a) and 1(b) show the total scattering efficiency of both systems. We see that an ideal Huygens dipole is reached in both cases near the designed wavelength, with a total efficiency at the peak around ~2.5 for the silver cluster and a higher value of ~5 for the silicon cluster. Figs. 1(d) and 1(e) shows that the scattered energy is radiated in the forward direction on a range of wavelength of about 100nm around the targeted wavelength.

For numerous applications, it is also interesting to maximize the forward scattering over a broad range of wavelengths. This can be realized by exploiting larger amount of mode all interfering destructively. Figure 1(c) and 1(f) present the optical properties of a cluster working as a broad Huygens multipole on a range of 1000 nm covering the visible and near infrared domain. The structure consists on a cluster of radius 300nm of 80 Si inclusions in water, of 50 nm in radius each. The volume fraction is f = 37%. Compare to the 13Si-cluster presented in figure 1(b), the radius of the full cluster has been increased leading to the excitation of much more multipole. At the same time, the volume fraction has been decreased, which induce a lower effective refractive index for the structure which in turn guarantee that electric and magnetic multipole of same order, up to the octupole, overlap two by two.

3. Metasurfaces applications: transparent devices with phase control and perfect absorber

Due to their directional scattering, Huygens sources open the possibility to manipulate transmitted wave, when organized into metalattices [12]. Functionalities such as focusing, beam deflection, vortex beam generation, holography,… are widely explore in the literature [13]. Such metasurfaces rely on the local spatial modulation of the phase, at the scale of a resonator, in order to shape the wavefront of the transmitted wave. At the same time, high transmission is conserved due to the absence of backscattering from the constituent meta-atoms.

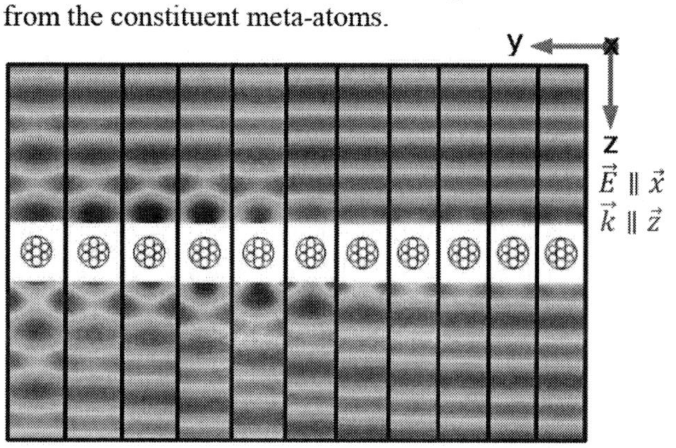

Figure 2. Beam shaping. Map of the electric field normalized to the incident field. Each box is a capture of a unit cell of an array of clusters with a given lattice spacing. Pitch takes value from 1585nm to 1305nm from left to right.

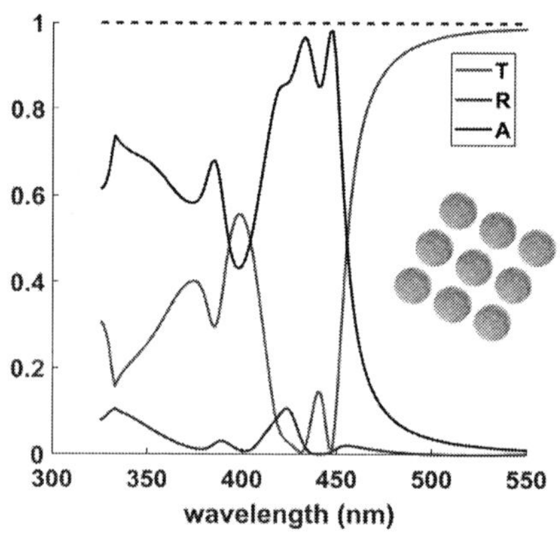

Figure 3. Perfect absorber. Absorption, transmission and reflection for an array of silicon Huygens clusters organized in a square lattice.

The mapping of the norm of the electric field presented in Figure 2, demonstrate that a cluster structure might be a suitable building block for such applications as it is possible to change the phase of the transmitted wave when varying the lattice spacing in an array of clusters. In this situation, we used silicon clusters at a wavelength of 1590nm as silicon exhibit no losses in the IR domain. The radius of the clusters are 390nm and the 13 Si inclusions composing the objects have a size of 129nm in radius. The full wave simulation were performed under COMSOL Multiphysics. The variation of phase in this situation cover a range of 1.7π when changing the pitch of the array from 1305nm to 1585nm, while the transmission is kept to value above 90% due to the overlapped resonances of the clusters at the working wavelength. Other numerical simulations not presented here indicates that it is also possible to obtain a variation of phase of 2π when varying the size of the clusters through a homothetic transformation.

Although high transmittance metadevices rely on lossless elements, it is also well known that lossy Huygens sources can be exploit to build perfect absorbers [14]. In our case, Silicon clusters in the visible domain are suitable candidates as the imaginary part of bulk silicon refractive index is quite high for wavelength below 500nm. Figure 3 present the transmission, reflection and absorption of an array of clusters designed as a Huygens dipole at the wavelength of 450nm. The cluster radius is 142nm and consist on 13 Si particles of radius 41 nm organized in a lattice with a pitch of 331nm. High intrinsic losses combines to Huygens source behavior allow us to obtain an absorption of nearly 100% at 450nm.

4. Conclusion

We proposed compact assemblies of particles into spherical clusters to fill the lack of Huygens sources with isotropic properties operating at optical frequencies. Both dipolar or broadband multipolar Huygens sources can be achieved with cluster of dielectric or metallic particles. The approach rely on the engineering of the refractive index of the structure rather than its geometrical shape. We believe the cluster structure is a rich system as it is scalable and its optical properties can be tuned by varying the nature, amount, size, and volume fraction of inclusions. We have as well explore ability of silicon cluster to control transmission, in both magnitude and phase, and absorption when organized into metalatices.

Acknowledgments

This work is supported by the LabEx AMADEus (ANR-10-LABX-42) in the framework of IdEx Bordeaux (ANR-10-IDEX-03-02), France.

References

[1] Person S, Jain M, Lapin Z, Saenz J J, Wicks G, and Novotny L 2013. Demonstration of zero optical backscattering from single nanoparticles. *Nano letters*, 13(4), 1806-1809
[2] Fu Y H, Kuznetsov A I, Miroshnichenko A E, Yu Y F, and Luk'yanchuk B 2013. Directional visible light scattering by silicon nanoparticles. *Nature communications*, 4, 1527.
[3] Luk'yanchuk B S, Voshchinnikov N V, Paniagua-Domínguez R, and Kuznetsov A I 2015. Optimum forward light scattering by spherical and spheroidal dielectric nanoparticles with high refractive index. *ACS Photonics*, 2(7), 993-999.
[4] Yang Y, Miroshnichenko A E, Kostinski S V, Odit M, Kapitanova P, Qiu M, and Kivshar Y S 2017. Multimode directionality in all-dielectric metasurfaces. *Physical Review B*, 95(16), 165426.
[5] Staude I *et al.* 2013. Tailoring directional scattering through magnetic and electric resonances in subwavelength silicon nanodisks. *ACS nano*, *7*(9), 7824-7832.
[6] Liu W, Zhang J, Lei B, Ma H, Xie W, and Hu H 2014. Ultra-directional forward scattering by individual core-shell nanoparticles. *Optics express*, 22(13), 16178-16187.
[7] Liu W, Lei B, Shi J, and Hu H 2016. Unidirectional superscattering by multilayered cavities of effective radial anisotropy. *Scientific reports*, 6, 34775.
[8] Dezert R, Richetti P, & Baron A 2017. Isotropic Huygens dipoles and multipoles with colloidal particles. *Physical Review B*, 96(18), 180201.
[9] Mackowski D W, & Mishchenko M I 1996. Calculation of the T matrix and the scattering matrix for ensembles of spheres. *JOSA A*, *13*(11), 2266-2278.
[10] Zhang S, Jiang R, Xie Y M, Ruan Q, Yang B, Wang J, and Lin H Q 2015. Colloidal Moderate-Refractive-Index Cu2O Nanospheres as Visible-Region Nanoantennas with Electromagnetic Resonance and Directional Light-Scattering Properties. *Advanced Materials*, *27*(45), 7432-7439.
[11] Sikdar D, Cheng W and Premaratne M 2015. Optically resonant magneto-electric cubic nanoantennas for ultra-directional light scattering. *Journal of Applied Physics*, *117*(8), 083101.
[12] Decker M, *et al.* 2015. High-efficiency dielectric Huygens' surfaces. *Advanced Optical Materials*, 3(6), 813-820.
[13] Staude I and Schilling J 2017. Metamaterial-inspired silicon nanophotonics. *Nature Photonics*, 11(5), 274.
[14] Alaee R, Albooyeh M, and Rockstuhl C 2017. Theory of metasurface based perfect absorbers. *Journal of Physics D: Applied Physics*, *50*(50), 503002.

All-dielectric metasurface filters for mid-infrared range

Zh O Dombrovskaya[1], **A V Rybina**[1], **A A Belov**[1,2] **and A N Bogolyubov**[1]

[1]Faculty of Physics, Lomonosov Moscow State University, Leninskie Gory, 119991, Moscow, Russia
[2]Peoples' Friendship University of Russia (RUDN University), Moscow 117198, Russia

dombrovskaya@physics.msu.ru

Abstract. In the present work, a metasurface composed of PbTe spherical particles and located on Ge substrate is considered. We show that it acts in two anti-reflective regimes: depending on geometrical parameters of the structure, it can be a narrow or a wide-range filter. The reason is that the incident radiation excites different dipole resonances of the metasurface. For high-index dielectric materials, we obtain conditions for the validity and accuracy of the dipole approximation and the locations of dipole magnetic and electric resonances.

1. Introduction

In thermal imagers, germanium (Ge) lenses are commonly used. In the actual range from 8 to 12 μm (long-wavelength mid-infrared range), this material is attractive due to its extremely low absorption and high index $n_{substr} \approx 4.0$. This results in a great reflection from the lens. To reduce the reflection, single-layer coatings are commonly applied. The latter is a thin film, for which the optimum refractive index equals to 2. However, such a material does not exist in nature. In practice, zinc selenide (ZnSe, $n \approx 2.4$) or commercial material Irtran-2 ($n \approx 2.2$) are used. Unfortunately, they do not provide complete suppression of the reflection from the structure at any wavelength from the considered range.

An alternative approach is to apply a coating based on the metasurface [1]. This tendency seems to be promising, because with the help of the metamaterial paradigm it is possible to obtain practically any effective refractive index of the coating. In this paper we consider an anti-reflective metasurface coating composed of lead telluride (PbTe, $n \approx 5.6$) spherical particles for the air-germanium interface. Optimization of the geometric parameters of the structure was carried out and zero reflection was obtained at the wavelength $\lambda_0 = 10$ μm.

2. Structure optimization

We start with a specification of the structure under study. A semi-infinite Ge substrate is illuminated by normally incident plane wave. On the substrate there is a square array with periodicity p of spherical dielectric scatterers with radius r. For the metaatoms PbTe material was chosen. It has one of the highest refractive indexes and also zero losses, therefore, it provides a significant field enhancement at the resonant wavelengths [2].

Our goal is to determine optimal geometrical parameters providing minimal reflection R_s from the whole structure (the metasurface and the substrate). Additional restrictions on the parameters are proposed in [3], [4]. They are associated with the fabrication process ($r \geq 50$ nm, $p \geq 4r$) and the applicability of the model (see the next paragraph).

The solution of the problem is reduced to multiple solutions of the direct problem [5], i.e., calculations of the characteristics of substrated metasurface. From one computation to another one, optimization parameters are directionally modified. It is advisable to use analytical models. Most of the modern models of metasurfaces are derived in the dipole approximation. Thereby, the question of its validity and accuracy arises.

2.1. Dipole response of spherical particles

Within the dipole approximation, each metaatom is replaced by a pair of electric and magnetic dipoles. In the present work, we verify the replacement basing on the dipole approximation (DA) validity criterion [6] (Figure 1). We show that for a high-index dielectric sphere, the dependence $\lambda/r_{max}(n)$ is linear in the mid-infrared range (Figure 2). Here, r_{max} is the maximal radius. We have also obtained the relations describing the positions of the electric (EDR) and magnetic (MDR) dipole resonances. They agree with experimental data [7]. Note that lines of MDR and EDR are located above black one (validity of the DA), that is, they satisfy the criterion.

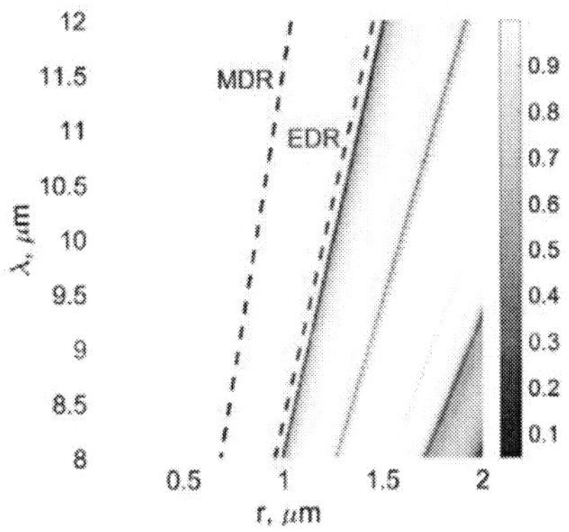

Figure 1. The validity of the dipole approximation for a PbTe spherical particle. When the criterion is fulfilled the colour is white.

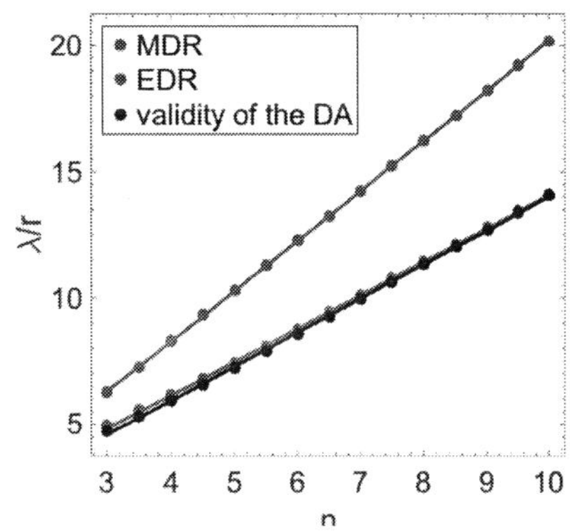

Figure 2. Line approximations of MDR ($\lambda/r_{max} = 1.98n + 0.37$), EDR ($\lambda/r_{max} = 1.32n + 0.87$) and the equation of the DA validity ($\lambda/r_{max} = 1.35n + 0.55$).

2.2. Results

The properties of the whole considered structure can be well described using the uncoupled-element model [8]. The Fresnel coefficients of the metasurface are calculated within the validity domain of the dipole approximation using the model of interacting induced dipoles [9].

The results of the optimization are depicted in Figure 3. The domain of the arguments is above the dashed line (technological constraint $p \geq 4r$). The markers show the positions of the obtained minima. Total reflectance compensation occurs when the magnitudes are matched and the phases of the electric and magnetic dipole moments differ by π. Figures 4-6 show the reflection spectra. The coloured lines correspond to the ranges in which the dipole approximation criterion is satisfied.

From the plots, one can see that the reflection minima can be caused by different reasons. 1) Tightly located non-resonant particles (Figure 4, magenta colour). 2) Interference between the substrate and the metafilm, the particle size is chosen to provide the first magnetic dipole resonance (Figure 5, blue colour). 3) Minimum occurring due to the magneto-electric interaction (Figure 6, red colour). Note that the latter case corresponds to the reflection compensation in the broad range. The width of the range with small reflection is determined by the positions of the dipole resonances.

For the sake of comparison, the reflection spectrum of a thin film made of Irtran-2 and placed on the same substrate is depicted (Figure 6, yellow colour). It is easy to see that using of this one-layer coating does not allow to achieve zero reflectance at 10 μm. However, it provides small reflectance in the entire range from 8 to 12 μm. On the other hand, the metasurface coating acts as a perfect high-quality filter (see blue curve) or a quite good broad filter (see red curve).

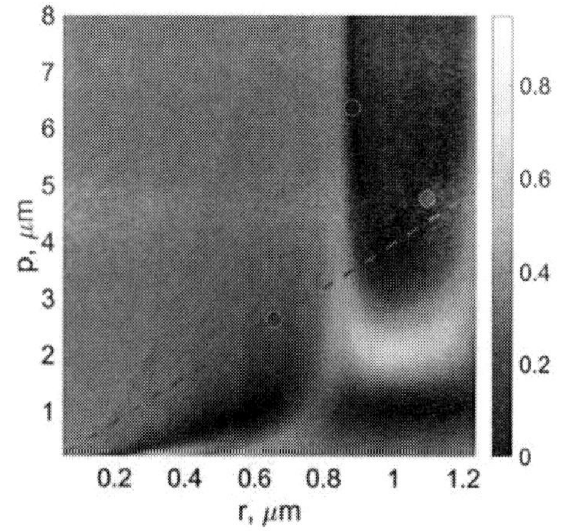

Figure 3. Reflectance $|R_s(p,r)|^2$ and results of the minimization of substrated metasurface at $\lambda_0 =$ 10 μm.

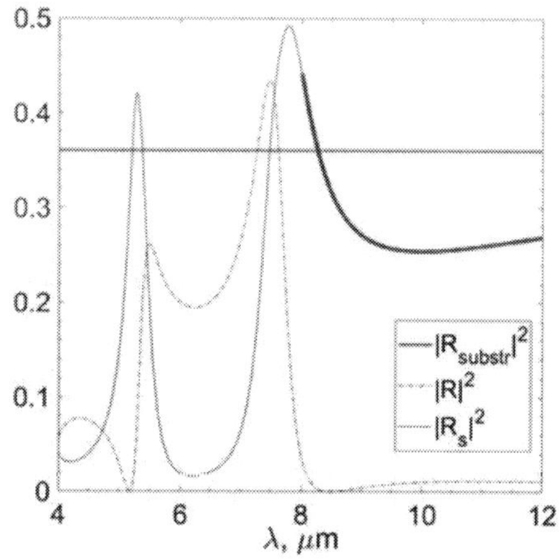

Figure 4. The reflection spectrum for the obtained optimal p and r at non-resonant case (magenta line). Other curves are indicated on the legend.

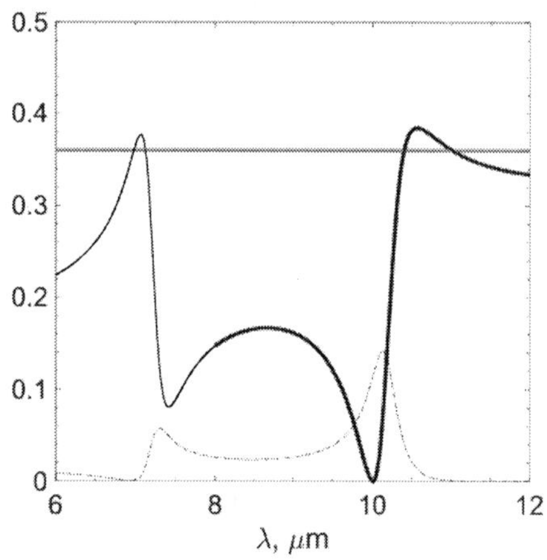

Figure 5. The reflection spectra for the obtained optimal p and r at MDR (blue line). Other curves are indicated on the legend (Figure 4).

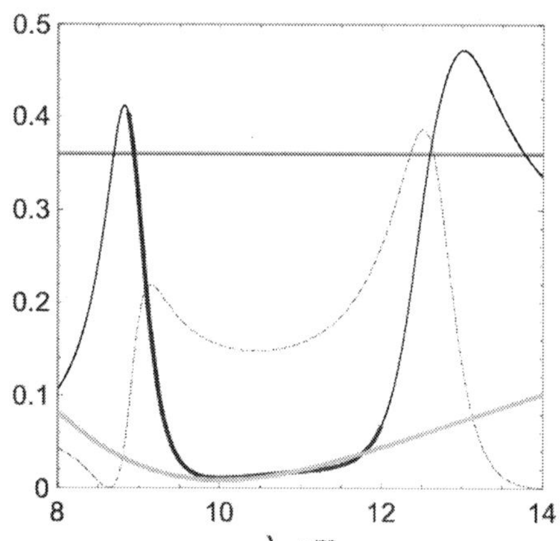

Figure 6. The reflection spectra for the obtained optimal p and r between EDR and MDR (red line). Yellow line corresponds to thin film coating. Other curves are indicated on the legend (Figure 4).

3. Conclusion

We consider the optimization of metasurface composed of PbTe spherical particles located on Ge substrate. The constrains on its geometric parameters are found. We show that such structure can be a perfect narrow or a good wide-range filter.

References

[1] Baryshnikova K V, Petrov M I, Babicheva V E and Belov P A 2016 *Sci. Rep.* **6** 22136
[2] Baranov D G, Zuev D A, Lepeshov S I, Kotov O V, Krasnok A E, Evlyukhin A B and Chichkov B N 2017 *Optica* **4** 814
[3] Dombrovskaya Zh O 2017 *Numerical Methods and Programming* **18** 455 [in Russian]
[4] Rybina A V, Dombrovskaya Zh O and Bogolyubov A N 2018 *Bulletin of the Russian Academy of Sciences: Physics* [submitted]
[5] Sveshnikov A G and Ilinskiy A S 1972 *Sov. Phys. Dokl.* **17** 527
[6] Dombrovskaya Zh O, Zhuravlev A V, Belokopytov G V and Bogolyubov A N 2016 *Phys. Wave Phenom.* **41** 96
[7] Kuznetsov A I, Miroshnichenko A E, Fu Y H, Zhang J and Luk'yanchuk B 2012 *Sci. Rep.* **2** 492
[8] Babicheva V E, Petrov M I, Baryshnikova K V and Belov P A 2017 *J. Opt. Soc. Am.* B **34** D18
[9] Evlyukhin A B, Reinhardt C, Seidel A, Luk'yanchuk B S and Chichko B N 2010 *Phys. Rev.* B **82** 045404

High-Q resonances supported by a single dielectric ridge on the surface of a slab waveguide

L L Doskolovich[1,2], E A Bezus[1,2] and D A Bykov[1,2]

[1] Image Processing Systems Institute of RAS — Branch of the FSRC "Crystallography and Photonics" RAS, 151 Molodogvardeyskaya st., Samara 443001, Russia
[2] Samara National Research University, 34 Moskovskoe shosse, Samara 443086, Russia

leonid@smr.ru

Abstract. Diffraction of slab waveguide modes on a dielectric ridge located on the waveguide surface is studied. Calculations, based on aperiodic rigorous coupled-wave analysis, demonstrate the existence of sharp resonant features in the reflectance and the transmittance spectra occurring at oblique incidence of TE-polarized waveguide mode on the ridge. Angular widths of the resonant peaks and dips vary from units to less than a thousandth of a degree. Using the effective index method, we explain the resonances by the excitation of cross-polarized modes of the ridge. The existence of high-Q resonances makes this structure promising for filtering, sensing and all-optical signal processing.

1. Introduction

In a wide class of planar (integrated) optoelectronic systems, optical processing is performed in a slab waveguide [1–6]. Such geometry corresponds to the "insulator-on-insulator" platform and is suitable for the creation of fully integrated optical devices. In this case, the processed signal corresponds to a superposition of slab waveguide modes with different propagation directions (in the case of spatial filtering) or with different frequencies (in the case of spectral filtering).

In this work, we show that a very simple structure consisting of a single subwavelength ridge on the surface of a slab waveguide has interesting resonant properties resulting from the excitation of a cross-polarized mode of the ridge. The discovered high-Q resonances supported by the presented planar structure make it promising for various applications including spatial (angular) and spectral filtering, transformation of optical signals and chemical or biosensing.

2. Diffraction of slab waveguide modes on a ridge

Let us study the diffraction of slab waveguide modes on the ridge located on the surface of a slab waveguide (figure 1). For the analysis, the following parameters were chosen: refractive index of the waveguide core layer and of the ridge $n_c = 3.3212$ (GaP at the wavelength $\lambda = 630\,\text{nm}$), refractive indices of the substrate and superstrate $n_{sub} = 1.45$ (fused silica) and $n_{sup} = 1$, respectively, waveguide thickness $h_c = 80$ nm, waveguide thickness at the ridge $h_r = 110$ nm. In this case, at $\lambda = 630$ nm the waveguide is single-mode both at TE- and TM-polarizations, and the effective refractive indices of the modes amount to $n_{\mathrm{wg,TE}} = 2.5913$ and $n_{\mathrm{wg,TM}} = 1.6327$, respectively. Effective refractive indices of the modes at $h_{\mathrm{r}} = 110$ nm equal $n_{\mathrm{r,TE}} = 2.8192$ and $n_{\mathrm{r,TM}} = 2.1867$.

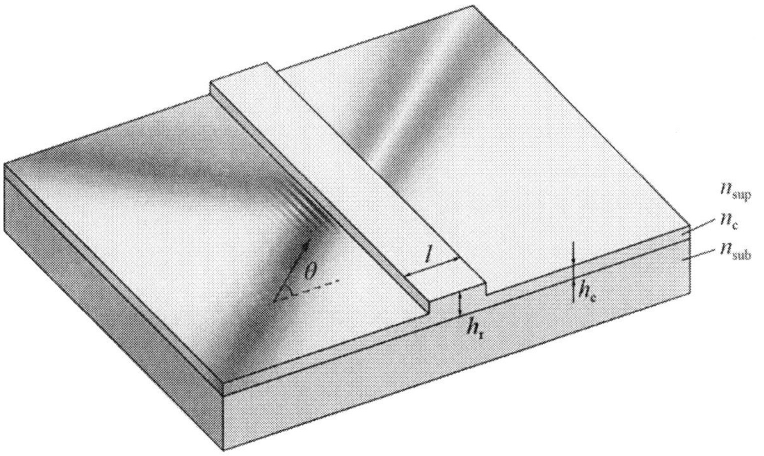

Figure 1. Geometry of the problem of diffraction of a waveguide mode on a dielectric ridge. One of the potential applications (spatial differentiation of the profile of an optical beam propagating in the waveguide) is schematically shown.

Figure 2 shows the reflectance $\left|R_{\mathrm{TE}}(\theta,l)\right|^2$ and the transmittance $\left|T_{\mathrm{TE}}(\theta,l)\right|^2$ of the ridge vs. the angle of incidence θ and the ridge length l in the case of an incident TE-polarized mode, where $R_{\mathrm{TE}}(\theta,l)$ and $T_{\mathrm{TE}}(\theta,l)$ are the complex reflection and transmission coefficients, respectively. The plots were calculated using an in-house implementation of the aperiodic rigorous coupled-wave analysis (aRCWA) technique [7]. RCWA, also called the Fourier modal method, is an established numerical technique for solving Maxwell's equations.

It is evident that several resonances (sharp reflectance maxima and transmittance minima) are present in the spectra of figure 2. Resonant reflectance peaks (transmittance dips) are located in the angle of incidence range $39.05° < \theta < 57.55°$, which is marked with dashed lines in figure 2. Let us explain these effects. In a general case, at oblique incidence of a TE-polarized guided mode on a ridge located on the waveguide surface, reflected and transmitted TE- and TM-polarized modes are generated, as well as a continuum of plane waves arising from "parasitic" scattering out of the guiding layer to the superstrate and substrate. However, if the angle of incidence exceeds a certain value $\theta_{\mathrm{cr,1}}$, the reflected and transmitted fields contain only the TE-polarized mode, and no out-of-plane scattering and polarization conversion occur [4–6]. Indeed, let us denote by $k_{x,\mathrm{inc}} = k_0 n_{\mathrm{wg,TE}} \sin\theta$, where $k_0 = 2\pi/\lambda$ is the wavenumber, the x-component of the wavevector of the incident TE-mode, which is parallel to the ridge boundaries. According to the boundary conditions for the Maxwell's equations, the $k_{x,\mathrm{inc}}$ component has to be conserved, i.e. it is the same for all the waves constituting the reflected and the transmitted fields, including the radiation scattered out of the waveguide. Therefore, at angles of incidence $\theta > \theta_{\mathrm{cr,1}} = \arcsin\left(n_{\mathrm{wg,TM}}/n_{\mathrm{wg,TE}}\right) = 39.05°$, reflected and transmitted TM-polarized modes become evanescent. At $\theta > \theta_{\mathrm{cr,1}}$, $k_{x,\mathrm{inc}}$ also exceeds the wave vector magnitudes of the propagating plane waves over and under the waveguide (in the regions with the refractive indices n_{sup} and n_{sub}, respectively) and therefore the incident TE-mode is not scattered out of the waveguide layer. Thus, at $\theta > \theta_{cr,1}$ the reflection and transmission coefficients corresponding to the TE-polarized mode satisfy the equality $|R_{TE}|^2 + |T_{TE}|^2 = 1$, since the materials of the structure are assumed to be lossless. The angle $\theta_{cr,1}$ is shown with a dashed line in figure 2, which corresponds to the upper boundary of the "resonance region". The lower boundary of this region corresponds to the cutoff angle of the TM-polarized mode in the ridge region, which amounts to $\theta_{cr,2} = \arcsin(n_{r,TM}/n_{r,TE}) = 57.55°$.

The presented analysis suggests that the resonances in figure 2 are associated with the excitation of TM-like eigenmodes of the ridge, which in this case acts as a leaky rib waveguide. Within the framework of the effective index method (EIM) [8], the TM-like modes of a rib waveguide can be approximately described by the dispersion relation of TE-polarized modes of a dielectric slab waveguide with the thickness l, in which the values $n_{r,TM}$ and $n_{wg,TM}$ are used as the refractive indices of the core layer and claddings, respectively.

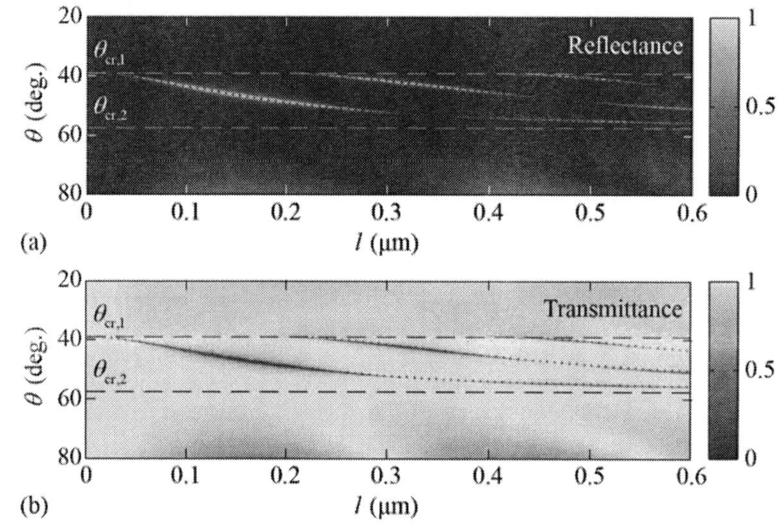

Figure 2. Reflectance (a) and transmittance (b) of the ridge vs. the ridge length l (horizontal axis) and the angle of incidence θ (vertical axis). Dashed lines show the boundaries of the resonance region and correspond to the cutoff angles of the TM-modes outside the ridge (at $h_c = 80$ nm, upper line) and in the ridge (at $h_r = 110$ nm, lower line). Dotted curves show the dispersion of the cross-polarized modes of the ridge.

The dispersion relation of a symmetric slab waveguide can be written in the form [9] $l = [\pi m + \arg r(\theta)]/(k_0 n_{r,TM} \cos\theta)$, where the integer m is the mode order, $r(\theta)$ is the complex reflection coefficient of a TE-polarized plane wave from the interface between the media with refractive indices $n_{r,TM}$ and $n_{wg,TM}$. A straightforward refinement of this EIM-based dispersion relation consists in the use of the reflection coefficient of the TM-polarized mode $r_{mode,TM}(\theta)$ from the interface between two waveguides with the thicknesses $h_r = 110$ nm and $h_c = 80$ nm instead of the plane-wave reflection coefficient $r(\theta)$. The dispersion curves $l = l(\theta)$ corresponding to this refinement are shown in figure 2 with dotted curves and are in an excellent agreement with the resonance locations. This confirms the hypothesis that the resonant features in the spectra of figure 2 are indeed due to the excitation of the cross-polarized modes of the rib waveguide.

Since in the resonant regime (at $\theta_{cr,1} < \theta < \theta_{cr,2}$) there is no out-of-plane scattering and polarization conversion in the reflected and transmitted radiation, the transmission coefficient strictly vanishes at the resonances [10, 11]. Moreover, at different ridge lengths the resonant peaks (dips) have different angular widths (different quality factor), which vary from units to less than a thousandth of a degree. This change in the Q-factor is caused by the interaction between the resonances of two types: Fabry-Perot resonance of the TE-mode and the guided-mode resonance of the TM-mode. When two types of resonances (two modes) interact, the so-called matrix Fabry-Perot resonances occur, which can have a very high Q-factor [12]. In fact, the Q-factor in the considered structure can reach infinity, i.e. the structure supports the so-called bound states in the continuum (BIC). A detailed theoretical study confirming the formation of robust BICs and high-Q resonances in the investigated structure will be the subject of a separate work.

3. Conclusion

In this work, we demonstrated that in the case of oblique incidence of TE-polarized modes of a slab waveguide on a dielectric ridge located on the waveguide surface, resonant changes in the reflectance and the transmittance spectra occur. The resonant peaks (dips) have different angular widths (different Q-factor), which can reach a thousandth of a degree or even less. Using a refinement of the effective index method, we explained these effects by the excitation of a cross-polarized mode of the ridge.

In our opinion, the presented planar structure can be used as an integrated optical spatial (angular) or spectral filter. Moreover, since the transmission coefficient of the ridge strictly vanishes at the resonances, the structure enables performing spatial optical differentiation in transmission (as it is schematically shown in figure 1) as well as spatial optical integration in reflection [13]. Finally, the existence of high Q-resonances and bound states in the continuum in the structure also makes it promising for chemical or biosensing. The authors believe that the presented configuration can also be extended to other platforms, e.g. Bloch surface waves propagating at the interfaces of photonic crystals.

Acknowledgments

This work was supported by Federal Agency of Scientific Organizations under Agreement 007-GZ/Ch3363/26 (theoretical analysis of the resonances of the ridge using the effective index method), and by Russian Science Foundation under Grant 14-19-00796 (implementation of the software tools for the rigorous simulation of the studied integrated structures).

References

[1] Haus H A 1984 *Waves and Fields in Optoelectronics* (Hemel Hempstead: Prentice-Hall)
[2] Calafiore G, Koshelev A, Dhuey S, Goltsov A, Sasorov P, Babin S, Yankov V, Cabrini S and Peroz C 2014 *Light: Sci. & Appl.* **3** e203
[3] Mossberg T 2001 *Opt. Lett.* **26** 414–6
[4] Doskolovich L L, Bezus E A, Golovastikov N V, Bykov D A and Soifer V A 2017 *Opt. Express* **25** 22328–40
[5] Doskolovich L L, Bezus E A and Bykov D A 2018 *Photon. Res.* **6** 61–5
[6] Hammer M, Hildebrandt A and Förstner J 2015 *Opt. Lett.* **40** 3711–4
[7] Silberstein E, Lalanne P, Hugonin J-P and Cao Q 2001 *J. Opt. Soc. Am. A* **18** 2865–75
[8] Kekatpure R D, Hryciw A C, Barnard E S and Brongersma M L 2009 *Opt. Express* **17** 24112–29
[9] Lifante G 2003 *Integrated Photonics: Fundamentals* (Chichester: Wiley)
[10] Bykov D A, Doskolovich L L, Golovastikov N V and Soifer V A 2013 *J. Opt.* **15** 105703
[11] Gippius N A, Tikhodeev S G and Ishihara T 2005 *Phys. Rev. B* **72** 045138
[12] Karagodsky V, Chase C and Chang-Hasnain C J 2011 *Opt. Lett.* **36** 1704–6
[13] Bezus E A, Doskolovich L L, Bykov D A and Soifer V A 2018 Spatial differentiation and integration of optical beams in a slab waveguide by a single dielectric ridge *Preprint* arXiv:1804.02267v1 [physics.optics]

The evidence of the role of surface plasmon polaritons in formation of femtosecond highly-regular laser-induced periodic structures on Cr films

A. V. Dostovalov[1,2], T. J.-Y. Derrien[3,4], F. Přeučil[3], T. Mocek[3], V. P. Korolkov[1,2], S. A. Babin[1,2] and N. M. Bulgakova[3]

[1] Institute of Automation and Electrometry of the SB RAS, 1 Acad. Koptyug Ave., 630090 Novosibirsk, Russia

[2] Novosibirsk State University, 2 Pirogova St., 630090 Novosibirsk, Russia

[3] HiLASE Centre, Institute of Physics, Czech Academy of Sciences, Za Radnicí 828/5, 25241 Dolní Břežany, Czech Republic

[4] Max Planck Institute for Structure and Dynamics of Matter (MPSD), Luruper Chaussee 149, 22761 Hamburg, Germany

dostovalov@iae.nsk.su, derrien@fzu.cz

Abstract. Cr films of controlled thickness deposited on glass substrates were irradiated by a high number of low-intensity femtosecond laser pulses below the Cr single-pulse damage threshold, producing periodic surface structures (LIPSS) of very high regularity via metal oxidation. To address the multiplicity of electromagnetic modes allowed for thin films, a rigorous numerical approach for modeling surface plasmon polaritons (SPP) in thin-film geometry has been developed. Three types of modes are predicted: the classical SPP with periodicities $\Lambda \sim \lambda$ (λ *is* laser wavelength) at air-film interface, and $\Lambda \sim \lambda/n$ at film-substrate interface (n is refractive index of substrate), and a propagation mode $\Lambda \sim \lambda/n_{Cr}$ where n_{Cr} is refractive index of film material. Experimentally observed LIPSS periods match well the predicted modes for the extreme cases, $\Lambda \sim \lambda/n$ at film thickness $h \leq 30$ nm and $\Lambda \sim \lambda$ at $h \geq 200$ nm, indicating respectively the dominance of SPP excited at the film-substrate interface and at the metal surface. For 30 nm < h < 200 nm, decreasing of the film thickness results in transition from the pure surface mode to its coupling with and finally domination of the electromagnetic wave excited at the film-substrate interface, providing evidence of the SPP mechanism of LIPSS formation under laser irradiation of thin metallic films.

1. Introduction

By scanning surfaces with high-repetition rate femtosecond laser pulses, highly regular periodic surface structures (LIPSS) can be formed at rates, which are competitive with existing industrial standards [1-4] in nanostructure demanding applications. The physical origin of the highly regular LIPSS was recently linked with the lossy properties of irradiated materials, where SPP decay length for propagation along the surface is small enough, thus limiting the interference of the multiple SPP excited at randomly distributed scattering centers in different parts of the laser-irradiation spot [4]. Whereas a number of lossy metallic materials enable the formation of highly regular LIPSS [1-4], low-loss plasmonic metals exhibit a lower quality LIPSS [4-6].

In this work, LIPSS were produced on Cr films of controlled thickness deposited on glass substrates by a high number of low-intensity femtosecond laser pulses below the Cr single-pulse damage threshold via metal oxidation (referred below as thermochemical LIPSS or TLIPSS). A systematic study of the role of film thickness on the TLIPSS period has been performed. The experimental results are analyzed based on an advanced thin-film SPP (TF-SPP) model, which accounts for the interference between SPPs excited at two film interfaces: with air and glass substrate. Simulations reveal that, depending on the film thickness, two possible electromagnetic modes can be responsible for TLIPSS formation: (1) SPP generated at the film-substrate interface for very thin films and (2) SPP excited on the film free surface for relatively thick films. The transition between these two modes suggested by the TF-SPP model and observed experimentally provides the evidence of the important role of SPP excitation upon laser irradiation of thin metallic films.

2. Experimental results

The experiments on TLIPSS formation were carried out on the setup described in detail in [7]. Femtosecond laser pulses with a central wavelength of 1026 nm, duration of 232 fs, and repetition rate of 200 kHz were focused by a lens with f = 35 mm providing the beam diameter ≈15 μm on the surface. The pulse energy E_p was in the range 100-110 nJ. The sample translation speed was set to 1 μm/s. Metal films were deposited on BK7 borosilicate glass substrates by the magnetron sputtering technique which provides controlled film thickness h in the range of 28-350 nm. Scanning electron microscopy (SEM) Hitachi TM3000 was used to inspect nanostructures formed on the sample surface.

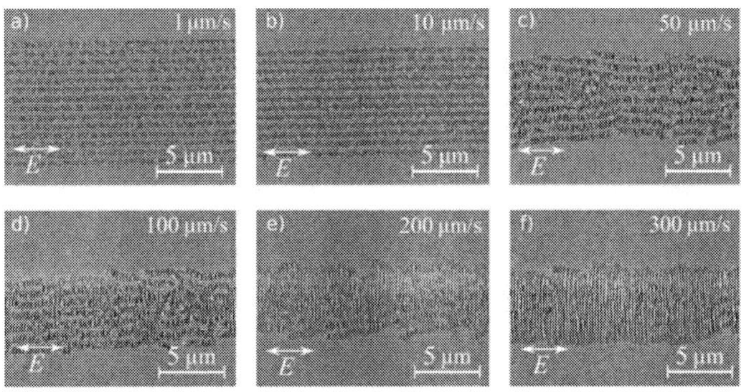

Fig. 1. The results of TLIPSS formation on chromium film with thickness of 28 nm at different scanning speeds: 1 μm/s (b), 50 μm/s (c), 100 μm/s (d), 200 μm/s (e), 300 μm/s (f).

The LIPSS produced on 28-nm Cr film at different scanning velocities v_s are presented in Fig. 1. Regular TLIPSS along the laser polarization direction with the period of 678 nm are formed at low v_s of 1 μm/s - 10 μm/s (Fig. 1a,b). They can be classified as low spatial frequency LIPSS (LSFL) [4]. Increasing v_s to 50 μm/s yields in reducing TLIPSS regularity (Fig. 1c). Under these conditions, the TLIPSS period slightly increases and the non-regular LIPSS on the top of TLIPSS with the direction perpendicular to polarization and the period of ~200-250 nm (high spatial frequency LIPSS or HSFL) becomes pronounced. Increasing v_s to 100-200 μm/s results in merging the adjacent ablative HSFL (Fig. 1 d, e). Upon merging, the HSFL regularity increases as clearly seen for v_s = 300 μm/s (Fig. 1f).

Comparison of TLIPSS produced on Cr films of different thicknesses is presented in Fig. 2. When increasing h to 70 nm (Figs. 1a and 2a), TLIPSS period increases to 842 nm (Fig. 2b) and to 909 nm on film with thickness of 125 nm (Fig. 2c), while the HSFL are more ordered as compared to thinner films and their average period for this h is 100 nm (Fig. 2d). The comparable period values of HSFL were observed in [8] at laser wavelength of 800 nm and 200 nm thick Ti film, but the orientation of structures in our case is different and perpendicular to the laser polarization direction.

3. Theoretical analysis

SPP properties in thin films can be studied in a simplified manner by solving an implicit equation based on the dielectric permittivities ε_1, ε_2, and ε_3, associated with the propagation wave-vectors k_1,

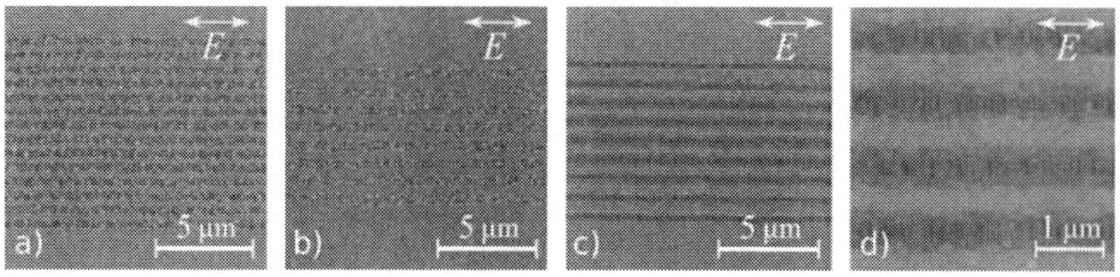

Fig. 2 The results of TLIPSS formation on chromium film with different thickness: 28 nm (a), 70 nm (b), 125 nm (c). The enlarged image of TLIPSS on chromium film with thickness of 125 nm (d).

k_2, and k_3, respectively [9]. The latter are defined by the light momentum conservation expressed as

$$k_j^2 = \beta^2 - \varepsilon_j \left(\frac{\omega}{c}\right)^2 \qquad (1)$$

where $j = 1$, 2 or 3 related to the film material (Cr in our case), air, and substrate respectively; c is light velocity in vacuum/air; ω is the laser frequency. The periods of possible modes of the surface electromagnetic waves resulting from the interference of the laser wave with the SPP wave is provided by the relation $\Lambda = 2\pi / \mathrm{Re}(\beta)$ where β is the complex wave number associated with the SPP modes. The periods can be calculated by numerically solving Eq. (1) with the following equation [10]

$$e^{-2k_1 t} = \left(\frac{k_1}{\varepsilon_1} + \frac{k_2}{\varepsilon_2}\right)\left(\frac{k_1}{\varepsilon_1} + \frac{k_3}{\varepsilon_3}\right) \Big/ \left(\frac{k_1}{\varepsilon_1} - \frac{k_2}{\varepsilon_2}\right)\left(\frac{k_1}{\varepsilon_1} - \frac{k_3}{\varepsilon_3}\right) \qquad (2)$$

Among the possible wave-numbers β, we consider the solutions which are associated with a distance characterizing SPP propagation (decay) length L_{SPP} expressed by $L_{\mathrm{SPP}} = \frac{l}{2\times\mathrm{Im}(\beta)}$ [10], which is comparable with the spot diameter (1 μm < d < 100 μm). Other modes of nanoscale periods are neglected as they are damped on distances much smaller than laser wavelength.

Figure 3 presents the comparison between the measured LIPSS periods and theoretical predictions. The classical SPP mode with the period $\Lambda \sim \lambda$ is characterized for chromium by L_{SPP} of the order of several micrometers [4]. For λ = 1026 nm, assuming a constant dielectric permittivity [11], the properties of SPP at the air/Cr interface are: L_{SPP} = 4.06 μm and Λ = 1026 nm. For the Cr/BK7 interface, using the optical data from [12]), one has Λ = 711 nm and L_{SPP} = 1.34 μm. These Λ values are marked by dashed horizontal lines in Fig. 3. The SPP modes at the air/Cr and BK7/Cr interfaces computed by the TF-SPP model are shown by solid lines, red and blue respectively, which are very close to the classical modes. Some increase in periodicity with decreasing the film thickness indicates preferable generation of asymmetric SPPs, according to [13].

Comparison of modeling with the experiment suggests that the laser light couples predominantly with the air/Cr asymmetric SPP mode when film thickness is larger than ~200 nm and with the Cr/BK7 mode when thickness is ≤ 30 nm. Note SPP decay depth $1/|k_1|$ is found ~32.8 nm, slightly larger than the penetration depth of the light in Cr at λ = 1026 nm (~23 nm). However, in the case when oxidation is active, SPP decay depth should increase. Then the SPP at the air/Cr surface can excite an SPP on the Cr/glass interface and the electromagnetic interaction between the two SPPs cannot be neglected [14]. If it is so, at 30 nm < h < 200 nm, a coupled mode arising from this interaction can drive the TLIPSS formation. With increasing film thickness, the interaction becomes weaker and the transition from Cr/glass to air/Cr mode gradually occurs that is not yet explained by the theoretical results. Returning to Fig. 1, we admit that the HSFL might be attributed to the guided mode in the film whose periodicity follows the tendency $\Lambda \sim \lambda/n_{\mathrm{Cr}}$. The periodicity can change due to partial oxidation that can be treated, e.g., in the frames of effective optical media. However, that still calls for further studies.

Conclusion

The thermochemical LIPSS were produced on Cr thin films of different thickness by applying femtosecond-laser scanning at pulse energy below the single-pulse damage threshold. A theoretical

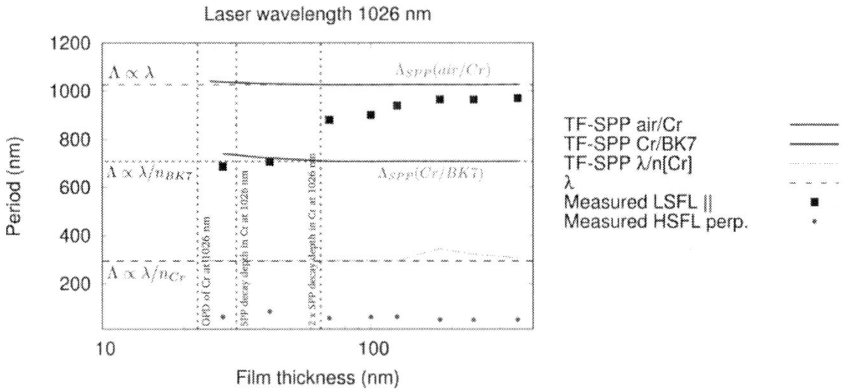

Fig. 3. Comparison between experimental results and theory for $\lambda = 1026$ nm. Measured LIPSS periods (see Fig. 2) are shown by black squares (LSFL) and red circles (HSFL). SPP modes computed by using TF-SPP model are shown by solid lines. Red line is for the SPP mode at the air/Cr interface and blue line is for the SPP mode at the Cr/BK7 interface. Horizontal black dashed lines correspond to λ and to the mode associated with light propagation toward the bulk λ/n with n to be refractive index of Cr. Vertical lines show the optical penetration depth in Cr (left) and the decay depth of SPP excited at the air/Cr interface (right).

analysis has been performed, which predicts the periods of the possible SPP modes of the films. For film thicknesses larger than 200 nm, the observed LSFL period matches with the air/Cr SPP mode while for thicknesses smaller than 50 nm it matches with the Cr/BK7 SPP mode. Although transition between the two cases is not yet directly described by modeling, these results provide evidence of the SPP mechanism of LIPSS formation under laser irradiation of thin films of lossy metals.

Acknowledgments
The research of A.V.D., V.P.K. and S.A.B. is financed by the Russian Foundation for Basic Research (Grant 16-32-60096). The research of T.J.-Y.D., F.P., N.M.B., and T.M. is financed by the European Regional Development Fund and the state budget of the Czech Republic (project BIATRI: CZ.02.1.01/0.0/0.0/15_003/0000445, project HiLASE CoE: No. CZ.02.1.010.00.015 0060000674, programme NPU I: project No. LO1602).

References

[1] B. Öktem, I. Pavlov, S. Ilday, et al., Nat. Phot. **7**, 897-901 (2013).
[2] A. Ruiz de la Cruz, R. Lahoz, J. Siegel, et al., Opt. Lett. **39**, 2491-2494 (2014).
[3] A. V. Dostovalov, V. P. Korolkov, S. A. Babin, Las. Phys. Lett. **12**, 036101 (2015).
[4] I. Gnilitskyi, T.J.-Y. Derrien, Y. Levy, et al., Sci. Rep. **7**, 8485 (2017).
[5] J. Xie, F. Wang, L. Jiang, et al., Las. Phys. **25**, 056103 (2015).
[6] A. Rodríguez, M. C. Morant-Miñana, et al., Appl. Surf. Sci. **351**, 135–139 (2015).
[7] A. V. Dostovalov, V. P. Korolkov, S. A. Babin, Appl. Phys. B **123**, 30 (2017).
[8] C. S. R. Nathala, A. Ajami, A. A. Ionin, et al., Opt. Exp. **23**, 5915-5929 (2015).
[9] B. Prade, J. Vinet, A. Mysyrowicz, Phys. Rev. B **44**, 24 (1991).
[10] S. A. Maier, Plasmonics, Fundamentals and Applications (Springer, 2007).
[11] P. B. Johnson, and R.-W. Christy, Phys. Rev. B **6**, 4370 (1972).
[12] I. J. Malitson, Opt. Soc. Am. **55**, 1205-1209 (1965).
[13] R. Charbonneau, P. Berini, E. Berolo, et al., Opt. Lett. **25**, 844-846 (2000).
[14] A.V. Zayats, I. I. Smolyaninov, A. A. Maradudin, Phys. Rep. **408**, 131–314 (2005).

IOP Conf. Series: Journal of Physics: Conf. Series **1092** (2018) 012026 doi:10.1088/1742-6596/1092/1/012026

Plasmon coupled super- and sub- radiance from dye shells

V C de Silva[1], M Moazzezi[1], Y V Rostovtsev[1] and V P Drachev[1,2]†

[1] Department of Physics and Advanced Materials Manufacturing Processing Institute, University of North Texas, Denton, TX 76203 USA.
[2]Skolkovo Institute of Science and Technology, Moscow, 121205, Russia.
†vladimir.drachev@unt.edu

Abstract. Coupled super- and sub- radiant emission of dye fluorescence near the plasmonic surface is shown to be an intrinsic property of the hybrid system. The bi-exponential decay is demonstrated for Au\SiO_2\ATTO655 core-shell nanoparticles, which persists even when averaging over a broad range of the coupling parameter. The superradiant and subradiant states are distinct and exclusive by the identical permutational symmetry of the atom-field interaction Hamiltonian of the two-level system. Our analytical approach shows that starting from two emitters, coupled for example via the plasmon modes, the symmetry breaking occurs, making detectable the simultaneous existence of the fast super-radiance and the slow sub-radiance. The relaxation in the sub-radiant system occurs mainly due to the interaction with the plasmon modes. Our theory shows that the relaxation leads to the population of the sub-radiant states by dephasing the super-radiant Dicke states giving rise to the bi-exponential decay in agreement with the experiments. Even if we have a group of many coupled emitters or many groups of coupled emitters with different coupling parameters, the averaging will affect the emission among the super-radiant fast transitions and the sub-radiant slow transitions and keep these two group distinct.

1. Introduction

If all atoms are initially inverted, the first spontaneously emitted photons trigger a buildup of a large macroscopic atomic polarization during the emission of a super-fluorescent pulse. The peak intensity and emission rate of such a super-fluorescent pulse is $\sim N$ times larger compared to that of N independent atoms [1], which in a classical picture corresponds to parallel dipole moments of all radiating atoms. However, there also exist the so-called subradiant states, in which the energy remains partially trapped in the atomic system since some of the atomic dipoles become antiparallel and the macroscopic polarization of the system is zero, thus they cannot couple to a radiation field. Numerical simulations by Temnov and Woggon [2] shows that a strong inhomogeneous broadening in the Dicke model leads to an intrinsically bi-exponential temporal evolution of the emission dynamics in which a fast superradiant decay is followed by a slow subradiant energy relaxation. Here, cooperative spontaneous emission is theoretically and experimentally investigated for an ensemble of emitters coupled to a core-shell structure.

2. Results and discussion

Simulations have been performed for the case of three atoms. We assume that three atoms are coupled to the same vacuum modes. The phase decay has two parts, one is coupling to the same bath (it describes the homogeneous broadening of coherence), and another part is coming from the coupling to the metal core (near fields, surface plasmons, etc.). Performing the simulations, we vary the coupling parameter. Without coupling to the plasmonic structure, the only single-exponential decay can be observed. The details will be published elsewhere. Once we have the coupling to the plasmonic structure, the results show the bi-exponential decays, $A_1e^{-t/\tau_1} + A_2e^{-t/\tau_2}$ as illustrated in Fig.1. One can see that an isolated atom has only one relaxation rate. When the system contains two or three atoms, the theory predicts the bi-exponential decay near the metamaterial structure that depends on the parameter of coupling to the structure. The decays related to the cooperative superradiant behavior of the atoms interacting with the same vacuum fields and related to the populations of the nondecaying states by local dephasing.

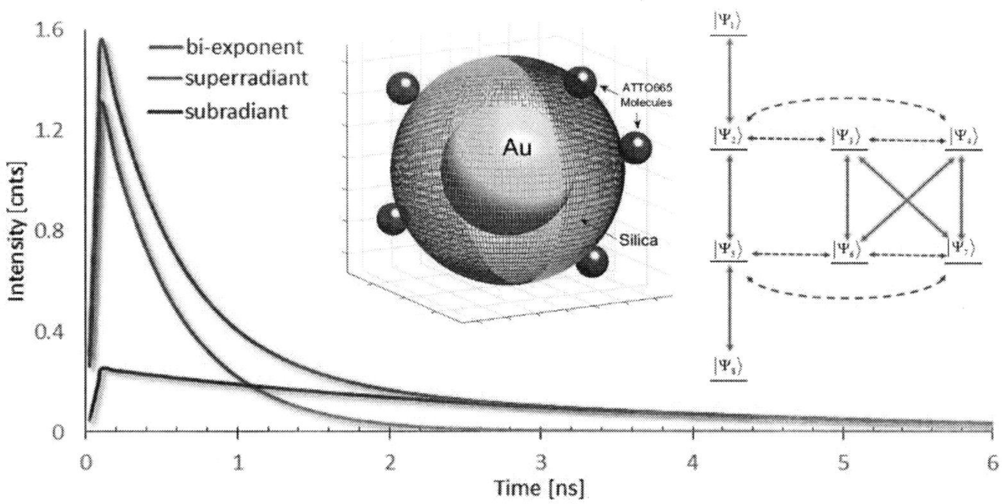

Figure 1. Bi-exponential emission decay containing supperradiant and subradiant pulses. (*Insert left*) The core-shell structure Au/SiO_2/ATTO655, (*Insert right*) The Dicke state basis for three emitters with the dipole-allowed transitions shown in blue lines; the mixing between states are in the dashed red lines.

To summarize all Dicke states can be exclusively divided into two groups. One set is the so-called super-radiant states. These states have transitions between each other from the most excited state to the ground state (Fig. 1 insert shown in blue). The second group of states is the sub-radiant states. They have weak transitions between each other. This second group have no ground state if they are decoupled from superradiant states. These sub-radiant states can trap some population since the trapped population cannot reach the ground state. The relaxation in the system occurs due to radiative coupling to the vacuum modes and due to the interaction with the phonon in the atomic matrix and the plasmonic structure (Fig. 1 insert shown in dashed red). The relaxation leads to the population of the subradiant states by dephasing the superradiant Dicke states giving rise to the bi-exponential decay that is observed in the experiments and shown in the theoretical section. Thus, the bi-exponential decay results from the cooperative atomic behaviour via the relaxation processes that transfer population from the super radiative states to the non-radiative states. Averaging over many groups of emitters coupled with different coupling parameter G preserves the bi-exponential behavior of the decay with almost the same ratio $\tau_2/\tau_1 \approx 5.5$ as the local ratio at specific G (see Fig,2a). This makes a strong evidence of the intrinsic exclusive properties of the coupling in each group of the coupled emitters. The ratio of amplitudes of two exponents, A2/A1 is slightly growing with G and always less than 0.25.
The results qualitatively agree with the experiments conducted with core-shell nanoparticles. The core-shell nanoparticles have a broad variation in the dielectric spacer thickness between Au core and

ATTO655 dye layer in the experiments, thus providing a case of average signal over broad range of coupling parameters. The theoretical averaging was done over the coupling parameter ranging from 0 to 14 and show distinct bi-exponential behavior.

Materials: Tetraethyl orthosilicate (TEOS), ethanol, (3-aminopropyl)trimethoxysilane and ammonium hydroxide were purchased from Sigma-Aldrich (St. Louis, MO, USA). SU8-2000 epoxy is purchased from Micro Chem. Au 20nm colloid is purchased from Ted Pella. ATTO655 is purchased from ATTO-TEC GmbH (Siegen, Germany).

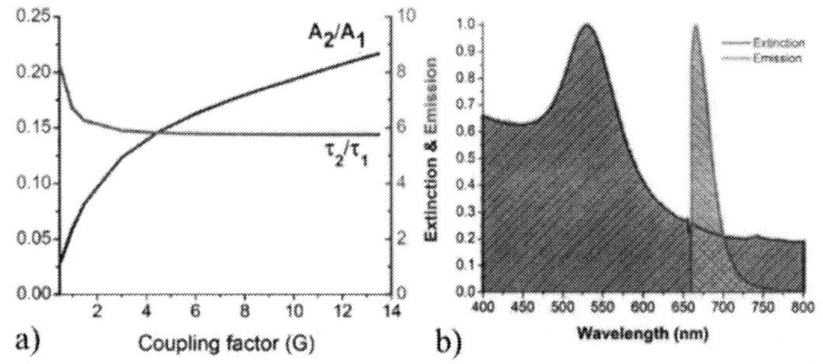

Figure 2. (a) Simulated amplitude ratio (left vertical scale) and decay time ratio (right scale) for three emitters versus coupling parameter G; (b) Experimental extinction and emission spectra of the core-shells on a substrate, excitation at 640 nm.

Synthesis of Au@SiO_2 core-shell is done under vigorous stirring of 1mL of Au 20 nm colloids concentration at 7x10^{11} particles/mL in 250 mL of 200 proof ethanol solution [3]. After the stirring done for 10 minutes, 4 mL of 30% ammonium hydroxide is added. Immediately after adding ammonium hydroxide, 100μL of TEOS were added to the solution. The solution was stirred for 2 hours and let the solution age without stirring at 4°C for 24 hours. After that, solution was centrifuged (3000rpm, 30 minutes) and washed with ethanol 4 times. Next, volume of 500μL of (3-aminopropyl)trimethoxysilane was added to the separated solution (10mL) and stirred vigorously for 24 hours and separated again as previously mentioned [3]. Preparation of fluorescent-coated Au@SiO_2 core-shell is done by dyeing (doping) method. One should add 4 mL of 6.2 μM ATTO655 (in ethanol) to 100μL of Au@SiO_2 (7x10^9 particles) and keep it in a shaker for 24 hours. After completion of the procedure, once again separate the particles using previously mentioned method and re-dispersed in 10% SU8.

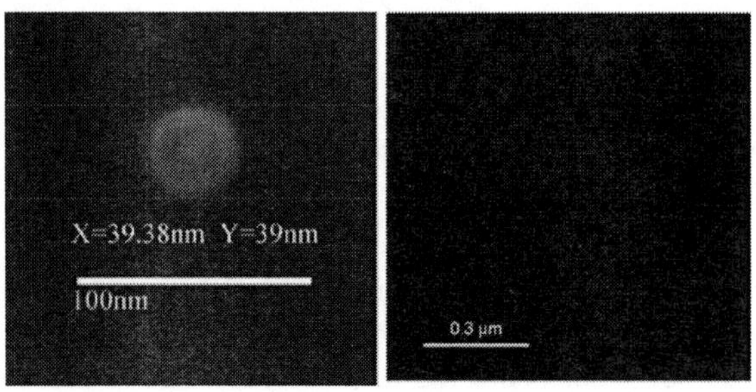

Figure 3. (*Left*) Scanning electron microscopy image, (*Right*) Fluorescence life time microscopy image.

Substrate preparation for optical time resolved confocal fluorescence spectroscopy was done as follows. Thin substrates (0.25mm thickness) were used in this experiment, and spin coated the final solution in 10% SU8 at different spin speeds from 500rpm to 4000rpm. Then let it dry for few hours and perform exticntion and emission spectroscopy shown in Fig. 2b. Fig. 3 shows scanning electron microscopy and fluorescence life time microscopy images (FLIM, Picoquant). The life time measurements show emission decay and then fitted with multiexponential function (representative decay measurements are in Fig.4). The characterization shows that dye doped core-shell nanoparticles are represented by broad distribution of particle size due to variation in the shell thickness and aggregate patterns. It results, consequently, in a variation in the coupling parameters. A high A2/A1 ratio is close to 0.5 in the case of core-shell Au/SiO2/ATTO655 nanoparticles in correspondence with average ratio obtained with the simulations. This bi-exponential decay is observed for very broad distribution in the geometry of the core-shell nanoparticles. Thus, the experiments show qualitative correspondence with

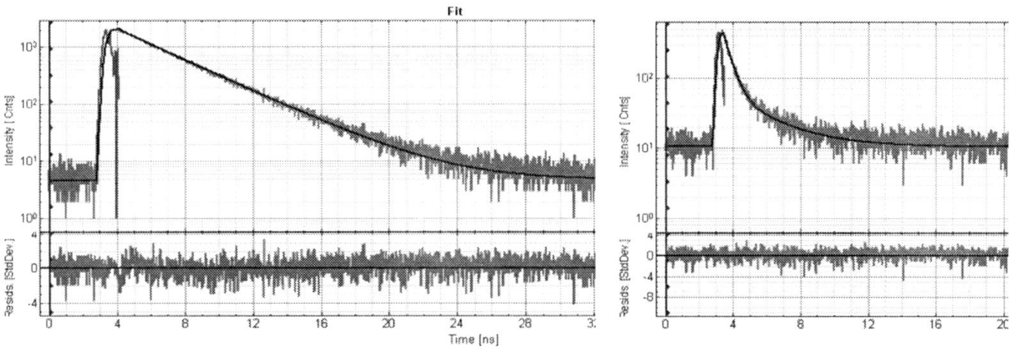

Figure 4. (*Left*) A single exponential emission decay for ATTO655 (313nM droplet 10% SU8) on a substrate, fitting 2.5 $e^{-t/3.2\ ns}$ kcnts with $\chi^2 = 1.006$; (*Right*) bi-exponential emission decay for Au@SiO_2 ATTO655 in 10% SU8 on a substrate 0.7 $e^{-t/0.44\ ns}$ + 0.06 $e^{-t/2.5\ ns}$ kcnts, $\chi^2 = 1.010$; green- raw data, black – fitting with the exponential reconvolution, red –instrumental reference function, green below - raw data minus fitted.

the theory. Namely, dye molecules emit superradiant and sub-radiant pulses when there is a coupling between them via plasmonic surface. ATTO655 dye molecules with solvents and on a glass substrate show single exponential fitting with very good χ^2 fitting parameters (χ^2 close to 1 means good accuracy). Dye dopped core-shell nanoparticles on a substrate always show bi-exponential decays even with the broad core-shell nanoparticle distribution in size.

3. Conclusions

The photoluminescence of the dye molecules has two components when coupled to the plasmonic nanostructure, a faster super-radiance followed by slower sub-radiance. Both times can be shorter than normal emitter. What is important is that the short pulse must have and has higher amplitude at the exponent. So short life time with high amplitude – these two properties cannot be separated. Indeed, the subradiant modes can be excited only through the coupling mechanism, sharing population from the superradiant modes. Thus the subradiant decay time is always longer and the amplitude is always less. Due to such an intrinsic link between two regimes the collective modes are always distinct and shown up in the form of fast and slow exponential decay.

Acknowledgments

This work was partially supported by Russian Ministry of Education and Science grant RFMEFI58117X0026.

References

[1] Dicke R H 1954 Coherence in spontaneous radiation process. *Phys. Rev.* **93**, 99 (1954).
[2] Temnov V V and Woggon U 2005 Superradiance and subradiance in an inhomogeneously broadened ensemble of two-level systems coupled to a low-*Q* cavity *Phys. Rev. Lett.* **95**, 243602.
[3] Xia X, Liu Y, Backman V and Ameer G A 2006 Engineering sub-100 nm multi-layer nanoshells. *Nanotechnology* **17**(21), 5435.

Visualization of spatial structure of optical modes in half-disk lasers

M S Dunaevskiy[1,2], P A Alekseev[1], A M Monahov[1], G S Sokolovskii[1,2], A N Baranov[3], R Teissier[3]

1. Ioffe Institute, Saint-Petersburg, 194021, Russia

2. ITMO University, Saint-Petersburg, 197101, Russia

3. Institute of Electronics and Systems, UMR 5214 CNRS – University of Montpellier, Montpellier, France

Mike.Dunaeffsky@mail.ioffe.ru

Abstract. The modal structure of half-disk lasers emitting in the middle IR-range was investigated. Scanning probe microscopy was used to map the intensity of optical radiation on the cleavages of half-disk lasers. The stability of optical modes in a half-disk lasers was studied by the geometric optics method of Poincare surface of section. It is shown that the deviation of the cleavage plane from the center of the disk leads to a rapid decrease in the number of stable spatial modes.

1. Introduction.

It is well known, that the high Q-factor of whispering gallery mode (WGM) disk lasers [1] (Figure 1a) provides relatively low current threshold values for laser emission. However, one of the main problems of these round-shaped lasers is the non-directionality of laser emission output. One of the approaches to improve directionality of WGM-lasers is the fabrication of half-disk resonators by cleaving the disk to two halves. Based on symmetry considerations, it is reasonable to assume that in an ideally split half-disk, the spatial structure of optical modes should be identical to that of the whole disk [2]. So, the areas of the most intense outcoming optical radiation must be at the very edge of the half-disk (Figure 1b). However, when splitting a disk, the cleavage plane does not pass exactly through the center (in this case it forms segment shaped resonator), which should lead to a perturbation of the spatial structure of optical modes (Figure 1c). The aim of this work was to study the spatial structure of optical modes in such segment shaped resonators. It is worth noting that this problem has not been previously studied in detail.

Recently, the authors of this work discovered the effect of a shift in the resonant frequency $\Delta\omega$ of an atomic force microscope (AFM) probe when it is illuminated [3]. This effect is due to the heating of the probe when the light is absorbed. In this case, the magnitude of the frequency shift is linearly proportional to the intensity of the light incident on the probe $\Delta\omega \sim I$. The model of light absorption related heating of the AFM-probe has been developed, which gives a satisfactory agreement with the experimental results [4]. It should be noted that in the paper [4] the absorption in traditional Si and Si_3N_4 AFM probes was calculated and it was shown that: 1) IR radiation is mainly absorbed in the upper metallization layer of the cantilever, 2) the absorption is significantly enhanced due to multiple light reflections in the "cantilever-surface" optical cavity, 3) heating by radiation can be enhanced by using cantilevers with a small cross-section and with low thermal conductivity, which slows the heat sink from the heated part of the cantilever, 4) the characteristic heating temperature of standard cantilevers by infrared radiation (P≈1 mW) is from fractions of Kelvin degrees to several Kelvin degrees. Based on this effect, a new probe method was developed to measure light-induced frequency shifts (LIFS) for the near-field mapping of the intensity distribution J(x,y) of the radiation of semiconductor laser structures (see Figure 1d) [3].

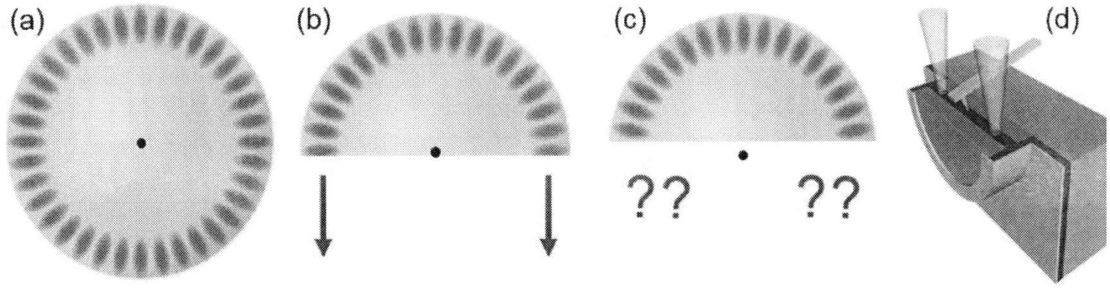

Figure 1. (a) – Scheme of WGM-modes in the disk-laser (b) - Scheme of WGM-modes in the half-disk laser, (c) - Scheme of unevenly chipped segment shaped laser, (d) – Scheme of light detection by AFM on the cleavage of half-disk laser.

2. Results and discussion.

By the above-mentioned LIFS method the near-field measurements were made on the cleavages of infrared (λ=2.1mkm) half-disk lasers (GaSb/GaAlSbAs/GaInSbAs) [3]. It was found that when the pump current exceeds the threshold value ($I > I_{th}$) the complex series of light-emitting spots (symmetric with respect to the center of the dissected disk) appear on the surface at some distance from the edges of the disk (see Figure 2).

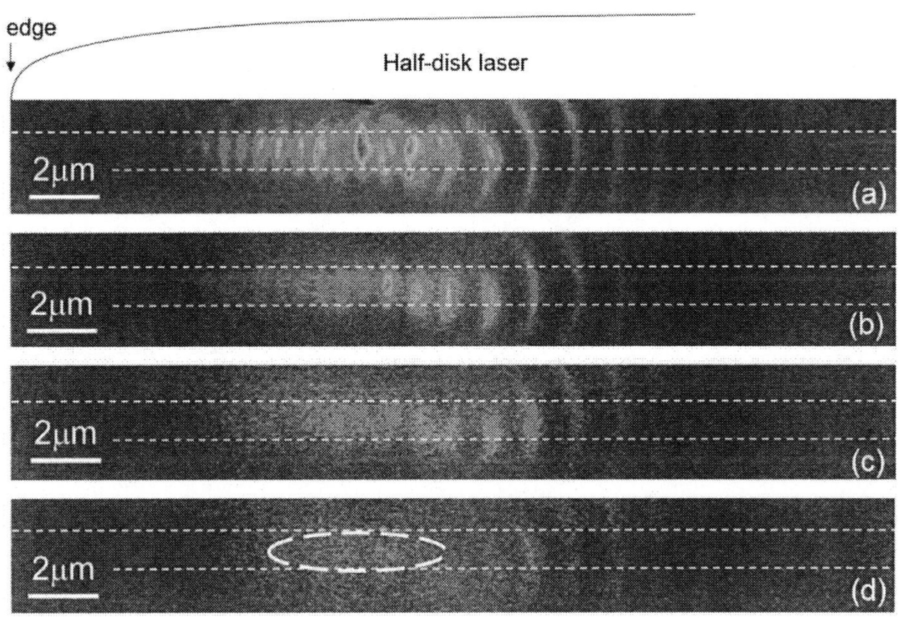

Figure 2. LIFS images J(x,y) measured near the edge of the half-disk laser obtained at different z-distances: (a) – 100nm, (b) – 400nm, (c) – 700nm, (d) – 1000 nm.

Figure 2 shows four LIFS images J(x,y) obtained near the edge of the half-disk and measured with different probe-surface z-distances (50 nm, 400 nm, 700 nm, 1000 nm). It can be seen that the intensity of a number of spots rapidly decreases with z-distance, that is, these spots have an evanescent nature. With an increase in the probe-surface distance up to 1000 nm, only the light emission region with a diameter of about 4 microns can be seen and this is the light that propagates as traveling waves and can be detected by an IR optical microscope. A number of half-disk lasers with diameters ranging from 100 microns to 300 microns were studied in a similar manner. For these structures, it has been experimentally found that the regions of maximum intensity of the modes that emit light are “moved away” from the edge of the half-disk by a significant distance ~30-50λ_m (λ_m - wavelength in a medium), which differs significantly from the case of the whole WGM-disk. In addition, it was found that the spatial position of the emitting spots differs significantly for half-discs chopped with different small deviations (δ/R~1-20 %) from the center of the disk. Also, it was possible to observe a series of evanescent spots, which makes it possible to obtain information about "sealed" modes that experience total internal reflection inside the laser cavity and do not emerge outward in the form of traveling waves.

Within the framework of geometric optics, it is possible to explain the observed phenomena. To describe the beams of closed (and unclosed) ray trajectories, we used the PSOS (Poincare surface of section) diagram method [5]. The PSOS-diagrams are constructed in the following way: (i) the optical ray is launched into the resonator and the following are then monitored: (ii) the positions of the points on the resonator boundary in which the ray undergoes reflections s_i (see Figure 3c), and (iii) the angles of reflection from the resonator boundaries θ_i (see Figure 3c). Then, for the whole set of rays in the resonator, a diagram is constructed in the coordinates (s; θ) – this is the PSOS-diagram. Stable trajectories in the PSOS-diagram will correspond to closed figures, and unstable trajectories will correspond to the "dust" of random points. Using the PSOS-diagram method, it is possible to determine in the 2D resonator the presence of stable modes and the so-called "chaotic" modes [6]. "Chaotic" modes correspond to trajectories of rays that do not close and move randomly in the resonator.

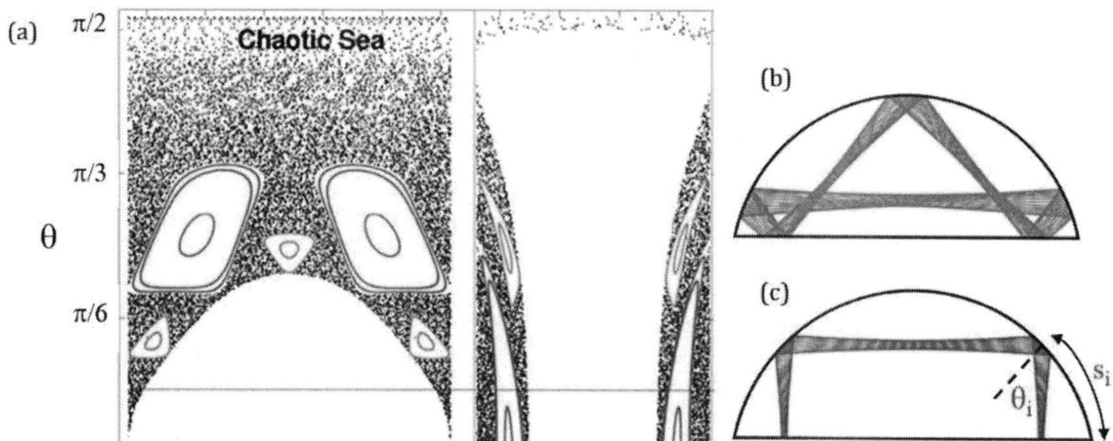

Figure 3. (a) PSOS-diagram of a half-disk cavity resonator cleaved with a 20% deviation from the center and (b) - Stable “sealed” mode (half of the triangle, m=3), corresponding to the green area in PSOS-diagram; (c) - Stable “emitting” mode (half of the rectangle, m=4), corresponding to the red area in PSOS-diagram.

It was observed, that with an increase in the δ/R deviation parameter (where δ is the deviation from the center of the half-disk) the portion of the "chaotic sea" in the PSOS diagram grows and only a small number of stable spatial modes remain (see Figure 3). It is also worth noting that as the deviation δ/R increases, the "chaotic sea" primarily devours those modes that have been located near the edge, and only modes that have been moved away from the edge remain stable. This important result qualitatively allows us to explain why the emission areas in unevenly chipped half-discs are moved away from the edge. The second important result that can be obtained by analyzing PSOS-diagrams is the presence of two categories among stable modes: (i) total internal reflection (TIR) modes, which correspond to closed ray trajectories that fall on all boundaries of resonator at angles greater than the TIR-angle and (ii) NonTIR-modes, which correspond to closed ray trajectories that fall on certain regions of the resonator boundaries at angles smaller than the TIR-angle. On the PSOS diagrams TIR-modes corresponds to "figures" entirely placed above the TIR-angle θ_{TIR}=arcsin(1/n), while NonTIR-modes correspond to figures located below the TIR-angle. It should be noted, that TIR-modes have a high Q-factor, they are "sealed" inside the resonator and cannot emit light outward in the form of a traveling wave. Outside the resonator, the TIR-modes can only appear in the form of exponentially decaying evanescent waves. On the other hand, Non-TIR modes can emit light outside of the resonator in the form of a traveling wave and this light can be detected by "far-field" optical microscopes.

It should be noted also, that the analysis of PSOS-diagrams makes it possible to estimate the angular divergence of the light beam emerging from the resonator [5]. In the simplest case, this can be done as follows: 1) one should choose a stable NON-TIR mode that contains an "island" located in the PSOS-diagram below the critical TIR-angle, 2) then the angular divergence of this beam inside the resonator is determined θ_{max}-θ_{min} (in the PSOS-diagram, this is simply the size of an "island" along the vertical axis of the incident angles θ), 3) then this value can be multiplied by the refractive index of the resonator (n), thus obtaining the angular divergence of the light beam emerging from the resonator. In segment-shaped resonators by changing the parameter δ the portion of the chaotic sea can be increased, thus reducing the size of stable modes "islands" on the PSOS-diagram. This reduces the angular divergence of the light beams and improves directionality of laser emission output.

Thus, analysis of PSOS-diagrams allows to determine the location of the regions of light emission and angular divergence of the emerging light beams for half-discs cleaved with a small deviation δ from the center. The presence of two effects: (i) the absorption by the "chaotic sea" of the modes located near the half-disc edge and (ii) the presence among the surviving stable modes of the nonradiating TIR-modes allows to explain why the light emission areas in unevenly cleaved discs are moved away from the edge.

Acknowledgements

M.S.D. acknowledges support from the Russian Foundation of Basic Research (project No. 17-52-16029).

References

[1] McCall S, Levi A, Slusher R, Pearton S and Logan R 1992 *Appl. Phys. Lett.* **60** 289
[2] Monakhov A, Sherstnev V, Astakhova A, Yakovlev Y, Boissier G, Teissier R and Baranov A 2009 *Appl. Phys. Lett.* **94** 51102
[3] Dunaevskiy M, Alekseev P, Baranov A, Monakhov A, Teissier R, Arinero R, Girard P and Titkov A 2013 *Appl. Phys. Lett.* **103** 53120
[4] Dunaevskiy M, Dontsov A, Alekseev P, Monakhov A, Baranov A, Girard P, Arinero R, Teissier R and Titkov A 2015 *Appl. Phys. Lett.* **106** 171105
[5] Nöckel J and Stone A 1998 *Nature* **385** 45
[6] Shinohara S, Harayama T, Fukushima T, Hentschel M, Sasaki T and Narimanov E 2010 *Phys.Rev.Lett.* **104** 163902

IOP Conf. Series: Journal of Physics: Conf. Series **1092** (2018) 012028 doi:10.1088/1742-6596/1092/1/012028

Circularly polarized thermal emission from chiral metasurface in the absence of magnetic field

S.A. Dyakov[1], **A.V. Ignatov**[1], **S.G. Tikhodeev**[2], **N.A. Gippius**[1]

[1]Skolkovo Institute of Science and Technology, Moscow, Russia

[2]Moscow State University, Moscow, Russia

E-mail: s.dyakov@skoltech.ru

Abstract. We report on the thermal radiation source which emits circularly polarized thermal radiation in the absence of an external magnetic field. The geometry of thermal source includes a photonic crystal slab waveguide with chiral morphology. We show that due to the absence of a mirror symmetry of such metasurface, the thermally generated electromagnetic waves are circularly polarized with the circular polarization degree of 0.87. Using the Fourier modal method we study the structure eigenmodes and analyze the field distributions in them.

1. Introduction

In recent years, the study of thermal emission of periodic materials attracted a great deal of attention from researchers due to its high potential for important applications in near- and far-field thermal management [1, 2]. Usually, the thermal radiation from bulk emitters is incoherent, broadband and isotropic. By changing the emitter symmetry one can control the angular emission diagram and the polarization of thermal radiation. In particular, a structure without a mirror symmetry can emit circularly polarized thermal radiation. The mirror symmetry can be broken down by applying an external magnetic field due to spin-orbit interaction of electrons. Another way to break down the mirror symmetry is creating a structure with a chiral surface morphology. Chiral metasurfaces were used in literature to obtain the circularly polarized photoluminescence of quantum dots [3]. The highest theoretically predicted circular polarization degree was as high as 99% while the corresponding experimental value was 81%. The circular polarization degree depends on the surface geometry and is a matter of theoretical optimization.

Here we use the chiral metasurface to generate the circularly polarized thermal radiation. We demonstrate a structure with artificial rotational anisotropy which radiates the circularly polarized thermal emission and discuss this effect in details.

2. Model structure and theory

In this paper, the thermal emitter consists of a KCl substrate capped by ZnSe waveguide with two-dimensional array of etched gammadions (Fig. 1a). The bottom surface of rectangles is covered by 20-nm thick layer of Si_3N_4. In this work we are focused on the 7–15 μm wavelength range. ZnSe and KCl are transparent in the middle infrared range and hence do not contribute to the thermal emission. In contrast, Si_3N_4 has a wide absorption band and hence is the source of thermal radiation in the structure. The emissivity is calculated using the Fourier

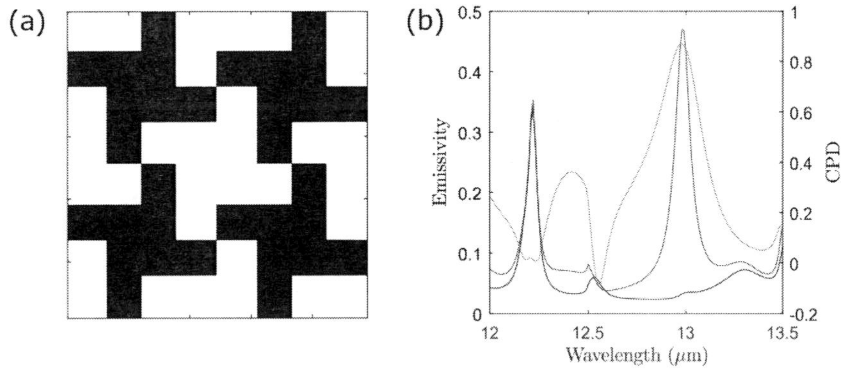

Figure 1. (a): Schematic of chiral metasurface. (b): LCP (blue) and RCP (red) emissivity spectra of the chiral metasurface. Circular polarization degree is shown by green.

modal method in the scattering matrix form [4]. To ensure the convergence of the method, in decomposition of electric and magnetic fields into Fourier series, we used 169 harmonics along with Li's factorization rules.

3. Results and discussions

To demonstrate the circularly polarized thermal emission, we calculate the emissivity spectra spectra of chiral metasurface in left circular polarization (LCP) and right circular polarization (RCP) as well as the circular polarization degree (CPD). We estimate the CPD as

$$\eta = \frac{I_{RCP} - I_{LCP}}{I_{RCP} + I_{LCP}} \tag{1}$$

As shown in Fig. 1b, in the displayed spectral range, the emissivity spectra are characterized by the peak at $\lambda \approx 13\,\mu$m which has different amplitudes in LCP and RCP. In the result, the circular polarization degree at this wavelength is non-zero and reaches the values of 0.87.

To understand the physical origin of the emissivity peaks, we study the structure eigenmodes by finding the poles of the scattering matrix. We also plot the fields distributions in the eigenmodes. Finally, we analyze the dependence of the circular polarization degree on the geometrical parameters of the metasurface and calculate the angular emission diagram of circularly polarized thermal radiation.

4. Conclusion

In conclusion, we have demonstrated that the chiral metasurface can emit the circularly polarized thermal radiation with the circular polarization degree as high as 85%. We attribute this effect to the shape anisotropy of the photonic crystal part of the structure.

References

[1] J.-J. Greffet, R. Carminati, K. Joulain, J.-P. Mulet, S. Mainguy, and Y. Chen, "Coherent emission of light by thermal sources," *Nature*, vol. 416, no. 6876, pp. 61–64, 2002.

[2] S. A. Dyakov, J. Dai, M. Yan, and M. Qiu, "Near field thermal memory based on radiative phase bistability of VO_2," *Journal of Physics D: Applied Physics*, vol. 48, p. 305104, 2015.

[3] S. V. Lobanov, T. Weiss, N. A. Gippius, S. G. Tikhodeev, V. D. Kulakovskii, K. Konishi, and M. Kuwata-Gonokami, "Polarization control of quantum dot emission by chiral photonic crystal slabs," *Optics letters*, vol. 40, no. 7, pp. 1528–1531, 2015.

[4] S. G. Tikhodeev, A. L. Yablonskii, E. A. Muljarov, N. A. Gippius, and T. Ishihara, "Quasiguided modes and optical properties of photonic crystal slabs," *Phys. Rev. B*, vol. 66, p. 045102, 2002.

Biexciton states in nanoheterostructures of dielectric quantum dots

Sergey I. Pokutnyi[a], Yuri N. Kulchin[b], Vladimir P. Dzyuba[b], Andrey V. Amosov[b]

[a]Chuiko Institute of Surface Chemistry of National Academy of Sciences of Ukraine, 17 General Naumov Str., Kyiv 03164, Ukraine
[b]InstituteofAutomation and Control Processes of FEB Russian Academy of Sciences, 5 Radio Str., Vladivostok 690950, Russia

Sergey Pokutnyie-mail: pokutnyi.serg@gmail.com

Abstract. It has been found that the formation of a biexciton in a nanoheterostructures (NHS) made up of aluminum oxide quantum dots (QD) synthesized in a dielectric matrix is of threshold character and can occur in a nanosystem where the distance D between the surfaces of QD is given by the condition $D_c^{(1)} \leq D \leq D_c^{(2)}$. We have shown that in such a NHS acting as "exciton molecules" (biexcitons consisting of spatially separated electrons and holes) are the quantum dots of aluminum oxide with excitons localizing over their surfaces. The position of the biexciton state energy band depends both on the mean radius of the quantum dots, and the distance between their surfaces, which enables one to purposefully control it by varying these parameters of the nanostructure.

1. Introduction

The studies of quasi 0-D nanosystems made up of spherical QD with the mean radius a in the range of 1 to 10 nm consisting of semiconductor (germanium, cadmium sulfide and selenide, gallium arsenide, zinc selenide) and dielectric (aluminum oxide) materials synthesized in dielectric (semiconductor) matrices attract considerable interest due to their unique photoluminescence properties, ability to effectively emit light in the visible and near infrared spectral range at room temperatures [1]. Optical and electro-optical properties of such quasi 0-D nanosystems are to a large extent governed by the energy spectrum of the spatially confined electron-hole pair (exciton) [1-3]. During investigation of the optical characteristics of nanosystems with Al_2O_3 QDs in experimental papers [1,2] it was found that the electron can be localized above the surface of the QD while the hole here moves in the volume of the QD.

In the present paper we show that the biexciton formation in a NHS made up of aluminum oxide quantum dots synthesized in a dielectric matrix is of threshold character and can occur in a nanosystem where the distance D between the surfaces of QD is given by the condition $D_c^{(1)} \leq D \leq D_c^{(2)}$. We also demonstrate that in such NHS acting as "exciton molecules" (or biexciton) are the quantum dots of aluminum oxide with excitons localizing over their surfaces. The position of the biexciton state energy band is shown to depend both on the mean radius of quantum dots, and the distance between their surfaces, which enables one to purposefully control it by varying these parameters of the NHS. It is established that at constant concentrations of biexcitons at temperatures T

below a certain critical temperature T_c due to the radiative annihilation of one of the excitons forming a biexciton one can expect a new spectral band of luminescence shifted relative to the exciton band by the biexciton binding energy E_B. This new luminescence band disappears at temperatures above T_c. At a constant temperature $T<Tc$ the growth of exciton concentration brings about weakening of the exciton band and strengthening of the biexciton band of luminescence.

2. Energy of the exciton quasimolecule ground singlet state

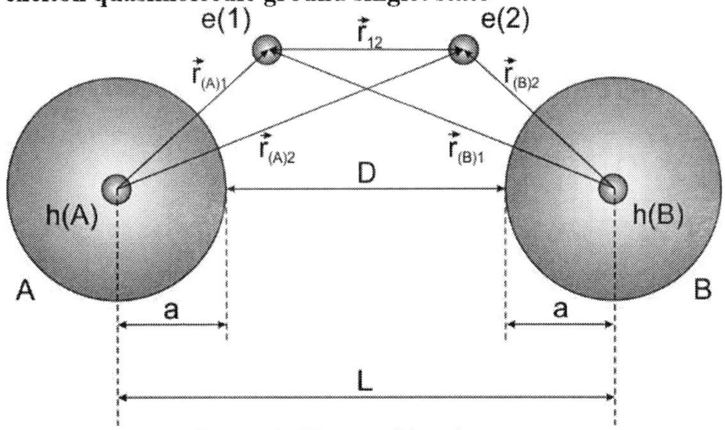

Figure 1. The considered system

Let us consider a model of a nanosystem composed of two dielectric (semiconductor) spherical quantum dots QD(A) and QD(B) of radius a synthesized in a dielectric matrix with permittivity ε_1 (D is the distance between the spherical surfaces of QD). QDs are of a dielectric (semiconductor) material with permittivity ε_2. For simplicity and without loss of generality let us assume that holes *h(A)* и*h(B)* with effective masses m_h are located at the centers of QD(A)and QD(B)whileelectrons *e*(1)and *e*(2)with effective masses $m_e^{(1)}$ are localized near the spherical surfaces of QD(A)and QD(B), respectively ($r_{A(1)}$ – the distance from the electron *e*(1)to the center of QD(A); $r_{B(2)}$– the distance from the electron *e*(2)to the center of QD(B)) (Figure 1) [4]. This assumption is justified by the fact that $m_e^{(1)}/m_h \ll 1$. Let us also infer that an infinitely high potential barrier exists on the spherical interface between the QD and the matrix so that the electrons cannot get into the quantum dot while the holes cannot escape from it.

Let us now use this model to consider the possibility of the formation of an exciton quasimolecule from spatially separated electrons and holes (the holes are located at the centers of QD(A) and QD(B) and electrons are localized near their spherical surfaces). Using adiabatic approximation and the effective mass approximation, the Hamiltonian of the exciton quasimolecule (of spatially separated electrons and holes) can be written in the form [4,5]:

$$\hat{H} = \hat{H}_{A(1)} + \hat{H}_{B(2)} + \hat{H}_{int}, \quad (1)$$

where $\hat{H}_{A(1)}$ and $\hat{H}_{B(2)}$ are the Hamiltonians of the excitons of spatially separated hole *h(A)* andelectron *e*(1) and hole *h(B)* and electron *e*(2), respectively. The contribution of the energy of polarization interaction with the surface of QD to the Hamiltonians of the excitons $\hat{H}_{A(1)}$ and $\hat{H}_{B(2)}$ can be, as a first approximation neglected [3,4]. In (1) In the first approximation we can neglect the contributions to the Hamiltonian $\hat{H}_{int}$ of the interaction energies of the electrons *e*(1) and *e*(2) and the holes $h(A)$ and $h(B)$ with polarization fields induced by these charge carriers on the surfaces of QD (A) and QD (B) [3,4]. Thus the Hamiltonian $\hat{H}_{int}$incorporates only the energies of Coulomb interaction of electron *e*(1) with hole *h(B),* and electron *e*(2) with hole *h(A),* as well as that between electrons *e*(1) and *e*(2), and holes *h(A)* and *h(B).* Under the assumption that the spins of the electrons

$e(1)$ and $e(2)$ are antiparallel let us write down the normalized wave function of the ground singlet state of the exciton quasimolecule as a symmetric linear combination of wave functions $\Psi_1(r_{A(1)}, r_{B(2)})$and $\Psi_2(r_{A(2)}, r_{B(1)})$[3,4]:

$$\Psi_s(r_{A(1)}, r_{A(2)}, r_{B(1)}, r_{B(2)}) = [2(1 + S^2(D, a))]^{-1/2}[\Psi_1(r_{A(1)}, r_{B(2)}) + \Psi_2(r_{A(2)}, r_{B(1)})], \quad (2)$$

where $S(D,a)$ is the overlap integral of single-electron wave functions. Assuming that the electrons $e(1)$ and $e(2)$ move independently from each other, let us represent the wave functions $\Psi_1(r_{A(1)}, r_{B(2)})$ and$\Psi_2(r_{A(2)}, r_{B(1)})$(2)as a product of single-electron wave functions $\varphi_{A(1)}(r_{A(1)})$and $\varphi_{B(2)}(r_{B(2)})$, as well as $\varphi_{A(2)}(r_{A(2)})$and $\varphi_{B(1)}(r_{B(1)})$,respectively [3,4]. Let us also represent the single-electron wave functions as variational functions of Coulomb type [3]. In the framework of the variational method, the energy of the exciton quasimolecule ground singlet state, as a first approximation, is given by the mean value of the Hamiltonian $\hat{H}(1)$ over the states described by the wave functions of the zeroth approximation Ψ_s (2). The total energy $E_0(\tilde{D}, \tilde{a})$ of the exciton quasimolecule ground singlet state takes the form [4,5]:

$$E_0(\tilde{D}, \tilde{a}) = 2E_{ex}(\tilde{a}) + E_{\text{B}}(\tilde{D}, \tilde{a}), \quad (3)$$

where $E_{\text{B}}(\tilde{D}, \tilde{a})$ is the binding energy of the ground singlet state of the exciton quasimolecule($\tilde{D} = (D/a_{ex}^{2D})$, the Bohr radius of the 2D exciton a_{ex}^{2D}), and $E_{ex}(\tilde{a})$ is the binding energy of the ground state of the exciton (consisting of spatially separated electron and hole) localized over the surface of QD, which was worked out in [3].

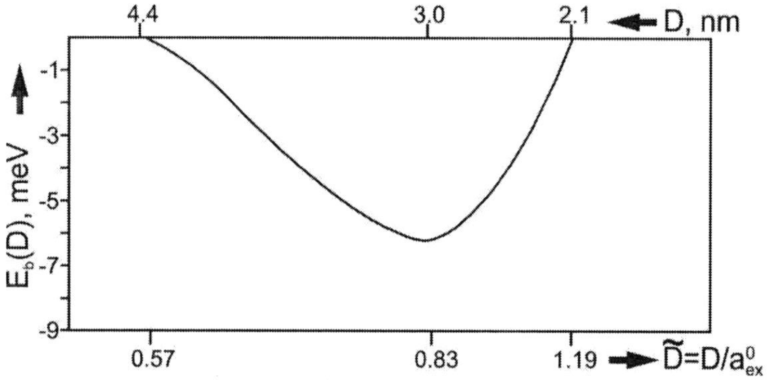

Figure 2. The dependence of the binding energy of the ground singlet state $E_{\text{B}}(D, \tilde{a}_1)$ of the biexciton (consisting of spatially separated electrons and holes) in a nanosystem made up of two spherical aluminum oxide quantum dots QD(A) and QD(B) with the mean radius $\tilde{a}_1 = 3.18$ nm, on the distance D between the surfaces of QD(A) and QD(B). Here $a_{ex}^0 = 0.35$nm is the Bohr radius of the two-dimensional exciton (consisting of spatially separated electron and hole).

Figure 2 presents the results of the variation calculations of the binding energy$E_{\text{B}}(\tilde{D}, \tilde{a})$ studied in [4] of the biexciton ground state in a nanosystem with QD of aluminum oxide of the mean radius $\tilde{a}_1 = 3.18$ nm, (permittivity $\varepsilon_2 = 10$, effective mass of the hole in QD $(m_h/m_0) = 6.2$), synthesized in a dielectric matrix of the vacuum oil VM-4 (permittivity $\varepsilon_1 = 1.96$,the effective mass of the electron in the matrix $(m_e^{(1)}/m_0)$was deduced in [6] and amounts to 0,537, the Bohr radius (5a) of the 2D exciton $a_{ex}^0 = 0.35$ nm). Authors of [1] studied nanostructures doped with x=0.003% to 1% concentrations of aluminum oxide. At concentrations of QD higher than $x \cong 0.6\%$ one needs to take into account the interactions of the charges localized over the surfaces of quantum dots. The variational method that we used for the calculation of the biexciton ground state binding energy$E_B(\tilde{D},\tilde{a})$ is applicable provided that it is much smaller than the binding energy of the exciton ground state $E_{ex}(\tilde{a})$, i.e. the following condition must be fulfilled:

$$\left(E_B(\tilde{D},\tilde{a})/E_{ex}(\tilde{a})\right) \ll 1 \quad (4)$$

The binding energy $E_B(\tilde{D},\tilde{a})$ of the biexciton ground state in a nanosystem with QD of aluminum oxide of the mean radius $\tilde{a}_1 = 3.18$ nm has a minimum $E_B^{(1)}(D_1,\tilde{a}_1) \approx -7.03$ meV (at the distance $D_1 \cong 2.93$ nm) (Fig. 2) ($E_B^{(1)}$ corresponds to the critical temperature $T_c \approx 81.6\ K$). As it follows form Fig.1, biexciton appears in the nanosystem at distances $D \geq D_c^{(1)} \cong 2.1$ nm between the surfaces of QD. The formation of such a biexciton („exciton molecule") is of threshold character and can occur in a nanosystem with quantum dots of the mean radius $\tilde{a}_1$, where the distance D between the surfaces of QD exceeds a certain critical value $D_c^{(1)}$. The existence of such distance $D_c^{(1)}$arises from quantum size effects in which the decrease in the energies of interaction of the electrons and holes entering into the Hamiltonian (Eq. (1)) of the "exciton molecule" with decrease of the distance D between the surfaces of the QD cannot compensate for the increase in the kinetic energy of the electrons and holes.
The binding energy of the exciton $E_{ex}(\tilde{a})$ amounts to $E_{ex}(\tilde{a}_1) \cong -100.8$ meV [3], with the energy of the biexciton ground state (3) taking the value $E_0(\tilde{D}_1,\tilde{a}_1) \approx -208.6$ meV. It should be emphasized that the criterion (4) of the applicability of the variational method for the calculation of the biexciton binding energy $E_B(\tilde{D},\tilde{a})$ is fulfilled ($\left(E_B^{(1)}(D_1,\tilde{a}_1)/E_{ex}(\tilde{a}_1)\right) \cong 0.07$). At larger distances D between the surfaces of QD: $D \geq D_c^{(2)} \cong 4.4$ nm, the biexciton breaks down into two excitons (consisting of spatially separated electrons and holes), localized over QD surfaces (Fig.1). Thus a biexciton can be formed in a nanosystem where $D_c^{(1)} \leq D \leq D_c^{(2)}$. Furthermore, a biexciton can exist only at temperatures lower than the critical temperature $T_c = 81.6$ K. In the aluminum oxide monocrystal with the binding energy E_B= 0.61 meV (which corresponds to the temperature 7.04 K) the biexciton binding energy $E_B^{(1)}$is almost 12 times as large. As follows form the results of the variational calculations, the major contribution to the biexciton binding energy is from the energy of exchange interaction of electrons and holes, which by far surpasses that from their Coulomb interaction (i.e. the ratio ≤ 0.11). Since the calculations of the biexciton ground state binding energy $|E_B(\tilde{D},\tilde{a})|$ in the nanosystem are variational, the values of $|E_B(\tilde{D},\tilde{a})|$ and $|E_B^{(1)}|$ can be somewhat underestimated.

3. Conclusions

Thus it has been shown that the biexciton formation in a NHS made up of aluminum oxide quantum dots (QD) of the mean radius $\tilde{a}_1$ is of threshold character and can occur in a nanosystem where the distance D between the surfaces of QD is given by the condition $D_c^{(1)} \leq D \leq D_c^{(2)}$. We have demonstrated that in such a NHS (with the mean radius $\tilde{a}_1 = 3.18$ nm and the distance between surfaces of aluminum oxide QD $D_1 = 2.93$ nm) at temperatures lower than the critical temperature T_c=81,6K (which corresponds to the biexciton binding energy E_B= 7.03 meV) an „exciton molecule" (a biexciton consisting of spatially separated electrons and holes) can be formed. Acting as "exciton molecules" in the NHS are aluminum oxide quantum dots with excitons (consisting of spatially separated electrons and holes) localizing over their surfaces. The interaction of the excitons with the QD surfaces ("intramolecular" interaction) is shown to be much stronger than that between quantum dots ("intermolecular" interaction).
We have also shown that the position of the biexciton band in the NHS depends both on the QD mean radius $\tilde{a}_1$ and the distance D_1 between QD surfaces. By varying these parameters, one can purposefully control the position of the energy band of biexciton states in the NHS. With increase in temperature above the threshold (T$\geq T_c$), a phase transition can occur from the biexciton to exciton state. It has been found that at a constant concentration of excitons (i.e. constant concentration of QD) and temperatures *T* below *Tc*, one can expect a new luminescence band shifted from the exciton band

by the value of the biexciton binding energy E_B. This new band disappears at higher temperatures ($T \geq T_c$). At a constant temperature below Tc, an increase in exciton concentration (i.e. in QD concentration) brings about weakening of the exciton luminescence band and strengthening of the biexciton one.

References

[1] V.A. Milichko, V.P. Dzuyba, Y.N. Kulchin 2013 Unusual nonlinear optical properties of SiO2 nanocomposite in weak optical fields *Appl. Phys. A* 11(1), 319 -322

[2] V.A. Milichko, V.P. Dzuyba, Y.N. Kulchin 2013 Anomalous optical nonlinearity of dielectric nanodispersions *Quantum Electronics* Volume 43, Issue 6, 2013, Pages 567-573

[3] S.I. Pokutnyi, Y.N. Kulchin, V.P. Dzyuba 2017 Spectroscopy of excitons in heterostructures with quantum dots *Journal of Applied Spectroscopy* 84(4), 603 –610.

[4] S.I. Pokutnyi, Y.N. Kulchin, V.P. Dzyuba, A.V. Amosov (2016) Biexciton in nanoheterostructures of dielectric quantum dots *Journal of Nanophotonics* 10(3), 036008-1 - 036008-8

[5] S.I. Pokutnyi 2017 Excitonic quasimolecules in nanosystems of quantum dots *Optical Engineering*, 56(9),091603-1 - 091603-7.

[6] Sergey I. Pokutnyi 1995 Size quantization of excitons in quasi-zero-dimensional structures *Physics Letters A.* 168, 433-436.

Acknowledgments

The work was supported by “Far East Program” of FASO Russia (grant № 18-3-023).

Modelling the electromagnetic field distribution in arbitrary inhomogeneous media using the Green's function approach

Vladimir P. Dzyuba[a], Andrey V. Amosov[a], Yuri N. Kulchin[a], Liudmila P. Liakhova[b]

[a]Institute of Automation and Control Processes of FEB Russian Academy of Sciences, 5 Radio Str., Vladivostok 690950, Russia
[b]FarEasern Federal University, 8 Suhanova Str., Vladivostok, 690950, Russia

e-mail: vdzyuba@iacp.dvo.ru

Abstract The equations of electric vector and its spectrum for the electromagnetic field in a stationary isotropic inhomogeneous medium with conductivity, magnetic and permittivity with its arbitrary dependence from coordinates are presented. The method of solving this equations is based Green's function together with the method of iterations. This approach is also applicable to optical frequencies.

1. Introduction

The methods for modelling electromagnetic fields in inhomogeneous media can be divided into two classes: the first is analytical one for finding the analytical solution of the related problem; the second is the modeling in computing environments, when the electromagnetic field distribution problem is solved by numerical methods. The second class contains a wide range of mathematical numerical methods (finite-difference time-domain, finite element method, etc.) which is used in computer programs for electrodynamic modeling [1, 2]. Among them there are direct universal methods for solving boundary value problems. The disadvantage of this methods is the high computational burden. Also should be noticed, that the solution of physical problems by numerical methods usually requires a number of mathematical approximations and significant physical simplifications of the media model. But this two classes can be combined, when exact analytical methods are used at the initial phase to simplify the problem and solve it by numerical methods further. Hereafter, we use this procedure for our problem.

2. Main part

Under inhomogeneous media one consider the media with electric conductivity σ, permittivity ε, permeability μ, current density $\vec{j}$ and speed of electromagnetic waves propagation c are the functions of media points coordinates. Thus the electric, and magnetic fields vectors, and the current density is a function of time and coordinates.

The wave equation for electric field vector is obtained from the Maxwell's equations by exclusion the magnetic field vector. Carrying out the transformation of the Maxwell's equations, it is necessary to take into account the relation:

$$\nabla(\nabla\cdot\vec{E}) = \left[\nabla\cdot\vec{J} - \vec{E}(\nabla\cdot\sigma)\right]\nabla\frac{1}{\sigma} + \frac{1}{\sigma}\nabla\cdot\left[\nabla\cdot\vec{J} - \vec{E}(\nabla\cdot\sigma)\right] \quad (1)$$

If $\nabla\cdot\vec{J} \neq 0$, then:

$$\nabla(\nabla\cdot\vec{E}) = \vec{E}\frac{(\nabla\cdot\sigma)^2}{\sigma^2} - [(\nabla\cdot\ln\sigma)\nabla]\vec{E} - (\vec{E}\cdot\nabla)(\nabla\cdot\ln\sigma) - (\nabla\cdot\ln\sigma)\times\nabla\times\vec{E} + (\nabla\cdot\vec{J}_{ext})\left(\nabla\cdot\frac{1}{\sigma}\right) + \\ + \frac{1}{\sigma}\nabla(\nabla\cdot\vec{J}_{ext}), \quad (2)$$

where $\vec{J}_{ext}$ is the external current density. Transforming the Maxwell's equations and taking into account (2) one can obtain the equation for the electric field vector in an inhomogeneous media with field source which electric charge flow from source volume isn't equal to zero:

$$\varepsilon\mu\frac{\partial^2}{\partial t^2}\vec{E} + \mu\sigma\frac{\partial}{\partial t}\vec{E} - \Delta\vec{E} - (\nabla\cdot\ln\mu\sigma)\times(\nabla\times\vec{E}) + \vec{E}\left(\frac{\nabla\cdot\sigma}{\sigma}\right)^2 - [(\nabla\cdot\ln\sigma)\nabla]\vec{E} - (\vec{E}\cdot\nabla)(\nabla\cdot\ln\sigma) = \\ -\mu\frac{\partial}{\partial t}\vec{J}_{ext} - (\nabla\cdot\vec{J}_{ext})\left(\nabla\cdot\frac{1}{\sigma}\right) - \frac{1}{\sigma}\nabla(\nabla\cdot\vec{J}_{ext}) \quad (3)$$

For absence the charges injection into the medium $\nabla\cdot\vec{J} = 0$. If the field is monochromatic, then the ratio of the first and the second members of (3) is equal $i\frac{\varepsilon}{\sigma}\omega$. Modulus of this ratio is small under high conductivity, low permittivity, and low-frequency value conditions. This corresponds to electromagnetic field quasistationarity in the media. In this case, one can neglect the term containing the second derivative with respect to time. Thus, the field propagation in the media mainly is of the diffusive character. When $\frac{\varepsilon}{\sigma}\omega \gg 1$the field propagation in the media is of the wave-type mainly. At present equation (3) is hard to solve correctly due to the lack of methods. Finding a solution that satisfies a given correctness can be done in two ways. The first is to use a numerical method. The second is to convert equation (3) to integral form using the Green's function for the Vector Helmholtz Equation for the electric field vector Fourier spectrum [3]. In the first case its require the σ, ε, μ value and solution will correspond only to this values. The general solution for arbitrary values of σ, ε, μ might not turn out to be feasible. In the second case its possible for σ, ε, μ as a coordinate function in the form of the Born-Neumann series. To accomplish this task, the electric field vector and current density can be written as Fourier spectrum $\vec{E}(\vec{r},\omega), \vec{J}(\vec{r},\omega)$. Let's consider conductivity and permittivity as a sum of constant and variable in space parts:

$$\sigma(\vec{r}) = \sigma_c + \sigma_1(\vec{r}),\ \mu(\vec{r}) = \mu_c + \mu_1(\vec{r}),\ \varepsilon(\vec{r}) = \varepsilon_c + \varepsilon_1(\vec{r}) \quad (4)$$

The constant terms can be defined as an average value along the field or current propagation track. Let's introduce the notations:

$$k^2(\vec{r}) = \left(\omega^2\varepsilon\mu - i\omega\mu\sigma - \left(\frac{\nabla\cdot\sigma_1(\vec{r})}{\sigma(\vec{r})}\right)^2\right)$$

$$\vec{f}_{ext}(\vec{r},\omega) = \left(\omega^2\varepsilon\mu - i\omega\mu\sigma - \left(\frac{\nabla\cdot\sigma_1(\vec{r})}{\sigma(\vec{r})}\right)^2\right) - \left(\nabla\vec{J}_{ext}(\vec{r},\omega)\right)\left(\frac{\nabla\sigma_1(\vec{r})}{\sigma^2(\vec{r})}\right) - \frac{1}{\sigma}\nabla\left(\nabla\cdot\vec{J}_{ext}(\vec{r},\omega)\right) \quad (5)$$

$$\vec{f}(\vec{r},\omega) = -\left(\frac{\nabla\mu_1(\vec{r})}{\mu(\vec{r})} + \frac{\nabla\cdot\sigma_1(\vec{r})}{\sigma(\vec{r})}\right)\times\left(\nabla\times\vec{E}(\vec{r},\omega)\right) - \left(\frac{\nabla\sigma_1(\vec{r})}{\sigma(\vec{r})}\nabla\right)\vec{E}(\vec{r},\omega) - \left(\vec{E}(\vec{r},\omega)\cdot\nabla\right)\left(\frac{\nabla\sigma_1(\vec{r})}{\sigma(\vec{r})}\right)$$

Using (5) and (4) for equation (3):

$$\Delta\vec{E}(\vec{r},\omega)+k^2(\vec{r})\vec{E}(\vec{r},\omega)=\vec{f}(\vec{r},\omega)+\vec{f}_{ext}(\vec{r},\omega) \tag{6}$$

In equation (6) the field source is not only the external currents $\vec{f}_{ext}(\vec{r},\omega)$ but also the media heterogeneity reradiating the external field. These sources are described by $\vec{f}(\vec{r},\omega)$. Its presence in (6) does not allow converting the equation to integral form. Nevertheless, by using the Green's function one can reformulate (6) in integral form in terms of the vector which solution can be found as the series of successive approximations. The problem of finding the Green's function of equation (6) in the analytical form is not solved for an arbitrary dependence of $k(\vec{r})$ on the coordinates. Using (4) one can formulate $k^2(\vec{r})$ as the sum of independent on coordinates ${k_c}^2$ and function on coordinates ${k_1}^2(\vec{r})$:

$$k^2(\vec{r})=(\omega^2\varepsilon_c\mu_c-i\omega\sigma_c\mu_c)+\omega^2\big(\varepsilon_c\mu_1(\vec{r})+\varepsilon_1(\vec{r})\mu_c+\varepsilon_1(\vec{r})\mu_1(\vec{r})\big)$$
$$-i\omega\big(\sigma_c\mu_1(\vec{r})+\sigma_1(\vec{r})\mu_c+\sigma_1(\vec{r})\mu_1(\vec{r})\big)-\left(\frac{\nabla\cdot\sigma_1(\vec{r})}{\sigma(\vec{r})}\right)^2={k_c}^2+{k_1}^2(\vec{r})$$

Thus equation (6) can be written as:

$$\Delta\vec{E}(\vec{r},\omega)+{k_c}^2\vec{E}(\vec{r},\omega)=\vec{f}_1(\vec{r},\omega)+\vec{f}_{ext}(\vec{r},\omega), \tag{7}$$

where $\vec{f}_1(\vec{r},\omega)=\vec{f}(\vec{r},\omega)-{k_1}^2(\vec{r})\vec{E}(\vec{r},\omega)$. For the vector fields, the Green's function is a tensor. In this case, the equation for electric field vector in inhomogeneous media breaks up into the system of three independent scalar equations for the $\vec{E^i}(\vec{r},\omega)$ or $\vec{E^i}(\vec{r},t)$projections on the axis of the orthogonal coordinate system. This simplifies the Green's function and it can be written as a vector function with terms are scalar function as $\vec{G}(\vec{r}-\vec{r}_1)=\sum_{i=1}^{3}\vec{n}_iG(\vec{r}-\vec{r}_1)$, where $\vec{n}_i$ are the unit vectors of the coordinate axes. Thus one can be written:

$$\vec{E}(\vec{r},\omega)=\textstyle\sum_{i=1}^{3}\vec{n}_i\int_\Omega G(\vec{r}-\vec{r}_1)\Big[\big(\vec{n}_i\cdot\vec{f}_{ext}(\vec{r}_1,\omega)\big)+\big(\vec{n}_i\cdot\vec{f}_1(\vec{r}_1,\omega)\big)\Big]dV \tag{8}$$

The integration is performed with respect to variable $\vec{r}_1$. There are three steps to solving (8):
1) Finding the solution of inhomogeneous Schrodinger equation using the Green's function method

$$\Delta\vec{E}_0(\vec{r},\omega)+{k_c}^2\vec{E}_0(\vec{r},\omega)=\vec{f}_{ext}(\vec{r},\omega)$$

$$\vec{E}_0(\vec{r},\omega)=\sum_{i=1}^{3}\vec{n}_i\int_\Omega G(\vec{r}-\vec{r}_1)\Big[\big(\vec{n}_i\cdot\vec{f}_{ext}(\vec{r}_1,\omega)\big)\Big]dV$$

2) Using this solution for $\vec{f}_1(\vec{r},\omega)$, and substituting the obtained result into equation (8) one can find the first solution approximation:

$$\vec{E}_1^i(\vec{r},\omega)=\int_\Omega G(\vec{r}-\vec{r}_1)\Big(f_1^i(E_0,\omega,\vec{r}_1)+f_{ext}^i(\omega,\vec{r}_1)\Big)dV$$

3) Substituting the first approximation into equation (8) for $\vec{f}_1(\vec{r},\omega)$ the second approximation for (8) can be obtained:

$$\vec{E}_2^i(\vec{r},\omega) = \int_\Omega G(\vec{r}-\vec{r}_1)\left(f_1^i(E_1,\omega,\vec{r}_1) + f_{ext}^i(\omega,\vec{r}_1)\right)dV$$

Integration is over infinite volume Ω and we shall take into account the field propagation velocity by using the inverse Fourier transform. In the real situation, the volume shall be such, that each volume point makes a notable contribution in the field as the second field source. Due to fast Green's function amplitude decrease and strong absorption of the electromagnetic field by the media, the integration domain may be small but contains the observation point $\vec{r}$. By acting in this way, we can achieve the required accuracy for the solution of calculating the electromagnetic field in an inhomogeneous medium problem. Finding the inverse Fourier transform of the solution approximation, using the expression (8), the electric field strength vector as a function of coordinates and time defines. For determining the magnetic field component one use Maxwell's equations and write it for magnetic and electric fields Fourier spectrum $\nabla \times \vec{E}(\vec{r},\omega) = -i\omega\vec{B}(\vec{r},\omega)$. Substituting (8) in this equation one can get:

$$-i\omega\vec{B}(\vec{r},\omega) = \int_\Omega \vec{e}\times\nabla\cdot G(\vec{r}-\vec{r}_1)\left[\left(\vec{f}_{ext}(\vec{r}_1,\omega)\right) + \left(\vec{f}_1(\vec{r}_1,\omega)\right)\right]dV$$

$$\vec{H}(\vec{r},\omega) = -i\frac{1}{\omega\mu(\vec{r})}\int_\Omega \vec{e}\times\nabla\cdot G(\vec{r}-\vec{r}_1)\left[\left(\vec{f}_{ext}(\vec{r}_1,\omega)\right) + \left(\vec{f}_1(\vec{r}_1,\omega)\right)\right]dV, \quad (9)$$

where $\vec{e}$ is the unit vector along the electric field vector. Expression (9) allows calculating the fields of spectra of magnetic field vector in inhomogeneous media from source or external field. To model the space-time distribution of the field in the media one can use the Green's space-time functions:

$$\vec{E}(\vec{r},t) = \vec{e}\int_{-\infty}^{\infty}dt_1\int_\Omega G(\vec{r}-\vec{r}_1,t-t_1)\left[\left(\vec{f}_{ext}(\vec{r}_1,t_1)\right) + \left(\vec{f}_1(\vec{r}_1,t_1)\right)\right]dV$$

$$\vec{H}(\vec{r},t) = -i\frac{1}{\omega\mu(\vec{r})}\int_{-\infty}^{\infty}dt_1\int_\Omega \vec{e}\times\nabla\cdot G(\vec{r}-\vec{r}_1,t-t_1)\left[\left(\vec{f}_{ext}(\vec{r}_1,t_1)\right) + \left(\vec{f}_1(\vec{r}_1,t_1)\right)\right]dV \quad (10)$$

It should be noticed, that inhomogeneous character of the media makes the electric and magnetic fields vectors cease to be always perpendicular to the direction of the field propagation, as in vacuum and homogeneous media.

3. Conclusions

Thus, the proposed method allows to convert the problem of electromagnetic field distribution in an arbitrary medium into the integration procedure, what is much simpler than the solution of the partial differential equations with variable coefficients.

References

[1] D. W. Prather 1999 Formulation and application of the finite-difference time-domain method for the analysis of axially symmetric diffractive optical elements J. Opt. Soc. Am. A. 16(5), 1131-1142.

[2] Ch. Caloz, T. Itoh 2006 Electromagnetic metamaterials: transmission line theory and microwave applications (the engineering approach). A John Wiley & Sons, Inc., 352.

[3] V. P. Dzyuba, A. M. Zapolsky 2000 The green function for the elastic half-space with a curvilinear boundary Doklady Earth Sciences, 372, 709-711.

Acknowledgments

The work was supported by "Far East Program" of FASO Russia (grant № 18-3-023).

IOP Conf. Series: Journal of Physics: Conf. Series **1092** (2018) 012031 doi:10.1088/1742-6596/1092/1/012031

Multi-functional RF coils for ultra-high field MRI based on 1D/2D electromagnetic metamaterials

Daniel Erni [1,2], Andreas Rennings [1], Jan Taro Svejda [1], Benedikt Sievert [1], Zhichao Chen [1], Thorsten Liebig [1], Juerg Froehlich [1,2,3]

[1] General and Theoretical Electrical Engineering (ATE), Faculty of Engineering, and CENIDE – Center for Nanointegration Duisburg-Essen, University of Duisburg-Essen, D-47048 Duisburg, Germany.

[2] Institute of Electromagnetic Fields (IEF), Department of Information Technology and Electrical Engineering, ETH Zürich, CH-8092 Zürich, Switzerland.

[3] Fields at Work GmbH, CH-8006 Zürich, Switzerland.

E-mail: daniel.erni@uni-due.de

Abstract. Electromagnetic metamaterials have already proven very valuable for the enhancement and molding of RF magnetic fields within ultra-high field MRI scanners at 7T. We report on our development of coil elements based on composite right-/left-handed (CRLH) 1D electromagnetic (EM) metamaterial transmission lines (*metalines*) operating in the zeroth order resonance (ZOR) to foster uniform RF magnetic field distributions along the scanner axis. Tailored EM metalines supporting full-wave or quarter-wave resonances are used either as metamaterial ring antenna or as dual-band coil elements for simultaneous $^{1}H/^{23}Na$ imaging. The EM metalines are key to the *MetaBore*, which is a fully adaptive RF field control scheme based on a periodic axial arrangement of conformal metamaterial ring antennas in the framework of high-field traveling-wave MRI. With the 2D EM metamaterials (*metasurfaces*) we realized high-impedance surfaces (HIS) in order to enhance the uniformity and directivity of the RF magnetic field from e.g. overlaid (elongated) dipole elements towards the probe volume. The designs include simulation studies of the overall multichannel coil systems, which are carried out with our home-made, open source electromagnetic 3D EC-FDTD solver *openEMS* supporting conformal cylindrical inhomogeneous meshing. Experimental verifications of our coils have been carried out within 7T MRI scanners (Siemens Magnetom).

1. Introduction

Since the seminal proposal of V. G. Veselago [1] in 1968, where for the first time the potential of properly tailoring electromagnetic (EM) material parameters had been laid out, electromagnetic metamaterials (MTMs) have evolved into one of the most powerful paradigms of modern engineering science. The possibility for dispersion engineering and meta-doping (e.g. the introduction resonant and radiative defects) has led to various novel EM applications encompassing e.g. compact leaky wave antennas, real-time analog signal processing, functional meta-surfaces for e.g. (spatial) transformation optics/electromagnetics, and miniaturized multi-functional metadevices [2] while covering most of the EM spectrum. The application of EM MTMs in ultra-high field (UHF) magnetic resonance imaging (MRI) has started midst of the noughties where mainly the lensing capabilities of MTM pads have been exploited for enhancing the B_1 efficiency in e.g. extremities. Although boosting MRI sensitivity via e.g. MTM surfaces (*metasurfaces*) has now improved into a mature technology [3], proper MTM-based coil designs for UHF MRI have only emerged in the wake of utilizing composite right-/left-handed (CRLH) MTM transmission lines (*metalines*) in modern RF engineering [4-7]. In the remainder of the paper we shall briefly summarize part of our research on the use of CRLH metalines in multi-functional coil design [4-13] for UHF MRI at 7T as well as on the field enhancement due to a HIS metasurface within an 8-channel coil design using optimized elongated dipole elements [15-16].

2. Coils based on CRLH metamaterial transmission lines (metalines)

2.1. Zeroth order resonance (ZOR) coil elements for 7T MRI

A common element in coil design based on CRLH metalines lies in the applicability of dispersion engineering as an additional degree-of-freedom. In coil elements operating in the zeroth order resonance (ZOR) the resulting uniform current (and $\boldsymbol{B}_1$ field) distribution (cf. figure 1) has been exploited for providing large longitudinal field of views (FoVs) [4, 5] up to 40cm [6] that are virtually scalable with the number of unit cells. A 4-channel coil design with frequency balanced CRLH unit cells [6, 7] yielded $|\boldsymbol{B}_1^+|_{max}/(SAR_{max})^{0.5} = -2.6$dB, an inter-element coupling $S_{ij} \leq -30$dB and a SAR_{10g} ~70% (cf. to microstrip dipole element) within an uniform octogonal phantom (where $|\boldsymbol{E}|^2$ directly maps to SAR).

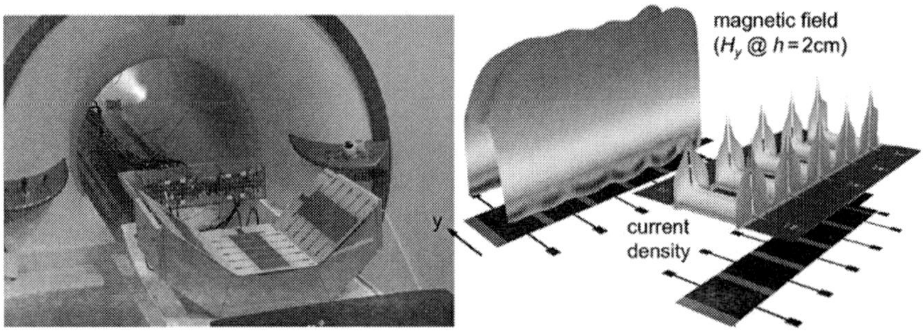

Figure 1. (left) CRLH ZOR coil element array; (right) FDTD simulations of the current density in the top metal layer and the resulting uniform $\boldsymbol{B}_1$ field distribution (H_y) 2cm above the metaline surface.

2.2. Dual-resonant coil elements for $^{23}Na/^{1}H$ 7T MRI

CRLH metalines are inherently dual-band due to the presence of both the right-handed and (folded) left-handed branch in the dispersion diagram. As shown in figure 2 a dual-band coil element has been realized based on a quarter-wave metaline resonator to target the Larmor frequencies of ^{23}Na and ^{1}H simultaneously for combined MRI at 7T [8, 9]. FDTD simulations provided (dimensionally adjusted) figures-of-merit $|\boldsymbol{B}_1^+|_{max}/(SAR_{max})^{0.5}$ = 2.7dB (298MHz) and 12.2dB (79MHz) [8]. First MRI scanner experiments with fast gradient-echo sequences on a cylindrical phantom with 4 ping-pong ball insets (200/400mMol/l NaCl in water respective in agar) have revealed a distinct sodium selectivity mainly for the inset with 400mMol/l in agar, which is due to the low SNR/high coil losses at 79MHz [8].

Figure 2. (left) Triple-layer MIM realization of the quarter-wave resonant CRLH dual-band coil element; (right) FDTD simulation of the combined current density on the top and bottom metal layer at the sodium frequency (79MHz) and the proton frequency (298MHz) with an equal λ/4 wave pattern.

2.3. *Full-wave resonant ring antenna coils for 7T traveling-wave MRI*

A full-wave resonance on a CRLH metaline, which is bent around to form a ring antenna (with 24 unit cells), provides a surface current distribution that has proven successful to excite the circularly polarized TE_{11} cylindrical waveguide mode within the context of traveling-wave MRI [10]. Using a longitudinally periodic arrangement of such ring antennas along the scanner bore allows – in the framework of an inverse problem solution for the excitation currents – for a distinct profiling of the $\boldsymbol{B}_1^+$ field along the patient's body. As displayed in figure 3 an accurate 15 cm wide profile (given the wavelength of the traveling waves of 3.6 m!) for *larynx* illumination has been achieved with a suppression of the unwanted $\boldsymbol{B}_1^-$ component (compared to $\boldsymbol{B}_1^+$) of 21dB while showing no SAR hot spots in the shoulder-neck region [11, 12]. It's worth noting that dispersion engineering underlying the CRLH ring antenna design allows the multiple ring antenna setting (i.e. the *MetaBore*) to fully conform to the inner surface of the cylindrical scanner bore providing an ergonomic as well as an electronically steerable «longitudinal shimming scheme» that may be conceptualized as an active cylindrical metasurface.

As proof-of-concept an in-vivo whole body 7T MRI study of a crab-eating macaque monkey has been carried out with a reduced setting of e.g. 2 CRLH ring antennas in Tx and Rx operation. Using FLASH (whole-body) and TSE (head) sequences well resolved images were retrieved with underlying $\boldsymbol{B}_1^+$ transmit efficiencies of 1.62μT/kW$^{0.5}$ (body) and 0.72μT/kW$^{0.5}$ (head) [13].

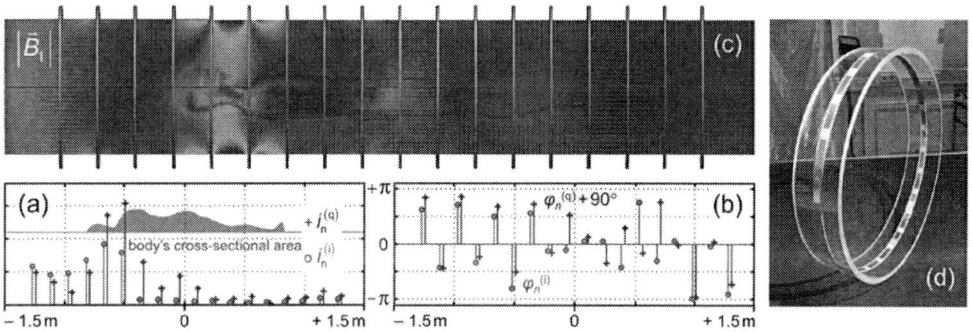

Figure 3. Simulated [18] traveling-wave profiling scenario for *larynx* illumination: (a) amplitudes and (b) phases of the 18 ring antenna's current excitations (width 1cm, pitch 15cm, diameter 64cm); (c) transversal RF magnetic field; and (d) CRLH ring antenna prototype (shown without ground plane); the body phantom is a cylindrically discretized version of *Duke*^{TM} from SPEAG's *Virtual Family*^{TM}.

3. Coils based on metamaterial high-impedance surfaces (HIS)

In order to enhance the uniformity and directivity of the RF magnetic field of e.g. elongated electric dipoles [15, 16] or series resonant loops [17] towards the probe volume, these coil elements were backed by an RF shield consisting of a 2D EM MTM (*metasurface*). The latter was realized as an uniplanar high-impedance surface (HIS) [14] where surface currents were efficiently suppressed yielding a realistic approximation to a perfect magnetic conductor (PMC). The rationale behind using HIS shields are the residual in-phase image currents that are in support of the currents in the overlaid coil elements providing an overall enhancement of the $\boldsymbol{B}_1$ field above the shielded coil. In this regard elongated dipole elements positioned 5mm over a HIS shield achieved a 26% enhancement of $\boldsymbol{B}_{1,max}$ compared to a perfect electrical conducting (PEC) shield in addition to a 40% broadening of the cross-sectional field profile [15]. The resulting 8-channel HIS dipole coil design (cf. figure 4) was based on extensive FDTD analyses (using conformal meshing) [18] and provided an improved $\boldsymbol{B}_1$ homogeneity within a cylindrical phantom compared to a PEC shielded realization at the expense (as expected from the profile broadening) of a 3dB increased inter-element coupling [16].

4. Concluding remark

The aim of the presented work is to show that dispersion engineering in 1D and 2D EM metamaterials can provide additional degrees-of-freedom for multi-functional UHF MRI coil design along two conceptual approaches; namely EM metalines as proper coil elements with tailored field distributions and/or multi-band features, and EM metasurfaces as additional measures for boosting MRI sensitivity.

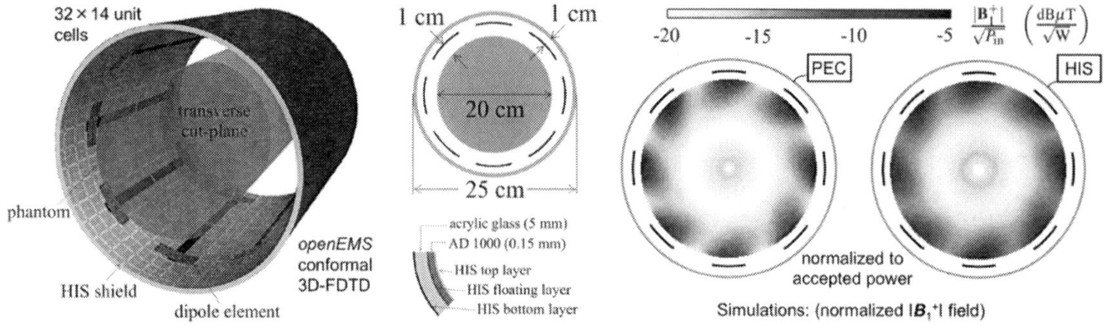

Figure 4. 8-channel HIS dipole coil: (left) FDTD simulation setup; (middle) coil dimensions with the simulated/experimental phantom; (right) simulated RF magnetic field (HIS vs. PEC ground planes).

Acknowledgments
We kindly acknowledge the Erwin L. Hahn Institute for Magnetic Resonance Imaging (ELH) of the university's hospital in Essen as well as at the Institute for Biometrics and Medical Informatics at the Otto-von-Guericke University in Magdeburg, both for their experimental support at their 7T MRI scanner systems (Siemens Magnetom).

References

[1] V. G. Veselago, *Sov. Phys. Usp.*, **10** (4), pp. 509-514, 1968.
[2] C- Caloz, *APSURSI 2016*, June 26- July 1, Fajardo, Puerto Rico, pp. 1299-1300, 2016.
[3] R. Schmidt, A. Slobozhanyuk, P. Belov, A. Webb, *Sci. Rep.*, **7** (1678), 2017.
[4] A. Rennings, J. Mosig, A. Bahr, C. Caloz, M. E. Ladd, D. Erni, *EuCAP 2009*, March 23-27, Berlin, pp. 3231-3234, 2009.
[5] A. Rennings, P. Schneider, C. Caloz, S. Orzada, *METAMATERIALS 2009*, Sept. 1-4, London, UK, pp. 126-128, 2009.
[6] A. Rennings, J. T. Svejda, K. Solbach, D. Erni, *MAGMA*, **26** (suppl 1), pp. 183-185, 2013.
[7] A. Rennings, J. Svejda, S. Otto, K. Solbach, D. Erni, *IEEE IMS 2013*, June 2-7, Seattle, USA, WE1E-1, 2013.
[8] J. T. Svejda, A. Rennings, D. Erni, *tm – Technisches Messen*, **84** (1), pp. 2-12, 2017.
[9] J. T. Svejda, D. Erni, A. Rennings, *MAGMA*, **29** (suppl 1), pp. S309, 2016.
[10] D. Erni, T. Liebig, A. Rennings, N. H. L. Koster, J. Froehlich, *33th IEEE EMBS 2011*, Aug. 30 - Sept. 3, Boston, MA, USA, pp. 554-558, 2011.
[11] H. Yang, T. Liebig, A. Rennings, J. Froehlich, D. Erni, *ICEAA 2013*, IEEE/URSI, Sept. 9-13, Torino, Italy, pp. 468-471, 2013.
[12] T. Liebig, J. T. Svejda, H. Yang, A. Rennings, T. Herrmann, J. Mallow, J. Bernarding, J. Froehlich, D. Erni, *ISMRM-ESMRMB 2014*, May 10-16, Milano, Italy, pp. 1358, 2014.
[13] T. Herrmann, *et al.*, *PLOS One*, **13** (1), paper e0191719, pp. 1-17, 2018.
[14] Z. Chen, K. Solbach, D. Erni, A. Rennings, *EuMC 2014*, Oct. 6-9, Rome, Italy, pp. 1576-1579, 2014.
[15] Z. Chen, K. Solbach, D. Erni, A. Rennings, *IEEE Trans. Microw. Theory Techn.*, **64** (3), pp. 972-983, 2016.
[16] Z. Chen, *Application of High Impedance Surfaces to Improve Radiofrequency Coil Performance for 7-Tesla Magnetic Resonance Imaging*, Ph.D. thesis, University of Duisburg-Essen, 2016.
[17] Z. Chen, K. Solbach, D. Erni, A. Rennings, *IEEE Trans. Microw. Theory Techn.*, **65** (3), pp. 988-997, 2017.
[18] T. Liebig, A. Rennings, S. Held, D. Erni, *Int. J. Numer. Model.*, **26** (6), pp. 680-696, 2013.

IOP Conf. Series: Journal of Physics: Conf. Series **1092** (2018) 012032 doi:10.1088/1742-6596/1092/1/012032

The effect of the strong spatial inhomogeneity of the electric field on plasmon rectification of terahertz radiation in graphene

D V Fateev [1], K V Mashinsky [1,2], H Qin[3], J D Sun[3], V V Popov [1,2]

[1] Kotelnikov Institute of Radio Engineering and Electronics of the Russian Academy of Sciences (Saratov Branch), Saratov 410019, Russia

[2] Saratov State University, Saratov 410012, Russia

[3] Key Laboratory of Nanodevices and Applications, Suzhou Institute of Nano-Tech and Nano-Bionics (SINANO),Chinese Academy of Sciences, Suzhou, 215123, China

FateevDV@yandex.ru

Abstract. The rectification of terahertz radiation in a spatially periodic graphene structure is studied. It is found that rectified current due to plasmonic electron-hole effect grows with spatial inhomogeneity of plasmon electric field in graphene.

1. Introduction

Research of terahertz (THz) wave rectification in graphene shows the increasing of rectified current in p-n graphene structures relatively to graphene having only n- or p-type of conductivity [1]. The p-n junctions in graphene may be created by both a chemical or electrical doping of graphene [2,3]. There were discussed several physical mechanisms for wave rectification in graphene structures, that are photothermoelectric effect [4], Drude heating of carriers by THz radiation [5] and nonlinear plasmonic effects [6].

In this paper, we discuss the effect of the strong spatial inhomogeneity of the electric field on the plasmon rectification of terahertz radiation in graphene structure with dual-grating gate.

2. Results and Discussion

We consider a graphene structure (figure 1) consisting of graphene located on the surface of a dielectric layer (SiO_2) of 300 nm thickness, which is located on a conductive substrate. Above the graphene is a metal dual-grating gate, which is separated from graphene by an insulating layer (Al_2O_3) of 26 nm thickness. The sub-grating electrodes are shifted relative to each other in the direction of periodicity in order to create an asymmetry of the unit cell of the periodic structure. The applied dc voltage between the electrodes of the grating gate and graphene, and between the conductive substrate and graphene, makes it possible to control the density of charge carriers in different sections of graphene. In this way, the Fermi energy of charge carriers in graphene can vary in such a way that graphene parts can have either electron and hole conductivity.

To describe the nonlinear dynamics of charge carriers in graphene, hydrodynamic equations, the continuity equation and the balance equation of the charge carrier momentum, are solved by the perturbation theory method [6]. The electric fields of plasmons in the graphene structure were calculated using a self-consistent electromagnetic approach based on the solution of integral equations formed with respect to oscillating currents on the conducting elements of the unit cell of the graphene structure [7]. The equilibrium spatial distribution of the Fermi energy of charge carriers in a unit cell of graphene was obtained by solving the electrostatic equations with taking into account the effect of the quantum capacity of graphene [8].

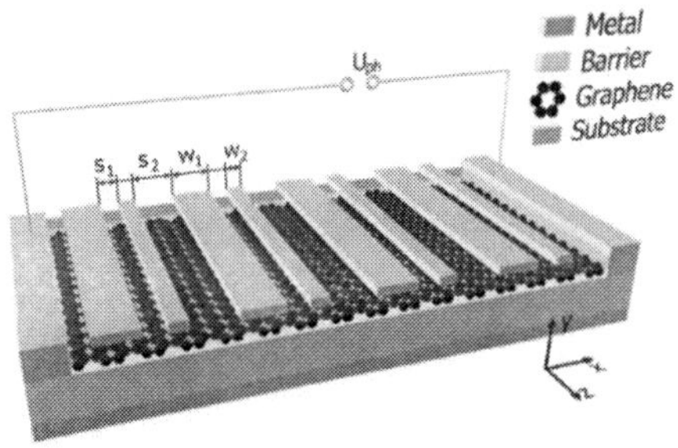

Figure 1. Schematic view of the dual-grating gate graphene structure.

The nonlinear hydrodynamic approach allows deriving the expression for calculation the rectified current in graphene structure [6]:

$$j_0 = -\frac{|e^3| v_F^2}{2\text{h}\pi\omega\gamma\left(\omega^2+\gamma^2\right)} \frac{1}{L} \int_{-\frac{L}{2}}^{\frac{L}{2}} \operatorname{sgn}\left(e(x)\right) \operatorname{Re}\left[\left(\omega - i\frac{\gamma}{2}\right) E_{x,\omega}(x) \frac{\partial \left(E_{x,\omega}(x)\right)^*}{\partial x}\right] dx \qquad (1)$$

where e is carrier charge ($e>0$ for holes and $e<0$ for electrons), $v_F = 10^6$ m/s is Fermi velocity, ω is the frequency of incident THz wave, γ is the momentum relaxation rate, L is the spatial period of the structure, and $E_{\omega,x}$ is in-plane electric field in graphene. The rectification of THz radiation in a periodic graphene structure can occur due to two different physical mechanisms. One physical mechanism is the rectification of THz radiation due to the effect of differential plasmon carrier drag in graphene [9]. The second rectification mechanism of THz radiation is the plasmon electron–hole ratchet effect. The rectification of THz radiation due to the plasmon electron–hole ratchet effect is possible only if different parts of graphene have different types of conductivity [6]. Such charge distribution in the unit cell of periodic graphene structure may be achieved by applying different dc potentials to different gates of the structure. Then the unit cell of graphene structure will have n-p-n junction. In such case the plasmon electron–hole ratchet effect dominates in THz frequency range. The rectified current due to plasmonic electron-hole ratchet is described by the formula

$$\begin{aligned} j_0^{rt} &= -\frac{|e^3| v_F^2}{\text{h}^2\pi\omega\left(\omega^2+\gamma^2\right)} \frac{4}{\gamma L}\left[\operatorname{sgn}\left(e_1\right) - \operatorname{sgn}\left(e_2\right)\right] \\ &\times \operatorname{Re}\left[\left(i\omega + \frac{\gamma}{2}\right) \sum_p \sum_{p \neq p'} E_{x,\omega,p} E_{x,\omega,p'}^* \frac{p'}{p-p'} \sin\left(\pi\left(p-p'\right)\frac{w_2}{L}\right) \exp\left(i\pi\left(p-p'\right)\frac{s_1-s_2}{L}\right)\right] \end{aligned} \qquad (2)$$

where $E_{\omega,x,p}$ is the spatial Fourier harmonics of the electric field $E_{\omega,x}$, with p being integer, e_2 is carrier charge in graphene under the gate w_2, and e_1 is the carrier charge in the rest of the graphene-structure unit cell. Besides geometrical asymmetry of the graphene structure, the resulted rectified current strongly depends on inhomogeneity of the plasmon electric field in graphene through the factor $\sum_{p}\sum_{p\neq p'} E_{x,\omega,p}E^{*}_{x,\omega,p'}$.

The calculations of the rectified current were performed for dual-grating gate graphene structure with parameters: $w_1 = 1$ μm, $w_2 = 0.25$ μm, $s_1 = 0.625$ μm, $s_2 = 0.125$ μm, the voltage at the back gate is $U_{bg} = 40$ V, the voltage at the subgrating w_1 $U_{w1} = 1$ V, and $\tau = 1$ ps.

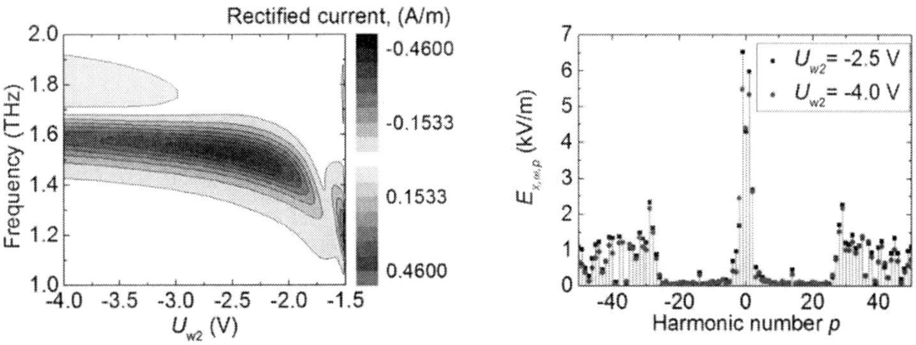

(a) (b)

Figure 2. (a) Dependence of the rectified current on frequency and dc gate voltage U_{w2} at the sub grating w_2. (b) Spectral composition of excited plasmon mode of the graphene structure at frequency 1.537 THz for different gate voltages U_{w2}.

The rectified current calculated within Eq. (1) grows near to the plasmon resonance in graphene structure (figure 2 a). The set of dc gate voltages U_{bg} and U_{w1} creates the electron conductivity in the structure unit cell except part of graphene under the gate w_2. Increasing dc voltage U_{w2} changes the spectral composition (figure 2 b) of excited plasmon mode of the graphene structure, which leads to growing the rectified current (figure 2 a) due to increasing the inhomogeneity of the electric field.

In conclusion, the excitation of the plasmon modes with strongly inhomogeneous electric field results in increasing rectification efficiency by plasmonic electron-hole ratchet effect in graphene.

This work was supported by the grant from the Russian Foundation for Basic Research (No. 17-52-53063, the grant from the Foundation for the Development of Theoretical Physics "BASIS", and National Natural Science Foundation of China through the grant 61611530708.

References

[1] Alonso-González P, Nikitin A Y, Gao Y, Woessner A, Lundeberg M B, Principi A, Forcellini N, Yan W, Vélez S, Huber A J, Watanabe K, Taniguchi T, Casanova F and Hueso L E 2017 *Nat. Nanotech.* **12** 31–6
[2] Williams J R, DiCarlo L and Marcus C M 2007 *Science* **317** 638–41
[3] Lohmann T, von Klitzing K and Smet J H 2009 *Nano Lett.* **9** 1973–9
[4] Koppens F H L, Mueller T, Avouris P, Ferrari A C, Vitiello M S and Polini M 2014 *Nat. Nanotechnol.* **9** 780–93
[5] Freitag M, Low T. and Avouris P 2013 *Nano Lett.* **13** 1644–9
[6] Fateev D V, Mashinsky K V and Popov V V 2017 *Appl. Phys. Lett.* **110** 061106
[7] Polischuk O V, Fateev D V and Popov V V 2017 *Semiconductors* **51** 11 1460–5
[8] Liu M-H 2013 *Phys. Rev. B* **87** 125427
[9] Olbrich P, Kamann J, Konig M, Munzert J, Tutsch L, Eroms J, Weiss D, Liu M-H, Golub L E, Ivchenko E L, Popov V V, Fateev D V, Mashinsky K V, Fromm F, Seyller T and Ganichev S D 2016 *Phys. Rev. B* **93** 075422.

IOP Conf. Series: Journal of Physics: Conf. Series **1092** (2018) 012033 doi:10.1088/1742-6596/1092/1/012033

Surface polaritons and Dyakonov-like surface waves in an isotropic-anisotropic nanocomposite system

I V Fedorin

Department of Materials for Electronics and Solar Cells, National Technical University "Kharkiv Polytechnic Institute", 61002, Kharkiv, Ukraine

fedorin.ilya@gmail.com

Abstract. The properties of surface polaritons as well as Dyakonov surface waves at the plane interface between a nanocomposite made of doped semiconductor spherical inclusions distributed in a transparent matrix and a nanocomposite made of the identical transparent matrix but with doped semiconductor inclusions of ellipsoidal shape are studied. Electrodynamic Maxwell-Garnett model is applied to simulate the effective resonance permittivity of both nanocomposites. The nanocomposite with spherical inclusions represents an isotropic medium, while the nanocomposite with ellipsoidal inclusions can be described as uniaxial material. It is shown, that by properly choosing the physical and geometrical parameters of the corresponding composites, like the volume fractions and materials of nanoinclusions, the effective control over surface plasmon-polaritons and Dyakonov-like surface waves properties is possible.

1. Introduction

Recently, the interest in study of physical properties of artificial structures of various physical nature has significantly increased. In particular, the special interest is devoted to the study of various types of nanocomposite media, metamaterials, photonic crystals, and hyperbolic media, which have a number of unusual properties that allow creating new materials with predetermined structural, electronic and optical characteristics. The properties of such structures can significantly differ from the properties of the materials which form the composites. In particular, by appropriately selecting the constituent materials, namely their concentration, geometrical dimensions and physical properties it is possible to achieve the bands of negative real parts of the complex dielectric and / or magnetic permeabilities in a certain frequency range, which results in a number of unusual physical phenomena, such as negative refraction, super-lenses and super-prism phenomena, to name but a few [1-6].

It is well known, that in the frequency range where one of the material parameters (permittivity and/or permeability) becomes negative the propagation of surface plasmon polaritons (SPP) along the plane interface is possible. The wave field of surface polaritons is localized in the near-surface region, the thickness of which on each side of the interface is of the order of the wavelength. In anisotropic media, e.g. in nanocomposites with anisotropic arrangement and shape of inclusions, the properties of the surface waves essentially depend on the direction of their propagation with respect to the anisotropy axes. It was shown that in such structures the so called Dyakonov surface waves (DSW) may be guided by the interface. Like plasmons polaritons, they exist at the interface of two different materials, and should feature similar excitation and detection properties. However, in opposition with SPPs, DSW have the peculiarity of possessing hybrid polarization [7-11].

The main goal of this paper is to study the properties of surface waves, both surface polaritons and Dyakonov-like hybrid modes, which take place at the interface between two composite porous materials, composed of doped semiconductor nanoparticles of spherical (isotropic nanocomposite) and ellipsoidal (uniaxial nanocomposite) shapes suspended in a dielectric matrix. The influence of geometrical as well as physical properties of artificial materials from both sides of the interface on the properties and types of supported surface modes is theoretically and numerically studied.

2. Geometry of the structure and basic equations

Let consider the plane interface between a nanocomposite (NC) semi-infinite media with doped semiconductor nanoinclusions of spherical shapes, which fills the half-space $x>0$ and adjacent to it NC semi-infinite media with doped semiconductor nanoinclusions of ellipsoidal shapes filling the space $x<0$ (Figure 1). NC's represent a non-conductive transparent matrix with permittivities $\varepsilon_{m1,2}$, which in the investigated frequency range does not depend on the frequency. In the volume of host material semiconductor nanoparticles with permittivities $\varepsilon_{i1,2}$ are evenly distributed. The dielectric permittivity of doped semiconductor nanoparticles can be describe using the Drude model [12]

$$\varepsilon_{i1,2} = \varepsilon_0 - \frac{\omega_p^2}{\omega^2 + i\gamma\omega}, \tag{1}$$

where ε_0 is the high-frequency relative permittivity, $\omega_p = \sqrt{4\pi e^2 N_e / m_{eff}}$ is the plasma frequency; e, N_e, m_{eff} are the charge, equilibrium concentration and effective mass of charge carriers, γ is the electron collision frequency.

We assume that the size of inclusions suspended in a dielectric matrix is much smaller than a wavelength and the electromagnetic field penetration depth in the material. In this case, if we neglect the interference effects on the inclusions and their volume fraction is as small as 1/3, it is generally known, that effective Maxwell–Garnett model can be applied to describe the optical properties of the nanocomposites under study [1]. Within the framework of this model the homogenization procedure can be applied and NC with spherical inclusions can be modeled as isotropic structure with the following effective dielectric permittivity [1]

$$\varepsilon_{nc1} = \varepsilon_{m1}\left[1 + \frac{f_1}{(1-f_1)/3 + \varepsilon_{m1}/(\varepsilon_{i1} - \varepsilon_{m1})}\right]. \tag{2}$$

For the NC with inclusions of ellipsoidal shape we assume that the rotation axes (and optical axis) of all nanoparticles coincide with the coordinate axis z (Figure 1). In this case, such NC has the properties of uniaxial crystal and its effective dielectric constant is described by the diagonal tensor with nonzero components $\varepsilon_{xx} = \varepsilon_{yy} = \varepsilon_{nc2}^{\perp}$ and $\varepsilon_{zz} = \varepsilon_{nc2}^{\parallel}$, which can be written as [3]

$$\varepsilon_{nc2}^{\perp,\parallel} = \varepsilon_{m2}\left[1 + \frac{f_2(\varepsilon_{i2} - \varepsilon_{m2})}{\varepsilon_{m2} + (1-f_2)(\varepsilon_{i2} - \varepsilon_{m2}) g_{\perp,\parallel}}\right], \tag{3}$$

where $f_{1,2}$ are the volume fractions of nanoparticles in the matrix; $g_{\perp,\parallel}$ are the depolarizing factors (form factors) that take into account the influence of the nanoparticles shape on the magnitude of the dipole moment induced on it. Remarkably, that in the case of spherical particles there is no form anisotropy, that is $g_{\perp,\parallel} = 1/3$ and equation (3) can be reduced to equation (2).

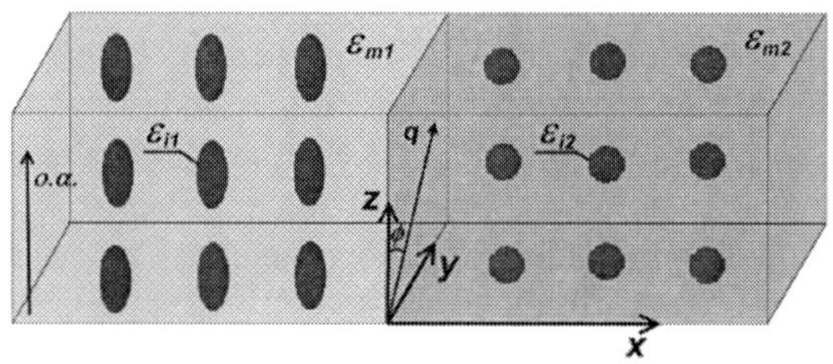

Figure 1. Schematic geometry of the system under study, consisting of a semi-infinite nanocomposite with spherical semiconductor inclusions ($x > 0$), and a nanocomposite with ellipsoidal semiconductor inclusions ($x < 0$) with optical axis (o.a.) along z-axis.

Let us neglect the losses in the semiconductor nanoparticles for analytical study, i.e. the damping constant $\gamma \to 0$. Let assume that the electromagnetic field amplitudes are evanescent, propagating from the interface at $|x| \to \infty$.

Using Maxwell equations, and applying boundary conditions at $x = 0$ for the tangential components of the electric and magnetic fields, it is possible to obtain the following dispersion equation for hybrid DSW [7-11].

$$\left(\chi_{nc1} + \chi_{nc2}^{E}\right)\left(\chi_{nc1} + \chi_{nc2}^{H}\right)\left(\varepsilon_{nc1}\chi_{nc2}^{H} + \varepsilon_{nc2}^{\perp}\chi_{nc2}^{E}\right) = \left(\varepsilon_{nc2}^{\parallel} - \varepsilon_{nc1}\right)\left(\varepsilon_{nc1} - \varepsilon_{nc2}^{\perp}\right)k_0^2\chi_{nc2}^{H}, \tag{4}$$

which provides a spectral map of allowed values of $\mathbf{q}$, which is the wave vector of DSW in the (k_y, k_z) plane. In (4) $\chi_{nc1}, \chi_{nc2}^{E}, \chi_{nc2}^{H}$ are the decay rates in the nanocomposites (in the uniaxial nanocomposite we have both extraordinary (H) and ordinary (E) waves), which are a real positive values, given by

$$\chi_{nc1} = \sqrt{k_z^2 + k_y^2 - k_0^2\varepsilon_{nc1}},\ \chi_{nc2}^{E} = \sqrt{k_y^2 + k_z^2\frac{\varepsilon_{nc2}^{\parallel}}{\varepsilon_{nc2}^{\perp}} - k_0^2\varepsilon_{nc2}^{\parallel}},\ \chi_{nc2}^{E} = \sqrt{k_y^2 + k_z^2 - k_0^2\varepsilon_{nc2}^{\perp}},\ k_0 = \omega/c, \tag{5}$$

Assuming that all decay rates, and relative permittivities are positive, it is clear that a hybrid surface waves localized at the interface exist only if $\varepsilon_{nc2}^{\parallel} > \varepsilon_{nc1} > \varepsilon_{nc2}^{\perp}$. As a consequence, a positive birefringence is mandatory to ensure a stationary solution of Maxwell's equations for DSW.

Thus, the conditions of DSW existence are mainly determined by the values of the permittivities of the contacting materials. Therefore, the properties of DSW can be effectively controlled by means of appropriate choosing of geometrical as well as physical parameters of the system under study, which is illustrated and discussed in the next section.

It should be also noted that dispersion equation for standard SPP propagating parallel or perpendicular to the optical axis ($k_y = 0$ or $k_z = 0$) have the standard form and the necessary condition of SPP existence require the negative values of relative permittivities of one of the contacting half-spaces.

3. Numerical calculations

Let us numerically analyze the permittivity components of nanocomposites under study (Figure 2) and angular range of DSW existence versus frequency (Figure 3) to identify the frequency ranges and angular properties of DSW. In the numerical calculations we use the following parameters of nanocomposites: n-InSb inclusions in both nanocomposites, with electron concentration $N_e = 8.65 \cdot 10^{14}\,\text{cm}^{-3}$, plasma frequency $\omega_p = 1.35 \cdot 10^{13}\,\text{s}^{-1}$, and high-frequency relative permittivity $\varepsilon_0 = 15.68$, suspended in a transparent optical glass with $\varepsilon_{m1} = \varepsilon_{m2} = 2.56$. Filling factor of semiconductor inclusions is $f_1 = f_2 = 0.05$.

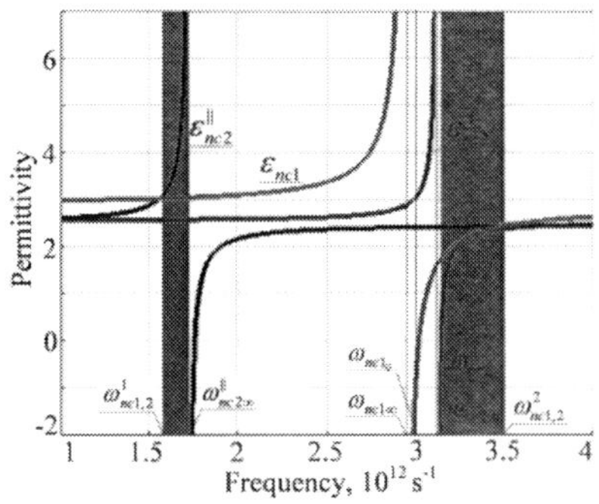

Figure 2. Nanocomposites permittivities as a function of frequency and DSW existence bands (grey areas).

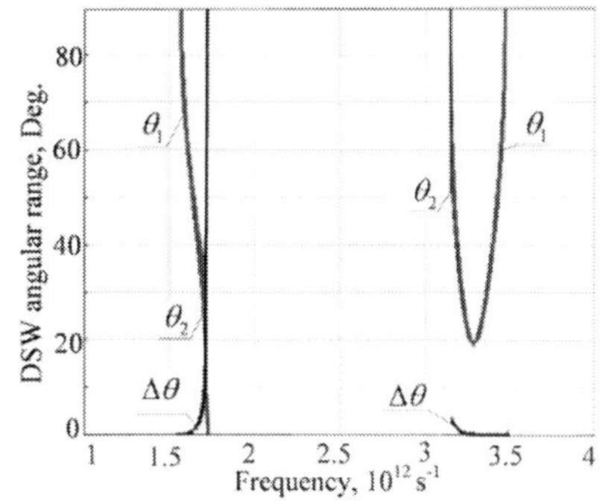

Figure 3. Angles of DSW existence (grey areas) versus frequency, $\Delta\theta = |\theta_1 - \theta_2|$.

As can be seen, there are two frequency bands where the condition for DSW existence is satisfied. First frequency band ($\omega^{1}_{nc1,2} < \omega < \omega^{\parallel}_{nc2\infty}$) occupies the frequency range between the frequency $\omega^{1}_{nc1,2}$, at which $\varepsilon^{\parallel}_{nc2} = \varepsilon_{nc1}$, and $\omega^{\parallel}_{nc2\infty}$ at which $\varepsilon^{\parallel}_{nc2}$ tends to infinite (in the absence of damping factor in the semiconductor inclusions). Second range of DSW existence arises in the frequency range $\omega^{\perp}_{nc2_0} < \omega < \omega^{2}_{nc1,2}$ between the frequency $\omega^{2}_{nc1,2}$ at which $\varepsilon^{\parallel}_{nc2} = \varepsilon_{nc1}$, and $\omega^{\perp}_{nc2_0}$ at which $\varepsilon^{\perp}_{nc2} \to 0$. It should be noted, that in the frequency bands where the components ε_{nc1} and/or $\varepsilon^{\parallel}_{nc2}, \varepsilon^{\perp}_{nc2}$ are negative, existence of surface plasmon polaritons (SPP) is possible under certain wavevector directions.

Fig. 3 represents the dependence of DSW existence angles versus frequency. As can be seen, the angular range of DSW existence varies with frequency from very small values up to very high values near to the boundary of first DSW existence band.

To conclude, in the current paper, the properties of hybrid-polarized surface waves that may propagate obliquely as well as surface polaritons at the boundary between two nanocomposites with semiconductor inclusion of different shapes are studied numerically and theoretically. The system is considered in the effective medium Maxwell-Garnett approximation, which implies significant limitations on the theory described in the paper. Nevertheless, it has been shown that by means of changing physical and geometrical parameters of the system under study it is possible to effectively control the properties of surface waves, namely to achieve several frequency bands, and wide angular range of surface waves existence. The influence of dissipation processes has been studied numerically. The results of current paper can found potential applications for chemical and biological sensors.

References

[1] Vetrov S, Bikbaev R, Rudakova N, Chen K-P and Timofeev I 2017 *J. Opt.* **19** 085103
[2] Baibak V, Fedorin I, Bulgakov A 2014 *Telecommunications and Radio Engineering* **73** 555
[3] Golovan L, Timoshenko V, Kashkarov P 2007 *Phys. Usp.* **50** 595
[4] Baibak V, Fedorin I, Bulgakov A 2013 *Progress in Electromagnetic Research* M **32** 229
[5] Fedorin I 2016 *Progress in Electromagnetic Research* B **67** 17
[6] Fedorin I, Baibak V, Bulgakov A 2014 *Eur. Phys. J. Appl. Phys.* **66** 20502
[7] Polo J, Nelatury S and Lakhtakia A 2007 *J. Opt. Soc. Am.* A **24** 2974
[8] Chen S, Shen Z and Wu W 2014 *J. Opt. Soc. Am.* A **31** 1923
[9] Darinskii A 2001 *Crystallography Reports* **46** 842
[10] Artigas D, Torner L 2005 *Phys. Rev. Lett.* **94** 013901
[11] Dyakonov M 1988 *Sov. Phys. JETP* **67** 714
[12] Bass F and Bulgakov A 1997 *Kinetic and Electrodynamic Phenomena in Classical and Quantum Semiconductor Superlattices* (New York: Nova Science Publishers) p 498

IOP Conf. Series: Journal of Physics: Conf. Series **1092** (2018) 012034 doi:10.1088/1742-6596/1092/1/012034

Intersubband plasmons induced negative refraction at mid-IR frequency in heterostructured semiconductor metamaterials

Mario Ferraro[1], Massimo Giudici[2], Angela Vasanelli[3], Miguel Montes Bajo[4], Julen Tamayo-Arriola[4], A. Hierro[4], Jean Michel Chauveau[1], Patrice Genevet[1]

[1]*Université Cote d'Azur, CNRS, CRHEA, rue Bernard Gregory, Sophia Antipolis 06560 Valbonne, France*

[2]*Université Cote d'Azur, CNRS, InPhyni, Route des Lucioles, Sophia Antipolis 06560 Valbonne, France*

[3]*Laboratoire Matériaux et Phénomènes Quantiques, Université Paris Diderot, Sorbonne Paris Cité, CNRS-UMR 7162, 75013 Paris, France*

[4]*ISOM and Dpto. Ing. Electrónica, Universidad Politécnica de Madrid, Avda. Complutense 30, 28040 Madrid, Spain*

Abstract. In this contribution, we report on negative refraction effect occurring in layered semiconductors. The origin of this effect is attributed to the presence of intersubband plasmon resonance induced by the electronic confinement of the electrons in thick quantum wells. Relying on the effective medium theory, we analyse both the in plane and the out of plane effective optical properties of highly doped ZnO/ZnMgO semiconductor material. We theoretically show and experimentally demonstrate that thicknesses and doping levels of each layer can be carefully chosen to feature strong intersubband transitions, leading to type 1 and type 2 hyperbolic response. Understanding the optical properties of intersubband materials in the frame of hyperbolic materials would shed new light on which physical mechanisms are controlling the radiative decay of intersubband plasmon excitations. This approach could be further utilized to designing efficient mid-IR sources.

1. Introduction

Metamaterials are artificial materials that feature uncommon physical properties. In optics the field of metamaterials has received tremendous interest over the last few years, with the demonstration of various unexpected and intriguing effects such as ultra-high refractive index and extraordinary optical activity. However optical metamaterials are mainly known owing to the fascinating possibility of realizing devices based on negative index of refraction, also called negative index materials (NIM). A great deal of attention on this domain has been attracted due to the potential of metamaterials in revolutionizing both fundamental and applicative research. Material with negative index of refraction was firstly been considered in a theoretical paper by Veselago in 1968. In this latter paper, it is shown that negative index can be obtained in a material presenting simultaneous negative dielectric permittivity and magnetic permeability, in brief by adjusting two and matching electric and magnetic resonances in the material. It implies that in such systems the absorption is quite relevant. In order to mitigate optical losses, lots of efforts have been made to reduce the absorption losses, considering for example gain material. A more efficient approach has been proposed in recent years. It consists on exploiting hyperbolic dispersion, i.e. opposite signs of permittivity along orthogonal in- and out- of plane directions. Affecting the optical response only along one direction, hyperbolic dispersion can be realized in anisotropic medium using a single optical resonance. In such a system, also dubbed hyperbolic metamaterial (HMM), one of the electric or magnetic resonance is replaced by leveraging on the material anisotropy. The conceptual and experimental simplification also comes with a drastic reduction of the optical losses. The main idea behind HMM is to make use of an alternating composition of both resonant and non-resonant materials. Peculiar attention has been given to semiconductor HMM [1-3] because of their functional and fabrication advantages, substituting the metallic resonant part by a highly doped semiconductor.

In this paper we present theoretical and experimental results confirming negative refraction in highly doped layered semiconductor materials and highlight the fact that this behaviour is essentially driven by the intersubband transitions (ISBT) in the quantum wells (QW) [4].

1 – Samples: materials and geometry

The physical system proposed in this paper is a stack of alternated layer of doped ZnO and undoped MgZnO forming a quantum wells along z direction grown on native ZnO substrate to reduce the dislocation density for sharper ISBT. The choice of ZnO is mainly related to the possibility of reaching very high doping level and because of its non-polar growth possibility, therefore avoiding the detrimental quantum confined Stark effect [5]. A scheme of the system is reported in Fig. 1a.

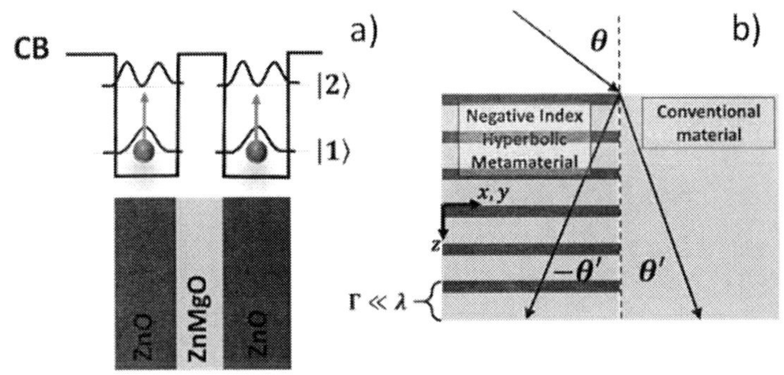

Figure 1. Left schematic represents the conduction band of ZnO/ZnMgO heterostructure. The electronic confinement in doped ZnO quantum wells produces intersubband transitions controlled by the quantum well thickness and the electron doping density. The right panel presents the two possible regimes. On the left (right) the material has negative (positive) index materials giving rise to negative (positive) refraction effect.

2 – Model

The problem was mathematically treated following Maxwell-Garnett theory, in the presence of both the QW and the barrier to obtain realistic material permittivity. Details on the model to account for phonon contribution, and anisotropic ISBT response will be discussed during the presentation. The main results are summarized in Fig. 2a and 2b that show both the in- and out-of-plane components of the real part of the effective medium permittivity. Fig. 2c summarizes the permittivity signs: the optical properties range from pure metal to pure dielectric passing through HMM type 1 and 2. We show that T1 HMM originates from the intersubband plasmon response of the material.

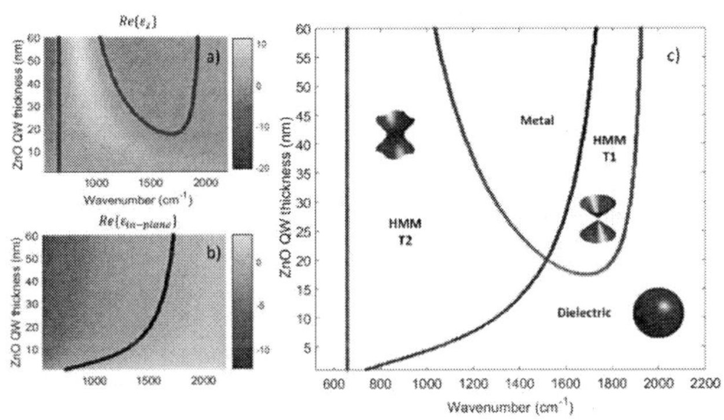

Figure 2. Hyperbolic metamaterial condition. (a,b) Real part of the effective permittivity out-of-plane (a) and in-plane (b) as function of the frequency and QW thickness. The curves show the sign inversions. (c) Optical behaviour of the system as function of frequency and QW thickness. It shows pure dielectric, metallic and HMM features depending on the parameters. The pure dielectric, HMM type 1 and type 2 zones are labelled with their correspondent isofrequency curve (sphere and hyperboloids respectively). The blank zones are metallic.

3 – Experimental results

The theoretical prediction were tested experimentally with two samples of respectively 35 and 4 nm thickness ZnO quantum wells. According to the figure 2, the former should exhibits negative refraction while the latter should behave as a conventional dielectric material in the 1400-1800 cm-1 wavenumber range. The samples are composed of 15 couples of QW-barrier having total thicknesses equal to 590 and 125 nm respectively.

Negative refraction have been detected by using a blade to stop either the positive refraction or the negative refraction depending on the sample orientation (Fig. 3a). According to the figure in presence of negative refraction a maximum in transmission for θ<0 is expected since the positive refraction is stopped. Symmetrically, a minimum is expected for θ>0. This result is confirmed by experimental spectra reported in Fig. 3b.

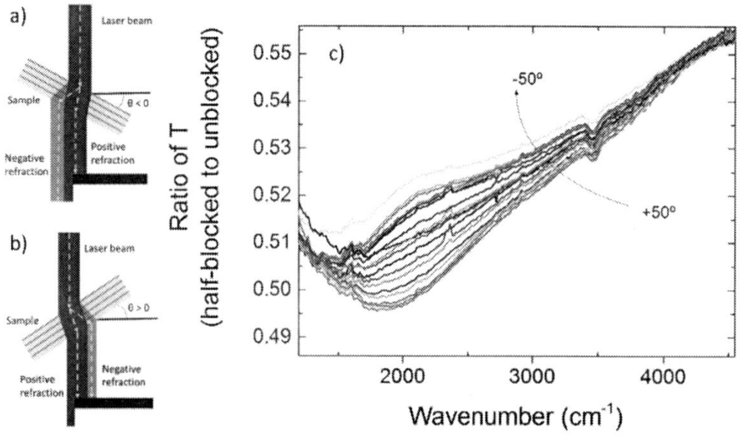

Figure 3. (a,b) Scheme of the experimental setup: stopping the positive and negative refracted laser beam respectively. (c) Experimental results: transmission ratio between half-blocked and unblocked beam. It shows a convexity inversion around 2000 cm^{-1} with the inclination angle θ according with the theoretical model.

4 – Conclusion

In conclusion we have experimentally demonstrated the possibility of reaching negative refraction due to intersubband transition in a layered semiconductor system. The experimental results are in agreement with the theoretical model. Although it is clear that intersubband designs are composed of several subwavelength layer of Drude-like materials, this work connects for the first time the concepts of intersubband plasmons with the photonic response of hyperbolic metamaterials. Due to the advanced photonic properties of HMM, we believe that these results could initiate new design for efficient mid-IR light emitting sources.

Funding. PG acknowledges financial support from the European Union's Horizon 2020 under the European Research Council (ERC) grant agreement No. 639109 (project Flatlight). JMC and AH acknowledge financial support from the European Union's Horizon 2020 Research and Innovation Program under grant agreement No. 665107 (project ZOTERAC).

References

[1] Gururaj V. Naik, Jingjing Liu, Alexander V. Kildishev, Vladimir M. Shalaev and Alexandra Boltasseva , Demonstration of Al:ZnO as a plasmonic component for near-infrared metamaterials, PNAS June 5, 2012. 109 (23) 8834-8838;

[2] Anthony J. Hoffman, , Aishwarya Sridhar, Phillip X. Braun, Leonid Alekseyev, Scott S. Howard, Kale J. Franz, Liwei Cheng, Fow-Sen Choa, Deborah L. Sivco, Viktor A. Podolskiy, Evgenii E. Narimanov, and Claire Gmachl, Midinfrared semiconductor optical metamaterials, Journal of Applied Physics 105, 122411 (2009);

[3] Anthony J. Hoffman, Leonid Alekseyev, Scott S. Howard, Kale J. Franz, Dan Wasserman, Viktor A. Podolskiy, Evgenii E. Narimanov, Deborah L. Sivco and Claire Gmachl, Negative refraction in semiconductor, Nature Materials volume 6, pages 946–950 (2007);

[4] Angela Vasanelli, Yanko Todorov, Carlo Sirtori, Ultra-strong light–matter coupling and superradiance using dense electron gases, C. R. Physique 17 (2016) 861–873;

[5] M. Montes Bajo, J. Tamayo-Arriola, M. Hugues, J. M. Ulloa, N. Le Biavan, R. Peretti, F. H. Julien, J. Faist, J. M. Chauveau, A. Hierro, Intersubband polarons in oxides, arXiv:1703.07743.

IOP Conf. Series: Journal of Physics: Conf. Series **1092** (2018) 012035 doi:10.1088/1742-6596/1092/1/012035

Wave phenomena and band-gap formation in layered phononic crystals with functionally graded interlayers and periodic arrays of cracks

Sergey I. Fomenko[1], **Mikhail V. Golub**[1], **Chuanzeng Zhang**[2]

E-mail: sfom@yandex.ru; m_golub@inbox.ru; c.zhang@uni-siegen.de
[1] Institute for Mathematics, Mechanics and Informatics, Kuban State University, Stavropolskaya 149, Krasnodar 350040, Russian Federation
[2] Department of Civil Engineering, University of Siegen, D-57068 Siegen, Germany

Abstract. Time-harmonic wave motion in a layered phononic crystals with internal inhomogeneities such as piezoelectric functionally graded interlayers and periodic arrays of cracks is considered. The complete band-gaps in layered piezoelectric functionally graded phononic crystal are revealed and analyzed. It is demonstrated that introduction of periodic cracks leads to the formation of extra band-gaps. Effect of formation of narrow band-gap exhibiting resonance nature is demonstrated within pass-band.

1. Introduction

Since 1990s considerable efforts have been made in the field of acoustic/elastic metamaterials and phononic crystals (PnCs) [1, 2]. PnCs can be used for elastic wave energy manipulation due to their unique properties like the effect of the signal absorption, negative refraction, wave resonances, wave localization, focusing etc. Accounting for these wave phenomena allows designing novel ultrasonic devices operating in broad frequency ranges, which are based on guided, surface or bulk waves.

Introduction of imperfections or cracks into composite structure provide an extra reflector that may lead to additional band-gaps or other unique physical phenomena, e.g. [3, 4]. Thus, if the periodicity of a layered PnC is disturbed and the cracks act as disorders, wave localization and resonance phenomena arise for some combinations of geometrical or material parameters [5]. At the same time, crack-like inhomogeneity is stress concentrator, which existence may cause failure of the structure. The problem of scattering of elastic waves in periodic composites by interlayers, made from functionally graded (FG) materials, and periodic arrays of crack-like inhomogeneities is considered in the present paper.

2. Configuration of phononic crystal

Plane wave propagation in a layered phononic crystal composed of N periodically situated unit-cells of thickness H between two identical half-spaces is considered. The unit-cell is a stack of layers infinite along axes Ox_1 and Ox_2. Two kinds of unit-cells of PnC shown in Figure 1 are studied: unit-cells with two FG interlayers and unit-cells with periodic array of cracks.

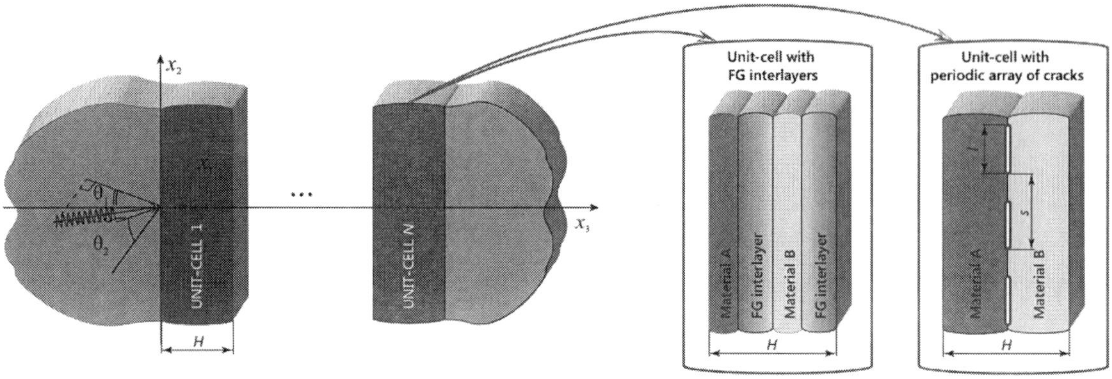

Figure 1. Two considered configurations of layered phononic crystals with inhomogeneities

Two homogeneous are assumed of piezoelectric material PVDF and silver, while two FG layers have the same thickness $H_{\rm FG}/2$, where material properties varies in accordance with power law given in [6]. The angles θ_1 and θ_2 denote azimuthal and radial angles of the incoming plane wave correspondingly. The second kind of PnC is a periodic set of unit-cells incorporating periodic arrays of interface cracks of widths $2l$, where cuts are situated at the distance s from each other. The material properties are assumed of lead and epoxy.

3. Layered phononic crystals with functionally graded interlayers

The solution of the problem for PnC with FG interlayers is based on the method of eigenvalue decomposition [7]. The mechanical displacements u_i and electrical potential φ are written as follows:

$$u_i = \sum_{n=1}^{4} u_{in}\lambda_n^{-N}, \quad \varphi = \sum_{n=1}^{4} \varphi_n\lambda_n^{-N}, \tag{1}$$

where λ_n are the eigenvalues of the unit-cell transfer matrix such as $|\lambda_n| \geq 1$. The semi-analytical formulae for the coefficients u_{in} and φ_n have been derived, the latter provides a numerically stable algorithm for wave-field calculation.

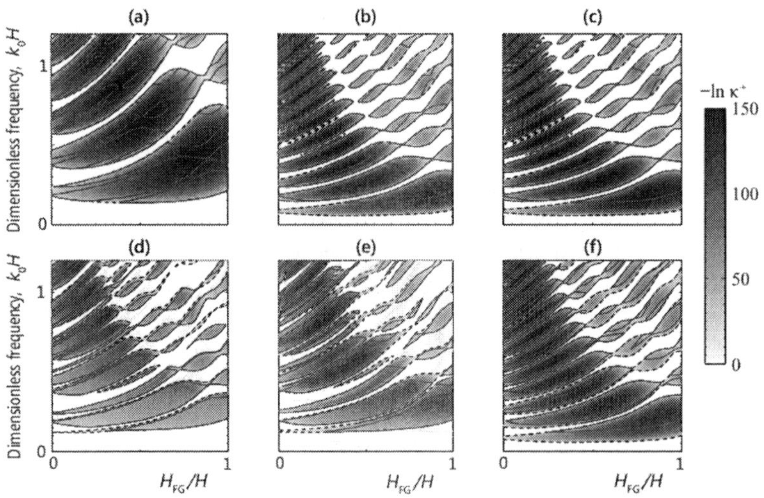

Figure 2. The influence of functionally graded layers thickness H_f on the transmission coefficient κ^+ for different types of incident waves: (a,d) – P-wave, (b,e) – SV-wave, (c,f) – SH-wave, and different angles of incidence: the first row (a,b,c) is for $\theta_1 = \theta_2 = 0°$, the second row (d,e,f) is for $\theta_1 = 45°$, $\theta_2 = 0°$

Considering a large amount of pairs of piezoelectric materials with FG interlayer, the authors have found that unit-cells made of PVDF (material properties are given in [8]) with by a very thin silver layer (aspect ratio between thicknesses 19 : 1) can be used in order to produce complete band-gaps providing full reflection for all angles of incidence. Moreover, the FG interlayers can be used to manipulate width of the band-gaps.

The influence of FG interlayers on wave transmission through the phononic crystal is shown in Fig. 2 for longitudinal (P), vertically polarized shear (SV) and horizontally polarized shear (SH) incident waves for normally and obliquely incident plane elastic waves. Here k_0 is the wavenumber of the transverse waves in the half-space. The coloured zones in the Figs. 2–3 factually correspond to band-gaps. The red solid lines are contour lines of the domains, where all eigenvalues $|\lambda_n| > 1$ in Eq.(1). In other words, these domains are classical band-gaps (or band-gaps of first kind in accordance with extended classification given in [7] for elastic anisotropic PnC). Other colored zones are the band-gaps of second kind or low-transmission pass-bands. For SV- and SH- incident waves the band-gaps of first kind are identical (see 2nd and 3rd column of Fig. 2). The band-gaps become wider and shift to high frequency side with increase of FGM portion.

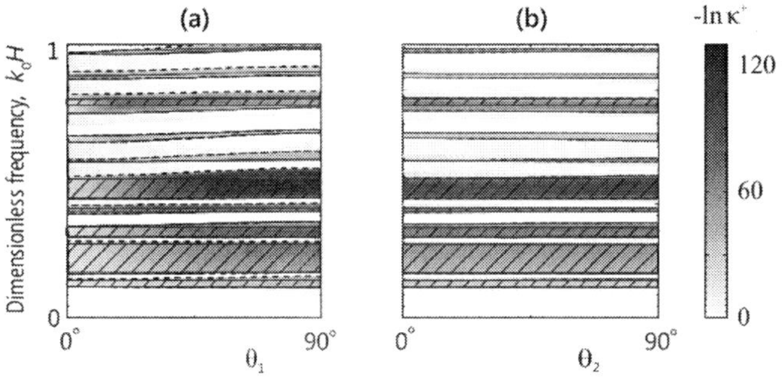

Figure 3. The influence of incidence angles of the longitudinal waves on band-gaps in functionally graded PnC ($H_f = 0.5H$); $\theta_2 = 0°$ in (a), and $\theta_1 = 45°$ in (b); shaded domains are full band-gaps

The full (complete) band-gaps can be observed in the considered piezoelectric PnC. These zones are indicated by shaded domains in Fig. 3 for dependencies of energy transmission coefficient on incident angles of P-waves.

4. Layered phononic crystals with multiple periodic arrays of cracks

Let us consider the propagation of plane elastic waves in a layered PnC with multiple periodic arrays of strip-like cracks as shown in Fig. 1. The total wave-field in such PnC is a sum of the incident field computed using the method described in the previous section and the scattered field, which is a superposition of the wave-fields scattered by all the arrays of cracks. The total wave-field satisfies the traction-free boundary conditions on all the crack-faces, this boundary condition leads to the unknown displacement jumps. The latter is determined using the boundary integral equation method and the integral approach [9]. Owing to the periodicity of the crack spacing s, only the region surrounding a reference crack in the centre of the Cartesian coordinate system needs to be considered. The crack density parameter $C = l/s$ is introduced to characterise periodic arrays of cracks.

Fig. 4 demonstrates the effect of crack introduction into the layered periodic structure on wave transmission. The figure depicts frequency dependence of the energy reflection coefficient κ^- and band-gaps marked by different colours for $l = 1$ ($C = 0.5$) and $l = 1.8$ ($C = 0.9$). First of all, it should be mentioned that presence of the cracks leads to the occurrence of extra band-gaps compared with PnC without cracks, the width of band-gap is wider the greater crack density C. The possibility of formation of narrow band-gaps for greater cracks can be observed in Fig. 4. These narrow band-gaps are very sensitive to the crack size l and exhibit resonance nature.

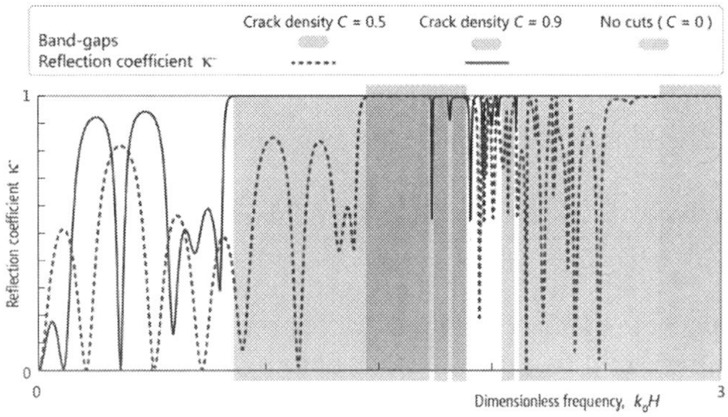

Figure 4. The energy reflection coefficient κ^- and band-gaps in the case of normally incident P-wave scattered by multiple periodic arrays of cracks with the crack density $(C = 0.5, 0.9)$, $s/H = 2$

5. Conclusion

The aim of this study is to investigate wave phenomena in layered PnCs and to demonstrate possible approach of an arrangement of unit-cells incorporating FG interlayers in layered PnCs for the formation and widening band-gaps. Presence of periodic arrays of cracks leads to the extra band-gaps, some of them are related to resonance wave transmission through PnC.

Acknowledgments

The authors are grateful to the support by Russian Foundation for Basic Research (joint project 18-501-12069 with German Research Foundation (DFG) ZH 15/29-1 and project 16-41-230352 with Krasnodar Region Administration).

References

[1] Deymier P 2013 *Acoustic Metamaterials and Phononic Crystals* (Springer-Verlag Berlin Heidelberg)
[2] Craster R and Guenneau S 2013 *Acoustic Metamaterials: Negative Refraction, Imaging, Lensing and Cloaking* (Springer-Verlag Berlin Heidelberg)
[3] Golub M V and Zhang C 2015 *Journal of Acoustical Society of America* **137** 238–252
[4] Biwa S and Ishii Y 2017 *The Journal of the Acoustical Society of America* **142** 2600–2600
[5] Golub M V 2016 *Journal of Applied Mechanics and Technical Physics* **57** 1190–1197
[6] Fomenko S I, Golub M V, Bui T Q, Zhang C and Wang Y S 2014 *International Journal of Solids and Structures* **51** 2491–2503
[7] Golub M V, Fomenko S, Alexandrov A, Zhang C, Wang Y S and Chen A L 2017 *Advanced Materials* (*Springer Proceedings in Physics*) (Springer) chap Elastic wave propagation in anisotropic and functionally graded layered phononic crystals: band-gaps, pass-bands and low transmission pass-bands, **175** 243–258
[8] Zabek D, Taylor J and Bowen C R 2016 *IEEE Transactions on Ultrasonics, Ferroelectrics, and Frequency Control* **63** 1681–1689 ISSN 0885-3010
[9] Glushkov E V, Glushkova N V, Golub M V and Zhang C 2009 *Acoustical Physics* **55** 8–16

Light scattering by resonant nanoparticles in a 2D lattice.

I M Fradkin[1,2], S A Dyakov[1], N A Gippius[1]

[1] Skolkovo Institute of Science and Technology, Nobelya Ulitsa 3, Moscow 121205, Russia

[2] Moscow Institute of Physics and Technology, Institutskiy pereulok 9, Moscow Region 141701, Russia

E-mail: Ilia.Fradkin@skoltech.ru

Abstract. Photonic crystals are wide-spread optical materials that are actively applied for manipulating light in various cases. However, it is still a challenge to calculate optical properties of photonic crystals, which include small deep-subwavelength resonant inclusions, in a fast way with acceptable precision. Here, we present a method for calculation of scattering matrix of a 2D lattice of resonant nanoparticles embedded into a homogeneous layer far from any interfaces. The proposed approach combines both specialized techniques for a description of high-gradient nanoparticles' near-field and theory, which accounts for the interactions between the nanoparticles. Our method makes it possible to increase the computational speed by several orders of magnitude in comparison with FEM and paves the way to a solution of several types of problems.

1. Introduction

Photonic crystal slabs and 2D photonic crystals are of great interest because of their unique optical properties, which appear in periodic structures and relative simplicity of fabrication. One of the most powerful methods for their description is Fourier modal method (FMM) [1-2], which is based on a decomposition of an electromagnetic field in terms of harmonics, which fulfill diffraction conditions. If the considered structure includes some tiny dielectric inclusions, then it is very likely that they might be just ignored because of negligible contribution in any observable quantity. Nevertheless, if these particles are plasmonic and have some resonances, then their influence becomes substantial and therefore very large number of harmonics are required to be taken into account. The requirement of FMM to calculate an inverse matrix, which size is proportional to the number of harmonics (N) leads to a cubic growth of algorithm complexity with N. Moreover, it is necessary to subdivide particles into many layers if they have a non-constant cross-section. These circumstances make FMM calculations for photonic crystals with small plasmonic nanoparticles almost impossible.

Near-field of small resonant particles can be efficiently calculated by such advanced methods as finite element method (FEM) or finite difference time domain (FDTD) method. However, straightforward utilization of these methods requires conducting calculations for a whole cell, which typically has relatively large dimensions. Thus, the problem again requires a huge amount of computational resources and time.

Problems of this type have been considered before. However, most studies are focused on some particular cases such as normal incidence of light and rectangular lattices [3-5]. The response of a single particle is often calculated analytically by Mie scattering theory [3] or conventional formulas for polarization of ellipsoids in electrostatic limit [4] or with dynamic corrections [5]. This strongly limits capabilities of application of these approaches and makes it necessary to develop a more general one.

Here, we develop a theory and corresponding programming code, which calculates scattering matrix of a 2D lattice of resonant nanoparticles in a homogeneous layer (see figure 1). This approach allows us to calculate scattering matrix of arbitrary, not necessarily rectangular lattice for any angle of incidence of external light. The main feature of our theory, which makes it faster than conventional ones, is a combination of methods aimed at near and far-field calculations and utilization of their advantages. Moreover, we introduce both dipole and resonant approximation, which simplify the model and allow to conduct all the calculations knowing only eigenmodes of a particle.

Figure 1. Schematic of a typical structure that might be calculated by the proposed theory. Plasmonic nanoparticles of a certain shape are embedded into a homogeneous layer far from interfaces.

2. Self-consistent equation on induced polarization and its solution

FMM is a powerful tool for operating with periodic structures till electromagnetic fields vary in space slowly. Therefore, it is not a surprise that it fails to cope with a high-gradient field of resonances. In this way, it seems reasonable to subdivide a problem into two parts.

The first problem is a determination of background field that polarizes certain particle. It consists of external field and field reradiated by adjacent particles. Thus, we obtain the following formula $\mathbf{E}_{\mathrm{b}}(\mathbf{r}_i) = \mathbf{E}_{\mathrm{ext}}(\mathbf{r}_i) + \sum_{j\neq i} \int \hat{G}(\mathbf{r}_i - \mathbf{r}_j)\mathbf{P}(\mathbf{r}_j)d^3\mathbf{r}_j$, where $\mathbf{P}$ is the induced in particles excessive polarization field and $\hat{G}$ is the dyadic Green function. Integration is conducted over the volume of particles and summation is conducting over all cells, except the considered one.

The second problem is a determination of polarization operator $\hat{\alpha}$ that shows what polarization is induced in a solitary particle under the influence of background field $\mathbf{P}(\mathbf{r}) = \hat{\alpha}[\mathbf{E}_{\mathrm{b}}(\mathbf{r})]$. It can be easily shown that this polarization field might be expressed in terms of full electric field $\mathbf{E} = \mathbf{E}_{\mathrm{b}} + \mathbf{E}_{\mathrm{s}}$, which includes both background and scattered parts. Indeed, $\mathbf{P}(\mathbf{r}) = (\varepsilon - \varepsilon_{\mathrm{b}})\mathbf{E}(\mathbf{r})/(4\pi)$, where ε and ε_{b} are permittivities of a particle and background medium respectively. Thus, this problem reduces to an issue of determination of full electric field, which may be solved in an efficient way. For simple geometries, it can be done analytically, whereas non-trivial cases can be resolved by FEM or any other appropriate numerical method. However, since we consider resonant nanoparticles, according to the theory of quasinormal modes (see figure 2) scattered field can be expressed as a superposition of eigenmodes $\mathbf{E}_{\mathrm{s}} = \sum_i A_i \tilde{\mathbf{E}}_i$ [6-7]. That is why all the Lorentz-like frequency dependences of a single particle polarization can be described by resonances' fields $\tilde{\mathbf{E}}_i$ that can be found, for instance, by FEM and corresponding amplitudes A_i that are given by theory [6].

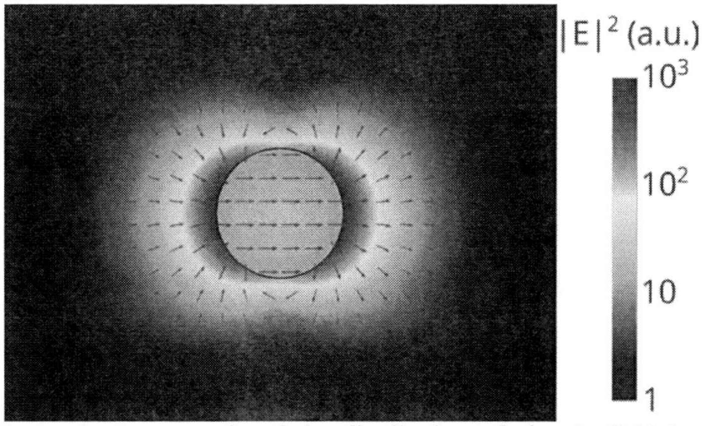

Figure 2. An example of the distribution of electric field in a resonant mode of plasmonic nanosphere. Red arrows correspond to the total current including its displacement part.

Thus, gathering these two observations we obtain self-consistent equation on induced polarization, which is a modified form of volume-integral equation method [8].

$$\mathbf{P}(\mathbf{r}_i) = \hat{\alpha}[\mathbf{E}_{\mathrm{ext}}(\mathbf{r}_i) + \sum_{j \neq i} \int \hat{G}(\mathbf{r}_i - \mathbf{r}_j)\mathbf{P}(\mathbf{r}_j)d^3\mathbf{r}_j] \quad (1)$$

The equation (1) fully describes the structure. If it included sum over all cells, then it might have been easily reduced to a system of linear equations by application of Fourier method. In order to do that we have implemented a special filter that is equal to zero at the zero point and smoothly raises up to the unity on a distance of a lattice period. Therefore, application of this filter helps to make the equation homogeneous, but absolutely identical to the initial one. Moreover, we have managed to compose such a filter, that it is still possible to make a Fourier transform of a filtered Green function analytically, which is very important for efficient calculations.

After reducing to the system of linear equations it might be additionally simplified for small particles, which corresponds to dipole approximation. Indeed, small particles reradiate the most part of energy through dipole emission [9], which means that we need to account for only 3 degrees of freedom. Finally, we obtain the following system of equations on a total background electric field:

$$\mathbf{E}_{\mathrm{b}} = \mathbf{E}_{\mathrm{ext}}(0) + \frac{2\pi^2 k_0^2}{S}\hat{g}_1\hat{\alpha}_{\mathrm{d}}\mathbf{E}_{\mathrm{b}}\,, \quad (2)$$

,where S is the area of a cell, k_0 is a wavenumber in the vacuum, $\hat{\alpha}_{\mathrm{d}}$ is the polarizability tensor of an isolated particle (which can be expressed in resonant form), $\hat{g}_1$ – is a complex matrix which is fully determined by Green function, geometrical properties of a lattice and wave vectors of external incident waves. Finally, $\mathbf{E}_{\mathrm{b}}$ is a total background field, which is a sum of an external one and a field induced by other particles. Also, it can be considered as $\mathbf{E}_{\mathrm{b}} = \hat{\alpha}_{\mathrm{d}}^{-1}\mathbf{D}$, where $\mathbf{D}$ is a full dipole moment of a particle in the central cell. Thus, by solving this equation numerically, which is not a challenging task, we obtain the value of induced dipole moment. After that, we can easily calculate reemitted field which together with transmitted one forms the scattered field of a lattice.

In order to illustrate the operation of the developed script, we consider a square lattice of gold nanospheres of 20 nm radius in a medium with $\varepsilon_{\mathrm{b}} = 10$. Period of the lattice is equal to 300 nm and gold is described by Drude model with the following parameters: $\omega_p = 1.34 \cdot 10^{16} s^{-1}$, $\gamma = 1.09 \cdot 10^{14} s^{-1}$, ε_∞ =8.5. The projection of an incident wave's wave vector on a lattice plane is directed along one of the lattice axis. From the figure 3, it is seen that coupling between nanoparticles leads to a significant bending of plasmonic resonance band.

IOP Conf. Series: Journal of Physics: Conf. Series **1092** (2018) 012036 doi:10.1088/1742-6596/1092/1/012036

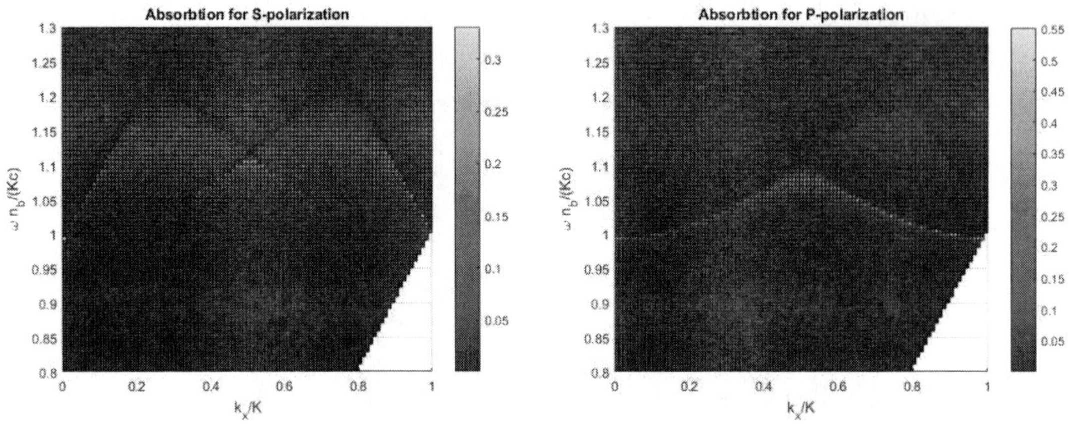

Figure 3. Dependence of light absorption in a lattice of gold spheres on frequency and projection of wave vector on a lattice axis for S and P polarizations.

3. Conclusion

In this work, we have constructed a theory, which makes it possible to calculate both polarization of resonant nanoparticles in a 2D lattice and their scattering matrix. In dipole approximation, this theory requires knowing of polarizability tensor, which might be calculated analytically, numerically using FEM or in any different way. Also, we propose to calculate this tensor in resonant approximation, which strongly accelerates all the calculations. The typical calculation time of scattering matrix turns out to be of an order of several milliseconds whereas full FEM calculations take at least several dozens of seconds, which is four orders of magnitude longer. Such speed paves the way to large-scale computations for purposes of design of optoelectronic devices and other photonic crystals based structures. Also, our method can be easily expanded on more useful and practical problems of a lattice placed on an interface between two layers and a lattice with a basis of several nanoparticles, which will be done in further studies.

References

[1] Tikhodeev S G, Yablonskii A L, Muljarov E A, Gippius N A and Ishihara T *Phys. Rev. B* **66** 045102

[2] Lifeng L 2003 *J. Opt. A: Pure Appl. Opt.* **5** 345

[3] Evlyukhin A B, Reinhardt C, Seidel A, Luk'yanchuk B S and Chichkov B N 2010 *Phys. Rev. B* **82** 045404

[4] Kosobukin V A 2012 *Phys. Solid State* **54** pp 2471-2480

[5] Auguié B and Barnes W L 2008 *Phys. Rev. Lett.* **101** 143902

[6] Sauvan C, Hugonin J P, Maksymov I S and Lalanne P 2013 *Phys. Rev. Lett.* **110** 237401

[7] Bai Q, Perrin M, Sauvan C, Hugonin J P and Lalanne P 2013 *Opt. Express* **21** 27371

[8] Novotny L, Hecht B. 2012 *Principles of Nano-optics* (New York: Cambridge University Press) chapter 16 pp 506-517

[9] Landau L, Lifshitz E 2013 *The Classical Theory of Fields* (Elsevier)

IOP Conf. Series: Journal of Physics: Conf. Series **1092** (2018) 012037 doi:10.1088/1742-6596/1092/1/012037

Second harmonic generation in nanoparticles with Mie resonances

K S Frizyuk[1,2], **D A Smirnova**[3], **A N Poddubny**[4], **M I Petrov**[1]

1. Centre for Nanophotonics and Metamaterials, ITMO University, Saint-Petersburg, Russia
2. Academic University, Saint-Petersburg, Russia
3. Institute of Applied Physics, Russian Academy of Sciences, Nizhny Novgorod 603950, Russia
4. Ioffe Institute, St Petersburg 194021, Russia

E-mail: k.frizyuk@metalab.ifmo.ru

Abstract. We study second harmonic generation (SHG) by high-index dielectric nanoparticles and possessing bulk quadratic nonlinearity of a tensorial form. The nanoparticles exhibit Mie resonances in infrared and optical regions. We theoretically show how the polarization structure of the Mie modes governs the SHG process, and demonstrate that the superposition of two resonances at the incident light wavelength and at the SH wavelength provides us the enhancement of SHG. We also perform the multipolar decomposition of second harmonic radiation, and show that it is governed by the selection rules, defined by the symmetry of the nanoparticle shape, incident wave polarization and crystalline symmetry of the lattice.

1. Introduction

The resonantly enhanced second harmonic generation (SHG) is well studied in metallic nanoparticles [1], where the enhancement of local electric field is achieved due to strong plasmonic resonances. The plasmonic nanostructrues emit nonlinear signal due to surface effects or by breaking the symmetry of the excited modes [4] as the bulk second order susceptibility tensor $\hat{\chi}^{(2)}$ is equal to zero [2] due to inverse symmetry of the lattice. All-dielectric nanophotonics has been intensively developed over the last years as an alternative to plasmonics, with pronounced Mie-type volume resonance instead of surface palsmon resonances. The nonlinear effects accompanying the light scattering, such as second- and third-harmonic generation, are also susceptive to the resonance properties of the dielectric nanostructures. At the same time, the variety of dielectric and semiconductor materials allows one to use of those with non-zero bulk second order susceptibility tensor $\hat{\chi}^{(2)}$. This allowed the researchers to observe strong enhancement of SHG in dielectric and semiconductor materials reported recently in number of papers [6, 9, 7, 8, 10]. However, to control the efficiency of SHG, it is vitally important to identify the mode coupling mechanism in a nonlinear process: eigen modes of a nanostructure at the fundamental frequency couple to the modes at the second harmonic (SH), with account for the specific tensor $\hat{\chi}^{(2)}$ structure. The symmetry of both fundamental modes and SH modes, multiplied over the crystalline symmetry, either allow or forbid the typical coupling channels. This work is devoted to theoretical studies of the SHG by spherical dielectric nanoparticle. According to Mie theory, subwavelength nanoparticle with high refractive index n has first-order resonances when $\lambda \sim nd$, where λ is incident light wavelength and d is nanoparticle characteristic size. We consider the enhancement of the SHG due to Mie mode excitation, and their coupling at the fundamental frequency and at the frequency fo the second harmonic.

2. Coherent and incoherent light emission

It is well known that the emission process is defined by the local density of the optical states (LDOS) or imaginary part of Green's function [3]. The Raman emission is dielectric nanostructures is one of the examples of such process [5]. At the same time the SH is a coherent non-elastic scattering process. One can derive the expression for the intensity of the SH emission basing on the Green's function properties:

$$I_{SH} = \frac{\omega_{SH}^3 \mu\mu_0}{2}\left[\int\int dV' dV'' \mathbf{P}^*(\mathbf{r}', \omega_{SH}, \omega) \Im\hat{\mathbf{G}}(\mathbf{r}', \mathbf{r}'', \omega_{SH}) \mathbf{P}(\mathbf{r}, \omega_{SH}, \omega)\right], \tag{1}$$

where ω_{SH} is the second harmonic frequency, $\mathbf{P}(\mathbf{r}, \omega_{SH}, \omega)$ is the induced polarization vector on SH double frequency. One can notice that here the intensity of SH process is defined by the cross density of states (CDOS) [11], which is defined by the imaginary part of dyadic Green function of dielectric sphere $\Im\hat{\mathbf{G}}(\mathbf{r}', \mathbf{r}'', \omega_{SH})$. According to Mie theory, exciting field is significantly enhanced at specific frequencies, which gives an enhancement to non-linear response respectively. When it's expanded into vector spherical harmonic series, coefficients before tensorial products of harmonics have resonant character. They have maximums at the same frequencies as Mie coefficients. Thus, if the ω_{SH} is resonant, SH signal will be enhanced due to Green's function poles even if incident light frequency ω is out of the resonance resonant. We are interested in the cases when both wavelengths are resonant.

3. Modelling the SHG emission

We consider SH field generated by a spherical dielectric nanoparticle with sizes of the order of hundreds nanometers in diameter with high refractive index $n \sim 2.5 - 3.5$. Such nanoparticles have Mie resonances in the optical range. We developed rigorous analytical approach to compute and predict nonlinear fields generated by such nanoparticles and applied it to SHG from AlGaAs (T_d crystalline symmetry) and $BaTiO_3$ (C_{4v} symmetry) nanoparticles assumed monocrystaline. We used electromagnetic dyadic Green's function approach in order to model the SH emission.

Nonlinear field at the frequency $2\omega = \omega_s$ outside the sphere can be expressed through the vector harmonic expansion, which give the following expression for the intensity of SHG:

$$\mathbf{E}(\omega_{SH}, \mathbf{r}) = \sum_{n=1}^{\infty}\sum_{m=0}^{n} E_0(D_{Memn}\mathbf{M}^{(1)}_{emn}(\omega_{SH}, \mathbf{r}) + D_{Momn}\mathbf{M}^{(1)}_{omn}(\omega_{SH}, \mathbf{r}) \tag{2}$$

$$+D_{Nemn}\mathbf{N}^{(1)}_{emn}(\omega_{SH}, \mathbf{r}) + D_{Nomn}\mathbf{N}^{(1)}_{omn}(\omega_{SH}, \mathbf{r})) \tag{3}$$

where $\mathbf{M}_{\substack{e\\o}mn}(\omega)$ and $\mathbf{N}_{\substack{e\\o}mn}(\omega)$ are vector spherical harmonics defined in [12].

As a basic example, we have considered $BaTiO_3$ nanosphere under non-resonant excitation, when the excitation wavelength is much longer than the lowest resonance wavelength. This provides the condition of magnetic dipolar resonance at the SH frequency. To achieve deeper understanding of the nonlinear signal properties, we decomposed the fields at the fundamental wavelength into the sum of magnetic $\mathbf{M}_{\substack{e\\o}mn}$ and electric $\mathbf{N}_{\substack{e\\o}mn}$ vector spherical harmonics defined in Ref. [12]. We performed the calculations and built multipolar decomposition of the SH intensity dependent from the orientation of crystalline lattice of the nanosphere (Fig. 2). We have observed the contribution of the dipolar multipoles in the total intensity on the wavelengths of magnetic dipolar resonance.

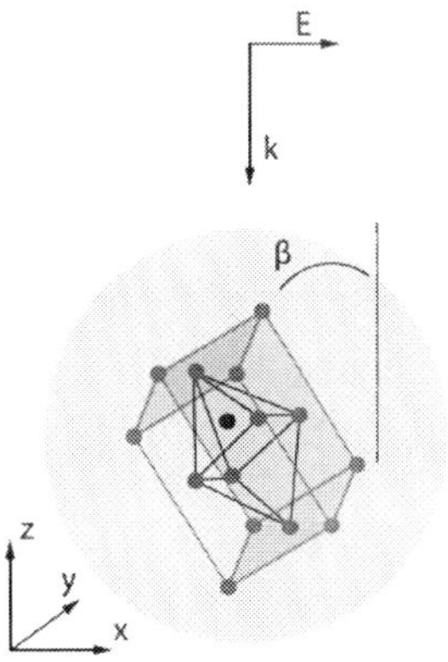

Figure 1. Geometry of the problem. We consider SHG from monocrystalline spherical nanoparticle with different lattice orientations, while incident field orientation is fixed.

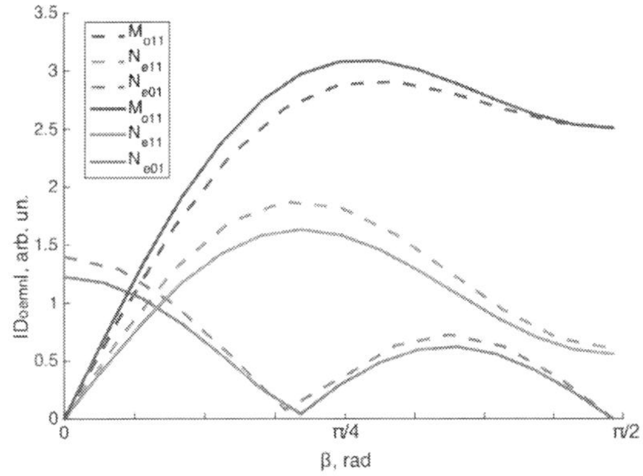

Figure 2. Angular dependence of $D_{{}^{e}_{o}mn}$ coefficients before Mie harmonics in SHG field. BT spherical monocrystalline nanoparticle, r = 120 nm, incident wavelength - 1200 nm. Dashed lines have been computed in COMSOL and solid lines show the analytical result

Although intensity is expected to be enhanced due to resonance behavior of Green's function, it can be noticed that the enhancement takes place only for particular lattice orientation. This effect is observed because the excitation of magnetic dipolar modes $\mathbf{M}_{{}^{e}_{o}mn}$ can be prohibited due to relative symmetry properties of the incident wave and crystalline lattice, Fig. 2 illustrates this case when $\beta = 0$

Using symmetry properties of the system, we developed simple rules to find if particular Mie-harmonics are presented in SH-field. We provide table of the selection rules when BT lattice-axis orientation is parallel to k-vector of incident field ($\beta = 0$) (table 1). We consider only dipolar terms in incident field. For the first case, magnetic dipoles are not presented in SH field, and can not be enhanced by resonant LDOS on the frequency of MD resonance.

Table 1. Selection rules for BT when $\beta = 0$. The magnetic dipoles are not presented in the SH field.

incident field	Second harmonic contains
$N_{e11} \times N_{e11}$	N_{e01}, M_{o22}, N_{e03}, N_{e23}, M_{o24}, N_{e05}, N_{e25}
$M_{o11} \times M_{o11}$	N_{e01} M_{o22} N_{e03} N_{e23}
$M_{o11} \times N_{e11}$	N_{e02} N_{e22} M_{o23} N_{e04} N_{e24}

As one can see from the selection rules, the total SH intensity is dependent on the relative orientation of the incident plane wave and the crystalline lattice.

The same rules can be applied for the other crystalline symmetries such as AlGaAs with T_d lattice group. We build the multipolar decomposition of the intensity dependence on nanoparticle radius for SHG by a spherical AlGaAs nanoparticle (Fig. 3). We see that the SH signal is enhanced when we have resonances at incident and SH wavelengths, but modulated by selection rules (not all spherical harmonics appear). In particular, all electric dipolar harmonics are prohibited for T_d symmetry when the incident field is parallel to crystalline lattice axis (see

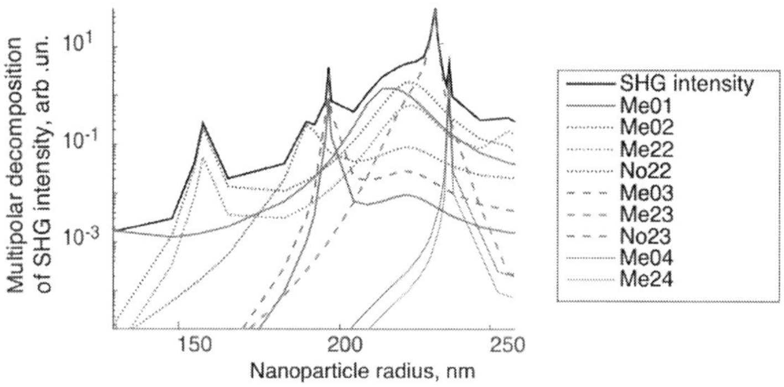

Figure 3. Multipolar decomposition of SHG intensity in AlGaAs nanoparticles of different radius. Pump wavelength - 1550nm.

Fig. 1).

4. Conclusion

We have analyzed the second harmonic generation from crystalline noncentrosymmetric nanoparticles, possessing second-order bulk nonlinearity. We derived the selection rules for typical crystalline systems and validated them with numerical modeling. We have calculated the dependence of the SH intensity on the orientation of the crystalline axes with respect to the incident light polarization. The maximum conversion efficiency was found to be achieved when both incident and SH wavelength are resonant, and resonant multipoles are not prohibited by the selection rules.

References

[1] J. Butet, P.-F. Brevet, and O. J. F. Martin, ÂOptical Second Harmonic Generation in Plasmonic Nanostructures: From Fundamental Principles to Advanced Applications,Â ACS Nano, vol. 9, no. 11, pp. 10545Â10562, 2015.
[2] R. W. Boyd, *Nonlinear optics* (Academic Press, San Diego, 2003).
[3] Novotny L., Hecht B. Principles of Nano-Optics.
[4] M. Celebrano et al., ÂMode matching in multiresonant plasmonic nanoantennas for enhanced second harmonic generation,Â Nat. Nanotechnol., vol. 10, no. 5, pp. 412Â417, Apr. 2015.
[5] K. Frizyuk,et al., Enhancement of Raman scattering in dielectric nanostructures with electric and magnetic Mie resonances Phys. Rev. B 97, 085414, Febr. 2018
[6] R. Camacho-Morales et al., ÂNonlinear Generation of Vector Beams from AlGaAs Nanoantennas,Â Nano Lett., vol. 16, no. 11, pp. 7191Â7197, 2016.
[7] S. Liu, G. A. Keeler, J. L. Reno, M. B. Sinclair, and I. Brener, ÂIII-V Semiconductor Nanoresonators-A New Strategy for Passive, Active, and Nonlinear All-Dielectric Metamaterials,Â Adv. Opt. Mater., vol. 4, no. 10, pp. 1457Â1462, 2016.
[8] L. Carletti, A. Locatelli, D. Neshev, and C. De Angelis, ÂShaping the radiation pattern of second harmonic generation from AlGaAs dielectric nanoantennas,Â ACS Photonics, vol. 3, no. 8, pp. 1500Â1507, 2016.
[9] F. Timpu, A. Sergeyev, N. Hendricks, and R. Grange, ÂSecond-Harmonic Enhancement with Mie Resonances in Perovskite Nanoparticles,Â ACS Photonics, vol. 4, no. 76Â84, 2017.
[10] S. V Makarov et al., ÂEfficient Second-Harmonic Generation in Nanocrystalline Silicon Nanoparticles,Â Nano Lett., vol. 17, no. 5, p. 3047, 2017.
[11] A. Caze, R. Pierrat, and R. Carminati, ÂSpatial Coherence in Complex Photonic and Plasmonic Systems,Â Phys. Rev. Lett., vol. 110, no. 6, p. 63903, Feb. 2013.
[12] C. F. Bohren and D. R. Haffman, Absorption and Scattering of Light by Small Particles. New York: Wiley, 1983.

IOP Conf. Series: Journal of Physics: Conf. Series **1092** (2018) 012038 doi:10.1088/1742-6596/1092/1/012038

Resonant Silicon Nanoparticles for Enhanced Light Harvesting in Halide Perovskite Solar Cells

A D Furasova[1], A A Zakhidov[1,2], A Di Carlo[3] and S V Makarov[1]

[1]Department of Nanophotonics and Metamaterials, ITMO University, St. Petersburg, 197101, Russia

[2] University of Texas at Dallas, Richardson, Texas 75080, USA

[3]C.H.O.S.E. (Centre for Hybrid and Organic Solar Energy), Department of Electronic Engineering, University of Rome-Tor Vergata, Rome, 00133, Italy

furasova@inbox.ru

Abstract. Improving performance and stability of lead halide perovskites solar cells (PSC) is one of the most important direction in modern photovoltaics. For this purpose, we added low-loss and chemically inert resonant silicon nanoparticles (NPs) between electron transport layer and perovskite layer that allows to increase the efficiency enhancement via improved light harvesting, photocurrent, open-circuit voltage and fill-factor. This results in a boost of the device efficiency up to 18.8%, being a record among the previously reported results on nanoparticles incorporation into $CH_3NH_3PbI_3$ ($MAPbI_3$) perovskite-based solar cells. Moreover, PSC with the silicon NPs shows efficiency improvement in time with respect to the cells without NPs. The proposed method is simple and can be implemented for PSCs with various compositions and architectures.

1. Introduction

Nowadays lead halide perovskites have a huge number of possible optoelectronic applications including diodes [1], laser technologies [2], photodetectors [3] but the main attention is focused on photovoltaics [4-6]. PSCs achieved remarkable power conversion efficiency (PCE) values exciding 22% [7] through a careful materials design and fabrication process optimizations. A lot of attempts were given to increase PCE by plasmonic NPs implementation into perovskite SC devices, which can play a role as either the absorber or increase charge transport, that makes them a promising method to engineer both optical and electronic device characteristics by making better use of incident light and engineering the carrier dynamics.

On the other hand, there is no any data about time stability of PSC contained chemical active plasmonic NPs. As shown in previous works [8], a metal part of PSC can interact with organic and perovskite layers under temperature higher that 30 °C. It is necessary to cover by additional inert shell plasmonic NPs, which not only includes complementary resources of device preparation, but also can reduces the NPs charge influence.

At the same time, Mie resonance silicon NPs recently has emerged as the powerful tool for various optical applications[9,10]. Due to their low cost, preparation simplicity, chemical inert properties and temperature resistance these NPs represent a viable and better alternative to noble metals for

IOP Conf. Series: Journal of Physics: Conf. Series **1092** (2018) 012038 doi:10.1088/1742-6596/1092/1/012038

photovoltaics application. In this work we incorporate Si NPs between mesoporous electron transport layer and $MAPbI_3$ to increase efficiency up to 18.8%.

2. Discussion

Figure 1 shows the device structure and thickness of each layer. Meso-superstructured PSC has a n-i-p architecture. A 50-nm TiO_2 compact layer is deposited on top of an electron collecting fluorine-doped tin oxide (FTO) electrode. A mesoporous TiO_2 (m-TiO_2) ETL has a 200-nm thickness, capped by photoactive $MAPbI_3$ perovskite (150/350 nm). The 400-nm hole transport layer (HTL), SPIRO-OMeTAD, doped with Bis(trifluoromethane)sulfonimide lithium salt, coats perovskite, and 80-nm gold layer, was deposited as top contact. The Si NPs, located between m-TiO_2 and perovskite, have size between 50 and 180 nm. This configuration allows to keep device morphology from thickness change or additional layers roughness.

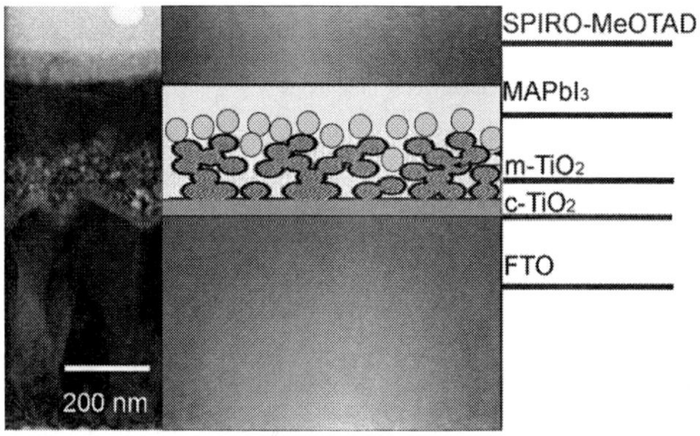

Figure 1. Cross-sectional bright-field STEM image and layer scheme of a complete solar cell. Bottom-up: FTO electrode, c-TiO_2 ETL, m-TiO_2 ETL, $MAPbI_3$ perovskite, SPIRO-OMeTAD HTL.

The best reference device showed an efficiency of 17.7%, when the best performance for device with Si NPs is 18.8%. The Figure 2 demonstrates the J-V scans for the best cells with the different design (with and without Si NPs). It is clearly seen that device with Si NPs has a well shape of J-V curve due to fill factor of 78.9% (for reference it is 77%), improving V_{oc} from 1.05 to 1.06 V and J_{sc} from 22.0 to 22.4 mA/cm^2.

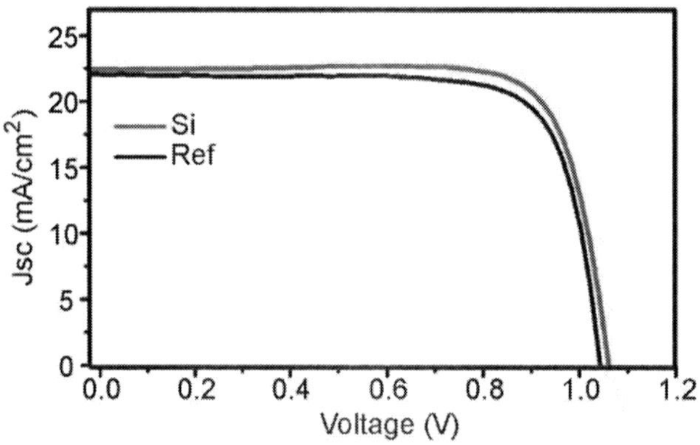

Figure 2. J-V curves for best cells with (orange) and without (black) incorporation of Si NPs.

According to the statistical data, measured for 34 cells and presented in figure 3, the average efficiency for standard devices is 17.2%, whereas for Si NPs contained this performance is 17.6%. It should be pointed, that we have observed that the Si NPs presence improves the statistic device performance after 4 days up to 17.9%, at the same time, reference cells lightly decrease it to 17.1%.

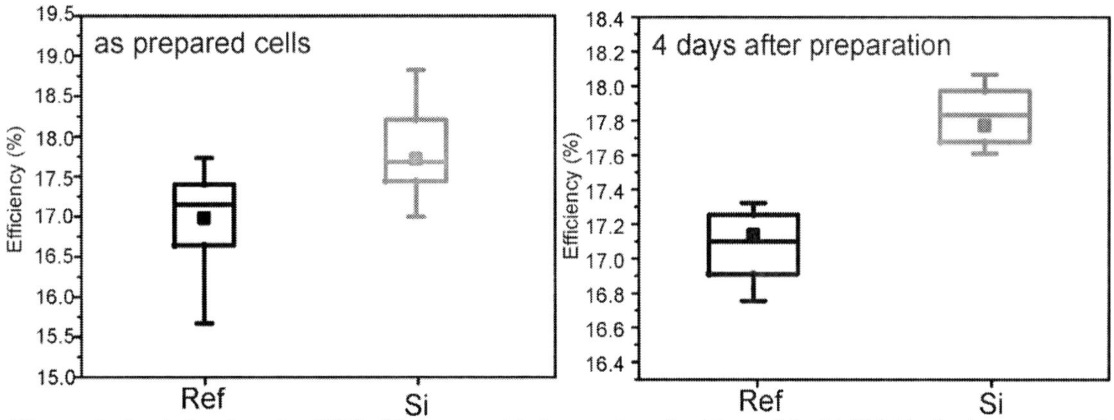

Figure 3. Statistics data for PSC efficiency with (orange) and without (black) Si NPs for just prepared cells (left) and after several days (right).

3. Conclusion

Si NPs present a novel approach to improve lead halide perovskite solar cells, being located at an interface of $MAPbI_3$ and m-TiO_2. The improved PSCs have demonstrated the efficiency of 18.8%, which are higher than those for the reference devices. Moreover, Si NPs show a unique property to improve devices after few days.

Acknowledgements
The work was supported by Russian Foundation for Basic Researches (grants 17-03-00621 and 16-29-05317)

References

[1] Sutherland B R, Sargent E H, 2016 *Nat. Photonics.* **10**, 295.
[2] Veldhuis S A, Boix P P, Yantara N, Li M, Sum T C, Mathews N, Mhaisalkar S G, 2016 *Adv. Mater.* **28**, 32.
[3] Dou L, Yang Y M, You J, Hong Z, Chang W-H, Li G, Yang Y, 2014 *Nat. Commun.* **5**, 5404.
[4] Manser J S, Christians J A, Kamat P V, 2016 *Chem. Rev.* **116**, 21.
[5] Grätzel M, 2014 *Nat. Mater.* **13**, 838.
[6] Correa-Baena J-P, Saliba M, Buonassisi T, Grätzel M, Abate A, Tress W, Hagfeldt A, 2017 *Science.* **358**, 6364.
[7] Saliba M, Matsui T, Domanski K, Seo J Y, Ummadisingu A, Zakeeruddin S M, Correa-Baena J P, Tress W R, Abate A, Hagfeldt A, Gratzel M, 2016 *Science* **354,** 6309.
[8] Domanski K, Correa-Baena J-P, Mine N, Nazeeruddin M K, Abate A, Saliba M, Tress W, Hagfeldt A, Gratzel M, 2016 *ACS Nano*, **10,** 6.
[9] Kuznetsov A I, Miroshnichenko A E, Brongersma M L, Kivshar Y S, Luk'yanchuk B, 2016 *Science*, **354**, 6314.
[10] Tiguntseva E, Chebykin A, Ishteev A, Haroldson R, Balachandran B, Ushakova E, Komissarenko F, Wang H, Milichko V, Tsypkin A, Zuev D, Hu W, Makarov S, Zakhidov A, 2017 *Nanoscale*, **9,** 12486.

Towards to the development of THz detectors based on carbon nanostructures

I. Gayduchenko[1], G. Fedorov[1,2], N. Titova[1], M. Moskotin[1], E. Obraztsova[3], M. Rybin[3] and G. Goltsman[1]

[1]Moscow State University of Education, 1/1 M. Pirogovskaya Str., Moscow, Russian Federation, 119991
[2]Moscow Institute for Physics and Technology (State University), 9 Institutskiy per., Dolgoprudny, Moscow Region, Russian Federation, 141701;
[3]Prokhorov General Physics Institute, Russian Academy of Sciences, 38 Vavilov Str., Moscow, Russian Federation, 119991

igorandg@gmail.com

Abstract. Demand for efficient terahertz radiation detectors resulted in intensive study of the carbon nanostructures as possible solution for that problem. In this work we investigate the response to sub-terahertz radiation of detectors with sensor elements based on CVD graphene as well as its derivatives – carbon nanotubes (CNTs). The devices are made in configuration of field effect transistors (FET) with asymmetric source and drain (vanadium and gold) contacts and operate as lateral Schottky diodes. We show that at 300K semiconducting CNTs show better performance up to 300GHz with responsivity up to 100V/W, while quasi-metallic CNTs are shown to operate up to 2.5THz. At 300 K graphene detector exhibit the room-temperature responsivity from R = 15 V/W at f= 129 GHz to R = 3 V/W at f = 450 GHz. We find that at low temperatures (77K) the graphene lateral Schottky diodes responsivity rises with the increasing frequency of the incident sub-THz radiation. We interpret this result as a manifestation of a plasmonic effect in the devices with the relatively long plasmonic wavelengths. The obtained data allows for determination of the most promising directions of development of the technology of nanocarbon structures for the detection of THz radiation.

1. Introduction

Increase in the sensitivity of the THz radiation detectors range can be achieved by reducing the size of the sensor element. Graphene is an almost ideal material for creating nanoscale structures for this purpose. One of the major advantages of graphene is its high carrier mobility and the associated large coherence length, so that the band structure is determined by the size quantization and can be controlled via the geometry of the structure [1].

Several successful configurations of detectors of THz radiation on the basis of graphene and nanotubes have been proposed recently [2-5]. Further improvement of this type of device is not possible without better understanding of the mechanisms that determine the magnitude of the response to radiation.

In this paper we present the results of our systematic studies of different configurations of detectors with sensor elements based on graphene as well as its derivatives – carbon nanotubes (CNTs) and graphene nanotubes (GNRs). The asymmetry means that different metals are used to contact the channel.

Detection of the radiation takes place through the rectification: due to asymmetry of the IV curve mean current under harmonically changing voltage is not zero. In case of a CNT rectification takes place due to transport through a Schottky barrier at V contact directly related to the nanotube's bandgap. In case of graphene the situation is more complicated. Transport of carriers through the p-n junction depends on the angle between the normal to the junction and the carrier momentum. For normally incident electrons there is no energy barrier and they shunt the nonlinear transport. At the same time current of the non-normally incident electrons is non-linear function of the applied voltage. This scenario is evidenced by the temperature dependence of the response of our structures to the radiation.

At room temperature our results show that semiconducting CNT (s-CNT) devices are preferable for frequencies up to 300GHz (cut off about 500 GHz), while quasi-metallic -CNT devices can be used as broad band detectors operating at frequencies up to at least 2.5 THz. At 300 K graphene detector exhibit the room-temperature responsivity from R = 15 V/W at f= 129 GHz to R = 3 V/W at f = 450 GHz.

At low temperatures (77K) we fan that graphene lateral Schottky diodes responsivity rises with the increasing frequency of the incident sub-THz radiation. We interpret this result as a manifestation of a plasmonic effect in the devices with the relatively long plasmonic wavelengths.

2. Experimental

For our experiments we fabricated asymmetric structures in configuration of field effect transistors (FET) with channel based on graphene or CNTs. The structures consist of a single layer graphene (SLG) or individual CNT on the SiO_2 substrate with moderately conducting silicon bottom layer (room temperature resistivity of 10 Ω·cm). The latter plays the role of the back gate. The source and drain contacts were made from different metals (vanadium and gold) and connected to the logarithmic spiral electrodes serving as an antenna coupling the incident radiation to the device (Fig. 1). The single layer graphene and single walled carbon nanotubes (SWNTs) used in our experiments as a FET channel were synthesized by chemical vapor deposition (CVD) . Detailed description of device fabrication and configuration is presented in our previous works [4-5].

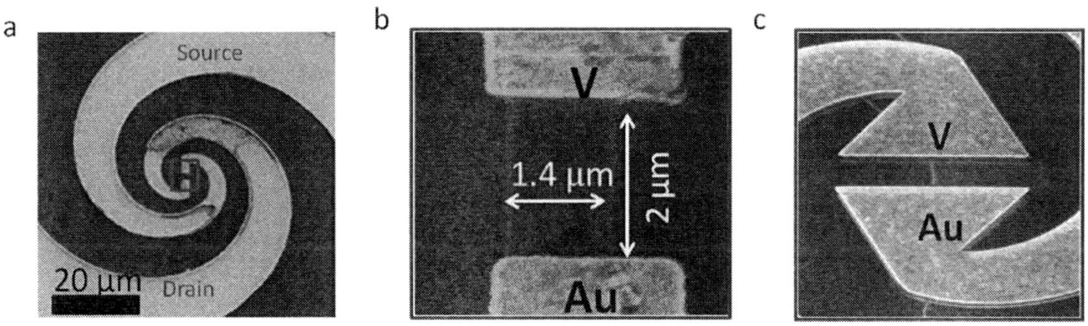

Figure 1. (a): Electron image of the device coupled to the radiation with a logarithmic spiral antenna. The red rectangle marks the transistor channel area; (b) Electron image of the device channel formed by graphene; (c): Electron image of the device channel formed by CNT

The terahertz radiation is provided by two backward wave oscillators (BWO) and laser. The power incident on the cryostat window is measured by a Golay cell. The first BWO is used as a source of radiation with a frequency of 129 GHz (Maximum power is 1mW). The second BWO is used for frequency range 265-450 GHz. Its maximum power was found to be about 400 µW. We also used a gas discharge laser operating on a 2.5 THz H_2O line. Its maximum output power is 200 µW.

First, we measured DC transport characteristics of our devices. Figure 2 (a-b) shows conductance of CNT devices as a function of gate voltage. The $G(V_G)$ curve shown on the Fig. 2(a) is typical for a semiconducting CNT (s-CNT). The transistor curve G(VG) on Fig. 2(b) is typical for a device with conductance channel formed with a quasi-metallic CNT (m-CNT)[4]. We note that the ON state conductance of m-CNT devices is much larger than that of s-CNT devices. This can be explained if we consider the Schottky barrier formed at the vanadium electrode that has the work function less than that of the CNT. The width of this barrier at the Fermi level is proportional to the nanotube's band gap. Thus the G(VG) curves confirm the existence of a Schottky barrier at the vanadium-CNT interface. In case of graphene (fig.2 (c)) device resistance in all cases goes up as the gate voltage is swept from -10 to10 V indicating p-doping of the graphene channel.

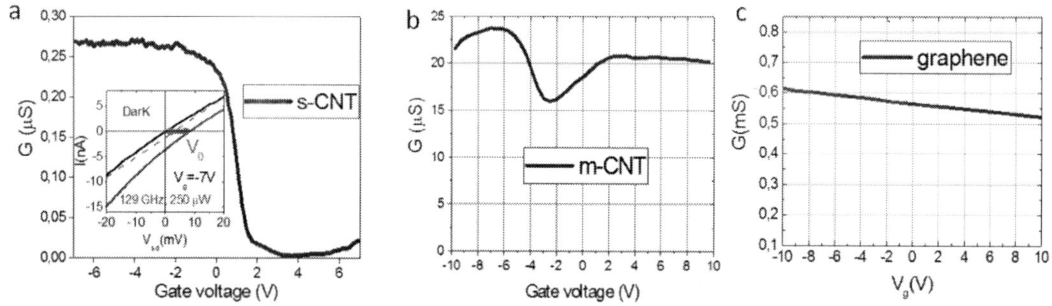

Figure 2. Transfer characteristics of semiconducting (a) , quasi-metallic (b) nanotubes and graphene (c) measured at room temperature

First, we characterize the room temperature response of our devices to radiation at the frequency of 129 GHz. For this we measured the devices IV curves with and without radiation. For all types of the devices under the radiation with a power P = 250 µW the IV curves shift to the right along the voltage axes, that clearly indicates the appearance of a DC voltage induced by the radiation equal to V_0, which we further refer to as response voltage V_{RESP}. The typical IV-curve of the s-CNT device measured at negative gate voltage with and without radiation is shown in the Figure 2.

Next, we measure the gate dependence of the response voltage $V_{RESP}(V_G)$ for all types of the devices. The results for CNT device are displayed in Fig. 3, with the responsivity defined as V_{RESP}/P, where P is a power incident on the device. Detection of the radiation takes place through combination of two mechanisms: photothermoelectric effect arises from nonuniform temperature distribution in FET channel rectification due Schottky barrier at V contact directly related to the nanotube's bandgap [4]. Figure 3 shows comparison of Seebeck coefficient (which is proportional to $G^{-1}*(dG/dV_G)$) and measured response. At negative and positive voltages measured signal isn't zero (opposite Seebeck coefficient), indicating additional contribution of diode response.

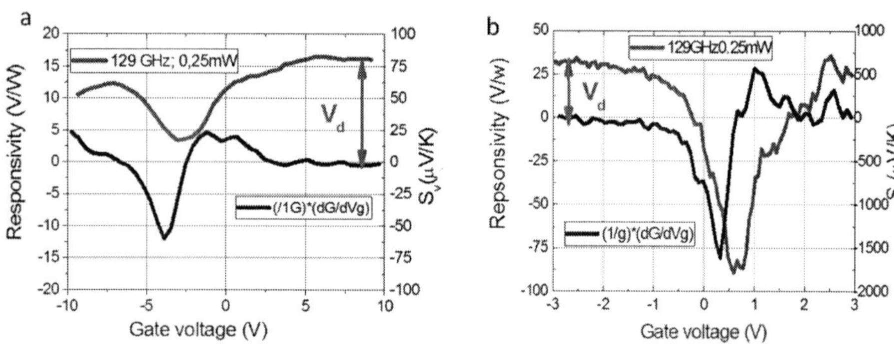

Figure 3. Seebeck coefficient and measured response as function of gate voltage for m-CNT (a) and s-CNT (b) devices.

Next measure CNT device response at different frequencies up to 2.5 THz. At room temperature our results show that semiconducting CNT (s-CNT) devices are preferable for frequencies up to 300GHz (cut off about 500 GHz), while quasi-metallic -CNT devices can be used as broad band detectors operating at frequencies up to at least 2.5 THz.

The results for graphene device are displayed in Fig. 4. As seen from the data, graphene device responsivity at 300 K and 77K is practically independent of the gate voltage. Figure 4(b) shows responsivity of graphene device at different frequencies. As seen from Fig. 4(b), at room temperature the responsivity decreases as the frequency is increased. While at 77 K the responsivity increases as the frequency increases within the range from 129 GHz to 450 GHz.

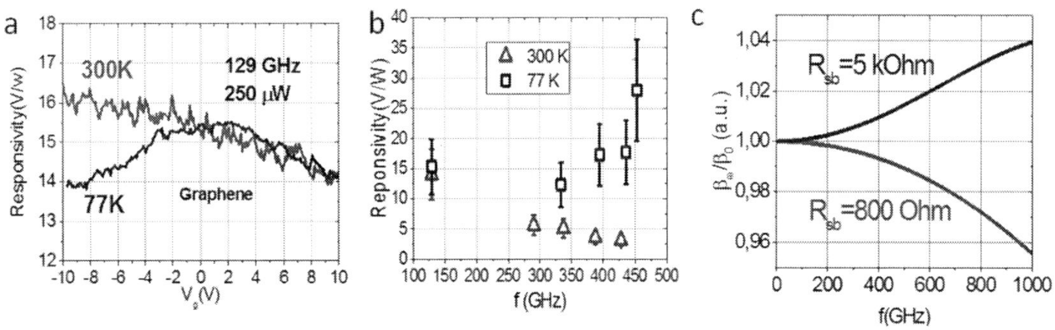

Figure 4. (a):Responsivity of graphene device as a function of gate voltage measured at 300K and 77K; (b): Responsivity of the graphene devices as a function of the radiation frequency measured at 300K and 77K; (c): Calculated frequency dependence of the rectified current normalized at its value at zero frequency [5].

The most important feature observed in our experiments is the qualitative change of the frequency dependence of the device responsivity upon a decrease of the temperature from 300 K to 77 K. To interpret obtained results we consider rectification on barrier formed on graphene/Vanadium interface for non-normally incident electrons and rake in account excitation of plasmons [5].

3. Summary

We show that at 300K semiconducting CNTs show better performance up to 300GHz with responsivity up to 100V/W, while quasi-metallic CNTs are shown to operate up to 2.5THz. At 300 K graphene detector exhibit the room-temperature responsivity from R = 15 V/W at f= 129 GHz to R = 3 V/W at f = 450 GHz. The minimum noise equivalent power, NEP, at 300K of our devices was nW/Hz$^{-0.5}$, which in order of magnitude corresponds to commercially available THz cameras [6].

We find that at low temperatures (77K) the graphene lateral Schottky diodes responsivity rises with the increasing frequency of the incident sub-THz radiation. We interpret this result as a manifestation of a plasmonic effect in the devices with the relatively long plasmonic wavelengths.

4. Acknowledgments

The research has been carried out with the support of the Russian Science Foundation (project No. 17-72-30036).

References

[1] Koppens et al, 2014, *Nat. Nanotechnol.* **9**, 780–793
[2] Vicarelli L., et al , 2012, *Nat. Mat.*, **11** 865
[3] Cai X. et al, 2014, *Nat. Nanotechnol,* **9**, 814–819
[4] Fedorov G. et al, 2018, *Phys. Status Solidi B*, **255**, 1700227
[5] Gayduchenko I. et al, 2018, *Nanotechnology*, **29**, 245204
[6] Dhillon S. S. et al 2017, *Journal of Physics D: Applied Physics,* **50**, 043001

Splitting of terahertz surface plasmons by polyimide films

V V Gerasimov[1,2], A K Nikitin[2,3], A G Lemzyakov[1], I A Azarov[2,4], I A Milekhin[2,4], B A Knyazev[1,2], E A Bezus[5,6], E A Kadomina[5] and L L Doskolovich[5,6]

[1] Budker Institute of Nuclear Physics of the Siberian Branch of the RAS, 11, Ac. Lavrentieva ave., Novosibirsk, 630090, Russia
[2] Novosibirsk State University, 2, Pirogova ave., Novosibirsk, 630090, Russia
[3] Scientific and Technological Center for Unique Instrumentation of the RAS, Moscow, 117342 Russia
[4] Rjanov Institute of Semiconductor Physics of the Siberian Branch of the RAS, 11, Ac. Lavrentieva ave., Novosibirsk, 630090, Russia
[5] Image Processing Systems Institute of RAS — Branch of the FSRC"Crystallography and Photonics" RAS, 151 Molodogvardeyskaya st., Samara, 443001, Russia
[6] Samara National Research University, 34 Moskovskoye shosse, Samara, 443086, Russia
v.v.gerasimov3@gmail.com

Abstract. Terahertz (THz) surface plasmons (SPs) can be effectively used in THz communication lines, information processing devices, spectroscopic measurements of submicron dielectric layers, and metal surface quality control. Such applications necessitate implementation of various optical schemes, including splitting of THz SPs on the guiding surface. This work is the first experimental demonstration of the fact that a polyimide kapton film of submillimetre thickness can effectively split THz SPs. Measurements of the transmission and reflection coefficients of SPs on gold-ZnS layer plane structures using monochromatic radiation of the Novosibirsk free-electron laser (λ=130 μm) have shown good accordance with theoretical calculations.

1. Introduction

Terahertz (THz) surface plasmons (SPs), a kind of surface TM-mode electromagnetic waves guided by a metal/dielectric interface, have specific features as compared with SPs of the visible and near-infrared ranges: their group velocity is very close to the speed of light, and the propagation length can be as large as tens of centimeters [1]. Due to these features, THz SPs can be effectively used in THz communication lines and information processing devices [2]. Besides, confinement and attenuation of THz SPs are sensitive to thin dielectric layers of thickness of microns to tens of nanometers deposited on the conducting surface [3]. This can be useful for spectroscopic measurements, especially for polymers and biological condensed layers in the THz range [4]. It was experimentally proved that SPs are totally reflected from plane mirrors placed on the SP guiding surface [5]. The reflection coefficient of THz SPs from plane mirrors depends on the tilt angle to the normal of the surface [6]. The width of the angular dependence is sensitive to the roughness of the surface, which can be used in metal surface quality control.

Such applications of THz SPs necessitate implementation of various optical schemes, including reflection, focusing, deflection, and splitting of THz SPs on the guiding surface. There are no papers about splitting of SPs. In this work, we present first experimental results on splitting THz SPs propagating along a metal-dielectric plane interface. The splitting is performed due to polyimide films placed on the SP track. Comparison with numerical calculations is discussed.

2. Experimental setup

As a source of terahertz radiation we used the monochromatic radiation of the Novosibirsk free electron laser, the wavelength tuned in the region from 30 to 240 μm [7]. In our experiments we used the wavelength $\lambda = 130$ μm. The SPs were launched by the end-fire coupling technique [8] on a 1/8 part of metalized (Au) glass cylinder (figure 1) with a curvature radius R=60 mm. To obtain the maximum coupling efficiency we covered the gold surface with zinc sulphite (ZnS) layer 1 μm thick. The cylinder was used for screening from parasitic bulk waves resulting from diffraction on its edge. The generated SPs proceeded from the coupling element to the adjoining plane sample. The SPs transmitted or reflected from the polyimide film were uncoupled on another cylinder set in position A or B. A polyimide film was placed on the sample at the angle θ to the SP track. The uncoupled SPs were detected with an optoacoustic detector (a GC-1D Golay cell with NEP 1.4×10^{-10} W/Hz$^{1/2}$). The detector was equipped with a metal diaphragm having a 2x10 mm rectangular slit oriented parallel to the plane of sample surface and cutting off parasitic bulk waves. The Golay cell was coupled to the lock-in amplifier SR-830 tuned to a 15 Hz chopping frequency.

The samples were 15.0×3.5×1.0 cm glass substrates metalized with a gold layer 0.3 μm thick by the magnetron sputtering technique. Optionally, the gold surface was coated with e-beam evaporated ZnS films 0-3 μm thick. As a polyimode film we used kapton 125 μm thick.

The film was sufficiently rigid, without surface bends. In the experiments we tried to make good contact between the film and sample surface, but it was not ideal. The gap could be 40-200 μm in each measurement.

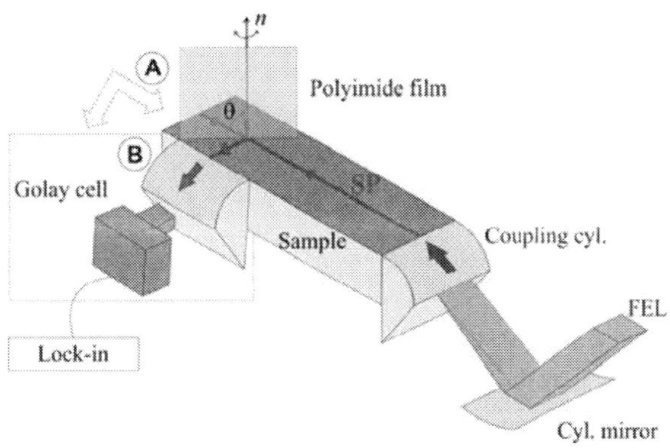

Figure 1. Experimental schema.

3. Experimental results and comparison with theory

The coefficient of SP transmission through the kapton film T vs. the incident angle θ was measured with the detector placed in position A (figure 1). The results for a sample with an Au-ZnS layer 0.45 μm thick are shown in figure 2 (a). On the same graph, the theoretical calculations of SP transmission are plotted (blue line). For calculations we used an in-house implementation of the aperiodic modification of the rigorous coupled-wave analysis (aRCWA) technique [9-10]. RCWA, also called the Fourier modal method, is an established numerical technique for solving Maxwell equations. In calculations we used dielectric permittivity values in THz range for the gold surface, $\varepsilon_{Au}\approx$-530+i*9010 (measured in our recent experiments with surface plasmon polaritons propagating along a bare gold surface), for ZnS, $\varepsilon_{Zns}\approx$8.7+i*0.059, and for kapton, $\varepsilon_{kap}\approx$3.5+i*0.13 film (measured with the Fourier-spectrometer Bruker Vertex 80 V).

The experimental and theoretical dependencies are in good agreement: with increasing θ the transmission gradually decreased. Additionally, on the same graph (figure 2, (a)) we plotted the transmission of s-polarised plane bulk wave through the kapton film (red points). The wave vector of the SPs was given as the wave vector of a plane wave. Both dependences almost coincide. For measuring the SP reflection coefficient R we moved the detector into position B (figure 1). Obtaining the angular dependence $R(\theta)$ using such schema is very difficult, and therefore we measured the reflection only for one angle $\theta = 45°$, which yielded $R = 0.28 \pm 0.03$. To compare this result with the theory we calculated $R(\theta)$ for different gaps between the kapton film and the sample surface (g = 0, λ/2, λ), see figure 2, (b). The gaps were introduced for taking the non-ideal contact between them into account. The theoretical dependencies show good agreement with the experiment. Adding the coefficients at $\theta = 45°$, it can be found $R+T \approx 0.8$, which is less than unity. It corresponds to the diffraction losses of SPs arisen from the non-ideal contact.

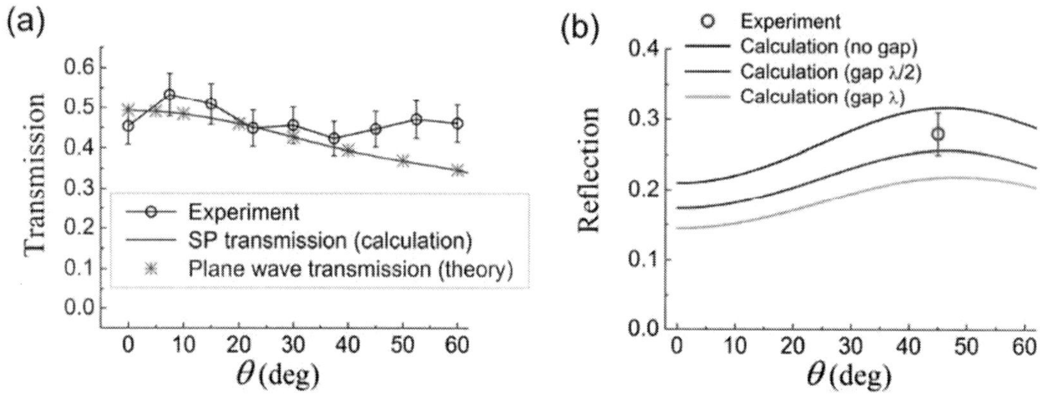

Figure 2. Experiments with sample with Au-ZnS layer 0.45 μm thick: (a) coefficient of SP transmission through kapton polyimide film T vs. incident angle θ (black circles: experiment, blue line: theoretical calculations; red points: transmission of plane wave); (b) reflection coefficient R vs. θ (red circle: experiment for $\theta = 45$; theoretical calculations for gap between kapton film and sample surface g = 0, λ/2, λ).

In addition, we measured the SP transmission coefficient for different ZnS thicknesses at $\theta = 45°$ (see figure 3). In the experiments the signal to noise ratio was low because of the high attenuation of SPs along the track. Comparison of the experimental results with the theoretical calculations showed qualitative agreement: with increasing ZnS thickness d, the transmission slightly decreased.

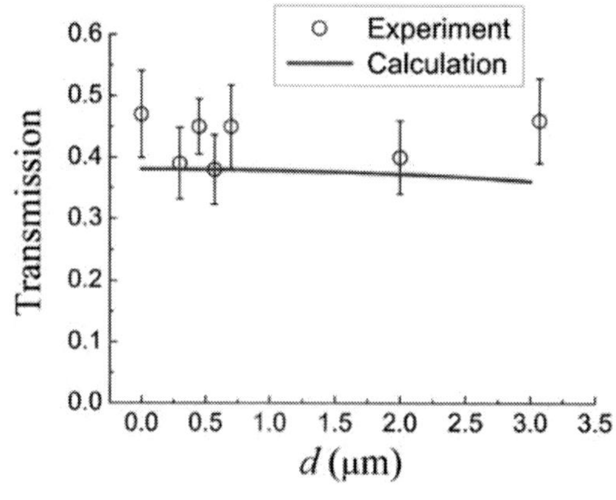

Figure 3. Coefficient of SP transmission through kapton polyimide film T vs. ZnS thickness d at $\theta = 45°$ (black circles: experiment; red line: theoretical calculations).

4. Summary

It has been demonstrated experimentally that a polyimide kapton film with submillimetre thickness can split THz surface plasmons (SPs). The coefficients of transmission T and reflection R of SPs on gold-ZnS layer plane structures were in a good accordance with the theoretical calculations. Besides, T and R almost coincide with the coefficients of plane bulk waves and depend on the gap between the kapton film and sample surface. The transmission of SPs slightly decreased when the gold was coated with the ZnS dielectric film of submicron thickness. Thus, rigid polyimide films can be used effectively as SP splitters in surface plasmon optical schemes.

5. References

[1] Jeon T-I and Grischkowsky D 2006 THz Zenneck surface wave (THz surface plasmon) propagation on a metal sheet *Appl. Phys. Lett.* **88** 061113
[2] Maier S A 2007 *Plasmonics: Fundamentals and Applications* (Springer Verlag, New York)
[3] Gerasimov V V, Knyazev B A, Lemzyakov A G, Nikitin A K and Zhizhin G N 2016 Growth of terahertz surface plasmon propagation length due to thin-layer dielectric coating *Journal of Optical Society of America B* **33** (11) 2196-2203
[4] Siegel P H 2004 Terahertz Technology in Biology and Medicine *IEEE Tranzactions on microwave theory and techniques* **52** (10) 2438-2447
[5] Gerasimov V V, Knyazev B A, Nikitin A K 2017 Reflection of terahertz monochromatic surface plasmon polaritons by a plane mirror *Quantum Electronics* **47** 65-70
[6] Gerasimov V V, Knyazev B A, Nikitin A K 2016 Wave-vector spectrum of monochromatic terahertz surface plasmon polaritons on real surfaces *Physics Procedia* **84** 157 – 164
[7] Kulipanov G N, *et al.* 2015 Novosibirsk free electron laser—facility description and recent experiments *IEEE Trans. Terahertz Sci. Technol.* **5** 798
[8] Stegeman G I, Wallis R F, Maradudin A A 1983 Excitation of surface polaritons by end-fire coupling *Optics Letters* **8**(7) 386-388
[9] Silberstein E, Lalanne P, Hugonin J-P and Cao Q 2001 Use of grating theories in integrated optics *J. Opt. Soc. Am. A* **18** 2865–2875
[10] Hugonin J P and Lalanne P 2005 Perfectly matched layers as nonlinear coordinate transforms: a generalized formalization *J. Opt. Soc. Am. A* **22** 1844–1849

Acknowledgments

The experiments were made using the equipment of The Siberian Synchrotron and Terahertz Radiation Center supported by The Ministry of Education and Science of The Russian Federation (project RFMEFI62117X0012) and by the Russian Science Foundation (grant No. 14-50-00080). Numerical calculations were supported by the Russian Foundation for Basic Research (project no. 16-29-11683). Authors thank the team of the Novosibirsk free electron laser for assistance in the experiments.

Using submicron dielectric coatings to reduce terahertz surface plasmon losses on metals

V V Gerasimov[1,2], A K Nikitin[2,3], B A Knyazev[1,2], A G Lemzyakov[1] and I A Azarov[2,4]

[1] Budker Institute of Nuclear Physics of the Siberian Branch of the RAS, 11, Ac. Lavrentieva ave., Novosibirsk, Russia, 630090

[2] Novosibirsk State University, 2, Pirogova ave., Novosibirsk, Russia, 630090

[3] Scientific and Technological Center for Unique Instrumentation of the RAS, 15, Butlerova str., Moscow, Russia, 117342

[4] Rjanov Institute of Semiconductor Physics of the Siberian Branch of the RAS, 11, Ac. Lavrentieva ave., Novosibirsk, Russia, 630090

v.v.gerasimov3@gmail.com

Abstract. Terahertz (THz) surface plasmons (SPs) are very promising for developing transmission lines and new photonic devices. Energy losses during propagation of SPs plays important role for these applications. We experimentally studied attenuation of SPs along plane and convex (curvature radius of 60 mm) gold surfaces coated with submicron dielectric (ZnS) layers using monochromatic radiation (λ=130 μm) of the Novosibirsk free-electron laser. The optimal dielectric thicknesses d^*, corresponding to the minimum energy losses of SPs, were found. The SP attenuation coefficients α were about 100 times higher for convex surfaces than for plane surfaces, which means intensive radiative losses on the surface bend. A ZnS coating with $d_1^*\approx 0.025$ μm on a plane gold surface reduced α by 20 % as compared with bare gold; coating with $d_2^*\approx 1$ μm on convex surface, three times.

1. Introduction

Currently, studies in the field of terahertz (THz) plasmonics are of great interest because using THz surface plasmons (SPs) is promising for development of transmission lines and new photonic devices [1]. Transmission and processing of information may be several orders faster in THz plasmonic devices as compared with the visible region and can excel the characteristics of existing technologies [2]. Besides, THz SPs propagate on plane metals over tens of centimeters [3], which is much longer than in the visible and near-infrared ranges [4].

Functioning of photonic devices depends much on the SP energy losses because of propagation of SPs along the conducting surface. It is well known that in the THz range the SP attenuation coefficient is about two orders less than the Joule losses calculated by the Drude model of metal conductivity [5-8]. There are two main reasons for this discrepancy. The most probable cause is the difference in the dielectric properties of real metal surface and respective bulk crystal [9]. Another explanation of the great attenuation of THz SPs can be the "radiative" losses of SPs on the roughness and intrinsic inhomogeneity of the surface. The characteristic dimensions of optical surface roughness are usually much less than the THz radiation wavelength. According to the theory [10], it means that the radiative losses (RLs) of SPs are expected to be negligible. However, in recent experiments we found [11] that RLs in the far infrared region could be significant. Also, in the last theoretical paper [12] the analytical solution for the propagation of SPs along a 2-D rough metal surface was obtained, from which it fallows that there is a roughness-induced energy flux from the surface reradiated into a free space. Estimated calculations showed that a 30 nm roughness can reduce SP propagation length in two times. It can be understood qualitatively if we take into account the small difference between the THz SP wave vector and that of plane waves in the free space, which may lead to intensive SP conversion into bulk waves even on tiny surface imperfections. The role of RLs is even greater on curved surfaces, where it depends on the curvature radius [13]. If a thin dielectric film coats a metal surface, the SP wave vector is increases, leading to the transformation of a surface wave on the imperfections and bends less possible, and as a result decreases the RLs. On the other hand, it is well known that a dielectric coating raises the SP Joule losses due to increase in the field portion of SPs carried inside the metal [14]. These two competitive processes can lead to existence of optimal dielectric thickness, corresponding to minimum total losses of SPs.

In this work, we present experimental results of measuring the energy losses of terahertz SPs propagating along plane and convex metal surfaces coated with submicron dielectric layers. Comparison with theoretical predictions is discussed.

2. Experiments on plane surfaces

As a source of terahertz radiation we used Novosibirsk free electron laser generated monochromatic radiation with the wavelength tuning in the region from 30 to 240 μm [15]. In our experiments we used the wavelength $\lambda = 130$ μm. SPs were launched by end-fire coupling technique [16] on 1/8 part of metalized (Au) glass cylinder (figure 1, (a)) with curvature radius R=60 mm. To obtain maximum coupling efficiency the gold surface was covered with zinc sulphite (ZnS) layer 1 μm thick. The cylinder was used to screen from parasitic bulk waves produced from diffraction on its edge. The generated SPs proceeded from the coupling element to the plane sample adjoining it. Samples were 100x120 mm size glass substrates metalized with a gold layer 0.3 μm thick using magnetron sputtering technique. Optionally the gold surface was coated with 0-1.1 μm thick ZnS films by e-beam evaporation. To change the path l which SPs run on the sample surface we used fixed and movable gold mirrors. SPs were uncoupled on the second cylinder and detected to the edge with the optoacoustic detector (a GC-1D Golay cell with NEP 1.4×10^{-10} W/Hz$^{1/2}$). The detector was equipped with metal diaphragm having a 2x10 mm rectangular slit (oriented parallel to the plane of sample surface), cutting off parasitic bulk-waves. The Golay cell was coupled to a lock-in amplifier SR-830 tuned at a 15 Hz chopping frequency.

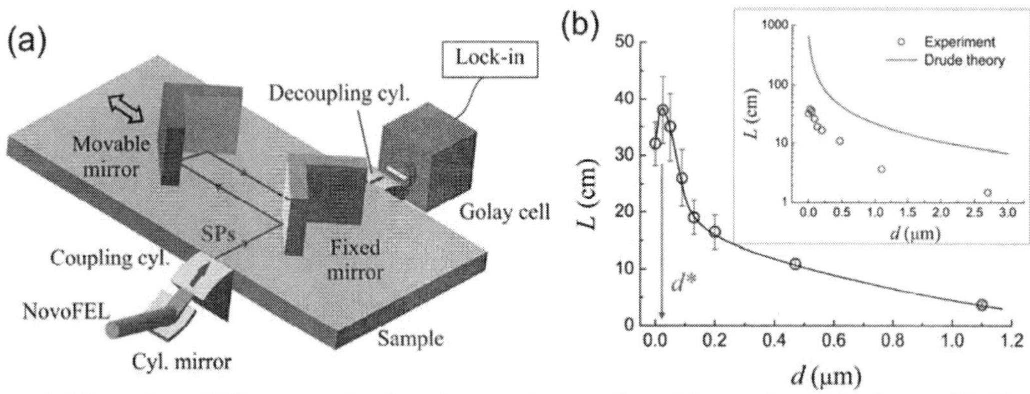

Figure 1. Measuring of SP propagation length on a plane surface: (a) experimental schema; (b) SP propagation length L vs. ZnS layer thickness d. In the inset – the comparison with Drude theory.

Using a motorized translation stage attached to the movable mirror, we scanned the SP intensity I vs. the run distance l on samples with different ZnS layer thicknesses d. Fitting I (l) with exponential decay functions we obtained SP propagation lengths L (the distance in which the intensity decreases by a factor of e). The results are shown in figure 1, (b). The propagation length increases with d, reaches the maximum at $d^* \approx 0.025$ μm, and gradually decreases. The maximum corresponds to minimum energy losses of SPs. The attenuation coefficient of SPs $\alpha = 1/L$ for d^* is about 20 % less than that for a bare gold surface. The existence of optimal dielectric thickness confirms our prediction about the existence of two competitive processes: the dielectric coating reduces the SP radiative losses and increases the Joule losses [11]. Note that the experimental value of the propagation length for bare gold ($L \approx 32$ cm) is about 20 times less than the Drude model predicts ($L_{Drude} \approx 660$ cm), see the inset in Figure 1b. This discrepancy gradually decreases as the ZnS layer thickness d goes up. One can assume that the main cause of this disagreement is the difference in the dielectric properties of a real metal surface and a respective bulk crystal. Taking for bare metal $L \approx 32$ cm and the decay length of SPs into the air ($\delta \approx 1.3$ mm [7]), one can obtain the dielectric permittivity of gold surface $\varepsilon Au \approx -7987+i*10036$. This value is significantly less than the Drude model predicts for bulk gold ($\varepsilon Drude \approx -101640+i*284090$ at $\lambda = 130$ μm).

3. Experiments on convex surfaces

We tested propagation of SPs along a cylindrical surface with a curvature radius R=60 mm (figure 2, (a)). A cylindrical mirror launched the SPs on the edge of the sample. The radiative losses emitted from the SP track were detected with a TPX lens and a Golay cell placed on an optical rail in a $2f$-$2f$ arrangement, where f=100 mm is the focal length. The intensity of SP radiative losses I_{rad} is proportional to the intensity of SPs I_{SP} and is expected to decay along the track as $I_{rad} \sim I_{SP} \sim \exp(-\alpha \cdot R \cdot \theta)$. Basing on this assumption, we recorded I_{rad} vs. deflection angle θ for different thicknesses d of the ZnS layer and found the propagation lengths L of the SPs (figure 2, (b)). The dependence $L(d)$ has a maximum at $d^* = 1$ μm, $L_{max} \approx 0.8$ cm, which is about 3 times larger than for a bare gold convex surface. Compared with plane samples, the attenuation coefficients of SPs on convex samples α are about 100 times larger due to high radiative losses on the surface band, and minimum energy losses take place at about 40 times higher dielectric thickness d^*.

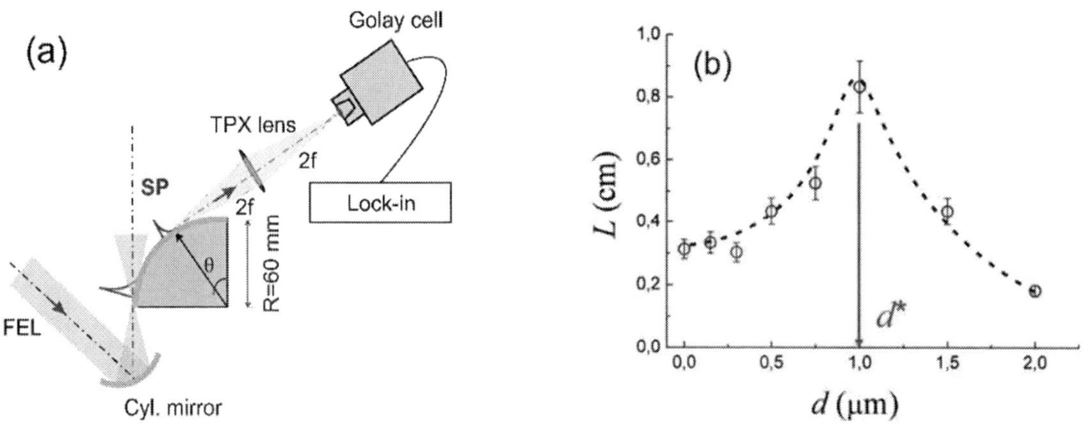

Figure 2. Measuring of SP radiative losses on a convex surface: (a) experimental schema; (b) $L(d)$.

Summary
We measured propagation lenses L of terahertz SPs along plane and convex (R=60 mm) gold surfaces coated with submicron dielectric (ZnS) layers. For both geometries the optimal dielectric thicknesses d^* corresponding maximal L (minimal energy losses) of SPs were observed. For convex surfaces the SP attenuation coefficients $\alpha = 1/L$ were about 100 times higher and d^* was 40 times larger, then for plane surfaces, corresponding to intensive radiative losses on the surface bend. Thus, a thin submicron dielectric coating can significantly reduce energy losses of THz SPs, which is very promising for developing of THz transmission lines and new photonic devices.

References

[1] Atwater H A 2006 The promise of plasmonics *Scintific American* **296** 56–63
[2] MacDonald K F, *et al.* 2008 Ultrafast active plasmonics *Nat. Photonics* **3** 55–58
[3] Jeon T-I and Grischkowsky D 2006 THz Zenneck surface wave (THz surface plasmon) propagation on a metal sheet *Appl. Phys. Lett.* **88** 061113
[4] Suárez A, Ferrando J, Marques-Hueso A, Díez R, Abargues R, Rodríguez-Cantó P J and Martínez-Pastor J P *Nanophotonics* **6**(5) 1109–1120
[5] Koteles E S and McNeill W H 1981 Far infrared surface plasmon propagation *Intern. J. Infrared Millim. Waves* **2** 361-371
[6] Schlesinger Z, Webb B C and Sievers A J 1981 Attenuation and coupling of far infrared surface plasmons *Sol. St. Comm.* **39** 1035-1039
[7] Gerasimov V V, Knyazev B A, Nikitin A K and Zhizhin G N 2011 A way to determine permittivity of real metal surfaces at terahertz frequencies *Appl. Phys. Letters* **98** 171912
[8] Gerasimov V V, Cherkassky V S, Knyazev B A, Kulipanov G N, Kotelnikov I A, Nikitin A K and Zhizhin G N 2013 Surface plasmon polaritons launched using a terahertz free electron laser: propagating along a gold-ZnS-air interface and decoupling to free waves at the surface tail end *Journal of Optical Society of America B* **30** 2182-2190
[9] Pandey S, Gupta B, Chanana A and Nahata A 2016 Non-Drude like behavior of metals in the terahertz spec tral range *Advances in Physics: X* **1** 176-193
[10] Raether H 1988 Surface plasmons on smooth and rough surfaces and on gratings *Springer Tracts in Modern Physics* **111** (Springer-Verlag)
[11] Gerasimov V V, Knyazev B A, Lemzyakov A G, Nikitin A K and Zhizhin G N 2016 Growth of terahertz surface plasmon propagation length due to thin-layer dielectric coating *Journal of Optical Society of America B* **33** (11) 2196-2203
[12] Kotelnikov I and Stupakov G 2016 Dispersion relation of a surface wave at a rough metal-air interface *Physical Review A* **94** 053847
[13] Hasegawa K, Noeckel J U, Deutsch M 2004 Surface plasmon polariton propagation around bends at a metal-dielectric interface *Appl. Phys. Lett.* **84** 1835
[14] Schlesinger Z and Sievers A J 1982 IR surface-plasmon attenuation coefficients for *Ge*-coated *Ag* and *Au* metals *Phys. Rev. (B)* **26**(12) 6444-54
[15] Kulipanov G N, *et al.* 2015 Novosibirsk free electron laser—facility description and recent experiments *IEEE Trans. Terahertz Sci. Technol.* **5** 798
[16] Stegeman G I, Wallis R F, Maradudin A A 1983 Excitation of surface polaritons by end-fire coupling *Optics Letters* **8**(7) 386-388

Acknowledgments

The experiments were made using the equipment of The Siberian Synchrotron and Terahertz Radiation Center supported by The Ministry of Education and Science of The Russian Federation (project RFMEFI62117X0012) and by the Russian Science Foundation (grant No. 14-50-00080). Authors thank the team of the Novosibirsk free electron laser for assistance in the experiments.

An analytically based computer model for evaluating effective elastic properties of nitride nanowire-polymer compositions using laser-generated surface acoustic waves

Evgeny Glushkov[1], Natalia Glushkova[1] and Olga Boyko[2]

[1] Kuban State University, Institute for Mathematics, Mechanics and Informatics,350080 Krasnodar, Russia

[2] Sorbonne University, Paris Institute of Nanosciences, 75252 Paris, France

E-mail: evg@math.kubsu.ru, Olga.Boyko@insp.jussieu.fr

Abstract. The investigation of physical and mechanical properties of nitride nanowires (NWs) is a topical field of modern nanotechnological research due to a wide range of promising applications. Normally, the NWs are embedded in a host polymer (e.g., hydrogen silsesquioxane – HSQ), and it becomes necessary to determine the effective properties of polymer-NW compositions rather than of separate wires. The central idea of the present research is the evaluation of the effective elastic moduli of the anisotropic composite HSQ-NW material via the minimization of the goal function that specifies the discrepancy between the measured and calculated characteristics of laser-generated surface acoustic waves. The experimental data are obtained on the basis of transient grating spectroscopy while the theoretical dispersion curves are calculated using the analytically based computer model for elastic guided wave excitation and propagation in multilayered elastic anisotropic half-spaces. The numerical examples illustrate the validation of the developed computer model against the experimental data acquired for sandwich microstructures with known elastic moduli as well as demonstrate the restoration of the latter in the case of HSQ-NW compositions with unknown mechanical properties.

1. Introduction

For many years the materials of the III-Nitrides group (GaN, AlN, InGaN, and others) have attracted the attention of researchers due to their exceptional properties that are very suitable for modern electronic and optoelectronic applications. From the early 2000s, the use of these materials in the form of nitride nanowires (NWs) has been intensively studied [1,2] due to their remarkable piezoelectric properties, which are very suitable for novel ultra-compact and efficient piezoelectric generators providing energy harvesting and autonomous operation of microelectronic devices. A review of research from single III-nitride nanowires to piezoelectric generators can be found in Ref. [3].

The electric output of a single piezoelectric NW depends on the law of its deformation in response to mechanical loading. Therefore, to maximize the output and optimize the design, a study of such laws is highly desirable. Meanwhile, the NWs are normally used in the form of arrays embedded in a host polymer (e.g., in hydrogen silsesquioxane – HSQ [4]), and the shape of their deformation is determined by the elastic properties of the polymer-NW composition as a whole. This fact was the incentive for the undertaken research aimed at developing non-destructive methods for evaluating elastic moduli of anisotropic HSQ-NW composites via the measurement of laser-generated surface acoustic waves (SAWs).

The preparation of the samples, the measurement procedure, and the experimental results are described in separate paper [5] submitted to the METANANO 2018 in parallel with this article. Here we present the mathematical and computer model used to restore sample parameters based on the measured data. To simulate the SAWs, we have used the integral transform technique and the Green's matrix approach yielding explicit integral and asymptotic representations for the wave fields generated in anisotropic multilayered structures by a given surface load [6,7]. Earlier, the evaluation of effective elastic properties has been performed for layered composite fiber-reinforced plastic plates using piezoelectrically induced guided waves and laser Doppler vibrometry [8].

2. Physical and mathematical framework

The experimental samples are sandwich compositions with an HSQ-NW interlayer of about micron thickness deposited on a silicon (Si[111]) substrate (half-space) and covered by a thin metallic layer which provides an electrical contact. The nanowires embedded into the HSQ polymer are vertically oriented across the interlayer, closing the circuit between the coating and the substrate. Thus, the effective elastic properties of the HSQ-NW composition can be simulated by a transversally isotropic medium with the horizontal plane of isotropy (x, y). Such materials are specified by five independent elastic moduli C_{ij} and density ρ.

To simulate the frequency spectra of the laser-generated SAWs, the boundary value problem (BVP) is formulated with respect to the complex displacement amplitude $\mathbf{u} = (u_1, u_2, u_3)$ of the time-harmonic oscillation $\mathbf{u}(\mathbf{x})e^{-i\omega t}$; $\omega = 2\pi f$ is angular frequency, f is frequency, $\mathbf{x} = (x, y, z)$ is a point in the Cartesian coordinate system; plane (x, y) and axis z are aligned with the sample's surface and the outward normal.

The solution to the BVP is obtained in terms of the Green's matrix $k(\mathbf{x})$ and the vector of load $\mathbf{q}$ applied to a surface area Ω, or equivalently, via their Fourier symbols $K(\alpha_1, \alpha_2, z) = \mathcal{F}_{xy}[k]$ and $\mathbf{Q}(\alpha_1, \alpha_2) = \mathcal{F}_{xy}[\mathbf{q}]$ [7-9]:

$$\mathbf{u}(\mathbf{x}) = \iint_{\Omega} k(x-\xi, y-\eta, z)\mathbf{q}(\xi, \eta)d\xi d\eta = \frac{1}{(2\pi)^2}\iint_{-\infty}^{\infty} K(\alpha_1, \alpha_2, z)\mathbf{Q}(\alpha_1, \alpha_2)e^{-i(\alpha_1 x+\alpha_2 y)}d\alpha_1\alpha_2.$$

Here $\mathcal{F}_{xy}$ is the Fourier transform with respect to x and y variables; the Fourier parameters α_1 and α_2 play the role of wavenumbers for the waves propagating along the (x, y) surface. The residues from the poles ζ_n of the matrix K elements describe traveling SAWs generated by the source load $\mathbf{q}$ [6,7].

3. Validation

To verify the applicability of the theoretical model developed, first, the samples with known material parameters have been chosen for the theory-to-experiment comparisons. Figure 1 gives an example of such comparisons for the test sample Ni/HSQ/Si with a rather thin nickel coating (the thickness $h_1 = 0.015$ μm) and an HSQ interlayer without NWs of thickness $h_2 = 1$ μm. The material parameters for all examples are collected in Table 1. Here and below we use the experimental data obtained by the authors of adjoined contribution [5]. In all figures, the experimental results are shown by circle markers.

The coloured level-line plot in Fig. 1, left depicts a surface plot of $|K_{33}|$ as a function of the wavelength $\lambda = 2\pi/\alpha_1$ in the x direction ($\alpha_2 = 0$) and frequency f. The element K_{33} of the matrix K describes the vertical displacement response of the multilayered substructure on a normal elementary loading (it is enough for laser generation and measurement). The areas of

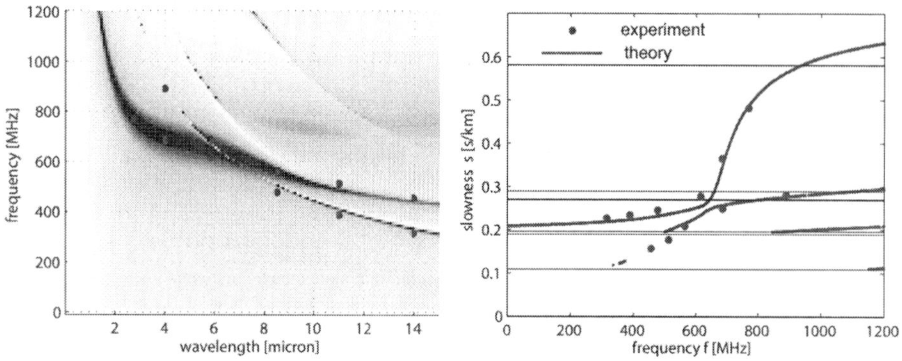

Figure 1. Test sample Ni/HSQ/Si: level-line surface $|K_{33}(\lambda, f)|$ (left) and SAW dispersion curves (right).

K_{33} maximal amplitudes correspond to the parameters of dominate SAWs. They yield peaks in the frequency spectra detected in experimental measurements for fixed wavelengths λ.

The ridges of K_{33} amplitude go along the lines specified by its real and nearly real poles $\alpha_1 = \zeta_n(f)$ that determine the dispersion characteristics of the traveling SAWs and leaky guided waves. The right subplot of Fig. 1 shows the corresponding dispersion curves in the frequency-slowness plane. The slowness $s = \alpha_1/\omega = 1/(\lambda f)$ is convenient for the analysis since its magnitude varies in a limited range specified by the frequency-independent body-wave slownesses of layer materials shown in the figures by thin horizontal lines.

4. Preliminary results

In real applications, a platinum coating is more suitable. However, the transition from the test sample to a Pt/HSQ/Si specimens with ten times thicker coatings ($h_1 = 0.15$ μm) considerably changed both theoretical curves and experimental markers that cease to be in agreement (e.g., Fig. 2, left). Nevertheless, via the variation by the HSQ polymer thickness h_2, which also could not be precisely controlled in the course of sample fabrication, it was achieved a match with the same material parameters as in the previous example and unexpectedly small thickness $h_2 = 0.25$ μm (Fig. 2, right).

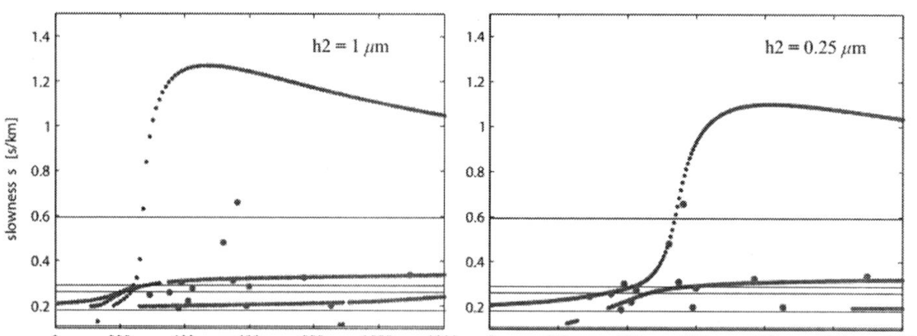

Figure 2. Slowness dispersion curves for Pt/HSQ/Si samples with the same HSQ properties as in the example above and different thicknesses: $h_2 = 1$ μm (left) and $h_2 = 0.25$ μm (right).

The presence of nanowires complicates elastic properties of HSQ-NW composites. It turned out hard to select them manually based on seemingly realistic guesses, even accounting for the

Table 1. Elastic moduli C_{ij} [GPa] and density ρ [10^3 kg/m^3]

	C_{11}	C_{33}	C_{12}	C_{13}	C_{44}	C_{66}	ρ
Ni	247	247	153	153	122	122	8.9
Pt	315	315	193	193	61	61	21.5
HSQ	11.89	11.89	5.95	5.95	2.97	2.97	1.0
HSQ-NW, guess	7.37	27.39	3.58	3.61	3.00	1.90	1.9
HSQ-NW, restored	20.77	105.50	10.08	0.	10.41	5.35	1.77
Si[111]	195	205	54.3	44.7	60.7	70.3	2.331

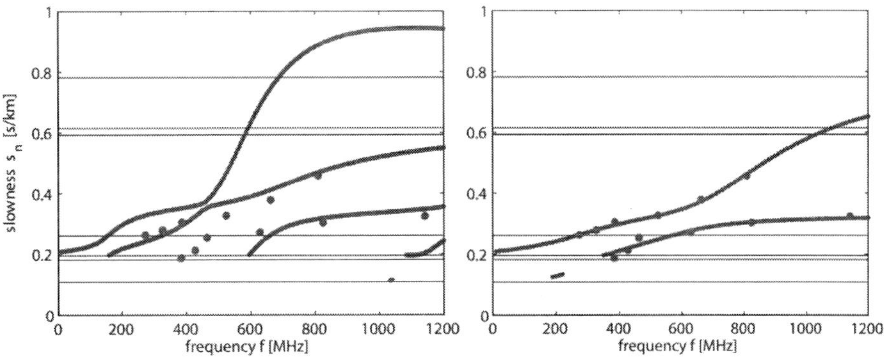

Figure 3. Nanowire specimen Pt/HSQ-NW/Si: guessed HSQ-NW parameters (left) and the adjusted ones (right).

known moduli of the NW material (GaN in the case, Fig. 3, left). To fit the curves to the experimental markers, we have minimized the goal function, which specifies their deviation, by varying the five independent elastic moduli C_{ij}, the density ρ, and the thickness h_2 of the HSQ-NW interlayer. Figure 3, right gives an example of such a fit reached at the set of HSQ-NW parameters shown in Table 1; the guessed and adjusted thicknesses are $h_2 = 1.1$ μm and $h_2 = 0.89$ μm, respectively.

Acknowledgments
The work is supported by the Russian Science Foundation (Project No. 17-11-01191).

5. References

[1] Qian F, Gradecak S, Li Y, Wen C Y and Lieber C M 2005 *Nano Letters* **5**(11) 2287
[2] Tchernycheva M, Lavenus P, Zhang H, Babichev A V, Jacopin G, Shahmohammadi M, Julien F H, Ciechonski R, Vescovi G. and Kryliouk O 2014 *Nano letters*, **14**(5) 2456
[3] Gogneau N, Jamond N, Chrétien P, Houzé F, Lefeuvre E and Tchernycheva M 2016 *Semicond. Sci. Technol.* **31**(10), 103002
[4] Liou H C, Pretzer J 1998 *Thin Solid Films* **335** 186
[5] O. Boyko, B. Bonello, E. Charron, L. Lu, M. Tchernycheva, N. Gogneau, 2018 Conf. Proc. METANANO (submitted).
[6] Glushkov E, Glushkova N and Eremin A 2011 *J. Acoust. Soc. Am.* **129** (5) 2923
[7] Glushkov E, Glushkova N and Zhang C 2012 *J. Appl. Phys.* **112** 064911
[8] Eremin A A, Glushkov E V, Glushkova N V and Lammering R 2015 *Composite Structures* **125** 449

Metamaterial filter design via surrogate optimization

Andrea Bacigalupo and Giorgio Gnecco

MUSAM and AXES Research Units
IMT School for Advanced Studies - Piazza S. Francesco, 19 - 55100 Lucca, Italy

{andrea.bacigalupo,giorgio.gnecco}@imtlucca.it

Abstract. Recently, an increasing research effort has been dedicated to analyse transmission and dispersion properties of periodic metamaterials containing resonators, and to optimize the amplitude of selected acoustic band gaps between consecutive dispersion curves in the Floquet-Bloch spectrum. Potential novel applications of this research are in the design of passive mechanical filters/diodes. The present work proposes a way to interpolate the objective functions in such band gap optimization problems, using Radial Basis Functions. The study is motivated by the high computational effort often needed for an exact evaluation of the original objective functions, when using iterative optimization algorithms. By replacing such functions with surrogate objective functions, well-performing suboptimal solutions can be obtained with a small computational effort. Numerical results demonstrate the feasibility of the approach.

1. Introduction

An increasing research effort has been recently dedicated to the analysis of transmission and dispersion properties of elastic waves propagating in periodic metamaterials containing one or more resonators per elementary cell [1,6]. These have demonstrated several improvements with respect to the associated periodic materials, which do not include the resonators. If properly designed, the geometrical and mechanical parameters of such metamaterials may allow the enhancement of specific properties. For instance, challenging issues arise in the optimization of properties of the dispersion curves for specific aims, such as opening, enlarging, closing or shifting band gaps in target acoustic frequency ranges, with potential novel applications like the design of fully customizable passive mechanical filters/diodes.

A promising approach to metamaterial design is based on formulating and solving suitable constrained nonlinear optimization problems, where the vector of optimization variables is made of design parameters such as the mass densities and radii of the resonators, the nonlinear constraints are of both mechanical and geometrical nature, and the nonlinear objective functions model any desirable property of the metamaterial, such as the amplitude of the band gap between two specific consecutive dispersion curves [2,3,4]. Unfortunately, in several cases, the application of classical iterative optimization algorithms [5] to solve such optimization problems is not easy since, at each iteration, the evaluation of the objective function can be computationally expensive (e.g., for each such evaluation, one may need to solve a sequence of eigenvalue subproblems, one for each value of the abscissa of the dispersion curves). Moreover, the associated computational effort typically increases with respect to the number of resonators per unit-cell, and to the number of design parameters to optimize [4].

By replacing the original objective functions with more-easily computable functions, surrogate optimization [10,15] can help in a first phase of the optimization, providing quickly good suboptimal solutions, to be possibly re-optimized locally in a second optimization phase, turning back to the original objective functions. In the paper, focus is made on the application of surrogate optimization based on Gaussian Radial Basis Function (RBF) interpolation to the optimization of the filtering properties of the tetrachiral periodic metamaterial considered in the recent work [2].

The paper is organized as follows. Section 2 motivates the need for interpolation of objective functions in band gap optimization problems. Section 3 describes the solution approach adopted here, based on the combination of Gaussian RBF interpolation with an iterative optimization algorithm and a quasi-Monte Carlo multi-start technique. Section 4 reports related numerical results and discusses them.

2. Need for objective function interpolation in band gap optimization problems

Each of the constrained nonlinear optimization problems considered in [2] consists in maximizing the (normalized) amplitude of the band gap between a specific pair of consecutive dispersion curves in the acoustic part of the Floquet-Bloch spectrum. Due to space constraints, the reader is referred to [2, Section 2 and Appendix] for details about the precise physical-mathematical model, which describes a tetrachiral periodic metamaterial. Here only the following computational issues are recalled, which arise from trying to solve numerically the optimization problems above (similar issues arise for other models, e.g., the ones considered in [3,4]). For each such problem, and for any given choice of the vector of design variables, one needs the next steps to compute the associated value of the band gap:

a) determination of the coefficients of a secular equation (obtained, e.g., by evaluation of a determinant of a suitable square matrix), and computation of the (square) roots of such a secular equation. This has to be repeated for every choice of the dimensionless wave vector on the (discretized) closed boundary of the irreducible Brillouin zone, which parameterizes the square matrix itself, together with the vector of design parameters;

b) determination of maximum/minimum values of two specific consecutive dispersion curves, and their comparison.

In the above, the computational effort is dominated by step a), and increases with the dimension of the matrix involved, and with the refinement of the discretization. It is worth mentioning that a symbolic computation of the coefficients of the secular equation does not help to reduce this effort, unless the square matrix is sparse (which is usually not the case), allowing in that case a significant reduction in the number of terms appearing in the resulting symbolic expressions. Otherwise, a numeric computation of the coefficients of the secular equation for each choice of the vector of design parameters is preferable. Hence, the only way to decrease significantly the computational effort consists in reducing the number of objective function evaluations using, e.g., function interpolation.

3. Proposed solution approach

To reduce the computational effort when solving numerically each band gap optimization problem, the following surrogate optimization approach is considered in this paper. Namely, the original objective function is evaluated exactly on a finite subset of the domain to which the vector of design variables belongs (training set), then the resulting values are interpolated. Hence, the band gap optimization is performed using not the original objective function, but its interpolant. The following choices characterize the approach considered here, and distinguish it from other possible approaches based on alternative machine-learning techniques (e.g., Support Vector Machine regression [7, Section 6.2]):

a) function interpolation is used instead of function approximation, since the objective function is known exactly at the training points;

b) for the interpolation, a mesh-free method is used, based on N_c strictly positive-definite RBF computational units [8], and on a quasi-Monte Carlo discretization [11] of the domain of design variables for the selection of the centers of the RBFs (including in the training set only points generated by a quasi-Monte Carlo sequence that also satisfy all the constraints of the original optimization problem). This choice is justified by the nice approximation error guarantees (expressed in the supremum norm) available for such an interpolation scheme (see, e.g., [8, Theorem 14.5]), when the objective function is sufficiently smooth, and the fill distance (a measure of how well the training points fill the domain) is sufficiently small. This last property is more easily obtained using a (mesh-free) quasi-Monte Carlo discretization rather than a (mesh-based) regular-grid discretization [11]. In the following, Gaussian RBFs with fixed width are used, and their common width is selected by leave-one-out cross-validation, following the approach described in [8, Section 17.1.3] and called Rippa method [12] therein. For the interpolation, the centers of the Gaussian RBFs are chosen as coincident with the training points. Hence, strict positive-definiteness of the RBFs guarantees non-singularity of the interpolation matrix, and uniqueness of the interpolant [8, Chapter 3];

c) for the optimization, a Sequential Linear Programming (SLP) algorithm including an adaptive trust region is used, employing the interpolant determined above as the surrogate objective function. This algorithm is similar to the one presented in [9] for a different application of optimization in materials science. At each iteration of the implemented algorithm, the original optimization problem is replaced by its linearization around the current vector of design variables. Changes in such variables are also limited by an additional adaptive trust region constraint, which depends on the quality of the previous linearizations. If the previous linear approximation is accurate, then the size of the trust region (which is centered on the current vector of design variables) is expanded (up to an upper bound on that size), otherwise it is reduced (up to a lower bound on it). In the paper, the SLP algorithm is applied for a fixed number N_i of iterations, although more sophisticated termination criteria may be also applied. Differently from [2,3,4], such an algorithm is used instead of the globally convergent version of the method of moving asymptotes (GCMMA) [13,14], since some preliminary simulations - whose results are not reported here, due to space constraints - have shown that, for the specific problem, the performance of the former is less dependent than the latter on the interpolation quality;

d) in a similar way as in [2,3,4], a quasi-Monte Carlo multi-start initialization approach, based on N_s different starting points for the iterative optimization algorithm, is used. In addition, a subset of the same training points used for the interpolation is selected also for the quasi-Monte Carlo multi-start initialization, since on such points there is no approximation error on the original objective function. Hence, at each initial iteration, the original objective function and the surrogate one assume the same value.

It is worth observing that the maximum value achieved by the Gaussian RBF interpolant on the whole domain does not coincide in general with the one assumed by the original objective function on the training set. For instance, when the constraints are not active at optimality, a necessary (but not sufficient) condition for having the same maximum values is that the gradient of the interpolant is equal to the all-zeros vector, for the elements of the training set associated with the maximum objective value on such a set. However, this additional condition, together with the interpolating conditions, forms a system of linear equations having typically no solution (instead, the interpolating conditions alone always determine a unique Gaussian RBF interpolant). A consequence of this is that the maximum value of the Gaussian RBF interpolant on the whole domain can be larger than the maximum value of the objective function on the training set, making the optimization of the Gaussian RBF interpolant potentially useful for the surrogate optimization of the original objective function.

4. Numerical results and discussion

In the following, numerical results are reported for the Gaussian RBF interpolant constructed as described in Section 3, by evaluating the original objective function on N_c =200 centers, generated according to a Sobol' sequence [11], and satisfying all the original constraints. Then, the SLP algorithm with the adaptive trust region is applied, using the first N_s=10 centers above as initial points. A variation of the optimization problem described in [2, Eq. (14)], with different bounds on the design parameters (only two of which are free), has been considered to get the numerical results presented in the following. Figure 1 illustrates, for each of the N_s repetitions of the optimization procedure above, the evolutions of the values assumed, respectively, by the surrogate and original objective functions during the iterations of the SLP algorithm (the latter values have been computed a-posteriori, for a validity check). Each repetition consists of N_i=100 iterations of the SLP algorithm with the adaptive trust region. The obtained numerical results are promising since, in each repetition, the values of the surrogate and original objective functions have evolved in a similar way (even in cases for which the quality of the interpolant may be low, possibly due to a small number of interpolation points).

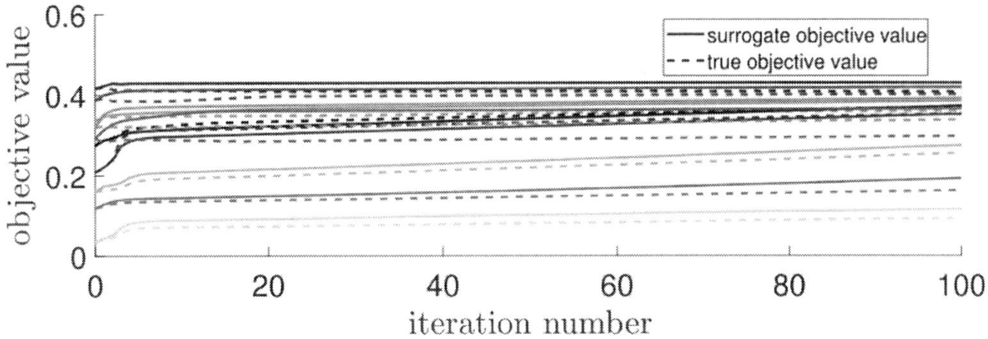

Figure 1. For each repetition of the optimization procedure: values of surrogate and original objective functions at each iteration of the SLP algorithm (same colors refer to the same repetition).

References

[1] A. Bacigalupo, L. Gambarotta, Simplified modelling of chiral lattice materials with local resonators, International Journal of Solids and Structures, 83, 126-141, 2016.

[2] A. Bacigalupo, G. Gnecco, M. Lepidi, L. Gambarotta, Design of acoustic metamaterials through nonlinear programming, 2nd International Workshop on Optimization, Machine Learning and Big Data (MOD 2016), Volterra (Italy), August 26th-29th, 2016. In Lecture Notes in Computer Science, 10122, 170-181, 2016.

[3] A. Bacigalupo, M. Lepidi, G. Gnecco, L. Gambarotta, Optimal design of auxetic hexachiral metamaterials with local resonators, Smart Materials and Structures, 25(5), article ID: 054009, 2016.

[4] A. Bacigalupo, G. Gnecco, M. Lepidi, L. Gambarotta, Optimal design of low-frequency band gaps in anti-tetrachiral lattice meta-materials, Composites Part B: Engineering, 115(5), 341-359, 2017.

[5] M.S. Bazaraa, H.D. Sherali, C.M. Shetty, Nonlinear Programming: Theory and Algorithms, John Wiley & Sons, 2006.

[6] D. Bigoni, S. Guenneau, A.B. Movchan, M. Brun, Elastic metamaterials with inertial locally resonant structures: application to lensing and localization, Physical Review B, 87, article ID: 174303, 2013.

[7] N. Cristianini, J. Shawe-Taylor, An Introduction to Support Vector Machines and Other Kernel-Based Methods. Cambridge University Press, 2000.

[8] G. E. Fasshauer, Meshfree Approximation Methods with MATLAB, World Scientific, 2007.

[9] C. Lin, Y.H. Lee, J.K. Schuh, R.H. Ewoldt, J.T. Allison, Efficient optimal surface texture design using linearization. In Advances in Structural and Multidisciplinary Optimization, Springer, 632-647, 2017.

[10] S. Koziel, L. Leifsson (eds.), Surrogate-Based Modeling and Optimization: Applications in Engineering, Springer, 2013.

[11] H. Niederreiter, Random Number Generation and Quasi-Monte Carlo Methods, SIAM, 1992.

[12] S. Rippa, An algorithm for selecting a good value for the parameter c in radial basis function interpolation, Advances in Computational Mathematics, 11, 193-210, 1999.

[13] K. Svanberg, The method of moving asymptotes - a new method for structural optimization, International Journal for Numerical Methods in Engineering, 24, 359-373, 1987.

[14] K. Svanberg, A class of globally convergent optimization methods based on conservative convex separable approximations, SIAM Journal on Optimization, 12, 555-573, 2002.

[15] S.M. Wild, C. Shoemaker, Global convergence of radial basis function trust-region algorithms for derivative-free optimization, SIAM Review, 55, 349-371, 2013.

Direct laser deposition as a method of biodegradable magnesium implant manufacturing

A S Gnedenkov[1,*], S L Sinebryukhov[1], D V Mashtalyar[1,2], S V Gnedenkov[1,2], E P Subbotin[3], A I Nikitin[3], K V Nadaraia[1,2] and Y N Kulchin[2,3]

[1] Institute of Chemistry of FEB RAS, 159 Pr. 100-letiya Vladivostoka, Vladivostok, 690022, Russia
[2] Far Eastern Federal University, 8 Sukhanova St., Vladivostok, 690950, Russia
[3] Institute of Automation and Control Processes of FEB RAS, 5 Radio St., Vladivostok, 690041, Russia

*E-mail: asg17@mail.com

Abstract. Magnesium can be used as a new biodegradable material for orthopaedic applications due to its outstanding properties including high biocompatibility, proper mechanical stiffness and biodegradability. Direct laser deposition (DLD) as a way of biomedical implant production has been used to make bulk pure Mg. Surface morphology, microstructure and mechanical properties of DLD sample have been studied. The relative porosity of DLD Mg sample has been determined.

1. Introduction

Nowadays, metals such as titanium, stainless steel, chromium–cobalt and its alloys are widely used as materials for orthopaedic applications [1]. At the same time, these implant materials have some shortcomings: difference in mechanical properties (microhardness, elastic modulus) between these biomaterials and natural bones can sufficiently damage the bone tissue [2]; the release of toxic elements may leads to the negative effect such as allergy or even cancer [3]. Currently, Mg has been proposed as a promising biomaterial, which possesses a good biocompatibility [4] and appropriate mechanical performance similar to the natural bone as compared to other metallic implant [5]. In addition, magnesium can continuously degrade in situ, which eliminates the necessity of secondary surgery as a result of Mg implant resorption in the human body [6].

Direct laser deposition (DLD) can be used to obtain implant materials. Polymers, ceramics, metals and their composites can be processed by DLD method. The DLD allows the layers of a powder to be successively melted by means of a laser beam heat. Each melted layer connects with the previous one until the final sample is formed.

In the field of personalized medicine, the magnesium based materials are very attractive to be used in the process of implant formation by means of such additive technique as DLD [7].

At the same time, Mg has some shortcomings such as high corrosion activity in the saline solutions, which leads to its too fast degradation [8]. Thereby, many scientific groups try to obtain new Mg alloys as well as to develop a new ways for treating and producing the personalized implant materials [9,10] as well as to protect Mg samples against intensive corrosion destruction [11-24]. Nevertheless, fast magnesium dissolution is the urgent problem in the implant surgery. Some researchers use laser treatment to reduce the magnesium alloy dissolution by means of Mg α-phase refinement and homogeneous element distribution [25-30]. It should be noted that DLD technique is basically used to obtain titanium alloys, steels but not magnesium alloys. Usually, scientists investigated the process of selective laser melting of pure Mg by single layer production [7]. There are a few studies dealt with the DLD of bulk Mg [31].

2. Experimental

Mg powder MPF-4 GOST 6001-79 (RusRedMet Research and Production Association) with following chemical composition: 0.05 wt. % Fe; 0.005 wt. % Cl; 0.1 wt. % H_2O; balance – Mg was used. The shape of particles of Mg powder is presented in figure 1a. The average size of Mg particles is 60–80 μm.

DLD process of Mg particles was realized by means of a laboratory robotic technological complex with KUKA robot KR-30 HA and a laser with 2.5 mm diameter beam. The Mg specimen was obtained in the isolated chamber suitable to work with inflammable materials, with strict control of the protective environment from process gases. Twenty-five layers were applied by DLD process to obtain Mg sample. The size of the samples was 30 mm × 15 mm × 6 mm.

X-ray diffraction (XRD) (Bruker D8ADVANCE) with Cu Kα radiation was applied to investigate the phase composition of the DLD Mg sample.

The optical microscope (Carl Zeiss Axiovert 40 MAT) was used to study the microstructure of the DLD specimen. Surface analysis was made using scanning electron microscopy (Carl Zeiss EVO 40) and energy dispersive spectroscopy (Shimadzu EDX-800).

Mechanical properties of the samples were determined by means of Shimadzu DUH-W201 ultramicrohardness tester at a load of 100 mN and a holding time of 5 s. Archimedean method was used to determine the density of the DLD sample.

3. Results and discussion

Figure 1b depicts the DLD Mg sample morphology. The specimen surface is rough with distinguished boundaries between DLD Mg layers. The presence of unmelted spheroidised particles can be found in the structure of the material.

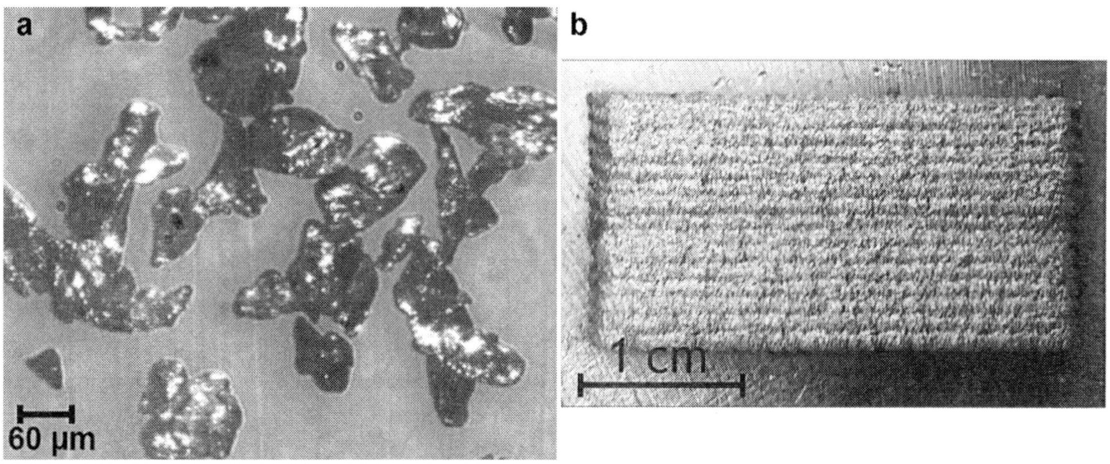

Figure 1. The images of Mg particles used for DLD (a) and the pure Mg specimen obtained by DLD process (b)

Analysis of the XRD measurements of the DLD Mg specimen indicates the presence of the only Mg phase (hexagonal structure) in the material composition.

Morphology of the specimen cross-section was studied by SEM analysis (figure 2a). These data indicate the presence of small cracks at the edge of each laser track. At the same time, the DLD sample has a well compactness structure. There are some micropores in the material structure but neighbor layers are well connected to each other, which is a result of a good metallurgical bond created by DLD. Caused by the Gaussian laser energy distribution, the molten pools with an elliptical bottom are aligned in layers, which reflects the overlapping layered peculiarity inherent in DLD method (figure 2a).

Analysis of the SEM image shows that the width track is equal to about (1.12 ± 0.09) mm and the penetration depth of molten pools is (0.37 ± 0.08) mm. Thereby, remelting process of the part of each layer of the specimen is realized. Multiple remelting processes are very important to reinforce the metallurgical bond between layers, as well as to determine the microstructure of DLD specimen. From figure 2a, where the detail information about laser scanning tracks is presented, the heterogeneous character of the sample microstructure can be seen. It should be emphasized that to obtain the high strength metallurgical bond, the length between two adjacent tracks must be shorter than the width of track [32]. Figure 2b shows some unmelted particles at the boundary of the track.

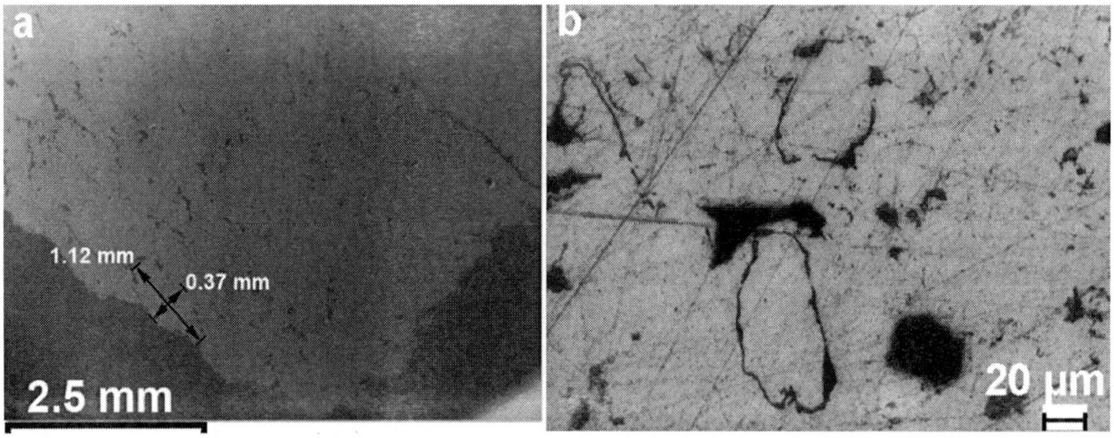

Figure 2. SEM (a) and optical (b) images of the cross-section of DLD Mg substrate

The sample microstructure can be described with fine grains because of fast solidification [33], which lead to the high relative density (93.3%) as well as to high mechanical properties of the DLD pure magnesium. The microhardness and elastic modulus of the DLD specimen is equal to about (0.49 ± 0.09) GPa and (30 ± 4) GPa, respectively. The specimen density is equal to 1.622 g/cm^3.

4. Conclusion

Analysis of the obtained results leads to the following conclusion.

DLD sample has a well compactness structure. There are some micropores in the material structure but neighbor layers are well connected to each other, which is a result of a good metallurgical bond created by DLD.

The DLD sample microstructure consists of α-phase of Mg.

The DLD specimen possesses a high mechanical properties and high relative density. The main strengthening mechanisms for DLD magnesium sample are solid solution strengthening and grain refinement.

Acknowledgments
This work was financially supported by the Basic Research Program of FEB RAS "Far East" (project №18-5-055).

References

[1] Al-Mobarak N A 2008 *Mater. Und Werkstofftechnik* **39** 486–91
[2] Geetha M, Singh A K, Asokamani R and Gogia A K 2009 *Prog. Mater. Sci.* **54** 397–425

[3] Okazaki Y and Gotoh E 2005 *Biomaterials* **26** 11–21
[4] Staiger M P, Pietak A M, Huadmai J and Dias G 2006 *Biomaterials* **27** 1728–34
[5] Chiu K Y, Wong M H, Cheng F T and Man H.C 2007 *Surf. Coat. Technol.* **202** 590–98
[6] Song G L and Song S Z 2007 *Adv. Eng. Mater.* **9** 298–302
[7] Chung C Ng, Savalani M and Man H C 2011 *Rapid Prototyping J.* **17** 479-90
[8] Agarwal S, Curtin J, Duffy B and Jaiswal S 2016 *Mater. Sci. Eng.* **68** 948–63
[9] Bettles C J 2008 *J. Mater. Eng. Perform.* **17** 297–301.
[10] Tandon R and Madan D 2014 *Powder Metall.* **57** 236–41
[11] Gnedenkov S V, Sinebryukhov S L, Mashtalyar D V, Egorkin V S, Sidorova M V and Gnedenkov A S 2014 *Corros. Sci.* **85** 52–9
[12] Gnedenkov S V, Sinebryukhov S L, Mashtalyar D V, Imshinetskiy I M, Gnedenkov A S, Samokhin A V and Tsvetkov Y V 2015 *Vacuum* **120** 107–14
[13] Gnedenkov S V, Sinebryukhov S L, Mashtalyar D V, Nadaraia K V, Gnedenkov A S and Bouznik V M 2016 *Corros. Sci.* **111** 175–85
[14] Gnedenkov A S, Sinebryukhov S L, Mashtalyar D V and Gnedenkov S V 2014 *Solid State Phenom.* **213** 143–8
[15] Gnedenkov A S, Sinebryukhov S L, Mashtalyar D V and Gnedenkov S V 2016 *Corros. Sci.* **102** 348–54
[16] Sun M, Yerokhin A, Bychkova M Y, Shtansky D V, Levashov E A and Matthews A 2016 *Corros. Sci.* **111** 753–69
[17] Ivanou D K, Yasakau K A, Kallip S, Lisenkov A D, Starykevich M, Lamaka S V., Ferreira M G S, Zheludkevich M L and Aparicio M 2016 *RSC Adv.* **6** 12553–60
[18] Gnedenkov A S, Sinebryukhov S L, Mashtalyar D V. and Gnedenkov S V. 2016 *Solid State Phenom.* **245** 89–96
[19] Mohedano M, Serdechnova M, Starykevich M, Karpushenkov S, Bouali A C, Ferreira M G S and Zheludkevich M L 2017 *Mater. Des.* **120** 36–46
[20] Gnedenkov A S, Sinebryukhov S L, Mashtalyar D V and Gnedenkov S V 2016 *Corros. Sci.* **102** 269–78
[21] Sinebryukhov S L, Gnedenkov A S, Mashtalyar D V and Gnedenkov S V 2010 *Surf. Coatings Technol.* **205** 1697–701
[22] Gnedenkov A S, Sinebryukhov S L, Mashtalyar D V and Gnedenkov S V 2012 *Phys. Procedia* **23** 98–101
[23] Gnedenkov A S, Sinebryukhov S L, Mashtalyar D V and Gnedenkov S V 2013 *Surf. Coatings Technol.* **225** 112–8
[24] Gnedenkov A S, Sinebryukhov S L, Mashtalyar D V, Gnedenkov S V and Sergienko V I 2017 *Corrosion Science and Technology-Korea* **16** 151-9
[25] Osakada K and Shiomi M 2006 *Int. J. Mach. Tools Manuf.* **46** 1188–93
[26] Liu C, Liang J, Zhou J, Wang L and Li Q 2015 *Appl. Surf. Sci.* **343** 133–40
[27] Cui Z-Q, Shi H-X, Wang W-X and Xu B-S 2015 *Trans. Nonferr. Met. Soc. China* **25** 1446–53
[28] Taltavull C, Torres B, Lopez A, Rodrigo P, Otero E, Atrens A and Rams J 2014 *Mater. Des.* **57** 40–50
[29] Coy A, Viejo F, Garcia-Garcia F, Liu Z, Skeldon P and Thompson G 2010 *Corros. Sci.* **52** 387–97
[30] Raman R S, Murray S and Brandt M 2013 *Surf. Eng.* **23** 107–11
[31] Hu D, Wang Y, Zhang D, Hao L, Jiang J, Li Z and Chen Y 2015 *Mater. Manuf. Process.* **30** 1298–304
[32] Wei K, Gao M, Wang Z and Zeng X 2014 *Materials Science and Engineering A.* **611** 212–22
[33] Song B, Dong S J, Zhang B C, Liao H L and Coddet C 2012 *Mater. Des.* **35** 120–25

Analogue simulation of two-body quantum dynamics with classical setup

M A Gorlach[1], M Di Liberto[2,3], A Recati[2], I Carusotto[2], A N Poddubny[1,4] and C Menotti[2]

[1] ITMO University, Saint Petersburg 197101, Russia
[2] INO-CNR BEC Center and Dipartimento di Fisica, Università di Trento, 38123 Povo, Italy
[3] Center for Nonlinear Phenomena and Complex Systems, Université Libre de Bruxelles, B-1050 Brussels, Belgium
[4] Ioffe Institute, Saint Petersburg 194021, Russia

E-mail: `m.gorlach@metalab.ifmo.ru`

Abstract. One-dimensional Bose-Hubbard models provide a variety of intriguing predictions including the formation of repulsively bound two-boson pairs, their collapse and revival and the realization of the two-boson topological states. While probing such systems in quantum regime is quite a challenging task, there is another appealing option to explore the properties of repulsively bound bosons based on mapping of the one-dimensional interacting system onto the two-dimensional classical model described by the standard coupled-mode equations. Adopting the latter approach, we develop the procedure to probe the energy of bound boson pairs as well as to perform the tomography of the bound pair quantum state in realistic dissipative systems.

1. Introduction

Since the theoretical prediction and experimental observation of repulsively bound boson pairs (doublons) in one-dimensional systems described by the Bose-Hubbard model [1] there is an immense interest in such strongly correlated two-boson states. Further studies revealed a number of their exciting properties including coherent destruction (collapse) [2], emergence of topological two-boson states [3, 4], interaction-induced topological order in two-dimensional systems [5] and rich interplay of local and non-local interaction mechanisms in extended Hubbard models [6, 7].

While the experimental studies of bound boson pairs remain quite challenging, there exists an alternative strategy based on the possibility to map a one-dimensional interacting system described by the Bose-Hubbard Hamiltonian onto a two-dimensional classical system obeying the usual coupled-mode equations [8]. Such two-dimensional systems have recently been realized in waveguide geometries thus paving a way to quantum simulation of few-body physics with photonic systems [9].

In this work, we analyze how one can extract doublon energies and a specific type of the doublon state (bulk or edge) from the experiments with such analogue quantum simulators.

2. Emulating bound boson pairs

We start by considering the dynamics of the two interacting bosons in the Su-Schrieffer-Heeger model illustrated schematically in Fig. 1(a). This system is described by the Bose-Hubbard

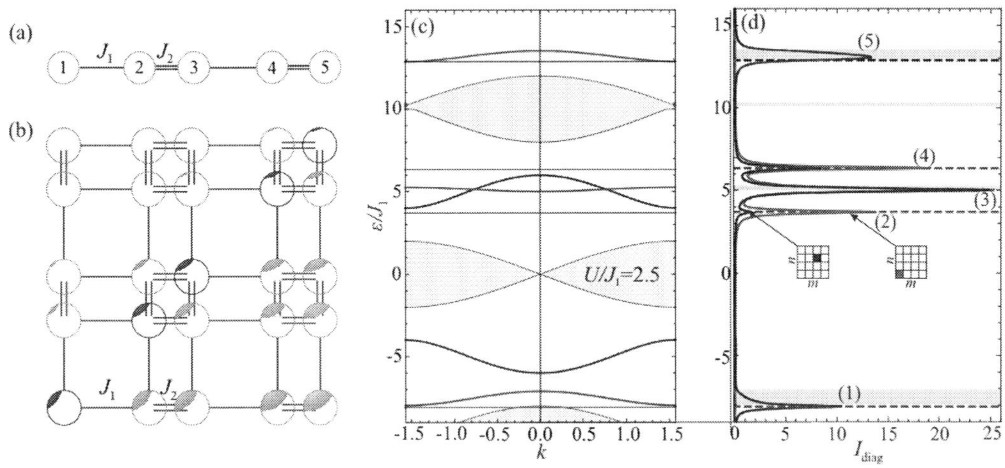

Figure 1. Illustration of the analogue simulation of quantum system with a classical setup. (a) One-dimensional two-body system described by the Bose-Hubbard Hamiltonian. The array starts from a weak link edge and finishes with a strong link edge. (b) Two-dimensional nearest-neighbor coupled setup emulating quantum two-body problem. (c) Spectrum of various two-body states possible in the system: green, blue and red bands designate scattering states, single-boson edge states and bulk doublon states, respectively. Horizontal black and magenta lines indicate doublon edge states localized at strong link and weak link edges, respectively. (d) Field intensity from the sites at the diagonals $m = n$ and $m = n \pm 1$. Light red regions show the boundaries of the doublon bands. Blue and green curves correspond to the excitation of the system from (3,3) and (1,1) sites, respectively. Insets illustrate the excitation protocol.

Hamiltonian

$$\hat{H} = \omega_0 \sum_m \hat{n}_m + U \sum_m \hat{n}_m (\hat{n}_m - 1) - \sum_m \left(J_1 \hat{a}^\dagger_{2m-1} \hat{a}_{2m} + J_2 \hat{a}^\dagger_{2m} \hat{a}_{2m+1} + \text{H.c.} \right) \qquad (1)$$

where $\hat{a}^\dagger_m$ and $\hat{a}_m$ are photon creation and annihilation operators for the m-th cavity, $\hat{n}_m = \hat{a}^\dagger_m \hat{a}_m$ is the photon number operator, $J_{1,2}$ are the tunneling constants, U is the interaction strength, ω_0 is the cavity eigenfrequency ($\hbar = 1$), and only one photon polarization state is considered. The analysis of the system [3, 4] demonstrates that this model is equivalent to a two-dimensional classical system [Fig. 1(b)] with nearest-neighbor coupling. The latter two-dimensional system can be realized using various technological platforms, for instance, arrays of polariton pillars, while its properties can be probed by exciting one or several pillars and measuring the intensity distribution in the array.

To detect doublon states, we compute the population of the three diagonal lines $m = n$ and $m = n \pm 1$ as a function of excitation frequency. The results of numerical calculation based on coupled mode theory are presented in Fig. 1(d). The excitation spectrum features several peaks with an amplitude that strongly depends on the choice of the excitation point. The intensity from the diagonal pillars $I_{\rm diag}$ quantifies the probability of two bosons co-localization in the same cavity in 1D quantum problem. Hence, we associate the peaks in the excitation spectrum of a 2D system with the doublon modes. However, a single excitation spectrum does not allow one to distinguish doublon bulk and edge states.

To draw such a distinction, we calculate the analogous excitation spectrum for a different choice of the excitation point, further away from the (1,1) corner of the system. Since doublon edge states decay away from the array edges, we expect that the corresponding peaks in the spectrum will be suppressed dramatically, while the amplitude of the peaks corresponding to bulk doublon modes will stay roughly the same. Based on this interpretation, we are able now to make a distinction between bulk doublon states associated with the peaks (1), (3), (5) and edge doublon states represented by the peaks (2) and (4). Comparing the obtained results with numerical calculation of the system eigenmodes, we verify the validity of our technique.

To summarize, our technique provides an experimentally feasible way to probe doublon bands and doublon edge states using simple photonic systems which uncovers interesting perspectives in analogue simulation of strongly correlated quantum states.

Acknowledgments

This work was supported by the Russian Science Foundation (Grant No. 18-72-00102). M.A.G. and A.N.P. acknowledge partial support by the Foundation for the Advancement of Theoretical Physics and Mathematics "Basis". M.D.L., A.R., I.C. and C.M. acknowledge funding from Provincia Autonoma di Trento, partly through the SiQuro project ("On Silicon Chip Quantum Optics for Quantum Computing and Secure Communications") and from the EU-FET Proactive grant AQuS, Project No. 640800. M.D.L. also acknowledges support from the ERC Starting Grant TopoCold.

References

[1] Winkler K, Thalhammer G, Lang F, Grimm R, Hecker Denschlag J, Daley A J, Kantian A, Büchler H P and Zoller P 2006 *Nature* **441** 853-856.
[2] Lin S, Zhang X Z and Song Z 2014 *Physical Review A* **90** 063411.
[3] Di Liberto M, Recati A, Carusotto I, Menotti C 2016 *Physical Review A* **94** 062704.
[4] Gorlach M A, Poddubny A N 2017 *Physical Review A* **95** 053866.
[5] Salerno G, Di Liberto M, Menotti C, Carusotto I 2018 *Physical Review A* **97** 013637.
[6] Gorlach M A, Poddubny A N 2017 *Physical Review A* **95** 033831.
[7] Di Liberto M, Recati A, Carusotto I, Menotti C 2017 *The European Physical Journal Special Topics* **226** 2751-2762.
[8] Longhi S, Della Valle G 2013 *Journal of Physics: Condensed Matter* **25** 235601.
[9] Mukherjee S, Valiente M, Goldman N, Spracklen A, Andersson E, Öhberg P, Thomson R R 2016 *Physical Review A* **94** 053853.

Glancing angle deposition of optical coatings: results of the full-atomistic simulation

F V Grigoriev*, V B Sulimov and A V Tikhonravov

Research Computing Center, M.V. Lomonosov Moscow State University, 119991, Moscow, Leninskye Gory, 1 build.4, Russian Federation

*Corresponding author, E-mail: fedor.grigoriev@gmail.com

Abstract. The earlier developed molecular dynamics approach is used to simulate the thin film glancing angle deposition. The deposition at glancing angles is used for the production of optical thin films with low refractive indices. Both low-energy and high-energy deposition processes are considered for the case of silicon dioxide thin films. It is revealed that glancing angle low-energy deposition forms silicon dioxide films with low densities of about 1,5 g/cm^3 and low refractive indices. Density profiles of growing films essentially depend on the initial conditions of the deposition process. The density of high-energy deposited films achieves 2,4 g/cm^3. This value is close to that obtained before in the case of normal incidence deposition.

1. Introduction

Glancing angle deposition (GLAD) is the promising technique for the production of optical thin films with low refractive indices [1]. In this technique angles between the directions of particles flow and the normal to the substrate exceed 80° so the deposited particles move almost parallel to the substrate. This leads to the formation of separated nanostructures which shapes depend on the deposition conditions [2]. These so-called "sculptured thin films" (STF) [3] have low densities and therefore have low refractive indices which ensure also low reflectance of the produced films [4].

Structural, mechanical and optical properties of GLAD-grown STF essentially depend on the deposition conditions: density of deposited particles flow, particle energy distribution, substrate temperature, and so on [1-4]. In the present work the earlier developed method for the full-atomistic simulation of thin film deposition process [5-7] is applied to study the GLAD-grown SiO_2 films structural properties. Film density profiles are obtained for the cases of high-energy and low-energy deposition. The porosity and morphology of simulated nanostructures are studied qualitatively using visualization techniques. It was found that in the case of low energy of deposited silicon atoms structural properties of the GLAD-grown and normal deposition grown SiO_2 films differ essentially. In particular density of the GLAD deposited films reduces noticeably compared with the normal deposited films. This density reduction is accompanied by the increase of concentration of pores having dimensions of up to several nanometers.

2. Simulation method

The simulation of thin films deposition is performed using the classical molecular dynamic (MD) method. DESIL force filed [5-7] is used for the calculation of the potential energy of interatomic interaction. Deposition process is organized as a step-by-step procedure described in [5]. At every deposition step the MD simulation is performed using periodic boundary conditions with *NVT* ensemble (constant number of atoms, volume and temperature). In the present work the following MD parameters are used: the deposition temperature is $T = 300$ K, the duration of one deposition cycle is 6 ps, the time step of MD modeling is 0.5 fs, the number of injected SiO_2 groups per one injection cycle is 25. Berendsen thermostat [8] is used to keep the simulation box temperature constant. Horizontal dimensions of the substrate are 10 nm and 11 nm, maximum value of film thickness is about 40 nm. Density and thickness of the substrate are 2.14 g/cm^3 and 4.5 nm, respectively. The electrostatic component of the interatomic energy is calculated using the Particle Mesh Ewald [9] method. MD simulation is performed using the GROMACS package [10]. The visual analysis is performed using Visual Molecular Dynamic (VMD) [11] program facilities.

All calculations have been performed using the supercomputers 'Lomonosov' and 'Lomonosov-2' of the Supercomputing Center of Lomonosov Moscow State University [12].

3. Results and discussion

Initial values of the deposited Si atoms velocities are specified as follows:

$$\upsilon_z = -\upsilon_0\cos\alpha;\ \upsilon_x = \upsilon_0 \mathrm{sn}\alpha\sin\beta;\ \upsilon_x = \upsilon_0\mathrm{sn}\alpha\cos\beta \qquad (1),$$

where α is the glancing angle (see Fig. 1), β angle is chosen in the interval from 0° to 90°, υ_0 denotes the initial value of the Si atoms kinetic energy, that is 10 eV for high-energy deposition and 0,1 eV for low-energy deposition. Coordinate axes are directed as shown in the right part of Fig. 2. Initial coordinates of the deposited silicon and oxygen atoms are specified as described in [5].

Results of the simulation are shown in Figs. 1, 2.

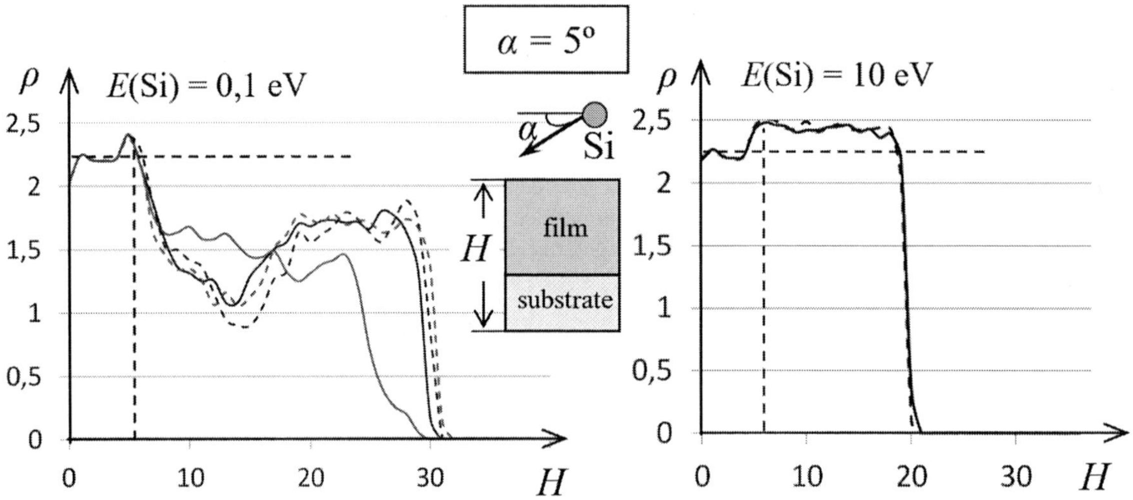

Figure 1. Density profiles of deposited films for the low- and high-energy deposition processes, ρ (g/cm^3) is the film density, *H*(nm) is the film thickness, α(grad) is the glancing deposition angle. The four curves at the left part and two curves at the right part differ by sets of random values of initial coordinates of silicon and oxygen atoms. The vertical dashed line denotes the upper boundary of the substrate, the horizontal dashed line denotes the experimental value of glassy SiO_2 density, that is 2,2 g/cm^3.

The density of the film fabricated using high-energy deposition is essentially higher than the density of low-energy deposited film. This density fluctuates near the average value of 2,4 g/cm^3 (the right part of Fig.1) which is close to the density of the film deposited at normal flux incidence° [7]. The low-energy deposition process produces films having fluctuating profiles (the left part of Fig.1). The profiles noticeably depend on the initial coordinates of deposited silicon and oxygen atoms. In contrary, in the case of high-energy deposition the two curves corresponding to different sets of initial coordinates are practically identical. The average value of density of low-energy deposited film is equal to 1,5 g/cm^3 which is much lower than the density of the film deposited at normal flux incidence [7]. The decrease of thin film refractive index as compared to that of fused silica can be estimated using the Gladstone–Dale equation [13] in the form of $\Delta n = 0{,}21 \cdot \Delta\rho$, where $\Delta\rho$ is the difference between film and glassy density. In our case $\Delta\rho = -0{,}7$ g/cm^3. So the variation of the refractive index in the case of low-energy deposition achieves $\Delta n = -0{,}14$. It is worth noting that the empirical character of the Gladstone–Dale equation limits it in the prediction of relationships between density and refractive index of structures with high porosity.

The atomistic structure of deposited films is shown in Fig. 2. Substrate atoms are painted as yellow (Si) and red (O), deposited atoms are painted as grey (Si) and blue (O). The constant density surface is shown in Fig. 2. Density at a certain point is calculated as a sum of atomic contributions exponentially reducing with the growth of distances between atoms and this point.

Accordingly to the density profiles (Fig. 1) the porosity of deposited films is higher in the case of deposition with $\alpha = 5°$ than in the case of normal incidence deposition. There are no pores in the film obtained using the high-energy deposition (Fig. 1, right part).

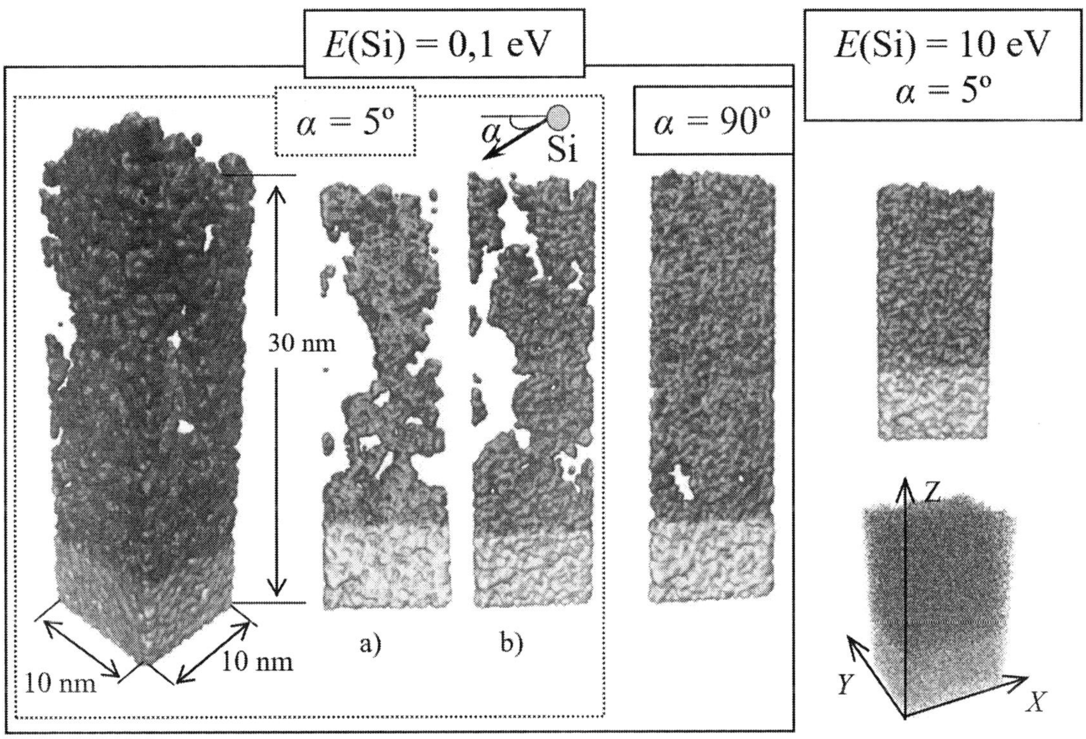

Figure 2. Atomistic structure of deposited films, α(grad) is the glancing deposition angle. The 3D film structure is shown at the left. Layers of atoms with x coordinates in the interval 4,5 nm < x < 5,5 nm are shown at the center and right side of the figure. Layers denoted as a) and b) differ by the initial sets of coordinates of deposited atoms.

4. Conclusions

The full-atomistic classical simulation of the glancing angle deposition of silicon dioxide films has been performed. Both low-energy and high-energy deposition processes are studied. Glancing angle deposition with low energy produces films with low density of about 1,5 g/cm^3 and with refractive index 1,32. Density profiles and visual representation of the low-energy deposited films show the essential dependence of their microscopic structure on deposition conditions.

The density of high-energy deposited films achieves 2.4 g/cm^3. This value is the same as that obtained earlier for the case of deposition with the atomic flux in the direction perpendicular to the substrate plane.

Acknowledgments

The work was supported by the Russian Science Foundation (grant number 14-11-00409).

References

[1] Kuo M L, Poxson D J, Kim Y S, Mont F W, Kim L K, Schuhert E F, Lin S Y 2008 *Optics Letters* **33** 2527

[2] Hawkeye M M, Brett M J 2007 *J. of Vac. Sci and Techn.* **25** 1317

[3] Robbie K, Brett M J 1997 *Technology A*, **15** 1460

[4] Trottier-Lapointe W, Zabeida O, Schmitt T, Martinu L 2016 *Applied Optics* **55** 8796

[5] Grigoriev F V, Sulimov A V, Kochikov I V, Kondakova O A, Sulimov V B, Tikhonravov A V 2015 *Int. J. of High Perf. Comp. Appl.* **29**, 184

[6] Grigoriev F V, Sulimov A V, Katkova E V, Kochikov I V, Kondakova O A, Sulimov V B, Tikhonravov A V 2016 *J. of Non-Cr. Sol.* **448** 1

[7] Grigoriev F V, Sulimov A V, Katkova E V, Kochikov I V, Kondakova O A, Sulimov V B, Tikhonravov A V 2017 *Applied Optics* **56** C87

[8] Berendsen H J C, Postma J P M, van Gunsteren W F, DiNola A, Haak J R 1984 *J. of Chem. Phys.* **81** 3684

[9] Darden T, York D, Pedersen L 1993 *J. Chem. Phys.* **98** 10089

[10] Abraham M J, Murtola T, Schulz R, Páll S, Smith J C, Hess B, Lindahl E 2015 *SoftwareX*, **1–2** 19

[11] Humphrey W, Dalke A, Schulten K 1996 *J. Molec. Graphics* **14** 33

[12] Sadovnichy V, Tikhonravov A, Voevodin V V, Opanasenko V 2013 *"Lomonosov": Supercomputing at Moscow State University. In Contemporary High Performance Computing: From Petascale toward Exascale,* (CRC Press) p 283

[13] Vedam K, Limsuwan P 1978 *J. Chem. Phys.* **69** 4772

IOP Conf. Series: Journal of Physics: Conf. Series **1092** (2018) 012047 doi:10.1088/1742-6596/1092/1/012047

Laser-printed plasmonic nanotextures with controlled porosity

S A Syubaev[1,2] , S O Gurbatov[1,2], A A Kuchmizhak[1,2,*]

[1] Institute of Automation and Control Processes, 5 Radio St., Vladivostok 690041, Russia

[2] Far Eastern Federal University, 8 Sukhanova St., Vladivostok 690950, Russia

alex.iacp.dvo@mail.ru

Abstract. We propose efficient approach for fabrication of various plasmonic nanostructures in the shape of the sub-micro rings and nanohillocks having tailored inner porosity. The fabrication approach combines the direct nanosecond-pulse ablation of the nitrogen-rich Au films covering glass or silicon substrates followed by a large-area etching with an Ar-ion beam. The average size of the nanopores inside the produced structures can be controlled via nitrogen doping level in the target metal films through tailoring the nitrogen gas concentration in the process of the film magnetron deposition, while the overall geometric shape of the produced plasmonic structures can be controlled by optimizing the thickness of the irradiated film as well as the laser irradiation parameters. Spherical-shape nanopores of variable size which depends on the initial nitrogen concentration in the irradiated film were directly revealed by preparing the cross-sectional cuts of the fabricated nanorings using focused ion beam milling. The proposed approach holds promise for easy-to-implement fast laser printing of the various nanostructures with the tailored porosity for chemo- and biosensing applications.

1. Introduction

Plasmonic nanostructures with pronounced porosity are of great interest for various applications ranging from the bio- and chemosensing to the photocatalysis and hydrogen storage [1]. In particular, the nanotextured surface of such porous strucures produces the strongly localized and enhanced electromagnetic near-fields (hot spots) which density can be controlled by tailoring the geometric dimensions and density of the nanopores. Fabrication of such porous plasmonic nanostructures is rather challenging procedure typically requiring implementation of several concequitive time-consuming and expensive fabrication processes. For example, nanostructures consisting of a two-component metal alloy (say, Au and Ag) were fabricated using the electron-beam lithography, while the subsequent chemical dealloying in the acid provides the way to remove one of the material from the alloyed nanotextures to produce nanopores [2].

Here, we demonstrate facile laser-based approach allowing fabrication of various porous nanotextures on the glass/silicon substrates. The proposed approach combines the direct single-pulse ablation of the nitrogen-doped gold/silver film followed by the large-scale soft etching with the argon-ion beam. The average size of the nanopores was found to correlate with the nitrogen doping layer of the deposited metal films controlled via mixing the nitrogen and argon discharge gases in the process of the magnetron deposition. The formation of the nanopores results from the boiling of the nitrogen-reach areas of the metal film irradiated with the laser pulse. Depending on the initial thickness of the irradiated film, the laser pulse can produce either a nanohillock or the nanoring allowing to tune the general geometry of the fabricated structures as well.

2. Results and Discussions

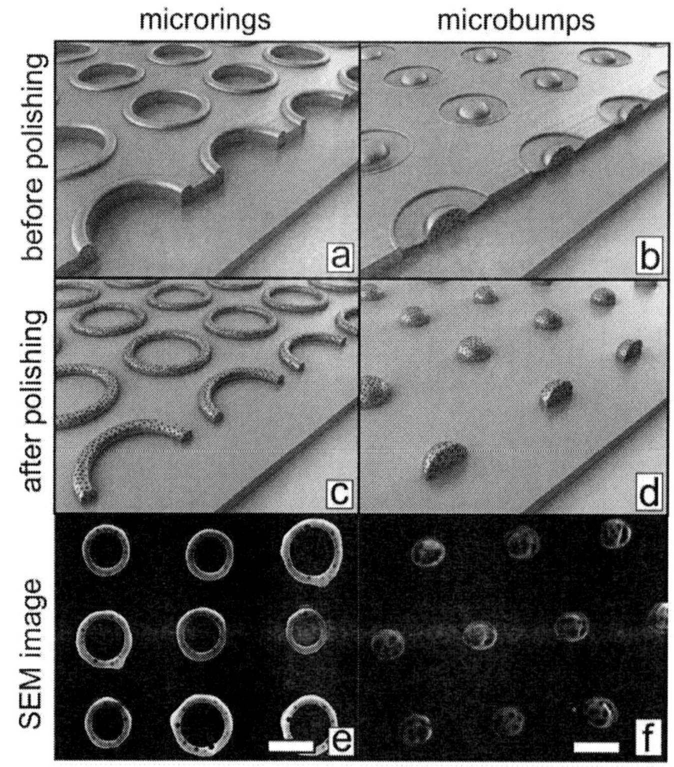

Figure 1. (a,b) Two-step fabrication of the porous nanorings and nanohillocks. Single nanosecond laser pulses first irradiate the surface of either "thin" (a) or "thick" Au film on a glass substrate. Depending on the Au film thickness, laser irradiation produces the through holes or the surface hillocks. The nanopores are mainly appear inside laser-modified areas. (c,d) Subsequent etching of the textured Au films with an accelerated defocused argon-ion beam results in removal of the surrounding unmodified metal film leaving the isolated plasmonic nanorings or nanohillocks on the substrate surface. Note that the etching process also reveals the initially hidden nanopores. (e,f) Normal-view SEM images showing arrays of porous nanorings and nanohillocks. Scale bar corresponds to 500 nm.

The proposed approach was adopted from [3,4] and contains two main fabrication steps. On the first step, we used the tightly focused (D_{opt}=1 μm) second-harmonic (532-nm) laser pulses generated by the Nd:YAG system (pulse duration of 7 ns, pulse repetition rate is 20 Hz) to locally ablate the Au film on a glass/silicon substrate (Fig.1(a,b)). We utilized the Au films with the thicknesses of 120 and 250 nm. Irradiation of the thinner film at the pulse above the ablation threshold [3] results in formation of the circular through holes with the rather thick resolidified rim. On the contrary, at similar applied energy the laser radiation is not able to melt the 250-nm thick Au film through all its thickness resulting in formation of parabola-shaped nanohillocks on the surface of the irradiated Au film. Subsequent etching of the textured Au films with an accelerated defocused argon-ion beam at etching rate of 1 nm/min results in removal of the surrounding unmodified metal film leaving the isolated plasmonic nanorings or nanohillocks on the substrate surface. Noteworthy, by shaping the focal-plane intensity distribution of the laser pulse, one could also expect to produce other nanostructures having various geometric shapes and sizes.

The argon is a common discharge gas used for magnetron deposition of the Au films. The ablation of the Au films evaporated in the Ar atmosphere and subsequient Ar-beam etching results in formation of smooth structures without any surface or subsurface pores inside as it was previously reported in [4]. In a shart contrast, the similar surface structures, nanorings and nanohillocks, produced from the Au film deposited in the nitrogen atmosphere demonstrate pronounced porosity (see Fig.1(e,f)). To study this feature in more details, using magnetron deposition we prepared a series of Au film evaporated at various mixing ratios of nitrogen and argon [5]. In all cases, the films have identical thickness of 120 nm. Our studies clearly shows the correlation between the initial concentration of the nitrogen in the evaporation chamber and the average size of the nanopores appeared in the fabricated microrings as it is illustrated by the series or SEM images (Fig.2(a-e)).

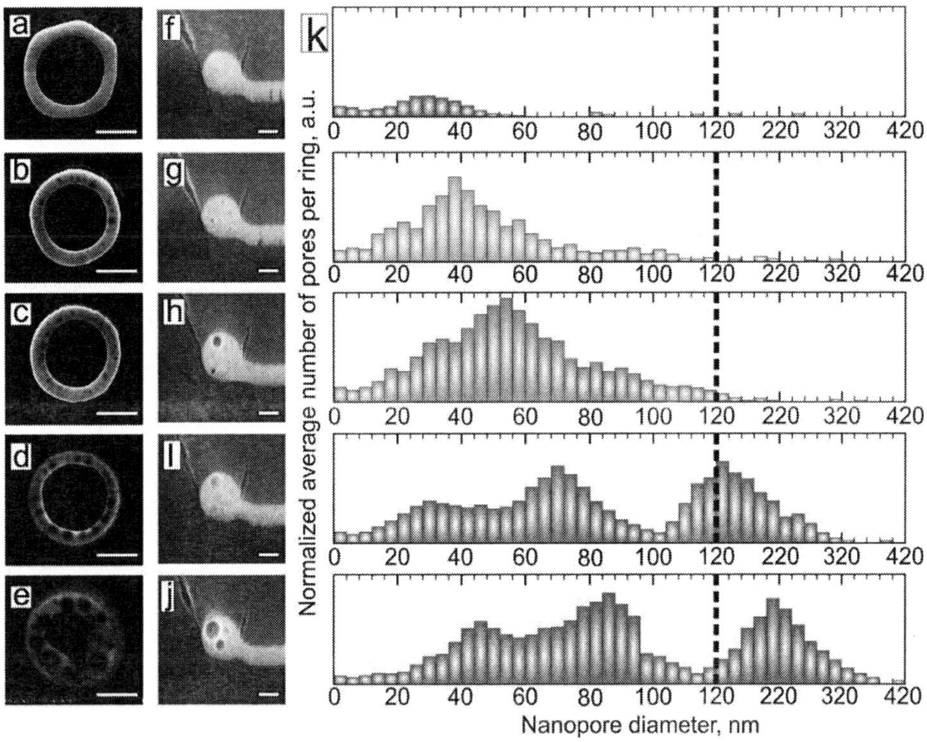

Figure 2. (a-e) Normal-view SEM images of the isolated 900-nm diameter nanorings and corresponding cross-sectional FIB cuts of the nanoring wall showing the position and the average size of the nanopores. (f-j) Corresponding cross-sectional FIB cuts (k) Averaged distribution of nanopore diameters inside the fabricated microrings. The statistics for each film type was accumulated over 50 microrings of the same size produced under similar experimental conditions

To reveal the location of the nanopores inside the fabricated nanorings we have prepared the cross-sectional cuts made across the ring wall using the focused ion-beam milling technique. For the nitrogen-doped Au films the formation of the spherical-shape nanopores can be identified in all the presented SEM images (Fig. 2(f-j)). The analysis of the SEM images indicates that the average diameter of the nanopores increases versus the concentration of the nitrogen in the Au film (Fig. 2(k)). Noteworthy, for weakly doped Au films (Fig.2(b,c)), the diameter distribution demonstrates the only size maxima, while for highly doped ones – several maxima in the diameter distribution can be identified (Fig. 2(k)). We believe this feature can be explained by merging the smaller nanopores into the bigger ones for nitrogen-rich Au films.

3. Conclusions

Here, we propose efficient two-step approach for rapid laser printing of the sub-micro rings and nanohillocks having tailored inner porosity. The presented approach combines the direct single-pulse ablation of the nitrogen-doped Au film followed by the large-scale soft etching with the argon-ion beam. The average size of the nanopores was found to correlate with the nitrogen doping layer of the deposited metal films controlled via mixing the nitrogen and argon discharge gases in the process of the magnetron deposition. This tendency was directly revealed using SEM inspection of the FIB cuts produced along the fabricated nanoring walls. The proposed facile laser-based approach with its potential to tune also the general shape of the printed nanotextures via optimization of the laser intensity distribution as well as target characteristics holds promise for easy-to-implement fast laser printing of the various nanostructures with the tailored porosity for chemo- and biosensing applications.

Acknowledgments
Authors acknowledge support from Russian Science Foundation 17-19-01325.

References

[1] Davis M 2002 *Nature* **417**, 813
[2] Hergert G, Vogelsang J, Schwarz F, Wang D, Kollmann H, Groß P, Lienau C, Runge E, Schaaf P 2017 *Light:Sci. Appl.* **6** e17075
[3] Kuchmizhak A, Gurbatov S, Vitrik O, Kulchin Yu, Milichko V, Makarov S, Kudryashov S 2016 *Sci. Rep.* **6** 19410
[4] Kuchmizhak A, Gurbatov S, Kulchin Y, Vitrik O 2015 *Opt. Commun.* **356** 1
[5] Syubaev S, Nepomnyashchiy A, Mitsai E, Pustovalov E, Vitrik O, Kudryashov S, Kuchmizhak A 2017 *Appl. Phys. Lett.* **111**, 083102

IOP Conf. Series: Journal of Physics: Conf. Series **1092** (2018) 012048 doi:10.1088/1742-6596/1092/1/012048

Numerical calculation and Cartesian multipole decomposition of optical pulling force acting on Si nanocube in visible region

E.A. Gurvitz, A. S. Shalin

ITMO University, 49 Kronverkskiy prospect, St. Petersburg, Russia, 197101

Corresponding author: egorgurvitz@gmail.com

Abstract. The multipole decomposition of optical force and scattering cross-section is considered for the two plane waves incident on Si nanocube. The obtained results show the high impact of a toroidal moment and high order multipoles in optical force, while they aren't represented main resonances in scattering cross-section.

1. Introduction

Since the first laser tweezers were introduced by A. Ashkin, the most popular and utilized principle for optical manipulations is pulling of an object to a maximum of the electric field that formed with tightly focused laser beams. Only laser tweezers allow to non-invasively control mechanical motion of organic and inorganic objects in nano- and microscales, that plays a crucial role for biology and medicine applications [3].

Optical tweezers are widely used in particle sorting [4], microfabrication [5] and with the developments of digital processing and 3D printing technologies (particularly two-photon photopolymerisation [6]) light trapped and controlled functional micro and nanoscale devices approaching new applications and developments [7,8]. Beside original classical approach for optical control in nano- and microscales there are very fast developing alternative approaches based on an interplay of an object with a substrate [9,10], manipulation using re-scattered light [11,12], optical binding [13,14], unusual optical beams (not a Gaussian or plane waves) [15,16]. One of the biggest breakthroughs in the optical tweezing become spatial light modulation allowing simultaneous control over tens trapping beam points and generate an arbitrary spatial distribution of electric field amplitude and its phase manipulation about 2π [17,18].

The opportunity to assemble arbitrary beam and fast development of all-dielectric based metamaterials [19,20] lead to necessity of analysis of complex systems and conditions. One of the tools allowing to perform an analysis is the optical force multipole decomposition [15]. In this work we performed numerical computation of Si nanocube in optical frequency range and multipole analysis of scattering cross-section and optical force acting on nanocube using Cartesian multipoles [15,19]. We are reporting two typical regions of the forces pulling an object in direction of the source. The first one is clearly explained by negative influence of electric and magnetic dipole interplay (cross-term) and the second one is frequently accompanied by contribution of electric toroidal moment and high-order multipoles cross-terms.

IOP Conf. Series: Journal of Physics: Conf. Series **1092** (2018) 012048 doi:10.1088/1742-6596/1092/1/012048

2. Results and discussion

Two plane waves forming Bessel Beam were directed on the Si nanocube as it is shown in Fig. 1. As it was reported [15] strongly inclined electric fields produce force overcoming radiation pressure allowing to obtain optical force pulling a particle back to the source.

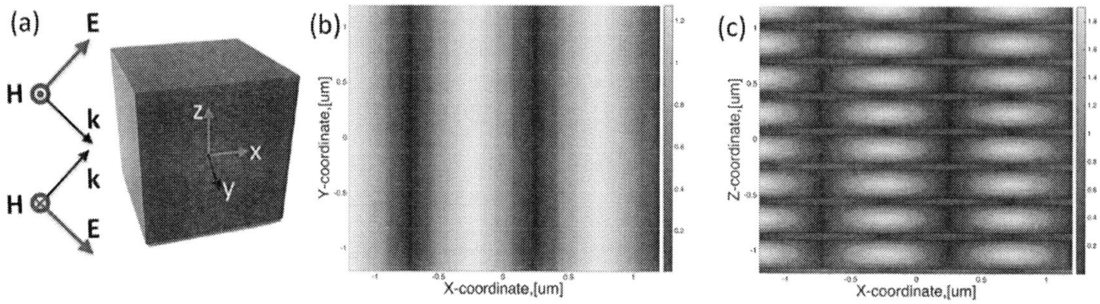

Figure 1. The field composed from two incident plane waves incident at angle of 70 deg to x-axis scattering on the Si nanocube with a=200[nm] (a). The XY and XZ cross-sections of normalized E-field amplitude for incident radiation with wavelength 600[nm] is illustrated in (b) and (c).

Using Cartesian multipole decomposition [15] of the electric field scattering cross-section and time-averaged optical force were obtained and shown in Fig.2. Due to the incident fields have the large angle of incidence the dominant optical response in scattering cross-section Fig2(a) and optical force Fig.2(b) is defined by magnetic dipole, less electric dipole; electric quadrupole and magnetic quadrupole

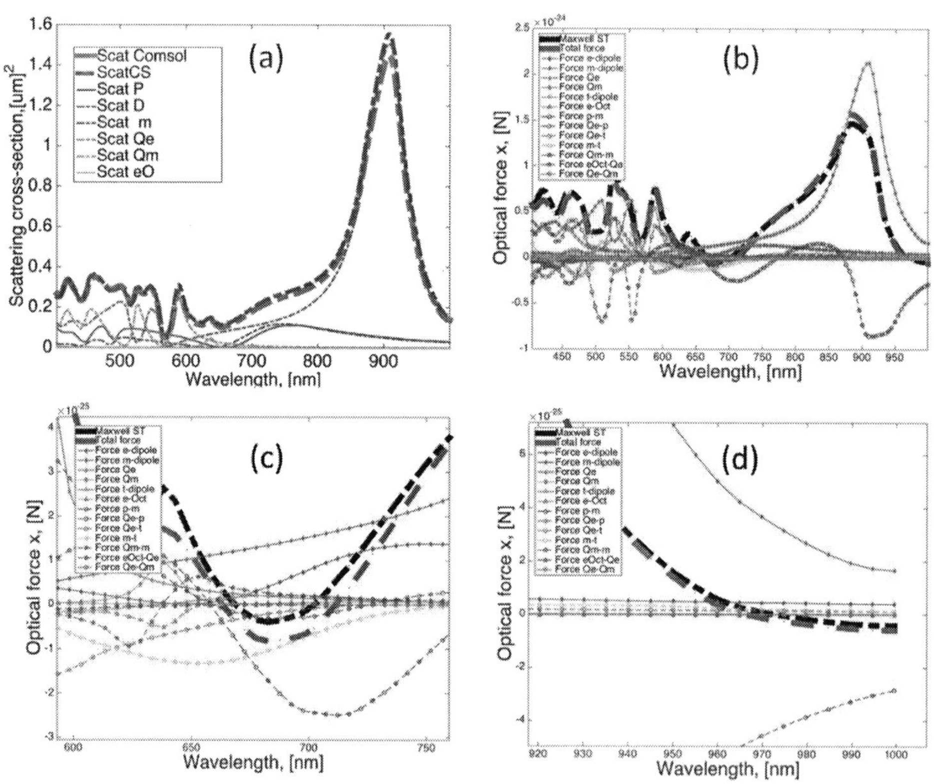

Figure 2. The multipole decomposition of scattering cross-section (a). The multipole decomposition of time-averaged optical force shown on panel (b) and panels (c,d) demonstrated regions, where optical pulling force is obtained.

moments are observed in short wavelength spectral region. The time-averaged optical force for the most part of spectral region coincided with the profile of scattering cross-section except the regions, where the interplay between the high order multipoles or toroidal one with electric and magnetic dipoles is observed. For the long wavelengths coupling between electric and magnetic dipoles leads to negative directed optical force Fig.4(d). That is the typical region for simple geometries (sphere, cylinder, cube, etc) of all-dielectric nanoparticle dependent on inclusions of incident radiation and less on the particle shapes. The second region presented in Figure 2(c) shows the negative force, caused by the interplay of high order multipoles with basic ones. The geometry and small changes of shapes more significantly cause the contribution to optical force. For the case of Si nanocube, we succeeded to forces consisting of electric and magnetic cross-term, magnetic toroidal cross-term and electric quadrupole toroidal cross-term overlapping at 680 [nm] and giving the substantial contribution to the negative force. The interplay of magnetic quadrupole and magnetic dipole, electric quadrupole and magnetic quadrupole for the chosen geometrical parameter and shape were shifted to the blue region and suppressed by pure magnetic dipole contribution.

In conclusion, it should be emphasized, that for the chosen medium and size of the nanoobject the optical force will have 2 typical regions defined by a simple interconnection of electric and magnetic dipoles and defined by several interconnections produced by high order multipoles. The analysis achieved using 3rd order Cartesian multipole decomposition for consideration of regions caused by multipoles larger order will produce difficulties, due to the very big number of interplays. The optical pulling force was achieved for all dielectric particle with low losses, that could be very helpful for further experimental studies

Acknowledgements

The authors acknowledge the financial support provided by Russian Fund for Basic Research (Projects No. 18-02-00414, 18-52-00005); Ministry of Education and Science of the Russian Federation (GOSZADANIE, Grant No. 3.4982.2017/6.7) and Government of Russian Federation, Grant 08-08. The calculations of multipole moments have been supported by the Russian Science Foundation Grant No. 16-12-10287

References

[1] Ashkin A., Dziedzic J. M., Bjorkholm J. E., and Chu S., 1986 Observation of a single-beam gradient force optical trap for dielectric particles *Opt. Lett.*, **11**, pp. 288.
[2] Ashkin, A., & Dziedzic, J. M. 1971 Optical levitation by radiation pressure *Applied Physics Letters*, *19*(**8**), pp. 283-285.
[3] Wang, M. D., Yin, H., Landick, R., Gelles, J., & Block, S. M. 1997 Stretching DNA with optical tweezers *Biophysical journal*, *72*(**3**), 1335-1346.
[4] MacDonald M. P., Spalding G. C., Dholakia K. 2003 Microfluidic sorting in an optical lattice *Nature* **426**, pp. 421.
[5] Agarwal, R., Ladavac, K., Roichman, Y., Yu, G., Lieber, C. M., & Grier, D. G. 2005 Manipulation and assembly of nanowires with holographic optical traps. *Optics Express*, 13(**22**), pp. 8906-8912.
[6] Maruo, S., Nakamura, O., & Kawata, S. 1997 Three-dimensional microfabrication with two-photon-absorbed photopolymerization *Optics letters*, *22*(**2**), pp. 132-134.
[7] Villangca, M. J., Palima, D., Bañas, A. R., & Glückstad, J. 2016 Light-driven micro-tool equipped with a syringe function *Light: Science & Applications*, *5*(**9**).

[8] Galajda P. and Ormos P. 2001 Complex micromachines produced and driven by light *Appl. Phys. Lett.* **78**, pp 249–25.
[9] Rodríguez-Fortuño, F. J., & Zayats, A. V. 2016 Repulsion of polarised particles from anisotropic materials with a near-zero permittivity component. *Light: Science & Applications*, 5(**1**), pp. 16022.
[10] Petrov, M. I., Sukhov, S. V., Bogdanov, A. A., Shalin, A. S., & Dogariu, A. 2016 Surface plasmon polariton assisted optical pulling force, *Laser & Photonics Reviews*, 10(**1**), pp. 116-122.
[11] Ivinskaya, A., Petrov, M. I., Bogdanov, A. A., Shishkin, I., Ginzburg, P., & Shalin, A. S. 2017 Plasmon-assisted optical trapping and anti-trapping. *Light: Science & Applications*, *6*(**5**), pp. 16258.
[12] Yue, L., Minin, O. V., Wang, Z., Monks, J. N., Shalin, A. S., & Minin, I. V. 2018 Photonic hook: a new curved light beam *Optics letters*, *43*(**4**), 771-774.
[13] Dholakia, K., & Zemánek, P. 2010 Colloquium: Gripped by light: Optical binding. Reviews of modern physics, 82(**2**), pp. 1767.
[14] Kostina, N., Ivinskaya, A., Sukhov, S., Bogdanov, A., Toftul, I., Nieto-Vesperinas, M., ... & Shalin, A. 2017 Optical binding via surface plasmon polariton interference. arXiv preprint arXiv:1708.05471.
[15] Chen, J., Ng, J., Lin, Z., & Chan, C. T. 2011 Optical pulling force. *Nature photonics*, *5*(**9**), pp. 531.
[16] Novitsky, A., Qiu, C. W., & Wang, H. 2011 Single gradientless light beam drags particles as tractor beams *Physical review letters*, *107*(**20**), pp. 203601.
[17] Liesener, J., Reicherter, M., Haist, T., & Tiziani, H. J. 2000 Multi-functional optical tweezers using computer-generated holograms *Optics Communications 185*(**1-3**), pp.77-82.
[18] Curtis, J. E., Koss, B. A., & Grier, D. G. 2002 Dynamic holographic optical tweezers. *Optics communications*, *207*(**1-6**), pp. 169-175.
[19] Jahani Saman & Jacob Zubin 2016 All-dielectric metamaterials, *Nature Nanotechnology* (**11**), pp. 23–36

Passive Decoupling Techniques in Ultra-High Field MRI

Anna Hurshkainen[1], **Constantin Simovski**[2], **Stanislav Glybovski**[1]
[1]Department of Nanophotonics and Metamaterials, ITMO University, 197101 St. Petersburg, Russia
[2]School of Electrical Engineering, Department of Electronics and Nanoengineering, Aalto University, 13000, FI-00076 AALTO, Finland

E-mail: a.hurshkainen@metalab.ifmo.ru

Abstract. In this work a comparison of two passive decoupling techniques for ultra-high field magnetic resonance imaging (MRI) is presented. An electromagnetic band-gap (EBG) structure and a single passive dipole are compared in terms of their capabilities to decouple two active dipole body array elements at 7 Tesla at an extremely short inter-element distance. Decoupling band as well as distribution of circularly polarized radiofrequency magnetic field inside a homogeneous phantom provided with these two techniques are compared.

1. Introduction

Ultra-high field (UHF) MRI (over 7 Tesla) is an extensively developed tool for medical diagnostics and preclinical biomedical research. UHF imaging provides enhanced spatial resolution of images, thanks to the improved signal-to-noise ratio (SNR) ensured by extremely high static magnetic field. A complex MRI system comprises multiple blocks each one affecting SNR of images. Moreover, SNR is known to depend on the static magnetic field intensity. A radiofrequency (RF) block of an MR scanner contains so-called coils – RF antennas used for nuclei excitation and MR signal detection. In low- and high-field MR scanners volumetric coils are mostly used for excitation since they provide homogeneous RF magnetic field distribution over the whole human body. At the same time, for reception local coils are used providing high SNR over the particular region of interest (ROI). In ultra-high field MRI this strategy of using volumetric coils for transmission and local dedicated coils for reception is not feasible anymore. Due to the elevated RF signal frequency (298 MHz at 7 Tesla for instance), the wavelength inside a high-permittivity human body becomes comparable to its size (around 10 cm at 7 Tesla), which causes interference artifacts - signal voids. Therefore dedicated surface coils are used in UHF for transmission as well, mostly combined in coil arrays providing RF transmit magnetic field homogeneity over targeted ROI via parallel transmission. In this technique constructive interference of RF signal, created by multiple local RF coils is provided by the input signal phase manipulation of each coil element. Importantly all the elements of a coil array should be isolated one from another, i.e. should have a negligible mutual impedance, which is mostly not the case due to short distances between neighboring antennas. The way of mutual coupling suppression in coil arrays to be chosen mostly depends on the type of antennas comprising an array. For the magnetic-loop coil mutual inductance compensation can be achieved simply by overlapping

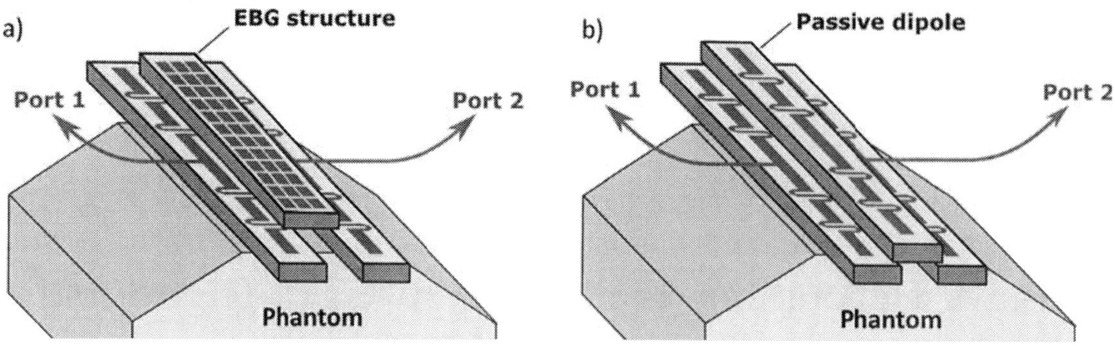

Figure 1. Two considered techniques for passive decoupling of dipole coil arrays operating at 7 Tesla: a) EBG structure ; b) passive dipole

neighboring elements of an array [1]. Other passive decoupling techniques are available for loop coils such as: passive resonant loop [2], stacks of miniaturized resonators comprising so-called magnetic wall structures [3] for mutual inductance reduction. In [4] decoupling of TEM elements of head coil array by stacked magnetic resonators was demonstrated.

In this work recently demonstrated [5],[6] methods of decoupling of dipole antennas in arrays for parallel transmission at 7 Tesla are compared. A relative frequency band of decoupling as well as the intensity and homogeneity of a created circularly polarized RF transmit magnetic field (B_1^+) are studied.

2. Mutual coupling reduction by EBG structure

A comprehensive analysis of using EBG structures for mutual coupling reduction of dipoles body array elements at 7 Tesla MRI was demonstrated in [5]. The employed EBG structure comprised a block of coupled periodically arranged unit cells of "mushroom" type: with a metal square patch connecting to the ground plane by a thin metal wire. Sub-wavelength operation of the EBG structure (containing 3×15 unit cells) was provided by an additional layer of patches introducing high capacity and tuning the unit cell at the operational frequency near 298 MHz while having a period of EBG structure equal to 20 mm. One of eigenmodes of the finite structure was used for decoupling. An experimental sample of the EBG structure was build and its decoupling capabilities were tested using two fractionated dipoles [7] located over the homogeneous liquid phantom with the permittivity of 45 and conductivity of 0.25 S/m. The axes of the dipoles were separated by the distance of 30 mm. S-parameters of two dipoles were measured by the vector network analyzer. Experimental setup is depicted in Fig. 1a and its detailed description is given in [5].

The frequency dependent reflection coefficient of two passive dipoles equipped with the EBG structure is depicted in Fig. 2a by the dashed line, while their transmission coefficient is plotted in Fig. 2b by the dashed line as well. Matching network was attached to both antennas to tune their input impedance in presence of the decoupling structure to 50 Ohm at the resonance frequency of 298 MHz. The transmission coefficient curve reveals that the relative decoupling band, where transmission coefficient is below -16 dB, is equal to 0.8 MHz. A single-frequency simulation of the above described setup was performed in CST Microwave Studio in order to evaluate the field created by the dipoles in the phantom in the presence of the EBG structure. Fig. 3a shows the transverse plane distribution of the circularly polarized magnetic field B_1^+ created in the transmit mode inside the homogeneous phantom normalized to 1W of accepted power.

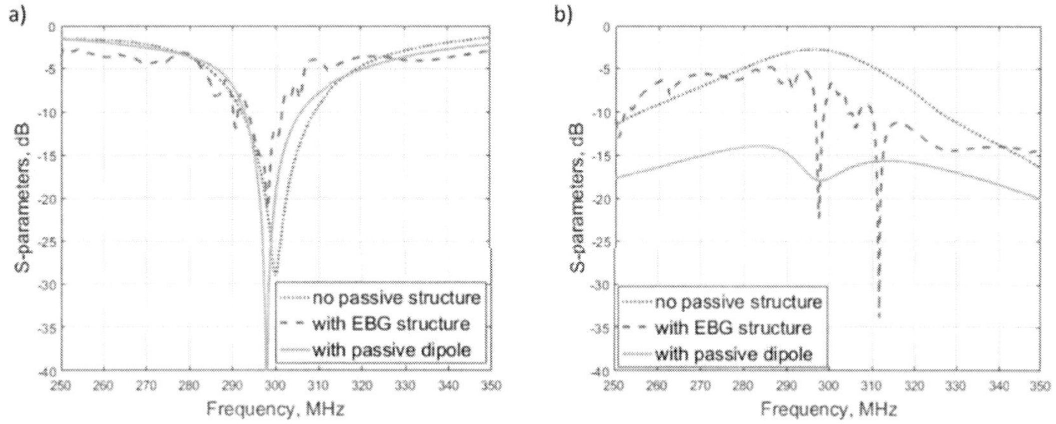

Figure 2. S-parameters of two active dipoles in the vicinity of passive decoupling structures: a) reflection coefficient; b) transmission coefficient

3. Mutual coupling reduction by a single passive dipole

In [6] another passive decoupling structure was studied analytically, numerically and experimentally. An analytical expression was derived using the electromotive force method determining the condition of full decoupling (for any combination of currents) of two active dipoles by a passive one, located in a middle between them. Using the aforementioned condition and analytical expressions for self- and mutual impedances of resonant dipoles, it was shown that full decoupling can be achieved within their resonance band. In this work we extended this study onto the pair of active fractionated dipoles equipped with the passive one located between them as depicted in Fig. 1b. This setup was studied numerically in CST Microwave Studio. S-parameters of the two dipoles in the frequency range of 250 – 350 MHz were calculated. The reflection coefficient of the dipoles in the presence of a passive one is depicted in Fig. 2a by the solid line, while the transmission coefficient is plotted in Fig. 2b by the solid line as well.

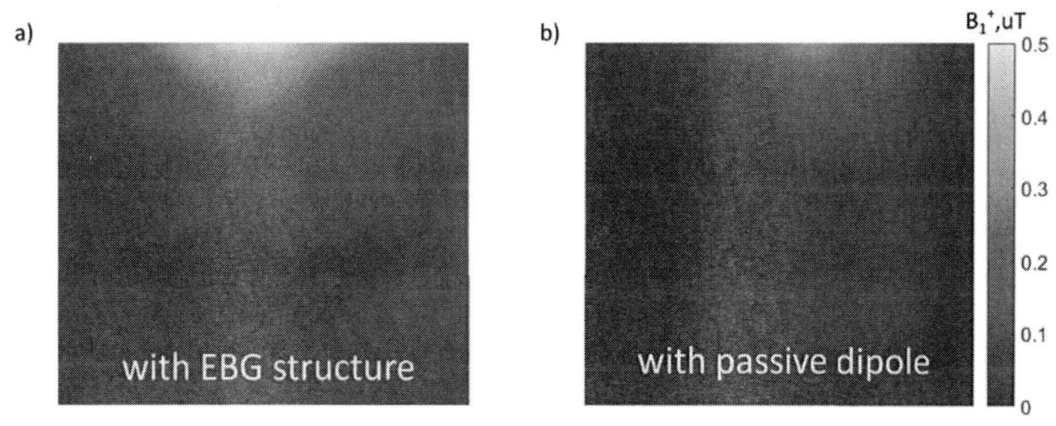

Figure 3. Transverse plane B_1^+ distribution normalized to 1W of accepted power of an active dipole in the vicinity of second active dipole and a passive decoupling element (area dimensions 350 by 200 mm, frequency 298 MHz) a) EBG structure b) dipole

Both active antennas were matched virtually at 298 MHz using the schematic toolbox of CST. Regarding to the S_{12} curve the relative decoupling band is 11 MHz, where the transmission coefficient stands below -16 dB. The field B_1^+ emitted by active dipoles at 298 MHz was also calculated and its transverse plane distribution normalized by 1W of accepted power is depicted in Fig. 3b.

4. Discussion and conclusion

Our comparative study of the two considered decoupling techniques revealed benefits and drawbacks in both of them. The relative decoupling frequency band determined by the level of S_{12} below -16 dB is 0.8 MHz for the EBG structure, while for the passive dipole approach it is 14 times broader. Though the band provided by both the techniques is sufficient for the application in MRI. The wider band of the single dipole can be more stable with respect to variation of the phantom properties. On the other hand the numerical study of the created field in the phantom revealed the benefit of the EBG structure in terms of its effect on the level and homogeneity of B_1^+. Since unit cells of the EBG structure are extremely miniaturized (much smaller than the operational wavelength) they do not produce high scattered radiative field and therefore do not create any destructive interferences of B_1^+ inside the phantom. This can be concluded from Fig. 3a, where the field produced by a dipole is similar to one produced in the absence of a decoupling structure. For the passive dipole the situation is opposite: the resonant passive dipole is an effective scatterer and creates destructively interfering fields in the bulk of the phantom reducing the efficiency of active dipoles. As can be seen in Fig. 3b this leads to a reduced level of B_1^+ and a distorted field distribution with worse homogeneity. Therefore choosing the decoupling strategy is a trade-off between the bandwidth stability of radiative elements and their transmission efficiency (magnetic field magnitude normalized by an accepted power).

In this work two passive decoupling techniques were comparatively studied: the EBG decoupling structure and the passive dipole decoupling technique. Their decoupling bands as well as their impact on the transmission efficiency of active dipoles were studied. Comparative study revealed that the choice of decoupling strategy for each particular case is a trade-off between these two aspects.

References

[1] Roemer P.B., Edelstein W.A., Hayes C.E., Souza S.P., and Mueller O.M., "The NMR phased array," *Magn. Reson. Med.*, vol. 16, pp. 192-225, 1990.

[2] Yan, X., Zhang, X., Feng, B., Ma, C., Wei, L., and Xue, R, "7T transmit/receive arrays using ICE decoupling for human head MR imaging," *IEEE Trans. Med. Imag.*, vol. 33(9), pp. 1781-1787, 2014.

[3] Connell, I. R., Gilbert, K. M., Abou-Khousa, M. A., and Menon, R. S, "Design of a parallel transmit head coil at 7T with magnetic wall distributed filters," *IEEE Trans. Med. Imag.*, vol. 34(4), pp. 836-845, 2015.

[4] Georget, E., Luong, M., Vignaud, A., Giacomini, E., Chazel, E., Ferrand, G., and Bonod, N, "Stacked magnetic resonators for MRI RF coils decoupling," *Journal of Magn. Reson.*, vol. 275, pp. 11-18, 2017.

[5] Hurshkainen A. A., Derzhavskaya T. A., Glybovski S. B., Voogt I. J., Melchakova I. V., Van den Berg C. A. T., and Raaijmakers A. J. E., "Element decoupling of 7 T dipole body arrays by EBG metasurface structures: Experimental verification," *J. Magn. Res.*, vol. 269, pp. 87–96, Aug. 2016.

[6] Mollaei M. S. M., Hurshkainen A., Glybovski S., and Simovski C., "Passive decoupling of two closely located dipole antennas," available in Arxiv:1802.07500.

[7] Raaijmakers, A. J., Italiaander, M., Voogt, I. J., Luijten, P. R., Hoogduin, J. M., Klomp, D. W., and van den Berg, C. A, "The fractionated dipole antenna: A new antenna for body imaging at 7 Tesla," *Magn. Reson. Med.*, vol. 75(3), pp. 1366-1374, 2016.

Acknowledgment

This work was supported by the Russian Science Foundation (Project No. 18-19-00482).

Properties of composite coatings formed by treatment of PEO-layers with fluoroparaffins

K V Nadaraia[1,2], D V Mashtalyar[1,2], I M Imshinetsky[1,*], A S Gnedenkov[1,2], S L Sinebryukhov[1], A N Minaev[2] and S V Gnedenkov[1,2]

[1]Institute of Chemistry of FEB RAS, 159 pr. 100-letiya Vladivostoka, Vladivostok, 690022, Russia
[2]Far Eastern Federal University, 8 Sukhanova St., Vladivostok, 690950, Russia

*e-mail: igorimshin@gmail.com

Abstract. The formation of protective multifunctional coatings on the MA8 magnesium alloy using the method of plasma electrolytic oxidation (PEO) and subsequent treatment with fluorocarbon materials has been suggested. A complex study of tribological characteristics and morphological features of the obtained composite coatings has been carried out. After the treatment of the PEO-coating with fluoroparaffins the wear resistance has increased by more than 10 times as compared to the base PEO-layer. Hydrophobic properties of the composite coatings have been studied. It was established the contact angle changed in the range from 122° up to 137° depending on the flouroparaffin type.

1. Introduction

The method of plasma electrolytic oxidation has been proven to be efficient at protective coatings formation on magnesium alloys. Due to a high adhesion to the substrate and high strength, such coatings improve sufficiently the part working parameters [1–6]. However, high surface roughness and high porosity reduce the protective characteristics of coatings. To eliminate the above disadvantages researchers suggest different approaches consisting in pores sealing with a chemically inert substance. In this regard, fluoropolymers are characterized by high chemical resistance and stability over time, and they can be considered as one of the promising materials [7–12]. The combination of the chemical stability of the PEO-layer and the chemical inertness of the fluoropolymer materials provides the composite polymer-containing coating with high protective properties. Fluoroparaffins are characterized by high plasticity, low melting point and low melt viscosity, therefore they are promising materials which do not require high temperatures to form composite coatings based on the PEO-layer.

The present work is devoted to studies of the properties of composite coatings on the magnesium alloy MA8 formed by PEO in the silicate-fluoride electrolyte and subsequent treatment with fluorocarbon materials.

2. Experimental part

The PEO-layer formed on MA8 magnesium alloy (wt. %: Mn – 1.3-2.2, Ce – 0.15-0.35, balance – Mg) was used as the base for the formation of the composite coating. The PEO process was carried out in a silicate-fluoride electrolyte ($Na_2SiO_3 \cdot 5H_2O$ – 15 g l^{-1}, NaF – 5 g l^{-1}). The frequency of the polarizing pulses was equal to 300 Hz, the duty cycle was equal to 50%. At the first stage, the anodic component was fixed in the galvanostatic mode at a current density of 0.5 A cm^{-2}, the cathode phase was set potentiostatically at –30 V. The duration of the first stage of the PEO was equal to 200 s, the anodic

voltage reached up to 300 V. At the second stage for 600 s the anodic component decreased potentiodynamically from 300 down to 200 V, and the cathodic one increased from –30 up to –10 V.
To obtain the composite coating, fluoroparaffins – perfluorinated accelerator powders (PAP) developed by Kirovo-Chepetsk Halopolymer Plant were used. Fluoroparaffins were agglomerated particles consisting of a mixture of perfluoroalkanes with different lengths of the carbon chain. To form the composite coating on the surface of the PEO-layer three commercially available fluoroparaffins PAP-90, PAP-110, PAP-180 (where numbers 90, 110, 180 indicate the melting temperature) were used. In accordance with used material the obtained composite coatings are identified below as CC (PAP-90), CC (PAP-110) and CC (PAP-180), respectively.
Fluoroparaffins were applied on the heated surface of the PEO-coated samples until complete melting. After then the spin-coating method at a rotation speed of about 200 rpm for 2-3 s was used to uniform distribution of the material over the surface. The heating temperature of the samples was chosen as 83-85 °C (PAP-90), 113-115 °C (PAP-110), and 183-185 °C (PAP-180).
The thickness of the formed composite layers was measured using the eddy-current thickness gauge VT-201 (KID, LLC, Russia). The surface morphology and microprofile investigation were carried out using non-contact optical surface profiler technique (OSP). OSP370 device installed on the M370 electrochemical workstation (Princeton Applied Research, USA) was used. Data were analysed using the Gwyddion 2.45 software.
Tribological tests were performed on the automated friction machine Tribometer (CSM Instruments, Switzerland) according to the "ball-on-plate" test scheme. A silicon nitride ball with a diameter of 10 mm was used as a counterbody. All studies were carried out under dry friction condition at a temperature of 25 °C and 10 N normal load, F_n. The sliding speed was equal to 50 mm s^{-1}. The experiment was stopped when the whole coating was abraded down to the metal substrate. The precision contact profilometer MetekSurtronic 25 (Taylor Hobson Ltd, England) was used to estimate the cross section area, S, of the wear track after tribological testing. Wear of the material, W, was calculated using equation $W = \Delta V_{sample}/(L \cdot F_n)$, where ΔV_{sample} is the sample volume loss during the test, L is the distance, and F_n is the applied normal load. The sample volume loss was calculated in accordance with $\Delta V_{sample} = S \cdot l$, where l is the track length.
The wettability of the obtained composite coatings was estimated by measuring the contact angle of the sessile droplet deposited on the sample surface using a DSA100 device (KRÜSS GmbH, Germany) with a corresponding software. Deionized water was used as the test liquid for measuring the contact angle. The volume of the drop was equal to 2 µl. The Young-Laplace method [13] was applied to calculate the contact angle.

3. Results and discussion

During the sample rotation due to the centrifugal force the fluoroparaffin was uniformly distributed over the specimen surface and seals the pores of the outer porous layer of the PEO-coating. The increase in the coating thickness as compared to the base PEO-layer was equal to about 5.4 µm (table 1).
Analysis of the surface roughness parameters shown in Table 1 made it possible to establish that fluoroparaffin presence on the PEO-layer surface leads to the increase of the composite coating roughness, regardless of the fluorocarbon material used. During the centrifugation process, the most part of the fluoroparaffins distributed nonuniformly over the heterogeneous layer surface due to the high melt viscosity. This fact was supported by the increase of the roughness parameters: R_t (the vertical distance between the highest and lowest points of the profile), R_a (the arithmetic average of the absolute values of the profile heights) and R_z (the average value of the absolute values of the heights of five highest profile peaks and the depths of five deepest valleys), and S_{real}/S_{pr} (the ratio of the calculated real surface area value to the area of its orthogonal projection), as well as the real surface area. A portion of the molten fluoroparaffin penetrates into the pores and seals them up.

Table 1. Surface roughness parameters of the base PEO-layer and composite coatings obtained using various fluoroparaffins

Coating type	PEO-coating	CC(PAP-90)	CC(PAP-110)	CC(PAP-180)
Thickness, μm	24±2	28.8±1.3	31±3	29±2
R_a, μm	0.71±0.11	0.92±0.12	0.93±0.14	0.96±0.14
R_z, μm	3.9±0.8	4.9±0.5	4.6±0.7	4.9±0.7
R_t, μm	4.9±0.9	6.2±0.8	6.0±1.3	6.9±0.9
S_{real}/S_{pr}	1.016	1.030	1.021	1.020

The scratch test data analysis confirms the positive effect of the fluoroparaffins in the composite coating on its mechanical properties (table 2). The increase of the load (L_{C2}) at which the coating adhesion strength alteration began, as well as the load (L_{C3}), at which scratching of the coating to the substrate occurred (plastic abrasion of the film to the metal) compared to the base PEO-coating was experimentally detected. The values of L_{C2} and L_{C3} for all composite coatings are practically identical. A slight difference is probably related to the nature of the fluoroparaffin used in the composite layer formation.

Table 2. The mechanical properties of the coatings

Coating type	L_{C2}, N	L_{C3}, N	Wear, mm^3 (N m)$^{-1}$	Number of cycles before coating abrasion
PEO-coating	8.4 ± 0.9	11.0 ± 1.5	$1.7\cdot10^{-3}$	2457
CC(PAP-90)	11 ± 3	12.6 ± 1.0	$1.5\cdot10^{-4}$	14377
CC(PAP-110)	10.3± 1.6	12.6 ± 0.6	$8.2\cdot10^{-5}$	41652
CC(PAP-180)	10.2 ± 0.4	11.3 ± 0.3	$8.5\cdot10^{-5}$	40560

The analysis of the tribological tests presented in figure 1 as a function of the friction coefficient versus the number of cycles revealed a significant fluoroparaffin effect on the wear time increase of the composite coating. The abrasion time increases from 5.8 up to 16.9 folds in comparison with the base PEO-coating (table 2). The fluoroparaffin in the porous layer works as a dry lubricant reducing the wear of the PEO-sublayer. For all samples with composite coatings monotonous abrasion is observed with friction coefficient increasing at the end of the experiment. Samples CC(PAP-110) and CC(PAP-180) have practically the same dependence of the friction coefficient on the number of cycles. During the wear test, there is a gradual friction coefficient increase in the range of about 0.1–0.2 and a sharp rise at the end of the experiment up to 0.4. The composite coatings obtained using fluoroparaffins PAP-110 and PAP-180 have the best wearproof. The wear of such coatings is almost by two orders of magnitude smaller than one of the PEO-layer.

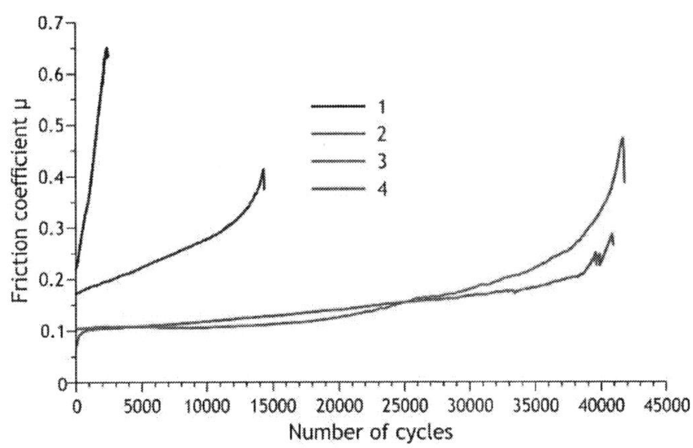

Figure 1. The friction coefficient dependence on the number of cycles for the samples with PEO-coating (1) and with composite coatings obtained with different fluoroparaffins: PAP-90 (2), PAP-110 (3), PAP-180 (4)

The wettability analysis of the composite coatings obtained using various fluoroparaffins indicates their hydrophobic properties (table 3) in contrast to the hydrophilic base PEO-coating. The presence of the fluoroparaffin layer on the PEO-coating enables one to give the hydrophobic properties to the composite coating, as indicated by the contact angle increase from 37° up to 122°–137°, depending on the used material (table 3). The maximum of the contact angle (137°) is registered for CC(PAP-90), the fluoroparaffins of which provide, probably, the minimum surface energy.

Table 3. The contact angles for the samples with various coating types

Coating type	Contact angle, °
PEO-coating	37±2
CC(PAP-90)	137±4
CC(PAP-110)	126±5
CC(PAP-180)	122±3

Conclusion
As a result of the performed studies, the method of the composite coatings formation on the surface of magnesium alloys by means of plasma electrolytic oxidation and subsequent application of fluoroparaffins is developed. The resulting coatings significantly increase the contact angle (from 122° up to 137° depending on the used fluoroparaffin). Fluoroparrafins sufficiently improve the tribological properties of composite coatings. Wear of composite coatings is reduced by two orders of magnitude as compared to the base PEO-layer.

Acknowledgments
This work was supported by the Grant of Russian Science Foundation (project №14-33-00009).

References
[1] Arrabal R, Matykina E, Hashimoto T, Skeldon P and Thompson G E 2009 *Surf. Coatings Technol.* **203** 2207
[2] Hussein R O, Nie X and Northwood D O 2013 *Electrochim. Acta* **112** 111
[3] Gnedenkov S V, Sinebryukhov S L, Mashtalyar D V, Imshinetskiy I M, Samokhin A V and Tsvetkov Y V 2015 *J. Nanomater.* **2015** ID154298
[4] Mashtalyar D V, Gnedenkov S V, Sinebryukhov S L, Imshinetskiy I M and Puz' A V 2017 *J. Mater. Sci. Technol.* **33** 461
[5] Han B, Yang Y, Deng H, Chen Y and Yang C 2018 *Int. J. Electrochem. Sci.***13** 5681
[6] Wang P, Wu T, Xiao Y T, Zhang L, Pu J, Cao W J and Zhong X M 2017 *Vacuum* **142** 21–8
[7] Castellanos A, Altube A, Vega J M, García-Lecina E, Díez J A and Grande H J 2015 *Surf. Coatings Technol.* **278** 99
[8] Martini C, Ceschini L, Tarterini F, Paillard J M and Curran J A 2010 *Wear* **269** 747
[9] Gnedenkov S V, Sinebryukhov S L, Mashtalyar D V, Nadaraia K V, Gnedenkov A S and Bouznik V M 2016 *Corros. Sci.***111** 175
[10] Gnedenkov S V, Sinebryukhov S L, Mashtalyar D V and Imshinetskiy I M 2015 *Surf. Coatings Technol.* **283** 347
[11] Eslamzadeh N, Ebrahimi-Kahrizsangi R, Karbasi S, Zarebidaki A and Gharavi F 2017 *J. Mater. Eng. Perform.* **26** 4255
[12] Imshinetskiy I M, Gnedenkov S V, Sinebryukhov S L, Mashtalyar D V, Samokhin A V and Tsvetkov Y V 2014 *Solid State Phenom.* **213** 125
[13] Li D and Neumann A W 1992 *J. Colloid Interface Sci.* **148** 190

Thermomechanical ablation under plasmonic field excited by ultrashort laser pulse. Part I: Plasmonics

A.I. Ignatov[1,2,3], **V.V. Zhakhovsky**[1,4], **A.M. Merzlikin**[1,2,3], **and N.A. Inogamov**[4,1]

[1] Dukhov All-Russia Research Institute of Automatics, 22 ul. Sushchevskaya, Moscow 127055, Russia

[2] Moscow Institute of Physics and Technology, 9 Institutskiy per., Dolgoprudny, Moscow Region, 141700, Russia

[3] Institute for Theoretical and Applied Electromagnetics of the Russian Academy of Sciences, 13 Izhorskaya ul., Moscow 125412, Russia

[4] Landau Institute for Theoretical Physics of the Russian Academy of Sciences, 1a prosp. Akademika Semenova, Chernogolovka, Moscow Region 142432, Russia

E-mail: nailinogamov@gmail.com

Abstract. Ultrashort laser pulse with duration $\tau_L \sim 1$ ps, e.g. acting in Kretschmann geometry, supports a standing plasmonic wave in a thin metal film deposited onto a surface of transparent dielectric prism during the pulse. It is very important that the standing plasmonic wave spatially redistributes energy of the 1st laser beam which excites a surface plasmon polariton (SPP), where SPP traps this energy, and laser energy incident on SPP (laser 2, region where standing wave locates). Thus a spatially periodic heating pattern is formed with dissipative hot spots around the antinodes and cold spots around the nodes of the standing wave. Distance between two subsequent nodes equals to plasmonic wavelength ~ 0.7 μm. Another important circumstance is the *ultrashort* duration of the non-homogeneous heating. It is shorter than the acoustic time scale $t_s = d_f/c_s$ defining mechanical response of the film to heating; here d_f is film thickness and c_s is a sound speed in metal. t_s is about a few tens of picoseconds for our range of thicknesses. This means that mechanical response of a film to the heating is retarded in time relative to the heating stage, which leads to the subsonic motion of material and large generated stresses because they are directly proportional to volume density of absorbed energy. Fast freezing of moving molten film fixes its deformed configuration. This opens a new way for fabrication of desirable structures at nanoscales. In the first part of our work we present a picture of standing wave. The second part given separately is devoted to the thermal and hydrodynamic response to the fast heating by the standing wave.

1. Introduction. Modification of surface properties by structuring can be done by different techniques listed in the paper II which follows this paper I. An important weakly studied approach using a joint action of ultrashort laser pulse and surface plasmon-polariton (SPP) wave is studied in our work divided on two parts. Here in the Part I we consider an electromagnetic (EM) configuration which creates a sequence of hot spots along a thin gold film mounted above the dielectric substrate. These hot spots are the antinodes of an EM standing wave. Below a scheme of such standing wave generated by interferences between a propagating SPP wave and a laser beam is considered. In the next part II we describe thermomechanical consequences of this joint fast thermal action onto a film. It is shown that this action results in appearance of ripples looking like periodic bending of a film.

IOP Conf. Series: Journal of Physics: Conf. Series **1092** (2018) 012051 doi:10.1088/1742-6596/1092/1/012051

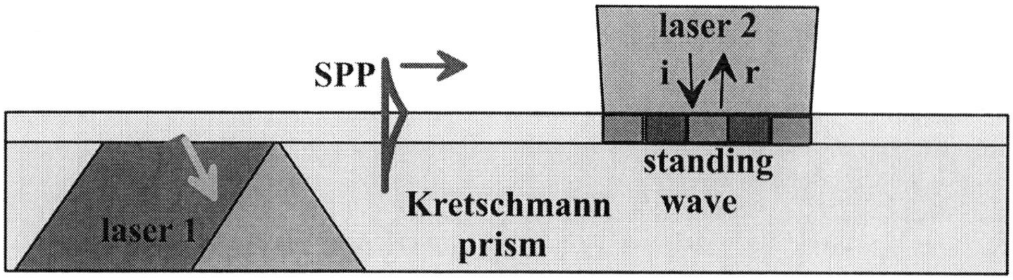

Figure 1. Generation scheme of nanostructured piece of a film with using plasmon-assisted management of dissipative spots location. The upper part of the Kretschmann prism with the gold film is shown. Laser 1 excites a surface plasmon polariton (SPP) wave. The SPP achieves the illuminated spot 2 where SPP together with laser beam 2 produces the standing wave. The standing wave generates the sequence of hot and cold intervals in the film. This sequence imprints into a final surface nanostructure.

2. Scheme of splitting of a laser spot 2 into a sequence of small hot spots. General scheme describing the new method of fine manufacturing is presented in figure 1. Briefly, (i) the laser beam 1 inclined at a resonance angle excites a SPP (surface plasmon polariton) according to the Kretschmann scheme [1]. The SPP propagates to the right side from the illuminated spot 1 in figure 1; (ii) there is the laser beam 2 consisting of incident and reflected waves (reflection from a film); (iii) the excited SPP interferes with the beam 2 forming a standing wave in the beam spot. The standing wave is schematized in figure 1 by the red and blue alternating motionless rectangles. The red ones relate to antinodes and places of high light dissipation power, while the blue ones correspond to the nodes. (iv) Later in time the motionless chain of the hot and cold rectangles imprints into geometry of a film forming designed nanostructure, see details in the Part II.

3. Hierarchy of three time scales. The problem concerning SPP-assisted structuring is difficult because formally the computational code has to include electrodynamic (ED), thermodynamic, and hydrodynamic subcodes combined consistently. We resolve these difficulties by separating physical processes according to their rates: fast, moderate, and slow, as described in Part II. The fastest one taking ~ 1 ps is EDs and the heating of film through its depth. Separation of a film from substrate, inflation of ripples, and freezing of them take longer time. Thus we can neglect deformations in description of the fastest process discussed here in Part I.

4. Electrodynamic simulation. Scheme of electrodynamic problem is shown in figure 1, and computation technique is partially presented in [2]. A gold film $d_{\mathrm{f}} = 80$ nm thick is deposited onto substrate made from fused silica. Two Gaussian $\lambda_{\mathrm{L}} = 785$ nm 2D laser pulses 1 and 2 are used. 2D means that in electrodynamic simulation the laser beams are infinitely long in the direction perpendicular to the plane of figure 1. Polarization vector E of electromagnetic field is in the plane of figure 1. Refractive indexes for this wavelength are $n = 0.149 + 4.78i$ [3] and $n = 1.454$ for gold and fused silica, respectively. The axis x of the reference frame in figure 1 is directed along a film in the plane of figure; the axis y is directed up; while the axis z is directed to us from the plane of figure 1 (right-hand triple). The level $y = 0$ corresponds to the contact boundary between gold and vacuum (or air). The level $y = -80$ nm is the film-silica contact.

The laser beam 2 in figure 1 is a 2D Gaussian beam limited in its width in plane (x, y) and unlimited in the direction z. The beam 2 comes from vacuum at normal angle of incidence. Beam e-folding radius ω_0^{inc} on an amplitude of electromagnetic field is $6\lambda_{\mathrm{L}}$. Intensity distribution of this beam is shown in figure 2(b).

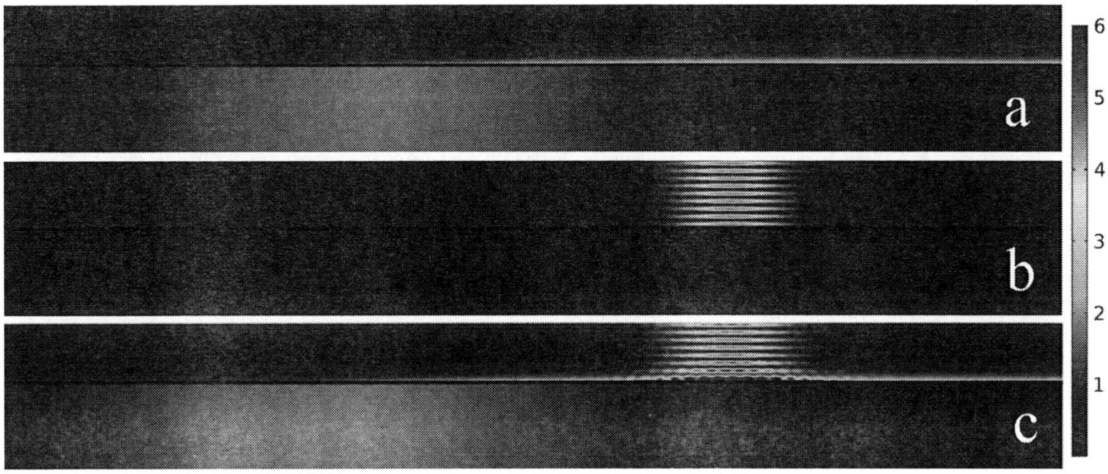

Figure 2. a) Intensity distribution of electric field for the Gaussian beam 1 (see figure 1) illuminating a film from the substrate (optically dense fused silica) side. The incident angle $\alpha = 44.8°$ corresponds to excitation of surface plasmon in the Kretschmann configuration. Electromagnetic field of the beam 1 is p-polarized. In this case a surface plasmon polariton (SPP) is excited at the upper boundary of a film shown in figure 1. The beam 1 is seen as the rather wide vertical light strip at the bottom left half of the substrate. The dark thin horizontal strip is the gold film dividing silica below and vacuum above. The bright thin strip above the film relates to averaged intensity of SPP propagating to the right side. Brightness and width in y-direction of this strip increases as it crosses the exciting beam 1, and after crossing the brightness and width remain constant at the considered distance. b) Distribution of intensity of electric field in the case of the beam 2 from figure 1. The beam comes normally from vacuum side. It is polarized in the plane of this figure. For the considered wavelength 785 nm the reflection coefficient of gold is high. Therefore the interference of incident and reflected waves produces a chain of bright horizontal intervals. Horizontal width of the intervals is defined by Gaussian beam width. Vertical distance between two successive intervals is $\lambda_L/2$. c) Time averaged intensity distribution if both laser beams 1 and 2 are present. The cases with beam 1 and beam 2 are shown in the above two panels separately. Now the SPP propagating from the spot 1 in figure 1 interferes with incident and reflected waves of the beam 2. The interference results in the standing wave with hot spots in its antinodes. These hot intervals are well seen as the bottom sequence of the short red intervals. Length of these intervals equals to the SPP wavelength. Dissipation in the skin layer of the film follows this sequence of the bottom red intervals. Thus the hot rectangles shown in figure 1 are produced. The color bar with arbitrary units is the same for all three panels (a), (b), and (c).

Figure 2(a) presents the intensity distribution for the gold film illuminated from the substrate side solely by the beam 1 shown in figure 1. The e-folding radius (at the illuminated surface) of decrease of amplitude of Gaussian beam when we go out from the beam axis is $\omega_0^{Kr} = 20\lambda$ in our ED simulation As was mentioned above, the beam 1 comes at the resonant angle $\alpha = 44.8°$ relative to a normal direction to the film. At these conditions (propagation from optically dense medium to metal film) according to the Kretschmann scheme [1] the SPP is excited at the metal-vacuum contact, because this angle corresponds to the resonance for excitation. The SPP wavelength for the consider pair metal-substrate and the chosen wave λ_L is equal to $\lambda_{spp} = 768$ nm, which is slightly less than $\lambda_L = 785$ nm.

In the right side of figure 2(a) at the upper film boundary we see the excited SPP wave

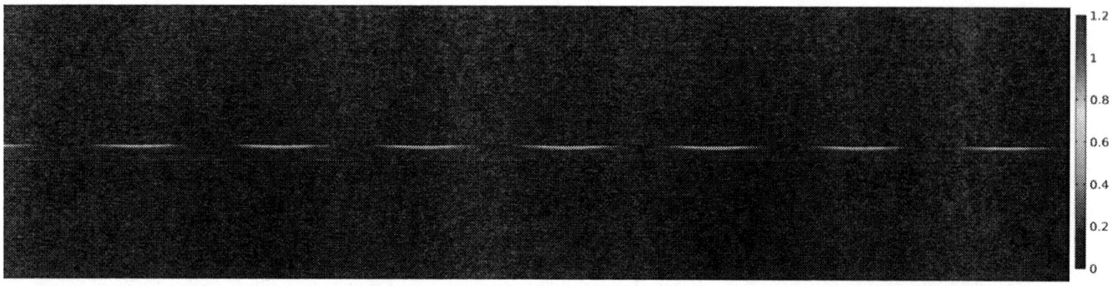

Figure 3. Spatial distribution of electromagnetic heating power inside the gold film. The fused silica is below the film, vacuum is above. The scheme of illumination by beams 1 and 2 is shown in figure 1. The vertical axis of the beam 2 is placed in the middle of this frame. The film thickness of 80 nm is ten times less than the SPP wavelength of $\lambda_{\rm spp} = 768$ nm. The more intense dissipation in the skin-layers near the upper and bottom boundaries is seen. The horizontal shift of the maximums of dissipation between the upper and the bottom boundaries is observed.

propagating in the right direction from the spot 1. The pattern of interference between the SPP and the beam 2 is clearly seen in figure 2(c). which presents intensity distribution of the total sum obtained from incident and reflected beams 1 and 2, together with SPP. In the presented here particular run of electrodynamic simulation the distance between axes of the two beams is 20 μm. To achieve the maximum visibility of the interference pattern in the spot 2 (see figures 1 and 2) the ratio of powers of the beams 1 and 2 was specially adjusted. To achieve a maximum the power of the beam 1 should be four times larger than power of the beam 2.

The calculated spatial distribution of local power $Q(x, y)$ of energy dissipation in metal per unit of volume for described above conditions is presented in figure 3. In the hydrodynamic and molecular dynamic simulations in Part II we use absorbed fluence $F_{\rm abs}(x)$. This function is $F_{\rm abs}(x) = \int Q(x, y)\, dy$, where the integral is taken through the film thickness. Heating at a picosecond time scale becomes uniform through the vertical intervals crossing the film. While spreading of heat along the SPP spatial period lasts much longer than the duration of the acoustic stage at which the partial or total separation of a film takes place. This is why we use the fluence $F_{\rm abs}(x)$ as the input data for the simulation of the second (or acoustic) stage described in Part II.

Conclusion

The complicated problem of how the "hot" plasmonics can be used for formation of surface nanostructures is studied. Solution is based on consecutive treatment of physical processes based on their time scales. Orders of magnitudes differ these scales from ~ 1 ps for EDs and heating through film thickness (subproblem #1 for the fastest processes) to 10–20 ps for sound propagation through the film (subproblem #2 for acoustic scale and separation from substrate) and to tens of ns for the slowest stage of inflation of ripples, their cooling and freezing due to lateral heat transfer (subproblem #3 for formation and freezing of ripples). Here the subproblem #1 is studied. The next two subproblems are discussed in Part II.

Authors INA and ZVV are supported by the RSF grant 14-19-01599.

References

[1] Kretschmann E and Raether H 1968 *Z. Naturforsch.* **23a** 2135–2136
[2] Ignatov A and Merzlikin A 2018 *Journal of the Optical Society of America B* **35** accepted
[3] Johnson P B and Christy R W 1972 *Phys. Rev. B* **6**(12) 4370–4379

Thermomechanical ablation under plasmonic field excited by ultrashort laser pulse. Part II

A.I. Ignatov[1,2,3], **V.V. Zhakhovsky**[1,4], **A.M. Merzlikin**[1,2,3], **and N.A. Inogamov**[4,1]

[1] Dukhov Research Institute of Automatics, 22 ul. Sushchevskaya, Moscow 127055, Russia
[2] Moscow Institute of Physics and Technology, 9 Institutskiy per., Dolgoprudny, Moscow Region, 141700, Russia
[3] Institute for Theoretical and Applied Electromagnetics of the Russian Academy of Sciences, 13 Izhorskaya ul., Moscow 125412, Russia
[4] Landau Institute for Theoretical Physics of the Russian Academy of Sciences, 1a prosp. Akademika Semenova, Chernogolovka, Moscow Region 142432, Russia

E-mail: `nailinogamov@gmail.com`

Abstract. Electrodynamic modeling of running SPP (surface plasmon-polaritons) and interference of the SPP wave with the beam 2 are presented in Part I of our work. Standing wave with alternating hot (antinodes) and cold (nodes) intervals appears thank to the interference. Electrodynamic simulations provide distribution of absorbed energy along the hot and cold intervals, which is utilized in this Part II. Here we discuss how can the ultrafast spatially alternating heating be imprinted into the periodic shape perturbations of a thin film atop a substrate. The Kretschmann's configuration (Kc) and additional (relative to Kc) laser beam #2 for periodic ripples of the film is used.

1. Introduction. Popular techniques of surface nanomodification utilize the ultrafast lasers, e.g. see [1, 2, 3]. Single or multiple laser actions with a number of shots $N > 1$ can be used to fabricate surface nanostructures. In the present work we consider optical lasers and single pulses; x-ray lasers need special analysis. Difference between single and multiple actions in the plane $(N, F_{\rm abs})$ is discussed in [4], see figure 11 in this Ref. Hereinafter $F_{\rm abs}$ is laser energy absorbed or dissipated in a target. In addition, structures formed by laser heating of a metal target through transparent liquid (laser ablation in liquid) have been studied. Different structures appear in the cases with large $R_{\rm L} \gg \lambda$ and small $R_{\rm L} \sim \lambda$ (tight focusing) illuminated spots; here $R_{\rm L}$ is the radius of surface spot illuminated by a beam with a Gaussian spatial distribution of laser intensity, the wavelength λ is $\sim \mu$m for the visible light.

Random nanostructures are formed in the case of large spots [4, 5, 6, 7]. Because we exclude from consideration the case of multiple illuminations of the same spot by $N > 1$ shorts, the LIPSS (laser induced periodic surface structures) are not studied here. The minimal spatial scales of the random surface nanostructures are determined by the surface tension, but not by the light wavelength λ. Diffraction-limited small spots are also used for nanostructuring, but the corresponding structures are very different from those which appear in large spots. Tight focusing produces individual structures – one per a small spot, and those structures are usually used as elements of a large array [3, 8, 9]. There are two methods of manufacturing the array of nanostructures. One of them uses frequent repetitions of shots simultaneously with shifting of surface. Thus every new shot comes to a fresh (previously unirradiated) spot, e.g. see figure 1 in [9]. In the second method a laser ray is split into a few subrays which interfere with each other thus producing simultaneously an array of small illuminated spots on the surface [3].

Above we list known methods of nanomodification of metal surface. The new method proposed here is based on a fine spatial redistribution of dissipated energy governed by surface plasmons. The scheme of the method of nanostructuring is given in figure 1 of the Part I.

2. Splitting the time scales. The problem stated above is too complicated to be numerically solved from its beginning to the end, because the development of a multi-physics computation code requires great efforts. To follow formation of nanostructures at a surface of metal film under the action of electromagnetic (EM) field, the electrodynamic (ED) and thermomechanical processes, both, have to be taken into account in one straight-through code. At the input of such code the target and a laser pulse characteristics should be given. At the output we expect to get a spatially deformed film which has been subjected to heating, melting, and shifting before freezing and forming the final nanostructure. Developing of such multi-physics code is work for future.

We solve the problem by splitting into three time scales. There are three codes and three stages at which the different processes corresponding to these three time scales are simulated. The output of the first code is the input for the second one, while the second output is the input for the third code.

The deformed surface is created from initially smooth plane film. We use a gold film with the thickness $d_f = 80$ nm, see Part I, deposited atop a dielectric prism. The first time scale ~ 1 ps is defined by duration of a laser pulse τ_L and duration needed for homogeneous heating the film through its thickness d_f; spread of dissipated heat along a spatial period of a plasmonic wave ~ 0.7 μm takes much longer times ($\sim$ nanoseconds) and relates to the third stage. We use electrodynamic (ED) code to describe excitation of plasmonic wave by the first laser beam, its interference with the second laser beam, formation of a standing wave, and dissipation in the standing wave, see figure 1 in Part I. The first stage is short relatively to the acoustic time $t_s = d_f/c_s = 27$ ps, where the sound speed $c_s = 3.1$ km/s. Mechanical interaction of a fast heated film with substrate lasts $\sim t_s$. This is the second stage which we solve using the second code: the one-dimensional (1D) two-temperature (2T) hydrodynamic code (2T-HD) [8]. Comparing the first and the second time scales we conclude that we can employ the ED code neglecting appearance of a gap between a film and substrate. The ED code gives distribution of thermal heating along the film; this distribution (obtained in Part I) serves as a starting state for thermomechanical simulations of the second and the third stages (these two stages are described in this Part II).

Only later at the long stage 3 a slow thermal cross-flow between the places where the hot antinodes and the cold nodes of the standing wave were located becomes significant in the energy balance. Due to thermal outflow outside the region occupied by the spot 2 of the laser beam 2 shown in figures in Part I leads to final freezing of the nanostructures. Freezing time should be less than tens nanoseconds to keep structure shapes formed by the ED imprint when the liquid film can move. If a longer time is required for cooling the structure will be smoothed out by capillary effects before freezing. Cooling rates impose strong limitation on the lateral size of the structured spot at the surface which should be less than few tens microns or we have to design alternative scheme of cooling with the usage of substrate with higher thermal conduction.

3. Thermomechanical response to plasmonic heating. For the stage 3 we employ the MD-MC code combining the molecular dynamics (MD) and Monte-Carlo (MC) methods [8]. Monte-Carlo subroutine allows to include electron heat conduction which is high in metal. Combined program follows material motion, melting, fast cooling and recrystallization. It should be emphasized that the non-equilibrium recrystallization with strong overcooling of liquid below melting temperature takes place at our spatiotemporal scales.

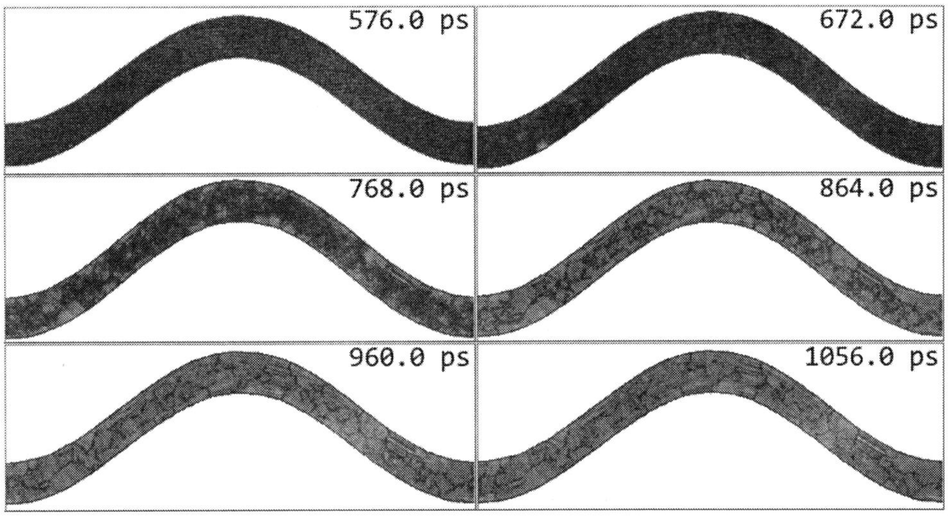

Figure 1. MD-MC simulated development of the SPP rippling. If the film remains in a liquid state, the height of sinusoidal-like perturbation will decrease later in time. Here inertia, surface tension, and the cooling rate are so that solidification proceeds at the stage of capillary stopping. The red color refer to the liquid phase, green – the solid state. Crystallization begins in strongly overcooled liquid gold, and the tiny crystallites of solid phase form gradually at the surface and in the bulk of gold.

In 2T-HD calculations we use the previously established (a) heat conduction coefficient valid in 2T and 1T states [10]; (b) electron-ion coupling parameter [10]; and (c) equation of state of gold [11], which is also valid for 2T and 1T states. 2T effects are significant because the duration of two-temperature relaxation 7–10 ps is comparable to acoustic time scale t_s for gold. The nonequilibrium 2T gold is softer than 1T one at the same density and total internal energy. Because in the 2T state a part of pressure is produced by excited electrons which have 2–3 times lower Gruneisen parameter [11]. This circumstance appreciably decreases expansion velocities at the same absorbed energy in comparison with the hypothetical case when gold always belongs to 1T state after absorption of plasmonic energy. Let's mention that equation of state used in the 2T-HD code describes well the first order phase transitions, melting and crystallization, which are important for our problem.

We use the 2T-HD code to determine the distributions of the temperature $T(x)$ and velocity $v(x)$ along a film, which are the output functions from the stage 2. To calculate those distributions the function $F_{\rm abs}(x)$ of dissipated fluence, which is the output function from the ED stage 1 and the input function for the stage 2 (2T-HD), is used.

Dynamics of one wavelength perturbation in a film under the periodic lateral boundary conditions was considered in a series of MD-MC simulations. Typical results are presented in figure 1. It shows evolution of a film after melting, separation from substrate, deformation under surface tension action, and solidification. Initial thickness of the gold film is 20 nm in our MD-MC simulations. Horizontal length (i.e. in x direction) of the rectangular box is $L = 200$ nm. Thickness of the simulation box in the z direction (perpendicular to the plane of figure 1) is 12 nm. Temperature of gold after fast melting is 1800 K. In this run we neglect the temperature difference along a simulation box because the main thermal process is not equilibration of temperature along a period of a SPP wave but cooling due to transfer of thermal energy outwards the hot spot 2, see Part I. In simulations we use the accurate EAM (embedded atom model) interatomic potential for gold [12] developed using the stress-matching method with the stress-strain curves obtained by DFT (density functional theory). This EAM potential gives the melting temperature of 1330 K close to the experimental melting point of 1337 K for gold.

Initially the velocity distribution $v(x) = v_0 \cos(2\pi x/L)$ with $v_0 = 50$ m/s is imposed in the run presented in figure 1. 2T-HD simulations show that for our range of dissipated energies the velocities of a film after separation are of the order of a few tens of m/s [8]. Capillary scale of speed is $v_{\rm cap} = 2\sqrt{\sigma/(\rho d_{\rm f})}$, where σ is the surface tension, ρ is the density of gold. This speed is 56 m/s for the film with $d_{\rm f} = 80$ nm and $\sigma = 1200$ erg/cm^2. In our MD-MC simulations $d_{\rm f} = 20$ nm and EAM potential has $\sigma \approx 600$ erg/cm^2. Then $v_{\rm cap} = 79$ m/s. MD-MC simulations show that to decelerate a film effectively and to prevent jetting the surface tension should be sufficiently strong to resist inertia. Then the dimensionless capillary normalized velocity $2v_0/v_{\rm cap}$ should be of the order of 1 or less [8]. In the run given in figure 1 the evolution of one SPP period of a film is followed in the coordinate system connected with a center of mass of this period. Momentum of the period is equal to zero in this system. This period moves outward in other coordinate system connected with the bulk of substrate . This corresponds to the few central bumps in the chain of bumps or periods of the SPP wave train covering the spot 2, see paper I. The bumps outside this central interval remains intact with substrate.

There are different situations depending on the dimensionless capillary normalized velocity and the rate of cooling, see [8]. In simulations we use the Langevin thermostat to reach the goal temperature 600 K with the characteristic cooling time 400 ps. The thermostat acts only on the z-component of atom velocity which is perpendicular to the plane of motion. Thus the thermostat does not decelerate motion in the x, y plane shown in figure 1. The thermostat provides approximately homogeneous cooling of gold inside the SPP period due to thermal energy transport outwards from the hot spot 2 which includes a train of the standing SPP wavelengths, see Part I.

The panels in figure 1 show evolution of phase states during cooling and freezing the film. Due to fast cooling the molten gold goes into a deeply supercooled state. Only at time of 590 ps an appreciable growth of nuclei of solid phase begins at temperatures ≈ 870 K, which is much below the equilibrium melting temperature of 1330 K for simulated gold. Nucleation starts like a explosive process driven by homogeneous appearance of multiple seeds of solid state everywhere in a film volume, because metastable liquid gold reaches a significant supercooling degree. Such homogeneous volume solidification takes few hundreds picoseconds to finally freeze gold and stop motion of the film.

Acknowledgements
Work of NAI and VVZ was supported by the RSF grant 14-19-01599.

References
[1] Domke M, Rapp S, Schmidt M and Huber H P 2012 *Opt. Express* **20** 10330–10338
[2] Ivanov D S, Kuznetsov A I, Lipp V P, Rethfeld B, Chichkov B N, Garcia M E and Schulz W 2013 *Applied Physics A* **111** 675–687
[3] Nakata Y, Miyanaga N, Momoo K and Hiromoto T 2013 *Appl. Surf. Sci.* **274** 27–32
[4] Inogamov N, Zhakhovsky V, Ashitkov S, et al 2015 *Engineering Failure Analysis* **47** 328 – 337
[5] Wu C and Zhigilei L V 2016 *J. Phys. Chem. C* **120** 4438–4447
[6] Zhakhovskii V V, Inogamov N A and Nishihara K 2008 *JETP Letters* **87** 423–427
[7] Fang R, Vorobyev A and Guo C 2017 *Light: Sci. Appl.* **6** e16256
[8] Wang X W, Kuchmizhak A A, Li X, Juodkazis S, Vitrik O B, Kulchin Y N, Zhakhovsky V V, Danilov P A, Ionin A A, Kudryashov S I, Rudenko A A and Inogamov N A 2017 *Phys. Rev. Appl.* **8**(4) 044016
[9] Kuchmizhak A, Vitrik O, Kulchin Y, et al 2016 *Nanoscale* **8**(24) 12352–12361
[10] Petrov Y, Inogamov N and Migdal K 2013 *JETP Lett.* **97** 20–27
[11] Petrov Y V, Migdal K P, Inogamov N A and Zhakhovsky V V 2015 *Applied Physics B* **119** 401–411
[12] Zhakhovskii V, Inogamov N, Petrov Y, Ashitkov S and Nishihara K 2009 *Appl. Surf. Sci.* **255** 9592 – 9596

IOP Conf. Series: Journal of Physics: Conf. Series **1092** (2018) 012053 doi:10.1088/1742-6596/1092/1/012053

Recent progress in emittance-controlled optical metasurfaces

Masanobu Iwanaga

National Institute for Materials Science (NIMS), 1-1 Namiki, Tsukuba 305-0044, Japan

E-mail: iwanaga.masanobu@nims.go.jp

Abstract. Emittance is one of the fundamental optical quantities characterizing optical metasurfaces. Optical metasurfaces with high emittance have been demonstrated to be very efficient for enhancing light emission of fluorescent molecules and rare-earth ions, strongly suggesting that the metasurfaces are useful to make light-emission materials more efficient. In particular, we found so far that stacked complementary metasurfaces are able to exhibit thousand-fold larger emission intensity in a quite uniform manner than non-enhancing bulk substrates like Si. We survey our recent progress in the high-emittance metasurfaces.

1. Introduction

Metasurfaces are very thin artificial single-layer or a-few-layer structures fabricated on substrates, and are attracting growing interest in their diverse functions such as ray control [1,2], anti-reflection [3], thermal emission [4–7], polarization and phase control [8–10], holograms [11], enhanced spectroscopy [12], hot electrons [13, 14], and so on.

Here, we focus on recent progress in emittance-controlled optical metasurfaces, and address the strategy and achievements to date. The basic idea is similar, in part, to thermal radiation. It is widely known that blackbody (that is, perfect absorber) is the most efficient thermal emitter because the study on blackbody, so-called Planck radiation, was a trigger of quantum physics. Due to reciprocity, light absorptance is equivalent to emittance [15]; therefore, perfect light absorber is an essential criterion to design ideal emitters. Indeed, metasurface thermal emitters were produced, which met the criterion [7].

To compare the blackbody and optical metasurfaces of artificially designed resonances, let us consider a simple situation. Input energy E_1 is supplied by external light and output energy $E_2(< E_1)$ comes out from the objects. Figure 1 shows schematic energy diagrams of (a) blackbody and (b) discrete electromagnetic (EM) resonant states in optical metasurfaces. Figure 1(c) shows emittance spectrum corresponding to Figure 1(b).

In the blackbody, the input energy is entirely consumed by inner relaxation such as phonon and inelastic scatterings in the constituent materials. Therefore, resultant thermal emission exhibits broad spectrum, reflecting low density of photonic state, and it is, in principle, difficult to selectively emit photons of the particular energy E_2. In contrast, the optical metasurface is designed to have a discrete energy level at the E_2, where the emittance is large (almost 100% in Figure 1(c)). Therefore, it is possible to selectively emit photons at the E_2. This scheme is intuitively understood in thermal emissions; however, this strategy has been proved to be valid for luminescence emission in the metasurfaces [16–19].

We briefly survey recent progress in emittance-controlled optical metasurfaces [16–20] in the next section: fluorescence (FL) enhancement in section 2.1, photoluminescence (PL) from rare-earth Er ions

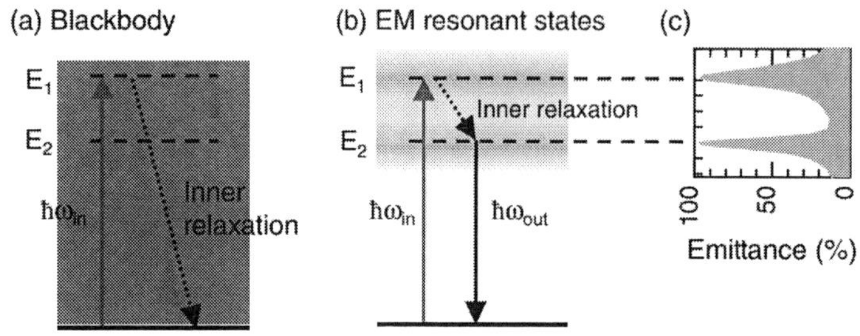

Figure 1. Schematics of energy diagrams of (a) blackbody and (b) EM resonant states of discrete levels in metasurfaces. (c) Emittance spectrum corresponding to (b).

in section 2.2, and unconventional superlinear PL responses from quantum dots coupling with plasmonic metasurfaces in section 2.3.

2. Recent progress

2.1. Fluorescence enhancement in high-emittance metasurfaces

Figure 2 shows a set of typical results for enhanced FL on a stacked complementary (SC) metasurface, which is illustrated in Figure 2(a) and consisted of SC Au layers and perforated silicon-on-insulator (SOI) layer [18]. FL molecules of IR783 (Sigma-Aldrich Inc., USA) were dispersed on the SC metasurface in a quite uniform manner.

Figure 2(b) shows enhanced FL-molecule signals with solid red and dashed green curves, which were measured on the SC metasurfaces with and without 10-carboxy-1-decanethiol (10-CDT) self-assembled monolayer (SAM) (Dojindo Laboratories, Japan), respectively. Reference FL signals are shown in Figure 2(c), which were measured on a flat Si wafer in an equivalent condition to Figure 2(b). From the comparison of the FL signals emitted from the SC metasurfaces and the reference, it is obvious that FL signals are highly enhanced on the SC metasurfaces. In Figure 2(b), for the Raman signals at about

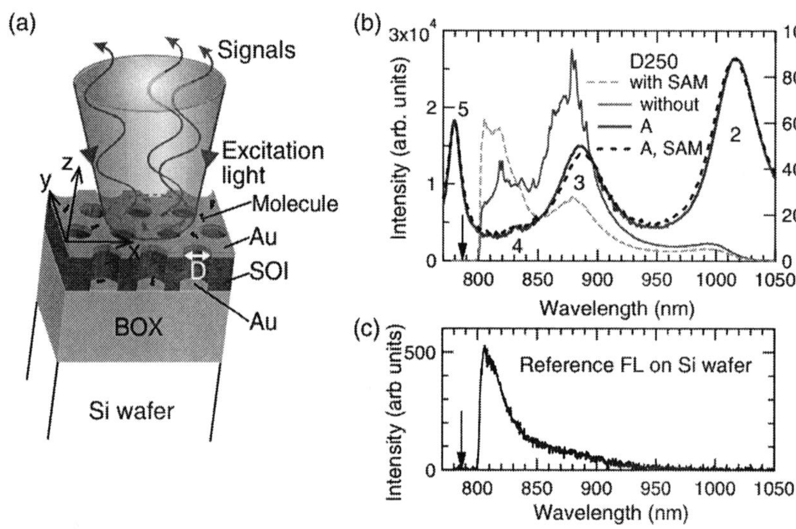

Figure 2. (a) Schematic of optical configuration on the SC metasurface. (b) Enhanced FL spectra (solid red and dashed green curves) measured on SC metasurfaces with and without the SAM, respectively. We note that, on the SC metasurface without the 10-CDT SAM, substantial Raman signals are seen as narrow peaks. (c) Reference FL spectrum of IR783 on Si wafer, measured in an equivalent condition to (b). Adapted with permission from [18].

880 nm that are estimated to be 10000 counts, the Raman signal in the reference spectrum is less than detection limit, which means less than 5 counts; therefore, enhancement factor of the Raman signal was evaluated to be more than 2000. Also, for the FL signal at 1000 nm in Figure 2(b), the enhancement factor was estimated to be 1400. Thus, the FL-intensity enhancing capability of the SC metasurfaces is prominent, being one of the best among the numerous plasmonic structures reported to date [21,22].

In particular, we stress that the enhanced FL and Raman signals appear at absorbance peaks, which are equivalent to emittance peaks due to reciprocity. Emittance is a macroscopic quantity of the metasurface. Thus, the experimental results support that highly enhanced optical signals from FL molecules are attained under emittance-controlled condition. Another successful example using a different FL molecule was reported on the SC metasurfaces [17], where enhancement factor was examined, in more details, in terms of electric-field intensity on the SC metasurfaces, lifetime of the FL, and the quantum yield of FL molecules.

2.2. Selective control of electric- and magnetic-dipole emissions from rare-earth ions

Figure 3(a) shows a schematic of one-dimensional periodic array of plasmon cavities, which are infinitely long along the y axis [19]. Figure 3(b) presents a set of EM-field intensity in the plasmonic cavity. Figure 3(c) shows reflection (solid green curve) and absorption (dashed black curve) spectra in the upper panel. The large absorption peak more than 90% was tuned to the emission wavelengths of Er ions around 1.53 μm. Figure 3(c) also shows electric- (red) and magnetic-field (blue) intensities in the lower panel. The intensities were taken at the positions A and B in Figure 3(b). Figure 3(d) shows scanning tunneling-electron-microscopy (STEM) images. Thin white lines indicated by red arrows represent Er-doped layer inside the plasmonic cavities. Red scale bars indicate 200 nm.

We set the Er-doped layers precisely at the maximum positions of electric- and magnetic-field intensities, and selectively enhanced electric-dipole (ED) and magnetic-dipole (MD) emissions in the Er ions [19]. The ED- and MD-emission-intensities were several-ten-fold enhanced, compared with Er-doped SiO_2 films under an equivalent experimental condition. From the experimental data, we succeeded in extracting ED and MD emission spectra; moreover, Purcell effect was observed for both ED and MD

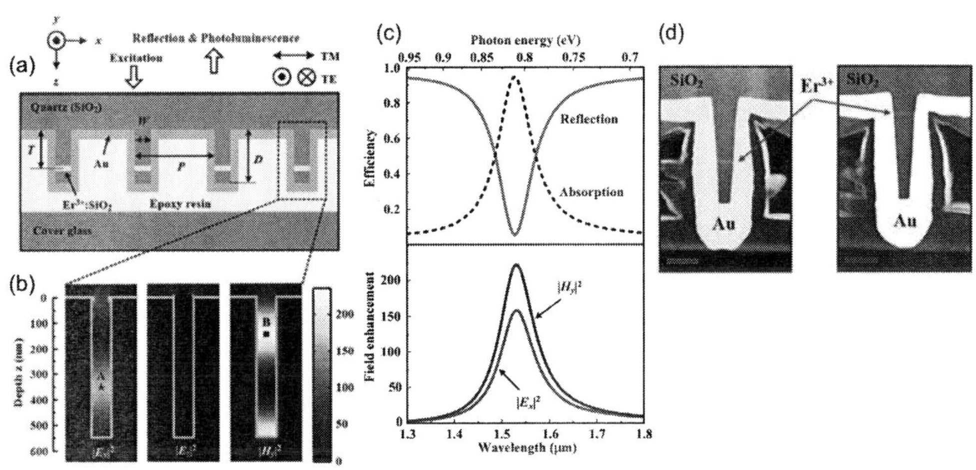

Figure 3. (a) Schematics of plasmonic cavity, in which Er ions are embedded. (b) EM resonant fields in the cavity. (c) Upper panel: reflection (green) and absorption (dashed black) spectra. Lower panel: electric- (red) and magnetic-field (blue) intensities at the positions A and B in (b), respectively. (d) STEM images of actual samples. Thin white layers are the Er-doped layers. Scale bars (red) indicate 200 nm. Adapted with permission from [19].

transitions.

Recently, it was reported that outmost surface of a dielectric metasurface was controlled by liquid crystal to obtain PL of constituent material in a more efficient way [23]. The control was active while the resultant PL-intensity enhancement was limited to be about two-fold.

2.3. Plasmon–quantum-dot coupling systems

The strategy in the preceding subsections is considered to be valid for other luminescent materials such as semiconductors. However, highly prominent results similar to the FL molecules and rare-earth ions have not yet reported to date. This probably means that detailed and relevant designs for the metasurfaces including semiconductors have not yet been obtained and that interplay between metasurfaces and semiconductors is more complicated. Indeed, recent reports have shown that, in coupling systems of plasmonic metasurfaces and quantum dots (QDs), unconventional superlinear PL dynamics takes place due to hot electrons [14, 24].

We mention that efficient extraction of PL from semiconductor QDs has been attempted in many other platforms such as photonic crystals [25, 26], which can be regarded as a kind of metasurfaces when they are located at the surface of bulk substrates. The PL was prominently (or 200∼1200-fold) enhanced at wavelengths of the large photonic density of states; broad PL spectrum coming from the Ge QDs was modified into PL spectrum of several narrow peaks [26]. In the configuration, there is no contribution from hot electrons, in contrast to the plasmon–QD coupling systems [14,24]; therefore, the enhancement mechanism of PL from the Ge QDs is ascribed to the photonic modes of large density of states.

3. Summary

We described our recent progress in emittance-controlled optical metasurfaces. It turns out that emittance, which is a macroscopic quantity featuring metasurfaces, plays a key role in significant enhancement effects on dipole emitters, irrespective of electric and magnetic components.

Acknowledgments

The author thanks Bongseok Choi for the contribution to FL- and rare-earth-emission enhancement experiment, and Takaaki Mano for the growth of III-V semiconductor quantum dots. This study was partially supported by JSPS KAKENHI Grant (number JP17H01066) and by HPCI System Research Project (IDs: hp170134, hp180107). The metasurfaces were fabricated at Nanofabrication Platform and Namiki Foundry in NIMS.

References

[1] Yu N, Genevet P, Kats M A, Aieta F, Tetienne J P, Capasso F and Gaburro Z 2011 *Science* **334**(6054) 333–337
[2] Aieta F, Genevet P, Kats M A, Yu N, Blanchard R, Gaburro Z and Capasso F 2012 *Nano Lett.* **12**(9) 4932–4936
[3] Spinelli P, Verschuuren M and Polman A 2012 *Nat. Commun.* **3** 692
[4] Hesketh P J, Zemel J N and Gebhart B 1986 *Nature* **324**(6097) 549–551
[5] Puscasu I and Schaich W L 2008 *Appl. Phys. Lett.* **92**(23) 233102
[6] Ye Y H, Jiang Y W, Tsai M W, Chang Y T, Chen C Y, Tzuang D C, Wu Y T and Lee S C 2008 *Appl. Phys. Lett.* **93**(3) 033113
[7] Miyazaki H T, Kasaya T, Iwanaga M, Choi B, Sugimoto Y and Sakoda K 2014 *Appl. Phys. Lett.* **105**(12) 121107
[8] Iwanaga M 2012 *Sci. Technol. Adv. Mater.* **13**(5) 053002
[9] Zhao Y and Alù A 2013 *Nano Lett.* **13**(3) 1086–1091
[10] Arbabi A, Horie Y, Bagheri M and Farao A 2015 *Nat. Nanotechnol.* **10**(11) 937–943
[11] Zheng G, Mühlenbernd H, Kenney M, Li G, Zentgraf T and Zhang S 2015 *Nat. Nanotechnol.* **10**(2) 308–312
[12] Iwanaga M 2016 *Plasmonic Resonators: Fundamentals, Advances, and Applications* (Singapore: Pan Stanford Publishing)
[13] Knight M W, Sobhani H, Nordlander P and Halas N J 2011 *Science* **332**(6030) 702–704
[14] Iwanaga M, Mano T and Ikeda N 2018 *ACS Photonics* **5**(3) 897–906
[15] Greffet J J and Nieto-Vesperinas M 1998 *J. Opt. Soc. Am. A* **15**(10) 2735–2744
[16] Iwanaga M, Choi B, Miyazaki H T, Sugimoto Y and Sakoda K 2015 *J. Nanomater.* **2015** 507656

[17] Choi B, Iwanaga M, Miyazaki H T, Sugimoto Y, Ohtake A and Sakoda K 2015 *Chem. Commun.* **51**(57) 11470–11473
[18] Iwanaga M, Choi B, Miyazaki H T and Sugimoto Y 2016 *Nanoscale* **8**(21) 11099–11107
[19] Choi B, Iwanaga M, Sugimoto Y, Sakoda K and Miyazaki H T 2016 *Nano Lett.* **16**(8) 5191–5196
[20] Iwanaga M and Choi B 2015 *Nano Lett.* **15**(3) 1904–1910
[21] Lakowicz J R, Ray K, Chowdhury M, Szmacinski H, Fu Y, Zhang J and Nowaczyk K 2008 *Analyst* **133** 1308–1348
[22] Bauch M, Toma K, Toma M, Zhang Q and Dostalek J 2014 *Plasmonics* **9**(4) 781–799
[23] Bohn J, Bucher T, Chong K E, Komar A, Choi D Y, Neshev D N, Kivshar Y S, Pertsch T and Staude I 2018 *Nano Lett.* **18**(6) 3461–3465
[24] Iwanaga M, Mano T and Ikeda N 2018 *Appl. Phys. A* **124**(2) 93
[25] Fujita M, Takahashi S, Tanaka Y, Asano T and Noda S 2005 *Science* **308**(5726) 1296–1298
[26] Yuan S, Qiu X, Cui C, Zhu L, Wang Y, Li Y, Song J, Huang Q and Xia J 2017 *ACS Nano* **11**(11) 10704–10711

IOP Conf. Series: Journal of Physics: Conf. Series **1092** (2018) 012054 doi:10.1088/1742-6596/1092/1/012054

Synchronized surface plasmon generation and scattering in organic light emitting diodes: measurements removing optical interference effect

T Ishiguro[1], K Kasahara[1], N Ikeda[2] and Y Sugimoto[2]
[1]Collegeof Science and Engineering, Ritsumeikan University,
1-1-1 Noji-higashi, Kusatsu, Shiga 525-8577, Japan
kasahara@se.ritsumei.ac.jp
[2]Center, National Institute for Materials Science (NIMS),
1-2-1 Sengen, Tsukuba, 305-0047, Japan

Abstract. It has been believed that the energy transfer of one exciton arising in an organic layer creates one SPP on a proximally-positioned metal surface. However, because excitons in the organic layer are close to one another, there is the possibility that multiple SPPs are generated in phase, entering an appearance of constructive interference among the scattered electromagnetic waves. This should lead to an increase in SPP scattering probability effectively. To investigate the possibility and light enhancement associated with this, we prepared a device consisting of Alq_3 on a quarter-wavelength dielectric multilayer film/glass as a reference instead of the usual Alq_3 /glass substrate. This reference device eliminated the influence of complicated optical interference, and made it possible to calculate SPP-mediated light enhancement accurately. The contributions of SPPs and non-radiative components to light emission were adjusted in the calculation as the results corresponded with the measured enhancement dependence on Alq_3 thickness. As a result, it was found that the SPP scattering rate was larger than the ordinary calculation based on the assumption that excitons create SPPs randomly. In-phase energy transfer from excitons to SPPs is supported by evidences obtained from this study.

1. Introduction

Surface plasmon polariton (SPP)-mediated light emission which occurs in organic light emitting diodes is of considerable practical concern. SPPs essentially become ohmic loss on the surface of a metal electrode. SPPs are essentially light waves arising from the strong interaction between electromagnetic fields and the free electron collective oscillations along the metal–dielectric interface. In addition to the intrinsic Ohmic losses of metals, scattering can be a major loss factor that limits the propagation length of SPPs. This scattering occurs at naturally formed submicrometre-scale roughness on the metal surface, and SPPs could be extracted as light again.

So far, we have investigated the possibility of the generation and scattering of SPPs in phase in organic light emitting diodes [1]. Due to excitons in the organic film being in close proximity to each other, SPPs can be generated simultaneously, and scattered synchronously, which should lead to an increase in the scattering probability of SPP due to the constructive interference of the electric fields of SPPs. In examining this possible phenomenon, the optical interference of fluorescence occurring in the organic film on a glass which was used as a reference was rather complicated, because emitted light from the organic layer entered the glass substrate, and it was difficult to calculate the influence. That is, optical interference occurring in the reference device stood in the way of the experiments. Consequently, ambiguity has remained concerning the probability of constructive SPP generation and scattering. In this study, we eliminated this obstacle by using a quarter-wavelength dielectric multilayer film (DMF) as a reference. It was found that the SPP generation in phase was probable in organic layers through the experimentally obtained enhancements and calculations.

2. Device fabrication and measurements of the enhancement

To avoid the optical interference effect, we deposited Alq_3 on a quarter-wavelength DMF/glass substrate as a reference. The DMF consisted of quarter-wavelength thick alternating SiO_2 and TiO_2 layers. The reflectivity of the DMF was set at almost 100 % at 530 nm where Alq_3 had the maximum intensity. We made Alq_3 /Ag /glass to investigate the SPP-mediated enhancement. Angle dependence of the luminescence was measured by rotating devices with the use of a 405-nm laser diode and a luminance meter. Florescence emitted up to a rotating angle of 15 degrees from the frontal direction was summed up with all devices. The coefficient of reflectance of the dielectric multilayer mirror could be assured to be -1 in the range of those angles. Input and reflection light power of the laser diode were measured, and the correction of the absorptance was performed for each measurement result. Following this step, light enhancement was acquired by dividing the total light output up to 15 degrees measured for Alq_3 on an Ag /glass by that of Alq_3 on a DMF/glass while the Alq_3 thickness was changed. At the same time, using an integration sphere, we measured the internal quantum efficiency of Alq_3 formed on frosted glasses. The surface of a glass substrate was treated with sand-blasting, and was unlevelled because the internal quantum efficiency of organic film on a flat glass was subject to an optical interference effect inside the organic layer. Measured enhancements are shown in figure 1. They were 1.2, 1.0, and 0.6 for 90 nm, 110 nm, and 145 nm, respectively, on Ag.

3. Results and discussion

3.1. Enhancement based on an ordinary picture that one exciton creates one SPP

To know the contribution of SPP-mediated emission, we calculated theoretical values corresponding to the measured enhancements. Thus, the sum of original radiation and that originating from the scattering of SPP created by energy transfer of an exciton was obtained first for the structure of an organic layer on a metal film.

In obtaining energy transfer from an exciton to radiation (b_r), SPP (b_{sp}), and lossy surface wave (b_{lsw}) in the organic layer/metal film structure, they are calculated from the following equations [2].

$$b_r(k_1, z) = \int_0^{k_1} W(k_1, k_{//}, z)dk_{//}, \tag{1}$$

$$b_{sp}(k_1, z) = \int_{k_1}^{2k_{sp}-k_1} W(k_1, k_{//}, z)dk_{//}, \tag{2}$$

$$b_{lsw}(k_1, z) = \int_{2k_{sp}-k_1}^{\infty} W(k_1, k_{//}, z)dk_{//}, \tag{3}$$

where k_1: wavenumber in the organic layer (in this case, wavenumber corresponding to 530 nm), $k_{//}$: x component of $k_1$that is parallel to the metal surface, $k_{sp} = (\frac{\omega}{c})\sqrt{\frac{\epsilon_1\epsilon_m}{\epsilon_1+\epsilon_m}}$ ($\epsilon_1$and ϵ_m are the permittivity of the organic material and metal, respectively), and W: dissipation of exciton energy per unit time, z: position of exciton in the direction normal to the metal surface, the origin of which was defined at the interface of the metal and organic layer, and d: thickness of the organic layer. Experimentally obtained light output in this study is given by eq. (4), where k_h is the value of $k_{//}$ corresponding to the maximum rotating angle (i.e.15 degrees).

$$b_{rh}(k_1, z) = \int_0^{k_h} W(k_1, k_{//}, z)dk_{//}. \tag{4}$$

With respect to the emission based on SPP scattering, we used the following eq. (5) instead of eq. (2), because the use of eq. (2) results in $k_{//}$ being larger than k_1, meaning that the wavelengths of SPP scattering-based radiation become shorter than the wavelength set in the enhancement measurements. Thus, it is necessary to calculate the contribution from SPPs to light output using eq. (5), and not by eq. (2).

$$b_{sph}(k_1, z) = \{1 - (1 - P_{scat})^N\} \times T_0 \times \int_{2\pi n/\lambda_{max}}^{k_1} W(k_1{}', k_{//}, z)dk_1{}', \tag{5}$$

In eq. (5), λ_{max} is the maximum emission wavelength, which we assumed 670 nm for Alq_3 in the calculation. The intensities of Alq_3 became comparatively weak at the wavelength compared to the peak intensity. The scattering rate of SPPs at the wavelength, P_{scat}, was calculated using the average height and full-width at half-maximum of a bump, and the average period between the bumps measured by AFM [2], which were 5 nm, 50 nm, and 100 nm, respectively. N is the number of scattering, which was obtained from the propagation distance of the SPP attenuating to $1/e$ due to ohmic loss divided by the measured average period. The propagation distance of Ag was calculated to be 4.2 μm at 530 nm. T_0 is the average transmissivity of light between 0 to 15 degrees emitted at $z = 0$ and appearing in the air. SPP scattering occurred at $z = 0$, and the extracted light in the air was assumed to be affected by the optical interference inside the organic film.

The portion of light emission over the original total exciton energy R_m is given by eq. (6).

$$R_m = \left(\frac{1}{d}\right) \int_0^d \frac{b_{rh}+b_{sph}}{(b_r+b_{sp}+b_{lsw})} dz \, . \tag{6}$$

Next, we calculated radiation for Alq_3 on the DMF/glass not having such losses as SPP generation and lossy surface wave. The portion of light emission over the original total exciton energy R_d is given by the radiation summing up to the maximum rotation angle to the total radiation, as is shown in eq. (7).

$$R_d = \left(\frac{1}{d}\right) \int_0^d \frac{b_{rh}}{b_r} dz. \tag{7}$$

Because the original exciton energy was the same with respect to two structures, the ratio of R_m and R_d becomes theoretical enhancement. Figure 1 shows the calculated values (solid diamonds), and we found that numerical calculations could not account for the measured enhancements. In the case of no existence of SPP-mediated light emission of non-radiative components, the aforementioned discussion holds true, even when the internal quantum efficiency of Alq_3 is not 100%.

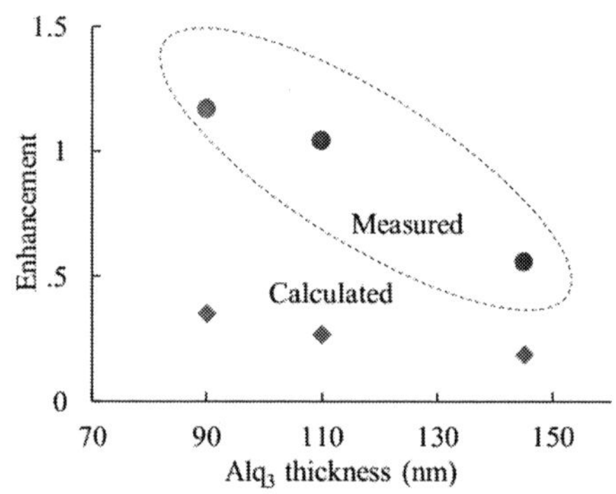

Figure 1. Measured and calculated enhancements of Alq_3 at 530 nm. Output light of Alq_3 on Ag emitted to a cone with the half-angle of 15 degrees was divided by that of Alq_3 on DMF. Original radiation and that originating from the scattering of SPP created by energy transfer of an exciton were taken into account in the calculation of R_m.

3.2. Calculation taken into account of possible generation of SPP in phase and scattering

It is understood that one SPP is created on a metal surface by energy transfer from one exciton in an organic layer. There is a possibility that multiple SPPs are generated in phase, leading to constructive interference occurrence among the scatted electromagnetic waves. The distance between the closest organic molecules is of the order of ~nm. It is possible that SPPs are generated simultaneously rather than separately, and the scattering rate of SPPs should become high compared with the case of individual SPP creation. Assume a system consisting of N-SPPs. The total electric field is expressed as $\sum_{j=1}^{N} E_j e^{i\theta_j}$ using individual phase θ_j of generated SPPs. When SPPs have the same phase, total scattering energy is proportional to $P_{scat}N^2E^2$ (i.e. $E = E_j$). This is $P_{scat}NE^2$ in the case of no phase synchronization, meaning that the substantive scattering probability results in N times the probability with out-of-phase in association with in-phase SPP creation.

In addition, we allowed for SPP-mediated light emission of non-radiative components, which can be converted to light after becoming SPPs, and by their scattering on the metal surface. We assumed that these two processes occurred in the experiments. In that case, b_{sph} and R_m is given by eqs. (8) and (9).

$$b_{sph}(k_1, z) = \{1 - (1 - XP_{scat})^N\} \times T_0 \times \int_{2\pi n/\lambda_{max}}^{k_1} W(k_1{}', k_{//}, z) dk_1{}' , \quad (8)$$

$$R_m = \eta \left(\frac{1}{d}\right) \int_0^d \frac{b_{rh} + b_{sph}}{(b_r + b_{sp} + b_{lsw})} dz + (1 - \eta) Y T_0 , \quad (9)$$

where η is the internal quantum efficiency, which was estimated to be 20% from an integration sphere used-measurements using Alq_3 on a sand-blasted glass as a sample. The process of energy conversion from non-radiative components to photon included an unknow factor. Namely, the probability of energy transfer from the non-radiative components to SPPs was unclear. Thus, the non-radiative component-to-SPP energy transfer rate times subsequent SPP scattering rate at bumps was represented by Y. R_d was the same as shown in eq. (7), because the internal quantum efficiency was multiplied for both b_{rh} and b_r in the integral in the right-hand side of eq. (7).

Figure 2 shows the calculated results (solid squares) assuming that $X = 10$, and $Y = 8 \times 10^{-3}$, and they favoured the measured results. $X = 10$ meant that 10 excitons created SPPs simultaneously. Here, a question arises as to what determines the number of in-phase excitons. The answer to the question might be in random lasers. Random lasers based on a suspension containing dyes and silver nanoparticles in methanol was reported by et al., the properties of which were dominated by an enhanced photon scattering mechanism induced by the surface plasmons [4]. Although we don't use such a suspension containing silver particles, multiple energy exchange between excitons and SPPs has the possibility of generating the local coherence.

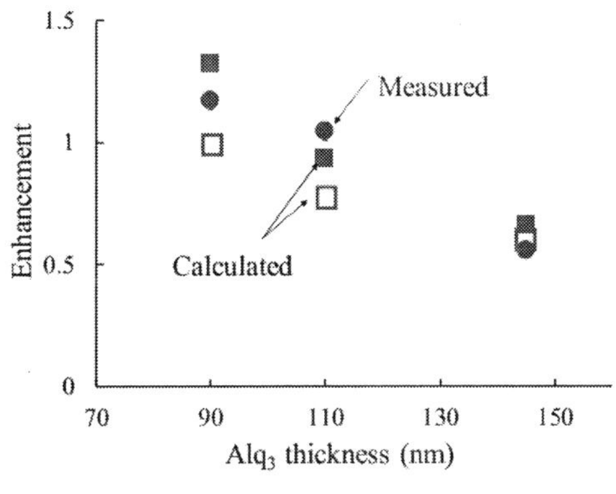

Figure 2. Measured and calculated enhancements of Alq_3 at 530 nm. It was assumed that SPPs were created synchronously from excitons, and that the non-radiative components were converted to light through SPP generation and scattering. Assuming that $X = 10$ and $Y = 8 \times 10^{-3}$ (solid squares), calculated data corresponded reasonably well with the experimental data. Open squares show calculated results on the assumption that $X = 1$ and $Y = 3 \times 10^{-2}$.

4. Conclusion

The contributions of SPPs and non-radiative components to light emission were adjusted as the result corresponded with the measured enhancement dependence on the Alq_3 film thickness. As a result, it was found that the SPP scattering rate was larger than the ordinary calculation based on the assumption that excitons create SPPs randomly. In-phase energy transfer from excitons to SPPs is supported by evidences obtained from this study.

References

[1] Mizoguchi Y, Yoneda T, Ishiguro T, Kasahara K, Ikeda N and Sugimoto Y 2017 *Proc. META2017* (Incheon-Seoul, South Korea) P-01-06
[2] Weber W H and Eagen C F 1979 *Opt. Lett.* **4** 236
[3] Sanchez-Gil J A and Maradudin A A 1999 *Phys. Rev. B* **160** 8359
[4] Dice G D, Mujumdar S and Elezzabi A Y 2005 *Appl. Phys. Lett.* **86** 131105

Electronic properties of quantum rings dressed by a high-frequency electromagnetic field

O. V. Kibis[1], **V. K. Kozin**[2], **I. V. Iorsh**[2], **I. A. Shelykh**[2,3]
[1]Department of Applied and Theoretical Physics, Novosibirsk State Technical University, Karl Marx Avenue 20, Novosibirsk 630073, Russia
[2]ITMO University, Saint Petersburg 197101, Russia
[3]Science Institute, University of Iceland, Dunhagi 3, IS-107, Reykjavik, Iceland

E-mail: oleg.kibis@nstu.ru

Abstract. It is shown theoretically that the strong coupling of quantum rings to a high-frequency off-resonant electromagnetic field (dressing field) substantially effects on their electronic characteristics. Particularly, we demonstrated that a linearly polarized dressing field decreases the spin splitting arisen from the Rashba spin-orbit interaction, whereas a circularly polarized field splits electron states corresponding to mutually opposite directions of angular momentum. Therefore, the dressing field can control spin and transport characteristics of the quantum rings and be potentially exploited in optoelectronic applications.

1. Introduction

Advances in laser physics and microwave technique achieved in recent decades have made possible the use of high-frequency fields as tools of flexible control of various atomic and condensed-matter structures (so called "Floquet engineering" based on the Floquet theory of periodically driven quantum systems). As a consequence, the properties of electronic systems driven by oscillating fields are actively studied to exploit unique features of composite states of field and matter. Particularly, electron strongly coupled to electromagnetic field — also known as "electron dressed by field" (dressed electron) — has become a commonly used model in modern physics. Recently, we elaborated the theory of dressed electrons in various 2D nanostructures, including quantum wells [1–3], graphene [4–6], etc. Developing this scientific trend, we present the theory of dressed electrons for semiconductor quantum rings (QRs) which can serve as a basis for prospective nanoelectronic applications [7].

2. Model

We will be to consider a semiconductor QR of the radius R, which lies in the (x, y) plane and is irradiated by a monochromatic electromagnetic wave propagating along the z axis. Since the frequency of the wave, ω, is assumed to be far from all resonant electronic frequencies, the wave cannot be absorbed by electrons and can be considered as a dressing field within the conventional Floquet theory. Introducing the electron-field interaction within the minimal coupling scheme, the Hamiltonian of dressed QR, $\hat{\mathcal{H}}(t)$, results from the "bare" Hamiltonian, $\hat{\mathcal{H}}_0(k_x, k_y)$, with the formal replacement $\mathbf{k} \to \mathbf{k} - (e/\hbar)\mathbf{A}(t)$, where $\mathbf{k} = (k_x, k_y)$ is the electron wave vector, $\mathbf{A}(t)$ is the time-dependent vector potential of the wave, and e is the electron charge. Solving the non-stationary Schrödinger problem with the time-dependent Hamiltonian, $\hat{\mathcal{H}}(t)$, within the recently elaborated approach [8–11], we can reduce the time-dependent Hamiltonian $\hat{\mathcal{H}}(t)$ to the effective time-independent Hamiltonian, $\hat{\mathcal{H}}_{\text{eff}}$, which describes stationary electronic properties renormalized by the dressing field. Considering the stationary Schrödinger problem with the effective Hamiltonian, $\hat{\mathcal{H}}_{\text{eff}}$, we can arrive at the renormalized electron energy spectrum of the dressed QR.

3. Results

It follows from the calculated electron energy spectra that electron states with mutually opposite directions of angular momentum, m and $-m$, are split by the circularly polarized dressing field. To clarify the physical nature of the splitting, it should be reminded that the degeneracy of these states takes place in the absence of the field due to the time-reversal symmetry (the Kramers theorem). Since the time reversal turns clockwise polarized photons into counterclockwise polarized ones and vice versa, the electron coupling to the circularly-polarized field breaks this symmetry and split the degenerated electron states [?]. As to a linearly polarized dressing field, it does not split doubly degenerated electron states with mutually opposite directions of angular momentum, m and $-m$, but renormalizes the electron effective mass and leads to the dependencies of the spin angle and the energy levels on the irradiation intensity. Particularly, the relatively weak irradiation can decrease the angle to tens percents of its initial value in the unirradiated QR. Since the modulation of spin orientation by various external actions lies in the core of modern spintronics, the found strong dependence of the spin polarization on the irradiation can be used in prospective ring-shaped spintronic devices operated by light.

4. Conclusions

We demonstrated that the key electronic characteristics of QRs — the structure of electron energy levels and the spin polarization of electrons — strongly depend on high-frequency off-resonant irradiation. Particularly, the field-induced splitting of electron energy levels as well as the modification of electron effective mass and the spin-orbit coupling appear. The irradiation-induced renormalization of electron energy spectrum can be observed in state-of-the-art optical experiments, whereas the light sensitivity of the spin orientation can be exploited in prospective spintronic devices operated by light.

Acknowledgments

The work was partially supported by RISE Program (project CoExAN), FP7 ITN Program (project NOTEDEV), Russian Foundation for Basic Research (project 17-02-00053), Rannis project 163082-051, and Ministry of Education and Science of Russian Federation (project 3.4573.2017/6.7, project 3.2614.2017/4.6, and project 14.Y26.31.0015).

References

[1] Morina S, Kibis O V, Pervishko A A and Shelykh I A 2015 *Phys. Rev. B* **91** 155312
[2] Pervishko A A, Kibis O V, Morina S and Shelykh I A 2015 *Phys. Rev. B* **92** 205403
[3] Dini K, Kibis O V and Shelykh I A 2016 *Phys. Rev. B* **93** 235411
[4] Kibis O V, Morina S, Dini K and Shelykh I A 2016 *Phys. Rev. B* **93** 115420
[5] Kibis O V, Dini K, Iorsh I V and Shelykh I A 2017 *Phys. Rev. B* **95** 125401
[6] Iorsh I V, Dini K, Kibis O V and Shelykh I A 2017 *Phys. Rev. B* **96** 155432
[7] Fomin V M 2014 *Physics of Quantum Rings* (Berlin: Springer)
[8] Kibis O V 2011 *Phys. Rev. Lett.* **107** 106802
[9] Sigurdsson H, Kibis O V and Shelykh I A 2014 *Phys. Rev. B* **90** 235413
[10] Hasan M, Iorsh I V, Kibis O V and Shelykh I A 2016 *Phys. Rev. B* **93** 125401
[11] Kozin V K, Iorsh I V, Kibis O V and Shelykh I A 2018 *Phys. Rev. B* **97** 035416

Orientational Kerr effect in liquid crystal ferroelectrics and modulation of partially polarized light

Alexei D. Kiselev[1], **Vladimir V. Kesaev**[2] **and Evgeny P. Pozhidaev**[2]

[1] Saint-Petersburg National Research University of Information Technologies, Mechanics and Optics, 49 Kronverksky Prospekt, St. Petersburg 197101, Russia

[2] Lebedev Physical Institute RAS, 53 Leninskiy Ave., Moscow 119991, Russia

E-mail: alexei.d.kiselev@gmail.com

Abstract. We study modulation of partially polarized light governed by the orientational Kerr effect in subwavelength-pitch deformed-helix ferroelectric liquid crystals. In our experimental setup based on the Mach-Zehnder interferometer, it is found that the electric field induced shift of the interference fringes crucially depends on the degree of polarization of the incident light. We show that the experimental data can be theoretically described in terms of the electrically dependent Pancharatnam relative phase. The electric field dependence of both the Pancharatnam and geometric phases is evaluated by comparing the experimental and theoretical results.

Orientational structures characterizing orientationally ordered anisotropic liquids known as the *liquid crystals* are extremely sensitive to both external fields and boundary conditions. This sensitivity underlines the modes of operation of liquid crystal (LC) spatial light modulators (SLMs) that have been widely utilized to modulate amplitude, phase, or polarization of light waves in space and time [1]. Ferroelectric liquid crystals (FLCs) are known to be characterized by very fast response time and represent most promising chiral liquid crystal material. Most of the FLC modes, however, are not suitable for phase-only modulation devices because their optical axis sweeps in the plane of the cell substrate producing undesirable changes in the polarization state of the incident light.

In order to get around the optical axis switching problem the system based on the orientational Kerr effect in a vertically aligned deformed helix ferroelectric LC (DHFLC) with subwavelength helix pitch was suggested in [2, 3] (a recent review of the physics of the DHFLC effect can be found in [4]). In the geometry of a uniform lying helix, phase modulation of light in DHFLC cells was also studied using the experimental technique based on a Mach-Zehnder two-arm interferometer [5, 6].

An important point is that all the above mentioned studies deal with the case of linearly polarized illumination where the incident beam is fully polarized. In our recent paper [7], we have found that, for unpolarized incident light, the electro-optic response is insensitive to the effect of electric-field-induced rotation.

In this paper, we present the results of investigation into the effects of the degree of polarization in the electro-optic behavior of the planar aligned DHFLC cells illuminated with partially polarized light and extend our theoretical analysis of the Pancharatnam and geometric

(Pancharatnam-Berry) phases to the general case of non-unitary evolution of mixed polarization states.

We consider a DHFLC layer of thickness D with the z axis normal to the bounding surfaces: $z = 0$ and $z = D$ (see Fig. 1). The geometry of a *uniform lying* DHFLC helix with subwavelength pitch where the helix (twisting) axis $\hat{\mathbf{h}}$ is directed along the x axis will be our primary concern. In this geometry depicted in Fig. 1 the equilibrium orientational structure can be described as a helical twisting pattern where FLC molecules align on average along a local unit director $\hat{\mathbf{d}} = \cos\theta\,\hat{\mathbf{h}} + \sin\theta\,\hat{\mathbf{c}}$, where θ is the *smectic tilt angle*; $\hat{\mathbf{h}} = \hat{\mathbf{x}}$ is the twisting axis normal to the smectic layers and $\hat{\mathbf{c}} \perp \hat{\mathbf{h}}$ is the c-director. The FLC director lies on the smectic cone depicted in Fig. 1(left) and rotates in a helical fashion about a uniform twisting axis $\hat{\mathbf{h}}$ forming the FLC helix. This rotation is described by the azimuthal angle around the cone ϕ that specifies orientation of the c-director in the plane perpendicular to $\hat{\mathbf{h}}$: $\hat{\mathbf{c}} = \cos\phi\,\hat{\mathbf{y}} + \sin\phi\,\hat{\mathbf{z}}$.

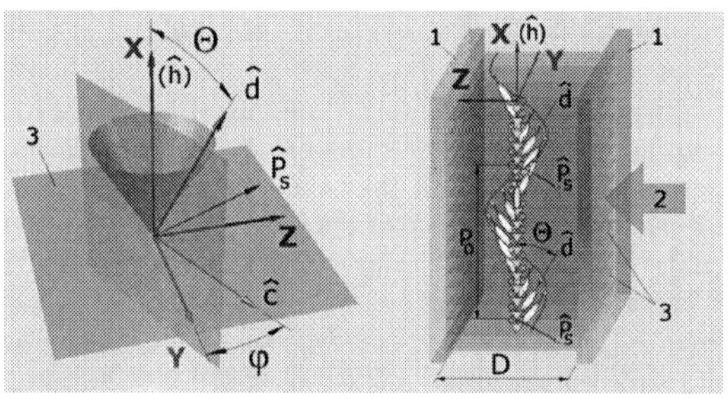

Figure 1. Geometry of smectic cone (left) and DHFLC cell (right). 1 – cell substrates; 2 – incident light beam; 3 – smectic layers.

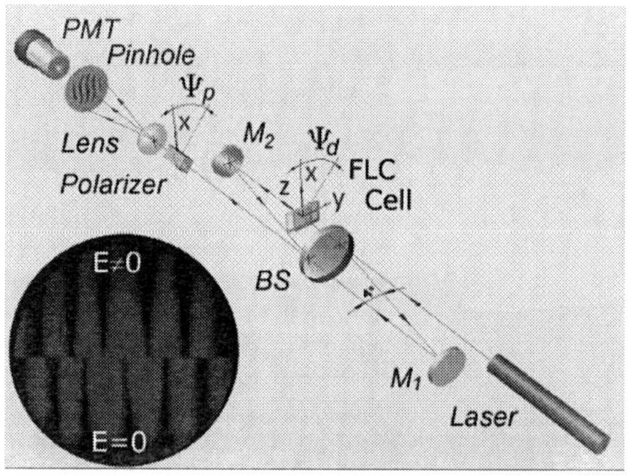

Figure 2. Experimental setup based on a Mach-Zehnder two-arm interferometer: BS is the beam splitter; M_1 and M_2 are the mirrors; PMT is the photomultiplier tube. The insert illustrates the electrically induced shift of the interference fringes.

Figure 1(right) illustrates the uniform lying FLC helix in the slab geometry with the smectic layers normal to the substrates, where $\mathbf{E} = E\,\hat{\mathbf{z}}$ is the applied electric field which is linearly coupled to the *spontaneous ferroelectric polarization*, $\mathbf{P}_s = P_s\hat{\mathbf{p}}$, directed along the *polarization unit vector*, $\hat{\mathbf{p}} = \hat{\mathbf{x}} \times \hat{\mathbf{c}}$.

According to Refs. [8, 9, 5], optical properties of such cells can be described by the effective dielectric tensor of a homogenized DHFLC helical structure. The zero-field ($\mathbf{E} = 0$) dielectric tensor is uniaxially anisotropic with the optical axis directed along the twisting axis $\hat{\mathbf{h}} = \hat{\mathbf{x}}$. The zero-field effective refractive indices of extraordinary (ordinary) waves, n_h (n_p), generally depend on the smectic tilt angle θ and the optical (high frequency) dielectric constants characterizing the FLC material (see, e.g., equation (8) in Ref. [5] giving the expressions for $\epsilon_h = n_h^2$ and $\epsilon_p = n_p^2$).

The electric-field-induced anisotropy is generally biaxial so that the dielectric tensor is characterized by the three generally different principal values (eigenvalues): $\epsilon_\pm = n_\pm^2$ and $\epsilon_z = n_z^2$. The in-plane principal optical axes (eigenvectors)

$$\hat{\mathbf{d}}_+ = \cos\psi_d\,\hat{\mathbf{x}} + \sin\psi_d\,\hat{\mathbf{y}}, \quad \hat{\mathbf{d}}_- = \hat{\mathbf{z}} \times \hat{\mathbf{d}}_+ \tag{1}$$

are rotated about the vector of electric field, $\mathbf{E} \parallel \hat{\mathbf{z}}$, by the azimuthal angle ψ_d. In the low electric field region, the electric field dependence of the angle ψ_d is approximately linear: $\psi_d \propto E$, whereas the electrically induced part of the principal refractive indices, $n_\pm$ and n_z, is typically dominated by the Kerr-like nonlinear terms proportional to E^2 (see, e.g., equations (10)–(18) in Ref. [5]). This effect — the so-called *orientational Kerr effect* — is caused by the electrically induced distortions of the helical structure. It is governed by the effective dielectric tensor of a nanostructured chiral smectic liquid crystal defined through averaging over the FLC orientational structure [2, 3, 8, 9]. In electro-optical studies of DHFLC cells, the

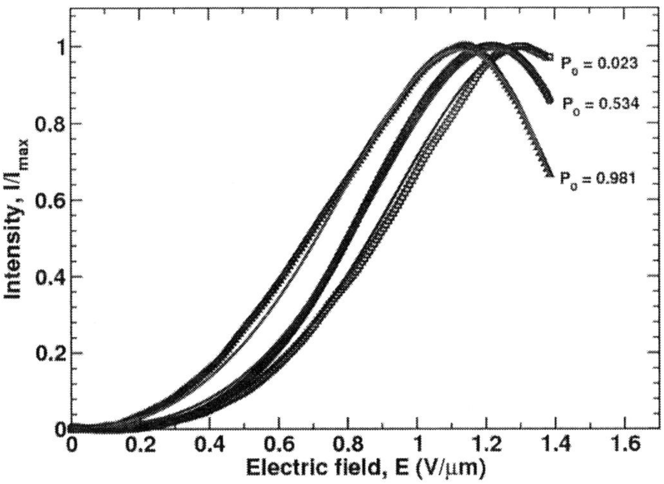

Figure 3. Normalized intensity of output wavefield measured as a function of the electric field at different values of the degree of polarization, P_0, for the DHFLC cell of thickness $D \approx 52$ μm filled with the FLC mixture FLC-587. Solid lines represent the theoretical curves. The parameters of the mixture are: $n_\perp \approx 1.47$ is the ordinary refractive index, $n_\parallel \approx 1.69$ is the extraordinary refractive index, $\theta \approx 33.5°$ is the smectic tilt angle, and $r_2 \approx 1.05$ is the biaxiality ratio.

orientational Kerr effect typically manifests itself in the electrically dependent transmittance of normally incident light passing through crossed polarizers [2, 3, 8, 9]. This effect was also experimentally studied in [5, 7] using the experimental setup shown in Fig. 2. This setup is based on a Mach-Zehnder two-beam interferometer where the FLC cell is placed in the path of the sample beam. Referring to Fig. 2, a beam splitter (BS) divides a collimated laser light into two beams, the reference and the sample beams, which, after reflection at the mirrors M_1 and M_2, are recombined at the semireflecting surface of the beam splitter (BS). The interfering beams emerging from the interferometer optionally pass through an output polarizer with the transmission axis azimuth ψ_p and then are projected by the lens on to a screen with a pinhole. Figure 3 shows the experimental results for the electric field dependence of the output light field intensity measured at different values of the degree of polarization in the DHFLC cell of thickness $D \approx 52$ μm filled with the FLC mixture FLC-587.

In our theoretical analysis, we consider the input lightwave field characterized by the 2×2 equal-time coherence matrix: $\mathbf{M}_0 = \langle \mathbf{E}_0 \otimes \mathbf{E}_0^* \rangle = I_0 \rho_0$, where $I_0 = \mathrm{Tr}\,\mathbf{M}_0$ is the intensity; ρ_0 is the polarization density matrix. In circular basis $\hat{\mathbf{e}}_\pm = (\hat{\mathbf{x}} \pm i\hat{\mathbf{y}})/\sqrt{2}$, the density matrix ρ_0 is given by

$$2\rho_0 = \boldsymbol{\sigma}_0 + P_0(\hat{\mathrm{s}}_0 \cdot \boldsymbol{\sigma}), \; \boldsymbol{\sigma}_0 = \mathrm{diag}(1,1), \; \boldsymbol{\sigma} = (\boldsymbol{\sigma}_1, \boldsymbol{\sigma}_2, \boldsymbol{\sigma}_3), \tag{2}$$

where $\boldsymbol{\sigma}_i$ are the Pauli matrices; $\hat{\mathrm{s}}_0 = (\sin(2\theta_0)\cos(2\phi_0), \sin(2\theta_0)\sin(2\phi_0), \cos(2\theta_0))$ is the normalized Stokes vector; ϕ_0 is the polarization azimuth; $-\pi/4 \le \pi/4 - \theta_0 \equiv \chi_0 \le \pi/4$ is the ellipticity angle; and $\epsilon_{\mathrm{ell}} = \tan\chi_0$ is the ellipticity. The transmission matrix of the DHFLC cell can be written in the similar form as follows

$$2\mathbf{T}_s(h) = (t_+ + t_-)\boldsymbol{\sigma}_0 + (t_- - t_+)(\hat{\mathrm{s}}_d \cdot \boldsymbol{\sigma}), \; t_\pm = |t_\pm| e^{i\Phi_\pm} = \frac{1 - \rho_\pm^2}{1 - \rho_\pm^2 \exp(2in_\pm h)} \exp(in_\pm h), \tag{3}$$

where $\rho_\pm = (n_\pm - n_{\rm m})/(n_\pm + n_{\rm m})$ is the Fresnel reflection coefficient; $n_{\rm m}$ is the refractive index of the ambient medium; $h = k_{\rm vac}D = \omega/cD$ is the thickness parameter; $\hat{\mathbf{s}}_d = (\cos 2\psi_d, \sin 2\psi_d, 0)$ can be regarded the normalized Stokes vector of a light beam linearly polarized along the optic axis $\hat{\mathbf{d}}_+$. The transmission and density matrices can be used as an input to derive the expression for the intensity of the total output lightwave and fit the experimental data presented in Fig. 3, where the resulting theoretical curves are shown as solid lines.

We can now apply the interferometry based approach formulated in Refs. [10, 11] to the case of plane wave which is normally incident on the DHFLC cell and is characterized by the polarization density matrix (2) and derive the expressions for the Pancharatnam and geometric phases, Φ_P and Φ_g, acquired by the light transmitted through the DHFLC cell. The result reads

$$\Phi_P = \Phi + \arg[\mathrm{Re}\,\tilde{F}_P + i\tilde{P}\,\mathrm{Im}\,\tilde{F}_P] = \arg\mathrm{Tr}[\mathbf{T}_s(h)\rho_0], \quad (4)$$

$$\tilde{F}_P = \cos\Delta\Phi\cos[\theta_\lambda(h) - \theta_\lambda(0)] + i\sin\Delta\Phi\cos[\theta_\lambda(h) + \theta_\lambda(0)], \quad (5)$$

$$\Phi_g = \arg[\mathrm{Re}\,\tilde{F}_g + i\tilde{P}\,\mathrm{Im}\,\tilde{F}_g],\ \tilde{F}_g = \tilde{F}_P \mathrm{e}^{-i\tilde{\Phi}_d}, \quad (6)$$

$$\tilde{\Phi}_d = \int_0^h \cos(2\theta_\lambda(s))\Delta\Phi_s'(s)\mathrm{d}s,\ \cos(2\theta_\lambda) = q_3/|\mathbf{q}|,\ |\mathbf{q}| = \sqrt{q_3^2 + q_0^2[1 - (\hat{\mathbf{s}}_d\cdot\hat{\mathbf{s}}_0)^2]} \quad (7)$$

where $\Phi = (\Phi_+ + \Phi_-)/2$ is the averaged phase shift; $\Delta\Phi = (\Phi_+ - \Phi_-)/2$ is the phase retardation; $\tilde{P} = \dfrac{\sqrt{p_+(0)p_+(h)} - \sqrt{p_-(0)p_-(h)}}{\sqrt{p_+(0)p_+(h)} + \sqrt{p_-(0)p_-(h)}}$, $p_\mu(0) = \dfrac{1+\mu P_0}{2}$, $p_\mu(h) = \dfrac{\mathrm{Tr}(\rho) + \mu|\mathbf{q}|}{2}$, $q_0 = P_0[\tau_+^2 - \tau_-^2]$, $q_3 = 2\tau_-\tau_+ + P_0[\tau_+^2 + \tau_-^2](\hat{\mathbf{s}}_d\cdot\hat{\mathbf{s}}_0)$, $\mathrm{Tr}(\rho) = \tau_+^2 + \tau_-^2 + 2P_0\tau_+\tau_-(\hat{\mathbf{s}}_d\cdot\hat{\mathbf{s}}_0)$, $\tau_\pm = (|t_+| \pm |t_-|)/2$.

In conclusion, we have studied electric-field-induced modulation of partially polarized light in the DHFLC with subwavelength helix pitch using the experimental technique based on the Mach-Zehnder interferometer. Such modulation occurs under the action of the voltage applied across the cell and manifests itself in the electrically dependent shift and contrast of the interference pattern. The effect of the degree of polarization is investigated theoretically and the theoretical results are in good agreement with the experimental data. Our analytical results show that, in contrast to the total Pancharatnam phase, the geometric phase is determined by the phase retardation and is independent of the averaged phase shift. In addition, the electric field dependence of the phases enters through the angle between the Stokes vectors $\hat{\mathbf{s}}_0$ and $\hat{\mathbf{s}}_d$ decribing the effect of electrically induced rotation of the optical axes. This effect is supressed in the limiting cases of unpolarized and circular polarized incident light.

We acknowledge financial support from RFBR under Grants 16-02-00441 a, 16-29-14012 ofi_m and 16-42-630773 p_a.

References

[1] Efron U (ed) 1995 *Spatial Light Modulator Technology: Materials, Applications, and Devices* (NY: Marcell Dekker)
[2] Pozhidaev E P, Kiselev A D, Srivastava A K, Chigrinov V G, Kwok H S and Minchenko M V 2013 *Phys. Rev. E* **87** 052502
[3] Pozhidaev E P, Srivastava A K, Kiselev A D, Chigrinov V G, Vashchenko V V, Krivoshey A V, Minchenko M V and Kwok H S 2014 *Optics Letters* **39** 2900–2903
[4] Kiselev A D 2017 *Journal of Molecular Liquids* ISSN 0167-7322 (in press) URL `https://doi.org/10.1016/j.molliq.2017.12.126`
[5] Kotova S P, Samagin S A, Pozhidaev E P and Kiselev A D 2015 *Phys. Rev. E* **92** 062502
[6] Kiselev A D, Kesaev V V and Pozhidaev E P 2017 *EPJ Web of Conferences* **161** 02010
[7] Kesaev V V, Kiselev A D and Pozhidaev E P 2017 *Phys. Rev. E* **95**(3) 032705
[8] Kiselev A D, Pozhidaev E P, Chigrinov V G and Kwok H S 2011 *Phys. Rev. E* **83** 031703
[9] Kiselev A D and Chigrinov V G 2014 *Phys. Rev. E* **90**(4) 042504
[10] Sjöqvist E, Pati A K, Ekert A, Anandan J S, Ericsson M, Oi D K L and Vedral V 2000 *Phys. Rev. Lett.* **85**(14) 2845–2849
[11] Tong D M, Sjöqvist E, Kwek L C and Oh C H 2004 *Phys. Rev. Lett.* **93**(8) 080405

IOP Conf. Series: Journal of Physics: Conf. Series **1092** (2018) 012057 doi:10.1088/1742-6596/1092/1/012057

Photoinduced electron transport in QDs-based hybrid structures with TiO_2 nanoparticles

E P Kolesova [1], O Cleary [2], Y K Gun'ko [1,2], V G Maslov [1] and A O Orlova [1]

[1] ITMO University, Saint-Petersburg, 197101, Russia
[2] Trinity College, Dublin, Dublin 2, Ireland

e.p.kolesova@gmail.com

Abstract. The multiexponential decay of QDs photoluminescence (PL) in hybrid structures based on QDs and TiO_2 nanoparticles has been investigated. We show that both fractions of QDs with long and short PL decay times should be regarded as an electron donor. From the point of view of QDs blinking theory, we estimate the efficiency of electron transfer from neutral and negatively charged QDs. We demonstrate that an efficiency of electron transfer from negatively charged QDs in spite of the effective Auger recombination is quite high. It suggests that a decrease in the Auger recombination efficiency will significantly improve the functionality of hybrid structures based on QDs, such as TiO_2/QDs structures.

Semiconductor quantum nanocrystals, due to their unique properties, are widely used in hybrid structures as a donor of energy or charge carrier. Considering quantum dots (QDs) as an effective donor of charge carriers, the most suitable acceptor is titanium dioxide nanoparticle (TiO_2) [1]. Hybrid structures TiO_2/QDs combine the ability to generate reactive oxygen species (ROS) caused by TiO_2 nanoparticles [3] with the ability to absorb external radiation of a wide spectral range caused by QDs, and the band structure of QDs makes possible to transfer the electron from QDs to TiO_2 nanoparticles. The realization of this process in TiO_2/QDs structures allows to obtain a system exhibiting photocatalytic activity under visible light that makes such structures promising for antibacterial therapy. Despite all of QDs advantages, a significant shortcoming of QDs is a multiexponential PL decay [4]. The presence of two or more components in PL decay of QDs indicates an existence of different QDs fractions characterized by different constants of the nonradiative relaxation channels in QDs. It can significantly affect on both QDs photoluminescence and electron transfer from QDs to TiO_2 nanoparticles. Thus, the investigation of influence of QDs multiexponential PL decay on electron transfer in TiO_2/QDs hybrid structures is an indispensable step for optimizing the parameters of hybrid structures. Until today, there is no universally accepted approach for interpreting this kind of QDs photoluminescence decay. But the blinking theory is the most frequently attracted theory to describe the multiexponential PL decay of QDs at room temperature. In the blinking theory, multiexponential decay is associated with the ability of QDs to randomly switch between states with neutral, negative or positive charges [5]. In the framework of this theory the presence of an additional electron leads to a doubling of the radiative constant due to a twofold increase in the probability of radiative recombination of the exciton, which is accompanied by a decrease in the characteristic decay time of QD photoluminescence. It is demonstrated that a reduction in the characteristic PL decay time is accompanied by a decrease in PL intensity of QDs as a result of effective Auger recombination with the participation of an additional electron [6]. Summarizing, in the blinking theory, a fraction with long PL decay time include neutral QDs (X^0), the presence of an additional electron in QD leads to formation of a negative trion (X^-) with shortened PL decay time, and the formation of a positive trion (X^+), which corresponds to QD with additional hole, further shortens their PL decay time. Obviously, the presence of an additional charge, especially for negative, in QDs can significantly impact on electron transfer efficiency, which is a key parameter of TiO_2/QDs hybrid structures functionality. In most of articles, where the patterns of electron transfer are studied, little attention has been paid to multiexponential PL decay of QDs. It should be pointed that only few articles relating electron transfer from QDs to TiO_2 nanoparticles have taken into account multiexponential PL decay of QDs and only neutral fraction of QDs is considered there [7].

The goal of our work is to identify the patterns of multiexponential PL decay of QDs in TiO_2/QDs hybrid structures and estimate the effect of QDs charge on electron transfer efficiency from QDs to TiO_2 nanoparticles.

For formation of TiO_2/QDs multilayer structures, the I type CdSe/ZnS QDs and TiO_2 spherical nanoparticles were used. The nanocrystals were consecutively deposited to a dielectric substrate by the modified Langmuir-Blodgett technique [8]. The ROS generation by TiO_2/QDs hybrid structures under visible radiation (530 nm) was investigated using a chemical sensor RNO [9]. It should be pointed that QDs are characterized by low efficiency of ROS generation in all spectral ranges [10] and TiO_2 nanoparticles are not capable of absorbing radiation from visible spectral range [11]. Taking these factors into account, the registered generation of ROS by TiO_2/QDs hybrid structures under visible radiation unambiguously testifies electron transfer from QDs to TiO_2 nanoparticles. In the TiO_2/QDs hybrid structures and dry layers of free QDs, the PL kinetic has been approximated by a biexponential function. In the framework of QDs blinking theory we consider it as contributions of neutral and negatively charged QDs in PL decay. Photoluminescence of positively charged QDs is strongly extinguished and, therefore, neglected in the present work (Table 1).

Table 1. PL decay times (τ), amplitudes (A) and quantum yields (φ) of QDs in dry layers and in TiO_2/QDs hybrid structures.

Structures	τ_1, ns	A_1, %	φ_1, %	τ_2, ns	A_2, %	φ_2, %
Layers of free QD	18	12	72	3.4	88	27
Hybrid structures TiO_2/QD	12	26	49	2.5	74	20

* $\varphi = k_r \cdot \tau$ [12].

The shortening of PL decay times (see Table 1) clearly demonstrates the formation of TiO_2/QD structures. In TiO_2/QDs structures, electron transfer competes with the radiative relaxation of photoexcitation in QDs, and, therefore, effective electron transfer from QDs to TiO_2 nanoparticles should be accompanied by quenching of QDs photoluminescence. Assuming that electron transfer is dominated in PL quenching of QDs in TiO_2/QD structures and the radiative constant (k_r) remains unchanged during the formation of hybrid structures [12], we have estimated the rate constants of electron relaxation processes in QDs (see Table 2). It is well known that the nonradiative Auger processes, whose effectiveness significantly increases with introducing an additional electron or hole in QDs, are the serious annoying pattern for core and core/shell quantum dots [13]. The high rate of Auger recombination ($k_{Auger} \sim 10^{10}$ s^{-1}) allows these nonradiative channels to compete effectively with both photoluminescence and charge transfer, thus greatly reducing the functionality of QDs-based hybrid structures.

Table 2. The rate constants of relaxation processes in QDs and in TiO_2/QDs hybrid structures.

State of QDs	k_r, $\times 10^7 s^{-1}$	k_{nr}+ k_{Auger}, $\times 10^7 s^{-1}$	k_{ET}, $\times 10^7 s^{-1}$	E_{ET}, %
Neutral QDs	4	1.5	3	37
Negatively charged QDs	8	22	10	25

*k_r – the radiative constant of CdSe QD [12], k_{nr}+ k_{Auger} – the sum of the nonradiative constant of QD and the rate constant of Auger recombination, k_{ET} – the rate constant of electron transfer from QD to TiO_2, E_{ET} – electron transfer efficiency estimated as $E_{ET}=k_{ET}/(k_r + k_{nr} + k_{Auger} + k_{ET})$.

Assuming that the nonradiative constant of QDs (k_{nr}) does not depend on QDs charge, we can suggest that the rate of Auger recombination (k_{Auger}) increases significantly with the appearance of an additional electron. Our estimation has shown that the rate of electron transfer (k_{ET}) from negatively charged QDs is almost 4 times greater than from neutral QDs. It clearly demonstrates that it is fundamentally wrong not to consider this QDs state in the analysis of electron transfer in QD-based hybrid structures. However electron transfer is not the dominant electron relaxation process for both fractions of QDs. The relatively low electron transfer efficiency for negatively charged QD with the high rate of this process is associated with a significant increase of Auger recombination efficiency. We are going to solve this problem using alloyed QDs as an electron donor in our future work. According to theoretical predictions, a smoothing of potential energy curve of alloyed QDs restores the restriction on Auger recombination [14]. It will significantly improve electron transfer efficiency and, as a result, TiO_2/QDs structures functionality.

References

[1] Costi R, Saunders A E and Banin U 2010 *Angewandte Chemie International Edition* **49**(29) 4878
[2] Talgorn E., Abellon R D, Kooyman P J., Piris J, Savenije T J, Goossens A and Siebbeles L D 2010 *ACS nano* **4**(3) 1723
[3] Nosaka Y and Nosaka A Y 2017 Chem. Rev. 117(17) 11302
[4] Schlegel G, Bohnenberger J, Potapova I and Mews A 2002 *Phys. Rev. Let.* **88** 137401
[5] Shimizu K T, Neuhauser R G, Leatherdale C A, Empedocles S A, Woo W K and Bawendi M G 2001 *Phys. Rev.* B **63**(20) 205316
[6] Galland C, Ghosh Y, Steinbruck A, Sykora M, Hollingsworth J A, Klimov V I and Htoon H 2011 *Nature* **479** 203
[7] Hamada M, Nakanishi S, Itoh T, Ishikawa M and Biju V 2010 *ACS nano* **4** (8) 4445
[8] Gromova Y A, Reznik I A, Vovk I A, Rackauskas S, Alaferdov A V, Orlova A O and Fedorov A V 2015 *MRS Online Proceedings Library Archive* **1787** 15
[9] Burns J M, Cooper W J, Ferry J L, King D W, DiMento B P, McNeill K, and Rose A L 2012 *Aquatic Sciences*, **74**(4), 683-734
[10] Ipe B I, Lehnig M and Niemeyer C. M. 2005 *Small* **1**(7) 706
[11] Khan M M, Ansari S A, Pradhan D, Ansari M O, Lee J and Cho M H 2014 *Journal of Materials Chem.* A **2**(3) 637
[12] Park Y S, Bae W K, Padilha L A, Pietryga J M and Klimov V I 2014 *Nano let.* **14**(2) 396
[13] Rea N, Campana S, Oosterbroek T, Perna R, Mereghetti S, Covino S and Mignani R 2004 *Phys. Rev. Lett.* 0408087
[14] Beane G A, Gong K and Kelley D F 2016 *ACS nano* **10**(3) 3755

Second harmonic generation in hyperbolic metamaterials

A.R. Pomozov, I.A. Kolmychek*, V.B. Novikov, A.P. Leontiev, K.S. Napolskii, T.V. Murzina

Moscow State University, 119991 GSP-1, Leninskie Gory, Moscow, Russia

*irisha@shg.ru

Abstract. The results of the synthesis, experimental studies, and numerical simulation of optical and nonlinear-optical properties of gold nanorods in porous anodic alumina templates are presented. Two features in the spectra of such structures in the vicinity of the wavelengths of 530 nm and 810 nm correspond to the epsilon-near-pole and epsilon-near-zero, respectively. Optical and nonlinear-optical properties of the structures in the spectral vicinity of these points are investigated. Experimentally observed features are consistent with the data of simulation.

1. Introduction

Hyperbolic metamaterials are composite artificial metal-dielectric nanostructured materials possessing the components of the permittivity and/or permeability tensor of opposite signs [1,2]. The key optical property of this class of the metamaterials is hyperbolic dispersion law, instead of elliptical dispersion in traditional materials. Moreover hyperbolic regime is directly related with two features of the spectra of the optical material's properties. The first one is a change in the sign of the real part of one of the components of the effective permittivity tensor when passing it through zero (Epsilon Near Zero, ENZ). The second one is appearance of a pole of effective dielectric constant (Epsilon Near Pole, ENP) associated with high and strong dispersive dielectric constant in narrow spectral region [3]. The most popular designs of hyperbolic media include ordered array of metallic nanorods in a dielectric matrix or film consisting of alternating metal and dielectric layers. A plenty of unusual effects have been already experimentally discovered and theoretically explained in hyperbolic metamaterials, such as negative refraction [4], enhancement of spontaneous emission [5] and Raman scattering [6], transformation of the evanescent field of the object (radiated or scattered) into propagating mode in the far field [7]. Optics of hyperbolic media attracts great attention of researchers that is associated with the broad possibilities of their application in nanophotonics, biosensing [8], imaging [9], and waveguides [10].

Second harmonic generation (SHG) is a widely known technique sensitive for the properties of nanostructures. Forbidden in the media with inversion symmetry, the SHG originates from the surfaces and interfaces, where the symmetry is broken [11]. Moreover the nonlinear-optical response enhances under the resonance excitation that makes the method applicable for the study of metallic plasmonic nanostructures [12]. The increase in the SHG intensity under the phase matching conditions provides the efficiency of the SHG technique for investigations of the media with unusual dispersion properties [13]. Thus, the study of the hyperbolic metamaterials by means of the nonlinear optical methods could discover fundamentally new effects, perspective for different applications [14]. At the same time, there is just a few experimental papers concerning the SHG in hyperbolic metamaterials [15].

In this paper complex experimental and numerical study of the optical and nonlinear-optical effects in arrays of gold nanorods in anodic alumina template is presented with the special accent on the ENZ and ENP spectral features.

2. Samples, experimental and simulation techniques

Arrays of metal nanorods were prepared by templated electrodeposition of gold into the pores of 50 μm thick anodic alumina films. To obtain alumina templates, two-step anodizing of high-purity aluminium (99.99%) in a 0.3 M H_2SO_4 solution at 25 V was used. According to the statistical analysis of the SEM images, the average distance between the centres of pores is 63 ± 5 nm, the pore diameter is 25 ± 5 nm, which corresponds to a porosity of about 14%. After the chemical dissolution of the remaining aluminium and the barrier oxide layer, 150 nm thick silver layer served as a current collector was formed on the bottom side of the alumina template by magnetron sputtering. The electrocrystallization of gold was carried out in a three-electrode electrochemical cell in the potentiostatic regime ($E_d = -0.95$ V) at room temperature from commercial electrolyte containing 0.05 M $[Au(CN)^2]^-$. The height of the rods was about 400 nm. Before carrying out optical measurements, the silver electrode was selectively dissolved in dilute HNO_3 solution. A scheme of the obtained structures is shown in Fig. 1.

Optical properties of the composite are defined by two components $\varepsilon_\perp$ and $\varepsilon_\parallel$ of effective dielectric tensor that corresponds to perpendicular and parallel rods' direction, respectively. To investigate linear optical properties of the metamaterial, transmittance of the composite film was measured. In experiment transmitted through the structure halogen lamp radiation was analyzed by a spectrometer. Because of the ENZ regime is governed by dielectric tensor component $\varepsilon_\parallel$, the measurements of the spectra were performed for p-polarized fundamental beam. That allowed revealing the ENZ corresponding transition from elliptic to hyperbolic dispersion law.

Nonlinear-optical set-up was based on a femtosecond Ti-Sapphire laser system, with the pulse duration of 60 fs and repetition rate 80 MHz, operating in the wavelength range from 750 up to 880 nm. The p-polarized fundamental beam was focused on the sample into a spot of about 30 μm in diameter. Reflected SHG passed through a necessary set of color filters and was detected by a photomultiplier. In order to account for the fluctuations of the laser parameters, the reference SHG channel with a crystalline quartz as a nonlinear element was used, the measured SHG from the sample was normalized to that in the reference channel.

The spectra of the permittivity tensor components were calculated within the effective medium theory that includes the dipole-dipole interaction of the neighboring nanorods, accordingly [16]. Nonlinear-optical response was calculated, taking into account strong anisotropy of the medium, the distribution of the vectorial electric field within the structure and neglecting the spectral dependencies of the second-order susceptibility tensor components.

3. Results and discussion

Figure 2 represents the transmission wavelength-angular spectrum of the sample. It reveals two minima centred at approximately 530 nm and 810 nm and that are absent in the case of bulky gold or alumina. The first resonance is associated with the so called *transverse* localized plasmon resonance that corresponds to the electronic oscillations in the direction perpendicular to the long axis of nanorods [16]. This resonant feature turned out to be independent on the angle of incidence. Long-wavelength minimum is related to the collective *longitudinal* oscillations of the electronic gas in metal nanorods. It vanishes for the normal incidence as there are no components of the fundamental field along the rods.

The spectrum of the SHG intensity, measured for the incident angle of 45°, is shown in Figure 3. The data demonstrate an enhancement of the quadratic signal more than two times in the spectral range centred at about 815-820 nm. It should be noted that in this spectral region the long-wavelength minimum for linear transmission was observed.

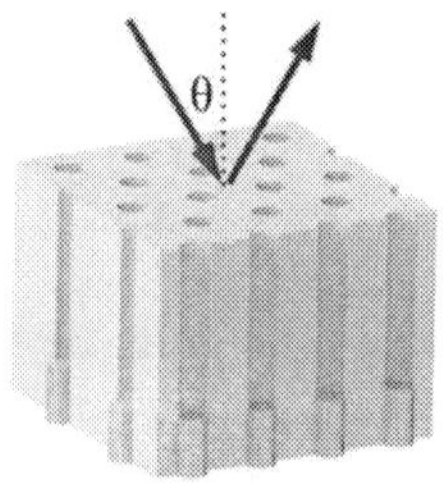

Fig.1.
Scheme of the samples.

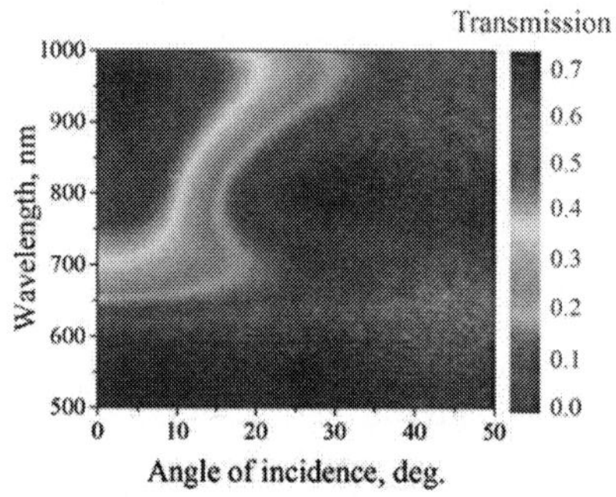

Fig. 2.
Transmission spectrum of the sample containing of Au nanorods in anodic alumina template.

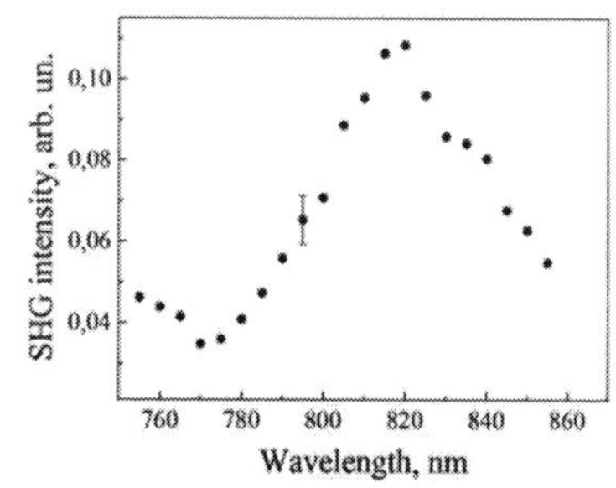

Fig. 3.
Experimental p-polarized SHG intensity spectrum recorded at incident angle of 45°.

The real parts of the permittivity tensor components perpendicular and parallel to the long axis of the rods ($\varepsilon_{\perp}$ and $\varepsilon_{||}$, respectively) were calculated in effective medium Maxwell-Garnett approach generalized to the case of an optically anisotropic system as described in [16], taking account of the morphology of Au/anodic alumina samples, i.e. the volume fraction of gold in the filled part of the template of about 14% and the length of the rods of about 400 nm. The obtained data are demonstrated in Figure 4. It can be clearly seen that at the wavelength of about 530 nm there is epsilon-near-pole point, where $\varepsilon_{\perp} \to \infty$. ENZ feature, corresponding $\varepsilon_{\parallel} \to 0$, is located near the wavelength 810 nm.

Figure 5 represents the calculated spectrum of the second harmonic intensity in the geometry corresponding to the experimental scheme, i.e. for the angle of incidence 45°, p-polarization of the fundamental and SHG radiation. The graph exhibits a strong amplification of the SHG response in the spectral vicinity of the wavelength of 810 nm. The origin of SHG enhancement is increasing of electric field in the structure caused by extremely small $\varepsilon_{\parallel}$ and continuity of normal component of electric displacement field.

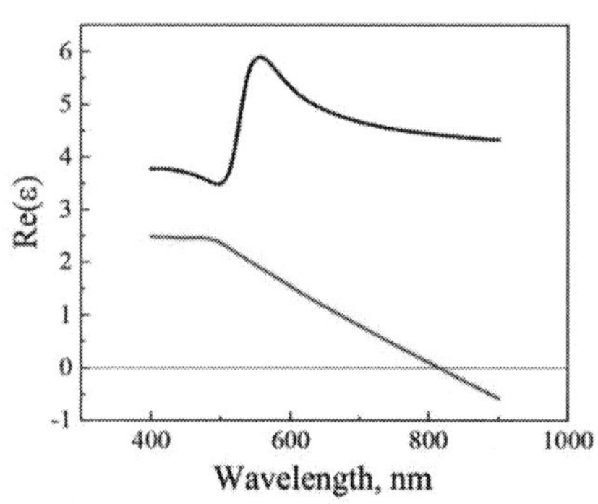

Fig. 4.
Simulated dispersion of the permittivity components:

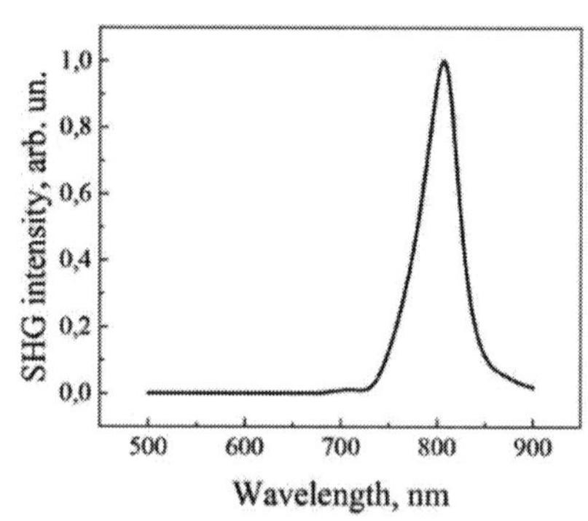

Fig. 5.
Simulated spectrum of the reflected p-polarized SHG

$Re(\varepsilon_{||})$ – red line, $Re(\varepsilon_{\perp})$ - black line. intensity. Angle of incidence is 45°.

Comparison of the obtained data allows one to conclude that the two features in transmission spectra at about 530 nm and 810 nm are associated with ENP and ENZ points, respectively. Thus, the hyperbolic dispersion regime is realized at the wavelengths above 810 nm. The SHG response is strongly enhanced in the spectral vicinity of the ENZ point that can be attributed to the local field amplification owing to the longitudinal plasmon resonance excitation. SHG experimental data stay in a good agreement with the results of numerical simulations. Small differences between them can be caused by some inaccuracies of effective medium approach and the fact that the permittivity of the bulk gold was taken for the simulation, whereas it can slightly differ from the corresponding value for electrodeposited metal.

4. Conclusion

Optical and nonlinear-optical properties of the array of gold nanorods in porous anodic alumina matrix were studied experimentally and numerically. Epsilon-near-pole and epsilon-near-zero points were detected as minima in transmission spectrum. Strong enhancement of the SHG signal was detected in the spectral vicinity of the epsilon-near-zero point owing to the local field enhancement within the structure.

Acknowledgments

The financial support of the President's grant MK-5704.2018.2 is greatly acknowledged. Some parts of the experiments were carried out using the scientific equipment purchased by M.V. Lomonosov Moscow State University Program of Development.

References

[1] A. Alu, M.G. Silveirinha, A. Salandrino, N. Engheta. Physical Review B **75**, 155410 (2007).
[2] A. Poddubny, I. Iorsh, P. Belov, Yu. Kivshar. Nature Photonics **7**, 948 (2013).
[3] P. Shekhar, J. Atkinson, Z. Jacob. Nanoconvergence **1** (14), 1 (2014).
[4] L. Ferrari, Ch. Wu, D. Lepage et al. Progress in Quantum Electronics **40**, 1 (2015).
[5] X. Ni, G.V. Naik, A.V. Kildishev et al. Applied Physics B **103**, (2011).
[6] S. Habouti, M. Matefi-Tempfli, C.-H. Solterbeck et al. J. Mater. Chem. **21**, 6269 (2011).
[7] L. Ferrari, J. Stephen, Th. Smalley et al. Nanoscale **9**, 9034 (2017).
[8] K.V. Sreekanth, M. ElKabbash, Y. Alapan et al. EPJ Appl. Metamat **4**, 1, (2017).
[9] T. Li, V. Nagal, D.H. Gracias, J.B. Khurgin. Optics Express **25** (12), 13588 (2017).
[10] N. Vasilantonakis, M. E. Nasir, W. Dickson et al. Laser Photonics Rev. **9** (3), 345 (2015).
[11] Y.R. Shen. The Principles of Nonlinear Optics. John Wiley & Sons, New York, 1984.
[12] V.L. Krutyanskiy, I.A. Kolmychek, E.A. Gan'shina et al. Phys. Rew. B **87**, 035116 (2013).
[13] Duncan, L. Perret, S. Palomba et al. Scientific reports **5**, 8983 (2015).
[14] W. Wu, L.Fan, W. Zang et al. Optics Express **25** (18), 21342-21348 (2017) .
[15] G. Marino, P. Segovia, A.V. Krasavin et al. Laser Photonics Rev. **1**, 1700189 (2018).
[16] R. Atkinson, W.R. Hendren, G.A. Wurtz et al. Phys. Rev. B **73**, 235402 (2006).

Whispering gallery modes in the two-photon luminescence of organic microspheres

I.A. Kolmychek[1,*], N.V. Mitetelo[1], K.D. Zhdanova[1], V.B. Novikov[1], E.A. Mamonov[1], M. Jyothi[2], D. Venkatakrishnarao[2], R. Chandrasekar[2], T.V. Murzina[1]

[1]Department of Physics, M. V. Lomonosov Moscow State University
[2]School of Chemistry, University of Hyderabad, Hyderabad 500046, India

*irisha@shg.ru

Abstract. Organic microstructures attract much attention due to their exciting properties that can be introduced by a proper choice of the composition, shape and alignment in an array. Here we apply the nonlinear optical microscopy for the studies of resonant properties of individual response of organic microspheres of various compositions. The observed spectral modulation of the two-photon luminescence spectra of such structures is caused by excitation of the whispering gallery modes (WGMs) that correspond to the two-photon luminescence spectrum of a particular microparticle specimen. Lasing and photobleaching effects induced by intense near-IR laser radiation are discussed.

1. Introduction

Intensive studies of microstructures of various design are stimulated by the modern trend in photonics for the search of effective, functional and tunable microelements. It was recognized that one of the perspective ways here is the usage of organic microstructures, which can be made via the self-assembling technique and reveal a number of attractive properties, such as a relatively easy way for the modification of the chemical composition, shape and dimensions, which govern their optical and nonlinear optical response[1]. The ability of this technique for the fabrication of axially symmetric structures allows for the principal realization of the excitation of whispering gallery modes (WGM) and relevant effects. It has been demonstrated that for the case of organic dyes, WGMs can be observed for microstructures of different shapes (μ-spheres, -hemispheres, -frustums, -rods) [2,3] and the quality factor up to several hundred can be achieved.

In this work, we discuss the nonlinear optical excitation of spherical organic microstructures, namely, the two-photon excitation of the photoluminescence and of the WGMs. The main benefit of this approach consists in the usage of a fundamental wavelength that corresponds to the transparency range of most organic specimen, thus providing a large penetration depth, while the irradiation of the resonating microstructures can be realized in the two-photon luminescence range. As the high intensity of the pump laser radiation is required in this case, a number of the nonlinear optical effects should be studied, including the multi-photon excitation and photoluminescence, photobleaching of the dyes, WGMs excitation and lasing.

2. Experimental section

The structures under study are the microspheres made by the molecular self-assembling technique described in detail elsewhere [4] and applied for the red dye (DCM). In this method, an acetonitrile solution of DCM dye (2 mg/2 ml) was sonicated for 30 sec and left undisturbed for 10 min, then a few drops of the solution was dispersed on a clean glass slide under slow solvent evaporation condition (self-assembly process) to form the microparticles. The evolution of spherical shape is associated with surface tension effect.

The nonlinear-optical (NLO) effects in individual DCM microspheres were studied by using the NLO microscopy set-up based on a femtosecond titanium-sapphire laser with the pulse duration of 60 fs and the central wavelength tunable in the range 760-860 nm. The pump beam passed through a dichroic mirror was focused on the edge on a microparticle by an objective with NA=0.7, which provided the optical resolution for the NLO signal of approximately 500 nm. The reflected NLO signal collected by this same objective was reflected by the dichroic mirror passed through a spectrometer and was detected by a PMT. A schematic view of the experiment is shown in Fig. 1a. It is worth noting that in these experiments the fundamental wavelengths correspond to the transparency region of the specimen.

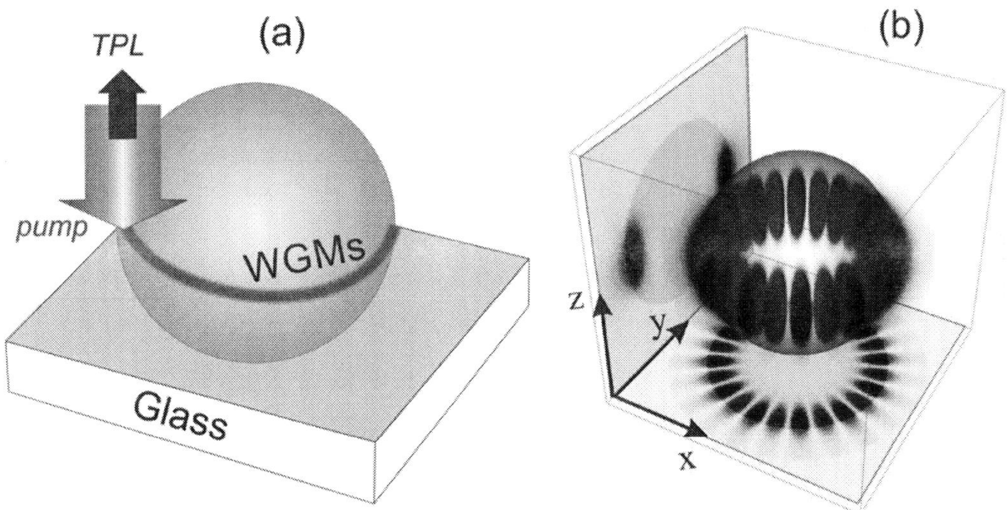

Figure 1. (a) Schematic view of the experiment of the WGMs excitation in the two-photon luminescence: the pump radiation excites the edge of a DCM microsphere, the TPL spectrum is registered in the reflection geometry in a nonlinear optical microscopy set-up. (b) Analytical calculations of the electric field distribution for a WGM with $m=l$.

3. Results and discussion

Figure 1b shows the results of the modeling of the electromagnetic field distribution in a typical DCM microsphere associated with the WGMs excitation for the mode with the angular numbers $m=l$. It can be seen that the field is localized in the equatorial plane, the number of antinodes being dependent on the sphere's diameter, material dispersion etc. So it is clear that in order to observe the resonant modes in the TPL spectra of a sphere, the signal should be detected from the edge of the structure.

This natural expectation was confirmed by the experimental studies of the spectra of the NLO response of individual microspheres. A typically observed spectrum is shown in Figure 2a. The following features should be noted here. First, the shape of the spectrum with the maximum at approximately 560 nm corresponds to the DCM dye. Second, a pronounced modulation of the TPL intensity can be seen, which is associated with the WGM excitation. While the absolute values of these maxima are not high, they correspond to a strong amplification of the TPL intensity and relevant electromagnetic field circulating within the cavity. According to our estimations, the quality factor of the WGMs is of the order of 200 and increases with spheres diameter.

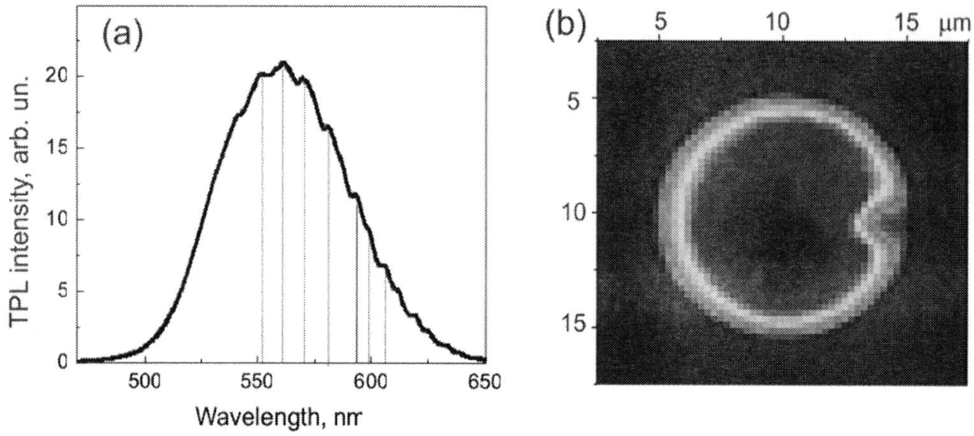

Figure 2. a) Two-photon luminescence spectrum of a DCM sphere of 7 um in diameter. (b) TPL map of a single DCM microsphere of approximately 10 microns in diameter.

First experiments on the power dependence of the intensity of definite WGMs show that under proper experimental conditions it demonstrated a fracture that should be associated with the lasing effect. At the same time, a confrontational effect that was observed in DCM structures is the photobleaching of the organic dye, which appears for the area illuminated by the pump beam. A typical TPL image of a microsphere after the illumination by an intense laser radiation (peak intensity of 300 GW/cm^2, which corresponds to the average power of 7 mW) is shown in Fig. 2 b. It can be seen that the right-hand side of the sphere reveals much less two-photon luminescence intensity due to this effect.

4. Conclusion

Summing up, we demonstrate that DCM-based organic microspheres support the excitation of whispering gallery modes' in the two-photon luminescence spectral range. The interplay between the photoluminescence and the photobleaching effects determine the luminescent properties of these organic microresonators excited by an intense laser radiation.

Acknowledgments
This work was financially supported by DST-New Delhi (Grant No. INT/RUS/RSF/P-05) and RSF-Moscow (Grant No. 16-42-02024). NVM thanks the financial support from the Foundation for the advancement of theoretical physics and mathematics "BASIS". MJ and DV thank the CSIR-New Delhi for fellowships.

References
[1] Kumar P 2010 Nanoscale Research Letters **5** 1367 ISSN 1556-276X
[2] Venkatakrishnarao D, Mohiddon M A and Chandrasekar R Advanced Optical Materials **5** 1600613
[3] Venkatakrishnarao D, Narayana Y S L V, Mohaiddon M A, Mamonov E A, Mitetelo N, Kolmychek I A, Maydykovskiy A I, Novikov V B, Murzina T V and Chandrasekar R Advanced Materials **29** 1605260
[4] Chandrasekhar N and Chandrasekar R 2010 Chem. Commun. **46**(17) 2915

Resonant excitation of spin waves in hybrid nanostructures via frequency combs

S A Kolodny, D I Yudin and I V Iorsh

ITMO University, Birjevaja line V.O., 14, Saint Petersburg, Russia, 199034

E-mail: `s.kolodny@metalab.ifmo.ru`

Abstract. Here we propose a new method of resonant optical excitation of spin waves in hybrid nanostructures via frequency comb. We consider a two-layer nanostructure consisting of an Au layer with a grating on the top and an yttrium iron garnet (YIG) thin film at the bottom. The plasmon evanescent fields induce time-independent magnetic polarization via the transverse magneto-optical Kerr effect (TMOKE) at the Au-YIG interface. Thus magnetic polarization in turn plays a role of a source for the excitation of the spin waves in YIG film. We show that by using the frequency comb as a source we can generate the magnetic polarization in resonance with the spin waves in YIG thus substantially increasing the magneto-optical response of the system. These results open new routes in the optical control of spintronics systems.

1. Introduction

Spin waves, or magnons, are propagating perturbations of magnetisation in magnetic materials [1]. Recently it has been shown that these waves can be applied as an alternative to charge current in applications of data processing and computation [2]. Excitation and detection of magnons can be accomplished optically by using pump-probe technique [3]. Magnon frequencies typically lie in GHz frequency band and thus the optical excitation of spin waves is inherently non-resonant. However, by exploiting transverse magneto-optical Kerr effect [4] and using the train of fs optical pulses as a sourse (i.e. a frequency comb), we can achieve resonant optical magnon excitation.

2. Optical resonant excitation of spin waves

In order to define the possibility of resonant optical excitation of magnons we considered a hybrid nanostructure consists of yttrium iron garnet ($Y_3Fe_5O_{12}$, YIG) thin film and Au plasmonic discrete grating on the top (see Fig. 1 (a)). YIG proved to be a good ferromagnetic material with cubic symmetry lattice. [5, 6]. The motions of magnetization in solid is described by the well-known Landau-Lifshitz-Gilbert equation (Eq. 1) [7]. The first term of equation describes a precession of magnetization around its equilibrium while the second is damping written in Gilbert form. We take Gilbert damping constant of YIG α equals to $3.2 * 10^{-4}$, however it can be lower as it is shown in Ref. [8]. M_0 is the equilibrium magnetization and H_0 is applied magnetic field.

$$\frac{d\vec{m}}{dt} = \gamma\mu_0[\vec{M}_0 \times \vec{H}_{eff}] + \frac{\alpha}{M_0}[\vec{M}_0 \times [\vec{M}_0 \times \vec{H}_{eff}]] \quad (1)$$

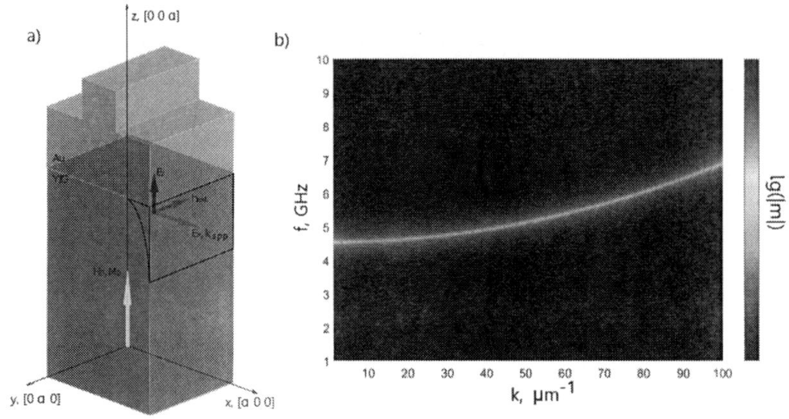

Figure 1. a) Elementary cell of the Au-YIG structure. b) Dispersion of spin waves inside YIG. Colorbar is given in decimal logarithmic scale.

H_{eff} is the effective magnetic field applied to magnetic material. In our calculation we use the following form:

$$\vec{H}_{eff} = \vec{h}_{ext} + \alpha_{ik}\frac{d^2\vec{m}}{dx_i^2} + \overline{N}^a\vec{m} - (Z_0 + Z_k)\vec{m} \tag{2}$$

Assuming the exciting field h_{ext} to be constant, we find the dispersion of magnons inside the yttrium of iron garnet (see Fig. 1 (a)). As it can be seen, the magnon resonance is located in the GHz region for the values of the plasmon wave vectors and it is very narrow because of the small value of the Gilbert damping.

To define the excitation field h_{ext}, we perform a numerical simulation in CST Microwave Studio using the eigenmode solver for infinite periodic structure (see Fig. 2). As it is shown on the Fig. 2 (a-b), the electric field inside the YIG layer has two phase shifted components of the field, E_x and E_z. The resulting magnetic field h_{ext} will be proportional to the following vector product (Fig. 2 (c)), according to TMOKE:

$$h_{ext} \sim [E \times E^*]. \tag{3}$$

Since the field in this case becomes time-independent, we propose using frequency combs to create a time and therefore frequency dependences. Thus, taking the repetition frequency of train of fs impulses equals to the resonant frequency of magnons in our structure, we can find the dispersion of the magnetic field (see Fig. 2 (d). Due to the periodicity of nanostructure and the discrete frequency dependence of fs impulses, the magnons dispersion has a discrete form with pronounced maxima. This allows us to find a dispersion of magnons $m(\omega, k)$ as $\vec{m}(\omega, k) = \hat{G}\vec{h}(\omega, k)$ (see Fig. 3 (a)). We can see that the dispersion has only one hot spot at 4.56 GHz with grating properties defined value of $|k|$ is $10\mu m^{-1}$, where the frequency of the exciting magnetic field coincides with the resonant frequency of the magnons in the yttrium iron garnet. So that is clearly seen the resonant mode of excitation. To check it we perform excitation using various frequencies in 1-10 GHz range for fixed wave vector $|k| = 10\mu m^{-1}$ with step 0.01 GHz because of narrow magnon resonance. In addition to the main resonance there are additional equidistant resonances at lower frequencies with a smaller magnitude of magnons. The calculations were carried out for the maximum normalized value of the exciting magnetic field equals to 1 a.u.. As a result, in the case of resonant excitation the magnitude of magnons is greater than it is in the case of nonresonance excitation. And if we take into account the high

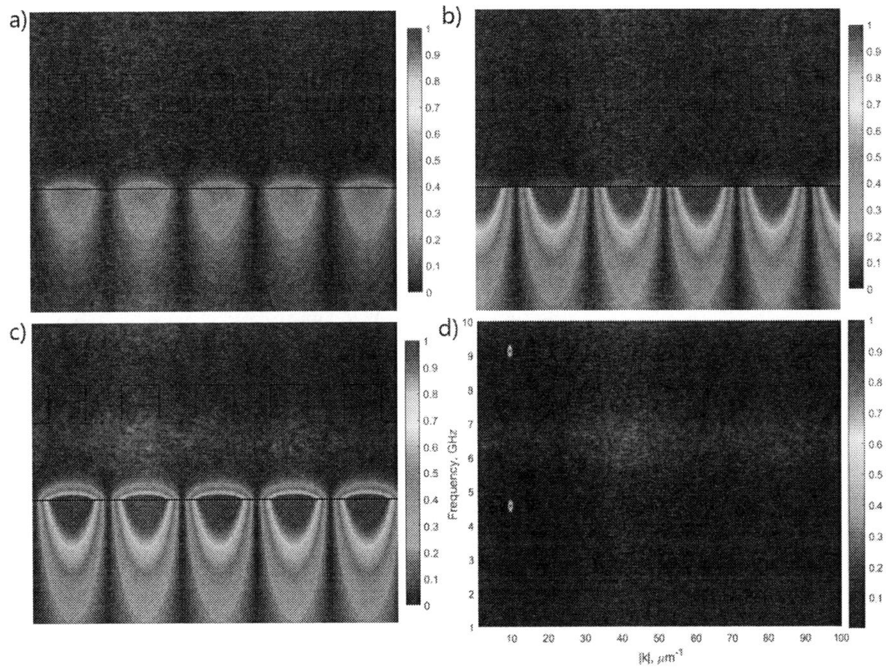

Figure 2. Fields distributions inside the infinite hybrid nanostructure: a) E_x, normalized by the maximum value of electric field, b) E_y, normalized by the maximum the maximum value of electric field and c) h_{ext}, normalized by the maximum value of magnetic field. d) Dispersion of h_{ext}, created by the train of fs optical pulses with repetition frequency 4.56 GHz.

localization of the field by the plasmon, it is obvious that excitation will occur with a higher efficiency at the same power of fs laser pulses.

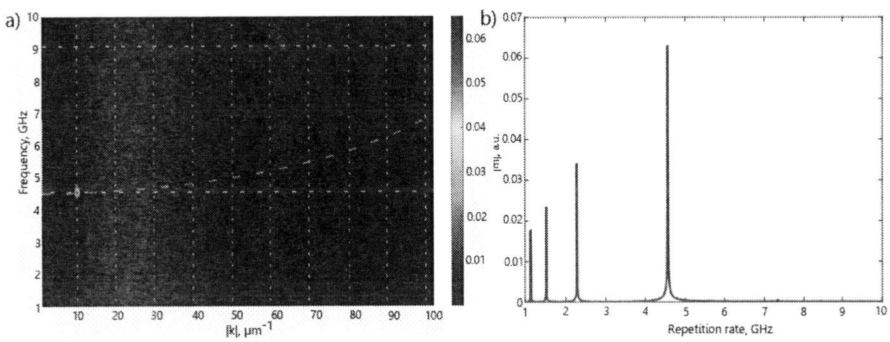

Figure 3. a) Dispersion of excited magnons in hybrid YIG-Au nanostructure. White dashed curve represents the dispersion line of spin waves in YIG, crossing points of pink dashed lines represents the dispersion of h_{ext}. b) Magnitude of excited spin waves vs repetition frequency of fs impulses.

3. Conclusions

In conclusion, we have theoretically investigated a possibility of resonant optical excitation of spin waves in YIG-Au hybrid nanostructure. We have demonstrated that by using plasmon fields and frequency combs as source the excitation of spin waves becomes more efficient in comparison with non-resonant case of excitation. We believe that the results lay the groundwork for further investigation of optically excited spin waves and can be applied for spintronic tasks.

Acknowledgements

We acknowledge the support from the Russian Academic Excellence Project 5-100, from Projects No. 3.1365.2017/4.6, No. 3.8884.2017/8.9, No. 3.4424.2017/HM and No. 3.2614.2017/4.6 of the Ministry of Education and Science of the Russian Federation, and from the Icelandic research fund (Rannis) under Project No. 163082-051.

References

[1] Phillips T G and Rosenberg H M 1966 *Reports on Progress in Physics* **29** 285
[2] Csaba G, dm Papp and Porod W 2017 *Physics Letters A* **381** 1471 – 1476 ISSN 0375-9601
[3] Kimel A, Kirilyuk A, Usachev P, Pisarev R, Balbashov A and Rasing T 2005 *Nature* **435** 655
[4] Belotelov V I and Zvezdin A K 2012 *Phys. Rev. B* **86**(15) 155133
[5] Pirro P, Brächer T, Chumak A, Lägel B, Dubs C, Surzhenko O, Görnert P, Leven B and Hillebrands B 2014 *Applied Physics Letters* **104** 012402
[6] Collet M, Gladii O, Evelt M, Bessonov V, Soumah L, Bortolotti P, Demokritov S, Henry Y, Cros V, Bailleul M *et al.* 2017 *Applied Physics Letters* **110** 092408
[7] Stancil D D and Prabhakar A 2009 *Spin waves* (Springer)
[8] Dubs C, Surzhenko O, Linke R, Danilewsky A, Brückner U and Dellith J 2017 *Journal of Physics D: Applied Physics* **50** 204005

IOP Conf. Series: Journal of Physics: Conf. Series **1092** (2018) 012061 doi:10.1088/1742-6596/1092/1/012061

Effect of grain orientation on properties of diamond/graphite metasurface fabricated by laser direct-write

M S Komlenok[1,2], **M A Dezhkina**[2], **A A Khomich**[1,3], **A S Orekhov**[4,5], **A S Orekhov**[5], **V G Ralchenko**[1,2], **S G Tikhodeev**[6,1], **V I Konov**[1,2]
[1]A.M. Prokhorov General Physics Institute, RAS, Moscow, 119991, Russia
[2]National Research Nuclear University MEPhI, Moscow, 115409, Russia
[3]V.A. Kotelnikov Institute of Radio Engineering and Electronics, RAS, Moscow region, Russia
[4]National Research Centre "Kurchatov Institute", Moscow, 123182, Russia
[5]A.V. Shubnikov Institute of Crystallography, Federal Scientific Research Centre "Crystallography and Photonics", RAS, Moscow, 119333, Russia
[6]M.V. Lomonosov Moscow State University, Leninskie Gory 1, Moscow, 119991, Russia

Abstract. We report on the correlation between the optical properties of the laser modified region with the crystallographic orientation of the irradiated surface. Graphitized structure is formed on CVD polycrystalline diamond surface after laser (KrF, τ = 20 ns, λ = 248 nm) irradiation. The thickness of the graphitized layer is analysed by transmission electron microscopy and by decreasing the intensity of the diamond line in the Raman spectra. It is found that the change of optical properties of modified layer depends on its thickness and the maximum thickness formed on the grain with orientation {111}. Simulation of transmittance of diamond/graphite metasurface confirms the variation of the graphitized layer thickness.

1. Introduction

Advances in technology of chemical vapour deposition (CVD) of polycrystalline diamond allow to obtain samples up to a millimeter thick and up to 100 mm in diameter. Different elements of photonics and electronics [1, 2] has been produced by laser induced graphitization of the surface [3-4] and bulk [5-7] of diamond. But the effect of crystal orientation of a diamond on the properties of modified area is not well studied. It is explained first of all by the use of surface graphitization of diamond for applications which does not need precise laser processing, for example for fabrication of two-dimensional diamond detectors with a size of one pixel exceeding 200 μm [8]. However for planar elements of photonics even for the THz range, when the lateral size of a single structure becomes comparable with that of a certain crystallite, the effect of crystal orientation becomes crucial. In this paper we investigate the correlation between the diamond grain orientation and the thickness of laser-induced graphitized layer by means of a Raman spectroscopy and examination of the cross section of the modified area with the help of transmission electron microscopy (TEM). Numerical modelling of transmittance of diamond/graphite metasurface confirms the variation of the thickness of graphitized layer.

2. Experimental

Experimental details of the laser direct-write technique, used to fabricate graphitized structures, and measurements of the transmittance spectra of the continuous layer (4x4 mm) and grid (width of 15 μm and period of 30 μm) can be found in our recent work [9]. A dual-beam scanning electron microscope FEI Versa FEG equipped by Hikari EBSD Camera was used for the morphology and grains orientation characterization. Diffraction patterns were acquired at 20 kV accelerating voltage and 4 nA beam current. The thickness of the graphitized layer on a diamond surface was analysed by TEM on a lamella prepared by the focused ion beam technique (FIB). Investigation of the laser induced graphitization of diamond surface was performed using a Raman spectrometer (Lab Ram HR800) in micro configuration. The probe beam wavelength was $\lambda_0 = 473$ nm.

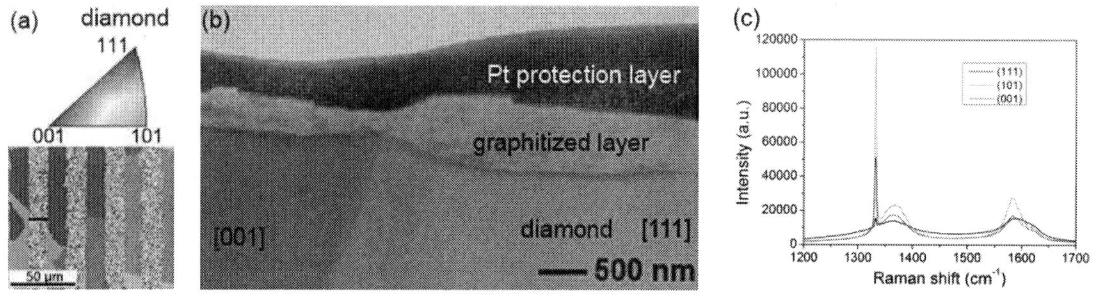

Figure 1. (a) EBSD map of crystallite orientation on the diamond surface with graphitized ribbons; **(b)** Bright field TEM image of graphitized layer cross-section on the diamond grains with different orientation; **(c)** Raman spectra of the modified area on diamond grains {111}, {101} and {001} orientation.

3. Results and discussion

To study the correlation of the properties of a laser graphitized diamond layer with the diamond grains orientation, we obtained a grain orientation map using electron backscatter diffraction (EBSD) of the sample surface (figure 1a). The indexed ribbons in figure 1a correspond to the diamond, and the graphitized ribbons look like non-indexed pixels. A TEM lamella for the top layer thickness analysis was cut from the area marked by black arrow (figure 1a) is presented in figure 1b. The area on the surface was chosen so that two grains with different orientations of {001} and {111} were represented on the cut. The boundary between the two crystallites is clearly visible in figure 1b. The difference in the thickness of the graphitized layer d_{nnn} at different grains is also evident: $d_{111} \approx 580$ nm on the orientation {111} and $d_{001} \approx 380$ nm on the orientation {001}.

The thickness of the graphitized layer on crystallites oriented along the three main planes in diamond was analyzed by Raman spectroscopy. The corresponding spectra are shown in figure 1c. The spectrum of the original diamond was also recorded, but not shown in the figure, because the intensity of diamond line, located at 1332.5 cm^{-1}, was much higher than the maximum value observed in the spectra for the graphitized surface. Intensity of the diamond peak in the spectra varies depending on the orientation of the crystallite. Similar result was observed in our recent work devoted to the diamond/graphite detectors of ionizing particles, but the orientation of diamond grains was unknown [10].

If we assume that the optical properties of the graphitized layer remain unchanged when passing from one diamond plane to another, but depend only on its thickness, then the ratio of the thicknesses d_n/d_{max} can be estimated from the ratio of the intensities of the diamond line I_{dn} in the modified region. It may readily be shown that this ratio is defined as the $\log_{I_{dmax}/I} I_{dn}/I$, where I is an intensity of the diamond line from untreated area. The minimum value of the intensity of the diamond line was recorded while passing through a graphitized layer on a crystallite with orientation {111}. Hence, the maximum thickness d_{111} of the modified layer is formed along the {111} planes. For the thicknesses of layers formed on crystallites with orientations {001} and {101}, the coefficients of 0.69 and 0.52 are obtained, i.e. $d_{001}/d_{111} = 0.69$ and $d_{101}/d_{111} = 0.52$. The result of the analysis of Raman spectroscopy correlates with the ratio of thicknesses of $d_{001}/d_{111} = 380/580 = 0.66$, obtained from electron microscopy.

So the thickness of graphitized layer depends on the grain orientation of the diamond and the question how to model optical properties of the diamond/graphite metasurface arises. The electromagnetic response of laser modified layer is unknown; we use Drude model to estimate it [9]. The only important parameters in modeling are thickness d and DC conductivity σ of the graphitized layer. The conductivity of the graphitized area with the size of 4x4 mm is controlled by the thickest part of the modified layer. So we used the maximum thickness of 580 nm for numerical simulation of the transmittance of continuous graphitized area on the diamond surface that is shown in figure 2a.

To calculate the optical properties of graphitized diamond structure with accounting for multiple reflections we employed the rigorous coupled waves analysis (RCWA) in form of the optical scattering matrix [11]. A good agreement between the simulated and measured results is observed and the best fit value is $\sigma = 305\ \Omega^{-1}\ \text{cm}^{-1}$. However, a modelling of transmittance of the graphitized grid, acting as a THz beam polarizer, with the same parameters ($d = 580$ nm and $\sigma = 305\ \Omega^{-1}\ \text{cm}^{-1}$) did not show a good agreement with the experiment. The comparison of measured and calculated transmission spectra in linear polarizations perpendicular (p-) and parallel (s-) to grid is presented in figure 2b. For p-polarization, the calculated transmissivity is only slightly higher than the experimental one, but for s-polarization the difference is significant. The best match is achieved by decrease of d to the value of 350 nm. In this case one can see a good agreement between the experiment and simulated results. This can be attributed to the fact that the DC conductivity of the ribbon is controlled by its smallest cross section. When the size of the certain grain is comparable with the width of the ribbon, a single diamond grain with the smallest thickness of graphitized layer is enough to control the conductivity.

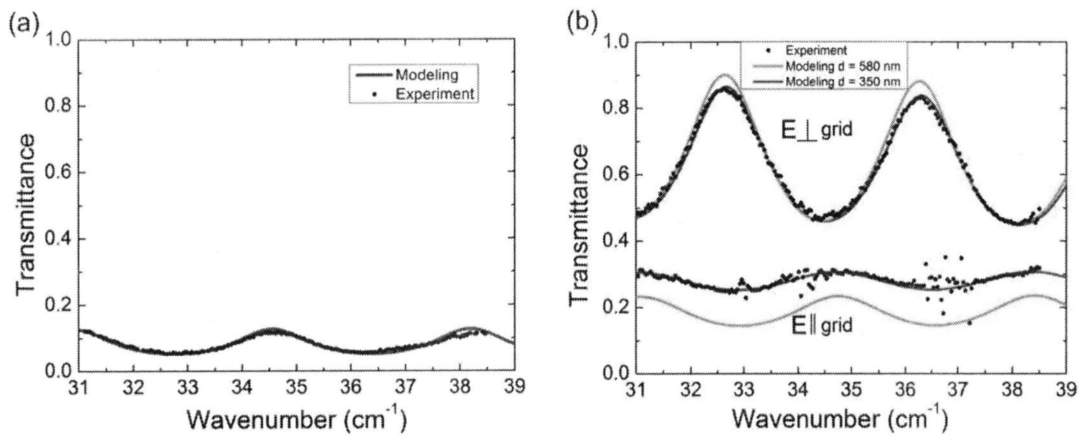

Figure 2. **(a)** Calculated (red curve) and experimental (black dots) transmittance spectra of the diamond plate with thickness of 578 μm and graphitized layer with thickness of 580 nm; **(b)** Calculated (curve) and experimental (black dots) transmittance spectra of periodic graphitized grid on the diamond surface for different polarizations of probe beam.

4. Conclusions

In conclusion, we have demonstrated the correlation between the grain orientation of CVD polycrystalline diamond and the thickness of laser-induced graphitized layer on it. Raman spectroscopy and STEM technique indicate that the maximum thickness of the modified layer is formed on the surface of the crystallite orientation {111}. Simulation of the diamond/graphite metasurface transmittance proved to be quite sensitive to the thickness of structures, and different thickness of graphitized layer was used for continuous area and grid to achieve a good agreement with experiment.

5. Acknowledgments

This work was supported by RFBR (16-32-60179), by the Competitiveness Enhancement Program of the MEPhI and by Federal Agency of Scientific Organizations under Agreement 007-GZ/Ch3363/26. The theoretical part of this work was supported by Russian Science Foundation (Grant No. 16-12-10538).

References

[1] Konov V I 2012 Laser in micro and nanoprocessing of diamond materials *Laser & Photonics Reviews* **6** 739-766

[2] Konov V I 2015 Carbon photonics *Quantum Electronics* **45** 1043

[3] Pacilli M, Allegrini P, Girolami M, Conte G, Spiriti E, Ralchenko V G, Komlenok M S, Khomic A A and Konov V I 2013 Polycrystalline CVD diamond pixel array detector for nuclear particles monitoring *Journal of Instrumentation* **8** C02043

[4] Pacilli M, Allegrini P, Conte G, Spiriti E, Ralchenko V G, Komlenok M, Bolshakov A, Khomich A A and Konov V 2014 Beta particles sensitivity of an all-carbon detector *Nuclear Instruments and Methods in Physics Research A* **738** 119-125

[5] Shimizu M, Shimotsuma Y, Sakakura M, Yuasa T, Homma H, Minowa Y, Tanaka K, Miura K and Hirao K 2009 Periodic metallo-dielectric structure in diamond *Optics Express* **17** 46-54

[6] Oh A, Caylar B, Pomorski M and Wengler T 2013 A novel detector with graphitic electrodes in CVD diamond *Diamond and Related Materials* **38** 9-13

[7] Lagomarsino S, Bellini M, Corsi C, Gorelli F, Parrini G, Santoro M and Sciortino S 2013 Three-dimensional diamond detectors: Charge collection efficiency of graphitic electrodes *Appl Phys Lett* **103** 233507

[8] Pacilli M, Allegrini P, Conte G, Spiriti E, Ralchenko V G, Komlenok M S, Bolshakov A P, Khomic A A and Konov V I 2013 Diamond-graphite pixel array for particles detection *Journal of Instrumentation* **8** C10013

[9] Komlenok M S, Lebedev S P, Komandin G A, Piqué A and Konov V I 2018 Fabrication and electrodynamic properties of all-carbon terahertz planar metamaterials by laser direct-write *Laser Physics Letters* **15** 036201

[10] Salvatori S, Oliva P, Pacilli M, Allegrini P, Conte G, Komlenok M, Khomich A A, Bolshakov A, Ralchenko V and Konov V 2017 Nano-carbon pixels array for ionizing particles monitoring *Diamond and Related Materials* **73** 132-136

[11] Tikhodeev S G, Yablonskii A L. Muljarov E A, Gippius N A, and Ishihara T, 2002 Quasiguided modes and optical properties of photonic crystal slabs *PhysicalReview B* **66**, 045102

Highly Reflective Mirror Based on Noble Metal 2D Babinet Metasurface

V V Kornienko[1,2,*], **A N Shaimanov**[1,2], **E M Khabushev**[1,3], **K M Khabarov**[1,3], **I A Rodionov**[1,4], **A V Baryshev**[1,3]

[1] Dukhov All-Russia Research Institute of Automatics (VNIIA), 22 Sushchevskaya str., Moscow 127055, Russia
[2] Faculty of Physics, Lomonosov Moscow State University, 1-2 Leninskie Gory str., Moscow 119991, Russia
[3] Moscow Institute of Physics and Technology, Dolgoprudny, Moscow Region 141700, Russia
[4] Bauman Moscow State Technical University, Vtoraya Baumanskaya str. 5, Moscow 105005, Russia

*E-mail: VladimirVKornienko@yandex.ru

Abstract. We study optical properties of plasmonic 2D nanostructures (metasurfaces) constructed by analogy with Babinet's principle – combinations of a perforated noble metal thin film and a complementary array of nanodiscs. Despite of large absorption in either the perforated films or the arrays of nanodiscs, the considered metasurfaces were highly reflective. The effect of increased reflectance is attributed to mutual cancellation of electric and 'magnetic' dipoles induced in antiphase, thus suppressing absorption; and a spacer in between the dipoles influences the spectral and incidence angle ranges for the effect observation. We show that a chosen metasurface can have an angle- and polarization-independent response in a wide spectral range.

1. Introduction

In optics, a basic formulation of Babinet's principle [1] is that the sum of a wave transmitted through a screen and a one transmitted through a complementary screen (e.g., a hole in a metal film and a metal disc of the same diameter) is exactly the incident wave. In the simplest case of a thin metal screen, a complementary screen is obtained via interchanging the metal and the holes in the initial screen configuration, i.e. via the structural inversion. Babinet's principle has been extensively used for designing the antennas with constant input impedance, studied in radio frequency. Subsequent modern applications in optics were due to the polarization-independent and spectrally-flat response of a self-complementary structure. At present, Babinet's principle is extended to absorbing and impedance surfaces, replacing the structural inversion with the impedance one in the latter case. Thin layers of substance structured at sub-wavelength level is commonly referred to as a metasurface (planar metamaterial) [2]. Babinet's principle in its initial formulation is valid for thin ($l << \lambda$) screens with large apertures in it ($d >> \lambda$) and should be clarified if a substance under study is structured at sub-wavelength level. Structuring the substance at sub-wavelength level results in peculiarities in the optical response. Extraordinary optical transmission of periodic metallic structures has been known since the early works by Ebbesen [3] and Pendry [4]. Modern designs of metasurfaces include checkerboard-type [5] and kagome [6] lattices. In [7], an effective lumped circuit model was developed for wavefront shaping on the basis of a complementary plasmonic structure.

In this paper we study the optical properties of plasmonic 2D Babinet-type nanostructures. We discuss the mechanism for increased reflectance and attribute it to the local field interaction between the grating plasmon resonances in the perforated metal film and those in ensembles of metal discs.

2. Model and numerical results

Let us consider two separated complementary 2D metasurfaces embedded into a thin dielectric film (BK7 glass) and a Babinet one, when adjoining them (Figure 1). Metasurface A is a periodic array of silver discs with the following parameters: lattice constants of $a = b = 350$ nm, a disc radius varied in a range of $r_{disc} = 30$–100 nm and a disc height of $h_{disc} = 30$ nm. Metasurface B is a perforated silver film: lattice constants of $a = b = 350$ nm, a hole radius varied in a range of $r_{hole} = 30$–100 nm and a film thickness of $h_{hole} = 30$ nm. Both metasurfaces A or B demonstrate a grating plasmon resonance (GPR)-modified spectra where interaction in the ensemble of discs or holes governs the spectral position of corresponding bands as shown in Figures 1(a) and 1(b). For the chosen $a = b = 350$ nm, these reflection spectra change with the radius of discs (or holes) as the effective refractive index of the metasurface changes [8]. For the sake of simplicity, the Babinet metasurface (***C***) was situated on a BK7 substrate (Figure 1(d)); the dielectric constant $\varepsilon_{BK7} = 1.51^2$ of the BK7 glass was used in the calculation models for all studied metasurfaces. Spacing dH between A and B varied from 10 to 100 nm. The total thickness of the considered metasurfaces was of the order of ~100 nm. Reflectivity of a Babinet-type structure is presented in Figure 1(c) for $dH = 20$ nm. We shall limit the spectral range to 600–1000 nm where GPR features are observed; spectra were calculated at normal and oblique incidence (α – the angle of incidence).

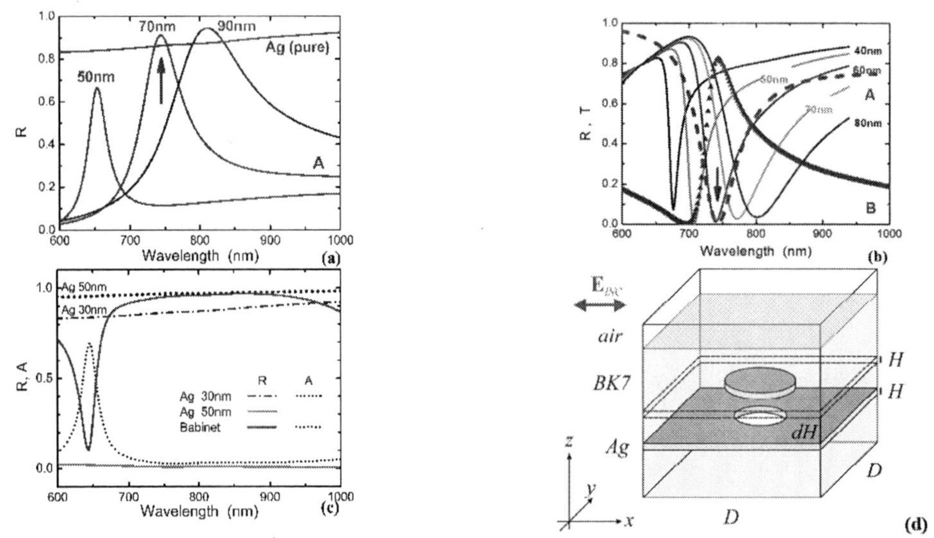

Fig. 1 **(a)** Normal-incidence reflection spectra for various disc radii for metasurface A. The reflectivity of a non-processed Ag film of the same thickness is shown with a red line. **(b)** Normal-incidence reflection spectra for various hole radii for metasurface B. Overlap takes place (shown with a red arrow) between transmission (2D metal disc array, dashed red curve) and reflection (perforated metal film) when $r_{disc} = 70$nm and $r_{hole} = 60$nm. Transmission spectrum for B ($r_{hole} = 60$nm). **(c)** Reflectivity (R) and loss ($A = 1 - R - T$) spectrum at normal incidence for a Babinet-type structure (metasurface ***C***) formed by a stacking A and B with $dH = 20$nm spacing in between. R and A for a non-processed Ag film are shown for the same (30nm, dash-dotted line) and increased (50nm, circles) thickness values. **(d)** General view of a Babinet-type structure.

When an array of nanodiscs is illuminated, the induced electric dipoles interact via the electromagnetic waves along the surface. The resonant wavelength (see Figure 1(a)) is governed by a diffraction grating condition at the order $m = \pm1, \pm2, \ldots$: $m\lambda_{GPR} = n_{EFF} D$. Here, λ_{GPR} is a vacuum wavelength, D is the period of the structure, and n_{EFF} is an effective permittivity of the medium, which can be calculated [9] in the frame of Maxwell-Garnett approximation. In the case of a perforated metal film, a number of mechanisms affect the resonance position [10] apart from the grating plasmon resonance; among them are the cavity modes and spoof plasmons. Arguably, the response of a single hole in a metal film can be described as a magnetic dipole [11]. Our modelling results (Figures 1(a) and 1(b)) suggest that metal discs of metasurface ***A*** can act as electric dipoles, while holes of metasurface ***B*** can be represented as magnetic dipoles and their transmission (T_A) and reflection spectra (R_B) can be quantitatively same, $T_A \approx R_B$. Note, also, that polarized light coupling to metasurfaces ***A*** and ***B*** goes differently at oblique incidence. The GPR band spectrally shifts in the *s*-polarized spectra of metasurface ***A*** and does not shift in the *p*-polarized ones as the angle of incidence changes [9]. One can observe the opposite behaviour in polarized spectra of metasurface ***B***; the GPR band shifts for the *p*-polarization. This is why we may expect an unusual polarization-independent optical response for metasurface ***C***.

Plot 1(b) illustrates that, taking the exactly complementary metasurfaces ***A*** and ***B***, the GPR features in their reflection and transmission spectra do not coincide. However, by simply changing r_{disc} which means influencing the effective refractive index, one can tune the spectral position of the transmission peak in the spectrum of metasurface ***A*** and adjust it to that in the spectrum of metasurface ***B***. It was found that the best overlap takes place when the holes were slightly smaller than the discs, $r_{hole} = 60$ nm and $r_{disc} = 70$ nm. Note that the reflection minimum for spectra of metasurface ***B*** is due to SPR excitation which is eventually associated with absorption. A reflection spectrum of metasurface ***C*** having $r_{hole} = 60$ nm, $r_{disc} = 70$ nm, H = 30 nm and dH = 30 nm is shown in Plot 1(c). The interaction between electric and 'magnetic' dipoles placed in the close vicinity unexpectedly resulted in an effect of increased reflectance for metasurface ***C***. The rise of reflection in the range of 700–900 nm can be associated with suppression of strength of plasmon near fields, diminishing light attenuation via two channels – surface plasmon polaritions on the surfaces of metasurface ***B*** and electric dipoles in the discs of metasurface ***A***. Thus, the reflectance from metasurface ***C*** was up to 10 percent larger than that of a conventional silver film with a thickness of 30 nm; the effective thickness of silver film is increased to 50 nm. Near-zero energy flow through the metasurface ***C*** occurs due to the fact that the electric dipole induced in ***B*** is in an antiphase [12] with the external field, as opposed to electric dipole in A. Near-field interaction between adjoined complementary metasurfaces ***A*** and ***B*** results in the cancellation of the transmitted wave.

We have observed several *dH*-dependent features in reflection spectra [Figure 2(a)]: (*i*) there is an optimal value of dH where the reflectance tends to maximum (see line 3), (*ii*) a wavelength range of the increased reflection shortens as dH increases, (*iii*) there is a band associated with resonant light coupling of the "disc–hole metaatom" in a range of 600-700 nm–an electric dipole resonace, which is not observed in spectra of metasurfaces ***A*** and ***B***, and (*iv*) a magnetic dipole resonance of the "disc–hole metaatom". Figure 2(b) illustrates robustness of the effect of increased reflection from metasurface ***C*** having dH = 30 nm. Here, the presented curves show a normalized variation of reflectance (ΔR/R) under variation of the BK7 refractive index ($\Delta n_{BK7} = 0.01$), hole ($\Delta r_{holes} = 2$ nm) and disc ($\Delta r_{disc} = 2$ nm) radii variation, and the variation of ΔR/R in *s*- and *p*-polarized spectra at for oblique incidence, $\Delta\alpha = 10°$. We attribute the increase in reflectivity to the manifestation of an interaction between two 'decoupled' sheets, one exhibiting electric dipole interaction properties while the other one manifesting the magnetic dipole ones. To validate this assumption, we varied numerically the gap distance *dH* between the layers. Increase in the gap distance *dH* leads to lowering in the reflection, as shown in Figure 2(a).

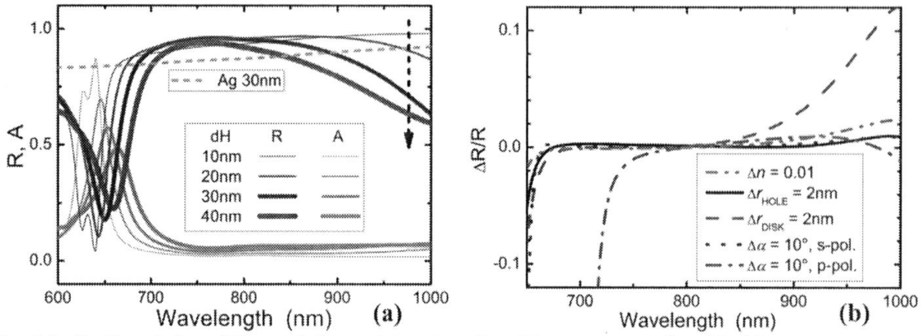

Figure 2. **(a)** Reflectivity (R) and loss ($A = 1 - R - T$) spectra of a Babinet-type structure for different spacing distances dH between complementary metasurfaces. Metal film thickness is 30 nm, r_{disc} = 70nm, r_{hole} = 60 nm. Increase in the line thickness (as shown with dashed arrow) corresponds to increase in dH from 10 nm to 40 nm in 10 nm steps. Dashed line shows the reflectivity of a 30nm-thick Ag film without (nano-) structuring. **(b)** Spectral changes in reflection spectra of a 'Babinet-type' structure due to small variations of structure parameters and interaction geometry

3. Summary

Near-field interaction between adjoined complementary metasurfaces – 2D arrays of discs and holes separated by a gap – results in the cancellation of the transmitted wave, thus significantly increasing reflectance as compared to that of the initial thin non-structured silver film. For such Babinet-type metasurface, the observed effect can be attributed to destructive interference of grating plasmon resonances originated from collective excitation of electric and 'magnetic' dipole modes of discs and holes. The resulting structure had a flat spectral response that was polarization-independent in a wide spectral range (~700-900nm). This spectral range was limited from the short/long wavelengths by bands associated by collective responses of "disc-hole metaatoms". When increasing the gap between the adjoined metasurfaces, the effect of increased reflectance was diminished, and the Babinet-type metasurface became absorbing and transmitting.

4. Acknowledgments

This work was supported by the Russian Foundation for Perspective Research, contract no. 7/004/2013-2018 of December 23, 2013.

References

[1] (new_01) Jackson J D 1999 *Classical Electrodynamics (3rd ed.)* (New York: J Wiley & Sons).
[2] (new_02) Decker M, Staude I, Falkner M, Dominguez J, Neshev D N, Brener I, Pertsch T and Kivshar Yu S 2015 *Advanced Optical Materials* **3** 813
[3] (new_03) Ebbesen T W, Lezec H J, Ghaemi H F, Thio T, and Wolff P A *Nature* **391** 667
[4] (new_04) Martin-Moreno L, Garcia-Vidal F J, Lezec H J, Pellerin K M, Thio T, Pendry J B and Ebbesen T W 2001 *Phys. Rev. Lett.* **86** 1114
[5] (new_05) Urade Y, Nakata Y, Nakanishi T and Kitano M 2015 *Phys. Rev. Lett.* **114** 237401
[6] (new_06) Nakata Y, Okada T, Nakanishi T and Kitano M 2012 Phys Rev B 85 205128
[7] (new_07) Liu T, Huang L, Hong W, Ling Y, Luan J, Sun Y and Sun W 2017 *Opt. Expr.* **25** 16332
[8] Yang S-H, Bandaru P R 2008 *Optical Engineering* **47** 029001
[9] Shaimanov A N, Khabarov K M, Merzlikin A M, Bykov I V and Baryshev A V 2017 *JETP* **124** 584
[10] Moreau A, Lafarge C, Laurent N, Edee K and Granet G 2007 *J. Opt. A: Pure Appl. Opt.* **9** 165
[11] Permyakov D V, Mukhin I S, Shishkin I I, Samusev A K, Belov P A and Kivshar Yu S 2014 *JETP Lett* **99** 724
[12] Janssen O T A, Urbach H P and Hooft G W 2006 *Opt. Expr.* **27** 11823

IOP Conf. Series: Journal of Physics: Conf. Series **1092** (2018) 012063 doi:10.1088/1742-6596/1092/1/012063

Temporal dynamics of a quantum emitter with multiple excited states in the vicinity of an anisotropic metasurface

D. F. Kornovan[1], I. D. Toftul[1], A. V. Chebykin[1], M. I. Petrov[1,2], and I. V. Iorsh[1]

[1] ITMO University , Nanophotonics and Metamaterials Department, Birjevaja line V.O., 14, 199034, St. Petersburg, Russian Federation

[2] University of Eastern Finland, Yliopistokatu 7, FI-80101 Joensuu, Finland

E-mail: d.kornovan@metalab.ifmo.ru

Abstract. In this work we focus on studying the temporal dynamics of a quantum emitter with few (3) excited states, which allows to observe an interplay between different spontaneous emission channels. We show that if the quantization axis is being rotated with respect to the normal of a metasurface it is possible to observe a difference in the transfer dynamics from one state to another and vice versa.

1. Introduction

The topic of anisotropic vacuum has been attracting a lot of attention since the first time it was discussed in [1] by G. Agarwal. The author proposed an idea that one can exploit the interference between different emission channels of a quantum emitter due to the anisotropy of the surrounding. An interesting manifestation of this effect is that it is possible to suppress the total emission rate due to destructive interference between the decay channels, which was studied in [2]. In this paper we want to study further the topic of coupling the orthogonal transitions of a single quantum emitter through the field modes of a metasurface.

2. Theoretical Framework and results

In this work we investigate the time evolution of a four level inverse tripod-type atom with a single ground state $|g\rangle$ and three excited states $|e_{-1}\rangle$, $|e_0\rangle$, $|e_+\rangle$ with the corresponding transition dipole moments $\mathbf{d_{-1}}$, $\mathbf{d_0}$, and $\mathbf{d_{+1}}$ as shown in Fig. 1, (a). These three dipole moment vectors form an orthonormal set in $\mathbb{R}^3$, which leads to the fact that the three corresponding excited states are decoupled in the isotropic environment. However, they might interact with each other via the modes supported by a photonic or plasmonic structure. From this perspective a metasurface presents an interesting system to consider, since it can be anisotropic in the interface plane. The optical properties of it can be characterized by the tensor of the effective surface conductivity σ, which can be chosen to be diagonal in some coordinate system. To describe the optical properties of a metasurface we use the effective conductivity described by [3]:

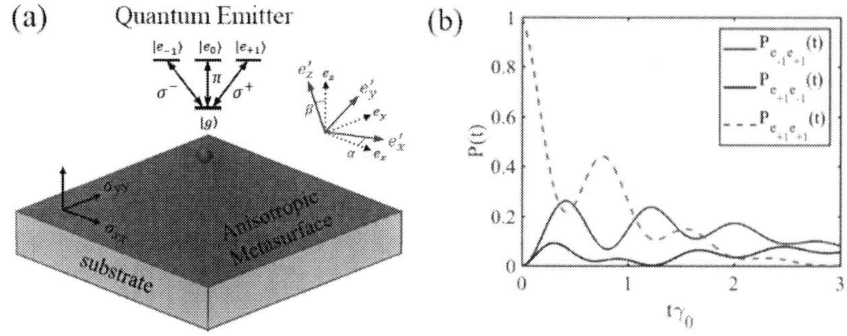

Figure 1. a) Schematic picture of the system: an atom with 3 excited states in the vicinity of an anisotropic metasurface. The dotted arrows represent the lab coordinate system, the blue solid arrows - the rotated on 2 Euler angles. b) Excitation transfer from e_1 to e_2 and vice versa along with the initially excited state population. The parameters are: $A_{xx} = A_{yy} = 1$, $\Omega_{xx} = 1$, $\Omega_{yy} = 3$, $\gamma_{xx} = \gamma_{yy} = 0.01$, $k_0 = 0.65\Omega_{xx}/c$, the rotation angles were $\alpha = \beta = \pi/4, \gamma = 0$. The substrate was considered as an infinite dielectric half-space with $\epsilon_2 = 4$. The atom-surface distance is $\Delta z = 0.05\, c/\Omega_{xx}$.

$$\sigma = \begin{pmatrix} \sigma_{xx} & 0 \\ 0 & \sigma_{yy} \end{pmatrix}, \quad \sigma_{jj} = A_{jj}\frac{ic}{4\pi}\frac{\omega}{\omega^2 - \Omega_{jj}^2 + i\gamma_{jj}\omega}, \tag{1}$$

where A_{jj} is the normalization constant, Ω_{jj} is the resonance frequency, γ_{jj} is the bandwidth of the corresponding resonance.

In order to solve for temporal dynamics of the emitter's states we employ the approach of the resolvent operator [4, 5], when it is possible to find the matrix elements of the evolution operator $\hat{U}(t,0)$ using:

$$\langle e_f|\hat{U}(t,0)|e_i\rangle = \int_C \frac{dz}{2\pi i} e^{-izt/\hbar}\langle e_f|\hat{G}(z)|e_i\rangle, \tag{2}$$

where $\hat{G}(z) = (z - \hat{H})^{-1}$ is the resolvent operator of the full Hamiltonian $\hat{H}$. Using this relation we can find probabilities for the atom to be in the excited state $|e_f\rangle$ at time t given that it was in the state $|e_i\rangle$ initially by simply finding $P_{e_f}(t) = |U_{e_f e_i}(t,0)|^2$.

We should note that here we work in the single excitation domain since we are studying the excited states evolution of a single atom. In this single excitation subspace of a Hilbert space the resolvent can be found to be:

$$\langle e_f|\hat{G}(z)|e_i\rangle = \langle e_f|\left[z - \hat{H}_0 - \hat{\Sigma}(z)\right]^{-1}|e_i\rangle \approx \left[(z - \hbar\omega_0)\,\delta_{e_f e_i} + 4\pi k_0^2 \mathbf{d}^*_{\mathbf{e_f}}\mathbf{G}(\mathbf{r_0},\mathbf{r_0},\omega_0)\mathbf{d}_{\mathbf{e_i}}\right]^{-1}, \tag{3}$$

here $\hat{H}_0$ is the unperturbed Hamiltonian and $\Sigma(z)$ is the level shift operator, δ_{ij} is the Kronecker delta. The level-shift operator in the latter expression is written within two approximations: near-resonant interaction of the states ($z = \hbar\omega_0$) and taking into account only proccesses of the second order [6], leading to a form of a classical dipole-dipole coupling constants: $\Sigma_{e_f e_i} = -4\pi k_0^2 \mathbf{d}^*_{\mathbf{e_f}}\mathbf{G}(\mathbf{r_0},\mathbf{r_0},\omega_0)\mathbf{d}_{\mathbf{e_i}}$. Therefore, in order to describe the interaction of the emitter with the electromagnetic modes of the field we only need to construct the Green's function of a metasurface, which can be done by following the procedure outlined in [7].

Now we want to rotate the transition dipole moments $\mathbf{d_q}$ with respect to the original coordinate system, in which σ tensor is diagonal. The rotation of dipole moments will alter the way excited states couple with each other, therefore, altering the time dynamics. Formally this can be done by first considering the level-shift operator matrix in the Cartesian coordinate system $\Sigma_{Cart}(\hbar\omega_0)$, which expresses the couplings of linear dipoles d_j, $j = x, y, z$. Then we can use the transformation matrix S, which transforms the Σ from the Cartesian coordinate system into the spherical tensor basis $\mathbf{d_q}$, $q = +1, 0, -1$. Finally, we can simply rotate the spherical tensors written in Cartesian coordinates with rotation matrix $M(\alpha, \beta, \gamma)$ (active representation, right-hand rule, $z'' - y' - z$ convention), where α, β, γ are the Euler angles:

$$\Sigma'_{Sph} = T^{-1}(\alpha, \beta, \gamma)\Sigma_{Cart}T(\alpha, \beta, \gamma),$$
$$T(\alpha, \beta, \gamma) = M(\alpha, \beta, \gamma)S. \tag{4}$$

After performing this, we can express the evolution operator matrix elements (2) through the matrix elements of $T(\alpha, \beta, \gamma)$, and the diagonal entries of Σ_{Cart} as:

$$U_{kl}(t, 0) = \sum_{j=1}^{3} T^*_{j,k}T_{j,l}e^{-i\Sigma_{jj}t/\hbar} = \sum_{j=1}^{3} C_j^{(kl)}e^{-i\Sigma_{jj}t/\hbar}, \tag{5}$$

Note that for Σ_{Cart} index j labels the Cartesian coordinates as $1 - x, 2 - y, 3 - z$, and for the case of spherical basis - $1 - (-1), 2 - (0), 3 - (+1)$.

Let us focus on the case, when we study how excitation is transferred between the two excited states: $|\mathbf{e_{+1}}\rangle \rightarrow |\mathbf{e_{-1}}\rangle$ and the inverse process $|\mathbf{e_{-1}}\rangle \rightarrow |\mathbf{e_{+1}}\rangle$. For them we can write explicitly the corresponding coefficients before the exponents:

$$C_1^{(13)} = -\frac{e^{2i\gamma}}{2}\left(cos(\alpha)cos(\beta) - isin(\alpha)\right)^2,$$
$$C_2^{(13)} = \frac{e^{2i\gamma}}{2}\left(cos(\alpha) - icos(\beta)sin(\alpha)\right)^2,$$
$$C_3^{(13)} = -\frac{e^{2i\gamma}}{2}sin^2(\beta), \tag{6}$$

and for the inverse process we have $C_j^{(31)} = (C_j^{(13)})^*$. One can notice from (6) that the Euler angle γ enters as an overall phase factor, therefore, the probabilities given by $|U_{kl}(t, 0)|^2$ are independent of it. This is natural since γ represents the rotation around the new quantization axis $\mathbf{e'_z}$ and such a rotation should not change the behavior of the system. It is also important to note that in a general situation, for arbitrary α and β, the coefficients $C_k^{(ij)}$ are complex. It means that when finding the probabilities $|U_{kl}(t, 0)|^2$, the $e^{-i\Sigma_{jj}t/\hbar}$ terms will interfere in such a way that the oscillating parts of the dynamics for $|e_{+1}\rangle \rightarrow |e_{-1}\rangle$ and $|e_{-1}\rangle \rightarrow |e_{+1}\rangle$ will have different initial phases as can be seen in Fig. 1, (b). Indeed,

$$P_{e_{-1}e_{+1}}(t) = \sum_{j} \xi^{(jj)}_{e_{-1}e_{+1}}e^{-\gamma_{jj}t} + \sum_{i<j} 2\xi^{(ij)}_{e_{-1}e_{+1}}cos\left(\left[\Delta_{ii} - \Delta_{jj}\right]t - \phi^{(ij)}_{e_{-1}e_{+1}}\right)e^{-(\gamma_{ii}+\gamma_{jj})t/2}, \tag{7}$$

where $\Delta_{ii} = \mathrm{Re}[\Sigma_{ii}]$, $\gamma_{ii} = -2\mathrm{Im}[\Sigma_{ii}]$, $\xi^{(ij)}_{e_{-1}e_{+1}}$ are some real-valued constants, $\phi^{(ij)}_{e_{-1}e_{+1}}$ are the initial phases of oscillations. Notice that both $\xi^{(ij)}_{e_{-1}e_{+1}}$ and $\phi^{(ij)}_{e_{-1}e_{+1}}$ depend only upon the Euler angles and have the following properties: $\xi^{(ij)}_{e_{-1}e_{+1}} = \xi^{(ij)}_{e_{+1}e_{-1}}$, but $\phi^{(ij)}_{e_{-1}e_{+1}} = -\phi^{(ij)}_{e_{+1}e_{-1}}$. The latter leads to a *phaseshift* in the oscillatory dynamics for the probabilities of the two processes under study. Also notice from (7) that in order to observe this effect the anisotropy of the structure is needed as if $\Delta_{ii} = \Delta_{jj}$, then the contributions from the phaseshifts can be simply absorbed in the definition of the constant $\xi^{(ij)}_{e_1e_3}$ and there will be no difference in the dynamics.

3. Conclusion

In our work we studied the dynamical behavior of an inverse-tripod type quantum emitter with three excited states interacting though the modes of an anisotropic metasurface. We demonstrated that if the quantization axis has an arbitrary orientation then it is possible to observe a non-reciprocity, which manifests itself in a phase delay in the oscillating part of the dynamics arising from the anisotropic properties of the structure.

4. Acknoledgements

This work was supported by the Russian Foundation for Basic Research (proj. # 18-32-00691). D. K. and M.P. acknowledge the support from Basis Foundation.

References

[1] Agarwal G S, 2000 Anisotropic vacuum-induced interference in decay channels, *Phys. Rev.* Lett., vol. **84**, p. 5500-5503.
[2] Hughes S and Agarwal G S 2017 Anisotropy-Induced Quantum Interference and Population Trapping between Orthogonal Quantum Dot Exciton States in Semiconductor Cavity Systems *Phys. Rev.* Lett., vol. **118**, p. 063601.
[3] Yermakov O Y, Ovcharenko A I, Song M, Bogdanov A A, Iorsh I V and Kivshar Yu S 2015 Hybrid waves localized at hyperbolic metasurfaces *Phys. Rev.* B, vol. **91**, p. 235423.
[4] Cohen-Tannoudji C, Dupont-Roc J and Grynberg G 2004 *Atom-Photon Interactions: Basic Processes and Applications*, Wiley-VCH Verlag GmbH & Co. KGaA.
[5] Kornovan D F, Petrov M I and Iorsh I V, 2017 Transport and collective radiance in a basic quantum chiral optical model, *Phys. Rev.* B, vol. **96**, p. 115162.
[6] Dung H T, Knöll L and Welsch D-G 2002 Resonant dipole-dipole interaction in the presence of dispersing and absorbing surroundings, *Phys. Rev.* A, vol. **66**, p. 063810.
[7] Lakhtakia A, 1992 Green's functions and Brewster condition for a halfspace bounded by an anisotropic impedance plane, *International Journal of Infrared and Millimeter Waves*, vol. **13**, p. 161-170.

IOP Conf. Series: Journal of Physics: Conf. Series **1092** (2018) 012064 doi:10.1088/1742-6596/1092/1/012064

Light-matter interaction between photonic bound states in the continuum and bright excitons in transition metal dichalcogenides

K. L. Koshelev[1,2], **S. K. Sychev**[1], **Z. F. Sadrieva**[1], **A. A. Bogdanov**[1], **I. V. Iorsh**[1]

[1]ITMO University, 197101 St. Petersburg, Russian Federation
[2]Nonlinear Physics Centre, Australian National University, Canberra ACT, Australia

E-mail: `ki.koshelev@gmail.com`

Abstract. Being motivated by recent achievements in the rapidly developing fields of optical bound states in the continuum (BICs) and excitons in monolayers of transition metal dichalcogenides, we analyze strong coupling between BICs in Ta_2O_5 periodic photonic structures and excitons in WSe_2 monolayers. We demonstrate that giant radiative lifetime of BICs allow to engineer the exciton-polariton lifetime enhancing it three orders of magnitude compared to a bare exciton.

1. Introduction

Monolayers of transition metal dichalcogenides (TMDCs) are a certain class of post-graphene two-dimensional materials [1], attracting vast research interest in recent years. TMDC are direct-gap semiconductors, exhibiting strong light-matter coupling [2], leading to the emergence of the new quasiparticles, exciton-polaritons [3] at room temperatures in structures comprising TMDC monolayer and an optical cavity.

Excitons-polaritons have been extensively studied in last two decades due to their fascinating fundamental properties. Strong coupling of TMDC excitons to light has been observed in the structures resembling the conventional microcavities, where the monolayer was sandwiched between two Bragg mirrors [4]. At the same time, since fabrication of high quality TMDC monolayers is based on the mechanical exfoliation techniques, it is quite technologically demanding. Thus, it would be extremely useful to realize high quality optical resonances without the requirement for the growth of the upper mirror.

In this paper we propose an alternative scheme for the realization of strong exciton-photon coupling in 2D materials without using mirrors which is beneficial both in terms of ease of realization and tunability. We focus on structures comprising of a TMDC and a photonic crystal slab (PCS). The idea behind that is the exploitation of the so-called optical bound state in the continuum (BIC) [5], supported by the PCS, as a high quality cavity mode. BICs in the periodic photonic structures originate due to destructive interference of the leaky modes supported by the PCS. They are characterized by extremely high quality factors.

The design of structure is shown in Fig. 1 - a WSe_2 flake is placed on top of an one-dimensional Ta_2O_5 PCS. We notice that strong light-matter coupling demands both high-Q

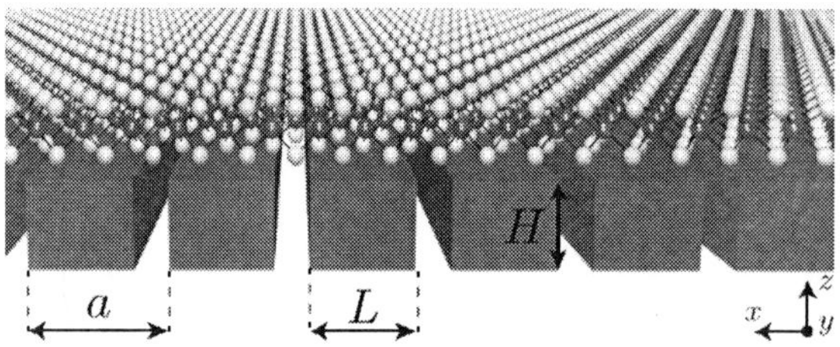

Figure 1. Isometric view of a PCS-TMDC structure. The refractive index of Ta_2O_5 bars is equal to 2.1. The TMDC made of WSe_2 is laid on top of the PhC slab.

photonic structures and long-living excitonic states. We address this problem by tuning the PCS shape and material parameters to the regime of bound states in the continuum providing a giant quality factor of resonator which is limited by surface roughness and finite size of the sample only.

We begin with analysis of the eigenmode spectrum of the PCS applying the guided-mode expansion (GME) method [6] widely used for characterization of photonic eigenmodes of periodic photonic structures.

We consider an air-suspended Ta_2O_5 one-dimensional grating with lower and upper air claddings being semi-infinite. The PCS consists of rectangular bars with height H and width L being spaced equidistant with a period of a (see Fig. 1). We put refractive index of Ta_2O_5 equal to 2.1 which is appropriate for the red band of the visible spectrum range. The calculations are performed for the PCS with $a = 1.03H$, $L = 0.90a$ and the eigenvalue problem is truncated by 101 plane waves and 8 guided modes of the effective waveguide kept in the expansion [6]. The spectrum of eigenfrequencies ω and inverse radiation lifetimes $\gamma = 1/(2\tau_{\text{rad}})$ of the PCS for in-plane wavevectors along the x direction of the first Brillouin zone is shown in Fig. 2(a,c) for TE-polarized and in Fig. 2(b,d) for TM-polarized modes, respectively. The BICs represent unusual leaky modes with $\gamma = 0$ and can be formed both at the center of the Brillouin zone (at-Γ BIC) and at specific points between the zone edge and center (off-Γ BIC).

Optical properties of WSe_2 monolayers are governed by very robust excitons with binding energies of the order of 500 meV [7]. We study A-type excitons representing bound states of electrons in the conduction band and holes in the upper subband of the valence band of $\mathbf{K}_+$ and $\mathbf{K}_-$ valleys [8]. We focus on bright excitonic states active for in-plane polarization of incident light, which represent, in general, a pair of valley-degenerate states with $\sigma_\pm$ polarization. In this case a linearly polarized pump excites a superposition of excitons with total polarization along the in-plane component of the electric field of light.

To investigate light-matter interaction between excitons in the TMDC monolayer and BICs in the PCS we apply the full quantum formalism for both photonic and excitonic states being an extension of the GME method [9]. Coupling between photonic and exciton modes is governed by the oscillator strength f and the overlap integral of mode profiles

$$V_{n,\mathbf{k}_\parallel} = -i\sqrt{f} \int\limits_{\text{cell}} d\mathbf{r}_\parallel \; \mathbf{e} \cdot \mathbf{E}^{\text{up}}_{n,\mathbf{k}_\parallel}(\mathbf{r}_\parallel) e^{-i\mathbf{k}_{\text{exc}}\mathbf{r}_\parallel}. \quad (1)$$

Here $\mathbf{e}$ is the unit vector of exciton polarization, $\mathbf{k}_\parallel$ and $\mathbf{k}_{\text{exc}}$ are in-plane wavevectors of light and exciton, respectively, and $\mathbf{E}^{\text{up}}_{n,\mathbf{k}_\parallel}$ is the electric field of n-th photonic mode at the upper surface of the PCS. Importantly, exciton-photon interaction is allowed only under condition $\mathbf{k}_{\text{exc}} = \mathbf{k}_\parallel + \mathbf{G}$, where $\mathbf{G}$ is the reciprocal lattice vector. We formulate the total Hamiltonian and reduce it to a non-Hermitian eigenvalue equation.

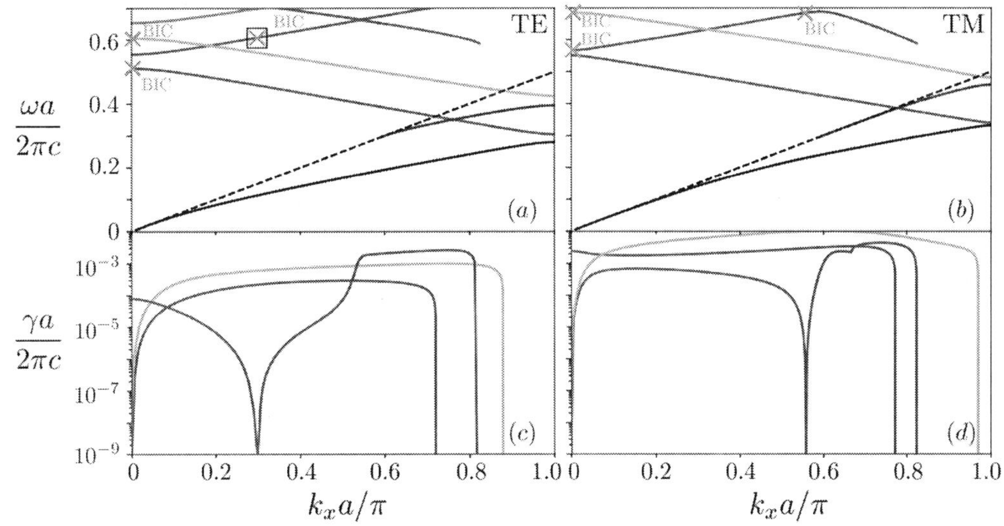

Figure 2. Eigenmode spectrum of air-suspended Ta_2O_5 one-dimensional grating. Band structure $\omega a/2\pi c$ for (a) TE-polarized and (b) TM-polarized modes, respectively. Black dashed lines represent the light lines $\omega = ck_x$. Dimensionless inverse radiation lifetime $\gamma a/2\pi c$ for (c) TE-polarized and (d) TM-polarized modes, respectively. BICs are marked by orange crosses.

We apply the procedure for characterization of band structure and damping rates of exciton-polaritons being formed in the vicinity of the BIC. At low temperatures of about 4 K energy of A-type bright exciton in WSe_2 is of order of $E_{\text{exc}} = 1.74$ eV [10] and its dispersion can be neglected at the scales of the problem. The radiative $\tau_{\text{exc,R}}$ and non-radiative $\tau_{\text{exc,NR}}$ lifetimes can estimated as 1 ps [11] and 1 ns [12]. For a bare exciton, radiative channel dominates and leads to the damping rate $\hbar/(2\tau_{\text{exc}}) = 0.33$ meV. However, when the TMDC is strongly coupled to the PCS, the radiative channel of exciton into the photonic system is enabled and it dominates with respect to direct radiation into free space [13]. It leads to renormalization of exciton radiative lifetime of exciton used as a parameter for eigenmode procedure. Therefore, in calculations we use non-radiative lifetime for the exciton damping rate $\hbar/(2\tau_{\text{exc,NR}}) = 0.33\ \mu$ eV. Oscillator strength depends on $\tau_{\text{exc,R}}^{-1}$ and is about 6.5×10^{-12} cm $\cdot$ eV^2. We focus on the off-Γ BIC marked with a black square in Fig. 2 which has TE polarization which leads to better coupling with in-plane excitons according to Eq. 1. We tune the BIC frequency to a resonance with E_{exc} adjusting the height of the PCS to the value of $H = 418$ nm.

The spectrum of energies $\hbar\omega$ and inverse lifetimes $\hbar/(2\tau)$ of upper and lower exciton-polariton branches (UP and LP, respectively) calculated by means of the GME is shown in Fig. 3(a,b). Figure 3(a) demonstrates strong coupling between the exciton and the off-Γ BIC which manifests itself as an avoided resonance crossing with Rabi splitting of order of 3 meV. The radiation losses of both exciton-polariton branches are shown in Fig. 3(b) in comparison with the losses of photonic and exciton modes. As it can be seen, for specific values of k_x the lifetime of polariton modes can exceed the bare exciton lifetime by almost three orders of magnitude and reaches 0.66 ns. Such giant enhancement is the special effect intrinsic to optical bound states in the continuum. As it was discussed before, direct radiation of excitons into free space is strongly suppressed due to coupling to the PCS. At the same time BICs are entirely uncoupled to the radiation continuum. In total, this leads to suppression of all possible mechanisms of radiation resulting in giant LP lifetime comparable with the non-radiative lifetime of a bare exciton.

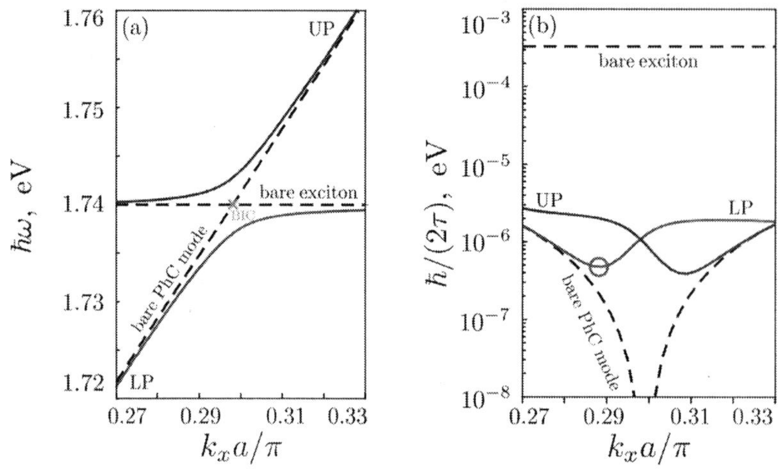

Figure 3. (a) Dispersion and (b) inverse lifetime of exciton-polaritons at the conditions of strong coupling between the TE-polarized photonic mode supporting an off-Γ BIC and in-plane polarized exciton with energy of 1.74 eV. State with maximal lifetime is marked by a green circle.

In conclusion, we have proposed an experimentally feasible scheme to achieve strong coupled exciton-photon system in a two-dimensional nanostructure comprising a TMDC monolayer and a periodic photonic nanostructure. Importantly, this scheme does not require the growth of Bragg mirrors, which substantially simplifies the fabrication. We believe, these findings open new avenues for the applications of strong light-matter coupling at the nanoscale.

Acknowledgments

The work has been supported by the Ministry of Education and Science of the Russian Federation (3.1668.2017/4.6), the Russian Foundation for Basic Research (16-37-60064) and the President of Russian Federation (Grant MK-403.2018.2).

[1] Bhimanapati G R, Lin Z, Meunier V, Jung Y, Cha J, Das S, Xiao D, Son Y, Strano M S, Cooper V R *et al.* 2015 *ACS Nano*
[2] Mak K F and Shan J 2016 *Nature Photonics* **10** 216–226
[3] Kavokin A V, Baumberg J J, Malpuech G and Laussy F P 2007 *Microcavities* vol 21 (Oxford University Press)
[4] Liu X, Galfsky T, Sun Z, Xia F, Lin E c, Lee Y H, Kéna-Cohen S and Menon V M 2015 *Nature Photonics* **9** 30–34
[5] Hsu C W, Zhen B, Stone A D, Joannopoulos J D and Soljačić M 2016 *Nature Reviews Materials* **1** 16048
[6] Andreani L C and Gerace D 2006 *Physical Review B* **73** 235114
[7] He K, Kumar N, Zhao L, Wang Z, Mak K F, Zhao H and Shan J 2014 *Physical review letters* **113** 026803
[8] Glazov M, Ivchenko E, Wang G, Amand T, Marie X, Urbaszek B and Liu B 2015 *physica status solidi (b)* **252** 2349–2362
[9] Gerace D and Andreani L C 2007 *Physical Review B* **75** 235325
[10] Wang G, Robert C, Suslu A, Chen B, Yang S, Alamdari S, Gerber I C, Amand T, Marie X, Tongay S *et al.* 2015 *Nature communications* **6**
[11] Palummo M, Bernardi M and Grossman J C 2015 *Nano letters* **15** 2794–2800
[12] Robert C, Lagarde D, Cadiz F, Wang G, Lassagne B, Amand T, Balocchi A, Renucci P, Tongay S, Urbaszek B *et al.* 2016 *Physical Review B* **93** 205423
[13] Poddubny A N, Belov P A and Kivshar Y S 2011 *Physical Review A* **84** 023807

IOP Conf. Series: Journal of Physics: Conf. Series **1092** (2018) 012065 doi:10.1088/1742-6596/1092/1/012065

Optical binding near plasmonic substrates

N A Kostina[1 a)], A N Ivinskaya [1], A S Shalin[1]

[1] ITMO University, St. Petersburg 197101, Russia

[a)] E-mail: nkostina@corp.ifmo.ru

Abstract. In this work we consider optical binding of two dipolar nanoparticles near plasmonic substrates, that introduces additional degree of freedom into the system and allows precise and versatile control over the particles' position. It is shown, that surface modes allow to overcome diffraction limit and increase stiffness of optical trapping by an order of magnitude.

1. Introduction

Through the last decades optical forces attracted a significant interest due to their applicability in biological, chemical and optoelectronical research. For ensembles of particles optical forces can lead to an optical binding effect [1], that is self-organization of the particles caused by light rescattering. One of the most important optical binding characteristics are locations of stable positions of the particles and stiffness of the optical trap. Usually, precision of the particles' positioning is restricted by incident light wavelength and is hard to vary. Also, for stable trapping one needs to improve trap stiffness what requires high intensity of incident field. It was shown earlier, that the presence of a substrate with surface [2] or volumetric modes [3] affects drastically optical force on a single particle, and helps to improve trapping, or, for certain conditions, obtain anti-trapping effects [4].

2. Calculation of the optical binding force near the substrate

In this work we investigate the simplest case of two dipolar nanoparticles placed above the substrate and illuminated by a plane wave, figure 1.

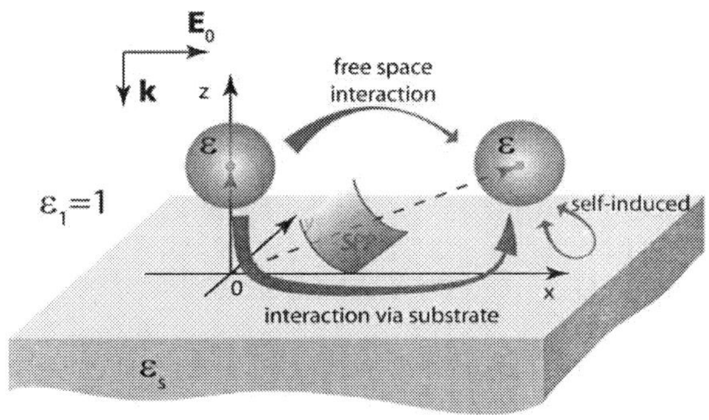

Figure 1. The scheme of the structure. Plane wave is polarized along x direction and incident along z. Origin of the coordinates coincides with the first particle's position, and the second has coordinates (x, y, z). Dielectric permittivity of the particles ε =3, radius R=15 nm.

The force acting on a small particle is described as follows:

$$\mathbf{F} = \frac{1}{2}\sum_i \mathrm{Re}(p_i^* \nabla E_i^{loc}), \tag{1}$$

where p_i is the ith component of particle induced dipole moment, and E_i^{loc} corresponds to the local field at the location of the particle, $i=x,y,z$.
General expression for local field at the location of one of the nanoparticles is obtained by using Green's function formalism:

$$\mathbf{E}_2 = \mathbf{E}_0 + \frac{k^2}{\varepsilon_1\varepsilon_0}\mathbf{G}_1\mathbf{p}_1 + \frac{k^2}{\varepsilon_1\varepsilon_0}\mathbf{G}_2\mathbf{p}_2. \tag{2}$$

Here, $\mathbf{p}_1$, $\mathbf{p}_2$ are induced dipole moments for the first and second nanoparticles correspondingly, $\mathbf{E}_0$ is incident field with wave vector k, and ε_0, ε_1 is vacuum permittivity and permittivity of the upper half-space. Green's functions $\mathbf{G}_1$, $\mathbf{G}_2$ for the first and second dipoles consist of the free space binding contribution, self-induced contribution through the substrate and multiple rescattering by the particles through the substrate.

3. Enhancement of optical binding

Figure 2 illustrates that surface plasmon polariton at the metal-dielectric interface enhances optical force by an order of magnitude in comparison with homogenous space. The distance between equilibrium positions of the particles can become smaller than diffraction limit due to dispersion relation of the plasmon. For example, the first equilibrium position at the substrate is around 140 nm, but it is approximately 400 nm for optical binding in a free space.

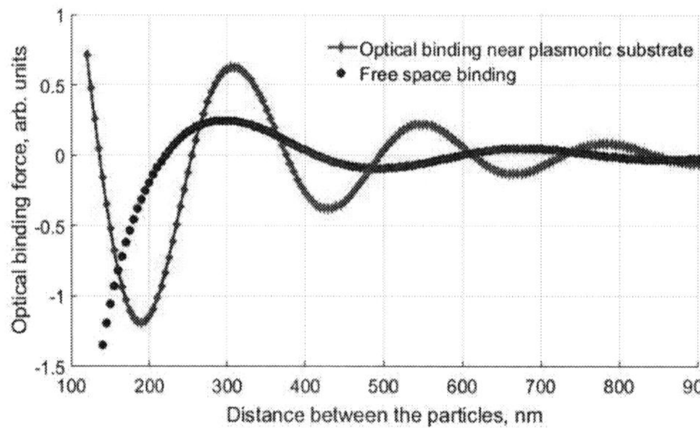

Figure 2. The optical binding force dependence on the distance between two dipoles. Blue line corresponds to the interaction in a free space, and red line in vicinity of the plasmonic substrate (silver with permittivity ε_s=-1.75+0.3i). For convenience the value for free space binding is multiplied by 10. The incident light wavelength is 350 nm.

Thus, in this work we analyze optical binding near to plasmonic substrates. It is shown that surface plasmon polariton interference plays crucial role in the interaction of nanoparticles. We demonstrate that two nanoparticles placed near metallic interface can be stably bounded at distances defined by surface plasmon polariton. While the classical diffraction limit prevents achieving deeply subwavelength arrangements, auxiliary nanostructures enable tailoring optical forces via additional interaction channels. In addition, the enhancement by one order of magnitude of the optical binding force is possible owing to the resonant surface plasmon polariton excitation. Moreover, stable configurations are formed along incident light polarization direction, while for binding in free space

IOP Conf. Series: Journal of Physics: Conf. Series **1092** (2018) 012065 doi:10.1088/1742-6596/1092/1/012065

they are mutually perpendicular. That makes possible control over a nanoparticle trajectory on subwavelength scales and opens opportunities for optical-induced organization of particles on a plasmonic substrate with different periods along the field polarization as well as perpendicular to it.

Acknowledgements
The authors acknowledge the financial support provided by Russian Fund for Basic Research (Projects No. 18-02-00414, 18-52-00005); Ministry of Education and Science of the Russian Federation (GOSZADANIE, Grant No. 3.4982.2017/6.7) and Government of Russian Federation, Grant 08-08.

References
[1] M. I. Petrov, S. V. Sukhov, A. A. Bogdanov, A. S. Shalin, A. Dogariu 2016, Laser Photonics Rev., **10**, 116–122
[2] K. Dholakia, P. Zemánek 2010, Reviews of modern physics **82**, 2, 1767.
[3] O. Kidwai, S. V. Zhukovsky, J. E. Sipe 2012, Physical Review A, **85**, 053842
[4] A. Ivinskaya, M. I. Petrov, A. A. Bogdanov, I. Shishkin, P. Ginzburg, A. S. Shalin, 2017, Light: Science & Applications, **6**(5), e16258.

Self-focusing does not occur for few-cycle pulses

S A Kozlov[1], A A Drozdov[1], M A Kniazev[1], D A Kislin[1], S Choudhary[2,3], and R W Boyd[2,3]

[1] ITMO University, 49 Kronverskiy pr., Saint-Petersburg, Russia, 197101
[2] University of Ottawa, Ottawa, Canada
[3] The Institute of Optics, University of Rochester, Rochester, NY 14627-0186, USA

E-mail: kozlov@mail.ifmo.ru

Abstract. A surprise of nonlinear optics of few-cycle waves, such as the disappearance of the phenomenon of high-power radiation self-focusing, is discussed. An expression is derived and analyzed for the conditions under which the concept of a critical power for self-focusing loses its physical meaning due to the dominance of the process of dispersion over diffraction.

1. Introduction

Self-focusing is a fundamental self-action effect in which a light beam comes to a focus as a consequence of the lens it induces in the nonlinear optical medium. The concept of a critical power is a key feature of conventional theories of self-focusing; the critical power determines when the self-induced focusing begins to dominate over diffractive spreading of the light beam. A large number of papers and books have been devoted to this nonlinear phenomenon from 1962 (see, for example, review [1] and references therein).

In the present work, we demonstrate that, for few-cycle wave packets with longitudinal dimension less the transverse size, the concept of critical power of self-focusing can lose its physical meaning because of the dominance of dispersion over diffraction. We present simple formulas for the estimation of the parameters of the field and medium under which self-focusing disappears and illustrate with numerical calculations the changes in the self-action phenomenon of light in this case.

2. Theoretical estimations

The normalized equation describing propagation of a linearly polarized paraxial quasi-monochromatic wave through a nonlinear media has well-known form [2]

$$\frac{\partial\tilde{\mathcal{E}}}{\partial z}+\frac{1}{L_w}\frac{\partial\tilde{\mathcal{E}}}{\partial\tilde{t}}+\frac{i}{L_{disp1}}\frac{\partial^2\tilde{\mathcal{E}}}{\partial\tilde{t}^2}-\frac{1}{L_{disp2}}\frac{\partial^3\tilde{\mathcal{E}}}{\partial\tilde{t}^3}-\frac{i}{L_{nl1}}\left|\tilde{\mathcal{E}}\right|^2\tilde{\mathcal{E}}+\frac{1}{L_{nl2}}\frac{\partial}{\partial\tilde{t}}\left(\left|\tilde{\mathcal{E}}\right|^2\tilde{\mathcal{E}}\right)=\frac{i}{L_{difr}}\tilde{\Delta}_\perp\tilde{\mathcal{E}},\quad(1)$$

where $\tilde{\mathcal{E}}$ is the normalized envelope of wave electric field, z is the beam propagation direction, $\tilde{t}$ is the normalized time, $\tilde{\Delta}_\perp=\frac{\partial^2}{\partial\tilde{x}^2}+\frac{\partial^2}{\partial\tilde{y}^2}$, $L_w=V\tau_0$. Dispersion lengths given by $L_{disp1}=\frac{2\tau_0^2}{\beta_2}$, $L_{disp2}=\frac{6\tau_0^3}{\beta_3}$; nonlinear lengths given by $L_{nl1}=\frac{1}{\gamma_1\mathcal{E}_0^2}=\frac{c}{\omega_0\Delta n_{nl}}$, $L_{nl2}=\frac{\tau_0}{\gamma_2\mathcal{E}_0^2}=\frac{c\tau_0}{\Delta n_{nl}}$

and diffraction length given by $L_{difr} = 2k_0 r_0^2$ characterize the influence on the character of light propagation of various physical phenomena: dispersion, nonlinearity and diffraction. Here $\mathcal{E}_0$ is the initial field amplitude, τ_0 and r_0 are the duration and radius of the beam at the entrance of nonlinear medium, $k(\omega) = \frac{\omega}{c} n(\omega)$ is the wave number, $V = \left(\frac{\partial k}{\partial \omega}\right)^{-1}_{\omega_0}$ is wave velocity, $\beta_2 = \left(\frac{\partial^2 k}{\partial \omega^2}\right)_{\omega_0}$, $\beta_3 = \left(\frac{\partial^3 k}{\partial \omega^3}\right)_{\omega_0}$, ω_0 is the central frequency of radiation, $k_0 = k(\omega_0)$, $\Delta n_{nl} = \frac{1}{2} n_2 \mathcal{E}_0^2 = n_2' I$, n_2 and n_2' are coefficients of nonlinear refractive index and I is the input wave intensity.

Let's consider a case

$$L_{nl1} = L_{difr}, \tag{2}$$

when the effect of medium nonlinear response and diffraction are close. In this case it is natural to introduce an expression for the self-focusing critical power $P_{cr} = (\pi r_0^2) \cdot I$ which is trivially deduced for (2) and takes the form:

$$P_{cr} = R_{cr} \frac{\lambda_0^2}{8\pi n_0 n_2'}, \tag{3}$$

where R_{cr} is the parameter of nonlinearity [3], $\lambda_0 = \frac{2\pi}{k_0}$ is the central wavelength of radiation and $n_0 = n(\omega_0)$ is the linear refractive index. This naively deduced expression for critical power (3) coincides with the well-known expression [1] to within a constant factor R_{cr}. This indicates the accuracy of our nonlinear and diffraction length estimates.

It is important, we note, that for

$$L_{disp1} < L_{difr} \tag{4}$$

nonlinearity will compete not with diffraction but with dispersion. Under these conditions, the concept of P_{cr} will begin to lose its original meaning. The inequality (4) can be expressed in terms of laboratory parameters as:

$$\frac{l_0}{D_0} < \sqrt{c\omega_0 n(\omega_0)\beta_2}, \tag{5}$$

where $l_0 = 2c\tau_0$ is longitudinal size of the wave packet and $D_0 = 2r_0$ is its transverse size.

For the simplest dispersion formula of $n(\omega) = N_0 + ac\omega^2$, where N_0 and a characterize the medium dispersion, the inequality (5) can be reduced to

$$\frac{l_0}{D_0} < \sqrt{6 N_0 \Delta n_{disp}}, \tag{6}$$

where $\Delta n_{disp} = ac\omega_0^2$ is the change of the refractive index at the central wavelength due to dispersion.

For example, for optical radiation with a central wavelength of 800 nm propagating through a fused silica with dispersion $n(\omega) = N_0 + ac\omega^2 - bc/\omega^2$ [4] we find that condition (5) becomes $l_0/D_0 < 0.18$. For terahertz radiation with a central frequency of 1.0 THz propagating through a stoichiometric MgO:$LiNbO_3$ crystal with $n(\omega) = N_0 + ac\omega^2$ [5] we have $l_0/D_0 < 0.87$ according to formulas (5) - (6). It means we can expect that wave packets with longitudinal dimension less than transverse size, the concept of crytical power of self-focusing can lose its physical meaning.

IOP Conf. Series: Journal of Physics: Conf. Series **1092** (2018) 012066 doi:10.1088/1742-6596/1092/1/012066

3. Numerical illustrations

We consider boundary conditions (the electric field of radiation at the entrance to nonlinear medium) in form of a Gaussian axisymmetric paraxial beam with a small number of oscillations ($N = \tau_0/T_0$)

$$E(0,r,t) = \mathcal{E}_0 \exp\left(-\frac{r^2}{{r_0}^2}\right)\exp\left(-\frac{t^2}{{\tau_0}^2}\right)\sin(\omega_0 t), \tag{7}$$

where $r = \sqrt{x^2+y^2}$, $T_0 = 2\pi/\omega_0$ is central period of electric field oscillations.

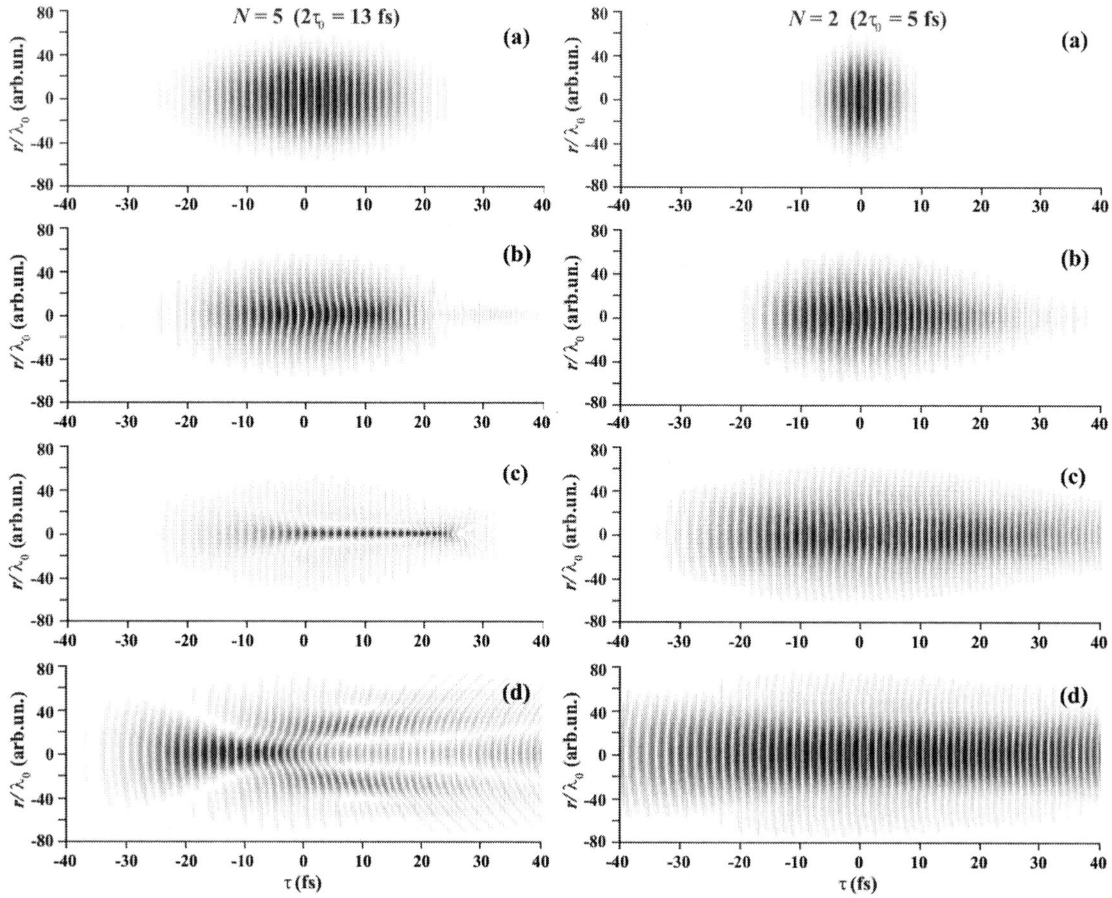

Figure 1. The spatiotemporal dynamics of the electric optical field with initially Gaussian transverse profile in bulk fused silica at distances (a) $z = 0$ mm, (b) $z = 1.0$ mm, (c) $z = 1.9$ mm, (d) $z = 3.0$ mm for two different input pulse durations: $2\tau_0 = 13$ fs (left column) and $2\tau_0 = 5$ fs (right column). Here $\tau = t - (c/N_0)z$ is retarded time.

Figure 1 demonstrates typical dependence of light field self-focusing dynamics on the number of oscillations N in the initial pulse. The typical changes in self-focusing dynamics are illustrated for ultrashort pulses of Ti:S laser for central wavelength of $\lambda_0 = 800$ nm with $N = 5$ ($2\tau_0 = 13$ fs) and $N = 2$ ($2\tau_0 = 5$ fs) propagated in bulk fused silica. Here initial pulse parameters are $I = 9\times10^{11}$ W/cm^2, $\lambda_0 = 800$ nm, $D_0 = 30\lambda_0$. We also used fixed ratio between initial power P_0 and critical power P_{cr} for self-focusing $P_0/P_{cr} = 2$. For the case $N = 5$ we get $l_0/D_0 = 0.17$, for $N = 2$ we have $l_0/D_0 = 0.07$. As one can see from Figure 1, reduction of N leads to decreasing of self-focusing efficiency because dispersion starts play a significant role in pulse evolution.

Acknowledgments
This work was financially supported by the "5-100-2020" Programme of the Russian Federation, Grant 08-08.

References

[1] Boyd R W, Lukishova S G and Shen Y 2009 *Self-focusing: Past and Present* (New York: Springer)
[2] Agrawal G 2013 *Nonlinear Fiber Optics* 5th ed (New York: Academic Press)
[3] Kandidov V P, Shlenov S A and Kosareva O G 2009 *Quantum Electronics* **39** 205
[4] Kozlov S A and Samartsev V V 2013 *Fundamentals of femtosecond optics* (Cambridge, UK: Woodhead)
[5] Drozdov A A, Kozlov S A, Sukhorukov A A and Kivshar Y S 2012 *Phys. Rev. A* **86**(5) 053822

Polyelectrolyte Parg/DS submicrocapsules functionalized by magnetite nanoparticles as effective MR contrast agents

A A Kozlova[1,4], S V German[2], M V Novoselova[2], P A Demina[1], V S Atkin[1], V V Zuev[3] and D A Gorin[1,2]

[1] Saratov State University, 83 Astrakhanskaya Str., 410012 Saratov, Russia
[2] Skolkovo Institute of Science and Technology, Nobelya Ulitsa 3, 121205 Moscow, Russia
[3] Saratov State Medical University, 112 Bolshaya Kazachia Str., 410012 Saratov, Russia

[4] Corresponding author: anastasia.kozlova245@yandex.ru

Abstract. Polyelectrolyte microcapsules functionalized by Fe_3O_4 nanoparticles have proved to be perspective magnetic resonance (MR) contrast agents. Here we report about submicron-sized capsules with a shell made of poly-L-arginine and dextran sulfate. Such polymers are biodegradable, thus they are suitable to use as components of drug delivery carriers. To functionalize submicrocapsules by magnetite freezing and layer-by-layer method were used. MR study of obtained samples was carried out, and signal intensity change dependencies on submicrocapsules concentration by mass amount of iron oxide (III) nanoparticles were shown. The sample with optimal MR contrast characteristics was found as well, and an ability to use it as a bimodal MR agent was demonstrated.

1. Introduction

One of the main task of nanotechnology scientific investigations is to create carriers that can be effectively used for controlled drug delivery. Such carriers should provide not only a high drug loading, but also its transfer, local distribution and an ability to control carrier degradation and thus encapsulated substance release [1].

Different types of containers have been developed for the current moment [2-4]. Nevertheless, polymer capsules of micron and submicron size possess better characteristics in the view of loaded substance volume, time and temperature stability [5]. Layer-by-layer assembly [6] allows us to obtain polyelectrolyte shells functionalized by various inorganic, magnetic nanoparticles. Some substances can be additionally loaded into an inner volume of capsules as well.

Adding of magnetite nanoparticles to polymeric capsules can allow to control them by an external magnetic field and to visualize by magnetic resonance imaging (MRI) [7].

In order to obtain carriers described above polyelectrolyte Poly-L-arginin/Dextran sulfate (Parg/DS) submicrocapsules were prepared. Those ones of micron size were investigated earlier [8]. They contained Fe_3O_4 nanoparticles in a shell or/and an inner volume. To investigate the MR signal intensity dependence on magnetite content, carriers obtained were studied by MRI.

2. Methods and experiments

Magnetite nanoparticles were prepared by chemical precipitation. 20 mL of 1M NaOH and 150 mL of water were poured into the reaction cell and then heated until 40°C. For further colloid formation di- and trivalent iron salts solution was prepared. 1.3 g of $FeCl_3$ and 0.48 g of $FeCl_2$ were dissolved in 25 mL of pure water. For magnetite colloid stabilization 25 mL of 6M citric acid solution were prepared. After heating of NaOH, iron salts solution were injected there under nitrogen pressure. Then citric acid was added to stabilize magnetite nanoparticles had formed. TEM (transmission electron microscopy) of Fe_3O_4 nanoparticles are presented in figure 1, a. The size was 13±5 nm, measured by DLS (dynamic light scattering) method (figure 1, b).

Polyelectrolyte Parg/DS submicrocapsules were formed by layer-by-layer assembly [6]. First, calcium carbonate cores were prepared by adding 0.4 mL of 0.5M sodium carbonate and 0.4 mL of 0.5M calcium chloride salts solutions to glycerin (4g) under constant mixing. After 1 hour submicron particles were washed three times by pure water, and then polymer shell were formed. Submicrocapsules were functionalized by magnetite nanoparticles that were loaded into an inner volume by freezing and in a shell by Lbl method (according to the protocol from [8]). The loading method "freezing" was described in the manuscript [9], which is currently under revision. After the shell formation calcium carbonate cores were dissolved under EDTA (ethylenediaminetetraacetic acid disodium salt dehydrate) solution treatment (0.2M in water).

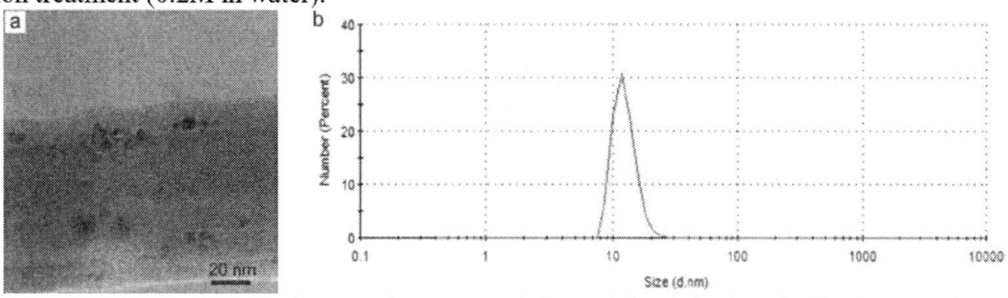

Figure 1. (a) TEM image of magnetite nanoparticles and (b) their size distribution, obtained from DLS.

Two samples of magnetic polymeric submicrocapsules were synthesized: with low/high magnetite concentration in an inner volume and with 1 layer of magnetite nanoparticles in a shell, Low_{core}+Lbl and $High_{core}$+Lbl, respectively. Fe_3O_4 aqueous solution with 1.6 mg/mL concentration was used to freeze magnetite into calcium carbonate cores, and 0.56 mg/mL solution was used for Lbl assembly. SEM (scanning electron microscopy) images of obtained submicrocapsules are presented in figure 2. The presence of magnetite into the submicrocapsules was confirmed by EDXS analysis.

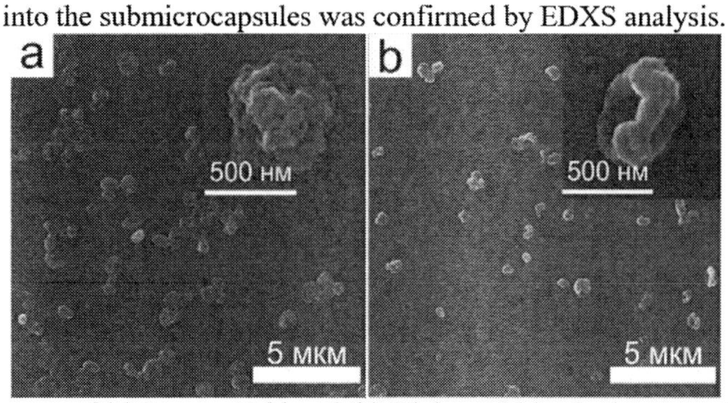

Figure 2. SEM images of (a) Low_{core}+Lbl and (b) $High_{core}$+Lbl submicrocapsules, functionalized by magnetite nanoparticles.

Evaluation of a mass fraction of magnetite nanoparticles adsorbed on a shell and loaded by freezing into submicrocapsules was carried out by colorimetric titration in terms of the iron ions (III) qualitative reaction with ammonium thiocyanate. Obtained polymer capsules with magnetite nanoparticles were dissolved in 1M sulfuric acid. The reaction was carried out by direct mixing of 2 ml of the investigated solution and 2 ml of 4M ammonium thiocyanate solution at rapid mixing, which results in formation of blood-red colour iron thiocyanate solution. Then a concentration of iron ions in the investigated solution was determined by colorimetric titration in terms of a standard iron (III) solution reaction with ammonium thiocyanate. The standard iron (III) solution was 1M sulfuric acid in which iron ions (III) were dissolved, its concentration was 0.1 mg/ml by iron amount.

The measurements of the magnetite nanoparticles size distribution were performed using a Zetasizer Nano ZS instrument (Malvern Instruments Ltd, UK).

TEM images of the magnetite nanoparticles were obtained with operating at 120 kV Libra-120 transmission electron microscope (Carl Zeiss, Germany). Investigated particles were deposited on a 300-mesh copper grid coated with formvar.

Scanning electron microscopy (SEM) was performed with MIRA II LMU (Tescan, Czech) microscope at operating voltage of 30 kV, in secondary and backscattering electron modes.

In vitro MRI study was carried out using a Philips Achieva 1.5 T high field MRI scanner equipped with a phased array coil. For obtaining T1-weighted images spin echo (SE) and gradient echo (FFE) sequences were applied, for T2-weighted images – turbo spin echo (TSE) sequence. The following parameters were used for conducting measurements: the repetition time (TR) was 488 ms and the echo time (TE) was 15 ms for the T1-weighted pulse SE sequence; the TR was 104 ms and the TE was 4.5 ms for the T1-weighted pulse FFE sequence; the TR was 3000 ms and the TE was 47.7 ms for the T2-weighted pulse TSE sequence. In order to increase MR signal in T1-weighted image the TR should be decreased, so tissues don't manage to relax after an external influence. At the same time, the T2 relaxation time TE decrease leads to low MR contrast in T2-weighted image [10].

3. Results and discussion

Low_{core}+Lbl and $High_{core}$+Lbl submicrocapsules were obtained by varying the number of freezing loading cycle. As a result, each sample contained 40 mg of $CaCO_3$ cores, 3.03±0.11 mg and 13.26±0.55 mg of magnetite nanoparticles, respectively. *In vitro* MRI study was carried for different samples' concentration to show the signal intensity dependence on the magnetite content in solution. For this T1 SE, T1 FFE and T2 TSE modes were used. MR images are performed in figure 3 and figure 4.

Dependences of MR signal intensity (SI) on submicrocapsules concentration by mass amount of magnetite are presented in figure 5 and figure 6 for Low_{core}+Lbl sample; in figure 7 and figure 8 for $High_{core}$+Lbl sample.

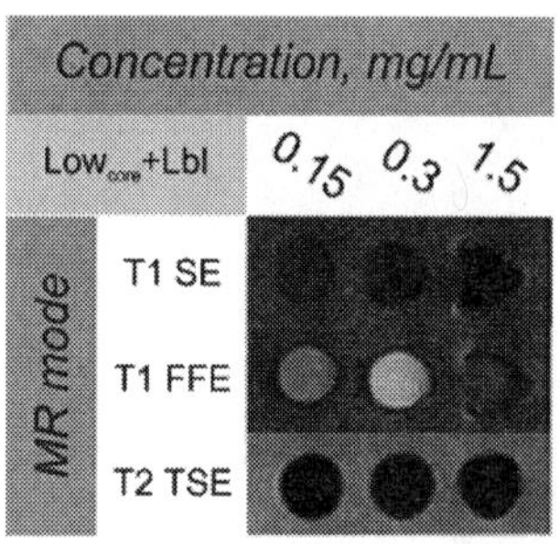

Figure 3. T1- and T2-weighted MR images of Low_{core}+Lbl submicrocapsules, concentration by magnetite mass amount

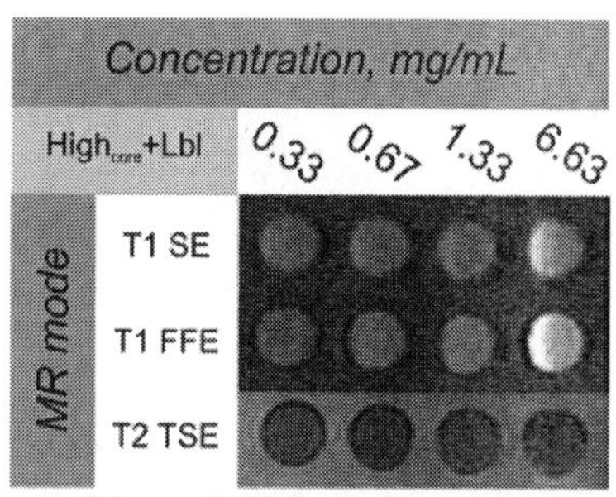

Figure 4. T1- and T2-weighted MR images of $High_{core}$+Lbl submicrocapsules, concentration by magnetite mass amount

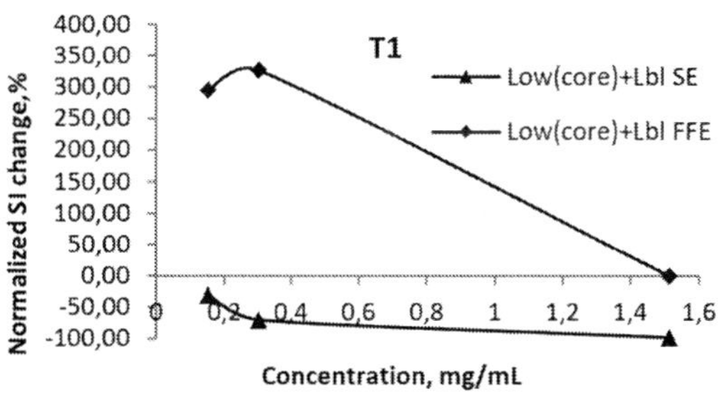

Figure 5. Low_{core}+Lbl normalized signal intensity (SI) dependence on concentration by magnetite mass amount for T1 mode

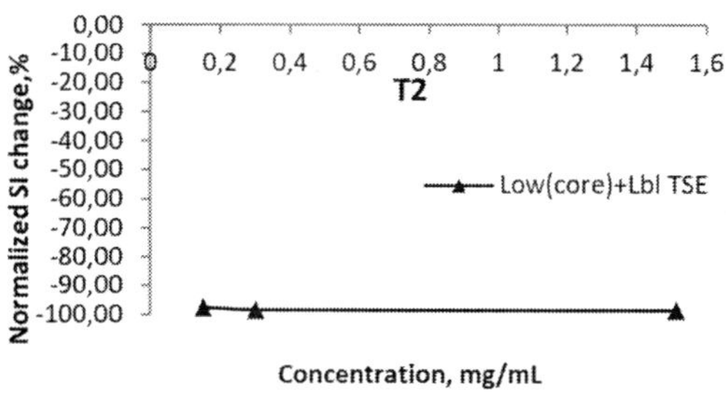

Figure 6. Low_{core}+Lbl normalized signal intensity (SI) dependence on concentration by magnetite mass amount for T2 mode

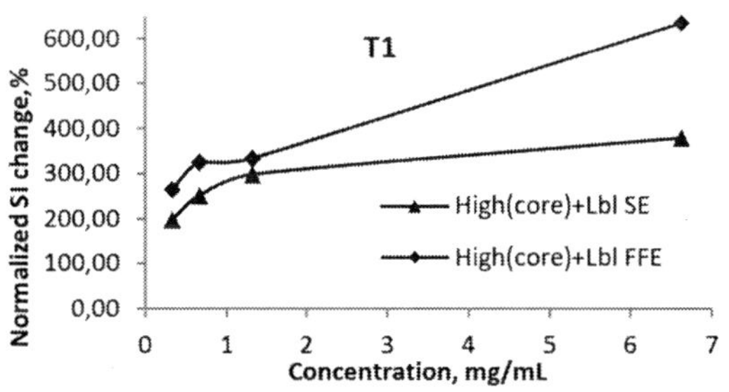

Figure 7. $High_{core}$+Lbl normalized signal intensity (SI) dependence on concentration by magnetite mass amount for T1 mode

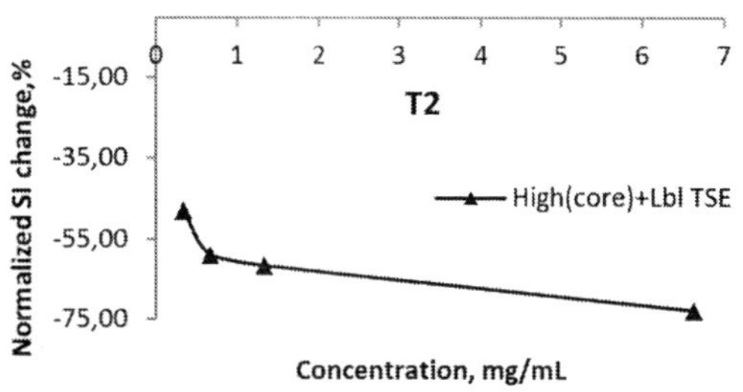

Figure 8. $High_{core}$+Lbl normalized signal intensity (SI) dependence on concentration by magnetite mass amount for T2 mode

The T1 SI change was about 300% for Low_{core}+Lbl, which is 2 times lower than for magnetite aqueous solution. But this sample possess high contrast for T2 mode. It is nearly -100% change, which leads to almost full decay of the signal.

The $High_{core}$+Lbl sample SI change from the data obtained is shown in figure 7 and figure 8. This sample possess high 300-600% T1 MR contrast for both SE and FFE sequences, and even the signal enhancement for the high concentration area where the signal usually decreases. The signal enhancement is near to that one for magnetite colloid. The curves observed are monotonous, do not have any additional peaks that provide the effective usage of the sample as T1 contrast agent. T2 mode was studied as well and provided 65% SI change. Having regard to the above, $High_{core}$+Lbl submicrocapsules can be effectively used as bimodal MR contrast agent. Therefore, due to the high magnetite concentration such carriers can be controlled by an external magnetic field, which is necessary for controlled drug delivery.

References

[1] Huyett L, Dassau E, Zisser H and Doyle F 2015 *Ind. Eng. Chem. Res.* **54** 10311–21
[2] Terentyuk G, Panfilova E, Khanadeev V, Chumakov D, Genina E, Bashkatov A, Tuchin V, Bucharskaya A, Maslyakova G, Khlebtsov and Khlebtsov B 2014 *Nano Res.* **7** 325–37
[3] Chen Y, Chen Y, Xiao D, Bose A, Deng R and Bothun G 2014 *Colloids Surfaces* B **116** 452–58
[4] German S, Navolokin N, Kuznetsova N, Zuev V, Inozemtseva O, Anis'kov A, Volkova E, Bucharskaya A, Maslyakova G, Fakhrullin R, Terentyuk G, Vodovozova E and Gorin D 2015 *Colloids Surfaces* B **135** 109–15
[5] Antipina M and Sukhorukov G 2011 *Advanced Drug Delivery Reviews* **63** 716–29
[6] Sukhorukov G, Donath E, Davis S, Lichtenfeld H, Caruso F, Popov V and Möhwald H 1998 *Polym. Adv. Technol.* **767** 759–67

[7] Rizzitelli S, Giustetto P, Cutrin J, Delli Castelli D, Boffa C, Ruzza M, Menchise V, Molinari F, Aime S and Terreno E 2015 *J. Control. Release* **202** 21–30
[8] German S, Bratashov D, Navolokin N, Kozlova A, Lomova M, Novoselova M, Burilova E, Zyev V, Khlebtsov B, Bucharskaya A, Terentyuk G, Amirov R, Maslyakova G, Sukhorukov G and Gorin D 2016 *Phys.Chem.Chem.Phys.* **18** 32238–46
[9] German S, Novoselova M, Bratashov D, Demina P, Atkin V, Khlebtsov B, Parakhonskiy B, Sukhorukov G and Gorin D *Scientific Reports* manuscript is under revision
[10] Westbrook C, Roth C and Talbot J 2011 *MRI in Practice 4th edition* vol 10–11(Oxford: Blackwell Publishing Ltd.)

Nonlinear Bound States in the Continuum in One-Dimensional Photonic Crystal Slab

S.D. Krasikov[1], **A.A. Bogdanov**[1], **I.V. Iorsh**[1]
[1]ITMO University, St. Petersburg 197101, Russia
E-mail: `s.krasikov@metalab.ifmo.ru`

Abstract. We explore the nonlinear optical response of the silicon one-dimensional photonic crystal slab supporting bound states in the continuum (BIC). We show the emergence of such nolinear effects as multistable behaviour, self-tuning of BIC and breaking of symmetry protected BIC. We define a class of the nonlinear solutions generated by the linear BIC state and analyze the modulation instability of the obtained solutions and the effect of the finite system size on the stability.

Bound states in the continuum [1] (BICs) are localized states whose energy lies within the radiation continuum. Such states can be considered as ultra high quality resonance modes and therefore can be used to enhance optical nonlinear effects [2, 3]. In particular, onset of multistable optical response is determined by the pump power density depending on the nonlinear susceptibility of the material and the quality factor of resonator. Hence, the threshold powers needed for the emergence of multistability in BIC-supporting structures are low and such structures may be used for the realization of all-optical switchers.

In our work [4], we show that BIC supporting systems indeed allow to achieve strong nonlinear response without cavity at low pump power. For that we consider 1D photonic crystal slab with Kerr-type nonlinearity and analyze the role of BIC in formation of multistable states. Also we analyze the stability of found states with respect to longitudinal perturbations.

Considered system, schematically shown in Fig. 1(a), can be treated as a slab waveguide with grating on the top side, so it can be described by the effective refractive index $n(z)$:

$$n(z) = n_0 + \delta n \frac{2}{\pi} \sum_{j=1}^{\infty} \frac{1}{j} \sin\left(\frac{j\pi w}{d}\right) \cos(jGz) + n_2 I. \tag{1}$$

where n_0 is the average refractive index of core layer in the absence of grating, δn is the modulation amplitude, $G = 2\pi/d$ is the reciprocal vector of the structure, d is the period of lattice, w is the width of the protrusions and $n_2 I$ is the term corresponding to the Kerr-type focusing nonlinearity. We consider the resonant excitation of the system, so it can be described by the coupled-mode theory within two-mode approximation [5, 6, 7]. Hence, the electric field can be written in terms of forward and backward propagating waveguide modes with amplitudes $E_\pm$ and wavevector β:

$$E(z,t) = \left[E_+(z,t)e^{i\beta z} + E_-(z,t)e^{-i\beta z}\right] e^{-i\omega_0 t} + c.c. \tag{2}$$

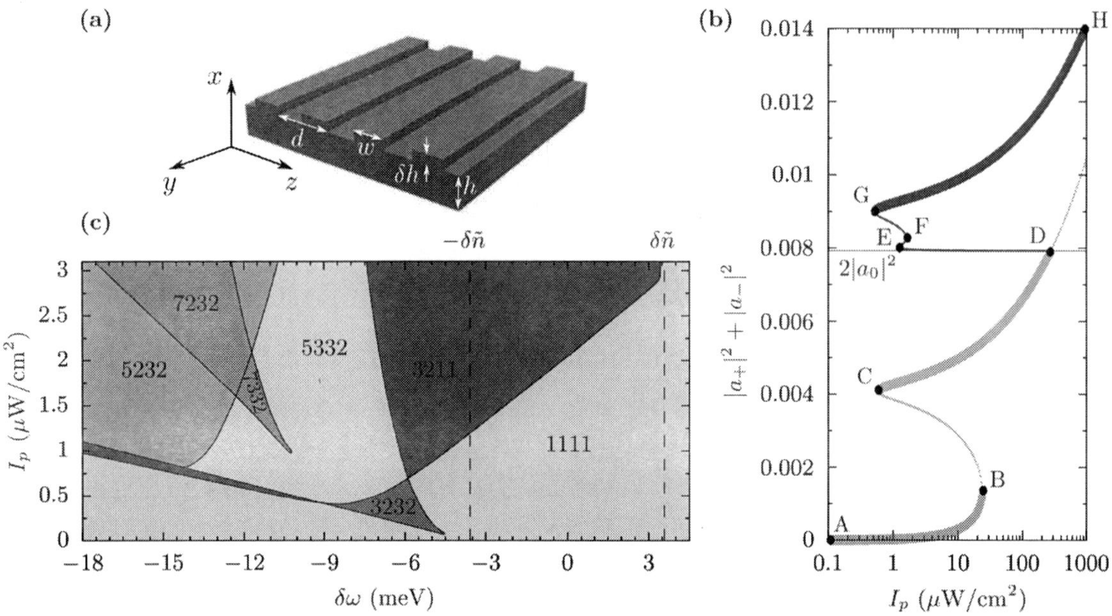

Figure 1. (a) Schematic picture of the considered system. (b) The total field as the function of pump intensity. Here $\delta\omega = -11.16$ meV. The green dots correspond to symmetric solutions and the blue dots correspond to asymmetric. The stable (unstable) solutions are shown by thick (tiny) dots. The red line corresponds to antisymmetric solutions $2|a_0|^2$. Labeled points are used to distinguish between different branches of solutions. (c) Phase diagram of the solutions. Numerical indices *abcd* have the following meaning: *a* shows the total number of solutions, *b* denotes how many of them are stable, *c* is the number of symmetric solutions and *d* is the number of stable symmetric solutions.

Assuming $\omega_0 = ck_0 = \beta c/n_0$, $\beta = G = 2\pi/d$ and using slow-varying approximation, we can get the resulting pair of the dimensionless coupled-mode equations with the pump and decay terms:

$$\frac{\partial \tilde{E}_\pm}{\partial \tilde{t}} = i\delta\tilde{n}\tilde{E}_\mp - \tilde{\gamma}(\tilde{E}_+ + \tilde{E}_-) + i\tilde{E}_\pm(|\tilde{E}_\pm|^2 + 2|\tilde{E}_\mp|^2) + \sqrt{\tilde{\gamma}\tilde{I}_p}e^{i\phi_0 - i\delta\tilde{\omega}\tilde{t}} \quad (3)$$

where $\tilde{I}_p = n_2\tilde{E}_p^2/2$ is pump intensity, $\delta\tilde{\omega} = \delta\omega/\omega$ is the detuning between the pump beam, coupling constant is $\gamma \approx k_0^2\delta\tilde{n}^2$ and the phase is $\phi_0 = -1/2i\log(-r_{slab} - t_{slab})$. The following normalized parameters were used:

$$\tilde{z} = \beta z; \quad \tilde{E}_\pm = E_\pm(n_2/n_0)^{1/2}; \quad \tilde{t} = \omega t; \quad \delta\tilde{n} = \delta n \sin(2\pi w/d)/(2\pi n_0). \quad (4)$$

For the linear case in the absence of pumping, system of equations (3) has symmetric solutions, corresponding to leaky modes, and antisymmetric solutions corresponding to the symmetry protected BIC:

$$\tilde{\omega} = \delta\tilde{n}; \qquad \tilde{E}_+ = -\tilde{E}_- \quad (5)$$

$$\tilde{\omega} = -\delta\tilde{n} - 2i\tilde{\gamma}; \qquad \tilde{E}_+ = \tilde{E}_-. \quad (6)$$

However, in the nonlinear case, amplitude of BIC depends on the frequency:

$$|\tilde{E}_\pm| = \sqrt{\frac{1}{3}(\delta\tilde{n} - \tilde{\omega})}. \quad (7)$$

Hence, there always exists a purely antisymmetric (and thus non-radiating) state at all the frequencies smaller than the frequency of the linear BIC ($\omega < \delta n$). This is the manifestation of the self-tuning of BIC. If we pump the sufficient energy density in the system, then the resonant frequency is tuned such that the BIC coincides with pump frequency, which can be regarded as a way to excite BIC [8, 9]

Considering inhomogeneous set of equations with solutions in the form $\tilde{E}_\pm = -ia_\pm e^{i\phi_0 - i\delta\omega t}$, we can obtain three classes of solutions. First of all, there are symmetric solutions $a_+ = a_- = a_s$, which generates a standard S-shaped bistability curve. The second class is asymmetric solutions, which have the form $a_\pm = a_0 e^{i\phi_\pm}$ such that

$$a_0 = \sqrt{\frac{1}{3}(\delta n - \delta\omega)}, \quad \tan\left(\frac{\phi_+ + \phi_-}{2}\right) = -\gamma/\delta n, \quad \cos\left(\frac{\phi_+ - \phi_-}{2}\right) = \frac{\sqrt{\gamma I_p}}{2a_0\sqrt{\gamma^2 + \delta n^2}}. \tag{8}$$

It is important to note, that a_0 does not depend on pump intensity and as $\tilde{I}_p$ approaches zero, the solution approaches the fully antisymmetric solution $\phi_+ - \phi_- = \pi$. Moreover, the solution becomes fully symmetric at $I_p = \frac{4}{3\gamma}(\delta n - \delta\omega)(\gamma^2 + \delta n^2)$. This type of solutions was termed Josephson-like current [10]. Despite the fact that for infinite system these solutions are unstable, they may be stable in the finite-size system as we will show later [see Fig. 2]. Finally, there is a class of asymmetric solutions such that $|a_+| \neq |a_-|$, and $\phi_+ = \pi + \phi_-$.

To analyze these solutions we have performed numerical calculations for a silicon slab waveguide in vacuum ($n_0 = 3.48$, $n_2 = 3 \cdot 10^{-18}$ m^2/W [11, 12]). The parameters are the following: depth of the grating is 10 nm, $h = 100$ nm, $\delta n \approx 0.0316$ and $w = d/4$. The lattice period d was chosen in such a way that $\beta = 2\pi/d$ at wavelength $\lambda = 1$ μm, material losses are accounted. Results of the calculations are shown in Fig. 1(b),(c). If $\delta\omega > \delta n$ than only single symmetric solutions can exist. Then, two additional asymmetric solutions can emerge in the region $-\delta n < \delta\omega < \delta n$. One of this solutions is stable and therefore the nonlinearity could result in breaking of the symmetry protected BIC and its transformation into symmetric solution. But when the detuning is less than $-\delta n$, a multistable behavior is observed with three stable solutions.

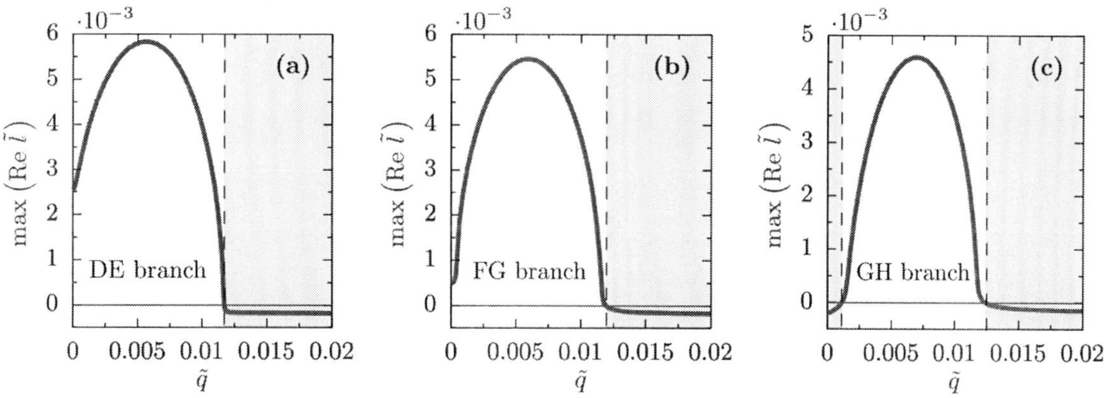

Figure 2. Stability analysis of solutions. For each plot $Ip \approx 0.015$ W/m^2 and $\delta\omega = -11.16$ meV. All other parameters are given in the main text. Solutions in Fig. 1 corresponds to the case of $\tilde{q} = 0$.

To analyze the stability of the obtained solutions substitute the electric field in the form

$$\tilde{E}_\pm(\tilde{z}, \tilde{t}) = \left(a_\pm + \epsilon f_\pm e^{i\tilde{q}\tilde{z}} e^{\tilde{l}\tilde{t}} + \epsilon g_\pm e^{-\tilde{q}\tilde{z}} e^{\tilde{l}^*\tilde{t}}\right) e^{-i\delta\tilde{\omega}\tilde{t}}, \tag{9}$$

to the initial differential equations and linearize them with respect to ϵ which is the amplitude of the small perturbation. Here $\tilde{q}$ is the wavevector of the perturbation, which is a parameter, and $\tilde{l}$ is the complex eigenfrequency. If at least one of calculated $\tilde{l}$ has positive real part, then the solution is unstable. We see that asymmetric and antisymmetric solutions are stable with respect to perturbations with wavevectors, larger than some finite critical value $\tilde{q}_{max}$, as it is shown in Fig. 2. Since only perturbations which have wavelength smaller than the system size may exist, the perturbations with $\tilde{q} < 1/N$ (where N is the number of periods) will decay. Thus, if the system size is less than $\tilde{q}_{max}$, the solutions could be stabilized.

To conclude, we have demonstrated that the nonlinear BIC can exist in 1D photonic crystal slab and we have shown how the existence of the BIC leads to the emergence of the multistable behavior in the structure. Also, we have shown that the solutions unstable in the infinite system may be stabile in the finite-size system. The simplicity of the structure and the moderate level of required pump intensities in BIC-supporting structures opens new avenues for the realization of all-optical switchers exploiting bound states in the continuum.

Acknowledgments
This work is supported by the Ministry of Education and Science of Russian Federation (Projects No. 3.1668.2017/4.6, No. 3.4573.2017/6.7, No. 3.2614.2017/4.6, No. 3.1365.2017/4.6, No. 3.8884.2017/8.9, and No. 14.Y26.31.0015). A.B. acknowledges the President of the Russian Federation (grant MK-403.2018.2) and BASIS foundation.

References
[1] Hsu C W, Zhen B, Stone A D, Joannopoulos J D and Soljačić M 2016 *Nature Reviews Materials* **1** 16048
[2] Bulgakov E N and Sadreev A F 2010 *Physical Review B* **81** 115128
[3] Bulgakov E N and Sadreev A F 2009 *Physical Review B* **80** 115308
[4] Krasikov S D, Bogdanov A A and Iorsh I V 2018 *Physical Review B* **97** 224309
[5] Kogelnik H 1975 *Integrated Optics* (Berlin, Heidelberg: Springer Berlin Heidelberg) chap 2, pp 13–81
[6] Sipe J E, Poladian L and de Sterke C M 1994 *Journal of the Optical Society of America A* **11** 1307–1320 ISSN 1084-7529
[7] Kogelnik H and Shank C V 1971 *Applied Physics Letters* **18** 152–154
[8] Bulgakov E, Pichugin K and Sadreev A 2015 *Optics express* **23** 22520–22531
[9] Pichugin K and Sadreev A 2016 *Physics Letters A* **380** 3570–3574
[10] Bulgakov E, Pichugin K and Sadreev A 2011 *Journal of Physics: Condensed Matter* **23** 065304
[11] Lin Q, Zhang J, Piredda G, Boyd R W, Fauchet P M and Agrawal G P 2007 *Applied physics letters* **91** 021111
[12] Bristow A D, Rotenberg N and Van Driel H M 2007 *Applied Physics Letters* **90** 191104

IOP Conf. Series: Journal of Physics: Conf. Series **1092** (2018) 012069 doi:10.1088/1742-6596/1092/1/012069

Enhanced magneto-optical Faraday effect in 2D magnetoplasmonic structures caused by orthogonal plasmonic oscillations

D M Krichevsky[1], A N Kuzmichev[2], D A Sylgacheva[1,2], A N Kalish[1,2] V I Belotelov[1,2]

[1] Lomonosov Moscow State University, 119991, Moscow, Russia

[2] Russian Quantum Center, 121353, Moscow, Russia

Abstract. In recent research we present results of transmission and Faraday angle rotation theoretical simulations and experimental study in magnetoplasmonic 2D structures based on perforated with subwavelength holes gold film on top of a smooth Bi-substituted yttrium iron garnet (BiYIG). It was shown that in the case of normal incident light it is possible to enhance Faraday angle rotation mainly through excitation of orthogonal plasmonic oscillations.

1. Introduction

Among different optical effects magneto-optics plays one of crucial role. This is caused by ability to control polarization and intensity of optical radiation in a wide range of frequencies. Metal-dielectric structures attract particular interest, which are able to enhance magneto-optical effects through plasmon-polariton oscillations exciting on the surface of the structure. At the same time nanoscale thickness of mentioned systems allows to increase optical transparency, which is important in device applications [1]. Perforated metallic films have been studied previously due to the extraordinary optical transmission (EOT) phenomenon existence [2]. By combining these structures with magnetic dielectrics it is possible to control Faraday effect enhancement through modifying EOT position. Moreover waveguiding modes of dielectric slab in such systems play an important role in Faraday effect enhancement [3].

In this paper we report on transmission and Faraday angle rotation experimental studies together with theoretical simulations performed using rigorous coupled-wave analysis [4] and transfer matrix method [5] of 2D magnetoplasmonic systems based on perforated gold layer and Bi-substituted yttrium iron garnet. The thickness of magnetic dielectric was extremely low to decrease an impact of waveguiding modes. It was shown that in such system orthogonal plasmonic oscillations in metal layer play primary role in Faraday rotation angle enhancement.

2. Results and discussions

The schematic representation of investigated structure is presented in figure 1. The structure consists of gold layer on the top of the system, perforated with round holes with a period of 570 nm. Uniform dielectric layer was deposited on the bottom, the sample was magnetized perpendicular to its surface. For experimental measurements and theoretical calculations normal incidence of the light was used.

IOP Conf. Series: Journal of Physics: Conf. Series **1092** (2018) 012069 doi:10.1088/1742-6596/1092/1/012069

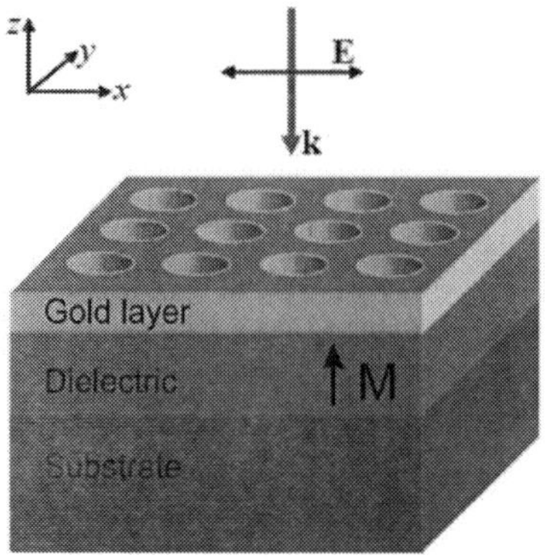

Figure 1. Schematic representation of considering structure.

In the case of nonmagnetic dielectic if the light is polarized along OX-direction (see figure 1) and its wavelength is equal to plasmon-polariton excitation wavelength there would be excited two plasmonic waves along OX-direction. This configuration will not excite plasmonic oscillations along OY-direction due to lack of required components of electromagnetic fields. But when dielectric slab is made of gyrotopic material (for instance, BiYIG) and is magnetized along OZ surface plasmon-polariton become polarized also in OY-direction. It makes possible to excite plasmonic waves orthogonal to incident light. These oscillations contribute into far field and increase Faraday rotation.

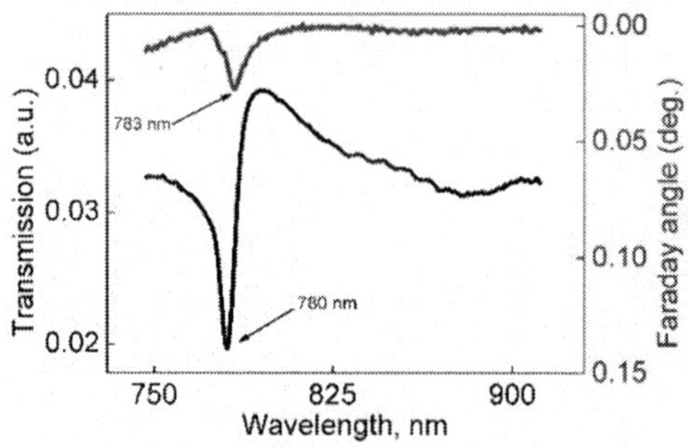

Figure 2. Transmission and Faraday rotation angle for developed system.

The idea was proved by theoretical calculations and experimental measurements for particular system. Measured transmission and Faraday rotation angle of mentioned structure are presented in figure 2. There is distinct Faraday angle enhancement (~5 times) near plasmon-polariton excitation wavelength (at 783 nm).

Acknowledgments

This work is supported by Russian Presidential Grant (MD-1615.2017.2) and Foundation for the advancement of theoretical physics BASIS.

References:

1. V. I. Belotelov, I. A. Akimov, M. Pohl, V. A. Kotov, S. Kasture, A. S. Vengurlekar, Achanta Venu Gopal, D. R. Yakovlev, A. K. Zvezdin, M. Bayer, Enhanced magneto-optical effects in magnetoplasmonic crystals, NATURE NANOTECHNOLOGY, 6, 370, 2011.
2. T.W. Ebbesen, H.J. Lezec, H.F. Ghaemi, T. Thio and P.A. Wolff, NATURE, 391, 667, 1998.
3. V.I. Belotelov, L.L. Doskolovich, A.K. Zvezdin, Extraordinary magneto-optical effects and transmission through metal-dielectric plasmonic systems, Physical Review Letters, 98, 077401, 2007.

4. M. G. Moharam, E. B. Grann, D. A. Pommet, T. K. Gaylord, Formulation of stable and efficient implementation of the rigorous coupled-wave analysis of binary gratings, JOSA A, 12, 1068, 1995.
5. J. Hamrle, Magneto-optical determination of the in-depth magnetization profile in magnetic multilayers, Université Paris-Sud XI, Orsay and Charles University in Prague: PhD thesis, 200 p., 2003.

IOP Conf. Series: Journal of Physics: Conf. Series **1092** (2018) 012070 doi:10.1088/1742-6596/1092/1/012070

Resonant pumping of polaritonic SSH chains

Yury Krivosenko, Ivan Iorsh, and Ivan Shelykh

E-mail: y.krivosenko@gmail.com
ITMO University, St. Petersburg 197101, Russia

Abstract. An optical-lattice realization of a Su-Schrieffer-Heeger chain is theoretically investigated. The topological state is controlled by the relative phase of the lasers constructing the lattice. The hopping amplitudes are calculated within the harmonic approximation to the optical potential in the vicinity of its minima. Finally, we examine polaritons loaded into adiabatically shaken optical lattices with damping and pumping and demonstrate that, in the case of resonant pumping, the system manifests different quantitative and qualitative behaviour for topologically trivial and non-trivial phases.

1. Introduction

A topological insulator (TI) is a state of matter that behaves as a typical insulator in the bulk but can host midgap states localized at its boundary [1] (the so-called *edge states*). The minimalistic model demonstrating topological properties was initially suggested by Su, Schrieffer, and Heeger (SSH) [2]. The system can be presented as a dimerized linear chain with each unit cell consinting of two sites with staggered intra- (t_1) and inter-cell (t_2) hopping amplitudes.

Due to the simplicity and rich properties, the SSH model and its different generalizations have been extensively studied during the recent years (in the field of electronics [3, 4] and waveguide photonics [5,6]). Periodically modulated (*Floquet-engineered* [7,8]) SSH chains have been investigated as well with respect to topological properties and phase transitions [9–11]. Floquet engineering is known to be achievable not only in the field of charged particles, but in the field of cold atoms physics as well. There, a possible realization of Floquet engineering is utilization of cyclically modulated (shaken) optical lattices (OL) [12–14].

In this research, we theoretically investigate the resonant pumping of the SSH chain modelled as an adiabatically shaken OL loaded with exciton-polaritons in microcavities. The topology is adjusted by tuning the relative phase of the lasers constructing the optical potential.

2. Theory

Behaviour of the SSH chain is described by the non-stationary Schrödinger equation

$$H_{\text{SSH}}\,|\psi(t)\rangle = i\,\partial_t|\psi(t)\rangle, \tag{1}$$

where the tight-binding SSH Hamiltonian in real space is [1]:

$$H_{\text{SSH}} = \sum_n t_1|n\rangle\langle n| \otimes \widehat{\sigma}_x + \sum_n t_2\left(|n+1\rangle\langle n| \otimes \frac{\widehat{\sigma}_x + i\widehat{\sigma}_y}{2} + \text{h.c.}\right). \tag{2}$$

Here, the external (n) and internal (α) states are separated by means of a tensor product: $|n,\alpha\rangle \to |n\rangle \otimes |\alpha\rangle$. $\hbar = 1$ is accepted throughout the paper, σ_i are Pauli matrices

To solve (1), $\psi(t)$ is conventionally sought in the form

$$|\psi(t)\rangle = \sum_n |n\rangle \otimes \begin{pmatrix} A_n(t) \\ B_n(t) \end{pmatrix} \tag{3}$$

(the explicit notation of the dependence on time is further omitted). Then, we insert (2) and (3) into (1) and move to a typically polaritonic domain: we add damping and pumping to the system (as in [15]) by means of the damping exponential factor γ (for all the sites) and by the pumping term $P \exp i(-\omega_\mathrm{p} t + \phi_0)$ placed into the equation relative to the pumped site $|n_0\rangle \otimes |\mathrm{C}\rangle$ (C equals A or B), whereupon we obtain the following set of equations

$$i\,\partial_t A_n = t_1 B_n + t_2 B_{n-1} - i\,\gamma A_n + \varepsilon A_n + \delta_{\mathrm{A,C}}\,\delta_{n,n_0}\,P \exp i(-\omega_\mathrm{p} t + \phi_0), \tag{4a}$$
$$i\,\partial_t B_n = t_1 A_n + t_2 A_{n+1} - i\,\gamma B_n + \varepsilon B_n + \delta_{\mathrm{B,C}}\,\delta_{n,n_0}\,P \exp i(-\omega_\mathrm{p} t + \phi_0), \tag{4b}$$

where P, ω_p, and ϕ_0 denote the amplitude, frequency, and relative phase of the pumping field, respectively, and ε, in the field of photonics, represents the cavity mode energy. Resonant pumping would mean the equality $\varepsilon = \omega_\mathrm{p}$.

Further, we utilize the fact that the optical potential profile and thus the topological phase can be controlled by the relative phase of the lasers forming the OL [16]. Consider an OL formed by three laser fields: $E\mathrm{e}^{ikx+i\alpha}$, $E\mathrm{e}^{-ikx}$, and $E\mathrm{e}^{3ikx}$, where α is the relative phase achieved by the frequency detuning ($\delta\omega$): $\alpha(t) = \delta\omega\, t$. The optical potential produced by them is

$$V_\mathrm{OL}(x,t) = V_0 \left|\mathrm{e}^{ikx+i\alpha(t)} + \mathrm{e}^{-ikx} + \mathrm{e}^{3ikx}\right|^2 = V_0\left(3 + 4\cos\kappa x\,\cos\alpha(t) + 2\cos 2\kappa x\right). \tag{5}$$

Here, $\kappa = 2k$, and V_0 is the dimensional constant. Within the potential, the intracell, Δx_1, and intercell, Δx_2, distances between the neighbour minima (which are attributed to the sites) are:

$$\Delta x_1(\alpha) = x_{n,\mathrm{B}} - x_{n,\mathrm{B}} = \frac{2\,\arccos\frac{\cos\alpha}{2}}{\kappa}, \qquad \Delta x_2(\alpha) = \frac{2\pi}{\kappa} - \Delta x_1. \tag{6}$$

In order to evaluate the hopping amplitudes $t_1 = t(x_{n,\mathrm{A}} \to x_{n,\mathrm{B}})$ and $t_2 = t(x_{n,\mathrm{B}} \to x_{n+1,\mathrm{A}})$, we use the harmonic approximation for the optical potential (5) in the vicinity of its minima. That leads to the magnitudes of the vibration frequency, ω, and the zero-vibrations amplitude, r_0:

$$\omega(\alpha) = \sqrt{\frac{2V_0\kappa^2\left(4-\cos^2\alpha\right)}{\mu}}, \quad r_0(\alpha) = \sqrt{\frac{1}{\mu\omega(\alpha)}} = \left[2V_0\mu\kappa^2\left(4-\cos^2\alpha\right)\right]^{-1/4}, \tag{7}$$

where μ is the oscillator reduced mass. The vibrations frequency ω remains the same for all the minima. The localized Wannier states are taken as the zero vibrational level harmonic wavefunctions centred at the corresponding minima. In this case, the hopping amplitudes can be calculated as $t_\mathrm{1eff} = t(\Delta x_1)$ and $t_\mathrm{2eff} = t(\Delta x_2)$ with

$$t(\Delta x_i) = \int_{-\infty}^{+\infty} \mathrm{d}x\; \chi_0(x+\Delta x_i)\,\frac{\mu\omega^2 x^2}{2}\,\chi_0(x) = \frac{\omega}{2}\cdot \mathrm{e}^{-\Delta_i^2}\cdot\left[\Delta_i^2 + \frac{1}{2}\right], \tag{8}$$

where $\Delta_i = \Delta_i(\alpha) = \Delta x_i(\alpha)/2r_0(\alpha)$ (see (6), (7)), and ω is a function of α too (7).

3. Results and discussion

The dependences of $\Delta x_{1(2)}$ (6) and ω (7) on α after a certain algebra disclose that

$$\Delta_1(\alpha) = \frac{\Delta x_1(\alpha)}{2r_0(\alpha)} = \arccos\left(\frac{\cos\alpha}{2}\right)\cdot\left[2V_0\mu\,\kappa^2\left(4-\cos^2\alpha\right)\right]^{1/4} = \Delta_2(\alpha-\pi), \tag{9a}$$

and consequently

$$t_2(\alpha) = t_1(\pi-\alpha). \tag{9b}$$

Here, we focus on the adiabatic regime of lattice shaking: $\delta\omega \ll t_1, t_2$. Fig. 1 represents the numerical solution of the adiabatic dynamic problem (1) for $\alpha = 0, \pi/4, \pi, 3\pi/4$ (see panels A, B, C, and D, respectively) at the zero initial conditions. The pumped site is the terminal one ($|0, A\rangle$). In the whole figure, green and red colours refer to topologically trivial and non-trivial chain configurations, respectively. Panels E and F display the delocalization of the excitation

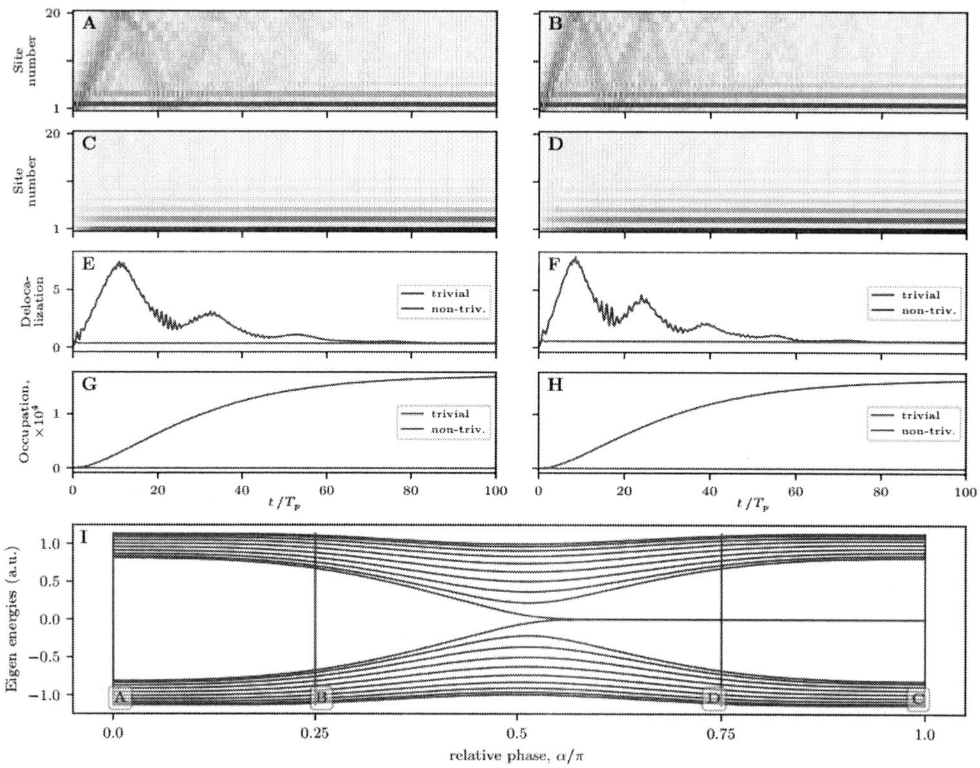

Figure 1. Adiabatic puming. A–D: evolution of the sites occupation of the SSH chain subject to the resonant pumping of the terminal (first) site for the values of relative phase $\alpha = 0$ (A), $\pi/4$ (B), π (C), and $3\pi/4$ (D). Darker colors correspond to greater values, the color normalization is power-law with the exponent 0.2. E and F: the delocalization of the excitation. G and H: the total occupations of the chain versus time (t). Panels E and G correspond to A and C, F and H – to B and D. I: the band structure dependent on α, the green and red labels serve to associate values α with relative A–D subplots.

(the dispersion), panels G and H – temporal dependences of the total occupation of the chain: A and C are summated into E, B and D – into F. On panel I, the band structure as function of α is demonstrated, the vertical lines and adjacent characters indicate the values α taken for the relative A–D figures. The parameters used for the simulations were: $\omega_p = \varepsilon = 1$ (resonant pumping), $V_0\kappa^2 = 2\mu = 0.5$, $\gamma = 0.0075$.

From this figure, we can see that the sites occupations of the chain differ both qualitatively and quantitatively for trivial and non-trivial phases. The non-trivial chain is pumped more intensively to a considerable extent (as there exist the edge state that can respond to and capture the pumping). For the same reason, the delocalization of the excitaion is (at time scales of several tens of pumping periods) is lower for the non-trivial chain but this difference finally vanishes. Moreover, the occupation pattern of the trivial chain (subplots A and B) clearly demonstrates the features of the solution corresponding to the no-pumping scheme with non-zero initial conditions when only the terminal state is initially occupied.

4. Conclusion

The main aim of the present study was to theoretically examine the topological properties of the Su-Schrieffer-Heeger chain subject to pumping. The chain was modelled by the condensate of non-interacting polaritons in the adiabatically shaken optical lattice. The topological phase of the chain was managed by the relative phases of the lasers constructing the lattice, the corresponding intersite hopping amplitudes were analytically calculated within the harmonic approximation to the local site potential. The chains were resonantly pumped at their terminal sites. We have shown that, under the considered conditions, the topologically trivial and non-trivial chains can exhibit perceptible behaviour.

Acknowledgments

The research project was supported by Russian Foundation for Basic Research (Project No. 16-02-01058), Rannis Project No. 163082-051, and Ministry of Education and Science of Russian Federation (Projects No. 3.2614.2017/4.6, No. 3.1365.2017/4.6, and No. 3.8884.2017/8.9).

[1] J.K. Asbóth, L. Oroszlány, A. Pályi. A short course on topological insulators: Band structure and edge states in one and two dimensions. *Lecture notes in physics* **919**, Springer (2016).

[2] W.P. Su, J.R. Schrieffer, and A.J. Heeger. Solitons in polyacetylene. *Phys. Rev. Lett.* **42**, 1698 (1979).

[3] L. Li, Z. Xu, and S. Chen. Topological phases of generalized Su-Schrieffer-Heeger models. *Phys. Rev. B* **89**, 085511 (2014).

[4] A.M. Marques and R.G. Dias. Multihole edge states in Su-Schrieffer-Heeger chains with interactions. *Phys. Rev. A* **95**, 115443 (2017).

[5] H. Schomerus. Topologically protected midgap states in complex photonic lattices. *Optics Letters* **38**, 1912 (2013).

[6] S. Weimann, M. Kremer, Y. Plotnik, et al. Topologically protected bound states in photonic parity-time-symmetric crystals. *Nature Materials* **16**, 433 (2017).

[7] N.H. Lindner, G. Refael, and V. Galitski. Floquet topological insulator in semiconductor quantum wells. *Nature Physics* **7**, 490 (2011).

[8] M.C. Rechtsman, J.M. Zeuner, Y. Plotnik, et al. Photonic Floquet topological insulator. *Nature* **496**, 196 (2013).

[9] V. Dal Lago, M. Atala, and L.E.F. Foa Torres. Floquet topological transitions in a driven one-dimensional topological insulator. *Phys. Rev. A* **92**, 023624 (2015).

[10] A. Gómez-León and G. Platero. Floquet-Bloch theory and topology in periodically driven lattices. *Phys. Rev. Lett.* **110**, 200403 (2013).

[11] Y. Hadad, A.B. Khanikaev, and A. Alù. Self-induced topological and edge states supported by nonlinear staggered potentials. *Phys. Rev. B* **93**, 155112 (2016).

[12] A. Eckardt, C. Weiss, and M. Holthaus. Superfluid-insulator transition in a periodically driven optical lattice. *Phys. Rev. Lett.* **95**, 260404 (2005).

[13] A. Eckardt. Colloquim: Atomic quantum gases in periodically driven optical lattices. *Rev. Mod. Phys.* **89**, 011004 (2017).

[14] H. Lignier, C. Sias, D. Ciampini, et al. Dynamical control of matter-wave tunneling in periodic potentials. *Phys. Rev. Lett.* **99**, 220403 (2007).
[15] A.A. Pervishko, T.C.H. Liew, V.M. Kovalev, I.G. Savenko, and I.A. Shelykh. Nonlinear effects in multi-photon polaritonics. *Optics Express* **21**, 15183 (2013).
[16] M. Atala, M. Aidelsburger, J.T. Barreiro, et al. Direct measurement of the Zak phase in topological Bloch bands. *Nature Physics* **9**, 795 (2013).

IOP Conf. Series: Journal of Physics: Conf. Series **1092** (2018) 012071 doi:10.1088/1742-6596/1092/1/012071

Elliptically polarized exciton-polariton condensate in a semiconductor microcavity with a chiral photonic crystal slab

V D Kulakovskii[1,1] A S Brichkin[1], N A Gippius[2], S G Tikhodeev[3], C Schneider[4], and S Höfling[4]

[1] Institute of Solid State Physics RAS, Chernogolovka, 142432, Russia,
[2] Skolkovo Institute of Science and Technology, Moscow 143026, Russia
[3] A.M. Prokhorov General Physics Institute RAS, Vavilova Str.38, Moscow 119991, Russia
[4]Technische Physik and Wilhelm-Conrad-Röntgen-Research Center for Complex Material Systems, Universität Würzburg, D-97074, Würzburg, Am Hubland, Germany

Abstract. We demonstrate a method for controling polarization of exciton-polariton condensate emission in a planar semiconductor microcavity based on modification of the electromagnetic mode structure in the microcavity. The modification is realized by fabrication of a chiral photonic crystal slab with partial etching of the upper Bragg mirror lowering the microcavity symmetry to C4. A degree of circular polarization as high as 70% has been demonstrated experimentally in the structure with optimized parameters without the use of a static magnetic field or birefringent wave plates.

1. Introduction

Exciton polaritons in semiconductor microcavities (MCs) are half-light-half-matter boson quasiparticles forming due to coupling of the optical and matter modes in the strong light-matter coupling regime [1]. Just like exciton and light, they are boson particles with an integer spin. Exciton-polariton systems have attracted broad interest during the last decade. Their properties are reliably controlled and probed by optical means as polaritons tunneling through Bragg mirrors turn into photons with the same statistics and vice versa. Owing to the small effective mass, about 10^{-8} times smaller than the atom mass, polaritons exhibit quantum coherent properties at relatively low densities of 10^{10}-10^{11} cm^{-2} at surprisingly high temperatures [2]. The condensation of polaritons at the low polariton (LP) band bottom in GaAs- and CdTe-based planar MCs is observed up to several tens of Kelvin [3-5] and up to room temperatures in GaN and ZnO-based MCs [6,7].

The total angular momentum J of a heavy-hole exciton-polariton in a quantum-well MC can have two possible projections on the growth axis of the structure, $Jz = +1$ (spin up) or $Jz = -1$ (spin down), corresponding to right and left circularly polarized emitted light, respectively. These states are degenerate due to high MC symmetry. The spontaneous symmetry breaking of the condensate manifests itself in spontaneous buildup of linear polarization of the emitted light [8]. In the absence of pinning, such polarization is randomly chosen by the system and changes stochastically from one

[1] kulakovs@issp.ac.ru

experiment to another [9]. Reduction of MC symmetry caused by magnetic field (lateral strain) results in splitting of the LP level into two sublevels with orthogonal circular (linear) polarizations.

In the present paper we investigate the possibility of controlling polarization of LP condensate emission in an unstrained planar semiconductor MC using modification of the MC electromagnetic mode structure by fabrication of a chiral photonic crystal slab at a zero magnetic field and show that the degree of circular polarization of LP condensate emission can be controlled by a very thin half-wave plate with a thickness of the order of the emitted light wavelength.

2. Experimental details.

AlAs/AlGaAs/GaAs high Q-factor MC was grown by molecular beam epitaxy on a (001)-oriented GaAs. It consists of a lower and an upper Bragg reflectors with 27 and 23 pairs of $AlAs/Al_{0.2}Ga_{0.8}As$ layers, respectively, and an active layer with three groups of four 13 nm GaAs quantum wells (QWs) separated by 4 nmAlAs barriers. A chiral layer is fabricated by electron-beam nanolithography and dry etching through top N_{etch}= 4.75 Bragg pairs of the upper mirror.

The schematics of the CPC are depicted in Fig. 1. It consists of a square lattice of rectangular pillars that have a broken in-plane mirror symmetry but possess a fourfold rotational axis and provides strong optical activity[10]. The vertical walls of nanopillars are normal to the [110] crystallographic directions. This structure has C4 point symmetry, and it is three-dimensionally chiral, because it does not have planes of mirror symmetry, including the horizontal one.

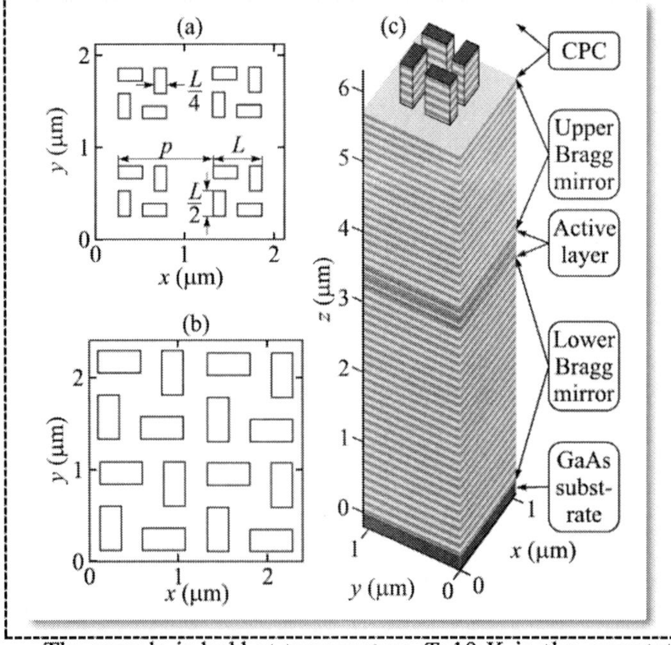

Figure 1. (a and b) The schematics of CPC composed of a square lattice of rectangular nanopillars in Samples A (a) and B (b), 2×2 periods are shown. (c) Schematics of the unit cell of Sample A. The nanopillars, composing the CPC, are etched through the top 4.75 (of 23) Bragg pairs of the upper Bragg mirror of a planar AlGaAs/AlAs MC with twelve GaAs QWs (three groups of four). Green and yellow colors represent AlAs and AlGaAs λ/4-layers, respectively. Red color represents the GaAs substrate, smoothing layers between Bragg pairs and active QWs.

The sample is held at temperature T=10 K in the cryostat. The excitation is carried out with a Ti-sapphire laser in the spectral range of the first reflection minimum of the MC. The laser spot has a diameter of about 10 μm. The emission is collected in an angle range of ±15°. It is dispersed by a monochromator and detected by a Si CCD camera. The polarization of the luminescence is analysed by a quarter wave retarder and linear polarizers feature.

3. Experimental results and discussion.

In [11], taking into account the reciprocity principle and the symmetry of the structure, we showed that a CPC slab with C4 symmetry can work as a waveplate due to Fabry-Perot interference between the vertically propagating modes in the slab and, for some optimal parameters, it can transmit light with one circular polarization and suppress almost entirely light with opposite polarization. No linear polarization is observed, because the oscillating dipoles are assumed to be randomly linearly polarized in the *xy* plane. To obtain a high degree of circular polarization of LP emission, the chiral structure was optimized on the basis of calculations of the frequency dependences of LP emission in the right- and left-circular polarizations using an optical scattering matrix and the modal Fourier method [11, 12].

Basing on the calculations, two periodic structures have been fabricated for experimental studies. They are shown schematically in Figs. 1(a) and 1(b). Sample A has period p=1060 nm and pillar feature size L=544 nm, and Sample B has p=1200 nm and L=960 nm. The height of the chiral layer is equal to 4.75 Bragg pairs. The horizontal size of chiral structure in each sample is 50 × 50 μm^2. Calculated degrees of circular polarization, ρ_c, of LP emission depend on the angle from the normal to the MC plane. The expected values of ρ_c are of ~0.75 and 0.6 for emission along the normal to the MC plane in the samples A and B, respectively.

In general, the used approximation is not completely valid for describing the condensate emission: the emission is calculated for oscillating point dipoles in the QW plane driven by external excitation, so that the intensities rather than the electromagnetic fields are summed up at the receiver. This approximation should work well below the threshold of LP condensation and is expected to be a reasonable starting point for optimization of the structures.

Figure 2 b shows circularly polarized emission spectra of LPs from a MC with a CPC at excitation density P slightly below and above the threshold of LP condensation for Sample A. The detuning of the photon and exciton modes is about -6 meV. Figure 2 b shows that with a 40% increase in P the LP emission intensity increases several-fold and the emission spectrum shows a threshold-like transition from a broad to narrow line with full width at half maximum (FWHM) of 0.35 meV. Comparison of the σ^+ and σ^- polarized spectra reveals that the transition is also attended with a large increase in the degree of circular polarization $\rho_c=(I^+-I^-)/(I^++I^-)$ where I^+ and I^- are the emission intensities in the right (σ^+) and left (σ^-) polarizations, respectively. The spectral dependences of ρ_c at P below and above P_{thr} are shown in Fig. 2a. It is seen that ρ_c increases from ~0.2 at P=0.7P_{thr} to ~0.65 at P= 1.05P_{thr}.

The change in the LP emission spectra with increasing pump density is shown in Fig. 3b. It is seen that with increasing pump density (i) the emission intensity increases superlinearly (ii) the emission line being shifted towards higher energies and slightly broadened. Underline that the blue shift of the emission line proves its polariton nature. Indeed, the shift indicates the growth of the repulsive

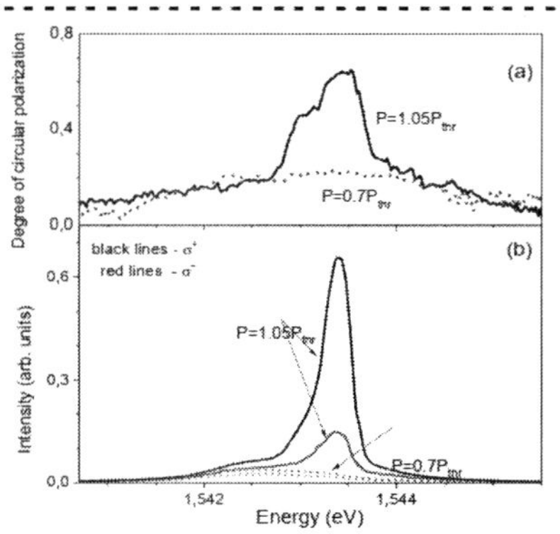

Figure 2. (a) The degrees of ρ_c and (b) circularly polarized emission spectra of LPs from MC with a CPC at P slightly below and above the threshold of LP condensation. The CPC parameters are p = 1060 nm and L = 544 nm.

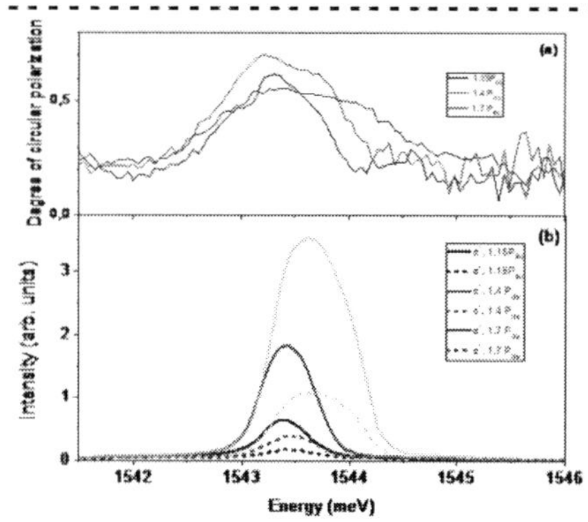

Figure 3. (a) The degrees of ρ_c and (b) circularly polarized emission spectra of LPs from MC with a CPC at several P above the threshold of LP condensation. The CPC parameters are p = 1060 nm and L = 544 nm.

interaction in the system under study, which is characteristic of LP systems and absent in light ones. Figure 3b shows that the degree of circular polarization changes relatively weakly with P, the maximum value of ρ_c~ 0.7 is observed at P = 1.4 P_{thr} and decreases somewhat with a further increase in P to ρ_c ~ 0.55 at P = 1.7 P_{thr}.

Finally, linear polarization along the (110) crystallographic direction of the AlGaAs and GaAs layers, ρ_{lin}~0.38. As a result the total degree of LP emission light polarization $\rho_\Sigma = (\rho_c^2 + \rho_{lin}^2)^{1/2}$ reaches 0.75. Note that the linear polarization is unexpected for the emission from structures with C4 symmetry. This can be related to possible (i) fabrication-induced deformation of the CPC structures, which could lead to a reduced structural symmetry or (ii) preferential linear polarization of the condensate state.

4. Conclusion.

We have demonstrated that fabrication of a thin, of the order of λ, CPC slab at the MC top can be used for controlling the degree of circular polarization of LP condensate emission without the use of a static magnetic field or birefringent wave plates. The magnitude of ρ_c in the fabricated structure with optimized parameters depends weakly on the condensate density and reaches 0.65-0.7 at P =1.05 – 1.5P_{thr}. In addition, it was found that the condensate emission also exhibits a partial linear polarization. The reason for the presence of linear polarization with the symmetry of the C4 structure is not yet clear and requires further research.

Acknowledgement This work is supported by Russian Science Foundation (Grant No. 16-12-10538) and State of Bavaria. We are grateful to M. Emmerling for preparing the nanopillars.

References

[1] Weisbuch C, Nishioka M, Ishikawa A and Arakawa Y 1992 *Phys.Rev. Lett.* **69** 3314
[2] Bramati A,. Modugno M, *Physics of Quantum Fluids* (Springer-Verlag, Berlin, 2013
[3] Kasprzak J, Richard M, Kundermann S, Baas A, Jeambrun P, Keeling J M J, Marchetti F M, Szymanska M H, Andre R, Staehli J L, Savona V, Littlewood P B, Deveaud-Pledran B, and Dang L S 2006 *Nature* **443** 409
[4] Balili R, Hartwell V, Snoke D, Pfeiffer L and West K 2007 *Science* **316** 1007
[5] Tsintzos S I, Pelekanos N T, Konstantinidis G, Hatzopoulos Z and Savvidis P G, 2008 *Nature* **453** 372
[6] Christopoulos S, von Högersthal G B H, Grundy A J D, Lagoudakis P G, Kavokin AV, Baumberg J J, Christmann G, Butté R, Feltin E, Carlin J F, Grandjean N 2007 *Phys. Rev. Lett.* **98** 12640
[7] Bhattacharya P, Frost T, Deshpande S, Baten M Z, Hazari A and Das A, 2014 *Phys. Rev. Lett.* **112** 236802
[8] Laussy F P, Shelykh I A, Malpuech G, and Kavokin A V 2006 *Phys. Rev.* B **73** 035315
[9] Read D, Liew T C H, Rubo Y G and Kavokin A V 2009 *Phys. Rev.* B **80** 195309
[10] Konishi K, Bai B, Meng X, Karvinen P, Turunen J, Svirko Y P and Kuwata-Gonokami M, 2008 *Opt. Express* 16 7189
[11] Lobanov S V, Weiss T, Gippius N A, Tikhodeev S G, Kulakovskii V D, Konishi K and Kuwata-Gonokami M 2015 *Opt. Lett.* **40** 1528
[12] Whittaker D M and I. S. Culshaw I S 1999 *Phys. Rev.* B **60**, 2610 .

IOP Conf. Series: Journal of Physics: Conf. Series **1092** (2018) 012072 doi:10.1088/1742-6596/1092/1/012072

Interlayer self-assembly of 2D antimony/multigrapnene nanocomposites

T V Kulikova*, A V Tuchin, D A Testov and L A Bityutskaya

Voronezh State University, 1, University Square, 394018, Voronezh, Russia

*e-mail: kaimt@mail.ru

Abstract. Paper studies conditions for the formation of 2D allotropes of layered precursors (antimony, graphite) and self-assembled composite structures based on them. The latter assemble from colloidal solution of the inter-layers. The above conditions have been studied both experimentally and theoretically within the framework of our research. The modes for the isopropanol/antimony self-activated colloidal solutions have been set up. These processes are accompanied by the long-term nonlinear hydrodynamic effects in a colloidal solution that correlate with periodic changes of solution particle sizes. It has been proved theoretically that the charge properties of multilayers of antimony are different, whereas the processes observed during the tests indicate the presence of 2D structures within different number of layers of the solution volume, ranging in type and magnitude of the charge. Two types of composite structures have been obtained: a multilayer multigraphene/antimony structure and a polymorphic multigraphene/nanofiber.

1. Introduction

Ever since the discovery of graphene [1,2], the world scientific community has multiplied its efforts to create similar 2D materials. The main trend of the modern materials science thereafter has become the creation and study of 2D materials properties, which would consist of one or two elements of the III - VI group of the periodic system, as well as the composites based on them [3]. The most persistent modifications of 2D materials are formed by the the nitrogen group. In contrast to graphene, it shows non-zero bandgap. Its electronic structure is sensitive to external deformations [4-6]. Antimony is distinguished as a heavy element with a strong spin-orbit interaction within this group, which allows us to observe topological effects, and also to form stable 2D allotropes [7,8].

Our purpose is to study the forming conditions for the composite structures in colloidal solutions of layered precursors (antimony, graphite).

2. Theoretical part

Self-assembly is a process in which a spontaneously ordered unit is formed by the individual components or constituents of a mixture due to the minimization of their total energy. Unlike molecular systems, the building blocks in colloidal solutions of layered precursors are represented by 2D allotropes. The colloidal solution of antimony corresponds to five principles of self-assembly formulated for molecular systems: availability of building blocks and their mobility, reversibility, the environment for the process, and intermolecular interactions [9]. However, it has a fundamental distinction – developed 2D surface. Quantum-chemical modeling of antimony colloidal solution was applied to study possible interactions of - mono- and multilayers' building blocks.

Quantum-chemical modeling of antimony layers fragments with elementary cells size of 5x5 passivated by hydrogen at their edges was carried out with the Density Functional Theory (DFT) applied. Antimonene is a corrugated bilayer, consisting of strained hexagons with antimony atoms located at their apexes. The distances between neutral atoms within the layer constitute R = 2.970 Å.

Sequential increase in the number of layers leads to qualitative change in distribution of the electron density. Atoms of the two-layer antimonene of the outer layers have an excess effective positive charge of 0.21 e, whereas the internal ones have -0.21 e of charge. The interatomic distance increases up to 2.978-2.979 Å. The minimum distance between the adjacent layers atoms is 3.351 Å.
The middle bilayer is almost neutral in the three-layer antimonene. Its effective charge is 0.009 e. This reveals the asymmetry of the charge distribution in the outer layers: the outer edges retain their positive charge of Qeff = 0.147 e, whereas the internal edges have a charge of -0.156 e. The interatomic distances in the middle layer comprise 2.978 Å, whereas in the outer layers they shrink to 2.977 Å, which matches the two-layer antimonene.
Layered growth of the structure determines the alignment of antimonene based on the increased number of interlayer bonds. However, the amount of reduced energy per layer decreases from E_{bsh} = 0.734 to 0.499 eV / unit cell layer for a two- and three-layer structure. The above result conforms to increased minimum distance between atoms of adjacent layers of 3.590 Å in the three-layer antimonene.
Thus, the redistribution of electron density in multilayer structures of antimony determines the following inducement of the excess positive charge on the outer edges of layers. The inner layers are neutral as it is the case of a single-layer antimonene, while the interlayer binding energy between the outer and inner layers is reduced by almost 1.5 times as compared to the two-layer structures in which there is no neutral layer observed. In general, consistent increase in the number of layers stabilizes multilayer antimony. However, it predetermines its easier stratification due to the lower interlayer binding energy and initial positive charge on the outer edge.

Table 1. The shortest intra-layer and interlayer distances; reduced energy and effective charge of antimonene, depending on the number of layers

n	R, Å	Rsh, Å	Eb, eV/ layer of the unit cell	Qeff, e
1	2.970	–	–	0
2	2.978			0.210
				-0.210
		3.351	0.734	
	2.978			-0.210
				0.210
3	2.977			0.147
		3.590	0.499	-0.156
	2.978			0.009
				0.009
		3.590	0.499	-0.156
	2.977			0.147

Thus, the charge in the layers allows us to develop special techniques and entire technology for self-assembling of planar 2D structures of layered precursors and the composites based on them.

3. Experimental part

The method of liquid-phase exfoliation [10] was applied to obtain planar 2D structures of antimony, graphite and composites based on them: the solution of isopropyl spirit and water, single crystal of antimony or highly oriented pyrolytic graphite was affected by the long ultrasonic impact. Dispersing process was carried out by ultrasound with a frequency of 22 kHz and 100 W of power. The solution was ultrasonically processed for 3 hours for graphite and 5 hours for antimony. In order to obtain antimony/carbon composites, the solutions were mixed and dispersed again for an hour.
Long-time nonlinear hydrodynamic effects were observed after dispersing of the colloidal solution of antimony:

- the formation of vortices in the given volume of colloidal solution, the number of which ranged depending on the concentration and volume of the solution;

- oscillatory change in the limpidity of the solution, occurring within 30 hours after the dispersion was over.

Long multi-hour dispersion of antimony colloidal solution leads to the formation of planar structures of various thicknesses and lateral sizes range from 0.5 to 10 μm. Antimony separation into mono-atomic layers is not observed as the studied modes were applied.

In case of graphite colloidal solution dispersion, delamination is quite efficient, and a multigraphene and polymorphic composite structure - multigraphene / nanofibers [11] - is formed. The composite is a set of graphene sheets with a nanofibres on their surface (Fig. 1a). The lateral dimensions of the composite are 7x2.5 μm in size, whereas the diameter of the fibers is about 50 nm.

As colloidal solution of antimony and graphite is dispersed, a multilayer composite 'multigraphene / antimony' structure with lateral dimensions of about tens of microns is formed (Fig. 1b).

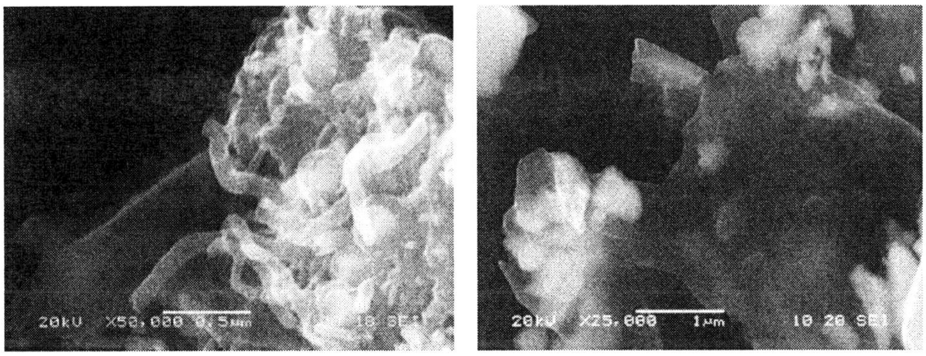

a b

Figure 1. SEM images of multigrafen / nanofibres composite (a) and multigrahen / antimony composite (b)

4. Discussion of results and conclusions

Paper studies consistencies of formation of planar 2D structures antimony as well as of graphite and composites based on the method of liquid-phase exfoliation of precursors. Several techniques were developed and particular modes of obtaining structures of different types have been optimized. Obtained multilayer structures were morphologically characterized.

The modes for self-activated colloidal solutions, followed by the long-time hydrodynamic effects related to the periodic change in particle size in solution were revealed experimentally. It has been proved theoretically that antimony multilayers range in their charge properties and all the processes observed experimentally indicate availability of 2D structures with a different number of layers in the solution volume, differing in the type and magnitude of the charge. Electrostatically active environment of colloidal solution creates conditions for self-assembly of both the planar structures of the precursor and the composite structures based on them.

Earlier we have demonstrated the possibility of self-assembly of core-shell structures based on antimony in a melt with a graphene-like composite shell [12-13].

As a result multilayered planar structures of antimony and graphite, as well as two types of composite structures were obtained: a multilayer multigraphene/antimony structure and a polymorphic-multigraphene/nanofiber.

References

[1] Novoselov K S 2011 *UFN* **181** 1299-1311 DOI: 10.3367 / UFNr.0181.201112f.1299
[2] Geim A K 2011 *UFN* **181** 1284-1298 DOI: 10.3367 / UFNr.0181.201112e.1284;
[3] Kulikova T V, Tuchin A V, Averin A A , Testov D A , Bityutskaya L A and Bormontov E N 2018 *Journal of Technical Physics (Zhurnal tekhnicheskoy fiziki)* **88(7)** 1025-1031;

[4] Yuanfeng Xu, Bo Peng, Hao Zhang, Hezhu Shao, Rongjun Zhang and Heyuan Zhu 2017 *Ann. Phys. (Berlin)* 1600152 DOI: 10.1002/andp.201600152;

[5] Shengli Zhang, Zhong Yan, Yafei Li, Zhongfang Chen and Haibo Zeng 2015 *Angew. Chem. Int. Ed.* **54(10)** 3112–3115 DOI: 10.1002/anie.201411246;

[6] Mingwen Zhao, Xiaoming Zhang and Linyang Li 2015 *Scientific Reports* 5:16108 DOI: 10.1038/srep16108;

[7] Bian G, Miller T and Chiang T C 2011 *Physical Review Letters* **107(3)** 036802 DOI: 10.1103/PhysRevLett.107.036802;

[8] Sung Hwan Kim, Kyung-Hwan Jin, Joonbum Park, Jun Sung Kim, Seung-Hoon Jhi and Han Woong Yeom 2016 *Scientific Reports* 6:33193 DOI: 10.1038/srep33193;

[9] Mansuri G A 2008 *Principles of nanotechnology* (Moscow: The scientific world (Nauchnyy mir)

[10] Gibaja C, Rodriguez-San-Miguel D, Ares P, Gomez-Herrero J, Varela M, Gillen R, Maultzsch J, Hauke F, Hirsch A, Abellan G and Zamora F 2016 *Angew. Chem. Int. Ed.* **55(46)** 14345–49 DOI: 10.1002/anie.201605298;

[11] Wei Fan, Longsheng Zhang and Tianxi Liu 2017 *Graphene-Carbon Nanotube Hybrids for Energy and Environmental Applications* (Springer Singapore) DOI: 10.1007/978-981-10-2803-8;

[12] Kulikova T V, Bityutskaya L A, Tuchin A V and Averin A A 2017 *Perspective materials (Perspektivnye materialy)* **3** 5-13

[13] Kulikova T V, Bityutskaya L A, Tuchin A V, Lisov E V, Nesterov S I, Averin A A and Agapov B L 2017 *Letters on materials* **7(4)** 350-354 DOI: 10.22226/2410-3535-2017-4-350-354

Fabrication of Bi_2Te_3 films on the surface of silica optical fibres by MOCVD

P. I. Kusnetsov, E. A. Savelyev, V.A Jitov, K. M. Golant

Kotel'nikov IRE RAS, Mokhovaya 11-7, Moscow, 125009, Russia

pik218@fireras.su

Abstract. Crystalline 2D bismuth and antimony chalcogenides, as well as graphene, can successively operate as saturable absorbers in various schemes of passively mode-locked fibre lasers. An attractive benefit of these saturable absorbers is their power to effectively operate at different wavelengths in the wideband from 1 to 2 microns. However, the problem is how to provide a reliable interaction between light in a fibre and the absorber. The main disadvantage of a standard solution to the problem essentially is to use exfoliated flakes dissolved in a coupling agent, which deposits additional heat that negatively affects the absorber. In the present research, thin crystalline films of a narrow-band semiconductor Bi_2Te_3 were deposited directly on silica optical fibres by means of metal-organic chemical vapour deposition (MOCVD). A ZnTe buffer layer was deposited first in a single growth run, providing the necessary condition for better adhesion of bismuth telluride film to the silica surface. Two types of fibre coatings were prepared and tested: the coating upon the cleaved end of the fibre and the one upon the lateral surface of the light-guiding fibre core. In the latter case, prior to deposition a fibre section of about 1 cm in length was uniformly thinned via chemical etching to a diameter of about 20 microns. Reflection and transmission of the growing films were monitored in situ during the deposition.

1. Introduction

Saturable absorbers based on atomically thin crystalline films of bismuth and antimony chalcogenides are used in the 1–2 μm wavelength range [1] despite the fact that the width of a forbidden band of these narrow-bandgap semiconductors does not fit the energy of photons in the wavelength range mentioned. This is possible due to the peculiarity of band structure and the optical absorption spectrum of thin films under consideration, which subjected to change under the impact of strong optical fields [2].

In practice, the manufacture of this type of saturable absorber is performed by mixing suspensions of nano-flakes exfoliated from the corresponding single crystal with a binding polymeric agent [3]. While relatively simple, this procedure results in the inevitable dispersion of flake sizes and the impairment of polymer-crystal interface states on the saturated absorber characteristics. In addition, high optical power density generates significant heat, which can lead to irreversible changes in the optical properties of an absorber.

An alternative way is the deposition of a thin semiconductor absorbing film directly on the surface of an optical waveguide. There are various technological methods suitable for this task ranging from molecular-beam epitaxy [4] to laser ablation [5] and magnetron sputtering [6]. Among them, we can highlight film deposition by metal-organic chemical vapour deposition (MOCVD) [7]. This process does not require ultra-high vacuum, allows the obtainment of homogeneous films, and easily vary their composition in a single deposition run.

In this paper, we present the results of the first application of MOCVD to the deposition of Bi_2Te_3 films on the surfaces of a silica optical fibre.

2. Experimental.

To obtain the thin-film coatings on a fibre, we used the MOCVD process outlined in [8]. The deposition took place in a horizontal tubular reactor heated by a resistance oven 10 cm in length to a temperature of 420–445 °C in the atmospheric pressure hydrogen flow. Trimethylbismuth ($BiMe_3$), diethyltellurium (Et_2Te) and diethylzinc ($ZnEt_2$) served as the precursors. The first two were used to deposit the Bi_2Te_3 absorbing film and the final one was given to obtain the ZnTe buffer and protective (top) layers. The 5–10 nm thick ZnTe buffer layer terminates open chemical bonds on the silica surface to strengthen the adhesion of the growing Bi_2Te_3 film to the surface.

Two types of silica fibre surfaces served as substrates for the deposition of films in our experiments. The first one was the cleaved end of a fibre. The second one was a lateral surface of a tapered section of standard single mode (SMF28) fibre. For taper manufacturing, the fibre was released from the polymer protective coating at a distance of 0.3–2 cm and placed in an etchant, the recipe of which is given in [9]. The diameter of the protection-free section of the fibre uniformly scaled down from 125 μm to approximately 20 μm during 100 min at a temperature of 27 °C. The fibre segments that were free of the protective polymer coating (12–15 cm in length) and accordingly prepared for deposition were housed in a hermetically sealed MOCVD reactor. Through the exteriorized fibre ends, one can continuously record changes in transmission and reflection spectra during the film growth on the fibre surfaces.

3. Results and discussion.

For deposition of the Bi_2Te_3 absorbing film, we used $BiMe_3$ at a partial pressure of 3×10^{-5} ATM and a 15-fold excess of tellurium in the vapour phase. A rather high mass flow of bismuth is necessary for a high-rate deposition. This step turned out to be a mandatory requirement for the formation of a thin film uniform coat.

Under these deposition conditions, the necessary (10–40%) change of the reflectivity for the cleaved fibre end due to film deposition took place in 3–13 sec. The EDX analysis indicated the stoichiometry of the deposited Bi_2Te_3 film. The Raman spectra of the obtained thin films in addition to vibration modes of Bi_2Te_3 contained intense LO modes of the ZnTe buffer layer. The growth of the taper films caused a decrease in fibre transmission during the first 7–10 seconds of deposition. The final transmittance depended on both taper diameter and length.

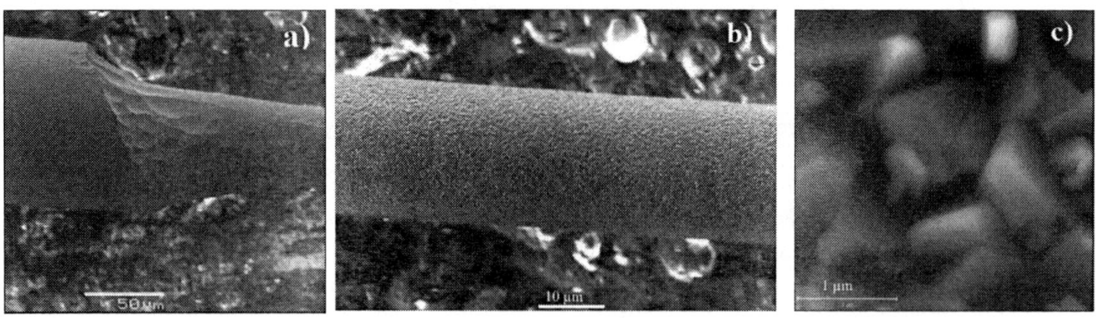

Fig.1 a), b) - SEM and c) - AFM images of fiber taper coated by a thick crystalline Bi_2Te_3 layer via MOCVD. Deposition time of the Bi_2Te_3 film amounts to 250 sec.

Figs. 1-a and -b show SEM images of the central part of the tapered fiber covered by a thick Bi_2Te_3 layer. The surface of this coating consists of large-size nano-flakes randomly oriented on the lateral surface of the fibre (Fig. 1c).

Conclusion.
The MOCVD technology has been employed for coating the cleaved ends of an optical fibre and/or lateral surface of the light-guiding core with a Bi_2Te_3 thin film saturable absorber. The developed technology is of immediate interest for the engineering of passively mode-locked fibres with a wavelength range of 1–2 μm and waveguide lasers.

Acknowledgements.
The authors are grateful to Dr. A.G. Temiryazev for AFM imaging.
This work was supported by the Russian Science Foundation (project no. 17-19-01057).

References
[1] Y. Wang, S. Liu, J. Yuan, P. Wang, J. Chen, J. Li, S. Xiao, Q. Bao, Y. Gao, J. He, Ultra-broadband Nonlinear Saturable Absorption for Two-dimensional $Bi_2Te_xSe_{3-x}$ Nanosheets, Scientific Reports, 6 (2016) 33070.
[2] V.V. Tugushev, E. Kulatov, K.M. Golant, Features of the electronic spectrum and optical absorption of ultrathin Bi2Se3 films, JETP Letters, 106 (2017) 422–428..
[3] M. Salim, R. Shaharuddin, M. Ismail, S. Harun, H. Ahmad, S.R. Azzuhri, Bi_2Te_3 based passively Q-switched at 1042.76 and 1047 nm wavelength, Laser Physics, 27 (2017) 125102.
[4] G. Wang, X. Zhu, J. Wen, X. Chen, K. He, L. Wang, X. Ma, Y. Liu, X. Dai, Z. Fang, J. Jia, Q. Xue, Atomically smooth ultrathin films of topological insulator Sb_2Te_3, Nano Research, 3 (2010) 874-880.
[5] B. Saha, P. Chaturvedi, A.K. Yadav, D. Saha, S. Ganguly, Pulsed laser deposition of highly oriented stoichiometric thin films of topological insulator Sb_2Te_3, Journal of Vacuum Science & Technology B, 34 (2016) 021806.
[6] S.-J. Joo, B.S. Kim, B.-K. Min, M.W. Oh, J.-E. Lee, B.K. Ryu, H.W. Lee, S.D. Park, Deposition of n-Type Bi_2Te_3 Thin Films on Polyimide by Using RF Magnetron Co-Sputtering Method, Journal of Nanoscience and Nanotechnology, 15 (2015) 8299–8304.
[7] L.D. Alegria, J.R. Petta, Controlled MOCVD growth of Bi_2Se_3 topological insulator nanoribbons, Nanotechnology, 23 (2012) 435601.
[8] P. I. Kuznetsov, V.O. Yapaskurt, B.S. Shchamkhalova, V.D. Shcherbakov, G.G. Yakushcheva, V.A. Luzanov, V.A. Jitov, Growth of Bi_2Te_3 films and other phases of Bi-Te system by MOVPE, Journal of Crystal Growth, 455 (2016) 122–128
[9] S. Ko, J. Lee, J. Koo, B.S. Joo, M. Gu, J.H. Lee, Chemical wet etching of an optical fiber using a hydrogen fluoride-free solution for a saturable absorber based on the evanescent field interaction, Journal of Lightwave Technology, 34 (2016) 3776–3784.

IOP Conf. Series: Journal of Physics: Conf. Series **1092** (2018) 012074 doi:10.1088/1742-6596/1092/1/012074

Numerical design of Au/Si core-shell nanoparticles

Larin A.O.[1*], Sun Y. [1], Zuev D. A. [1]

[1]Departure of Physics and Engineering, ITMO University, St. Petersburg 197101, Russia

*artem.larin@metalab.ifmo.ru

Abstract. Resonant metal-dielectric nanostructures of the core-shell type can provide unusual optical properties originated from spectral overlapping of electric and magnetic resonances of the nanostructure components. Here we numerically investigate the geometry and scattering cross section spectra of a core-shell nanostructure, where the core is gold, and the shell is crystalline silicon. We study spherical and cylindrical types of the nanostructures and search optimal geometrical parameters in the terms of spectral overlapping of the resonances. The obtained geometry parameters can be utilized for the first experimental realization of the metal-dielectric core-shell nanostructures.

1. Introduction.

Resonant plasmonics and dielectric nanoparticles have unique optical properties: surface enhancement Raman scattering [1], Kerker effect [2] and Fano resonance [3, 4], generation of high harmonics [5], etc. The unification of these types of nanoparticles in one structure gives combination of the properties of dielectric and metallic components as well as more freedom in obtaining of novel optical properties in such nanosystem [6].

In this paper, we present the results of analytical and numerical analysis of metal-dielectric nanoparticles with a core of gold and shell of silicon. Recently, the broadband unidirectional scattering was demonstrated theoretically in spherical metal-dielectric core shell nanoantenna [2]. Here, we examined the spherical as well as cylindrical shapes of the core-shell nanostructure to obtain spectral overlapping electric (ED) and magnetic (MD) dipole resonances of gold and silicon components respectively. The materials and size of cylindrical nanoantenna were set to obtain geometrical parameters which can be achieved by standard nanofabrication methods.

2. Results and discussion.

At first we have investigated the spherical core-shell nanoparticle. The description of the core is also similar to that one in problems for a homogeneous sphere in the field of an incident plane monochromatic wave; this method does not spend much time on the calculations. Scattering on spherical nanoparticles was modeled analytically by the Mie theory [7]. The field inside the core depends only on the first-order Bessel function. In the shell, the field is dependent on the Bessel functions of the first and second order [7]. Data on dielectric constants were taken from Ref. [8, 9]. We have investigated different values of the inner radius of the nanostructure in the range from 10 to

100 nm with the outer radius remaining constant of 150 nm. The simulation showed the dependence of the position and intensity of the scattering cross section of the electric dipole of a metal core on its internal radius. The spectrum was chosen, in which the point of the maximum value of the ED scattering cross section of the metal core corresponds to the value of the MD scattering cross section of the dielectric shell, in order to fulfill the first Kerker condition (figure 1). This is possible for inner and outer radius of the particle of 37.4 nm and 150 nm, respectively, at a wavelength of 1169 nm.

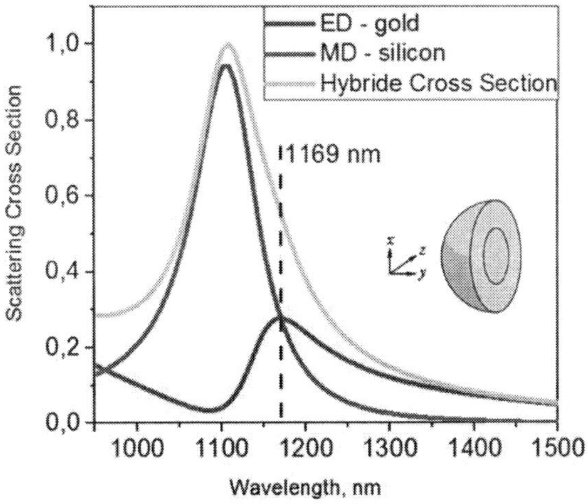

Figure 1. Scattering cross section for a spherical core-shell nanoparticle in the air calculated on the basis of the Mie theory (shell radius of 150 nm, core radius of 37.4 nm). A dashed line designates the wavelength of the incident radiation at which unidirectional scattering is observed.

Developed model for a spherical nanoparticle were used to simulate a cylindrical structure, which can be fabricated by standard nanolithography techniques. To solve this problem, we used the FTDF method from the CST MicroWave studio. Features of geometry are the following: the half height of the cylinder corresponds to its radius, which was applied for both core and shell of the nanoparticle. This configuration maximizes the cylindrical shape to a spherical one. In addition, a plane monochromatic wave is directed along the axis of the cylinder, where the geometric cross section of the object gives a projection like that one for a spherical core-shell nanoparticle. The resulting spectrum of the scattering cross section for a designed cylindrical structure is shown on the figure 2.

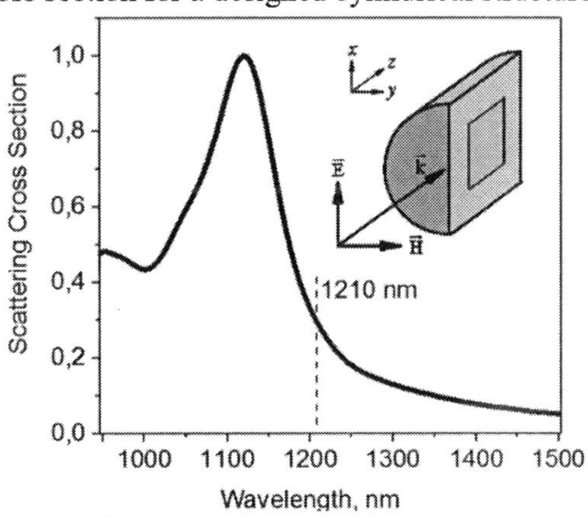

Figure 2. Scattering cross section for cylindrical core-shell nanoparticle in the air calculated on the basis of FTDT method (radius of a shell is 136 nm, radius of a core is 14.7 nm). The dashed line designates the wavelength of the incident radiation at which unidirectional scattering and azimuthally symmetry are observed. The inset shows the modeled structure and direction of propagation of the incident wave.

The size of the outer radius of the cylinder was chosen in such a way that the silicon nanoparticles of two types of the shape without core would have the same values of the resonant frequency for the MD resonance of the silicon. For this structure, the Kerker condition is satisfied at the wavelength of 1210 nm. The fulfillment of the Kerker condition is represented in the form of directional diagrams of the scattered light (figure 3). The smallest lobe beam width was obtained with an outer radius of cylinder of 136 nm and an inner one of 14.7 nm. The first figure (on the left side, figure 3) demonstrates the absence of backward scattering, the second (on the right side, figure 3) a presence of an azimuthally symmetry.

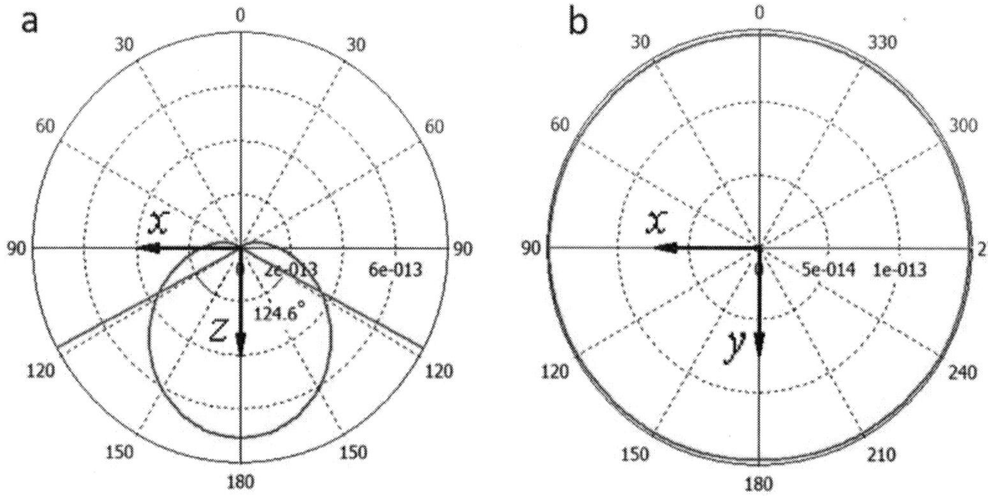

Figure 3. Scattering diagrams for a cylindrical core-shell nanoparticle: the diagram (on the left side) in the Oxz plane and (on the right side) in the Oxy plane. The coordinate system for the nanoparticle is indicated on the insets.

The carried out research demonstrates possibility of realization of unidirectional scattering in cylindrical hybrid nanostructure. The opportunity to adjust the optical properties is made by selecting materials for the core and shells, the particle size, and the relationship between the core and shell dimensions.

3. Conclusion.

In this work we have studied hybrid metal-dielectric core-shell nanoparticle to search conditions of resonances spectral overlapping (ED for gold core and MD for silicon shell) for unidirectional scattering realization. The geometry and material parameters were found for cylinder nanostructure which can be fabricated by existing nanolithography methods. The experimental realization of hybrid core-shell nanostructure and following experimental study of optical properties of such nanostructure can provide a set of interesting effects after the appropriate adaptation of the geometry of the structure and selection of materials: enhancement of Raman scattering, high order harmonic generation, etc. Moreover, application of silicon as a shell makes it possible to work with biological objects.

4. Acknowledgments

The Ministry of Education and Science of Russian Federation (Project 2.2267.2017/4.6) as well as Russian Foundation for Basic Research (Projects 17-02-00538, 16-29-05317). D. Zuev acknowledges the scholarship of the President of the Russian Federation.

References

[1] Jiangcai W and et. al. 2017 Surface Plasmon Resonance Sensors on Raman and Fluorescence Spectroscopy *Sensors* **17** 2719 – 2730

[2] Liu W and et. al. 2012 Broadband Unidirectional Scattering by Magneto-Electric Core Shell Nanoparticles *ACS Nano* **6(6)** 5489-5497

[3] Luk'yanchuk B and et.al. 2010 The Fano resonance in plasmonic nanostructures and metamaterials *Nature Material* **9** 707 – 715

[4] Limonov M F and et.al. 2017 Fano resonances in photonics *Materials* **9** 543 – 554

[5] Wang H and et.al. 2015 Janus magneto-electric nanosphere dimers exhibiting unidirectional visible light scattering and strong electromagnetic field enhancement *ACS Nano* **9(1)** 436–448

[6] Toshihiko S and et al. 2017 Efficient third harmonic generation from metaldielectric hybrid nanoantennas *Nano letters* **17(4)** 2647–2651

[7] Bohren C F and Huffman D R 1998 Absorption and Scattering of Light by Small Particles (Canada: A Wiley-Inter-science publication) A Potpourri of Particles: Coated sphere

[8] Johnson P B and Christy R W 1972 Optical Constants of the Noble Metals *Phys. Rev. B* **6** 4370–4379

[9] Rakiand A D and et.al. 1998 Optical properties of metallic films for vertical-cavity optoelectronic devices *Appl. Opt.* **37** 5271–5283

IOP Conf. Series: Journal of Physics: Conf. Series **1092** (2018) 012075 doi:10.1088/1742-6596/1092/1/012075

Surface acoustic wave modulation of single photon emission from GaN/InGaN nanowire quantum dots

S Lazić[1], E Chernysheva[1], A Hernández-Mínguez[2], P V Santos[2] and H P van der Meulen[1]

[1] Departamento de Física de Materiales, Instituto 'Nicolás Cabrera' and Instituto de Física de Materia Condensada (IFIMAC), Universidad Autónoma de Madrid, 28049 Madrid, Spain
[2] Paul-Drude-Institut für Festkörperelektronik, Hausvogteiplatz 5-7, 10117 Berlin, Germany

lazic.snezana@uam.es

Abstract. On-chip quantum information processing requires controllable quantum light sources that can be operated on-demand at high-speeds and with the possibility of in-situ control of the photon emission wavelength and its optical polarization properties. Here, we report on the dynamic control of the optical emission from core-shell GaN/InGaN nanowire (NW) heterostructures using radio frequency surface acoustic waves (SAWs). The SAWs are excited on the surface of a piezoelectric lithium niobate crystal equipped with a SAW delay line onto which the NWs were mechanically transferred. Luminescent quantum dot (QD)-like exciton localization centers induced by compositional fluctuations within the InGaN nanoshell were identified using stroboscopic micro-photoluminescence (micro-PL) spectroscopy. They exhibit narrow and almost fully linearly polarized emission lines in the micro-PL spectra and a pronounced anti-bunching signature of single photon emission in the photon correlation experiments. When the nanowire is perturbed by the propagating SAW, the embedded QD is periodically strained and its excitonic transitions are modulated by the acousto-mechanical coupling, giving rise to a spectral fine-tuning within a ~1.5 meV bandwidth at the acoustic frequency of ~330 MHz. This outcome can be further combined with spectral detection filtering for temporal control of the emitted photons. The effect of the SAW piezoelectric field on the QD charge population and on the optical polarization degree is also observed. The advantage of the acousto-optoelectric over other control schemes is that it allows in-situ manipulation of the optical emission properties over a wide frequency range (up to GHz frequencies).

1. Introduction

The implementation of semiconductor heterostructures on a nanowire (NW) platform provides a scalable miniaturized architecture for nanoscale photonic and optoelectronic devices and facilitates their integration into modern on-chip technologies. However, to fully exploit the potential of NW-based devices, appropriate techniques are required to allow in-situ control and manipulation of their fundamental physical properties on a nanoscale. Compared to other control schemes, which typically require selective doping and sophisticated nanofabrication of electrical contacts on individual NWs with sub-micrometer dimensions, the acousto-optoelectric approach based on radio-frequency (rf) surface acoustic waves (SAWs) has emerged as a powerful contactless tool for reliable manipulation of the optical emission characteristics from small-scale NW heterostructures, including NW quantum dots

(QDs). Moreover, because the SAWs propagate at the speed of sound, high operating frequencies (from several tens of megahertz up to several tens of gigahertz) can be achieved, which are usually inaccessible by other control schemes.

Most reports on SAW-modulated NW QD structures to date are limited to III-arsenide material systems. In this contribution, we review our experiments on the effect of SAW-induced strain and piezoelectric fields on the photon emission dynamics from wurtzite III-nitride dot-in-a-nanowire heterostructures. Applying SAW-techniques to group III-nitrides is vital, as their wide bandgap, large exciton binding energy and band offsets make them an ideal candidate for high-power and high-temperature device applications.

2. Methods

A schematics of the sample configuration is displayed in Fig. 1(a). All optical experiments are performed on a 128° Y-cut lithium niobate ($LiNbO_3$) substrate photo-lithographically patterned with an acoustic delay line consisting of two interdigital transducers (IDTs). The IDTs are designed to generate SAWs with an acoustic wavelength λ_{SAW}=11.67 μm, corresponding to a room-temperature resonant frequency of f_{SAW}~332 MHz and temporal period of T_{SAW}~3 ns. Information on the rf reflection and transmission characteristics of the SAW-chip can be found in [1]. The epitaxially grown semiconductor NWs containing a pencil-like GaN core, an InGaN nanoshell and an external conformal GaN cap were mechanically transferred onto the area between the two IDTs. The intrawire InGaN nanoshell hosts QD-like exciton localization centers induced by facet-dependent indium incorporation and composition fluctuation. More details on the NW growth and QD formation mechanism are given elsewhere [2,3].

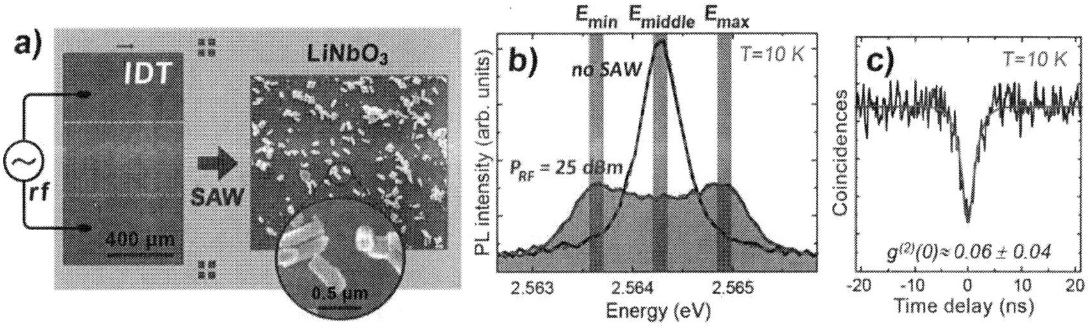

Figure 1. (a) Layout of the SAW-chip formed by IDTs patterned on the surface of a $LiNbO_3$ crystal containing GaN/InGaN NWs dispersed on the SAW propagation path. Inset: Magnified view of the optically probed NWs. (b) Time-integrated micro-PL spectrum of a single NW QD without (unshaded trace) and with (shaded trace) SAW applied. The rectangles labeled E_{min}, E_{middle} and E_{max} marke the spectral detection windows used in Fig. 2(a) (see text). (c) Autocorrelation histogram of the unperturbed (i.e. no SAW) emission peak in (b).

The effect of the SAW on the optical emission from NW QDs formed on the topmost polar *c*-facet of the InGaN nanoshell was characterized by low-temperature (T~10 K) spatially, time- and polarization-resolved micro-PL spectroscopy. The experiments were performed under direct optical excitation of single dispersed NWs. To monitor the acousto-mechanic and acousto-electric effects on the probed NW QD emission dynamics we employ a stroboscopic micro-PL technique with either time-integrated or time-resolved detection [1,4]. The photon emission statistics was analyzed using a Hanbury-Brown and Twiss (HBT) setup, which delivers a histogram of the photon coincidences as a function of the arrival time delay at the two HBT detectors [1]. When triggered by the laser pulses, the same setup also yields time-resolved PL traces.

3. Results and discussions

Figure 1(b) presents low-temperature micro-PL spectrum of one of the optically excited NW QDs exhibiting strong electromechanically coupling to the propagating SAW [5]. In the absence of a SAW, the spectrum (black trace) recorded under continuous wave (cw) optical excitation at 442 nm shows a sharp emission line due to the ground-state exciton (X) transition [1]. Its autocorrelation histogram in Fig. 1(c) reveals a pronounced signature of single photon emission: an anti-bunching dip at zero-time delay (τ=0) with the background-corrected $g^{(2)}(0)$ value of 0.06±0.04 [1]. When the SAW is applied, it couples to the probed NW QD through the deformation potential mechanism and, consequently, periodically modulates the X's emission energy [4]. Averaged over time, this modulation gives rise to an apparent splitting of the X emission line (cf. Fig. 1(b), red trace). Due to the time-integrated detection and the unlocked excitation scheme (i.e. cw excitation), the recorded data provides an averaged picture of the SAW-governed carrier recombination. In this way, we can assess the full bandwidth of the SAW-driven periodic spectral tuning of the QD optical transitions, which for the highest SAW intensity at P_{rf} =+25 dBm reaches a maximum value of ΔE =1.55 meV [4,5].

By collecting the photons within a narrow energy window (cf. Fig. 2(a)), the SAW-driven spectral tuning forces the X emission energy to move in and out the selected detection range and, consequently, changes in the photon count rate are detected at one of the HBT detectors within each SAW cycle. As seen in Fig. 2(a), light collected at the highest (E_{max}) and lowest (E_{min}) energy of the SAW-induced dynamic spectral modulation is of opposite phase and is emitted at the SAW frequency. In contrast, at the central wavelength (E_{middle}) the emission occurs at twice the acoustic frequency as the X peak moves past this energy twice per acoustic cycle. This spectral detection filtering enables the photon output to be clocked at the acoustic frequencies and can thus be employed to control the photon emission time. Such acoustically regulated control of the photon emission time provides an efficient mechanism for deterministic on demand generation of single photons without the need for a pulsed laser. By increasing the SAW frequency, high repetition rate of the emitted photons can be achieved.

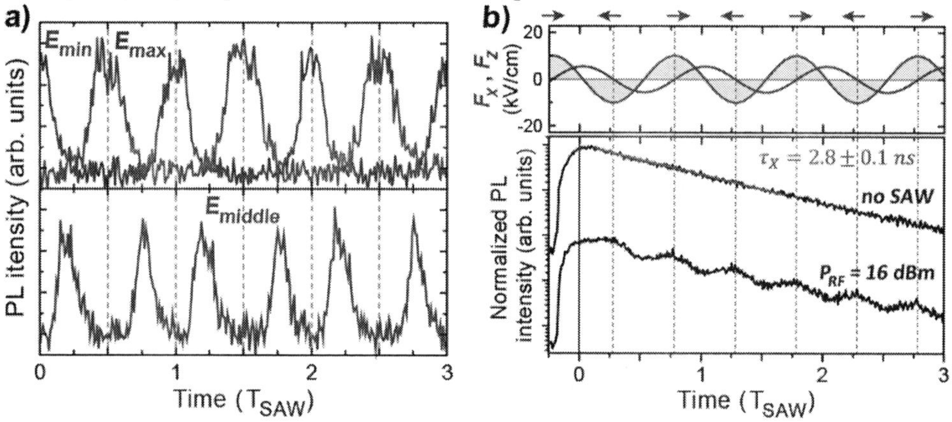

Figure 2. (a) Time-resolved PL intensity of the probed NW QD emission recorded under stroboscopic detection at the spectral positions marked by the corresponding rectangles labeled E_{min}, E_{middle} and E_{max} in Fig. 1(b). (b) Lower panels: PL decay curves of the NW QD emission with no SAW applied (upper trace, showing an exciton lifetime of τ_X =2.8±0.1 ns) and under SAW excitation with P_{rf} =+16 dBm (lower trace). The traces are vertically offset for clarity. Upper panels: time dependence of the SAW longitudinal F_X (red shaded curves) and transverse F_Z (blue curves) electric field components at the position of the probed NW QD.

To better understand the underlying dynamics of the excitonic transition in the acoustically modulated QD confinement potential, we registered the time evolution of its PL under SAW-synchronized pulsed optical excitation and time- and spectrally resolved detection conditions. The lower panel in Fig. 2(b) presents the temporal decay at a moderate acoustic power of the PL, collected over the entire spectral modulation bandwidth of the SAW-split X line in Fig. 1(b). The measured time-resolved PL profiles reveal the dependence of the SAW-mediated QD emission dynamics as a function of the local phase of the SAW. Contrary to the unperturbed PL transients (upper trace in lower panel of Fig. 2(b)), the stroboscopic time-resolved measurement shows weak oscillations in the PL emission intensity during the decay. For a sufficiently slow decay of the order of the acoustic period (i.e. $\tau_X \approx T_{SAW}$), these PL intensity oscillations appear twice per SAW cycle, yielding multiple characteristic signatures in the X

PL time-transient. These increased PL intensity signatures coincide with the acoustic phases marked with vertical dashed lines and horizontal red arrows in Fig. 2(b). As seen from the numerical modelling (see Ref. [4] for more details) of the temporal evolution of the longitudinal F_X (i.e. along SAW propagation direction) and transverse F_Z (i.e. perpendicular to the substrate surface) components of the SAW-induced electric field for P_{rf} =+16 dBm, these marked SAW phases correspond to the maximum and minimum values of F_X. Considering that the studied QD is, essentially, a potential trap induced by indium content fluctuations within the topmost polar InGaN nanoshell section, it acts as an optically active deep trap for electrons and holes photo-excited in the InGaN region. We, thus, relate the observed PL intensity oscillations to the photo-generation in a continuum of states of the InGaN nanoshell region surrounding the QD, where the SAW's F_X induces, at phases corresponding to its maximum magnitude, simultaneous transfer of photo-excited electrons and holes into the energetically lower QD states [4]. The transverse electric field component F_Z, on the other hand, has negligible effect on the charge population of the SAW-modulated QD potential owing to its small contribution compared to the strong built-in electric field acting along the axis of wurtzite III-nitride NWs [5].

The effect of the SAW piezoelectric field on the polarization degree of the QD optical emission is also observed and is presented by the substantial reduction (up to 30%, not shown) of the initially high linear polarization degree (~80%) with increasing the SAW intensity. For our NW QDs, previous measurements on the same batch as the one studied here have shown that the SAW fields can also tune the biexciton binding energy [5]. As the biexciton binding energy is determined by electron-hole Coulomb interactions in the QD, this shows that the SAW fields can as well affect various aspects of the QD band structure, such as band energy separation or mixing. We, hence, attribute the observed behavior to the SAW-induced change in the QD valence band structure (unpublished results). In fact, a significant change in the linear polarization degree has been predicted in Ref. [6] as a function of the ratio between the in-plane QD anisotropy parameter (Ξ_{6a}) and the in-plane component of the spin-orbit interaction (Δ_2). Actually, in wurtzite III-nitride QDs, as a result of the small energy separation between the two uppermost valence bands (i.e. small Δ_2), even a small in-plane anisotropy is able to produce large values of the polarization degree. In addition, it has been shown that the spin-orbit coupling can be effectively manipulated in InAs QDs by an external electrical bias [7]. Thus, for the underlying mechanism of the observed reduction in the photon polarization degree we propose the modification of the spin-orbit coupling by the SAW piezoelectric field.

In summary, we have demonstrated the SAW-induced fine-tuning of the III-nitride NW QDs emission energy and combined it with the spectral detection filtering to control the photon output timing. We have also shown that the SAW piezoelectric field can be employed to control the NW QD charge population and its optical polarization degree. The latter is particularly relevant for quantum communication and information processing applications for which real-time control of the photon polarization state is required.

References

[1] S. Lazić, A. Hernández-Mínguez and P. V. Santos, Semicond. Sci. Technol. **32**, 084002 (2017).

[2] Ž. Gačević, N. Vukmirović, N. García-Lepetit, A. Torres-Pardo, M. Müller, S. Metzner, S. Albert, A. Bengoechea-Encabo, F. Bertram, P. Veit, J. Christen, J. M. González-Calbet and E. Calleja, Phys. Rev. B **93**, 125436 (2016).

[3] Ž. Gačević, M. Holmes, E. Chernysheva, M. Müller, A. Torres-Pardo, P. Veit, F. Bertram, J. Christen, J. M. González Calbet, Y. Arakawa, E. Calleja and S. Lazić, ACS Photonics **4**, 657 (2017).

[4] S. Lazić , E. Chernysheva, A. Hernández-Mínguez, P. V. Santos and H. P. van der Meulen, J. Phys. D: Appl. Phys. **51**, 104001 (2018).

[5] S. Lazić, E. Chernysheva, Ž. Gačević, H. P. van der Meulen, E. Calleja, and J. M. Calleja Pardo, AIP Advances **5**, 097217 (2015).

[6] R. Bardoux, T. Guillet, B. Gil, P. Lefebvre, T. Bretagnon, T. Taliercio, S. Rousset and F. Semond, Phys. Rev. B: Condens. Matter Mater. Phys. **77**, 235315 (2008).

[7] Y. Kanai, R. S. Deacon, S. Takahashi, A. Oiwa, K. Yoshida, K. Shibata, K. Hirakawa, Y. Tokura and S. Tarucha, Nature Nanotech. **6**, 515 (2011).

Novel Optimized Hybrid Terahertz Photoconductive Antennas

S.I. Lepeshov[1], A.A. Gorodetsky[1,2,4], N.A. Toropov[1], T.A. Vartanyan[1], A.E. Krasnok[3], P.A. Belov[1], and E.U. Rafailov[4]
[1]ITMO University, St. Petersburg, Russia
[2]Department of Chemistry, Imperial College London, London, UK
[3]Department of Electrical and Computer Engineering, The University of Texas at Austin, Austin, USA
[4]Aston Institute of Photonic Technologies, Aston University, Birmingham, UK

Abstract. We demonstrate the results of on THz emission from hybrid THz photoconductive antenna loaded with silver nanoantenna. The results of experimental investigations are in a good agreement with numerical simulations presented in our recent work. The conversion efficiency reveals over 5-fold improvement at certain frequencies, if compared with similar photoconductive antenna without silver nanoparticles, while previous results for this type of antenna barely exceeded 2-fold conversion efficiency gain. We propose a cost-effective fabrication procedure to realize such hybrid THz antennas with optimized plasmonic nanostructures via thermal dewetting process, which does not require any post processing and makes the proposed solution very attractive for applications.

1. Introduction

Terahertz photoconductive antennas (PCAs) are one of the most promising sources of pulsed THz radiation [1,2]. The principles of THz PCAs operation are based on the effect of ultrafast variations of surface photoconductivity of a semiconductor substrate under fs-laser irradiation: after exciting the gap between the electrodes with a fs-laser, the concentration of charge carriers increases sharply for a short period of time, and THz pulse generation occurs. Since the generated signal is a short single pulse hundreds of fs in duration, its spectrum spreads over several octaves in THz frequency range. The broad bandwidth is ideal for spectroscopic investigations of organic molecules [3], material science [4] as well as wireless THz transmitters and receivers [5]. However, conventional THz PCAs have a rather low conversion efficiency that prevents data transmission over long distances, they do not provide a large signal-to-noise ratio, and require ultrafast laser pump at wavelengths around λ = 800 nm, usually provided by expensive and bulky Ti:Sapphire lasers. The low conversion efficiency is mainly related to two predominant factors: the photocarrier screening effect [6], and the low absorption coefficient of the surface layer of the semiconductor substrate. To overcome these limitations, optical nanoantennas have been proposed to be placed in the gap of THz PCA [7,8]. The nanoantenna-based PCAs have been called hybrid terahertz-optical PCAs. It has been shown that this solution provides high fs-laser pump absorption, shorter photocarrier lifetimes and excellent thermal efficiency. Despite the fact that many nanoantenna designs have been studied for enhancement of the THz generation from PCA, an optimized geometry has not yet been proposed.

Here, we present the results of our recent experimental investigations of a log-periodic THz photoconductive antenna coupled to a silver (Ag) nanoantenna array arranged in the gap of a THz antenna [9]. We obtain an enhancement of optical-to-THz conversion efficiency 2-fold larger in comparison with previously reported results [10]. As a byproduct, we propose a cost-effective fabrication procedure allowing to produce such hybrid THz antennas with optimized plasmonic nanostructures via thermal dewetting process.

2. Sample fabrication

First, we fabricate the hybrid PCAs with optimized Ag nanoantenna design proposed in Ref. [9]. The PCAs were fabricated on semi-insulating GaAs substrates containing self-assembled InAs quantum-dots (QDs). Such QD based antennas have been recently demonstrated to generate effectively both pulsed and CW THz radiation [11,12]. Since the 800 nm wavelength pump is used, and the carriers are generated in the whole volume of the GaAs matrix, the dots serve only as carrier lifetime shorteners, in a similar way as defects in low temperature grown GaAs [11]. The log-periodic shape of electrodes has been selected for the realization of the THz antenna due to the fact that such design provides a broadband radiative spectrum. The latter circumstance allows studying the THz antenna operation in the wider-frequency range. The log-periodic antenna electrodes have been deposited onto the QD-based substrate by UV optical lithography. The fabricated THz antennas have a smallest gap of 8 μm, and overall diameter of 1.8 mm. Then, a 20-nm silver film has been deposited onto the substrate surface of one of the fabricated antennas. Upon thermal dewetting process caused by heating, the silver film has been transformed into a disordered array of spheroid nanoparticles. The size of the resulting nanoparticles depends on the thickness of the silver film and can further be changed by laser pump [13]. The sizes of the fabricated nanoparticles proximately correspond to calculated optimal ones and the average distance between nanoparticles is 280 nm. The dashed oval in Fig. 1(a) shows the location of the resulting nanoantennas on the radius-center distance map. The SEM pictures of the produced hybrid THz antenna are shown in Fig. 1(a).

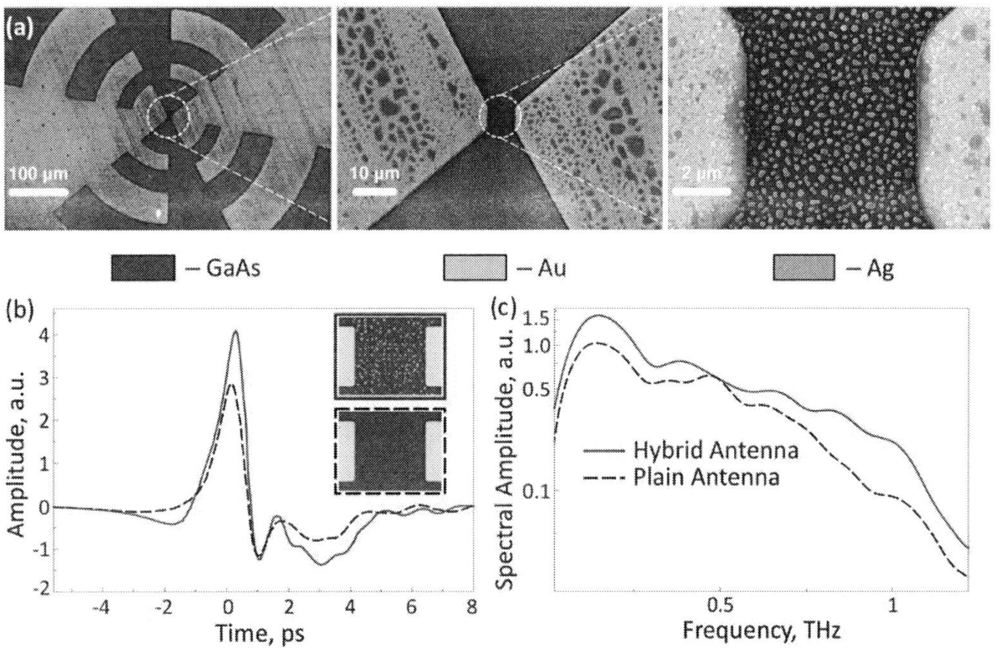

Figure 1. (a) The typical SEM images of the fabricated log-periodic THz PCA, its central part and its gap filled with silver nanoantennas. (b) Time domain THz field profiles and (b) their corresponding spectral amplitudes measured for hybrid (red) and plain (black) PCAs.

3. Experimental results

Then, we perform experimental investigations of hybrid PCAs with optimized Ag nanoantennas and compare the obtained results with previous studies. For experimental investigation, a standard THz time-domain spectroscopic (TDS) system has been used. THz-TDS is pumped with Sprite-XT (M Squared Ltd.) femtosecond Ti:sapphire laser that delivers pulses of 120 fs duration at 80 MHz repetition rate with central wavelength of 800 nm. As a THz detector, a LT-GaAs photoconductive antenna made by Teravil Ltd. was used, and the beam was guided between the transmitter and detector by two off-axis parabolic mirrors. To estimate the effect of silver nanoantennas, signals from standard and nanoantenna-enhanced THz antennas were measured for similar pumping and bias conditions. The experimental results are summarized in Fig. 1(b,c).

Fig. 1(b) shows the time domain THz signals generated with the log-periodic PCA with and without Ag spheroidal nanoantennas. The corresponding amplitude spectra, obtained using the Fourier transform, are shown in Fig. 1(c). It can be seen that the antenna demonstrates an uneven enhancement across the spectrum. The enhancement reaches its maximum around 1 THz, and the enhancement value corresponds to the 5-fold theoretically derived value of the power absorption enhancement. Lower effect at other frequencies and negligible enhancement at 0.5 THz can be associated with the change in the THz antenna impedance induced by highly conductive silver in the gap [9]. The results for THz PCAs with nanoantennas demonstrate over 5-fold increase in comparison with the case of nanoantenna absent. This is 3 times higher than the results that have been previously reported [10].

4. Conclusion

In conclusion, we have demonstrated an unprecedented enhancement of the photoconductive antenna operation efficiency by optimized spheroidal Ag plasmonic nanoantennas. The resulting hybrid PCA demonstrates over 5-fold increase in the generated THz signal around 1 THz, and over 2-fold increase in overall generated THz power. Moreover, we have proposed the cost-effective fabrication procedure to realize such hybrid THz antennas with optimized plasmonic nanostructures via thermal dewetting process, which does not require any post processing and makes the proposed solution very attractive for applications. We believe that our results may be useful for many relevant applications, requiring compact and effective room-temperature THz sources, including spectroscopy, biological sensing, security imaging, and ultrafast data transmission.

Acknowledgments
This work was supported by Russian Foundation for Basic Research (Projects 16-07-01166a and 18-07-01492a).

References

[1] Auston D H, Cheung K P and Smith P R Picosecond photoconducting hertzian dipoles 1984 *Appl. Phys. Lett.* **45** 284–586.

[2] Jepsen P, Cooke D and Koch M Terahertz spectroscopy and imaging – Modern techniques and applications 2011 *Laser&Photonics Reviews* **5** 124-166

[3] Fischer B, Walther M and Jepsen P Far-infrared vibrational modes of DNA components studied by terahertz time-domain spectroscopy 2002 *Physics in medicine and biology* **47** 3807.

[4] Grischkowsky D, Keiding S, Van Exter M and Fattinger C Far-infrared time-domain spectroscopy with terahertz beams of dielectrics and semiconductors 1990 *JOSA B* **7** 2006-2015.

[5] Akyildiz I, Jornet J and Han C Terahertz band: Next frontier for wireless communications 2014 *Physical Communication* **12** 16-32

[6] Loata G, Thomson M, Löffler T and Roskos H Radiation field screening in photoconductive antennae studied via pulsed terahertz emission spectroscopy 2007 *Applied Physics Letters* **91** 232506.

[7] Krasnok A, Maksymov I, Denisyuk A, Belov P, Miroshnichenko A, Simovskii C and Kivshar Yu Optical nanoantennas 2013 *Phys. Usp.* **56** 539-564

[8] Lepeshov S, Gorodetsky A, Krasnok A, Rafailov E and Belov P Enhancement of terahertz photoconductive antenna operation by optical nanoantennas 2016 *Laser&Photonics Reviews* **11** 1600199

[9] Lepeshov S, Gorodetsky A, Krasnok A, Toropov N, Vartanyan T A, Belov P, Alú A and Rafailov E U Boosting Terahertz Photoconductive Antenna Performance with Optimised Plasmonic Nanostructures 2018 *Scientific Reports* **8** 6624

[10] Park S, Choi Y, Oh Y and Jeong K Terahertz photoconductive antenna with metal nanoislands 2012 *Optics Express* **20** 25530-25535.

[11] Leyman R, Gorodetsky A, Bazieva N, Molis G, Krotkus A, Clarke E and Rafailov E Quantum dot materials for terahertz generation applications 2016 *Laser&Photonics Reviews* **10** 772-779

[12] Fedorova K, Gorodetsky A and Rafailov E Compact All-Quantum-Dot-Based Tunable THz Laser Source 2017 *IEEE Journal of Selected Topics in Quantum Electronics* **23** 1-5
[13] Toropov N, Gladskikh I, Parfenov P and Vartanyan T Fabrication and laser-assisted modification of the Ag particles ensembles supporting quadrupole plasmon oscillations 2017 *Optical and Quantum Electronics* **49** 154

Strong Coupling in Si Nanoparticle Core - 2D WS_2 Shell Structure

S. Lepeshov[1], A. Krasnok[2], O. Kotov[3], and A. Alu[2]
[1] ITMO University, St. Petersburg, Russia
[2] Department of Electrical and Computer Engineering, The University of Texas at Austin, TX 78712, USA
[3] N.L. Dukhov Research Institute of Automatics, Moscow 127055, Russia

Abstract. Two-dimensional semiconducting transition metal dichalcogenides (TMDCs) are attractive materials for optoelectronics. In this paper, a strong coupling regime in Si nanoparticle covered by single layer WS_2 system at the magnetic optical Mie resonance is observed. This regime is achieved owing to the symmetry of the magnetic dipole Mie mode, polarized along the TMDC surface, and low dissipative losses of the Si nanoparticle. A Rabi splitting energy exceeding 110 meV is obtained.

1. Introduction

Single-layer transition metal dichalcogenides (1L-TMDCs) are an emerging class of 2D semiconductors that enable attractive optoelectronic applications, such as photodetection and light harvesting [1-3], phototransistors and modulation [4,5], light-emitting diodes and lasers [6-8]. 1L-TMDC is formed by a hexagonal layer of transition metal atoms (Mo, W) hosted between two hexagonal lattices of chalcogenide atoms (S, Se). In the atomic monolayer limit, transition metal dichalcogenides are particularly interesting, because their bandgaps become direct enabling enhanced interaction with light [9]. However, for practical applications, it is necessary to further increase the coupling strength between elementary excitations of 1L-TMDC (excitons and trions) and light. For this purpose, a variety of optical structures and devices has been proposed and studied [10]. One of the types of such structures providing enhanced photon-exciton interaction at nanoscale are plasmonic nanoparticles and nanoantennas. These nanoscale objects made of noble metals (gold, silver) along with 1L-TMDC enable the strong coupling regime [11], pronounced Fano resonance [12] and plasmon-exciton resonance energy transfer [13]. But the typical plasmonic materials, gold and silver, have finite conductivities at optical frequencies leading to the inherent dissipation of electromagnetic energy. For many applications, heat generation at the nanostructure is detrimental, since the behaviour of the excitonic system can be modified dramatically with temperature. Recently developed all-dielectric nanoantennas made of high-index materials (Si, GaP, GaAs) with insignificant dissipation losses in visible range can be more efficient alternatives their plasmonic counterparts in the TMDC-nanoantenna composites [14]. Dielectric nanoantennas support not only electric but also magnetic Mie resonances providing control of both electric and magnetic light components [15].

Here, we present our recent studies [16] of the resonant coupling in a single Si nanoparticle-monolayer WS2 core-shell nanostructure. We predict a strong coupling regime for Si nanoparticles (SiNPs) covered by 1L-WS_2 at the magnetic optical Mie resonance with a Rabi splitting energy exceeding 110 meV. The strong coupling regime is achieved owing to the symmetry of the magnetic Mie mode, which allows a strong interaction with atomically thin materials.

2. Results and discussion

A core-shell spherical particle with a Si core and a monolayer 1L-WS_2 shell is a simple model configuration that can be analysed analytically. First, by using Mie theory we calculate separately the scattering cross sections of a SiNP and a 1L-WS_2 empty shell with atomic thickness (h ≈0.7 nm). The dielectric permittivities of c-Si and 1L-WS_2 have been taken from Refs. [17,18] for ambient temperature T=300 K. For SiNPs, we obtain the strong electric (ED) and magnetic (MD) dipole and weak magnetic quadrupole (MQ) resonances (Figure 1(a)) whose positions depend on the particle radius. By choosing the radius of SiNP to be R=75 nm, we obtain the magnetic dipole resonance at λ≈615 nm (~2 eV), at which the electrically neutral A exciton (X^0) resonance in WS_2 monolayers takes place. The scattering cross section of the WS_2 shell is two orders of magnitude less than that of SiNP of same radius (Figure 1(b)), but it demonstrates a sharp peak (X^0), corresponding to the A exciton in 1L-WS_2.

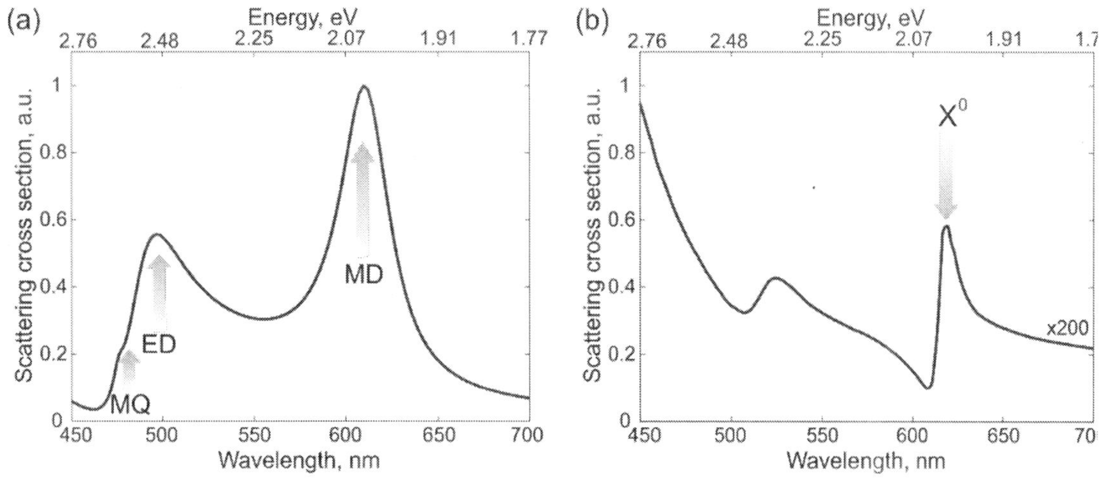

Figure 1. Scattering cross sections of (a) Si nanoparticle of radius R = 75 nm and (b) WS_2 shell with thickness of h = 0.7 nm.

Then, we use Mie theory generalized for core-shell spherical particles [19] for the calculation of the scattering cross section of the SiNP-WS_2 core-shell system. We use the parameters fitted earlier to achieve the maximum overlap of the magnetic dipole Mie resonance of SiNP and the X^0 exciton peak of the 1L-WS_2 monolayer. Interaction of these two resonances gives a strong Fano-like resonance at λ≈615 nm, Figure 2(a), for which the coupling constant g can be extracted using coupled mode theory (CMT). In the harmonic oscillator approximation, the scattering cross section of our system with two coupled oscillators is given by:

$$\sigma_{\mathrm{scat}} \propto \left| \frac{\omega^2 \tilde{\omega}_{\mathrm{ex}}^2}{\tilde{\omega}_{\mathrm{md}}^2 \tilde{\omega}_{\mathrm{ex}}^2 - g^2 \omega^2} \right|^2, \tag{1}$$

where $\tilde{\omega}_{\mathrm{ex}}^2 = \omega^2 - \omega_{\mathrm{ex}}^2 + i\gamma_{\mathrm{ex}}\omega$ and $\tilde{\omega}_{\mathrm{md}}^2 = \omega^2 - \omega_{\mathrm{md}}^2 + i\gamma_{\mathrm{md}}\omega$ are the harmonic oscillator terms for the exciton and magnetic dipole resonances, respectively, with ω_{ex}, ω_{md} and γ_{ex}, γ_{md} being the corresponding resonance frequencies and dissipation rates (full width at the half maximum). After fitting the curve shown in Figure 1(c), we obtain that the coupling constant $g \approx 58$ meV, whereas the dissipation rates equal $\gamma_{\mathrm{ex}} \approx 33$ meV and $\gamma_{\mathrm{md}} \approx 84$ meV.

It is known that in order to achieve strong coupling, the system should satisfy the criterion $\hbar\Omega > (\hbar\gamma_{\mathrm{ex}} + \hbar\gamma_{\mathrm{md}})/2$, where $\hbar\Omega$ is the Rabi splitting energy. The coupling strength and Rabi splitting energy are related to each other via $g = \hbar\Omega / 2$. The total loss $\hbar\gamma_{ex} + \hbar\gamma_{md}$ is 117 meV, whereas the Rabi splitting energy $\hbar\Omega$ equals 116 meV, which satisfies the criterion $\hbar\Omega > (\hbar\gamma_{ex} + \hbar\gamma_{md}) / 2$.

Figure 2(b) shows the scattering cross section of the SiNP-WS_2 core-shell nanostructure as a function of wavelength and SiNP radius. These results confirm the expected anti-crossing behaviour of scattering at the intersection of excitonic (X^0) and MD resonances with high-energy and low-energy polariton branches.

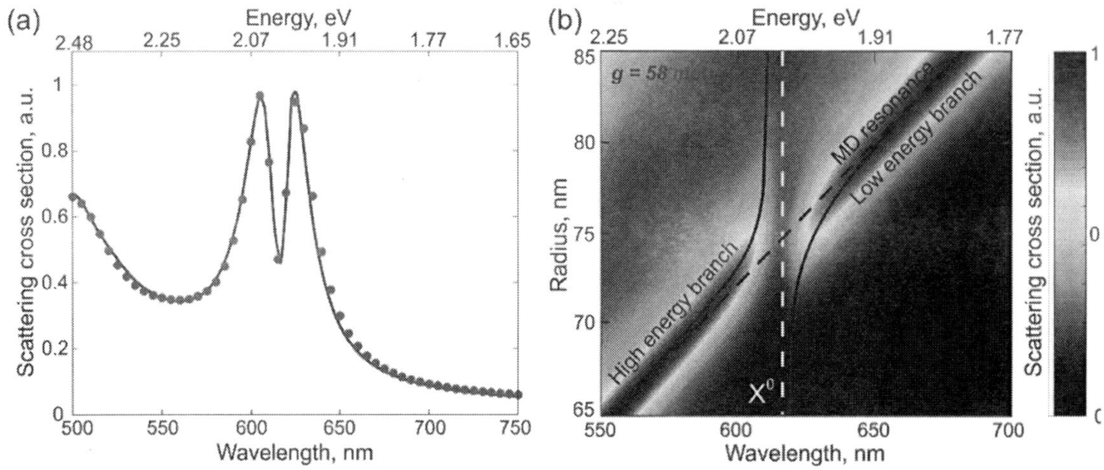

Figure 2. (a) Scattering cross sections of Si-WS_2 core-shell system with SiNP R = 75 nm in air. Red dots show the fitting by coupled mode theory (CMT). (b) Scattering cross section of Si-WS_2 core-shell system as a function on wavelength and SiNP radius.

3. Conclusion

We have addressed the issue of resonance coupling in single Si nanoparticle-monolayer TMDC (WS_2) structures. We have theoretically predicted a strong coupling regime with a Rabi splitting energy exceeding 110 meV for a Si NPs covered by 1L-WS_2 at the magnetic optical Mie resonance due to the symmetry of the mode and low dissipative losses. Our findings pave the way to a novel class of low-loss nanophotonic, optoelectronic and quantum optical devices based on dielectric resonant nanoparticles coupled with 2D excitonic materials.

Acknowledgments

This work was supported by Russian Foundation for Basic Research (Project 18-37-00336).

4. References

[1] Lopez-Sanchez O, Lembke D, Kayci M, Radenovic A, Kis A, Ultrasensitive Photodetectors Based on Monolayer MoS2 2013 *Nat. Nanotechnol.* **8** 497–501

[2] Britnell L, Ribeiro R M, Eckmann A, Jalil R, Belle B D, Mishchenko A, Kim Y-J, Gorbachev R V, Georgiou T, Morozov S V, Grigorenko A N, Geim A K, Casiraghi C, Neto A H C, Novoselov K S, Strong Light-Matter Interactions in Heterostructures of Atomically Thin Films 2013 *Science* **340** 1311–4

[3] Perea-Lõpez N, Elías A L, Berkdemir A, Castro-Beltran A, Gutiérrez H R, Feng S, Lv, R, Hayashi T, Lõpez-Urías F, Ghosh S, Muchharla B, Talapatra S, Terrones H, Terrones M, Photosensor Device Based on Few-Layered WS2 Films 2013 *Adv. Funct. Mater* **23** 5511–7

[4] Yin Z, Li H H, Jiang L, Shi Y, Sun Y, Lu G, Zhang Q, Chen X, Zhang H, Single-Layer MoS2 Phototransistors 2012 *ACS Nano* **6** 74–80

[5] Sun Z, Martinez A, Wang F, Optical Modulators with 2D Layered Materials 2016 *Nat. Photonics* **10** 227–38

[6] Withers F, Del Pozo-Zamudio O, Mishchenko A, Rooney A P, Gholinia A, Watanabe K, Taniguchi T, Haigh S J, Geim A K, Tartakovskii A I, Novoselov K S, Light-Emitting Diodes by Band-Structure Engineering in van Der Waals Heterostructures 2015 *Nat. Mater.* **14** 301–6

[7] Liu C H, Clark G, Fryett T, Wu S, Zheng J, Hatami F, Xu X, Majumdar A, Nanocavity Integrated van Der Waals Heterostructure Light-Emitting Tunneling Diode 2017 *Nano Lett.* **17** 200–5

[8] Ye Y, Wong Z J, Lu X, Ni X, Zhu H, Chen X, Wang Y, Zhang X, Monolayer Excitonic Laser 2015 *Nat. Photonics* **9** 733–7

[9] Zhao W, Ghorannevis Z, Chu L, Toh M, Kloc C, Tan P-H, Eda G, Evolution of Electronic Structure in Atomically Thin Sheets of WS2 and WSe2 2013 *ACS Nano* **7** 791–7

[10] Krasnok A, Lepeshov S, Alú A, Nanophotonics with 2D Transition Metal Dichalcogenides 2018 *arXiv preprint:1801.00698*

[11] Baranov D G, Wersäll M, Cuadra J, Antosiewicz T J, Shegai T, Novel Nanostructures and Materials for Strong Light–Matter Interactions 2018 *ACS Photonics* **5** 24–42

[12] Abid I, Chen W, Yuan J, Bohloul A, Najmaei S, Avendano C, Péchou R, Mlayah A, Lou J, Temperature-Dependent Plasmon-Exciton Interactions in Hybrid Au/MoSe2 Nanostructures 2017 *ACS Photonics* **4** 1653–60

[13] Wang M, Li W, Scarabelli L, Rajeeva B B, Terrones M, Liz-Marzán L M, Akinwande D, Zheng Y, Plasmon–trion and Plasmon–exciton Resonance Energy Transfer from a Single Plasmonic Nanoparticle to Monolayer MoS2 2017 *Nanoscale* **9** 13947–55

[14] Baranov D G, Zuev D A, Lepeshov S I, Kotov O V, Krasnok A E, Evlyukhin A B, Chichkov B N, All-Dielectric Nanophotonics: The Quest for Better Materials and Fabrication Techniques 2017 *Optica* **4** 814

[15] Evlyukhin A B, Novikov S M, Zywietz U, Eriksen R L, Reinhardt C, Bozhevolnyi S I, Chichkov B N, Demonstration of Magnetic Dipole Resonances of Dielectric Nanospheres in the Visible Region 2012 *Nano Lett.* **12** 3749–55

[16] Lepeshov S, Wang M, Krasnok A, Kotov O, Zhang T, Liu H, Jiang T, Korgel B, Terrones M, Zheng Y, Alú A, Tunable Resonance Coupling in Single Si Nanoparticle-Monolayer WS2 Structures 2018 *ACS Applied Materials & Interfaces* Just Accepted Manuscript DOI: 10.1021/acsami.7b17112

[17] Li Y, Chernikov A, Zhang X, Rigosi A, Hill H M, van der Zande A M, Chenet D A, Shih E-M, Hone J, Heinz T F, Measurement of the Optical Dielectric Function of Monolayer Transition-Metal Dichalcogenides: MoS2, MoSe2, WS2, and WSe2 2014 *Phys. Rev. B* **90** 205422

[18] Vuye G, Fisson S, Nguyen Van V, Wang Y, Rivory J, Abelès F, Temperature Dependence of the Dielectric Function of Silicon Using in Situ Spectroscopic Ellipsometry 1993 *Thin Solid Films* **233** 166–70

[19] Aden A L, Kerker M, Scattering of Electromagnetic Waves from Two Concentric Spheres 1951 *J. Appl. Phys.* **22** 1242–6.

IOP Conf. Series: Journal of Physics: Conf. Series **1092** (2018) 012078 doi:10.1088/1742-6596/1092/1/012078

Coherently enhanced wireless power transfer: theory and experiment

S. Li[1], **A. Krasnok**[1,2], **D. Baranov**[3,4], **A. Generalov**[5] **and A. Alù**[2]

[1] ITMO University, St. Petersburg 197101, Russia

[2] Department of Electrical and Computer Engineering, The University of Texas at Austin, Austin, Texas 78712, USA

[3] Department of Physics, Chalmers University of Technology, 412 96 Gothenburg, Sweden

[4] Moscow Institute of Physics and Technology, Dolgoprudny 141700, Russia

[5] Department of Electronics and Nanoengineering, Aalto University, 02150 Espoo, Finland

E-mail: `akrasnok@utexas.edu`

Abstract. Extraction of electromagnetic energy by an antenna to a coupled waveguide is the crucial part of wireless power transfer. Efficiency of this process is usually defined by the coupling strength between the antenna and the outcoupling waveguide or cable. We show that there is an additional possibility to improve the receiving efficiency by coherent excitation of the outcoupling waveguide by a backward propagating guided mode with a specific amplitude and phase. This additional wave creates a special interference picture in the system and result in increased amount of energy extracted to the waveguide from the free space. We develop a simple analytical model predicting this effect, demonstrate it in numerical FDTD simulations, and verify in microwave experiment.

1. Introduction

An antenna is one of the most important elements in wireless technology, including communications and power transfer [1]. The first antennas were invented at the time of the discovery of electromagnetic waves by H. Hertz in 1888 and, since then, this technology has been progressing continuously. Nowadays, various types of antennas are commonly used for radio, microwave, THz and optical frequencies, where they have become irreplaceable elements for quantum optics and interconnections on a chip [2, 3]. While wireless communications are well established, wireless power transfer (WPT), proposed in the beginning of 20th century by N. Tesla [4], has been experiencing a rebirth in recent years, caused by demonstrations that the WPT efficiency, i.e., the ratio of energy received by an antenna over the total amount of emitted energy, can be drastically enhanced in the so-called near-field WPT regime, when the power is transferred via resonant coupling [5].

In this work, we introduce a conceptually different approach to realize robust WPT systems, relying on the local control of the wave impedance offered by interference phenomena. Specifically, we demonstrate that there is a possibility to improve the receiving efficiency of an antenna by *coherent excitation* of the outcoupling waveguide with a backward propagating signal of specific amplitude and phase, Fig. 1. This signal creates a specific interference pattern in the system that results in optimal wave impedance at the feed location, maximizing the energy transferred to the receiving antenna from free space. We develop an illustrative analytical model

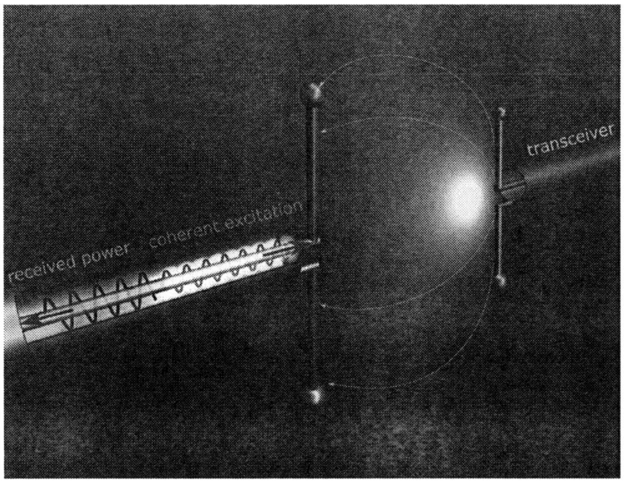

Figure 1. A schematic of coherently enhanced wireless energy transfer. The incident radiation of transceiver excites the antenna, which couples to waveguide creating a forward-propagating waveguide mode. Coherent excitation of the outcoupling waveguide with a backward propagating guided mode with specific amplitude and phase can retune the matching condition and largely improve the WPT performance.

predicting this effect, demonstrate it in full-wave numerical simulations, and in a microwave experiment.

2. Theoretical model

To demonstrate the effect of enhancing WPT's efficiency, we develop a theoretical model on the basis of temporal coupled mode theory [6, 7], which can be applied for both near-field and far-field WPT systems. The system of a waveguide-coupled antenna is schematically presented in Fig. 1. The dipole antenna is excited by an external field s_F created by a transceiver, which can be either free-propagating or evanescent. The antenna couples to a waveguide with amplitude $s_{\rm w}^-$, which carries the out-coupled energy to the receiver. To model the structure, we will assume that the antenna has a single resonance at the operating frequency ω_0. Moreover, for the sake of simplicity, we reduce the direct scattering matrix,which reflects the direct pathway between input and output channels, to a scalar c, assuming that the coupling of the antenna with free-space modes leading to additional radiation losses is taken into account by τ_F. The resulting amplitude of reflected mode that carries the transferred energy can be calculated via following equation:

$$s_{\rm w}^- = c s_{\rm w}^+ + \kappa_{\rm w} \frac{\kappa_{\rm w} s_{\rm w}^+ + \kappa_F s_F}{i(\omega - \omega_0) + 1/\tau}. \quad (1)$$

where $s_{\rm w}^+$ is the amplitude of the input mode, τ is the mode damping time, $\kappa_{\rm w} = \sqrt{2/\tau_{\rm w}}$ and $\kappa_F = \sqrt{2/\tau_F}$, $\tau_{\rm w}$ is the antenna decay time into the waveguide mode, τ_F is the antenna decay time into radiation in free space. In order to study this dependence we plot the *energy balance* in transferred energy [Fig. 2(a)], defined as the total received energy $|s_{\rm w}^-|^2$ minus the energy spent in the auxiliary signal $|s_{\rm w}^+|^2$ that is sent from the waveguide port to the load, as a function of the auxiliary power $|s_{\rm w}^+|^2$ for different values of the relative phase $\varphi = Arg(s_{\rm w}^+/s_F)$. The shaded region marks all achievable values of the energy balance for phase from 0 to 2π, indicating that for a suitable combination of phase and amplitude it is possible to achieve perfectly matched regime, even after subtracting the energy carried by the back-propagating signal. In particular,

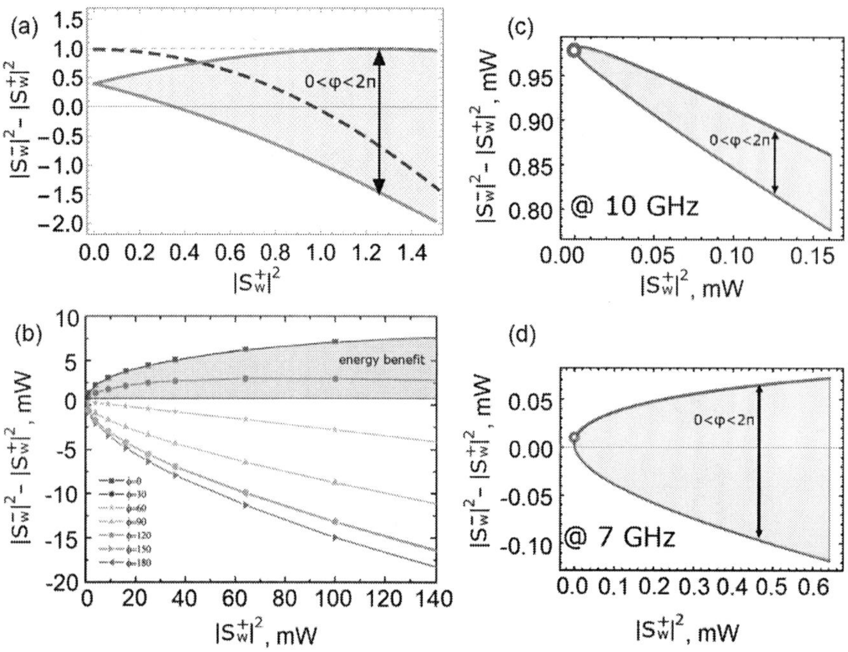

Figure 2. (a) Analytically calculated energy balance for a resonant matched (gray dashed curve) and resonant mismatched antenna (color solid curves). (b) Numerical calculations of energy balance as a function of the auxiliary signal intensity for different relative phase values. (c,d) The net extracted energy $|s_w^-|^2 - |s_w^+|^2$ vs power of the additional signal $|s_w^+|^2$ for two different frequencies. The shaded area corresponds to all values of the relative phase between the external (default or main) signal and the auxiliary mode. The power of 1 mW is fed to port 1 in both cases.

for a certain phase difference the amount of extracted power (purple curve) reaches the ideal value of a resonant critically coupled antenna.

3. Numerical simulations and experiment

To verify the predictions of the analytical model, we carried out FDTD simulations of coherently enhanced outcoupling by using the CST Microwace Studio. In simulation, a coaxial cable was coupled to a non-resonant dipole antenna with 5 cm arm-length, which is excited by microwave radiation from the open end of a rectangular waveguide at 1.36 GHz. A distance between the antenna and the rectangular waveguide is about 40 cm. The results obtained by numerical calculation are shown in Fig. 2(b) and confirm the theoretical predictions. We note that the relative phase denoted here as ϕ is calculated for the $\mathrm{Arg}(s_w^+)$ and $\mathrm{Arg}(s_F)$ taken at different points (at the dipole antenna center and at the rectangular waveguide end, respectively) and, hence, depends on their relative distance. We would like to note here that the value s_F we used is a constant. However, when the coupling between the transmitter and receiver is sufficient, the coherent wave s_w^+ may influence on the matching conditions of the system transmitting antenna-its transmission line. It would lead to decreasing of s_F and therefore to decreasing of s_w^- and WPT efficiency.

Finally, we experimentally verified this concept by using a microwave two-port system. Each coaxial cable is connected to a vector network analyzer and terminated with a waveguide to coaxial adapter (WCA) that transforms electrical currents to propagating electromagnetic radiation. The cutoff frequency of WCAs is around 8 GHz, which manifests itself in high reflection (s_{11}) at frequencies below 8 GHz. Thus, the WCA operates as a mismatched antenna for frequencies below 8 GHz and an almost matched antenna for frequencies above 8 GHz. The port 1 creates free space radiation s_F, a part of which (s_w^-) then receives by port 2. The port 2 is excited by the auxiliary signal (s_w^+) of variable amplitude and phase. The resulting s-parameter spectra allow to observe coherently enhanced energy transfer. The dependence of the net extracted energy $|s_w^-|^2 - |s_w^+|^2$ versus power of the additional excitation $|s_w^+|^2$ for the whole set of relative phases $0 < \phi < 2\pi$ presented in Fig. 2(c) and (d) highlights the different behavior of the system in various spectral regions. At 10 GHz, where reflection by each WCA back to the coaxial cable is very low, excitation with the auxiliary signal barely improves the energy transfer. At 7 GHz, however, the situation is strikingly different: high reflection by the WCA enables order of magnitude enhancement of the transferred power, Fig. 2(d). This enhancement becomes possible due to a small transmission from port 1 to port 2 even at 7 GHz, which can be increased by interference with the auxiliary signal.

4. Conclusion

In conclusion, we have shown that coherent signals sent from the receiving port of a WPT system can largely enhance and control the power transfer efficiency. This additional signal creates a tailored interference in the system, modifying the local wave impedance at the antenna load, thus enabling conjugate matching and critical coupling even if the antenna itself is largely mismatched, resulting in increased amount of energy extracted to the waveguide from free space. We have developed an illustrative analytical model predicting this effect and demonstrated it in numerical simulations and in a microwave experiment. Our approach of coherently enhanced WPT can be applied for the development of efficient wireless power transfer systems with robust operation in rapidly changing environments, as common in practical situations and setups.

5. Acknowledgements

We acknowledge the support from the Russian President's Scholarship, from the Grant from the President of the Russian Federation (MK-381.2017.2), and from Russian Foundation for Basic Research (RFBR) (16-37-60076 mol_a_dk).

References

[1] Balanis C A 1997 *Antenna Theory: Analysis and Design* (New York: John Wiley & Sons)
[2] Agio M and Alù A 2013 *Optical Antennas* (Cambridge: Cambridge University Press)
[3] Novotny L and Hecht B 2012 *Principles of Nano-Optics* (Cambridge: Cambridge University Press)
[4] Brown W C 1984 *IEEE Transactions on Microwave Theory and Techniques* **32** 1230
[5] Kurs A, Karalis A, Moffatt R, Joannopoulos J D, Fisher P and Soljacic M 2007 *Science* **317** 83
[6] Haus H 1984 *Waves and Fields in Optoelectronics* (New Jersey: Prentice Hall)
[7] S Fan W S and Joannopoulos J D 2003 *JOSA A* **20** 569

IOP Conf. Series: Journal of Physics: Conf. Series **1092** (2018) 012079 doi:10.1088/1742-6596/1092/1/012079

Numerical simulation of electron energy loss spectroscopy accounting for nonlocal effect in plasmonic nanoparticles

Ivan Lopushenko

PhD student, *Lomonosov Moscow State University*, Faculty of Physics, Chair of Mathematics, Leninskye Gory 1/2, Moscow, Russian Federation

E-mail: lopushenko.ivan@physics.msu.ru

Abstract. Interaction between an electron in straight uniform motion and a nearby plasmonic nanoparticle deposited in homogeneous medium is examined numerically in no-recoil approximation. To account for the nonlocal effect occurring in the plasmonic medium Generalized Nonlocal Optical Response scattering problem statement is considered. Discrete Source Method is employed to construct the solution. Electron energy loss probability is computed as a function of incident field frequency and obtained surface plasmon resonance energy is compared to local response simulations. It is shown that surface plasmon resonance peak frequency blueshift can be resolved within the proposed approach.

1. Introduction

An increasing demand in accurate characterization of solid microstructures, such as plasmonic nanoparticles (NPs), requires precise experimental and theoretical techniques. For this reason electron-beam spectroscopies, such as electron energy-loss spectroscopy (EELS), have recently gained much attention in nanophotonics research and have proven to be a powerful tool for studying surface plasmons (SPs) related to certain applications in sensing, waveguiding, plasmonic nanoantennas, optical trapping, and others [1, 2]. As experimental EELS setup requires very expensive instrumentation, numerical simulations seem to be a promising and cheap way to obtain EEL spectra for detailed analysis of specimens relevant parameters. In order to ensure simulation accuracy an adequate and reliable electromagnetic modeling tool is required.

2. Electron energy loss spectroscopy model

Let us consider the following scattering problem geometry (figure 1): an electron propagating in vacuum D_0 with velocity v passes a spherical homogeneous plasmonic particle D_i of radius r in the order of ten nanometers. We hereby assume that electron does not pass through the particle (absolute value of the impact parameter $\mathbf{b}$ is larger than r) and that it is represented by point charge e in uniform straight motion along z axis. Analytical expressions for the incident electromagnetic field due to given electron beam are available and can be efficiently used in the simulation procedures [3–5]. Moving point charge induces an electromagnetic field scattered back from the sample, which interacts with the electron causing characteristic energy losses.

Consequently, the scattered electromagnetic field has to be calculated in order to compute the energy loss probability P of the incident electron.

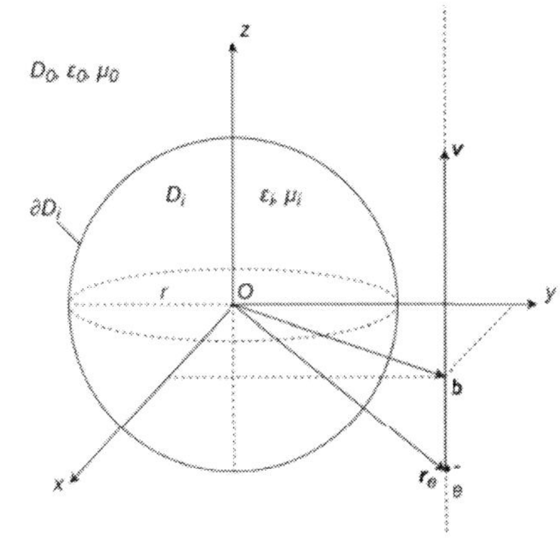

Figure 1. Mathematical model of the EELS configuration

Assuming that the electron energy loss is small compared to the initial kinetic energy of the incident electron (0.5eV - 50eV in the valence-loss region against initial 50keV -300keV), let us consider that the electron velocity v remains constant. This simplification is known as *no-recoil approximation*. Under the given circumstances a sufficient condition on electron loss probability P can be written in the form of expression that connects P to the scattered electromagnetic field $Esca$ [3–5]:

$$P(\omega) = \frac{e}{\pi\hbar\omega}\int_{-\infty}^{+\infty} Re\left\{\mathbf{E}_{sca}(\mathbf{r}_e,\omega)e^{-i\omega z/v}\right\} d\mathbf{r}_e. \qquad (1)$$

3. Generalized Nonlocal Optical Response

Numerical simulations of the electromagnetic scattering problems are traditionally based on either classic or quantum electromagnetic models, with quantum-based simulations being more precise but at the same time very expensive computationally. In turn, many classic high-performance techniques exist that ensure good accuracy of the solution for relatively large plasmonic particles. However, the studies of SPs in small metal nanoparticles with size less than 10nm demand more accurate yet efficient simulation tools due to nonlocal effects occurring in the plasmonic medium [6]. Despite the quantum nature of these effects it has been shown that they could be efficiently accounted for within the scope of classic approaches via incorporation of the *Generalized Nonlocal Optical Response* (GNOR) theory into the electromagnetic model [7]. It has also been shown that account for nonlocal effects could drastically influence scattering characteristics and so efficient GNOR-enabled numerical techniques are required to accurately resolve the properties of nanoscaled plasmonic systems [8, 9].

Within GNOR model material's dielectric permittivity is represented as a scalar function with nonlocal supplement. In this case the internal field inside the plasmonic NP ceases to be purely transversal and acquires longitudinal component [7]. Key feature of the longitudinal component is that corresponding wave number significantly differs from the wave number of the transversal field. This leads to fast oscillations of bessel functions employed to construct the solution which makes it more difficult to implement an accurate and efficient computational scheme.

Moreover, as longitudinal component is an additional unknown parameter, scattering problem statement has to be modified. In particular, new boundary condition is required to ensure the uniqueness of the solution to the Maxwell system. We would generally follow the approach featured in works [7–9] and introduce the electric field normal component discontinuity condition in order to write down the appropriate problem statement.

4. Discrete Sources Method

In this work GNOR-enabled scheme of the Discrete Sources Method (DSM) is applied to solve the problem of plasmonic nanoparticle excitation with electron beam. The proposed method is known to be one of the most flexible techniques for boundary value scattering problems analysis, in particular due to its many remarkable features including *a posteriori* error estimation, efficient usage of computational resources and capability of rigorously resolving the nonlocal effect influence on particle properties [10, 11].

Within DSM both scattered field everywhere in D_0 and total internal field inside the nanoparticle D_i are constructed as finite linear combinations of the fields resulting from discrete sources (DS) located inside the particle. The internal DS fields are chosen to analytically satisfy Maxwells equations in D_i. In order to account for the longitudinal component of the internal electromagnetic field (and, consequently, for nonlocal effects) corresponding DS analytically satisfying GNOR governing equations are also introduced [8, 9]. External DS fields in turn satisfy both Maxwells equations in D_0 and the required infinity conditions, and so does the constructed solution for the scattered field. Then unknown amplitudes of all discrete sources are determined from the only conditions not satisfied analytically: from boundary conditions enforced at the nanoparticle surface.

In this work we would use the hybrid DSM scheme which employs electric and magnetic dipoles to construct the scattered field solution and internal discrete sources of the same order to approximate the internal field. The sources are distributed along the z axis inside D_i. Unknown DS amplitudes are computed via the Generalized Point-Matching Technique and surface residual minimization approaches in least square sense [8], thus ensuring boundary conditions and allowing to numerically obtain scattered field necessary for determining electron energy loss probability via (1).

5. Numerical simulations

We would consider a silver nanosphere in homogeneous medium with diameter d and electron beam with impact factor $\mathbf{b} = (b_x, b_y)$, where $b_x = d + 0.5\text{nm}$, $b_y = 0$. Corresponding material data is obtained from [7]. In particular, we would employ metal plasma frequency in $\hbar\omega_p = 8.99$ eV, Drude damping rate $\hbar\gamma = 0.025$ eV, Fermi velocity $v_F = 1.39 \cdot 10^6\text{ms}^{-1}$ and electron diffusion constant $D = 3.61 \cdot 10^{-4}\text{m}^2\text{s}^{-1}$. Besides, we would consider $\varepsilon_0 = \mu_0 = \mu_i = 1$.

In order to validate the proposed approach let us analyze how NP diameter d influences the position of plasmon resonance peak in local and GNOR approximations. For this purpose we would conduct series of numerical experiments varying d from 4nm to 15nm and comparing resonance energy $\hbar\omega_0$ obtained from plane wave (PW) scattering and from EELS computations. For the plane wave scattering simulations we would employ both DSM and analytical solution, and for the electron energy loss spectroscopy we would employ the proposed DSM approach. Corresponding results are presented on fig. 2.

As seen on the figure, significant blueshift of the plasmon resonance peak is predicted by GNOR-enabled computations with decrease of diameter d, which does not occur at all in the local response approximation. Moreover, DSM is shown to be capable of resolving this blueshift in both PW scattering and electron energy loss probability computations. It should be also noted that these results are in good agreement with available experimental data, where blueshift of plasmon resonance peak is also observed [7].

Generally, DSM allows examining particles and structures of various shapes [11]. For this reason the obtained results provide a promising outlook on further development of the proposed approach to nonspherical particles and dimers, which is relevant to applications.

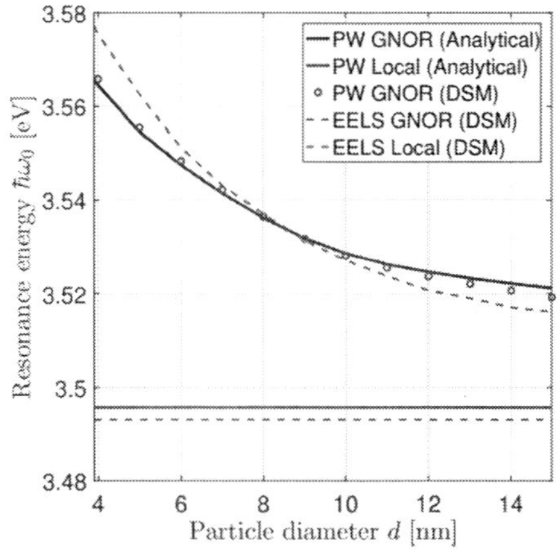

Figure 2. Dependence of the resonance energy $\hbar\omega_0$ on the NP diameter d. *Solid lines*: analytically obtained results; *line of dots*: results obtained from plane wave (PW) scattering analysis via DSM; *dash lines*: results obtained from EEL spectra analysis via DSM.

6. Conclusion

The problem of electron beam excitation of small plasmonic nanoparticles with diameter less than 15nm was considered. GNOR-enabled Hybrid Discrete Sources Method scheme was successfully applied to construct the solution to the corresponding boundary value problem and to computation of electron energy loss probability spectra. Numerical simulations confirm that proposed approach is capable of resolving surface plasmon resonance blueshift observed for small silver nanoparticles from EEL spectra.

Acknowledgments

This work is supported by the German-Russian Interdisciplinary Science Center (G-RISC) funded by the German Federal Foreign Office via the German Academic Exchange Service (DAAD), project M-2017b-7.

References

[1] Thomas S, Matyssek C, Hergert W, Arnold M, Kiewidt L, Karamehmedović M and Wriedt T 2015 *Plasmonics* **11** 865–874
[2] Guillaume S O, de Abajo F J G and Henrard L 2013 *Physical Review B* **88**
[3] Kiewidt L, Karamehmedović M, Matyssek C, Hergert W, Mdler L and Wriedt T 2013 *Ultramicroscopy* **133** 101–108
[4] Geuquet N and Henrard L 2010 *Ultramicroscopy* **110** 1075–1080
[5] de Abajo F J G 2010 *Reviews of Modern Physics* **82** 209–275
[6] de Abajo F J G 2008 *The Journal of Physical Chemistry C* **112** 17983–17987
[7] Raza S, Bozhevolnyi S I, Wubs M and Mortensen N A 2015 *Journal of Physics: Condensed Matter* **27** 183204
[8] Eremin Y A and Lopushenko I V 2017 *Moscow University Computational Mathematics and Cybernetics* **41** 165–172
[9] Lopushenko I V 2017 *Plasmonics: Design, Materials, Fabrication, Characterization, and Applications XV*
[10] Eremin Y A and Lopushenko I V 2016 *Moscow University Computational Mathematics and Cybernetics* **40** 1–9

[11] Wriedt T and Eremin Y (eds) 2018 *The Generalized Multipole Technique for Light Scattering* (Springer International Publishing)

Surface Engineering of Two-Dimensional Hydrogenated Silicon Nanosheets for Tailored Applications

A. Lyuleeva[a], A. Rivadeneyra[a], T. Helbich[b], P. Holzmüller[b], B. Rieger[b], P. Lugli[c], M. Becherer[a]

[a] Chair of Nanoelectronics, Technical University of Munich, Theresienstr. 90/I, 80333 München, Germany

[b] Catalysis Research Center/Wacker-Chair of Macromolecular Chemistry, Technical University of Munich, Garching, Germany

[c] Faculty of Science and Technology, Free University of Bozen-Bolzano, 39100 Bolzano, Italy

Email: alina.lyuleeva@mytum.de

Abstract. Approaching the predicted limits by Moor's law, which come with the shrinking size of electronics, science needs to find alternatives for the fabrication of novel devices. Within the last decades, more and more hybrid materials have been developed based on 0D- 1D- and also 2D nanomaterials. This development is also driven by the search for fast, cheap, reproducible and environmental-friendly low-cost devices. Hydrogenated silicon nanosheets (SiNSs) consist of just a buckled monolayer of sp^3-hybridized silicon atoms, which are decorated with hydrogen atoms. The sheets show green and consistent photoluminescence at ~510 nm and are predicted show semiconducting behaviour. SiNSs may be used as an alternative for not only bulk silicon, but also already known two dimensional materials. Well-studied and established nanosilicon surface chemistry offers solid ground for precise surface engineering, leading to control of the nanomaterial for reproducible device fabrication. New applications and processing techniques can thus be developed, which lead to facile fabrication of enhanced sensors and (opto)electronics. In this regard, research pursues additionally the combination of organic materials' properties with those of the nanomaterials, building the basis for new devices. One example is the combination of the widely used organic semiconductor poly(3-hexylthiophene-2,5-diyl) with the precisely modified SiNSs. Again, surface engineering plays the key role in the hybrid materials' synthesis and subsequent application. This gives the opportunity to tune the properties of the hybrid material in the desired way. In this context, we present different ways of surface engineering with the modification of SiNSs as an example for the subsequent application in various (opto)electronic devices (*e.g.*, SGFETs and photonic sensors).

1. SiNSs as Novel Two-dimensional Material

With the discovery of graphene by Novoselov *et al.*, a new area of nano-based applied science has been discovered. Following this example, further two-dimensional (2D) analogues[1] such as *e.g.*, silicene, germanene[2], or transition metal dichalcogenides[3] have been synthesized and characterized.

The synthesis of silicene[4] and its hydrogenated form silicane[5][6] turned out to be particularly interesting, as silicon is the main element in the semiconductor industry. Starting with the chemical exfoliation of calcium ions from the crystalline $CaSi_2$ Zintl phase, nano-sized silicane, also referred to as layered polysilane, can be liberated. These slightly buckled free standing 2D silicon sheets (SiNSs), ranging from hundreds of nanometers to micrometers size, are nearly planar. With the small dimensions of SiNSs, new properties and possibilities arise, of which some examples are the green photoluminescence (PL) and the semiconducting behavior of the material. As already shown with the pioneer graphene, the band gap can be modified and Fermi level tuning can be performed *via* external influences on the surface.[7][8] This precise surface engineering of the nanomaterial can be either covalent or non-covalent. In the case of SiNSs, none of the methods leads to a significant deformation, or destruction of the surface, due to the sp^3-hybridized nature of the material. For covalent functionalization, the H atom is available on the surface, hence there is no need to introduce defects as with graphene. Physical strain[9], the underlying substrate[10], external electrical fields[11] and surface functionalization[12] are predicted to have an influence on physical properties of the nanomaterial and open the ability for the manipulation of the band gap. This approach can be used for surface engineering, leading to the tailored design of the surface, which is most suitable for the desired needs. Thus, new devices like SiNS-based photonic or humidity sensors can be fabricated. These could eventually replace the already widely used bulk silicon, which is limited by its features, such as the indirect energy band gap[13], lack of transparency and flexibility.

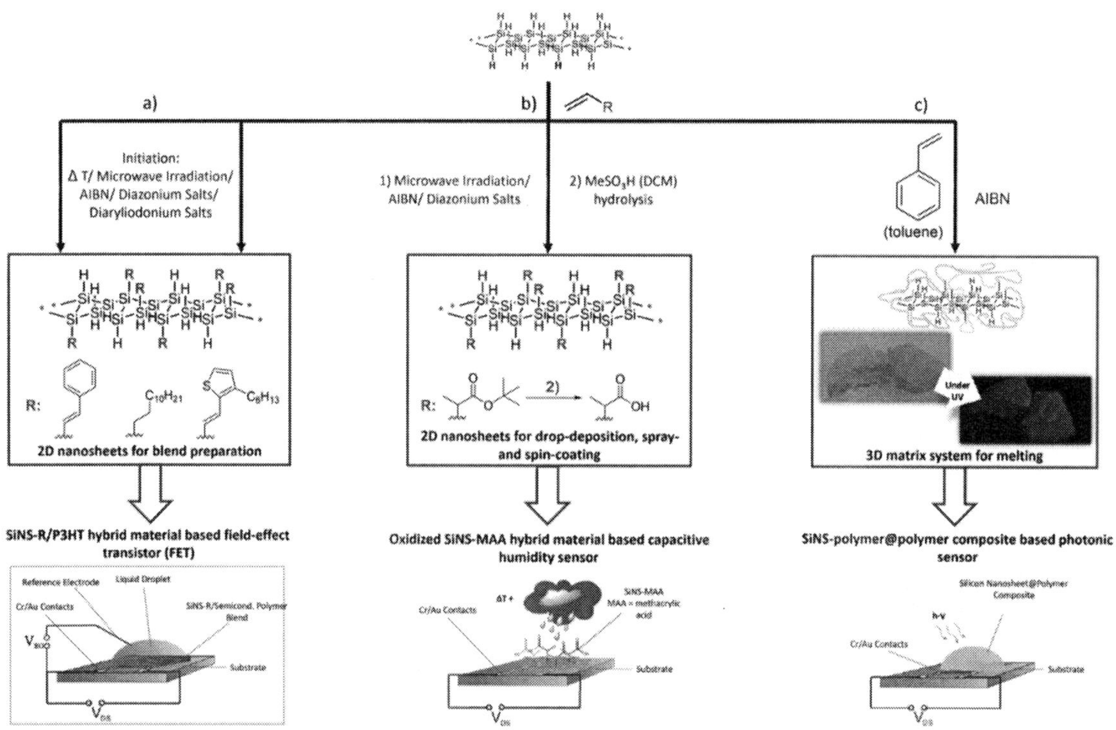

Figure 1: Surface engineering *via* functionalization of SiNSs using hydrosilylation on the surface of the Si-based 2D material. Different initiators can be used for the radical reaction, which lead to the functionalization with a variety of molecules. The modification introduces an additional property to the nanomaterial, such as additional π-electrons to the system (a), hydrophilicity of the surface (b) and encapsulation in a polymer (c). This step not only helps to protect the sensitive material from external influences, but also to control it for subsequent processing and application in novel nanomaterials-based electronics.

2. Surface Engineering of Nanomaterials

In general, to enable the use of these nanomaterials in ambient conditions and for facile big scale fabrication, they need to be controlled and stabilized.[14] Especially the 2D SiNSs tend to agglomerate right after their exfoliation, due to their high surface to volume ratio. Additionally, like most of the silicon-based materials, they show high affinity to oxygen and can therefore be destroyed by high energy light (UV light) and reaction with air-oxygen.[15] As a result, the formation of silicon dioxide takes place, which is an insulator with a wide energy band gap[16]. SiNSs can be stabilized *via* adjusted surface functionalization, which was already described in our previous works.[15][17] A variety of organic functional groups, such as dodecene, styrene or *tert*-butylmethacrylate (^{t}BMA) were attached to the surface, using thermal energy, diazonium salts[15], or AIBN[17] as an initiator for a hydrosilylation reaction. This surface modification is a significant increase in the lifetime of SiNSs. Additionally, the nanomaterial can be controlled more easily, due to the possibility to prepare various dispersions, which can be used for subsequent device fabrication. Different material deposition techniques, such as spray deposition or spin coating can be applied. As a result, reproducible and ultra-low-cost conductive patterns *via* inkjet printing technique[18] can be possible in the future, which is currently under further investigation in our group.

Hydrosilylation reaction and simultaneous radical polymerization has also been used for the synthesis of a composite SiNSs-polymer@polymer, which consists of functionalized and at the same time in a polymer matrix embedded SiNSs. This three-dimensional network is formed in a one-step reaction, offering both materials' properties. Thus, the features of the polymer open up even more new possibilities for application. As we could already show in our previously published work[19], new photosensitive devices can be built based on these composites. As an example, polystyrene has been used to embed and functionalize SiNSs, forming SiNS-PS@PS. The resulting composite can be melted on top of pre-patterned gold electrodes, forming an optoelectronic sensor. In this case, both rigid and flexible substrates can be used. Very sensitive optical response can be detected using this device, while exposing it to a light source with wavelengths lower than 450 nm.

Combining the high surface to volume ratio and the surface engineering of SiNSs with functional groups, we went further and realized the fabrication of the first oxidized SiNSs (OSiNSs)-based capacitive humidity sensors. First, ^{t}BMA was used as functional molecule for the modification of the SiNSs' surface. The *tert*-butylic ester group of the methacrylic acid can be *e.g.,* exposed to saponification.[20] This lead to the formation of methacrylic acid groups (MAA), covering the surface of the SiNSs and thus resulted in a facile transformation of the properties of the hybrid material. These hydrolysis reactions only take place with the functional groups of the organic molecules on the SiNSs' surface. Thus, the Si-Si-bond splitting can be excluded. The hydrophilic nature of the herewith synthesized hybrid material is crucial for the device to be able to work as a humidity sensor. In an additional step, the hybrid material was oxidized, in order to form highly porous OSiNS-MAA structure, which can be penetrated by surrounding water molecules. These can interact with the hybrid material and influence its dielectric behavior, resulting in a change in the capacitance. For the measurements, the humidity at a constant temperature was changed from around 25% relative humidity (RH) to up to 95 %RH. In general, this modification method shows an additional way for properties tuning of the hybrid material. A variety of further applications without destroying the conductive core can be developed.

In this work, not only an established way for a facile synthesis of hybrid materials, such as functionalized SiNSs and SiNS-based polymer composites, is highlighted, but also a detailed characterization of the surface with a variety of analysis techniques is presented. Furthermore, first steps towards optoelectronic applications are described. A facile integration of the modified and in a polymer matrix embedded SiNSs into the already well-known fabrication methods and the first prototypes of a highly sensitive photonic sensor are presented. Additionally, the successful fabrication of a SiNSs-based humidity sensor is described. This enables new possibilities in the field of capacitive sensors, whereby the first steps towards response to surrounding molecules, such as water at a certain temperature, are studied.

References

[1] J. A. G. Sheneve Z. Butler, Shawna M. Hollen, Linyou Cao, Yi Cui, *ACS Nano*, vol. 7, no. 4, pp. 2898–2926, 2013.
[2] K. L. Low, W. Huang, Y.-C. Yeo, and G. Liang, *IEEE Trans. Electron Devices*, vol. 61, no. 5, pp. 1590–1598, 2014.
[3] H. X. Luan, C. W. Zhang, F. Li, and P. J. Wang, *Phys. Lett. Sect. A Gen. At. Solid State Phys.*, vol. 377, no. 39, pp. 2792–2795, 2013.
[4] P. Vogt, P. De Padova, C. Quaresima, J. Avila, E. Frantzeskakis, M. C. Asensio, A. Resta, B. Ealet, and G. Le Lay, *Phys. Rev. Lett.*, vol. 108, no. 15, p. 155501, Apr. 2012.
[5] O. Böhm, J.; Hassel, *Z. anorg. allg. Chem.*, vol. 160, p. 152, 1927.
[6] S. Yamanaka, H. Matsu-ura, and M. Ishikawa, *Mater. Res. Bull.*, vol. 31, no. 3, pp. 307–316, 1996.
[7] K. S. Mali, J. Greenwood, J. Adisoejoso, R. Phillipson, and S. De Feyter, *Nanoscale*, vol. 7, no. 5, pp. 1566–1585, 2015.
[8] M. Steenackers, A. M. Gigler, N. Zhang, F. Deubel, M. Seifert, L. H. Hess, C. H. Y. X. Lim, K. P. Loh, J. A. Garrido, R. Jordan, M. Stutzmann, and I. D. Sharp, *J. Am. Chem. Soc.*, vol. 133, no. 27, pp. 10490–10498, 2011.
[9] F. Li, R. Lu, Q. Yao, E. Kan, Y. Liu, H. Wu, Y. Yuan, C. Xiao, and K. Deng, *J. Phys. Chem. C*, vol. 117, no. 25, pp. 13283–13288, 2013.
[10] M. Niu, D. Cheng, and D. Cao, *Sci. Rep.*, vol. 4, p. 4810, 2014.
[11] Z. Ni, H. Zhong, X. Jiang, R. Quhe, G. Luo, Y. Wang, M. Ye, J. Yang, J. Shi, and J. Lu, *Nanoscale*, vol. 6, pp. 7609–18, 2014.
[12] M. J. S. Spencer, T. Morishita, M. Mikami, I. K. Snook, Y. Sugiyama, and H. Nakano, *Phys. Chem. Chem. Phys.*, vol. 13, no. 34, pp. 15418–22, Sep. 2011.
[13] Y. Sun and H. H. Wang, *Adv. Mater.*, vol. 19, no. 19, pp. 2818–2823, 2007.
[14] Y. Sugiyama, H. Okamoto, T. Mitsuoka, T. Morikawa, K. Nakanishi, T. Ohta, and H. Nakano," *J. Am. Chem. Soc.*, vol. 132, no. 17, pp. 5946–7, May 2010.
[15] T. Helbich, A. Lyuleeva, I. M. D. Höhlein, L. Scherf, J. Kehrle, T. F. Fässler, and P. Lugli, *Chem. - A Eur. J.*, p. Chem. Eur. J. 10.1002/chem.201505134, 2016.
[16] B. El-Kareh, *Fundamentals of Semiconductor Processing Technology*. 1995.
[17] T. Helbich, A. Lyuleeva, T. Ludwig, L. M. Scherf, T. F. Fässler, P. Lugli, and B. Rieger, *Adv. Funct. Mater.*, pp. 6711–6718, 2016.
[18] A. Albrecht, A. Rivadeneyra, A. Abdellah, P. Lugli, and J. F. Salmerón, *J. Mater. Chem. C*, vol. 4, no. 16, pp. 3546–3554, 2016.
[19] A. Lyuleeva, T. Helbich, M. A. Naretto, A. Falco, F. Hegmann, B. Rieger, and P. Lugli, *Journal of Physics D: Applied Physics*, 10019, 2016.
[20] L. H. Hess, A. Lyuleeva, B. M. Blaschke, M. Sachsenhauser, M. Seifert, J. A. Garrido, and F. Deubel, *ACS Appl. Mater. Interfaces*, vol. 6, no. 12, pp. 9705–9710, 2014.

Spectrally tunable illumination system based on acousto-optic diffraction of light

A S Machikhin[1,2], **A B Kozlov**[1,3], **D D Khokhlov** [1,2], **V E Pozhar** [1,4], **V I Batshev** [1,4]
[1] Laboratory of acousto-optic spectrometry, Scientific and Technological Center of Unique Instrumentation of Russian Academy of Sciences, 15 Butlerova, Moscow, 117342, Russian Federation
[2] Department of Electrical Engineering and Introscopy, Institute of Automatics and Computer Engineering, National Research University "Moscow Power Engineering University", 14, Krasnokazarmennaya street, Moscow, 111250, Russian Federation
[3] POLYUS Research Institute of M.F. Stelmakh Joint Stock Company, Building 1, 3 Vvedenskogo St., Moscow, 117342, Russian Federation
[4] Department of optic-electronic devices, Bauman Moscow State Technical University, 5, 2-ya Baumanskaya street, Moscow, 105005, Russian Federation

E-mail: machikhin@ntcup.ru

Abstract. Spectrally tunable illumination is widely used for spectral imaging, spectroscopy, spectral-domain optical coherence tomography and other applications. Most of swept-wavelength light sources are based on mechanical change of narrow-band filters or on a supercontinuum laser coupled with a tunable filter. Such commercially available sources suffer either from a limited number of spectral channels or from a spatial noise caused by parasitic interferences. Here, we describe an alternative solution, which is free from the mentioned disadvantages and provides speckle-free uniform illumination. It is based on acousto-optic (AO) diffraction of a wide-band light. AO tunable filters (AOTFs) are fully software-controlled solid-state devices and do not require mechanical scanning. The selected wavelength of light is determined by the period of the volume phase grating induced in the crystal by an ultrasound wave, i.e. by the frequency of the electric signal applied to the piezo-transducer that excites ultrasound wave. Due to high spectral resolution, small entrance pupil diameter and small angular aperture of AOTFs, its utilisation together with wide-band light sources is problematic even with a specific optical coupling to optimize the light-energy and aberration parameters of the filtered light. In this paper, we show that efficiency of this approach may be radically improved by acoustic frequency modulation.

1. Introduction

In spectroscopy, there are two techniques of spectral tuning: by means of the illumination arm or the detection arm, i.e. with the spectral element location before and behind the inspected object. Though the second scheme is used more often, the first one is sometimes preferable, for example, in order to limit the radiation flow upon the object. Nowadays, spectrally tunable illumination is widely used for spectral imaging, spectroscopy, spectral-domain optical coherence tomography and other applications[1-3].

[1] To whom any correspondence should be addressed.

Modern spectral tunable illumination systems can be divided into two classes: laser sources and combination of wideband source with a tunable filter. Despite the outstanding characteristics like high spectral brightness, high spectral resolution, and high speed of sweeping, lasers are not suitable for many applications because of speckle structure. Most of swept-wavelength light sources are based on mechanical change of narrow-band filters or on a supercontinuum laser coupled with a tunable filter. Such commercially available sources suffer from a limited number of spectral channels and from a spatial noise caused by parasitic interferences, correspondently.

Here, we describe an alternative solution, which is free from the mentioned disadvantages and provides speckle-free uniform illumination. It is based on acousto-optic (AO) diffraction of a wide-band light. Acousto-optical tunable filters (AOTFs) are electronically-driven solid-state devices and do not require mechanical scanning. The selected wavelength of light is determined by the period of the volume phase grating induced in the crystal by an ultrasound wave, i.e. by the frequency of the electric signal applied to the piezo transducer.

It should be noted, that effective coupling of AOTFs with wide-band light sources is a challenge due to narrow angular aperture and rather small entrance pupil diameter of the filters[4-5]. Additionally to these losses, the significant decrease of the output light intensity results from narrow-band filtering. In this paper, we show that efficiency of this approach may be improved significantly by controllable broadening of the spectral bandwidth by means of acoustic frequency modulation.

2. Experimental setup

Spectrally tunable illumination system (figure 1) comprises wide-range light source, AOTF and all the necessary optical elements. The incandescent lamp or light-emitting diode can be used as an optical source. The coupling optics provide light beam shaping in respect to cross-section as well as to AOTF field-of-view. The output optics enable adjusting to the studied sample or to the working optical system.

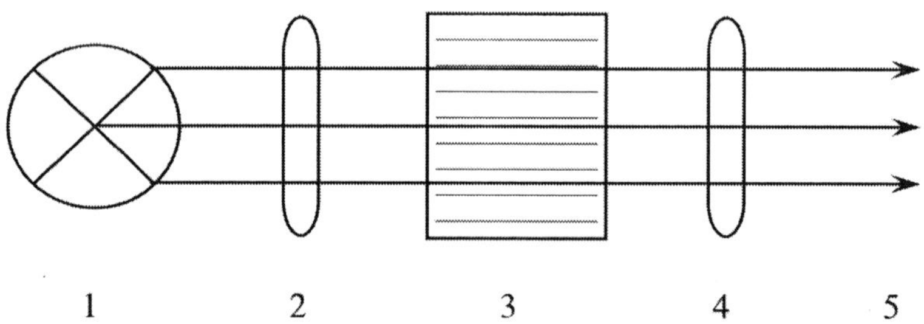

Figure 1. Spectrally tunable illumination system.
1 – wideband source, 2 – coupling optics, 3 – AOTF, 4 – output optics,
5 – spectral tunable radiation.

The specific feature of the AOTF (figure 2, *a*) is two different operation techniques: single-frequency (classical) and transmission window broadening modes. In the first modes, the single-frequency signal is supplied to AO crystal cell, thus producing homogeneous grating in crystal. Therefore, the bandwidth written in optical frequency units ($\nu \equiv 1/\lambda$) is determined by the acoustic grating width $\Delta\nu \sim 1/D$. In the second mode, the frequency is changed rapidly, so the grating becomes inhomogeneous across the light beam, and therefore different parts of grating diffract different spectral components. In this way, the transmission window of AOTF is broaden (see figure 3 below). For modulation, we applied sawtooth-like periodical frequency variation (figure 2, *b*) to keep the window motionless. Varying the limits of acoustic frequency sweeping (f_{min}, f_{max}) we can control AOTF bandwidth and spectral location.

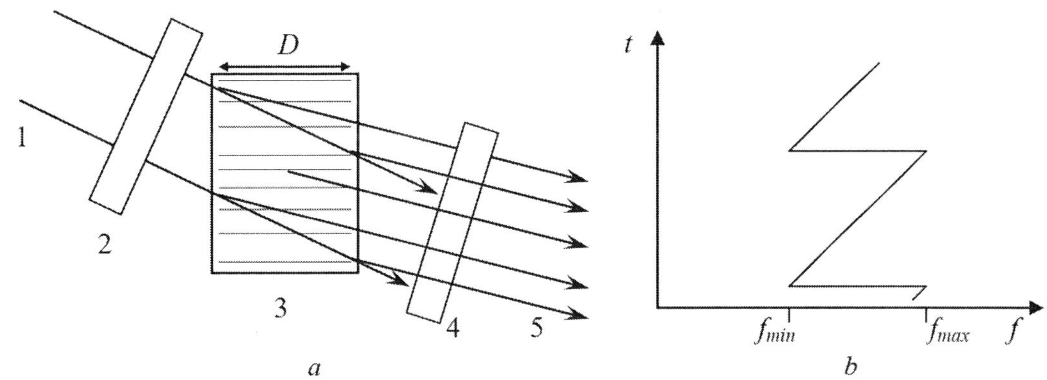

Figure 2. AOTF optical diagram (*a*) and modulation principle (*b*).
1 – input light beam, 2 – input and output crossed polarizer, 3 – AOTF,
5 –output spectral selected beam, *f*(*t*) – acoustic frequency variation (modulation).

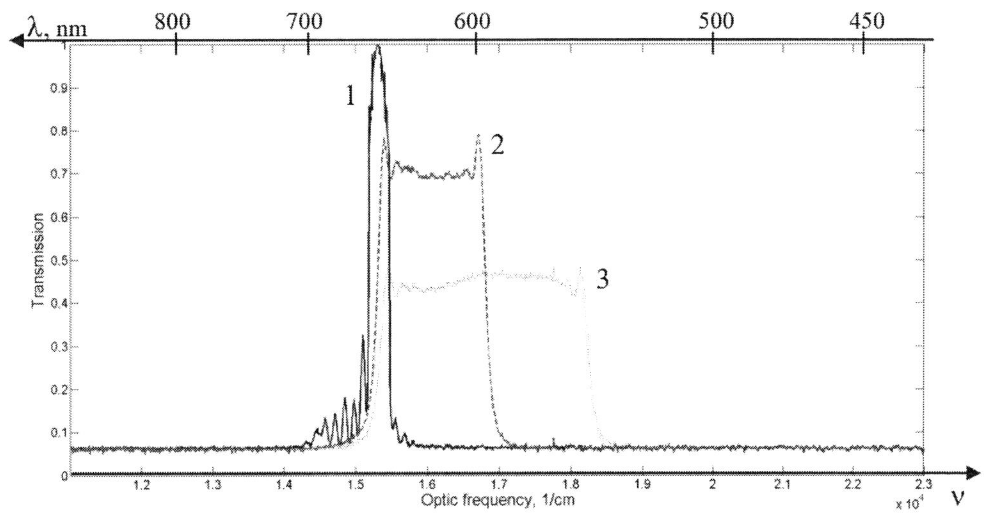

Figure 3. Transmission windows of the system.
1 – single frequency (narrow-band) mode,
2, 3 – broaden mode with different acoustic frequency sweep

3. Results

The basic characteristics of spectrally tunable illumination system are following:

spectral range	- 450-850 nm,
bandwidth	- from 60 cm^{-1} upto 500 cm^{-1}
spectral tuning	- computer-controlled addressing (less 1 ms)
AOTF	- non-collinear wide-angle TeO_2

The developed illumination system provides fast random spectral access over the spectrum with a function of bandwidth control.

4. References

[1] Demtröder W 1981 *Laser Spectroscopy* Tunable Coherent Light Sources (Berlin: Springer, Heidenberg) **5** 321-74
[2] Chinn S R, Swanson E A, Fujimoto J G 1997 *Optics Letters* Optical coherence tomography using a frequency-tunable optical source **22** 340-2

[3] Park J, Lee M, Grossberg M D, Nayar S K 2007 *Proc. of the 11th IEEE International Conference on Computer Vision* Multispectral imaging using multiplexed illumination
[4] Batshev V, Machikhin A, Pozhar V, Khokhlov D 2017 *Proc. SPIE* Design of an optical system for coupling a wideband light source with an acousto-optical tunable filter **10466** 1046639
[5] Machikhin A, Batshev V, Polschikova O, Khokhlov D, Pozhar V, Gorevoy A 2017 *Proc. SPIE* Conjugation of fiber-coupled wide-band light sources and acousto-optical spectral elements **10592** 105920I

Acknowledgments
The Russian Science Foundation (project #17-19-01355) financially supported the work.

Femtosecond laser nanostructuring of reinforcement bars surface for improvement of its interaction with concrete

N V Makarova[1,2], S V Makarov[3]

[1] Institute of Automation and Control Processes FEB RAS, Vladivostok, 690041, Russia
[2] Far Eastern Federal University, Vladivostok, 690000, Russia
[3] ITMO University, Saint Petersburg, 197101, Russia

E-mail: maknat@bk.ru

Abstract. Femtosecond laser nanostructuring of metal surfaces is a powerful tool for improvement of their various properties. Namely, increase of optical absorption ("black metals"), superhydrophobity, surface hardening, as well as improvement of tribological properties have been reported previously. Here we show the improvement of bond slip of nanostructured reinforcement bars with concrete. We demonstrate strong dependence of straining force and pulling force on the regime of surface nanostructuring. In particular, two-fold increase of straining force and pulling force can be achieved at the highest fluence in the regime of "black metal". Our finding has high practical importance for advanced civil engineering.

1. Introduction

Laser-induced nanoscale surface structures can be generated in a simple single-step process and allow a wide range of surface functionalization, including colorization [1-3], the control of surface wetting properties [4-7], friction management [8-10], and surface hardening [11]. The reduction of friction and wear by nanostructures has attracted the industrial attention, because state-of-the-art laser processing costs are around 0.1 €/cm^2 [10].

Despite various materials have been used for fs-laser nanostructuring, steel is a key material for modern industry, including civil engineering. The low carbon steel possessing sufficient strength and elasticity is widely used as reinforcement for concrete constructions. Its main disadvantage is the low corrosion resistance. In turn, cold-drawn wires made of stainless steel with high corrosion resistance are widely used in mechanical engineering and medicine, whereas its applications in civil engineering are limited due to less preferable tribological properties. In this regard, fs-laser nanostructuring would solve this problem. Indeed, experimental studies of femtosecond laser irradiation of metal surfaces by pulsed laser ablation have shown that direct laser interaction in air ambient leads to oxidation or carbonization of the surface [11]. These effects of loosening the surface of the metal, which can improve the interaction of reinforcing bars with concrete.

In this work, we study experimentally, for the first time to our best knowledge, the improvement of bond slip of the fs-laser nanostructured reinforcement bars with concrete.

2. Results

Surface of a sample was nanostructured by a Yb^{+3} laser (TeMa, Avesta Project) with a center wavelength of 1050 nm, pulse duration 150 fs and energy up to 100 nJ in the TEM_{00} mode. The target surface was irradiated in the scan regime at a pulse repetition rate of 10 kHz formed by pulse picker (Avesta Project). Focusing of the laser beam was performed by a lens with 5 cm focal length. Laser power was varied by a Glan prism coupled with half-wave plate. We studied three regimes of laser nanostructuring: near-threshold, 1.5 times exceeding of threshold power, 3 times exceeding of threshold power. Threshold of nanostructuing was determined with optical microscope, when visible damage occurs at some value of laser power, which was measured by means of powermeter (Ophir). Our samples represented bars of cold-drawn wire with 4-mm in diameter made of stainless steel 12X18H10T GOST18143-72 (ASTM: 321H AISI).

The adhesion of the bars to concrete was determined by standard pulling tests. On the surface of 6 rods, laser surface treatment was applied at the different modes of nanostructuring. Optical images are shown in Fig.1a. The areas of nanostructuring were created from 4 opposite sides of the rods, the length of each of the area was L = 3.25 mm, the width was around 2 mm (Fig. 1a). All samples were compared with the bars without nanostructures on surface.

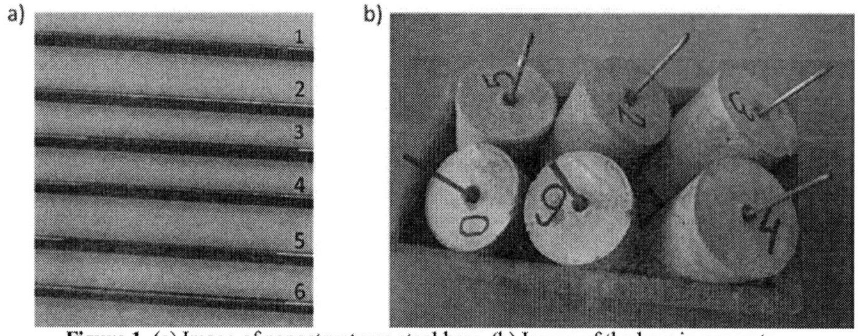

Figure 1. (a) Image of nanostructures steel bars. (b) Image of the bars in concrete.

After keeping the fabricated bars for 5 years (possible shelf life in the warehouse) at normal humidity and temperature of 15-25 degrees Celsius, reinforced concrete samples with their use were made. The samples were cylindrical in diameter 55 mm, height 65 mm. In the exit points of the rods, notches were made to exclude surface cracking of the concrete (Fig. 1b). The concrete used for making samples was fine-grained at the age of 10 days; the uniaxial compression strength was 12.4 MPa.

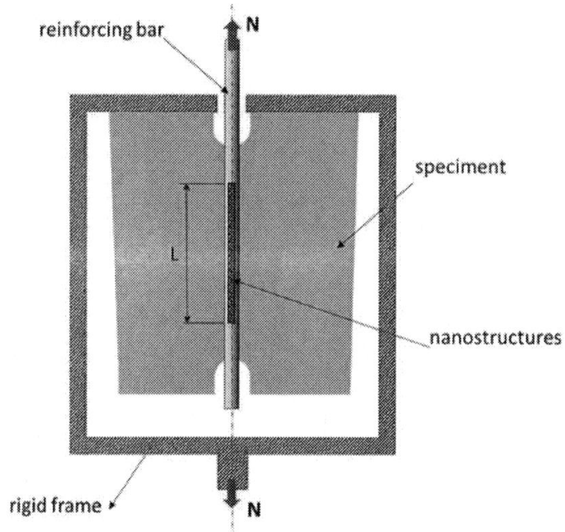

Figure 2. Schematic illustration of the principle of reinforcement bars testing.

The testing device was a rigid metal frame with a hole for the rod (Figure 2). The tests were carried out using a universal testing machine (Shimazu AGS-X 10 kN). The loading speed was 2 mm/min. The values of the straining force (breaking of adhesion) and full pulling of bar (inclusion of friction forces into work) were fixed. The results of the tests are given in Table 1.

Table 1. Dependence of straining force and pulling force on the regime of fs-laser nanostructuring of stainless steel bars.

Nanostructuing regime	Straining force, N	Pulling force, N
without nanostructures	226	424
near-threshold sample #1	120	131

near-threshold sample #2	217	235
$1.5P_{th}$ sample #3	248	317
$1.5P_{th}$ sample #4	295	540
$3P_{th}$ sample #5	447	784
$3P_{th}$ sample #6	355	627

3. Discussion and conclusion

An analysis of the results showed that, the samples corresponded to near-threshold regime of nanostructuring showed a lower slip bonding strength as compared with control sample without nanostructures. Also, there was significant reducing of frictional forces for these samples. The samples where the incident power was slightly higher than nanostructuring threshold exhibited almost the same characteristics as compared with the reference samples. However, the work of friction forces increased significantly. The samples corresponded to the "black metal" regime (3-fold exceeding of threshold) showed considerable increase of both straining forces and total pulling force as compared to the control unstructured sample. According to the previous works [12,13], the third regime of "black metal" corresponds to formation of a bimodal relief with microscale cones covered by nanostructures.

To conclude, we have shown that the control of laser processing regimes allows to obtain the surface with improved characteristics for reinforced concrete structures, which is stable in time.

Acknowledgements

This work was financially partially supported by Government of Russian Federation (Grant 08-08).

References

[1] Vorobyev A, Guo C 2008 *Appl. Phys. Lett.* **92** 041914
[2] Dusser B, Sagan Z, Soder H, Faure N, Colombier J, Jourlin M, Audouard E 2010 *Opt. Express* **18** 2913–2924
[3] Ionin A, Kudryashov S, Makarov S, Seleznev L, Sinitsyn D, Golosov E, Golosova O, Kolobov Y and Ligachev A 2012 *Appl. Phys. A* **107**(2) 301-305
[4] Zorba V, Stratakis E, Barberoglou M, Spanakis E, Tzanetakis P, Fotakis C 2008 *Appl. Phys. A* **93** 819–825
[5] Vorobyev A, Guo C 2012 *Science of Advanced Materials* **4** 432–438
[6] Saltuganov P, Ionin A, Kudryashov S, Rukhadze A, Gavrilov A, Makarov S, Rudenko A, Zayarny D 2015 *Journal of Russian Laser Research* **36**(1), 81-85
[7] Kolobov Y, Smolyakova M, Kolobova A, Ionin A, Kudryashov S, Makarov S, Saltuganov P, Zayarny D, Ligachev A 2014 *Las. Phys. Lett.* **11**(12), 125602
[8] Yasumaru N, Miyazaki K, Kiuchi J 2008 *Appl. Surf. Sc.* **254** 2364
[9] Bonse J, Koter R, Hartelt M, Spaltmann D, Pentzien S, Höhm S, Rosenfeld A, Krüger J 2014 *Appl. Phys. A* **117** 103–110
[10] Bonse J, Koter R, Hartelt M, Spaltmann D, Pentzien S, Höhm S, Rosenfeld A, Krüger J 2015 *Appl. Surf. Sci.* **336** 21–27
[11] Kolobov YR, Korneeva EA, Kuz'menko IN, Skomorokhov AN, Kudryashov SI, Ionin AA, Makarov SV, Kolobova A, Manokhin S, Betekhtin V, Kadomtsev A 2018 *Technical Physics* **63**(3) 385-390
[12] Ionin A, Kudryashov S, Makarov S, Rudenko A, Seleznev L et al. 2013 *Appl. Surf. Sci.* **284** 634-637
[13] Ionin A, Kudryashov S, Makarov S, Rudenko A, Seleznev L et al. 2014 *Appl. Phys. A* **116** (3), 1133-1139

IOP Conf. Series: Journal of Physics: Conf. Series **1092** (2018) 012083 doi:10.1088/1742-6596/1092/1/012083

Metamaterials-inspired resonator for wireless power transfer systems

A. Markvart[1], M. Song[1], S. Kosulnikov[1], S. Glybovski[1], P. Belov[1], C. Simovski[2,1], P. Kapitanova[1]
[1]Department of Nanophotonics and Metamaterials, ITMO University, 197101 Saint-Petersburg, Russia
[2]School of Electrical Engineering, Department of Electronics and Nanoengineering, Aalto University, P.O. Box 15500, 00076 Aalto, Finland

E-mail: a.markvart@metalab.ifmo.ru

Abstract.
Metamaterial-inspired resonator for a wireless power transfer system is proposed and studied numerically. The resonator consists of two orthogonal wire layers separated by a short distance. To reduce the resonant frequency from several hundreds of MHz it is placed in a background dielectric material with high permittivity. By tuning the permittivity of the background material, the radiation losses can be significantly reduced resulting in a high Q-factor of 900. The wireless power transfer efficiency greater than 80% from the wire resonator to a small loop receiver has been numerically demonstrated at the frequency of 19 MHz. The large area of the proposed resonator (50 cm × 50 cm) offers a possibility to charge multiple receivers simultaneously.

1. Introduction

Wireless power transfer (WPT) through magnetic resonance coupling is thought to be a popular and safe way for charging consumer electronic devices [1, 2]. Traditionally the system utilizes a transmitter and receiver operating at the same resonant frequency which are coupled through the magnetic fields. Two main parameters to be optimized to improve the WPT efficiency are the Q-factors and the couping coefficient between the transmitter and receiver [2]. Usually, metallic coils or planar spiral resonators are used as a transmitter and receiver. Their Q-factors are normally limited due to the Ohmic loss and do not exceed 500 [3–5]. Metamaterials are artificial periodic structures which can realize unique electromagnetic (EM) properties that does not exist in nature. A variety of metamaterials have been both theoretically [8–10] and experimentally [11, 12] verified to improve the couping coefficient between the transmitter and receiver in a WPT system. Wire medium is a volumetric metamaterial which has been extensively investigated in recent years for controlling the propagation of the EM waves and has been used to realize subwavelength imaging [6]. The metasurface (2D metamaterial) composed of an array of wires has been proposed to enhance the uniformity of the magnetic field in magnetic resonance tomography [7]. In this work, we propose a metamaterials-inspired resonator which is composed of two orthogonal wire layers separated by a short distance placed in a background material with high permittivity. We perform numerical studies to show that a high Q-factor of the resonator can be achieved with the optimization of the background material permittivity.

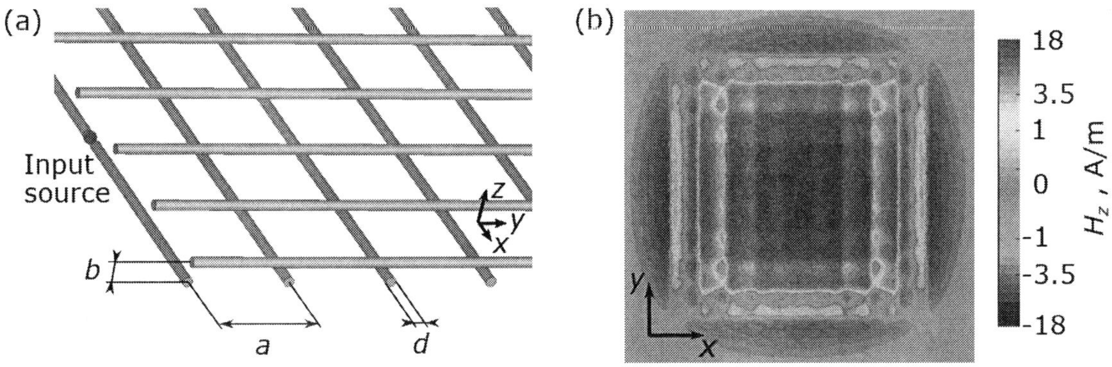

Figure 1. (a) Design of the metamaterials-inspired resonator. (b) The z-component of magnetic field distribution on the plane 12 mm above the resonator formed by 10×10 wires.

Moreover, the magnetic field distribution over the resonator is quasi-uniform which can be used to charge several devices placed on it simultaneously. Finally, the WPT efficiency from the proposed transmitter resonator to a loop receiver at the operational frequency of 19 MHz is studied.

2. The Resonator Design and Results of Numerical Simulations

The metamaterials-inspired resonator under study is demonstrated in figure 1(a). In principle the proposed resonator structure can be consisted of two orthogonal layers formed by an arbitrary number of wires with an arbitrary length L, periodicity a, and diameter d, which are separated by a spacing b. Taking into consideration the practical resonator size, e.g. for a WPT charging pad use, we chose 10×10 copper wires with $a = 5.55$ cm, $b = 6$ mm, $d = 2$ mm and $L = 50$ cm. We performed numerical simulations in CST Microwave Studio 2017. First, the eigenmode solver was used to analyze the modes which exist in such resonator. We observed a series of eigenmodes corresponding to certain current directions in the wires. For WPT application, a uniform magnetic field over the whole area of the resonator is preferable for charging multiple receivers. The fundamental mode of the resonator meets this requirement. Next, to excite the fundamental mode of the resonator, a discrete port with 50 Ohm impedance was embedded in to a gap of a lateral wire, as shown in Fig.1(a). The frequency domain solver of CST Microwave Studio 2017 was used to numerically study the input admittance and the field distribution of the resonator. From the simulated Y-parameter a resonant peak at $f = 150$ MHz was observed. The magnetic field distribution simulated at this frequency is demonstrated in figure 1(b), in which a quasi-uniform magnetic field is generated in the central area of the resonator. Thus, one may conclude that the fundamental mode is excited in the resonator. The uniformity of the magnetic field over the resonator could be favorably utilized to charge multiple small receivers. The classical 3 dB method to determine the Q-factor of the resonator from the Y-parameter spectrum was used and the Q-factor of the resonator operating at the fundamental mode was obtained as $Q = 13$. The low value of the Q-factor can be explained by the strong radiation losses of the resonator.

It is absolutely clear, that the current resonator design can not be used in a WPT system due to the low Q-factor. Moreover, the operational frequency determined by the wires length is beyond 100 MHz. With respect to the AirFuel Alliance standard the resonator system shall operate at 6.78 MHz. So, for the practical application of the proposed design it is important to reduce the resonant frequency and radiation losses. To solve these problems, we propose to immerse the resonator into a high permittivity background material. As an initial study,

IOP Conf. Series: Journal of Physics: Conf. Series **1092** (2018) 012083 doi:10.1088/1742-6596/1092/1/012083

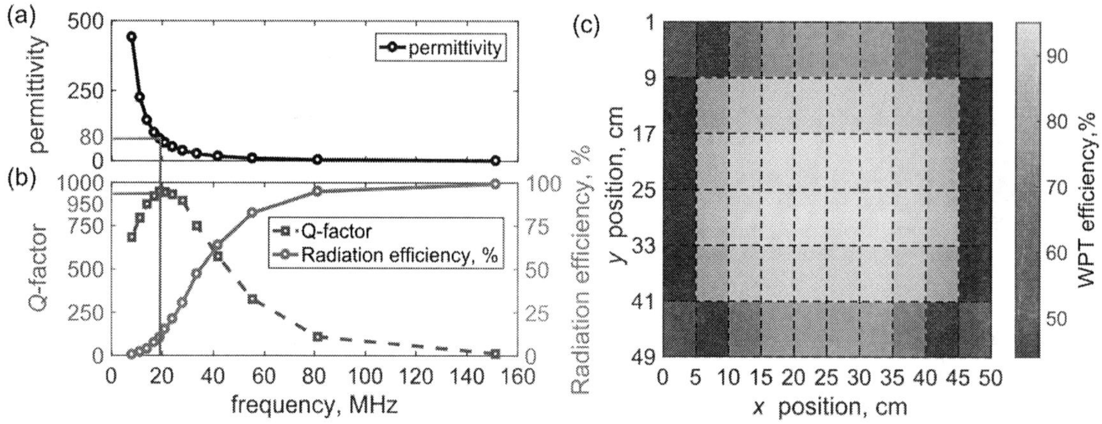

Figure 2. (a) The permittivity of the background material as a function of the corresponding resonant frequency. (b) Q-factor and the radiation efficiency as a function of the resonant frequency. (c) Calculated WPT efficiency as a function of the location of the receiving loop. Grid dotted lines represent the position of the receiving loop.

the resonator was situated inside a lossless dielectric material with the overall dimensions of 50 cm × 50 cm × 2 cm. The resonant frequency can be tailored by tuning the permittivity of the dielectric material as plotted in figure 2(a). The resonant frequency decreases as the permittivity of the dielectric material increases. At the same time, we observe the radiation efficiency also decreases (see figure 2(b)). It can be explained by the fact that the physical dimension in terms of wavelengths became much smaller leading to the reduced of the radiation efficiency. However, as the permittivity increases, the Q-factor does not increase monotonically. There exists a maximal Q-factor $Q = 950$ at $f = 19$ MHz corresponding to an optimal permittivity of $\varepsilon = 80$ as shown in figure 2(b). This is because of the overall capacitance between the wire layers increases significantly for the high permittivity values, which reduces the Q-factor. Water is a natural material characterized by a permittivity of $\varepsilon = 78.4$ at the frequency of 19 MHz. As a further investigation, the loss of the water is taken into consideration. The numerical study for Q-factor was conducted when the water conductivity was set as $\sigma = 5.55 \times 10^{-6}$ S/m, which resulted in a slightly reduced Q-factor from 950 to 900. However, the obtained Q-factor value is still higher then metallic coils or planar spiral resonators. Finally, the embedding of the metamaterials-inspired resonator inside the water helped to significantly improve the Q-factor, decrease the resonant frequency keeping the resonator dimensions the same (50 cm × 50 cm × 2 cm).

Now we proceed to investigate the WPT efficiency of the optimized metamaterials-inspired resonator design. For generality, we employ a copper rectangular loop with the dimensions of 5 cm × 8 cm. A 50 Ohm discrete port was embedded in a loop gap as a load. The loop size corresponds to a typical dimensions of a mobile phone. The receiver is placed in $x - y$ plane 5 mm above the dielectric surface of the resonator. The WPT efficiency for different positions of the receiver above the transmitter was calculated via the simulated S-parameters of the WPT system as $\eta = |S_{21}|^2$. Note that this equation is valid only when the impedance matching conditions at both ports are satisfied, i.e. $|S_{11}| = |S_{22}| = 0$ at the resonant frequency. Thus, additional matching networks at both ports were employed for all the receiver positions. The calculated WPT efficiency is shown in figure 2(c). In the central area of the metamaterials-inspired resonator, the WPT efficiency of more than 80% is obtained, whereas the efficiency of 50% is still achievable as the receiver approaches the edges.

3. Conclusion

The design of metamaterial-inspired resonator for planar WPT system has been proposed. The quasi-uniform magnetic field generated in the resonator has been numerically demonstrated. To improve the resonator's Q-factor the structure has been optimized by tuning the background material permittivity. The optimized design of the metamaterials-inspired resonator with the dimensions of 50 cm $\times$ 50 cm $\times$ 2 cm offered a Q-factor of 900 at the resonant frequency of 19 MHz. The power transfer from the metamaterial-inspired resonator to the metallic loop has been numerically investigated. The WPT efficiency of more than 80% in the central area of the resonator has been demonstrated. The further optimization of the resonator to meet the AirFuel Alliance standard operational frequency of 6.78 MHz can be achieved by adjusting the wire length.

4. Acknowledgment

The work was supported by the Russian Science Foundation (Project No. 17-79-20379).

References

[1] Kurs A, Karalis A, Moffatt R, Joannopoulos J D, Fisher P and Soljacich M 2007 *Science* **317** 83–86
[2] Song M, Belov P and Kapitanova P 2017 *Appl. Phys. Rev.* **4** 021102
[3] Tierney B B and Grbic A 2014 *IEEE Trans. Antennas and Propag.* **62** 3310–20
[4] Jolani F, Yu Y and Chen Z 2014 *IEEE Antennas Wireless Propag. Lett.* **13** 1648–51
[5] Klein A and Katz N 2011 *Curr. Appl. Phys.* **11** 1188–91
[6] Simovski C R, Belov P A, Atrashchenko A V and Kivshar Y S 2012 *Adv. Mater.* **24** 4229–48
[7] Slobozhanyuk A P, Poddubny A N, Raaijmakers A J, van den Berg C A, Kozachenko A V, Dubrovina I A, Melchakova I V, Kivshar Y S and Belov P A 2016 *Adv. Mater.* **28** 1832–38
[8] Zhu L, Luo X and Ma H 2016 *Appl. Phys. Lett.* **109** 024103
[9] Zhao Y and Leelarasmee E 2014 *Microwave Opt. Technol. Lett.* **56** 867–75
[10] Urzhumov Y and Smith D 2011 *Phys. Rev.* B **83** 205114
[11] Wu Q, Li Y H, Gao N, Yang F, Chen Y Q, Fang K, Zhang Y W and Chen H 2015 *Europhys. Lett.* **109** 680005
[12] Yoo Y J, Yi C, Hwang J S, Kim Y J, Park S Y, Kim K W, Rhee J Y and Lee Y P 2016 *Sci. Rep.* **6** 33416

IOP Conf. Series: Journal of Physics: Conf. Series **1092** (2018) 012084 doi:10.1088/1742-6596/1092/1/012084

Conjugate impedance matched metamaterials: Physical modeling and wave interaction effects

Stanislav I. Maslovski[1], Hugo R. L. Ferreira[1], Iurii O. Medvedev[1], Nuno G. B. Brás[2], Abdelghafour Abraray[3], Telmo Fernandes[3], Henrique J. A. Silva[1], and António L. Topa[2]

[1]Instituto de Telecomunicações – Coimbra
Universidade de Coimbra – Pólo II, 3030–290 Coimbra, Portugal

[2]Instituto de Telecomunicações – Lisboa
Instituto Superior Técnico, Av. Rovisco Pais 1, 1049–001 Lisboa, Portugal

[3]Instituto de Telecomunicações – Leiria
Morro do Lena, Alto do Vieiro, 2411–901 Leiria, Portugal

E-mail: stas@co.it.pt

Abstract. Conjugate impedance matched metamaterials are shown to be effective traps for electromagnetic waves. Objects made of such materials are able to receive radiation even when it is not directly incident on their surface. Here, we develop methods of physical modeling of such objects and investigate interactions of conjugate impedance matched superabsorbers with passing electromagnetic radiation. We study realizations of such metamaterials with meshes of loaded transmission lines and develop a theory of electromagnetic wave propagation and absorption in such media. Peculiar wave propagation, wave trapping and absorption effects in metamaterial black holes and wormholes are demonstrated. Possible modifications under the goal of optimizing absorption while minimizing complexity of the involved metamaterials are discussed. Conjugate-impedance matched superabsorbers may find applications as efficient harvesters of electromagnetic radiation, and as novel antennas and sensors.

1. Introduction

From wave optics, we know that the scattering and absorption cross sections of resonant particles can be much greater than those of non-resonant bodies with the same dimensions [1, 2]. For instance, absorption in subwavelength metallic particles exhibiting plasmonic resonance can be orders of magnitude higher than the same for a black-body type absorber of a comparable physical size. Effectively, such resonant particles are able to collect the incident wave power from an area much bigger than the physical cross section of the particles.

The same physical principle of optimal resonant absorption is used when designing compact receiving antennas. From the theory of wire antennas it is known that a short wire dipole (with length much smaller than half wavelength) is a rather ineffective receiver unless it is loaded with complex impedance $Z_{\rm load}(\omega) = Z^*_{\rm dip}(\omega)$, where $Z^*_{\rm dip}(\omega)$ is the complex-conjugate of the input impedance of the dipole antenna at the frequency ω. Such a conjugate matched load compensates for the excess reactance of the short dipole antenna, tunes it in resonance with the incident field, and provides for the maximum of the received power.

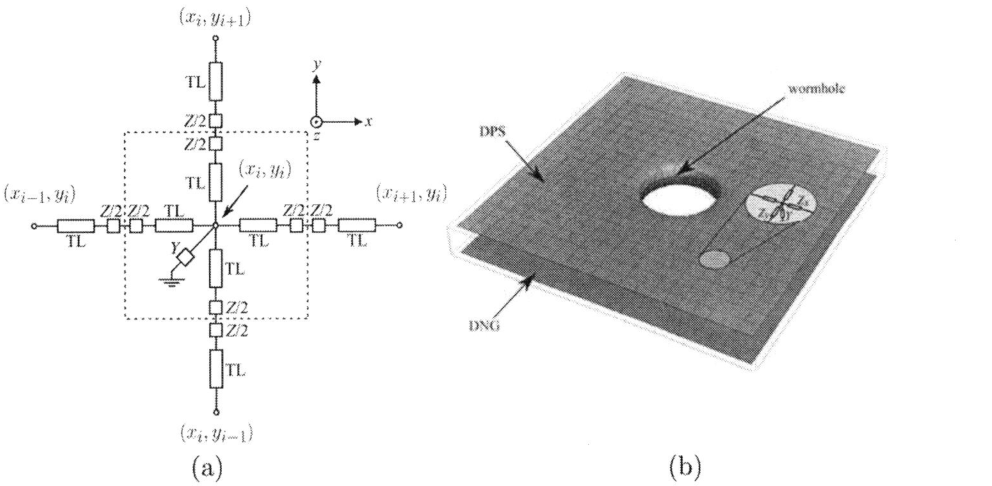

Figure 1. (a) Circuit diagram of the loaded 2D transmission line (TL) mesh. The unit cell of the structure is indicated with a dashed square. (b) Schematic representation of a TL-based wormhole structure formed by two electrically connected two-dimensional DPS and DNG domains. The two separate domains can be realized with unloaded (DPS) and loaded (DNG) TL meshes. The two TL meshes are electrically connected at the circumference $r = a$ of the wormhole neck.

If an object supports higher order multipolar resonances (in addition to the main electric dipolar mode) at the same frequency ω, its absorption cross section can be made higher than that of a resonant dipole. In fact, it can be shown [3]–[5] that there is no ultimate upper limit on the effective absorption cross section of a resonant object when more and more multipolar modes of the object pile up at the same resonant wavelength. With this principle, an exotic object — metamaterial black hole — formed by a medium with simultaneously negative permittivity and permeability can be constructed to posses arbitrarily large absorption cross section, theoretically, independently of the physical size of the object. The required condition is that double-negative (DNG) metamaterials with arbitrarily small loss and arbitrarily large ranges of permittivity and permeability values are attainable. Such metamaterial superabsorbers can be used as efficient harvesters of electromagnetic radiation at microwave frequencies or as super-Planckian radiative heat emitters at infrared and optical frequencies.

2. Physical modeling of superabsorption effect in two dimensions

From our previous study [3] we know that a spherical metamaterial body with radius a and radially dependent isotropic complex permittivity $\varepsilon(r) = \varepsilon'(r) - j\varepsilon''(r)$ and permeability $\mu(r) = \mu'(r) - j\mu'(r)$ satisfying $\varepsilon'(r)/\varepsilon_0 = \mu'(r)/\mu_0 = -a^2/r^2$, $|\varepsilon''(r)/\varepsilon'(r)| = |\mu''(r)/\mu'(r)| = \tan\delta \to 0$, has (theoretically) infinite effective absorption cross-section $\sigma_{\text{abs}}(\omega) \to \infty$ at the frequency ω, independently of the physical radius of the body a. Here, ε_0 and μ_0 stand for the permittivity and the permeability of the surrounding space, e.g., free space. For objects made of the materials with finite values of the loss tangent: $\tan\delta > 0$, and a limited variation range of the relative material parameters when $r \to 0$, the absorption cross section is finite, but still it can be large as compared to the physical dimensions of the body, even when $a \gg \lambda$, where λ is the radiation wavelength. An analogous result can be obtained as well in the case of a cylindrical body with the radius a and the anisotropic radially dependent material parameters [4].

Unfortunately, with the facilities that are currently at our hands a practical demonstration of the superabsorption effect in three dimensions appears to be rather difficult. Therefore, here

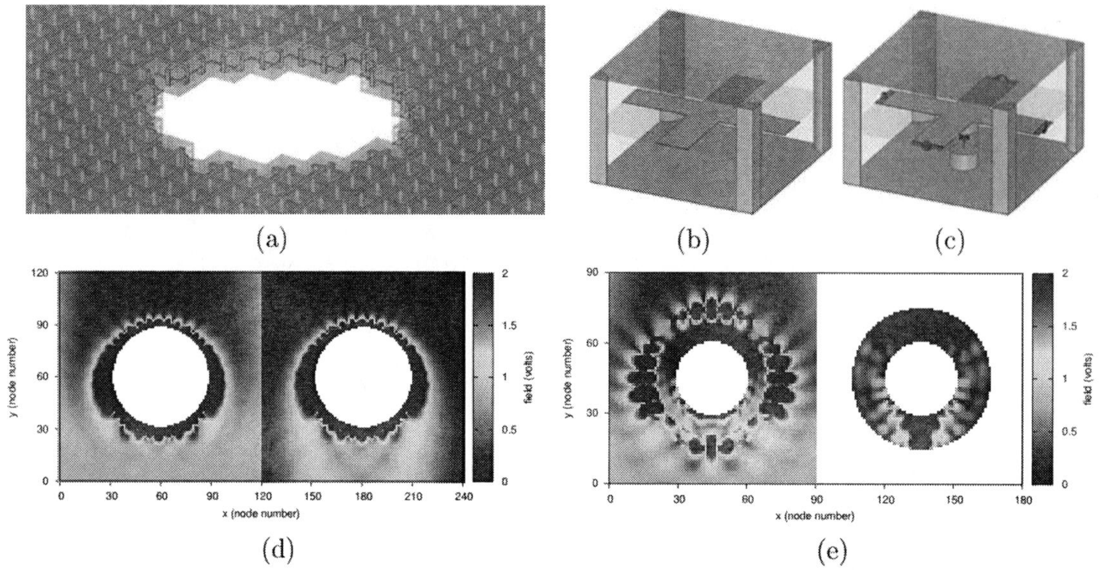

Figure 2. Panel (a) shows a realization of the wormhole structure shown in Figure 1 (b) with the DPS and DNG unit cells depicted in panels (b) and (c), respectively. The DPS unit cell is formed by an unloaded crossing of two symmetric strip lines. The DNG unit cell is formed by two crossed stripline segments loaded at the crossing by a lumped inductor connected to the ground (which realizes Y), and by four lumped capacitors at the edges of the unit cell (which realize $Z/2$). The lumped elements are shown in blue. (d) Distributions of the nodal voltage for the wormhole structure under plane wave incidence as functions of the geometrical coordinates $x < 120$ (the DPS domain) and $x > 120$ (the DNG domain) and y expressed in units of the unit cell size d. (e) The field distribution for the ring object confined to the region $r < R_{\rm obj} = 30d$ (where the region $x < 90$ represents the DPS domain with the object and $x > 90$ corresponds to the DNG domain) with optimized values of the characteristic impedance and the propagation factor and with a greatly reduced number of DNG cells. In all these plots, there are no unit cells in the white colored regions, and, respectively, there is no wave propagation in these regions.

we aim at physical modeling of this effect in a setup with reduced dimensionality, namely in two dimensions. Many limitations of volumetric metamaterials can be overcome when using two-dimensional metamaterials (2D) realized with meshes of loaded transmission lines (TL). The theory of such 2D metamaterials has been developed in a number of well-known works. There is also a possibility to extend such concepts to 3D (e.g., Ref. [6]). The unit cell of a generic 2D TL-based metamaterial is shown in Figure 1 (a). The electromagnetic waves in such structures are represented by waves of electric currents and voltages in the transmission line segments. By selecting proper loads, a TL-based metamaterial can be made to support forward or backward waves, and thus operate effectively as a double positive (DPS) medium with $\varepsilon'_{\rm eff} > 0$ and $\mu'_{\rm eff} > 0$, or as a DNG medium with $\varepsilon'_{\rm eff} < 0$ and $\mu'_{\rm eff} < 0$.

In this work we show that in the 2D case the superabsorption effect can be modeled by a wormhole structure [7] composed of two separate TL meshes: The unloaded DPS mesh and the loaded DNG mesh, electrically connected at the circumference of the wormhole. This structure is schematically shown in Figure 1 (b). The DPS and DNG domains in the top and bottom halves of this structure have to be conjugate impedance matched for all spatial harmonics of the incident field. In the practical realization of such a structure we employ meshes of strip lines [Figure 2 (a)] with the unit cells shown in Figure 2 (b,c). These unit cells fill the DPS and the

DNG domains shown in Figure 1 (b) by the golden and the blue surfaces, respectively. These DPS and DNG networks are laid atop one another and are separated by the common ground plane. A wave propagating in the DPS network towards the wormhole, after passing through the connections at the circumference of the wormhole, propagates in the DNG region in the outward direction and is absorbed at the edge of the DNG mesh by matched loads.

Simulation results obtained with the frequency-domain transmission line matrix (FDTLM) method are shown in Figure 2 (d). In this case, the wormhole structure is formed by the DPS and DNG domains occupying an area of 120×120 cells with each cell being a square of size $d \times d$ and a wormhole of radius $R_{\rm WH} = 30d$, with $d = 5$ mm. There are no cells at the middle of the domains where $r < R_{\rm WH}$. Respectively, in this middle region (the white area in the figure) there is no propagation. In this case, the numerically calculated value of of the absorption cross section is $\sigma_{\rm abs}/(2R_{\rm WH}) = 1.49$, i.e., this metamaterial object performs about 50% better than a simple black body type absorber of the same radius. Note that this result is achieved for an object with a rather large electrical size $\beta_0 R_{\rm WH} = 30\beta_0 d = 9.45$, which means that the circumference of the object is on the order of 10 wavelength.

In order to reduce the number of DNG unit cells in the structure, we perform a numerical optimization to confine all DNG cells within the region $r < R_{\rm obj}$. The optimization establishes optimal variation profiles for $\beta_0 d$ (the electric length) and Z_0 (the TL impedance) inside the object. We consider an object with $R_{\rm obj} = 30d$, which is formed by 14 concentric rings of DNG cells. The inner radius of the smallest ring is $r = R_{\rm WH} = 16d$, at which point the wormhole starts. In the DNG domain the region $R_{\rm WH} < r < R_{\rm obj}$ is filled with the cells whose electric parameters match to the parameters of the last ring. The field distribution for the optimized ring structure is depicted in Figure 2 (e).

3. Conclusions

In this work, physical modeling of the superabsorption effect in the conjugate impedance matched metamaterials has been considered. The effect has been modeled with 2D TL-based metamaterial structures. We have shown that, in this model, a finite-size conjugate impedance matched superabsorbing object can be equivalently represented with the wormhole structure formed by two electrically connected DNG and DPS domains. We have also found that the superabsorption effect can be observed in nonuniform structures comprising fewer DNG cells. Namely, such an effect can be observed for a metamaterial ring object that fits entirely within a region of finite radius $r < R_{\rm obj}$, while $\beta_0 R_{\rm obj} \gg 1$. This is especially interesting for applications of the metamaterial superabsorbers as efficient harvesters of electromagnetic radiation which absorb more energy from an incoming wave than what is incident directly on their surface.

References

[1] C. F. Bohren and D. R. Huffman, *Absorption and Scattering of Light by Small Particles* (Weinheim: Wiley-VCH Verlag GmbH & Co. KGaA), 2007.
[2] S. Tretyakov, Maximizing absorption and scattering by dipole particles, *Plasmonics,* vol. 9, pp. 935–944, 2014.
[3] S. Maslovski, C. Simovski, S. Tretyakov, Overcoming black body radiation limit in free space: metamaterial superemitter, *New J. Phys.,* vol. 18, p. 013034, 2016.
[4] C. A. Valagiannopoulos, J. Vehmas, C. R. Simovski, S. A. Tretyakov, S. I. Maslovski, Electromagnetic energy sink, *Phys. Rev. B,* vol. 92, p. 245402, 2015.
[5] S. I. Maslovski, C. R. Simovski, S. A. Tretyakov, Conjugate-impedance matched metamaterials for super-Planckian radiative heat transfer, *Proc. SPIE 9883, Metamaterials X,* p. 98830O, 2016.
[6] P. Alitalo, S. Maslovski, S. Tretyakov, Three-dimensional isotropic perfect lens based on *LC*-loaded transmission lines, *J. Appl. Phys.,* vol. 99, p. 064912, 2006.
[7] S. I. Maslovski, H. R. L. Ferreira, I. O. Medvedev, N. G. B. Bras, Superabsorbing metamaterial wormhole: Physical modeling and wave interaction effects, arXiv:1806.07818 [physics.class-ph], 2018.

Planar plasmonic optics: from basic elements to quantum generator

V I Balykin and P N Melentiev

Institute of Spectroscopy, Troitsk, Moscow 108840, Russia
Corresponding author's e-mail: balykin@isan.troitsk.ru

Abstract. The key advantage of plasmonics is in pushing our control of light down to the nanoscale. It is possible to envision lithographically fabricated plasmonic devices for future quantum information processing or cryptography at the nanoscale in two dimensions. Here we demonstrate the development of the basic elements of planar plasmonic optics: plasmonic optics media, focusing and reflecting plasmonic elements, plasmonic interferometer, plasmonic autocorrelator and planar plasmonic quantum generator.

1. Introduction

Plasmon optics, like all other types of optics (photon, electron, atomic) - solves the problem of controlling the appropriate types of waves. For effective control of plasmon waves, it is necessary to create appropriate control elements: the medium in which the wave (surface plasmon – polaritons, SPP) propagates, focusing elements (lenses, mirrors), beam splitter, interferometers, detector and SPP wave generator. We report development of the basic elements of plasmon optics, including:

- Propagation medium for SPP waves (single-crystal metal surfaces with a large propagation length of SPP waves);
- Nanolocalized sources of SPP waves in the visible and ultra-violet wavelength region;
- mirrors for SPP waves;
- SPP interferometer;
- SPP wave detector;
- Plasmonic nanolaser.

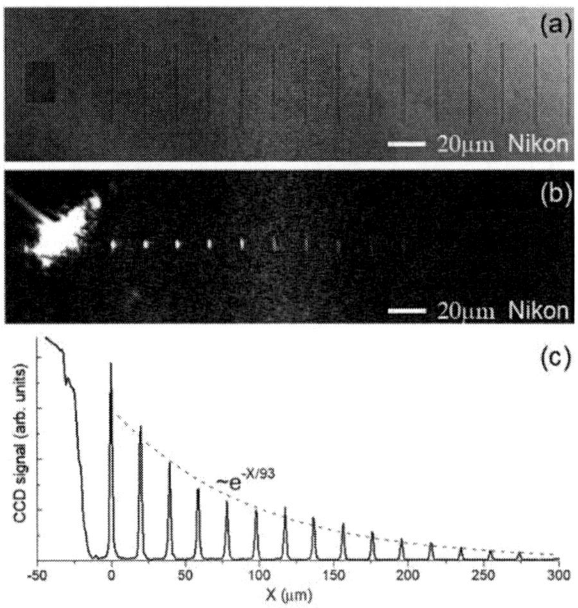

Figure 1. Optical microscopy of SPP waves: (a) optical image of a Ag film surface with nanostructures created by an ion-beam lithography, (b) optical image of Ag film surface when SPP excited by a laser radiation, (c) the cut image of figure 1b along the SPP propagation direction; the dotted curve - the approximation by exponential curve with the characteristic decay length of 93 μm [1].

2. Propagation medium

The choice of material for creating plasmonic nanostructures is decisive in achieving its best properties. We show [1] that the newest developed method of epitaxial growth of silver films makes it possible to realize in nanoplasmonics a propagation medium for SPP with losses corresponding to the theoretical values. The SPP on the surface of such nanofilms (780 nm) is characterized by a record long propagation length equals to 200 μm. This value is more than twice larger than the best known experimental results. Fig. 1 show the optical image of a Ag film surface with a nanostructures created by a focused-ion-beam lithography. The nanostructures used to excite SPP and to measure SPP propagation length on the Ag surface (figure 1).

3. Nanolocalized sources of SPP waves

We proposed and developed a new element of nanoplasmonics - a hybrid nanostructure: Split - Hole Resonator (SHR) [2], which has the following unique properties: (1) record high efficiency of third harmonic generation and multiphoton luminescence [2-4]; (2) a record high sensitivity to radiation polarization, with an extinction coefficient of the order of 4×10^4 [5]. We have shown that the presence of a strong localized plasmon resonance of the SHR nanostructure and the use of a femto laser pulse (two periods of light wave oscillations) allows the realization of an effective nanolocalized light source of the 2-nd and 3-rd harmonics in ultra-violet (UV) spectral region [4]. The amplitude of the near-field radiation at the frequency of the 3-rd harmonic is 0.6% of the field amplitude at the fundamental frequency, which is currently the maximum achievable value.

4. Focusing of SPP waves

Reflection and focusing of the SPP waves opens a unique possibility of controlling the propagation characteristics of the SPP waves. A method for controlling the SPP wave propagation was proposed and implemented, using SPP wave reflection from a "SPP wave mirror", created in the form of a nanoslit located on the surface of a metal nanofilm. The realization of reflection by means of a "SPP wave mirror" allows not only the control of propagation direction of the SPP wave, but also to focus it in a diffraction-limited spot [6].

Due to losses in the metal, the propagation length of the SPP wave on the surface of the metal is severely limited and this is a great limitation for numerous SPP applications. We have shown that the focusing of SPP allows a significant increase in the effective propagation length of SPP wave. A significant increase in the propagation length of the SPP wave makes it possible to realize a high-speed data transmission line at a distance of the order of 1 mm at a wavelength of 780 nm. It is shown that the information transfer rate of this transmission line is as high as 3.5 THz. The large propagation length of the SPP wave and high information transfer rate open up new possibilities for using SPP waves **in** information transfer applications [7].

5. SPP wave detector

The SPP wave detector is based on the nanostructuring of a metal surface. A technique for detecting SPP waves by its scattering on nanoparticles (nano-grooves) created by the method of ion lithography on a metal surface is developed [8]. The scattering cross section greatly decreases with the particle size, while the absorption is linearly dependent on the volume. Thus, the scattering is more sensitive and preferable for SPP detectors. The method of scattering detection is applicable to extremely clean surfaces of high optical quality. The scheme of the SPP wave detection on the surface of single-crystal gold or silver films based on nanogrooves was realized. Nanogrooves make it possible to "visualize" the process of propagation of a SPP wave on the surface of a metallic nanofilm. The use of nanogrooves as local detectors allows us to determine the propagation length of a SPP wave. The obtained value of the propagation length of the SPP wave is in good agreement with the measurement data obtained with the help of a SPP interferometer.

IOP Conf. Series: Journal of Physics: Conf. Series **1092** (2018) 012085 doi:10.1088/1742-6596/1092/1/012085

6. SPP interferometer

In experiments on nanoplasmonics, it would be ideal to measure the duration of a laser pulse directly in the object plane of the sample. We present results of our experimental investigations on the creation of a SPP tilted slit–grove (TSG) interferometer [9], which was prepared in a metal gold film. The possibility of monitoring the pulse duration in the interval from 6 to 50 fs is shown.

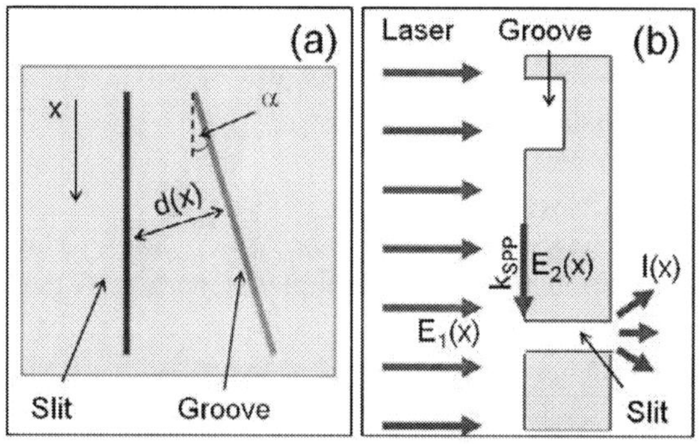

Figure 2. Scheme of SPP interferometer [9].

Figure 2 shows the schematic image of a TSG interferometer that was prepared in a gold film by a slit and a groove tilted at a certain angle to the slit. When measuring the laser pulse duration, the interferometer was illuminated by the laser beam perpendicularly to the film plane. The scattering of the laser light by the groove leads to the excitation of a SPP wave with wave vector k_{sp}, which is perpendicular to the groove. The SPP wave propagating from the groove in the direction toward the slit is partially *scattered* by the slit and creates a near field in the region of the nanoslit. Because of the occurrence of angle between the slit and the groove, the distance traveled by the SPP wave from the groove to the slit varies linearly along the slit axis. This leads to a dependence of the field amplitude on coordinate and to a similar dependence of its phase. The occurrence of the angle between the slit and the groove gives rise to interference of the near fields of the laser radiation in the slit. The SPP interferogram obtained by excitation of SPP interferometer with laser radiation is shown on figure 3.

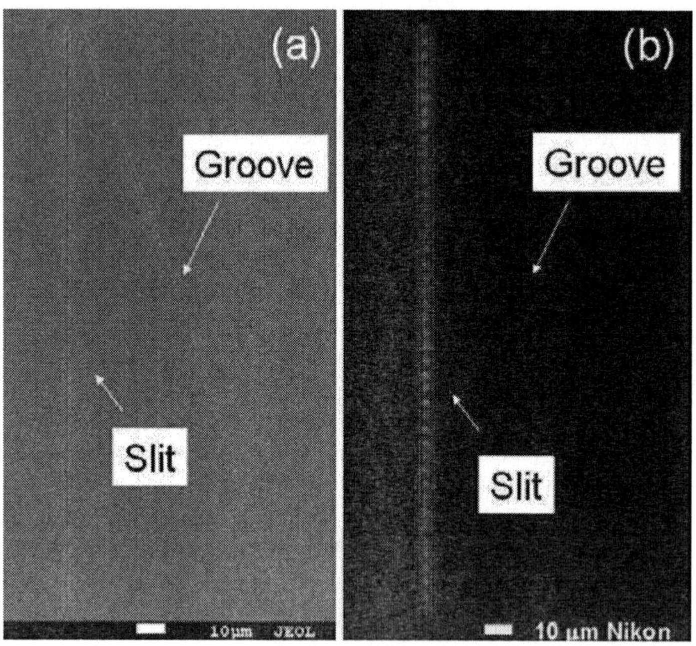

Figure 3. (a) SEM image of SPP interferometer, (b) SPP wave interferogram obtained by excitation of an interferometer with laser radiation of a wavelength 800 nm [9].

7. The plasmonic nanolaser

We design [10] a planar plasmonic nanolaser, extending the idea and implementation of plasmonic lasers reported in [11]. It is a planar waveguide made of a layer of liquid gain medium deposited on a plasmonic crystal (see figure 4). The plasmonic crystal is formed by an array of nanoholes (175 nm diameter and a pitch of 565 nm) created in a 100-nm-thick silver film deposited onto a quartz substrate. Dye molecules are optimal for plasmonic lasers because the dye molecules are capable of nearly 100% yield and allow for the creation of an optically homogeneous medium with high gain. We use a liquid gain medium: solution of dye molecules R101. The use of a liquid gain medium offers the advantages over a solid gain medium by fixing the bleaching problem. Also, the employment of microfluidics make plasmonic intracavity spectroscopy suitable for the lab-on-chip technology.

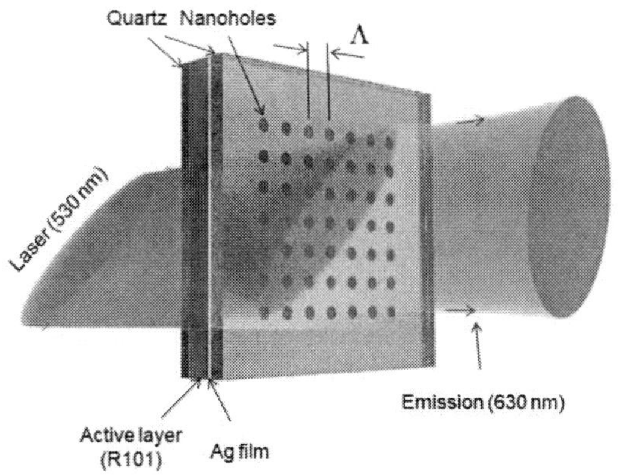

Figure 4. Scheme of plasmonic nanolaser: Planar waveguide made of a liquid gain medium layer (a solution of R101 dye in dimethyl sulfoxide (DMSO)) deposited on a plasmonic crystal [10].

The lasing occurs at pumping intensities above 500 kW/cm. The spectrum of the plasmonic laser for two values of pumping are shown in figure 5. At intensities below 500 kW/cm^2, the spectral contour is very similar to that of usual R101 dye luminescence. However, at higher pumping intensities, the luminescence contour changes dramatically, and a very narrow spectral line with the half-width of 1.7 nm emerges at 628 nm. Such a narrow line width indicates that we deal with lasing.

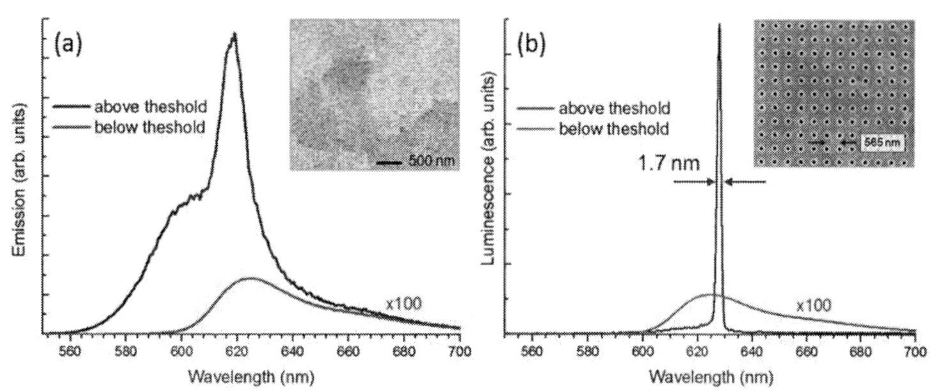

Figure 5. (a) The spectrum of dye luminescence in the quartz/silver/DMSO/quartz system without nanoholes; (b) the luminescence spectrum of the plasmonic laser. The insets show electronic microscope images of the samples [10].

References

[1] Baburin A, Kalmykov A, Kirtaev R, Negrov D, Moskalev D, Ryzhikov I, Melentiev P, Rodionov I and Balykin V 2018 *Opt. Mat. Express* sent
[2] Melentiev P, Afanasiev A, Kuzin A, Baturin A and Balykin V 2013 *Opt. Express* **21** 13896-13905
[3] Melentiev P, Afanasiev A, Kuzin A, Zablotskiy A and Balykin V 20115 *Opt. Express* **23** 11444
[4] Melentiev P, Afanasiev A, Kuzin A, Gusev V, Kompanets O, Esenaliev R and Balykin V 2016 *Nano Lett.* **16** 1138-1142
[5] Melentiev P, Afanasiev A, Kuzin A, Baturin A and Balykin V2013 *Opt. Lett.* **38** 2274
[6] Melentiev P, Kuzin A, Negrov D and Balykin V2018 *Plasmonics* accepted
[7] Melentiev P, Kuzin A, Negrov D, Klimov V and Balykin V 2018 *Nano Lett.* sent
[8] Melentiev P, Kuzin A and Balykin V 2017 *Quant. Electron.* **47** 266 – 271
[9] Melentiev P, Kuzin A, Gritchenko A, Kalmykov A and Balykin V 2017 *Opt. Commun.* **382** 509
[10] Melentiev P, Kalmykov A, Gritchenko A, Afanasiev A, Balykin V, Baburin A, Ryzhova E, Filippov I, Rodionov I, Nechepurenko I, Dorofeenko A, Ryzhikov I, Vinogradov A, Zyablovsky A, Andrianov E and Lisyansky A 2017 *Appl. Phys. Lett.* **111** 213104
[11] Meng X, Liu J, Kildishev A and Shalaev V 2014 *Laser & Phot. Rev.* **8** 896-904

Silicon nanoparticles for nonlinear frequency conversion of mid-IR radiation

Elizaveta V. Melik-Gaykazyan

Lomonosov Moscow State University, Faculty of Physics, Moscow 119991, Russia

melik@nanolab.phys.msu.ru

Abstract. Here a new way to use silicon nanoparticles for medium-infrared photonics is suggested. This contribution is focused on solving the fundamental problem of optical interaction between subwavelength silicon resonators and mid-IR laser radiation. In this spectral range the coefficient of two-photon absorption, which is the main mechanism for limiting the efficiency of nonlinear optical processes in silicon, is close to zero. The study provides answers to the fundamental questions about high-efficient nonlinear-optical transformations of four-wave mixing, and will serve as the basis for new nonlinear devices of integrated silicon nanophotonics.

1. Introduction

High-index nanoparticles are in the focus of many research groups specialized in photonics and nanophotonics. This interest appeared quite recently: it was observed that both electric and magnetic multipoles were excited in nanoparticles (spheres and disks) made out of high-index materials in the visible and near-IR spectral ranges [1-3], what is explained by Mie scattering theory [4].

The significant enhancement of the third-harmonic generation in a metasurface of closely packed silicon nanodisks comparing with non-structured area was detected [5] in the spectral vicinity of a magnetic dipolar resonance; it became the first work considering the study of nonlinear optical properties of all-dielectric nanostructures with optical magnetism. Moreover, optically isolated nanoparticles also reveal the effective up-conversion process at the magnetic dipolar Mie-type resonance [6] in contrast to the electric dipolar mode of a particle.

In order to increase the efficiency of nonlinear processes several nanostructure patterns have been studied including the combination of particles into oligomers, quadrumers [7], which leads to the excitation of collective optical Fano-type modes and 100-times enhancement of the third-harmonic-generation signal due to the redistribution of local electromagnetic field inside a silicon nanoparticle cluster.

The experimentally achieved nonlinear frequency conversion efficiency reaches values up to 10^{-5} in the consideration of the third-harmonic generation in silicon Mie-resonant particles; the main mechanism limiting the four-wave-mixing process efficiency is the two-photon absorption effect in silicon. The two-photon absorption coefficient is about 2.5 cm/GW in the near-IR spectral range for silicon [8, 9], whereas the coefficient becomes close to zero for a medium-infrared part of the spectrum, the multi-photon absorption parameters are also known for crystalline silicon [10].

The nonlinear optical process efficiency, particularly, the third-harmonic generation process in the case of silicon Mie-resonators, has not been studied for the mid-IR pump yet, as well as the excitation of optical modes in silicon nanoparticles at triple pump frequencies, whilst the research of three-wave-mixing conversions in silicon nanoresonators with multipolar modes at the double pump frequency from the visible spectral range is conducted [11].

2. Idea

The two-photon absorption coefficient of silicon becomes close to zero for the wavelengths from 2.1 microns and longer, i.e. in the mid-IR spectral range. In order to keep the nanoscale of Mie-type resonator sizes it is suggested to consider nanoparticles supporting the excitation of electric and magnetic multipolar Mie-type optical modes at the frequencies of nonlinear optical response – at the triple pump frequencies.

3. Methods

Numerical simulations of the third-harmonic output from a silicon nanodisk on a glass substrate are performed by the finite-difference time-domain method with the help of the *FDTD Solutions* software by *Lumerical Solutions, Inc*. The material of the disk is considered as nonlinear one by adding the value of its third-order nonlinear susceptibility. Each spectral point is a simulation result for the specific central pump frequency of an external radiation source modeled by a plane wave pulse; this procedure also enables to take into account the dispersion of a silicon nonlinear susceptibility, its tensor nature is ignored. The optical signal at the vicinity of a third-harmonic generation wavelength is detected by a two-dimensional monitor and, then, integrated as a function of these two spatial coordinates and frequency.

The sizes of considered silicon disks provide the excitation of multipolar resonances at the triple pump frequencies which correspond to the spectral range from 650 to 800 nm.

4. Discussion

Performed calculations reveal the strong enhancement of the third-harmonic generation signal from silicon nanoparticles at the vicinity of the magnetic dipolar resonance of a silicon nanodisk when the optical mode is excited at the triple pump frequency. The maximum value of normalized nonlinear signal is observed to be shifted in the blue area of the spectrum with respect to the Mie-type resonance due to dispersion dependencies of optical material parameters taken into account. These spectral features of the nonlinear response qualitatively coincide with the experimental findings reported previously for a silicon disk metasurface with spectrally overlapped electric and magnetic dipolar modes [12] and explain them by the model chosen for the numerical simulations and described above.

The proposed approach provides the resonant efficiency enhancement of the four-wave-mixing process in silicon nanoparticles excited by the mid-IR radiation as well as overcomes the fundamental conversion efficiency limit due to the two-photon absorption effect in silicon. These findings open a new avenue for novel nonlinear photonic devices.

Acknowledgements

A financial support provided by the Russian Foundation for Basic Research according to the research project No. 18-32-01038 is acknowledged.

References

[1] Kuznetsov AI, Miroshnichenko AE, Fu YH, Zhang J, Lukyanchuk B 2012 Magnetic light *Scientific Reports* **2** 492

[2] Evlyukhin AB, Novikov SM, Zywietz U, Eriksen RL, Reinhardt C, Bozhevolnyi SI, Chichkov BN 2012 Demonstration of magnetic dipole resonances of dielectric nanospheres in the visible region *Nano Letters* **12** 3749

[3] Staude I, Miroshnichenko AE, Decker M, Fofang NT, Liu S, Gonzales E, Dominguez J, Luk TS, Neshev DN, Brener I, Kivshar Y 2013 Tailoring directional scattering through magnetic and electric resonances in subwavelength silicon nanodisks *ACS Nano* **7** 7824

[4] Mie G 1976 Contribution to the optics of turbid media, particularly of colloidal metal solutions *London: Royal Aircraft Establishment, Library Translation 1873, RAE-Lit-Trans-1873* 72

[5] Shcherbakov MR, Neshev DN, Hopkins B, Shorokhov AS, Staude I, Melik-Gaykazyan EV, Decker M, Ezhov AA, Miroshnichenko AE, Brener I, Fedyanin AA, Kivshar YS 2014 Enhanced third-harmonic generation in silicon nanoparticles driven by magnetic response *Nano Letters* **14** 6488

[6] Melik-Gaykazyan EV, Shcherbakov MR, Shorokhov AS, Staude I, Brener I, Neshev DN, Kivshar YS, Fedyanin AA 2017 Third-harmonic generation from Mie-type resonances of isolated all-dielectric nanoparticles *Philosophical Transactions of the Royal Society A* **375** 20160281

[7] Shorokhov AS, Melik-Gaykazyan EV, Smirnova DS, Hopkins B, Chong KE, Choi D-Y, Shcherbakov MR, Miroshnichenko AE, Neshev DN, Fedyanin AA, Kivshar YS 2016 Multifold enhancement of third-harmonic generation in dielectric nanoparticles driven by magnetic Fano resonances *Nano Letters* **16** 4857

[8] Bristow AD, Rotenberg N, Van Driel HM 2007 Two-photon absorption and Kerr coefficients of silicon for 850-2200 nm *Applied Physics Letters* **90** 191104

[9] Garcia H, Kalyanaraman R 2006 Phonon-assisted two-photon absorption in the presence of a dc-field: the nonlinear Franz–Keldysh effect in indirect gap semiconductors *Journal of Physics B* **39** 2737

[10] Gai X, Yu Y, Kuyken B, Ma P, Madden SJ, Campenhout JV, Verheyen P, Roelkens G, Baets R, Luther-Davies B 2013 Nonlinear absorption and refraction in crystalline silicon in the mid-infrared *Laser & Photonics Reviews* **7** 1054

[11] Makarov SV, Petrov MI, Zywietz U, Milichko V, Zuev D, Lopanitsyna N, Kuksin A, Mukhin I, Zograf G, Ubyivovk E, Smirnova DA, Starikov S, Chichkov BN, Kivshar YS 2017 Efficient second-harmonic generation in nanocrystalline silicon nanoparticles *Nano Letters* **17** 3047

[12] Melik-Gaykazyan EV, Shorokhov AS, Zubyuk VV, Kroychuk MK, Shcherbakov MR, Dolgova TV, Fedyanin AA, Choi D-Y, Neshev DN, Kivshar YS 2016 Second and third harmonic generation in silicon metasurfaces with spectrally overlapped electric and magnetic dipolar resonances *Zhurnal Prikladnoi Spectroscopii* **6** 136

IOP Conf. Series: Journal of Physics: Conf. Series **1092** (2018) 012087 doi:10.1088/1742-6596/1092/1/012087

Surface enhanced IR absorption by cortisol molecules

I. A. Milekhin[1,2], S. A. Kuznetsov[1,3], A. G. Milekhin[1,2], T. A. Duda[1], O. P. Cherkasova[4], E. E. Rodyakina[1,2,] A. V. Latyshev[1,2]
[1]Novosibirsk State University, Pirogov 2, 630090 Novosibirsk, Russia
[2]Rzhanov Institute of Semiconductor Physics, Lavrentiev Ave. 13, 630090 Novosibirsk, Russia
[3]Rzhanov Institute of Semiconductor Physics, Novosibirsk Branch "TDIAM", Lavrentiev Ave. 2/1, Novosibirsk 630090, Russia
[4]Institute of Laser Physics, Novosibirsk, Russia

E–mail of the corresponding author: mia2994@gmail.com

Abstract. We report on the study of surface enhanced infrared absorption (SEIRA) by steroid hormone cortisol molecules with different concentrations deposited on linear Au nanoantenna arrays. Localized surface plasmon resonances (LSPRs) arise in the nanoantennas under the influence of external electromagnetic radiation. LSPR frequency depends mainly on the nanoantenna length and can vary from the visible to the terahertz range. We establish the ratio between the structural parameters of nanoantenna arrays and LSPR frequencies based on the results of 3D electrodynamic simulations. Using nanolithography we fabricate nanoantenna arrays having LSPR frequencies close to the frequencies of the most intense absorption modes of steroid hormone cortisol. We deposit cortisol molecules onto the surface of nanoantenna arrays by drop– coating. SEIRA spectra of the nanoantenna arrays make it possible the determination of the presence of cortisol and establishing the sensitivity limit of this method. Thus, we show possibility of SEIRA application for cortisol concentration analysis.

1. Introduction

Among the variety of organic compounds, steroid hormones are of great importance, since they affect many physiological processes in humans and animals. Cortisol is the main glucocorticoid hormone produced both in the adrenal gland and in some other types of tissue, to regulate blood glucose level. Violation in the cortisol production leads to the development of diabetes, Cushing's syndrome and other pathologies [1]. Therefore, the determination of cortisol concentrations in blood and tissue is an important and urgent task.

The typical IR spectroscopy has a relatively low sensitivity for detecting the vibrational modes in organic or inorganic molecules which is however not sufficient for biological sensors applications. The idea of SEIRA is based on specially designed Au nanoantenna arrays providing high electromagnetic field in the immediate vicinity of the plasmon nanostructure under resonance excitation when the frequency of the incident electromagnetic radiation coincides with the eigenfrequency of the electronic oscillations in the nanoantennas and the LSPR arises [2]. As a result, the local electromagnetic field near the edges of the nanoantennas is enhanced, which makes it possible to detect a small amount of substances [3] placed in the local field [4–6]. The linear nanoantenna array consists of metal rods (for example, Au, Al, Ag) with a width of about 100 nm, the characteristic length which can vary in the range from tens of nanometers to several micrometers. As a rule, nanoantenna arrays are fabricated on dielectric (SiO_2, CaF_2, Al_2O_3) or semiconductor (Si) substrates using nanolithography. In our work, we successfully employed SEIRA for detection of cortisol with a low concentration.

2. Experimental

The optimum structure values were determined by electrodynamic simulations using 3D full wave simulations in the software ANSYS Electromagnetics Suite 18 [8] to set up the plasmon resonance to a specific wavelength in the mid infrared spectral range.

Homogeneous arrays of linear Au nanoantennas 3×3 mm^2 in size with structural parameters determined from the calculations were formed on pure Si (001) substrates by focused ion beam as described earlier [7]. The width (height) was chosen to be 60 nm (50 nm), and the antenna length was varied within 2–15 μm. Figure 1 demonstrates a typical scanning electron microscopy (SEM) image of nanoantenna array using the same Raith-150 system at 10 kV acceleration voltage, 30 μm aperture, and 6 mm working distance. Structural parameters determined from the SEM image coincide with the calculated ones with the experimental accuracy of about 10 nm.

Cortisol was extracted from human blood plasma according to the method [9] and extract was dissolved in 24 μl of a mixture of ethanol–water (80:20 v/v) and dropped onto the substrate surface from solutions. The drop volume was 2 μl.

IR transmission spectra of Au nanoantenna arrays with deposited cortisol molecules were measured in the spectral range from 600 to 5000 cm^{-1} using a Bruker Vertex 80v Fourier spectrometer. For further estimations, the ratio of the IR transmittance spectra corresponding to light polarization along the axis of the nanoantenna arrays and perpendicular to it was calculated and analyzed.

3. Results and discussions

The experimentally determined LSPR wavelength depends linearly on the antenna length in a wide spectral range (Fig. 2).

Figure 1. Representative SEM image of a linear Au nanoantenna array.

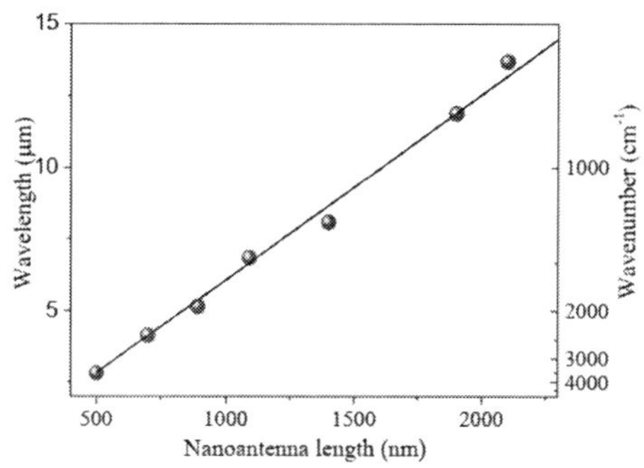

Figure 2. LSPR energy in nanoantenna arrays fabricated on bare Si surfaces as a function of the nanoantenna length.

Cortisol molecule has two carbonyl groups, which are responsible for spectral features in the range of 1600–1700 cm^{-1} [11]. These modes have a strong intensity in IR absorption and were used for further enhancement by SEIRA technique. SEIRA spectra of cortisol covered by nanoantennas with specific lengths providing the LSPR mode (deep minima near 1500 cm^{-1}) with the frequency close to that IR absorption bands of cortisol molecules to achieve the maximum absorption enhancement are shown in Figure 3 [10].

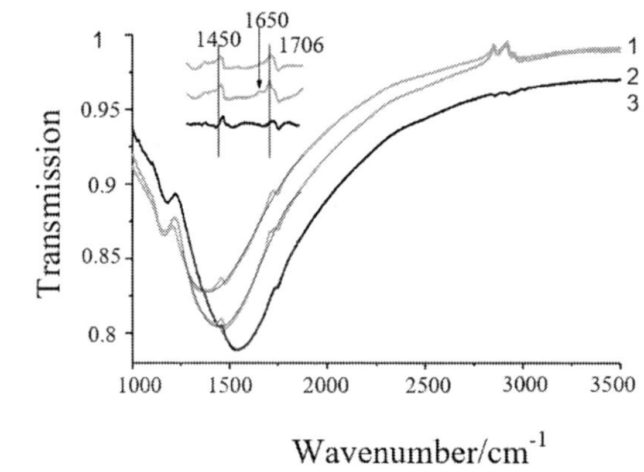

Figure 3. IR spectra of nanoantenna arrays after dropping 8 μl of blood plasma extract with different quantities of cortisol: 1 – low concentration of cortisol (2 μg/μl); 2 – high concentration of cortisol (11 μg/μl); 3 – without cortisol. Inset to the figure shows the fragments of the same spectra with subtracted background.

As you can see from the figure nanoantennas covered by cortisol with different concentration reveal similar IR response that indicates exceeding measurement limit. With decreasing the concentration of cortisol allows to determine the sensitivity limit of cortisol detection which amounts to 10 ng/μl. Thus, a possibility of application of SEIRA for analysis of steroid hormone cortisol in blood plasma was shown.

4. Conclusion

We employed the SEIRA detection of cortisol molecules extracted from human blood and deposited on linear Au nanoantenna arrays. Nanoantennas with optimal structural parameters providing the LSPR frequencies overlapping the characteristic vibrational modes of cortisol were fabricated. We measured IR response by cortisol with different concentration deposited on Au antenna arrays and determined the detection limit of 10 ng/μl.

5. Acknowledgment

This work has been supported by Russian Foundation for Basic Research (project 18–32–00551 mol –a).

References

[1] Cherkasova O P., Activity of 11β– hydroxysteroid dehydrogenase of rat kidney and liver in inherited stress– induced arterial hypertension, Biochemistry (Moscow) **2007**;1:172–175. doi: 10.1134/S1990750807020126.

[2] Schatz, G. C.; Young, M. A.; Van Duyne, R. P. Electromagnetic Mechanism of SERS. In Surface–Enhanced Raman Scattering: Physics and Applications; Kneipp, K.; Moskovits, M.; Kneipp, H., Eds.; Topics in Applied Physics, Vol. 103; Springer: Berlin, Germany, **2006**; pp 19–45.

[3] Nanoantenna: Plasmon–Enhanced Spectroscopies for Biotechnological Applications, ed. Chapelle, M. L. de la and Pucci, A. CRC Press.Taylor & Francis Group: Boca Raton, **2013**. 405 p.

[4] Hartstein, A.; Kirtley, J. R.; Tsang, J. C. *Phys. Rev. Lett.* **1980**, 45, 201– 204.

[5] Osawa, M. Surface–Enhanced Infrared Absorption, Kawata S. (Ed.): Near– Field Optics and Surface Plasmon Polaritons, *Topics Appl. Phys.* **2001**, 81, 163–187, Springer– Verlag Berlin Heidelberg 2001.

[6] Orendorff, C. J.; Gole, A.; Sau, T. K.; Murphy, C. J. *Anal. Chem.* **2005**, 77 (10), 3261–3266.

[7] Yeryukov, N.A.; Milekhin, A.G.; Sveshnikova, L.L.; Duda, T.A.; Rodyakina, E.E.; Sheremet,

E.S.; Ludemann, M.; Latyshev, A.V. ; Zahn, D.R.T. *Thin Solid Films* **2013**, 543, 35–40.
[8] ANSYS, Inc.; url: https://www.ansys.com/products/electronics.
[9] Cherkasova O. P., Biochemistry (Moscow), **2007,** Vol. 1, No. 2, pp. 172–175.
[10] Milekhin, A.G., Cherkasova, O.P., Kuznetsov, S.A., et al, Beilstein Journal of Nanotechnology, **2017**, 8 (1), p. 975– 981.
[11] Minaeva V. A., Minaev B. F., Baryshnikov G. V.,et al. Study of IR and Raman spectra of steroid hormones cortisone, cortisone acetate and cortisol by quantum–chemical method of densityfunctional theory // Cherkassy University Bulletin:Chemical sciences, **2011**, vol. 195, part II, 60–83.

IOP Conf. Series: Journal of Physics: Conf. Series **1092** (2018) 012088 doi:10.1088/1742-6596/1092/1/012088

A new tool for imaging and reconstruction of optically cleared invertebrate animals

V A Milichko[1], V A Dyachuk[1,2]

[1] Department of Nanophotonics and Metamaterials, ITMO University, St. Petersburg 197101, Russia

[2] Department of Neuroscience, Karolinska Institutet, Stockholm, Sweden

slava.dyachuk@ki.se

Abstract. The invertebrate animal group reaches more than 90% of all living organisms and includes more than 1.3 million described species. In zoological and morphological studies, big size animals and their deep anatomy investigations are limited by examination of serially sectioned material using different microscopy techniques. Whole-mount visualization such huge biological samples are problematic using the classical protocol of immunostaining and microscopy techniques. Here we demonstrate advantages and disadvantages of DMSO-based immune-staining protocols and better hand optical clearing of invertebrates for the study of inner organs and nervous system anatomy, in particular, that could be useful for studying of morphology in several groups of animals.

The ability to detect the whole animal body rather than presented in a set of series of histological sections is essential to make easy the deep analysis of separated organ [1-4] and even organ system (blood, nervous, and muscle systems) avoiding the loos information that may can occur in a result of slicing of the tissues. Here, we developed whole-mount immunostaining protocol with the optical clearing of the soft oyster body by organic solvents BABB improving the penetration depth. The present study was focused on detection of neurotransmitters of central and peripheral nervous systems that we were able to see through using modified optical clearing protocol.

We used an adult of bivalve oyster *Crassostrea gigas* to study of inner organ morphology and neuroanatomy. Using polyclonal antibodies against the neuronal markers Phe-Met-Arg-Phe (FMRFamide), 5-hydroxytryptamine (5-HT), and acetylated tubulin to identify neural elements and their distribution in the soft oyster body. Here, we present an approach that allows imaging of whole fixed adult oyster after modified immunostaining procedure together with the deep analysis of neuroanatomy on cryosections. Used these approaches allow us to describe the deeply nervous system and innervation of several inner organs of adult animals.

Used methods facilitate the morphological descriptive analysis of biological objects and together with immunostainings allow to investigate several inner organs and organ systems without destroying of dissections and serial sectioning.

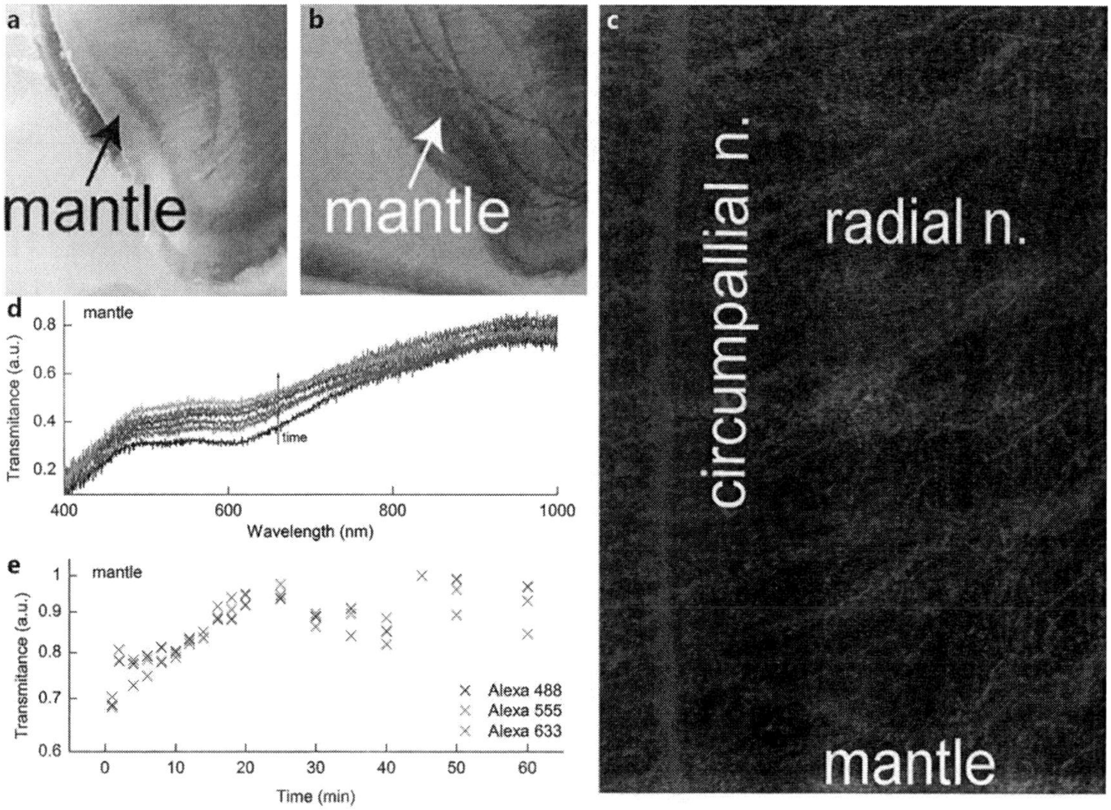

Figure 1. (a,b) Optical images of uncleared and cleared oyster body. (c) The 5-HT-ergic nervous system of adult oyster *Crassostrea gigas*. The red colour is coming from AlexaFlour 555. (d,e) Evolution of the transmission spectrum of oyster mantle under the clearing process in time. Alexa 488, Alexa 555, and Alexa633 indicate the corresponding wavelength at which the dyes luminescent and the transmission increases (e).

References

[1] Chung K, Wallace J, Kim S Y, Kalyanasundaram S, Andalman A S, Davidson T J, Mirzabekov J J, Zalocusky K A, Mattis J, Denisin A K, Pak S, Bernstein H, Ramakrishnan C, Grosenick L, Gradinaru V and Deisseroth K 2013 *Nature* **497** 332.

[2] Dodt H U, Leischner U, Schierloh A, Jährling N, Mauch C P, Deininger K, Deussing J M, Eder M, Zieglgänsberger W and Becker K 2007 *Nature Methods* **4** 331.

[3] Hama H, Kurokawa H, Kawano H, Ando R, Shimogori T, Noda H, Fukami K, Sakaue-Sawano A and Miyawaki A 2011 *Nature Neuroscience* **14** 1481.

[4] Tomer R, Ye L, Hsueh B and Deisseroth K 2014 *Nature Protocols* **9** 1682.

Acknowledgments

The authors wish to thank Prof. Y. S. Kivshar for mentoring this project. V. A. Milichko acknowledges Russian Science Foundation (Grant 17-19-01637, "All-dielectric and hybrid resonant integrated biophotonics") on optical measurements; V. A. Dyachuk acknowledges the President Scholarship (СП-2890.2016.4) on work with invertebrate animals.

Light induced heating of Ge nanoparticle covered by BSA

A A Krasilin[1], D A Zuev[1], V A Dyachuk[1,2], V A Milichko[1]

[1] Department of Nanophotonics and Metamaterials, ITMO University, St. Petersburg 197101, Russia

[2] Department of Neuroscience, Karolinska Institutet, Stockholm, Sweden

v.milichko@metalab.ifmo.ru

Abstract. We report on the changes in conformation of a bovine serum albumin (BSA) adsorbed on surface of single all-dielectric (germanium) nanoparticles following light-induced heating to 640 K. This protein is considerably more resistant to heat when adsorbed on the nanoparticle in air than when in solution or the solid state. These observations will facilitate biocompatible usage of all-dielectric nanoparticles in medicine.

All-dielectric nanophotonics [1] is a new paradigm in optics where the multitude of Mie-type optical resonances govern various linear and nonlinear optical properties of dielectric nanoparticles [1,2]. All-dielectric nanoparticles provide optical field localization and concentration of light into biomolecules [3] that interact poorly with light. Such nanoparticles are very promising candidates for bioimaging [4], biosensing [5] and heat therapy [3,6,7]. However, few available reports on biophotonic applications of all-dielectric nanoparticles have not addressed important information regarding the structure of biomolecules adsorbed on their surface.

To create all-dielectric (germanium) nanoparticles, we employed laser-induced forward transfer technique described in detail elsewhere [8]. The nanoparticles were printed by femtosecond laser onto the acceptor substrate (60-nm gold film on fused silica) in air as shown in figure 1. The crystallinity of single nanoparticles was confirmed by confocal Raman spectroscopy [7]. The peak observed at 301 cm^{-1} with corresponding full width at half maxima of 4.7 cm^{-1} was indicative of crystallinity rather than an amorphous state. BSA can be easily adsorbed onto oxidized germanium nanoiparticles (figure 1) via long-range electrostatic and short-range hydrophobic interactions [9].

The specific sizes of nanoparticles allowed resonant interaction with light, while the non-zero imaginary part of the permittivity promoted light absorption and conversion of light to heat [3,6]. Since heat influences the lattice dynamics of solids, the local temperature of the nanoparticle-biomolecule complex can be determined from the Raman spectrum [3,7].

As shown in figure 2, the dependency of the intensities of the S-S (508 cm^{-1}) and amide I (1650 cm^{-1}) modes on local temperature was similar to that for the protein unfolding in heated solution. However, in our case the slope was less steep, and the residual bonds appeared above 400 K. We speculate that this is due to the steric hindrance of BSA adsorbed on the nanoparticle surface in air [10] as a consequence of strong interactions via electrostatic and hydrophobic forces. The intensity of all modes decreased slowly for the protein adsorbed on germanium nanoparticle. This leads us to speculate that BSA on the surface is more resistant to heat-induced changes in conformation. It should be knotted that BSA in solution or solid state is completely denatured at 350-390 K, when adsorbed on germanium nanoparticles the corresponding temperatures is 450 K (figure 2).

Here we study the changes in the conformation of BSA adsorbed to the surface of single all-dielectric crystalline nanoparticles (germanuim) following light-induced heating to 640 K [7]. Monitoring of S-S and amide I modes via confocal Raman spectroscopy, we revealed that the protein demonstrates the resistance to light-induced heating. This is an important finding for safe utilization of all-dielectric nanoparticles in biology and medicine.

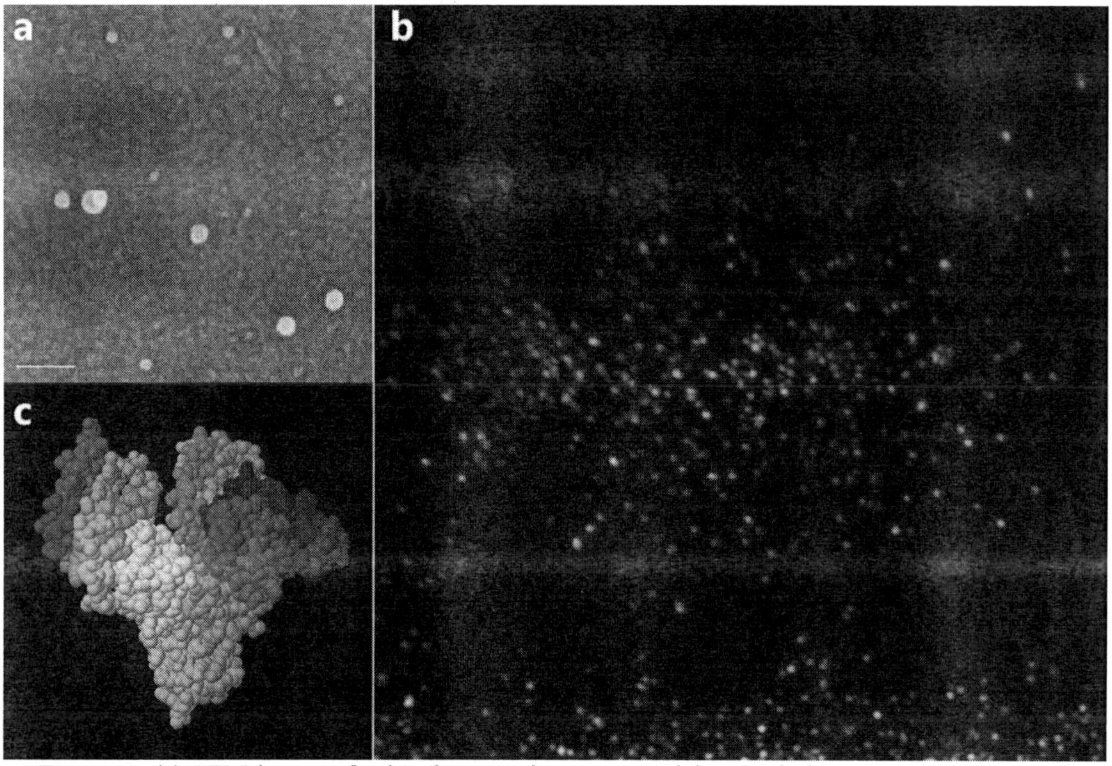

Figure 1. (a) SEM image of printed germanium nanoparticles. Scale bar, 150 nm. (b) Dark field image of germanium nanoparticles onto gold substrate. (c) Schematic illustration of BSA (from https://www.rcsb.org/structure/3V03).

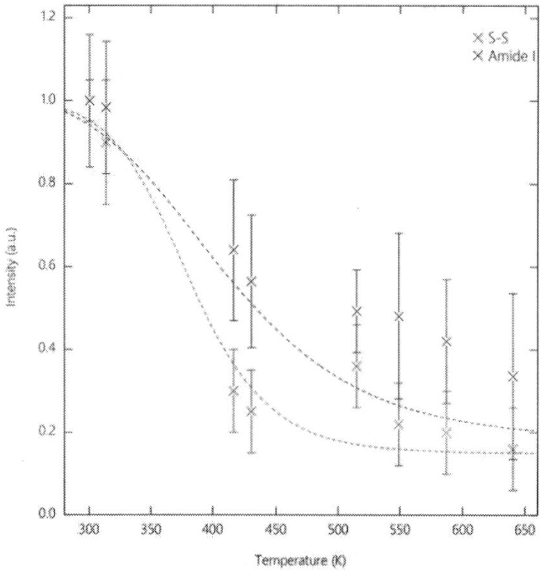

Figure 2. The dependency of the intensities of the S-S (508 cm^{-1}) and amide I (1650 cm^{-1}) modes on the local temperature of the germanium nanoparticle. Experimental values are represented by crosses, the dashed lines represent the fit of these values to the analytical expression described in ref. [7].

References

[1] Kuznetsov A I, Miroshnichenko A E, Brongersma M L, Kivshar Y S and Luk'yanchuk B 2016 *Science* **354** aag2472.

[2] Makarov S V, Petrov M I, Zywietz U, Milichko V, Zuev D, Lopanitsyna N, Kuksin A, Mukhin I, Zograf G, Ubyivovk E, Smirnova D A, Starikov S, Chichkov B N and Kivshar Y S 2017 *Nano Lett.* **17** 3047.

[3] Milichko V A, Zuev D A, Baranov D G, Zograf G P, Volodina K, Krasilin A A, Mukhin I S, Dmitriev P A, Vinogradov V V, Makarov S V and Belov P A 2018 *Laser Photon. Rev.* **12** 1700227.

[4] Pantazis P, Maloney J, Wu D and Fraser S E 2010 *Proc. Natl. Acad. Sci. USA* **107** 14535.

[5] Bontempi N, Chong K E, Orton H W, Staude I, Choi D Y, Alessandri I, Kivshar Y S and Neshev D N 2017 *Nanoscale* **9** 4972.

[6] Zograf G P, Petrov M I, Zuev D A, Dmitriev P A, Milichko V A, Makarov S V and Belov P A 2017 *Nano Lett.* **17** 2945.

[7] Krasilin A A, Volodina K, Sukhova A A, Petrov M I, Zuev D A, Dyachuk V A and Milichko V A 2018 *J. Biophotonics* e201700322.

[8] Dmitriev P A, Makarov S V, Milichko V A, Mukhin I S, Gudovskikh A S, Sitnikova A A, Samusev A K, Krasnok A E and Belov P A 2015 *Nanoscale* **8** 5043.

[9] Kubiak-Ossowska K, Tokarczyk K, Jachimska B and Mulheran P A 2017 *J. Phys. Chem. B* **121** 3975.

[10] Rocker C, Potzl M, Zhang F, Parak W J and Nienhaus G U 2009 *Nature Nanotechnol.* **4** 577.

Acknowledgments

The authors wish to thank Prof. Y. S. Kivshar for mentoring this project. This work was supported financially by Russian Science Foundation (Grant 17-19-01637, "All-dielectric and hybrid resonant integrated biophotonics").

Nanometer scale metal-organic framework nanoparticles for optical application

A Nominé[1], S A Sapchenko[2], J Ghanbaja[1], S Bruyère[1], A S Zalogina[3], P Trofimov[3], A A Krasilin[3], V A Milichko[3]

[1] Université de Lorraine, Institut Jean Lamour, UMR CNRS 7198, Nancy F-54011, France

[2] Nikolaev Institute of Inorganic Chemistry, Siberian Branch of the Russian Academy of Sciences, 3 Acad. Lavrentiev Ave., 630090 Novosibirsk, Russia

[3] Department of Nanophotonics and Metamaterials, ITMO University, St. Petersburg 197101, Russia

v.milichko@metalab.ifmo.ru

Abstract. We report on the laser assisted technique to fabricate nanometer scale derivatives from Cd-based metal-organic frameworks. The nanoparticles obtained have an average size of hundreds of nanometres and represent an amorphous phase with preserved composition. The interaction between light and the nanoparticles promote the optical response (polarization dependent scattering) within the visible range.

Metal-organic frameworks (MOFs) represent a special class of molecular crystals that uniquely combine the physical and chemical properties of both organic and inorganic components. In particular, this synergy provides such crystals with a whole series of nonlinear optical properties like generation of higher harmonics [1,2], topological optical states [3], exciton energy transport [4], laser generation [5], optical recording of information [6], birefringence [7] and many others. The potential for the use of such crystals as functional materials in optics lies in the high nonlinear-optical response of organic components as opposed to inorganic nonlinear crystals, as well as the stability of the structure to high-intensity optical radiation in comparison with purely organic nonlinear polymers due to high crystallinity of MOF. However, for the modern optics (optical data processing and storage), the strategy to miniaturize the functional elements is a common strategy and is a big challenge for metal-organic compounds. In this regards, to fabricate MOF nanoparticles (NPs) and achieve an efficient and rapid control of their optical properties is still a fundamental issue in physics and chemistry to be overcome.

Here we report on the nanostructuring of metal-organic frameworks by a femtosecond laser post-processing. The nanoparticles obtained provide enhanced interaction with radiation due to Mie's optical resonances, which, in turn, can effectively control the linear and nonlinear optical response of MOFs at the nanometer scale.

Since the pioneer report [8], the strategy to nanostructuring MOFs is developed by different chemical approaches: the template method [9], microwave synthesis [10], etching method [11], self-assembly [12] and spray drying [13]. The use of the approaches leads to the fabrication of the NPs and nanocrystals of MOFs, hollow MOF microcapsules, thin MOF films, meso-structures and MOF photonic crystals [14], and nanometer scale MOF derivatives like carbon-based and metallic nanoparticles [15,16]. The main issue to be solved is the improvement of the functional chemical properties of metal-organic frameworks, namely the catalytic, filtration, sorption, sensing and capacitive properties, as well as targeted properties for contrasting in MRI [8]. However, recently proposed laser-assisted technique [17] currently can be applied for different MOFs to fabricate the NPs with different composition and structures allowing the intriguing optical properties.

Following this strategy we fabricate MOF NPs (figure 1a,b,e) from [Cd(sdc)(C4)]·DMF [18]. The nanoparticles obtained demonstrate spherical shape with an average diameter of 500 nm. Transmission electron microscopy and EDX analysis revealed the amorphous inner structure, while the chemical composition is still the same. We also studied the NPs by dark field microspectroscopy: The interaction of light with the NPs provide the scattering of light due to the excitation of Mie type resonances and supports the different color of the NPs at different polarization (*s* and *p*) of light being interesting for nanophotonic application (figure 1c,d).

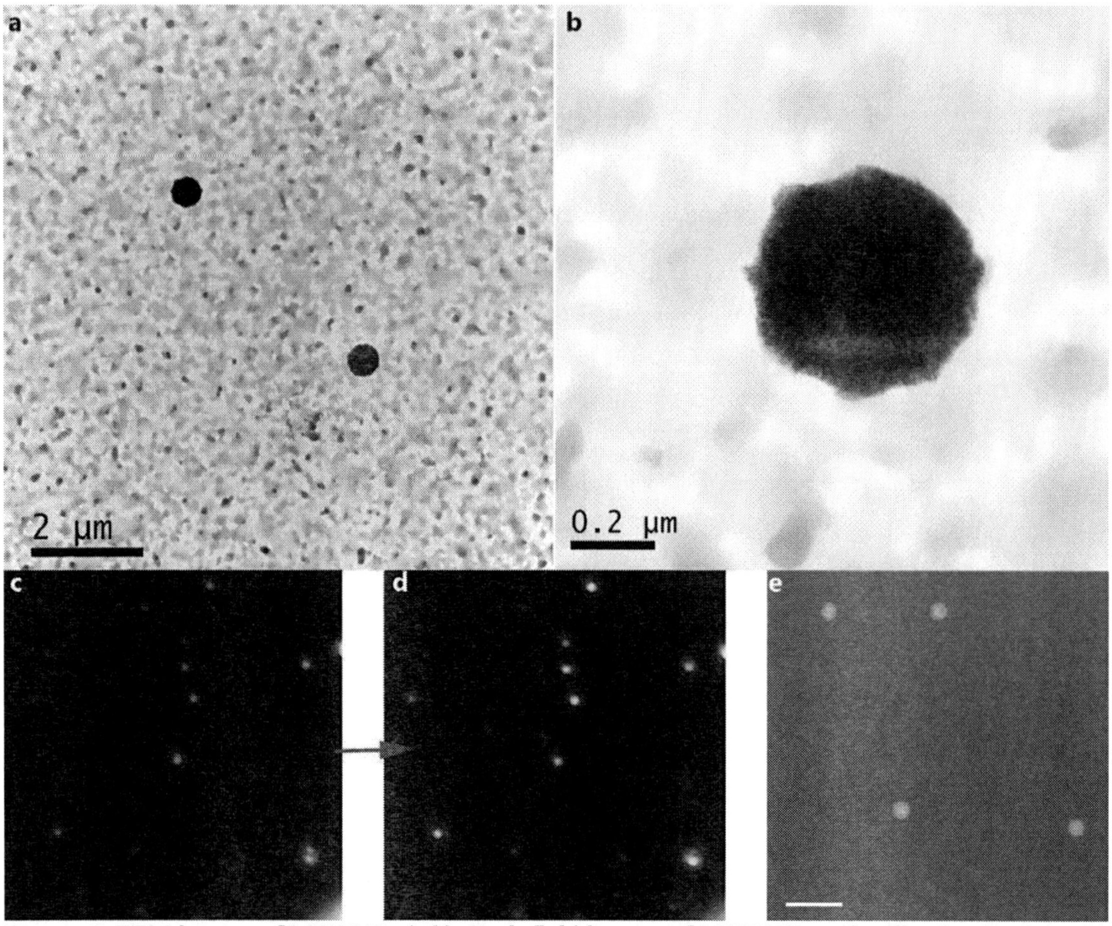

Figure 1. TEM images of MOF NPs (a,b). Dark field images of MOF NPs under illumination by *p* (c) and *s* (d) polarized light. Changing color comes from excited Mie type resonances. (e) SEM image of MOF NPs. Scale bar, 1 μm.

References

[1] Evans O R and Lin W 2002 *Acc. Chem. Res.* **35** 511.
[2] Mingabudinova L R, Vinogradov V V, Milichko V A, Hey-Hawkins E and Vinogradov A V

2016 *Chem. Soc. Rev.* **45** 5408.
[3] Wang Z F, Su N and Liu F 2013 *Nano Lett.* **13** 2842.
[4] Son H J, Jin S, Patwardhan S, Wezenberg S J, Jeong N C, So M, Wilmer C E, Sarjeant A A, Schatz G C, Snurr R Q, Farha O K, Wiederrecht G P and Hupp J T 2013 *J. Am. Chem. Soc.* **135** 862.
[5] Medishetty R, Nalla V, Nemec L, Henke S, Mayer D, Sun H, Reuter K and Fischer R A 2017 *Adv. Mater.* **29** 1605637.
[6] Milichko V A, Makarov S V, Yulin A V, Vinogradov A V, Krasilin A A, Ushakova E, Dzyuba V P, Hey-Hawkins E, Pidko E A and Belov P A 2017 *Adv. Mater.* **29** 1606034.
[7] Vinogradov A V, Milichko V A, Zaake-Hertling H, Aleksovska A, Gruschinski S, Schmorl S, Kersting B, Zolnhofer E M, Sutter J, Meyer K, Lönnecke P and Hey-Hawkins E 2016 *Dalton Trans.* **45** 7244.
[8] Rieter W J, Taylor K M L, An H, Lin W and Lin W 2006 *J. Am. Chem. Soc.* **128** 9024.
[9] Reboul J, Furukawa S, Horike N, Tsotsalas M, Hirai K, Uehara H, Kondo M, Louvain N, Sakata O and Kitagawa S 2012 *Nat. Mater.* **11** 717.
[10] Guo W, Sun W, Lv L P, Kong S and Wang Y 2017 *ACS Nano* **11** 4198.
[11] Hu M, Ju Y, Liang K, Suma T, Cui J and Caruso F 2016 *Adv. Func. Mater.* **26** 5827.
[12] Ameloot R, Vermoortele F, Vanhove W, Roeffaers M B, Sels B F and De Vos D E 2011 *Nat. Chem.* **3** 382.
[13] Carné-Sánchez A, Imaz I, Cano-Sarabia M and Maspoch D 2013 *Nat. Chem.* **5** 203.
[14] Avci C, Imaz I, Carné-Sánchez A, Pariente J A, Tasios N, Pérez-Carvajal J, Alonso M I, Blanco A, Dijkstra M, López C and Maspoch D 2018 *Nat. Chem.* **10** 78.
[15] Pachfule P, Shinde D, Majumder M and Xu Q 2016 *Nat. Chem.* **8** 718.
[16] Wang X X, Hwang S, Pan Y T, Chen K, He Y, Karakalos S, Zhang H, Spendelow J S, Su D and Wu G 2018 *Nano Lett.* doi: 10.1021/acs.nanolett.8b00978.
[17] Milichko V A, Zalogina A, Mingabudinova L R, Vinogradov A V, Ubyivovk E, Krasilin A A, Mukhin I, Zuev D A, Makarov S V and Pidko E A 2017 *AIP Conference Proceedings* **1874** 030025.
[18] Barsukova M O, Sapchenko S A, Kovalenko K A, Samsonenko D G, Potapov A S, Dybtsev D N and Fedin V P 2018 *New J. Chem.* **42** 6408.

Acknowledgments
The authors wish to thank Prof. T. Belmonte (CNRS, Institut Jean Lamour), Prof. V. P. Fedin (Nikolaev Institute of Inorganic Chemistry SB RAS), Prof. E. A. Pidko (TU Delft), and Dr. D. A. Zuev (ITMO University) for fruitful discussion. This work was supported by the Ministry of Education and Science of the Russian Federation (project No 14.587.21.0050 with unique identification RFMEFI58718X0050).

IOP Conf. Series: Journal of Physics: Conf. Series **1092** (2018) 012091 doi:10.1088/1742-6596/1092/1/012091

Second harmonic generation from metal-organic framework derivatives

P Trofimov[1], S A Sapchenko[2], A Zalogina[1], V A Milichko[1]

[1] Department of Nanophotonics and Metamaterials, ITMO University, St. Petersburg 197101, Russia

[2] Nikolaev Institute of Inorganic Chemistry, Siberian Branch of the Russian Academy of Sciences, 3 Acad. Lavrentiev Ave., 630090 Novosibirsk, Russia

v.milichko@metalab.ifmo.ru

Abstract. We report on second harmonic generation (SHG) in Cd-based metal-organic framework (MOF) derivatives. The MOF derivatives have an average size of 100 nm and represent the MOF composition in an amorphous phase. Intriguing, despite the latter, the nonlinear optical properties (SHG) of single MOF crystal and its derivatives is the same.

Metal-organic frameworks (MOFs) represent a unique class of crystalline materials that combine the physical and chemical properties of both organic and inorganic building blocks [1]. In physics, this synergy provides MOFs with a series of nonlinear optical properties as generation of higher harmonics [2,3], birefringence [4], lasing [5], optical recording of information [6] and many others. The utilizing of MOFs as functional materials in optics is due to the high nonlinear-optical response of organic components, as well as the stability of the structure to extremal condition. However, for the nanophotonics, efficient and stable nonlinear optical response of nanoparticles and MOF derivatives is still a fundamental issue in physics and chemistry. Here we use a laser-assisted technique [7] to fabricate MOF derivatives. The latter provide enhanced interaction with radiation and represent stable nonlinear optical response within the visible range at the nanometer scale.

The use of different approaches to fabricate MOF derivatives [8,9] is aimed basically to improve the functional chemical properties of metal-organic frameworks, namely the catalytic, filtration, sorption, sensing and capacitive properties. However, the recently proposed laser-assisted technique to fabricate MOF derivatives can be applied for different composition and provide optical response.

Here we fabricate MOF derivatives (figure 1) from [Cd(dmf)(sdc)(C6)]·DMF [10] with use of laser-assistant technique [7]. The derivatives obtained demonstrate spherical shape with an average diameter of 100 nm. Transmission electron microscopy revealed the amorphous inner structure, while the chemical composition is still the same. The interaction of light with such nanoparticles provide the generation of second optical harmonics being interesting for nanophotonic application of MOF derivatives in biology and imaging (figure 1).

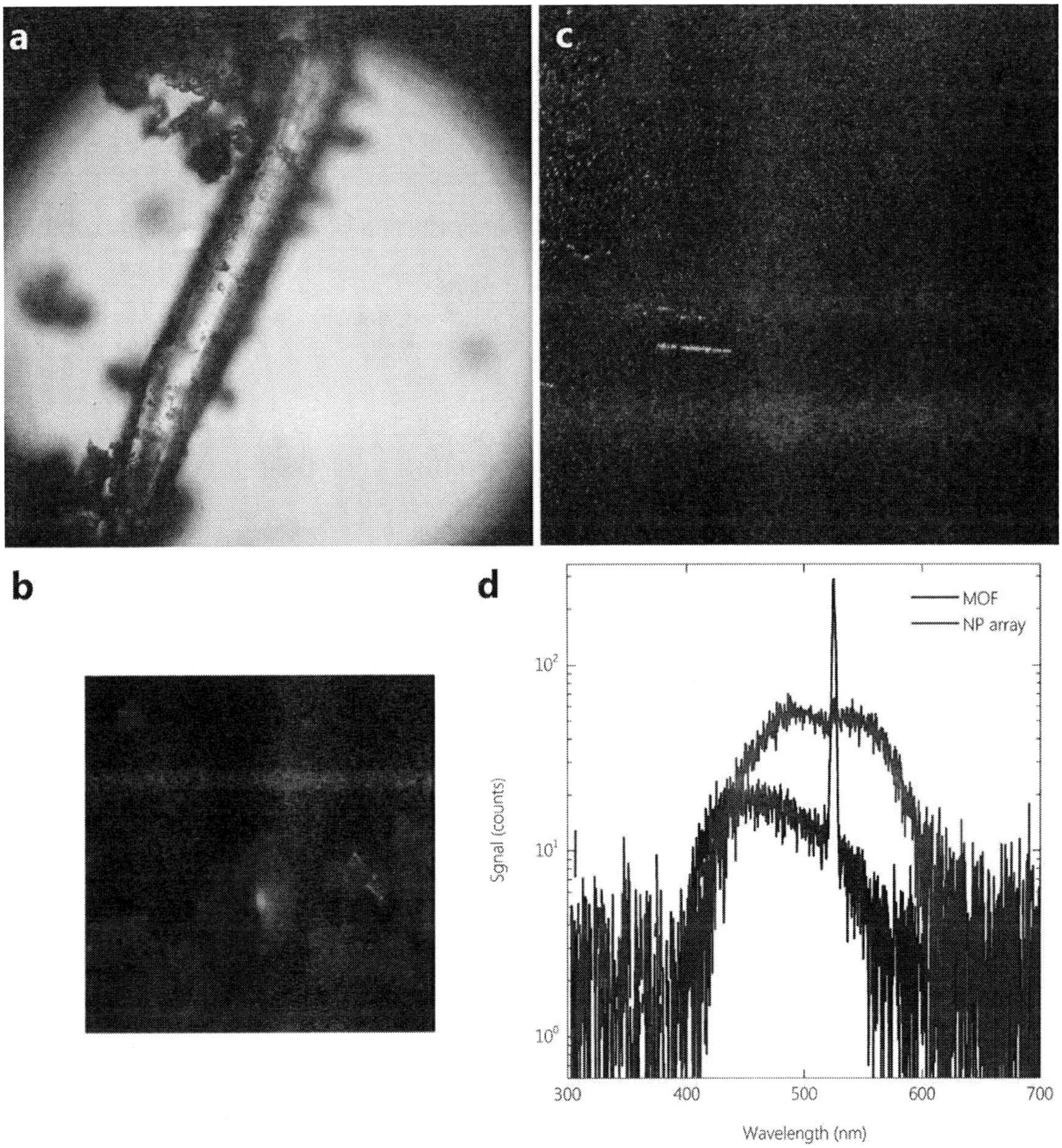

Figure 1. (a) Optical image of single crystal MOF in transmission geometry. (b) SHG from single crystal MOF. (c) Dark field images of MOF derivatives under illumination of white light. Small colored spots are derivatives with an average size of 100 nm. (d) SHG spectrum from single crystal MOF and an array of its derivatives.

References

[1] Zhou H C, Long J R and Yaghi O M 2012 *Chem. Rev.* **112** 673.
[2] Markey K, Krüger M, Seidler T, Reinsch H, Verbiest T, De Vos D E, Champagne B, Stock N and van der Veen M A 2017 *J. Phys. Chem. C* **121** 25509.
[3] Mingabudinova L R, Vinogradov V V, Milichko V A, Hey-Hawkins E and Vinogradov A V 2016 *Chem. Soc. Rev.* **45** 5408.
[4] Vinogradov A V, Milichko V A, Zaake-Hertling H, Aleksovska A, Gruschinski S, Schmorl S, Kersting B, Zolnhofer E M, Sutter J, Meyer K, Lönnecke P and Hey-Hawkins E 2016 *Dalton*

Trans. **45** 7244.
[5] Medishetty R, Nalla V, Nemec L, Henke S, Mayer D, Sun H, Reuter K and Fischer R A 2017 *Adv. Mater.* **29** 1605637.
[6] Milichko V A, Makarov S V, Yulin A V, Vinogradov A V, Krasilin A A, Ushakova E, Dzyuba V P, Hey-Hawkins E, Pidko E A and Belov P A 2017 *Adv. Mater.* **29** 1606034.
[7] Milichko V A, Zalogina A, Mingabudinova L R, Vinogradov A V, Ubyivovk E, Krasilin A A, Mukhin I, Zuev D A, Makarov S V and Pidko E A 2017 *AIP Conference Proceedings* **1874** 030025.
[8] Ji W, Xu Z, Liu P, Zhang S, Zhou W, Li H, Zhang T, Li L, Lu X, Wu J, Zhang W and Huo F 2017 *ACS Appl. Mater. Interfaces* **9** 15394.
[9] Zhang H, Nai J, Yu L and Lou X W 2017 *Joule* **1** 25.
[10] Barsukova M O, Sapchenko S A, Kovalenko K A, Samsonenko D G, Potapov A S, Dybtsev D N and Fedin V P 2018 *New J. Chem.* **42** 6408.

Acknowledgments
The authors wish to thank Dr. A. Nominé, Dr. J. Ghanbaja, Dr. S. Bruyère (CNRS, Institut Jean Lamour) for TEM support; Prof, E. A. Pidko (TU Delft), Dr. A. A. Krasilin (ITMO University) and Dr. K. Kovalenko (Nikolaev Institute of Inorganic Chemistry SB RAS) for fruitful discussion. This work was supported by the Russian Foundation for Basic Research (Grant No. 16-37-60073 mol_a_dk) and by Grant of the President of the Russian Federation (MK-2952.2017.2).

IOP Conf. Series: Journal of Physics: Conf. Series **1092** (2018) 012092 doi:10.1088/1742-6596/1092/1/012092

Gain of terahertz plasmons in a double-layer graphene

M Yu Morozov[1], **I M Moiseenko**[1, 2]
[1]Kotelnikov Institute of Radio Engineering and Electronics (Saratov Branch), Russian Academy of Sciences, Saratov 410019, Russia
[2]Physics Department, National Research Saratov State University, Saratov 410012, Russia
Email: mikhail.yu.morozov@gmail.com

Abstract. Amplification of terahertz plasmons in a pair of parallel active graphene monolayers is studied theoretically. It is shown that symmetric mode increment of plasmons in the double-layer graphene may be greater than that in a single graphene layer by approximately two times of magnitude.

1. Introduction

Graphene, being natural two-dimensional material with zero band-gap [1], appears as a perfect platform for terahertz (THz) radiation amplification [2]. Justification of possibility of possessing negative conductivity of graphene at THz frequencies [3] along with the possibility of excitation of highly confined graphene plasmons [4] resulted in proposals of THz graphene lasers [5] and plasmonic amplifiers [6–8]. Double-layer graphene structure consists of two parallel graphene monolayers with a narrow dielectric gap (barrier layer) between them. Electromagnetic fields of plasmons propagating in these layers interact with each other giving rise to a single unified plasmon in the pair of parallel graphene monolayers surrounded by dielectric claddings [9]. In this paper, we study the amplification of THz plasmons in the double-layer graphene.

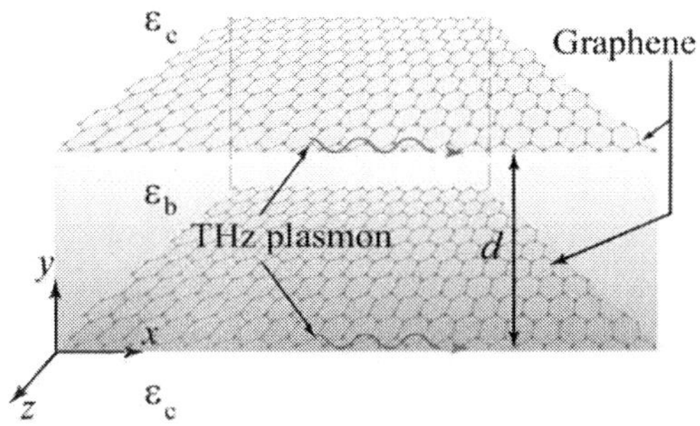

Figure 1. Schematic view of the structure under consideration.

2. Theoretical model

The structure under consideration is schematically shown in Fig. 1. We consider a symmetric structure with two coupled graphene layers with identical inverted population of free charge carriers in each graphene layer which are separated by a dielectric gap (barrier layer) of thickness d and covered by dielectric claddings. Dielectric permittivity of surrounding materials, ε_c are equal for the top and bottom claddings while the value of dielectric permittivity of the barrier layer, ε_b, differs from ε_c.

Representing the spatio-temporal dependence of the plasmon fields in the form $\propto \exp(-i\omega t + ik_x x \mp k_{yc} y)$ in claddings (minus and plus sign stands for the top and bottom cladding, respectively) and $\propto \exp(-i\omega t + ik_x x)\left[A_{0b}^{+} \exp\left(k_{yb} y\right) + A_{0b}^{-} \exp\left(-k_{yb} y\right)\right]$ (A_{0b}^{+} and A_{0b}^{-} are the amplitudes of electric or magnetic fields of the forward and the counter waves) in the barrier layer, we write the dispersion relation as [9]

$$\tanh\left(k_{yb} d\right) = -\frac{2\Psi}{1+\Psi^2} \tag{1}$$

with

$$\Psi = \frac{k_{yb}}{\varepsilon_b}\left[\frac{\varepsilon}{k_{yc}} + i\frac{\sigma(\omega)}{\varepsilon_0 \omega}\right], \quad k_{yj} = \sqrt{k_x^2 - \varepsilon_j \omega^2 / c^2}, \tag{2}$$

where $j = \mathrm{b}, \mathrm{c}$ stands for the barrier layer and claddings, respectively, k_x and ω are the wavevector and frequency of the plasmon, ε_0 is the electric constant, $\sigma(\omega)$ is the dynamic graphene conductivity [6]. In a symmetric structure the dispersion equation (1) splits into two branches:

$$\tanh\left(k_{yb} d/2\right) = -1/\Psi \tag{3}$$

corresponding to the symmetric mode and

$$\tanh\left(k_{yb} d/2\right) = -\Psi, \tag{4}$$

defining the antisymmetric mode. These modes are termed according to the tangential to graphene component of the plasmon electric field distribution across the symmetry plane of the structure under consideration, which is symmetric for symmetric mode and antisymmetric for the other one.

3. Results and discussion

In Fig. 2, one can see the symmetric mode increment (gain) of plasmons defined as $\alpha = -2\,\mathrm{Im}\,k_x$ as a function of the barrier layer thickness *d*. With decreasing the barrier layer thickness *d*, the plasmon wavelength monotonically increases (see curve *2* in Fig. 2) leading to increasing the localization length $L_{\mathrm{loc}} = 1/\mathrm{Re}\,k_x = \lambda/2\pi$ of plasmon electric field. The dependence of symmetric mode increment on barrier layer thickness *d* has an extremum (see curve *1* in Fig. 2). Let us define the physical reasons of such behaviour. The power density, released from graphene layers $P = -\mathrm{Re}\,\sigma \left|E_{x\mathrm{gr}}\right|^2$, where $\left|E_{x\mathrm{gr}}\right|$ is the tangential to graphene component of the plasmon electric field in graphene layers and plasmon energy flux in the *x* direction $W = 1/2 \int_{-\infty}^{\infty} \mathrm{Re}\left(E_y \cdot H_z\right) dy$ obey to a simple relationship representing the energy conservation law: spatial variation of energy flux W is defined by the value of power density released from graphene layers, P, i.e. $dW/dx = P$. Taking into account that P and W are proportional to $\exp(-2\,\mathrm{Im}\,k_x)$, the symmetric mode gain α could be rewritten as $\alpha = -2\,\mathrm{Im}\,k_x = P/W$. In other words optical mode gain is power density, released from graphene layers at fixed plasmon energy flux $W = 1\ \mathrm{W/m}$.

With decreasing the barrier layer thickness *d*, the power density P initially increases due to constructive interference of the plasmon fields in both graphene monolayers. At large separation between graphene layers the plasmon energy flux W (that we assume to be fixed) is mainly concentrated in the barrier layer because the dielectric permittivity value of barrier layer is bigger than that of the claddings (see curve *1* in Fig. 3). At small barrier layer thicknesses the plasmon energy flux W is mainly concentrated in claddings (see curve *2* in Fig. 3). Thus, the decreasing of P is caused by weaker localization of the plasmon electric field in the vicinity of graphene layers at small barrier layer thicknesses due to concentration of the plasmon energy flux in the claddings with small value of the dielectric permittivity.

In conclusion, we have studied the gain of the symmetric plasmon mode in the double-layer graphene heterostructure. It is shown that symmetric mode increment of plasmons in the double-layer graphene may be greater than that in a single graphene layer by approximately two times of magnitude. Amplified plasmons can be used in low-loss interconnects and active elements in the THz plasmonic graphene nanocircuits.

The authors are grateful Russian Science Foundation for the support of the work through Grant No. 18-72-00009.

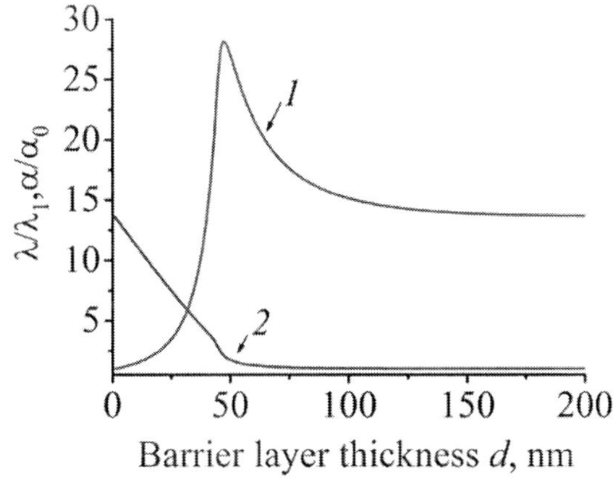

Figure 2. The symmetric mode increment of plasmons (curve *1*) and plasmon wavelength (curve *2*) versus barrier layer thickness *d*.

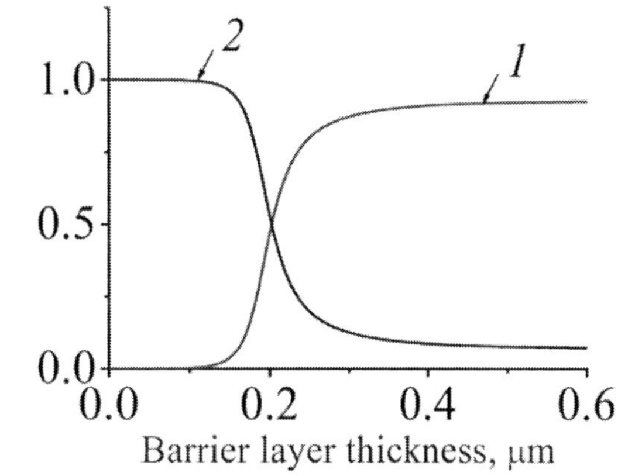

Figure 3. The plasmon energy flux *W* in the barrier layer (curve *1*) and in claddings (curve *2*) versus barrier layer thickness *d*.

References

[1] Novoselov K S, Fal'ko V I, Colombo L *et al.* 2012 *Nature* **490** 192
[2] Grigorenko A N, Polini M, and Novoselov K S 2012 *Nature Photonics* **6** 749
[3] Ryzhii V, Ryzhii M, Otsuji T 2007 *Journal of Applied Physics* **101** 083114
[4] Koppens F H L, Chang D E, and Garcıa de Abajo F J 2011 *Nano Lett.* **11** 3370
[5] Aleshkin V Ya, Dubinov A A, Ryzhii V 2009 *JETP Lett.* **89** 63
[6] Dubinov A A, *et al.* 2011 *J. Phys.: Condens. Matter* **23** 145302
[7] Popov V V *et al.* 2012 *Phys. Rev. B* **86** 195437
[8] Morozov M Yu *et al.* 2015 *Appl. Phys. Lett.***106** 061105
[9] Gan C H, Chu H S, and Li E P 2012 *Phys. Rev. B* **85** 125431

IOP Conf. Series: Journal of Physics: Conf. Series **1092** (2018) 012093 doi:10.1088/1742-6596/1092/1/012093

Unit cell for 1-bit reconfigurable transmitarray with circular polarization

I V Munina, V V Kirillov, A P Verevkin and P A Turalchuk

Microelectronics and Radio Engineering Department,
St. Petersburg Electrotechnical University "LETI", Saint Petersburg, Russia

IVMunina@etu.ru

Abstract. The unit cell structure for a transmitarray with circular polarization is presented. The unit cell operates at 5.8 GHz. The dependence of the phase shift on the unit cell parameters is investigated. The realization of the reconfigurable unit cell providing two out-of-phase states is under consideration. The proposed unit cell can be promising for a design of the transmitarray with electronic beamsteering capability.

1. Introduction

Increasing number of users of modern wireless communication systems and rising the data throughput creates new challenges in the transceiver design. An antenna array as a key component requires to provide wide angle electronic beam steering capability for reliable user access. For increasing data throughput, the operational frequencies go up to mm-waves and for overcoming path losses the antenna systems have to provide high gain. The traditional realization of an analog phased antenna array requires a lot of phase shifters, that makes the structure more complicated and increases the system cost. Another approach to a high-gain antenna design is using dielectric lenses [1]. However, the beam steering in such structures is realized by switching the feed antenna position, that allows to provide beamsteering in a very narrow angular sector. Also, the disadvantage of the dielectric lenses is the volumetric implementation. In general, there is a challenge to design low cost, low loss and easy implemented antenna system.

In the last years, it became popular the planar design of the beamforming network such as a transmitarray, a reflectarray and a frequency selective surface. The beamforming is usually realized in dielectric lenses by changing the feed position. The other way is to incorporate the tunability in side of unit cell of such systems. The transmitarray antenna is typically formed by half wavelength unit-cells and a source placed at a focal distance F. The unit cells generate a phase distribution across the array and form a plane wave front. If the phase shift is controlled electronically in each unit cell, a reconfigurable antenna is obtained with beam-forming and beam-steering capabilities. The different types of a realization of the reconfigurable transmitarray by using varactor diodes and PIN-diodes are shown in [2-4].

In this paper, a novel type of unit cell for 1-bit reconfigurable transmitarray with circular polarization operating at 5.8 GHz is proposed. The unit cell is designed as two defected patches excited by U-shaped resonators connected by vias. Changing the position of U-shaped strip excitation by using two SPDT switches allows changing the phase shift from 0° to 180° and providing beam steering capability of the transmittarray.

2. The unit-cell architecture

In the transmitarray, the radiation from the source focuses due to forming a special parabolic phase distribution along the aperture for compensating phase advances from the feed. The phase distribution can be formed by varying the size parameters of unit cell. The required phase distribution forming the plane wave defines by a focal distance between the feed and the transmitarray and by the size of the aperture of the transmitarray. The beamsteering capability is provided by changing the initial phase distribution along the aperture by changing the phase delay independently in each unit cell. In [5] it was shown that 1-bit transmitarray with 180° resolution is capable to realize wide angle scanning and even polarization manipulation.

2.1. Unit cell structure

The proposed unit cell is formed by two patches with square defects. Adding the defects in the patch structure allows increasing the way of current flow and in a such way to decrease the size of the patches. Each defected patch is capacitively loaded by two U-shaped half wavelength resonators (Figure 1-(c)). Each U-shaped resonator of receiving part is connected to U-shaped resonator of transmitting part by via hole. The pairs of U-shaped resonators are placed in orthogonal planes. The structure can be excited by plane wave with linear polarization (TE or TM) or circular polarization. The location of the U-shaped resonators is shown in Figure 1(b). In turn, receiving and transmitting layers are divided by ground plane. The metallic patterns are etched on Rogers RO4003 (ε=3.5, tanδ=0.0018) substrate with a thickness of 0.51 mm. The two layers of bonding film Rogers RO4003C (ε =3.38, h=0.1mm) are used between the layers of RO4003.

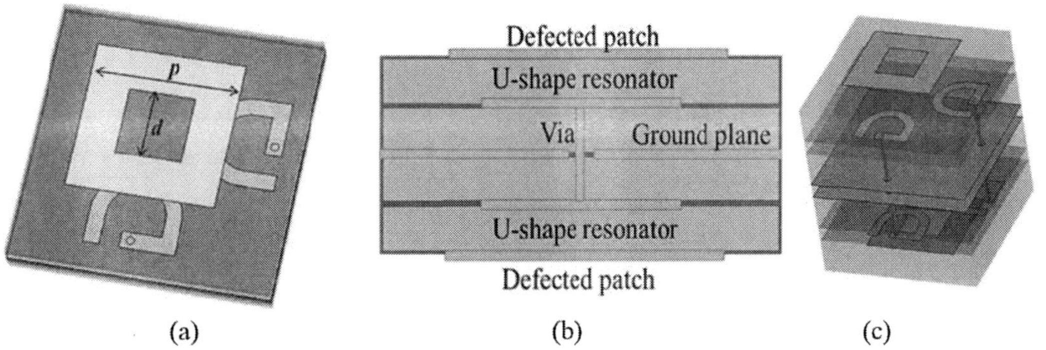

(a) (b) (c)

Figure 1. Details of the proposed unit-cell architecture: 3D view (a), side view (b) and layer-by-layer view (c).

The phase distribution along the transmit array can be formed by varying the geometry of unit cell. We consider two parameters: the size of the patch p and the size of the defect in the patch d. It is important to note that the phase variations are limited by preserving the operational passband. At the graphs in Figure 2 the dependences of the magnitude and the phase of transmission coefficient of d,

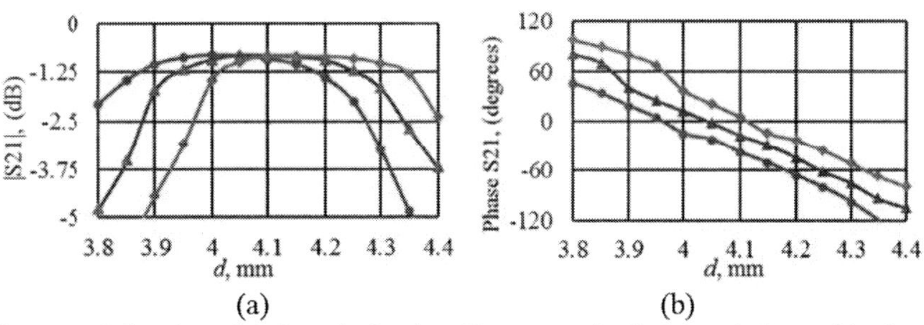

(a) (b)

Figure 2. Influence of the size of defect d of unit cell on magnitude (a) and phase (b) of transmission coefficient for patch size p=12.2 mm (red line); 12.25mm(blue line); 12.3mm (green line) at 5.8 GHz.

for p = 12.2 mm (red curve); 12.25 mm(blue curve); 12.3 mm (green curve), are presented at 5.8 GHz. Both graphs show that the transmission coefficient is very sensitive to size changing. For the patch size 12.25 mm on condition preserving the magnitude of the transmission coefficient better than -1 dB the phase changes from 12.45° to -28° for the defect size from 4 mm to 4.2 mm. The further increasing of the defect size leads to shifting the operational bandwidth of the unit to the higher frequencies.

2.2. 1-bit reconfigurable unit cell

In the proposed unit cell, the 180°phase resolution can be realized. For that, it is necessary to place two RF SPDT switches between the two ends of the U-shaped resonators on the receiving side of the transmitarray (Figure 3). The switch state defines the direction of the current flow in the U-shaped resonator and the patch on the receiving part. The state, when the U-strips are connected from the same side ends is called 0° state, because the current flowing in the patches are codirectional. For the opposite case, the current flows in the patches in inverse direction, and the phase shift changes on 180°. The results of full wave EM simulation of the reconfigurable unit cell (p = 12.3 mm, d = 4.2 mm) are shown in Figure 4. In the passband of the structure the phase difference between phases in the two states is equal 180° (Figure 4 (b)). Independent control of the switch states of the each unit cell of the transmitarray allows to form different phase distribution and provides beamsterring capability.

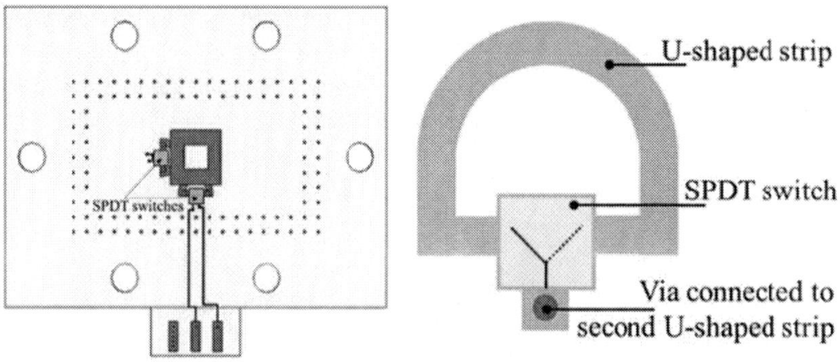

(a) (b)

Figure 3. The test structure of the reconfigurable unit cell (a) and switch position (b)

For the experimental study, the unit cell will be placed inside the rectangular waveguide. As a part of the unit cell the RF switch SKY13585-679LF will be used. It is an ultra-miniature SPDT switch providing a high-isolation, low dc power consumption from SKYWORKS. The insertion loss of the switch at 5.8 GHz is 0.5 dB and the isolation is lower than 20 dB, respectively.

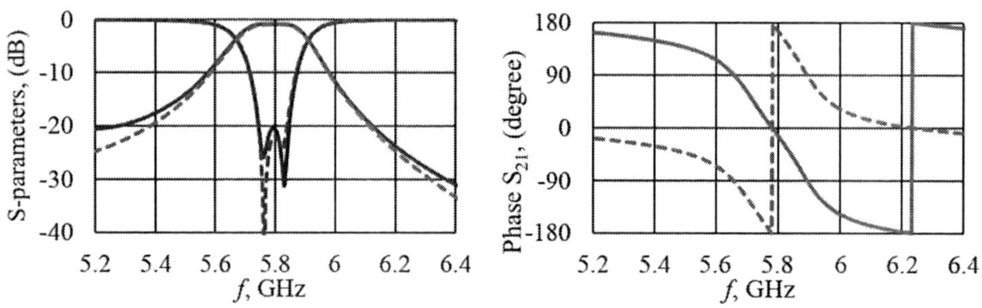

(a) (b)

Figure 4. Simulated frequency response of 0° (solid lines) and 180° (dashed lines) phase states for a single unit-cell: reflection coefficient (blue) and transmission coefficient (red) (a), phase of transmission coefficient (b). Periodic boundary conditions and normal incidence of wave have been considered.

3. Conclusion
In this paper the multilayer structure of the reconfigurable unit cell for 1-bit transmit array is proposed. The simulated results of the phase of transmission coefficient for 2 states are shown. The scenario for experimental results validation is suggested. The unit cell will be tested as a part of a transmit array with electronic beamsteering. As a future research the design will be scaled to mm wave range.

Acknowledgment
The authors would like to thank Dr. Dmitry Kozlov for fruitful discussions of the work. The work was supported by Russian Science Foundation (project 17-79-20374).

References
[1] Artemenko, Alexey, et al. "Millimeter-wave electronically steerable integrated lens antennas for WLAN/WPAN applications." *IEEE Transactions on Antennas and Propagation* **61.4** (2013): 1665-1671.
[2] Huang, Cheng, et al. "Using reconfigurable transmitarray to achieve beam-steering and polarization manipulation applications." *IEEE Transactions on Antennas and Propagation* **63.11** (2015): 4801-4810.
[3] Frank, Martin, Robert Weigel, and Alexander Koelpin. "Design of a 24 GHz reconfigurable transmitarray element with continuous phase range." *Antennas and Propagation (EUCAP), 2017 11th European Conference on. IEEE*, 2017.
[4] Lau, Jonathan Y., and Sean V. Hum. "A wideband reconfigurable transmitarray element." *IEEE Transactions on Antennas and Propagation* **60.3** (2012): 1303-1311.
[5] Di Palma, Luca, et al. "Circularly-polarized reconfigurable transmitarray in Ka-band with beam scanning and polarization switching capabilities." *IEEE Transactions on Antennas and Propagation* **65.2** (2017): 529-540.

IOP Conf. Series: Journal of Physics: Conf. Series **1092** (2018) 012094 doi:10.1088/1742-6596/1092/1/012094

Enhancement of the intensity magneto-optical effect in magnetophotonic metasurfaces

A I Musorin, M G Barsukova, V S Snigirev, A S Shorokhov and A A Fedyanin

Quantum Electronics Department and Quantum Technology Center, Faculty of Physics, Lomonosov Moscow State University, Moscow 119991, Russia

E-mail: `fedyanin@nanolab.phys.msu.ru`

Abstract. Intensity magneto-optical response enhancement is shown numerically in magnetophotonic metasurfaces. Spectral overlapping of magnetic and electric dipole Mie resonances leads to increase of the effect and to modification of the resonance line shape.

1. Introduction

The field of optical metasurfaces is one of the most rapidly developing area of the modern photonics nowadays [1, 2]. These subwavelength structures are two-dimensional analogues of metamaterials. They allow one to control an amplitude and a phase of light opening new ways for creation of flat optical devices [3]. A lot of studies are devoted to develope an active metasurface, which can be controlled by an external stimulus [4]. Magnetic field can be one of it. Its influence has been recently shown on the transmittance of the array of hybrid Ni/Si nanoparticles when the magnetic dipole (MD) Mie resonance is excited [5, 6]. In this research, the next step is realized by studying magneto-optical response enhancement due to the hybridization of magnetic and electric dipole (ED) modes in magnetophotonic metasurfaces.

2. Sample

The sample under study is a 2D dense array of silicon nanodisks covered with 5-nm-thick nickel nanodisks. The sample is designed to support excitation of electric and magnetic dipole Mie resonances in the near-IR region. The nanoparticles are embedded into a dielectric matrix with the refractive index of 1.5. Height of Si disk is 200 nm. Two diameters are examined: 275 and 325 nm. The lattice constants are 475 and 525 nm, respectively. The wavelength-dependent dielectric constants of silicon and nickel [7] are taken into account as well as the dispersion of non-diagonal elements of the permittivity tensor of nickel [8].

3. Numerical calculations

The numerical calculations are based on the finite-difference time-domain technique. Optical and magneto-optical spectra are simulated. The intensity magneto-optical effect is examined for transmittance in the Voigt geometry. Configuration of a wave vector, a linear polarization and a direction of external magnetic field should satisfy the right-hand rule. Then, variation of the transmittance induced by the magnetic field can be defined as

$$\delta = \frac{T(H) - T(0)}{T(0)}, \tag{1}$$

where T(H) and T(0) are transmittance of the sample in case of existence and absence of the magnetic field, respectively. Figure 1 shows optical (black curve) and magneto-optical responses (red curve) spectra for the structures under study. The excitation of separate MD and ED modes leads to the resonant increase of the magneto-optical signal (see Fig. 1 (a)). Enlarging of the disks results in spectral shift of these resonances (see Fig. 1 (b)). It leads to the case when phases of resonances interfere constructively causing significant magneto-optical enhancement up to 20%.

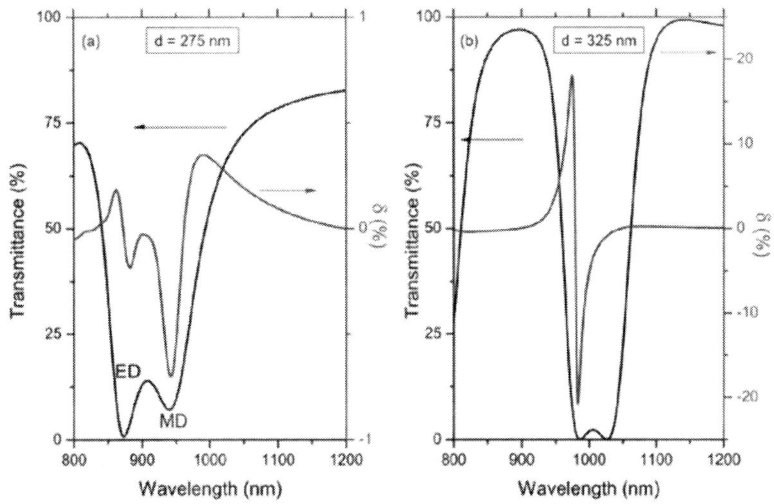

Figure 1. Optical (black curves) and magneto-optical (red curves) spectra of metasurfaces with disk diameters (a) d = 275 nm and (b) d = 325 nm.

4. Conclusions

We demonstrate that magnetophotonic metasurfaces can modulate transmittance of the sample up to 20% by external magnetic field. These results can be used for the further optimization of active metasurfaces.

Acknowledgments

The authors are grateful to Y.S. Kivshar, D.N. Neshev and B.S. Lukyanchuk for fruitful discussions. This work was partly supported by Russian Ministry of Education and Science (grant 14.W03.31.0008, calculation of optical spectra). Magneto-optical calculations were performed as a part of the President of the Russian Federation for leading scientific schools project (grant No. NSh-6862.2018.2), supported by Russian National Technology Initiative and Russian Foundation for Basic Research (grant No. 18-52-50021).

References

[1] Yu N and Capasso F 2014 *Nat. Mater.* **13** 139
[2] Kuznetsov A I, Miroshnichenko A E, Brongersma M L, Kivshar Y S and Luk'yanchuk B 2016 *Science* **354** aag2472

IOP Conf. Series: Journal of Physics: Conf. Series **1092** (2018) 012094 doi:10.1088/1742-6596/1092/1/012094

[3] Yu Y F, Zhu A Y, Paniagua Domnguez R, Fu Y H, Luk'yanchuk B and Kuznetsov A I 2015 *Nat. Mater.* **9** 412
[4] Decker M and Staude I 2016 *J. Opt.* **18** 103001
[5] Barsukova M G, Shorokhov A S, Musorin A I, Neshev D N, Kivshar Y S and Fedyanin A A 2017 *ACS Photonics* **4** 2390
[6] Musorin A I, Barsukova M G, Shorokhov A S, Neshev D N, Kivshar Y S and Fedyanin A A 2018 *J. Magn. Magn. Mater.* **459** 165
[7] Palik E D 2012 *Handbook of Optical Constants of Solids* (Boston: Academic Press)
[8] Krinchik G S and Artemev V A 1968 *Sov. Phys. JETP* **26** 1080

Corrosion properties of composite coatings on AMg3 alloy surface

K V Nadaraia[1,2], D V Mashtalyar[1,2], V S Egorkin[1,2], I E Vyaliy[1], I M Imshinetsky[1], S L Sinebryukhov[1] and S V Gnedenkov[1,2]

[1] Institute of Chemistry FEB RAS, 159 Pr. 100-letiya Vladivostoka, Vladivostok 690022, Russia

[2] Far Eastern Federal University, 8 Sukhanova St., Vladivostok 690091, Russia

Corresponding author: nadaraiakv@mail.ru

Abstract. Method of the composite polymer-containing coatings formation on the AMg3 aluminum alloy by plasma electrolytic oxidation (PEO) with the subsequent application of a polymer from the suspension of superdispersed polytetrafluoroethylene is presented. The optimal parameters such as a multiplicity of the application, the temperature and the time of heat treatment were determinate in order to obtain coatings with the highest protective properties. Composite layers significantly improve the anticorrosion properties of aluminum alloy, reducing corrosion current density by 2 orders of magnitude in comparison with the base PEO-coating. This makes it possible to significantly reduce the possibility of corrosion damage of the protective coating during operation, and, consequently, improve its reliability.

1. Introduction

Aluminum alloys as structural and functional metals have a number of technical and economic advantages, providing expediency of their use in aircraft, automotive and electronics industries [1–2]. The mechanical properties of a number of aluminum alloys are not inferior to the strength of steels, which makes it possible to reduce the metal consumption of the structure, facilitates the transportation of equipment, installation and repair work, etc.

However, aluminum alloys have a low corrosion resistance in some aggressive media [3–4]. One of the most promising methods of aluminum alloys protection is the application of coatings on their surface by plasma electrolytic oxidation (PEO), which allows to form anticorrosive coatings on the surfaces of metals and alloys [5–7]. Moreover, the convoluted and rough surface of the PEO-coating allows to form the composite polymer-containing layers in order to provide additional protection to products [8, 9]. Particular interest such polymer component of composite coatings (CC) are fluoropolymer compounds due to their unique resistance to various aggressive media.

Earlier, we developed methods of introducing polymers into the porous part of the PEO-coating, which made it is possible to purposefully form composite protective layers using the suspensions of superdispersed polytetrafluoroethylene on a magnesium alloy [10, 11]. The use of the PEO-method and fluoropolymer materials with directed formation of composite coatings provides a number of advantages in comparison with other approaches. This expands the range of practical application of metals and alloys. In particular, improving of the electrochemical characteristics of the surface layers of the material is extremely useful for aluminum alloys.

In this paper, original methods for the formation of protective composite polymer-containing layers on the AMg3 aluminum alloy are presented.

2. Materials and methods

For the formation and investigation of CC the AMg3 aluminum alloy samples were used. Samples were in the form of rectangular plates with the size of 15 mm × 30 mm × 1 mm. Prior to oxidation all samples were mechanically treated with grinding paper with a sequential reduction of grain size down to 14 μm in order to remove burrs, outgrowths, and various defects from the metal surface, washed with distilled water and degreased with ethanol. This is necessary for standardizing the surface of the samples. As an electrolyte for PEO was used following aqueous solution: 0.6 g l^{-1} NaF, 5 g l^{-1} $C_4H_4O_6K_2 \cdot 0.5H_2O$, 10 g l^{-1} $Na_2MoO_4 \cdot 2H_2O$, 10 g l^{-1} $Na_2B_4O_7 \cdot 10H_2O$, and 10 g l^{-1} $Na_3PO_4 \cdot 12H_2O$. A 15 % suspension of superdispersed polytetrafluoroethylene (SPTFE) in isopropanol was used to prepare a composite polymer-containing coating.

Formation of the PEO-coatings was carried out in a bipolar mode, including galvanostatic and potentiodynamic stages. During the PEO process, in the cathodic part at all stages the current density was equal to –0.1 A cm^{-2}. The anodic component was potentiostatically stabilized at a value of 30 V at the first stage. At the second stage, the anodic voltage was increased from 30 to 420 V. Then, at the third stage, the anodic component decreased from 420 to 230 V. The oxidation time was equal to 45 min. The PEO-layers were coated with a polymer by immersion in a suspension of SPTFE followed by heat treatment. The polymer was applied one- (CC-1x), two- (CC-2x), and threefold (CC-3x). Most homogeneous layer was obtained by the slow immersion, exposure to the solution for 10–15 s, and rapid withdrawal. An applied layer was predried at air for 20 min at 25 °C. Next step was a heat treatment at 350 °C during 15 min.

The surface morphology of formed coatings was investigated with the use of an Evex Mini-SEM scanning electron microscope (SEM) (Evex Analytical Instruments, USA).

The electrochemical properties of the coatings were studied using the VersaSTAT MC electrochemical system (Princeton Applied Research, USA). The measurements were carried out in a three-electrode cell at room temperature in a 3 wt. % NaCl solution. The niobium mesh covered with platinum was used as a counter electrode. The saturated calomel electrode (SCE) was used as a reference electrode. The exposed area of the sample was equal to 1 cm^2. The samples were kept in a solution for 30 min prior the start of the electrochemical tests to achieve steady state. The potentiodynamic polarization measurements were carried out at a scan rate of 1 mV s^{-1}.

To describe the experimental dependence of the current density I on the potential E, the Levenberg-Marquardt method was used (1) [12]:

$$I = I_C\left(10^{\frac{E-E_C}{\beta_a}} + 10^{-\frac{E-E_C}{\beta_c}}\right) \tag{1}$$

This approach allows one to obtain the most precise calculated values of such parameters as corrosion potential E_C, corrosion current density I_C, the slope of the cathodic polarization curve β_c, and the slope of the anodic polarization curve β_a.

The polarization resistance R_P was determined in a separate experiment by potentiodynamic polarization of the sample in the potential range $\Delta E = E_C \pm 20$ mV at a scan rate of 0.167 mV s^{-1}, in which the linear dependence $I = f(E)$ is observed.

R_P values was calculated according to the equation (2):

$$R_P = \frac{\Delta E}{\Delta I} \tag{2}$$

3. Results and Discussion

The analysis of the SEM data indicates the developed morphology of the base PEO-layer (figure 1a). Pores and microdefects, which are a consequence of the sharp cooling of the microdischarge zone after its attenuation, are presented on the surface of the sample. After a single application of SPTFE, particles of this material appear on the surface of the PEO-coating; however, no significant changes in the morphology of the surface layer are observed (figure 1b). With an increase in the multiplicity of the treatment with organofluorine material up to two and three times, pores and defects are sealed, and a formation of the uniform polymer film is observed (figure 1c, d). After a twofold treatment with organofluorine material, an increasing of the thickness of the polymer-containing coating is observed (from 19 to 31 μm). It should be noted that there is no significant difference in morphological structure between polymer-containing layers with double and triple application of SPTFE.

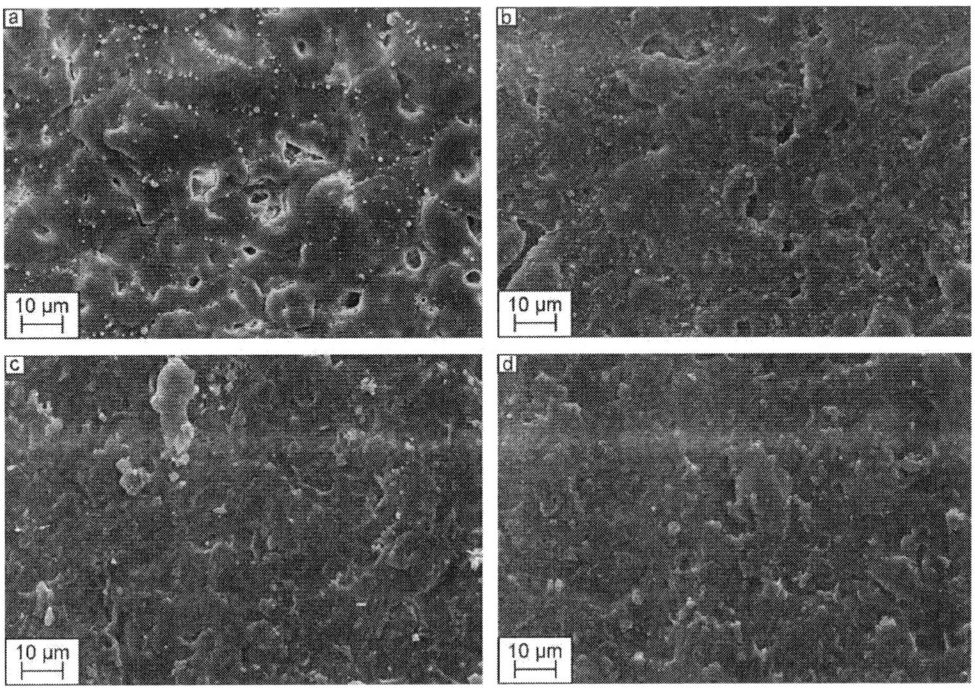

Figure 1. SEM-images of PEO-coating (a) and composite coatings with different multiplicities of treatment: CC-1x (b), CC-2x (c), CC-3x (d).

Analysis of the experimental data obtained by the potentiodynamic polarization (table 1) indicates an increase in the protective characteristics of the formed polymer-containing layers in comparison with the base PEO-coating and the uncoated material. At a single treatment of the PEO-layer with a fluoropolymer material, the corrosion current density I_C is reduced by more than 28 times (table 1). Further, with increasing of the treatment multiplicity up to two, the corrosion current density I_C is reduced, and the polarization resistance R_P is increased by one order of magnitude (table 1). The highest protective characteristics is observed for CC-3x: for this coating the I_C values decreased by more than 4 and 2 orders of magnitude in comparison with the uncoated and PEO-coated materials, respectively.

Table 1. Corrosion properties of studied samples

Sample	E_C (V vs. SCE)	I_C (A cm^{-2})	R_P (Ω cm^2)
Without coating	–0.76	8.3×10^{-6}	0.5×10^{3}
With PEO-coating	–0.63	5.4×10^{-8}	1.0×10^{5}
CC-1x	–0.64	1.9×10^{-9}	2.5×10^{7}
CC-2x	–0.64	1.4×10^{-10}	3.2×10^{8}
CC-3x	–0.61	2.3×10^{-11}	1.6×10^{9}

Such improvement in the protective properties of composite layers is associated with sealing of the pores and blocking of the defects in the PEO-coating with a polymer, as well as forming a uniform PTFE layer on the sample surface (figure 1). The higher protective characteristics of CC-3x in comparison with CC-2x results a more complete penetration of the polymer into the pores of the coating and increase of the thickness of composite layer from 31 to 35 μm with an increase in the number of the fluoropolymer material applied.

4. Conclusion

The method of composite polymer-containing coating formation on the AMg3 aluminum alloy by PEO and subsequent treatment with polymer have been developed. The protective properties of composite coatings are significantly higher than those of the base PEO-layer: the corrosion current density is decreased by more than 2 orders of magnitude. The presence of the chemically stable PTFE in the composition of polymer-containing coating increases the samples resistance to corrosion. The formed composite coatings have prospects for using in automobile and aircraft manufacture, electronics, and medicine.

Acknowledgments
This work was supported by the grant of the Russian Science Foundation (project no. 14-33-00009).

References

[1] Park C H, Jeong H Y, Eom G S, Son H T, Yu H and Park S S 2014 *Korean J. Met. Mater.* **52** 53–74
[2] Aguirre-De la Torre E, Afeltra U, Gomez-Esparza C D, Camarillo-Cisneros J, Perez-Bustamante R and Martınez-Sanchez R 2014 *J. Mater. Eng. Perform.* **23** 581–7
[3] Xie H-J, Cheng Y-L, Li S-X, Cao J-H and Cao L 2017 *Trans. Nonferrous Met. Soc. China* **27** 336–51
[4] Fatimah S, Kamil M P, Kwon J H, Kaseem M and Ko Y G 2017 *J. Alloys Compd.* **707** 358–64
[5] Barati N, Yerokhin A, Golestanifard F, Rastegari S and Meletis E I 2017 *J. Alloys Compd.* **724** 435–42
[6] Barati N, Meletis E I , Golestani Fard F, Yerokhin A, Rastegari S and Faghihi-Sani M A 2015 *Appl. Surf. Sci.* **356** 927–34
[7] Egorkin V S, Gnedenkov S V, Sinebryukhov S L, Vyaliy I E, Gnedenkov A S and Chizhikov R G 2018 *Surf. Coat. Technol.* **334** 29–42
[8] Gnedenkov S V, Sinebryukhov S L, Mashtalyar D V, Nadaraia K V, Kiryukhin D P, Buznik V M, Kichigina G A and Kushch P P 2015 *Russ. J. Inorg. Chem.* **60** 975–86
[9] Srinivasan P B, Scharnagl N, Blawert C and Dietzel W 2010 *Surf. Eng.* **26** 354–60
[10] Gnedenkov S V, Sinebryukhov S L, Mashtalyar D V, Nadaraia K V, Gnedenkov A S and Bouznik V M 2016 *Corros. Sci.* **111** 175–85
[11] Mashtalyar D V, Nadaraia K V, Sinebryukhov S L and Gnedenkov S V 2017 *J. Mater. Sci. Technol.* **33** 661–7
[12] Shi Z and Atrens A 2011 *Corros. Sci.* **53** 226–46

Lateral radiative forces exerted by evanescent fields along a hyperbolic metamaterial slab

I S Nefedov[1], **C A Valagiannopoulos**[2], **A S Shalin**[1]

[1]Laboratory "Nanooptomechanics", ITMO University, St. Petersburg, 197101, Russia
[2] Department of Physics, Nazarbayev University, Qabanbay Batyr Ave 53, Astana 010000, KZ.

E-mail: isnefedov@gmail.com

Abstract. We show and investigate the optical forces acting on a particle in the vicinity of a planar waveguide which is filled with hyperbolic material and supports propagation across its plane (two-dimensional). The anisotropy axis of its medium lies in plane of the waveguide. In contrast to commonly considered pushing or pulling forces, acting in one-dimensional guiding structures, in the case of two-dimensional wave propagation, the angles between the momentum and the total energy flow may take any value around the circle. Accordingly, evanescent fields out of the slab exert lateral radiative forces on a nanoparticle oriented parallel to momentum being controllably different from the total energy flow direction. This provides a flexibility in manipulation by nanoparticles by employing suitably engineered hyperbolic structures.

1. Introduction

Laser beam manipulation of nanoparticles has found applications in bio-physics, micro-fluidics, and optomechanical devices. In literature, conservative (gradient) forces and non-conservative pushing and pulling forces exerted by electromagnetic fields have been mainly discussed. Especially the pulling forces have attracted particular interest due to its amazing property to act against the source of electromagnetic energy [1]. In this context, hyperbolic metamaterials (HMMs) provide additional flexibility for controlling light propagation and enable construction of a platform supporting strong radiative pulling forces [2]. However, in most of the papers devoted to such effects in HMMs, the forces are studied *within* a nanostructured metamaterial [2, 3] hampering their experimental observation. Importantly, many works present the outcome of numerical simulations without explaining the reason for the appearance of the pulling forces in hyperbolic metamaterials.

In [4], the effect of the pulling force is explained in terms of partial energy flows propagating inside the HMM and outer space. Namely, the energy flow integrated over outer space, which is vacuum or isotropic dielectric, is always directed as the momentum. On the contrary, the energy flow integrated over the hyperbolic material cross-section, propagates in the opposite direction. Therefore, the total energy flow, i.e. integrated over full cross-section which includes HMM and external region, may be directed as the momentum or oppositely depending on the sign of waveguide dispersion. If one extends the aforementioned logic in the case of two-dimensional wave propagation, the angle between momentum and the total energy flow can take all possible values. Indeed, the energy flow in outer isotropic space is always parallel to the momentum, while the lateral radiative force on a neutral nanoparticle may be directed at an arbitrary angle with respect to the total energy flow or the group velocity.

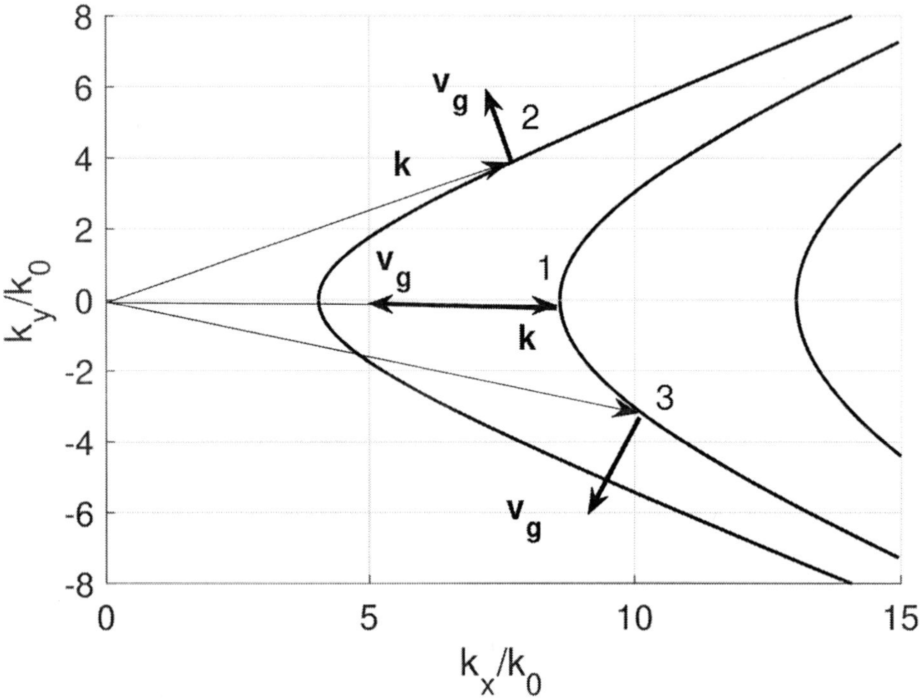

Figure 1. Isofrequency contours for three lowest modes corresponding to the same frequency in plane of normalized wavevector components $(k_x/k_0, k_y/k_0)$. Thin arrows show directions of the wavevector $\mathbf{k}$ and thick arrows show directions of the group velocity $\mathbf{v_g}$.

2. Hyperbolic waveguide description and hybrid waves

A periodic metal-dielectric structure with the layers lying in the (xz)-plane is considered as the hyperbolic metamaterial. Gold is assumed to be a metal with its relative permittivity defined by Drude formula in the optical range [5]. The components of the effective permittivity tensor are expressed as in [6], with a fraction ratio of gold $f = 0.05$ and under considered parameters: $\varepsilon_{xx} < 0$, $\varepsilon_{zz} < 0$, $\varepsilon_{yy} > 0$. The external medium is vacuum, while the thickness of the HMM slab is taken equal to 250 nm.

In contrast to the case of one-dimensional wave propagation where the structure is isotropic in the (xy)-plane [4], the waves here are hybrid i.e. cannot be classified as TM and TE waves. Dispersion equation for hybrid waves in anisotropic slab, based on full-wave theory, has been derived using 4×4 transfer matrix and solved numerically.

3. Waveguide dispersion and directions of momentum and energy flow

Fig. 1 illustrates waveguide dispersion in form of isofrequency contours on the map of the normalized wave vector components $(k_x/k_0, k_y/k_0)$, where $k_0 = 2\pi/\lambda$ is the wavenumber in vacuum, calculated at the frequency ≈231 THz ($\lambda = 1.3\,\mu$m). In contrast to unbounded HMM there exists a countable number of modes with different field distribution along the the waveguide thickness, i.e. the z-axis. We show only three isofrequency loci to the lowest modes. High-order modes correspond to larger transverse components of the wave vector k_x, k_y, see [7]. Thus, several modes can be excited simultaneously by an external source, like a dipole that may make difficult their inentification. However, the modes with large wavevectors have small propagation length and their influence disappears far from the source.

The vector of group velocity $\mathbf{v_g}$ is perpendicular to the isofrequency curve at a point corresponding to the wavevector $\mathbf{k}(k_x, k_y)$ and directed towards the increasing frequency. The group velocity has the same direction as the total energy flow or the Poynting vector integrated over whole cross-section. One can observe that in point 1 (corresponding to $k_y = 0$), the directions of the group velocity (the total energy flow) and momentum $\mathbf{k}$ are opposite. It means that evanescent fields of these modes exert pulling forces on a particle placed nearby the HMM slab. At the same time, in point 2 the direction of the momentum (and the Poynting vector in outer space and the radiative force) forms an angle $\approx 90°$ with the group velocity. Finally, at point 3, this angle is sharp tending to 180° when $k_y \to 0$.

The radiative force, exerted on a dipolar particle, can be written as [8]:

$$\langle\mathbf{F}\rangle = \frac{1}{4}\mathrm{Re}\{\alpha\}\nabla|\mathbf{E}|^2 + \sigma\frac{1}{2}\mathrm{Re}\left\{\frac{1}{c}\mathbf{E}\times\mathbf{H}^*\right\} + \sigma\frac{1}{2}\mathrm{Re}\left\{i\frac{\varepsilon_0}{k_0}(\mathbf{E}\cdot\nabla)\mathbf{E}^*\right\}, \quad (1)$$

where c is the speed of light and ε_0 is the permittivity of vacuum. The symbol α is used for the polarizability of a particle; in the case of a sphere with radius r filled with material of relative permittivity ε is written as:

$$\alpha = \frac{\alpha_0}{1 - i\alpha_0 k_0^3/(6\pi\varepsilon_0)} \quad , \quad \alpha_0 = 4\pi\varepsilon_0 r^3\frac{\varepsilon-1}{\varepsilon+2}, \quad (2)$$

with $\sigma = k_0\mathrm{Im}\{\alpha\}/\varepsilon_0$. The first term in (1), related to the gradient forces, causes the attraction of the particle towards the slab interface due to the z-dependence of evanescent fields. In the second term, the lateral component of the Poynting vector exerts the force in (xy)-plane along the waveguide interface. The third term of (1) equals to zero for the evanescent field in a lossless isotropic medium as shown in [9]. Thus, the non-conservative dipolar force is proportional to the corresponding Poynting vector and it acts in the same direction.

In conclusion, we have demonstrated that not only pushing and pulling non-conservative forces can be observed in vacuum nearby anisotropic waveguide, but also forces directed at any angle to the group velocity (or, equivalently, the total energy flow).

Acknowledgement

This work was partially supported by no. 090118FD5349 NU Small Grant project: *"Super transmitters, radiators and lenses via photonic synthetic matter"* and no. 20162031 NU ORAU Grant entitled: *"Structured light for nonlinear and topological photonics"*. Funding from MES RK state-targeted program BR05236454 is also acknowledged.

The authors acknowledge the financial support provided by Russian Fund for Basic Research (Projects No. 18-02-00414, 18-52-00005); Ministry of Education and Science of the Russian Federation (GOSZADANIE, Grant No. 3.4982.2017/6.7) and Government of Russian Federation, Grant 08-08.

References

[1] Chen J, Ng J, Lin Z, and Chan C T 2011 *Nat. Phot.* **5** 531
[2] Shalin A S, Sukhov S V, Bogdanov A A, Belov P A, and Ginzburg P 2015 *Phys. Rev. A* **91** 063830
[3] Bogdanov A A, Shalin A S, and Ginzburg P 2015 *Sc. Rep.* **5** 15846
[4] Nefedov I and Shalin A S 2017 *Phys. Status Solidi RRL* 1700219
[5] Etchegoin P G, Le Ru E G and Meyer M 2006 *The Journal of chemical physics* 125 164705
[6] Cortes C L and Jakob Z 2013 *Phys. Rev. B* **88** 045407
[7] Nefedov I and Tretyakov S 2011 *Phys. Rev. B* **84** 113410
[8] Albaladejo S, Marqués M I, Laroche M and Sáenz J J 2009 *Phys. Rev. Lett.* 102 113602
[9] Nefedov I S and Rubi J M 2017 *Sc. Rep.* **7**, 6155

Coupled-mode theory of field transfer processes in surface plasmon resonance structures

D V Nesterenko[1,4,*], S Hayashi[2,3] and Z Sekkat[3,5,6]
[1]Image Processing Systems Institute RAS - Branch of the FSRC "Crystallography and Photonics" RAS, Samara, Russia
[2]Kobe University, Kobe, Japan
[3]Moroccan Foundation for Science, Innovation and Research (MAScIR), Rabat, Morocco
[4]Samara National Research University, Samara, Russia
[5]Osaka University, Suita, Japan
[6]Mohammed V University in Rabat, Rabat, Morocco
[*]E-mail address: dtnesteren@gmail.com
This work was supported by the Federal Agency of Scientific Organizations (agreement # 007-GZ/Ch3363/26) and by the Russian Foundation for Basic Research (project #18-07-00613).

Abstract. We study the response of the metal-dielectric planar structures supporting surface plasmon polariton modes to an external electromagnetic field. Analysing the field transfer processes within the coupled-mode theory, we demonstrate the role of non-resonant and resonant responses in the appearance of Fano line shapes in spectra of the metal-dielectric interfaces, Otto and Kretschmann ATR configurations.

1. Introduction

Surface plasmon polariton (SPPs) modes supported by metal-dielectric interfaces can be excited using high-index prisms in the attenuated total reflection (ATR) configurations in planar plasmonic structures [1, 2]. The optical properties of surrounding media determine the resonance characteristics of SPP modes and can be characterized by the analysis of resonances in ATR spectra. Experimental and theoretical angle-scan ATR spectra in the Kretschmann and Otto configurations exhibit strongly asymmetric resonance line shapes. However, the SPP ATR resonances have not been considered as asymmetric ones yet. The asymmetry in the resonance profiles was first explained by Fano as an interference of responses from discrete and continuum states of physical systems [3]. To explain the origins of asymmetric resonance line shapes in the ATR spectra of the planar plasmonic structures, we apply coupled-mode (CM) theory to analyze the field transfer processes and provide a deeper insight into the general Fano-like phenomena of mode coupling. The fundamental mechanisms of field transfer and coupling in the planar plasmonic structures are crucial for analysis and design of general resonance structures that support modes of different nature.

2. Coupled-mode theory at metal-dielectric interfaces

The CM approach in optics was implemented for analysis of Fano resonances in terms of interaction of electromagnetic fields [4, 5]. We consider the generation of evanescent field at the metal-dielectric interface by an external wave. In the case of near-field enhancement by resonant excitation of an SPP mode, the overall response from the interface includes non-resonant and resonant components [6]. The three field transfer processes are distinguished in the formation of these response components. The outcoupling of the non-resonant component of evanescent field, resonant coupling to and outcoupling from the SPP mode to the resonant component are schematically demonstrated in figure 1.

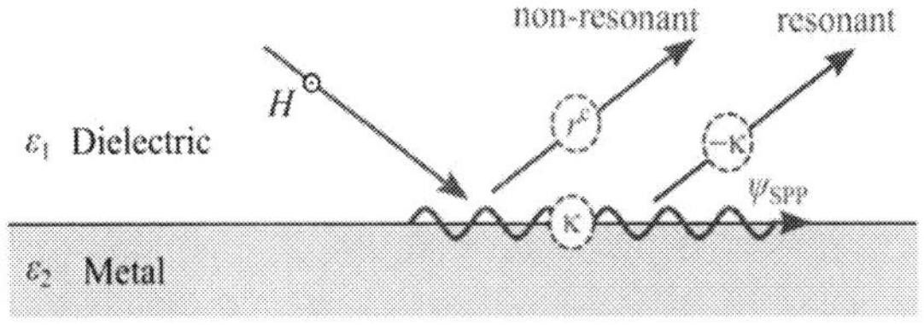

Figure 1. Field transfer processes of coupling between the external wave and an SPP mode at the metal-dielectric interface.

For the coupling process of an evanescent-field with a SPP mode $\psi_{SPP} \propto \exp(i\gamma_{SPP}x)$ propagating along the interface in a two-layer metal-dielectric system, the dynamic equation for ψ_{SPP} induced by incoupling of an evanescent wave H can be written as

$$\frac{1}{i\gamma_{SPP}}\frac{d\psi_{SPP}}{dx} = \psi_{SPP} + \kappa H, \tag{1}$$

where κ is a coefficient of coupling between the mode and external waves.

The total outcoupled evanescent wave H_r is composed of a component generated by the non-resonant outcoupling of H with the continuum coefficient r^c of near-field enhancement and a component resulted from the resonant outcoupling of the mode ψ_{SPP} with coupling coefficient $-\kappa$. Expression for H_r can be represented as

$$H_r = r^c H - \kappa\psi_{SPP}. \tag{2}$$

Assuming arbitrary propagation constant α along x axis for the incoupling wave, $H \propto \exp(i\alpha x)$, the spectral dependence of the mode amplitude ψ_{SPP} is represented by a Lorentzian

$$\psi_{SPP} = \frac{\gamma_{SPP}}{\alpha - \gamma_{SPP}}\kappa H. \tag{3}$$

The overall near-field enhancement coefficient $r \equiv H_r / H$ can be obtained by substituting Eq. (3) into Eq. (2) as the interference of the non-resonant and resonant components

$$r = r^c - \frac{\gamma_{SPP}}{\alpha - \gamma_{SPP}}\kappa^2. \tag{4}$$

These components are indistinguishable in the overall response registered by a detector. Eq. (4) can be represented in a general form of a Fano formula

$$r = r^c \frac{\alpha - \gamma_{SPP}\left(1 + \kappa^2 / r^c\right)}{\alpha - \gamma_{SPP}}. \tag{5}$$

Here, $\kappa^2\gamma_{SPP}$ is defined as the complex amplitude of a single SPP mode response, the relation of the resonance amplitude to the continuum amplitude $\kappa^2\gamma_{SPP} / r^c$ determines a shift of the zero parameter of the function r from γ_{SPP}. All the considered parameters κ, γ_{SPP}, and r^c are the intrinsic parameters of the two-layer structure that depend on the permittivities of both layers.

Based on Eq. (5), we can conclude that the field enhancement spectra $|r|^2$ in the metal-dielectric interfaces are described by the Fano line shape. The appearance of the non-resonant and resonant components in the near-field enhancement coefficient can be also demonstrated analytically by the rigorous electromagnetic theory.

3. Coupled-mode theory in ATR geometries

A metal-dielectric structure considered in the previous section can be coupled to the prism by the spacer layer in two ways. In the well-known Otto and Kretschmann configurations, the spacer layer is a dielectric and a metal, respectively. Excitation of SPP modes at a metal-dielectric interface is performed by evanescent waves generated by a high-index prism under ATR conditions. These configurations can be represented by a three-layer structure, in which the prism and outer layer are denoted as semi-infinite layers L_0 and L_2, respectively; the spacer layer of a finite thickness h separates the layers L_0 and L_2 and is denoted as L_1 as depicted in figure 2. Therefore, the spacer layer L_1 is a dielectric in the Otto configuration and a metal in the Kretschmann configuration.

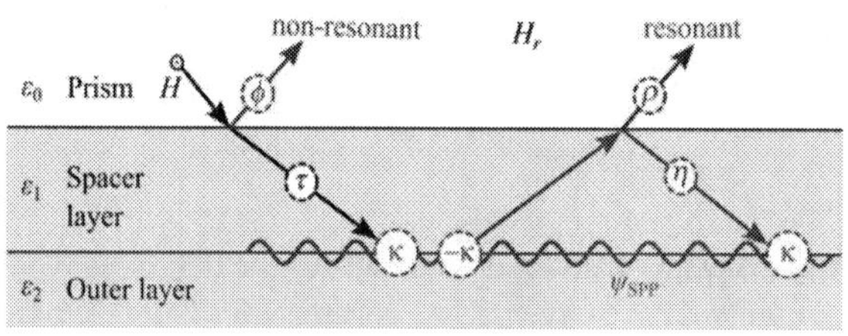

Figure 2. Field transfer processes of SPP mode excitation in three-layer structures.

We develop CM theory for the field transfer processes in three-layer resonance systems. Neglecting the non-resonant reflection from the metal-dielectric interface, the four field transfer processes can be outlined. As schematically demonstrated in figure 2, an incident propagating plane wave H is split onto a beam ϕ non-resonantly reflected from the prism/spacer layer interface and an evanescent wave in the spacer layer. The generated evanescent field of the complex amplitude τ at the spacer layer/outer layer interface resonantly incouples to a SPP mode ψ_{SPP} with coupling coefficient κ.

The excited SPP mode ψ_{SPP} resonantly outcouples into the evanescent field with coefficient $-\kappa$. The generated evanescent wave exponentially decays with distance from the metal-dielectric interface. At the prism/spacer layer interface, the decayed evanescent field generates two waves. One is a wave propagating in the prism. The amplitude of the propagating wave is proportional to the decay coefficient ρ. The second wave is an evanescent wave, which reflects back to the spacer layer/outer layer interface and couples again to ψ_{SPP}. The amplitude of this evanescent wave at the spacer layer/outer layer interface is proportional to the decay coefficient η. Due to the resonant outcoupling, the reflected wave ρ and the evanescent wave η, which is coupled back to the SPP mode, have resonant behavior.

Based on the field transfer processes, we can write generalized CM equations to describe the field transfer dynamics for the outgoing reflected wave H_r represented as a sum of the resonant and non-resonant components

$$\frac{1}{\mathrm{i}\gamma_{\text{SPP}}}\frac{d\psi_{\text{SPP}}}{dx} = \psi_{\text{SPP}} - \eta\kappa^2\psi_{\text{SPP}} + \tau\kappa H\,, \tag{6}$$

$$H_r = \phi H - \rho\kappa\psi_{SPP}. \tag{7}$$

The amplitude of ψ_{SPP} is obtained from Eq. (6) as a Lorentzian function:

$$\psi_{SPP} = \frac{\gamma_{SPP}}{\alpha - \gamma_{SPP}\left(1-\eta\kappa^2\right)}\tau\kappa H. \tag{8}$$

The reflection coefficient $r \equiv H_r / H$ is obtained by substitution of Eq. (8) into Eq. (7)

$$r = \phi\frac{\alpha - \gamma_{SPP}[1-(\eta - \rho\tau/\phi)\kappa^2]}{\alpha - \gamma_{SPP}(1-\eta\kappa^2)}. \tag{9}$$

The position and depth of the SPR dip are determined by the zero parameter $\gamma_{SPP}[1-(\eta - \rho\tau/\phi)\kappa^2]$. In the three-layer system, the mode self-coupling $-\eta\kappa^2$ in the spacer layer leads to the shift of $-\eta\kappa^2\gamma_{SPP}$ in the pole and zero values in respect to its intrinsic value γ_{SPP}. The weaker field decay in the spacer layer and stronger coupling degree result in a larger shift of the propagation constant. In addition to the influence of self-coupling $-\eta\kappa^2$, the resonance dip parameter in the three-layer system is perturbed by the coupling of external fields with the mode and non-resonant reflection from the structure.

The reflection coefficient r can be also expressed as the interference of the non-resonant reflection from the prism interface and a resonant component arising from the SPP excitation that results in a sum of a Lorentzian function and a continuum

$$r = \phi - \frac{\gamma_{SPP}}{\alpha - \gamma_{SPP}(1-\eta\kappa^2)}\rho\tau\kappa^2. \tag{10}$$

Eqs. (9) and (10) represent Fano formulas, and the reflectivity $|r|^2$ exhibits an asymmetric Fano line shape. Therefore, the presented model of the field transfer processes can be effectively used to describe the asymmetric resonance responses in the ATR structures.

4. Conclusions

The general expressions obtained by the CM theory for the spectral responses of the two- and three-layer planar plasmonic structures demonstrate that the structures acts like non-tunable and tunable resonators, respectively. In both cases, the appearance of resonant and non-resonant components in the response results in the Fano interference, where the origins of non-resonant component are different from those of the resonant one. The resonance features of the asymmetric Fano line shape in spectra of the considered structures are determined by the coupling strength between the SPP mode and the external field. In the case of the three-layer structure, the spacer layer allows the self-interaction of the SPP mode that modifies the position and line shape of the Fano resonances. The obtained results may open a new avenue in studies on field transfer processes in matter.

References

[1] Otto A 1968 *Z. Phys.* A **216** 398
[2] Kretschmann E and Raether H 1968 *Z. Naturforsch.* A **23** 2135
[3] Fano U 1935 Il Nuovo Cimento **12** 154
[4] Haus H A 1984 *Waves and fields in optoelectronics* (Prentice-Hall, Englewood Cliffs; London)
[5] Fan S, Suh W, and Joannopoulos J D 2003 J. Opt. Soc. Am. A 20, 569
[6] Nesterenko D V, Hayashi S and Sekkat Z 2018 *Phys. Rev.* B **97** 235437

Highly stable silver nanoparticles for SERS applications

S M Novikov[*,1,2], **V N Popok**[3], **A B Evlyukhin**[1,4], **M Hanif**[5], **P Morgen**[6], **J Fiutowski**[7], **J Beermann**[1], **H-G Rubahn**[7] **and S I Bozhevolnyi**[1]

[1]Centre for Nano Optics, University of Southern Denmark, Campusvej 55, DK-5230 Odense, Denmark.

[2]Laboratory of Nanooptics and Plasmonics, Moscow Institute of Physics and Technology, 9 Institutsky Lane, Dolgoprudny 141700, Russia.

[3] Department of Physics and Nanotechnology, Aalborg University, Skjernvej 4A, DK-9220 Aalborg, Denmark.

[4] ITMO University, 49 Kronversky Ave., 197101 St. Petersburg, Russia.

[5] Interdisciplinary Nanoscience Center (iNANO), Aarhus University, Gustav Wieds Vej 14, 8000 Aarhus, Denmark.

[6] Department of Chemical Engineering, Biotechology and Environmental Technology, University of Southern Denmark, Campusvej 55, DK-5230 Odense, Denmark.

[7] Mads Clausen Institute, University of Southern Denmark, NanoSYD, Alsion 2 DK-6400 Sønderborg.

* novikov-s@yandex.ru

Abstract. Plasmonics sensor configurations utilizing localized plasmon resonances in silver nanostructures typically suffer from rapid degradation of silver in ambient atmospheric conditions. In this work, we report on the fabrication and detailed characterization of ensembles of *monocrystalline* silver nanoparticles (NPs), which exhibit a long-term stability of optical properties under ambient conditions without any protective treatments. Ensembles with different densities (surface coverages) of size-selected NPs (diameters from 12.5 to 24 nm) on quartz substrates are fabricated using the cluster-beam technique and characterized by linear spectroscopy, surface-enhanced Raman scattering microscopy as well as transmission electron, helium-ion and atomic force microscopies. It is found that the fabricated ensembles of monocrystalline silver NPs preserve their plasmonic properties (monitored with optical spectroscopy) and strong field enhancements (revealed by surface enhanced Raman spectroscopy) at least 5 times longer as compared to chemically synthesized silver NPs with similar sizes. The obtained results are of high practical relevance for the further development of sensors, resonators and metamaterials utilizing plasmonic properties of silver NPs.

1. Introduction

Light interaction with nanostructures, especially those fabricated of noble metals, gives rise to various fascinating optical phenomena [1]. One of the current research directions in nano-optics is the development of metal nanostructures that efficiently interconvert propagating and localized optical fields and thereby facilitate the generation of strongly enhanced local electrical fields. Field enhancements (FEs) occur due to resonantly excited surface plasmons, representing collective electron oscillations in metals coupled to electromagnetic fields in neighboring dielectrics [2]. Strong FEs are extremely important for practical applications for example in sensing and catalysis as well as in surface-enhanced Raman scattering (SERS) [3, 4]. Noble metal nanostructures are also very promising for the formation of surface plasmon resonators, which have demonstrated enormous enhancement of the fluorescence of quantum emitters, thus facilitating the development of ultra-bright and stable single-photon sources. In addition, they are can be used as real-time sensors, plasmonic solar cells, building blocks for light-energy guiding devices and metamaterials with unique optical properties, *etc.* Silver exhibiting excellent plasmonic properties[5] can advantageously be used in the majority of the above-mentioned applications [5-6].

There are many strategies for the fabrication of plasmonic structures using silver. Different approaches such as particle formation by evaporation or sputtering, high-fluency ion implantation, chemical and photo reduction, production of colloid nanoparticles (NPs), *etc.* were suggested. Many of them allow formation of nanomaterials with an attractive efficiency. However, rapid oxidation /sulfidation from the ambient atmosphere dramatically decreases all bonuses of this metal [7, 8] and causes complications in terms of practical applications. For example, the sensors containing Ag usually require storage in inert gases and the operational time of such devices in the ambient atmosphere is rather limited. One existing solution is to make a protection layer, which can prevent the Ag structures from the degradation due to reactions with environmental species. However, these treatments are not always applicable or rather complex to be used on an industrial scale. A possible approach to solve this problem could be the formation of very pure particles with perfect crystalline structure. They should be more stable against the above-mentioned degradation phenomena. A very promising approach for preparation of such NPs is aggregation of metal atoms into clusters from very pure sources in vacuum. These clusters can be collimated into beams, size selected and deposited or implanted on/in different substrates, which provides great capabilities for controlling the structure and properties of materials on the nanoscale [9]. In this work, we utilize a cluster beam deposition technique based on magnetron sputtering [10] as a method of fabrication of ensembles of monocrystalline Ag NPs with well-controlled sizes. Due to these properties they demonstrate excellent stability of plasmon band intensity (it decreases only for about 20% after 30 days) and strong FE which is also preserved over rather long time period at room temperature and ambient atmosphere. We anticipate that the obtained results are very promising for sensor applications.

2. Experimental section.

2.1. Ag NPs by a cluster beam deposition technique

A magnetron sputtering NPs apparatus (MaSCA) [10] utilizing a commercial source, NC200U from Oxford Applied Research, was used for cluster production from silver targets of 99.99% purity. In the source, the target material was sputtered by an Ar plasma and was condensed into NPs of various sizes in the aggregation region with the help of He as a buffer gas. Thereafter, the NPs were expanded through a nozzle and collimated into a beam. NPs from a beam were selected using an electrostatic quadrupole mass selector (EQMS). Only charged particles were selected and steered toward deposition on 10x10 mm^2 quartz substrates in the vacuum with a background pressure of approximately 5×10^{-7} mbar. The size-selection was performed at two different EQMS voltages of 300 and 1600 V. According to earlier studies, these voltages correspond to mean particle diameters of 12.5±1.1 and 24.0±2.0 nm, respectively. Particles of these sizes are further referred to as small and large NPs. The size-selected silver NPs were deposited in a so-called soft-landing regime securing that the cluster kinetic energy is much smaller than the cohesive energy of atoms leading to no or very small distortion of the particles shape, which is supposed to be close to spherical in a free cluster beam. Samples were prepared with several different particles coverages (surface densities of NPs) ranging between approximately 300-600 particles/μm^2 for small NPs (samples DN1-DN3) and 70-300 particles/μm^2 for large NPs (samples DN4-DN6). Figure 1 shows HIM image and linear spectra of the fabricated samples.

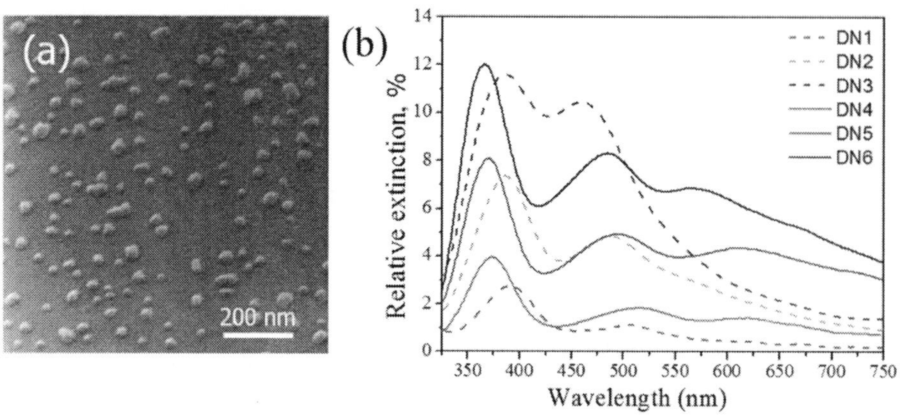

Figure 1. (a) HIM image of deposited large silver NPs (24 nm). (b) Experimental extinction spectra for the ensembles of small (dashed lines) and large (solid lines) Ag NPs with three different surface densities.

2.2. SERS characterizations

In order to estimate the applicability of the samples for SERS, the SERS characterizations are performed. The measurements are carried out on the samples kept in the air for 8 days after the fabrication. All samples are covered by an ethanol 10^{-6} M solution of Crystal Violet (CV) dye and dried under ambient conditions. As shown elsewhere, the oxidation of silver NPs or other reactions with species present in air typically lead to significant decrease or completely disappearing SERS signal on a relatively short time scale of a few days [11-12]. The typical SERS spectra are shown in Figure 2 and they demonstrate considerable or even high intensity depending on particle size and surface coverage. SERS images show rather homogeneous signal distribution across the samples except a few fluctuations in intensity related to cluster diffusion and aggregation around natural surface defects of quartz. As can be seen from the spectra presented in Figure 2a, the SERS intensity increases with increasing cluster density from DN1 to DN3 and from DN4 to DN6.

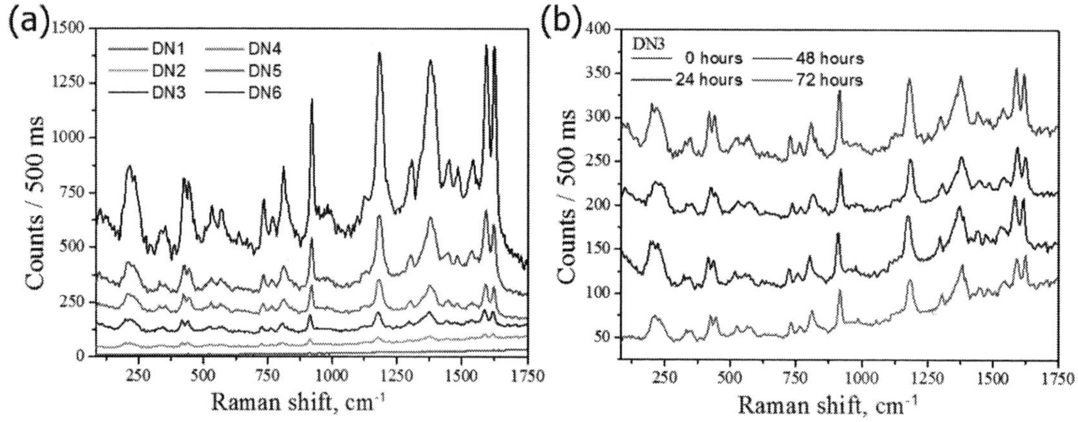

Figure 2. SERS spectra (a) for the samples DN1-DN6 obtained 8 days after the deposition of small and large Ag clusters on the glass substrate (particles surface density increases from DN1 to DN3 for the small NPs (12.5 nm) and from DN4 to DN6 for the large NPs (24 nm)), (b) for the sample with cluster surface density DN3 and different time durations after coverage by analyte.

This is an obvious tendency because every cluster represents the so-called hot spot, i.e. the place with enhanced local field. Hence, an increase in cluster density gives higher intensity. At the same time, the intensity for the samples with large NPs tends to be much higher compared to the samples with small ones due to the factor of two differences in radius between the small and large NPs yielding 8 times difference in the effective volume for individual particles. Thus, more dye molecules surrounding large clusters become excited and contribute to SERS signal. It is also known that the maximal gain can be expected for specific (efficient) particle size and distances between them. However, study of this phenomenon was not among the scopes of current work. In order to monitor evolution of spectra after covering the silver NPs by the analyte, we performed the measurements each 24 hours. These SERS spectra (see Figure 2b) shows only near 15% decrease of intensity after 72 hours, thus, demonstrating great stability of the cluster ensembles with respect to SERS in ambient atmospheric conditions.

The stability for SERS application of Ag NPs obtained by cluster beam deposition technique is compared with that for the NPs obtained by chemical synthesis. For the comparison of the stability of our clusters for SERS application, we used colloidal silver nanoparticles with size 21±6 nm. Several samples were prepared using these colloids by drop-casting on the glass substrate. Each sample consists on 3-4 drops located separately on one glass slide. The samples were covered by CV 3, 24, 32, and 48 hours after the drop-casting, respectively; each with the same concentration as used for clusters, and dried under ambient conditions. The level of SERS signals is found to be strong only for the samples covered by the dye 3 and 24 hours after the drop-casting of NPs. The mapping shows homogeneously strong SERS signals across the whole area selected for the scan. For the sample covered by CV 32 hours after the drop-casting, the intensity of spectral peaks is significantly decreased and they nearly disappear for the sample covered by the dye 48 hours after the particle deposition. The mapping made for these two samples shows that the SERS signal is present only on some parts of the scanned area. In order to monitor stability of colloidal NPs *after covering by the analyte*, we performed the measurements each 24 hours, similar to that carried out for the NPs deposited from cluster beams. The SERS spectra show nearly 3 times decrease of intensity after first 24 hours, thus, demonstrating much faster degradation of FE compared to the cluster beam-made particles. To compare a signal increase in SERS, with that in ordinary Raman keeping the same experimental parameters, the analytical enhancement factor (EF expression) is used [13]. The average EF are estimated to be ~0.05×10^4 (DN1), ~0.46×10^4 (DN2), ~1.41×10^4 (DN3), ~0.25×10^5 (DN4), ~0.52×10^5 (DN5) and ~1.39×10^5 (DN6), respectively. We would like to stress that in this case we have used non-resonant dye, whereas for resonant cases the enhancement estimation would be one-to-two orders of magnitude higher. The demonstrated stability of plasmonic properties is caused by the monocrystalline structure of silver NPs revealing strong resistance to oxidation and other reactions with atmospheric gases or typical contaminants (for example, sulfurous compounds) during the storage in ambient atmosphere at room temperature. In addition, it is worth mentioning that the purpose of this work [14] is not optimization of structures to get the highest SERS signal and finding the minimal resolvable concentration of an analyte but rather to estimate the stability and applicability of the prepared cluster ensembles. A simple change of substrate type, for example, substituting quartz by Au or Ag, which allow to excite surface plasmon modes, could significantly increase the signal. The reported results are of high relevance for the fabrication of sensors, resonators and metamaterials utilizing plasmonic properties of silver NPs. This work will be followed by further research in order to optimize parameters of the fabricated ensembles of NPs. In future work we are going to work on the optimization of cluster sizes and surface density in order to reach all benefits from the proposed method of fabrication and combine the current technique with lithography. Furthermore, it is worth mentioning that the samples demonstrate several resonances covering almost all visible spectral intervals and one can vary the spectral position by changing the cluster size and surface coverage.

Acknowledgment. The authors gratefully acknowledge financial support from the Russian Science Foundation Grant No. 16-12-10287.

References

[1] Gramotnev D K, Bozhevolnyi S I 2010 *Nature Photonics* **4** 83.
[2] Maier S. *Plasmonics: Fundamentals and Applications.* Springer 2007 New York.
[3] Beermann J, Novikov SM, Albrektsen O, Nielsen M G, Bozhevolnyi S I 2009 *J. Opt. Soc. Am. B.* **26** 2370.
[4] Samal A K, Polavarapu L, Rodal-Cedeira S, Liz-Marzán L M, Pérez-Juste J, Pastoriza-Santos I 2013 *Langmuir* **29** 15076.
[5] Brazhe NA, Evlyukhin A B, Goodilin EA, Semenova AA, Novikov S M, Bozhevolnyi S I, Chichkov B N, Sarycheva A S, Baizhumanov A A, Nikelshparg E I, et al. 2015 *Sci. Rep.* **5** 13793.
[6] Vishnupriya S, Chaudhari K, Jagannathan R, Pradeep R. 2013 *Part. Part. Syst. Char.* **30** 1056.
[7] Zhang X, Hicks E M, Zhao J, Schatz G C, Van Duyne R P 2005 *Nano Lett.* **5** 1503.
[8] Cao W, Elsayed-Ali H E, 2009 *Mater. Lett.* **63** 2263.
[9] Popok V N, Samela J, Nordlund K, Campbell E.E.B. 2010 *Phys. Rev. B.* **82** 201403.
[10] Hartmann H, Popok V N, Barke I, V. von Oeynhausen and Meiwes-Broer K-H 2012 *Rev. Sci. Instrum.* **83** 073304.
[11] Han Y, Lupitskyy R, Chou T M, Stafford C M, Du H, Sukhishvili S. 2011 *Anal. Chem.* **83** 5873.
[12] Qi H, Alexson D, Glembocki O, Prokes S M, 2010 *Nanotechnology.* **21** 215706.
[13] Le Ru E C, Blackie E, Meyer M, Etchegoin P G 2007 *J. Phys. Chem. C* **111** 13794.
[14] Novikov S M, Popok V N, Evlyukhin A B, Hanif M, Morgen P, Fiutowski J, Beermann J, Rubahn H-G, Bozhevolnyi S I 2017 *Langmuir* **33** 6062.

All-optical switching of pulse transmission in disordered resonant media

Denis V Novitsky[1,2] **and Alexander S Shalin**[1]
[1]ITMO University, Kronverksky Prospekt 49, 197101 St. Petersburg, Russia
[2]B. I. Stepanov Institute of Physics, National Academy of Sciences of Belarus, Nezavisimosti Avenue 68, 220072 Minsk, Belarus

E-mail: dvnovitsky@gmail.com

Abstract. Using numerical calculations of Maxwell-Bloch equations, we propose a scheme for all-optical transmission switching based on interaction between ultrashort copropagating pulses in disordered resonant media. The physical mechanism of this nonlinear effect is revealed as well as its dependence on the amplitudes of pulses and the interval between them.

Switching and modulation of optical response are the basic effects needed for controlling light signals and construction of tunable optical devices. A number of mechanisms and materials are proposed to realize ultrafast switching. For example, photonic crystals can be used as all-optical modulators based on the band-gap shift effect. Such shift can be reached due to nonlinear response of solid materials (usually semiconductors, such as GaAs [1] or doped metal oxides [2]) and is characterized by picosecond relaxation times. Semiconductors are also proposed as a key element for compact tunable metasurfaces where switching is a result of shifting the Mie-type resonances [3]. Another class of perspective materials is the transparent conductive oxides, which claim to replace metals in tunable and switchable plasmonic devices [4].

In this paper, we propose a new mechanism for all-optical transmission switching based on disorder-induced interaction of copropagating coherent pulses in resonant media. It is known that in the uniform resonant medium (disorder parameter $r = 0$), such pulses eventually transform into standard 2π solitons and interact elastically, i.e., they do not lose any energy as a result of collision [5]. We show that the presence of random variations of active particles density ($r \neq 0$) results in effective inelastic interaction between the pulses when portion of light energy gets absorbed by the medium. This light trapping depends on the strength of disorder, which is a necessary ingredient for the effect to occur. Launching either one or two pulses inside disordered resonant medium, one will obtain strongly different levels of transmission having a means for realizing all-optical switching.

The system considered in this paper consists of a background dielectric doped with active (two-level) atoms, which initially are in the ground state. Light propagation in such media is described with the semiclassical Maxwell-Bloch equations for the dimensionless electric-field amplitude $\Omega = (\mu/\hbar\omega)E$ (normalized Rabi frequency), complex amplitude of the atomic polarization ρ, and difference between populations of ground and excited states w [5]:

$$\frac{d\rho}{d\tau} = il\Omega w + i\rho\delta - \gamma_2\rho, \quad (1)$$

$$\frac{dw}{d\tau} = 2i(l^*\Omega^*\rho - \rho^* l\Omega) - \gamma_1(w-1), \tag{2}$$

$$\frac{\partial^2\Omega}{\partial\xi^2} - n_d^2\frac{\partial^2\Omega}{\partial\tau^2} + 2i\frac{\partial\Omega}{\partial\xi} + 2in_d^2\frac{\partial\Omega}{\partial\tau} + (n_d^2-1)\Omega = 3\epsilon l\left(\frac{\partial^2\rho}{\partial\tau^2} - 2i\frac{\partial\rho}{\partial\tau} - \rho\right), \tag{3}$$

where $\tau = \omega t$ and $\xi = kz$ are the dimensionless time and distance, μ is the dipole moment of the quantum transition, $\hbar$ is the reduced Planck constant, $\delta = \Delta\omega/\omega = (\omega_0 - \omega)/\omega$ is the normalized frequency detuning, ω is the carrier frequency, ω_0 is the frequency of the atomic resonance, $\gamma_1 = 1/(\omega T_1)$ and $\gamma_2 = 1/(\omega T_2)$ are the normalized relaxation rates of population and polarization respectively, and T_1 (T_2) is the longitudinal (transverse) relaxation time. The dimensionless parameter $\epsilon = \omega_L/\omega = 4\pi\mu^2 C/3\hbar\omega$ is responsible for the light-matter coupling, where C is the density of two-level atoms and ω_L is the normalized Lorentz frequency. Quantity $l = (n_d^2+2)/3$ is the local-field enhancement factor due to the polarization of the background dielectric with refractive index n_d by the embedded active particles. Further we numerically solve Eqs. (1)–(3) using the finite-difference time-domain (FDTD) approach adapted to this particular problem [6, 7].

We consider the system in which the density of active particles experiences periodical random variations along the light propagation direction, so that the strength of light-matter coupling at some point z of the medium is given by [7]

$$\omega_L(z) = \omega_L^0[1 + 2r(\zeta(z) - 0.5)], \tag{4}$$

where ω_L^0 is the constant determined by the average density of two-level particles, $\zeta(z)$ is the random number uniformly distributed in the range [0; 1], and r is the parameter of disorder strength. Such disordered resonant medium can be considered as a multilayer (total thickness L) consisting of slabs (thickness δL) with different density of active atoms. The transition to light localization in such structures was studied in [7].

The parameters used for calculations agree with both rare-earth atoms and semiconductor quantum dots as the active particles. In particular, we take the relaxation times $T_1 = 1$ ns and $T_2 = 0.1$ ns, the Lorentz frequency $\omega_L^0 = 10^{12}$ s^{-1}, the frequency detuning $\delta = 0$, the full thickness of the medium $L = 1000\lambda$, the period of random density variations $\delta L = \lambda/4$, and the background refractive index $n_d = 1$. Incident light pulses have the central wavelength $\lambda = 0.8$ μm and the Gaussian envelope $\Omega = \Omega_p \exp(-t^2/2t_p^2)$, where the pulse duration is $t_p = 50$ fs. The peak Rabi frequency Ω_p is measured in the units of $\Omega_0 = \lambda/\sqrt{2\pi}ct_p$ corresponding to the pulse area of 2π.

Let us consider two pulses copropagating in the disordered resonant medium. Interactions of copropagating pulses can be realized due to the fact that pulse speed depends on its intensity: more powerful pulses move faster. So, if the first pulse has lower amplitude than the second one, they can collide and interact after a certain propagation length. The main result of this paper is shown in Fig. 1, where the dependencies of transmission, reflection and their sum on disorder strength r are shown for the copropagating pulses with amplitudes $\Omega_{p1} = \Omega_0$ and $\Omega_{p2} = 1.5\Omega_0$ and the interval $\Delta t = 10t_p$ between peaks at the entrance. We compare the case of two simultaneously propagating pulses with the case of single pulses. This latter case is used to calculate transmission and reflection for the situation with *apriori* no interaction between pulses; namely, transmitted or reflected energy of two non-interacting pulses can be calculated as $E_{12} = (E_1\Omega_{p1}^2 + E_2\Omega_{p2}^2)/(\Omega_{p1}^2 + \Omega_{p2}^2)$, where E_1 and E_2 are the data from the single pulse propagation calculations.

It is clearly seen from Fig. 1 that the resulting transmission and reflection for the pulses with and without interaction are the same in the uniform medium ($r = 0$) and are sharply different in

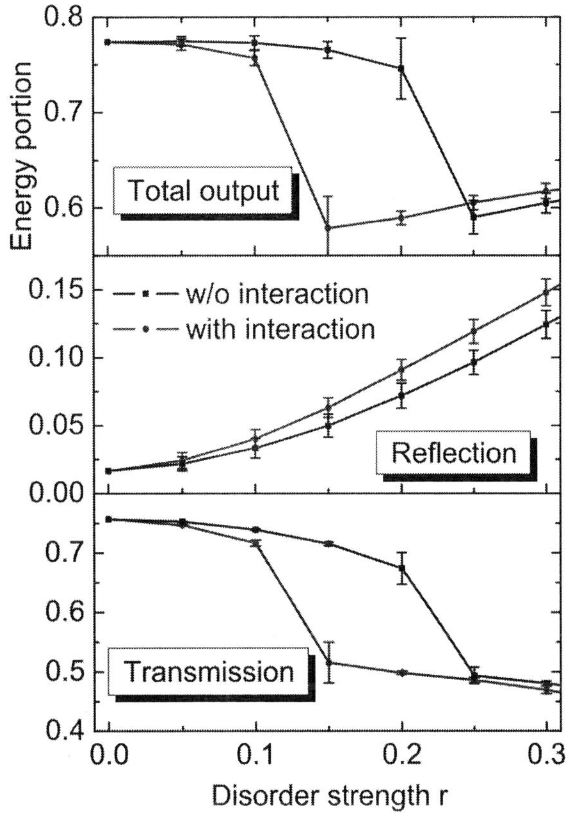

Figure 1. Average transmitted, reflected and total energy of pulses as a function of the disorder parameter r. The cases of interacting co-propagating pulses and single pulses with the amplitudes propagating without interaction are shown. The amplitudes of the pulses are $\Omega_{p1} = \Omega_0$ and $\Omega_{p2} = 1.5\Omega_0$; the interval between pulses is $\Delta t = 10t_p$. Energy averaged over 100 realizations was calculated for the time interval $700t_p$ and was normalized on the input energy. The error bars show the unbiased standard deviations for the corresponding average values.

the presence of disorder ($r > 0$). In other words, disorder induces effective interaction between copropagating pulses, which now collide inelastically and lose a portion of their energy due to interaction, in sharp contrast to the ordered case. The reason is that the medium is not uniform anymore, so that even relatively weak reflections due to varying densities of active particles can give rise to the coupling between the pulses.

This additional loss can be used to control transmission of light through the system. Indeed, after the collision, the pulses have lower intensity and, hence, move slower. They become more vulnerable to trapping inside the disordered medium. As a result, transmission and total output demonstrate a drop at rather low disorder, i.e., the threshold of localization is shifted to lower r: this threshold occurs at $r > 0.2$ for the interaction-free propagation, whereas light is localized already at $r > 0.1$ for the interacting copropagating pulses. The situation considered can be treated as a peculiar scheme of all-optical transmission switching based on disorder-induced pulse-pulse interactions: launching the second (control) pulse significantly modifies transmission of light through the medium in comparison to the case of single (signal) pulse.

We have also analyzed the influence of pulse amplitude, interval between pulses, and background dielectric on the efficiency of transmission switching. Calculations show that as we decrease the amplitude of the second pulse from $\Omega_{p2} = 1.5\Omega_0$ down to $1.1\Omega_0$, transmission drops from 0.5 to almost 0.3, so that the ratio of transmission for the pulses without and with interaction grows from approximately 1.5 up to 2. This means that the pulses with close amplitudes (e.g., $\Omega_{p1} = \Omega_0$ and $\Omega_{p2} = 1.1\Omega_0$) are better suited for transmission switching. The reason is that these pulses move with closer velocities and therefore have more time to interact than colliding pulses with strongly differing velocities.

As to the interval Δt between pulses, according to our calculations, transmission increases as the interval decreases, i.e., pulses collide too soon after launching and lose smaller portion of energy. On the contrary, increase in Δt first lowers transmission, which at $\Delta t \geq 20t_p$ keeps practically the same value. Thus, efficiency of transmission modulation is high in the wide range of intervals between pulses.

Finally, to examine the possibility of transmission switching in a more realistic set-up, we have simulated propagation of pulses in the disordered resonant medium with the background refractive index $n_d = 1.5$. The distinction of this situation is that there is additional reflection at the entrance and exit interfaces due to refractive index discontinuity. This reflection can strongly influence the results even in the ordered case, since a reflected portion of the faster (control) pulse can interact with the slower (signal) one. This situation resembles the well-known case of counterpropagating pulses which interact inelastically and lose large portion of energy due to collision [5]. In order to suppress this possibility and save the copropagating geometry, the system can be supplemented with the caps of pure background material of proper thickness (40λ) from both sides of the medium. This is enough to obtain almost interaction-free propagation of pulses at $r = 0$, so that the curves obtained in this case are generally similar to those shown in Fig. 1 and corroborate the presence of the transmission switching effect for non-vacuum background.

The work was supported by the Russian Foundation for Basic Research (Projects No. 18-02-00414 and 18-32-00160). Numerical simulations of light interaction with resonant media were supported by the Russian Science Foundation (Project No. 17-72-10098).

References

[1] Ermolenko M V, Buganov O V, Tikhomirov S A, Stankevich V V, Gaponenko S V and Shulenkov A S 2010 *Appl. Phys. Lett.* **97** 073113.
[2] Paternò G M, Iseppon C, D?Altri A, Fasanotti C, Merati G, Randi M, Desii A, Pogna E A A, Viola D, Cerullo G, Scotognella F and Kriegel I 2018 *Sci. Rep.* **8** 3517.
[3] Shcherbakov M R, Liu S, Zubyuk V V, Vaskin A, Vabishchevich P P, Keeler G, Pertsch T, Dolgova T V, Staude I, Brener I and Fedyanin A A 2017 *Nat. Commun.* **8** 17.
[4] Ferrera M and Carnemolla E G 2018 *J. Opt.* **20** 024007.
[5] Novitsky D V 2011 *Phys. Rev.* A **84** 013817.
[6] Novitsky D V 2009 *Phys. Rev.* A **79** 023828.
[7] Novitsky D V 2018 *Phys. Rev.* A **97** 013826.

IOP Conf. Series: Journal of Physics: Conf. Series **1092** (2018) 012100 doi:10.1088/1742-6596/1092/1/012100

Light dynamics in $\mathcal{PT}$-symmetric multilayers: Phase transition, nonreciprocity, and propagation direction locking

Denis V Novitsky[1,2], Alina Karabchevsky[3], Andrei V Lavrinenko[4], Alexander S Shalin[2] and Andrey V Novitsky[4,5]

[1]B. I. Stepanov Institute of Physics, National Academy of Sciences of Belarus, Nezavisimosti Avenue 68, 220072 Minsk, Belarus
[2]ITMO University, Kronverksky Prospekt 49, 197101 St. Petersburg, Russia
[3] Ben-Gurion University of the Negev, David Ben-Gurion Blvd, P.O.B. 653, 84105 Beer-Sheva, Israel
[4]DTU Fotonik, Technical University of Denmark, Ørsteds Plads 343, DK-2800 Kongens Lyngby, Denmark
[5]Department of Theoretical Physics and Astrophysics, Belarusian State University, Nezavisimosti Avenue 4, 220030 Minsk, Belarus

E-mail: dvnovitsky@gmail.com

Abstract. We report the numerical study of $\mathcal{PT}$-symmetry breaking in one-dimensional structures with resonantly absorbing and amplifying layers described by the Maxwell-Bloch equations. Temporal dynamics of light interacting with such a structure is governed by strong field amplification and subsequent saturation of loss and gain above the exceptional point resulting in the lasing-like regime with powerful pulses generated both in reflection and transmission. In this regime, we predict and investigate the uniqueness of phase transition due to saturation, transmission nonreciprocity, and the direction locking of the transmitted and reflected radiation.

Optical $\mathcal{PT}$-symmetric structures and the effects connected to $\mathcal{PT}$-symmetry breaking is one of the hot topics in modern photonics [1, 2, 3]. In one-dimensional (1D) geometry, optical $\mathcal{PT}$-symmetric structure can be realized as a photonic-crystal-like system composed of alternating loss and gain layers. Such a multilayer should have proper spatial variation of the complex permittivity satisfying the necessary condition $\varepsilon(z) = \varepsilon^*(-z)$, i.e., the real part of the permittivity is an even function of the coordinate, whereas the imaginary part is an odd function. There is a number of phenomena associated with violation of the $\mathcal{PT}$ symmetry, such as sharp change in polarization response of the system [4], enhanced sensitivity to external perturbations near exceptional point [5, 6], new effects of lasing [7, 8] and anti-lasing [9].

In this work, we consider a 1D structure in which both loss and gain materials are described as a two-level resonant medium. This allows us to describe self-consistently dynamics of field and material parameters and to shed new light on change of system's temporal response at the exceptional point where the phase transition of $\mathcal{PT}$-symmetry breaking occurs. The system considered in this work is a periodic planar structure composed of $2N$ alternating loss and gain layers. It is illuminated by normally incident monochromatic light of frequency ω. We describe both loss and gain in a similar manner, using the model of a homogeneously-broadened two-level medium. Then light interaction with the structure is given by the Maxwell-Bloch equations for microscopic polarization ρ, population difference w, and electric field amplitude A [10],

$$\frac{d\rho}{d\tau} = il\Omega w + i\rho\delta - \gamma_2\rho, \tag{1}$$

$$\frac{dw}{d\tau} = 2i(l^*\Omega^*\rho - \rho^* l\Omega) - \gamma_1(w - w_{eq}), \tag{2}$$

$$\frac{\partial^2\Omega}{\partial\xi^2} - n_d^2\frac{\partial^2\Omega}{\partial\tau^2} + 2i\frac{\partial\Omega}{\partial\xi} + 2in_d^2\frac{\partial\Omega}{\partial\tau} + (n_d^2 - 1)\Omega = 3\alpha l\left(\frac{\partial^2\rho}{\partial\tau^2} - 2i\frac{\partial\rho}{\partial\tau} - \rho\right), \tag{3}$$

where $\tau = \omega t$ and $\xi = kz$ are respectively the dimensionless time and distance, $\Omega = (\mu/\hbar\omega)A$ is the normalized Rabi frequency, $k = \omega/c$ is the wavenumber in vacuum, c is the speed of light, $\hbar$ is the reduced Planck constant, μ is the dipole moment of the quantum transition, $\alpha = \omega_L/\omega = 4\pi\mu^2 C/3\hbar\omega$ is the normalized Lorentz frequency, C is the concentration of active atoms, $\delta = (\omega_0 - \omega)/\omega$ is the normalized detuning of laser frequency from the atomic resonance, $\gamma_1 = 1/(\omega T_1)$ and polarization $\gamma_2 = 1/(\omega T_2)$ are the normalized relaxation rates expressed by means of the longitudinal T_1 and transverse T_2 relaxation times, $l = (n_d^2 + 2)/3$ is the local-field enhancement factor, and n_d is the real-valued refractive index of the background medium.

We describe both gain and loss materials with the same Maxwell-Bloch equations (1)-(3) using the equilibrium population difference w_{eq} as a key parameter governing the level of pumping. Indeed, in the stationary approximation the resonant medium can be characterized by the effective permittivity $\varepsilon_{eff} \approx n_d^2 + 3il^2\omega_L T_2 w_{eq}$ [11]. Since gain and loss correspond to negative and positive w_{eq}, it is straightforward to obtain a $\mathcal{PT}$-symmetric structure composed of alternating layers with balanced loss (ε_{eff+}) and gain (ε_{eff-}), where $\varepsilon_{eff\pm} \approx n_d^2 \pm 3il^2\omega_L T_2|w_{eq}|$. The necessary condition $\varepsilon(z) = \varepsilon^*(-z)$ is obviously fulfilled for such a structure, providing even (odd) function of z for the real (imaginary) part of the permittivity. It can be also shown that $\mathcal{PT}$ symmetry approximately follows directly from the Maxwell-Bloch equations, if the steady state is established.

Equations (1)–(3) are solved numerically using the FDTD approach developed earlier [12]. For calculations, we use semiconductor doped with quantum dots as an active material characterized by the following parameters: $n_d = 3.4$, $\omega_L = 10^{11}$ s^{-1}, $T_1 = 1$ ns, and $T_2 = 0.5$ ps. We also assume the exact resonance $\delta = 0$. The multilayer structure contains $N = 20$ unit cells with both loss and gain layers having the same thickness $d = 1$ μm.

We first explore the response of the system in $\mathcal{PT}$-symmetric phase, i.e., below the exceptional point. Our numerical simulations of monochromatic light propagation are in good agreement with the transfer-matrix calculations performed in the stationary approximation with the effective permittivities given above. This agreement is due to the rapid establishment of stationary transmission and reflection example of which is shown in Fig. 1(a). It is also seen that owing to reciprocity of the system, the transmission of oppositely propagating (forward and backward) waves is the same ($T_{LG} = T_{GL} = T$), but the reflection is different ($R_{LG} \neq R_{GL}$). In particular, we are able to reproduce the anisotropic transmission resonances characteristic for 1D $\mathcal{PT}$-symmetric structures when $T = 1$ and one of the reflection coefficients goes to zero [13].

$\mathcal{PT}$-symmetry breaking occurs when the pumping parameter reaches a certain threshold value, which can be determined in the stationary approximation as an exceptional point where eigenvalues of the scattering matrix cease to be unimodular, whereas eigenvectors coincide [13]. For the structure parameters given above, the position of this phase transition can be found within the transfer-matrix method as $w_{eq} \approx 0.222$. However, as the pumping parameter approaches the exceptional point, the stationary approximation becomes inapplicable: the

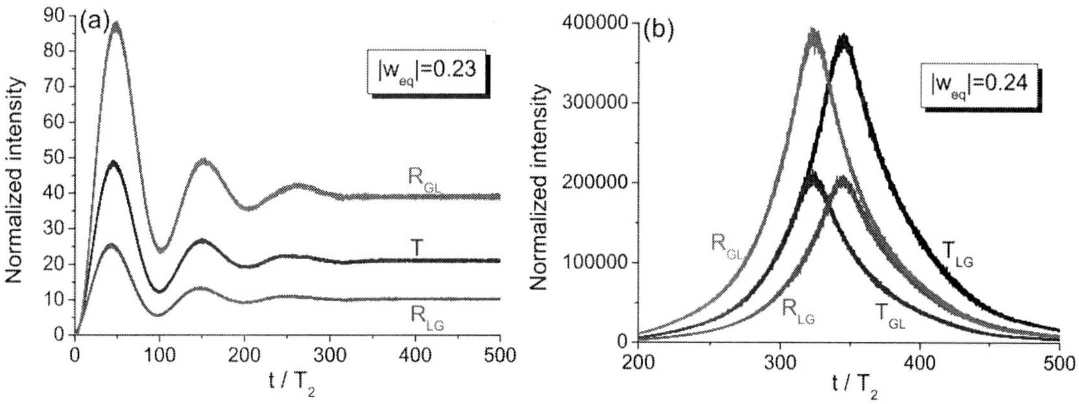

Figure 1. Temporal dynamics of the reflected (R) and transmitted (T) intensity for the pumping parameter (a) $|w_{eq}| = 0.23$ and (b) $|w_{eq}| = 0.24$. The two directions of wave propagation are denoted with subscripts LG and GL originating from the order of layers in the unit cell of the structure.

discrepancy between the results given by the transfer-matrix method and numerical simulations of the Maxwell-Bloch equations sharply increases. This means that near the phase transition point we should solve the differential equations in time domain in order to correctly predict the response of the system.

Above the exceptional point, a strong amplification of the electromagnetic field inside the structure occurs, since loss and gain are not balanced anymore. This enhancement is limited by saturation that develops in the resonant medium as the field becomes strong enough. Indeed, whereas population difference preserves its initial value $w(t) = w_{eq}$ in the $\mathcal{PT}$-symmetric phase below the exceptional point, it starts changing due to saturation above the exceptional point. The resulting temporal profiles of transmitted and reflected intensity have the form of powerful pulses due to the rapid release of energy stored in the gain layers [see Fig. 1(b)]. Therefore, the $\mathcal{PT}$-symmetry-broken phase can be called the lasing-like regime (in analogy with [11]), which can be treated as a dynamical feature of this phase state.

Saturation is also the reason for the irreversibility of $\mathcal{PT}$-symmetry breaking: since the necessary condition $\varepsilon(z) = \varepsilon^*(-z)$ is not fulfilled anymore due to changing population difference, the return of the system with a further increase of the pumping parameter back into the PT-symmetric state as predicted in the stationary approximation is now impossible.

In the lasing-like regime, the structure becomes nonreciprocal, since transmission is no more symmetric ($T_{LG} \neq T_{GL}$) as clearly seen in Fig. 1(b). Moreover, the intensities of the pulses escaping the system do not depend on the direction of the incident light: almost the same pulses are emitted from the gain and loss ends of multilayer after reversing the input light direction ($T_{LG} = R_{GL}$ and $T_{GL} = R_{LG}$). This novel unusual feature due to $\mathcal{PT}$-symmetry breaking can be called the propagation direction locking of radiation transmitted and reflected by the structure. Saturation is not enough to explain this effect as evidenced by calculations for relatively powerful incident wave when $\mathcal{PT}$ symmetry is broken at every value of the pumping parameter. Therefore, the $\mathcal{PT}$-symmetry breaking is necessary for the locking effect to occur. This phenomenon can be viewed as a possible basis for peculiar all-optical diodes and transistors.

The work was supported by the Belarusian Republican Foundation for Fundamental Research (Projects No. F16K-016 and F16R-049), the Russian Foundation for Basic Research (Projects No. 18-02-00414, 18-52-00005 and 18-32-00160), Ministry of Education and Science of the Russian Federation (GOSZADANIE, Grant No. 3.4982.2017/6.7), Government of Russian

Federation (Grant 08-08), and the Israeli Ministry of Trade and Labor Kamin Program (Grant No. 96717). Numerical simulations of light interaction with resonant media were supported by the Russian Science Foundation (Project No. 17-72-10098). The calculations of field distributions have been supported by the Russian Science Foundation (Project No. 16-12-10287). Partial financial support of Villum Fonden (DarkSILD project) is acknowledged.

References

[1] Zyablovsky A A, Vinogradov A P, Pukhov A A, Dorofeenko A V and Lisyansky A A 2014 *Phys. Usp.* **57** 1063.
[2] Feng L, El-Ganainy R and Ge L 2017 *Nat. Photon.* **11** 752.
[3] El-Ganainy R, Makris K G, Khajavikhan M, Musslimani Z H, Rotter S and Christodoulides D N 2018 *Nat. Phys.* **13** 11.
[4] Lawrence M, Xu N, Zhang X, Cong L, Han J, Zhang W and Zhang S 2014 *Phys. Rev. Lett.* **113** 093901.
[5] Hodaei H, Hassan A U, Wittek S, Garcia-Gracia H, El-Ganainy R, Christodoulides D N and Khajavikhan M 2017 *Nature (London)* **548** 187.
[6] Chen W, Özdemir S K, Zhao G, Wiersig J and Yang L 2017 *Nature (London)* **548** 192.
[7] Feng L, Wong Z J, Ma R-M, Wang Y and Zhang X 2014 *Science* **346** 972.
[8] Hodaei H, Miri M-A, Heinrich M, Christodoulides D N and Khajavikhan M 2014 *Science* **346** 975.
[9] Wong Z J, Xu Y-L, Kim J, O'Brien K, Wang Y, Feng L and Zhang X 2016 *Nat. Photon.* **10** 796.
[10] Novitsky D V 2011 *Phys. Rev.* A **84** 013817.
[11] Novitsky D V, Tuz V R, Prosvirnin S L, Lavrinenko A V and Novitsky A V 2017 *Phys. Rev.* B **96** 013856.
[12] Novitsky D V 2009 *Phys. Rev.* A **79** 023828.
[13] Ge L, Chong Y D and Stone A D 2012 *Phys. Rev.* A **85** 023802.

Creation of metamaterials from layered quantum structures based on graphene

A.K. Nukhov[1,2,3]

[1]Moscow Institute of Physics and Technology, Institutskyi 9, 141700 Dolgoprudnyi, Russia
[2]Daghestan Institute of Physics after Amirkhanov, Yaragskogo str., 94, 367003,Makhachkala, Russia,
[3]Dagestan State University, Gadzhiyev Street, 43a , 367000, Makhachkala Russia,

nukhov1984@mail.ru

Abstract. The paper considers low-dimensional electronic systems interacting with graphene, and the possibility of creating on their basis artificially nanostructured layered metamaterials with given optical properties for various nanophotonic devices and for the elemental base of nanoelectronics.

1. Introduction

The use of graphene gives new opportunities in the development of nanostructured metamaterials with unique optical properties. Calculation and analysis of electronic characteristics of graphene and low-dimensional systems based on it require development of new computational algorithms and the use of powerful computing systems to model and create such nanostructured metamaterials

Earlier, electronic states of low-dimensional systems (quantum dots, films) interacting with graphene were examined and the theoretical possibility of controlled influence on electronic states of these systems by means of electric and magnetic fields was demonstrated [1, 2]. These effects are directly related to the optical properties of these low-dimensional systems. The possibility of varying the optical properties of low-dimensional systems is one of the opportunities for the development of promising technologies for the artificial nanostructuring of metamaterials.

2. Methods and results

Let us consider a system of quantum dot–graphene-monolayer–SiO2/+n+Si-substrate (Fig. 1). The consideration is based upon the Anderson–Newns model [1-4]. As can easily be shown, the density of states (DOS) in the epitaxial layer of graphene near the Dirac point can be expressed as follows:

$$\rho_g(\varepsilon) = \frac{\varepsilon}{\pi \upsilon_F^2 \mathrm{h}^2}, \qquad (1)$$

where ε is the energy variable, υ_F is the Fermi velocity (for electrons in graphene, υ_F ~ 108 cm/s), and the subscript "g" refers to graphene. The spectral function for a QD in the Anderson–Newns model can be written as follows:

$$A(\varepsilon) = \frac{\Gamma + \pi |V|^2 \rho_g(\varepsilon)}{(\varepsilon - E_n)^2 + \left(\Gamma + \pi |V|^2 \rho_g(\varepsilon)\right)^2}, \qquad (2)$$

where E_n is the energy spectrum of QD electrons, V is the hybridization potential, Γ - damping of one-electron excitations due to collisional processes in the QD. The DOS in the QD is expressed as

$$\rho(\varepsilon)=\sum_{n=1}^{n_F}\frac{\Gamma+\alpha\Omega}{\left(\varepsilon_0\Omega-E_n\right)^2+\left(\Gamma+\alpha\Omega\right)^2}\quad, \tag{3}$$

where $\alpha=|V|^2\varepsilon_0/\upsilon_F^2\hbar^2$, $\Omega=\varepsilon/\varepsilon_0$ is the dimensionless energy, ε_0 is a certain characteristic energy, and n_F is the number of states under the Fermi level. A QD can be considered as a three dimensional potential well with infinitely high walls. Then, the energy spectrum of QD electrons can be written as $E_n=\frac{\pi^2\hbar^2n^2}{2mL^2}=E_1n^2$, where L is the linear size of the QD. The integration of DOS given by formula (3) from 0 to $\Delta\Omega_F=\left(E_F-\varepsilon_F\right)/\varepsilon_0$ yields the following expression for the charge transfer in the system under consideration:

$$\langle n\rangle_\pm=\pm\frac{\alpha}{\left(1+\alpha^2\right)^2}\sum_{n=1}^{n_F}\left\{\frac{1}{2}\ln\left[\frac{\left(\varepsilon_0\Delta\Omega_F-E_n\right)^2+\Lambda}{E_n^2+\Lambda}\right]+\right.$$
$$\left.+\frac{E_n}{\sqrt{\Lambda}}\left[arctg\left(\frac{\varepsilon_0\Delta\Omega_F-E_n}{\sqrt{\Lambda}}\right)+arctg\left(\frac{E_n}{\sqrt{\Lambda}}\right)\right]\right\}, \tag{4}$$

where the plus and minus signs refer to the cases of charge transfer from graphene to QD ($\varepsilon_F > E_F$) and vice versa ($\varepsilon_F < E_F$), respectively; $\Lambda=E_n^2\left(1-1/\left(1+\alpha^2\right)^2\right)$, ε_F is the Fermi energy of electrons in graphene, and $E_F=E_{n_F}$ is the Fermi energy of electrons in the QD. The ε_F value is nonzero due to the applied bias voltage V (Fig. 1). Moreover, by varying this voltage, one can control the Fermi energy level. Using the applied voltage of different polarity, it is possible to dope graphene with either electrons or holes [4], thus controlling both the number and type of charge carriers in graphene and, hence, the amount of transferred charge. The Fermi level depends on the applied bias voltage V as follows [4]:

$$\varepsilon_F=\upsilon_F\hbar\sqrt{\frac{\xi V}{ed}} \tag{5}$$

where d is the substrate thickness, e is the electron charge, and ξ is the dielectric permittivity of the substrate. Substituting expression (5) into formula (4), one can readily determine the transferred charge as a function of the applied voltage (Fig. 2).

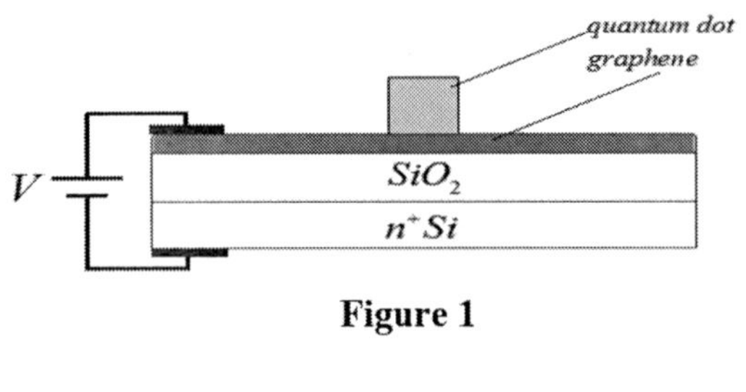

Figure 1

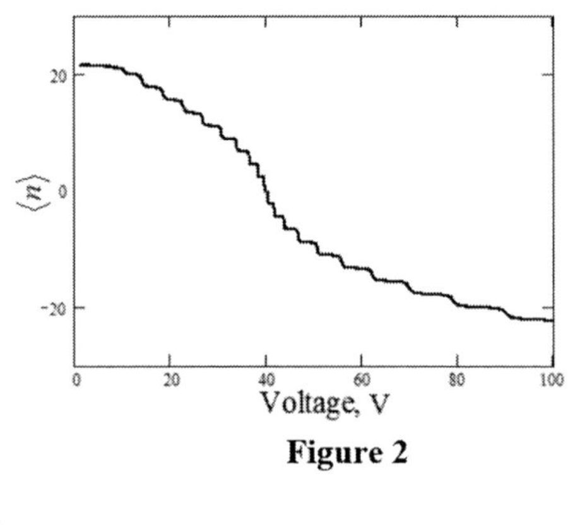

Figure 2

3. Discussion and conclusions

One of the ways to create metamaterials is the formation of layered structures of dielectrics and semiconductors. Such structures allow to control the electric current, but they are very power-consuming (10 kW per 1 cm^2). Firstly, it is energetically unprofitable; secondly, it is very difficult to find materials capable of withstanding such power. This problem can be solved by creating intermediate layers where the electron can live longer, but it is necessary to think what materials these layers can be made of, how this barrier will be overcome, and how electrons will be transported to these layers. In this paper we consider layered metamaterials based on graphene.

Firstly, graphene is a truly two-dimensional structure. Secondly, thanks to the Dirac fermions, graphene occupies an intermediate position between metals and semiconductors. Electrons in graphene can move with zero effective mass, what leads to long free paths. All this makes graphene an ideal candidate for use as an intermediate layer in such layered metamaterials. Moreover, graphene can be used as an insulating layer between a semiconductor and an intermediate layer, where the electron will stay in the free state for longer. This possibility follows from the work [1], where it is shown that with the help of graphene, as an intermediate material, , it is possible to control both the quantity and the type of charge carriers passing between the two systems applying an external electric field of different polarity and magnitude. This effect becomes possible due to the fact that the Fermi level in graphene depends on the applied external electric field. In this case graphene will serve as a transit material between these layers. All this makes graphene an excellent candidate for use in structuring metamaterials of nanoscale.

It can also be noted that the development of technologies for creating metamaterials based on graphene in the form of layered structures is a great step towards improving the so-called planar metamaterials. They are two-dimensional metamaterials, which, in particular, are used to cover solar panels in order to increase their efficiency.

Acknowledgments
This work was supported by the Ministry of Education and Sciences of the Russian Federation (16.7162.2017/8.9).

References

[1] Z. Z. Alisultanov, R.P. Meilanov, A.K. Nukhov. Tech. Phys. Lett., 2012, V.38, N.6, pp. 552–554.
[2] Anderson P.W. // Phys. Rev. 1961. V. 124. N. 1. P. 41-53.
[3] Newns D.M. // Phys. Rev. 1969. V. 178. N. 3. P. 1123.
[4] Davydov S.Yu. // JETP Letters. 2011. V. 37. N. 10. pp. 64-67.

Saturation of fluorescence from NV centers in Mie-resonant diamond particles

Dmitry V. Obydennov[1], **Daniil A. Shilkin**[1,2], **Evgeny V. Lyubin**[1], **Maxim R. Shcherbakov**[1,3], **Evgeny F. Ekimov**[4], **Oleg S. Kudryavtsev**[5], **Igor I. Vlasov**[5,6], **Andrey A. Fedyanin**[1]

[1] Quantum Electronics Department and Quantum Technology Center, Faculty of Physics, Lomonosov Moscow State University, Moscow 119991, Russia
[2] Center for Functionalized Magnetic Materials, Immanuel Kant Baltic Federal University, Kaliningrad 236041, Russia
[3] School of Applied and Engineering Physics, Cornell University, Ithaca NY 14853, USA
[4] Institute for High Pressure Physics, Russian Academy of Sciences, Troitsk 142190, Russia
[5] Prokhorov General Physics Institute, Russian Academy of Sciences, Moscow 119991, Russia
[6] National Research Nuclear University MEPhI, Moscow 115409, Russia

E-mail: obydennov@nanolab.phys.msu.ru

Abstract. Diamond particles with color centers have found numerous applications in quantum optics, bioimaging and nanosensing. Here, we report on an experimental study of the optical Mie resonances of subwavelength diamond particles and their effect on the fluorescence of nitrogen-vacancy centers. We employ single-particle scattering spectroscopy to reveal the fundamental optical resonances of individual diamond particles and measure the fluorescence saturation curves using continuous-wave excitation to evidence the difference in the spontaneous emission rate of the color centers in different particles. A simple analytical model has been used to predict the dependence of the saturation intensity on the particle size for a free-standing spherical diamond. Experimentally, we have shown that the magnetic dipole and quadrupole modes are excited in the fluorescence spectral range for particles of approximately 250–350 nm in size and that the saturation intensity for such particles is by an order of magnitude higher than for a smaller one, which is in agreement with our calculations. The results demonstrate that the efficiency of diamond-based sensors and single-photon sources can be improved by simply choosing particles of a proper size.

1. Introduction

Diamond particles with color centers are extensively used in quantum optics, bioimaging and nanosensing [1, 2]. To efficiently handle the fluorescence of the color centres, hosting particles were proposed to be placed into properly designed resonators. In this way, the emission can be enhanced and directed into a desired direction. However, the fact that the diamond particle itself can be resonant is usually omitted from consideration. High-index nanoparticles exhibit a set of optical resonances, the first of which is excited when the particle size is approximately equal to the wavelength in the particle material [3]. As they reveal themselves in the scattering spectra, these resonances are also referred to as Mie resonances. The high refractive index of diamond does allow for the excitation of Mie resonances in subwavelength diamond particles, although no evidence of this had been shown until our recent work on scattering spectroscopy of single nanodiamonds [4]. The effects of these resonances on the luminescence of color centers in submicrometer diamond particles have been repeatedly discussed in the literature [5, 6, 7, 8, 9]; however, a direct experimental evidence of this relationship is still lacking.

In this contribution, we report on an experimental study of the fluorescence from nitrogen-vacancy (NV) color centers in Mie-resonant subwavelength diamond particles. First, we employ single-particle dark-field scattering spectroscopy to reveal the fundamental Mie resonances of individual diamonds. Second, we measure the fluorescence saturation curves using continuous-wave excitation to characterize the influence of the excited resonances on the spontaneous emission rate. Finally, we calculate the dependence of the saturation intensity in a free-standing spherical diamond on its diameter and compare the resulted dependence with the experimental observations.

2. Results and discussion

The studied diamonds were synthesized by the HPHT technique from a mixture of adamantane ($C_{10}H_{16}$) and adamantane carbonitrile ($C_{11}H_{15}N$) in a ratio of 50:1 at a temperature of 1800 K and a pressure of 8–9 GPa (see [10] for more details). For further study, diamonds were dispersed on a cover glass from a water suspension. Four diamond particles of subwavelength dimensions were chosen for the fluorescence experiments. Scanning electron micrographs of the chosen particles are shown in Figure 1a.

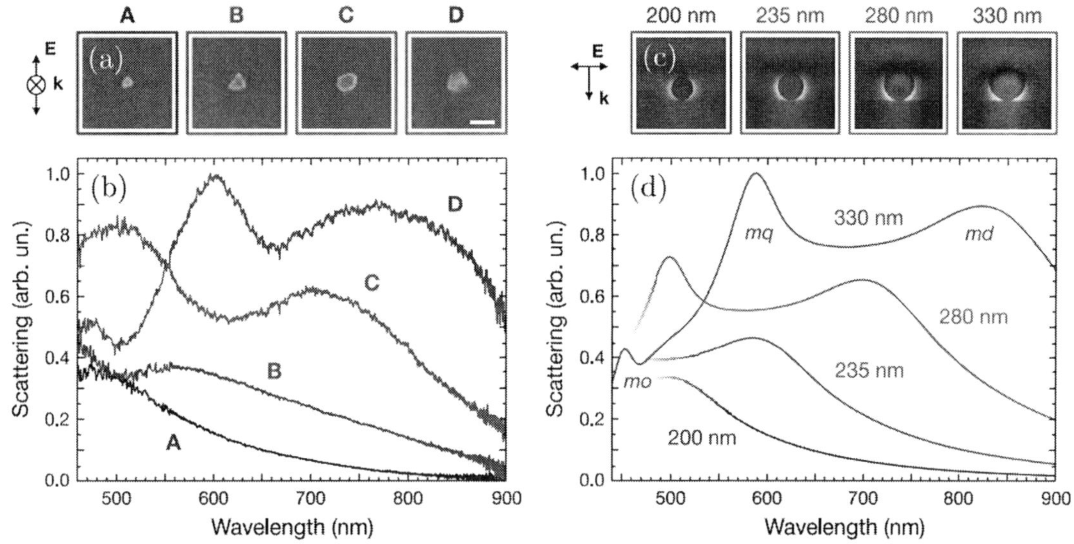

Figure 1. (a) Scanning electron micrographs of the studied diamond particles (scale bar is 500 nm). The orientation of the laser polarization in the fluorescence experiments is shown on the left. (b) Non-polarized scattering spectra of the studied particles. All spectra are normalized to the maximum value obtained for particle 'D'. (c) FDTD-calculated electric field amplitude in spherical diamonds of different size located on a glass substrate under plane wave illumination with a wavelength of 660 nm. The simulation geometry is shown on the left. (d) FDTD-calculated spectra of scattering into the experimental collection aperture for spherical diamonds of different size located on a glass substrate. For a 330-nm sphere, the peaks corresponding to the excitation of the magnetic dipole (md), quadrupole (mq) and octupole (mo) modes are denoted.

The optical resonances of the diamond particles were charactirezed by single-particle scattering spectroscopy performed as described elsewhere [4]. The non-polarized scattering spectra for the four studied particles are shown in Figure 1b. With increasing particle size, the resonances shift to the red side of the spectrum and the absolute scattering increases. Both numerical and analytical calculations have been carried out to provide an insight into the nature of the observed resonances. Figures 1c,d demonstrate the results of numerical finite-difference time-domain (FDTD) simulations of scattering by spherical diamonds of different size located on a glass substrate. The shown scattering spectra were calculated for the experimental collection aperture. Additionally performed analytical calculations have shown that the observed peaks are associated with the excitation of the magnetic dipole, quadrupole and octupole modes.

For the fluorescence experiments, the radiation of a multimode DPSS laser with a wavelength of 532 nm was filtered by a single-mode polarization-maintaining optical fiber and focused into a 3-μm spot on the sample. The fluorescence signal was collected in transmittance geometry by a water-immersion objective lens with a numerical aperture of 1.2, filtered from the pump and directed into an Ocean Optics USB4000 spectrometer with a resolution of 1 nm. The obtained spectra, with the dark noise subtracted, are shown in the top row of Figure 2. The brightest fluorescence is observed for the largest particle that is at least partly due to a bigger number of color centers contained. With increasing optical power, the saturation effect takes place as shown in the bottom row of Figure 2. The given dependences were obtained by integrating the registered fluorescence signal over the wavelengths of 610–710 nm. The background signal shown by the gray open circles was collected from empty regions of the substrate in the vicinity of the studied diamonds. For each particle, the saturation curve was approximated by the following law: $K = K_\infty P/(P + P_{\rm sat}) + \alpha P$, where the coefficient α was predetermined by fitting the background signal with a proportional dependence. The obtained values of the saturation power are given in Figure 2 above the saturation plots.

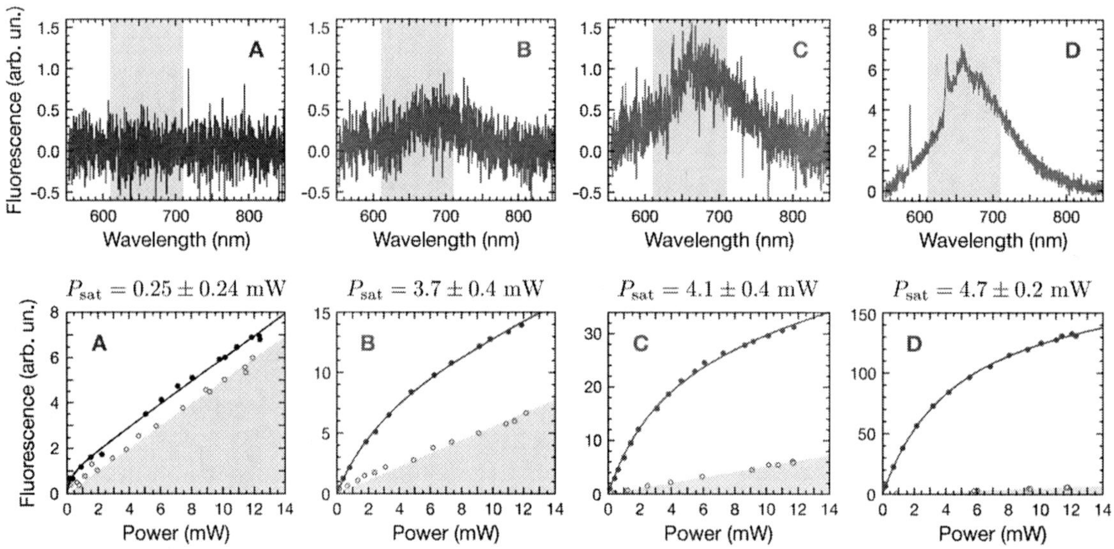

Figure 2. The top row presents the fluorescence spectra measured at a 532-nm pump power of 10 mW with the dark noise signal subtracted. The bottom row presents the fluorescence saturation curves obtained by integrating the fluorescence signal over the wavelengths of 610–710 nm. The background signal measured in the vicinity of the corresponding particles is shown by gray open circles. The obtained values of the saturation power are given above the saturation plots.

The excitation of Mie resonances in a fluorescent particle affects the saturation process in two ways. First, if the particle is resonant at the laser wavelength, the pump field inside the particle is enhanced that leads to a faster saturation. Second, if the particle is resonant at the fluorescence wavelength, the spontaneous emission rate is increased; this phenomenon is known as the Purcell effect and leads to an increase of the saturation intensity. We use equations from [11, 12] to analytically calculate the field enhancement E and the Purcell factor F_P in a spherical particle. We then suppose that the saturation intensity for a point source is proportional to the Purcell factor divided by the squared field enhancement at the pump frequency. In the case of multiple emitters, as relevant for our experiment, the saturation intensity I_sat can be roughly determined by averaging this ratio over the sphere volume:

$$I_\mathrm{sat} \propto \frac{1}{V} \int \frac{F_\mathrm{P}|_{\lambda=660\,\mathrm{nm}}}{E^2|_{\lambda=532\,\mathrm{nm}}} \, dV. \tag{1}$$

Here, the Purcell factor F_P is averaged over all possible orientations of the emitting dipole. Note that although the substrate and nonradiative transitions can significantly modify the excited-state lifetime [13], we omit these effects from consideration expecting that they do not strongly change the dependence on size in the considered range.

The resulted dependence of the saturation intensity on the sphere diameter is shown in Figure 3 by the gray line. To compare it with the experimental results, the size of the studied particles was quantified based on the scattering spectra. For this purpose, the measured scattering spectra were approximated by numerically calculated ones using the least squares fitting technique with the diameter and a scale factor as estimated values. The resulted experimental dependence of the saturation power on the effective particle size is shown in Figure 3 by the color circles. The error bars for the particle size correspond to the two-fold increase of χ^2 compare to the best fit point. From comparison with the theoretical curve one can see that the more than an order-of-magnitude variance of the saturation power can be explained by the excitation of Mie resonances in the diamonds.

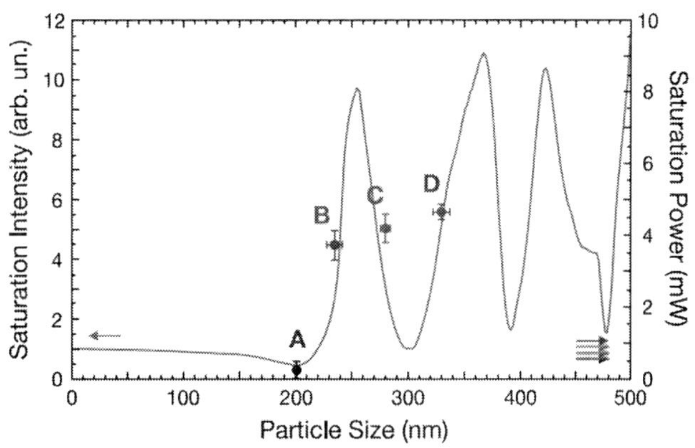

Figure 3. Gray line shows the saturation intensity averaged over the particle volume calculated for a dipole source emitting a wavelength of 660 nm in a free-standing diamond sphere as a function of the sphere diameter. Color circles show an experimental dependence of the saturation power on the effective particle size determined from the scattering spectra.

3. Conclusion

In conclusion, we have experimentally demonstrated the effect of Mie resonances of individual subwavelength diamond particles on the saturation of fluorescence from their NV centres. The results demonstrate that the efficiency of diamond-based sensors and single-photon sources can be improved by choosing particles of a proper size.

References

[1] Aharonovich I, Greentree A D and Prawer S 2011 *Nat. Photonics* **5** 397
[2] Chen X and Zhang W 2017 *Chem. Soc. Rev.* **46** 734–760
[3] Kuznetsov A I, Miroshnichenko A E, Brongersma M L, Kivshar Y S and Luk'yanchuk B 2016 *Science* **354** aag2472
[4] Shilkin D A, Shcherbakov M R, Lyubin E V, Katamadze K G, Kudryavtsev O S, Sedov V S, Vlasov I I and Fedyanin A A 2017 *ACS Photonics* **4** 1153–1158
[5] Greffet J J, Hugonin J P, Besbes M, Lai N, Treussart F and Roch J F 2011 *arXiv* 1107.0502
[6] Inam F A, Gaebel T, Bradac C, Stewart L, Withford M J, Dawes J M, Rabeau J R and Steel M J 2011 *New J. Phys* **13** 073012
[7] Castelletto S and Boretti A 2011 *Opt. Lett.* **36** 4224–4226
[8] Savelev R, Zalogina A, Kudryashov S, Ivanova A, Levchenko A, Makarov S, Zuev D and Shadrivov I 2018 *J. Phys.: Conf. Series* vol 961 p 012007
[9] Zalogina A S, Savelev R S, Ushakova E V, Zograf G P, Komissarenko F E, Milichko V A, Makarov S V, Zuev D A and Shadrivov I V 2018 *Nanoscale* **10** 8721–8727
[10] Ekimov E A, Kudryavtsev O S, Mordvinova N E, Lebedev O I and Vlasov I I 2018 *ChemNanoMat* **4** 269–273
[11] Bohren C F and Huffman D R 2008 *Absorption and scattering of light by small particles* (John Wiley & Sons)
[12] Chew H 1988 *Phys. Rev. A* **38** 3410
[13] Inam F A, Grogan M D W, Rollings M, Gaebel T, Say J M, Bradac C, Birks T A, Wadsworth W J, Castelletto S, Rabeau J R and Steel M J 2013 *ACS Nano* **7** 3833–3843

Acknowledgments

This work was partially supported by the Russian Ministry of Education and Science (# 14.W03.31.0008), the Russian Foundation for Basic Research (# 18-02-00880, spectroscopy experiments), the Russian Science Foundation (# 14-12-01329, synthesis and characterization of the diamonds; # 15-12-00065, calculations), and the Russian National Technology Initiative.

Functional metasurfaces based on water

Mikhail Odit[1], Andrei Saynskyi[1], Irina Munina[2], and Pavel Belov[1]

[1]Department of Nanophotonics and Metamaterials, ITMO University, 197101 Saint-Petersburg, Russia

[2] Microwave Microelectronics Laboratory, St. Petersburg Electrotechnical University LETI, Prof. Popov str. 197376 St. Petersburg, Russia

E-mail: odit@mail.ifmo.ru

Abstract. Here we report results on numerical investigation of the all-dielectric metasurfaces based on water. Two different functionalities of the water-based metasurfaces are shown: focusing an anomalous reflection of the electromagnetic wave. Metasurfaces are comprised of dielectric bianisotropic resonators and operate at microwave frequency.

1. Introduction

The metasurfaces made of high-index materials attracted a lot of attention due to their high efficiency in wave manipulation, diversified functionality, relatively simple fabrication and compact size [1]. Moreover all-dielectric metasurfaces exhibit no Ohmic losses and are fully compatible with standard industrial semiconductor technology. All-dielectric metasurfaces are usually made of sub-wavelength dielectric particles exhibiting Mie resonances. Combination of different Mie modes and Fabry-Perot resonances make it possible to use all-dielectric metasurfaces in controlling emitted radiation in both forward and backward directions, near-field engineering, enhancement of linear and nonlinear effects [2, 3].

In order to form homogeneous metasurface with closely packed sub wavelength resonant particle,s high-permittivity material should be used for resonators. While semiconductor is most common material for optical range [4], in microwave region high-index ceramics is usually used. It was recently shown that even usual water can be used as a constituent material for microwave metamaterials [5]. Water-based metasurfaces were shown be able to act as microwave absorbers [6, 7], tunable filter [8], exhibit torroidal response [9].

In this work we demonstrate water-based reflective metasurfaces with two different functionalities: a metasurface for focusing the reflected incident wave (focusing metamirror) and a metasurface anomalously reflecting normally incident waves (reflective metamirror).

2. Focusing metamirror

The focusing metamirror is composed of the dielectric bianisotropic particles arrange in a square array (Fig. 1). The shape of each particle has a form of cylinder of the height H and radius R with a notch inside having the depth h and the radius r (see inset to Fig. 1). The dielectric properties of the particles are the same as for distilled water at the temperature 25°: $\varepsilon = 78$, $\mu = 1$, electric conductivity $\sigma = 1.59$ [S/m]. The period of the array is $a = 15$ mm. Note that period should be less than half of the operating wavelength to avoid undesired reflection.

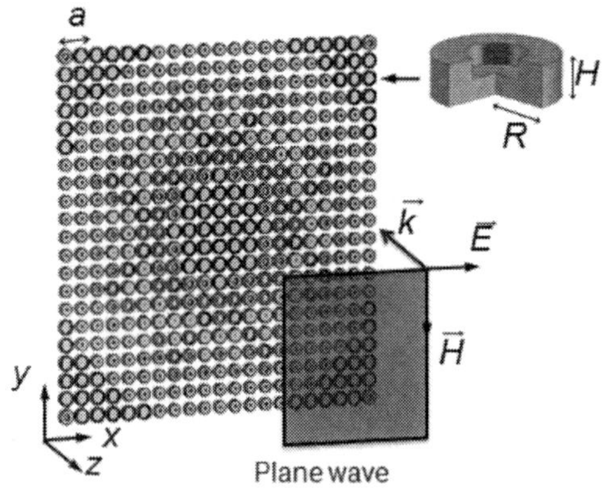

Figure 1. Focusing metamirror comprised of the array of 21x21 bianisotropic dielectric water particles. Inset is showing a quarter cut-off of a single particle.

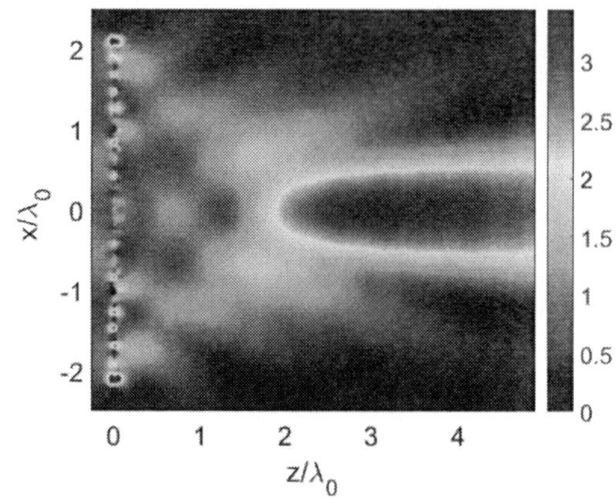

Figure 2. $|E_x|$-field component in the yz-plane in the middle of the metamirror. Focal spot is clearly visible at the distance 3λ from the metamirror which is positioned in the xy plane at $z = 0$.

Metasurface achieves focusing functionality by incorporating parabolic profile of the reflection phase on its surface providing constructive interference of the reflected field at a focal point in front of the metasurface:

$$\varphi(r_0) = \varphi_0 + \frac{2\pi}{\lambda}\sqrt{r_0^2 + F^2}, \tag{1}$$

where φ is a desired phase shift, r_0 is the distance from the lens center to the point at the surface of the lens, F is focal distance, λ is the operational wavelength and φ_0 is an arbitrary constant phase.

In order to get required phase shift the dimensions or permittivity of the particles should be varied. For given particle geometry the most convenient is to change the outer R and the inner r radii while keeping height constant. Required phases can be calculated by numerical simulation of the single particle with imposed unit cell boundary conditions. The final geometry of the each particle is defined by two conditions: minimum difference between required and obtained phase and maximum value of the reflection coefficient. The operation frequency of the was chosen to be 4 GHz. Corresponding calculated dimensions are $H = 5.2$ mm and $h = 2.8$ mm with the radii varying in the range $r = 0.5 \div 6.4$ mm and $R = 6 \div 7.4$ mm.

Finite size metasurface consisting of 21x21 particles has been then numerically simulated with the time integral equation solver of CST Microwave Studio. Open boundaries have been imposed from all directions and plane wave has been chosen as the excitation source. The performance of the designed optical metamirror is demonstrated in Fig. 2 where $|E_x|$-field component of the reflected incident field is plotted. The reflected field has a maximum at a distance 3λ from the metasurface.

3. Reflective metamirror

Next we have designed a metamirror to reflect normally incident plane wave to an chosen angle $\vartheta = 45°$ from the normal. From the generalized Snell's law [10] we can define the angle ϑ_r of the reflected wave for the case of normally incident wave and the periodic structure with the period a:

$$\sin\theta_r = \frac{\lambda_0}{aN} \tag{2}$$

where N is the number of cells within the range of wavelength, and λ_0 is the wavelength in vacuum. According to Eq. (2), the reflection angle ϑ_r can be tuned only by changing the number of elements N and the unit period a. To design the metasurface similar bianisotropic elements as in previous section have been used arranged with the period $a = 15$ mm. Correspondingly 7 elements with different dimensions were chosen to provide required set of phases changing from $0°$ to $360°$ (Fig. 3). An array consisting of 9 supercells has been numerically simulated under TE-polarised plane wave excitation. To reduce calculation time periodic boundaries have been imposed along y-direction. The results in a form of the real value of H_z field component is shown in Fig. (4) for the frequency $f = 4$ GHz. The white arrow represents the reflected wave which is equal to $\vartheta_r = 45^circ$ corresponding to the designed value.

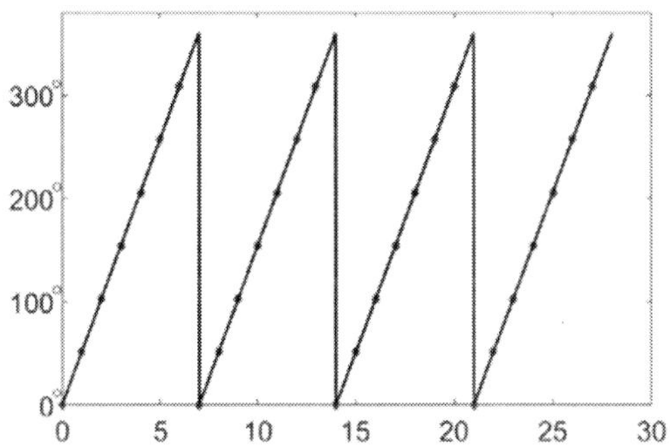

Figure 3. Distribution of the reflection phase over metasurface (for 3 supercells).

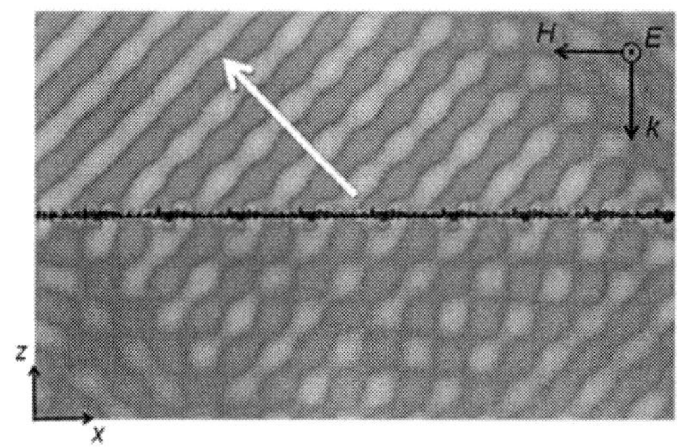

Figure 4. $z-$component of the magnetic field of the reflection metasurface (9 supercells) in the xz plane at $z = 0$.

4. Conclusion

It was shown by numerical simulation that simple distilled water can be used as a material to design all-dielectric metasurfaces. Two common features has been demonstrated including focusing and anomalous reflection of the reflected waves. There are no fundamental limitations in using water as a dielectric for different types of metasurfaces. This material opens an opportunity to design out-of-band transparent tunable cheap large-scale microwave metasurfaces.

Acknowledgments

The work was supported by the Russian Foundation for Basic Research (Project No. 18-37-00486).

References

[1] Glybovski S et al. 2016 *Phys. Rep.* 634 1-72
[2] Kruk S and Kivshar Y 2017 *ACS Photonics* 4(11) 2638-2649
[3] Yang Y, Miroshnichenko A E, Kostinsk S V, Odit M A, Kapitanova P et al. 2017 *Phys. Rev. B* 95 165426
[4] Kamali S M, Arbabi E, Arbabi A and Faraon A 2018 *Nanophotonics* 7(6) 10411068
[5] Andryieuski A, Kuznetsova S M, Zhukovsky S V, Kivshar Yu S and Lavrinenko A V*Sci Rep* 5 13535
[6] Song Q, Zhang W, Wu P, Zhu W, Shen Z, et. al. 2017 *Adv. Opt. Mat.* 5(8) 1601103
[7] Xie J et al. 2018 *Opt. Expr.* 26 5052-5059
[8] Odit M, Kapitanova P, Andryieuski A, Belov P and Lavrinenko A V 2016 *Appl. Phys. Lett.* 109 011901

[9] Stenishchev I V and Basharin A A, 2017 *Sci. Rep.* 7 9468
[10] Yu N et al. 2011 *Science* 334 333337

Optical coupling between resonant dielectric nanoparticles and dielectric waveguides probed by third harmonic generation microscopy

K I Okhlopkov[1], A A Ezhov[1], P A Shafirin[1], A S Shorokhov[1], N A Orlikovsky[2], M R Shcherbakov[1,3] and A A Fedyanin[1]

[1] Quantum Electronics Department and Quantum Technology Center, Faculty of Physics, Lomonosov Moscow State University, Moscow 119991, Russia

[2] Bauman Moscow State Technical University, Moscow 105005, Russia

E-mail: okhlopkov@nanolab.phys.msu.ru

Abstract. We report on the observation of optical coupling between a Mie-resonant subwavelength silicon nanodisk and a silicon waveguide, revealed through third harmonic generation microscopy. Changing the gap between the nanodisk and the waveguide, we observe third harmonic intensity modulation of up to 4.5. We attribute this strong modulation to modifications of the magnetic dipole resonance and local electric fields inside the nanodisk, as supported by full-wave simulations.

1. Introduction

Nanoparticles have important significance for nanophotonics as they possess the strong resonances in the visible and near-infrared spectral ranges. Recently, subwavelength all-dielectric Mie-resonant nanoparticles have been proposed as a promising analogue of plasmonic nanoparticles due to high refractive indices, low Ohmic losses and CMOS-compatible materials [1]. These features make them perfect candidates for integrated photonic devices, such as waveguides and nanoantennas. Moreover, the excitation of strong magnetic dipole (MD) resonance leads to significant enhancement of local electric field inside all-dielectric nanoparticles that enhances nonlinear-optical effects like third-harmonic generation (THG) [4].

Among the studies on single-particle scattering [2, 3], there are a plenty of works devoted to optical coupling between dielectric nanoparticles formations in dimers [5] and high-order oligomers [6]. However, the investigations of optical coupling between all-dielectric Mie-resonant nanoparticles and elements of integrated nanophotonics is lacking. In this contribution, we observe optical coupling between single subwavelength silicon nanoparticles and a silicon waveguide by studying THG as a function of distance between them.

2. Result and discussion

2.1. Sample design

We fabricated samples of "silicon nanodisk—waveguide" pairs out of a silicon-on-insulator wafer with the top silicon and buried SiO_2 layers thicknesses of 280 nm both. The width of waveguide

[3] Present address: School of Applied and Engineering Physics, Cornell University, Ithaca, 14853 NY, USA.

was chosen to be 435 nm. The pairs have two varied parameters: the nanodisk diameter and the gap between the nanodisks and waveguides. The diameters of nanodisks were chosen to be 380, 430 and 480 nm, and the distances between nanodisks and waveguides were 105, 185 and 320 nm.

2.2. Experimental method

To probe optical coupling between nanodisks and waveguide, THG microscopy was realized using a confocal laser scanning microscope (CLSM) complemented by a femtosecond Er^{3+} fiber laser with a pulse duration of 120 fs, a repetition rate of 70 MHz, a central wavelength of laser spectrum of 1545 nm and a width of 40 nm. The "silicon nanodisk—waveguide" pairs were illuminated by the laser pulses tightly focused to a waist of 3 μm in diameter, with the electric component of the incoming field orthogonal to the waveguides. The scattered third harmonic signal was collected by a CLSM objective and was detected by a CLSM photomultiplier.

The most significant THG enhancement, with respect to the THG from the silicon substrate, was observed for nanodisks with diameters of 480 nm, for which its MD resonance lied at the central wavelength of the pump radiation, 1545 nm. The maximum enhancement of THG signal is up to 26. Moreover, THG signal is highly sensitive to the distance between the nanodisk and the waveguide, leading to a significant modulation of THG signal by up to 4.5 in the case of the gap size changing from 185 to 105 nm (blue circles in Fig.1(a)).

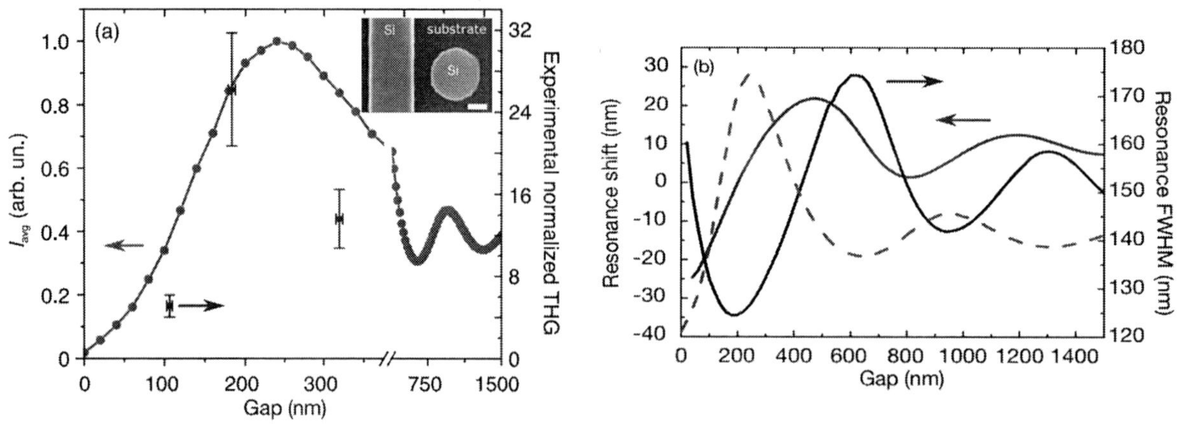

Figure 1. (a) Experimental THG enhancement from resonant (D = 480 nm) nanodisks as a function of the gap between the nanodisk and the waveguide (blue squares) and normalized local electric fields, given by Eq.(1) (connected red squares). The inset shows a scanning electron micrograph of the sample with parameters D = 515 nm and L = 165 nm. The scale bar is 200 nm. (b) MD resonance parameters as a function of gap L, with its central position plotted in blue, and full width at half maximum plotted in black. The dashed line is the same as the red curve from panel (a).

2.3. Numerical calculations

In order to study how the waveguide affects the local fields within the nanodisk, electric field strength distribution $\mathbf{E}(\mathbf{r})$ was calculated and following quantity was plotted in Fig.1(a):

$$I_{\text{avg}} = \int\limits_{\text{disk}} |\mathbf{E}(\mathbf{r})|^6 dV. \quad (1)$$

Together with our experimental observations, this non-monotonic distance dependence means that the non-resonant waveguide can enhance field localization in nanodisk and constructively amend its THG signal. To explain this dependence, we calculated full-width at half-maximum (FWHM) and the central wavelength (plotted in Fig.1(b)) of the MD resonance of nanodisk. The maximum of local electric fields (dashed red line) is achieved at the minimum of FWHM of 125 nm at the gap $\approx$ 200 nm. At the same time, the minimum of local electric fields is achieved at the maximum of FWHM of 175 nm at the gap $\approx$ 600 nm. It may be concluded that the main contribution to the alternating local fields of the nanodisk is provided by the dependence of the MD resonance radiative decay constant. Together with the dependence of FWHM and the central wavelength of MD resonance on the gap, the measured third harmonic represents a manifestation of optical coupling between the nanodisk and the waveguide.

Acknowledgments

This work was partly supported by Russian Ministry of Education and Science (grant 14.W03.31.0008, calculation of local fields). THG experiments were performed with the support of the grant of the President of the Russian Federation for leading scientific schools (grant No. NSh-6862.2018.2) and Russian Foundation for Basic Research (grant No. 18-32-01073). Nonlinear FDTD calculations were supported by Russian Science Foundation (grant No. 18-12-00475). Scanning electron microscopy was supported by Russian National Technology Initiative.

[1] Kuznetsov A I, Miroshnichenko A E, Brongersma M I, Kivshar Y S and Lukyanchuk B S 2016 *Science* **354** 2472
[2] Evlyukhin A B, Novikov S M, Zywietz U, Eriksen R L, Reinhardt C, Bozhevolnyi S I and Chichkov B N 2012 *Nano Lett.* **12** 3749-55
[3] Kuznetsov A I, Miroshnichenko A E, Fu Y H, Zhang J and Lukyanchuk B S 2012 *Sci. Rep.* **2** 492
[4] Shcherbakov M R, et al 2014 *Nano Lett.* **14** 6488-92
[5] Bakker R M, Permyakov D, Yu Y F, Markovich D, Paniagua-Domínguez R, Gonzaga L, Samusev A, Kivshar Y S, Lukyanchuk B S and Kuznetsov A I 2015 *Nano Lett.* **15** 2137-42
[6] Miroshnichenko A E and Kivshar Y S 2012 *Nano Lett* **12** 6459-63

IOP Conf. Series: Journal of Physics: Conf. Series **1092** (2018) 012105 doi:10.1088/1742-6596/1092/1/012105

Generation of surface plasmon-polariton pairs by a nonlinear nanoparticle

N A Olekhno[1], **M I Petrov**[1,2] **and I V Iorsh**[1]

[1]Department of Nanophotonics and Metamaterials, ITMO University, 199034 St. Petersburg, Russia

[2]Department of Physics of Condensed Matter, St Petersburg Academic University, 194021 St. Petersburg, Russia

E-mail: `nikita.olekhno@metalab.ifmo.ru`

Abstract. In the present paper, we study theoretically the generation of surface plasmon-polariton pairs in the process of spontaneous parametric downconversion of light by a nonlinear nanoparticle located near the metal-dielectric interface. A nanoparticle made of GaAs is considered as a particular example. The particle is assumed to be much smaller than the pump radiation wavelength and considered as a point source. We show that spatial correlations between the generated plasmons can be understood as a result of an interference of electric fields produced by dipole sources that are induced in the nanoparticle by the pump photon. A directional two-plasmon generation in certain geometries is observed.

1. Introduction

Spontaneous parametric downconversion (SPDC) is the main method for the generation of entangled photon pairs [1]. In this process, a medium with a non-zero second-order nonlinear susceptibility $\chi^{(2)}$ absorbs a pump photon at the frequency ω_{pump} and emits two photons, the idler and the signal one, with frequencies ω_i and ω_s such that $\omega_{pump} = \omega_i + \omega_s$ [1]. Entangled states of light are a subject of primary interest in the area of quantum optics. Their wavefunction can not be represented as a product of one-particle wavefunctions, thus there are correlations between the photons of the pair. Entangled pairs play a fundamental role in quantum mechanics, and are suitable for various applications such as quantum cryptography, sensing and metrology [1].

The generation of quantum states with nanosystems attracts considerable attention in the areas of nanophotonics and quantum optics. For example, the generation of entangled photon pairs in semiconductor waveguides of subwavelength cross sections [2], as well as by quantum dots [3] or plasmonic nanoantennas [4, 5] has been considered. The generation of plasmon-polariton pairs in nanostructures attracts considerable attention both experimentally and theoretically [6] from the time the entanglement of plasmons was first observed [7]. However, plasmonic systems do not possess an intrinsic nonlinearity due to the centrosymmetric lattice structure of noble metals. Thus, they act as electric field amplifiers, or perform the nonlinear generation only due to a non-symmetric geometry of resonant modes. To this end, a study of the generation of photon and plasmon-polariton pairs by dielectric nanoparticles and nanoantennas with $\chi^{(2)}$ nonlinearity is of considerable interest.

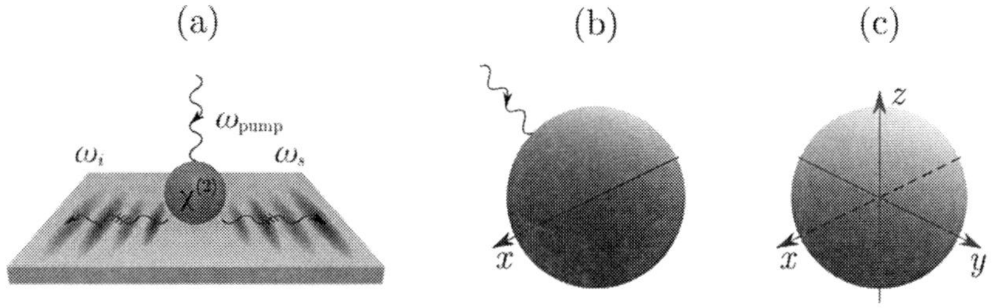

Figure 1. (a): A sketch of the considered geometry. The nanoparticle is shown with grey, the metallic half-space is shown with yellow. The electric field of the pump wave is parallel to the x-axis, and the pump wave vector is perpendicular to the interface. (b): Initial polarization of a nanoparticle in the x-direction. (c): Induced dipoles in the z- and y-directions.

In the present work, we consider the generation of plasmon pairs with a nanoparticle made of a nonlinear material (figure 1a). A nanoparticle is assumed to be small enough so that the following condition is satisfied:

$$d \ll \lambda/n, \tag{1}$$

where d is the typical size of a nanoparticle, λ is the pump radiation wavelength and n is the refractive index of the nanoparticle material. Thus, a nanoparticle can be treaten theoretically as an electric dipole. Hereafter we consider the second-order nonlinear susceptibility tensor for GaAs: $\chi^{(2)}_{xyz} = \chi^{(2)}_{zxy} = \chi^{(2)}_{yzx} = \chi^{(2)}_{yxz} = \chi^{(2)}_{zyx} = \chi^{(2)}_{xzy} = 4d_{36}$, with other matrix elements being zero [8].

2. Theoretical description of the nonlinear generation

We apply the general theoretical framework of paper [6]. Within this approach, spatial correlations between photons or plasmons of the generated pair are described via the two-particle amplitude

$$T_{is}(\mathbf{r}_i, \mathbf{r}_s) = \int \mathbf{d}_{i,\alpha} G_{\alpha\beta}(\mathbf{r}_i, \mathbf{r}_0, \omega_i) \Gamma_{\beta\gamma}(\mathbf{r}_0) G_{\gamma\delta}(\mathbf{r}_0, \mathbf{r}_s, \omega_s) \mathbf{d}_{s,\delta} d\mathbf{r}_0, \tag{2}$$

where the subscript i denotes the first (idler) photon with the frequency ω_i, and the subscript s denotes the second (signal) one with the frequency ω_s. The dipole moments $\mathbf{d}_i$ and $\mathbf{d}_s$ of idler and signal detectors define the polarizations of the detected photons. In the equation above, $\hat{G}(\mathbf{r}, \mathbf{r}_0, \omega)$ is the matrix of dyadic Green's function of the system, and $\hat{\Gamma}(\mathbf{r})$ is the generation matrix: $\Gamma_{ij}(\mathbf{r}) = \chi^{(2)}_{ijk}(\mathbf{r}) E^{pump}_k(\mathbf{r}) \mathrm{e}^{-i\omega_{pump}t}$, where $\chi^{(2)}_{ijk}$ is the second-order nonlinear susceptibility tensor, and E^{pump} is the pump field which causes nonlinear generation. When a pump photon interacts with a nanoparticle, the latter polarizes homogeneously (figure 1b). This polarization field acts as the actual electric field responsible for the nonlinear generation in the nanoparticle. The integration in Eq.(2) is performed over the volume of the nonlinear system which generates photon pairs. The probability of simultaneous detection of idler and signal photons of given polarizations at the given positions $\mathbf{r}_i$ and $\mathbf{r}_s$ is related to the two-photon amplitude as $W(\mathbf{r}_i, \mathbf{r}_s) = (2\pi/\hbar)\delta(\hbar\omega i + \hbar\omega_s - \hbar\omega_{pump})|T_{is}(\mathbf{r}_i, \mathbf{r}_s)|^2$ [6].

Since the condition (1) is satisfied, the nanoparticle can be described as a point electric dipole. The Green's function of the point dipole located near the metal-dielectric interface consists of the free space Green's function G_0 and the reflected part G_{ref}: $G(\mathbf{r}, \mathbf{r}_0, \omega) = G_0(\mathbf{r}, \mathbf{r}_0, \omega) + G_{ref}(\mathbf{r}, \mathbf{r}_0, \omega)$. The free space part has the following form [9]:

$$G_{0\alpha\beta}(\mathbf{r}, \mathbf{r}_0, \omega) = \frac{\exp(ik_0 R)}{4\pi R}\left[\left(\delta_{\alpha\beta} - \frac{R_\alpha R_\beta}{R^2}\right) - \frac{1 - ik_0 R}{k_0^2 R^2}\left(\delta_{\alpha\beta} - 3\frac{R_\alpha R_\beta}{R^2}\right)\right], \tag{3}$$

where $\mathbf{R} = \mathbf{r} - \mathbf{r}_0$, $R = |\mathbf{R}|$, and $k_0 = \omega/c$. The reflected part allows only an integral representation [9, 10]:

$$G_{ref}(\rho, \varphi, z) = \frac{ik_0}{8\pi} \int_0^\infty \hat{M}(s, \rho, \varphi) \exp(is_{1z}z) ds. \tag{4}$$

In the relation above, we use the polar representation with $\rho = \sqrt{(x - x_0)^2 + (y - y_0)^2}$, $\varphi = \mathrm{arctg}((y - y_0)/(x - x_0))$ and $z = z_{i,s} + z_0$. The normalized wavenumber is $s = \sqrt{k_x^2 + k_y^2}/k_0$, and $s_{1z} = k_1 z/k_0$. The index 1 denotes the upper half-space filled with a dielectric. Elements of the matrix $\hat{M}$ that are relevant for the following example have the form $M_{zz} = 2\pi J_0(s\rho) r_p(\omega) s^3/s_{1z}$ and $M_{yz} = -M_{zy} = -2\pi i s^2 r_p(\omega) J_1(s\rho) \sin(\varphi)$, with $J_0(s\rho)$ and $J_1(s\rho)$ being the zero order and the first order Bessel functions of the first kind, respectively, and r_p being the Fresnel reflection coefficient for p-polarized waves. For numerical calculations we assume that the permittivity of the dielectric is $\varepsilon_d = 1$, and the permittivity of the metal is $\varepsilon(\omega) = 1 - \omega_p^2/(\omega^2 + i\gamma\omega)$, where the plasma frequency of the metal $\omega_p = 7 \cdot 10^{15}\ s^{-1}$, and $\gamma = 0.01\omega_p$.

The generation matrix can be represented via the dyadic decomposition: $\Gamma = \sum_{\alpha,\beta} C_{\alpha\beta} |e_\alpha\rangle \langle e_\beta|$, where α and β take values x, y and z, and coefficients $C_{\alpha\beta}$ are defined by $\chi^{(2)}$ and the pump field. Thus, the amplitude T can be represented as a sum of products of two components of Green's functions having form $\langle e_i| G |e_\alpha\rangle \langle e_\beta| G |e_s\rangle$, which corresponds to the interference of electric fields from induced dipole sources.

3. Generation of surface plasmon-polariton pairs

Figure 2 shows the two-photon detection probability $W(\mathbf{r}_i, \mathbf{r}_s)$ for the generation of plasmon-polaritons by a GaAs nanoparticle located at the position $x_0 = y_0 = 0$, $z_0 = 10$ nm. In the case when the pump field is polarized along the x-axis, the two induced sources are dipoles parallel to the y- and z-axis, correspodingly (figure 1b,c). Indeed, $\Gamma \propto \chi^{(2)} |e_x\rangle \propto |e_y\rangle \langle e_z| + |e_z\rangle \langle e_y|$. The z-dipole emits plasmons in all directions in the (xy) plane with the same efficiency, whereas the y-dipole does not emit in the direction of the x-axis. Combinations of the corresponding Green's functions define the spatial structure of W. It is seen that if one detector is located at the angle $\varphi = \pi/2$ or $\varphi = -\pi/2$ with respect to the pump field, the probability of detection of the second plasmon is maximal in the same direction and vanishes in the opposite direction (figure 2a,b). In the limiting case when one of the detectors is located along the x-axis, the second one can detect a photon in the upper or in the lower half-planes with the same probability (figure 2c). Thus, by changing the orientation of detectors with respect to the nanoparticle one can tailor the correlation between the generated plasmons up to the case when both plasmons are emitted in one direction.

In macroscopic crystals, the efficiency of the nonlinear generation is maximal when the phase matching condition is satisfied. The latter reads as [8]

$$\mathbf{k}_{pump} = \mathbf{k}_i + \mathbf{k}_s, \tag{5}$$

where $\mathbf{k}_{pump}$, $\mathbf{k}_i$ and $\mathbf{k}_s$ are wave vectors of the pump, idler and signal photons, respectively. As seen from figure 2, this condition is not satisfied. Thus, in our case of SPDC by a nanoparticle, the phase matching condition is replaced by some kind of geometrical selection rules for eigenmodes of the considered system.

Both of the generated surface plasmon-polaritons share the same polarizations due to the structure of the Green's function G_{ref}. In particilar, at an arbitrary point near the interface, the electric field of plasmon-polaritons has non-vanishing components only in the vertical $\mathbf{e}_z$ and radial $\mathbf{e}_\rho$ directions. Hence, the considered generation of plasmon pairs is analogous to the SPDC Type 1 process [1].

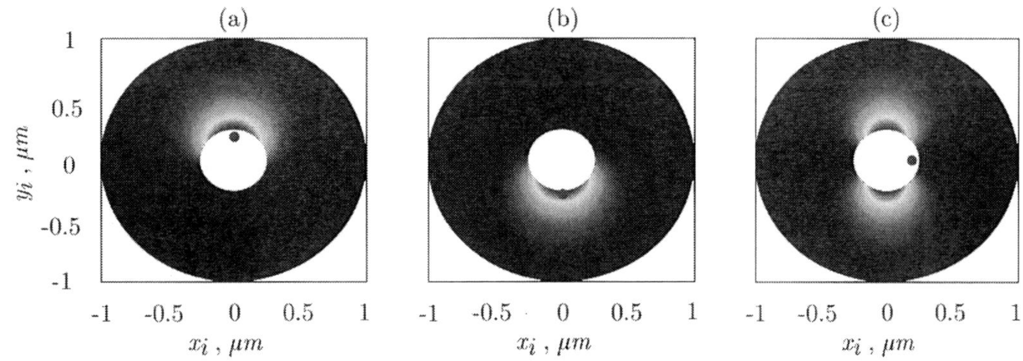

Figure 2. (a): The two-photon detection probability $W(\mathbf{r}_i, \mathbf{r}_s)$ for $\omega_i = \omega_s = 4.5 \cdot 10^{15}$ s^{-1}, $\rho_s = 300$ nm and $\varphi_s = \pi/2$. Vertical positions of the nanoparticle centre z_0 and detectors $z_{i,s}$ are $z_0 = z_i = z_s = 10$ nm. The measured polarization of the electric field is directed along the z-axis. (b): The same as (a), but with $\rho_s = 400$ nm and $\varphi_s = -\pi/2$. (c): The same as (a), but with $\varphi = 0$.

4. Conclusion

In the present work, we describe the generation of surface plasmon-polariton pairs by a GaAs nanoparticle in the process of spontaneous parametric downconversion. Spatial correlations between generated plasmons can be understood in terms of an interference between the electric fields generated by induced electric dipole sources. A number and orientations of these induced dipoles are defined by the second order nonlinear dielectric susceptibility tensor of the nanoparticle material, as well as by the direction of the pump electric field. In the simple case, when two of the principal axes of the nanoparticle are parallel to the metal-dielectric interface, one can observe an emission of both plasmons in the same direction for a linearly polarized pump. The efficiency of the generation increases significantly, if the frequencies of the generated plasmons are close to the the surface plasmon resonance frequency $\omega_p/\sqrt{2}$.

Acknowledgments

We are grateful to A.N. Poddubny and A.A. Sukhorukov for fruitful discussions. The work is supported by the Russian Foundation for Basic Research (proj. №18-32-01052 and 18-02-01206). NAO acknowledges the support of the Foundation for the Advancement of Theoretical Physics and Mathematics "BASIS" (№18-1-5-124-1) and the FASIE.

References

[1] Shih Y 2003 *Rep. Prog. Phys.* **66** 1009–44
[2] Solntsev A S, Sukhorukov A A 2017 *Reviews in Physics* **2** 19–31
[3] Orieux A, Versteegh M A M, Jons K D and Ducci S 2017 *Rep. Prog. Phys.* **80** 076001
[4] Poddubny A N, Ginzburg P, Belov P A, Zayats A V and Kivshar Y S 2012 *Phys. Rev.* A **86** 033826
[5] Straubel J, Sarniak R, Rockstuhl C and Slowik K 2017 *Phys. Rev.* B **95** 085421
[6] Poddubny A N, Iorsh I V and Sukhorukov A A 2016 *Phys. Rev. Lett.* **117** 123901
[7] Altewischer E, van Exter M P and Woerdman J P 2002 *Nature* **418** 304–6
[8] Boyd R W 2008 *Nonlinear Optics (3rd ed.)* (San Diego: Academic Press)
[9] Novotny L and Hecht B 2006 *Principles of Nano-Optics* (Cambridge University Press)
[10] Sinev I S et al 2017 *Laser Photonics Rev.* **11** 1700168

IOP Conf. Series: Journal of Physics: Conf. Series **1092** (2018) 012106 doi:10.1088/1742-6596/1092/1/012106

Perforated all-dielectric anapole metamaterials

Anar K Ospanova (1, 2, 3), Ivan V Stenishchev (1, 2, 3) and Alexey A Basharin (1, 2, 3)
(1) National University of Science and Technology (MISiS), Department of Theoretical Physics and Quantum Technologies, 119049, Moscow, Russia
(2) National University of Science and Technology (MISiS), The Laboratory of Superconducting metamaterials, 119049, Moscow, Russia
(3) Politecnico di Torino, Department of Electronic and Telecommunications, 10129, Torino, Italy

Abstract. In this work, we propose a novel type of perforated silicon metamaterials, exhibiting anapole mode in optics. The interplay of electric and toroidal dipole moments leads to transparency of the system at resonant frequency due to destructive interference between them. This type of metamaterials is demanding for practical nanophotonics application due to one-step simple fabrication and fascinating properties inherent to anapole mode excitation.

Metamaterials are composite media with unnatural optical properties. Among others, dielectric metamaterials are widely used due to their low dissipative losses and scalability in optical frequency range. In this rate, silicon metamaterials are mechanically, electrically, optically and thermal robust and comprehensive for experimental investigation [1, 2]. Novel toroidal and anapole metamaterials highly evaluated due to their field localization and transparency properties [3-7]. Recent materials shows transparency effects in optical frequency range [8]. However, the excitation of toroidal and anapole response requires sophisticated 3D toroidal topology, particularly in optical frequency range.

1 The structure of the metamaterial

The unit cell of metamaterial (called metamolecule) represents the cluster of four neighboring holes in square silicon slab of 100 nm thickness and 200 nm length. The radius of holes is 45 nm, period of clusters is 200 nm and center-to-center distance between holes is 55 nm (Fig. 1). Linearly polarized electromagnetic wave illuminates metamaterial from the top and, according to Mie theory, induces displacement currents along **E** component. These displacement currents are responsible to electromagnetic response of the structure, called Mie resonance. According to geometry of the metamolecule, displacement currents **j** take the shape of loops and generate magnetic vortex **m** inside them. Such electromagnetic configuration generates toroidal dipole **T** moment co-directional with incident wave [3, 4].

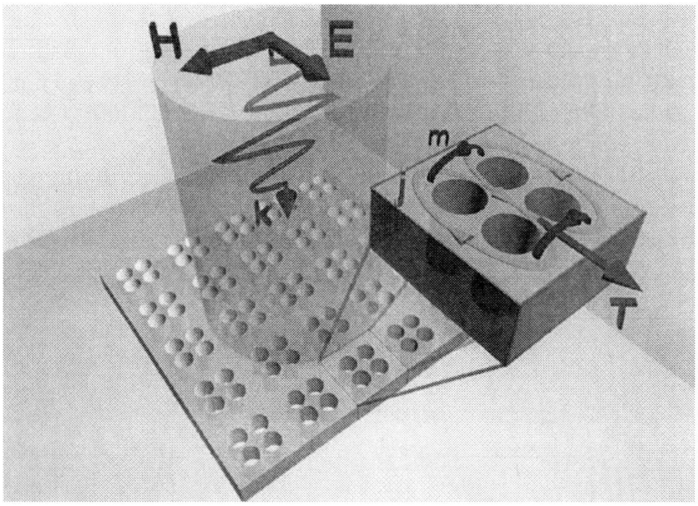

Fig. 1. Illustration of the proposed optical metamaterial supplemented with current and mode distribution within each metamolecule. Inset shows toroidal mode excitation; here **m** stands for magnetic dipole moment, **j** – displacement current loops, **T** – toroidal dipole moment.

2 Results

We simulate metamaterial by commercial version of CST Microwave Studio.The electromagnetic response of metamaterial exhibits sharp resonant response at f=566.5 THz that corresponds to full transmission |T|=1 (Fig. 2 (a)). Field maps provides an absolute value and cutplane in the direction of wave propagation of both electric and magnetic component at resonant frequency f=566.5 THz (Fig. 2 (b)). Electric field strongly concentrates between pairs of holes so that underpins formation of displacement current loops **j**. These currents create magnetic fields **m** circulated above and below holes (Figure 2(b)). Moreover, magnetic field strongly concentrates on the top and bottom facets of metamolecule and rotates clockwise. This featured colocation of electric and magnetic fields within metamolecule induce toroidal response **T** within metamolecule oscillating up and down on the direction of electric component **E**.

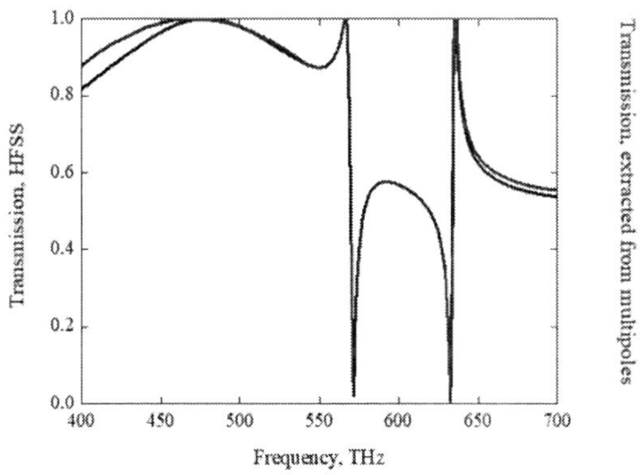

Fig 2 (a). Simulated and reconstructed transmission spectrum for metamolecule. Sharp transparency peak corresponds to f=566.5 THz.

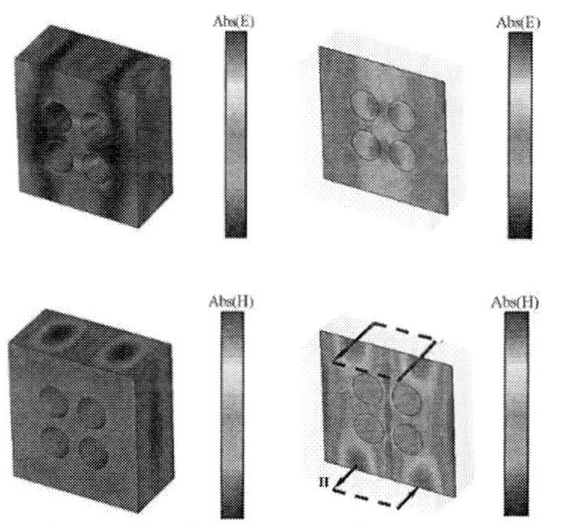

Fig. 2 (b) Field maps of absolute value of electric field with cross-section and absolute value of magnetic field intensities with cross-section.

Such featured electric and magnetic fields distribution along with low radiation losses are inherent to anapole mode excitation. To confirm this assumption, we provide multipole decomposition of displacement currents up to second order multipoles: electric **P** and magnetic dipoles **M**, toroidal dipole **T**, electric **Q**e, magnetic **Q**m and toroidal **Q**t quadrupoles illustrated in Fig. 3. Thus, close to resonant frequency there is compelling electric dipole **P** and toroidal dipole **T** which at resonant frequency f=566.5 THz come into oscillation of $\mathbf{P} = ik\mathbf{T}$ so that establish anapole mode, resulting in full transparency due to nonradiating configuration of anapole mode [3, 4]. Rather high value of magnetic dipole **M** as well as electric **Q**e, magnetic **Q**m and toroidal **Q**t quadrupoles introduces losses to transmission spectrum, however, reconstructed transmission spectrum perfectly matches simulated transmission results (see fig. 1).

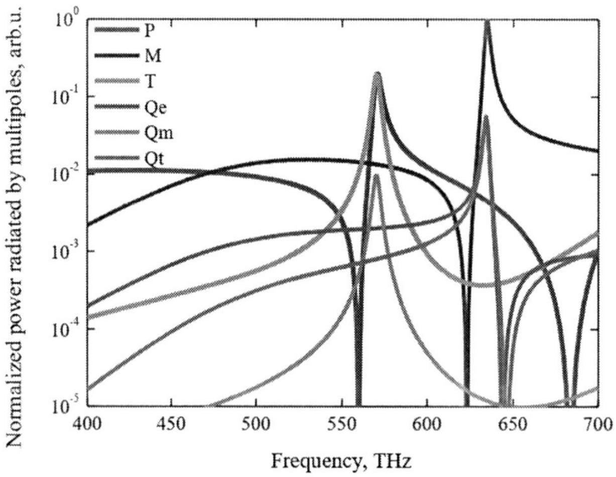

Fig. 3. Normalized power radiated by metamaterial of near-field distribution of metamolecule up to second order multipoles.

3 Conclusion

In this paper, we demonstrated for the first time anapole mode sustaining all-dielectric metamaterial in visible spectral range. Furthermore, this type metamaterial advantages in fabrication, i.e. in time and price among other known nanoscale metamaterials. Our sample can be simply made by perforating holes in solid silicon slab. Our metamaterial paves the way for advanced optical devices on the base of all-dielectric metamaterials demonstration subtle sensing, nonradiative data transfer, Aharonov-Bohm effect and other tempting applications in nanophotonics.

References

[1] Kuznetsov A I, Miroshnichenko A E, Brongersma M L, Kivshar Y S and Luk'yanchuk B 2016 *Science* **354** 6314.

[2] Decker M and Staude I 2016 *Journal of Optics* **18** 10.

[3] Fedotov V A, Rogacheva A V, Savinov V, Tsai D P and Zheludev N I 2013 *Sci Rep-Uk* **3** 2967

[4] Papasimakis N, Fedotov V A, Savinov V, Raybould T A and Zheludev N I 2016 *Nat Mater* **15** 263-271.

[5] Miroshnichenko A E, Evlyukhin A B, Yu Y F, Bakker R M, Chipouline A, Kuznetsov A I, Luk'yanchuk B, Chichkov B N and Kivshar Y S 2015 *Nat Commun* **6** 8069.
[6] Basharin A A, Chuguevsky V, Volsky N, Kafesaki M and Economou E N 2017 *Phys Rev B* **95** 035104 .
[7] Basharin A A, Kafesaki M, Economou E N, Soukoulis C M, Fedotov V A, Savinov V and Zheludev N I 2015 *Phys Rev X* **5** 011036.
[8] Wu P C, Chun Y L, Savinov V, Chung T L, Chen W T, Huang Y W, Wu P R, Chen Y H, Liu A Q, Zheludev N I and Tsai D P 2018 *ACS Nano* **12** (2) 1920-1927.

Optical Interaction Effects in Nanoplasmonic Antennas at Visible Wavelengths

Soroush Abedi and Tavakol Pakizeh*

Faculty of Electrical Engineering, K. N. Toosi University of Technology, Tehran 1631714191, Iran. *t.pakizeh@kntu.ac.ir

Abstract. In this article we study the effects of nanoelements' optical interactions on the properties and response of the nanoplasmonic antennas. Here, we report on some simple, ultra directive, and compact optical antenna designs working based on a hybridized plasmon mode. By enhancing the electromagnetic (EM) coupling between the nanoparticles, we achieve new optimum designs. Using two, three, or four closed nanoelements, the nano-antenna shows high directionality and gives us the opportunity to have a compact, color-switched, and spectrally-tunable optical nanoantenna. Such design are experimentally realized by HCL – a bottom-up nanofabrication based on colloidal self-assembly. Finally, in order to design a highly directive nanoantenna we present a new highly directive and packed Yagi-Uda (P-YU) nano-antenna.

1. Introduction

Nano-antennas are properly designed for manipulating the emission pattern of quantum emitters in addition to localizing freely propagating waves from specific direction. Plasmonic resonances lead us to beat the diffraction limit, therefore using metallic nanoparticles due to the plasmon modes properly suited as nano-antenna.

In order to achieve higher forward (FW) to backward (BW) directivity (G_{FB}) we study interaction effects of nanoelements. Here we theoretically calculate a significant G_{FB} for various designs. We have the capability for estimating the optical response of the nano-antenna by modeling each particle as a dipole by using quasi-electrostatic theory, coupled dipole approximation (CDA). The optical properties of the system can be found by solving the coupled-dipole equation $P_i(\omega) = \alpha_{MWi}(\omega)\left[E_0(r,\omega) + \sum_{j=1 j\neq i}^{n}(A_{ij}(r,\omega)P_j(\omega))\right]$, where E_0 is the incident-field and P_i and α_{MWi} are the self-consistently determined electric dipole moment and the modified polarizability of the particle i, respectively [1]. One can define A_{ij} by: $A_{ij} = \frac{e^{ikr_{ij}}}{r_{ij}}k^2\left(\widehat{r_{ij}}\widehat{r_{ij}} - I_3 + \frac{(ikr_{ij})}{r_{ij}^2}(3\widehat{r_{ij}}\widehat{r_{ij}} - I_3)\right)$; where $k = 2\pi/\lambda$ is the wave number, r_{ij} is the separation between dipoles i and j, $\widehat{r_{ij}}$ is the unit vector along r_{ij}, and I_3 is the 3×3 identity matrix. The term A_{ij} contains three important components which are related to the near-field ($1/r^3$), mid-field ($1/r^2$) and far-field ($1/r$) terms of each dipole's E-field.

2. Effects of hybridized plasmon modes

The hybridized plasmon modes are rooted in the near-field interaction of nano elements which are closed enough. By using these modes and considering the terms effecting on the interaction of the nano-antenna elements we can manipulate the pattern of the nano-antenna.

A simple nano-antenna consisting of a pair of interacting nanoparticles acts as a unidirectional ultra-compact optical antenna. A nanoplasmonic structure containing a pair of Au nanodisks (diameter D=90 nm, thickness T=20 nm) with the surface-to-surface distance, d=10 nm, is such that a strong near-field coupling and plasmon hybridization occurs [2]. Due to this, as it is shown in Fig. 1(a), by exciting the nano-antenna by a Hertzian-dipole placed at point P (see inset in Fig. 1(a)), its FW-BW directionality, as defined by $G_{FB} = 10 \times \log_{10}(S_{FW} / S_{BW})$, where S_{FW} and S_{BW} are electromagnetic powers radiated to the farfield in the FW- and BW-direction, reaches ~18 dB in a narrow spectral range around 665 nm, whereas the single nanodisk antenna radiates almost equally in the FW and BW directions. At the wavelength around 665 nm the excited nanodisk oscillates with a wavelength close to the antiphase mode thus the system only radiates in one direction which is a unique unidirectionality [2].

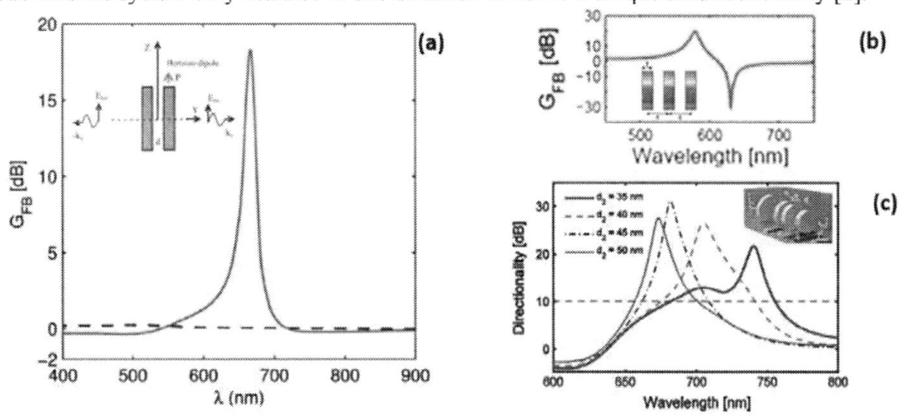

Fig. 1 Hybridized plasmon mode's effect on the response of simple nanoantennas. G_{FB} of (a) a pair of Au nanodisks with surface-to-surface distance (d= 10 nm) which are excited by a local emitter at position P [2], (b) asymmetric three nanodisks with equal distances (d= 40 nm) [3], and (c) four nanodisks for different d_2= 35 (blue solid-curve), 40 (dashed-curve), 45 (dash-dot curve), and 50 (green solid-curve) [4].

Moreover by adding the third nanodisk to the proposed nano-antenna, it is found that the emission of a local light source, feeding the first disks, can be remarkably directed and switched based on its color between two opposite directions (Fig. 1(b)) [3]. This switchable high directionality is based on the strong plasmonic interaction of the three asymmetric nanodisks and resulting hybridized modes of the nano-antenna. Interestingly, it is possible to obtain flexible control over the desired spectroscopic features of the proposed nano-antenna. Adding the forth nanodisk addresses the spectral demands in comparison to the simpler designs with two or three elements with more degree of freedom [4]. Relaying on EM coupling of the disks, the introduced nano-antenna offers tunable spectral performance as a function of the separating distances.

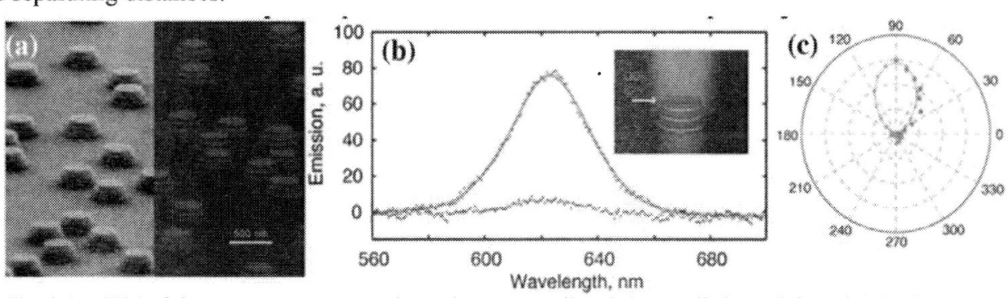

Fig. 2 (a) SEM of the nano-antennas on a glass substrate revealing their overall shape (left) and their elements (right, backscattered electrons image), (b) The directionality function of the nanoantenna is probed by the CdSe QDs with excitation at 532 nm and emission at 625 nm, coupled to the element I: forward – red dots and solid fit, backward – blue dots and dashed fit. The emission spectra are accompanied by (C) emission' polar plots, with dotted orange as experimental emission, and dotted blue – as modeled with FDTD [5].

As depicted in Fig. 2, these nano-antennas are easily produced in a macroscopic array format using affordable bottom-up nanofabrication – hole-mask colloidal lithography (HCL) [5]. In Fig. 2(a) there is an array of the nano-antenna, followed by their sequential increase in size. Each element is a metal nanodisk, supporting a LSPR, which is separated from the neighboring elements by a dielectric spacer. QDs are embedded into the dielectric spacers between the elements to preserve the robustness of the design with the outlook for possible applications in integrated nanophotonics. Directional emission or reception is realized into free-space, which is faced by the smallest element of the device [5].

3. Packed Yagi-Uda nano-antenna

Different designs of optical nano-antennas analogous to their microwave counterparts have been proposed such as Yagi-Uda (YU) [6], log-periodic [7]. One of these most actionable designs is a set of metallic nanoparticles in the case of the YU nano-antenna. Plasmonic nanorods, nanospheres, and nanodisks are used as elements in YU scheme containing a feed and several parasitic elements. The feed is the only active element in YU configuration. By adding some parasitic elements around the feed element one can steer feeds emission in a specific direction.

While in YU nano-antenna the feed element has an omnidirectional emission [6], using a unidirectional feed has a great impact on G_{FB} of the nano-antenna. Therefore we revise the former feed element designs and consider a designed feed with a unidirectional pattern including two resonance elements. A pair of optically close-enough nanorods, with one of them being assumedly excited by a quantum emitter modeled as a single electric-dipole positioned 2 nm above it, efficiently excites the hybridized plasmon modes. Figure 2(a) represents the spectral response of the YU and the new packed YU nano-antenna (P-YU) [8].

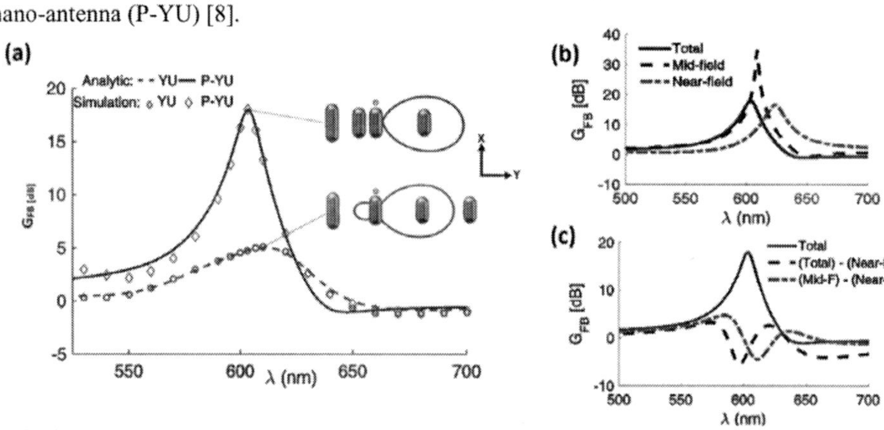

Fig. 3 (a) Theoretical expectation by the CDA method for G_{FB} of YU (red dot) and P-YU (blue diamond) are compared with the simulation results of G_{FB} for YU (red dashed curve) and P-YU (blue solid curve). Insets clearly show radiation patterns of the nano-antennas in ZY-plane [8]. (b) The effects of near-field and mid-field terms of induced dipoles on the G_{FB} of P-YU nano-antenna. (c) The effect of omitting the near-field term of induced dipoles on the G_{FB} of P-YU nano-antenna.

3.1. Near-field and far-field characteristics of P-YU

Using two closed nanorods instead of a single one as a feed element in P-YU nano-antenna increases the optical interaction between the elements [8]. By studying the three components of A_{ij} we can shed lights on the effect of these components on G_{FB} of the P-YU nano-antenna. As depicted in Fig. 1(b), by employing the CDA method and only considering the near-field (ignoring other terms of A_{ij}), the response of the nano-antenna is almost unchanged. Moreover omitting the near-field term of A_{ij} the response of the nano-antenna lessens significantly (Fig. 3(c)). Due to this, the geometrical parameters of the nano-antenna which influence the near-field of it have great impact on the nano-antenna response. Thus, by changing the length and distance between the P-YU elements specially the nanorods in feed, the directivity of the P-YU nano-antenna will be dramatically affected.

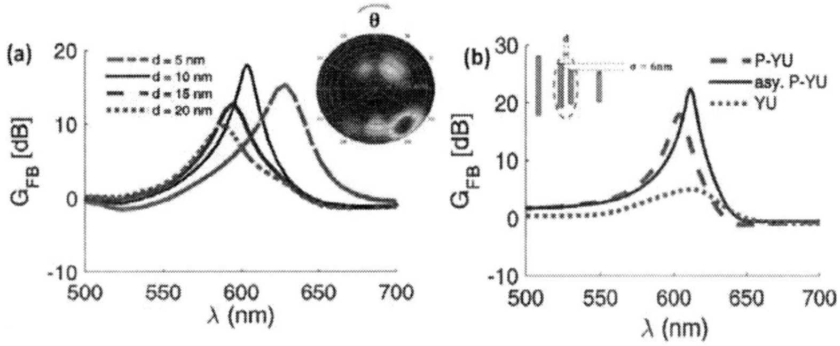

Fig. 4 Effects of varying the parameters of the interacting plasmonic feed's nano elements on G_{FB}. (a) G_{FB} of P-YU nano-antenna for various distance between the nanorods in feed (d= 5, 10, 15, 20 nm), inset shows the substrate (n= 1.5) effect on the direction of the P-YU's pattern (b) Analytical results for G_{FB} of the YU, P-YU and P-YU with asymmetric feed while the length of nanorods in feed differ by an amount 2δ (as shown in the inset).

3.2. *Optical interaction effects in P-YU*

By employing the CDA method, suitably verified by the full-wave simulations, we proved that the near-field interaction of P-YU's nano elements significantly affect the radiation pattern of the nano-antenna. In this section we change the near-field distribution of *E*-field around the feed by using changing the length and the distance of nanorods. Figure 4(a) shows that by changing the distance between the nanorods in feed package from 5 to 20 nm the directivity of the nano-antenna changes. Thus changing the distance between the nanorods in feed (*d*) positively effects on the nano-antenna response. The optimum distance between the nanorods is 10 nm. As it is shown in Fig. 4(b) by using one nanorod longer than the other we can improve G_{FB}. This is due to the fact that the longer element acts as reflector. The proposed feed improves the P-YU's response of the nano-antenna due to its dependency to the near-field distribution of *E*-field.

To summarize, we have shown that relaying on EM coupling by employing two, three or four closed nano elements a high directional, ultra compact, and switchable nanoantenna with significant bandwidth can be designed and fabricated. By using these compact and unidirectional systems instead of a single nanorod as a feed of conventional YU nano-antennas, G_{FB} will dramatically increases due to the stronger optical interaction of nanorods employed in the feed. It is shown that the optical interaction of the nano-antenna is really sensitive to the distance between the nanorods and the length of the feed elements. Thus employing interacting plasmonic nanoparticles in our designs opens new avenues for engineering highly directive and compact nano-antennas.

References

[1] T. Pakizeh, Unidirectional radiation of a magnetic dipole coupled to an ultracompact nanoantenna at visible wavelengths. *JOSA* B 29, 2446-2452 (2012).
[2] T. Pakizeh and M. Käll, Unidirectional ultracompact optical nanoantennas, *Nano Lett.* 9, 2343-2349 (2009).
[3] S. A. Lavasani and T. Pakizeh, Color-switched directional ultracompact optical nanoantennas. *JOSA* B 29, 1361-1366 (2012).
[4] H. Nouri, and T. Pakizeh, Spectrally-tunable directionality of compact optical nanoantennas. *Plasmonics* 8, 1633-1641 (2013).
[5] L. Tong, T. Pakizeh, L. Feuz, and A. Dmitriev, Highly directional bottom-up 3D nanoantenna for visible light. *Scientific Reports* 3, 2311 (2013).
[6] A. G. Curto, et al., Unidirectional emission of a quantum dot coupled to a nanoantenna. *Science* 329, 930–933 (2010).
[7] R S Pavlov, A G Curto, N F van Hulst, *Opt. Commun.* 285, 3334 (2012).
[8] S. Abedi and T. Pakizeh, "Packed Yagi–Uda nano-antennas using a unidirectional feed at visible wavelengths", *Opt. Lett.* ,42, 4788-4791 (2017).

Interaction of electromagnetic waves with VO_2 nanoparticles and films in optical and millimetre wave ranges: Prospective for nano-photonics, nano-antennas, and sensors

V.V. Koledov[1] , V.G. Shavrov[1] , N.V. Shahmirzadi[2] , T. Pakizeh[2] , A.P. Kamantsev[1] , D.S. Kalenov[1] , M.P. Parkhomenko[1] , S.V. von Gratowski[1] , A.V. Irzhak[3,4] , V.M. Serdyuk[5] , J.A.Titovitsky[5] , A.A. Komlev[6] , A.E. Komlev[6] , D.A. Kuzmin[7] , I.V. Bychkov[7], P. Yupapin[8]

[1]Kotelnikov Institute of Radioengineering and Electronics of Russian Academy of Sciences, Mokhovaya 11-7, Moscow 125009, Russia. [2]Faculty of Electrical Engineering, K.N. Toosi University of Technology, Tehran 1631714191, Iran. [3]National University of Science and Technology MISiS, Leninskii pr. 4, Moscow 119049, Russia. [4]Institute of Microelectronic Technology and High Purity Materials of Russian Academy of Sciences, Ac. Osipyana 6, Chernogolovka, Moscow Region 142432, Russia. [5]Institute of Applied Physical Problems, Belarusian State University, Kurchatova 7, Minsk 220045, Belarus. [6]Saint-Petersburg Electrotechnical University "LETI", Prof. Popova 5, St.-Petersburg 197376, Russia. [7]Chelyabinsk State University, Br. Kashirinykh 129, Chelyabinsk 454001, Russia. [8]Faculty of Electrical & Electronics Engineering, Ton Duc Thang University; 19 Nguyen Huu Tho Street, Tan Phong Ward, District 7, Ho Chi Minh City, Vietnam.

victor_koledov@mail.ru, t.pakizeh@kntu.ac.ir

Abstract. Recently great interest in the field of nano-optic is attracted to nano-antennas in both basic and applied research. Nano-antennas are represented by metallic nanoparticles and holes in thin films with arrays of the holes. The usual nano-antennas have fixed functionality, but it will be very interesting to get tunability of nano-antennas controlled by external field. Materials with phase transitions are candidates for the development of tuneable nanoantennas. In this report metal-dielectric phase transition in VO_2 micro- and submicron particles, films and holes are studied experimentally and theoretically. The complex dielectric permittivity of VO_2 is varied drastically with temperature due to structural phase transition, accompanying metal-insulator transition (MIT) at critical temperature (T_c=340 K). In this work, VO_2 film on glass substrate were prepared and investigated in MM wave range (27–37 GHz). Then submicron holes arrays were formed on VO_2 films by FIB milling and their optical Raman spectra were studied. The MM wave response of nanosized VO_2 films reveals strong anomalies near MIT. The holes and arrays show strong change of the Raman spectra at the wavelength 530 nm under heating by laser radiation. The optical and MM wave electromagnetic properties of homogeneous VO_2 nanospheres embedded in the air are studied theoretically. Size effects on the optical properties of the VO_2 nanosphere are investigated. In VO_2 nanosphere, converting into metallic phase by heating leads to formation of a localized surface plasmon resonance (LSPR) which red shifts by increasing dimension.

1. Introduction

In the last decades a real breakthrough has been made in the field of nano-photonics, nano-plasmonics, NAs, and metasurfaces (optical nanomaterials). One of the main problems facing the wide application of these devices is that the properties of such devices are fixed at time, and cannot be tuned. Controllable, or as they are called in these areas "active" (not to be confused with energy-generating) devices would not only be much more in demand, but could create many new scientific and technical opportunities. Recently developed devices like nano-optic, nano-plasmonic, nano-antennas, optical metamaterials/metasurfaces for many applications need the operations either at different frequency bands or different radiation characteristics (such as pattern or polarization) it will be optimal that the devices of nano-photonic, will have multiple characteristic. Thus one of the challenges in this area is to find mechanisms that can provide the control of the properties of functional materials and the devices based on these phenomena. There are many types of reconfigurable antenna that according to its function are divided into frequency reconfigurable antenna, polarization reconfigurable antenna, pattern reconfigurable antenna and hybrid reconfigurable antennas [1, 2]. The different tuning mechanisms have been studied in the literature can be classified into three general categories: (1) Physical tuning (shift, deformation, microfluidic); (2) material tuning (liquid crystal, ferromagnetic/ferroelectric, semiconductors, phase change materials); (3) circuit tuning (open/short, semiconductor devices, MEMS, Non Foster, varactor). One of the most promising mechanisms of tuning is using of phase transition materials, such as VO_2. In this report the phenomena that gives the possibility to create nano-photonic devices, such as nano-antennas in wide frequency range from optics to millimeter range in VO_2 thin films and NAs are studied.

2. VO_2- phase change material

One of the most promising phase change material is VO_2 [3-5]. Single crystals of VO_2 have high transparency and low absorption due to a low concentration of free carriers (~ 10^{18} cm^{-3} at $f < 6.7$ THz). At a temperature of $T_c = 340$ K [3,5] VO_2 undergoes metal-isolator phase transition (MIT) at which a monoclinic crystallographic structure is transformed into a tetragonal crystalline structure [3-5]. Under the illumination with a laser pulse with a dose of 0.15 mJ/cm^2, switching occurs for 8 ps [8]. Under the dose of 2.0 mJ/cm^2 for pumping, the switching time lies in the range of several hundred femtoseconds [4-5]. The transition is accompanied by a sharp change in the optical properties, IR, THz and MM wavelenths [6]. The goal of this report is study of VO_2 films properties in the MM wave range (8 mm) ranges, to make prototypes of nanoantennas (NAs), micro-holes, and slots (see Fig.1), as well as to study their interaction with electromagnetic wave in the visible range using Raman scattering methods, and to study theoretically the optical characteristics of VO_2 nanospheres.

2.1. Raman spectra of NAs based on VO_2

The spectra obtained from the structure of two adjacent holes in the VO2 film and the surface of the film after subtraction of the background and normalization show the follows: the relative intensity of the 620 cm-1 peak obtained in the region of the bridge between the holes is much smaller than that of the analogous peak in the spectrum, obtained on the film. See Figs.1 and 2. There is also a shift to smaller wave numbers at the peak obtained at 614 and 622 cm^{-1}. On the spectrum obtained after irradiation of the VO_2 film usig laser of 20 mW, there are additional peaks 146, 284, 531, 703 and 997 cm^{-1}, compared with the unirradiated one. The origin of these peaks may be due to photoinduced phase instability of the material [7].

2.2. MM wave range properties of the VO_2 films

Fig. 3 shows the reflection coefficient of VO_2 in the range 27-37 GHz at different temperatures in the vicinity of MIT in thin VO_2 film: below MIT at 331 K, near MIT at 340 K and above MIT at 345 K. Also with this method the hysteretic behavior of the reflection coefficient of the VO_2 film is shown for direct and inverse MIT, while the shape of the curve depends strongly on the frequency (Fig. 11). To explain the frequency dependence of the reflection coefficient, a theoretical model has been proposed within the framework of the Drude theory [9].

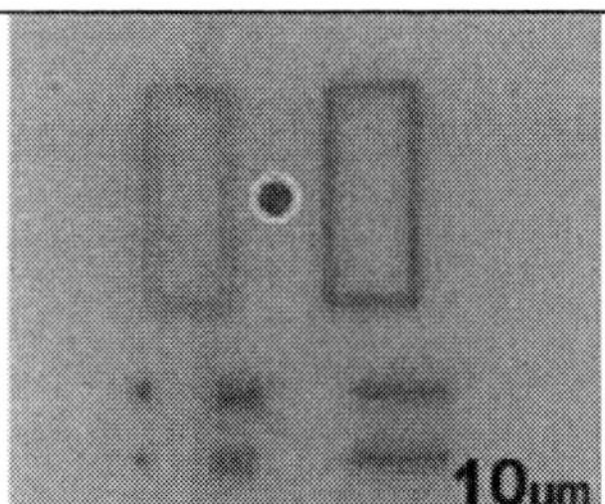

Fig. 1. The holes, hole matrixes, and slots produced by FIB milling in VO_2-370 nm thick film on SiO2 substrate.

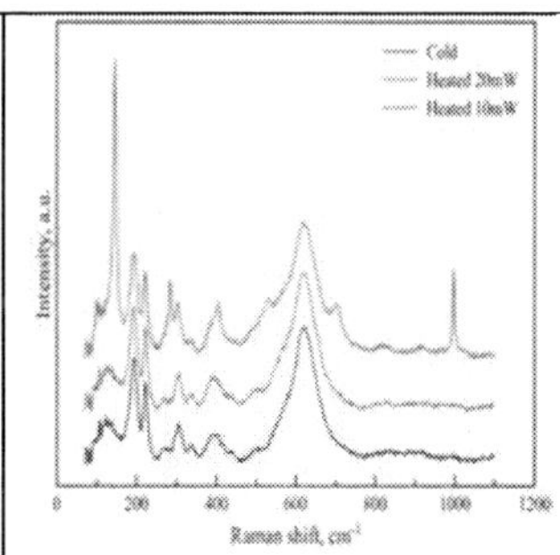

Fig. 2. Raman spectrum of VO_2 film before (cold), and after laser irradiating by 10 mW laser power (heated 10mW) and after laser irradiating with 20 mW laser power (heated 20mW).

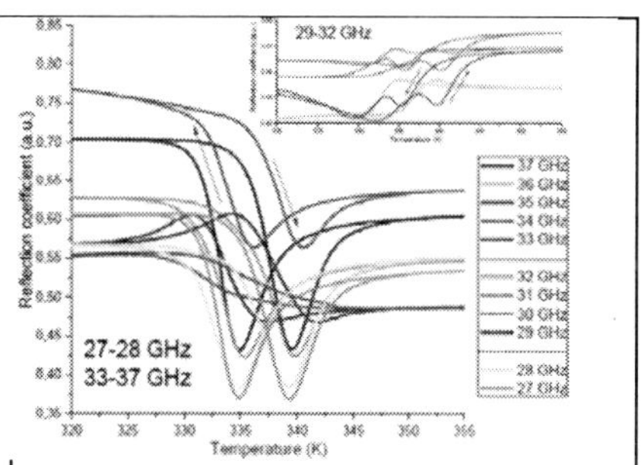

Fig. 3. Temperature dependence of the reflection coefficient, recorded on a panoramic reflectometer at different frequencies from 27 to 37 GHz, showing strong hysteretic anomalies near MIT.

IOP Conf. Series: Journal of Physics: Conf. Series **1092** (2018) 012108 doi:10.1088/1742-6596/1092/1/012108

2.3. Properties of VO2 nanosphere in optical spectrum

The optical properties of VO_2 nanosphere and also the temperature, size and surrounding medium effects on these properties are explored theoretically by exploiting dipole approximation (DA) [10], modified long wavelength approximation (MLWA) [11] and Mie theory [10]. For this purpose, the optical constant of VO_2 are extracted from [12] in two different temperature: i) 300 °K , in which the VO_2 nanosphere acts as an insulator nanosphere, ii)355 °K , where the VO_2 nanosphere manifests metallic properties. These two temperatures are far from the critical temperature (T_C) in which insulator-metal transition occurs in VO_2. In the temperatures near the T_C the insulating and metallic domains coexist, therefore the effective medium theory elucidates the optical constants of VO_2 [13]. In the metallic phase of VO_2, the nanosphere experiences the LSPR. So there is an enhancement in optical properties of VO_2 nanosphere which is disappeared in insulator case. See Fig. 5 and 6. By increasing the radius of nanosphere, the LSPR wavelength slightly redshifts. There is a peak in scattering and absorption spectra of VO_2 nanosphere in insulator state, in visible range corresponding combined modes. According to the obtained results, it is obvious that by increasing the radius of nanosphere, the DA method lose its accuracy.

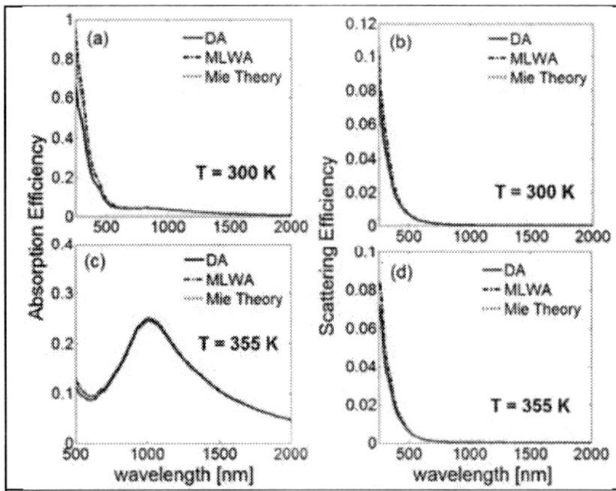

Fig. 4. Optical properties of 20 nm VO_2 nanosphere, (a) and (b) at room temperature, (c) and (d) above T_C

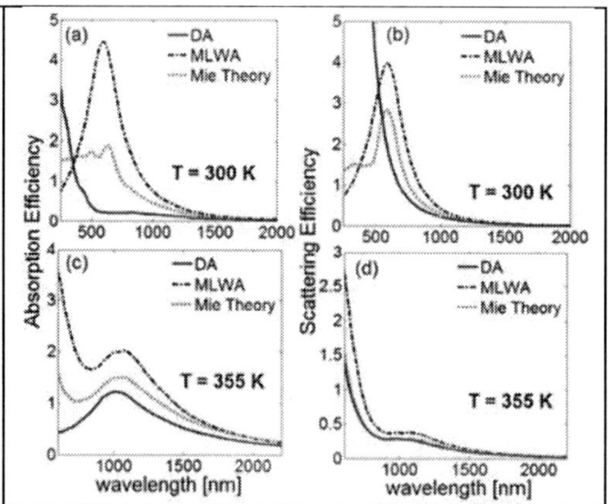

Fig.5. Optical properties of 100 nm VO_2 nanosphere, (a) and (b) at room temperature, (c) and (d) above T_C

The optical constant of VO_2 is shown in Fig.6. According to this data, the real part of VO_2 optical constant is near zero at LSPR wavelengths of VO_2 nanosphere with different radiuses. This is in contrary to LSPR condition (Fröhlich condition) [14]. It seems that the imaginary part of VO_2 has tremendous effect on the LSPR wavelength. This effect is examined in Fig.6. The extinction spectrum of 20 nm nanosphere approaches to Fröhlich condition as the value of imaginary part for VO_2 is lowered by 0.5, 0.25 and 0.1 factors. It can be concluded that the principle of resonance around 1000 nm is the LSP which is influenced severely by the imaginary part of the dielectric constant.

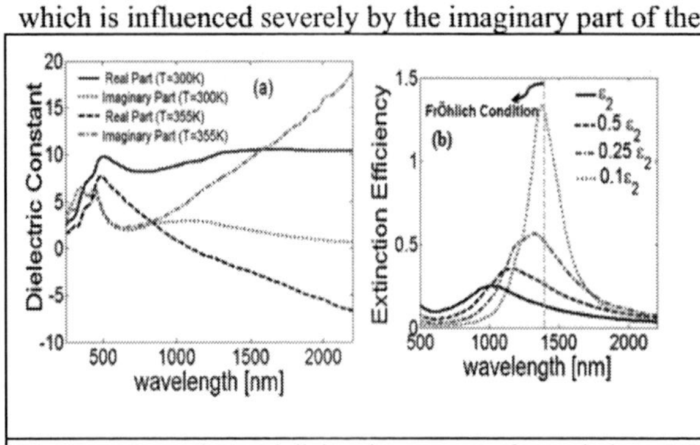

Fig. 6. (a) Experimental dielectric constant of VO_2, below and above the T_C, extracted from [12], (b) Effects of imaginary part of VO2 dielectric constant on spot of LSPR wavelength.

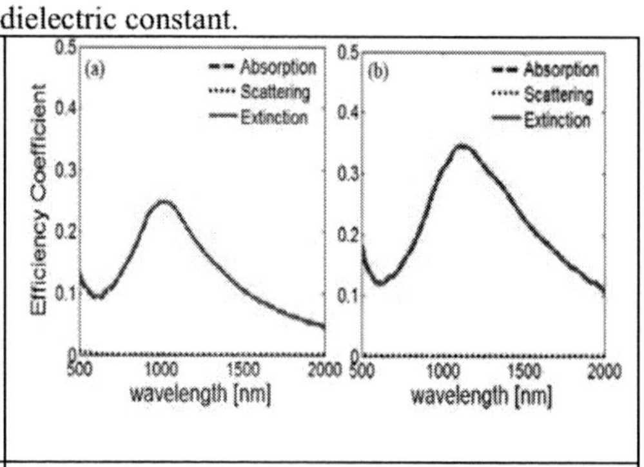

Fig.7. Effect of surrounding media on optical properties of VO_2 nanosphere in metallic phase, (a) in air, (b) in a medium with ε =2.25.

The effects of surrounding media ($\varepsilon = 2.25$) on the LSPR wavelength of 20 nm VO_2 nanosphere are investigated in Fig.7. The LSPR wavelength red shifts by increasing the permittivity of surrounding medium as expected from theory [10].

3. Conclusion

The results of the work are as follows:

1) VO_2 film on glass substrate were prepared and investigated in MM wave range (27–37 GHz). The MM wave response of nanosized VO_2 films reveals strong anomalies near MIT.

2) The submicron holes arrays were formed on VO_2 films by FIB milling and their optical Raman spectra were studied. The holes and arrays show strong change of the Raman spectra at the wavelength 530 nm caused by laser radiation heating.

3) The optical and MM wave electromagnetic properties of homogeneous VO2 nanospheres embedded in the air are studied theoretically. Size effects on the optical properties of the VO2 nanosphere are investigated. In VO_2 nanosphere, converting into metallic phase by heating leads to formation of a localized surface plasmon resonance (LSPR) which red shifts by increasing dimension.

Acknowledgments

This work was supported by grant RFBR 1757-560002 and INSF grant No. 86003683.

References

[1] Haupt, R. L., & Lanagan, M. (2013). Reconfigurable antennas. IEEE Antennas and Propagation Magazine, 55(1), 49-61.

[2] Turpin, J. P., Bossard, J. A., Morgan, K. L., Werner, D. H., & Werner, P. L. (2014). Reconfigurable and tunable metamaterials: a review of the theory and applications. International Journal of Antennas and Propagation, 2014.

[3] Y. Zhao, J. H. Lee, Y. Zhu, M. Nazari, Ch. Chen, H. Wang, A. Bernussi, M. Holtz, and Zh. Fan. Structural, electrical, and terahertz transmission properties of VO2 thin films grown on c-, r-, and m-plane sapphire substrates. Journal of Applied Physics. 2012, Vol. 111, No. 5, P. 053533.

[4] J. S. Kyounga, S. B. Choib, H. S. Kima, B. J. Kimc, Y. H. Ahnd, H. T. Kimc, D. S. Kim. Nanoresonator Enabled Ultrafast All - optical Terahertz Switching Based on Vanadium Dioxide Thin Film. AIP Conference Proceedings, 2011, Vol. 1399, No. 1, P. 1027-1028.

[5] Ch. Chen, Ya. Zhu, Y. Zhao, J. H. Lee, H. Wang, A. Bernussi, M. Holtz, and Zh. Fan. VO2 multidomain heteroepitaxial growth and terahertz transmission modulation. Applied Physics Letters, 2010, Vol. 97, No. 21, P. 211905.

[6] C. Chen, Y. Zhu, Y. Zhao, J.H. Lee, H. Wang, A. Bernussi, M. Holtz, Z. Fan. VO2 multidomain heteroepitaxial growth and terahertz transmission modulation. Appl. Phys. Lett., 2019, Vol. 97, P. 211905.

[7] S. Kumar, F. Maury, N. Bahlawane. Electrical Switching in Semiconductor-Metal Self-Assembled VO2 Disordered Metamaterial Coatings. Scientific Reports, 2016, Vol. 6, P. 37699.

[8] Zhan H., Astley V., Hvasta M. The metal-insulator transition in VO2 studied using terahertz apertureless near-field microscopy. Applied Physics Letters. 2007, Vol. 91, No. 16, P. 162110.

[9] I.V. Bychkov, V.A. Golunov, D.S. Kalenov, A.P. Kamantsev, D.S. Kuchin, V.V. Koledov, D.A. Kuzmin, V.V. Meriacri, S.V. von Gratowski, M.P. Parkhomenko, A.V. Mashirov, V.G. Shavrov. The intrinsic radiation and electromagnetic wave reflection coefficient in the range of 8 mm of $Ni_{2,14}Mn_{0,81}GaFe_{0,05}$ and Ti-Ni alloys in the temperature interval near the phase transitions of the 1st and 2nd order. Journal of Radioelectronics. 2014, No. 12, Available on the web http://jre.cplire.ru/jre/dec14/27/abstract_e.html

[10] C.F. Bohren, D.R. Huffman, "Absorption and scattering of light by small particles", John-Wiley 2008.

[11] K.L. Kelly, E. Coronado, L.L. Zhao, et al., "The opticalproperties of metal nanoparticles: the influence of size,shape, and dielectric environment", J. Phys. Chem. B,Vol. 107, No. 3, pp. 668-677, 2003.

[12] H.W. Verleur, A.S. Barker Jr, C.N. Berglund, "Optical properties of VO2 between 0.25 and 5 eV", Phys. Rev.,Vol. 172, No. 3, pp. 788-798, 1968.

[13] Jepsen, Peter Uhd, Bernd M. Fischer, Andreas Thoman, Hanspeter Helm, J. Y. Suh, René Lopez, and R. F. Haglund Jr. "Metal-insulator phase transition in a V O 2 thin film observed with terahertz spectroscopy." Physical Review B 74, no. 20 (2006): 205103.

[14] S.A. Maier, 2007. "Plasmonics: fundamentals and applications",Springer 2007.

Analysis of power confinement in Trench-Assisted Fiber for Sensing Applications

V Palodiya, Prakash Pareek
Department of Electronics and Communication Engineering,
Sreenidhi Institute of Science and Technology, Yamnampet, Ghatkesar, Hyderabad, Telangana, India-501301.
Department of Electronics and Communication Engineering,
Vaagdevi Engineering College (Vaagdevi Group of Colleges),
Bollikunta,Warangal,Telengana, India-506005.

E-mail: vikrampalodiya@gmail.com

Abstract. In this paper, we have the analysis on the effect of power confinement in trench assisted fiber. In single mode trench-assisted optical fibers at a certain value of trench width, all power propagates through the cladding. Single mode operation in the fiber is realized by the small diameter of the core and low mode field diameter gives tight confinement. The optimized design parameters are obtained by introducing trench between inner cladding and the outer cladding layer. The analysis shows that the relative refractive index difference in trench affects the properties of optical fiber for sensing applications.

Keywords: trench-assisted optical fiber, mode field diameter, power confinement.

1. Introduction

Bend insensitive fibers (BIF) are exciting engineers and scientists because of the possibility of their use in fiber-to-the-home (FTTH) applications and thereby reducing a power penalty caused by sharp bends in optical fibers used in the Broadband communications networks [1]. Broadband communications and access networks, including FTTH have been evolving rapidly. BIF, which has a low bending loss, has become particularly attractive as the FTTH service has expanded [2, 3]. There have been many reports on a BIF that employ such techniques as reducing the mode field diameter (MFD) [4], depressing cladding, adding a low index trench.

When talking about trench-assisted fiber (TAF), bend-insensitive optical fiber can be used in FTTH applications [5, 6]. The trench can reduce both macro-bending loss [7] and micro-bending loss [8]. It performs perfectly in following the ITU-T G.657 standard to achieve very low bending loss. Besides reducing bending loss, the low refractive index trench affects some other parameters in the optical fiber communication system such as chromatic dispersion [9, 10], mode field diameter [11] and even crosstalk in multi-core fiber [12]. Theoretically, an ideal optical fiber used in optical communication system should have these following characteristics: small MFD, tight power confinement and low bending loss. These characteristics, which guarantee the signal quality in the FTTH, can be optimized by trench-assisted structure. Although there have been many discussions about the TAF, we will study how different parameters of the trench affect power confinement, mode field diameter and other properties of optical fiber respectively. In this paper, we have focused on optical field distribution in the fundamental mode, effective mode field diameter and power confinement for various applications.

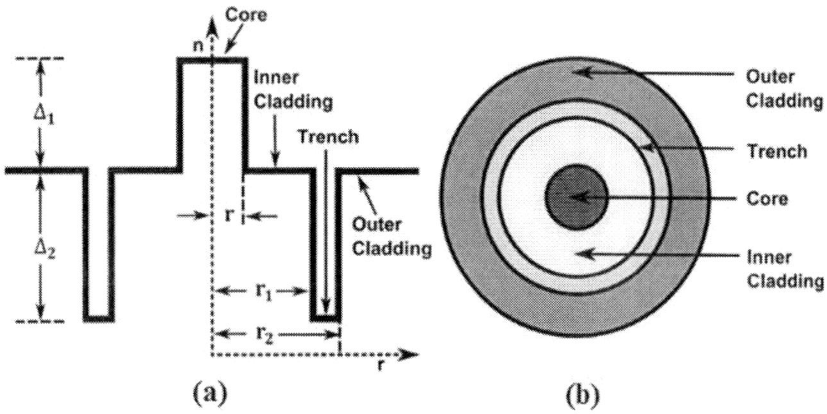

Fig 1: (a) Trench assisted fiber profile and (b) Cross sectional view of TAF.

2. Optimized Objects of Trench-Assisted Fiber

Based on the reported proposal [13], we have designed TAF with a Fluorine trench, inner cladding and outer cladding layout of this work. If layers of cladding are set inside the fiber, the cutoff wavelengths of the interior cores tend to be longer than that of the exterior cores [14]. This is due to the tight confinements in the interior cores, which are caused by the trench structure deployed around the core. Moreover, excessive MFD will also depend on the trench cladding [15]. Thus, we only arrange one layer of cores with a ring layout for the TAF. The design concept model is shown in Fig. 1. (a) Refractive index profile and (b) Cross sectional view. It illustrates the schematic of a fiber with trench-Index and stand for core radius. The different parameters associated with the proposed structure can be represented as follow: the distance between the center of the core and the inner circumference of trench denoted by r_1, the distance between the center of the core and outer circumference of trench denoted by r_2, the thickness of the trench layer denoted by $r_2 - r_1$, the relative refractive-index difference between core and cladding denoted by Δ_1, the relative refractive-index difference between cladding and a trench denoted by Δ_2. Table 1 shows design parameters of proposed TAF given below

Table 2.1. Structural parameters of Trench-Assisted Fiber

Parameter	Unit	Model-I	Model-II
R	μm	4	4
r_1	μm	4	8
r_2	μm	8	24
$r_2 - r_1$	μm	4	16
Δ_1	%	0.35~0.70	0.35~0.70
Δ_2	%	-0.20 ~-0.250	-0.25 ~-0.30

IOP Conf. Series: Journal of Physics: Conf. Series **1092** (2018) 012109 doi:10.1088/1742-6596/1092/1/012109

3. Simulated Results and Discussion

Although the trench structure can reduce the bending loss significantly and optimize the mode field diameter, it also affects the optical field distribution and power confinement for the fundamental mode. Therefore, we will have a direct insight into the variation of the mode area and power distribution in fundamental mode with trench width. We have study effect on fundamental mode and optical power change according to change in trench properties. Fig. 2. shows the highest field in the center part of the fundamental mode and power confinement in center core. This result shows a tight power confinement gives low mode field diameter, low bending loss. These characteristics show TAF use in broadband and FTTH application.

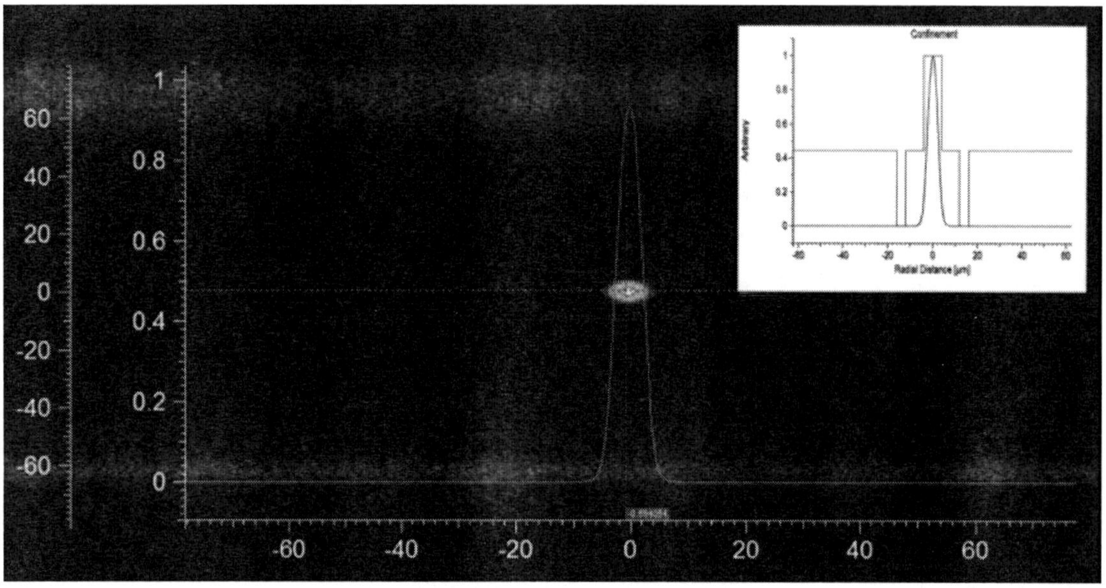

Fig 2: Optical field distributions in fundamental LP_{01} Mode and power confinement in center of the core.

In Fig. 3. shows the highest field on the surface of the fundamental mode and maximum power confinement in an outer cladding of the fiber. This result shows a power confinement in cladding gives this fiber sensitive to stress and strain, low bending loss. These characteristics show TA-F use in sensing application.

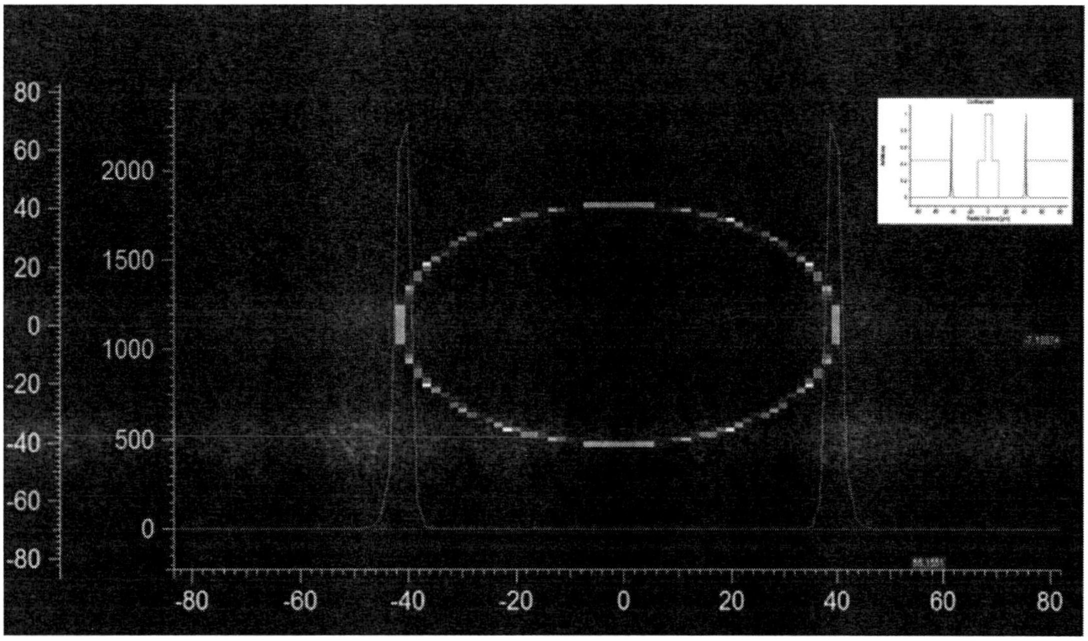

Fig 3: Optical field distributions in fundamental LP_{01} Mode and power confinement in the outer cladding.

4. Conclusion

To conclude our work, we have proposed a model of TA-F for sensing applications. The trench-index profile illustrates the advantages of the bend insensitive fiber for broadband and FTTH applications. The model shows the maximum power confinement in outer cladding. Hence TA-F second model used in sensing applications. We have studied the effect of different trench parameters on optical field distribution, bending loss, mode field diameter and power confinement of the fiber, theoretically. Based on the numerical analysis, it can be observed that important properties such as optical field distribution, power confinement and bending loss are sensitive to trench thickness. On the other hand, if the trench, thickness increase the MFD decrease with the acceptable bend loss and supporting single mode operation. The designed TA-F has low MFD of about 8.7 μm at 1550 nm wavelength, as well as low bending losses. It can show potential application in the field of sensing throughout the entire spectrum range of 1.3 μm to 1.6 μm wavelengths.

5. References

1. Watekar P, Ju S and Han W 2008 *Optics Express* **16** 1180-1187.
2. Himeno K, Matsuo S, Ning Guan and Wada A 2005 *Journal of Lightwave Technology* **23** 3494-3499.
3. Palodiya, V, Raghuwanshi, S K 2015 *Optica Applicata* **45** 341—353.
4. Tangonan G, Hsu H, Jones V and Pikulski J 1989 *Electronics Letters* **25** 142-149.
5. de Montmorillon L A, Matthijsse P, Gooijer F, Achten F, Molin D, Montaigne N and Maury J 2006 *55th IWCS/Focus* 342–347.
6. Palodiya, V, Raghuwanshi, S K 2015 *International Conference on Photonics Solutions* 9659 96590A.
7. Cattelan S, Ruzzier M, and Travagnin M 2009 Macrobending *58th IWCS/IICIT* 258-263.
8. Sillard P, Richard S, de Montmorillon L A, and Bigot-Astruc M 2010 *ECOC*, 1 - 3.

9. Ghafoori Fard H, Dehesht A, Rostami A and Nejati S 2008 *Optics Communications* **281** 5779-5787.
10. Palodiya V and Raghuwanshi S 2016 *Journal of Optical Communications* 37.
11. Kim J, Moon D S, and Eom J B 2003 *SPIE* 191 191–198.
12. Tsuchida Y, Saitoh K and Koshiba M 2005 *Optics Express* **13** 4770.
13. Raghuwanshi S, Kumar V and Talabattula S 2012 *Progress In Electromagnetics Research B* **36** 249-265.
14. Takenaga K, Arakawa Y, Tanigawa S, Guan N, Matsuo S, Saitoh K, and Koshiba M 2011 *OFC/NFOEC* 1 - 3.
15. Takenaga K, Arakawa Y, Sasaki Y, Tanigawa S, Matsuo S, Saitoh K and Koshiba M 2011 *Optics Express* **19** B543.

Effect of high index sphere size on effective refractive index of random nanocomposites

Andrey V. Panov

Institute of Automation and Control Processes, Far East Branch of Russian Academy of Sciences, 5, Radio st., Vladivostok, 690041, Russian Federation

E-mail: andrej.panov@gmail.com

Abstract. The linear and nonlinear optical properties of random composites are investigated with the use of the three-dimensional finite-difference time-domain (FDTD) simulations of light propagation. An effective linear refractive index of the mixture as a function of the GaP spherical inclusion size is constructed. This dependence is non-monotonic reaching a maximum just below the first optical resonance. It is shown that the effective Kerr third-order susceptibility of the nanocomposite grows by orders near the optical resonance.

The optical resonances of high refractive index nanoparticles have been studied for more than a century [1]. Recently, particular attention has been given to unusual properties of the high index particle structures: induced transparency [2], controllable strong anisotropic scattering [3], switchable invisibility [4]. It was supposed that these particles show enhanced optical nonlinearity due to the field concentration at the optical resonances [5].

The small dielectric spheres with the high refractive index have certain modes at which optical resonances are displayed. The resonance inside the particles arises from the self-maintaining oscillations of the electric and magnetic fields. The magnetic dipole resonance is most pronounced and has the smallest size of the particle when [6]

$$nkr = \pi, \tag{1}$$

where n is the refractive index inside the sphere, $k = 2\pi/\lambda$ are the wavenumber and the wavelength in vacuum, r is the radius of the sphere. At these resonances, the scattering from the particle has the Fano profile as a result of the interference between the resonant Mie scattering and the non-resonant background scattering [5].

The optical properties of the nanoparticle systems are usually treated by effective medium theories, for example, linear [7, 8] or nonlinear, [9]. In these theories, the effective refractive index of the mixture containing the inclusions is calculated. Generally, the effective medium theories assume the uniform field distribution inside inclusions that is $nkr \ll 1$.

In this work, the effective nonlinear refractive index of the nanocomposite containing identical spherical inclusions is examined as a function of the particle size by means of the FDTD modeling with the MEEP solver [10]. The simulations were conducted for the randomly arranged systems of the particles with the sizes from small compared with the wavelength to the magnetic dipole resonance radius and larger. The modeling was done using the method described in Ref. [11]. This method utilizes the phase change of the Gaussian beam transmitted through the studied sample. This phase change is computed for different intensities of the Gaussian beam enabling one to estimate the real parts of both the linear

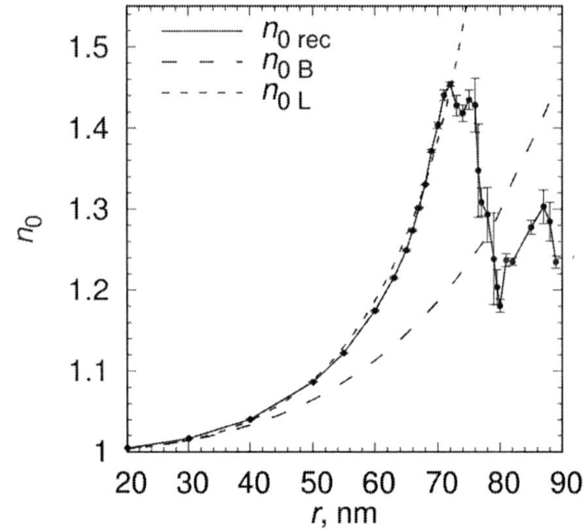

Figure 1. Dependence of the effective linear refractive index of the random nanocomposite on the sphere radius r for a fixed number of the particles (1000). The effective refractive index $n_{0\,\mathrm{rec}}$ is calculated using FDTD modeling, $n_{0\,\mathrm{B}}$ and $n_{0\,\mathrm{L}}$ are results of the effective medium theories [7, 8].

refractive index and third-order susceptibility of the sample. All the simulations were conducted for the wavelength of 532 nm.

The studied samples represent the system of the disjoint spheres of the same radius r randomly arranged in space. Two different arrangements of a thousand of the particles inside the sample were randomly generated, the size of the spheres was varied for each configuration. The gallium phosphide (GaP) was selected as a material of the inclusions because of the high value of the real part of linear refractive index $n_{0\,\mathrm{in}} = 3.49$ at $\lambda = 532$ nm [12] and moderately low extinction coefficient (0.0026). In this work, only the real part of the refractive index was used since MEEP cannot simulate simultaneously lossy and nonlinear medium. The third-order optical susceptibility of the inclusions $\chi^{(3)}_{\mathrm{in}} = 2 \times 10^{-10}$ esu [13] was applied to the computations.

The size of the computational domain for the FDTD simulations was $4 \times 4 \times 30$ μm with the space resolution of 5 nm. The size of the sample with the nanoparticles was $4{\times}4{\times}0.8$ μm, the distance between the Gaussian beam source and the studied sample was 0.9 μm. The beam radius w_0 at the beam waist was 1.1 μm.

Fig. 1 depicts the computed values of the linear effective refractive index $n_{0\,\mathrm{rec}}$ of the samples as a function of the sphere radius. The error bars show the standard deviations of $n_{0\,\mathrm{rec}}$. For comparison, the curves calculated using effective medium theories [7, 8] are given. It should be underlined that this effective medium theories are unusable for most range in Fig. 1. Moreover, the approximation in Ref. [8] is elaborated for the cubic lattice of the spheres. As can be seen from the figure, the dependence of $n_{0\,\mathrm{rec}}$ on the sphere size is non-monotonic: $n_{0\,\mathrm{rec}}$ has a maximum below the magnetic dipole resonance at $r = 76.1$ nm (Eq. 1), then it drops and rises again. In particular, for the particles with the sizes under the first optical resonance, the computed effective refractive index coincides with one predicted by the Lewin approximation [8].

The results of the calculations of the effective third-order susceptibility $\chi^{(3)}_{\mathrm{rec}}$ are presented in Table 1.

r, nm	f, %	$n_{0\,\mathrm{rec}}$	$n_{0\,\mathrm{B}}$	$n_{0\,\mathrm{L}}$	$\chi^{(3)}_{\mathrm{rec}}$, esu	$\chi^{(3)}_{\mathrm{RZPA}}$, esu
20	0.365	1.00464 ± 0.00002	1.00434	1.00380	$(2.0 \pm 0.7) \times 10^{-15}$	1.51×10^{-15}
30	1.18	1.01636 ± 0.00008	1.01424	1.01442	$(1.0 \pm 0.2) \times 10^{-14}$	5.35×10^{-15}
40	2.70	1.0405 ± 0.0002	1.03320	1.03908	$(4.4 \pm 0.5) \times 10^{-14}$	1.44×10^{-14}
50	5.07	1.0864 ± 0.0005	1.06468	1.08926	$(1.13 \pm 0.07) \times 10^{-12}$	3.52×10^{-14}
60	8.44	1.174 ± 0.001	1.11336	1.18693	$(1.12 \pm 0.05) \times 10^{-12}$	8.63×10^{-14}
65	10.5	1.249 ± 0.001	1.14620	1.26733	$(2.6 \pm 0.2) \times 10^{-11}$	1.38×10^{-13}
70	12.9	1.403 ± 0.004	1.18618	1.38453	$(5.0 \pm 1.2) \times 10^{-09}$	2.25×10^{-13}
80	18.6	1.180 ± 0.008	1.29352	1.88407	$(3.0 \pm 1.8) \times 10^{-09}$	6.47×10^{-13}

Table 1. The linear refractive indexes and nonlinear susceptibilities $\chi^{(3)}$ of random mixtures reconstructed with the FDTD modeling and predicted by the effective medium theories for different radii of spheres r. Here f is the volume fraction (concentration) of inclusions, $n_{0\,\mathrm{rec}}$ is the reconstructed linear refractive index, $n_{0\,\mathrm{B}}$ and $n_{0\,\mathrm{L}}$ are ones calculated using the Bruggeman [7] and Lewin [8] formulas, $\chi^{(3)}_{\mathrm{rec}}$ is the third-order susceptibility retrieved in this work, $\chi^{(3)}_{\mathrm{RZPA}}$ is calculated using [9].

It was not possible to compute $\chi^{(3)}_{\mathrm{eff}}$ for the resonance range owing to large errors. The effective $\chi^{(3)}$ calculated using Ref. [9] shows the largest value among other approximations [11]. It is of great importance that $\chi^{(3)}_{\mathrm{rec}}$ is much larger than the prediction of the nonlinear effective medium theory when the sphere size approaches the optical resonance. Near the magnetic dipole resonance, $\chi^{(3)}_{\mathrm{rec}}$ of the nanocomposite even exceeds the third-order susceptibility of bulk GaP. This is due to the giant field concentration inside the spheres near the resonance [5].

To sum up, the dependence of the nonlinear refractive index of the random nanocomposite on the size of the spherical inclusions is investigated numerically. The linear effective refractive index of the mixture falls at the magnetic dipole resonance. The Kerr third-order susceptibility of the composite abruptly increases near this resonance.

Acknowledgements

The results were obtained with the use of IACP FEB RAS Shared Resource Center “Far Eastern Computing Resource” equipment (https://www.cc.dvo.ru). This work was partly supported by “Far East Program” of FASO Russia (grant No. 18-3-023).

[1] Debye P 1909 *Ann. Phys. (Leipzig)* **335** 57–136
[2] Evlyukhin A B, Reinhardt C, Seidel A, Luk'yanchuk B S and Chichkov B N 2010 *Phys. Rev. B* **82**(4) 045404
[3] Fu Y H, Kuznetsov A I, Miroshnichenko A E, Yu Y F and Lukyanchuk B 2013 *Nat. comm.* **4** 1527
[4] Rybin M V, Samusev K B, Kapitanova P V, Filonov D S, Belov P A, Kivshar Y S and Limonov M F 2017 *Phys. Rev. B* **95**(16) 165119
[5] Tribelsky M I and Miroshnichenko A E 2016 *Phys. Rev. A* **93**(5) 053837
[6] van de Hulst H C 1981 *Light scattering by small particles* (Dover Publications)
[7] Bruggeman D A G 1935 *Ann. Phys. (Leipzig)* **416** 636–664
[8] Lewin L 1947 *Proc. Inst. Electr. Eng.* **94**(27) 65–68(3)
[9] Rukhlenko I D, Zhu W, Premaratne M and Agrawal G P 2012 *Opt. Express* **20** 26275–26284
[10] Oskooi A F, Roundy D, Ibanescu M, Bermel P, Joannopoulos J D and Johnson S G 2010 *Comput. Phys. Commun.* **181** 687–702
[11] Panov A V 2018 *Opt. Lett.* **43** 2515–2518
[12] Aspnes D E and Studna A A 1983 *Phys. Rev. B* **27**(2) 985–1009
[13] Kuhl J, Rhee B K and Bron W E 1985 *Time-Resolved Vibrational Spectroscopy* ed Laubereau A and Stockburger M (Berlin, Heidelberg: Springer Berlin Heidelberg) pp 30–34

Impact of Sn content on frequency response of GeSn SQWIP

Prakash Pareek[1], **Vikram Palodiya**[2]

[1] Department of Electronics and Communication Engineering, Vaagdevi Engineering College, Bolikunta, Warangal, Telengana, India-506005
[2] Department of Electronics and Communication Engineering, Sreenidhi Institute of Science and Technology, Hyderabad, Telengana, India-501301

ppareek1@gmail.com

Abstract. In this paper a detailed theoretical investigation of Sn content dependent frequency response in strain balanced SiGeSn/GeSn single quantum well infrared photodetector (SQWIP) is made. The rate equation in the quantum well is solved to obtain photo generated current density in frequency domain. Impact of Sn content in the well on the frequency response, and bandwidth is studied. The result show that a large bandwidth can be achieved at higher Sn content of active region i.e. quantum well in the device. This is mainly due to the consideration of strain balancing condition in the structure of proposed SQWIP. On using proper choice of Sn content in well, this structure may be used in CMOS compatible photonic integrated circuit.

1. Introduction

In last few years, III-V quantum well infrared photodetectors (QWIP) are used in ground and space-based applications such as night vision, temperature detection, early warning systems, navigation, flight control systems, weather monitoring, as well as security and surveillance [1]. However, the broad use of these III-V QWIPs is restricted due to their high cost and incompatibility to silicon which is very important to realize electronic photonic integrated circuit [2]. In recent years, emergence of genuine direct band gap group IV alloy i.e. GeSn, paved the path of monolithic QWIP [3,4]. Therefore study as well as modeling of GeSn QWIP is very crucial before its fabrication to reengineering its properties. In this process, strain also come into play obviously, due to a large lattice mismatch between Ge and Sn. Thus, strain balanced structure is used to protect the active GeSn well layer from excessive strain during fabrication [5]. In a strain balanced system, strain in well region can be minimized by adjustment of lattice constant of barrier and buffer layers. Due to its simple structure, a single quantum well structure is most appropriate for a better understanding of various physical aspects.

In this context, authors had already proposed a strain balanced GeSn based single QWIP (SQWIP) and reported a higher absorption coefficient for GeSn active layer [6]. Thus in present paper, extending our earlier work [6], effect of Sn content of well on frequency response in strain balanced $Si_{0.09}Ge_{0.8}$ $Sn_{0.11}/Ge_{0.83}Sn_{0.17}$ QWIP is studied. In this work, we derived an analytical expression of frequency dependent current density in QWIP using rate equation in quantum well.

2. Model Description and theoretical formulation

The cross section of the model considered in our analysis is shown in Fig 1. It consists of $Ge_{0.83}Sn_{0.17}$ quantum well layer sandwiched between two wide bandgap $Si_{0.09}Ge_{0.8}Sn_{0.11}$ barriers. The Ge and Sn content of well and barrier are optimized to facilitate low direct bandgap well layer as well as wide bandgap barrier for quantum carrier confinement [6].

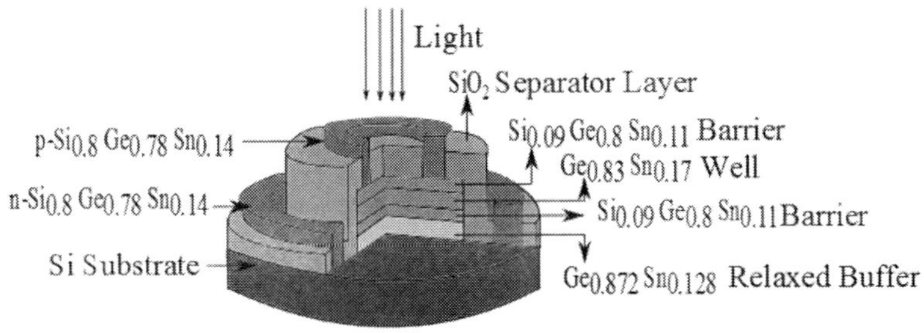

Fig. 1. Cross section of proposed QWIP

In this work, bias voltage is assumed to be large enough so that the carriers are drift across barriers at saturation velocity. Carrier transport mechanism is an important process which needs to be addressed in order to obtain frequency response in detector. Fig.2 shows a simple schematic of GeSn QWIP for calculating the frequency dependent photocurrent. Let the light of appropriate wavelength incident from p-side of the P-i-N detector. After incident of light, electron hole pairs are generated. These photogenerated electrons and holes move in opposite direction under influence of electric field. Diffusion of photogenerated carriers can be neglected due to a large electric field across depleted region. This movement of photogenerated carriers induces a current density in the structure. Let this current density due to moving electron, J_{em}(considering electron first). Actually, J_{em} depends upon the distribution function of uncollected electrons in the active region of the QWIP structure [7]. The distribution function includes the effect of decaying nature of light intensity. So, from the figure 2, expression of current density due to moving electrons is given by[7],

$$J_{em} = Kqv_e\left[1 - \exp(-\alpha(w_d - v_e t)\right] \quad (1)$$

where K is a constant which depends upon the intensity of input light. w_d is quantum well layer thickness, v_e is saturation velocity of electron and α is absorption coefficient of active GeSn well layer as computed in our previous work [6].

The transport mechanism of photogenerated carriers in quantum well can be fully understood by using rate equation in quantum well. The rate equation for electrons in conduction band can be written as[7],

$$\frac{\partial n_{QW}}{\partial t} = \frac{J_{em}}{q} - n_{QW}R \quad (2)$$

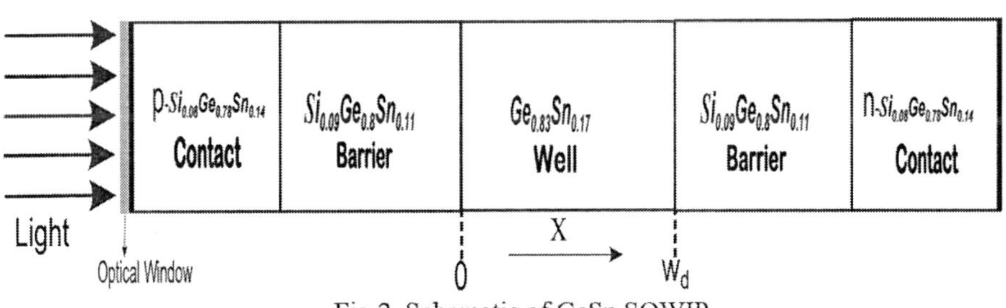

Fig.2. Schematic of GeSn SQWIP

where, J_{em} =current density due to moving electron, n_{QW}=electron sheet concentration in quantum well in cm^{-2}, J_{em}=Current density at quantum well due to moving carriers, r_{esc}=rate of thermionic emission in quantum well, r_r=rate of recombination of carrier in quantum well, R= $r_{esc}+r_r$ i.e. total rate of thermionic escape, tunneling and recombination of carriers. The rate equation involves the processes of carrier capture and escape from quantum well. The two main mechanisms of carrier escape from quantum well are considered in our study which are tunneling emission and thermionic emission. Thus the carriers which are injected from the contact, some get captured in quantum well; some escape from quantum well due to thermionic emission and tunneling, reach the contact at the other end and extracted by strong electric field [8].Now, the rate equation in quantum well can rewritten as:

$$\frac{\partial n_{QW}}{\partial t} = K.v_e\left(1-\exp(-\alpha\left(w_d - v_e t\right)\right) - n_{QW}\left(r_{esc} + r_{tunneling} + r_{recombination}\right) \quad (3)$$

On using Laplace transform to solve the above expression

$$n(j\omega) = K.\left[\left(\frac{1-R.\ \exp(-\alpha w_d)}{\left(\frac{\alpha v_e}{R}+1\right)}\right).\left(\frac{1}{\alpha v_e - R - j\omega}\right)\right] \quad (4)$$

Same calculation is done for obtaining hole current density i.e. p(s). Thus the frequency dependent current density in SQWIP can now be written as

$$J(j\omega) = n(j\omega)qv_e + p(j\omega)qv_h \quad (5)$$

3. Results and discussion

Before calculating frequency response, some appropriate approximations are made for material parameters like saturation velocity, recombination time for Sn for which exact values are not reported in the literature. To understand the effect of Sn content on the frequency response of GeSn QWIP, Fig. 3 is plotted. The figure shows the frequency response for different Sn content at zero bias. It is clearly observed from the plot, with increasing Sn content in quantum well layer, 3 dB bandwidth increases. The above statement become more clear when bandwidth is extracted from this figure and plotted as a function of applied bias and is shown in Fig. 4. In figure, bandwidth increases with bias due to decrement of effective band height which further increases thermionic emission [7]. Moreover, thermionic emission is more dominant here due to consideration of room temperature (300K). The figure also depicts a significant increase in the bandwidth on increasing Sn content. This is due to the strain balanced nature of the structure.

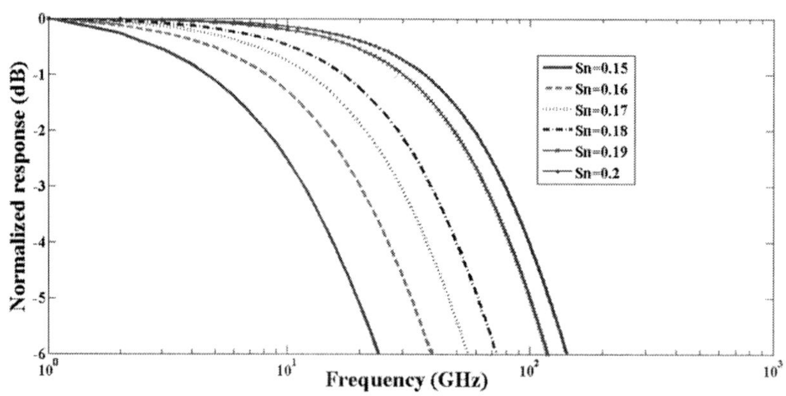

Fig. 3. Plot of frequency response of SQWIP for different Sn content in the well

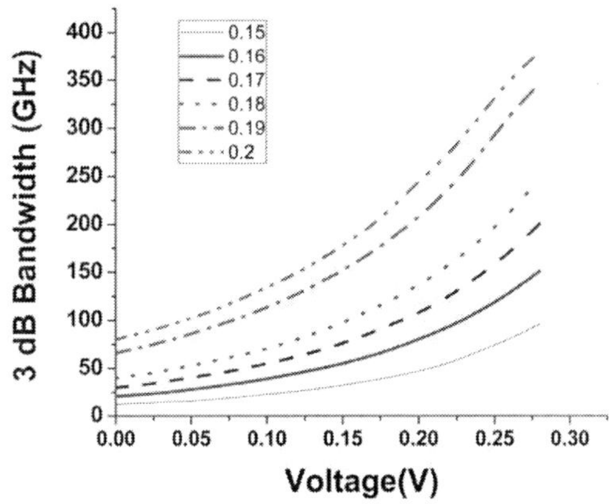

Fig.4. Plot of bandwidth of GeSn SQWIP as a function of different Sn contents

When Sn content is increased then the barrier thickness is reduces due to strain balanced condition. Due to decrease of barrier thickness, transit time of the carriers reduces. This causes increase in 3 dB bandwidth. The increment is more for higher Sn content in well. It is worth to be mentioned here at this critical juncture that increasing Sn content beyond 20% causes serious fabrication dislocations [9]. Therefore proper optimization of Sn should be done before actual implementation of GeSn SQWIP.

4.Conclusion

In this work, we have theoretically studied the frequency response in strain balanced SiGeSn/GeSn QWIP. The alloy composition of well, barrier and buffer layers are chosen to ensure type I direct band gap quantum well as well as strain balanced structure. The frequency dependent photogenerated current density is obtained using rate equation in quantum well including various physical aspects involved in QWIP physics. The nature of frequency response is investigated under variation of Sn content in well. The 3dB bandwidth which is obtained from this frequency response is also plotted as a function of Sn content in active layer as well as applied bias. A higher bandwidth up to 200 GHz is obtained at a bias of 0.28V for considered Sn content (0.17). The bandwidth increases with Sn content in well due to strain balanced condition but Sn content should be constrained below 20 % due to intolerable strain. The content of Sn in active well layer can play a decisive role in the performance of QWIP. Thus the proposed QWIP is a potential candidate to be used in cheap and commercial CMOS based EPICs.

5. References

[1] Downs Chandler and Vandervelde Thomas E 2013 *Sensors* **13** 5054
[2] Lo G Q, Ang K W, Liow T Y, Fang Q, Zhang J, Zhu S Y, Song J F, Xiong Y Z, Ren F F, Yu M B and Kwong D L 2010 *ECS Transactions* **28** 3
[3] Kouvetakis J, Menedez J. and Chizmeshya A V G 2006 *Ann. Rev. of Material Research* **36**, 497
[4] Zheng Jun, Wang Suyuan, Liu Zhi, Cong Hui, Xue Chunlai, Li Chuanbo, Zuo Yuhua, Buwen Cheng and Wang Qiming 2016 *App. Phys. Letter* **108** 033503
[5] Daukes Ekin, Kawaguchi K. , and Zhang 2002 *J. Crystal growth and design* **2** 287
[6] Pareek Prakash and Das Mukul K 2016 *Optical and Quantum Electronics* **48** 1(11)
[7] Das N R and Deen M J 2001 *IEEE Journal of Quan. Elect.* **37** 1574
[8] Zhou G. and Runge P. 2014 *IEEE Journal of Quan. Elect.* **50** 220
[9] Maczko H S, Kudrawiec R and Gladysiewicz M 2016 *Scientific Reports* **6** 34082

IOP Conf. Series: Journal of Physics: Conf. Series **1092** (2018) 012112 doi:10.1088/1742-6596/1092/1/012112

Graphene electrochemistry: edge vs. basal plane sites

S V Pavlov[1,2] and S A Kislenko[1]

[1] Joint Institute for High Temperatures of the Russian Academy of Sciences, Izhorskaya 13 Bldg 2, Moscow 125412, Russia
[2] Moscow Institute of Physics and Technology, Institutskiy Pereulok 9, Dolgoprudny, Moscow Region 141700, Russia

E-mail: sergey.v.pavlov@phystech.edu

Abstract. This work is focused on the topic of electrocatalytic activity of graphene edges. DFT calculations and quantum mechanical theory of charge transfer is used for the investigation of heterogeneous electron transfer kinetics by the example of the oxygen reduction reaction, which is important for electrochemical applications. It is shown that the rate of electron transfer at zig-zag edges is ~4 times larger than far from edges. Electron transfer occurs in nonadiabatic regime all over the graphene cluster at distances greater than 5Å. The highest occupied molecular orbital of the graphene cluster makes the main contribution to the electron transfer. Electrocatalytic effect of zig-zag edges is explained by preferential localization of the highly occupied molecular orbital on edges.

1. Introduction

Investigation of the heterogeneous electron transfer (ET) at the graphene surface is highly relevant today due to the perspectives of graphene use in electrochemical energy-storage devices [1]. Electrochemical properties of defective graphene is of particular interest by virtue of the fact that point and linear defects show high electrocatalytic activity [2, 3]. This effect is potentially useful for the development of cheap and catalyst-free graphene-based electrodes for electrochemical applications.

Kinetics of the redox processes at the graphene edge (linear topological defect) is mostly studied but controversial field. It was stated that the rate of ET at the edges is faster than at basal plane of graphite and sidewall of nanotubes [4, 5], and the latest delicate experiment on a single-layer graphene sheet confirmed it [3]. But on the other side, this effect was not reproduced and disputed by others [6, 7].

Important data for interpretation of this electrocatalytic effect was obtained theoretically in Fujita's work where it was shown the existence of the electronic states, which are strongly localized on zig-zag edges and have the energies near the Fermi level [8] (confirmed experimentally in [9]). According to the Gerischer model such states have the highest contribution to the ET rate. In our previous works, we have shown that besides the surface electronic properties, the rate of redox processes also depends on the electrode/electrolyte interface structure [10]. In particular, the distributions of the redox-active components and the barriers for their adsorption are different at the edge and the basal plane sites [11-14], which is another possible factor determining the catalytic activity.

Despite of many works on the subject, there is a lack of quantitative calculations of the ET kinetics on the graphene edges, which could allow to separate contributions of different factors to the redox processes. The aim of this work is quantitative analysis of the effect of surface electronic properties on the ET rate by calculating the electronic transmission coefficient near and far from zig-zag edge. The transmission coefficient represents the probability of the ET from the surface to the reactant in solutions when donor and acceptor levels are aligned. The calculations have been performed for the oxygen reduction reaction $O_2 + e \rightarrow O_2^-$, which is one of the key processes in fuel cells and metal-air batteries.

2. Computational details

The calculations were carried out for the graphene fragment $C_{150}H_{30}$ with zig-zag edges terminated by hydrogen atoms (figure 1). We studied the electron transfer from the $C_{150}H_{30}$ molecule to the O_2 molecule located above different points near and far from edge perpendicularly to the surface. At the same time, the distance from the oxygen molecule to the graphene surface was scanned.

Since we used cluster model, the surface has a discrete energy spectrum. In this case the rate constant of ET equals [10, 15-18]:

$$k = \sum_{i=1}^{N} \frac{\omega_{eff}}{2\pi} \kappa_e^{(i)} \exp\left(\frac{(\lambda+\varepsilon_F-\varepsilon_i-e_0\eta)^2}{4\lambda kT}\right) \quad (1)$$

where $\kappa_e^{(i)}$ is the electronic transmission coefficient from i-th orbital of the surface to the acceptor molecular orbital; λ is the reorganization energy of solvent; ε_F is the Fermi level of electrode, i.e., the highest occupied molecular orbital (HOMO) of the cluster; e_0 and η are the electron charge and the overpotential respectively, ω_{eff} is the polarization frequency. Since the oxygen being in triplet state $3\sum_g^-$ has two degenerate single occupied molecular orbitals (SOMOs), we calculated maximum of coupling between cluster's orbitals and both of SOMOs. We used typical values of the reorganization energy and the polarization frequency, which are equal to 50 kcal/mol and 10^{13} s^{-1} respectively [10].

We used the Landau-Zener theory [19-21] for the transmission coefficient calculations, which can be estimated from the following equation in the nonadiabatic limit [10, 16]:

$$\kappa_e^{(i)} \approx \frac{2\pi}{\hbar\omega_{eff}} {H_{if}}^2 \sqrt{\frac{\pi}{\lambda kT}} \quad (2)$$

where H_{if} is the electronic coupling matrix element: $H_{if} = \int \psi_i \hat{H} \psi_f d\Omega$. To calculate H_{if} we used method based on the linear correlation between the electronic coupling matrix element and the overlap between corresponding orbitals: $S_{if} = \int \psi_i \psi_f d\Omega$ [22]. Idea of this approach originates from Hückel theory [23].

$$H_{if} \approx CS_{if} \quad (3)$$

Constant C was obtained from the model problem of the electron transfer from benzene to the oxygen molecule, comparing the overlap integral of the benzene HOMO and the oxygen SOMOs and the electronic coupling matrix element calculated within the framework of the generalized Mulliken-Hush (GMH) [24].

Density functional theory (DFT) level was used to obtain the molecular orbitals of graphene cluster and oxygen molecule. We used B3LYP hybrid functional and 6-31G basis set. Configuration of the graphene cluster was built using Atomic Simulation Environment (ASE). Visualization of the electronic density were obtained using VESTA [25].

3. Results and discussion

Since there is a difference between the Fermi level and the level of i-th molecular orbital in exponential part of the equation (1), the HOMO has the highest contribution to the kinetics of heterogeneous electron transfer. The next energy level (HOMO-1) is 0,4 eV lower than the HOMO level. In case of zero overpotenial this leads to a decrease of the contribution from this orbital by 3 orders of magnitude. So, we can ignore all the orbitals below the HOMO when analyzing ET kinetics. Figure 1 depicts the distribution of the electronic density from the HOMO of the graphene cluster. One can see preferential localization of the electronic density on zig-zag edges of graphene.

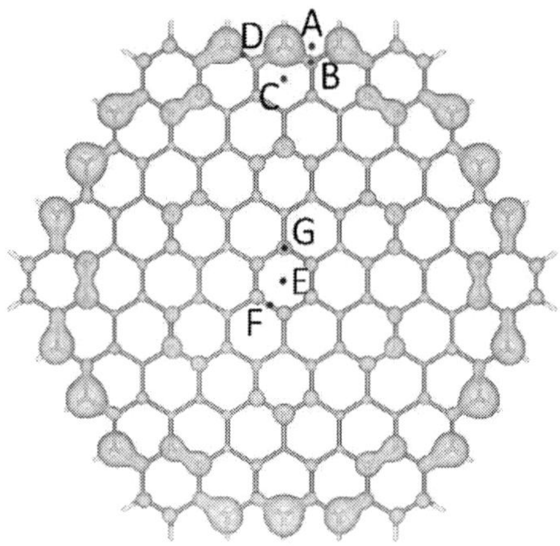

Figure 1. Distribution of the electronic density from HOMO

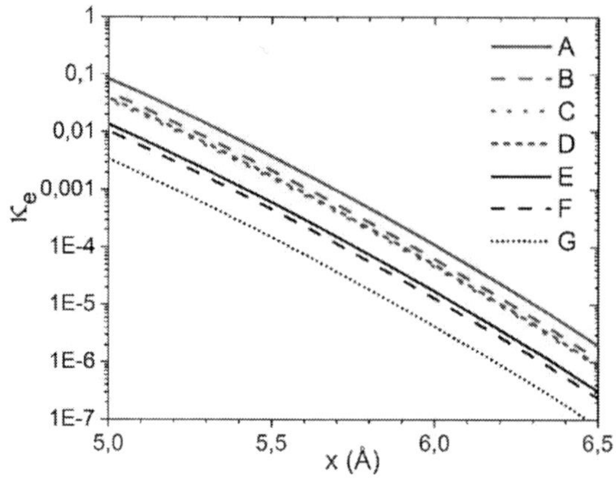

Figure 2. Electronic transmission coefficient as a function of the distance to the graphene surface opposite points A-G depicted in figure 1

Figure 2 shows the electronic transmission coefficient as a function of the distance to the graphene surface opposite the characteristic points, depicted in figure 1. The distance was scanned from 5 to 6.5Å. One can observe the exponential decay of the transmission coefficient with distance, which agrees with previous theoretical studies on other redox systems [10] and experimental observations [26]. Transmission coefficient values are much smaller than 1 indicating that ET occurs in the nonadiabatic regime all over the graphene cluster at such distances. The highest electronic transmission coefficient is observed in point A – the midpoint of the segment connecting the outer carbon atoms of zig-zag graphene edge. The highest transmission coefficient over the basal plane of the graphene cluster is observed in the hollow position (point E). The maximum value at the edge exceeds the maximum value at basal plane sites by 4 times. For remaining points (B, C and D) at the edge, the values of the transmission coefficient also exceed interior values.

Our results clearly support the dominate thinking in this field that zig-zag edges exhibit electrocatalytic properties. This effect is in agreement with many experimental works [3-5]. As for other works where the effect was not confirmed we propose the following possible explanations [6, 7]. It may be associated with 1) variety graphene edges morphology in experiments such as armchair edges, which does not have localized states in contradiction to zig-zag ones [8, 9]; 2) defects on the edges, in the presence of which local states are suppressed [9, 27]; 3) and various functional groups terminating edges, which can also strongly influence the electronic properties of graphene [28].

4. Conclusion

In this work we investigated the effect of zig-zag graphene edge on the kinetics of the heterogeneous electron transfer. It was observed, that the highest occupied molecular orbital has the greatest contribution to the kinetics of electron transfer. Distribution of electronic density from the HOMO of the graphene cluster was calculated. We observed the preferential localization of the electronic density on zig-zag edges of the graphene. The electronic transmission coefficient representing the probability of the electron transfer from the graphene to the oxygen molecule was calculated opposite the characteristic point at the edge and the basal plane sites of the graphene cluster. The 4 times increase in the maximum of transmission coefficient at the edge compared to the basal plane was found. We observed the catalytic effect of the zig-zag edge of graphene for the heterogeneous electron transfer on the oxygen molecule as one of the key stages of the oxygen reduction reaction in fuel cells and metal-air batteries.

Acknowledgements
The work is supported by grant No. 14-50-00124 from the Russian Science Foundation. The research was carried out using supercomputers at Joint Supercomputer Center of the Russian Academy of Sciences (JSCC RAS) and the equipment of the shared research facilities of HPC computing resources at Lomonosov Moscow State University. The authors acknowledge the Supercomputer Centre of JIHT RAS for providing computing time.

References

[1] Raccichini R, Varzi A, Passerini S and Scrosati B 2015 *Nat. Mater.* **14** 271-9
[2] Zhong J, Zhang J, Jin X, Liu J, Li Q, Li M, Cai W, Wu D, Zhan D and Ren B 2014 *J. Am. Chem. Soc.* **136** 16609-17
[3] Yuan W, Zhou Y, Li Y, Li C, Peng H, Zhang J, Liu Z, Dai L and Shi G 2013 *Sci. Rep.* **3** 2248
[4] Banks C, Davies T, Wildgoose G and Compton R 2005 *Chem. Commun.* **0** 829-41
[5] Ambrosi A, Bonanni A and Pumera M 2011 *Nanoscale* **3** 2256-60
[6] Patel A, Collignon M, O'Connell M, Hung W, McKelvey K, Macpherson J and Unwin P 2012 *J. Am. Chem. Soc.* **134** 20117-30
[7] Belova A, Kwabi D, Yashina L, Shao-Horn Y and Itkis D 2017 *J. Phys. Chem.* C **121** 1569-77
[8] Fujita M, Wakabayashi K, Nakada K and Kusakabe K 1996 *J. Phys. Soc. Jpn.* **65** 1920-3
[9] Niimi Y, Matsui T, Kambara H, Tagami K, Tsukada M and Fukuyama H 2006 *Phys. Rev.* B **73** 085421
[10] Nikitina V, Kislenko S, Nazmutdinov R, Bronshtein M and Tsirlina G 2014 *J. Phys. Chem.* C **118** 6151-64
[11] Kislenko S and Pavlov S 2017 *High Energy Chem.* **51** 51-5
[12] Pavlov S and Kislenko S 2018 *J. Phys.: Conf. Ser.* **946** 012028
[13] Pavlov S and Kislenko S 2016 *Phys. Chem. Chem. Phys.* **18** 30830-6
[14] Kislenko S 2018 *J. Phys.: Conf. Ser.* **946** 012029
[15] Dogonadze R, Kuznetsov A and Marsagishvili T 1980 *Electrochim. Acta* **25** 1-28
[16] Kuznetsov A and Ulstrup J 1999 *Electron transfer in chemistry and biology: an introduction to the theory* (John Wiley & Sons Ltd.)
[17] Levich V 1970 *Kinetics of reactions with charge transfer, Physical Chemistry, an Advanced Treatise* vol. Xb, ed H Eyring, D Henderson, and W Jost (New York: Academic Press)
[18] Nazmutdinov R, Tsirlina G, Petrii O, Kharkats Y and Kuznetsov A 2000 *Electrochim. Acta* **45** 3521-36
[19] Landau L 1932 *Phys. Z. Sowjetunion* **2** 46
[20] Landau L 1932 *Phys. Z. Sowjetunion* **1** 88
[21] Zener C 1932 *Proc. R. Soc. London, Ser.* A: **137** 696-702
[22] Gajdos F, Valner S, Hoffmann F, Spencer J, Breuer M, Kubas A, Dupuis M and Blumberger J 2014 *J Chem. Theory Comput.* **10** 4653-60
[23] Hoffmann R 1963 *J. Chem. Phys.* **39** 1397-412
[24] Cave R and Newton M 1996 *Chem. Phys. Lett.* **249** 15-9
[25] Momma K and Izumi F 2011 *J. Appl. Cryst.* **44** 1272-6
[26] Nikitina V, Rudnev A, Tsirlina G and Wandlowski T 2014 *J. Phys. Chem.* C **118** 15970-7
[27] Huang B, Liu F, Wu J, Gu B and Duan W 2008 *Phys. Rev.* B **77** 153411
[28] Kan E, Li Z, Yang J and Hou J 2008 *J. Am. Chem. Soc.* **130** 4224-5

Plasmonic Macroscopic Structures: from linear assemblies to 3D structured super-crystals

Nicolas Pazos-Perez[1,*], and Ramon Alvarez-Puebla[1,2]

[1]Department of Physical Chemistry and Inorganic and EMaS, Universitat Rovira i Virgili, Spain

[2]Institució Catalana de Recerca i Estudis Avançats (ICREA), Spain

*corresponding author, e-mail: nicolas.pazos@urv.cat

Abstract. Confinement controlled drying in combination with spin coating is presented as a novel fast and inexpensive approach to produce organized arrays of plasmonic nanoparticles. The presented method is quite versatile allowing the production of many different structures, from simple linear assemblies with a defined periodicity, width, and high to complex 3D micro-structured super-crystals with define morphology and dimensions which, can be produced either as continuous or as isolated assemblies. Moreover, this methodology is completely up-scalable allowing micro-structuration over large areas in the macro scale. Furthermore, their plasmonic collective behavior can be tuned by controlling the size and shape of the nanoparticles and the structures independently. Obtained organized structures have been proved to be very effective for SERS sensing showing very good reproducibility among big areas.

1. Introduction

Great control over the chemical synthesis of plasmonic nanoparticles and thus, their optical response has been achieved during the last decade. However, for many technological applications,[1-8] their controlled organization over extended areas and the control of their collective behavior is crucial. Long-ranged ordered macroscale materials using metal nanoparticles (NPs) as building blocks[5, 9] are very attractive because they preserve their nanoscale properties, meanwhile new properties arise due to the interaction between their individual constituents.[10-12] Unfortunately, the fabrication of such structures is not trivial due to the small size of the building blocks, and was mostly restricted to expensive and time-consuming lithographic techniques.[13-14] Therefore, bottom-up approaches based on the self-assembly of colloidal particles[12, 15-16] known as super crystals are in the focus of many researches as a promising alternative.[6, 8, 17-18]

In this work, we report novel methods to produce arrays of nanoparticles, either as continuous or as isolated super-crystals. These structures are made of plasmonic nanoparticles in a new fast and inexpensive approach based on the drying confinement of colloidal particles combined with spin coating. Allowing the production of periodical arrays (linear assemblies and super-crystals) of organized nanoparticles over extended areas. Furthermore, this technic allows the production of 3D complex morphologies that can be either connected between them or discontinuous. Additionally, the plasmonic properties of these structures were exploited for the generation of ultrasensitive and ultrafast sensors based on surface-enhanced Raman scattering (SERS) spectroscopy.[12]

2. Results and discussion

One of the crucial points for the successful generation of densely packed and well-ordered three-dimensional super-crystals is the quality of the nanoparticles. They have to be highly pure and monodisperse. Moreover, since a high quantity of NPs are required to form the crystals, they must be stable at very high concentrations meanwhile having low surfactant concentrations which will disturb the crystal structure.

2.1. Assembly process steps

The assembly process can be divided in two main steps: 1) Template production; 2) Nanoparticles confinement.[5, 9, 12]

2.1.1. Template production. Two types of templates were produced either to make linear arrays or super crystals. In both cases, the elastomer polydimethylsiloxane (PDMS) was chosen because their surface properties can be easily changed from hydrophobic to hydrophilic by plasma treatment. *Wrinkled templates:* The cross-linked PDMS elastomer was cut into stripes. These stripes were stretched uniaxially in a custom-made apparatus to a strain of 125% of their initial length. The stretched substrate was oxidized a concrete time in an oxygen plasma. After that, the sample was relaxed and wrinkles with a wanted spacing were obtained (controlling the plasma exposure time, the period of the wrinkles can be tuned). The hydrophilic wrinkled substrates were used directly after plasma treatment (**Figure 1** shows a schematic representation of the process). *Templates with complex morphologies:* Replica molding was used to topographically duplicate Si-masters with various topographies. Two replica molding steps were performed to obtain the negative and positive replicas of the Si-masters. The 1st replicas (negative), were prepared by casting PDMS on the Si-masters and curing them. To Produce the exact same morphologies of the masters (positive replicas), a 2nd replica molding step was performed using the negative replicas as templates. However, to avoid that the 1st PDMS replica cross-link and merge irreversibly with the new PDMS casted, the 1st PDMS replica needs to be silanized before use.

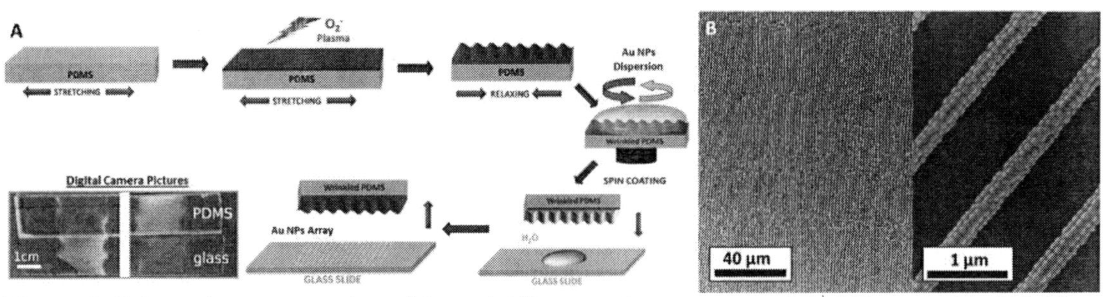

Figure 1. Schematic representation of the wrinkling template process and the NPs assembly (A). Two representative SEM images of Au NPs assembled into periodic lines(B).

2.1.2. Nanoparticles confinement and assembly. Two different but similar approaches were used to organize the NPs. *I) On wrinkled PDMS:* To assembly the NPs into lines, first a highly concentrated drop of the NPs solution was spin coated onto the wrinkled PDMS to confine the particles on the groves. After that, to transfer the particles to a substrate, a drop of water was placed on a glass slide and the wrinkled PDMS was placed on top of the wet surface and left undisturbed until water evaporates. Next, the PDMS was removed leaving the particles organized on the substrate. Remarkably, this method leads to a highly homogeneous patterning over cm^2. **Figure 1** shows a schematic representation of the wrinkling process and particle assembly together with some representative SEM images. *II) On PDMS replicas with complex morphologies:* In order to confine the NPs wetting contrast in a defined region was used. To do that, a PDMS piece with a hole in the middle was placed on the structured surface of the PDMS replica and treated with O_2 plasma. After removing the PDMS on top a hydrophilic region on the hydrophobic PDMS surface was created. Because of the wetting contrast, the NPs droplet remains confined in the hydrophilic region without spreading. After that, the NPs are allowed to sediment. A humidity chamber was used to avoid solvent evaporation. After sedimentation, continuous super crystals can be obtained removing the system from the humidity chamber and letting it dry. If discrete crystals are desired, the solvent containing surfactant is removed via spin-coating and two additional cleaning steps are performed using water to remove the excess of NPs on top. The crystals were then peeled out with adhesive tape and transfer to the desired surface (see **Figure 2**).

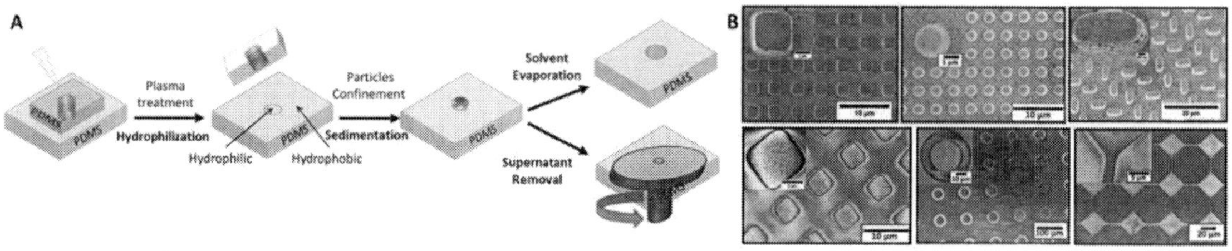

Figure 2. Schematic representation of the NPs assembly process using PDMS replicas with complex morphologies (A). Two representative SEM images of continuous (up) and discrete (down) crystals made of Au NPs(B).

2.2. *SERS enhancing properties*

The colloidal crystals were cleaned with O_2 plasma to remove all residual organics from the metallic surfaces.[19] A Raman active molecule as Benzenethiol (BT) was selected as the Raman probe and deposited on the crystals through gas phase. After characterizing crystals with different morphologies, two major features were found: I) All 3D microstructured crystals exhibit larger intensities than the non-structured ones. II) The absolute intensity increases as the wide of the structures increase independently of the crystal geometry, and the enhancements are similar for similar crystal wide (see **Figure 3**).

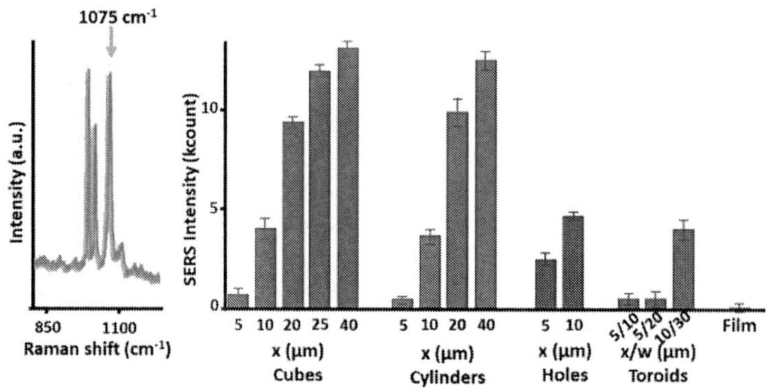

Figure 3. Benzenethiol SERS spectra (left). Comparison of the absolute SERs intensity of 3D super crystals with different morphologies. Showing an increase as the wide (x) of the structures increases for all crystal geometries. Similar enhancements were found for similar wide (x) for different morphologies.

3. Conclusions

In summary, we have developed an innovative cheap, easy, and fast process to produce homogeneous macroscale assemblies of NPs. This technique allows the production of linear assemblies and microstructured super crystals with complex geometries and 3D features over very large areas using plasmonic nanoparticles as building blocks. Permitting also the production of either continuous or discrete structures. The proposed process, the plasma drop confinement is very convenient and innovative, because of its simplicity and versatility for macroscopic areas. The obtained plasmonic super crystals also show a very high and uniform SERS enhancement.

References

[1] R. F. Aroca, R. A. Alvarez-Puebla, N. Pieczonka, S. Sanchez-Cortez, J. V. Garcia-Ramos, *Advances in Colloid and Interface Science* **2005**, *116*, 45-61.

[2] S. Lal, S. Link, N. J. Halas, *Nat Photon* **2007**, *1*, 641-648.

[3] O. D. Velev, S. Gupta, *Advanced Materials* **2009**, *21*, 1897-1905.

[4] M. L. Brongersma, V. M. Shalaev, *Science* **2010**, *328*, 440-441.

[5] N. Pazos-Perez, W. Ni, A. Schweikart, R. A. Alvarez-Puebla, A. Fery, L. M. Liz-Marzan, *Chemical Science* **2010**, *1*, 174-178.

[6] R. A. Alvarez-Puebla, A. Agarwal, P. Manna, B. P. Khanal, P. Aldeanueva-Potel, E. Carbó-Argibay, N. Pazos-Pérez, L. Vigderman, E. R. Zubarev, N. A. Kotov, L. M. Liz-Marzán, *Proceedings of the National Academy of Sciences* **2011**, *108*, 8157-8161.

[7] R. A. Álvarez-Puebla, *The Journal of Physical Chemistry Letters* **2012**, *3*, 857-866.

[8] M. Alba, N. Pazos-Perez, B. Vaz, P. Formentin, M. Tebbe, M. A. Correa-Duarte, P. Granero, J. Ferré-Borrull, R. Alvarez, J. Pallares, A. Fery, A. R. de Lera, L. F. Marsal, R. A. Alvarez-Puebla, *Angewandte Chemie International Edition* **2013**, *52*, 6459-6463.

[9] M. Mueller, M. Tebbe, D. V. Andreeva, M. Karg, R. A. Alvarez Puebla, N. Pazos Perez, A. Fery, *Langmuir* **2012**, *28*, 9168-9173.

[10] O. Hess, J. B. Pendry, S. A. Maier, R. F. Oulton, J. M. Hamm, K. L. Tsakmakidis, *Nat Mater* **2012**, *11*, 573-584.

[11] N. Pazos-Perez, T. Borke, D. V. Andreeva, R. A. Alvarez-Puebla, *Nanoscale* **2011**, *3*, 3265-3268.

[12] M. Tebbe, S. Lentz, L. Guerrini, A. Fery, R. A. Alvarez-Puebla, N. Pazos-Perez, *Nanoscale* **2016**, *8*, 12702-12709.

[13] C. M. Soukoulis, M. Wegener, *Nat Photon* **2011**, *5*, 523-530.

[14] O. Benson, *Nature* **2011**, *480*, 193-199.

[15] O. D. Velev, A. M. Lenhoff, *Current Opinion in Colloid & Interface Science* **2000**, *5*, 56-63.

[16] C. Burda, X. Chen, R. Narayanan, M. A. El-Sayed, *Chemical Reviews* **2005**, *105*, 1025-1102.

[17] N. Pazos-Perez, F. J. Garcia de Abajo, A. Fery, R. A. Alvarez-Puebla, *Langmuir* **2012**, *28*, 8909-8914.

[18] M. Tebbe, P. Cherepanov, E. V. Skorb, S. K. Poznyak, J. G. d. Abajo, A. Fery, D. V. Andreeva, R. A. A. Puebla, N. Pazos-Perez, *Particle & Particle Systems Characterization* **2014**, *31*, 1134-1140.

[19] M. Tebbe, M. Maennel, A. Fery, N. Pazos-Perez, R. A. Alvarez-Puebla, *The Journal of Physical Chemistry C* **2014**, *118*, 28095-28100.

IOP Conf. Series: Journal of Physics: Conf. Series **1092** (2018) 012114 doi:10.1088/1742-6596/1092/1/012114

Effect of high-frequency pumping on thin-film topological insulators

A A Pervishko[1], D Yudin[1], and I A Shelykh[1,2]

[1] Department of Nanophotonics and Metamaterials, ITMO University, Kronverksky Pr. 49, St. Petersburg 197101, Russia

[2] Science Institute, University of Iceland, Dunhagi-3, Reykjavik IS-107, Iceland

E-mail: anas0005@e.ntu.edu.sg

Abstract. Thin-film topological insulators (TIs) with the thickness below 10 nm are characterized by the presence of a gap in the spectrum of topologically protected surface states. This unusual behavior, also known as anomalous finite size effects, is associated with the hybridization between the states propagating along the opposite boundaries of thin-slab TI. In this work we consider a bismuth-based TI and show how an intense high-frequency linearly polarized light can be used for the tuning of the value of a gap. We also theoretically predict the effect of band inversion in the spectrum of the surface states under external electromagnetic pumping.

1. Introduction

The interest towards the studying of topological matter represents the mainstream in modern condensed matter physics [1, 2, 3]. Among such systems are the recently predicted [4, 5, 6] and experimentally realized [7, 8] three-dimensional topological insulator (TI), a structure which behaves as an insulator in the bulk but has gapless conducting electronic modes at the two-dimensional boundaries. Despite enormous progress TIs still remain an extremely active area of research due to their unique properties and potential applications [9, 10]. As an example of a strong three-dimensional TI one can consider Bi_2Se_3 that has the gapful spectrum of surface states, if the thickness of a structure is below 10 nm due to the interference between top and bottom states and gapless for larger thicknesses [11, 12, 13].

The idea of utilizing off-resonant pumping to drive low-dimensional electronic systems to non-trivial topological phases has recently been pushed forward [14, 15, 16]. Nevertheless, almost no attention was paid to the idea of using high-frequency electromagnetic radiation to the finite-size systems despite their potential impact on the realization of all-optical control in topological systems. In this paper we show that high-frequency pumping leads to renormalization of the bare Hamiltonian of a three-dimensional TI by studying as an example a thin slab of Bi_2Se_3 within the paradigm of Brillouin-Wigner perturbation theory [15], and demonstrate that the anomalous finite size effects in the spectrum of surface states are sensitive to the intensity of the driving.

2. Effective Hamiltonian

We introduce the coordinate system so that the z axis to be normal to the surface plain of the TI. The $\mathbf{k}\cdot\mathbf{p}$ Hamiltonian of Bi_2Se_3 in the vicinity of the Γ−point with no external pumping can be written as

$$H = \mathcal{E}_{\mathbf{k}} + \begin{pmatrix} \mathcal{M}_{\mathbf{k}} & A_1 k_z & 0 & A_2 k_- \\ A_1 k_z & -\mathcal{M}_{\mathbf{k}} & A_2 k_- & 0 \\ 0 & A_2 k_+ & \mathcal{M}_{\mathbf{k}} & -A_1 k_z \\ A_2 k_+ & 0 & -A_1 k_z & -\mathcal{M}_{\mathbf{k}} \end{pmatrix}, \qquad (1)$$

where the three-dimensional wave vector $\mathbf{k} = (k_x, k_y, k_z)$ specifies the electron states, $\mathcal{E}_{\mathbf{k}} = C + D_1 k_z^2 + D_2 k_\perp^2$, $\mathcal{M}_{\mathbf{k}} = B_0 - B_1 k_z^2 - B_2 k_\perp^2$, with $k_\pm = k_x \pm i k_y$ and $k_\perp^2 = k_x^2 + k_y^2$, while the parameters A_1, A_2, B_0, B_1, B_2, D_1, and C have to be fit to the results of ab-initio simulations [8].

Assume that the TI is pumped with a time-periodic electromagnetic field $\mathbf{E} = \mathbf{E}_0 \sin\omega t$ with the amplitude E_0 and the frequency ω propagating along the z axis. Restricting to linearly polarized electromagnetic field (along y direction) we take account of the external field by virtue of Peierls substitution $k_y \to k_y - eE_0\cos(\omega t)/(\hbar\omega)$. To find the renormalized parameters of the Hamiltonian (1) we use the high-frequency expansion in the form of Brillouin-Wigner perturbation theory. Performing the expansion up to terms proportional to $1/\omega^2$ we arrive at the effective Hamiltonian

$$H_{\text{eff}} = \mathcal{E}_{\mathbf{k}} + \begin{pmatrix} \tilde{\mathcal{M}}_{\mathbf{k}}\hat{\tau}_z + \tilde{A}_1 k_z \hat{\tau}_x & (\tilde{A}_{2x}k_x - i\tilde{A}_{2y}k_y)\hat{\tau}_x \\ (\tilde{A}_{2x}k_x + i\tilde{A}_{2y}k_y)\hat{\tau}_x & \tilde{\mathcal{M}}_{\mathbf{k}}\hat{\tau}_z - \tilde{A}_1 k_z \hat{\tau}_x \end{pmatrix} \qquad (2)$$

with $\tilde{A}_1 = A_1(1-a-b)$, $\tilde{A}_{2x} = A_2(1-a-b)$, $\tilde{A}_{2y} = A_2\left(1 - 2aB_0B_2/A_2^2 + 7b\right)$, and $\tilde{\mathcal{M}}_{\mathbf{k}} = \tilde{M} - \tilde{B}_1 k_z^2 - \tilde{B}_2 k_\perp^2 - 2aB_2k_y^2$, where we have set $a = [A_2\gamma/(\hbar\omega)]^2$, $b = \left[B_2\gamma^2/(4\hbar\omega)\right]^2$, $\gamma = eE_0/(\hbar\omega)$, $\tilde{B}_i = (1-a)B_i$, and $\tilde{M} = \tilde{B}_0 - (2-a)B_2\gamma^2/4$. In the expression (2) we have also defined a set of Pauli matrices $\hat{\tau}_x$ and $\hat{\tau}_z$. As one can clearly see the effect of renormalization in (2) brings is a certain anisotropy to the bare Hamiltonian (1).

3. Band structure of surface states

Consider a three-dimensional TI of the thickness L in the z direction, yet being infinite along x and y. To calculate the spectrum of surface states of the TI subjected to high-frequency pumping one has to solve the Schrödinger equation with the renormalized Hamiltonian (2), on condition that boundary conditions $\psi(x, y; z = 0, L) = 0$ are imposed, meaning that the wave function should be zero at the surface of the TI. Solving the eigenvalue problem with the Hamiltonian (2) we arrive at the dispersion relation in the form

$$\sum_{\pm}\left[\frac{\tanh(\lambda_\pm L/2)}{\tanh(\lambda_\mp L/2)} - \frac{\lambda_\pm}{\lambda_\mp}\left(1 \pm \frac{D_1^2 - B_1^2}{D_1^2 - \tilde{B}_1^2}\frac{\sqrt{R}}{\tilde{A}_1^2}\right)\right] = 0 \text{ with } \lambda_\pm = \left[\frac{F \pm \sqrt{R}}{2(D_1^2 - \tilde{B}_1^2)}\right]^{1/2}, \qquad (3)$$

where $F = (l_+ + l_-)D_1 + (l_+ - l_-)\tilde{B}_1 - \tilde{A}_1^2$, $R = F^2 - 4(D_1^2 - \tilde{B}_1^2)[l_+l_- - (\tilde{A}_{2x}k_x)^2 - (\tilde{A}_{2y}k_y)^2]$, and $l_\pm = C \pm \tilde{\mathcal{M}}_{\mathbf{k}} \mp \tilde{B}_2 k_\perp^2 - \varepsilon$.

The expression (3) implicitly determines the spectrum of the surface states $\varepsilon_{\mathbf{k}}(L)$ as a function of the layer thickness L. Solving this equation numerically we find that an external field allows to reduce the value of the energy gap. To model the properties of Bi_2Se_3 we utilize the following parameters [8]: $C = 6.8 \times 10^{-3}$ eV, $B_0 = 0.28$ eV, $A_1 = 2.2$ eV Å, $A_2 = 4.1$ eV Å, $B_1 = 10$ eV Å^2, $B_2 = 56.6$ eV Å^2, $D_1 = 1.3$ eV Å^2, and $D_2 = 19.6$ eV Å^2. The phase diagram that shows an interplay between the thickness of a thin-film TI and the intensity of the driving field in terms of the gap value is plotted in Fig. 1.

As one can see the external electromagnetic field not only leads to the suppression of anomalous finite size effects, but also may be used to induce band inversion in the spectrum of surface states. In particular, for a given layer thickness, slightly below 7 nm, the system undergoes topological phase transition along the black line in Fig. 1 where the gap collapses from topologically non-trivial phase (yellow area) to topologically trivial state (blue area).

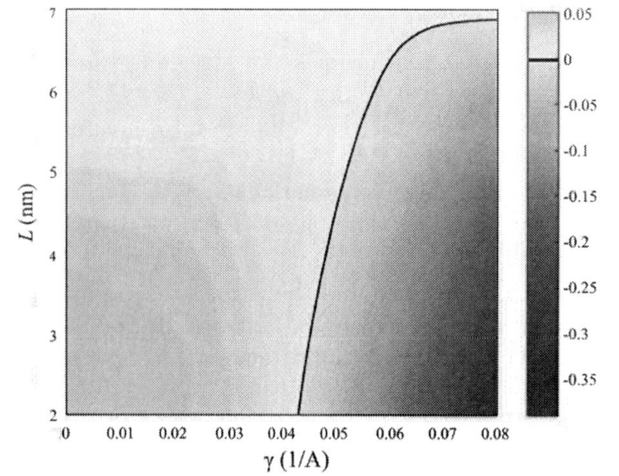

Figure 1. Phase diagram showing how the value of the gap at the Γ−point in the spectrum of surface states of TI depends on parameters of the pumping. For a given layer thickness one can notice the tendency towards band inversion induced by light upon increasing field intensity (blue area on the right) and passing through gapless region marked with a black solid line.

4. Conclusions

We have investigated anomalous finite size effects in a thin-film TI subjected to an intense off-resonant linearly polarized electromagnetic radiation. The high-frequency expansion reveals that the gap in the spectrum of surface states is quite sensitive to the parameters of the field, and can be suppressed upon increasing the intensity of the field within experimentally accessible range. Moreover, a further increase of the intensity can also lead to the light-induced band inversion.

Acknowledgments

The research leading to these results received funding from the Russian Science Foundation under Project 17-12-01359.

References

[1] Thouless D J, Kohmoto M, Nightingale M P and den Nijs M 1982 *Phys. Rev. Lett.* **49** 405
[2] Wen X-G 1995 *Adv. Phys.* **44** 405
[3] Hasan M Z and Kane C L 2010 *Rev. Mod. Phys.* **82** 3045
[4] Kane C L and Mele E J 2005 *Phys. Rev. Lett.* **95** 146802
[5] Fu L and Kane C L 2007 *Phys. Rev. B* **76** 045302
[6] Roy R 2009 *Phys. Rev. B* **79** 195322
[7] Xia Y, Qian D, Hsieh D, Wray L, Pal A, Lin H, Bansil A, Grauer D, Hor Y S, Cava R J and Hasan M Z 2009 *Nature Phys.* **5** 398
[8] Zhang Y, Liu C-X, Qi X-L, Dai X, Fang Z and Zhang S-C 2009 *Nature Phys.* **5** 438
[9] Vobornik I, Manju U, Fujii J, Borgatti F, Torelli P, Krizmancic D, Hor Y S, Cava R J and Panaccione G 2011 *Nano Lett.* **11** 4079
[10] Pesin D and MacDonald A H 2012 *Nature Mater.* **11** 409
[11] Linder J, Yokoyama T and SudbøA 2009 *Phys. Rev. B* **80** 205401

[12] Zhang Y, He K, Chang C-Z, Song C-L, Wang L-L, Chen X, Jia J-F, Fang Z, Dai X, Shan W-Y et al 2010 *Nature Phys.* **6** 584
[13] Lu H-Z, Shan W-Y, Yao W, Niu Q and Shen S-Q 2010 *Phys. Rev. B* **81** 115407
[14] Kitagawa T, Oka T, Brataas A, Fu L and Demler E 2011 *Phys. Rev. B* **84** 235108
[15] Mikami T, Kitamura S, Yasuda K, Tsuji N, Oka T and Aoki H 2016 *Phys. Rev. B* **93** 144307
[16] Hasan M, Yudin D, Iorsh I, Eriksson O and Shelykh I 2017 *Phys. Rev. B* **96** 205127

Electron hydrodynamic instabilities stimulated by asymmetry

Aleksandr S. Petrov **and D. Svintsov**

Laboratory of 2D Materials' Optoelectronics, Moscow Institute of Physics and Technology, Dolgoprudny 141700, Russia

E-mail: aleksandr.petrov@phystech.edu

Abstract. Direct current in confined two-dimensional (2d) electron systems can become unstable with respect to the excitation of plasmons. Numerous experiments and simulations hint that structural asymmetry somehow promotes plasmon generation, but a constitutive relation between asymmetry and instability has been missing. We provide such relation in the present paper and show that bounded perfect 2d electron fluids in asymmetric structures are unstable under arbitrarily weak drive currents. To this end, we develop a perturbation theory for hydrodynamic plasmons and evaluate corrections to their eigenfrequency induced by carrier drift, scattering, and viscosity. We show that plasmon gain continuously increases with degree of plasmon mode asymmetry until it surrenders to viscous dissipation that also benefits from asymmetry.

1. General overview

In high-mobility semiconductors, the electrons at room temperature demonstrate a fluid-like (hydrodynamic) behaviour [1]. Several remarkable manifestations of electron hydrodynamics include negative resistance due to viscous flow [2], higher-than-ballistic conduction of constrictions [3], and strong violations of Wiedemann-Franz law [4]. An intriguing yet weakly explored area of electron hydrodynamics is related to turbulence, i.e. instabilities of electron drift. A strong effort in this field was triggered by the statement that HD electron flow in a confined gated two-dimensional electron system (2DES) is unstable with respect to the self-excitation of plasmons under certain boundary conditions, now known as Dyakonov-Shur (DS) instability [5]. An observable outcome of instability in confined 2DES is the electromagnetic emission due to radiative plasmon decay [6].

The boundary conditions providing instability in original work of DS assumed zero alternating current at the drain, which is hardly fulfilled in real transistors. A successful demonstration of instability-induced THz emission in an ordinary HEMT [7] in spite of boundary condition problem stimulated dozens of instability proposals in related systems, including ungated [8] and partly gated transistors [9], Corbino discs [10], and edges of 2DES [11]. Both experiments [12, 13] and theories [9, 14] for numerous plasmonic structures hinted that it is structural asymmetry that drives the instability, but not a specific boundary condition. However, the link between asymmetry and instability remained elusive so far, contrary to the well-established symmetry constraint for the inverse process, photodetection.

In this paper, we provide a constitutive relation between asymmetry of confined 2DES-based plasmonic resonators and stability of their plasmon modes. To this end, we develop an analogue of quantum-mechanical perturbation theory for hydrodynamic equations in confined 2DES and express the growth/decay rates of plasmons through their field distributions in the absence of drift and dissipation. We show that the necessary condition for instability is the asymmetry of plasmon field: the mode gain turns to zero for even and odd field distributions, but is generally nonzero for modes without certain parity. The asymmetry in confined 2DES can be induced by inequivalence of source and drain contacts (which is the case of famous DS instability), asymmetry of gating environment, non-uniform doping, or all of them. The developed theory can be used for design of plasmonic oscillators with high gain and/or low threshold current.

2. Perturbation theory for electron hydrodynamics

Within this section, we shortly discuss the main points of our theory and its possible applications. A more detailed analysis can be found in [15].

Rewriting linearized hydrodynamic equations of a 2d charged fluid in an operator representation we managed to construct an analogue of the quantum-mechanical perturbation theory for an arbitrary transistor structure. It leads to the following expression for the plasmon eigenfrequency correction $\delta\Omega$:

$$\delta\Omega_\lambda = \frac{i}{\Pi}\left\{j_0\left[K(L) - K(0)\right] - Q_{loss}\right\}, \tag{1}$$

where $K(x) = mu_\lambda(x)^2/2$ is the local kinetic energy in a plasmon mode, u_λ is the velocity variation, m – electron effective mass, λ enumerates plasmonic modes,

$$\Pi = e^2 \int\limits_0^L \mathrm{d}x\mathrm{d}x' n_\lambda(x) G(x,x') n_\lambda(x') \tag{2}$$

is the potential energy of interacting charge density fluctuations, n_λ stands for electron density variation, G is the non-local Green's function of the electrode system, the integration is performed over the transistor channel of length L and

$$Q_{loss} = \frac{1}{2}\int\limits_0^L \mathrm{d}x \left\{ \frac{e^2 n_0}{m\Omega^2 \tau_\mathrm{p}} E_\lambda^2 - u_\lambda \partial_x [\eta \partial_x u_\lambda] \right\} \tag{3}$$

is the energy loss due to viscous friction and momentum non-conserving scattering, where e denotes the elementary charge, $n_0(x)$ – electron density profile in the channel, Ω – plasmon eigenfrequency, τ_p – momentum scattering rate, E_λ – electric field in the channel, η – dynamic viscosity of the electron fluid.

Equation (1) is the main point of our theory. We readily observe that the origin of plasmon growth, $\mathrm{Im}\,\delta\Omega > 0$, is the excess of kinetic energy entering the mode at source over the energy lost at the drain. The necessity for asymmetry to achieve wave growth now becomes apparent. Indeed, for zero-order functions with certain parity $u^2(L) = u^2(0)$, and only the loss term retains in (1). For modes without parity, the compensation of ingoing and outgoing energy can appear only accidentally. Therefore, the plasmon modes of any asymmetric nanostructure with perfect electron fluid ($\eta = 0$, $\tau_{sc}^{-1} = 0$) are unstable with respect to arbitrary weak drive current. This property hallmarks the instabilities of confined hydrodynamic plasmons from Smith-Purcell [16] and beam [17] instabilities in extended systems. The latter typically develop above the threshold drift velocity order of plasmon phase velocity.

The need for asymmetry becomes even more apparent from Fig.1, top, where we show the dependencies of plasmon growth rate for a partly gated transistor structure (shown on the inset) at different gate lengths and density ratios. The orange line shows the growth rates of

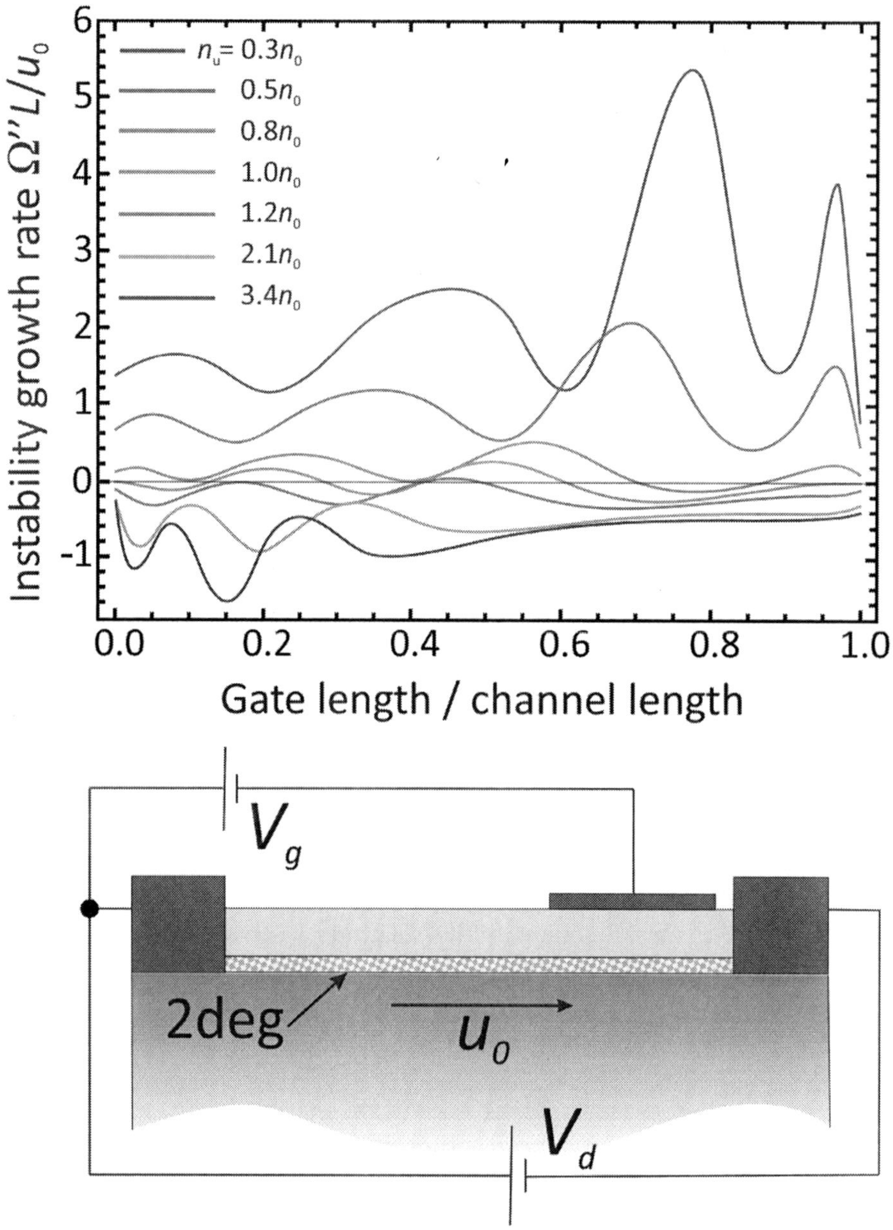

Figure 1. Top: calculated instability growth rates for the third plasmon mode in a partly gated FET (shown in inset) with different gate lengths and carrier densities. The growth rates are normalized by u_0/L, where u_0 is the drift velocity at the drain (same for all curves), the density at the drain n_0 is also fixed. The instability benefits if the drift is directed from low- to high-density region and is especially pronounced if low-density region is ungated. A structure with uniform density (orange line) also supports instabilities due to the asymmetry of electrical environment. Bottom: spatial distribution of plasmon potential in partly gated FET for two modes: (1) with almost symmetric in-plane distribution of potential and zero gain (2) with highly asymmetric in-plane potential and high gain.

plasmon modes in FET with *uniform* carrier density and metal gate adjacent to drain side. It demonstrates an oscillatory behaviour with varying gate length with a relatively small maximum growth rate $\sim u_0/L$, where u_0 is the drift velocity in the ungated region. Addition of extra density asymmetry, $n_g \neq n_u$, results in much more pronounced instability (for $n_g > n_u$) or more pronounced current-induced plasmon decay (for $n_g < n_u$).

Discussion

With relation (1), it becomes possible to study the 2d plasmon instabilities without going into complicated simulations of turbulent flows [14]. All necessary ingredients to judge on the possibility of gain are the plasmon mode profiles $\varphi_0(x)$ ("zero-order functions") that can be obtained with commercial electromagnetic simulators. Our perturbative expressions for plasmon eigen frequency allowed to reveal the competition between current-driven gain and viscosity both of which increase with the degree of asymmetry. Moreover, they allowed us to find an "optimal degree of asymmetry" for which the plasmon gain is highest or instability threshold current is the lowest. There exists a universal lower bound of Reynolds number for development of plasmon instability in 2DES with fixed voltage drop between source and drain, which equals $\mathrm{R}_{\min} = 2\sqrt{3}$.

References

[1] Gurzhi R 1968 *Physics-Uspekhi* **11** 255–270
[2] Bandurin D A, Torre I, Kumar R K, Ben Shalom M, Tomadin A, Principi A, Auton G H, Khestanova E, Novoselov K S, Grigorieva I V, Ponomarenko L A, Geim A K, Polini M 2016 *Science* **351** 1055–1058
[3] Kumar R K, Bandurin D A, Pellegrino F M D, Cao Y, Principi A, Guo H, Auton G H, Ben Shalom M, Ponomarenko L A, Falkovich G, Watanabe K, Taniguchi T, Grigorieva I V, Levitov L S, Polinin M and Geim A K 2017 *Nat. Phys.* **13** 1182–1185
[4] Crossno J, Shi J K, Wang K, Liu X, Harzheim A, Lucas A, Sachdev S, Kim P, Taniguchi T, Watanabe K, Ohki T A, Fong K C 2016 *Science* **351** 1058–1061
[5] Dyakonov M and Shur M S 1993 *Phys. Rev. Lett.* **71**, 2465
[6] Chaplik A 1985 *Surf. Sci. Rep.* **5** 289
[7] Knap W, Lusakowski J, Parenty T, Bollaert S, Cappy A, Popov V and M. Shur 2004 *Appl. Phys. Lett.* **84** 2331
[8] Dyakonov M and Shur M S 2005 *Appl. Phys. Lett.* **87** 111501
[9] Petrov A S, Svintsov D, Rudenko M, Ryzhii V and Shur M S 2016 *Int. Jour. High Speed Electron. Syst.* **25** 1640015
[10] Sydoruk O, Syms R and Solymar L 2010 *Appl. Phys. Lett.* **97** 263504
[11] Dyakonov M 2008 *Semicond.* **42** 984
[12] El Fatimy A, Dyakonova N, Meziani Y, Otsuji T, Knap W, Vandenbrouk S, Madjour K, Theron D, Gaquiere C, Poisson M, Delage S, Prystawko P and Skierbiszewski C 2010 *J. Appl.Phys.* **107** 024504
[13] Boubanga-Tombet S, Yadav D, Knap W, Popov V V and Otsuji T 2018 Plasmonic instabilities and terahertz waves amplification in graphene metamaterials *arXiv preprint* arXiv:1801.04518
[14] Koseki Y, Ryzhii V, Otsuji T, Popov V V and Satou A 2016 *Phys. Rev.* B **93** 245408
[15] Petrov A S, Svintsov D 2018 *arXiv preprint* arXiv:1802.03994
[16] Mikhailov S 1998 *Phys. Rev.* B **58** 1517
[17] Krasheninnikov M and Chaplik A 1980 *Sov. Phys. JETP* **79** 555

Direct imaging of isofrequency contours in all-dielectric optical metasurface

Pidgayko Dmitry[1], Ivan Sinev[1], Dmitry Permyakov[1], Anton Samusev[1], Stanislav Sychev[1], Andrey Bogdanov[1], Andrey Lavrinenko[2], Viktoriia Rutckaia[3], Joerg Schilling [3]

[1] Department of Nanophotonics and Metamaterials, ITMO University, 49, Kronverksky pr., 197101, Saint-Petersburg, Russia
[2] Department of Photonics Engineering, Technical University of Denmark, 2800 Kongens Lyngby, Denmark
[3]ZIK SiLi-nano, Karl-Freiherr-von-Fritsch-Strasse 3 D - 06120 Halle (Saale), Germany

E-mail: d.pidgayko@metalab.ifmo.ru

Abstract. We study the properties of optical surface waves in all–dielectric anisotropic metasurface represented by an array of silicon slabs. We employ back focal plane microscope with solid immersion lens for direct imaging of isofrequency contours of surface waves of the metasurface to reveal both elliptic and hyperbolic–like regimes and reconstruct the dispersion curves for transverse and longitudinal directions.

1. Introduction

Metasurfaces are two–dimensional subwavelength–structured media that exhibit unusual optical properties and offer control over light propagation, reflection, and refraction unachievable in conventional media. In comparison with bulk metamaterials, metasurfaces are fully compatible with modern planar fabrication technology, preserving most of the functionalities of three–dimensional metamaterials.

Currently, most of the research devoted to metasurfaces is focused on their interaction with far field. This includes different phenomena, such as negative refraction, manipulation over wave front, phase and polarization of light etc. [1] Strongly localized surface states of metasurfaces deserve special interest, since dispersion law of such waves lays under the light line, so they propagate along the structure without leaking to the far field. It makes metasurfaces very promising for light manipulation in modern optoelectronic and alloptical circuits.

Direct experimental observation of unusual regimes of surface waves propagation in metasurfaces has been demonstrated in only a few papers. In particular, positive-to-negative refraction transition, diffractionless propagation and other effects were demonstrated in visible–frequency plasmonic metasurface. [2] Hyperbolic dispersion of surface waves wavefront was achieved by applying a nanoimaging technique to a 2D–material–based metasurface in midinfrared region. [3] Both the shape of the wavefronts and the propagation direction of surface waves are determined by group velocity $\mathbf{v}_g = \mathrm{grad}_{\mathbf{k}}\omega$, which is aligned normal to the isofrequency contour. Therefore, a method for direct imaging of equal–frequency contours of surface waves would be an important tool for experimental studying of surface waves. On the other hand, since plasmonic structures have high intrinsic losses, which imposes significant restrictions on the propagation length of localized modes, employment of the all–dielectric platform for engineering the dispersion of surface waves in the visible and near–infrared regions is preferable.

In our work, we propose all-dielectric metasurface with elliptic and hyperboliclike equal–frequency contours of surface waves in the visible to near–infrared spectral region. We resort to Back Focal Plane microscopy (BFP) with Solid Immersion Lens (SIL) in order to directly image the isofrequency contours. Spectral dependence of isofrequency contours obtained experimentally allows fast extraction of the dispersion law of surface waves in arbitrary direction.

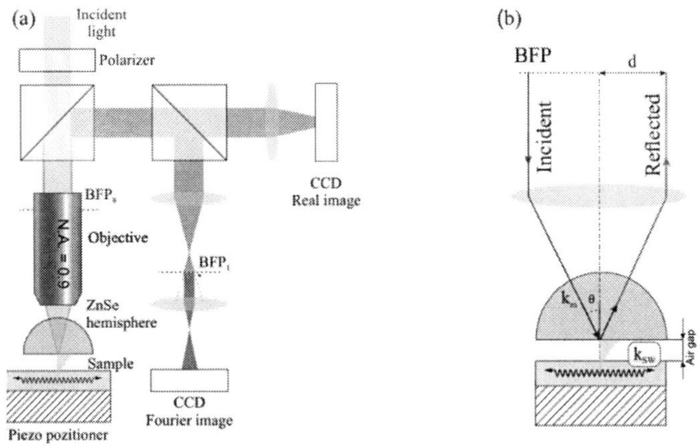

Figure 1. (a) Layout of the SIL–BFP setup.(b) Surface wave excitation and BFP imaging scheme. Surface waves are excited via frustrated total internal reflection in Otto geometry. In BFP of the objective distance d is proportional to k_{lat}/k_0 therefore reflectivity map in k - space is observed.

2. Methods

2.1. Experimental setup

Since the dispersion law of surface waves resides under the light line, special techniques are necessary to excite such states. For this purpose we apply Otto geometry with SIL based on ZnSe hemisphere (Fig.1 a). In the setup, the light passes through the objective and focuses on the flat surface of SIL. [4] At a certain angle and wavelength the lateral wave vector of the incident light coincides with the wave vector of surface wave which leads to its excitation via frustrated total internal reflection (Fig.1 b). The surface wave excitation is manifested as a dip in the reflectivity map for the respective k_x, k_y point in BFP of the objective. The full reflectivity map obtained in SIL–BFP setup for particular excitation wavelength thus illustrates the isofrequency contours of the surface waves excited in the sample. Maximum available angles and hence the wave vectors to be excited are determined by refractive index of SIL and numerical aperture of the objective: for ZnSe hemisphere refractive index is n $\approx$ 2.5 and N.A. = 0.9 for our objective (Mitutoyo, M Plan Apo HR, 100, 0.9 NA). The excitation efficiency depends on the air gap which is controlled with high accuracy by piezopositioner.

2.2. Metasurface design and simulation method

We propose a simple design of all–dielectric anisotropic metasurface. (Fig 2 a,b) Metasurface parameters are chosen in such a way that the target isofreqency contour regimes (elliptic and hyperbolic-like) are observed in 650 — 1100 nm range. Due to anisotropy, the optical properties of surface waves depend on the polarization of indecent light (transverse and parallel to bars).

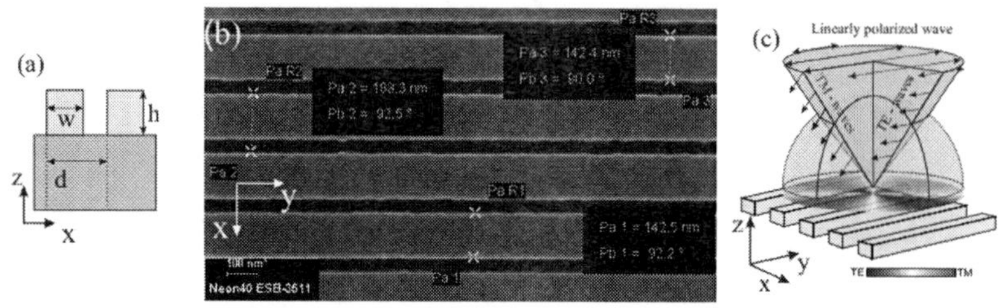

Figure 2. (a) Layout of all - dielectric the metasurface. We set w = 140 nm, d = 190 nm, h = 75 nm. Bar material is c - Si, substrate material is SiO_2 (b) SEM - image of metasurface. (c) Distribution of polarization in experiment. For our metasurface incident light polarization can be directed parallel or perpendicular to the bars.

To simulate the metasurface reflectivity in Otto geometry with different wavelength and polarization we used Fourier Modal Method (FMM). In particular, FMM makes it possible to obtain k – space reflectivity at BFP.

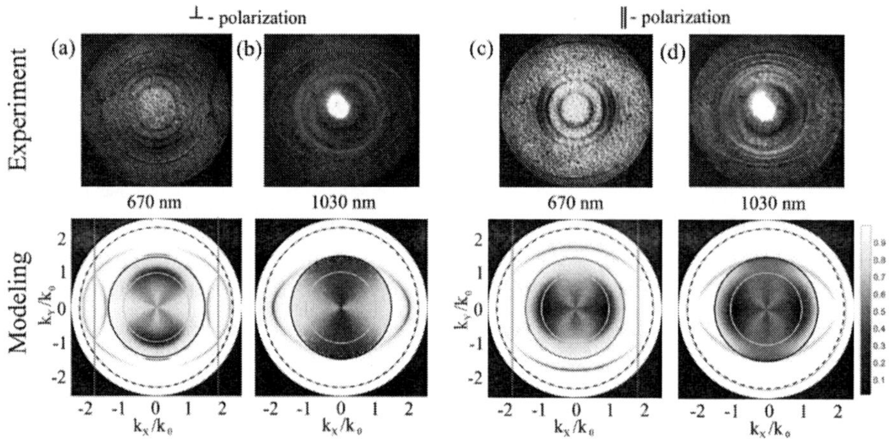

Figure 3. (a-d) Experimental and modeled images for metasurface reflectivity in BFP for transverse (a, b) and longitudinal (c, d) polarizations. In modeled images gray and black circles correspond to lateral wave vectors in free space and in substrate respectively, dashed curve meet the maximum wave vectors $n_{ZnSe}\cdot$ N.A. In(a, c) vertical blue lines means first Brillouin zone boundary. For transverse polarization hyperbolic - like (a) and elliptic (b) contours are observed. For longitudinal polarization(c, d) only elliptic curves are observed.

3. Results and discussion

The metasurface composed of an array of silicon slabs has been fabricated on a SOI substrate used focused ion beam milling.

Both in numerical simulations and experiment we observe that our metasurface can support substantially different regimes of surface wave depending on the wavelength and polarization of incident beam. For polarization parallel to the slabs, only the waves with elliptic isofrequency contours are excited, see Fig 3 (c, d). On the contrary, for transverse polarization we observe both hyperbolic–like and elliptic isofrequency contours of surface waves (Fig 3 a, b).

The origin of hyperbolic–like regime can be explained as follows. With the decrease of the wavelength elliptic curve crosses first Brillouin zone (BZ) boundary (blue vertical lines in Fig.

3 a, Modelling) and is translated according to the Bloch theorem from the second BZ to the first BZ forming hyperbolic–like contour characterized by negative curvature of the isofrequency contour.

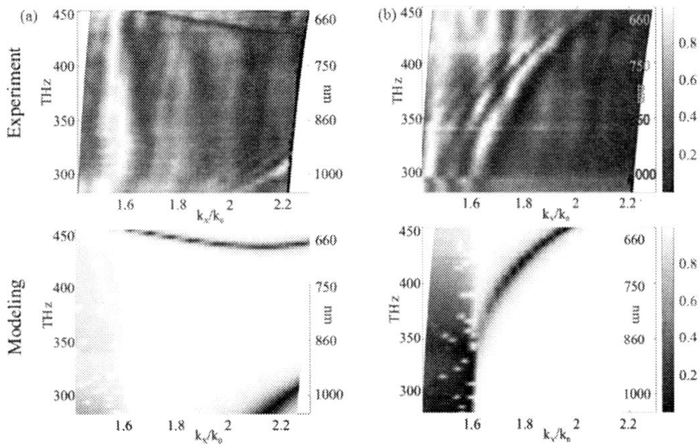

Figure 4. (a, b) Restored dispersion for metasurface. (a) In x direction wide band gap is observed. (b) In y direction ordinary dispersion curve is observed.

To determine wavelength range for all regimes and to reconstruct the dispersion curves for surface waves in longitudinal and transverse directions we measure the spectral dependence of SIL–BFP maps by tuning the excitation wavelength.

As a result, in longitudinal direction we see ordinary dispersion curve (Fig. 4 b, d), whilst in transverse direction we observe a wide band gap for surface waves (Fig. 4 a, c).

4. Conclusion

We have studied optical properties of surface waves supported by all–dielectric anisotropic metasurface represented by an array of silicon slabs. We resorted to BFP–SIL method to directly image isofrequency contours of surface waves in metasurface at different wavelengths and polarization of incident beam. We have shown that such a metasurface can support elliptic and hyperbolic–like regimes of isofrequency contours. The whole set of experimental data allows to reconstruct the dispersion curves of surface waves propagating in arbitrary direction.

5. Acknowledgements

This work was supported by the Ministry of Education and Science of the Russian Federation (Zadanie No.3.1668.2017/4.6, No.3.8891.2017/8.9).

References

[1] Nanfang Yu and Federico Capasso. Flat optics with designer metasurfaces. *Nature materials*, 13:139150, 2014.

[2] Alexander A. High, Robert C. Devlin, Alan Dibos, Mark Polking, Dominik S. Wild, Janos Perczel, Nathalie P. de Leon, Mikhail D. Lukin, and Hongkun Park. Visible-frequency hyperbolic metasurface. *Nature*, 522:192196, 2015.

[3] Peining Li, Irene Dolado, Francisco Javier Alfaro-Mozaz, Flix Casanova, Luis E. Hueso, Song Liu, James H. Edgar, Alexey Y. Nikitin, Sal Vlez, and Rainer Hillenbrand. Infrared hyperbolic metasurface based on nanostructured van der waals materials. *Science*.

[4] D. V. Permyakov, I. S. SinevS., K. Sychev, A. S. Gudovskikh, A. A. Bogdanov, A. V. Lavrinenko, and A. K. Samusev. Visualization of isofrequency contours of strongly localized waveguide modes in planar dielectric structures. *JETP Letters*, 107:1014, 2018.

Formation on magnesium alloy MA8 bioactive coatings containing nanosized hydroxyapatite

A B Podgorbunsky[1], K V Nadaraia[1, 2], I M Imshinetsky[1], S L Sinebryukhov[1] and S V Gnedenkov[1, 2]

[1] Institute of Chemistry FEB RAS, 159 Pr. 100-letiya Vladivostoka, Vladivostok 690022, Russia
[2] Far Eastern Federal University, 8 Sukhanova St., Vladivostok 690091, Russia

Corresponding author: pab@ich.dvo.ru

Abstract. Calcium hydroxyapatite (HAp) in the form of rounded nanoparticles 100–150 nm in diameter was produced by chemical synthesis. Coatings were formed on the surface of the MA8 magnesium alloy by plasma electrolytic oxidation in an electrolyte containing HAp nanoparticles. The modes of oxidation, electrolyte concentration and oxidation time were optimized, it allows to obtain coatings with high adhesion to the substrate. The morphology of the coatings, the phase and elemental composition were estimated.

1. Introduction

Applied conventional metal implants are made of titanium and stainless steel. This materials possess a high strength, wear resistance and corrosion resistivity However the stresses arising between the bone and the titanium implant due to the difference in the modulus of elasticity (~105 GPa and 27 GPa, respectively [1, 2]) leads to osteopenia and osteoporosis [3–6]. Furthermore, the inert of titanium cause the occurrence of inflammatory processes at long-term implantation [7, 8], which, in turn, often leads to revision surgery and extracting the implant [9, 10]. This factor plays a special role in maxillofacial surgery, as a secondary intervention can lead to an increase of scarring and the impossibility of using aesthetic surgery.

Currently, magnesium and its alloys are widely used in aviation, development of high-tech devices and other areas of human activity. In the past decades, there has been an explosion of interest to magnesium alloys as implant material due to their physiological and mechanical properties. Advantages of magnesium are its plasticity, low density, high strength-weight ratio [11, 12].The application of magnesium in orthopedics can be regarded as a revolutionary event for biodegradable materials, in particular, metals. The elastic modulus of Mg (45 GPa) is the closest to bone indexes in comparison with the metals used today in implantation medicine. In addition, the possibility of controlled dissolution of the magnesium implant [13, 14] eliminates the need for secondary surgical intervention, thereby reducing the health risks and costs of healing. Recent evidence indicates that magnesium corrosion products can be beneficial to the patient's health, in contrast to the products of the destruction of other metals [15].

Hydroxyapatite (HAp) is the basic mineral component of bones (50% of the total bone mass) and teeth (95% in enamel). As a component of coating HAp may promote the growth of a new bone [16].The brittle nature of the ceramic restricts the clinical orthopedic and dental applications despite ones excellent biocompatibility. In addition to providing the protective function of internal organs and levers for muscle action, human bones also provide a natural reservoir for cells and mineral ions that play an important role in maintaining the biochemical balance within the body. For example, calcium, whose level in the body is closely monitored and regulated by a process called homeostasis is an important element involved in muscular action and nerve conduction [17, 18]. Thus, the development of a composite material based on a magnesium alloy coated with HAp acting as a protective / bioactive layer [14, 16, 18] is a unique finding for many tasks of implantology and orthopedic surgery.

In the present study, a coating with incorporated nanoparticles of calcium hydroxyapatite on the surface of the MA8 magnesium alloy has been formed. The composition, morphology, and structure of the obtaining layers on the surface of a magnesium alloy were studied.

2. Materials and methods

In the present paper, calcium hydroxyapatite was obtained by maturation (7 days) in aqueous solution (0.1M $Ca(OH)_2$ and 0.1M H_3PO_4) followed by filtration, drying and annealing of the resulting precipitate at 800°C for 3 h. The procedure for obtaining HAp is similar to [19], slightly increased addition time of orthophosphoric acid. The single-phase nature of the obtained HAp was confirmed by X-ray diffraction using Bruker D8 ADVANCE diffractometer with CuK_α–radiation.

The coatings were formed on the surface of a MA8 magnesium alloy (mas. %): 1.5–2.5 Mn, 0.15–0.35 Ce, balance – Mg). For the deposition of coatings, the method of plasma electrolytic oxidation (PEO) was used. Preliminary for the surface standardization samples of Mg alloy (15x20x1 mm^3) underwent mechanical treatment by grinding paper of different grain sizes (600, 800, and 1200), cleaned and degreased with acetone in an ultrasonic bath for 90 s. The electrolyte containing sodium fluoride (5 g/l) and silicate (15 g/l) was selected to treat samples using the PEO-method. The obtained hydroxyapatite nanopowder with the particle size of 50–100 as a component of the electrolyte for PEO was used to produce the composite coating. PEO-coatings on the MA8 magnesium alloy were formed at the concentration of 10 g/l HAp particles in the working electrolyte. Since at the plasma electrolytic oxidation the coating growth proceeds, as a rule, during the sample anodic polarization, to ensure the maximal incorporation of particles into the coating, an anionic surfactant (sodium dodecylsulfate) was used as a stabilizer of the dispersed system. Its concentration in the final electrolyte was equal to 0.5 g/l. Formation of coatings occurred in the two-stage bipolar PEO-mode as described in [20]. After PEO process, the samples were cleaned with de-ionized water to remove the remained electrolyte and then dried in air. All the samples were sealed by waterproof silica gel, just leaving an exposed area of 1 cm^2 for the electrochemical tests.

The electrochemical properties of the coatings were studied using the VersaSTAT MC electrochemical system (Princeton Applied Research, USA). The measurements were carried out in a three-electrode cell at room temperature in a 0.9% wt. % NaCl solution. The niobium mesh covered with platinum was used as a counter electrode. The saturated calomel electrode (SCE) was used as a reference electrode. The exposed area of the sample was equal to 1 cm^2. The samples were kept in a solution for 30 min prior the start of the electrochemical tests to achieve steady state. The potentiodynamic polarization measurements were carried out at a scan rate of 1 mV s^{-1}.

3. Results and Discussion

To reduce the number of agglomerates and increase the dispersion the HAp powder was subjected to milling in a ball mill in isopropyl alcohol medium. X-ray diffraction patterns of the compound after milling in isopropyl alcohol medium (fig. 1) showed no differences in the phase composition of compounds. The particle size of the HAp lies in the range of 50–100 nm, as evidenced by the SEM data (fig 2).

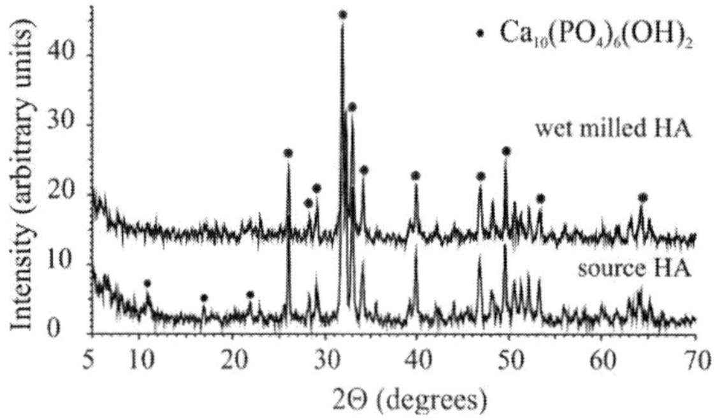

Fig. 1. XRD patterns of HAp powder and one subjected to wet-milling.

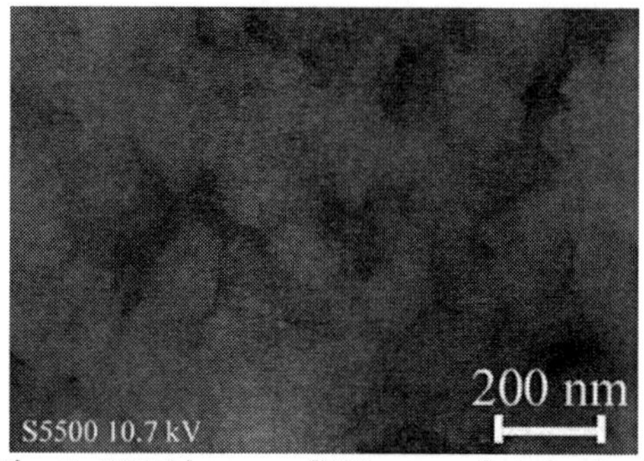

Fig. 2. SEM image of HAp powder obtained from an aqueous solution after wet milling.

Analysis of data obtained by the XRD method for formed coating indicates that there is no HAp on the surface of the coating in an amount sufficient to identify. The element analysis of cross-section obtained on the MA8 alloy indicates the presence of Mg, Na, O, F, Si, C, Ca and P elements (fig. 3). In the near-surface layer of the coating mainly contain Ca and P elements (fig. 4).

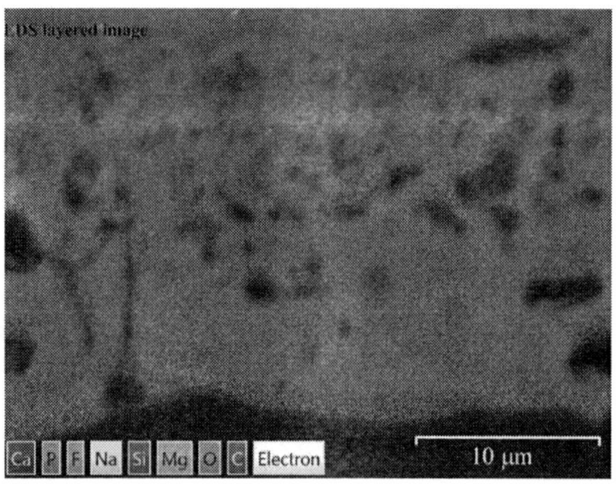

Fig. 3. Cross sectional morphology of PEO/HA sample with distribution of elements.

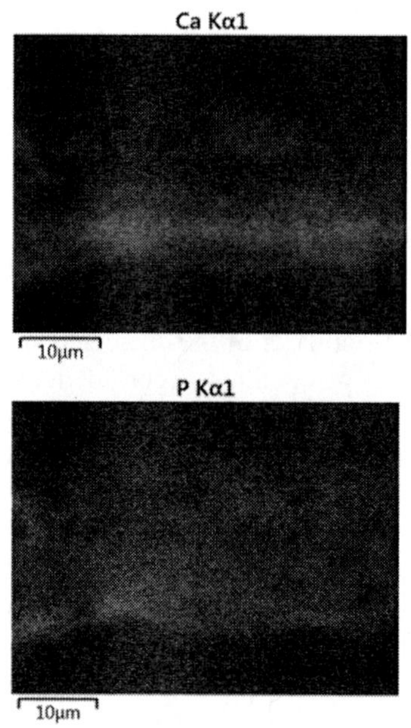

Fig. 4. Corresponding scanning maps of Ca and P distribution.

The data obtained by the EDX method make it possible to describe the structure of the coating formed on the alloy surface. The inner layer adjacent to the surface of the metal consists predominantly of MgO and MgF – dense, substantially nonporous layer with a thickness of ~ 5-10 µm. Further, in the thickness of the coating, pores appear, on the walls of which higher concentrations of Ca and P (fig 4).

Electrochemical studies indicate that coatings formed on the surface of magnesium alloy MA8 have a protective properties. Moreover, the coating formed with the use of calcium hydroxyapatite nanoparticles makes it possible to enhance the protective properties of the PEO coating against corrosion (table 1). The polarization resistance increased by more than threefold in comparison with the PEO coating and by two orders of magnitude compared to the uncoated alloy. The suggested mechanism for increasing the protective properties of a coating formed in an electrolyte containing nanoparticles is reducing the porosity due to their filling with HAp nanoparticles. In addition, the chemical stability of nanoparticles in the oxidation process is of great importance. As SEM data shown the HAp particles do not form agglomerates and do not appear in the structure of cross-section at any given magnification. Near the surface, the concentration of the elements increased, reaching a maximum in the near-surface layer (~ 10 μm).

Table 1. Electrochemical parameters for samples of alloy MA8 with and without coatings

Sample	E, V	$I_{cor.}$, A cm^2	R_p, Ω·cm^2	$\|Z\|_{f\to 0\ Hz}$, Ω·cm^2
MA8	-1.56	$1.42\cdot10^{-5}$	$2.16\cdot10^{3}$	$2.0\cdot10^{3}$
MA8 PEO	-1.45	$6.70\cdot10^{-7}$	$1.03\cdot10^{5}$	$1.1\ 10^{5}$
MA8 PEO/HA	-1.46	$2.40\cdot10^{-7}$	$3.79\cdot10^{5}$	$2.6\cdot10^{5}$

4. Conclusion

Calcium hydroxyapatite was synthesized by maturation from an aqueous solution in the form of a round shaped nanoparticles. The presented method of formation of composite coatings on magnesium alloy MA8 demonstrated flexibility of the application regime, improved resistance to corrosion, successful filling of pores in PEO coating with calcium hydroxyapatite. Compared to the uncoated metal the corrosion current density for the sample PEO/HA are reduced by more than two orders of magnitude, and impedance module at low frequencies increased by more than two orders of magnitude (up to 10^5 Ω·cm^2). The abovementioned factors indicate that the PEO/HA composite is promising in surface modification of biodegradable magnesium-based implants for orthopedic application.

Acknowledgments
This work was supported by the grant of the Russian Science Foundation (project no. 14-33-00009).

References

[1] International A. Standard Specification for Unalloyed Titanium, for Surgical Implant Applications. ASTM F67-06 Medical Device Standards and Implant Standards.
[2] C. Elias, J. Lima, R. Valiev, M. Meyers. Miner Met Mater Soc. 60(3) 46–49 (2008).
[3] T. Terjesen, K. Apalset. J. Orthop. Res. 6(2) 293–299 (1988).
[4] Terjesen, Acta Orthop. Scand. 55(1) 69 (1984).
[5] K. Wang, Mater. Sci. Eng. A. 213(1–2) 134–137 (1996).
[6] T. Karachalios, C. Tsatsaronis, G. Efraimis, et al., J. Arthroplasty. 19(4) 469–475 (2004).
[7] N. Taki, J. Tatro, J. Nalepka et al., J. Orthop. Res.. 23(2) 376–383 (2005).
[8] B. Warme, et al., J. Biomed. Mater. Res. B Appl. Biomater. 71(2) 360–366 (2004).
[9] Y.H. Kim, J.S. Kim, S.H. Cho, J. Arthroplasty. 14 538–548 (1999).
[10] J.J. Callaghan, et al., J. Bone Joint. Surg., 80A 704–714 (1998).
[11] B. Mordike, T. Ebert, Mater. Sci. Eng. A, 302(1) 37–45 (2001).
[12] S.V. Gnedenkova, et al., Surf. Coat. Technol. 204(14) 2316–2322 (2010).
[13] S.V. Gnedenkov et al., Russ. J. Inorg. Chem., 61(2) 135–141 (2016).
[14] S.V. Gnedenkov et al., Russ. J. Inorg. Chem., 61(4) 424–428 (2016).
[15] R.M. Touyz, Frontiers in Bioscience, 9 1278–1293 (2004).
[16] Gnedenkov, S.V., Sharkeev, Y.P., Sinebryukhov, S.L., et al., Corros Rev., 34(1-2) 65–83 (2016).
[17] G.A. Rodan, Bone homeostasis, The National Academy of Sciences, 95 13361-13362 (1998).
[18] G.E. J. Poinern, et al., American Journal of Biomedical Engineering, 2(6) 218–240 (2012).
[19] M. Akao, H. Aoki, K. Kato, J. Mater. Sci, 16 809–812 (1981).
[20] D.V. Mashtalyar, S.V. Gnedenkov, et al. // J. Mater. Sci. Technol. 33(5) 461–468 (2017).

Bound state in the continuum supported by a low refractive index contrast waveguide in a woodpile structure

L.Yu. Pogorelskaya,A.A. Bogdanov, K.B. Samusev, A.D. Sinelnik, Z.F. Sadrieva

The International Research Centre for Nanophotonics and Metamaterials, ITMO University, St. Petersburg 197101, Russia

E-mail: lydia.pogorelskaya@metalab.ifmo.ru

Abstract. In this work we demonstrate that the bound state in the continuum emerges in low refractive index periodic structures. We analyse properties of woodpile structure fabricated from polymer material with refractive index

1. Introduction

In traditional resonators, for example, those that employ the whispering gallery mode effect, high refractive index contrast between the waveguide and the claddings material is essential to obtain high quality factor. We propose a resonator with Q-factor, which does not depend on refractive index contrast because it is based on bound states in the continuum (BICs). These states remain localized due to destructive interference rather than total internal reflection. BICs are states with energies lying within the radiation modes continuum (above the light line in optical systems), in theory they have infinite quality factor. BICs were first described theoretically by von Neumann and Wigner in 1929 [1] for artificial quantum system; active research on BICs in optical systems is conducted actively in the last ten years both theoretically and experimentally. There are theoretical examples of a BIC-based waveguide with low refractive index on a high index thin membrane [2] and of BIC existing in fiber Bragg gratings with vanishing index contrast [3]. In our work we analyze BIC supported by woodpile structure fabricated from low refractive index material.

2. Results and discussion

We consider the woodpile structure composed of waveguide layer with higher effective refractive index ($\sim$1.4) on top of several layers with lower effective index ($\sim$1.2) on SiO2 substrate (Fig. 1).

If we consider such waveguide suspended in air, symmetry protected BIC is observed in Γ-point, the zeroth diffraction channel is forbidden due to destruction interference. However, higher order diffraction channels (that are not forbidden) may be opened in high index substrate, thus enabling leakage and destroying BIC [4]; we use layers with low effective refractive index to isolate waveguide from the substrate. The design of the structure was developed numerically using Fourier modal method (FMM). The model calculates reflection from the structure. Then we extract Q-factors from reflection spectra. Samples were realized experimentally using 3D laser lithography (direct laser writing).

Acknowledgments

The numerical model was developed with a support of RFBR according to the research project 18-32-00976 (mol a). Treatment of the numerical results and fabrication of the sample were carried out with a support of Russian Science Foundation (17-12-01581).

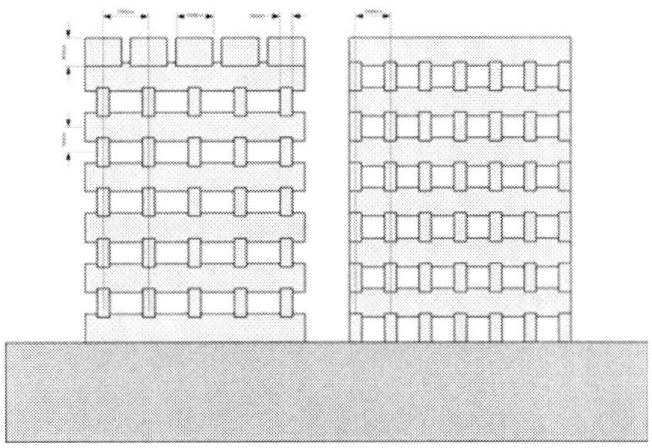

Figure 1. Schematic images of side views of the woodpile structure.

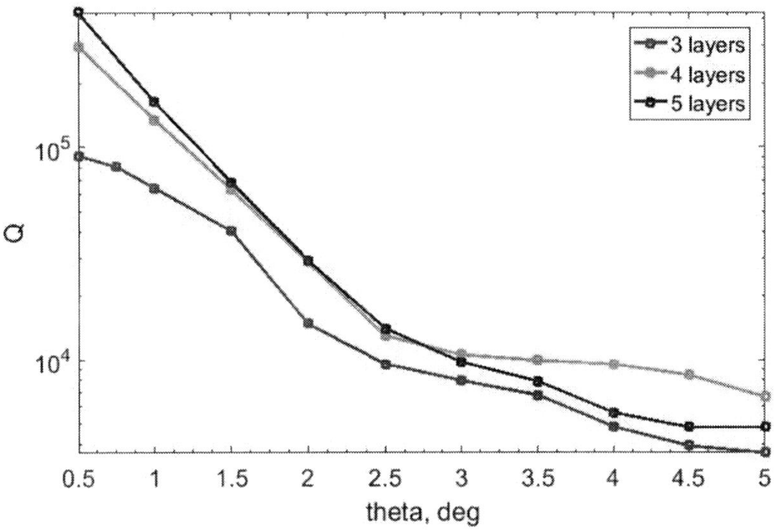

Figure 2. Q-factor of the woodpile structure calculated using FMM.

References

[1] von Neumann J and Wigner E 1929 Uber merkwurdige diskrete Eigenwerte *Phys.Zeitschrift.* **30** 465–7

[2] Zou C L 2015 Guiding light through optical bound states in the continuum for ultrahigh-Q microresonators *Laser Photonics Rev.* **9(1)**, 114–119

[3] Gao X, et al 2017 Bound states in the continuum in low-contrast fiber Bragg gratings preprint arXiv:1707.01247

[4] Sadrieva Z F, et al 2017 Transition from optical bound states in the continuum to leaky resonances: role of substrate and roughness, Photonics **4(4)** 723–727

Double frequency plasmonic amplification of terahertz radiation in a periodical double-layer graphene

O V Polischuk[1], I M Moiseenko[1,2], M Y Morozov[1] and V V Popov[1,2]
[1]Kotelnikov Institute of Radio Engineering and Electronics (Saratov Branch), Russian Academy of Sciences, Saratov 410019, Russia
[2] Saratov State University, Saratov 410012, Russia

Abstract. The plasmon amplification spectrum of terahertz radiation in a double-layer graphene nanoribbon array is theoretically studied. It is shown that this graphene structure exhibits strong plasmon response and giant amplification at vicinity of the anticrossing regime between optical and acoustic plasmon modes at room temperature.

1. Introduction

Graphene, a two-dimensional monolayer of graphite, has received a great deal of interest recently due to its unique electronic properties stemming from a linear (Dirac-type) gapless carrier energy spectrum. Graphene is also remarkable for high free-carrier mobility (temperature independent carrier mobility about 250000 $cm^2/V\cdot s$ was observed in multilayer epitaxial graphene on 4H-SiC substrate [1, 2]). Such carrier mobility corresponds to 1 ps carrier scattering times for the quasi-Fermi energy about 40 meV at room temperature and concentration of nonequilibrium electrons (holes) about 1.2×10^{11} cm^{-1}. Carrier scattering times longer than 1 ps were observed recently in optically pumped graphene [3, 4].

Graphene exhibits strong plasmonic response at terahertz (THz) frequencies due to both high density and small "relativistic" effective mass of the charge carriers [5, 6]. Stimulated emission of near-infrared and THz photons from population inverted graphene was recently observed [7]. As compared with the stimulated emission of the electromagnetic modes (photons), the stimulated emission of plasmons by the interband transitions in the population inverted graphene exhibits a much higher gain due to a small group velocity of the plasmons in graphene and strong confinement of the plasmon field in the vicinity the graphene layer [8].

Double-layer graphene structure consists of two parallel graphene layers with a narrow dielectric barrier slab between them. The electromagnetic fields of plasmons propagating in these layers interact with each other giving rise to a single unified plasmon in the pair of parallel graphene monolayers surrounded by dielectric claddings [9]. Double-layer graphene structure is actively studied due to their fundamental importance and practical applications.

2. Theoretical model

In this paper, we consider the amplification of a THz wave by the stimulated generation of resonant plasmons in a patterned double-layer population inverted graphene structure. Consider two graphene microribbon arrays separated by a thin dielectric barrier slab and placed on the semi-infinite dielectric substrate (see inset on figure 1). External THz wave is incident normally to the plane of the structure along y-direction with the polarization of the electric field $E_x^{(0)}$ across the graphene strips. The optical problem of interaction between THz wave and the double-layer graphene structure has been solved by using a self-consistent electromagnetic approach and integral equations method. The system of two coupled integral equations for the surface current density $j_x^{(w)}(x,y)$ on the graphene microribbon in both planes (at $y=d$ and $y=0$) are

$$\begin{aligned} j_x^{(w)}(x,d)+\int_0^w j_x^{(w)}(x',d)G_m^{(1)}(x,x',d)\,dx'+\int_0^w j_x^{(w)}(x',0)G_m^{(2)}(x,x',d)\,dx' &= \sigma_{w,d}Z^{(1)}E_x^{(0)}\delta_{m0}, \\ j_x^{(w)}(x,0)+\int_0^w j_x^{(w)}(x',d)G_m^{(1)}(x,x',0)\,dx'+\int_0^w j_x^{(w)}(x',0)G_m^{(2)}(x,x',0)\,dx' &= \sigma_{w,0}Z^{(2)}E_x^{(0)}\delta_{m0}, \end{aligned} \quad (1)$$

where $G_m^{1(2)}(x,x',d)=\sigma_{w,d}\frac{1}{L}\sum_m Z_{m,d}^{1(2)}\exp\left(iq_m(x-x')\right)$, $G_m^{1(2)}(x,x',0)=\sigma_{w,0}\frac{1}{L}\sum_m Z_{m,0}^{1(2)}\exp\left(iq_m(x-x')\right)$ are the kernels of the integral equations, $Z_{m,d}^{1(2)}, Z_{m,0}^{1(2)}$ are the explicit algebraic coefficients, calculated from the Maxwell equations. The response of optically pumped graphene is described by the complex dynamic surface conductivity $\sigma_{w,d(0)}$ [8]. The system of Fredholm integral equations of the second kind (1) was solved numerically by the Galerkin method. In the results, the induced electric and magnetic fields at the any point of the structure can be found.

Because the period of the structure is much shorter than the electromagnetic wavelength, only zero-order spatial Fourier harmonics are the travelling waves emitted into the ambient medium and substrate while all higher-order Fourier harmonics are evanescent fields decaying for $y\to\pm\infty$. Therefore, the reflection and transmission coefficients in the far-field zone can be calculated as $R=\left|E_{x0}(d)\right|^2/\left|E_x^{(0)}\right|^2$ and $T=\left|E_{x0}(0)\right|^2\sqrt{\varepsilon_s}/\left|E_x^{(0)}\right|^2$, respectively, where $E_{x0}(0)$ and $E_{x0}(d)$ are the zero-order spatial Fourier harmonics of the x-component of the induced electric field in graphene planes $y=0$ and $y=d$. The absorption/amplification coefficient is calculated as $A=1-R-T$ ($A>0$ in the absorption regime and $A<0$ in the amplification regime).

3. Results and discussion

Figure 1(a) shows the calculated spectrum of the plasmon resonances as a function of the dielectric barrier layer thickness values and frequency at room temperature. Calculations were performed for a small value of the quasi-Fermi energy equal to 20 meV.

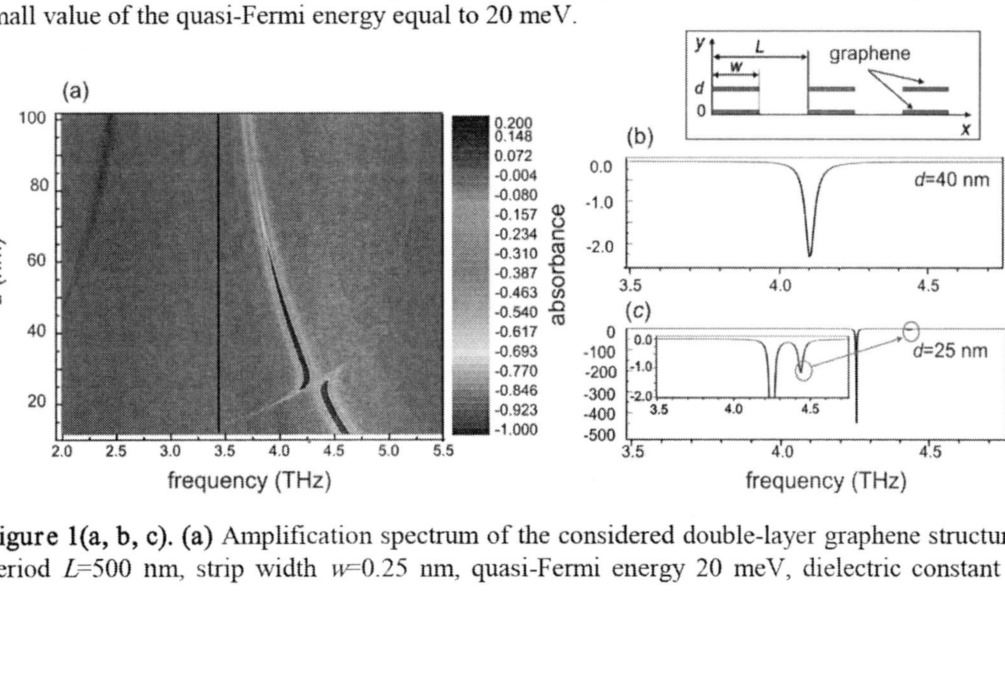

Figure 1(a, b, c). (a) Amplification spectrum of the considered double-layer graphene structure with period L=500 nm, strip width w=0.25 nm, quasi-Fermi energy 20 meV, dielectric constant of the substrate and barrier layer 11.7. The electron scattering time in graphene is 1 ps for temperature 300K. Schematic view of the double-layer graphene nanoribbon array and the coordinate system is shown in the inset at the top right. **(b)** Amplification coefficient of the considered graphene structure with d=40 nm. **(c)** Amplification coefficient of the considered graphene structure with d=25 nm. The insert in (c) shows the fragment for the lower values of the ordinate-axis.

However, such a low pumping is sufficient for the interband transitions in graphene prevail over Drude losses. Consequently a real part of graphene conductivity becomes negative $\mathrm{Re}(\sigma_{w,d(0)}) < 0$, that corresponds to the energy gain (negative absorbance) in graphene. Solid black line on Figure 1(a) corresponds to transparent inverted graphene $\mathrm{Re}(\sigma_{w,d(0)}) = 0$, and the real part of the inverted graphene conductivity is positive (negative) to the left (right) of this line.

The system under consideration supports the optical and acoustic plasmon modes. The figure 1(a) shows (from left to right) the fundamental acoustic, fundamental optical and first higher acoustic modes. The resonant frequencies of the optical and acoustic modes vary in the opposite way with the change of the thickness of the insulating dielectric layer d, which makes possible the anticrossing regime. The maximum amplification is achieved in the vicinity of the anticrossing regime (figure 1(c)) due to interaction of the optical and acoustic plasmon modes.

4. Conclusions

It is shown that THz amplification has a broad dynamic range and a wide range of dielectric barrier layer thickness values. Largest amplifications are achieved at lower dielectric barrier layer thickness values at the two resonant frequencies in the vicinity of the anticrossing regime, even for small quasi-Fermi energy 20 meV at room temperature.

Acknowledgments

This work was supported by the Russian Foundation for Basic Research (Project No. 16-02-00814).

References

[1] Orlita M, Faugeras C, Plochocka P, Neugebauer P, Martinez G, Maude D K, Barra A-L, Sprinkle M, Berger C, de Heer W A, and Potemski M 2008 *Phys. Rev. Lett.* **101** 267601

[2] Sprinkle M, Siegel D., Hu Y, Hicks J, Tejeda A, Taleb-Ibrahimi A, Le Fèvre P, Bertran F, Vizzini S, Enriquez H, Chiang S, Soukiassian P, Berger C. de Heer W A, Lanzara A, and Conrad E H 2009 *Phys. Rev. Lett.* **103** 226803

[3] Dawlaty J M, Shivaraman S, Chandrashekhar M, Rana F and Spencer M G 2008 *Appl. Phys. Lett.* **92** 042116

[4] Otsuji T, Boubanga Tombet S A, Satou A, Fukidome H, Suemitsu M, Sano E, Popov V, Ryzhii M and Ryzhii V 2012 *J. Phys. D: Appl. Phys.* **45** 303001

[5] Chen J, Badioli M, Alonso-González P, Thongrattanasiri S, Huth F, Hillenbrand R, and Koppens F 2012 *Nature* **77** 487

[6] Fei Z, Rodin A S, Andreev G O, Bao W, McLeod A S, Wagner M, Zhang L M, Zhao Z, Thiemens M, Dominguez G, Fogler M M, Castro Neto A H, Lau C N, Keilmann F, and Basov D N 2012 *Nature* **82** 487

[7] Boubanga-Tombet S, Chan S, Watanabe T, Satou A, Ryzhii V and Otsuji T 2012 *Phys. Rev. B* **85** 035443

[8] Dubinov A A, Aleshkin V Y, Mitin V, Otsuji T and Ryzhii V 2011 *J. Phys.: Condens. Matter* **23** 145302

[9] Gan C H, Chu H S, and Li E P 2012 *Phys. Rev. B* **85** 125431

IOP Conf. Series: Journal of Physics: Conf. Series **1092** (2018) 012120 doi:10.1088/1742-6596/1092/1/012120

Ab-initio study of electronic properties of a two-dimensional array of carbon nanotubes

N Yu Senkevich[1], R G Polozkov[1], S Morina[1,2], E D Cherotchenko[1], M E Portnoi [1,3] and I A Shelykh[1,2]

[1]ITMO University, Saint Petersburg, 197101, Russia
[2]Science Institute, University of Iceland, Dunhagi 3, IS-107, Reykjavik, Iceland
[3]School of Physics, University of Exeter, Stocker Road, Exeter EX4 4QL, United Kingdom

e-mail: polozkov@corp.ifmo.ru

Abstract. The equilibrium geometry and electronic band structure of a planar array of carbon nanotubes are studied with the use of the Quantum Espresso code - a plane-wave realisation of the density functional theory (DFT). The many-electron correlations and van der Waals corrections are taken into account. The optimal distance between nanotubes in the array corresponding to the minimum of the total energy of the system is found. A strongly anisotropic hyperbolic dispersion is demonstrated for low-energy charge carriers in an array of quasi-metallic (15,0) carbon nanotubes with the optimal inter-tube separation governed by van der Waals forces.

1. Introduction

Graphene-based nanomaterials have attracted great attention due to their unique electronic and mechanical properties. Graphene sheet rolled into a cylindrical shape forms a structure called carbon nanotube (CNT) [1]. CNTs manifest drastically different electronic properties from being a metal or a very narrow-gap semiconductor (quasi-metal) to a mid-gap semiconductor depending on their chiral vector - the way they are rolled.

Ab-initio methods are widely used for modeling electronic properties of single nanotubes with different chirality and more complex systems constructed on the basis of CNTs [2,3]. Moreover, such structures are extensively studied experimentally. For example, results of experimental studies of self-assembled horizontally aligned carbon nanotube arrays were recently published [4]. Such structures show great potential as building blocks for transparent displays, nanoelectronics, quantum lines, field emission transistors, super-strong tethers, aeronautics and astronics materials, and even space elevators. Recently, a giant optical absorption peak corresponding to inter-subband transitions has been observed in a film of highly-aligned for the excitation polarized normal to the tubes [5].

In this paper, we have estimated using DFT calculations the optimal distance between zigzag quasi-metallic (15,0) CNTs assembled in a perfect two-dimensional (2D) array, see Fig.1. We have calculated the electronic band structure of the corresponding CNT array for the bands closest to the Fermi level.

2. Computational details

The ab-initio calculations of the band structure and optimal distance between CNTs in the array were performed employing an ultra-soft pseudo-potential and a plane-wave basis in the Quantum Espresso package [6]. Quantum Espresso is based on the first principles density functional theory which allows the calculation of ground state properties of rather complex systems.

At the first stage we made an optimization of the spatial geometry of a planar array of (15,0) zigzag CNTs using the well-known generalized gradient approximation (GGA) for exchange-correlation potential - Perdew, Burke and Ernzerhof (PBE) functional [7] with and without the van der Waals (vdW) correction. The vdW interaction was included by the use of the method of Grimme (PBE-D2) [8]. The calculations were done for different inter-tube separations.

At the second stage, the band structure of the (15,0) CNT array was calculated using the PBE-D2 approach for an optimal distance *d*, which was found in the first stage. In order to get accurate results, the plane-wave cut-off was set to a high value of 800 eV, for which the structural relaxations and the electronic energies fully converged. A Gaussian smearing for the occupations was used with a smearing width of 0.01 eV. We performed a convergence test with respect to the number of k-points in the Monkhorst-Pack grid $10\times10\times1$.

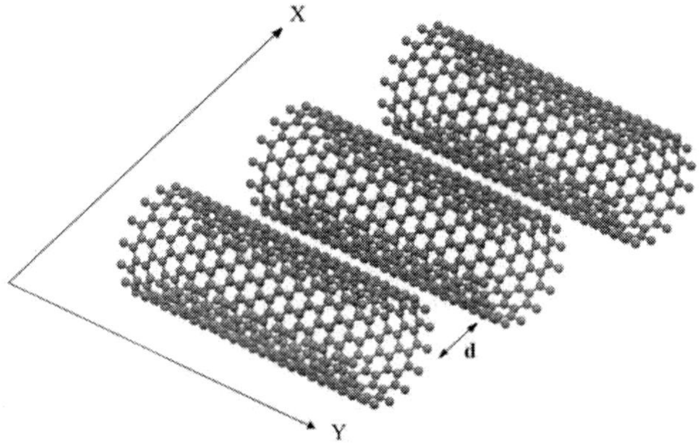

Figure 1. Horizontally aligned array of carbon zigzag nanotubes (15,0) separated by distance d between walls of tubes. The diameter of (15,0) CNT is 22.2 a.u.

3. Results

In this section we present the results of our calculations of the optimal inter-nanotube distance *d* and the electronic band structure of an array composed of (15,0) CNTs separated by this optimal distance.

To model a planar array, we used a hexagonal lattice with a separation of 60 atomic units (a.u.) between layers and 60 carbon atoms per unit cell. A diameter of an (*m*,*n*) CNT can be computed in a.u. as $D = 1.48\sqrt{m^2+n^2+\mathrm{mn}}$, where *n* and *m* are the integer components of the chiral vector [9]. For a (15,0) CNT the diameter is equal to *22.2* a.u.

3.1. Optimal distance

The calculated dependence of the total energy of the unit cell of the (15,0) CNT array on distance *d* between the carbon nanotube walls for PBE and PBE-D2 approaches is shown in Fig. 2.

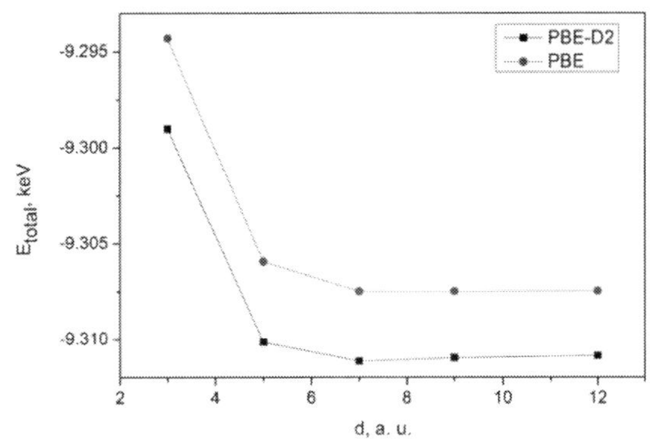

Figure 2. Dependence of the total energy on the distance between the nanotube walls obtained with PBE-D2 approach (lower curve) and PBE exchange-correlation potential (upper curve).

The optimal distance between nanotubes should correspond to the minimal value of the total energy of the system. It is interesting that the PBE approach (upper curve in the Fig. 2) is not enough to obtain the minimum in the dependence of the total energy on the distance. Taking into account the vdW correction (lower curve in the Fig. 2) leads to the appearance of the minimum and the optimal distance between nanotubes, respectively.

Finally, the results obtained with the PBE-D2 approach can be approximated by the Lennard-Jones potential (see Fig.3): $V(d) = \Delta E\left[\left(\frac{d_0}{d}\right)^{12} - 2\left(\frac{d_0}{d}\right)^{6}\right]$, where $\Delta E = 0.41$ eV is the depth of the Lennard-Jones potential, and d_0 corresponds to the total energy minimum.

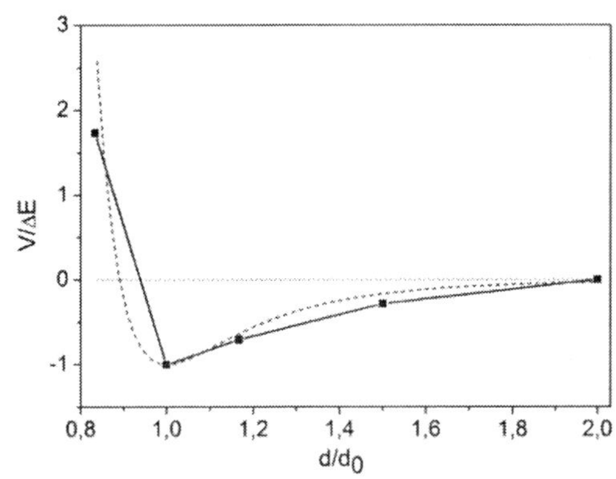

Figure 3. Dependence of total energy of the system (solid curve) and Lennard-Jones potential V with corresponding parameters (dashed curve) in units of depth of the potential well ΔE=0.41 eV of CNTs array unit cell on the normalized distance between nanotube walls (d_0=6 a.u.) obtained with PBE-D2.

3.2. Electronic band structure

The calculated band structure of a (15,0) CNT array along tube axis for $k_x = 0$ is shown in Fig. 4 (a). It is well known that an isolated (15,0) zigzag carbon nanotube has a small band gap, which is equal to 0.029 eV [10], and all its bands are doubly degenerate. It can be seen from Fig. 4 (a) that for the array of these nanotubes the degeneracy is removed, and the band gap for $k_y = 0$ and $k_x = 0$ is equal to 0.12 eV.

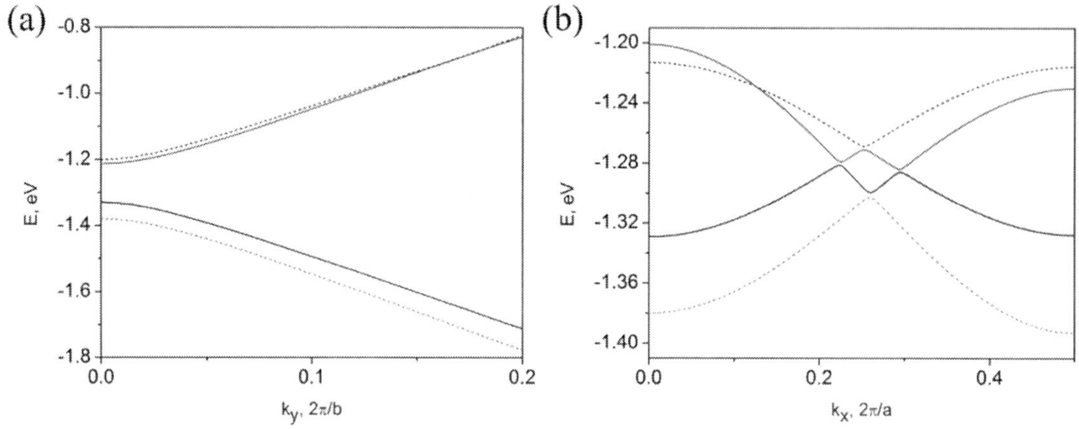

Figure 4. Electronic band structure of an array of (15,0) zigzag nanotubes separated by d=6 a.u. obtained with the PBE-D2 approach: (a) along the tube axis; (b) in the direction normal to the tube axis. The wave vectors are given in reciprocal space units.

For a 2D array of nanotubes, the electronic dispersion in the direction normal to the tube axis can also be calculated, see Fig. 4 (b). As can be seen from this figure, the band gap collapses away from the $k_x = 0$ point and the considered 2D material is in fact a metal. Finally, the two-dimensional dispersion for the bands closest to the Fermi level has saddle points, as can be seen from the Fig.4 (b). It means that the considered array represents a hyperbolic material for charge carriers.

4. Conclusion

We have found that a planar array of (15,0) zigzag CNTs separated by an optimal distance governed by van der Waals forces has a strongly anisotropic hyperbolic dispersion for its low-energy charge carriers. Thus, we have predicted a new type of hyperbolic van der Waals material with a rich energy spectrum promising interesting physical effects and optoelectronic applications.

5. Acknowledgements

This work was supported by the Icelandic Research Fund (Grant No. 163082-051), Horizon 2020 RISE project CoExAN (Grant No. 644076), mega-grant No 14.Y26.31.0015 and Project 3.2614.2017/4.6 of the Ministry of Education and Science of Russian Federation. This work also was financially supported by the Government of Russian Federation (Grant 08-08).

References

[1] Iijima S 1991 *Nature* **354** 56–58
[2] Charlier J-C, Blase X *et al* 2007 *Rev. Mod. Phys.* **79** 677
[3] Osadchy A *et al* 2013 *J. Nanoelectron. Optoelectron.* **8** 91
[4] Zhang R, Zhang Y *et al* 2017 *Chem. Soc. Rev.* **46** 3661-3715
[5] Yanagi K *et al* 2018 *Nature Communications* **9** 1121
[6] Giannozzi P *et al* 2009 *J.Phys.:Condens.Matter* **21** 395502
[7] Perdew J *et al* 1996 *Phys. Rev. Lett.* **77** 3865
[8] Grimme S 2006 *J. Comp. Chem.* **27** 1787
[9] Saito R, Dresselhaus G and Dresselhaus M S 1998 *Physical Properties of Carbon Nanotubes*, Imperial College Press
[10] Matsuda Y, Tahir-Kheli J, Goddard W 2010 *J. Phys. Chem. Lett.* **1** 2946–2950

IOP Conf. Series: Journal of Physics: Conf. Series **1092** (2018) 012121 doi:10.1088/1742-6596/1092/1/012121

Terahertz transitions in narrow-gap carbon nanotubes and graphene nanoribbons

V A Saroka[1,2], R R Hartmann[3] and M E Portnoi[1,4,5]

[1]School of Physics, University of Exeter, Stocker Road, Exeter EX4 4QL, UK
[2]Institute for Nuclear Problems, Belarusian State University, Minsk 220030, Belarus
[3]Physics Department, De La Salle University, 2401 Taft Avenue, Manila, Philippines
[4]ITMO University, St. Petersburg 197101, Russia

[5]Corresponding author: m.e.portnoi@exeter.ac.uk

Abstract. We show that the curvature effects in quasi-metallic carbon nanotubes and edge effects in narrow-gap graphene nanoribbons not only open band gaps in the THz range but also result in giant enhancement of the transition probabilities across these gaps. This makes these nanostructures perspective candidates for sources and detectors of THz radiation.

1. Introduction
Creating reliable, portable, tunable sources and detectors of terahertz (THz) radiation is one of the most challenging tasks of contemporary applied physics. One of the recent trends in bridging the so-called THz gap is to use carbon-based nanostructures [1]. Several original schemes utilizing the unique electronic properties of carbon nanotubes (CNTs) and graphene for THz application were brought forward by our group [2-6]. These schemes include THz generation by hot electrons in quasi-metallic CNTs, frequency multiplication in chiral-nanotube-based superlattices controlled by a transverse electric field, tunable THz radiation detection and optically-pumped emission in metallic CNTs in a strong magnetic field and using graphene p-n junctions for sub-wavelength polarization-sensitive THz detection. In the current presentation, we focus on direct interband dipole transitions in narrow-gap CNTs and graphene nanoribbons (GNRs).

2. Narrow-gap carbon nanotubes and graphene nanoribbons
The main conclusions of the reported work is that the curvature effects in quasi-metallic single-walled CNTs and edge effects in gapless graphene nanoribbons not only open bang gaps, which are typically in the THz range, but also result in giant enhancement of the probability of optical transitions across these gaps. Moreover, the matrix element of the velocity operator for these transitions has a universal value when the photon energy coincides with the band gap energy Eg. Namely, the absolute value of this matrix element for band-edge transitions is equal to graphene's Fermi velocity v_F. Upon increasing the excitation energy the absolute value of the transition matrix element first rapidly decreases (for photon energies remaining in the THz range but exceeding $2Eg$ it is reduced by three orders of magnitude) and thereafter it starts to increase proportionally to the photon frequency. A similar effect occurs in an armchair CNT with a band gap opened and controlled by a magnetic field applied along the nanotube axis [4-6]. As can be seen from Fig. 1, there is a direct correspondence between armchair graphene nanoribbons (AGNRs) and single-walled zigzag CNTs.

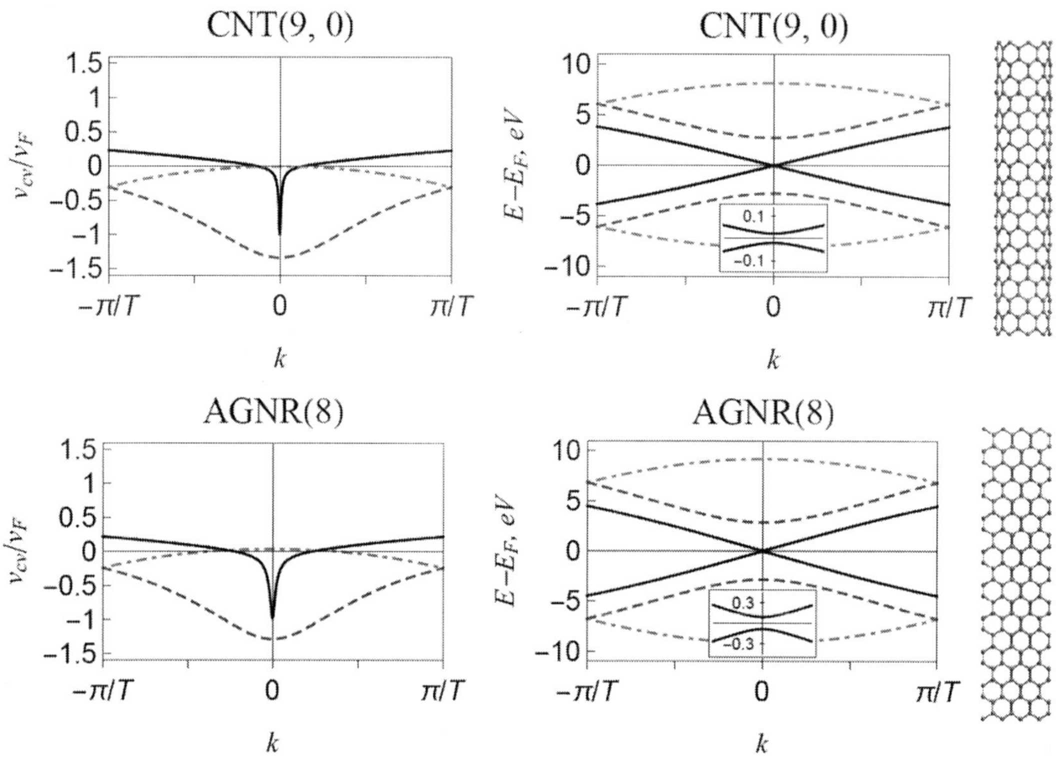

Figure 1. Energy spectra and interband transitions velocity matrix elements for CNT (9,0) (upper panel) and AGNR (8) (lower panel) as a function of momentum in the first Brillouin zone.

The described sharp photon-energy dependence of the transition matrix element together with the van Hove singularity at the band gap edge of the considered quasi-one-dimensional systems make them promising candidates for active elements of coherent THz radiation emitters. The effect of Pauli blocking of low-energy interband transitions caused by residual doping can be suppressed by creating the population inversion using high-frequency (optical) excitation. Excitonic effects, which are known to dominate the optical properties of semiconductor CNTs, are of less importance in narrow-gap CNTs and ANGRs where the exciton binding energy is proportional to the bandgap [7] and dark excitonic states become irrelevant.

This work was supported by the EU H2020 RISE project CoExAN, FP7 ITN NOTEDEV and visitors program at ITMO.

References

[1] R. R. Hartmann, J. Kono, and M. E. Portnoi, Nanotechnology, **25**, 322001 (2014).
[2] O. V. Kibis, D. G. W. Parfitt, and M. E. Portnoi, Phys. Rev. B **71**, 035411 (2005).
[3] O. V. Kibis, M. Rosenau da Costa, and M. E. Portnoi, Nano Lett. **7**, 3414 (2007).
[4] M. E. Portnoi, O. V. Kibis, M. Rosenau da Costa, Superlattices Microstruct. **43**, 399 (2008).
[5] M.E. Portnoi, M. Rosenau da Costa, O.V.Kibis, I.A.Shelykh, Intern. J. Modern Phys. B **23**, 2846 (2009).
[6] R.R.Hartmann and M.E.Portnoi, IOP Conf. Ser.: Mater. Sci. Eng. **79**, 012014 (2015).
[7] R.R.Hartmann, I.A.Shelykh, and M.E.Portnoi, Phys. Rev. B **84**, 035437 (2011).

IOP Conf. Series: Journal of Physics: Conf. Series **1092** (2018) 012122 doi:10.1088/1742-6596/1092/1/012122

Cesium lead mixed-halide perovskites in polymer matrix

A.P. Pushkarev[1], V.I. Korolev[1], S.S. Anoshkin[1], F.E. Komissarenko[1], S.V. Makarov[1], A.A. Zakhidov[1,2]

[1]ITMO University, St. Petersburg 197101, Russia

[2]The Alan G. MacDiarmid NanoTech Institute, University of Texas at Dallas, Richardson, TX, USA

anatoly.pushkarev@metalab.ifmo.ru

Abstract. Inorganic perovskites of $CsPbX_3$ (X = Cl, Br, I) type are excitonic semiconductors for optoelectronics and nanophotonics showing outstanding optical properties that can be controlled via their chemical composition tuning. Pure chloride, bromide and iodide phases of such materials exhibit intense photoluminescence with FWHF = 12-30 nm (FWHM – full width at half maximum) peaked at 420, 525 and 710 nm, respectively. In the meanwhile, mixed-halide perovskites are aimed to fill these gaps and demonstrate emission in the whole visible spectrum. The conventional method of formation of mixed-halide perovskites is mixing of different metal halides in organic solvents affording a purpose solution for further casting and annealing. This method has a serious disadvantage that stems from the inequivalent solubility of the precursors. Herein, we report an easy way of $CsPbCl_{3-x}Br_x$ ($0<x<3$) formation by means of Chemical Vapor Anion Exchange (CVAE) method. The thin films consisting of $CsPbBr_3$ grains embedded into polymer matrix were treated by HCl vapor to give mixed-halide phases emitting light in the 420-525 nm range.

1. Introduction

Since the breakthrough reports dedicated to utilization of $MAPbI_3$ (MA - methyl ammonium) material for efficient photoconversion in thin film mesoscopic solar cells [1,2], researchers working in the fields of organic electronics and nanophotonics have been discovering new applications of lead halide perovskites [3]. To date, the most promising applications, with exception of photovoltaics, are perovskite LEDs [4] and nanolasers [5] and in contrast to light harvesting devices they have to possess a crucial feature – broad spectral tuning of luminescence. To achieve wide-range light emission from lead halide perovskites the approach of variation in their chemical composition can be exploited. For instance, one-sort-halide $APbX_3$ (A = methyl ammonium (MA^+), formamidinium (FA^+), Cs^+; X = Cl^-, Br^-, I^-) materials reveal featured luminescence with FWHM = 12-30 nm in deep blue (415-425 nm), green (525-535 nm) or near infra-red (710-800 nm) spectral region. In the meanwhile, $APbX_3$ (X = $Cl_{1-y}Br_y$, $Br_{1-y}I_y$; $0<y<1$) mixed-halide perovskites are aimed to fill these gaps and demonstrate emission in the whole visible spectrum. The conventional method of mixed-halide perovskites production is mixing of different metal halides in organic solvents giving specific ink for further usage. This method suffers of the inequivalent solubility of the precursors (e.g. CsCl is extremely poor soluble in common perovskite solvents, namely, dimethyl sulfoxide and dimethyl formamide). To overcome this hindrance the "soft-kind" perovskite crystal lattice can be modified by halogen anion substitution when the material reacts with HX (X = Cl, Br, I) vapor [6].

Herein, we report an easy way of $CsPbCl_{3-x}Br_x$ (0<x<3) formation by means of Chemical Vapor Anion Exchange (CVAE) method. The thin films consisting of $CsPbBr_3$ grains embedded into polyethylene oxide (PEO) matrix were treated by HCl vapor for 7-300 s at 120 °C to give mixed-halide phases emitting light in the 420-525 nm range. Fluorescent microscopy images of the films confirmed the existence of individual phases. Remarkably, time dynamics of PL of the samples revealed no changing in peak position. On the contrary, the pure $CsPbCl_{3-x}Br_x$ films showed very fast light induced spectral changes caused by segregation phenomenon [7]. Therefore, the strategy of dispersion of perovskite materials in polymer matrix can be utilized for stabilization of mixed-halide phases.

2. Results

To produce polymer-perovskite films $PbBr_2$ (99.999%), CsBr (99.999%), N,N-dimethylformamide (DMF, anhydrous, 99.8%), dimethyl sulfoxide (DMSO, anhydrous, 99.8%) and poly(ethylene oxide) (PEO, average Mw 70 000) were used as received from commercial suppliers. Perovskite solution was prepared by dissolving PbBr2 (36.7 mg, 0.1 mmol) and CsBr (21.2 mg, 0.1 mmol) in 1 ml of DMSO. The mixture was shaken for 10 min to give a clear solution. PEO (10 mg) was dissolved in 1 ml of DMF by stirring for 24 h at 70 °C. The perovskite and PEO solutions were mixed in 1:3 ratio to afford polymer-perovskite ink. The ink was stirred at 70 °C for 30 min and then filtered by using 0.45 μm syringe filter with PTFE membrane. Glass substrates were cleaned with detergent water, acetone, and isopropanol for 5 min with sonication, then treated with oxygen plasma at 100 W power for 3 min. The organic-inorganic composite films were spin-casted onto the substrates at 1500 rpm for 1 min and further annealed at 110 °C for 3 min on a hotplate. Solution preparation and film deposition were carried out inside N_2-filled glovebox.

Chemical vapor rehalogenation of the PEO:$CsPbBr_3$ films was done at ambient conditions. The sample was placed in a glass Petri dish and preheated on the hotplate at 120 °C before the treatment. A glass low form weighing bottle was preheated in an oven at 120 °C as well. 3 mg of HCl azeotrope (37%) was dropped near the sample, and then it was covered by the hot glass weighing bottle. After a certain amount of time the system was dismantled and the sample was heated up to 200 C to eliminate structural defects of the perovskite crystal lattice. Thus, exposure of the samples to HCl vapor for 7, 18, 60, 152, and 300 s at 120 °C yielded $CsPbCl_{3-x}Br_x$ ($0 < x \leq 3$) grains dispersed in PEO matrix that exhibited PL peaks at 504, 485, 465, 445, and 421 nm, respectively (Fig. 1).

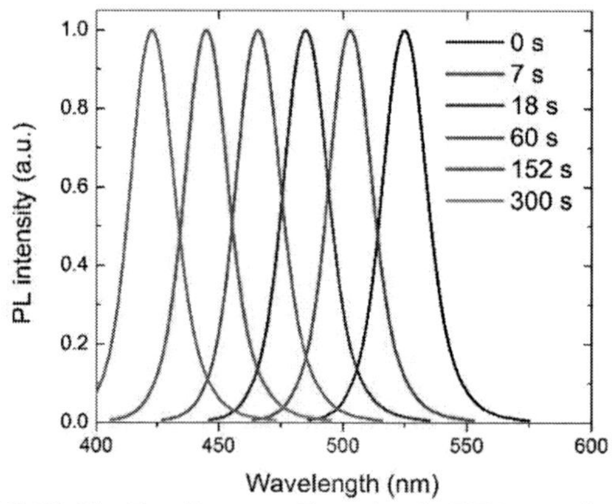

Figure 1. PL spectra of $CsPbCl_{3-x}Br_x$ ($0 \leq x \leq 3$) grains in PEO matrix obtained by exposure of $CsPbBr_3$-containing films to HCl vapor for 0-300 s time intervals.

We studied the films obtained by using a fluorescent microscope and found out the anion exchange occurs evenly over the whole area of the sample. Fig. 2 demonstrates images of initial film containing $CsPbBr_3$ grains (a) emitting light at 525 nm and $CsPbCl_{3-x}Br_x$ ones (b) with PL peaked at 485 nm. SEM images of the samples showed that small perovskite grains of 120-200 nm size unite in the larger clusters of 500-800 nm size.

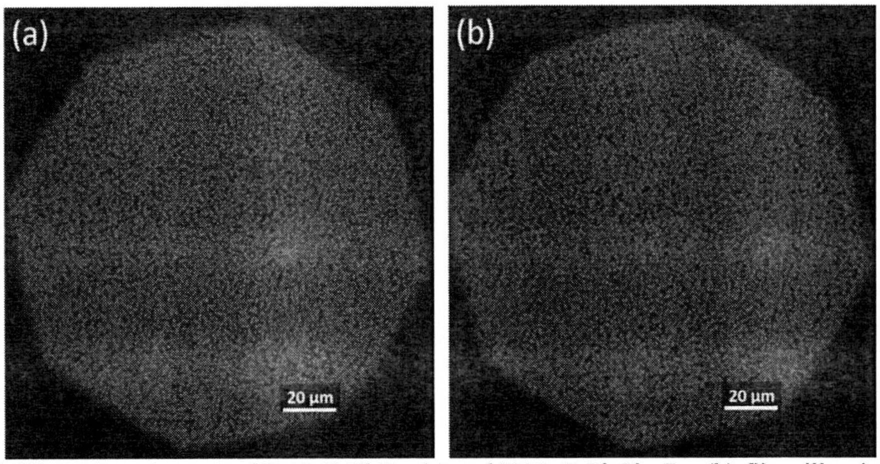

Figure 2. Fluorescent images of PEO:$CsPbBr_3$ (a) and PEO:$CsPbCl_{3-x}Br_x$ (b) films illuminated by 360 nm UV-lamp obtained with 50x objective.

Inorganic perovskites of $CsPbX_3$ (X = Cl, Br, I) type are excitonic semiconductors for optoelectronics and nanophotonics showing outstanding optical properties that can be controlled via their chemical composition tuning. However, phase instability of the mixed-halide perovskites is known as their main issue. Illumination of such materials by intense UV light or placing them in the strong electrostatic field leads to the migration of halogen anions resulting in formation of the grains enriched by only one sort of halogen atoms [6]. As a result of structural modification, the PL band of the materials undergoes spectral shift depending on duration of their expousure to light or field. This effect is called segregation. Surprisingly, the polymer-perovskite films obtained did not exibit any segregation under severe UV light for 1000 s (Fig. 3a). To check out $CsPbCl_{3-x}Br_x$ phase stability a polymer-free film was tested at the same conditions. The latter sample consisting of small tightly bound nanoparticles with the sizes less than 100 nm demonstrated extremely fast segregation in the 0.5-2 s time range (Fig. 3b). We assume the reason of PEO:$CsPbCl_{3-x}Br_x$ composite phase stability lies in the formation of relatevely large perovskite grains having surface free energy exceeding the volume one. Dominant contribution of the surface free energy to Gibbs free energy of the considered objects is supposed to be responsible for the mixed-halide material phase stabilization [8].

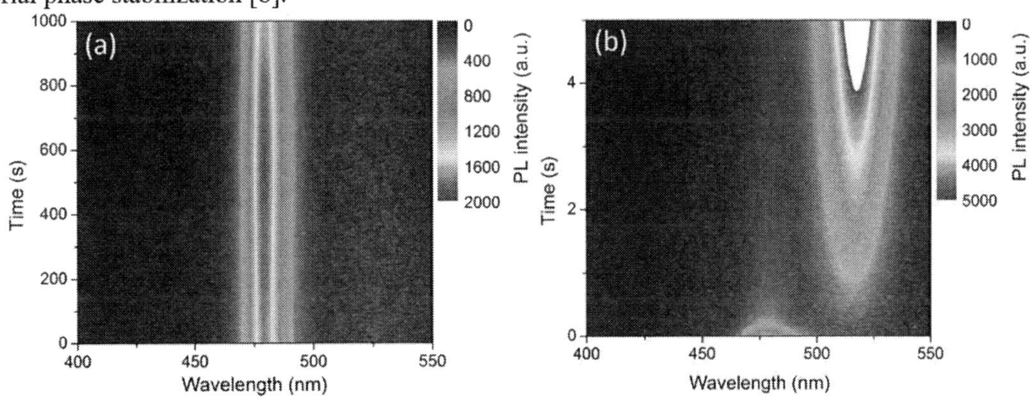

Figure 3. PL time evolution for PEO:$CsPbCl_{3-x}Br_x$ film (a) and polymer-free sample (b).

IOP Conf. Series: Journal of Physics: Conf. Series **1092** (2018) 012122 doi:10.1088/1742-6596/1092/1/012122

3. Conclusion

In summary we studied new method of inorganic perovskite crystal lattice modification. $CsPbCl_{3-x}Br_x$ ($0 < x \leq 3$) phases emitting light in the 421-504 nm spectral range were obtained by reaction of the solid $CsPbBr_3$ material with HCl vapor. Fluorescent microscopy of the treated films confirmed the anion exchange occurs evenly over the whole area of the samples. Furthermore, the polymer-perovskite films obtained did not suffer of segregation effect even when illuminated by severe UV light for 1000 s. We believe the dominant contribution of the surface free energy to Gibbs free energy of the dispersed in polymer matrix perovskite grains is supposed to be responsible for the mixed-halide material phase stabilization. As the light initiated and electrostatic field induced segregation effects in mixed-halide perovskites have the similar nature our findings might be useful for the futher progress in fabrication of polymer-perovskite LEDs emitting in the blue spectral region.

4. Acknowledgement

Synthesis of the perovskite-based materials was financially supported by the Russian Foundation for Basic Research (grant no. 18-33-00669). Optical characterization of the materials was financially supported by the Russian Science Foundation (grant no. 17-73-20336).

References

[1] Lee M M, Teuscher J, Miyasaka T, Murakami T N, Snaith H J 2012 *Science* **338** 643
[2] Kim H S, Lee C-R, Im J-H, Lee K-B, Moehl T, Marchioro A, Moon S-J, Humphry-Baker R, Yum J-H, Moser J E, Grätzel M, Park N-G 2012 *Sci. Rep.* **2** 591
[3] Tiguntseva E Y, Zograf G P, Komissarenko F E, Zuev D A, Zakhidov A A, Makarov S V, Kivshar Y S 2018 *Nano Lett.* **18** 1185
[4] Li J, Shan X, Bade S G R, Geske T, Jiang Q, Yang X, Yu Z 2016 *J. Phys. Chem. Lett.* **7** 4059
[5] Zhu H, Fu Y, Meng F, Wu X, Gong Z, Ding Q, Gustafsson M V, Tuan Trinh M, Jin S, Zhu X-Y 2015 *Nature Mater.* **14**, 636
[6] He X, Liu P, Zhang H, Liao Q, Yao J, Fu H 2017 *Adv. Mater.* 1604510
[7] Hoke E T, Slotcavage D J, Dohner E R, Bowring A R, Karunadasa H I, McGehee M D 2015 *Chem. Sci.* **6** 613
[8] Wang Q, Zheng X, Deng Y, Zhao J, Chen Z, Huang J 2017 *Joule* **1**, 371

IOP Conf. Series: Journal of Physics: Conf. Series **1092** (2018) 012123 doi:10.1088/1742-6596/1092/1/012123

Quantum effects amplification due to energy dissipation in parametrically driven qubit-cavity systems

S.V. Remizov[1,2], **A.A. Zhukov**[1,3], **D.S. Shapiro**[1,2], **W.V. Pogosov**[1,4], **Yu.E. Lozovik**[1,5,6]

[1]N.L. Dukhov All-Russia Research Institute of Automatics, 127055 Moscow, Russia
[2]V.A. Kotel'nikov Institute of Radio Engineering and Electronics, Russian Academy of Sciences, 125009 Moscow, Russia
[3]National Research Nuclear University (MEPhI), 115409 Moscow, Russia
[4]Institute for Theoretical and Applied Electrodynamics, Russian Academy of Sciences, 125412 Moscow, Russia
[5]Institute of Spectroscopy, Russian Academy of Sciences, 142190 Moscow region, Troitsk, Russia
[6]Moscow Institute of Physics and Technology, 141700 Moscow region, Dolgoprudny, Russia

E-mail: sremizov@gmail.com

Abstract. A qubit system coupled to a resonator may be treated as a quantum metamaterial. Quantum nature of such a system results in large variety of non-trivial effects. In particular, we study a parametrically driven system of qubits coupled to a single-mode resonator and show that the energy dissipation in the resonator can enhance population of the excited state of the qubits and generate the entanglement between qubits. The energy dissipation in the qubit subsystem can enhance the number of photons generated from vacuum. We demonstrate that these effects survive in the steady state as well. The optimal parameters for these two cases of dissipation-induced quantum effects amplification are shown to be results of fine balance of different processes in this hybrid system.

1. Introduction

A key feature of a quantum metamaterial is a coherent quantum dynamics. One of the possible quantum object which can be used as an element of such a metamaterial is a superconducting qubit [1]. It has prominent perspectives from experimental point of view due to its high degree of flexibility and tunability [2], and especially due to the possibility of dynamical changing of the coupling strength [3, 4, 5]. Thus, the investigation of parametrically driven systems is important for fundamental science and applications.

2. Hamiltonian and basic equations

We study a hybrid system of qubits coupled to a single-mode quantum resonator. The Hamiltonian of the system reads as

$$\mathcal{H}(t) = \omega a^{\dagger} a + \sum_j \epsilon_j \sigma_j^{+} \sigma_j^{-} + \sum_j g(t)(\sigma_j^{+} + \sigma_j^{-})(a^{\dagger} + a), \tag{1}$$

where $a^\dagger$ and a are photon creation and annihilation operators, σ_j^+, σ_j^- are Pauli operators acting in the subspace of j-th qubit degrees of freedom, and $g(t)$ is dynamically tunable coupling constant. The last term of the Hamiltonian is related to the interaction of qubits and photons in the resonator. It contains two contributions. The first one is rotating wave term, $V_1 = \sum_j g(t)(\sigma_j^+ a + \sigma_j^- a^\dagger)$. The second one is counter-rotating wave term, $V_2 = \sum_j g(t)(\sigma_j^+ a^\dagger + \sigma_j^- a)$. V_2 term can be essential for correct description of the system's dynamics when the coupling constant is periodically modulated [6, 7, 8].

To take into account the decoherence we solve numerically the Lindblad equation

$$\partial_t \rho(t) - \Gamma[\rho(t)] = -i[\mathcal{H}(t), \rho(t)],$$

where $\rho(t)$ is time-dependent full density matrix of the systems. We assume that the bath is Markovian so that the decoherence depends on the rates of energy dissipation in the resonator κ, in the qubit γ, and on the pure qubit dephasing rate γ_φ:

$$\Gamma[\rho] = \kappa\left(a\rho a^\dagger - \frac{1}{2}\{a^\dagger a, \rho\}\right) + \gamma\left(\sigma^-\rho\sigma^+ - \frac{1}{2}\{\sigma^+\sigma^-, \rho\}\right) + \gamma_\varphi\left(\sigma^z\rho\sigma^z - \rho\right).$$

3. The effect of energy relaxation on qubit entanglement

For the simplicity we study in this section two identical qubits coupled to the resonator. We assume that initially qubits and resonator are in their ground states, $|\downarrow\downarrow, 0\rangle$. The effect of coupling constant modulation is strongest when parametric resonance takes place, i.e. the modulation frequency is twice the resonator frequency [9, 10]. To further simplify the description we separate fast and slow degrees of freedom and perform time averaging over the fast ones. This procedure eliminates fast oscillations with the frequency of the photon field which have small amplitude. Only oscillations with much smaller frequencies of the order of the Rabi frequency survive. In the limit $|g(t)| \ll \omega$, the system's dynamics is controlled by two Fourier components [9, 10], $\langle g(t)\rangle_t \equiv p$ and $\langle g(t)\exp(-2i\omega t)\rangle_t \equiv q$. The only energy scale of the problem is defined by the coupling strength g_0 and we represent p and q as $p = g_0\theta$, $q = g_0(1-\theta)$, where $0 \le \theta \le 1$. Parameters p and q control the coupling strength in V_1 and V_2 channels respectively.

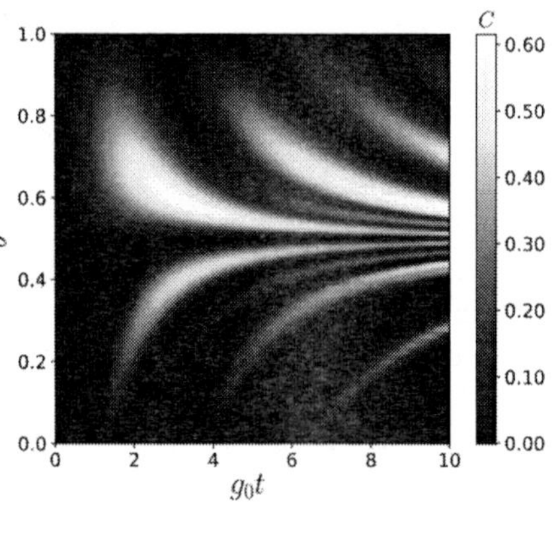

Figure 1. Decoherence-free time evolution of quantum concurrence.

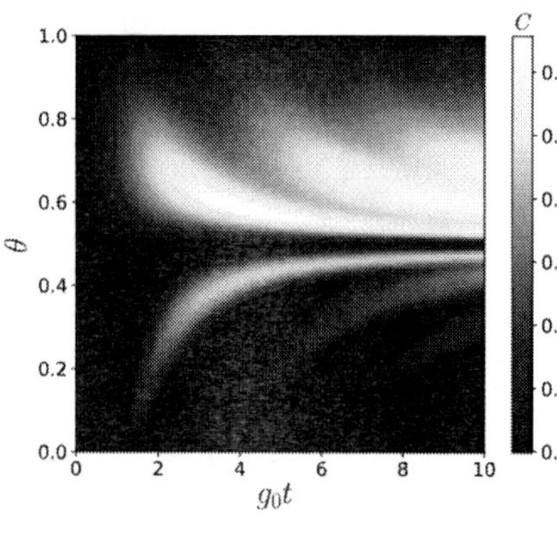

Figure 2. Time evolution of quantum concurrence at $\kappa = 0.2g_0$, $\gamma = 0$.

We start from the case of zero decoherence. Fig. 1 demonstrates the time evolution of quantum concurrence C, which we calculate with use of a standard procedure [11]. The time evolution of C reveal oscillations with the Rabi frequency. Relatively large maximum value of

C results from the resonance between excitation energies of qubits and the resonator frequency. As one can expect, non-zero value of the relaxation in the resonator blurs the picture as it is shown in Fig. 2. The effect of non-zero qubit relaxation is also quite expected — the concurrence oscillations become smaller with time and finally concurrence disappears. Surprisingly, turning on both relaxations simultaneously can stabilize concurrence, so that it remains non-zero in steady state even for non-zero qubit relaxation. This effect takes place only when $\theta > 1/2$, as it is illustrated in Fig. 3, 4.

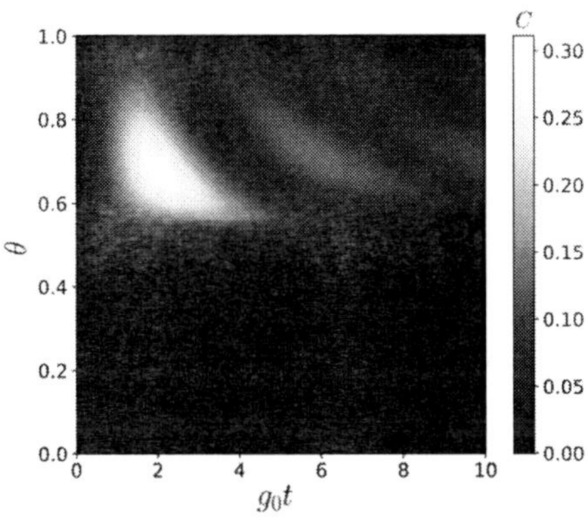

Figure 3. Time evolution of quantum concurrence at $\kappa = 0.1g_0$, $\gamma = 0.2g_0$.

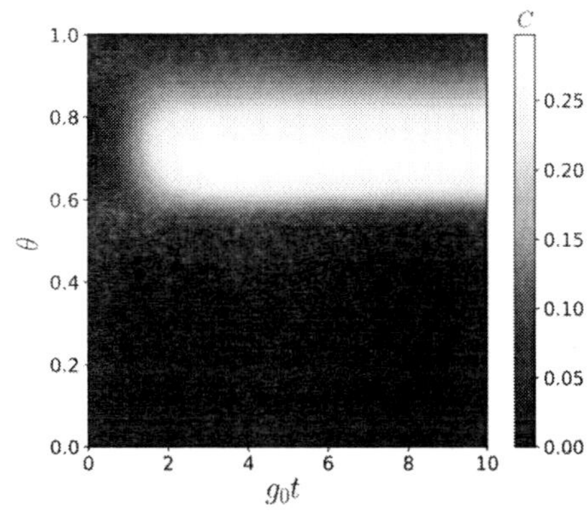

Figure 4. Time evolution of quantum concurrence at $\kappa = g_0$, $\gamma = 0.2g_0$.

Steady state analysis shows that there is an optimal value of resonator relaxation rate κ as it is illustrated in Fig. 5. A non-zero κ allows to create a kind of a circle consisting of three processes. At first, the system is excited from the lowest energy state $|\downarrow\downarrow, 0\rangle$ to the state $|\uparrow\downarrow, 1\rangle + |\downarrow\uparrow, 1\rangle$ via V_2 process. Then states $|\uparrow\uparrow, 0\rangle$ and $|\downarrow\downarrow, 2\rangle$ are produced by V_1 process and $|\uparrow\uparrow, 0\rangle$ state contributes to C. Non-zero κ leads to the decay of $|\downarrow\downarrow, 2\rangle$ state to the initial state $|\downarrow\downarrow, 0\rangle$ and thus closes the cycle. This explains positive role of κ for the quantum concurrence stabilization. For complete analysis the other processes should be taken into account as well. In particular, the decay of $|\uparrow\downarrow, 1\rangle + |\downarrow\uparrow, 1\rangle$ state may suppress the occupation of $|\uparrow\uparrow, 0\rangle$ state via V_1 process. Thus, quantum concurrence appears as a result of a fine balance between different processes.

As seen in Fig. 6, in general the behavior of the mutual information I is close to the one of C. Significant difference exists in the region $\theta \leq 1/2$, where $C = 0$, but $I > 0$. It means that in this case the qubits are correlated but not entangled.

In steady state zero is the optimal qubit relaxation rate for C due to the energy dissipation destroys quantum effects in the same channel. As a result, the less qubit relaxation rate the larger the steady state quantum concurrence.

4. The effect of energy relaxation on occupation numbers

For the simplicity in this section we study single qubit coupled to the resonator. It is parametrically excited from its ground state by the counter-rotating term of the Hamiltonian. Results for qubit occupation number in the steady state, n_q, are provided in Fig. 7. It demonstrates a non-trivial behavior — the largest effect of parametric excitation achieved for some optimal value of κ. This feature can be treated as a consequence of fine balance of counter-rotating processes and energy relaxation in resonator. The influence of γ is quite predictable — it suppress the population of qubit's excited state.

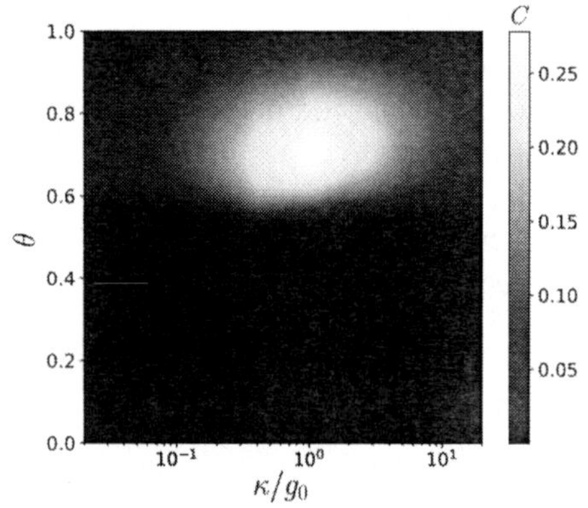

Figure 5. Steady state quantum concurrence at $\gamma = 0.2g_0$.

Figure 6. Steady state mutual information at $\gamma = 0.2g_0$.

The mean photon number in the steady state, n_{ph}, shows tricky behavior. The maximum value of n_{ph} is achieved at $\theta = 0.5$ and $\gamma = 0$. This regime indicates that the smaller the energy dissipation, the stronger the quantum effects. In addition, there is another local maximum of n_{ph} at $\theta = 1$ and at finite γ. These two maxima are connected by a kind of an arc with locally enhanced n_{ph}. As κ grows and approaches g_0, this local maximum become a global one. Thus, the enhancement of photon generation from vacuum due to the energy dissipation in the qubit subsystem starts to dominate at non-zero threshold of resonator relaxation $\sim g_0$. As one may expect, the increase of κ lead to suppression of mean photon number.

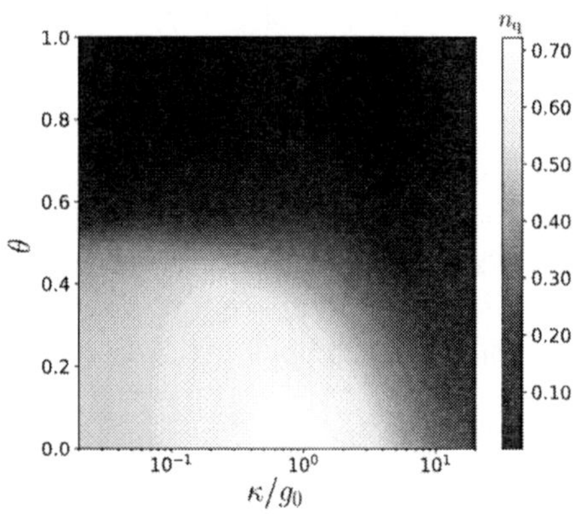

Figure 7. Steady state population of qubit's excited state at $\gamma = 0.2g_0$.

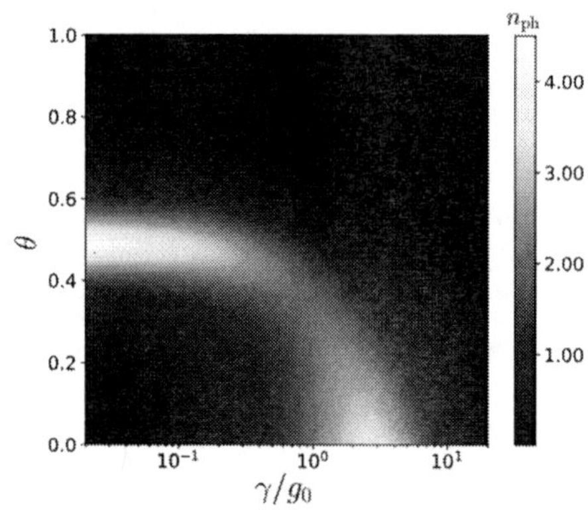

Figure 8. Steady state mean photon number at $\kappa = 0.2g_0$.

We also analyze the influence of pure dephasing γ_φ on our results. Generally, it suppresses the amplification of quantum effects, but this suppression is negligible as long as condition $\gamma_\varphi \lesssim \gamma$ is valid.

5. Conclusion

We studied theoretically the dynamics of parametrically driven system of qubits coupled to a single-mode resonator. We demonstrated that the energy dissipation does not always lead to the suppression of quantum effects. Instead, energy dissipation in one subsystem can greatly increase quantum effects in the other subsystem. Thus, non-zero value of the resonator decay rate can stabilize quantum concurrence and quantum mutual information of two qubits and allows them to survive even in the steady state despite of non-zero qubit relaxation rate. Non-zero qubit relaxation rate, in turn, leads to the amplification of photon generation from vacuum. This effect is more pronounced when coupling constant modulation is strong.

Our results may be useful for further research and development of quantum metamaterials.

6. Acknolegements

D. S. S. acknowledges a support by the Fellowship of the President of Russian Federation for young scientists (fellowship no. SP-2044.2016.5) and the Russian Science Foundation (contract no. 16-12-00095). Yu. E. L. acknowledges a support from RFBR (project no. 17-02-01134).

References

[1] Macha Pascal, Oelsner Gregor, Reiner Jan-Michael, Marthaler Michael, André Stephan, Schön Gerd, Hübner Uwe, Meyer Hans-Georg, Il'ichev Evgeni and Ustinov Alexey V 2014 *Nature Communications* **5** 5146
[2] Peng Z H, de E G S, Tsai J S and Astafiev O V 2016 *Nature Communications* **7** 12588
[3] Hoffman A J, Srinivasan S J, Gambetta J M and Houck A A 2011 *Phys. Rev. B* **84**(18) 184515
[4] Srinivasan S J, Hoffman A J, Gambetta J M and Houck A A 2011 *Phys. Rev. Lett.* **106**(8) 083601
[5] Zeytinoğlu S, Pechal M, Berger S, Abdumalikov A A, Wallraff A and Filipp S 2015 *Phys. Rev. A* **91**(4) 043846
[6] Zhukov A, Shapiro D, Remizov S, Pogosov W and Lozovik Y 2017 *Physics Letters A* **381** 592–596
[7] Dodonov V V 1998 *Phys. Rev. A* **58**(5) 4147–4152
[8] De Liberato S, Gerace D, Carusotto I and Ciuti C 2009 *Phys. Rev. A* **80**(5) 053810
[9] Shapiro D S, Zhukov A A, Pogosov W V and Lozovik Y E 2015 *Phys. Rev. A* **91**(6) 063814
[10] Zhukov A A, Shapiro D S, Pogosov W V and Lozovik Y E 2016 *Phys. Rev. A* **93**(6) 063845
[11] Wootters W K 1998 *Phys. Rev. Lett.* **80**(10) 2245–2248

IOP Conf. Series: Journal of Physics: Conf. Series **1092** (2018) 012124 doi:10.1088/1742-6596/1092/1/012124

Reduced Graphene Oxide Nanostructures by Light: Going Beyond the Diffraction Limit

Raul D. Rodriguez*[1], Ma Bing[1], Alexey Ruban[1], Sergey Pavlov[1], Ammar Al Hamry[2], Varnika Prakash[3], Munis Khan[2], Gennadiy Murastov[1], Ashutosh Mukherjee[2], Zoheb Khan[2], Suhail Shah[2], Anna Lipovka[1], Olfa Kanoun[2], Surinder K. Mehta[3], and Evgeniya Sheremet[1]

[1]Tomsk Polytechnic University, 30 Lenin Ave, 634050 Tomsk, Russia

[2]Chemnitz University of Technology, D-09107 Chemnitz, Germany

[3]Panjab University, Chandigarh, India

*corresponding author, E-mail: raulmet@gmail.com

Abstract. Graphene oxide (GO) offers excellent possibilities that are recently demonstrated in many applications ranging from biological sensors to optoelectronic devices. The process of thermal annealing aids in removing the oxygen-containing groups in GO, making GO more graphene-like, or the so-called reduced graphene oxide (rGO). Thermal reduction can also be achieved by intense light. Here, we demonstrate a scalable, inexpensive, and environmentally friendly method to pattern graphene oxide films beyond the diffraction limit of light using a conventional laser. We show that contrary to previous reports, non-linear effects that occur under high intensity conditions of laser irradiation allow the fabrication of highly conductive carbon nanowires with dimensions much smaller than the laser spot size. The potential of this method is illustrated by the fabrication of several devices on flexible and transparent substrates, including hybrid plasmonic/rGO sensors.

1. Introduction

Similar to graphene, graphene oxide (GO) has become a hot-topic in the 2D materials community. However, in comparison to graphene, GO has low electrical conductivity due to the presence of sp^3 as well sp^2 hybridized carbon atoms. GO electrical conductivity as well as its chemical properties can be drastically enhanced by applying excess energy that removes the oxygen-containing functional groups. In this way, the sp^2 hybridization of graphene is partly restored making reduced graphene oxide (rGO) an electrical conductor. Light irradiation can be considered a form of thermal annealing [1], in particular when using a laser to generate laser-reduced graphene [2]. Besides light, there are other ways to produce rGO from GO films, that includes, an electron beam or a heated atomic force microscopy (AFM) tip. In the latter cases, it is possible to obtain submicrometer-sized patterns of rGO that can be exploited in the fabrication of micro and nanosystems including biomedical applications [3]. However, patterning by an electron beam is not scalable or inexpensive. The same goes for patterning by AFM; it is time-consuming and definitively non-scalable limiting the development of devices. In this contribution to NANOMETA 2018, we aim at developing and demonstrating an inexpensive, scalable, and faster way to produce rGO patterns from GO films deposited on any arbitrary surface. Our method shows the additional advantage of creating conductive patterns of rGO with spatial dimensions below the laser spot size.

2. Materials and Methods

Graphene oxide commercially available from "Graphenea" was used to obtain GO films deposited by drop-coating. Different substrates were used: a rigid substrate (glass slips) and two flexible ones (PDMS and cellophane).

Different laser irradiation systems were investigated to produce the rGO patterns in lines, as shown schematically in Figure 1. A ytterbium pulsed laser engraving machine was used with 1064 nm excitation wavelength along with following parameters: laser spot 80 µm, speed rate 400 mm/s, frequency 90 kHz, pulse duration 50 ns at 5% nominal power. The second system was a portable laser engraving system with 1 W laser excitation output at 405 nm wavelength. The third method investigated involved two solid-state lasers at 785 nm and 532 nm. The Raman spectroscopy experiments were performed in a DXR™2xi Raman Imaging Microscope in the backscattering geometry with a 532 nm laser focused on the sample by a 50x LWD objective.

Spread resistance atomic force microscopy experiments were performed with a NTEGRA Prima SPM using monolithic tips made of Au. The current maps and localized IV curves were obtained in contact mode. The macro-scale conductivity was evaluated using a homemade voltage-amperometer device. After contacting the ends of rGO lines the voltage-current characteristics were recorded for samples obtained under different irradiation conditions. The (non-scaled) sketch of the laser irradiation to make conductive lines and the configuration for resistance measurements are shown in Figure 1.

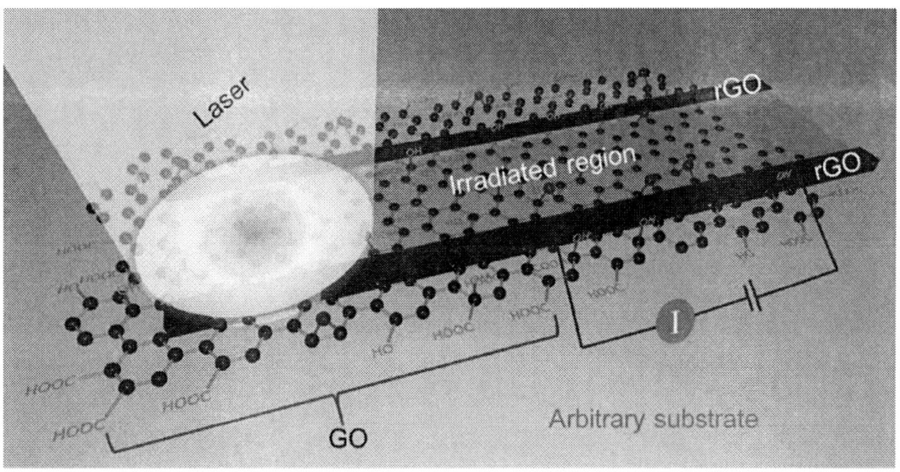

Figure 1: Schematics of the laser reduction of graphene oxide in patterned lines and the I-V characterization

3. Results and Discussion

The Figure 2a shows the conductivity map for a laser-induced pattern made on a GO film. This pattern was obtained using conventional laser annealing conditions with a 532 nm laser focused by a LWD objective (NA=0.5), and power of 8 mW. The Figure 2b is the topography map showing that there are no morphological modifications related to the laser-illuminated pattern. The Figure 2c shows the IV curves locally obtained with a metallic AFM tip used as electrode. In agreement with the conductivity map in Figure 2a, we see from the IV curve that the GO region is not conductive at all. The two other IV curves shown in Figure 2c, labelled “Conventional” and “Beyond” correspond to patterns obtained under different laser irradiation parameters, mild illumination under 532 nm, and high laser power illumination under 405 nm, respectively.

In several control samples made of linear patterns we systematically observed that: 1) there was no material remaining in the patterned lines, the laser ablated down to the plastic substrate (see Figure 3), and 2) the patterned lines showed an unexpected high electrical conductivity. To reconcile this apparent contradiction, we took a closer look at the laser-irradiated regions with Raman spectroscopy. The results obtained from dozens of spectra measured in the lines and on the pristine GO films show spectral changes that can be attributed to a modification of the carbon at the edges of the pattern (see Figure 2d and Figure 3a). Current sensing atomic force microscopy results allowed us to spatially visualize the regions responsible for the conductivity of patterns obtained under the highest laser power. These results (not shown) demonstrate that the edges were responsible for the electrical conductivity. This allowed us to conclude that by working in an extreme regime of high laser power, much higher than any reported until now in laser-reduced GO, highly confined regions of rGO could be obtained. The creation of these highly conductive edges occurred at the interface between the ablated GO regions where only the bare substrate remained, and the pristine GO at the other side of the edge. It can also be seen from the Raman microscopy image in Figure 3b that the laser spot size represented by the ablated area (blue color) is much wider in size than the thin green-bluish edges made of rGO.

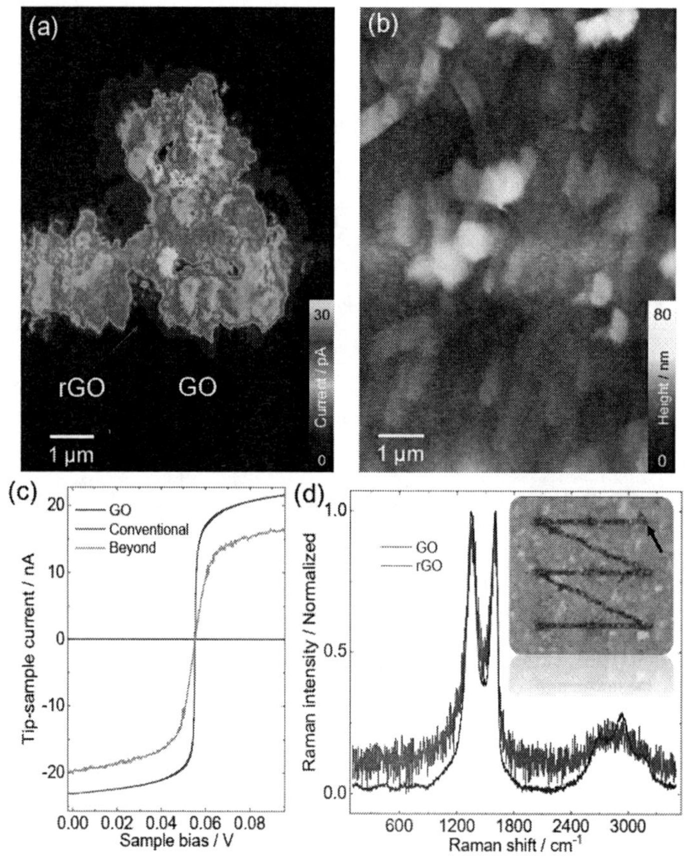

Figure 2: (a) Current sensing and (b) topography images obtained by conductive atomic force microscopy and (c) the electrical IV characterization. (d) The Raman spectra of GO and rGO; the D, G, and second order phonons show the effect of laser reduction. The inset shows a microscopy image of the laser-reduced pattern with an arrow indicated the spot imaged by AFM.

The generality and power of our method are demonstrated by the fabrication of different kinds of devices including flexible electrochemical sensors with plasmonic nanoparticles for dual-channel sensing and electronic circuits for lightings. Furthermore, this method opens the door to the creation of tunable optoelectronic devices by controlling the reversibility of the reduction process *via* an external bias. This latter is an exciting part of our ongoing research that will be briefly discussed as future perspective.

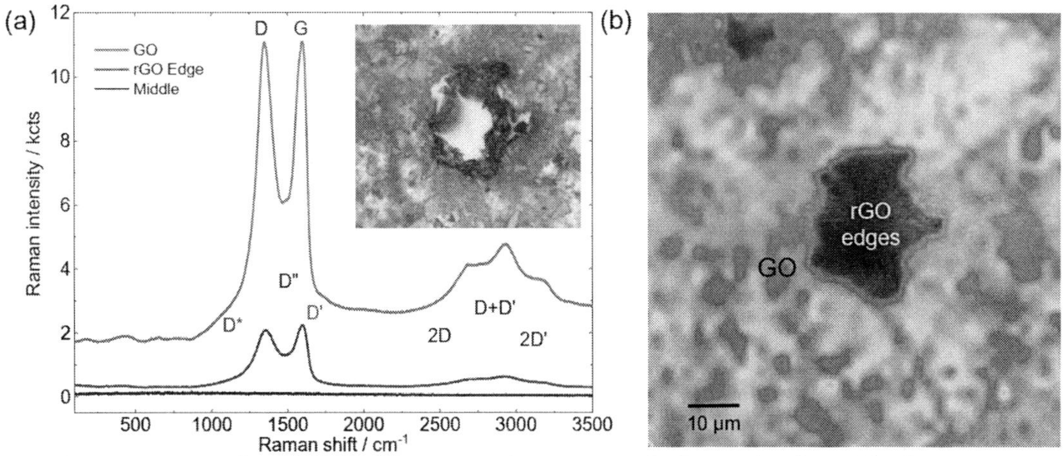

Figure 3: (a) Raman spectra from the main three regions of interest around a laser-reduced spot. The changes in intensity ratios between different peaks and the peak widths demonstrate the effects of reduction. The inset shows the optical microscopy image of the single laser irradiated spot on a GO film deposited on Au. (b) Raman intensity map for the G peak at 1590 cm^{-1}. The blue region at the middle of the spot shows no presence of GO or rGO film as also shown in the spectrum in (a).

4. Conclusion

Our research has tremendous implications in the inexpensive and large-scale fabrication of graphene-based optoelectronic devices with dimensions much smaller than the laser spot size circumventing the diffraction limit of light.

Acknowledgements

The research is carried out at Tomsk Polytechnic University within the framework of Competitiveness Enhancement Program. This work was performed in the context of the EU COST Action MP1302 Nanospectroscopy.

References

[1] A. Al-Hamry, H. Kang, E. Sowade, V. Dzhagan, R.D. Rodriguez, C. Müller, D.R.T. Zahn, R.R. Baumann, O. Kanoun, Tuning the reduction and conductivity of solution-processed graphene oxide by intense pulsed light, Carbon N. Y. 102 (2016). doi:10.1016/j.carbon.2016.02.045.

[2] Y. Zhang, L. Guo, S. Wei, Y. He, H. Xia, Q. Chen, H.B. Sun, F.S. Xiao, Direct imprinting of microcircuits on graphene oxides film by femtosecond laser reduction, Nano Today. 5 (2010) 15–20. doi:10.1016/j.nantod.2009.12.009.

[3] L. Dedelaite, R.D. Rodriguez, E. Andriukonis, M. Hietschold, D.R.T. Zahn, A. Ramanavicius, Surfaces functionalized by graphene oxide nanosheets for single cell investigations, Sensors Actuators, B Chem. (2018). doi:10.1016/j.snb.2017.08.187.

IOP Conf. Series: Journal of Physics: Conf. Series **1092** (2018) 012125 doi:10.1088/1742-6596/1092/1/012125

Porous germanium with Ag nanoparticles formed by ion implantation

A.M. Rogov[1], Y.N. Osin[1], V.I. Nuzhdin[2], V.F. Valeev[2], A.L. Stepanov[2]

[1] Kazan Federal University, 420008, Kazan, Russia
[2] Kazan Physical-Technical Institute, Russian Academy of Sciences, 420029, Kazan, Russia
Corresponding author: *alexeyrogov111@gmail.com*

Abstract. A novel approach is proposed and tested for the synthesis of thin porous PGe layers with Ag nanoparticles based on low-energy high-dose ion implantation of single-crystal *c*-Ge. To demonstrate a successes of this technique, an Ag^+-ion implantation of a polished *c*-Ge substrates with an energy of 30 keV at various doses of $1.0 \cdot 10^{16}$ - $1.5 \cdot 10^{17}$ ion/cm^2 and a current density of 5 μA/cm^2 was performed. Various analytical methods such as scanning electron and probe microscopy, as well as EDX analysis and electron backscattered diffraction were applied for observation of PGe formation of a spongy structure consisting of a network of intersecting Ge nanowires. At the ends of the nanowires, the synthesis of Ag nanoparticles were detected. It was also found that the formation of pores during Ag^+-ion implantation was-accompanied by efficient spattering of the Ge surface.

[1] Introduction

Porous germanium (PGe), which for first time was mentioned in the work [1], is widely used nowadays in optoelectronics and in the field of solar cell elements creation [2] as alongside with porous silicon (PSi). Multiple technologies for the creation of PGe layers were developed, for example the electrochemical treatment of single-crystal c-Ge in highly concentrated electrolytes [3], plasma-stimulated chemical vapor deposition [4], spark discharge method [5], thermal annealing of GeO_2 ceramic films in a hydrogen atmosphere [6] and others.

An effective nonchemical method of fabrication of nanoscale thin PGe layers on the c-Ge surface is a high-dose implantation with various ions in a vacuum. The first direct evidence of the PGe formation by such technological approach was obtained by electron-microscopic observations, while c-Ge was irradiated by Ge^+-ions at energy of 50 keV [7]. The porous structure was recorded, as shown in the review [8], for as amorphous (a-Ge) as c-Ge substrates by low-energy high-dose (> 1 MeV) implantation with variety of ions. It should also be mentioned some publications [9, 10] in which the Ag^+-ion implantation at high energies (100 MeV) and low doses in the range of $5.0 \cdot 10^{12}$ - $2.0 \cdot 10^{14}$ ion/cm^2, was used to create PGe layers.

This paper is also devoted to the creation of PGe layers by the ion implantation, but the aim is to synthesize Ag nanoparticles simultaneously with the PGe structure (Ag:PGi). For this purpose, the low-energy (<100 keV) in dose range ($1.0 \cdot 10^{16}$ - $1.5 \cdot 10^{17}$ ion/cm^2) Ag+ -ion implantation of c-Ge was applied. Interest in PGe structures with noble metal nanoparticles showing plasmon resonance properties is due to their possible use in various applications: to increase the absorptivity in solar cells [11], to improve photoconductivity [12], to generate electron-hole pairs [13] and so on.

[2] Experimental

Polished c-Ge substrate with thickness of 0.5 mm was used to obtain a structured surface Ag:PGe layer. For this purpose the implantation of c-Ge was carried out by Ag^+ ions with an energy of 30 keV at various irradiation doses of $1.0 \cdot 10^{16}$ -$1.5 \cdot 10^{17}$ ion/cm^2 and a current density of 5 $\mu A/cm^2$ with the ion accelerator ILU-3 at the room temperature of the substrate. Observation of morphology of the sample surface and energy-dispersive X-ray (EDX) spectral microanalysis of implanted Ge was carried out with a self-emission high-resolution scanning electron microscope (SEM) Merlin (Carl Zeiss) ~~and~~ which was equipped with Aztec X-Max spectrometer (Oxford Instruments). Structure of surface sample layer was characterized by electron backscattered diffraction (EBSD) patterns using HKL NordLys detector (Oxford Instruments).

In order to obtain the distribution of Ag nanoparticle ~~we used~~ scanning probe microscope (SPM) Dimension FastScan (Bruker) was used. This microscope is known for quantitative nanomechanical mapping feature which allows to obtain information on mechanical properties of the sample surface, such as adhesion, Young's modulus etc, alongside with classic atomic force measurements in a single-pass scanning.

[3] Results and discussion

SEM images of the Ge surface implanted with Ag^+ ions presented with various scales are shown in Fig. 1 and 2. Unlike the virgin polished c-Ge substrate, the morphology of the irradiated surface is seen as the highly developed open porous spongy structure (Fig. 1). It's also important to note that similar structure of PGe was formed by implantation of c-Ge, for example, by Bi+-ion with an energy of 30 keV [8], but it differs significantly from the columnar type PGe formed by implantation with a lighter Ge+ ion at low energies [13, 14]. In principal, the possibility of creating pores by implantation with Ge+ self-ions indicates that the formation of pores is happened not due the presence of an impurity, but because of some specific energy conditions during irradiation [18], which could be assumed also for our case Ag:PGi too, although heavy implants such as Ag could stimulate the appearance of a generally spongy structure.

Based on the SEM and SPM microphotographs (fig. 1,2,3), we also assume the following mechanism of distribution of Ag in PGe: small dose of $1.0 \cdot 10^{16}$ ion/cm^2 is characterized with small Ag nanoparticles (bright dots on fig 2a and 3a) but not a very effective spattering of Ge, the increase of the dose leads to better spattering and formation of spongy structure and at the same time the size of Ge nanowires in this structure also increases and starting from the dose range of around $1.0 \cdot 10^{16}$ ion/cm^2 these nanowires are decorated with Ag nanoparticles (fig 2b and 3b). The effect of the rise if the size of the nanowires can be explained in terms of covering initial Ge nanowires with Ag "film".

Acknowledgements

This work was supported by the Russian Science Foundation No. 17-12-01176.

References

[1] T.M. Donovan, K. Heinemann, Phys. Rev. Lett. 27 (1971) 1974-1976.

[2] E.G. Rojas, J. Hensen, J. Carstensen, H. Föll, R. Brendel, ESC Transactions. 33 (2011) 95-102.

[3] G. Flamand, J. Poortmans, K. Dessein, Phys. Stat. Sol. C 2 (2005) 3243-3247.

[4] J. Shieh, H.L. Chen, T.S. Ko, H.C. Cheng, T.C. Chu, Adv. Mat. 16 (2004) 1121- 1124.

[5] Kartopu G., S.C. Bayliss, R.E. Hummel, Y. Ekinci, J. Appl. Phys. 95 (2004) 3466-3472.

[6] C. Jing, C. Zhang, X. Zang, W. Zhou, W. Bai, T. Lin, J. Chu, Sci. Technol. Adv. Mater. 10 (2009) 65001-1 – 65001-6.

[7] I.H. Wilson, J. Appl. Phys. 53 (1082) 1698-1705.

[8] R. Böttger, K.-H. Heinig, L. Bischoff, B. Liedke, S. Facsko, Appl. Phys. A 113, (2013) 53-59.

[9] T. Som, J. Ghatak, O.P. Sinha, R. Sivakumar, D. Kanjilal, J. Appl. Phys. 103 (2008) 123532-1 - 123532-5.

[10] S. Hooda, S.A. Khan, B. Satpati, D. Strange, D. Buca, M. Bala, C. Pannu, D. Kanjilal, D. Kabiraj, Appl. Phys. A 122 (2016) 227-1 – 227-7.

[11] A. Polman, Science 322 (2008) 868-869.

[12] H. Atwater, A. Polman, Nature Mat. 9 (2010) 205-213.

[13] A. Cavalcoli, B. Fraboni, G. Impellizzeri, L. Romano, E. Scavetta, M.G. Grimaldi, Microporous and Mesoporous Mater. 196 (2014) 175-178.

[14] J.F. Ziegler, J.P.Biersack, U. Littmark, The Stopping and Range of Ions in Solids. Pergamon Press: New York., 1985.

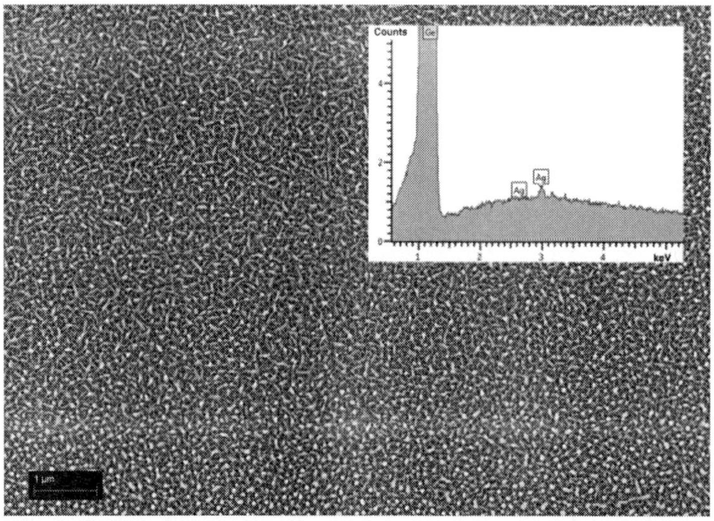

Fig. 1 SEM image of the Ge surface implanted with Ag+ ions

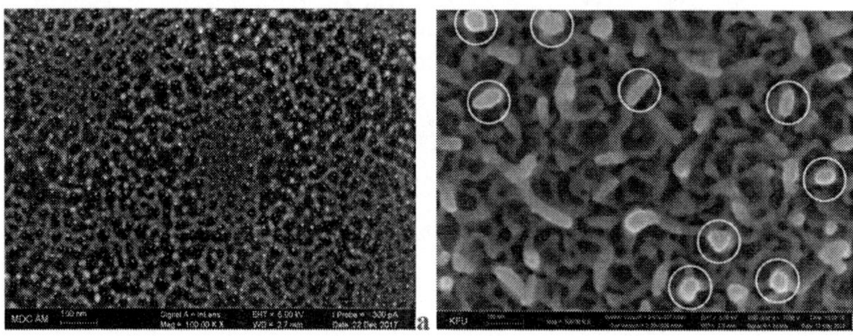

a b

Fig. 2 SEM images of the Ge surface implanted with Ag+ ions a. dose 1.0$^{\cdot 16}$ ion/cm^2; b. dose 1.5$^{\cdot 17}$ ion/cm^2

IOP Conf. Series: Journal of Physics: Conf. Series **1092** (2018) 012125 doi:10.1088/1742-6596/1092/1/012125

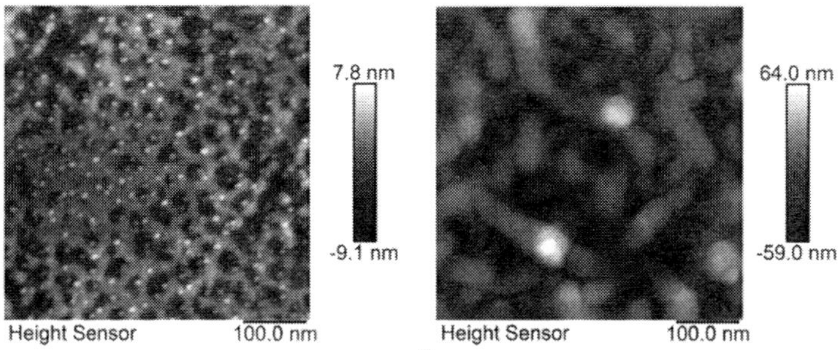

a b

Fig. 3 **AFM images of the Ge surface implanted with Ag+ ions a. dose $1.0^{\cdot16}$ ion/cm^2; b. dose $1.5^{\cdot17}$ ion/cm^2**

IOP Conf. Series: Journal of Physics: Conf. Series **1092** (2018) 012126 doi:10.1088/1742-6596/1092/1/012126

HgCdTe based quantum well heterostructures for long-wavelength lasers operating in 5 - 15 THz range

V V Rumyantsev[1,2], N S Kulikov[1,2], A M Kadykov[1], M S Zholudev[1,2], M A Fadeev[1], V. Ya. Aleshkin[1,2], S Winnerl[3], N N Mikhailov[4,5], S A Dvoretskii[4], V I Gavrilenko[1,2], M. Helm[3] and S V Morozov[1,2]

[1] Institute for Physics of Microstructures of RAS, 603950,Nizhny Novgorod, Russia

[2] Lobachevsky State University, 603950, Nizhny Novgorod, Russia

[3] Institute of Ion Beam Physics and Materials Research, Helmholtz-Zentrum Dresden-Rossendorf, D01328 Bautzner Landstraße, Dresden, Germany

[4] A.V. Rzhanov Institute of Semiconductor Physics, Siberian Branch of Russian Academy of Sciences, 630090 Novosibirsk, Russia

[5] Novosibirsk State University, 630090, Novosibirsk, Russia

E-mail: rumyantsev@ipmras.ru

Abstract. Stimulated emission (SE) at wavelengths up to 20.3 μm (14.7 THz) is demonstrated from HgCdTe based quantum well heterostructures. The mitigation of Auger processes resulting from «graphene-like» energy spectrum in structures under study is exemplified. By studying the carrier lifetime via time resolved photoconductivity and transmission pump-probe measurements we show that further increase in SE wavelength is feasible and discuss the promising routes towards the 5 – 15 THz frequency domain, where quantum cascade lasers are scarce and HgCdTe lasers may be competitive to the prominent emitters.

1. Introduction

The problem of developing terahertz (THz)/long-wavelength emitters is one of the hottest topics in the applied physics. For many spectroscopy related applications, it would be favorable to employ compact semiconductor lasers as the sources of long wavelength radiation. Quantum cascade lasers (QCLs) demonstrate remarkable performance in the spectral range from 1 THz to 5THz and above 15 THz [1]. However, the majority of the QCLs exploit GaAs and InP semiconductors, in which lattice absorption becomes too strong at frequencies lower than 15 THz. GaN QCL are tackling the 5 – 15 THz "gap" from the low-frequency side, but their figures of merit are yet to be improved [2]. The interband lasers are a straightforward alternative, but it requires narrow-gap materials, in which the non-radiative Auger recombination is expected to be very effective. The spectral range 5 – 15 THz is now partly covered only with the lead salt diode lasers, which provide emission wavelengths up to 46 μm [3]. The factor that mitigates the Auger recombination in PbSnSe(Te) is the symmetry between the dispersion laws in the conduction band and in the valence band. It can be shown that for the energy-momentum conservation laws to be fulfilled, the summarized kinetic energy of the three particles involved in the Auger process has to be over a certain threshold energy E_{th} [4] that depends on carrier energy spectra. For the certain types of carrier spectra, e.g. for the Dirac relativistic energy spectrum, it is not possible to fulfill the energy-momentum conservation laws in Auger process at all, i.e. E_{th} is infinite. Note that the particular case of massless Dirac fermions is realized in graphene, which has been studied intensively as a promising candidate for THz lasers and sensors. However, there is still some debate concerning the effectiveness of the Auger recombination in graphene structures because the marginal case of relativistic spectrum with zero bandgap and linear dispersion law requires rigorous consideration [5]. As for lead salt lasers, their figures of merit are limited by the growth technology: there are formidable problems in manufacturing quantum well (QW) structures for PbSnSe(Te) solid solutions, and the residual carrier density remains high $\sim 10^{17}$ cm^{-3} [3].

There are alternative semiconductor systems that allow one to come close to the "graphene-like" energy-momentum law, but leave a finite bandgap. As it has been shown in Ref. [6] one of such systems is HgTe/CdHgTe QW heterostructures. Unlike graphene, HgCdTe based QW allow tailoring the bandgap by changing the QW width and Cd content in it. Modern molecular beam epitaxy (MBE) delivers high quality HgCdTe epitaxial structures not only for native CdZnTe substrates, but for "alternative" GaAs substrates as well. HgCdTe epitaxial structures grown on GaAs(013) substrates have demonstrated remarkable figures of merit during our previous studies of photoconductivity (PC) and photoluminescence (PL) in the very long wavelength infrared (λ = 15 – 30 μm) range [7, 8]. In HgCdTe QW structures with bandgap ~ 60 -- 80 meV we observed the increase of carrier lifetimes with the pumping intensity up to several microseconds due to saturation of Shockley-Read-Hall (SRH) recombination centers, inferring that optical gain can be obtained under reasonable pumping intensity [7]. However, previously lasing in HgCdTe has been studied only in the short-wavelength part of the mid-infrared range (2 – 5 μm) [9]. In 1993, Arias et al. achieved the emission wavelength of 5.3 μm at 45K in diode laser that did not exploit QWs in its design [10]. As for the HgCdTe lasers with QWs, only emission near $\lambda \sim 2$ μm and $\lambda \sim 3$ μm was demonstrated [9] under optical pumping. In this work, we report stimulated emission (SE) from HgCdTe structures at wavelengths up to $\lambda \sim 20$ μm, and discuss the underlying physics that controls the Auger recombination in such structures. We demonstrate that there is a considerable reserve for HgCdTe based lasers to enter the 5 – 15 THz frequency domain, where no QCLs are available yet, and HgCdTe emitters might appear as a better alternative than the lead salt lasers.

2. Results and discussion

Structures under study were MBE-grown on semi-insulating GaAs(013) substrates with ZnTe and CdTe buffers using in situ ellipsometric control of the layer content and thickness [11]. Ex situ characterization of structures was performed by measuring interband photoconductivity (PC) and photoluminescence (PL) spectra at different temperatures. The temperature dependence of the bandgap was extracted from the PC and PL spectra and then compared to the energy spectra, calculated in the framework of Kane 8x8 Hamiltonian axial model, allowing us to determine the width and Cd content for each QW.

Structure#1 contains ten 12 nm thick $Hg_{0.87}Cd_{0.13}Te/Cd_{0.65}Hg_{0.35}Te$ QWs and structure#2 -- five 5.4 nm thick $Hg_{0.91}Cd_{0.09}Te/Cd_{0.6}Hg_{0.4}Te$ QWs. The structures were not intentionally doped; the residual carrier density obtained from the Hall measurements is several 10^{10} cm^{-2} and the typical dislocation density determined from the etch pits density was of ~ 10^{6} cm^{-2}. The structures were designed so as to effectively confine light to in-plane direction, therefore the "active" region (5 -- 10 QWs) was grown inside a thick (several micron) waveguide. Due to a quite specific growth direction (013), naturally cleaved facets do not form the Fabri-Perot resonator: the angle between the sample's surface and the lateral edge planes is far from 90^0. Thus, the SE studied in this work results from single-pass amplification.

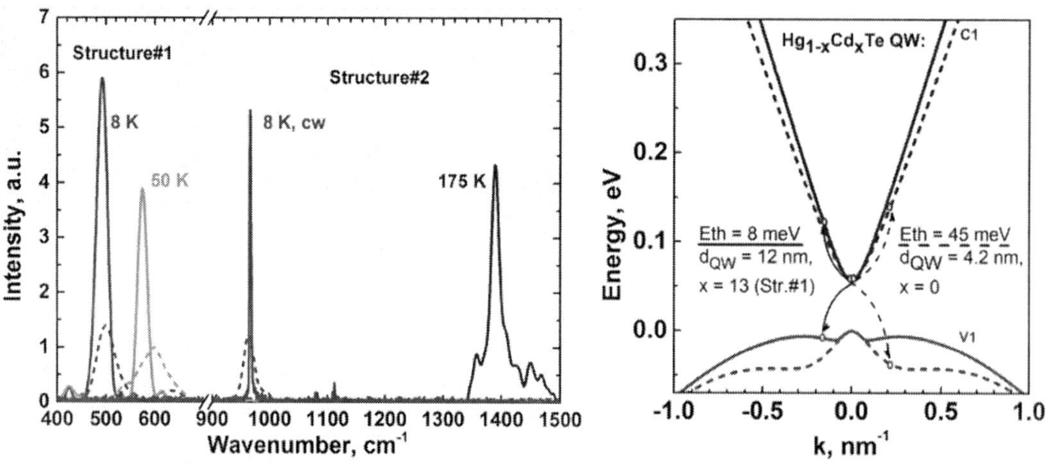

Fig. 1 Stimulated emission spectra for structures under study at different temperatures (solid curves). SE was obtained under pulsed pumping with 2300 nm wavelength and 10 kW/cm^2 intensity for Structure#1 (T = 8K, T = 50K) and 65 kW/cm^2 for Structure#2 at 175K. SE at cw excitation with 7W/cm^2 intensity at 900 nm wavelength is presented for Structure#2 at 8K. Dash curves show PL spectra obtained with the same cw pumping source at 5 W/cm^2 intensity for Structure#1 (T = 8K, T = 50K) and 1 W/cm^2 for Structure#2 at 8 K. Fig. 2 Calculated energy spectrum of Structure#1 (solid lines) and HgTe/CdHgTe QW with the same bandgap (dash lines), T = 0. Threshold energies and carrier configuration are shown for Auger CCHC process.

Fig. 1 shows the SE spectra for two samples under study at 8K and "critical" temperature T_{max}, above which no SE could be observed. The intensity of emission demonstrates clear threshold dependence on pumping power, typical for the onset of SE. Another telltale sign of the SE is PL line narrowing: it can be seen that the SE line FWHM is ~ 1.5 meV immediately above the threshold, and with increase in pumping intensity broadens up to ~ 4 meV, staying however less than the FWHM of PL spectra under continuous wave (cw) excitation and much less than typical width of PL from single HgCdTe QWs under pulsed excitation with the same intensity, which is ~ 10 meV [7].

The corresponding thresholds are 5 kW/cm^2 and 120 W/cm^2 for SE wavelength ~20 μm (15 THz) and ~10 μm (30 THz). However, these values are obtained for "below barrier" excitation, when the wavelength of the pumping radiation is longer than the cutoff wavelength of the interband transitions in the waveguide material and QW barriers. In this case one can presume that non-equilibrium carriers are generated only in QWs. Taking for estimation the QW absorption as 1 % one can find that corresponding carrier density in each QW should be $n_{th} = 1.4\times10^{11}$ cm^{-2} for $\lambda_p = 2.3$ μm and $I_{th} =$ 0.12 kW/cm^2 pumping intensity. This n_{th} value agrees well with our previous calculations: in Ref. [12] the possibility of radiation amplification in narrow gap HgTe QWs was stated for excess carrier densities above 1×10^{11} cm^{-2}. It also allows one to estimate the equivalent threshold current density for the same $N_{well} = 5$ QWs placed into a *p-n* junction as $J_{th} = eN_{well}n_{th}/\tau_{pulse}$ (*e* is the elementary charge), which for $\tau_{pulse} = 10$ ns gives $J_{th} = 11$ A/cm^2. This value is at least the order of magnitude less than that of lead salt lasers operating at the same temperature and emission wavelength and practically equal to threshold current density of InAs-based interband QCL emitting near 10.4 μm at 80 K [13]. In particular, the threshold is low enough to obtain SE under cw excitation: Fig 1. shows the corresponding spectrum, measured with cw pumping at 900 nm wavelength. In this case, pumping radiation generates electron-hole pairs in the barriers and as a result, the threshold is as low as 5 W/cm^2 and the SE line FWHM is only 0.5 meV ($\Delta\lambda/\lambda \sim 0.004$).

Obviously, the threshold grows with increase in the SE wavelength and the maximum "operating" temperature T_{max} gets lower. To understand the energetic scale that determines the critical temperature T_{max} consider the energy spectrum of Structure#1, presented in Fig.2. As can be seen from Fig. 2, for large k-vectors the hole effective mass increases dramatically and the dispersion laws in the valence band and the conduction band are no longer quasi-symmetrical. When holes reach larger-mass region the Auger recombination becomes efficient. Indeed, we have calculated the threshold energies E_{th} for CCHC Auger process (the energetic threshold for CHHH process is high compared to CCHC process, therefore the latter is the most important) and Fig. 2 shows the corresponding configuration of carriers for Auger process at the threshold. When compared to the critical temperature T_{max}, observed in the experiment, for Structure#2 under study we get $T_c \approx E_{th}/2$ (the same relation appears to be valid for all structures under study), corresponding to ~ 25% of holes having kinetic energy above the threshold in case of Bolzmann distribution law (as can be seen the portion of electron kinetic energy is the threshold energy in neglegible). However, the detrimental impact of side maxima in the valence band can be reduced in QW from pure HgTe. Fig. 2 shows the energy spectrum for HgTe/CdHgTe QW with the same bandgap as Structure#1. It can be seen that larger-mass region in the valence band shifts to lower energies, increasing the threshold energy dramatically. As it will be shown elsewhere, changing the sign of the QW strain (by adding Zn to barrier layers, for instance) can additionally suppress the Auger recombination.

Carrier lifetime measurements agree well with the abovementioned experimental data. In particular, the calculated threshold energy E_{th} is 29 meV for Structure #2, which gives $T_{max} \sim E_{th}/2 \sim$ 168 K, close to experimental value. Indeed, calculations demonstrate that radiative lifetime dependence on carrier density is enough to describe well the experimental PC and PL dynamics at 77K, showing that Auger recombination is of minor importance. For Structure#1 the PC decay measurements at 77K give carrier lifetime of ~ 60 ns for carrier density ~ 10^{11} cm^{-2}, while the calculated radiative lifetime is an order of magnitude higher, explaining the absence of SE at 77K. Obviously, carrier lifetimes are expected to decrease for QW structures with narrower bandgap. The sub-nanosecond carrier lifetimes have been explored via the pump-probe measurements of a sample's transmission using a free electron laser [14]. For HgCdTe QW with 20 meV, corresponding to 4.8 THz frequency, the carrier lifetime is no less than 100 ps for carrier density of 10^{11} cm^{-2} that is sufficient to achieve population inversion. One can estimate a threshold pumping intensity of 10 kW/cm^2 for an optically pumped laser exploiting such QWs as an active media. Thus, HgCdTe QWs should be able to provide amplification of radiation in 5 – 15 THz range as well.

The work was supported by the Russian Foundation for Basic Research (#16-32-60172) and Ministry of Education and Science of the Russian Federation MK-4399.2018.2. The work was done using equipment of Center "Physics and technology of micro- and nanostructures" at IPM RAS.

References

1. M.S. Vitiello *et al.*, Opt Express, 2015. **23**(4): p. 5167-82.
2. M.F. Anwar *et al.*, Proc. SPIE, 2015. **9483**: p. 948304.
3. K.V. Maremyanin *et al.*, Semiconductors, 2016. **50**(12): p. 1669-1672.
4. V.N. Abakumov *et al.*, *Nonradiative Recombination in Semiconductors*. 1991, North-Holland: Elsevier Science Publishers.
5. G. Alymov *et al.*, arXiv 2018: p. 1709.09015.
6. B.A. Bernevig *et al.*, Science, 2006. **314**(5806): p. 1757-61.
7. S.V. Morozov *et al.*, Applied Physics Letters, 2014. **105**(2): p. 022102.
8. V.V. Rumyantsev *et al.*, Semiconductors, 2013. **47**(11): p. 1438-1441.
9. J. Bleuse *et al.*, Journal of Crystal Growth, 1999. **197**(3): p. 529-536.
10. J.M. Arias *et al.*, Semiconductor Science and Technology, 1993. **8**(1S): p. S255.
11. S. Dvoretsky *et al.*, Journal of Electronic Materials, 2010. **39**(7): p. 918-923.
12. S.V. Morozov *et al.*, Semiconductors, 2012. **46**(11): p. 1362-1366.
13. Z. Tian *et al.*, Electronics Letters, 2012. **48**(2): p. 113-114.
14. S. Ruffenach *et al.*, APL Materials, 2017. **5**(3): p. 035503.

IOP Conf. Series: Journal of Physics: Conf. Series **1092** (2018) 012127 doi:10.1088/1742-6596/1092/1/012127

Effect of finite lateral size of dielectric grating on optical bound state in the continuum

Z. F. Sadrieva, A. A. Bogdanov

Department of Nanophotonics and Metamaterial, ITMO University, St. Petersburg, Russia

E-mail: zfsadrieva@corp.ifmo.ru

Abstract. Optical bound states in the continuum (BIC) are localized states with energy lying above the light line and having infinite lifetime. Any losses taking place in real systems result in transformation of the bound states into resonant states with finite lifetime. In this work, we analyze properties of BIC in CMOS-compatible one-dimensional photonic structure based on silicon-on-insulator and silicon nitride wafers. We reveal that finite lateral size of samples could destroy both off-Γ BIC and at-Γ BIC turning them into resonant states due to in-plane symmetry breaking.

1. Introduction

Localization of light in cavities (waveguides) implies that there is no coupling of cavity modes with modes of free space forming a continuum spectrum, i.e. there is no leakage losses. It follows from the Maxwell's equations straightforwardly that energies of localized states lie under the light line of the surrounding space. Optical bound states in the continuum (BICs) are a remarkable exception for this general rule. The BICs are spatially bounded (localized in one, two or even three dimensions [1, 2]) eigenstates of an optical system with quality factor that could, in principle, approach infinity despite lying within the light cone of the surrounding space, i.e. in the continuum spectrum of radiative modes. The concept of BIC was firstly proposed by von Neumann and Wigner in 1929 for electron placed in a specific potential [3]. More recently, another examples of electronic BICs have been predicted theoretically in atomic and molecular systems [4] and in artificial systems [5]. In optics, the term bound state in the continuum first appeared around 2008 [6] although this phenomenon has been studied earlier [7]. Experimental observation of optical BIC followed only in 2011 [8].

The main advantage of the bound optical states is the high quality factor which can be of the order of 10^7 and is limited only by the losses in the material and by the roughness of the photonic structures [9]. High quality, CMOS-compatibility and simplicity of geometry and design make the resonators based on the optical bound states in the continuum attractive for wide range of applications beginning with lasers and sensors and ending with nonlinear optical devices.

The next significant feature is feasibility to excite the bound optical states lying inside the light cone by an ordinary plane electromagnetic wave without using any auxiliary devices or tricks such as prisms, diffraction gratings or surface irregularities which are necessary to excite the states outside the light cone. This advantage offers great possibilities of technical applications.

In theory, BICs have infinite high quality factor, however, for real systems it is limited because of material absorption, technological imperfections, roughness, finite lateral size of samples and leakage into the substrate. Here, we analyze influence of the finite lateral size on frequency and Q-factor of BICs.

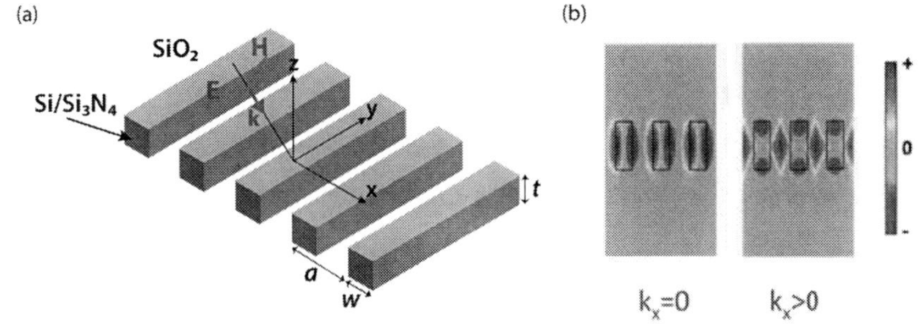

Figure 1. (a) Schematic image of a sample. (b) Electric field $|E_y|$ distribution for TE-like mode $\mathbf{E} \sim |E_y|e^{-ik_x x}$.

2. Results and Discussion

The design of the photonic structure under study is shown in Fig. 1(a). It consists of rectangular bars made of crystalline silicon or silicon nitride surrounded by fused silica. Since the array is periodic in the x-direction, the eigenstates of the structure are characterized by the frequency f, the wave vector along the bars k_y and by the Bloch quasi-wave vector k_x restricted to the first Brillouin zone. It has been shown that such one-dimensional photonic structures supports BICs either if wave vector k_y or quasi-wave vector k_x is zero, i.e. an excitation wave goes along the x- or y-direction. For the proposed design, BIC occurs when the k_y component of wave vector is zero. Hence, in this case, the spectrum of the structure consists of pure TE and TM modes, see Fig. 1(b).

Table 1. Position of BIC for photonic structure composed by bars with height about 200 nm. The incident angel α is calculated in both air and silica.

(period,width) nm	f, THz (nm)	polarization	$\alpha_{air(SiO2)}$ (deg)
(500,200)	241.2 (1243)	TM	65 (39)
(600,270)	204.6 (1462)	TM	62 (38)
(600,300)	197.2 (1520)	TM	74 (42)
(600,240)	213.9 (1402)	TM	52 (33)
(700, 315)	186.6 (1607)	TM	49 (32)
(700,280)	193.8 (1546)	TM	43 (28)
(750,300)	185 (1621)	TM	40 (27)
(750,300)	186 (1612)	TM	39 (26)
(800,360)	172.6 (1734)	TM	41 (27)
(800,360)	171,3 (1750)	TM	42 (28)
(900,405)	161.9 (1851)	TM	33 (22)
(900,405)	158.9 (1887)	TM	36 (24)

We consider the gratings fabricated from both silicon and silicon nitride. We numerically solved Maxwells equations by finite element method using COMSOL Multiphysics software.

One can see from Fig. 2 and Tab. 1 that there are many designs of grating supporting TE and TM-like BICs. We will choose one the most suitable for fabricating and experiment.

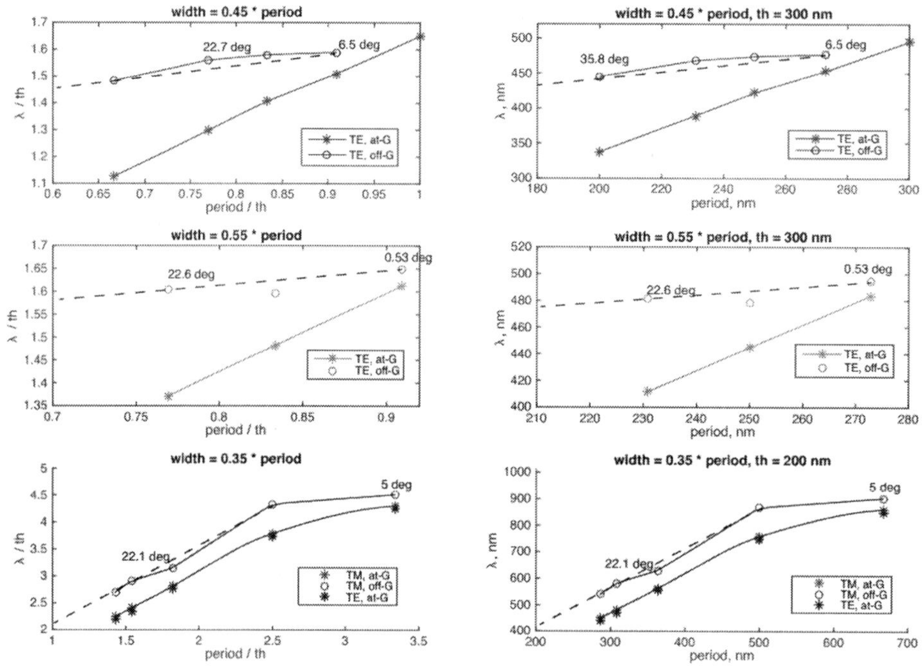

Figure 2. Position of BICs supported by silicon nitride grating.

Also, we studied the influence of finite lateral size, i.e. finite number of periods, on frequency and Q-factor of both at-Γ and off-Γ BICs. In finite quasy-periodic array bloch wave vector ceases to vary continuously whithin the first Brillouin zone but with step $\Delta k = \pi/aN$, where N is a number of periods a of the structure. Figure 3 shows dispersion curves and Q-factors. One can see that both at-Γ and of-Γ BICs keep the frequencies while the Q-factor drops significantly for off-Γ BIC. We suppose that the at-Γ BIC is less sensitive to finite lateral size due to protection by rotational symmetry. The finite structure has lost it's rotational symmetry so the BIC transforms in resonant state with finite lifetime. But the more periods include the array the higher is it's Q-factor.

To sum up, we have studied one-dimensional periodic structures. We confirm that the BIC transforms into resonant state with finite Q-factor due to finite lateral size of the structure. The obtained results provide useful guidelines for practical implementations of structures supporting optical bound states in the continuum and could find a number of applications in optical communications, on-chip photonics, laser physics and sensing.

Acknowledgments

Development of the numerical model was carried out with a support of Russian Science Foundation (17-12-01581). Numerical calculations and treatment of the results were provided with a support of RFBR according to the research project 18-32-00976 (mol a).

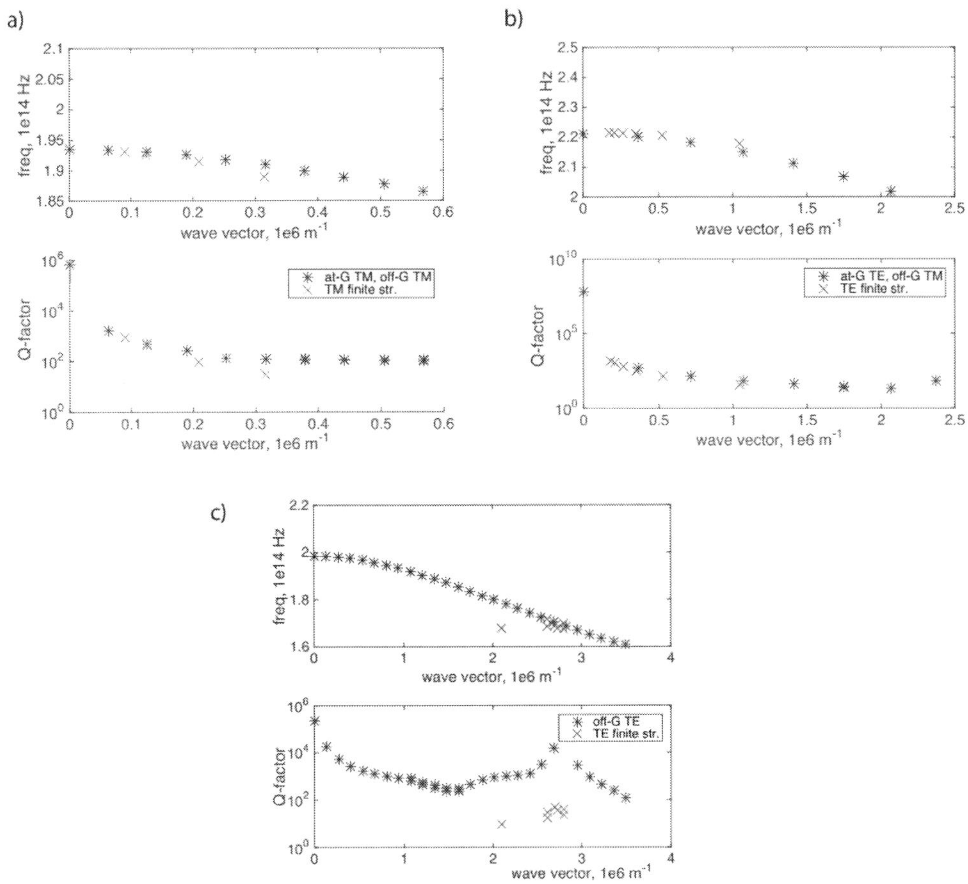

Figure 3. Calculated band structure and quality factor of silicon nitride grating supporting (a) TM-like BIC, (b,c) TE-like BIC.

References

[1] Alu A and Engheta N 2005 Achieving transparency with plasmonic and metamaterial coatings *J. Phys. Rev.* E **72** 1–9
[2] Monticone F and Alu 2014 Embedded photonic eigenvalues in 3D nanostructures *J. Phys.Rev. Lett.* **112** 1–5
[3] von Neumann J and Wigner E 1929 Uber merkwurdige diskrete Eigenwerte *Phys.Zeitschrift.* **30** 465–7
[4] Stillinger F H ,Herrick D R, Laboratories B, Hill M, Bound states in the contimuun *Phys. Rev.* A 1975 **11** 446–54
[5] Bulgakov E N, Pichugin K N, Sadreev A F, Rotter I Bound states in the continuum in open Aharonov-Bohm rings *JETP Lett.* 2006 **84** 430–5
[6] Bulgakov E N, Sadreev A F Bound states in the continuum in photonic waveguides inspired by defects. *Phys. Rev.* B 2008 **78** 075105
[7] Ochiai T and Sakoda K, Dispersion relation and optical transmittance of a hexagonal photonic crystal slab *Phys. Rev.* B 2001 **63** 125107
[8] Plotnik Y, Peleg O, Dreisow F, Heinrich M, Nolte S, Szameit A, Segev M, Experimental Observation of Optical Bound States in the Continuum *Phys. Rev. Lett.* 2011 **107** 183901
[9] Sadrieva Z, et al, Transition from optical bound states in the continuum to leaky resonances: role of substrate and roughness *Acs Photonics* 2017 **4(4)** 723–727

IOP Conf. Series: Journal of Physics: Conf. Series **1092** (2018) 012128 doi:10.1088/1742-6596/1092/1/012128

Topological interface states due to spontaneous symmetry breaking in a chain of anharmonic oscillators

Roman S. Savelev[1], **Maxim A. Gorlach**[1], **Alexander N. Poddubny**[1,2]
[1]ITMO University, Saint Petersburg 197101, Russia
[2]Ioffe Institute, Saint Petersburg 194021, Russia
E-mail: r.savelev@metalab.ifmo.ru

Abstract. In this work, we propose a one-dimensional system of coupled anharmonic oscillators with nonlinear coupling and study a spontaneous formation of interface excitations in the linearized spectrum of such periodic array. We show that during its evolution such system can undergo a topological transition from the disordered and topologically trivial phase into the nontrivial one. The topological transition is accompanied by the formation of the interface state in the spectrum of linearized excitations of the equilibrium state. Employing the technique based on the calculation of mean chiral displacement we demonstrate that the in-gap interface state exhibits a topological nature and is analogous to that in the well-known Su-Schrieffer-Heeger model. Our results demonstrate the formation of topological interface state induced by spontaneous symmetry breaking mechanism and open novel possibilities for designing various nonlinear electronic, photonic and polaritonic systems with disorder-robust control of information flows.

Topological edge or interface states of electrons [1], light [2, 3, 4] and sound [5] have recently received much attention due to their prospects for realization of disorder-robust one-way transport of information. Presently, the interest is shifting towards topological states in nonlinear and interacting systems promising higher tunability and richer fundamental physics [6, 7, 8, 9, 10, 11, 12, 13, 14]. However, there is still no clear recipe to realize a nonlinear system with edge or interface states between topologically distinct domains appearing due to spontaneous symmetry breaking. This work addresses an interesting question raised in this regard: is it possible to realize a topological phase transition from a trivial to a gapped nontrivial phase accompanied by appearance of the edge or interface states between topologically distinct domains?

To this end, we consider a system depicted in Fig. 1(a), which is based on the array of identical anharmonic oscillators with double-well on-site potential and nonlinear coupling between the nearest neighbors. The entire array is described by potential function

$$U = \sum_{n=1}^{N}(a_2 y_n^2 + a_4 y_n^4) + \sum_{n=1}^{N-1}\left[b_2(y_n - y_{n+1})^2 + b_4(y_n - y_{n+1})^4\right], \tag{1}$$

where y_n is the displacement of the nth oscillator, a_2, a_4 and b_2, b_4 are on-site and inter-site force constants, respectively, and N is the number of oscillators. The evolution of bulk oscillators are

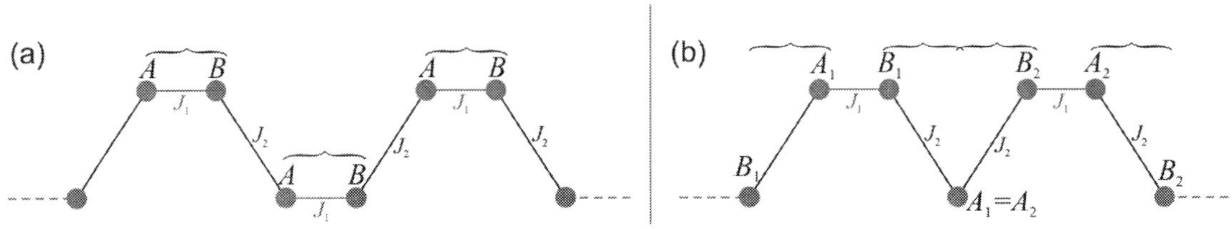

Figure 1. (a) Scheme of an infinite chain in the tetramer state. (b) Scheme of the possible interface configurations between two semi-infinite chains in different tetramer phases. The brackets indicate the unit cell choice in left and right domains.

then described with the following equation of motion:

$$\ddot{y}_n+2\gamma\dot{y}_n = -2a_2y_n-4a_4y_n^3-2b_2(y_n-y_{n-1})-2b_2(y_n-y_{n+1})-4b_4[(y_n-y_{n-1})^3+(y_n-y_{n+1})^3], \quad (2)$$

where γ term describes the friction. Equations of motion for the first and Nth oscillators can be obtained from (2) by omitting the brackets with y_{n-1} and y_{n+1}, respectively. The equilibrium states of the system $y_n^{(0)}$ can be found from the system of equations (2) with zero left side. Our analysis reveals that one of the stable static states of such system is the tetramer stationary state in the form $y_{4n}^{(0)} = y_{4n+1}^{(0)} = -y_{4n+2}^{(0)} = -y_{4n+3}^{(0)} = v_T = \pm\sqrt{-\dfrac{a_2+2b_2}{2(a_4+8b_4)}}$.

The interest in the tetramer state stems from the behaviour of small oscillations $z_n(t)$ near this equilibrium state. The system of linear equations for amplitudes z_n can be written down in the frequency domain, after substituting $y_n(t) = y_n^{(0)} + z_n\exp(-i\omega t)$ into the equation (2) and linearizing it. A linear dispersion of an infinite chain then reads:

$$\begin{cases}(\omega^2-\omega_0^2)A+(J_1+J_2e^{-ik})B=0,\\ (J_1+J_2e^{ik})A+(\omega^2-\omega_0^2)B=0,\end{cases} \quad (3)$$

where $\omega_0^2 = \omega_R^2 + J_1 + J_2$, $\omega_R^2 = 2(a_2+6a_4v_T^2)$, $J_1 = 2b_2$, $J_2 = 2b_2 + 48b_4v_T^2$, A, B are the amplitudes of the two oscillators in the unit cell [see Fig. 1(a)], and k is the normalized Bloch wavenumber. The interaction constants for the tetramer state differ by $|J_{n1} - J_{n2}| = 48b_4v_T^2$, which makes this model similar to the Su-Shrieffer-Heeger one [15]. Consequently, the solution of this system has a gap in linear dispersion.

In general, the edge oscillators have different static displacements than the bulk ones. However, it can be shown that the tetramer state still persists at the edges, if the following condition is fulfilled: $b_2/a_2 = 4b_4/a_4$. This condition along with the constraints imposed by the requirement of existence and stability of the tetramer solution defines the range of possible values of the force constants. In our calculations we have chosen $a_2 = -20$, $b_2 = -2$, $a_4 = 1$ and $b_4 = 1/40$.

In constrast to the linear Su-Shrieffer-Heeger model, such nonlinear system does not possess edge states, which is due to the nonlinear detuning of the resonance frequencies of oscillators. Therefore, we focus on the interface states which can exist even in the nonlinear system at the boundary between two different tetramer phases [16] [see Fig. 1(b)]. In order to probe the emergence of the expected topological interface state, we analyze the dynamics of $N = 80$ oscillators by directly solving full dynamic equations with small friction term $\gamma = 0.03$ included for convergence. Initial displacements of the oscillators were deviated from the static tetramer state with the boundary at the 40th oscillator [the scheme is shown in Fig. 1(b)] by a random value in the range $(-\delta_y, \delta_y)$, with maximum deviation $\delta_y = 0.3|v_0|$ (the static displacement was

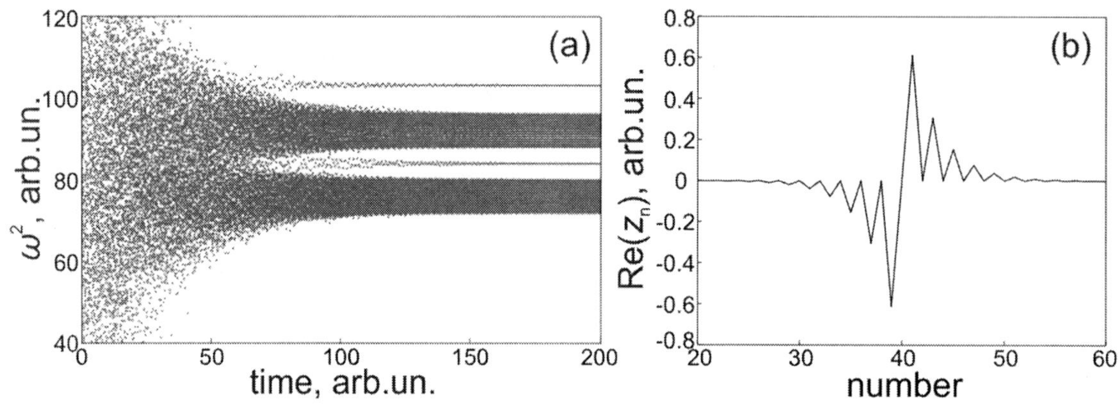

Figure 2. (a) Spectrum of small oscillations during time evolution of the array of 80 oscillators. (b) The distribution of small oscillations for the interface mode.

$|v_0| \approx 3.16$). Similarly, initial velocities were defined by random values in the range $(-\dot{\delta}_y, \dot{\delta}_y)$, with $\dot{\delta}_y = 1.5$. At each moment of time t the calculated $y_n(t)$ were considered as stationary displacements, and the spectrum of small oscillations was evaluated. In this way we recovered the evolution of the spectrum presented in Fig. 2(a). It is seen that during the evolution the spectrum of the system becomes gapped, and two edge states appear, the in-gap state and the state above the allowed bands. We are interested in the former state, corresponding to the topological zero-energy state in the SSH model [15]. The calculated displacement distribution depicted in Fig. 2(b) confirms that this state is indeed localized at the interface.

Since in the considered system the elements of the interaction matrix that describes the dispersion of the linearized system are random in the beginning of system evolution (due to random initial conditions) and converge to stationary values only in the limit of $t \to \infty$, at each moment of time the system is not periodic, and consequently the topological properties of the system cannot be derived from direct calculation of the Zak phase. Thus, we employ the technique based on the calculation of mean chiral displacement [17], which allows for determining of the topological phase of the not necessarily periodic systems. For a given system one can calculate the MCD of a freely evolved state $\boldsymbol{\Psi}$ as a function of time delay τ as follows:

$$\mathrm{MCD}(t,\tau) = \sum (\mathbf{\Gamma n})\boldsymbol{\Psi}(\tau + t), \tag{4}$$

where $\mathbf{\Gamma}$ is the matrix of the chiral operator, $\mathbf{n}$ is the matrix of the position operator, so that $\mathbf{\Gamma n} = \mathrm{diag}(\ldots, -1, 1, 0, 0, 1, -1, 2, -2, \ldots)$; and the vector of displacements $\boldsymbol{\Psi} = [\ldots, y_1, y_2, \ldots]^T$ is localized in an arbitrary unit cell for $\tau = 0$. For the SSH model the value of MCD at large times τ converges to either 0 or 0.5 depending on the choice of the unit cell, which corresponds to the values of the Zak phase $\gamma = 0$ or $\gamma = \pi$, respectively [17]. However, for non-periodic systems time dependence of the MCD might not converge to any certain value. Such behaviour can be observed in Fig. 3, where we plotted the limit values of MCD at $\tau = \infty$ as a function of time t. At small times all oscillators possess relatively large random displacements, and consequently the calculated values of the MCD are also random, i.e. the system does not exhibit nontrivial topological properties. However, at large times $t \gtrsim 150$ the values of $\mathrm{MCD}(\tau \to \infty, t)$ converge to 0 and 0.5 for the left and the right sides of the chain, respectively, indicating the formation of topologically different phases and consequently the topological origin of the interface state.

To summarize, our findings prove that nonlinear systems can dynamically switch from the disordered regime to the regime with non-zero-frequency topological edge states due to the spontaneous symmetry breaking mechanism. We believe that the fundamental link between

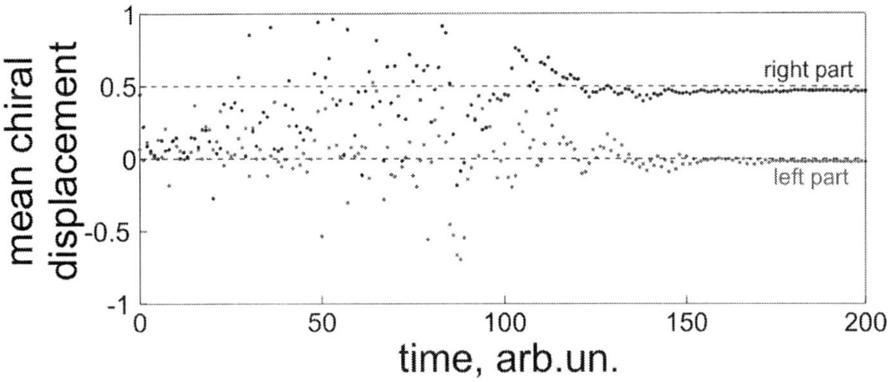

Figure 3. Mean chiral displacement calculated for the left (red dots) and right (blue dots) tetramer domains, see Fig. 1(b).

spontaneous symmetry breaking and dynamical topological states demonstrated here on a simple example of mechanical system is much more general being applicable to a wide variety of nonlinear electronic, photonic and polaritonic systems.

Acknowledgments

This work was supported by the Russian Foundation for Basic Research, according to the research project No. 16-37-60092 mol_a_dk. R.S. acknowledges the support by Grant from the President of the Russian Federation (MK-381.2017.2). ANP has been supported by the "Basis" foundation.

References

[1] Bernevig B and Hughes T 2013 *Topological Insulators and Topological Superconductors* (Princeton University Press)
[2] Lu L, Joannopoulos J D and Soljačić M 2014 *Nat. Photonics* **8** 821–829
[3] Lu L, Joannopoulos J D and Soljačić M 2016 *Nat. Phys.* **12** 626–629
[4] Khanikaev A B and Shvets G 2017 *Nat. Photonics* **11** 763–773
[5] Huber S D 2016 *Nat. Phys.* **12** 621–623
[6] Hadad Y, Khanikaev A B and Alu A 2016 *Phys. Rev. B* **93** 155112
[7] Hadad Y, Vitelli V and Alu A 2017 *ACS Photonics* **4** 1974–1979
[8] Bardyn C E, Karzig T, Refael G and Liew T 2016 *Phys. Rev. B* **93**(2) 020502
[9] Solnyshkov D D, Nalitov A V and Malpuech G 2016 *Phys. Rev. Lett.* **116**(4) 046402
[10] Solnyshkov D D, Bleu O, Teklu B and Malpuech G 2017 *Phys. Rev. Lett.* **118** 023901
[11] Leykam D and Chong Y D 2016 *Phys. Rev. Lett.* **117**(14) 143901
[12] Zhou X, Wang Y, Leykam D and Chong Y D 2017 *New J. Phys.* **19** 095002
[13] Di Liberto M, Recati A, Carusotto I and Menotti C 2016 *Phys. Rev. A* **94** 062704
[14] Gorlach M A and Poddubny A N 2017 *Phys. Rev. A* **95** 053866
[15] Shen S Q 2012 *Topological Insulators. Dirac Equation in Condensed Matters* (Springer)
[16] Blanco-Redondo A, Andonegui I, Collins M J, Harari G, Lumer Y, Rechtsman M C, Eggleton B J and Segev M 2016 *Phys. Rev. Lett.* **116**(16) 163901
[17] Cardano F, D'Errico A, Dauphin A, Maffei M, Piccirillo B, de Lisio C, Filippis G D, Cataudella V, Santamato E, Marrucci L, Lewenstein M and Massignan P 2017 *Nat. Commun.* **8** 15516

The influence of co-doping agents on bismuth clustering in silica synthesized by SPCVD

E.A. Savelyev,[1,*] O.V. Butov,[1] V.O. Yapaskurt,[1,2] and K.M. Golant[1]
[1]Kotel'nikov Institute of Radioengineering and Electronics of RAS, 11-7, Mokhovaya Str., Moscow 125009, Russia
[2]Department of Petrology, Geological Faculty, Moscow State University, Vorobievy Gory, Moscow 119899, Russia
*swoopermsu@yandex.ru

Abstract. We investigated the variation of near-infrared (NIR) luminescence spectra of bismuth-doped silica-based glasses caused by exposure to intense deep UV (193 nm wavelength) laser radiation. Bismuth-doped optical waveguides consisting of silica with additives of Al and P as the light-guiding core, synthesized via surface-plasma chemical vapour deposition (SPCVD) technology, served as the samples in our experiments. Spectra of steady-state NIR luminescence excited by 808, 904, and 970 nm wavelength laser diodes were measured in a wavelength band of 1-2 μm at temperatures of 105 and 300 K. We found that exposing the glass to intense ArF excimer laser radiation leads to NIR luminescence integral intensity and spectrum profile changes. The observed changes are discussed in the assumption that the UV laser exposure causes photoinduced rearrangement of the glass network in the vicinity of bismuth inclusions. It is expected that the latter are present in the glass network in the nature of separate ions, atoms, as well as 3D clusters.

1. Introduction

Currently, a complete example of the active centre responsible for near-infrared (NIR) luminescence of bismuth [1] is absent. The majority of relevant papers consider the origin of this luminescence in terms of the presence of bismuth ions, molecules, dimers and nano-clusters (Bi^0, Bi^+, Bi^{5+}, BiO, Bi_2^-, Bi_2^{2-}, Bi_2) in the host glass [1–5]. Therein bismuth clusters can be semiconductors with a bandgap depending on their size and shape [6].

A stronger acidity of phosphorus oxide (P_2O_5) as compared to silicon dioxide [7] leads to a stabilization in the P-doped silica subvalent states of bismuth, including the clusters [8]. From this point of view, aluminium additive to the silica has to prevent the formation of clusters due to the amphoteric nature of this metal. In such a way, phosphorus and aluminium dopants have to act in all the contrary way with regard to the probability of the formation of bismuth clusters in silica.

As a rule, the bandgap value of a nonmagnetic semiconductor decreases with respect to temperature leading to a redshift of the recombination luminescence spectrum, although there are instances where a blueshift can occur [9]. For this reason, the recording of the luminescence spectra at different temperatures could give additional information about the origin of the active bismuth centres.

Another way to gain further insight on the origin of NIR luminescence in Bi-doped silica-based glasses could be to examine the dependence of the luminescence spectrum on the excitation wavelength. For interband transitions, the spectral position of the luminescence emission peak weakly depends on the excitation wavelength, whereas for atoms and ions with strong non-uniform broadening in the luminescence spectrum, the emission peak must blueshift with an increase of the pump photon energy.

In addition to identifying the size of the active bismuth centres, there is an important problem on how to recognize their neighbours in silica with variations in co-doping. If the bismuth NIR-luminescence centres are linked to inherent silica defects, an external impact that destroys or forms these defects would change the intensity and the fine structure of bismuth luminescence as well. Deep UV laser radiation is one such impact, which destroys a variety of defects in silica that contain different dopants.

2. Samples and experimental technique

We prepared individual samples constituting cylindrical silica rods of about 4 mm in diameter. Each rod contained a doped silica core of approximately 700 μm in diameter activated by bismuth. For this purpose, we synthesized a layer of amorphous silicon dioxide containing bismuth and other co-dopants via surface-plasma chemical vapour deposition (SPCVD) on the inner surface of a reference silica tube 20 mm in diameter and 2 mm wall thickness. The tube with the deposited layer was consolidated in a rod by means of profusion rotation in a flame of the moving oxygen-hydrogen burner.

These cylindrical optical waveguides with bismuth–doped cores as the experimental units were pulled out from thus obtained rods. We have investigated the following three Bi-doped P or Al co-doped silica samples: S.BiP (phosphorous content of about 2.5 at.%), S.BiAl1 (aluminium content of about 2.4 at.%) and S.BiAl2 (aluminium content of about 3.2 at.%). NIR luminescence was excited by means of three different laser diodes at wavelengths of 808, 904, and 967 nm.

Pump radiation from laser diodes was delivered to a flat end of the sample core with the help of a pure silica core multimode fibre. Luminescence light through the lateral surface of the sample via the other multimode fibre was delivered to the DDS-30 (Lomo™) spectrometer equipped with the PDA10DT (Thorlabs™) InGaAs photoreceiver.

A portion of samples was irradiated by the ArF excimer laser radiation (193 nm wavelength). The exposure dose was as high as 5 kJ/cm^2 for each sample.

Measurements at a low (~105 K) temperature were carried out with the help of a homemade optical cryostat. In this case, pump light from the diode lasers was delivered to the active core of the sample by means of a collimator-lens assembly.

3. Results and discussion

We found that the NIR luminescence band for the S.BiP sample is centred at the wavelength of 1180 nm (1.05 eV) for both 904 and 970 nm pumping. Such a performance of the luminescence spectrum is inherent to the recombination luminescence of semiconductors. However, in the case of the S.BiAl2 sample, the peak position redshifts with an increase of the pump wavelength. This suggests a rather point-like nature of the active bismuth centres intrinsic in the silica sample with an aluminium additive.

Fig.1 depicts steady-state luminescence spectra obtained at temperatures of 300 K and 105 K. One can see that the cooling influences the NIR luminescence spectra of active bismuth centres in an unexpected way. The temperature reduction of the S. BiP sample pumped at the wavelength of 904 nm leads to an increase of luminescence intensity in the neighbourhood of 1.1 eV (Fig. 1b). This indicates that 3D bismuth clusters are the centres responsible for NIR luminescence. This luminescence is a consequence of the indirect radiative recombination process accompanied by phonon excitation. Pump absorption in this process takes place via a direct electron transition and weakly depends on the temperature. Conversely, with the S.BiAl2 sample, one can observe only a narrowing of the luminescence spectrum and no shifting in the position of the spectral peak (Figs. 1a and 1c). This suggests not only a point-like nature of this type of bismuth centre, but a lack of or an extremely small magnitude of the Stark splitting of the excited state energy level as well. The $^3P_1 \rightarrow {}^3P_0$ transition observed for Bi^+ ions [3] fully fits these requirements.

Fig. 2 illustrates the impact of the UV laser exposure on the spectra of NIR luminescence. The most significant observable changes are in the spectra of the sample S.BiAl1 excited at a wavelength of 808 nm. These changes can be explained by the recharging of AlO_4^- defects followed by the formation of neutral AlO_4 aggregates [10]. Therefore, one can suggest the presence of AlO_4^- in the neighbouring environment of the bismuth centre responsible for NIR luminescence in silica glass with aluminium additives.

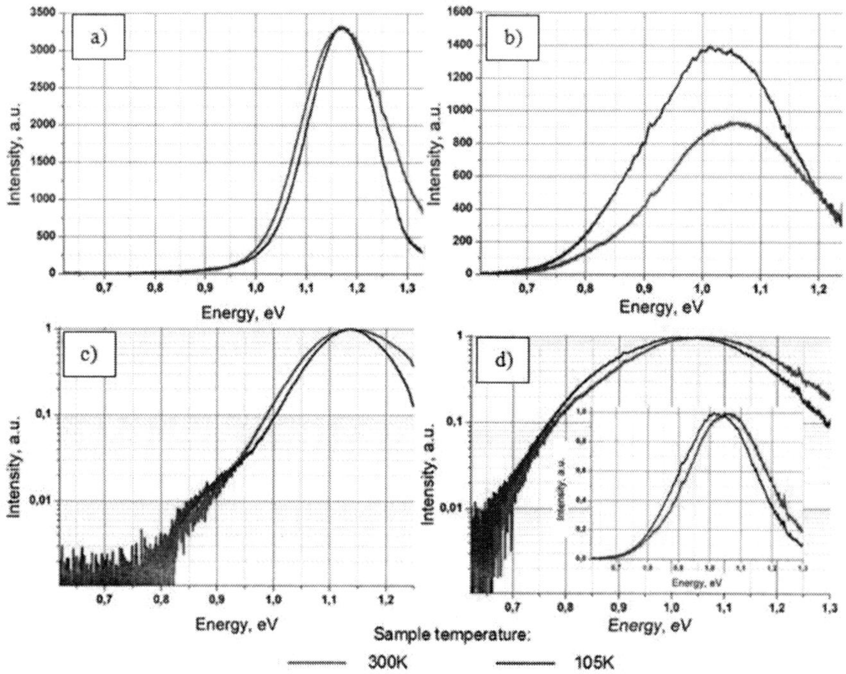

Fig.1. Steady-state NIR luminescence spectra of different samples (S.BiAl2 - a), c); S.BiP - b), d)) at temperatures of 300K and 105K excited by laser diodes at wavelengths of 904 nm - a), b), d) and of 970 nm - c).

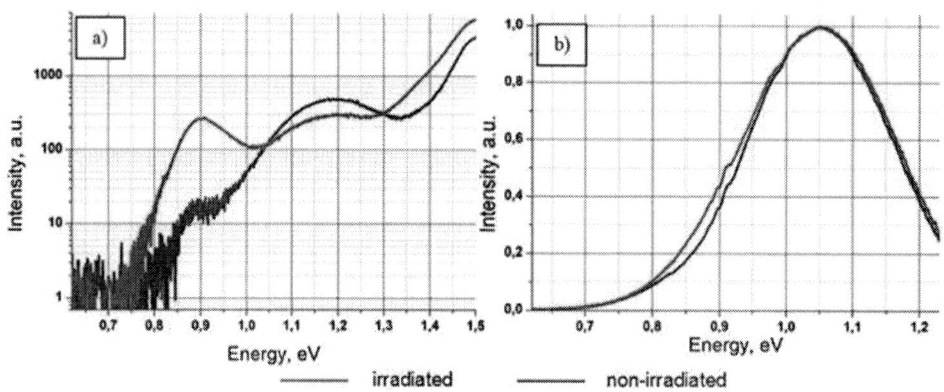

Fig.2. Changes in the steady-state NIR-luminescence spectra, as the result of glass exposure to the intense UV laser radiation; a) - S.BiAl1 excited at the wavelength of 808 nm, b) - SBi.P excited at the wavelength of 970 nm.

4. Conclusions

The results of our experiments with bismuth-doped silica glasses manufactured by the surface-plasma chemical vapour deposition technology are as follows. The addition of aluminium in the glass network shifts the equilibrium towards the formation of interstitial Bi^{+} ions environed by AlO_4^{-}. Meanwhile, the addition of phosphorus most likely favours the formation of 3D bismuth clusters. These two interstitials contribute to the NIR luminescence observed in these glasses.

Acknowledgements
This work was supported by the Russian Basic Research Foundation (Project 16-07-00371).

References

[1] Fujimoto Y and Nakatsuka M 2001 Infrared Luminescence from Bismuth-Doped Silica Glass 2001 *Jpn. J. Appl. Phys.* **40** L279
[2] Peng M, Zollfrank C and Wondraczek L 2009 Origin of broad NIR photoluminescence in bismuthate glass and Bi-doped glasses at room temperature *J. Phys.: Condens. Matter* **21** 285106
[3] Meng X, Qiu J, Peng M, Chen D, Zhao Q, Jiang X, and Zhu C 2005 Near infrared broadband emission of bismuth-doped aluminophosphate glass *Opt. Express* **13** 1628
[4] Sokolov V O, Plotnichenko V G, and Dianov E M 2008 Origin of broadband near-infrared luminescence in bismuth-doped glasses *Opt. Lett.* **33** 1488
[5] Peng M, Qiu J, Chen D, Meng X, and Zhu C 2005 Superbroadband 1310 nm emission from bismuth and tantalum codoped germanium oxide glasses *Opt. Lett.* **30** 2433
[6] Yuan K, Chen H, Kuang A L, Miao Y, and Xiong Z H 2008 Density-functional study of small neutral and cationic bismuth clusters Bi_n and Bi_n^{+} (n=2–24) *J. Chem. Phys.* **128** 094305
[7] Kozhina E L and Shultz M M 1999 Thermodynamic properties of sodium-containing glass forming oxide melts *Ceram. Silik.* **44** 91
[8] Sun H, Sakka Y, Shirahata N, Matsushita Y, Deguchi K, and Shimizu T 2013 NMR, ESR, and Luminescence Characterization of Bismuth Embedded Zeolites Y *J. Phys. Chem. C* **117** 6399
[9] Li Q, Xu S J, Xie M H and Tong S Y 2005 Origin of the 'S-shaped' temperature dependence of luminescent peaks from semiconductors *J. Phys. Condens. Matter* **17** 4853
[10] Trukhin A N, Teteris J, Fedotov A, Griscom D L, Buscarino G 2009 Photosensitivity of SiO_2–Al and SiO_2–Na glasses under ArF (193 nm) laser *J. Non-Cryst. Solids* **355** 1066

IOP Conf. Series: Journal of Physics: Conf. Series **1092** (2018) 012130 doi:10.1088/1742-6596/1092/1/012130

Sensing and photocatalytic activity of zinc sulfide quantum dots based on charge transfer process to analyte molecule

A A Sergeev[1,2], K A Sergeeva[2], A A Leonov[1,2], S S Voznesenskiy[2]

[1]Institute of Automation and Control Processes, Far East Branch the Russian Academy of Sciences, 5 Radio Street, Vladivostok, 690041, Russia
[2]Far Eastern Federal University, 8 Sukhanova Street, Vladivostok, 690950,Russia

E-mail: sergeev@iacp.dvo.ru

Abstract. We demonstrated the efficientcharge transfer between manganese doped zinc sulfide quantum dots (QDs)and an analyte molecule, methaneor methylene blue dye, by means of photoluminescence properties of ZnS QDs. In the presenceof methane QDs arefound to exhibit non-uniform increasing of luminescence intensity due to specific charge transfer from analyte to QD defect states. It is revealed that charge transfer leads to destruction of methylene blue dye molecule which makes possible to use the QDs for organic dye removal in aqueous solution. The limit of methane detection is 100 ppm.The photocatalytic rate of methylene blue degradation is 0.033 s^{-1} which is almost two orders higher than Degussa P25 rate.

1. Introduction

Semiconductor nanocrystals that are quantum confined in three dimensions, or quantumdots, have received much attention as low-cost active materials for highly efficient optoelectronicdevices [1] andsolar-to-fuel conversion applications, namely, photocatalysis, because of their unique size-dependent properties andcost-effective solution processing. These applications involve QDs as chargeseparation centers, and a critical and fundamental step is the electron or hole transfer from photoexcited QDs to analyte molecule acting like electron or hole acceptors through a nanoscale interface [2].A lot of emission centers in QDs result in strong dependence of electron transfer process inside the QD to its surface states, which makes QDs very attractive for implementation in luminescent sensors based on electron transfer process from QD to analyte molecule. Depending on specific of QD-to-analyte charge transfer analyte presence can initiate enhancement or quenching of QDs luminescence [3]. For example, if the QD is close to a molecule exhibiting a LUMO level at lower energy than that of the QD conduction band edge, a photoinduced electron transfer from the QD to the electron acceptor can occur and the luminescence of the QD is consequently quenched. On the other hand,in the presence of an electron donor exhibiting aHOMO level that is less positive than the valence bandpotential, electron transfer from the donor to the holesassociated with the valence band occurs resulting in decreasing of non-radiative transitions in QDs and, correspondingly, enchanting QD luminescence.

Due to high redox nature of QDs such charge transfers took place only for analyte molecules with strong chemical bond. Otherwise, photogenerated charge carriers can destruct analyte chemical bonds leading to photoinduced (or photocatalytic) degradation of analyte molecule [4].

Here we report the successful realization of photoinduced charge carrier transfer between QDs and analyte molecule. Depending on analyte nature charge transfer results in formation of luminescent sensor response in analyte presence or photocatalytic degradation of analyte molecule.

2. Materials and Methods

The quantum dots were prepared using aqueous solutions of zinc and manganese acetates (Zn/Mn ratio from 0.01/0 to 0.01/0.0004 mol/l), glutathione, sodium hydroxide and sodium sulfide. Under vigorous stirring, the solution of zinc/manganese acetate was introduced into a solution of glutathione. Then a solution of NaOH was added and stirred in a water bath at 80 °C for 30 minutes. A solution of sodium sulfide was added dropwise to the mixture for 30 minutes. The resulting mixture was heated at 90 °C for an hour.

Fluorescence spectroscopy study was performed with Fluorolog 3 fluorescence spectrometer (Horiba, Japan) equipped with continuous 450 W Xenon lamp source and pulse 150 W Xenon lamp source.Study of sensor characteristics was carried out in a special sealed cell (filled with aqueous QDs solution) coupled with spectrometer sample chamber. The methane level was preset by bubbling 10% argon-methane gas mixture with a flowrate of 100 ppm of methane per minute.

The photocatalytic degradation of methylene blue (MB), a typical conjugated aromatic dye, was chosen as a model reaction to evaluate ZnS:Mn^{2+} photocatalyst activity. MB is a weakly luminescent dye with two PL excitation (PLE) bands at 1.85 and 4.27 eV, PL maximum is close to 1.81 eV. The energy of the exciting photons was 4 eV that, on the one hand, corresponds to the ZnS QD band gap, and on the other hand, it ensures the excitation of the MB luminescence. Taking into account that PL intensity of methylene blue is directly proportional to its concentration, the simultaneous excitation of the QDs and the dye allowed us to observe the process of dye degradation in real time by decreasing the PL band at 1.81 eV.

The rate of photocatalytic decomposition was determined as

$$-\ln\left(\frac{C}{C_0}\right)=kt \qquad (1),$$

where C_0 is the initial MB concentration; C is the MB concentration at time t, and k is the rate of catalytic decomposition [5].

3. Results and Discussion

Fig. 1 shows the room temperature PL spectra of Mn^{2+}-doped ZnS QDs under 4 eV excitation.

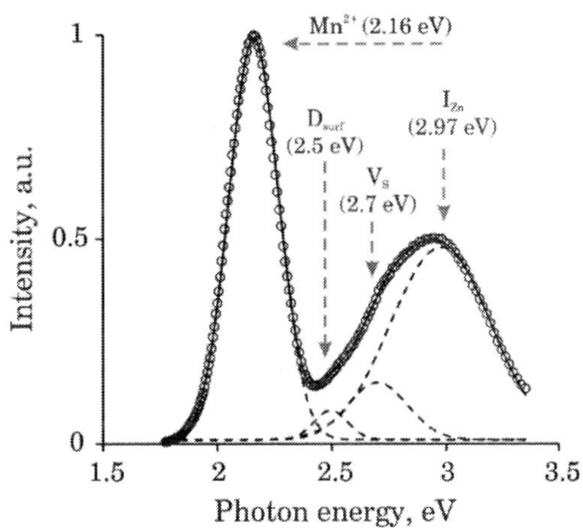

Figure 1. PL spectra of Mn^{2+} doped (0.05 at.%) ZnS QDs in glutathione. The solid line shows the resulting curve, the dashed lines indicate Gaussian components.

In undoped ZnS QDs the only one emission band at 3.1 eV is present.However, all the doped samples show broad emission band with maxima at 3.1 and 2.11 eV. Deconvolution of emission band at 3.1 eV by Gauss functions revealed three luminescence centers corresponded to interstitial zinc ions (I_{Zn}, E_{max} = 2.97 eV), sulfur vacancies (V_S, E_{max} = 2.7 eV) and surface defects (D_{surf}, E_{max} = 2.5 eV). Such luminescence centers are well-known for ZnS QDs and are in agreement with otherreports [3].Emission band at 2.11eV is related to $^4T_1 \rightarrow {}^6A_1$ transition within $3d^5$ configuration of Mn^{2+} ion. Increasing of Mn^{2+} concentration leads to the PL intensity decreasing at 3.11 eV band. The intensity of Mn^{2+} related emission increases up to dopant concentration of 2 at.% after that the concentration quenching occurs. Because both ZnS- and Mn^{2+} related PL bands took place, one can conclude that Mn^{2+} ions are incorporated in ZnS crystal lattice [3, 6].

To reveal the possibility of sensor response realization due to QD-to-analyte energy transfer a methane as analyte sample was choosen. Methane has 4 bonding and 4 antibonding molecular orbitals (MO) with energy levels varies from -18.79 eV to 3.9 eV. Thefour bottom methane MOs are all occupied and have energy lower than the non-bonding energy level (0 eV). The highest four methane molecular orbitals are empty, so it is possible to put an electron in one of these MOs, resulting in energy transfer. For studied ZnS:Mn QDs this process can occur by direct electron transfer to methane LOMO level from QD conduction band or through electron transfer from QD bandgap intrinsic defects including Mn^{2+} energy levels.

Under exposure even relatively low methane concentrations QD luminescence changessignificantly(Fig. 2). Bubbling by 100 ppm of methane through QD solution leads to increasing ofboth ZnS- andMn^{2+}-emission bands. Herewith, ZnS-relatedPL band shifts to the region corresponded to sulfur vacancies and interstitial zinc ion emission. Mn^{2+} PL band, as well as V_S and I_{Zn}bands,demonstrates ≈1.5-2 times amplitude increasing whilst D_{surf} emission band increases its amplitude up to 4 times (Fig. 4).

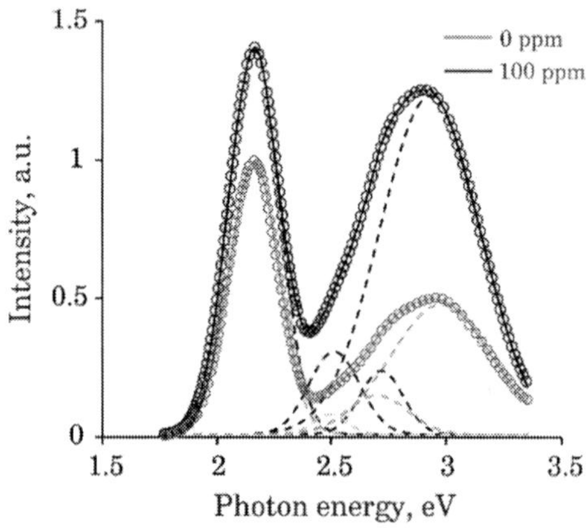

Figure 2. PL spectra of Mn^{2+} doped (0.05 at.%) ZnS QDs under 100 ppm methane exposure (black curves) compared to initial PL spectra (grey curves).

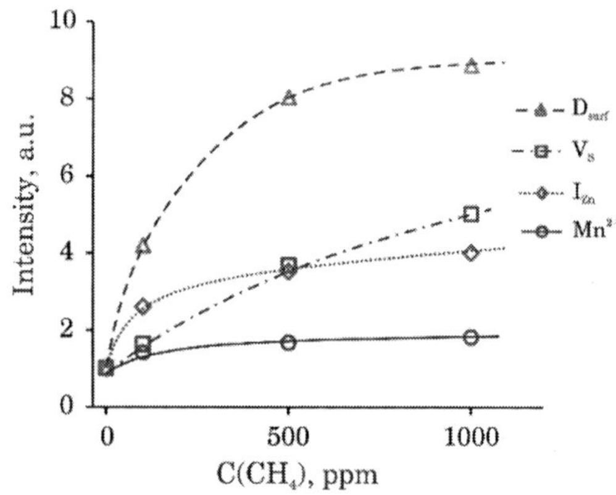

Figure 3. Intensity of QDs emission bands under exposure to different methane concentrations. Intensities was normalized to Mn^{2+} band at initial state.

Considering the fact that QD solution was bubbled by argon-methane gas mixture the observed luminescence increase may be caused by argon presence too. To exclude QD response to argon exposure an optical response to pure argon exposure of the same concentration was investigated. The≈2 times increasing for all PL bands was detected. This fact allows us to suggest that QD sensor response to argon-methane gas mixture exposure involves two processes: (I) QD passivation by argon molecules resulting in gradual luminescence increasing for all emitted centers; (II) specific reaction of QD surface to methane molecules resulted in selective sensor response formation.

It was found that the photocatalytic decomposition rate of methylene blue depends on the concentration of the Mn^{2+} (Fig. 4). For manganese-doped ZnS QDs, the decomposition of the most

MB molecules occurs within 200 seconds after the onset of irradiation. At the same time, a gradual increase in the dopant concentration leads to a reduction in the degradation time of MB. A minimum time achieved is 120 s for the sample with 0.125 at.% of Mn^{2+}. A further increase in the manganese concentration leads to an increase of photocatalytic decomposition time of methylene blue, presumably due to a change in the recombination rate of photoexcited charge carriers noted by a number of authors [7] and which is of great importance in the photocatalytic material development.

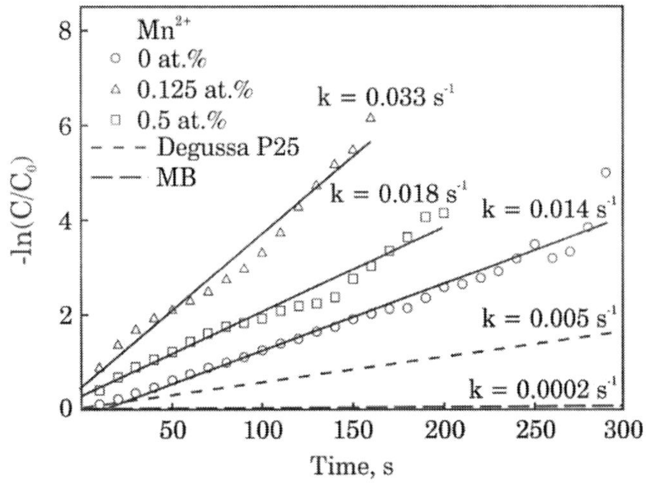

Figure 4. Photocatalytic degradation rate of MB in the presence of $ZnS:Mn^{2+}$ QDs in comparison with commercial photocatalyst Degussa P25..

The deviation from the linear dependence of the dye concentration logarithm can be explained as the processes of sorption of the dye molecules by the surface of the quantum dot. The efficiency of photocatalytic decomposition is affected by both the sorption probability of closely located MB molecule by the surface of the quantum dot and the rate of photocatalytic decomposition. Thus, at a sufficiently high concentration of the MB (at a time from 0 to 75 s), the sorption and photocatalytic decomposition processes take place quite rapidly. With a decrease in the concentration of methylene blue, the probability of a close location of the quantum dot and the dye molecule decreases that causes non-linear dependence of dye concentration logarithm in 100 s.

A comparison of the obtained results of the photocatalytic activity of $ZnS:Mn^{2+}$ QDs with the available passport data of the reference photocatalyst Degussa P25 demonstrate the high efficiency of proposed nanocrystals.

Acknowledgments
The work is supported by Russian Foundation of Basic Research (project No. 18-32-00135 mol_a).

References

[1] Bera D, Qian L, Tseng T K and Holloway P H2010 *Materials*3 2260.
[2] Dong Y, Choi J, Jeong H K and Son D H 2015 *J. Am. Chem. Soc*, **137** 5549.
[3] Freeman R and Willner I 2012 *Chem. Soc. Rev.***41** 4067.
[4] Joicy S, Saravanan R, Prabhu D, Ponpandian N andThangadurai P 2014 *RSC Adv.***4** 44592.
[5] Yang C, Dong W, Cui G, Zhao Y, Shi X, Xia X, Tang Band Wang W 2017*RSC Adv.***7** 23699.
[6] Wang Y, Liang X, Ma X, HuY, Hu X, Li X and Fan J 2014*Appl. Surf. Sci.***316** 54.
[7] Qu YandDuan X 2013 *Chem. Soc. Rev.***42** 2568.

Tractor beams at metamaterial substrates

A Ivinskaya[1], N Kostina[1], M I Petrov[1], A A Bogdanov[1], S Sukhov[2], P Ginzburg[3], and A S Shalin[1,*]

[1]"Nano-Optomechanics" Laboratory, Department of Nanophotonics and Metamaterials, ITMO University, Birzhevaja line, 14, 199034, St. Petersburg, Russia

[2]CREOL, The College of Optics and Photonics, University of Central Florida, Orlando, Florida, USA.

[3]School of Electrical Engineering, Tel Aviv University, Ramat Aviv, Tel Aviv 69978.

alexandesh@gmail.com

Abstract. Optical forces acting on nanoobjects can be enhanced, inverted or cancelled out by the presence of carefully chosen environment. Here we consider metamaterial substrate which can modify optical forces through the excitation of surface waves and volumetric hyperbolic modes. Both types of interaction channels will be discussed in the present work where we show new effect of pulling forces on subwavelength dielectric particles above multilayer hyperbolic substrates.

1. Introduction

A special class of metamaterials - hyperbolic metamaterials - can support both volumetric modes and surface waves and are especially interesting for optomechanical applications. Hyperbolic metamaterials based on vertically aligned metal nanorods were assessed as a platform for flexible optomechanical manipulation [1], [2]. While a set of peculiar effects including tractor beams were predicted by employing homogenization approach [1], consideration of near-field interactions within actual unit cells of the nanorod arrays require more detailed numerical analysis [2].

In this work we focus on metamaterial multilayer substrates for optomechanical applications. For a particular substrate, surface waves, being the key for effects such as tractor beam [3], anti-trapping in Gaussian beam [4], anisotropic binding [5], have many in common if certain peculiarities in the excitation process and interaction mechanism are omitted. Leaving out these differences and concentrating on the main features, we study a basic setup exceptionally sensitive to surface waves and modes in the structure, that is a particle over the metamaterial substrate illuminated by a plane wave.

2. Formalism

The force on a dipole particle with polarizability α near semi-infinite substrate, Fig. 1, can be written as [6]

$$F_x = \frac{1}{2}\mathrm{Re}(\alpha \mathbf{E}^{tot}\partial_x \mathbf{E}^{0*}) + |\alpha|^2 \omega^2 \mu_0 \,\mathrm{Im}(\mathrm{E}_x^{tot}\mathrm{E}_z^{tot*})\,\mathrm{Im}(\partial_x \mathrm{G}_{xz}), \qquad (1)$$

where $\mathbf{E}^{tot}$ is a local field at the particle location, $\mathbf{E}^0$ is incident field, $\mathbf{G}$ is Green's function. Green's function defines structure response to incident field and enters both expression for total field $\mathbf{E}^{tot}$ and the force $\mathbf{F}$.

Figure 1. Optomechanically manipulated subwavelength dielectric particle ($\varepsilon = 3$, $R = 15$ nm, $z = 15$ nm) next to anisotropic (hyperbolic) metasurface. Plane wave of p-polarization is incident at 35°.

Green's function encapsulates the response from the auxiliary layered structure, which manifest itself via reflection coefficients r_s and r_p for s and p polarizations:

$$\partial_x G_{xz} = \frac{1}{8\pi k_1^2} \int_0^\infty r_p k_\rho^3 \, \mathrm{e}^{-2ik_{1z}z} \, \mathrm{d}k_\rho \tag{2}$$

Hyperbolic materials can support both surface and bulk waves. Volumetric waves with hyperbolic dispersion have non-resonant features, whereas existence of surface modes can be identified via observation of poles in the reflection coefficient. Those coefficients for the case of the uniaxial crystal with an optical axis, pointing perpendicular to the interface, (ε_x, ε_z are permittivities along main axes) are given by

$$r_p = \frac{\varepsilon_x k_{1z} - \varepsilon_1 k_{2z}^p}{\varepsilon_x k_{1z} + \varepsilon_1 k_{2z}^p},$$
$$k_{2z}^p = \left(\varepsilon_x k_0^2 - \frac{\varepsilon_x}{\varepsilon_z} k_\rho^2\right)^{0.5}, \quad k_{1z} = \left(\varepsilon_1 k_0^2 - k_\rho^2\right)^{0.5}, \tag{3}$$

where k_0 is a wave vector of incident radiation in vacuum, ε_1 and $k_1 = \sqrt{\varepsilon_1} k_0$ are the permittivity and the wave vector in the upper half-space (air is considered hereafter) with components k_x, k_y, k_{1z} and $k_\rho = (k_x^2 + k_y^2)^{0.5}$ is a transversal wave vector. The branch of the square root solution in Eq. (2) should be chosen with an imaginary part of k_{2z}^p (wave vector in the substrate) positively defined for a wave to decay away from the interface. At the same time for idealized lossless hyperbolic materials, having different signs of permittivities along the two main axes and of the crystal $\varepsilon_x', \varepsilon_z'$, wave vector k_{2z}^p acquires real part for $k_\rho = k_{cr} \geq k_1$. It implies that at this condition evanescent waves scattered by the particle transforms into propagating volume modes inside the hyperbolic substrate.

3. Multilayer substrates for negative transversal force

For a setup with a particle over substrate excited by a plane wave, we start by overview of arbitrary uniaxial anisotropic substrates to identify dielectric parameters when strong response of optical forces takes place compared to free space. While certain types of dielectric permittivity tensors are hardly accessible, hyperbolic materials are very promising and are given by feasible multilayer geometries. Metal-dielectric multilayers are studied to find suitable geometric configurations. For hyperbolic isotropic and multilayered materials an emphasis is made on the difference in optical forces caused by surface waves and volumetric modes in such structures. Figure 2 shows that surface and volume modes in substrates with hyperbolic dispersion can lead to pulling forces over a broad wavelength range.

The material composition of the first layer (either metal or dielectric) of multilayer substrate has a crucial impact on the pulling force. This effect is solely related to the realization of the metamaterial and, of course, cannot be observed in the studies of the homogenized substrate. Those three different scenarios are considered hereafter. Fig. 2(a) demonstrates the value of the pulling force as the function of wavelength of the incident radiation (the silver filling factor is fixed at f1 = 0.2). The values of the force for the homogenized case typically lie in between those of multilayers with either dielectric (light blue lines) or metal (dark blue lines) top layer. If metal is set as an outer layer, maximum force amplification is achieved. At the longer wavelengths (infrared) regimes of metal or dielectric top layer can give an order of magnitude difference in force values. As expected, multilayers with smaller period (closely resembling the homogenized description) have broadband spectra while larger periods give sharper features.

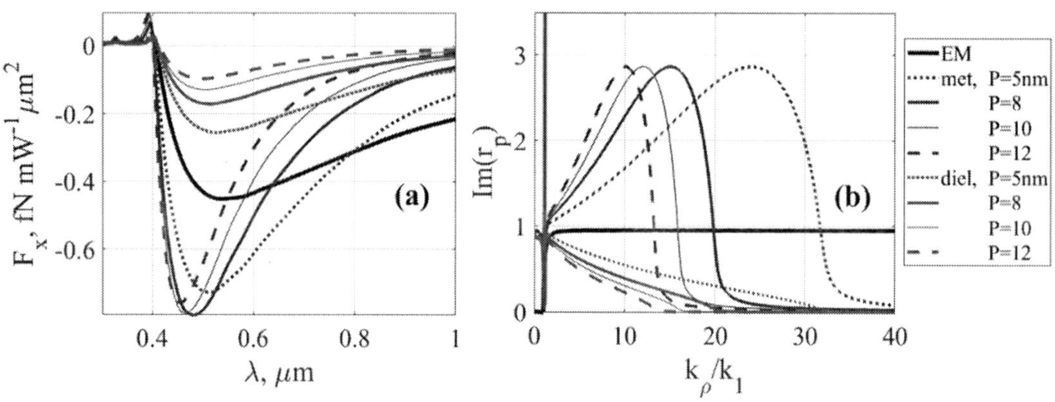

Figure 2. (a) Transversal force F_x acting on the particle, as the function of the wavelength. (b) Imaginary part of the reflection coefficient for semi-infinite Ag-polymer (n=1.05) multilayer composite as the function of the transverse wave vector. Dark (light) blue lines correspond to the case, when the top layer, interfacing the air, is metal (dielectric). Black line is obtained within the effective medium approximation. A set of multilayer periods is specified in the legend. Other parameters are: R = 15 nm, ε = 3, z = 15 nm, the structure is illuminated by a plane wave, incident at 35°.

4. Conclusions

The optical force acting on a nanoparticle near a planar substrate is governed by incident light and excitation of surface and volume modes of the substrate. Plane wave results in pulling forces towards the source for certain types of anisotropic substrates and multilayers. For multilayer substrate, top metal layer interfacing the air introduces additional enhancement of value of the pulling force.

Acknowledgements

The authors acknowledge the financial support provided by Russian Fund for Basic Research (Projects No. 18-02-00414, 18-52-00005); Ministry of Education and Science of the Russian Federation (GOSZADANIE, Grant No. 3.4982.2017/6.7) and Government of Russian Federation, Grant 08-08.

References

[1] Shalin A S, Sukhov S V, Bogdanov A A, Belov P A, and Ginzburg P, Optical pulling forces in hyperbolic metamaterials, Physical Review A, 91, 063830 (2015).
[2] Bogdanov A A, Shalin A S, Ginzburg P, Optical forces in nanorod metamaterial, Scientific Reports, 5, 15846 (2015).
[3] Petrov M I, Sukhov S V, Bogdanov A A, Shalin A S, and Dogariu A, Surface plasmon polariton assisted optical pulling force, Laser Photonics Rev. 10, 116-122, (2016).
[4] Ivinskaya A, Petrov M I, Bogdanov A A, Shishkin I, Ginzburg P and Shalin A S, Plasmon-assisted optical trapping and anti-trapping, Light: Science and Applications, 6, e16258 (2017).
[5] Kostina N, Ivinskaya A, Petrov M I, Bogdanov A A, Shishkin I, Ginzburg P and Shalin A S, Optical binding via surface plasmon polariton interference, in press.
[6] Novotny L and Hecht B, *Principles of Nano-Optics*, 2nd ed. Cambridge University Press, 2012.

IOP Conf. Series: Journal of Physics: Conf. Series **1092** (2018) 012133 doi:10.1088/1742-6596/1092/1/012133

All-Dielectric anti-reflective metasurface at visible frequencies

H. K. Shamkhi, A. S. Shalin

ITMO University, Kronverkskiy prospect 49, 197101, St. Petersburg, Russia

Corresponding author: h.shamkhi@optomech.ifmo.ru.

Abstract. We analytically study the broadband antireflection of a square array of dielectric pillars with a glass substrate. The presented condition exploits the optical response of the nanoparticles to achieve complete cancellation of the reflected field by means of destructive interference.

1. Introduction

The incident electromagnetic wave is partially reflected when encountering the interface between two different refractive indices mediums while the rest transmitted into the second medium. Suppressing the reflected waves is of essential interest to several optics devices such as solar-cells[1], light detectors[2], etc. Conventionally an optimum layer of quarter-wavelength dielectric antireflection coating ARC has been used to eliminate the wavelength at specific wavelength through the destructive interference. However $\lambda/4$ coating requires regularly nonexistence materials with a suitable refractive index also the needed layer thickness increase dramatically for longer wavelengths area of applications. During the earlier decades, the rapid advancement of metasurfaces grants an immense degree of freedom in engineering light-matter interactions at the sub-wavelength scale, thereby realizing tunable effective optical properties by altering its geometrical profile[3].

In this work, we analytically and numerically investigate the condition of the broadband

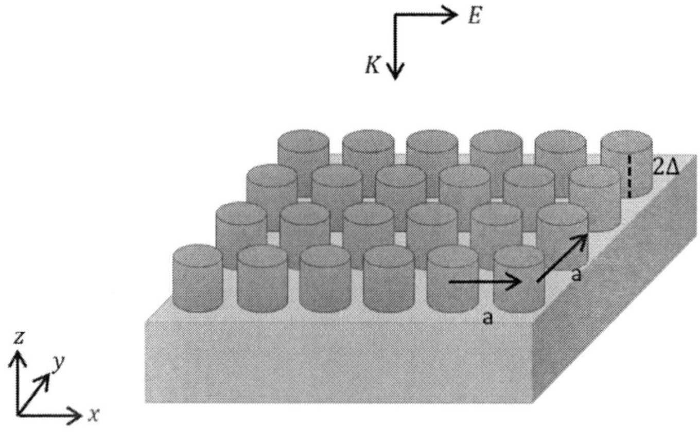

Figure 1. Schematic of the pillars array.

antireflection state in the visible spectrum region for a square array of silica pillars on a glass substrate. The layer antireflection is comparable to the traditional quarter-wave antireflection, but with two improvements—it does not require refractive index matching of the dielectric substrate and the overall antireflection coating thickness is thinner.

2. Theoretical background

We study the optical antireflection of a pillars layer infinitely extended on a substrate with a relative dielectric constant $\boldsymbol{n_d}$. The particle array is located in the $\boldsymbol{xy}$- plane with periodicity in the $\boldsymbol{x}$ and $\boldsymbol{y}$ directions. The structure is irradiated by an external optical plane wave with electric and magnetic fields $\boldsymbol{E_{inc}}\mathbf{exp}(\boldsymbol{ikr})$ and $\boldsymbol{H_{inc}}\mathbf{exp}(\boldsymbol{ikr})$, respectively with time dependent term omitted. The wave vector $\boldsymbol{k}$ is given in the free space medium, $\boldsymbol{r}$ define point in space, see figure 1.

The particles are approximated as a point electric and magnetic dipoles with electric and magnetic polarizabilities $\boldsymbol{\alpha_p}$ and $\boldsymbol{\alpha_m}$, respectively. The total reflected field can be presented as the vector sum of the directly scattered waves by the collective array dipoles, whose wave structure, in turn depends on the geometrical design of the system and fields reflecled from the interface[4]:

$$\boldsymbol{E}_{ref} = \mathrm{r}e^{ik\Delta}\boldsymbol{E}_{inc} + \alpha_p(\overleftrightarrow{g}_0 + e^{ik\Delta}\overleftrightarrow{g}_{ref}).\boldsymbol{E}_{eff} + \frac{\alpha_m}{c}(\overleftrightarrow{f}_0 + e^{ik\Delta}\overleftrightarrow{f}_{ref}).\boldsymbol{H}_{eff} \tag{1}$$

with

$$\boldsymbol{E}_{eff} = (1 + re^{2ik\Delta})\boldsymbol{E}_{inc} + \alpha_p(\overleftrightarrow{g} + \overleftrightarrow{g}_{ref}).\boldsymbol{E}_{eff} + \frac{\alpha_m}{c}(\overleftrightarrow{f} + \overleftrightarrow{f}_{ref}).\boldsymbol{H}_{eff} \tag{2}$$

$$\boldsymbol{H}_{eff} = (1 + re^{2ik\Delta})\boldsymbol{H}_{inc} + \alpha_m(\overleftrightarrow{g} + \overleftrightarrow{g}_{ref}).\boldsymbol{H}_{eff} - c\alpha_p(\overleftrightarrow{f} + \overleftrightarrow{f}_{ref}).\boldsymbol{E}_{eff} \tag{3}$$

Where $\boldsymbol{E}_{eff}$ and $\boldsymbol{H}_{eff}$ are the effective electromagnetic field at each particle center. For the case of incident wavelength larger than the lattice period, $\overleftrightarrow{g}_0$ and $\overleftrightarrow{f}_0$ Green tensors describe the scattered electric field in the far-field approach where evidently only the particle at the center contributes[5]. r, c and Δ are the free space - substrate reflection, speed of the light and the height of the point dipole from the substrate, respectively. The tensors $\overleftrightarrow{g}_{ref}$ and $\overleftrightarrow{f}_{ref}$ define the scattered fields which reflected by the substrate. These waves are only evanescent beams and dose not contribute to the far filed reflection nor interact with dipoles of the particles[6][7]. Under normal incidence and because of the symmetry of the array, the coupling between the electric and magnetic dipoles also disappears and imply $\overleftrightarrow{f} = 0$ [5]. Finally we derive the condition of the antireflection by inserting equations (2) and (3) into equation (1) and considering s-polarized incident light as follow:

$$R_l = r\frac{e^{i(k\Delta - \pi)}}{1 + re^{2i(k\Delta)}} \tag{4}$$

with

$$R_l \approx \frac{ik}{2a^2}\left(\frac{1}{\alpha_p - g^{xx}} - \frac{1}{\alpha_m - g^{yy}}\right) \tag{5}$$

Where R_l is the array reflection and a is the array periodicity. The tensor components g^{xx} and g^{yy} calculated by using the method of dominate part extraction and Possion summation[8].

3. Results and discussions

Following the condition presented in equation (4); in order to achieve full suppression, one needs to carefully tune the array and the particle geometrical parameter to compensate both the amplitude and the phase on the two sides of the equation. The aspect ratio of the pillars particles proposed to be height =70 nm and radius=30 nm of silica material with refractive index $n = 4$. The reflection spectra (blue line) plot are shown in figure 2(a). With periodicity $a = 100\ nm$,the low reflection region cover

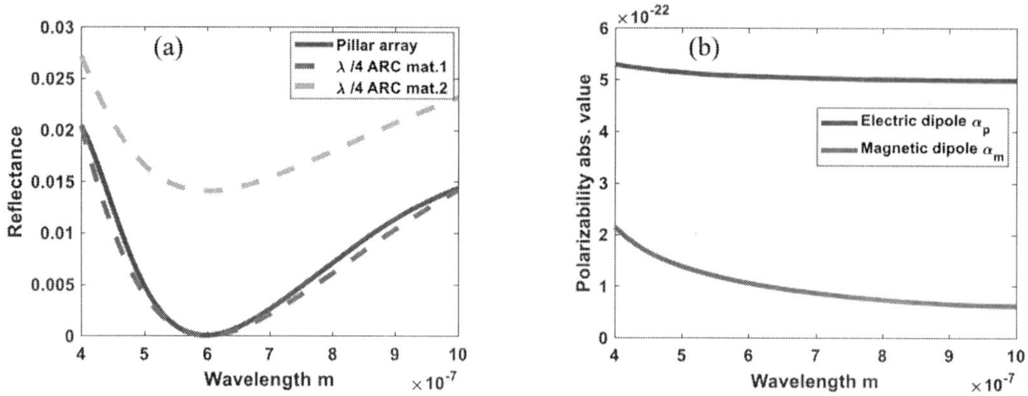

Figure 2. (a) The reflectance of the pillars layer with periodicity $100\,nm$ (blue line), the ARC optimum $\lambda/4$ antireflection with refractive index $n = 1.23$ mat.1 and $n = 1.38$ mat.2 and thickness $t = 122\,nm$. (b) The electric and magnetic dipole polarizability for material with constant refractive index $n_p = 4$.

wide band of the visible spectrum where it is deigned to reach zero at 590 nm wavelength. The comparison with the reflection spectrum of the quarter- wave antireflection coating (dashed lines) designed to suppress the reflection optimally at the same central minima shows similar wavelength response for the first material (mat.1) and much better for the second one (mat.2).

Figure 2(b) shows the net difference between the electric and magnetic material polarazabilites where it slowly changes over wavelength leading to expand the antireflection bandwidth. The position of the reflection minimum is impacted mainly by the height of the pillared particle. For instance increasing the pillared height leads to red shift the minima and vice versa while increasing layer periodicity enhances the antireflection without shifting the minima.

Acknowledgements

The authors acknowledge the financial support provided by Russian Fund for Basic Research (Projects No. 18-02-00414, 18-52-00005); Ministry of Education and Science of the Russian Federation (GOSZADANIE, Grant No. 3.4982.2017/6.7) and Government of Russian Federation, Grant 08-08. The calculations of multipole moments have been supported by the Russian Science Foundation Grant No. 16-12-10287

References

[1] Voroshilov P M, Simovski C R, Belov P A and Shalin A S 2015 Light-trapping and antireflective coatings for amorphous Si-based thin film solar cells *J. Appl. Phys.* **117** 203101

[2] Spinelli P, Verschuuren M A and Polman A 2012 Broadband omnidirectional antireflection coating based on subwavelength surface Mie resonators *Nat. Commun.* **3** 692

[3] Li X, Pu M, Ma X, Guo Y, Gao P and Luo X 2018 Dispersion engineering in metamaterials and metasurfaces *J. Phys. D. Appl. Phys.* **51** 54002

[4] Baranov D A, Dmitriev P A, Mukhin I S, Samusev A K, Belov P A, Simovski C R and Shalin A S 2015 Broadband antireflective coatings based on two-dimensional arrays of subwavelength nanopores *Appl. Phys. Lett.* **106**

[5] Evlyukhin A B, Reinhardt C, Seidel A, Luk'Yanchuk B S and Chichkov B N 2010 Optical response features of Si-nanoparticle arrays *Phys. Rev. B - Condens. Matter Mater. Phys.* **82** 1–12

[6] De Abajo F J G 2007 Colloquium: Light scattering by particle and hole arrays *Rev. Mod. Phys.* **79** 1267–90

[7] Novotny L and Hecht B 2006 *Principles of Nano-Optics* (Cambridge: Cambridge University Press)
[8] Swiecicki S D and Sipe J E 2017 Periodic Green functions for 2D magneto-electric quadrupolar arrays: explicitly satisfying the optical theorem *J. Opt.* **19** 95006

IOP Conf. Series: Journal of Physics: Conf. Series **1092** (2018) 012134 doi:10.1088/1742-6596/1092/1/012134

Wood's anomaly for plasmonic biosensor based on 1D magnetooptical nanostructure

A N Shaimanov[1,2], N A Orlikovsky[1,4], E M Khabushev[1,3], A V Zverev[1,4], A A Pishimova[1,4], G V Sharonov[1,5], G M Yankovskii[1], I A Rodionov[1,4], A V Baryshev[1,3]

[1]Dukhov All-Russia Research Institute of Automatics, 127055 Moscow, Russia

[2]Moscow State University, 119992 Moscow, Russia

[3]Moscow Institute of Physics and Technology, 141700 Dolgoprudny, Moscow Region, Russia

[4]Bauman Moscow State Technical University, 105005 Moscow, Russia

[5]Shemyakin–Ovchinnikov Institute of Bioorganic Chemistry, 117997 Moscow, Russia

Abstract. We demonstrate spectra of slabs of plasmonic 1D nanostructures and show their ability to detect a specific binding of low-density lipoproteins. Optical spectra of the slabs exhibiting a spectrally sharp resonant peak have been analyzed numerically to demonstrate responses of biosensors under study. We show that the sensitivity to biomolecular binding can be considerably increased by utilizing magnetooptical materials as constituent element of plasmonic 1D nanostructures.

1. Introduction

For detecting and monitoring diseases at early stages and in development, when investigating concentrations of biomarkers in the interval from negligible to reference values, highly sensitive, reliable diagnostic approaches are necessary [1,2]. Optical biosensing techniques–using surface plasmon resonance (SPR) [3-5]–are up to date the most commonly used in healthcare and science. The way to increase the sensitivity is utilization of the magnetic layer in the plasmonic structure to measure a magnetooptical (MO) response instead of the transmittance one [6].

In the present work, we demonstrate 500-μm-sized slabs of plasmonic 1D nanostructures with and without magnetic layers for detecting low density lipoproteins (LDL). We experimentally and theoretically study their optical spectra, elucidate main dependencies between structural parameters of optical resonances, give an explanation on resonant light coupling. For the nanostructures having optimal designs, we measured and numerically analyzed optical and MO responses in the transmission geometries when modeling biomolecular binding for LDL with a low concentration.

2. **Fabrication, calculation model and experiment**

The fabricated 1D pattern had periods of D = 600, 625, 650 and 675 nm, a height and a width of the hydrogen silsesquioxane(HSQ) ridge were set parameters for all samples d_{HSQ} = 80 nm and h_{HSQ} = 65 nm correspondingly.

Magnetoplasmonic 1D nanostructure (sample B) had the following parameters: period D=500 nm, height and width of the bismuth-substituted yttrium iron garnet(Bi:YIG) ridge were d_R=60 nm and h_R=100 nm, correspondingly, Bi:YIG film's thickness $d_{Bi:YIG}$ = 135 nm.
The transmission spectra were measured for polarized light by a spectrometer (Ntegra Spectra, NT-MDT) in the wavelength range of 700–950 nm. The measurements were carried out for the E_x-polarized light (see Fig. 1).
For COMSOL Multiphysics models of sample A and B were set up for one unit cell of the 1D grating, with the Floquet boundary conditions along the *x* axis describing the periodicity and being infinite along the *z* axis [Fig. 1]. Experimental differential spectra illustrating biomolecular binding were modeled for water conditions by changing the refractive index (*n*) and extinction coefficient (*k*) of a covering approx. 20-nm-thick biolayer (the size of LDL molecules), which was 10-nm-lifted from the slab surface (the size of fuctionalizing CS_2/ApoB biolayer).

3. Discussion

The main optical features (the Wood anomaly) in zero-order transmission spectra of the series of the samples follow the period of the 1D grating as shown in Fig.2 (a). The asymmetrical Fano contour has a maximum of λ_F discussed in Ref. [7] that was red-shifted from a wavelength of $\lambda_R = n_{PBS}D$ (the Rayleigh anomaly) and a minimum at $\lambda_{SPR} = n_{eff}D$, where n_{PBS} and n_{eff} are the refractive index of PBS and effective refractive index for the SPR wave, and D is the period of the 1D grating. The spectra show a similar tendency–a slight decrease of the intensity of the Fano peak. These spectra were calculated numerically. The calculated spectra were in a good agreement with the experimental ones; see solid lines and stars for samples A having D = 650 nm and dashed line for sample B having D = 500 nm. Plots (b) in Figs. 2 illustrate responses ΔT for optimized slabs. For considered samples A and B, the maximum of ΔT =0.01 (sample A) and ΔT = 0.005 (sample B) was obtained. The optical response of sample A was twice as much as response of sample B.
We have found that, by introducing Bi:YIG layer, the sensitivity can be significantly enhanced when measuring the MO response in the vicinity of the Fano peak. Indeed, a change in the angle of Faraday rotation $\Delta\theta = 3^0$ was demonstrated for sample B with optimized parameters. The mechanism of enhanced polarization rotation is discussed in detail in Ref. [8]. Suppression of the incident $\mathbf{E}_x$ wave and decoupling the born via magneto-optical activity $\mathbf{E}_z$ wave having a phase shift is a distinctive feature for the plasmon-enhanced MO response in the nanostructures under present study.
It is worth noting that the demonstrated experimental and calculated optical response allows us to make the next conclusions: (*i*) there are optimization parameters for a certain design of the grating; they are connected with each other and, correspondingly, influence the Q-factor of the Fano peak defining the sensitivity of the sensor and (*ii*) a significant increase in sensitivity can be achieved by introducing a magnetic component.

4. Conclusion

Optical spectra of various submillimeter-sized lithography-made 1D gratings have been studied and competitive levels of sensitivity to target biomolecules have been experimentally demonstrated. Our numerical studies demonstrated that the sensitivity can be increased when measuring magnetooptical response of plasmonic 1D nanostructures fabricated from a noble metal and a magnetooptical material.

References

[1] J.D. Wulfkuhle, L.A. Liotta, E.F. Petricoin, Proteomic applications for the early detection of cancer, Nat. Rev. Cancer 3 (2003) 267–275.
[2] P.J. Nestor, P. Scheltens, J.R. Hodges, Advances in the early detection of Alzheimer's disease, Nat. Rev. Neurosci. 5 (2004) 34–41.
[3] C.R. Taitt, J.P. Golden, Y.S. Shubin, L.C. Shriver-Lake, K.E. Sapsford, A. Rasooly, F.S. Ligler, A portable array biosensor for detecting multiple analytes in complex samples, Micro. Ecol. 47 (2004) 175–185.
[4] A. Patton, M.C. Mullenix, S.J. Swanson, E. Koren, An acid dissociation bridging ELISA for detection of antibodies directed against therapeutic proteins in the presence of antigen, J. Immunol. Methods 304 (2005) 189–195.

[5] A. Szabo, L. Stolz R.Granzow, Surface plasmon resonance and its use in biomolecular interaction analysis (BIA), Curr. Opin. Struct. Biol. 5 (1995) 699–705.
[6] D. O. Ignatyeva, G. A. Knyazev, P. O. Kapralov, G. Dietler, S. K. Sekatskii, and V. I. Belotelov, Magneto-optical plasmonic heterostructure with ultranarrow resonance for sensing applications, Sci. Rep. 6, (2016) 28077.
[7] K.L. Lee, J.B. Huang, J.W. Chang, S.H. Wu, P. K. Wei, Ultrasensitive Biosensors Using Enhanced Fano Resonances in Capped Gold Nanoslit Arrays, Sci. Rep. 5 (2015) 8547.
[8] A.V. Baryshev, A. M. Merzlikin , Tunable plasmonic thin magneto-optical wave plate. J. Opt. Soc. Am. B **33,** (2016) 1399.

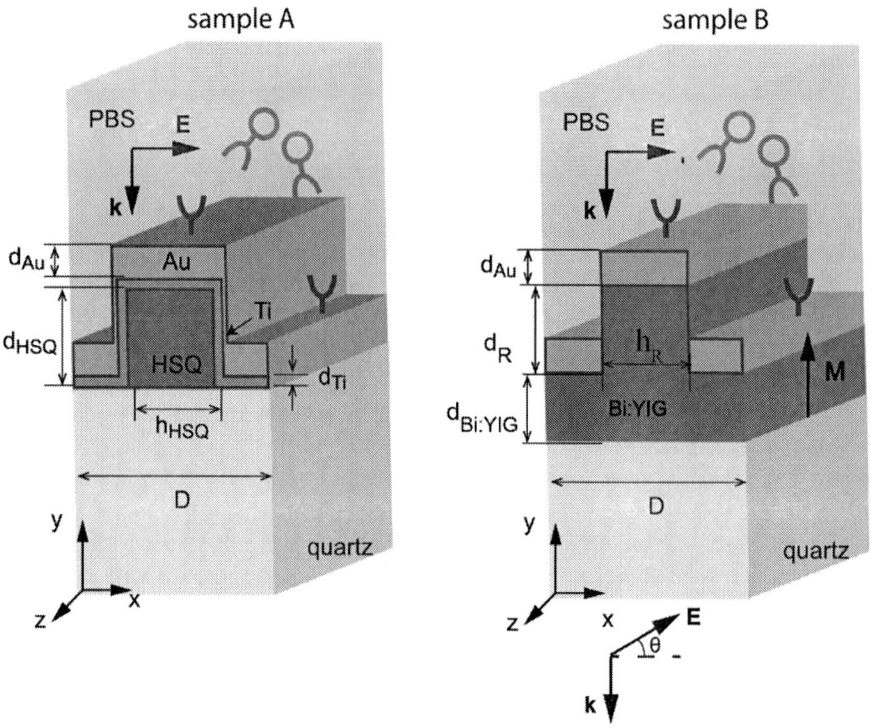

Fig. 1. A sketch of the unit cells of experimental sample A and the model of sample B.

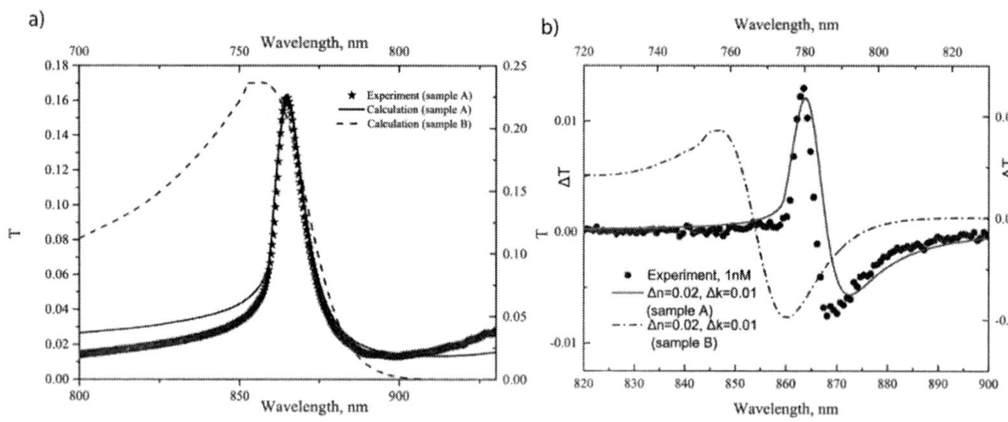

Fig. 2. (a) – Deferential transmission spectra for sample A (D = 650 nm) when detecting LDL with a concentration of 0.5 ng/μl (circles). Solid line shows fits of experiment (sample A) by modeling a 20-nm thick layer of binding molecules in water, changing optical constants of the layer). Dashed line shows magnetoplasmonic structure (sample B) by modeling a 20-nm thick layer of binding molecules in water, changing optical constants of the layer. (b) – Calculated (solid line) and measured (stars) spectra for sample A.

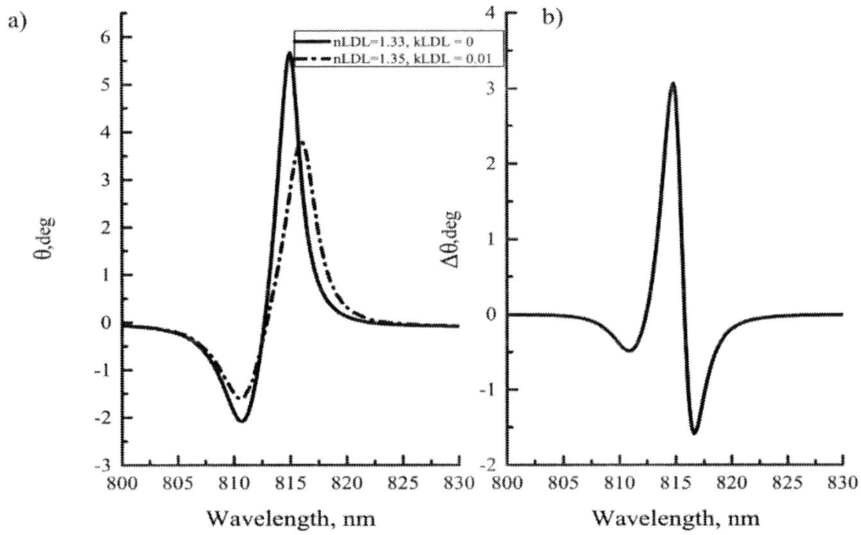

Fig. 3 Magnetooptical response of sample B with an optimized structure (D = 500 nm and h_R =100 nm). (a) – Faraday rotation spectra illustrating biomolecular binding for LDL; (b) – a differential Faraday rotation spectrum.

IOP Conf. Series: Journal of Physics: Conf. Series **1092** (2018) 012135 doi:10.1088/1742-6596/1092/1/012135

Non-classical plasmon states generation in nonlinear spaser systems

M. Yu. Gubin[1,2], **A. V. Shesterikov**[1], **S. N. Karpov**[1], **and A. V. Prokhorov**[1]

[1] Department of Physics and Applied Mathematics, Vladimir State University named after A. G. and N. G. Stoletovs, Vladimir, 600000 Russia
[2] Moscow State Pedagogical University, Moscow, 119991, Russia

E-mail: avprokhorov33@mail.ru

Abstract. The article is focused on the investigation of features of quantum dynamics for localized plasmons in spaser systems consisting of metal nanoparticles and semiconductor quantum dots. The non-classical plasmon states generation in a three-particle spaser system with nonlinear plasmon-exciton interaction is predicted.

1. Introduction

Chains of near-field coupled nanoparticles (NPs) and quantum dots (QDs) are of significant interest as a platform for quantum computing [1, 2]. The bunching (antibunching) effects can be observed for emitted photons from self-assembled QD structures [3]. On the other hand, if QDs are coupled with shared NP [4, 5], direct transmission of quantum correlations from bound excitons to plasmons can be achieved. One of the main advantages of such correlated plasmons is the possibility of simpler external control and manipulation of their carriers – NPs. In particular, this addressing can be carried out using the epifluorescence microscopy technique for single quantum objects [6].

In this paper we propose a model of three nanoobjects (NP-QD-NP) coupled by nonlinear dipole-dipole interactions in the presence of an external magnetic field. The nonlinear regime of this ensemble corresponds to the two-quantum processes of the QD biexciton decay in the case $|\delta| > \Omega_{1,2}, |\Delta|$, where $\delta = \bar{\omega} - \omega$ and $\Delta = \bar{\omega} - \omega_p$, $\bar{\omega}$ is the spasing frequency, and $\Omega_{1,2}$ are the Rabi frequencies of dipole-dipole interactions between the QD and NPs. As a result of this nonlinear process, one can expect the appearance of strongly correlated plasmon pairs.

In the technical framework, the presented spaser systems can be used to generate nonclassical states of the electromagnetic field at the nanoscale. Such systems can be integrated in the individual plasmonic waveguides [7] and plasmonic circuits for quantum information processing [8, 9].

2. Entangled plasmon generation in the nonlinear regime of a three-particle spaser

Nonlinear regimes of interaction between NP and QD can be realized, firstly, in the presence of a two-photon pump in the system [10, 11] and secondly, under the condition that the coupling energy between two electron-hole pairs is of the same order of magnitude as the internal coupling energy of a single pair. In this situation the coupled states of two electron-hole pairs (biexcitons

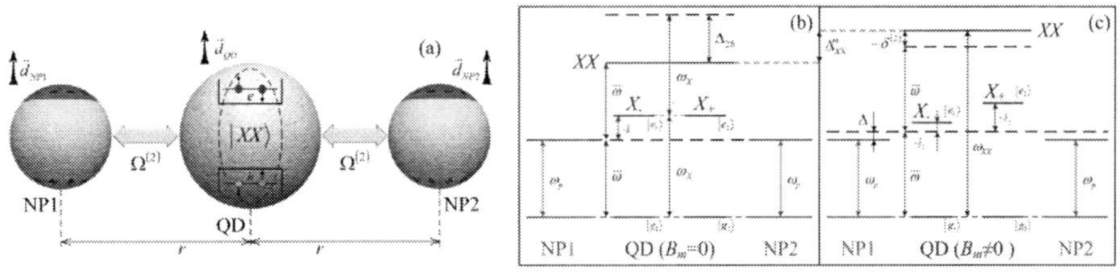

Figure 1. (a) A model of a three-particle spaser consisting of two NPs and single QD (NP-QD-NP), in which the generation of entangled plasmons is realized due to the QD's biexciton states decay. The mapping of exciton and plasmon energy levels is in the absence of (b) and in the presence of (c) external magnetic field ($\Delta_{XX}^{m} = 2kB_m^2$).

of QDs) can appear [10]. The energy of biexciton state XX differs from the double energy of the exciton X by the biexciton binding energy Δ_{2b} (see Fig. 1).

We choose such parameters of the spaser for which a cascade process is not realized in the system, but a pure two-quantum transition occurs [12]. In the conditions corresponding to scheme in Fig. 1 this regime is realized due to far-off-resonant interaction in the case $|\delta| > \Omega_{1,2}$. Such a process can lead to the generation of nonclassical states of plasmons and it is described by the following Hamiltonian [13]:

$$H = \hbar\omega_{p1}\hat{c}_1^+\hat{c}_1 + \hbar\omega_{p2}\hat{c}_2^+\hat{c}_2 + \frac{\hbar\omega_{XX}}{2}D + \hbar\Omega^{(2)}\left(\hat{c}_1\hat{c}_2\hat{S}^+ + \hat{c}_1^+\hat{c}_2^+\hat{S}\right), \quad (1)$$

where the last term in the brackets comprises the annihilation operator $\hat{S}$ of the biexciton XX state and the creation operators $\hat{c}_1^+$ and $\hat{c}_2^+$ of a pair of plasmons, whose energies differ slightly for the different intermediate levels X_+ (X_-) with frequencies ω_{X_+} (ω_{X_-}). The basis states $|g\rangle_1 = |1S\,(h)\,,\, m_s = -1/2\rangle$, $|g\rangle_2 = |1S\,(h)\,,\, m_s = +1/2\rangle$, $|e\rangle_1 = |1S\,(e)\,,\, m_s = +1/2\rangle$, $|e\rangle_2 = |1S\,(e)\,,\, m_s = -1/2\rangle$ for electrons and holes of QDs in Fig. 1 differ in the m_s values. The parameter $\Omega^{(2)} = \frac{\Omega_1\Omega_2}{2|\delta|}$ is the effective two-quantum Rabi frequency. Parameters $\delta = \bar{\omega} - \omega_X$ and $\delta^{(2)} = 2\bar{\omega} - \omega_{XX}$ are the effective detunings, where $\omega_{XX} = \omega_{X_-} + \omega_{X_+} - \Delta_{2b}$ is the biexciton frequency [14]. We assume $\omega_{X_-} = \omega_{X_+} = \omega_X$. We suppose that both NPs are identical, i.e. $\omega_{p1} = \omega_{p2} = \omega_p$, $\Delta = \bar{\omega} - \omega_p$, and the interaction between distant NPs placed at distance $2r$ can be ignored in the problem. The orientation of the dipole moments of the nano-objects in Fig. 1 corresponds to the nonradiative regime of the spaser, therefore the Rabi frequency of interaction between the QD and NP is $\Omega_1 = \Omega_2 = \frac{\mu_{QD}}{r^3}\sqrt{\frac{\omega_p a_{NP}^3}{4\pi\hbar\varepsilon_0}}$, where a_{NP} is the NP radius, ω_p is the plasmon resonance frequency of NPs, μ_{QD} is the dipole moment of the interband transitions in QD, ε_0 is vacuum permittivity, $\hbar$ is the Planck constant.

We use an approximation of slowly varying amplitudes, i.e. we assume that $\hat{c}_1 = \hat{c}_1 e^{-i\bar{\omega}t}$, $\hat{c}_2 = \hat{c}_2 e^{-i\bar{\omega}t}$ and $\hat{S} = \hat{S}e^{-2i\bar{\omega}t}$. Then the system of Heisenberg-Langevin equations corresponding to (1) can be represented in the following form:

$$\dot{\hat{c}}_1 = i\left(\Delta + \frac{i}{\tau_c}\right)\hat{c}_1 - i\Omega^{(2)}\hat{c}_2^+\hat{S} + \hat{F}_{c1}, \quad (2a)$$

$$\dot{\hat{c}}_2 = i\left(\Delta + \frac{i}{\tau_c}\right)\hat{c}_2 - i\Omega^{(2)}\hat{c}_1^+\hat{S} + \hat{F}_{c2}, \quad (2b)$$

$$\dot{\hat{S}} = i\left(\delta^{(2)} + \frac{i}{\tau_S}\right)\hat{S} + i\Omega^{(2)}\hat{c}_1\hat{c}_2 D + \hat{F}_S, \quad (2c)$$

$$\dot{D} = 2i\Omega^{(2)}\left(\hat{c}_1^+\hat{c}_2^+\hat{S} - \hat{S}^+\hat{c}_1\hat{c}_2\right) - \frac{D - D_0}{\tau_D} + \hat{F}_D, \quad (2d)$$

where the decay rates of different plasmon modes are assumed to be equal each other, i.e. $\tau_{c1} = \tau_{c2} \equiv \tau_c$. We also assume that the pump D_0 has a rate of $1/\tau_D$ for the exciton mode.

The amplitude of plasmons for a stationary solution (2) takes a form:

$$\bar{c}_i = \frac{e^{i\phi_i}}{2}\sqrt{\frac{\tau_c}{\tau_D}\left(D_0 - \bar{D}\right)}. \tag{3}$$

These stationary solutions are defined up to the plasmon phase ϕ_i. The amplitude of exciton mode takes the form

$$\bar{S} = -\frac{i\Omega^{(2)}\tau_c}{4B_1\tau_D}\bar{D}\left(D_0 - \bar{D}\right)e^{i(\phi_1+\phi_2)}. \tag{4}$$

Finally, the corresponding stable solution for population imbalance takes the form:

$$\bar{D} = \frac{D_0}{2} - \frac{1}{2}\sqrt{D_0^2 + \frac{16\tau_D}{{\Omega^{(2)}}^2\tau_c}\left(\Delta\delta^{(2)} - \frac{1}{\tau_c\tau_S}\right)}. \tag{5}$$

Initializing our system in the absence of a magnetic field, we assume $\bar{\omega}_0 = \omega_{XX0}/2$, where $\omega_{XX0} = \omega_{XX}|_{B_m=0}$. Then we get $\delta_0^{(2)} = 0$ and the required significant value of detuning is $\delta_0 = -\Delta_{2b}/2$ and additional condition $\Delta = 0$. Then the frequency of the transition in the QD can be expressed as $\omega_{X0} = \omega_p + \Delta_{2b}/2$. We choose $\lambda_p = 520$ nm for gold and $\Delta_{2b} = 2.881 \cdot 10^{14}$ s^{-1} (0.19 eV) for the CdSe QD [15, 16]. We determine the size of the QD $D_{QD} = 4.635$ nm to satisfy the conditions of two-quantum transition in the scheme. The dipole moment of corresponding transition in the QD will take new value $\mu_{QD} = 0.303 \cdot 10^{-28}$ C $\cdot$ m.

The simulation parameters correspond to $\tau_c = 5 \cdot 10^{-12}$ s, $\tau_S = 4 \cdot 10^{-11}$ s and frequency detuning to $\delta_0 = -1.441 \cdot 10^{14}$ s^{-1}. The distance between the NP and QD equals $r = 5$ nm. For a NP radius $a_{NP} = D_{QD}/2$, the single-plasmon Rabi frequency is equal to $\Omega_1 = \Omega_2 = \Omega = 1.534 \cdot 10^{13}$ s^{-1}, while the two-plasmon Rabi frequency equals $\Omega^{(2)} = 1.634 \cdot 10^{12}$ s^{-1}. These values of the Rabi frequencies approximately correspond to the study [17], where the NP-QD coupling factor is $1.516 \cdot 10^{12}$ s^{-1}.

The action of the magnetic field on the spaser system occurs at the spin moment of the electron and hole, which leads to a Zeeman splitting of the exciton energy. Then the frequencies of the excitons X_+ (X_-) will take the form $\omega_{X_-} = \omega_{X0} - \alpha B_m$ ($\omega_{X_+} = \omega_{X0} + \alpha B_m$), where $\alpha = g^F\mu_B/\hbar$ [18] and g^F is the Lande g-Factor, $\mu_B = 9.27 \cdot 10^{-24}$ J/T. The parameter g^F depends on the QD radius [19]. For very small QDs, this parameter almost coincides with $g_e^F = 2$ for the free electron and decreases to $g_{CdSe}^F = 0.68$ [20] for the CdSe bulk semiconductor.

However, the resulting biexciton frequency ω_{XX0} and the effective detuning $\delta_0^{(2)}$ do not change due to Zeeman splitting [18]. This is due to the fact that the Zeeman shifts for X_+ and X_- compensate each other. If, however, the diamagnetic shift in the QD is taken into account, then the exciton and biexciton energies take the forms $\omega_{X_-} = \omega_{X0} - \alpha B_m + kB_m^2$ ($\omega_{X_+} = \omega_{X0} + \alpha B_m + kB_m^2$) and $\omega_{XX} = \omega_{XX0} + 2kB_m^2$, where $k = \frac{e^2a_{ex}^2}{4\hbar\mu^*}$. The parameters $\mu^* = \left(\frac{1}{m_e} + \frac{1}{m_h}\right)^{-1}$ and $a_{ex} = \sqrt{\frac{1}{2}\left(r_e^2 + r_h^2\right)}$ are determined by the mass and radius of the exciton [18], where r_e and r_h are the effective radiuses of the electron and hole respectively.

Then the corresponding detunings will take the forms $\delta_{1,2} = \bar{\omega} - \omega_{X_-,X_+}$ and $\delta^{(2)} = 2\bar{\omega} - \omega_{XX0} - 2kB_m^2$, where $\bar{\omega}$ determines the new spaser frequency in the system, taking into account the magnetic field. Expression for $\bar{\omega}$ can be obtained in the next form:

$$\bar{\omega} = \frac{\tau_c\omega_p + \tau_S\omega_{XX}}{\tau_c + 2\tau_S}. \tag{6}$$

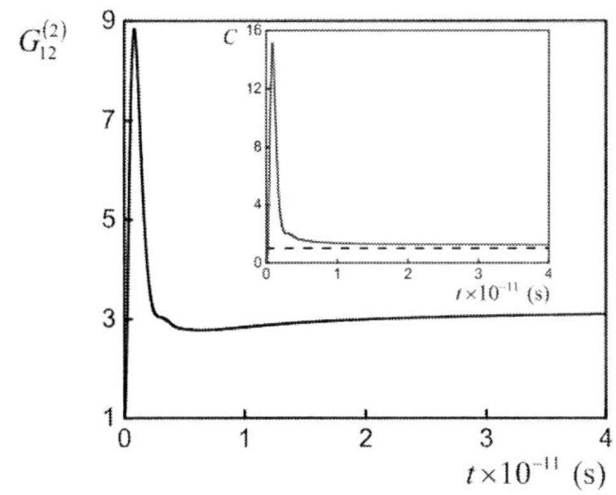

Figure 2. The time dependence of cross-correlation function $G_{12}^{(2)}$ (solid blue line) for plasmons in nonlinear NP-QD-NP spaser. Inset: time dependence of C parameter (solid red line) and the unity value of this parameter $C = 1$ (dashed black line) indicating on the violation of Cauchy-Schwarz inequality for generated plasmons in region above it.

Taking into account the solution (6) and in the presence of magnetic field with magnitude $B_m = 5$ T, the spaser frequency becomes equal to the value $\bar{\omega} = 3.625 \cdot 10^{15}$ s^{-1} ($\delta_1 = -1.433 \cdot 10^{14}$ s^{-1} and $\delta_2 = -1.448 \cdot 10^{14}$ s^{-1}, $\Delta = 8.678 \cdot 10^{10}$ s^{-1}, $\delta^{(2)} = -1.085 \cdot 10^{10}$ s^{-1}). The frequency detunings δ_1 (δ_2) satisfy to inequality $|\delta_{1,2}| > \Omega, \frac{1}{\tau_c}, \frac{1}{\tau_S}$ for the nonlinear regime in the spaser. Then the corresponding Rabi frequencies will take the values $\Omega_1^{(2)} = 1.642 \cdot 10^{12}$ s^{-1}, $\Omega_2^{(2)} = 1.625 \cdot 10^{12}$ s^{-1}.

In this paper we analyze the dynamics of the parameter

$$G_{12}^{(2)}(t,\tau) = \frac{\langle \hat{c}_1^+(t)\,\hat{c}_1(t)\,\hat{c}_2^+(t+\tau)\,\hat{c}_2(t+\tau)\rangle}{\langle \hat{c}_1^+(t)\,\hat{c}_1(t)\rangle\langle \hat{c}_2^+(t+\tau)\,\hat{c}_2(t+\tau)\rangle}, \tag{7}$$

which corresponds to the cross-correlation function and is a criterion for establishing correlations between plasmonic modes $\hat{c}_1$ and $\hat{c}_2$. In particular, the condition $G_{12}^{(2)} \equiv G_{12}^{(2)}(t,0) > 1$ is associated with the intermode plasmon bunching. Moreover, the violation of the Cauchy-Schwarz inequality

$$C \equiv \frac{G_{12}^{(2)}(t,0)^2}{g_1^{(2)}(t,0)\,g_2^{(2)}(t,0)} \leq 1$$

indicates the nonclassical character of the correlations between plasmonic modes, where $g_i^{(2)}(t,0) = G_i^{(2)}(t,0)+1$, $i = 1,2$. This is the necessary condition for the generation of entangled plasmons in the spaser system. In order to study the dynamics of the correlation parameters, we have derived a self-consistent system of equations for bi-linear combinations on the basis of plasmonic and excitonic modes. The value of parameter $G_{12}^{(2)}$ in such stationary conditions reaches the level 3.15, which demonstrates strong intermode bunching between modes $\hat{c}_1$ and $\hat{c}_2$. Moreover, the maximum value of C parameter is $C_{\max} = 15.15$ and the stationary regime value is $C_{\text{st}} = 1.2$, which demonstrates the nonclassical character of correlations between plasmon modes in spaser system, see inset in Fig. 2. This corresponds to the case when the initial entanglement between plasmons $\hat{c}_1$ and $\hat{c}_2$ is completely absent. Thus, the main result of our simulation is a demonstration of the development of quantum correlations between two localized plasmonic modes and the possibility for formation of an entangled state during the process of biexciton state decay in a nonlinear NP-QD-NP spaser system.

3. Conclusion

We have proposed a novel mechanism for the control of plasmons' quantum properties in the spaser, where a pair of NPs is coupled by nonlinear near-field interactions with a single QD (NP-QD-NP spaser).

We have optimized the system parameters providing the stationary regime of the entangled plasmons generation. Further development of this work may be aimed on a complex simulation of nonlinear collective processes [21] with non-classical states of surface plasmon-polaritons (SPP) [22] and SPP waves structures [23].

Acknowledgments

We would like to thank A.B. Evlyukhin for helpful discussions. This work was supported by the Russian Foundation for Basic Research Grant No. 17-42-330029 and the Grant of MSPU No. 0-1067.

References

[1] González-Tudela A, Martín-Cano D, Moreno E, Martín-Moreno L, Tejedor C and García-Vidal F J 2011 *Phys. Rev. Lett.* **106** 020501
[2] Struchalin G I, Pogorelov I A, Straupe S S, Kravtsov K S, Radchenko I V and Kulik S P 2016 *Phys. Rev. A* **93** 012103
[3] Becher C, Kiraz A, Michler P, Imamoğlu A, Schoenfeld W V, Petroff P M, Lidong Zhang and Hu E 2001 *Phys. Rev. B* **63** 121312(R)
[4] Andrianov E S, Pukhov A A, Dorofeenko A V, Vinogradov A P and Lisyansky A A 2015 *Quantum Electronics* **45** 245–249
[5] Gubin M Y, Shesterikov A V, Karpov S N and Prokhorov A V 2018 *J. Phys.: Conf. Ser.* **951** 012034
[6] Naumov A V, Gorshelev A A, Vainer Yu G, Kador L and Köhler J 2011 *Phys. Chem. Chem. Phys.* **13** 1734–1742
[7] Reinhardt C, Evlyukhin A B, Cheng W, Birr T, Markov A, Ung B, Skorobogatiy M and Chichkov B N 2013 *J. Opt. Soc. Am. B* **30** 2898–2905
[8] Fedyanin D Yu, Yakubovsky D I, Kirtaev R V and Volkov V S 2016 *Nano Lett.* **16** 362–366
[9] Lemke C, Schneider C, Leißner T, Bayer D, Radke J W, Fischer A, Melchior P, Evlyukhin A B, Chichkov B N, Reinhardt C, Bauer M and Aeschlimann M 2013 *Nano Lett.* **13** 1053–1058
[10] Levy R, Hönerlage B and Grun J B 1979 *Phys. Rev. B* **19** 2326
[11] Perlin E Yu 2000 *Optics and Spectroscopy* **88** 390–396
[12] Catalano I M, Cingolani A and Minafra A 1973 *Phys. Rev. B* **8** 1488
[13] Toor A H and Zubairy M S 1992 *Phys. Rev. A* **45** 4951
[14] Gauthier D J 2003 *Progress in Optics* vol 45 ed Wolf E (Amsterdam: Elseivier) chap 4, pp 205–272
[15] Patton B, Langbein W and Woggon U 2003 *Phys. Rev. B* **68** 125316
[16] Magarian K A, Fedyanin V V, Karimullin K R, Vasilieva I A and Klimusheva G V 2013 *J. Phys.: Conf. Ser.* **478** 012007
[17] Otten M, Shah R A, Scherer N F, Min M, Pelton M and Gray S K 2015 *Phys. Rev. B* **92** 125432
[18] Kuther A, Bayer M, Forchel A, Gorbunov A, Timofeev V B, Schäfer F and Reithmaier J P 1998 *Phys. Rev. B* **58** R7508(R)
[19] Pryor C E and Flatté M E 2006 *Phys. Rev. Lett.* **96** 026804
[20] Gupta J A, Awschalom D D, Al L Efros and Rodina A V 2002 *Phys. Rev. B* **66** 125307
[21] Golovanova A V, Gubin M Y, Gladush M G and Prokhorov A V 2016 *Bull. Russ. Acad. Sci. Phys.* **80** 808–813
[22] Shesterikov A B, Gubin M Y, Gladush M G and Prokhorov A V 2017 *JETP* **124** 18–31
[23] Dzedolik I V and Pereskokov V 2016 *J. Phys.: Conf. Ser.* **737** 012006

IOP Conf. Series: Journal of Physics: Conf. Series **1092** (2018) 012136 doi:10.1088/1742-6596/1092/1/012136

FLUORESCENT PROPERTIES OF GD-DOPED ZNO NANONPOROUS NETWORKS & ITS APPLICATION IN OPTICAL BIOSENSING

Shulga A. [1], Butusov L.A. [1,2], Boruleva E.A.[1], Chudinova G.K.[2,3], Sheshko T.F.[1], Kurilkin V.V.[1], Kochneva M.V. [1]

[1] Peoples' Friendship University of Russia (RUDN University), Miklukho-Maklaya str. 6, Moscow, Russia, 117198
[2] Institute of General Physics. A. Prokhorov RAS, Vavilov str. 38, Moscow, Russia, 119991
[3] National Research Nuclear University "MEPhI", Kashirskoye sh., 31, Moscow, Russia, 115409

E-mail: aleksandra-box@mail.ru

Abstract. This research presents a study of the fluorescent properties of new materials based on gadolinium-doped zinc oxide nanoporous networks obtained by sol gel method on the surface of microcrystalline silicon. The effect of co-doping of different concentrations of Gd ions on the emission properties of ZnO nanoparticles has been investigated. Emissions of such biomolecules as protein, amino acids and porphyrin and its detection limits were studied for the purpose of practical application of Gd-doped ZnO nanonporous networks as an element of an optical biosensor technology.

1. Introduction

Recently, scientists have attracted special attention to nanoparticles, which have multifunctionality, especially nanoparticles that combine several properties, for example semiconducting, electro-optical and luminescent [1]. Such materials contribute to a significant expansion of their application and zinc oxide is one of them.

Zinc oxide is a thermally and chemically stable material, while it is biocompatible. On the basis these properties zinc oxide is in use at biological systems and biosensors.

The doped structures, such as [5-7] studied the effect of doping with cadmium, magnesium and sulfur on the structural characteristics of ZnO films by an x-ray diffraction method and observed an increase in the lattice parameter with increasing impurity concentration. There is also important research on doping semiconductors with gadolinium to increase or decrease photoluminescent properties of ZnO [8]. Areas of application of such materials are very diverse. Consequently, it is widely used in fabrication of various optoelectronic devices such as transparent conducting oxide, solar cells or flat panel displays. In some cases, for the selective determination of a biomolecule, it is necessary to reduce the intrinsic fluorescence of zinc oxide, which is possible due to the introduction of the dopant. Besides semiconducting properties, ZnO nanomaterials also exhibit various desirable traits for biosensing such as high catalytic efficiency, strong adsorption capability, and high isoelectric point (IEP; ~9.5) which are suitable for adsorption of certain proteins by electrostatic interaction. Furthermore, high surface area, good stability, low toxicity, and high electron transfer capability also make them promising nanomaterials for biosensors [9].

Due to its unique optical, acoustic and electrical properties, zinc oxide has found application in gas sensors, varistors, and devices for generating surface acoustic waves. Being transparent in a wide spectral range, ZnO has a high resistance to irradiation, it is chemically etched and relatively cheap, which makes it attractive for use in microelectronics [2,3]. Doping of ZnO films with some elements leads to an increase in the width of the forbidden band, an increase in the activation energy of the donor centers and their stability [4]. The authors [5-7] studied the effect of doping with cadmium, magnesium and sulfur on the structural characteristics of ZnO films by an x-ray diffraction method and observed an increase in the lattice parameter c with increasing impurity concentration. Doping elements of the III group - aluminum, gallium, and indium - allows to obtain n-ZnO films with high conductivity. Doping elements of the V group made it possible to obtain ZnO films of p-type conductivity. To successfully create light-emitting devices based on ZnO in the form of p-n junctions or optoelectronic devices on heterostructures using ZnO, careful studies of the properties of doped ZnO films are necessary. The most convenient is gadolinium because its excitation wavelength coincides with the substrate materials.

In this article we consider the behavior of nanosized ZnO doped Gd films with different weight ratios of dopant and zinc oxide, for example, interactions with biomolecule such as myoglobine (MB) of various concentrations.

2. Experiment

The preparation of materials based on zinc oxide is currently carried out using various techniques having both advantages and disadvantages. In the present work the sol-gel method was used, one of the most effective methods for the formation of films whose surface is structured at the nanolevel. (nanosize). Materials of ZnO-Gd (III), which have not been studied before, have been obtained. The change in their spectral characteristics has been studied with varying concentrations of dopant-gadolinium ion – 7,8,12 stoichiometrically percentage (wt%). These films are the basis of an optical biosensor device with fluorescent detection, designed to investigate biomacromolecules in concentrations of 10^{-15}M and even lower by fluorescence spectroscopy method.

Zn^{2+}/Gd^{2+} + C_2H_5OH→ (≈60∘C→sirring) → ZnO/Gd_2O_3 (sol) → (monoethanolamine) → ZnO/Gd_2O_3 (sol/gel) → depositing the seed layer of the sol onto a substrate → dried in oven at 140∘C to obtain oxide films

Synthesis of undoped and Gd doped ZnO was carried out using analytical grade zinc acetate $[(CH_3COO)_2Zn]$, gadolinium nitrate $[Gd(NO_3)_3]$ and monoethanolamine (C_2H_7NO) in as-received condition. In the synthesis process, a required amount of zinc nitrate was completely dissolved in deionized water and a required amount of aqueous C_2H_7NO solution was added drop by drop to the aqueous zinc nitrate. The solution was stirred and maintained at room temperature for 40 min, and then kept at 60 °C for 2 h. until complete dissolution of the white precipitate For maturing the solution was kept at ambient temperature (22 ± 2)∘C for 2-7 days. After applying the seed layer sol thickness of 70 nm on a silicon substrates placed in a muffle furnace for drying at a temperature of 150 ∘C for 10 min, then annealed at a temperature of 500∘C for 2 hours. The process of deposition and drying and annealing were repeated until the desired coating thickness. For the synthesis of $Zn_{1-x}Gd_xO$ (from 0,01 to 0,33) NPs, a calculated amount of gadolinium nitrate was mixed with zinc nitrate solution. The required amount of aqueous C_2H_7NO solution was added drop by drop to the homogenous mixture to get a white precipitate. Further, a similar procedure adopted for the preparation of undoped ZnO.

3. Results and discussion

In this research we obtained thin films of a hybrid material of zinc oxide doped with different percentage by mass (7,8,12 wt%). Gadolinium is making a significant contribution to its own fluorescence sensor. We investigated the possibilities of fluorescence analysis of thin films of myoglobin protein using these sensor variations. Figure 1A shows the emission spectra (λ_{ex}=280nm) of myoglobin in concentrations from 10^{-4} to 10^{-12} mg / ml on the ZnO sensor surface with 7% of dopant. We note that the bathochromic and hypsochromic shifts of the fluorescence maxima did not exceed 3-5 nm. We observe a fluorescence decreasing with decreasing protein concentration up 10^{-8}

IOP Conf. Series: Journal of Physics: Conf. Series **1092** (2018) 012136 doi:10.1088/1742-6596/1092/1/012136

mg/ml of MB, then there is a linear dependence area and fluorescence ignition from 10^{-12} mg / ml of MB.

Figure 1B shows the fluorescence maximum values of all protein concentrations for three sensory surfaces doped with different amounts of phosphor. All three have a common feature - the ignition of fluorescence from 10^{-12} MB.

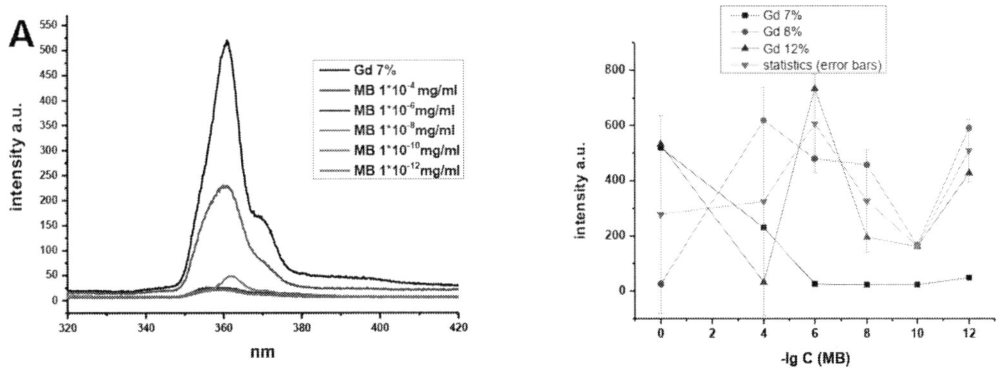

Fig.1 A) florescense speactra of Myoglobine's different concentrations (10^{-4}-10^{-12} increament of dilution x100) on the surface of ZnO-Gd doped (7 wt%) (λ_{ex}=280nm) B) fluorescence response of the protein (MB) (λ_{ex}=280nm) at different concentrations on three diferent surfaces of ZnO (modified by 7,8 and 12 weight percent of Gd)

In this case, the most significant are plates with 8 wt% of the Gd, which is almost 24 times higher than the base point without a protein, and 1.5 times exceeds 12 wt% of the Gd. The explanation of these phenomena is the task of our future research. Preliminary studies with a scanning electronic microscope using the example of 7% Gd are presented in Figure 2, these structures are nanoporous networks with a fairly large grain size of 8 μm, the average width of such filaments is 20-30 nm, while the length can reach several tens of μm.

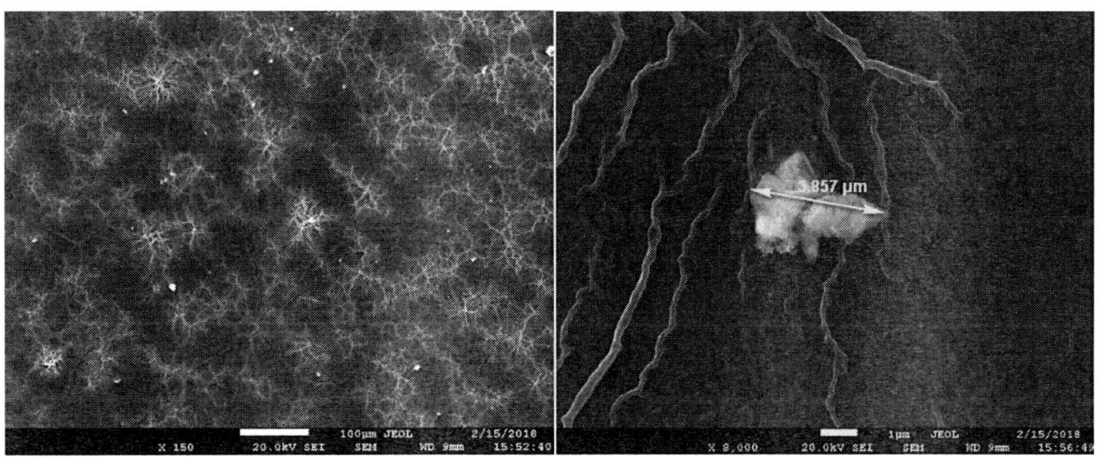

Fig.2 SEM images of ZnO modified by 7 weight percent of Gd

The excitation wavelength of 280nm corresponds to the excitation of a fragment of the molecule of myoglobin, the amino acid of tryptophan with emission in the region of 340-360nm, while we observe small "shoulders" in the region of 350-360nm associated with partial excitation of zinc oxide

nanostructures (see electronic transition in this region). In addition to tryptophan, the protein contains a porphyrin fragment, which, however, was not determined at concentrations below 10^{-6} mg / ml (λ_{ex}= 420nm λ_{em}= 680-710 nm).
The results obtained indicate good prospects for using gadolinium-doped zinc oxide structures as a fluorescent optical sensor for detecting biomacromolecules in the near ultraviolet region, apparently a quenching of the fluorescent response appears in the visible light region, which can also be used in specific cases.

4. Acknowledgments

This publication was prepared with the support of the "RUDN University Program 5-100".

5. References

[1] Y.Chen,T.Yao.Mater.Sci.Eng.B, 75, 190 (2000).

[2] D.C.Look.Mater.Sci.Eng.B,80, 383 (2001).

[3] D.C. Look, D.C. Reynolds, C.W. Litton, R.L. Jones,D.B. Eason, G. Cantwell. Appl. Phys. Lett.,81 (10), 1830 (2002).

[4] Y.W. Heo, Y.W. Kwon, Y. Li, S.J. Pearton, D.P. Norton. Appl.Phys. Lett.,84, 3474 (2004).

[5] Y.-Z. Yoo, Zheng-Wu Jin, T. Chikyow. Appl. Phys. Lett., 81(20), 3798(2002).

[6] T. Makino, Y. Segawa, M. Kawasaki, A. Ohtomo, R. Shiroki,K. Tamura, T. Yasuda. Appl. Phys. Lett.,78 (9), 1237 (2001).

[7] A. Ohtomo, M. Kawasaki, T. Koida, K. Masubuchi,H. Koinuma, Y. Sakurai, Y. Yoshida, T. Yasuda, Y. Segawa.Appl. Phys. Lett.,72 (19), 2466 (1998).

[8] Palvinder Kaur, Sanjeev Kumar, Chi Liang Chen, Ying-Ya Hsu, Ting-Shan Chan, Chung-Li Dong, C.Srivastava, Anupinder Singh, S.M.Rao. Investigation on structural, magnetic and electronic structure of Gd-doped ZnO nanostructures synthesized using sol-gel technique. Appl. Phys. A (2016) 122; 161.

[9] Kumar SA, Chen SM. Nanostructured zinc oxide particles in chemically modified electrodes for biosensor applications. Anal Lett. 2008; 41:141–58.

Decoupling of Dipole Antennas by a Split Loop

Constantin Simovski[1,2], Masoud Sharifian Mazraeh Mollaei[1], Stanislav Glybovski[2] and Anna Hurshkainen[2]
[1]Dept. Electronics and Nano-Engineering, Aalto University, P.O. Box 11000, FI-00076, Aalto, Finland.
[2]ITMO University, Dept. Nano-Photonics and Metamaterials, Kronverkski 49, 197101, St. Petersburg, Russia.
E-mail: konstantin.simovski@aalto.fi

Abstract. The decoupling of two very closely located resonant dipoles by a similar dipole located in the middle between them has been recently demonstrated. An approximate analytical model of this decoupling was built, validated by exact numerical simulations and confirmed experimentally. In this work we show that the similar decoupling can be achieved using another scatterer, namely a split-loop resonator may nicely replace the dipole. This replacement allows us to broaden the operation band of the antenna system nearly twice.

1. Introduction

A complete passive decoupling of two resonant dipole antennas 1 and 2 separated by an arbitrary gap d (the minimal value of d is restricted only by the requirement $d \gg r_0$, where r_0 is the wire cross section radius) was suggested and studied theoretically and experimentally in [1]. The decoupling is achieved by placing a similar resonant but passive dipole 3 (half-wave straight wire) in the middle of the gap. Then the electromotive force (EMF) induced by dipole 1 in dipole 2 is compensated by a part of the electromotive force induced in dipole 2 by scatterer 3. Similarly, scatterer 3 when excited by dipole 2 compensates the EMF induced by dipole 2 in dipole 1. This decoupling is complete, meaning that the power flux from dipole 1 to dipole 2 (and vice versa) is substituted by the flux from these dipoles to the scatterer for whatever relations between currents in dipoles 1 and 2. They both can be active, one of them can be active whereas the other can be loaded at the center by a lumped load or can be shortcut, they remain decoupled. However, this decoupling is approximate in the meaning that the power flux between dipoles 1 and 2 is suppressed not completely. For practical applications it is enough to reduce the mutual power transmittance by 10-12 dB. However, the bandwidth of antennas may suffer of the presence of scatterer 3. This is the case of [1], when the band of the resonant lossless matching (when the antenna circuits are tuned at the decoupling frequency) shrunk seven times due to passive dipole 3; and the bandwidth of the decoupling regime was as narrow as the resonance band. The purpose of the present study is to find a decoupling scatterer for two dipole antennas separated by a gap $d < \lambda/30$ as compact and efficient as the half-wave straight wire but more broadband. We will show that it can be achieved using an elongated split loop resonating at the same frequency as dipoles 1 and 2. We called this scatterer split-loop resonator (SLR).

2. Theoretical prerequisites

Now let us prove that decoupling of dipoles 1 and 2 located in free space is possible with an SLR symmetrically located between them. Since the loop contour C comprises the gap g we may consider the SLR as a wire scatterer. The method of induced EMFs is applicable to our SLR, as well as it was applicable to the dipole of our previous work [1]. Therefore, the condition of the complete decoupling expressed by formula (12) of work [1]

$$Z_{13}{}^2 = ZZ_M \tag{1}$$

remains valid for the structure whose side view is shown in Fig. 1. Here Z_M is mutual impedance between dipoles 1 and 2, Z is the self-impedance of our SLR, and $Z_{13} = Z_{23}$ is its mutual impedance with antenna 1 or antenna 2.

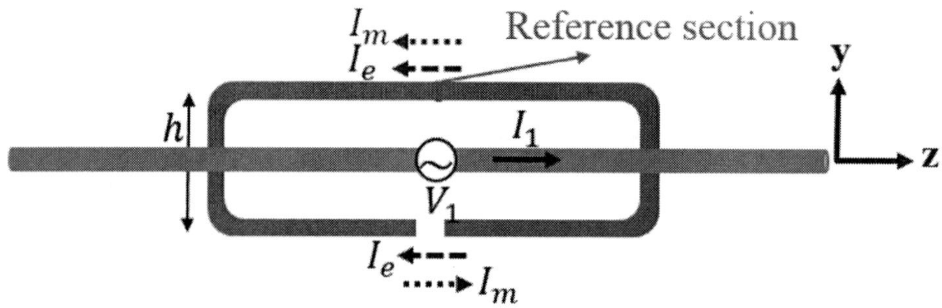

Figure 1. The side view of the structure comprising two active dipoles 1 and 2 (not seen behind) driven by arbitrary external voltages V_1 and V_2 and the passive SLR 3 between them whose self-impedance Z_{33} and mutual impedance $Z_{13} = Z23$ with the dipoles 1 and 2 refers to the Reference section. Current I_3 induced in the SLR is the sum of the electric $I_e = 0.5I_0 f_e(z)$ and magnetic $I_m = \pm 0.5 I_0 f_m(z)$ modes. At $(y = +h/2,\ z = 0)$ $I_m = I_e = I_0/2$, at $(y = -h/2,\ z = 0)$ $I_m = -I_e = -I_0/2$.

A primary source V_1 in the center of dipole 1 induces in our SLR 3 two current modes – an electric one I_e symmetric and an antisymmetric magnetic one I_m with respect to the plane $y = 0$. If the current in the reference section of the SLR i.e. at point $(y = +h/2,\ z = 0)$ is denoted as I_0 both modes have the same amplitude $I_0/2$ at this point, whereas they mutually cancel each other at the gap that can be approximated by point $(z = 0, y = -h/2)$. The distribution of the electric dipole mode along the SLR is similar to that in a straight wire. Contrary to the electric mode, the magnetic one is maximal at the vertical sides of the loop. This is so because these sides are shortcuts if our SLR is considered as a two-wire line. This model of the loop results in the following approximation:

$$f_e(z) \equiv \frac{I_e(z)}{I_0/2} = \frac{\sin k\left(\frac{L_l}{2} - |z|\right)}{\sin \frac{kL_l}{2}}, \quad f_m(z) \equiv \frac{I_m(z)}{I_0/2} = \pm \frac{\cos k\left(\frac{L_l}{2} - |z|\right)}{\cos \frac{kL_l}{2}}. \tag{2}$$

Sign plus corresponds to the top side of the loop $(y = +h/2)$, sign minus – to the bottom side $(y = -h/2)$. We analytically calculate the mutual impedance Z_{13} between dipole 1 and SLR 3 applying the general formula of the induced EMF method:

$$Z_{13} = \frac{1}{I_1} \int_C E_{13}(l) f_3(l)\, dl, \tag{3}$$

IOP Conf. Series: Journal of Physics: Conf. Series **1092** (2018) 012137 doi:10.1088/1742-6596/1092/1/012137

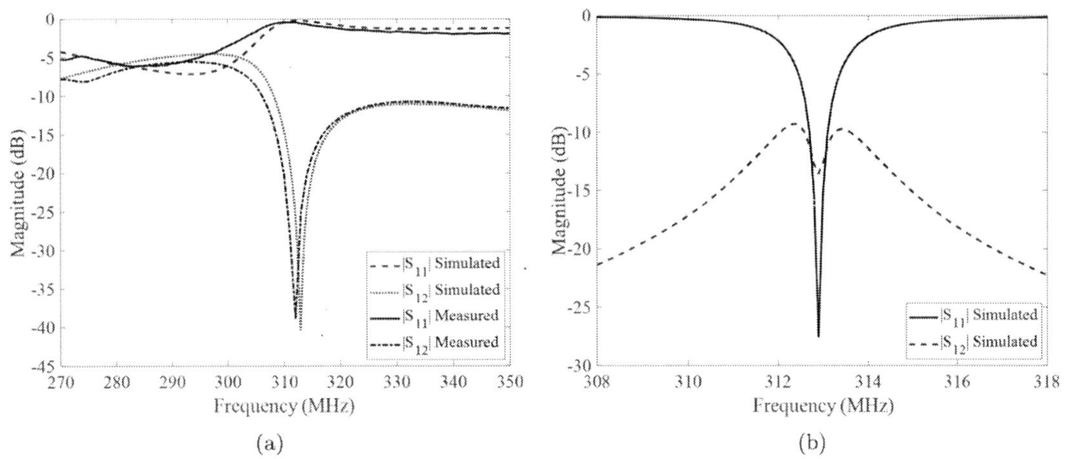

Figure 2. Frequency dependencies of S_{11} and S_{12} for the system of our dipoles 1 and 2 decoupled by our SLR 3. (a) – mismatched antennas, (b) – matched antennas.

where I_1 is the current at the center of dipole 1, $E_{13}(l)$ is the tangential component of the electric field produced by this primary current at a point l of the wire contour C of scatterer 3, and $f_3(l) = f_e(l) + f_m(l)$ is the current distribution in scatterer 3. Decomposition of the current induced in 3 onto the electric and magnetic modes allows us to split the right-hand side of (3) into electric and magnetic mutual impedances formed by the coupling of the primary current I_1 with the electric and magnetic modes, respectively. The analytical calculation of the integrals was done using the saddle point method. In this integration we essentially use the assumption that the SLR and the dipoles are resonant (and their resonance bands overlap).

In the case $h \ll d$ Eq. (1) – decoupling condition – transforms into a simple relation from which we find the detuning γ

$$\beta\gamma = \left(\eta k L_l^2/dR_0\right) - 1 \tag{4}$$

and check is this detuning really corresponds to the resonance band of both dipole 1 (or 2) and SLR 3. Choosing as an example $L_w = 500$ mm and $r_0 = 1$ mm (then the resonance band of dipoles 1 and 2 centered by the resonance frequency can be specified as 290-310 MHz) we fit the resonance band of the SLR to that of the dipoles when $h = 10$ mm and $L_l = 290$ mm. For $d = 3$ cm (in this case $h = d/3$) and $L_l = 29$ cm (4) yields $\gamma \approx 0.0423$ that implies the decoupling at the upper edge of the resonance band – at 312.8 MHz. Of course, the model is approximate and it is an approximate decoupling, however, in our terminology it is complete since should be observed for whatever relations of currents in the active dipoles.

3. Validation

Our experimental and numerical results are presented in Fig. 2. The complete decoupling implies that it is achieved for arbitrary relations between currents in the antennas. It is enough to inspect the decoupling for two different relations, corresponding to the mismatched and matched dipole antennas. In both mismatched and matched regimes (Figs. 2(a) and 2(b), respectively), minima of S_{12} were simulated at 312.8 MHz that is the indication of the complete decoupling. Due to the difficulty of the tunable matching circuit, we measured the S-parameters only for the mismatched structure. Our measurements agree very well with simulations and can be considered as a confirmation of the theory.

Our simulations for the matched case show that the insertion of SLR 3 decreases S_{12} at 312.8 MHz by 10 dB (from -4 dB corresponding to the reference structure [1] to -14 dB). This is probably sufficient for many applications. The operational band of the decoupled system can be defined as the minimal one of two bands – that of the matching (the band where $S_{11} \leq -15$ dB using a lossless matching circuit) and that of the decoupling (the band where $S_{12} \leq -10$ dB). In these definitions both bands of the matching and decoupling are equal to 1.3 MHz. This band is much wider than that offered by a decoupling dipole in [1] and this broadening is the main practical result. It follows from the fact that the extra mismatch due to the presence of the SLR at the distance $d/2$ from our antennas is not as high as the extra mismatch due to the presence of the dipole scatterer. We have not compared the simulation results of decoupling by passive SLR and dipole because the decoupling frequency is not the same for these cases; however, the enhancement of operating band is clear from comparison of S_{11} in Fig. 2(a) with S_{11} in Fig. 3 of paper [1].

4. Conclusions

We theoretically and experimentally showed the complete passive decoupling of two very closely located resonant dipoles achieved by adding a passive scatterer different from the similar dipole. Decoupling is granted by a long and narrow split loop, having the resonance in the same frequency band as the dipole antennas. The usefulness of this technical solution is the enlarged operation band of the antenna system.

References

[1] M. S. M. Mollaei, A. Hurshkainen, S. Glybovski, and C. Simovski, "Passive decoupling of two closely located dipole antennas," submitted to *IEEE Trans. Antennas Propag.*, available in arxiv: 1802.07500.

Acknowledgment

This work was supported by the Russian Science Foundation (Project No. 18-19-00482).

IOP Conf. Series: Journal of Physics: Conf. Series **1092** (2018) 012138 doi:10.1088/1742-6596/1092/1/012138

Light Trapping In Thin-Fim Solar Cells: Gain in Spite of Defects

Constantin Simovski[1,2], Mikhail Omelyanovich[1] and Victor Ovchinnikov[1]

[1]Dept. Electronics and Nano-Engineering,
Aalto University, P.O. Box 11000, FI-00076, Aalto, Finland.
[2]ITMO University, Dept. Nano-Photonics and Metamaterials, Kronverkski 49, 197101, St. Petersburg, Russia.

E-mail: konstantin.simovski@aalto.fi

Abstract. In this work, we report a nanostructure that grants a significant enhancement in optical efficiency to a thin-film solar cell based on amorphous silicon. The fabrication technique is very cheap and the nanostructure can be easily and rapidly prepared on a large area. This is possible by price of numerous mm-sized defects. In the areas of these defects the optical efficiency suffers. However, the enhancement of optical efficiency in the non-defective regions is so significant that the presence of defects still allows the gain granted by our light-trapping structure compared to a standard anti-reflective coating.

1. Introduction

While the light trapping in epitaxial solar cells was successfully achieved by texturing the photovoltaic (PV) layer, no one light-trapping structure (LTS) was commercialized for amorphous thin-film solar cells (TFSCs) dedicated to flexible substrates and having a lossy bottom electrode. In the literature, the idea of the resonant light trapping in such solar cells still dominates. Nowadays, from plasmonic nanostructures researchers switch to all-dielectric ones. However, their resonances are narrowband, whereas solar spectrum is ultra-broad. To cover the visible light range with a multi-resonant structure is possible but the cost of such the structure will be huge. Arrays of lenses represent a cheaper solution. However, all known attempts to trap the light using them resulted in a very modest enhancement for organic solar cells and in the absence of enhancement for cells based on a-Si. In our group, we have developed non-resonant all-dielectric nanostructures granting a significant gain to a-Si solar cells. However, in our experiments, we used a very cheap technology based on self-assembly. As a result, our LTSs strongly suffered of defects. Removal of the defective regions made our solar cells impractically small. In this paper, we report a nanostructure that grants a so significant enhancement in optical efficiency that the presence of defects still allows the gain compared to a standard anti-reflective coating (ARC). We report a significant gain obtained for 5 cm large samples based on a-Si.

IOP Conf. Series: Journal of Physics: Conf. Series **1092** (2018) 012138 doi:10.1088/1742-6596/1092/1/012138

2. Theoretical prerequisites

In work [1] we paid attention to the fact that a 300 nm thick layer of a-Si that corresponds to the maximal internal quantum efficiency of the corresponding solar cell is sufficiently thick for total absorption of solar light per one passage if the light is concentrated by a simple focusing of the incident wave beam. The absence of the enhancement is explained by a non-favorable distribution of the photocurrent in the bulk of a PV layer. Really, a lens collects the sunlight into a focal spot, whereas the most of the PV layer remains dark. Therefore, the drop of quantum efficiency compensates the gain in optical efficiency. We have suggested to locate an array of densely packed one-micron spheres of silica on top of the transparent conductive oxide (Al-doped ZnO nano-layer). In these structure the effect of focusing is cascaded for the oblique incidence. Then wave beam incident on one sphere penetrates into adjacent spheres and after total internal reflections and enters the PV layer. The part of the incident wave beam that transmits into the PV layer after one passage through the sphere and the part that transmits into it through the adjacent sphere both experience the focusing effect called photonic nanojet. This is not a usual focusing since the diameter of the sphere is comparable with the wavelength and the geometrical optics does not work. The formation of the nanojet corresponds to the constructive interference of spherical harmonics in the transmitted light. It is inherent only to transparent spheres whose diameter is 3-10 times larger than the wavelength [2]. Cascade focusing results in appearing the additional nanojets. For large angles of incidence – 45° and more – the whole volume of the PV layer turns out to be filled by nanojets. Since the light in nanojets is concentrated, the absorption in the PV layer is enhanced compared to the plane-wave regime. And what is important, for large incidence angles the parasitic transmission of light is prevented without the damage for the photocurrent distribution.

Similar structures had been known but in available works they operated either as optical cavities allowing only a narrow-band light trapping or as ARCs granting no light trapping. In the first case the diameters of the spheres are adjusted with the precision of 1-2 nm and in the second case – with the precision of dozens of nm. However, in both cases these spheres have submicron diameters. In our case 1 micron is exactly the optimum between the anti-reflective operation and the light trapping. For larger diameter the reflection losses grow due to the parasitic scattering. For lower diameter the photonic nanojet is not formed. Though for the normal incidence when the cascade focusing is absent our LTS yields to a standard ARC, shown in Fig. 1(a), the gain granted by it for large incidence angles is high. The solar cell operates not only in the midday. Its optical efficiency and photocurrent should be both averaged over the angles of incidence of sunlight corresponding to the daytime. Therefore the impact of the oblique incidence is important. For TFSCs based on a-Si even with 400 nm thick PV layer we experimentally obtained a 17% daytime-averaged gain in the photocurrent compared to the case when the same TFSC was covered by an optimized flat anti-reflective coating [3].

However, this design solution suffered of a serious drawback. The deposition of silica spheres in [3] was achieved by the self-assembly of the solar particles on a moving surface of the conductive oxide. The spheres transfer from the colloid to the surface forming a nice densely-packed monolayer. This method is much cheaper than the spin-coating and colloidal microspheres are also cheap, that looks very promising. However, a defect-free monolayer of spheres cannot be implemented in this way on a cm scale. Several mm-sized defects where the self-assembly failed and the spheres were isolated turned out to be unavoidable. In every square cm of our TFSC there were up to 10 such defects. They result in the strong backscattering from the initial 5 cm large samples. Scattering losses due to the defects compensated or even overcompensated the enhancement granted by the regular part of our LTS. Therefore, in work [3] we have removed the defective domains – the reported samples of TFSCs had millimeter sizes. Such TFSCs are suitable in order to report the enhancement granted by our LTS, however, they are impractically small and have no chances to commercialization.

This drawback urged us to search different mechanism of light trapping. We found this mechanism in the structure of notches implemented on top of the conducting oxide as shown in Fig. 1(b). Every notch is filled by a semi-diamond of silica and operates like a spherical microlens forming the photonic nanojet. Then the layer of the transparent oxide with densely packed notches becomes both all-angle anti-reflective coating and all-angle LTS without damaging its main function of the transparent electrode for the photocurrent collection. Simultaneously the conical shape of the silica protrusion is much more favorable in the defective areas because it promises lower scattering than the spherical shape. The backscattering cross section of our new microlens is nearly twice lower than that of the same notch comprising a silica sphere and is lower by an order of magnitude than that of a sphere located on the flat surface.

For the LTS described in our work [3] the negative impact of the defective areas was critical. In the new LTS the negative impact of the isolated notches with semi-diamond inclusions, in accordance, to simulations is almost absent.

3. Results

In Fig. 1 one can see the difference in the spatial distributions of the electromagnetic flux density in the reference structure with flat anti-reflective coating and in the suggested structure with patterned Al-doped ZnO. In the first case we observe the plane-wave transmission with the standard decay rate. In the second case we see the collimated wave beam composed under the notch. This color map fully confirms our initial insight of the light-trapping mechanism. Let us stress that this beam collimation is not a resonant effect. In Fig. 1 we show the distributions of the electromagnetic flux at four occasionally chosen wavelengths and in all cases observe the beam collimation.

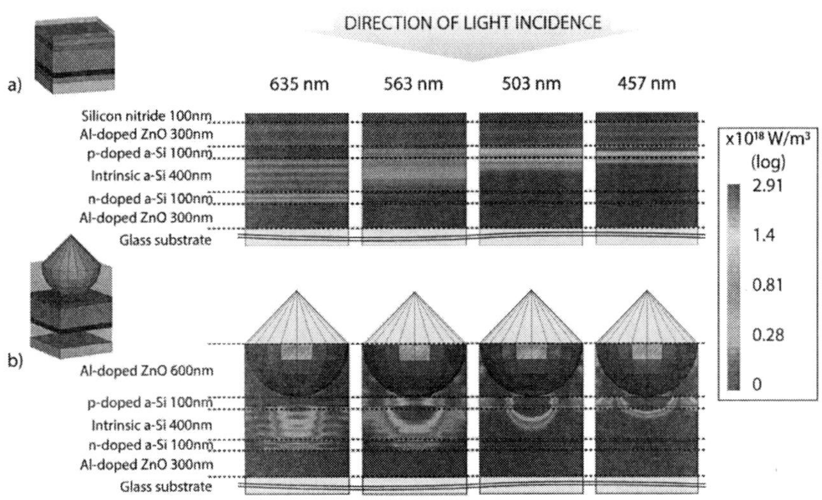

Figure 1. Comparison of the electromagnetic flux density in the reference (a) and suggested (b) structures at 4 wavelengths (normal incidence). Photonic nanojet is seen for the suggested LTS at all wavelengths.

As to the fabrication technology, it is also affordable: we use the Atomic Layer Deposition (ALD) and ion beam etching (IBE) in the small copy of an industrial machine. This technique is much cheaper than the ion beam lithography based on focused ion beams. It can process any samples up to wafer size, at any angle in uniform broad beam of Ar ions. Etching rate is 20 -

50 nm/min. We prepare a monolayer of colloidal particles on top of a PV layer using the self-assembly. Next, we deposit a layer of Al-doped ZnO by ALD. Al-doped ZnO was successfully deposited on the amorphous silicon substrate in the gaps between the spheres. ALD is done with the ratio of steps 19:1 (19 steps of ZnO to 1 step of Al2O3 deposition). At the last stage we utilize IBE in order to remove the top half of the structure for 10 minutes and 30 sec. The notches filled with the silica semi-diamond particles automatically result from the IBE – conical nano-protrusions do not require any special technique. The optimal size of the silica sphere for our LTS is 1.2 μm. Such spheres in colloidal solutions are as available and cheap as one-micron spheres.

Thorough measurements with several samples impinged by obliquely incident light of the solar simulator confirmed our theoretical expectations. In spite of defective regions which cover 30% of the area of our 5 cm large sample, the daytime-averaged enhancement of the PV absorption granted by our LTS is more than double compared to the bare solar cell and exceeds by 50% the enhancement offered by to the standard ARC (without defects). The absorption coefficient of the PV layer in presence of the LTS exceeds 70% and for the incidence angle 60° approaches to unity.

4. Conclusions

We have introduced a new LTS suitable for a-Si TFSCs combined with the top-electrode and does not require an additional ARC on its top. Our LTS is not resonant, does not contain lossy inclusions and grants the all-angle operation with a noticeable increase of the PV absorption compared to the flat anti-reflective coating located on top of the same solar cell. The structure can be fabricated in the affordable way that makes it attractive for industrial adaptation. An only expensive step is the application of the ALD machine. However, ALD though still considered as a rather expensive technology in the PV community, is developing and has good perspectives in the mass production of solar cells. Therefore, we believe that our finding is really useful and may give a revitalizing pulse to the development of a-Si solar photovoltaics.

References

[1] C.R. Simovski, A.S. Shalin, P.M. Voroshilov, and P.A. Belov, *J. Appl. Phys.* **114**, 103104 (2013).
[2] A.V. Itagi and W.A. Challener, *JOSA A* **22**, 2847 (2005).
[3] M. Omelyanovich, V. Ovchinnikov, and C. Simovski, *J. Optics* **17**, 025102 (2015).

IOP Conf. Series: Journal of Physics: Conf. Series **1092** (2018) 012139 doi:10.1088/1742-6596/1092/1/012139

Scattering of light from disordered photonic structures

A D Sinelnik[1 a)], **K B Samusev** [1,2], **M V Rybin**[1,2], **S Y Lukashenko**[1] **and M F Limonov**[1,2]

[1] ITMO University, St. Petersburg 197101, Russia
[2] Ioffe Institute, St. Petersburg 194021, Russia

[a)] E-mail: artem.sinelnik.95@mail.ru

Abstract. We study theoretically and experimentally the scattering of light from ordered and disordered photonic structures. As an example, we consider a thin slab of a woodpile-type photonic crystal and introduce random fluctuations in the rod orientation, revealing a crossover from the Laue diffraction to the formation of speckle patterns. We observe a nontrivial interplay between order and disorder when the orientational disorder is added only in one direction of the square woodpile structure. Namely, ordered sets of rods produce disordered patterns and, vice versa, disordered sets of rods produce ordered patterns. We explain this effect theoretically and experimentally by the fact that light scatters only from the intersection points of different rods.

1. Introduction
The macroscopic optical properties of a random medium essentially depend on the different length scales involved in the system on the microscopic level [1]. When both the size of the scatterers and the average distance between them are smaller than the wavelength, the random medium behaves as a homogeneous effective medium called as 3D metamaterials or 2D metasurfaces. When the refractive index of the medium displays fluctuations on length scales comparable or larger than the wavelength, undergoes a series of scattering events that randomizes the direction and phase of propagating waves. For the ordered photonic crystal, a laser light diffraction experiment demonstrates a Bragg diffraction pattern. For the disordered photonic structures including photonic glasses, a laser light diffraction experiment demonstrates a speckle pattern [2]. Laser speckle is an interference pattern produced by light scattered from different parts of the disordered object. The intensity at any point on the image is determined by the algebraic addition of all the wave amplitudes arriving at the point [3].

In this study, we measured experimentally optical diffraction from fabricated ordered and disordered woodpile-type structures and found unusual interplay between order and disorder. We found that the first-order diffraction patterns from ordered set of rods become randomized while diffraction patterns from disordered set of rods continue to be bright and sharp.

2. Samples preparation
The aim of this work was the synthesis of ordered and disordered woodpile-type photonic structures using the direct laser writing technique [4,5]. The structure was created using a hybrid organic-

inorganic material based on zirconium propoxide with an Irgacure 369 photo-initiator. The polymerization was performed with a 50 fs TiF-100F laser. The perfect woodpile structure is built *xy*-layer by *xy*-layer, four layers make up the lattice period *c* along the *z* axis. The building block that composes the structures is a square rod with the height *c*/4. In the first layer, the rods are parallel to each other with *a* period along the *x* axis, in the second layer the same rods with the same *a* period are parallel to each other along the *y* axis.

The disordered woodpile structures were fabricated as follows. Each individual rod in the layer was turn about its center (along *x*- or *y*-axis) by random angle α_i with respect to the ordered state ($\alpha_i = 0$ for the rods is the ordered structure). We employed function $\sigma = p\frac{\pi}{4}$ $(0 \le p \le 1)$for the normal distribution and for the uniform distribution $-\alpha_{max} \le \alpha_i \le \alpha_{max}$ with $\alpha_{max} = p\frac{\pi}{4}, 0 \le p \le 1$. All structures have the lattice parameters varied in the range of $0.5\mu m \le a \le 2.0\mu m$, the number of layers along the *z*-axis N was ranging from two layers (a metasurface) to 10 layers (ordered or disordered thin film). The correspondence of the resulting materials to the designed structures was confirmed by scanning electron microscopy (SEM).

3. Light scattering from ordered photonic structures

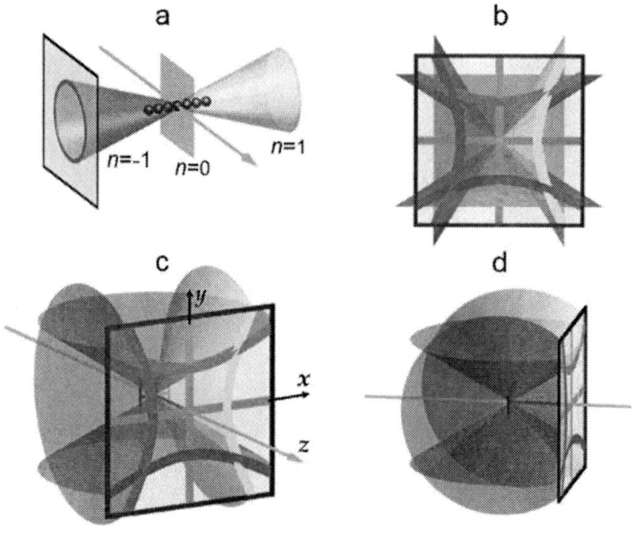

Figure 1. (a) Schematic of the zero-order (n = 0) and first-order (n = ± 1) Laue diffraction from the horizontally oriented chain of periodic scatterers. (b-c) Schematic of the zero- and first-order Laue diffraction from the ordered 2D structure with square symmetry in the case $a_x = a_y$. Diffraction patterns on a flat screen are shown by thick lines. Scattered light is shown by different colors for clarity.

The ordered woodpile structure can be considered as a structure composed of two sets of mutually orthogonal chains along the *x*- and *y*-axes. We use the Born approximation for the analysis of light scattering from low-contrast ordered 2D structures. For the 2D structures with the square lattice symmetry, the position of each scatterer is determined by the 2D vector $\mathbf{r}_i = \mathbf{a}_1 n_1 + \mathbf{a}_2 n_2$, where $\mathbf{a}_1$ and $\mathbf{a}_2$ stand for the basis mutually perpendicular vectors ($\mathbf{a}_1 \cdot \mathbf{a}_2 = 0$) of the square woodpile-like lattice ($a_1 = a_2 = a$). First, we consider the scattering from one-dimensional linear chain of scatterers.

The appearance of the Laue diffraction maxima in the case of the linear chain and normal incidence is described by the simple formula:

$$\theta_s = \cos^{-1}(n\lambda/a) \qquad (1)$$

where $a = |a_1|$, λ is the wavelength of incident light, and θ_s is the angle of light scattering from the chain between vectors a_1 and the wave vector of the scattered waves k_s. Note that the inverse cosine function is only defined in the interval from -1 to 1. Therefore the equation (1) defines the diffraction selection rules in relation to the ratio between λ and a. The zero-order diffraction (n = 0) is observed for any ratio between λ and a in the plane perpendicular to the axis a since the angle of light scattering becomes $\theta_s = 90°$, figure 1 (a). A pair of diffraction cones of the n-th order appears at $a > n\lambda$. For $\lambda < a < 2\lambda$, one can distinguish in the light scattering additionally to the plane (n=0) a pair of cones (n=1) that has the axes of symmetry coinciding with **a** and the apex angle of scattering $\theta_{s1} = \cos^{-1}(\lambda/a)$, figure 1 (a). For $2\lambda < a < 3\lambda$, one can observe the plane and two pairs of cones with $\theta_{s1} = \cos^{-1}(\lambda/a)$ (n=1) and $\theta_{s2} = \cos^{-1}(2\lambda/a)$ (n=2) and so on.

4. Disorder-induced light scattering

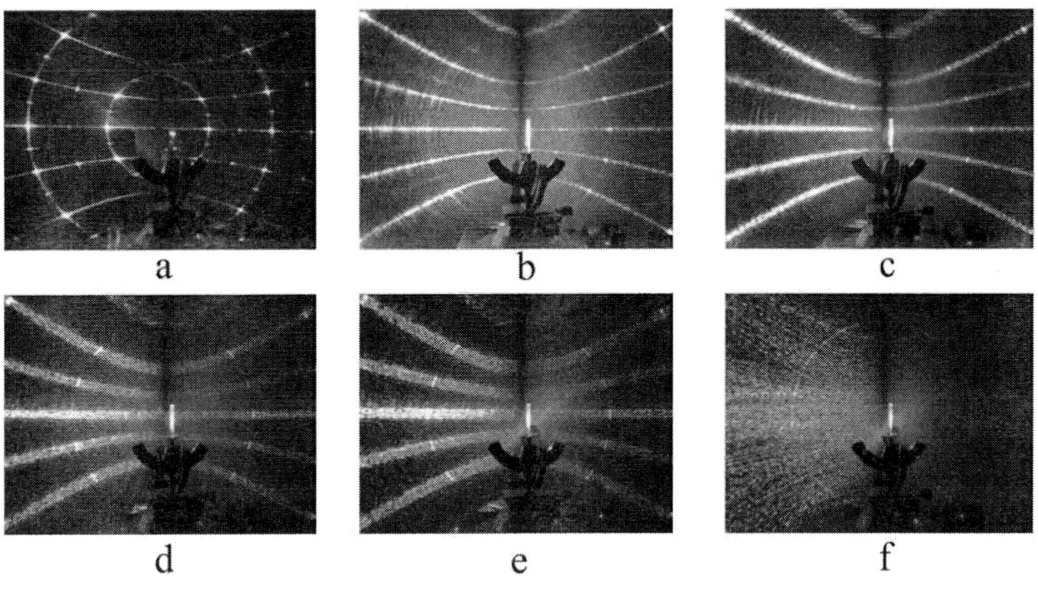

Figure 2. Diffraction pattern evolution for anisotropic woodpile thin slab depending on the disordered parameter p for the normal distribution. The anisotropic structures with disorder in rods oriented along vertical *y*-axis. $p= 0$ (a), $p =0.01$ (b), $p =0.04$ (c), $p =0.08$ (d), $p =0.1$ (e), $p =0.5$ (f). All structures have external size in the *xy* plane of 100x100μm, the lattice parameters of the ordered sample $a_x = a_y = 2\mu m$, the number of layers along the *z*-axis N=4. The patterns are observed on a flat screen positioned perpendicularly to the sample, as shown in figure 1 (a). $\lambda = 0.53\mu m$.

Using scattering geometry show in figure 1 (a), we investigated the patterns observed on a flat screen positioned perpendicular to the sample. Figure 2 demonstrates the diffraction patterns from the anisotropic woodpile with disorder in rods oriented along vertical *y*-axis. The n-order diffraction patterns depend on the interscatterers distance accordingly to the Laue equation $\theta_s = cos^{-1}(n\lambda/a)$. Therefore for the randomized parameter a, the angle θ_{sn} becomes uncertain. With increasing of the disordered parameter p, arcs and circles become more and more randomized and a granular distribution of light intensity appears at strong disorder $p = 0.5$.

To summarize, in woodpile disordered structures, the zero-order and higher-order diffraction patterns show different behavior as a function of the disorder parameter p. The zero-order diffraction is not significantly modified for small and intermediate disorder, but the higher-order diffraction can be affected strongly. For anisotropic woodpile structures when orientation disorder is introduced only in one direction of square woodpile, we found an interesting effect. When disorder increases, the higher-order patterns from ordered set of rods becomes randomized and finally hardly observed while higher-order patterns from disordered set of rods continue to be bright and sharp. The reason for such effect is that the light scatters not from the whole dielectric rod but from intersection points of different rods. This conclusion can be considered as general features of light scattering in dielectric photonic structures.

Acknowledgements
We acknowledge a support of the Ministry of Education and Science of the Russian Federation (Project 3.1500.2017/4.6), the Presidium of RAS (Program 32, Nanomaterials) and the Russian Foundation for Basic Research (Grant 18-02-00427).

References

[1] M. F. Limonov and R. D. L. Rue 2012 *Optical Properties of Photonic Structures: Interplay of Order and Disorder*, (New York CRC Press, Taylor & Francis Group).
[2] P. D. García, R. Sapienza, and C. López, Photonic Glasses: A Step Beyond White Paint, Adv. Mater. 22, 12 (2010)
[3] J.W. Goodman 1984 *Statistical properties of laser speckle patterns Laser Speckle and Related Topics 2nd ed* (Berlin: Springer)
[4] M. V. Rybin, K. B. Samusev, S. Yu. Lukashenko, Y. S. Kivshar, M. F. Limonov 2016 *Sci. Rep.* **6**, 30773
[5] A.D. Sinelnik, M.V. Rybin, S.Y. Lukashenko, M.F. Limonov, K.B. Samusev 2017 Phys. Rev. A, **95**, 063837

IOP Conf. Series: Journal of Physics: Conf. Series **1092** (2018) 012140 doi:10.1088/1742-6596/1092/1/012140

2π steering of surface plasmon polaritons with silicon nanoantennas

Ivan Sinev[1], Andrey Bogdanov[1], Filipp Komissarenko[1,2], Ivan Mukhin[1,2], Anton Samusev[1], Ivan Iorsh[1], and Andrei Lavrinenko[3]

[1]ITMO University, Kronverksky pr. 49 197101 St. Petersburg, Russia
[2]St. Petersburg Academic University, Khlopina st. 8/3 194021 St. Petersburg, Russia
[3]Technical University of Denmark, 2800 Kongens Lyngby, Denmark

E-mail: i.sinev@metalab.ifmo.ru

Abstract. We experimentally demonstrate full-angle control over the direction of excitation of surface plasmon polaritons on thin gold film with a single silicon nanosphere. Upon oblique excitation of the nanosphere with circularly polarized light beam, the directivity pattern of the excited surface waves exhibits rapid wavelength-dependent steering, which is the result of combined action of chiral response of the nanoantenna driven by mutual interference of its magnetic and electric dipole moments and chirality of the incident wave.

1. Introduction

obust spin momentum locking for surface and guided electromagentic waves allows for their directional excitation with circularly polarized light or chiral antennas [1, 2]. This opens up great prospects in many areas from optomechanics to quantum nanophotonics [3, 4, 5]. Recently, it was shown that dielectric nanoantenna provides chiral response with strong spectral dependence due to the interference of electric and magnetic dipole momenta for specific excitation conditions [6]. Notably, the effect does not require elliptical polarization of the pump beam or the geometric chirality of the nanoantenna, and thus realizes the concept of directional coupling of surface waves *beyond* spin-momentum locking [7]. Since the chiral response due to interference of dipoles is induced normally to the plane of incidence, silicon nanoantenna provides resonant switching between forward and backward excitation of surface plasmon polariton modes within a narrow spectral band.

On the other hand, excitation of nanoantenna with circularly polarized light should provide switching of the excited SPP from left to right with respect to the plane of incidence depending on the helicity of light [8, 9]. Here, we show that by utilizing both these effects, full control over the direction of excitation surface wave can be achieved, and provide first experimental evidence of plasmonic beam steering for silicon nanosphere on metal substrate (Fig. 1a).

2. Experiment

Single silicon nanoparticles were fabricated using laser ablation of amorphous silicon films [10] and transferred to thin (40 nm) gold film using nanomanipulations under electron beam [11].

To demonstrate the effect of SPP directivity switching experimentally, we used the setup for leakage radiation microscopy combined with Fourier plane imaging optics[12]. SPP was launched

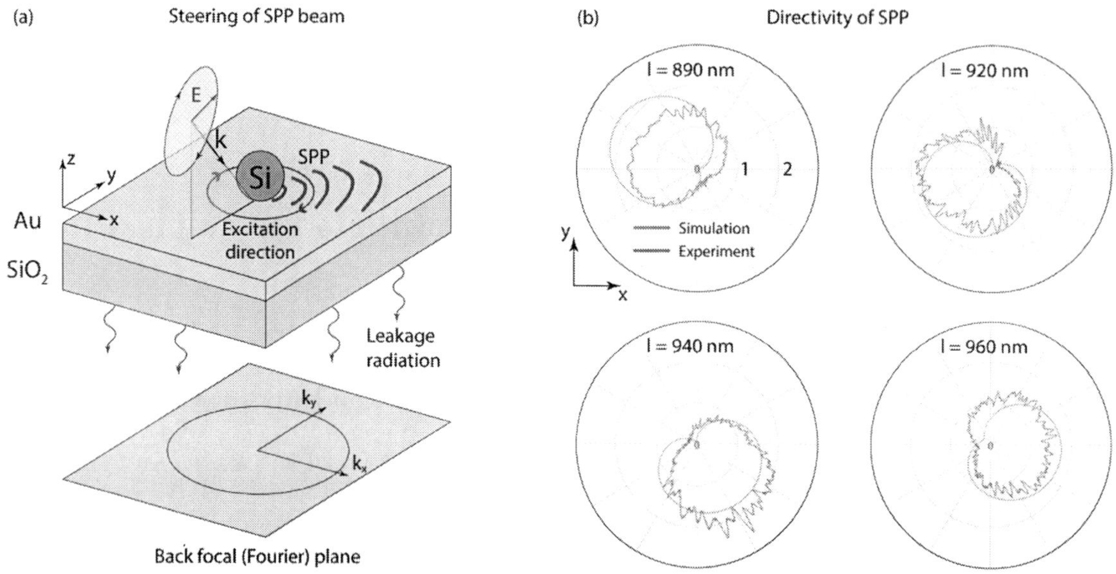

Figure 1. (a) Sketch of the experiment. SPP is excited silicon nanoantenna illuminated by obliquely incident circularly polarized wave. The directivity is measured using leakage radiation microscopy setup combined with Fourier plane imaging optics. (b) Measured and calculated directivity patterns of SPP at different wavelengths.

from the sphere by exciting it with a polarized laser beam with tunable wavelength incident at $\approx$25 degrees to the substrate normal and mildly focused with a 5 cm achromatic doublet lens. Spectral scanning was realized with Fianium SC400-6 supercontinuum source with acousto-optical tunable filter, which provided beam with central wavelength within 650-1100 nm range. The SPP leakage radiation present due to relatively small thickness of gold layer was collected from the bottom with an oil immersion objective (Zeiss 100x, NA=1.46). In the Fourier imaging optical channel, the incident beam was filtered with a beamstop to avoid camera overexposure.

We probe a 310 nm silicon nanosphere with circularly polarized excitation of different helicity and reveal the spectral evolution of the surface plasmon polariton (SPP) directivity patterns using leakage radiation microscopy setup combined with Fourier plane imaging optics [12]. The measured data show that within the spectral range where p-polarized light induces forward-to-backward switching of surface plasmon polariton from the nanoantenna [6], circularly polarized excitation enables wavelength-dependent steering of the directivity pattern. The measured SPP directivity patterns for several characteristic wavelengths are shown in Fig. 1b. The experimental data are in great agreement with the results of analytical calculation based on Green's function approach [13]. We show that by using both helicities of the incident wave, full 2π coverage of the direction of SPP excitation is possible.

3. Conclusions

In summary, we have experimentally demonstrated full-angle surface plasmon polariton beam steering with a single high-index dielectric nanoparticle. We showed that for particular angle of incidence, the combination of chiral response of dielectric antenna, which enables directional launching of surface waves within the plane of incidence, and helicity of the incident wave provides wavelength-dependent tuning of the surface wave excitation direction, thus realizing plasmonic beam steering. The experimental demonstration of the effect was carried out via leakage radiation microscopy combined with Fourier plane imaging optics, which allowed to reconstruct full spectral dependence of the SPP directivity pattern for both helicities of the incident beam. Our findings have important practical implications for on-chip optical communications and surface photonics.

Acknowledgments
This work was supported by the Ministry of Education and Science of the Russian Federation (Zadanie No.3.1668.2017/4.6). Ivan Sinev acknowledges the support of Russian Foundation for Basic Research, grant no. 18-32-00527.

References

[1] Lin J, Mueller J B, Wang Q, Yuan G, Antoniou N, Yuan X C and Capasso F 2013 *Science* **340** 331–334
[2] Mitsch R, Sayrin C, Albrecht B, Schneeweiss P and Rauschenbeutel A 2014 *Nature communications* **5** 5713
[3] Rodríguez-Fortuño F J, Engheta N, Martínez A and Zayats A V 2015 *Nature communications* **6** 8799
[4] Coles R, Price D, Dixon J, Royall B, Clarke E, Kok P, Skolnick M, Fox A and Makhonin M 2016 *Nature communications* **7** 11183
[5] Petrov M I, Sukhov S V, Bogdanov A A, Shalin A S and Dogariu A 2016 *Laser & Photonics Reviews* **10** 116–122
[6] Sinev I S, Bogdanov A A, Komissarenko F E, Frizyuk K S, Petrov M I, Mukhin I S, Makarov S V, Samusev A K, Lavrinenko A V and Iorsh I V 2017 *Laser & Photonics Reviews* **11**
[7] Picardi M F, Zayats A V and Rodríguez-Fortuño F J 2018 *Physical review letters* **120** 117402
[8] Rodriguez-Fortuño F J, Marino G, Ginzburg P, O'Connor D, Martínez A, Wurtz G A and Zayats A V 2013 *Science* **340** 328–330
[9] Rodriguez-Fortuño F J, Barber-Sanz I, Puerto D, Griol A and Martínez A 2014 *ACS Photonics* **1** 762–767
[10] Dmitriev P, Makarov S, Milichko V, Mukhin I, Gudovskikh A, Sitnikova A, Samusev A, Krasnok A and Belov P 2016 *Nanoscale* **8** 5043–5048
[11] Denisyuk A I, Komissarenko F E and Mukhin I S 2014 *Microelectron. Eng.* **121** 15–18 ISSN 0167-9317
[12] Drezet A, Hohenau A, Koller D, Stepanov A, Ditlbacher H, Steinberger B, Aussenegg F, Leitner A and Krenn J 2008 *Materials Science and Engineering: B* **149** 220–229
[13] Miroshnichenko A E, Evlyukhin A B, Kivshar Y S and Chichkov B N 2015 *ACS Photonics* **2** 1423–1428

IOP Conf. Series: Journal of Physics: Conf. Series **1092** (2018) 012141 doi:10.1088/1742-6596/1092/1/012141

Ultrafast dynamics of light scattering in resonant GaAs nanoantennas

Viacheslav S. Snigirev[1, a)], Alexander S. Shorokhov[1], Irina V. Soboleva[1], Dmitriy N. Gulkin[1], Vladimir O. Bessonov[1] and Andrey A. Fedyanin[1]

[1]Quantum Electronics Department and Quantum Technology Centre, Faculty of Physics, Lomonosov Moscow State University, Moscow 119991, Russia
[a)]Corresponding author: snigirev@nanolab.phys.msu.ru

Abstract. Active dielectric nanostructures have become one of the most popular trend lines in the modern dielectric nanophotonics due to great opportunities, which are offered by their numerous practical applications. Particularly, the concept of an optical switcher in nanophotonic circuitry can be realized by so-called all-optical mechanism, which could be carried out by a nanoantenna, such as an asymmetric dimer comprised of resonant nanoparticles made of material with high refractive index. In this case all-optical switching can be defined as the change of the trajectory of the femtosecond laser pulse after scattering on the nanoantenna, which was pumped by the other strong laser pulse. The realization of such effects would be a significant advance on the path to a novel technology. In this paper, we numerically demonstrate all-optical switching in the cylindrical asymmetric GaAs nanodimer, with the relaxation time of 10 ps and the deflection angle of the probe pulse of 7° for the relatively low intensities of the pumping ($\sim 15\ \frac{GW}{cm^2}$).

1. Introduction

Optical metamaterials represent an intensively developing sector of modern nanophotonics. Aiming at widening the range of their performance, they have passed a long way from plasmonic to dielectric nanostructures [1—3]. Dielectrics, apart from low dissipative losses at optical frequencies, provide great opportunities in terms of the scattering properties. Indeed, the magnitude of magnetic dipole (MD) modes in dielectric nanoresonators is comparable with the magnitude of electric dipole (ED) modes and for specific geometries can even exceed it [4]. The interference of the modes in the far field results in various emission patterns of the scattered light. Unidirectional scattering, formed by the interaction of two dipole modes under certain conditions [4], is a particularly relevant example of this interference.

The construction of the conventional dielectric nanoresonators is not designed to provide for the change of the system's configuration. For the case of unidirectional scattering that means that light of the same wavelength is always scattered in the same direction. Therefore, these nanostructures are passive and their optical properties cannot be externally manipulated. The development of active nanostructures can solve this problem.

Semiconductors can also be chosen as a material for a nanostructure. This class of materials preserves all advantages of dielectrics and is characterized by smaller values of the band gaps. The latter aspect is important for the effective free-carrier injection – phenomenon that leads to the changes of optical properties of the material due to varying concentration of electrons and holes in semiconductor.

The injection can be carried out by optical means. For instance, the optical response of the nanostructure can be changed by the strong laser pumping pulse. As a result, the trajectory of the probe pulse will be modulated, indicating this change. This effect is called all-optical switching. It was numerically demonstrated for an asymmetric Si-nanodimer, composed of two spherical nanoparticles: for the intensity of a pumping impulse $I_{peak} = 40\ \frac{GW}{cm^2}$ the deflection angle of the probe pulse was $\Delta\varphi = 20°$ [5]. It is worth noticing that the considerable change of optical properties can be achieved for lower intensities of the pumping pulse. The measurements of the change in refractive index of the direct-gap semiconductor (GaAs) metasurface demonstrated the change of the reflection coefficient $\Delta R = 0.35$ for smaller peak intensities $\sim 10\ \frac{GW}{cm^2}$ [6].

Taking into account the recent advances in the field of active photonic nanostructures, an idea of a new investigation emerges – experimental observation of all-optical switching in the asymmetric dimer. Experimental character dictates the choice of a new geometry (it is convenient to take disks instead of spheres) and of a new material (GaAs). This structure has never been considered before in the context of all-optical switching, therefore, numerical investigation of light scattering in it becomes attractive, paving the way to the future experimental studies.

2. Results and discussion

The calculations of dynamics of light scattering represent a complex task and should be conducted in several steps. The first one is the optimization of the sample's geometry.

2.1. *Geometry*

The idea of the experiment is to transform initially asymmetric scattering indicatrix of the nanoantenna to a symmetric shape by pumping the sample with a femtosecond laser pulse. The symmetrization happens because of the different quantity of the injected free-carriers and, consequently, different changes of the refractive n and extinction k indices of the substance. That is why there must be the pronounced difference in the absorption cross-sections of two nanodisks in the vicinity of the pulse central wavelength ($820\ nm$). At the same time, the resonance of the MD mode should be located in the same interval of the spectral domain. Moreover, the dimer must be entirely illuminated, and that fact also adds the condition on the distance between two cylinders.

To meet all the requirements, that were mentioned before, the following parameters were chosen (Fig. 1(a)): $r_s = 95\ nm$ and $r_j = 80\ nm$ (the radii of the larger r_s and smaller r_j disks); $h = 200\ nm$ (the height of the disks); $d = 450\ nm$ (the distance between the axes of symmetry of two nanoresonators).

2.2. *Change of optical characteristics*

The next step was the development of the physical model, which explains the change of n and k indices. The quantum description of the electron-hole transitions in semiconductors points out two effects, responsible for the change of optical properties – band filling and band-gap shrinkage. In addition, the classical Drude effect, describing optical properties of the electron gas in metals, should be also considered. The detailed analytical description of some phenomena can be found in literature [7]. To supplement the existing model and, hence, to get the accomplished analytical description, we added a formula for the change of the extinction coefficient, deduced on the basis of the classical Drude theory of metals [8]. The concentrations of electrons N and holes P are included as parameters in the analytical expressions for Δn and Δk. In further calculations, we assume $N = P$ (quasineutrality).

2.3. *Dynamics of free-carrier concentration and light scattering*

The free-carrier concentration dynamics $N(t)$ can be found as the numerical solution of the Cauchy problem. The differential equation, applied for these calculations, contains terms, representing recombination of the electron-hole plasma, generated by the pumping pulse, and the process of pumping as well. The information about coefficients for the recombination processes was taken from the literature [6, 9]. The term related to the pump process also was introduced in the equation. The calculations show that the typical time scales are $\sim 0.1\ ps$ (for the pumping rate) and $\sim 10\ ps$ (for the relaxation rate) for the peak intensities of the pumping pulses around $15\frac{GW}{cm^2}$ (Fig. 1(b)), what is equivalent to the values of the injected free-carrier concentration $\sim 10^{19} cm^{-3}$.

By putting the calculated dynamics of concentration as a function of time in the expressions for Δn and Δk, the dynamics of material's dispersion data was obtained. In this way, it becomes possible to simulate the dynamics of light scattering in the asymmetric cylindrical nanodimer, which was carried out by means of 3D finite-difference time-domain simulations [10]. The results prove that the direction of the maximal intensity of the scattered radiation changes on 7° for the pumping pulse with $I_{peak} = 15\frac{GW}{cm^2}$ (Fig. 1(c)) with characteristic time scale of 10 ps.

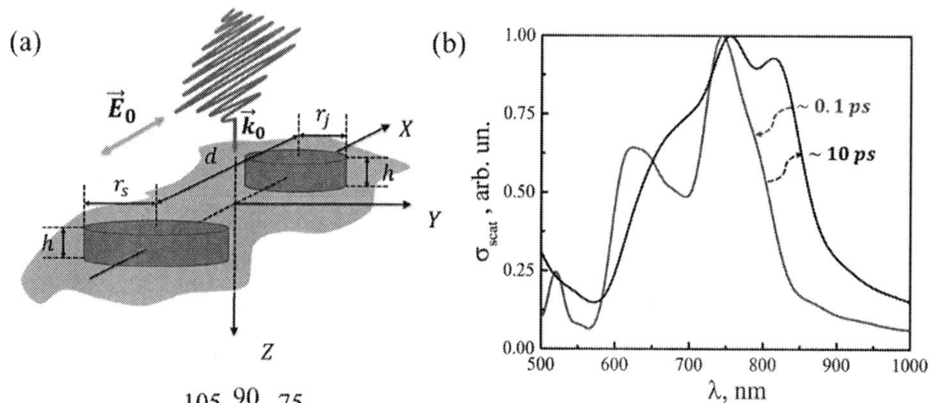

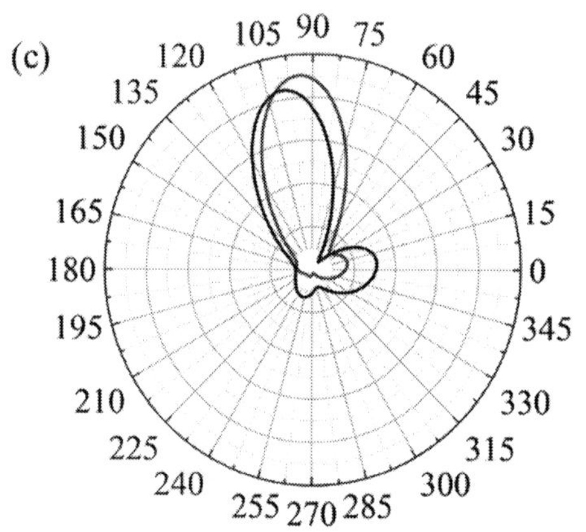

Figure 1. Ultrafast dynamics of light scattering in the asymmetric cylindrical GaAs nanodimer. (a) The configuration of the sample and the incident radiation. (b) Scattering cross-sections before pumping of the sample by a laser pulse with peak intensity $\boldsymbol{I_{peak} = 15\frac{GW}{cm^2}}$ (black curve) and right after it (red curve). (c) The scattering indicatrix for the cases of the unexcited (black curve) and excited (red curve) dimer.

3. Conclusion

We have numerically demonstrated all-optical switching in the asymmetric cylindrical GaAs nanodimer. The deflection angle of 7° was numerically demonstrated for realistic experimental conditions.

It is remarkable that, pursuing the particular aim to calculate light scattering in the asymmetric dimer, we built the general model of the dispersion dynamics that could be used as a foundation for further numerical investigations of various active nanostructures. The general model for the material dispersion dynamics was developed and introduced to the numerical simulations by FDTD approach, which could be used as a foundation for further numerical investigations of various active nanostructures based on direct-gap semiconductors.

IOP Conf. Series: Journal of Physics: Conf. Series **1092** (2018) 012141 doi:10.1088/1742-6596/1092/1/012141

4. Acknowledgements
This work was supported by the Russian Foundation for Basic Research (grant №17-02-01286) and in part by Russian National Technology Initiative.

References
[1] A. Femius Koenderink et al., *Science*, **348**, 516 (2015).
[2] Arseniy I. Kuznetsov et al., *Science,* **354**, aag2472 (2016).
[3] Isabelle Staude et al., *Nature Photonics*, **11**, 274 (2017).
[4] Yuri Kivshar et al., *Optics & Photonics News*, 1047-6938 (2017).
[5] Denis G. Baranov et al., *Las. & Phot. Rev.*, **6**, 1009–1015 (2016).
[6] Maxim R. Shcherbakov et al., *Nat. Commun.*, **8**, 17 (2017).
[7] Bennett, B. R., et al., *IEEE J. of Quantum Electron.*, **26**, 1, 113-121 (1990).
[8] Neil W. Ashcroft, N. David Mermin, *Solid State Physics* (1976).
[9] U. Strauss et al., Appl. Phys. Lett. **62**, 55 (1993).
[10] Lumarical Solutions Inc., https://www.lumerical.com/tcad-products/fdtd.

A Radiofrequency Coil Based on Hybridized Modes Two Resonant Dipoles

G. Solomakha[1], S. Glybovski[1], R. Abbdedaim[3], C. Simovski[2], I. Melchakova[1]

[1]Department of Nanophotonics and Metamaterials, ITMO University, 197101 Saint Petersburg, Russia
[2]School of Electrical Engineering, Department of Electronics and Nanoengineering, Aalto University, P.O. Box 15500, 00076 Aalto, Finland
[3]Fresnel, France

E-mail: `g.solomakha@metalab.ifmo.ru`

Abstract. We propose to use a resonator, consisting of mutually coupled dipoles as a radiofrequency coil with two orthogonal channels for 7 Tesla MRI in parallel trancieve configuration. In this work we show by numerical simulation that the two dipoles being driven symmetrically and anti-symmetrically produce similar field patterns to a dipole and a loop coil. It is shown that this equivalence is achieved for a certain distance between the dipoles.

1. Introduction

Operation of the proposed RF-coil is based on excitation of hybridized eigenmodes [?] in a pair of resonators (dipoles in our case). A single dipole is a resonant antenna with a linear current distribution which creates the RF-field distribution mostly described by a dipole electric moment. When used as a RF-coil for body imaging and put horizontally along the MRI bore direction, the dipole moment is parallel to a surface of a subject (i.e. a human body). In contrast, a loop antenna when used as a surface coil for body imaging, can be considered in terms of the field in the subject similarly to a magnetic dipole, directed normally to the surface, when used as MRI-coi. Therefore, a small dipole and a small surface loop produce mutually orthogonal field patterns and may be combined by centering at one position in an array with no mutual coupling. Indeed a pair of straight metal wires may possess an a antisymmetric quadrupole mode (also called odd mode), for which the wires support anti-parallel electric currents. These currents at electrically short distances produce magnetic field similarly to a continuous loop (e.g. a circular or a rectangular one). The same pair of wires may also support currents flowing parallel (even or dipole mode). In this case an electric dipole moment is produced and the near-field patters distantly enough from the dipoles are equivalent to ones of a single dipole. The similarity between the corresponding field patterns is schematically shown in Fig. 1 (a,b,c).

IOP Conf. Series: Journal of Physics: Conf. Series **1092** (2018) 012142 doi:10.1088/1742-6596/1092/1/012142

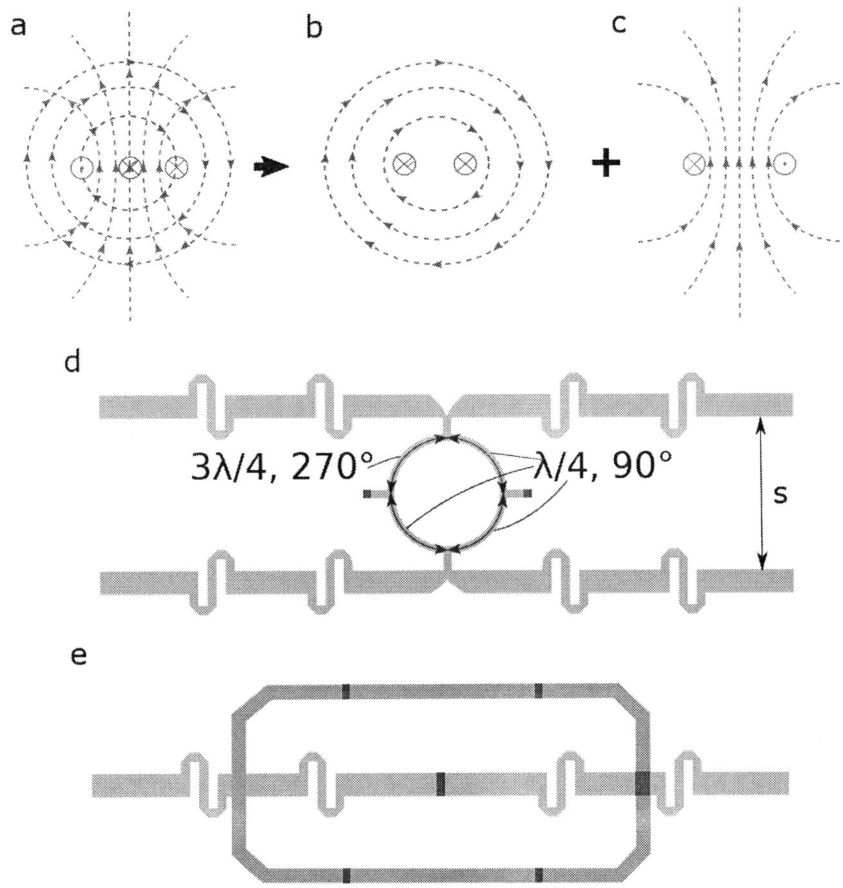

Figure 1. Analogy in the magnetic field created by the combination of a dipole and a surface loop (a) with one produced by the dipole mode (b) and the quadrupole mode (c) of a pair of coupled dipoles (dashed lines represent lines of magnetic field, while circles represent directions of currents of coils). Schematic layout of the proposed coil (d) consisting of two dipoles antennas. Schematic layout of the reference coil (f) consisting of loop and dipole antennas. Feeding points are shown by the red rectangles, lumped capacitors - by the the blue rectangles.

2. Design and simulation

The proposed coil is composed of two identical 300-mm-long fractionated dipole antennas [1] placed parallel to each other with a separation d between them as shown in Fig. 1 (d). The configuration of the reference design is depicted at Fig. 1 (e). The reference coil consists of a single fractioned dipole and rectangular surface loop with four distributed lumped capacitors and with size 89×190 mm^2[2]. The dipoles of the proposed coil have two ports at their middle splits. The even mode can be excited when the ports are driven in-phase, while the odd one is excited by out-of-phase ports. This dual-mode excitation can be realized by using an appropriate 180 hybrid, also known as a rat-race coupler[3].

Numerical simulation software (CST Studio Suite, CST, Darmstadt, Germany) was used to parametrically adjust the separation d between the dipoles by parametric optimization and calculate the electromagnetic fields, produced by the proposed dual-mode array element in pelvis-shaped homogeneous phantom with electrical properties of ε=61.8, ς=0.87 S/m. The goal of the parametric optimization was to obtain the same profiles of the radiofrequency magnetic fields created by both modes of the proposed coil as ones of the reference coil. All channels were

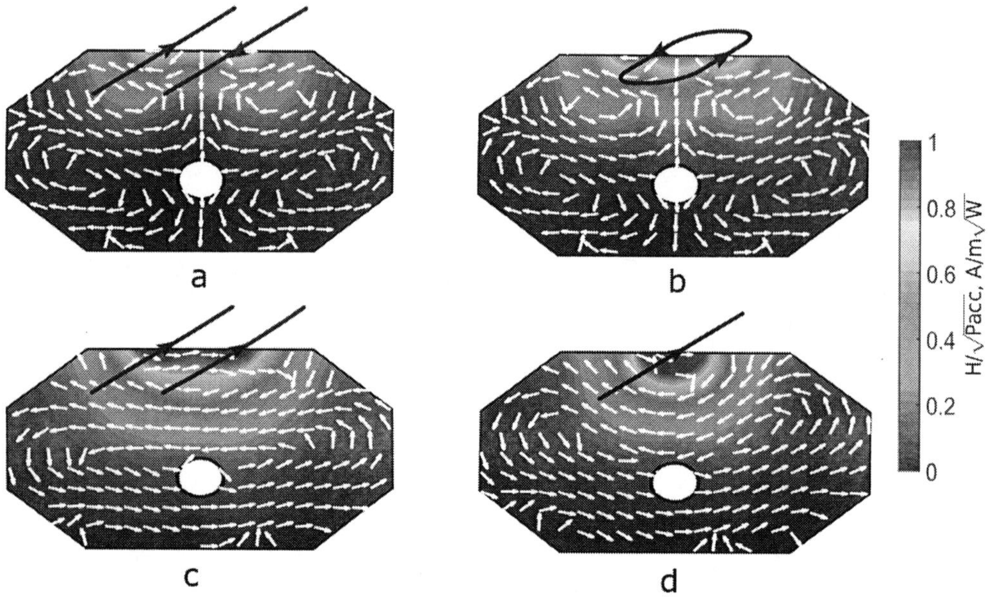

Figure 2. Simulated magnetic RF-field distributions of proposed and reference coils. Color plot demonstrates distribution of B_1^+ magnitude normalized by square root of accepted power. Arrows depict directions of H-field vector in the phantom : (a) odd (quadrupole) mode port of the proposed coil is driven; (b) loop of the reference coil is driven; (c) even (dipole) mode port of the proposed coil is driven; (d) dipole port of the reference coil is driven.

manually matched in CST DS module using lumped elements. The aim of the simulations was to confirm identity of field distributions provided by the proposed coil and dipole and loop coils.

3. Results

The magnetic field distribution normalized by the square root of the accepted power $\sqrt{P_{acc}}$ in the transversal plane in the center of the homogeneous phantom is presented in Fig. 2 both for the proposed coil (with the distance between the dipoles corresponding to the optimized case) and the reference one. The color plots demonstrate the magnetic field magnitude distribution, while the vector plots show the moment directions of the magnetic field vector. That plots demonstrate that the magnetic field distributions of the proposed coil are very similar to ones of the reference coil, especially at the depth of the phantom. As it is seen from Figure 2 (c) the magnetic field of the quadrupole (odd) mode of the proposed coil has the dominating vertical component. These are the same properties as of the surface loop part of the reference coil. From comparison of the Figures 2 a and b as well as Figures c and d, the optimal distance between the dipoles of the proposed coil provides equivalence of the proposed and the reference coils in terms of the RF-field patterns created when both of their channels are fed. Magnetic field magnitude vs. the depth profiles for different distances between the dipoles of the reference coils are compared in Figure 3 with the same curves of the dipole and loop antennas of the reference coils. As one can conclude, the corresponding curves for the proposed and the reference coils fit each other the best for the certain value $d = 6$ cm.

4. Conclusion

In this work, we propose a novel coil for 7 Tesla MRI based on hybridized modes of two coupled resonators. The results have shown an equivalence between the proposed coil and the dipole-

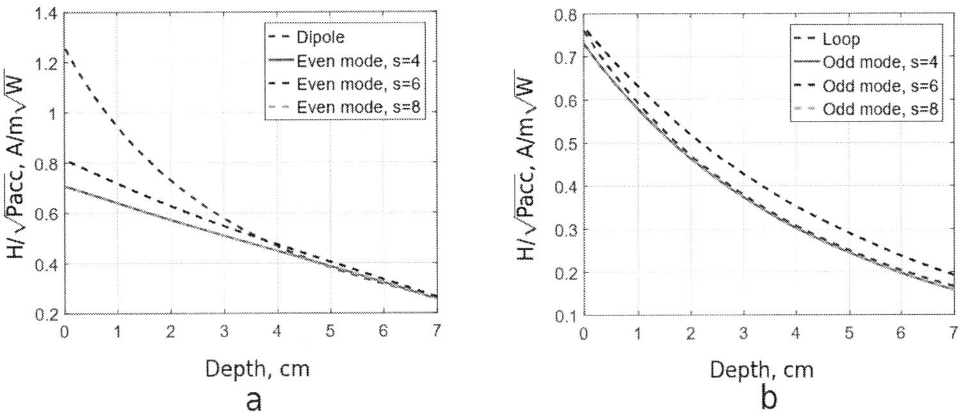

Figure 3. Simulated profiles of magnetic field magnitude vs. depth into a phantom normalized by the square root of accepted power for the even mode of proposed coil v.s. dipole of reference coil (a) and odd mode of proposed coil vs. loop channel of reference coil (b). The depth is measured from surface of phantom along the maximum of radiation inside the phantom.

loop combination in terms of magnetic field pattern created in a phantom. From the presented numerical simulation on can conclude that the proposed coil works similarly to the reference one when the distance between the two dipoles is optimized.

5. Acknowledgment

This work was supported by the Ministry of Education and Science of the Russian Federation (project No. 14.587.21.0041 with the unique identifier RFMEFI58717X0041) and European Union Horizon 2020 research and innovation programme under the grant agreement No. 736937.

References

[1] Raaijmakers AJE, Italiaander M, Voogt IJ, Luijten PR, Hoogduin JM, Klomp DWJ, et al. The fractionated dipole antenna: A new antenna for body imaging at 7 Tesla. Magnetic Resonance in Medicine. 2016;75(3):1366–1374.

[2] Ertrk MA, Raaijmakers AJE, Adriany G, Uurbil K, Metzger GJ. A 16-channel combined loop-dipole transceiver array for 7 Tesla body MRI. Magnetic Resonance in Medicine. 2017;77(2):884–894.

[3] David M. Pozar Microwave Engineering, 4th Edition Wiley, 2011

IOP Conf. Series: Journal of Physics: Conf. Series **1092** (2018) 012143 doi:10.1088/1742-6596/1092/1/012143

Influence of the crystalline structure of metal films on the performance of plasmonic biosensors

Y V Stebunov[1,2], D I Yakubovsky[1], V G Leiman[1], A V Arsenin[1,2] and V S Volkov[1,2,3]

[1] Moscow Institute of Physics and Technology, Institutsky 9, 141700 Dolgoprudny, Russia
[2] GrapheneTek, 7 Nobel Street, Skolkovo Innovation Center, 143026, Russia
[3] SDU Nano Optics, Mads Clausen Institute, University of Southern Denmark, Campusvej 55, DK-5230, Odense, Denmark

stebunov@phystech.edu

Abstract. The development of plasmonic biosensors requires careful consideration of various factors associated with the deposition of thin metal films. The quality of metal films is mostly determined by their polycrystalline structure, where smaller crystallite sizes produce higher optical losses. In present work we obtained the dependence of optical losses in thin gold and copper films on their thicknesses. These experimental results are used for the analysis of the performance of plasmonic biosensors, which demonstrate an increase in the limit of detection of up to 50% and 35% for gold and copper films, respectively.

1. Introduction

Nowadays, plasmonic biosensing has diverse applications in such areas as biochemical and pharmaceutical research, personal healthcare, veterinary, and monitoring of food quality and environment [1-3]. The performance of current plasmonic biosensors limits the range of possible applications and strongly depends on the used plasmonic materials and their optical quality. Here, we consider plasmonic biosensors based on surface plasmon resonance (SPR) excited in thin metal films by means of the Kretschmann configuration [4]. Various factors influence the optical quality of thin metal films such as growth kinetics for a particular substrate, deposition regimes, annealing, and others, because metals films with thicknesses below 100 nm deposited on dielectric substrates are mostly polycrystalline [5]. Their crystalline structure determines optical losses in metal due to electron scattering on crystallite boundaries [6-8]. Here, we study optical losses in thin gold and copper films and analyze their influence on SPR biosensing.

2. Deposition and characterization of thin metal films

Gold and copper films were deposited on silicon wafers with 285-nm-thick thermally-grown SiO_2 layer using electron-beam evaporation system Nano-Master NEE-4000. The films were deposited under vacuum conditions ($1\text{-}3\times10^{-6}$ Torr) and at a deposition rate of about 0.2 A/s. Dielectric functions of metal films were measured using spectroscopic ellipsometer VASE produced by J.A. Woollam Co in the wavelength range 300 – 1800 nm. For ellipsometry modelling, we used copper and gold dielectric functions represented by Drude Model:

$$\varepsilon = \varepsilon_\infty - \frac{\omega_p^2}{\omega^2 + i\Gamma\omega}, \quad (1)$$

where the dielectric function at an infinite frequency ε_∞, the plasma frequency ω_p, and the damping rate Γ are fitting parameters. The changes in a film thickness only slightly influence the plasma frequency and the permittivity at the infinite frequency, while considerable changes in damping factors of metals respond for their varying optical properties due to the different crystalline structures of the deposited films (Figure 1) [9]. For analytical calculations, we take the plasma frequency equaling 1.55×10^{16} and 1.26×10^{16} 1/s for gold and copper films, respectively. The corresponding permittivity at the infinite frequency equals 5.5 and 5.4, respectively.

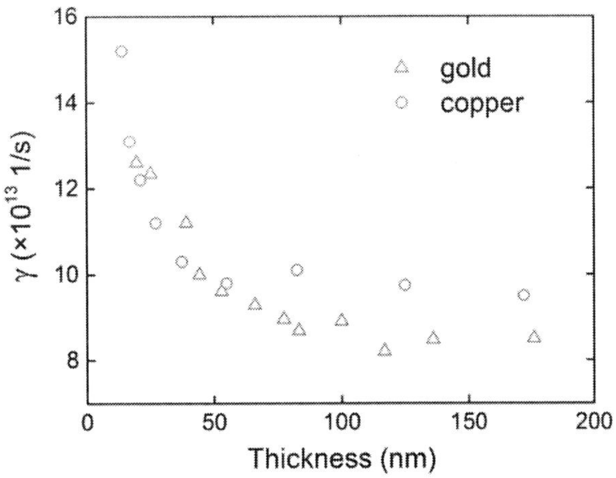

Figure 1. The damping factors for the gold and copper thin films of different thickness.

3. Plasmonic biosensing

To investigate the influence of the optical quality of metals on the performance of plasmonic biosensors, we analyse the SPR excitation in thin gold and copper films with a thickness of 47 nm. The excitation of SPR is considered according Kretschmann configuration. Using the transfer matrix model [10], we analyzed the reflection from a multilayer structure composing the following layers: 1) the glass layer with refractive index (RI) 1.523; 2) 47-nm-thick metal films; and 3) the top aqueous layer with RI of 1.33. The dielectric functions of metals are presented by the Drude model (1) with the varying damping rate in the range from 8.2×10^{13} to 12.6×10^{13} 1/s for gold films and from 9.5×10^{13} to 13.1×10^{13} 1/s for copper films. Figure 2(a) shows the SPR reflectivity curves for the gold films with different damping rates at the wavelengths of 880, 1035, 1315, and 1550 nm. The performance of SPR biosensing is characterized by such parameters as the full-width at half-maximum (FWHM) α_{FWHM} of SPR angular curve and the sensitivity to RI changes S_{RI} defined by the ratio of the shift of SPR angle $\Delta\alpha$ to the corresponding RI change of the media Δn above the metal film:

$$S_{RI} = \frac{\Delta\alpha}{\Delta n}. \quad (2)$$

Both these parameters depend on the optical properties of metal films. However, variation in the damping rate results in altered optical losses in metal films determined by the α_{FWHM} while producing negligible effect on the S_{RI}. The detection limit of SPR biosensors is proportional to the figure of merit defined as the ratio of S_{RI} to α_{FWHM}. Figure 2(b) shows α_{FWHM} and FOM obtained for gold films in dependence on the damping rate of gold. For the values of Γ considered in calculations, we obtain the improvement in FOM of up to 50%. The analogous results on the SPR excitation were obtained for copper films, which have promising applications in SPR biosensing (Figure 3) [11-12]. For the considered damping rates of copper films, the FOM can be improved up to 35%.

IOP Conf. Series: Journal of Physics: Conf. Series **1092** (2018) 012143 doi:10.1088/1742-6596/1092/1/012143

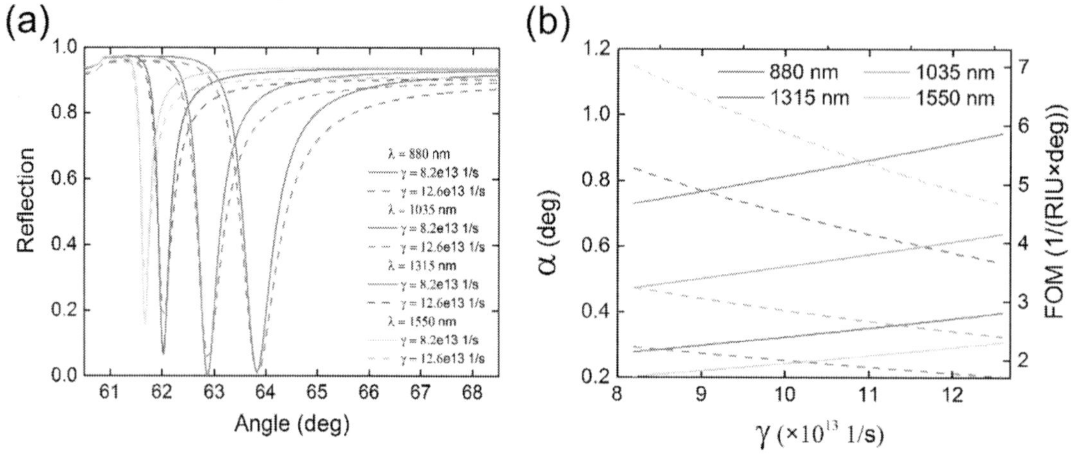

Figure 2. (a) SPR angular reflectivity curves for gold films characterized by different damping factors and at different wavelengths. (b) Full-width at half-maximum of SPR reflectivity peaks and figure of merit for SPR biosensing based on thin gold films characterized by different damping factors.

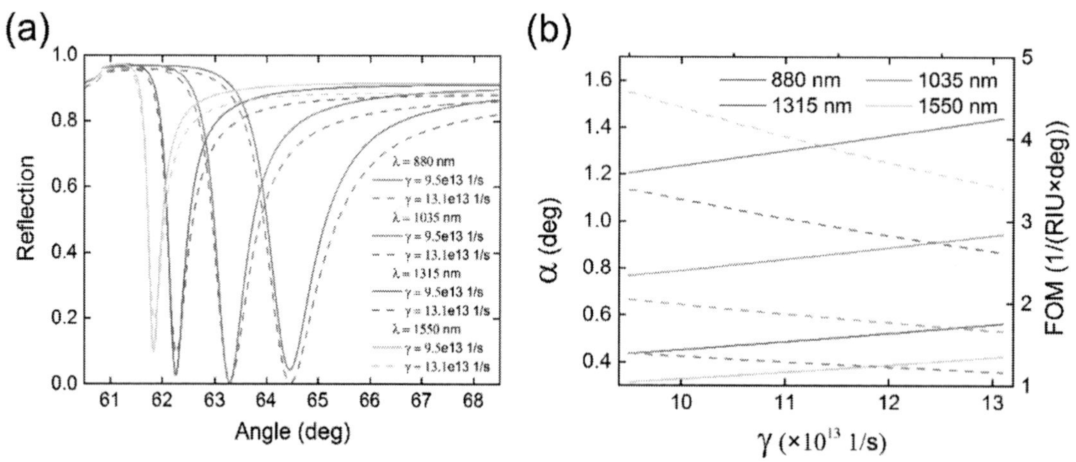

Figure 3. (a) SPR angular reflectivity curves for copper films characterized by different damping factors and at different wavelengths. (b) Full-width at half-maximum of SPR reflectivity peaks and figure of merit for SPR biosensing based on thin copper films characterized by different damping factors.

4. Conclusions

The performance of plasmonic biosensors depends on the optical quality of metal structures used for SPR excitation. In the case of thin metal films, a number of factor determines their optical properties, which is mostly reflected in their crystalline structure formed during the deposition process. In present study, we used the Drude model for the description of dielectric functions of gold and copper. The variances in damping rates for gold and copper films were obtained using ellipsometric measurements. Thereafter, using the obtained optical parameters, we analyzed the SPR excitation in thin metal films with application to optical biosensing. For the experimentally obtained variances in the damping rate of gold and copper, the difference in the FOM characterizing biosensing efficiency can reach up to 50% and 35%, respectively. Due to this, the development of plasmonic biosensors as well as other plasmonic and optoelectronic devices based on thin metal films requires the optimization of film deposition process in order to reach the best performance.

Acknowledgments
AA, DY and VL thank for support RFBR research projects No. 16-07-00837-a and 16-29-03432-ofi_m. YS thanks for support RFBR research projects No. 18-37-00302-mol_a and 18-07-01339-a and the Scholarship of the President of the Russian Federation (SP-4754.2018.5).

References
[1] Cooper M A 2002 *Nat. Rev. Drug Discovery* **1** 515.
[2] Karlsson R J 2004 *J. Mol. Recognit.* **17** 151.
[3] Homola J 2008 *Chem. Rev.* **108** 462.
[4] Kretschmann E 1971 *Z. Phys. A: Hadrons Nucl.* **241** 313.
[5] McPeak K M Jayanti S V Kress S J P Meyer S Iotti S Rossinelli A Norris D J 2015 ACS Photonics 2 326.
[6] Sotelo J Ederth J Niklasson G 2003 Phys. Rev. B 67 195106.
[7] Yakubovsky D I, Fedyanin D Yu, Arsenin A V, Volkov V S 2017 *AIP Conf. Proc.* **1874** 040057.
[8] Yakubovsky D I, Arsenin A V, Stebunov Y V, Fedyanin D Y, Volkov V S 2017 *Opt. Express* **25** 25574.
[9] Yakubovsky D I, Fedyanin D Y 2017 Proc. META'17 Incheon - Korea The 8th Int. Conf. on Metamaterials, Photonic Crystals and Plasmonics (Seoul), ed S Zouhdi, J Rho, and H Lee p 1474.
[10] Stebunov Y V, Aftenieva O A, Arsenin A V, Volkov V S 2015 *ACS Appl. Mater. Interfaces* **7** 21727.
[11] Volkov V S, Stebunov Y V, Yakubovsky D I, Fedyanin D Yu, Arsenin A V 2017 AIP Conf. Proc. 1874 020008.
[12] Stebunov Y V, Yakubovsky D I, Fedyanin D Yu, Arsenin A V, Volkov V S 2018 *Langmuir* **34** 4681.

The zoo of nonconservative optical forces

S V Sukhov[1,2]

[1] CREOL, The College of Optics and Photonics, University of Central Florida, 4304 Scorpius Str., Florida 32816, USA
[2] Kotel'nikov Institute of Radio Engineering and Electronics of Russian Academy of Sciences (Ulyanovsk branch), 48/2 Goncharov Str., 432071 Ulyanovsk, Russia

E-mail: ssukhov@knights.ucf.edu

Abstract. Optical forces are usually described as conservative ones originating from intensity gradients in optical tweezers. However, the fundamental optical action on matter is nonconservative. In contrast to gradient forces, the spectrum of action of nonconservative forces is much wider: they can propel, pull, rotate objects or move objects along complicated trajectories. Different manifestations of nonconservative optical forces will be reviewed and their dependence on the specific spatial properties of optical fields will be discussed. New developments relevant to the nonconservative optical forces such as negative forces (tractor beams) and transversal forces will also be discussed.

1. Introduction

The first indication that light can exert a force came from comet tails, observed to point away from the Sun. Back in the 17-th century, Kepler predicted that those tails were driven by the pressure of sunlight, and this conjecture still holds true for the many comet tails which consist of dust. Since then, sustained experimental attempts tried to test the concept of light pressure, the simplest manifestation of *nonconservative* optical force (NCF). The invention of lasers with their high intensity and directional beams produced new interest in optical forces. However, the main attention shifted to trapping particles at lasers' focal spot, the mechanism that relies on *conservative* gradient forces.

In contrast to gradient forces, the spectrum of action of nonconservative forces is much wider: they can propel, pull, rotate objects or move objects along complicated trajectories [1]. In contrast to conservative traps, nonconservative forces can produce mass manipulation of particles [2] or exert forces on macroscopically large objects (such as solar sails). However, the full range of applications of nonconservative forces still needed to be discovered.

2. Manifestations of nonconservative optical forces

Although different theoretical models have been employed to describe the optical forces acting on matter, they all can be traced to a common source: the exchange of momentum between radiation and matter: photons carry momentum and momentum conservation laws apply whenever a beam of light changes its direction due to refraction, reflection or scattering. But even with this simple interpretation and with a plane wave serving as an illumination source, NCFs can have very unexpected manifestations. For example, when particles are located at the interface of two media with different dielectric properties, the mechanical action of illumination may acquire additional features due to the transformation of linear momentum at the interface. In particular, the momentum of electromagnetic wave increases when light propagates into more optically dense medium. One interesting consequence is the appearance of attractive optical forces for surface-bound particles that were observed experimentally for droplets of oil on water surface [3]. The presence of *plasmonic* interfaces can generate other interesting effects. For instance, if a small particle is located near an interface and it is illuminated by a linearly polarized field, it can manifest a rotating dipole which appears because of the interference between the incident and the reflected waves. This rotating dipole excites asymmetrically propagating surface plasmon-polariton waves as shown in Figure 1. As the wavenumber of surface plasmons is larger than the one in free-space, the momentum of scattered wave effectively increases because of the surface plasmon wave propagating in the forward direction. This excess of linear momentum leads to the appearance of a negative force acting on the object ("plasmonic tractor beam") [4]. Besides negative, the lateral force can also exist in the case of circularly polarized plane wave excitation of particles next to the dielectric or plasmonic interfaces [5].

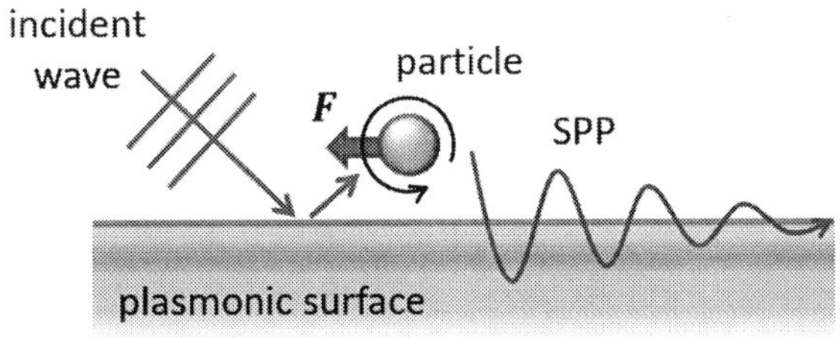

Figure 1. The schematics of negative force $\boldsymbol{F}$ appearance for a particle located next to the plasmonic interface. The interference of incident and reflected waves creates rotating dipole moment, which excites asymmetric surface plasmon-polariton (SPP) wave resulting in a recoil force acting on the particle.

The simplest extension of the case of a single plane wave, two-wave interference, brings a whole new range of optomechanical phenomena. Notably, in this case, the Poynting momentum density may not coincide with the canonical momentum. This can lead to transversal optical forces which point perpendicular to the direction of propagation of both waves, an effect that was detected experimentally [6] (Figure 2).

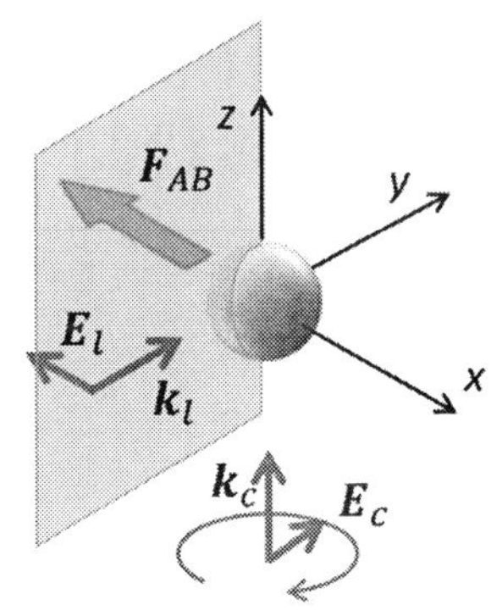

Figure 2. Scattering of two orthogonal waves (linearly l and circularly c polarized for this particular example) results in a nonconservative force $\boldsymbol{F}_{AB}$ acting on a scattering particle in a direction perpendicular to both wavevectors $\boldsymbol{k}_l$ and $\boldsymbol{k}_c$. The entire geometry resembles the setting of the Aharonov-Bohm (AB) experiment [6].

In the case of two or more overlapping waves (structured field), the interference between different multipoles can increase forward scattering and induce NCF acting in a direction opposite to the overall energy flow creating, so called, tractor beam [7]. Another example of structured field are Airy and other curved beams, which allow transport of particles along curved trajectories [8].

3. Conclusion

Nonconservative optical forces manifest themselves at all possible scales starting from atomic and up to cosmic dimensions. Even in the context of man-made devices, the nonconservative optical forces play a significant role from large-scale machines such as the LIGO interferometer down to the scales of micro-particles. There is a variety of effects ranging from the simple radiative pressure in a plane wave to the unexpected existence of negative and transversal forces acting on complex particles placed in structured fields. Nonconservative optical forces accompany a number of intriguing phenomena such as light momentum transformation in dielectric media, weak values of momentum near vortices and in evanescent fields or the elusive Belinfante momentum in structured fields. Understanding the intricacies of these nonconservative forces still raises fundamental questions and may open new possibilities for mechanical manipulation that go beyond the direct transfer of linear momentum from the electromagnetic field to a material body.

References

[1] Sukhov S, Dogariu A 2017 Non-conservative optical forces. *Reports on Progress in Physics* **80** 112001.
[2] Kajorndejnukul V, Sukhov S, Dogariu A 2015 Efficient mass transport by optical advection. *Scientific Reports* 5:14861.
[3] Kajorndejnukul V, Ding W, Sukhov S, Qiu C W, Dogariu A 2013 Linear momentum increase and negative optical forces at dielectric interface. *Nature Photonics* **7** 787.
[4] Petrov M I, Sukhov S V, Bogdanov A A, Shalin A S, Dogariu A 2016 Surface plasmon polariton assisted optical pulling force. *Laser & Photonics Reviews* **10** 116.
[5] Sukhov S, Kajorndejnukul V, Naraghi R R, Dogariu A 2015 Dynamic consequences of optical spin–orbit interaction. *Nature Photonics* **9** 809.
Rodríguez-Fortuño F J, Engheta N, Martínez A, Zayats A V 2015 Lateral forces on circularly polarizable particles near a surface. *Nature Communications* 6:8799.
[6] Sukhov S, Kajorndejnukul V, Broky J, Dogariu A 2014 Forces in Aharonov–Bohm optical setting. *Optica* **1** 383.
Fardad S, Salandrino A, Samadi A, Heinrich M, Chen Z, Christodoulides D N 2016 Scattering detection of a solenoidal Poynting vector field. *Optics Letters* **41** 3615.
[7] Dogariu A, Sukhov S, Sáenz J 2013 Optically induced'negative forces'. *Nature Photonics* **7** 24.
[8] Ang A S, Karabchevsky A, Minin I V, Minin O V, Sukhov S V, Shalin A S. 2018 'Photonic Hook' based optomechanical nanoparticle manipulator. *Scientific Reports* **8** 2029.

IOP Conf. Series: Journal of Physics: Conf. Series **1092** (2018) 012145 doi:10.1088/1742-6596/1092/1/012145

Absorption and Purcell factor control in hybrid oligomers

Y Sun and D Zuev

Department of Nanophotonics and Metamaterials, ITMO University, Saint-Petersburg, Russia

Corresponding author: yali.sun@metalab.ifmo.ru

Abstract. The control of optical properties of nanoobjects is cornerstone of modern nanophotonics. Here we numerically demonstrate, that absorption and Purcell factor can be modified in oligomer composed on hybrid metal-dielectric nanostructures by the shape modification of metal components of the nanostructures. We also reveal broadband enhancement of Purcell factor for discrete dipole placed inside the oligomer. The achieved results can be applied in optical filters, quantum optics as well as biophotonic applications.

1. Introduction

Oligomers consisting of a central particle of varying size, encircled by nanoparticles of constant size are widely studied for the interaction between optical dipole mode excited in the central nanostructure and collective mode of the encircled nanostructures [1-2]. Recently we have demonstrated that Fano resonance in oligomer composed on metal-dielectric nanostructures can be controlled fs-laser [3]. This method implies modification of metal components of hybrid nanostructures by the beam of fs-laser operating in continuous mode.

In this paper, we numerically study the opportunity of absorption as well as Purcell factor control for discrete dipole placed inside the oligomer changing the shape of plasmonic components of hybrid nanostructures.

2. Results and Disscusion

The hybrid oligomer studied consists of a central hybrid nanoparticle, encircled by six nanoparticles of different size. Fig. 1(a) shows the gold nanoparticles in the shape of disks located on a truncated silicon cones forming the hybrid oligomers. The bottom and upper diameter of central nanocone are 150 nm and 60 nm. The bottom and upper diameter of hexamer nanocone are 190 nm and 60 nm. The height of central cone, hexamer cone and nanodisk are 200 nm, 200 nm and 20 nm, respectively. The diameter of all the nanodisks equals to the bottom diameter of their corresponding nanocones. The distance between the central nanoparticle and the hexamer is 50 nm. In previous experiments on the fs-laser modifications of oligomers [3], the gold nanodisks are firstly melted to nanocups, and then reshaped as nanospheres with the continuous increase of power. As the gold nanoparticles are surrounded by a very thin chromium film, the volume of the gold nanoparticles stays the same within the reshaping process. Nevertheless, we can roughly estimate the geometry of nanocups and nanospheres as well as study absorption and Purcell factor tunning.

We first look into the absorption properties of the hybrid oligomers. We assume the incident plane wave is propagating along the axis of symmetry of the nanoparticles. The absorption spectrum for hybrid oligomers with gold nanoparticles of disk shapes is demonstrated on the Fig. 2(b).

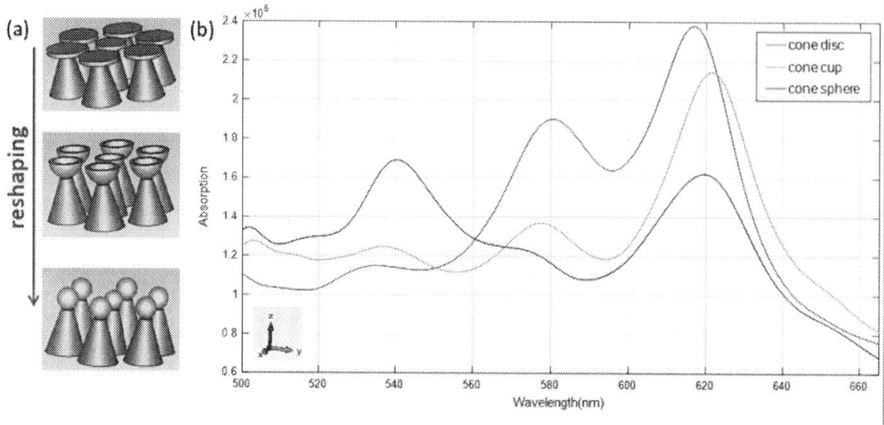

Fig. 1 (a) Schematic of hybrid oligomers with modification of the gold nanoparticles shape; (b) tuning of absorption properties with different gold nanoparticles shape

Modification of gold nanoparticle shape has an influence on the absorption spectra over the whole wavelength range. Obviously, at resonance around 540 nm, absorption increases above 50 % after the modification. Besides, at resonance around 580 nm, absorption decreases above 40 % after the modification. Moreover, at resonance around 620 nm, we can slightly tune the absorption resonance with modification of the gold nanoparticles shape. In this case, absorption property can be dramatically (about 60%) tuned during this process of modification.

Besides absorption property, we also investigate Purcell factor with different modified degree of the gold nanoparticles, as depicted on the Fig. 2.

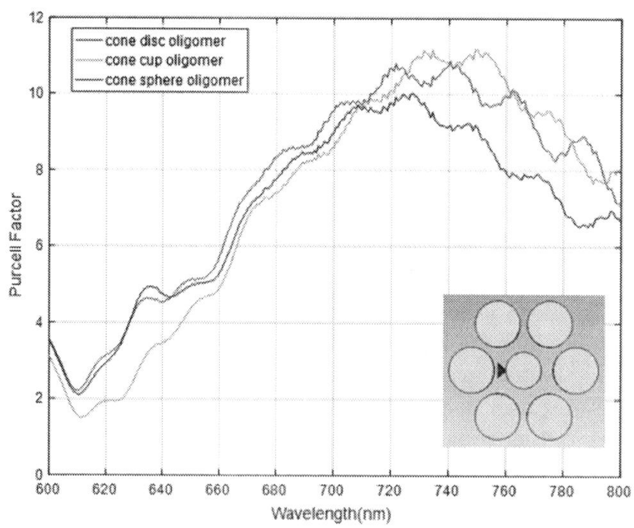

Fig. 2 Influence of the gold nanoparticle shape on the Purcell factor in hybrid oligomers

The inset shows the position of the discrete dipole placed in the middle between the central nanoparticle and hexamer. The length of the dipole is 20 nm. As can be seen from the figure, the increasing Purcell factor can be achieved in a wide spectral range. Moreover, the resonance of Purcell factor is also shifted with modification of the gold nanoparticles shape.

3. Conclusion

In this paper, we demonstrate that absorption spectrum as well as Purcell factor enhancement can be engineered by the change of geometry of the gold component. We reveal, that change of the gold nanoparticle shape from disk to sphere can modify absorption up to 60% at certain wavelength. We also show, that discrete dipole placed inside the oligomer has high enhancement of Purcell factor in a wide spectral range. These facts make the hybrid oligomer a promising system for different nanophotonic applications.

4. Acknowledgments

This work was supported by the Ministry of Education and Science of Russian Federation (Project 2.2267.2017/4.6) as well as Russian Foundation for Basic Research (Projects 16-37-60101, 16-29-05317). Y Sun acknowledges the support from the China Scholarship Council (201706160135).

References

[1] Miroshnichenko A E, Kivshar Y S. Fano Resonances in All-Dielectric Oligomers[J]. Nano Letters, 2012, 12(12):6459.

[2] Chong K E, Hopkins B, Staude I, et al. Observation of Fano Resonances in All - Dielectric Nanoparticle Oligomers[J]. Small, 2014, 10(10):1985-1990.

[3] Lepeshov S, Krasnok A, Mukhin I, Zuev D, Gudovskikh A, Milichko V, Belov P, and Miroshnichenko A. *Fine-Tuning of the Magnetic Fano Resonance in Hybrid Oligomers via fs-Laser-Induced Reshaping*[J]. Acs Photonics, 2017, 4(3).

Magnetoplasmonic crystals for detection of short wavelength spin waves

D A Sylgacheva[1,2] *, A N Kalish[1,2], V I Belotelov[1,2]
[1]Faculty of Physics, Lomonosov Moscow State University, Moscow 119991, Russia
[2]Russian Quantum Center, Skolkovo, Moscow 143025, Russia

*sylgacheva.darjja@physics.msu.ru

Abstract. Presently the coherent excitation of spin ensembles using plasmonic structures is of prime research importance since it potentially can expand the functionalities of magnonics. Optical visualisation of spin waves of small wavelengths (50 - 300 nm) is not possible by light due to the diffraction limit. However, plasmonic gratings allow it. In this work, gold gratings on magnetic films are investigated, and the optical properties of the plasmonic crystals were studied by simulations and experiments of the transmittance and magneto-optical effects.

1. Introduction

To date, the latest achievements of science and technology have led to the possibility of creating new functional magnetic structures of various design, which have modified optical properties, with characteristic dimensions in the order of several tens of nanometers. Such nanostructures allow control the characteristics of optical radiation at a scale smaller than the wavelength of light. The effectiveness of this control is achieved mainly by nanostructuring materials [1], which enables to create a medium with predetermined optical properties and enhance the optical and magneto-optical effects by only optimizing the geometric structure.

2. Plasmonic grating of the subwave period

Magnetostatic spin waves are waves of magnetization in a continuous magnetic medium having a rather wide range of wavelengths. Of special interest for modern quantum and magnon technologies are spin waves, in which the wavelength is less than 1 micron. Such waves can be detected using a grid of microwires [2], using the effect of electromagnetic induction. However, this method is not always convenient, since it does not allow rearranging the grating properties and changing the sensitivity to spin waves of different types and different spectral ranges. Therefore, a more promising method of excitation and detection is a magneto-optical method based on recording the precession of magnetization due to the magneto-optical Faraday effect and the transverse Kerr effect. However, this method has a spatial resolution constraint associated with the diffraction limit, because of which registration of spin waves with a submicron wavelength is difficult to perform. In this work we present a method for solving such a problem by using a plasmonic grating deposited on a magnetic dielectric in which spin waves propagate. The excitation of plasmonic resonances leads to an increase in the sensitivity of the magneto-optical technique and can make it possible to record spin waves whose wavelength is much less than the diffraction limit.

Electromagnetic simulations of the optical properties of the structure in the presence of a spin wave was carried out by the method of rigorous coupled-wave analysis (RCWA), adapted for the case when the gyration vector proportional to the magnetization vector and specifying the off-diagonal components of the permittivity tensor changes laterally in space according to a harmonic law. In this case, the period of the structure is a multiple of the period of gyration variation in space and the corresponding directions of the periodicity coincide. The direction of the gyration vector can be arbitrary, which makes it possible to calculate the magneto-optical effects in any configuration.
Due to the plasmonic grating, waveguide modes and surface plasmon polaritons are excited in the structure under consideration, which leads to resonances in the transmission spectra, the Faraday angle, and the magneto-optical transverse Kerr effect. Resonances in these spectra are optimized by selecting the parameters of the grating: the height, the width of the air gap and the period.
The dependencies of the angle Faraday and the transverse Kerr effect at a fixed resonance wavelength on the relative position of the slit and the maximum of the gyration demonstrate a harmonic appearance analogous to the dependence of the gyration in the film on the coordinate. Moreover, the extrema of these dependences of the Faraday angle and the transverse Kerr effect arise when the extremum of the gyration coincides with the center of the gap. With increasing k waves number, that is, with an increase in the number of spin wavelengths within the gold grating period, the value of magneto-optical effects decreases, that is, the recording of shorter spin waves is limited by the sensitivity level of the spectrometer or detector.
The investigations carried out show that when spin waves propagate through a magnetic film with a plasmonic coating, precession of the magnetization can be recorded by measuring the dependence of the Faraday angle or the transverse Kerr effect on time. At the same time, when observed at a wavelength corresponding to the resonance of the magneto-optical effect, the amplitude of the detected signal in the plasmonic crystal exceeds by more than 10 times the amplitude of the signal when viewed with a focused laser beam in a magnetic film without plasmonic coating.
Thus, it has been demonstrated that the application of a plasmonic coating to a magnetic film makes it possible to obtain information about the amplitude, wavelength, and frequency of the spin wave propagating in the material much more effectively by a magneto-optical method. It is important to note that there is no need to focus the laser beam into a spot close to the diffraction limit, which greatly simplifies the experimental procedure.

3. Acknowledgments
The work was supported by the Russian Presidential Grant MD-1615.2017.2.

References
[1] Stanciu C D, Hansteen F, Kimel A V, Kirilyuk A, Tsukamoto A, Itoh A, Rasing T 2007 *Phys. Rev. Lett.* **99** 047601
[2] Yu H, Kelly O D, Cros V, Bernard R, Bortolotti P, Anane A, Brandl F, Heimbach F, Grundler D 2016 *Nature communications* **7** 11255

Tuning chirality of laser-printed plasmonic nanoneedles via tailored spiral-shape pulses

S A Syubaev[1,2,*], A A Kuchmizhak[1,2], A P Porfirev[3]

[1] Institute of Automation and Control Processes, 5 Radio St., Vladivostok 690041, Russia
[2] Far Eastern Federal University, 8 Sukhanova St., Vladivostok 690950, Russia
[3] Samara National Research University, Moskovskoe shosse, 34, Samara 443086, Russia

trilar@bk.ru

Abstract. Chiral mass transfer on the surface of plasmonic-active metals appeared upon their ablation with vortex laser pulses was recently found to be driven by a helical-shape temperature and corresponding surface tension gradients rather than optical angular momentum transfer from the incident beam. Here, we demonstrate that by properly designing and tailoring the spiral-shape intensity pattern used for direct single-pulse laser ablation, the chirality of produced nanoneedles can be controlled in a wide range of parameter. Such optimization of the laser intensity pattern governing the helical movement of the transiently molten metal allows to produce nanostructures with controlled chirality suited for various nanophotonics and biosensing applications.

1. Introduction

Optical vortex (OV) [1], a singular optical beam characterized by a helical wavefront (or an orbital angular momentum, OAM, *l*) and a circular polarization (or a spin angular momentum, SAM), has refreshed research interest in the area of laser-material processing [2]. Such interest can be generally attributed to the specific character of the OV-matter interaction. Particularly, an OV laser pulses were demonstrated to "twist" irradiated matter producing a helical relief on the surfaces of various materials as transition metals [3,4], semiconductors [5] and azo-polymers [6]. For transition metals, the appearance of such helical-shape nanorelief was initially described in term of the direct OAM transfer from the OV pulse to a transient material melt [3]. Meanwhile, recent experiments provided alternative explanation of this phenomenon based on spiral intensity distribution emerging as a result of constructive interference of the incident donut-shape OV beam and its reflected replica [4]. Further studies performed for plasmonic-active metal films ablated with specially designed spiral-shape intensity patterns having both zero OAM and zero SAM confirmed the suggestion made [7]. Additionally, it was qualitatively demonstrated that the chirality of the produced nanoneedles is affected by a number of spirals in the designed intensity distribution.

In this study, we designed and fabricated several diffraction optical elements (DOEs) generated various intensity patters having variable number of intensity spirals. The chirality of the obtained nanoneedles defined as a number of turns per height unit was found to depend on the spiral intensity pattern.

2. Experiment

The spiral intensity patterns were obtained with the various binary spiral axicon with the phase given by:

$$\tau(r,\varphi)=\begin{cases}0, \cos(k\alpha_0 r+m\varphi)\geq 0,\\ \pi, \cos(k\alpha_0 r+m\varphi)<0,\end{cases}$$

where (r, ϕ) are the polar coordinates, $\alpha_0 = \lambda/d$ is the so-called numerical aperture of the axicon with the period d; m is the topological charge of the OV, $k = 2\pi/\lambda$ is the laser wavenumber at the wavelength λ (Figure 1a). The transmission function of the binary spiral axicon can be represented as a superposition of transmission functions of the scattering and collecting spiral axicons with the opposite topological charges. In this case, the total OAM of the generated focal-plane intensity distribution will be zero.

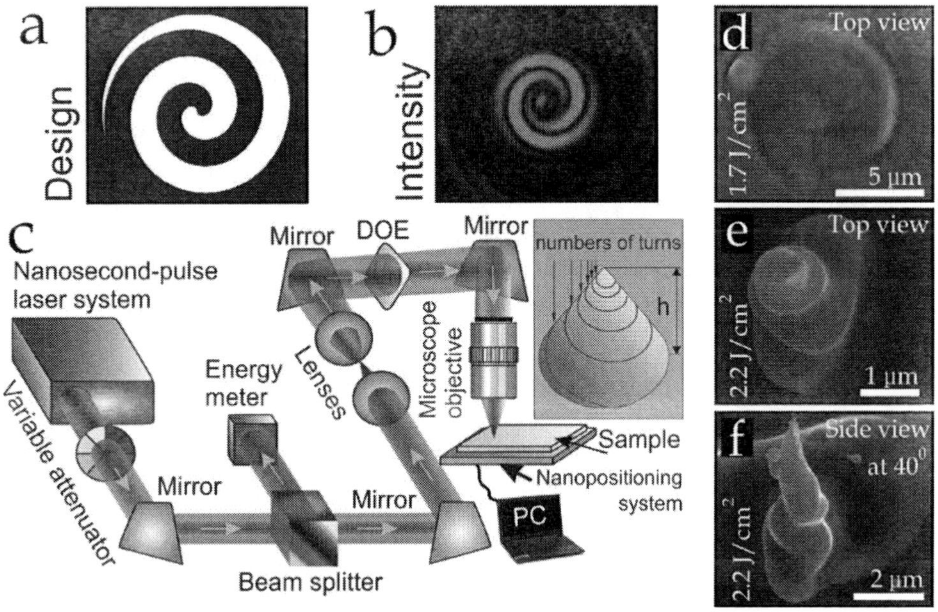

Figure 1. a) Designed phase pattern of the binary spiral axicon (white-color area denotes the phase of π rad, while the black one - 0 rad). b) Focal-plane intensity profile generated with the designed DOE. c) Experimental setup for laser nanotexturing. The inset explains the definition of the nanoneedle chirality. d-f) SEM images of the twisted nanoneedle printed with the double-spiral beam (Figure 1b) on the surface of a 500-nm thick Ag film at NA=0.42 and different incident fluences.

Single-pulse printing of twisted nanoneedles with the zero-OAM spiral intensity pattern was performed with a microfabrication setup (Figure 1c), using second-harmonic pulses of a Q-switched Nd:YAG nanosecond laser. The Gaussian laser beam was directed through a beam collimator and the transmissive DOE to generate a beam with a spiral-shape intensity pattern. This pattern was projected onto a sample surface by microscope objective with numerical aperture of 0.42. We used thick silver films with a thickness of 500 nm, which were deposited onto silica glass substrates, using e-beam evaporation. In the process of nanotexturing, the samples were mounted onto a three-dimensional motorized translation stage with a movement resolution of 150 nm. All structures were produced under single-pulse irradiation.

3. Results and discussions

SEM images in Figure 1(d-f) illustrate the surface modification produced at laser fluence slightly above the ablation threshold of the irradiated 500-nm thick film as well as a chiral nanoneedle formed at higher fluence of 2.2 J/cm^2. As seen, the produced structures are twisted in the clockwise direction, coinciding with the chirality of the double-spiral intensity pattern used (Figure 1b). Upon increasing the applied fluence, the geometric shape of the nanoneedle typically evolves us follows [7]. The formation of the protrusion starts in the center of the irradiated area via accumulation of the molten material. At further increase of the fluence, the height of this protrusion gradually grows. For a certain height, the liquid helical nanoneedle becomes hydrodynamically unstable decaying via ejection of the droplets. Taking into account this evolution, we further analyze the nanoneedles with the maximal height (aspect ratio), which can be observed at moderate fluences according to our previous studies. For such high-aspect ratio needles, we found direct correlation of their helical shape and the number of spirals in the designed intensity pattern used for single-pulse ablation.

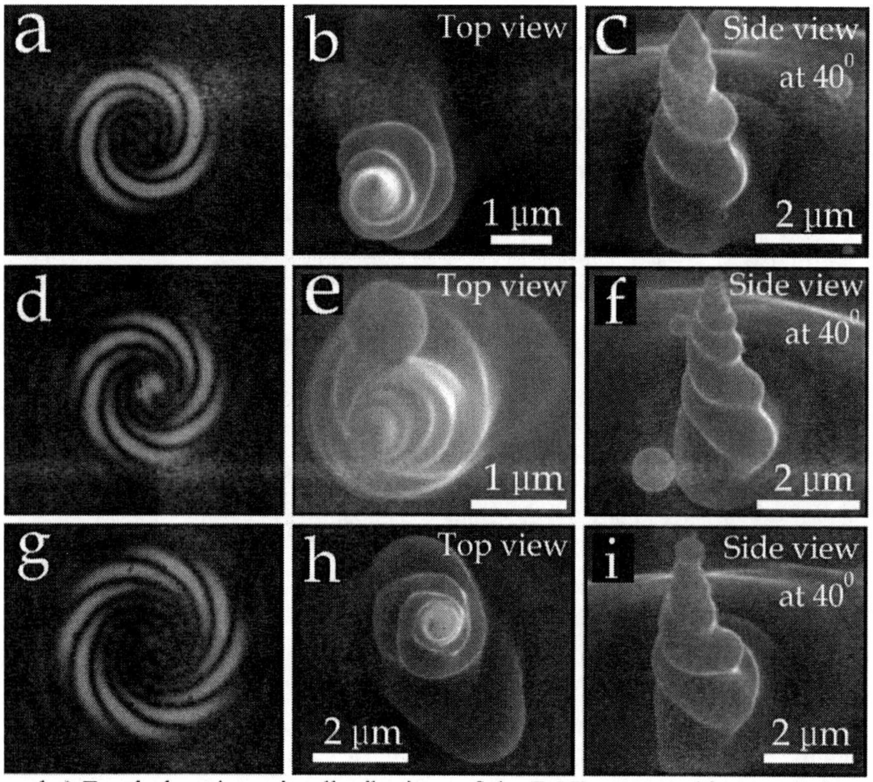

Figure 2. a,d,g) Focal-plane intensity distributions of the DOE-generated beams designed to have 4, 6, and 8 spirals. (b-i) Top-view and side-view SEM images of the twisted nanoneedles produced on the surface of the 500-nm Ag film using the generated spiral-shape beams.

For these experiments, we have designed and produced several DOEs generating beams with a lateral intensity distribution having various number of spirals (Figure 2(a,d,g)). Presented SEM image clearly illustrate that the chirality of the needles produced with multi-spiral intensity pattern significantly increase being compared to the needles produced by simple double-spiral pattern (see Fig.1(e,f)). To further quantitatively analyze the obtained results, we classified the chirality of the high-aspect-ratio nanoneedles as a number of spiral turns normalized on the structure height (see inset in Figure 1(c)). Analyzing the geometry of the produced structures from this point of view, the chirality was found to increase gradually with the number of intensity spirals reaching its maximum value of 2.4 μm^{-1} for six-spiral intensity pattern (Figure 3).

It is worth noting that using 8-spirals pattern results in decrease of the nanoneedle's chirality. Apparently, it is due to the fact that on the target surface, the material doesn't resolve spirals individually and they become overlapped owing to thermal blurring of the intensity distribution governed by strong lateral thermal conductivity in the silver film irradiated by the laser pulse.

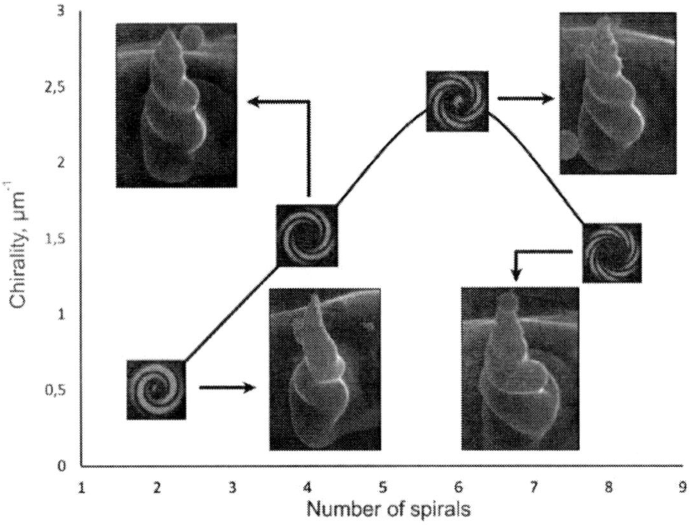

Figure 3. Chirality of the laser-printed nanoneedles versus the number of the spirals in the designed intensity pattern. The statistics for each intensity pattern having different number of intensity spirals is accumulated over 50 twisted nanoneedles produced under similar experimental conditions.

4. Conclusions

Herein, we showed that the chirality of the laser-printed nanoneedles can be tailored by precisely tuning the spiral-shape lateral intensity distribution used for single-pulse laser printing. Specifically, the chirality was found to increase gradually with the number of intensity spirals, reaching maximum value of 2.4 μm^{-1} for six-spiral intensity pattern. Further increase of spiral number in the intensity pattern decreases the chirality of the laser-printed nanoneedles, apparently owing to thermal blurring of the intensity distribution governed by strong lateral thermal conductivity in the silver film irradiated by the laser beam. Presented flexible and high-performing approach is promising for making functional nanotextures with controllable chirality.

Acknowledgments

Authors acknowledge support from Russian Science Foundation 17-19-01325.

References

[1] Bliokh K, Rodriguez-Fortuno F, Nori F and Zayats A V 2015 *Nat. Photonics* **9** 796

[2] Nivas J J J, Shutong H, Anoop K K, Rubano A, Fittipaldi R, Vecchione A, Paparo D, Marrucci L, Bruzzese R and Amoruso S 2015 *Opt. Lett.* **40** 4611

[3] Toyoda K, Takahashi F, Takizawa S, Tokizane Y, Miyamoto K, Morita R and Omatsu T 2013 *Phys. Rev. Lett.* **110** 143603

[4] Syubaev S, Zhizhchenko A, Kuchmizhak A, Porfirev A, Pustovalov E, Vitrik O, Kulchin Y, Khonina S and Kudryashov S 2017 *Opt. Express* **25** 10214

[5] Takahashi F, Miyamoto K, Hidai H, Yamane K, Morita R and Omatsu T 2016 *Sci. Rep.* **6** 21738

[6] Masuda K, Nakano S, Barada D, Kumakura M, Miyamoto K and Omatsu T 2017 *Opt. Express* **25** 12499

[7] Syubaev S, Porfirev A, Zhizhchenko A, Vitrik O, Kudryashov S, Fomchenkov S, Khonina S and Kuchmizhak A 2017 *Opt. Lett.* **42** 5022

IOP Conf. Series: Journal of Physics: Conf. Series **1092** (2018) 012148 doi:10.1088/1742-6596/1092/1/012148

Amplified spontaneous emission in polymer films doped with CdSe/CdS/ZnS colloidal quantum dots

Peter N.Tananaev[a),1], **Georgiy M. Yankovskii**[1], **Alexander V. Baryshev**[b),1,2]

[1]All-Russia Research Institute of Automatics, 127055, Moscow, Russia
[2]Moscow Institute of Physics and Technology, 141700, Dolgoprudny, Moscow Region, Russia

[a)] corresponding author: peternt@gmail.com
[b)] baryshev@vniia.ru

Abstract. Cd- and Pb-based colloidal quantum dots (QDs) and nanoplatelets are perspective materials for fabrication of active layers in photonic and plasmonic devices due to high values of optical gain (hundreds to 1200 cm^{-1}) under sufficient pump conditions [1, 2]. To obtain an optical gain, these nanocrystals are usually deposited densely and close-packed. Several drawbacks like scattering and difficulties associated with fabrication of such layers limit their use. We report successful aggregate free doping of polymer layers with homogeneously dispersed CdSe/CdS/ZnS QDs and provide description of two compositions that can accommodate up to 30-50 mass. % of QDs. The compositions can be polymerized with usual acrylate photoinitiation process or via thiol-ene polymerization process. QDs are well dispersed before and after heat- and photopolymerization, form no aggregates and, thus, the formed layers are almost free from scattering losses. Amplified spontaneous emission (ASE) in QD doped polymers was observed with 532 nm wavelength laser excitation in the variable stripe length (VSL) configuration.

1. Introduction

Plasmonic nanostructures can be used to localize electromagnetic radiation and to govern plasmon-polariton wave propagation [3]. Active plasmonics concept uses active medium with sufficient gain characteristic to mitigate intrinsic losses of metals (a usual plasmonic material) and improve characteristics of devices like waveguides [4] and spasers [5,6,7]. Quantum dots are especially good for such active media because of having large transition dipole moment; optical gain is proportional to its square and is only linearly proportional to volume concentration of excited quantum emitters [8]. Films of specially designed colloidal quantum dots (QDs) and other semiconducting nanocrystals (like colloidal quantum wells, nanorods etc) demonstrate optical gain coefficients over 1000 cm^{-1} and are successfully used to obtain spaser devices [7]. Compared to dyes with limited gains below 100 cm^{-1} [9] they also show improved photostability.

The most used are cadmium chalchogenide colloidal quantum dots, emitting in the visible. Since the first works on QDs lasing [10] synthetic chemistry have been improved; Cd-based core-shell QDs with CdSe cores demonstrate near unity single exciton quantum yields and improved biexciton and multiexciton quantum yields with non-radiative Auger recombination largely suppressed [11, 12]. The latter was the main factor limiting QDs lasing to femto- and picosecond pump regime. Most recent works demonstrate CW lasing and electrical pumping of Cd-based QDs layers [13].

QDs thin layers are usually made by spin-coating, dip coating or any other suitable method that can provide densely packed nanocrystals film. Some usual complications are: heat management, fluorescence quenching due to Förster resonant energy transfer and unacceptable optical and mechanical properties of such films. One can also mention large losses of QD material during deposition.

Another rarely used approach is to disperse quantum dots in polymer matrix, which leads to lower optical gain, but improves all abovementioned issues if done properly. Since Cd-based QDs are not easily dispersed in common polymers like PMMA [14], we developed several liquid compositions where QDs were first dispersed evenly and randomly, and then these were used to obtain nano- to micrometer thick films via photopolymerization or heat-induced polymerization. Absence of significant aggregation lowers optical scattering and associated losses. Since these compositions are applied in a liquid form and have no solvent evaporation step (100% material conversion), the method can be scaled up to more environmental friendly and ecological production.

2. Materials and methods

2.1. Materials and chemical processes description

Mercaptopropyltrimethoxysilane (MPTMS), Diphenyl(2,4,6-trimethylbenzoyl)phosphine oxide (TPO), 2-ethylhexyl acrylate (EHA), 1,6-hexanedioldiacrylate (HDDA), oleyilamine (OLA) are from Sigma-Aldrich. Deionized water, toluene, methanol, butanol, isopropanol (IPA), tetrahydrofuran (THF), hexane and other solvents (Sigma-Aldrich) were used as received.

CdSe/CdS/ZnS (QD630) quantum dots in toluene are synthesized according to [11] in NIIPA, Dubna, Russia. Their photoluminescence (PL) maximum is 630 nm and quantum yield is about 60%. QDot 655® CdSe/CdS/ZnS quantum dots (Invitrogen) 4-6 mg are supplied in 4 ml decane. All of the QDs were first precipitated from solvent with 1:3 IPA:methanol solvent mix, centrifuged and redispersed in hexane for subsequent processing. All quantity of QDot655 was heated with 100 uL OLA under nitrogen flow and then reprecipitated two times with IPA:methanol and methanol and redispersed in hexane.

2.1.1. First polymer composition. Hexane dispersions were transferred to glass microvials and evaporated, then 5% TPO (photoinitiator) solutions in EHA were added and heated to 90 °C to obtain QDs dispersions with 1- 20 mass. % concentrations of QDs. Obtained acrylate dispersions were deposited on clean quartz substrates via spin-coating and photopolymerized under UV LED 380 nm under nitrogen to obtain 100-200 nm thin films. Alternatively 10 uL of dispersions were dropped on quartz substrates, covered by hydrophobized glass, cured under UV LED light; glass cover was easily removed afterwards and from 3- to 30-μm-thick films were obtained; thickness was regulated via amount of liquid dispersion and applied pressure.

2.1.2. Second polymer composition. QD630 100 mg was redispersed in 15 ml toluene and reacted with 30 uL MPTMS in 5 ml toluene for 3 hours under inert atmosphere. Then 900 uL of 1% water in THF was added and left stirring for 3 days. The resulting QDs with modified surface (thiol groups through intermediate SiO_2 layer) were cleaned through evaporation of excess solvent and precipitation/redispersion with hexane and IPA. TPO 5% in HDDA were added to obtain 10-30% dispersions; 50% dispersion was obtained via careful evaporation of intermediate solvent. These dispersions can be polymerized by heating to 120 °C in inert atmosphere or by UV cure analogous to described above; arcylate and thiol-acrylate radical polymerization reactions are involved.

2.2. Characterization

The thickness of all obtained samples was measured via white light profilometry and AFM where necessary. Aggregation of QDs was analyzed from photoluminescence maps obtained by NTegra microscope (excitation with 532 nm wavelength laser). The most informative were measuring of photoluminescence and observation of ASE by utilizing the VSL method. In these experiments, a Nd:YAG laser second harmonic (532 nm, 13 ns, 10 Hz) was focused into a narrow line 200 μm x 10 mm, 800 kW/cm^2. Emission was collected via a 20x objective and fed into a monochromator via fiber waveguide. The collected light was analyzed by scanning in the 590-700 nm range with a 0.3 nm resolution and was measured at the exit slit of the monochromator via a Si detector.

3. Results and discussion

Visual examination and optical microscopy of all obtained samples shows high degree of their homogeneity. PL maps show absence of aggregates with sizes above 0.5 μm and low variation of PL intensity (<20%) over areas as large as 20 μm x 20 μm. For the first composition, some aggregation becomes prominent at 20 mass % concentration of both quantum dot types. The second composition yields better results even at 30 mass % for both QDs types and some visible aggregation at around 50 mass % for modified QDot655.

When the samples were exposed to laser radiation, emission collected from the end of the pump stripe showed characteristic signs of ASE – narrowing of photoluminescence spectra. The pronounced narrowing up to 4 nm is achieved for different types of samples when the stripe length exceeds 2 mm. For example, 8 μm thick sample containing 10% QDot655 under pump has a spectra shown on Figure 1. The onset of ASE could be attributed by the narrowing of the FWHM of the spectrum to 7 nm (line 3) at a stripe length L of 8 mm, contrary to the wide spectra achieved when the stripe length is less than 2 mm (lines 1 and 2). At the same time, samples with detected aggregation demonstrate typical QDot655 PL spectra with FWHM of 25-28 nm.

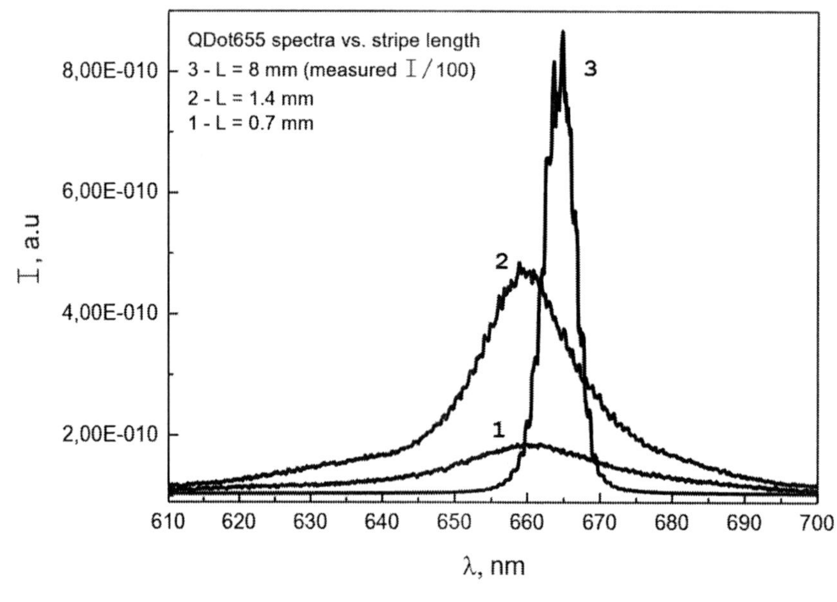

Figure 1. Dependence of QDs PL spectra on stripe length (VSL method).

4. Conclusions

The fabricated compositions employed in this work allowed us to obtain layers with homogeneously dispersed quantum dots in concentration range 1-50%. Methods that are reported here can be utilized if one is limited in amount of quantum dots for experiments or if 100% material conversion is desired during making of active layer. Absence of aggregation leads to low scattering losses, which allowed us to observe large magnitudes of optical amplification of spontaneous emission. Optical peculiarities and magnitudes of the modal and material optical gain values for different samples are to be reported.

5. Acknowledgments

This study was carried out with support from Russian Foundation for Advanced Research Projects (contract № 7/004/2013-2018 dated 23.12.2013).

References

[1] M.M.Adachi, F.Fan et al. Nature Communications, 2015, V.6, 8694
[2] S.Yakunin et al. Nat Commun. 2015, 6: 8056
[3] Krasavin A.V., Zayats A.V., Zheludev N.I. J. Opt. A: Pure Appl. Opt., 2005, V.7, pp.85–89.
[4] *Maier S.A.* Plasmonics: Fundamentals and Applications. Springer-Verlag US, 2007. 224 p.
[5] Bergman D.J., Stockman M.I. Phys. Rev. Lett., 2003, V.90, 027402-1-4.
[6] Stockman M.I. Nature Photonics, 2008, V.2, June, p.327.
[7] S.J.P.Kress et al. Sci Adv. 2017 Sep; 3(9): e1700688.
[8] A.V.Dorofeenko, A.A.Zyablovsky, A.A.Pukhov, A.A.Lisyansky, A.P.Vinogradov. Physics-Uspekhi, V.55, N.11, p.1080.
[9] I. Gozhyk et al. Phys.Rev.B, 2015, 92, 214202
[10] V.I.Klimov et al. Science. 2000, Oct 13; 290(5490):314-7.
[11] Chen Ou et al. Nature Mater., 2013, V.12, pp.445–451
[12] M.Nasilowski, P.Spinicelli, G.Patriarche, B.Dubertret. Nano Lett., 2015, 15 (6), pp 3953–3958
[13] J.Lim, Y.-S.Park, V.I.Klimov. Nature Materials, 2018, V.17, pp. 42–49
[14] P.N.Tananaev et al., Journal of Optical Technology (in Russian), 2018, V.3 №85

Low-threshold lasing in ZnO microtetrapods

A P Tarasov[1,2,4], Ch M Briskina[1], V M Markushev[1], A M Opolchentsev[3] and A S Lavrikov[3]

[1] Kotelnikov Institute of Radio-Engineering and Electronics of Russian Academy of Sciences, 11-7 Mokhovaya, Moscow, Russia, 125009
[2] Moscow Institute of Physics and Technology (State University), 9 Institutskiy per., Dolgoprudny, Moscow Region, Russia, 141701
[3] Federal Scientific Research Centre "Crystallography and Photonics" of Russian Academy of Sciences, 59 Leninskiy pr., Moscow, Russia, 119333

[4] Corresponding author's e-mail address: tarandrew17@gmail.com

Abstract. Lasing of ZnO microtatrapods in the spectral range of exciton recombination was studied at room temperature. Samples were produced by a method of pyrolytic carbothermal synthesis. With nanosecond pumping in Al-doped ZnO sample the record low value of the lasing threshold was observed. The threshold pumping energy was 0.13 μJ/cm^2. The dependences of the integrated intensity on the pumping level were obtained. As expected, they appeared to be linear in the lasing region. It was shown that with pumping level increase the intensities of laser lines is red shifted, i.e. long wavelength lines become more intensive. This is suggested to be related to the band-gap renormalisation as a result of possible heating or electron-hole plasma formation. The possible nature of the laser resonator in the tetrapod was discussed. Most likely that the resonator is formed at the bottom of the tetrapod leg where its cross-section is hexagonal. The appearance of the whispering gallery modes is possible in such resonator.

Zinc oxide (ZnO) is a direct semiconductor with a band-gap energy of about 3.3 eV. High exciton binding energy (59 meV) allows registration of the exciton recombination radiation in near-UV up to temperature 500°C. ZnO is known to be able to crystallize in many different forms, from which tetrapod-shape morphology is one of the most attractive.

Among all possible properties of such structure, lasing appearance (particularly, in the spectral range of the exciton radiation) due to existence of local resonators is of our great interest. A number of studies devoted to tetrapod-based microlasers are presented in the literature [1,2] but many problems remain to be solved. Namely, one can single out the achievement of the low lasing threshold and clear understanding what resonator participates in lasing and where it is located.

In this work we studied the lasing in ZnO microtetrapods at room temperature. The samples were produced by a method of pyrolytic carbothermal synthesis which consists in the following. The decalcified paper filter that served as a porous cellulose currier was impregnated with Zn acetate water solution (40 g/*l*). Then, thermal treatment at high temperature (maximum temperature was 1050°C, rate was 4.2 deg/min) was conducted. The holding at constant temperature was 15 - 50 min. The syntesis was perfomed in high purity alundum crucibles in air flow with a rate of 0.5 - 9 *l*/min. The subsequent cooling up to a room temperature was provided during ~ 20 h. Some samples were doped with Al by pouring the metal powder on the paper filters before the synthesis. Doping level was 5 at.%. The examples of the photograph of the crucible with ZnO tetrapods after the synthesis and the SEM image of ZnO tetrapod are presented in figure 1.

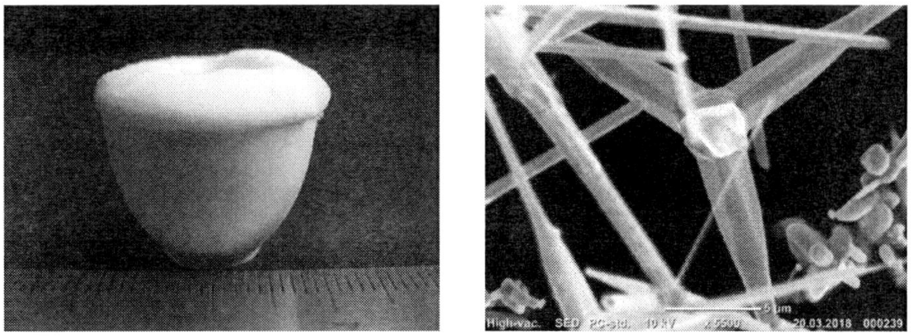

(a) (b)

Figure 1. a) The photograph of the crucible with ZnO tetrapods; b) The microphotograph (SEM) of the ZnO microtetrapod.

The excitation of the samples radiation was performed by the 3rd harmonic of a Nd:YAG laser (355 nm). Pulse duration and repetition rate were ~10 ns and 15 Hz, respectively. The pumping energy was varied from 30 nJ to several μJ. The laser beam was focused into a spot on the sample using a short-focus lens. Radiation of the samples was registered with the use of the monochromator MDR-206 in couple with the Peltier-cooled CCD camera Videoscan-285. The registered spectral range was 35 nm (from 373 to 408 nm). In order to register stable good-quality signal, the accumulation was used (50-200 pulses).

In the spectra of all samples studied in this work we observed the permanence of laser lines positions and their relative intensity ratios from shot to shot at the constant pumping level. Also the spectra shape before and after the accumulation did not change. All of this indicates that observed lasing was caused by the optical modes of tetrapods and it was not random lasing.

In figure 2(a) the spectra at two different pumping levels of one of the studied samples are shown. For this sample, the dependence of the integrated intensity (the integral of the spectrum over the wavelength) on the pumping level was obtained (figure 2(b)). Here, one can see that the slope in the range from zero pumping to the threshold (~1000 nJ) is noticeably lower than that at higher pumping levels.

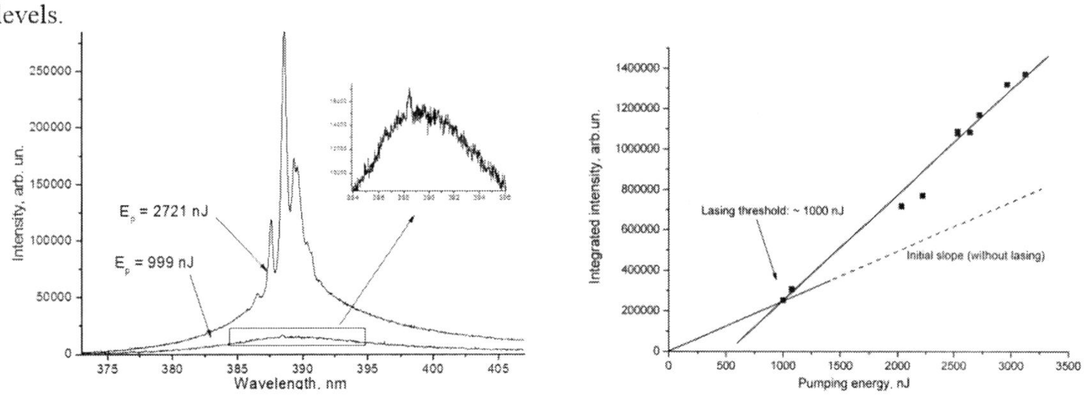

(a) (b)

Figure 2. a) The exciton radiation spectra at two different pumping levels: at the threshold (pumping energy E_p = 999 nJ), above the threshold (E_p = 2721 nJ); b) The dependence of the integrated intensity on the pumping level.

Studying the radiation of Al-doped sample (ZnO:Al) we observed significantly lower lasing threshold than one for the previous-discussed sample. The spectra of this sample at pumping levels near the lasing threshold are shown in figure 3(a). Here, the threshold is just ~ 40 nJ. One can see that the most intensive laser line is located at 386.8 nm.

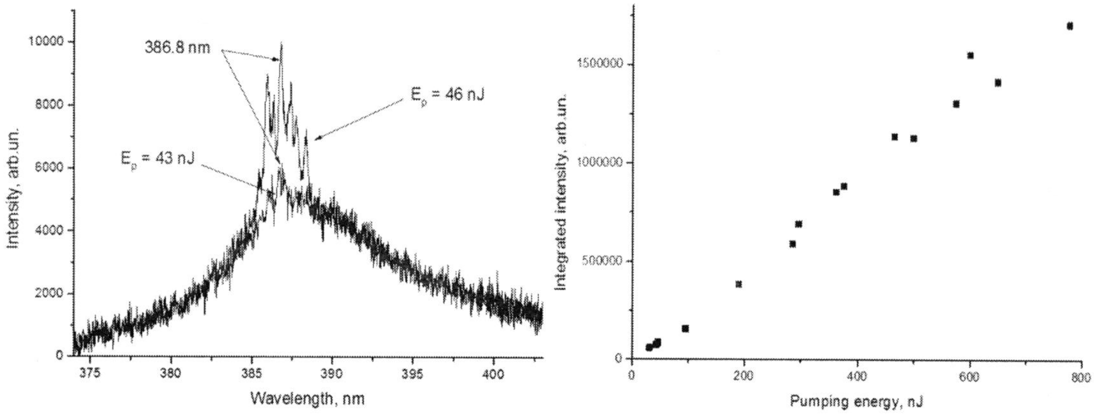

(a) (b)

Figure 3. a) The exciton radiation spectra of the ZnO:Al sample at pumping levels near the threshold; b) The dependence of the integrated intensity on the pumping level for ZnO:Al sample.

The dependence of the integrated intensity on the pumping level for ZnO:Al sample is shown in figure 3(b). Since the estimated size of a pumping spot on the sample is ~200 μm, for the threshold pumping energy of ~ 40 μJ, we obtain the energy density ~ 0.13 mJ/cm^2 and the power density ~ 13 kW/cm^2. As we know, it is the record low value. For example, in Ref.[3] and Ref.[4] where ZnO tetrapods were produced by a similar method the thresholds were 0.55 mJ/cm^2, and 0.815 mJ/cm^2, correspondingly. Here, it is important to note that comparing the threshold values in the units of energy density for different samples is correct only in case of equal (or at least close) durations of pumping pulses. In other cases it is more correct to compare them in the units of power density.

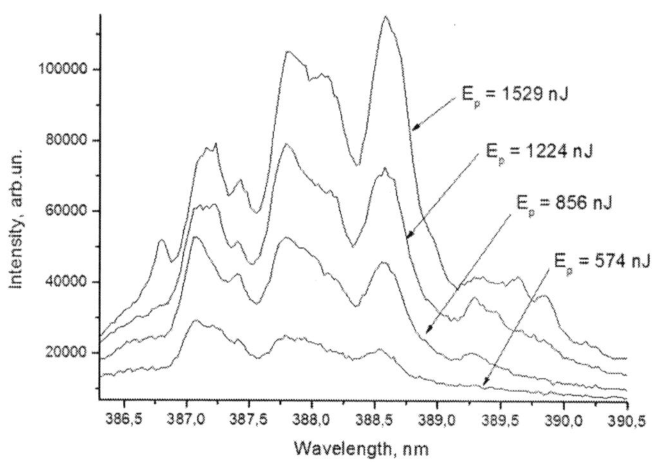

Figure 4. The lasing spectra of the ZnO:Al sample at different pumping levels.

In figure 4 the lasing spectra (in range from 386.3 nm to 390.5 nm) registered in a wide range of pumping levels are presented. It is seen that at low pumping levels the short wavelength laser lines are the most intensive ones. With the increase of the pumping level the intensity ratio increases towards long wavelength lines. Such relative change of the intensities speaks to the dependence of the gain medium on the pumping level. We suggest that it is related to the band-gap renormalization as a result of possible electron-hole plasma formation, which leads to the redshift of the exciton luminescence. Also such shift can appear due to heating of a sample area under the pumping. We will carefully check this in future work.

The nature of the laser resonator in the tetrapod is a crucial point. To our opinion, two possible resonators can exist in the tetrapod. The first one is formed by the pair of tetrapod legs, which can provide closed trajectories of photons [5]. Since in the present case the cross-section along the tetrapod legs changes its shape and size, most likely that the resonator is formed at the bottom of the leg where the cross-section is hexagonal (see figure 5). The appearance of the whispering gallery modes (WHM) is possible in such resonator. For example, such situation was observed in [6] where lasing in microtowers shaped similar to tetrapod legs in the present case was studied.

Figure 5. The SEM image of the tetrapod with a hexagonal cross-section in its leg.

To summarize, low-threshold lasing was obtained in Al-doped ZnO tetrapods produced by a method of carbolytic carbothermal synthesis. In further work the effect of Al-doping on the threshold value will be studied in detail.

We also observed the relative change in the intensity of the laser lines with the increase of the pumping level, which is probably related to the band-gap renormalization. It is suggested that a hexagonal part of the tetrapod leg forms a possible resonator with WGM.

References

[1] Modi G 2015 *Adv. Nat. Sci.: Nanosci. Nanotechnol.* **6** 03300
[2] Zalamai V V, Ursaki V V, Tiginyanu I M, Burlacu A, Rusu E V, Klingshirn C, Fallert J, Sartor J and Kalt H 2010 *Appl. Phys. B* **99** 215
[3] Li L E and Demianets L N 2008 *Opt. Mater.* **30** 1074
[4] Demyanets L N, Li L E, Lavrikov A S and Nikitin S V 2010 *Crystallogr. Rep.* **55** 142
[5] Markushev V M, Ursaki V V, Ryzhkov M V, Briskina Ch M, Tiginyanu I M, Rusu E V and Zakhidov A A 2008 *Appl. Phys. B* **93** 231
[6] Markushev V M, Ryzhkov M V, Briskina Ch M, Borodkin A A, Rumyantsev S I, Shen W Z, Xu D H and Lyaskovskii V L 2012 *J. of Russ. Laser Res.* **33** 122

IOP Conf. Series: Journal of Physics: Conf. Series **1092** (2018) 012150 doi:10.1088/1742-6596/1092/1/012150

Periodic structures, high impedance and semitransparent surfaces in antennas for centimeter and millimeter precision of positioning with the Global Navigation Satellite Systems

D V Tatarnikov, A V Astakhov, A P Stepanenko and A A Generalov

Topcon Positioning Systems, Derbenevskaya nab. 7/22, Moscow, Russia

DTatarnikov@topcon.com

Abstract. In designing GNSS high-precision positioning antennas it is desirable to achieve П-shaped pattern which is homogeneous in the top semi-sphere up to the local horizon and has rapidly decreasing gain in the downwards directions. Such requirements as minimal loss factor and compactness are reasonably applied. Patch antennas with substrates formed of periodic metal structures have been developed, the structures substituting common dielectrics. High impedance surfaces are employed in design of ground planes. Convex and flat ground planes are discussed and presented; the antenna with a large flat impedance ground plane has allowed to achieve the level of positioning error below 1mm rms (root mean square) in real time. The antenna serves as a reference for research purposes. Semitransparent surfaces have been considered as means to obtain a sharp drop of the gain (cutoff) while crossing a local horizon. Plane and concave structures have been treated, and potentials for cutoff are also shown. Practical helical travelling wave and backfire antennas have been realized, with the 1.5mm-rms-positioning error being achieved. In addition to positioning, the antennas are considered as sensors in atmospheric research.

1. Introduction

High precision positioning with the Global Navigation Satellite Systems (GNSS) is achieved [1] in differential mode when a moving object (rover) is positioned with respect to a base station. Path propagation delays of signals from all satellites in view are measured by base station and rover receivers, base station readings are transmitted to a rover in real time, the difference in propagation delays of each signal is calculated at the rover. RTK (Real Time Kinematic) algorithm is applied to resolve ambiguities in carrier phases and to estimate clock offsets, this allows to achieve cm – precision of positioning. The signals of the satellites are grouped into the so-called the Upper L-band and the Lower L-band with frequencies of (1545…1610) and (1160….1300)MHz respectively, the signals being right-hand circular-polarized (RHCP). The error contribution that is unique to a site is known as ground multipath. The error originates from unavoidable reflections of satellite signals from a terrain underneath the antennas. To mitigate the multipath, it is desirable to achieve a reasonable approximation to П-shaped antenna pattern which is homogeneous in the top semi-sphere up to the local horizon and has rapidly decreasing gain in the downwards directions. Such requirements as minimal antenna noise figure and compactness are applied.

The paper reflects the developments done in Moscow Technology Center of Topcon Corp. (Japan). For a broader view of GNSS antenna technology the reader is referred to [2,3].

2. Periodic Structures as Patch Antennas Substrates.

With the common patch antennas the bandwidth is proportional [4] to the substrate thickness. To avoid the use of thick RF ceramics in L-band, light weight metal structures have been developed. In figure 1, a parallel-plate waveguide is partially filled with a structure of metal ribs located at $-b < z < 0$. Then, in the case of $b, d << \lambda$ (with λ being a free space wavelength) the slowdown factor β of the wave propagation in x direction in the domain $0 < z < d$ is $\beta = \sqrt{\varepsilon_{eff}} \approx \sqrt{1 + b/d}$ [5]. Here ε_{eff} is equivalent to effective dielectric permittivity of the substrate. The antenna structure of circular polarization is shown in figure 2. Next, with TM_{10} cavity mode of a patch antenna, the electric field intensity in the area near the antenna center of symmetry is small. Removing pins of the central area has led to the patch stack with substrates in the form of a capacitive frame. At figure 3; the frame is a structure of interlapping metal pins. To estimate ε_{eff} one is to mention that for a linear polarized patch antenna the slowing down performance of the structure is a product of two factors: the propagation of a wave through a chain (figure4.a) and the capacitive load to radiation impedance (figure4.b); estimates [1] have shown that ε_{eff} on the order of 4…6 is achievable with the common manufacturing process.

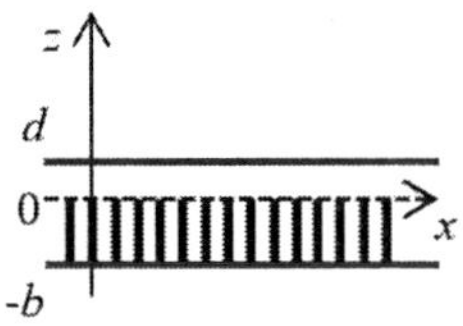

Figure 1. Parallel-plate waveguide filled with a structure of ribs.

Figure 2. Dual band patch antenna with comb substrates.

Figure 3. Dual band patch antenna with capacitive frame.

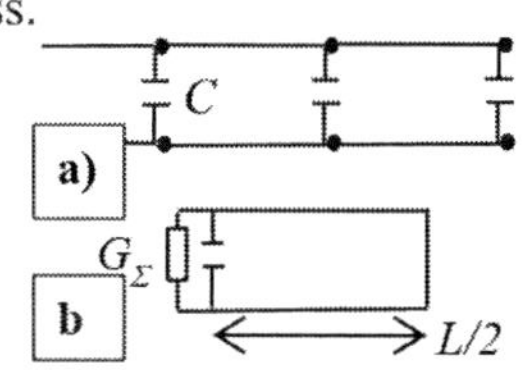

Figure 4. a) Equivalent chain circuit of longitudinal structure, b) capacitive load to radiation impedance G_Σ; L -resonant patch size.

3. High Capacitive Impedance Ground Planes.

For antenna gain reduction in the directions underneath the antenna, a flat metal ground plane is a common arrangement. Potentials of the approach are limited: in 2-dimensional approximation the field intensity underneath the ground plane is proportional to $\sqrt{\lambda / L}$. Here L is the ground plane size. Opposite, with high capacitive impedance ground planes in the same approximation the gain decreases as $(\lambda / L)^{3/2}$. The latter is equivalent to antenna pattern null formation in the horizon direction for substantially large L. To increase the said gain and thus to improve tracking of low elevated satellites by a receiver, a convex impedance structure [7] has been developed (figure 5). To achieve a high capacitive surface impedance and broadband functionality, a straight pins structure is utilized. As discussed in [7], as the radius of curvature decreases, the gain in the directions underneath stays almost constant within a reasonable range of radii while the gain for low elevation satellites improves rapidly. Figure 6 illustrates the antenna pattern of the developed structure in comparison [1] to a common Choke Ring ground plane antenna originally designed by Jet Propulsion Laboratory (JPL) of the USA. For correct understanding of the plots one is to mention that the carrier phase multipath error is proportional [1] to antenna down-up ratio in the form of $F(-\theta^e)/F(\theta^e)$. Here F is the antenna gain pattern and θ^e is the elevation angle with respect to a local horizon.

Flat impedance ground plane of large L has been employed to approximate the Π-shaped pattern. At figure 7, the ground plane is 3m in diameter. Within the physical optics approximation [1,8], the down-up ratio of the antenna is setup by the ground plane size while the gain for low elevations is defined by the distance in between the pins structure and the antenna element in the center. With the antenna, it is for the first time that the error of positioning below 1mm root mean square (rms) has been demonstrated.

IOP Conf. Series: Journal of Physics: Conf. Series **1092** (2018) 012150 doi:10.1088/1742-6596/1092/1/012150

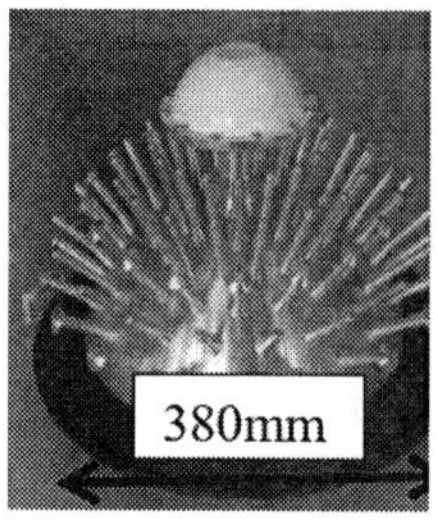

Figure 5. Convex Impedance Ground Plane Antenna.

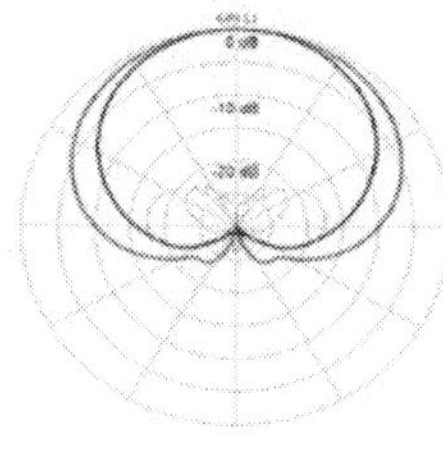

Figure 6. Antenna directivity patterns. Red – antenna of figure 5, Blue – JPL Choke Ring ground plane. RHCP(LHCP) – solid (dashed) lines.

Figure 7. GNSS antenna with large impedance ground plane.

4. Semitransparent surfaces.

Semitransparency is understood as obeying boundary conditions for electrically thin sheet in the form $\vec{n}_0 \times (\vec{H}^+ - \vec{H}^-) = \vec{E} / Z_g$. Here $\vec{n}_0$ is the unit vector normal to the sheet, $(\vec{H}^+ - \vec{H}^-)$ is the discontinuity of the magnetic field intensity on the two sides of the sheet, $\vec{E}$ is the electric field intensity, Z_g is the complex grid impedance that characterizes the sheet. The sheets of interest are grids of strip conductors with embedded impedances [9]. To achieve a Π-shaped pattern, radiation from a parallel plate waveguide with semitransparent walls has been considered [10], the profile of Z_g has been synthesized and dimensions have been evaluated. The approach has led to practical implementation in the form of quadrifilar helix antennas shown in the next Section. Another implementation [11] is a concave screen (figure 8). At the figure, the perfectly conducting portion is shown by a solid line and the semi-transparent by a dashed line. The arrangement allows to achieve a sharp drop of the gain while crossing the desirable light-shadow boundary. For large radius *b* of a screen, the screen is synthesized analytically within the geometrical optics approximation, for smaller radii a numerical optimization procedure has been applied. Figure 9 illustrates the patterns that are achievable for initially cardioid pattern of the source. The 20-db drop of the gain is realized by a screen with radius of 2 wavelength.

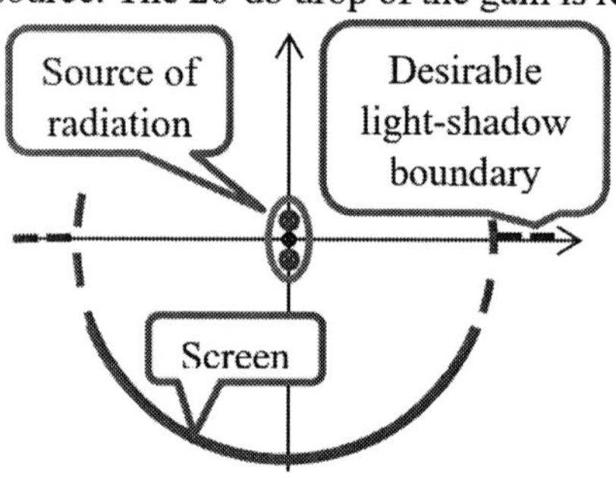

Fig.8 Geometry of concave semi-transparent screen.

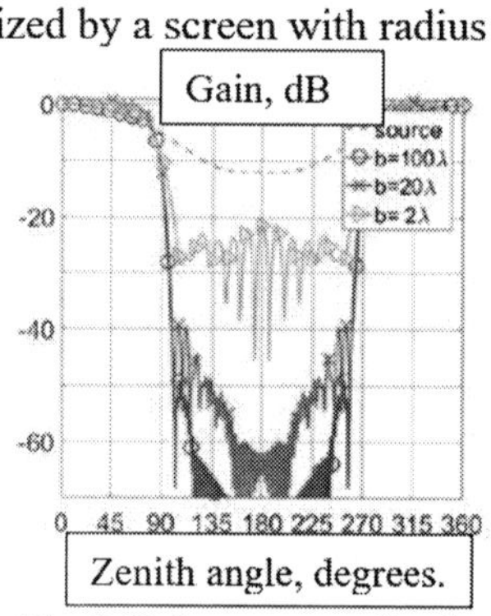

Fig.9 Gain patterns for the concave screens

5. Helix antennas for mm precision of positioning.

Helix antennas with embedded impedances (figure 10a) and of backfire type with alternating winding angle (figure 10b) have been developed [10,12]. Type of the gain pattern achieved is shown in figure 11. Real time error in positioning is illustrated in figure 12. Here, the multipath contribution falls below thermal noise, besides the noise the remaining error is estimated at 1mm rms. The result is valid for comparatively short distances between base station and rover. As the distance increases, the contribution of tropospheric gradients becomes noticeable. The technology is of potential interest for tropospheric slant delays estimates at precision of about 0.5mm rms.

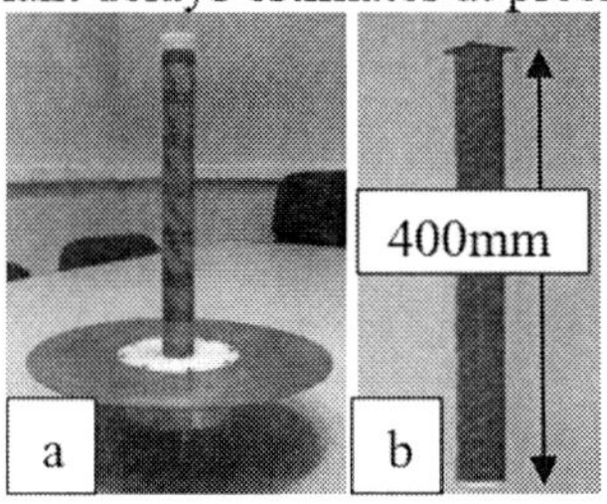

Fig.10 Quadrifilar helix antennas with embedded impedances (a) and alternating winding angle (b).

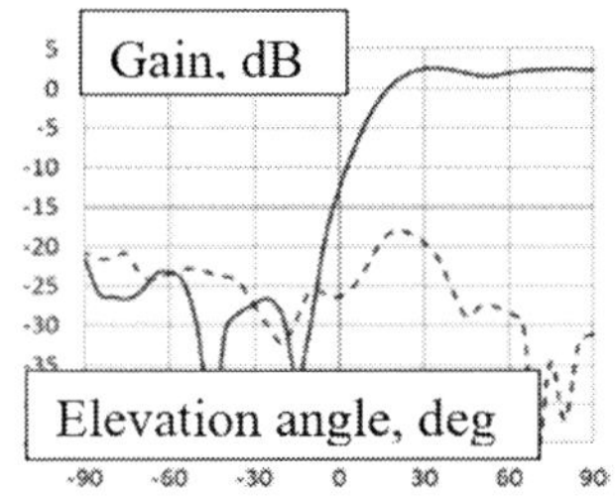

Fig.11 Gain pattern of helix antennas, RHCP (LHCP) – solid (dashed) lines.

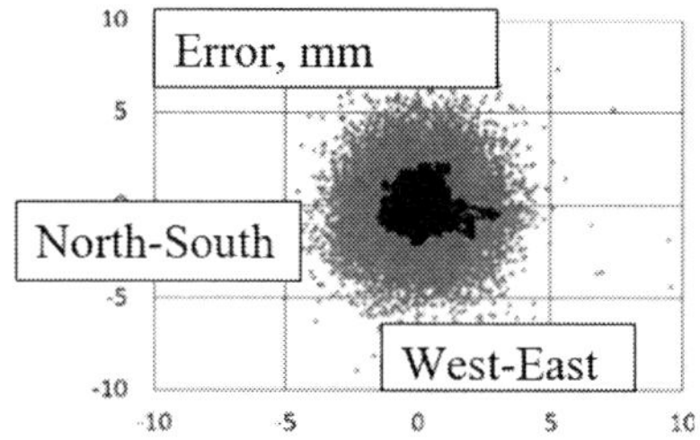

Fig.12 Error in real time positioning. Grey – noise floor, black – the remaining multipath error.

6. Conclusions

Practical antennas for high precision positioning that utilize periodic structures of various kinds have been developed. The structures form artificial dielectrics, high impedance and semitransparent surfaces. П-shaped antenna patterns with a 20 dB drop of the gain within +/-10 degrees angular sector about the local horizon have been realized. Precision of positioning of 1mm rms has been demonstrated.

References

[1] Leick A, Rapoport L and Tatarnikov D, GPS Satellite Surveying, 4-th Edition , Wiley, New York, 2015

[2] Rao B R, Kunysz W, Fante P and McDonald K *GPS/GNSS Antennas*, Artech House, Norwood, 2013

[3] Chen X, Parini C G, Collins B, Yao Y and Rehman M U *Antennas for Global Navigation Satellite Systems*, Wiley, New York, 2012

[4] Garg R, Bhartia P, Bahl I and Ittipiboon A *Microstrip Antenna Design Handbook*, Artech House, Norwood, 2001.

[5] Silin R A and Sazonov V P, *Slow-Wave Structures*, National Lending Library for Science and Technology, Boston Spa Eng., 1971

[6] Veitsel A V, Veitsel V A, Tatarnikov D V "Appratus for High Precision Positioning with the Global Navigation Satellite Systems. High Precision Antennas. Special Methods to Improve the Accuracy of Positioning.", Moscow, MAI-PRIN, 2010 (in Russian)

[7] Tatarnikov D, Astakhov A and Stepanenko A "GNSS Reference Station Antenna with Convex Impedance Ground Plane: Basics of Design and Performance Characterization", Proc. of the Int. Tech. Meeting of the Institute of Navigation ION ITM 2011, pp. 1240-1245

[8] Tatarnikov D and Astakhov A "Large Impedance Ground Plane Antennas for mm-Accuracy of GNSS Positioning in Real Time", Proc. of Progress in Electromagnetic Research Symposium PIERS 2013, Stockholm, Sweden, pp.1825-1829

[9] Tatarnikov D "Semi-Transparent Ground Planes Excited by Magnetic Line Current", *IEEE Trans. AP*, 60(6), pp. 2843-2852, 2012

[10] Tatarnikov D, Chernetsky I, Travelling Wave Antennas with Semitransparent Surfaces for Forming a Cutoff Pattern, PIERS Proceedings, pp.1168 - 1171, July 6-9, Prague, 2015

[11] Tatarnikov D, Generalov A, Concave Semitransparent Screen for Cutoff of the Fields in the Shadow Domain http://jre.cplire.ru/jre/jan18/7/text.pdf (in Russian)

[12] Tatarnikov D, Stepanenko A, Astakhov A, Moderately compact helix antennas with cutoff patterns for millimeter RTK positioning, *GPS Solutions, v.20, 2016, pp. 587-594*

IOP Conf. Series: Journal of Physics: Conf. Series **1092** (2018) 012151 doi:10.1088/1742-6596/1092/1/012151

The grating of micro-holes as a chemosensor with rhodamine 6G molecules

T T H Nguyen[1,3,4,*], S I Kudryashov[1,2,h], P A Danilov[1], A A Ionin[1], R A Khmelnitskii[1], A A Rudenko[1] and I N Saraeva[1]

[1]Lebedev Physical Institute, Leninskiy prospect 53, 119991 Moscow, Russia
[2]ITMO University, Kronverkskiy prospect 49, 197101 St. Peterburg, Russia, Russia
[3]Moscow Institute of Physics and Technology, Institutskiypereulok 9, 141701 Dolgoprudny, Moscow Region, Russia
[4]Ha Tinh University, No 447, street 26/3, precinct Dai Nai, city Ha Tinh, province Ha Tinh, Viet Nam.
*Corresponding author: E-mail: trang.nguyenthihuyen@htu.edu.vn

Abstract. Chemo-sensing of the produced 4×4 mm^2 gratings of micro-holes (approx. half-million of micro-holes) on the 30-nm thick silver film on the IR-transparent CaF_2 substrate was investigated in terms of their resonant EOT-enhanced IR-transmission as surface-enhanced IR-absorption (SEIRA). Selective IR absorption at 1261 cm^{-1} enhanced by 455 times, was demonstrated for rhodamine 6G molecules. The reference absorption lines were taken near 2900 cm^{-1}, where the IR radiation is freely channeling through the microholes, indicating the reference substrate coverage by the dye molecules for its relative internal calibration.

1. Introduction

Ultrasensitive detection of critical chemical or biological agents is demanding nowadays for counterfeiting and medical applications. For the last decade, surface-enhanced Raman scattering (SERS), utilizing different surface-nanotextured platforms, demonstrated its brilliant achievements in detection of single molecules [1]. Meanwhile, this nonlinear-optical method relying upon nanoscale near-field enhancement of intense electromagnetic (EM) fields of localized plasmons in the UV-near-IR range [2], in general, has its intrinsic drawbacks, such as critical nanocontamination (poisoning) of sensing nanostructured surfaces, luminescence background, degradation of sensing nanoelements under elevated thermal loads in CW laser fields. As a result, alternative sensing principles - surface-enhanced IR absoprtion/reflection (SEIRA/R) - were previously developed [3-6], invoking nanoscale enhancement of IR-range fields by localized surface plasmons in metallic surface nanoislands and nanoshells [7], or their micro-scale enhancement by delocalized surface plasmon-polaritons (SPP) in diverse regular micropatterns [8]. These micro-structured sensing platforms provide an opportunity for luminescence background-free, broadband or spectrally selective IR-range sensing of functional groups - fingerprints of explosives, narcotics. Comparing to elaborate, multi-step nanofabrication processes, diverse miscroscale surface patterns - e.g. arrays of microholes in thin films of plasmonic metals - can be readily directly printed with a high throughput by focused single nano- or femtosecond laser pulses [9-12], coming at (sub) MHz repetition rates [13].

In this work we report high-rate femtosecond laser fabrication of a number of SEIRA sensors on a thin silver film deposited on an IR-transparent fluorite substrate and FT-IR characterization of their sensing capabilities, using rhodamine 6G dye as a model analyte.

2. Experimental

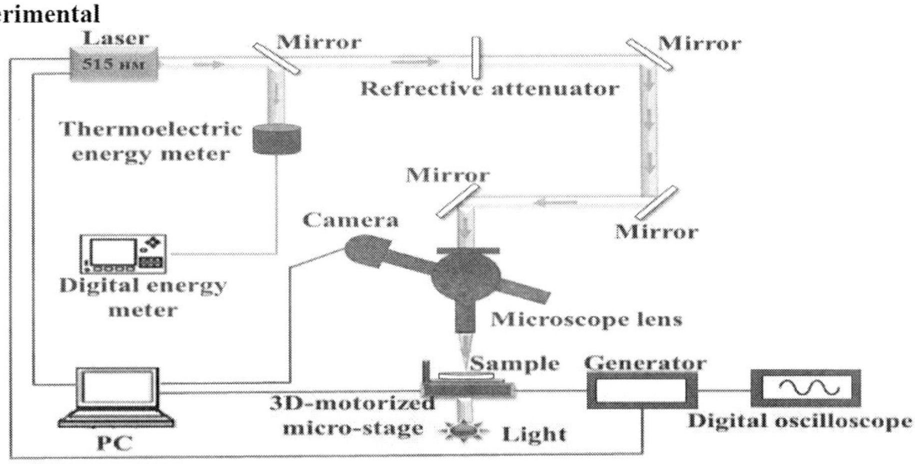

Figure 1. The schematic femtosecond laser pulses for fabrication microholes in thin metallic film

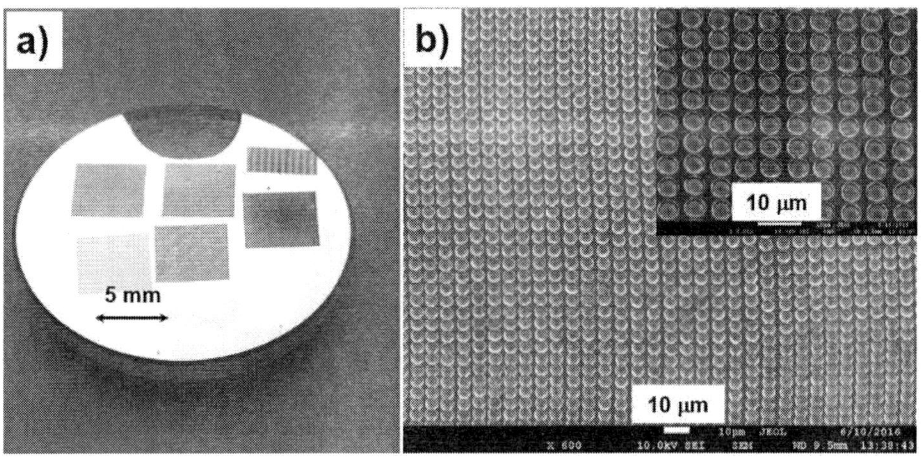

Figure 2. (a) Optical image of the 11-mm wide CaF_2 substrate with the top 30-nm thick Ag-film and a number of micro-hole arrays (typical square – 4×4 mm^2). b) Top-view SEM image of the array with diameter D = 4 µm and period P = 6 µm (inset: its magnified view).

We used a silver (Ag) film with thickness 30 nm, deposited onto a 1-mm thick CaF_2 substrate by magnetron sputtering in argon. Laser radiation was focused 515 nm, 220 fs TEM_{00}-mode with an energy of 85 nJ on the surface of the sample in air through the objective of an optical microscope with the numerical aperture 0.25 (Fig. 1). Holes in the film located on a three-dimensional motorized holder with computer control and linearly scanned at the scan speeds of 750 µm.s^{-1} in a number of lines with a shot-to-shot shift of 6 µm and interline spacing of 6 µm to produce on its frontal surface 4 mm × 4 mm wide large regular arrays of millions microholes with a 4 µm large diameter (Fig. 2). We used Rhodamine 6G for research surface enhanced IR absorption of sensors. The arrays of microholes characterized by means of scanning electron microscopy (SEM, JEOL 7001F). IR unpolarized transmittance of the arrays was measured in vacuum in the near-mid IR range of spectral wavenumbers υ = 400-5000 cm^{-1}, using a FT-IR spectrometer V-70 (Bruker).

3. Results and Discussion

First, the actual, apparently, inhomogeneous R6G surface coverage was considered, accounting for the different hydrophilicity of metallic and CaF_2 spots on the CaF_2 substrate and the grating. Such internal calibration was performed, considering the waveguide-like transmission through the array in the mid-IR range of 2800-3000 cm^{-1} at the transmittance level ≈0.08 (8%) in the clean grating and ≈0.069 (6.9%) in the grating with the R6G monolayer. The corresponding transmittance dips of R6G bands are 0.11% (≈2850 cm^{-1}), 0.35% (≈2920 cm^{-1}) and 0.2% (≈2960 cm^{-1}) on the CaF_2 substrate, representing combination CH-stretching modes of the main xanthene ring (XR), phenyl, ethyl and methyl groups with the lower-wavenumber external group modes (EGM), but with the overall wave-numbers less than those of C-H stretching overtones (Fig.3b). On the grating, these R6G absorption bands demonstrated, on average, much – almost 7-fold – lower signals of 0.04% (≈2850 cm^{-1}), 0.03% (≈2920 cm^{-1}), while the R6G absorbance at 2960 cm^{-1} unexpectedly appears even two-fold higher (0.4%). Comparing to the 13-fold stronger IR-transmission through the CaF_2 substrate – 92% vs 7% (this grating), such 7-fold lower R6G absorbance on the grating effectively means its 2-fold, on average, enhancement of R6G absorbance, with the corresponding 26-fold enhancement of the absorption band at 2960 cm^{-1}. Generally, such enhancement could occur due to plasmonic (spectrally homogeneous over the narrow range of 2800-3000 cm^{-1}) or chemical (R6G-Ag-comlexing, spectrally-selective for the different R6G bonds) factors, which will be discussed below during the analysis of the characteristic absorption bands. Meanwhile, taking the minimal enhancement conditions at ≈2920 cm^{-1}, 13-fold difference in the R6G absorbance on the CaF_2 and the grating could be expected over the entire IR spectrum at the absence of the relevant plasmonic and chemical enhancement factors.

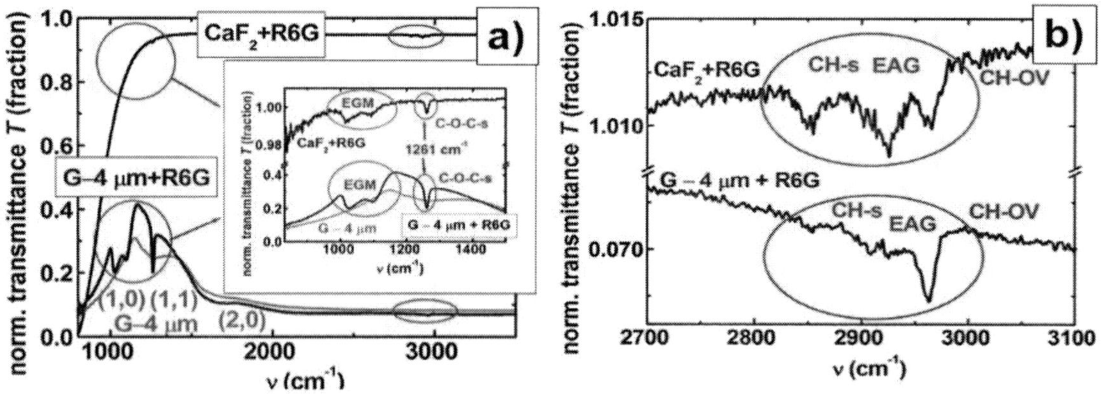

Figure 3. (a) Normalized FT-IR transmission spectra of the CaF_2 substrate with the R6G monolayer, of the clean array and of the array with the R6G monolayer. Inset: the magnified view of their normalized low-ν transmittance with the assignment of the R6G absorption bands on the CaF_2 substrate and the array. b) Magnified view of their normalized high-ν transmittance in the overtone region with the assignment of the composition R6G absorption bands.

In the striking contrast, the several main characteristic absorption bands of R6G in the range of 900-1300 cm^{-1} (the strongest one at 1261 cm^{-1}) appear strongly – by 20-40 times – enhanced, comparing to its non-enhanced IR-absorption bands on the CaF_2 substrate (Fig.3a). Specifically, in the original FT-IR spectrum (Fig. 3a) and in the normalized one (inset in this figure) of the R6G monolayer on the CaF_2 substrate there are well-known strong broad bands at ≈1020 cm^{-1} (EGM, 0.6%) and ≈1090 cm^{-1} (phenyl group, 0.4%), while rather narrow one at ≈1262 cm^{-1} (XR C-O-C stretching, 0.3%). These bands of the polar EGM ethyl/methyl-amino and XR C-O-C groups, as well as of the phenyl group emerge at the visible absence of very strong IR-bands of EGM C-H deformation (1305 cm^{-1}), non-identified EGM vibrations at 1500 cm^{-1} and 1528 cm^{-1}, and skeletal stretching vibrations of the external phenyl group (1606 cm^{-1}). Obviously, on the polar (ionic) CaF_2 substrate should benefit IR-absorption mainly for the corresponding polar (amino-, C-O-C, C-O and C=O) groups and to less extent for the polarizable (conjugated xanthene ring, phenyl) groups, inducing minor "red" spectral shifts due to small bond weakening (cfg., in this work the XR C-O-C stretching band at ≈1262 cm^{-1} versus its normal appearance at ≈1269 cm^{-1}, but no spectral shifts for the Raman-inactive EGM-absorption band at 1020 cm^{-1} and the IR/Raman phenyl group band at 1090 cm^{-1}).

Hence, the general high-wavenumber 13-fold reduction in R6G absorbance in the EOT region of 900-1300 cm^{-1} is (4-6)-fold enhanced by the EOT effect and then by 60-100 times enhanced by the chemical effects on the silver surface, resulting in the *analytical* enhancement of 20-40 times for the specific polar XR and EGM bands at ≈1025 cm^{-1}, ≈1100 cm^{-1} and ≈1262 cm^{-1}, as compared to the CaF_2 substrate in Fig.3a. Moreover, the evaluated total enhancement factors (~300-400 times, as the ratio of the R6G absorbance on the grating in the ranges of 900-1300 cm^{-1} and 2800-3000 cm^{-1} – similarly to the CaF2 substrate) are comparable to the previously reported values ~10^2-10^3 (the theoretically evaluated enhancement up to ~10^4), where no chemical enhancement effects were accounted for.

4. Conclusion

Our research has demonstrated high mid-IR sensitivity and ultralow limit of background-free detection of the easy-to-fabricate microhole-array SEIRA sensor on the 30 nm thick silver film, as well as, for the first time, the important opportunity for its relative internal calibration in the near-IR spectral range.

Acknowledgments

This work was supported by the grant of the RAS Presidium program, the Russian Foundation for Basic Research (grant nos. 15-02-08400, 16-32-00880), as well as by the Government of the Russian Federation (Grant 074-U01) for ITMO University.

References

[1] Kneipp K, Wang Y, Kneipp H, Perelman L T, Itzkan I, Dasari R R and Feld M S 1997 *Phys. Rev. Lett.* **78** 1667
[2] Klimov V 2014 *Nanoplasmonics* (Boca Raton, FL: CRC Press)
[3] Hartstein A, Kirtley J R and Tsang J C 1980 *Phys. Rev. Lett.* **45** 201
[4] Zayarny D A et al 2014 *JETP Lett.* **100** 295
[5] Baikova T V et al 2016 *Laser Phys. Lett.* **13** 075602
[6] Danilov P A et al 2016 *Laser Phys. Lett.* **13** 055602
[7] Kundu J, Le F, Nordlander P and Hallas N J 2008 *Chem. Phys. Lett.* **452** 115
[8] Williams S M, Stafford A D, Rodriguez K R, Rogers T M and Coe J V 2003 *J. Phys. Chem. B* **107** 43
[9] Kulchin Yu N, Vitrik O B, Kuchmizhak A A, Nepomnyashchii A V, Savchuk A G, Ionin A A, Kudryashov S I and Makarov S V 2013 *Opt. Lett.* **38** 1452–4
[10] Danilov P A et al 2014 *Quantum Electron.* **44** 540
[11] Zayarny D A et al 2015 *JETP Lett.* **101** 394
[12] Kuchmizhak A A, Ionin A A, Kudryashov S I, Makarov S V, Rudenko A A, Kulchin Yu N, Vitrik O B and Efimov T V 2015 *Opt. Lett.* **40** 1687
[13] Ionin A A, Kudryashov S I, Makarov S V, Levchenko A O, Rudenko A A, Saraeva I N, Zayarny D A, Nathala C R and Husinsky W 2015 *Laser Phys. Lett.* **13** 025603

IOP Conf. Series: Journal of Physics: Conf. Series **1092** (2018) 012152 doi:10.1088/1742-6596/1092/1/012152

High impedance MRI tunnel analysis with integral-equation methods

D. Tihon, H. Bui Van and C. Craeye
ICTEAM institute, Université catholique de Louvain, Place du Levant 3, 1348, Louvain-la-Neuve, Belgium
E-mail: denis.tihon@uclouvain.be

Abstract. In MRI devices, the antennas used to generate the RF field are generally shielded from the rest of the device in order to limit the interference. A proper design of this shielding using metasurfaces is a promising way to improve the performances. In this paper, we propose a method to rapidly simulate metasurfaces whose behaviour is approximated using surface impedance boundary conditions. The method is based on the Method of Moments (MoM) and focuses on the rapid simulation of MRI devices for many different surface impedances, which is crucial for design and optimization. The results are validated using the CST commercial software.

1. Introduction

Magnetic Resonance Imaging (MRI) is an active field of research gathering researchers from many different areas, given the wide variety of challenges that it offers. Generally, in order to decouple the RF antennas problem from the rest of the device, a shielding is used. Most of the times, for simplicity reason, this shielding consists of a conducting plate. However, this shielding presents two major drawbacks. First, it greatly reduces the radiation efficiency of dipole-like antennas located close to it, thereby creating SAR hot-spots and reducing the penetration depth and the homogeneity of the magnetic field [1, 2, 3]. Second, such conducting boundary may support surface modes that may enhance coupling between antennas [3].

In [1, 2], the authors propose to use a High Impedance Surface (HIS) for the shielding. Such surface improves both the field penetration and homogeneity within the body. In [3], the HIS is also used to decouple closely packed antennas. In [4] the study is extended to a wider range of surface impedances. Such surface impedances may be implemented using well-designed metasurfaces.

In the first steps of the design of such a metasurface, its properties are generally modelled through an equivalent surface impedance [4, 7]. One crucial point is the ability to rapidly solve the electromagnetic problem for a given distribution of surface impedance. Methods to simulate such impenetrable sheets with boundary conditions described using surface impedances are known for long [5, 6, 7, 8].

This paper aims at providing a rapid tool to simulate such surface impedance using the Method of Moments (MoM). The method focuses on the rapid simulation of MRI shields for many different impedance distributions. The MoM simulation can be decomposed into two parts: the computation of the MoM matrices and the solution of the resulting system of equations. In the formulation presented here, the impedance matrices are computed only once. Then, these impedance matrices can be reused for any surface impedance distribution. The method is easy to implement provided that one already has some code providing the classical MoM impedance matrices, and the computational overhead is negligible.

2. Description of the method

The MoM is based on the equivalence principle. Using equivalent electric (**J**) and magnetic (**M**) currents on the interface between homogeneous media, the electromagnetic problems on both sides of the interface are decoupled. The fields inside one material only depend on the sources located within this material and on the unknown equivalent currents on its boundary. In order to find the value of these unknown currents, boundary conditions are applied on the interfaces. In this paper, the Poggio-Miller-Chu-Harrington-Wu-Tsai (PMCHWT) formulation is used, so that the boundary conditions imposed for penetrable bodies correspond to the continuity of the tangential electric and magnetic fields [10].

In order for the computation to be tractable by a computer, the problem is discretized. The unknown currents are approximated as a weighted sum of predefined Basis Functions (BF) and the boundary conditions are imposed on average over some Testing Functions (TF). Doing so, the problem of finding the continuous currents distribution on the interfaces between different media is reduced to the problem of finding the unknown weights one has to attribute to the BF in order to find the electric and magnetic currents distribution. The system of equation to be solved then reads:

$$\underline{\underline{Z}} \begin{pmatrix} \mathbf{x}_J \\ \mathbf{x}_M \end{pmatrix} = - \begin{pmatrix} \mathbf{b}_E \\ \mathbf{b}_H \end{pmatrix} \tag{1}$$

with

$$\underline{\underline{Z}} = \sum_m \begin{pmatrix} \underline{\underline{Z}}_m^{EJ} & \underline{\underline{Z}}_m^{EM} \\ -\underline{\underline{Z}}_m^{EM} & \underline{\underline{Z}}_m^{EJ}/\eta_m^2 \end{pmatrix}, \tag{2}$$

$\underline{\underline{Z}}_i^{EJ}$ ($\underline{\underline{Z}}_i^{EM}$) whose (i, j) entry corresponds to the electric field on TF i generated by unitary electric (magnetic) currents on BF j through medium m, $\mathbf{x}_J$ ($\mathbf{x}_M$) the unknown coefficients one has to apply to the BF to find the equivalent electric (magnetic) currents, $\mathbf{b}_E$ ($\mathbf{b}_H$) the incident electric (magnetic) fields on the testing functions and η_m the impedance of medium m [10].

Now consider an infinitely thin impenetrable slab whose reaction to incident fields is described through Impedance Boundary Conditions (IBC) that links the total tangential electric and magnetic fields on each of its two sides:

$$\mathbf{E}_{t,1}^{\text{tot}} = Z_1 \hat{\mathbf{n}}_1 \times \mathbf{H}_{t,1}^{\text{tot}} \tag{3}$$

$$\mathbf{E}_{t,2}^{\text{tot}} = Z_2 \hat{\mathbf{n}}_2 \times \mathbf{H}_{t,2}^{\text{tot}} \tag{4}$$

with Z_i the impedance on side i of the slab, $\mathbf{E}_{t,i}^{\text{tot}}$ ($\mathbf{H}_{t,i}^{\text{tot}}$) the total electric (magnetic) field on side i of the slab at position $\mathbf{r}$ and $\hat{\mathbf{n}}_i$ the outer normal to the surface i of the slab. Note that since the slab is considered infinitely thin, both sides of the slab converge to a single surface, so that one position $\mathbf{r}$ may be on either side of the slab.

Using these boundary conditions and testing them on a given set of testing functions, considering that the fields vanish inside the slab and, following a reasoning similar to [5], one finally finds that equivalent currents on that slab are found imposing

$$\left[\underline{\underline{Z}} - \begin{pmatrix} \frac{Z_1 Z_2}{Z_1 + Z_2} \underline{\underline{G}} & \frac{Z_1 - Z_2}{Z_1 + Z_2} \underline{\underline{H}} \\ \frac{Z_1 - Z_2}{Z_1 + Z_2} \underline{\underline{H}} & \frac{1}{Z_1 + Z_2} \underline{\underline{G}} \end{pmatrix} \right] \cdot \begin{pmatrix} \mathbf{x}_J \\ \mathbf{x}_M \end{pmatrix} = - \begin{pmatrix} \mathbf{b}_E \\ \mathbf{b}_H \end{pmatrix} \tag{5}$$

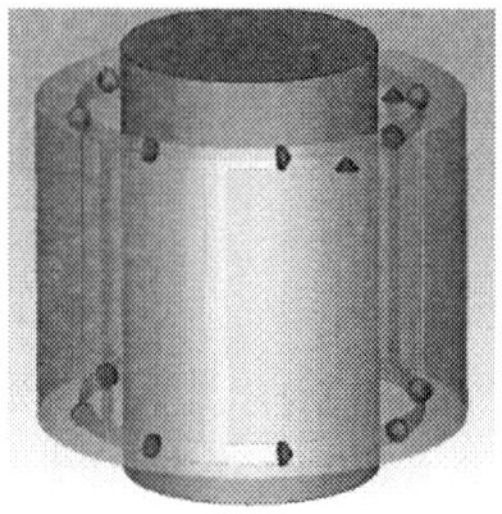

Figure 1. Geometry of the birdcage antenna studied. The blue arrows correspond to lumped capacitors, while the red arrows correspond to the two ports.

with

$$G_{ij} = \iint \mathbf{f}_j^B(\mathbf{r}) \cdot \mathbf{f}_i^T(\mathbf{r}) dS, \tag{6}$$

$$H_{ij} = \frac{1}{2} \iint \left(\hat{\mathbf{n}}_1 \times \mathbf{f}_j^B(\mathbf{r})\right) \cdot \mathbf{f}_i^T(\mathbf{r}) dS, \tag{7}$$

with $\mathbf{f}_j^B$ ($\mathbf{f}_i^T$) being the jth BF (i^{th} TF). Note that the above formula is only valid for constant surface impedance. If the impedance is varying over the surface, the factors involving the impedance in (5) should be included within the integral in (6) and (7). It is important to notice that the overhead for the computation of the $\underline{\underline{G}}$ and $\underline{\underline{H}}$ matrices is very small since the matrices are largely sparse and that the integrand is generally smooth. Moreover, if the impedance of the surface is constant by part and the meshing is conformal with respect to these parts, the $\underline{\underline{G}}$ and $\underline{\underline{H}}$ matrices only need to be computed once.

3. Numerical validation

In order to validate our results, we simulated a full 7T MRI device made of the 8-legs birdcage within a shielding surface presented in [9] (see Fig. 1). In order to mimic the presence of a body, a dielectric cylinder has been put inside the birdcage, with a relative permittivity $\varepsilon_r = 61$ and a conductivity of $\sigma = 0.8$ S/m. The birdcage is made of perfectly conducting strips with lumped capacitors of 11 pF located on the upper and lower rings, halfway between each leg. It is excited by two 50 Ω ports linking the top of two legs to the upper ring. These legs are chosen to be separated by a $\pi/2$ angle in order to efficiently excite the birdcage. The radius of the shielding is 10 cm and its height is 15 cm. The birdcage is made of 1 cm wide PEC conducting sheets. Its height and radius are 15 cm and 8 cm, respectively. The phantom is 20 cm high and its radius is 6 cm.

The 8-fold symmetry of the structure has been exploited in the simulations; details about the code used (symmetries, frequency sweep, etc.) can be found in [9] where the same geometry has been simulated for PEC shielding. In order to validate the presented method, the S parameter of the two ports have been computed for a very-low impedance surface ($Z_1 = 0.01\,\Omega$ and $Z_2 = 0.01\,\Omega$) and for high-impedance inner surface ($Z_1 = 1e5\,\Omega$ and $Z_2 = 0.01\,\Omega$), and the results compared with CST [11]. For CST, perfect electric conductor (PEC) and perfect magnetic conductor (PMC) have been used to mimick these extreme values of the impedance. The results can be seen in Figure 2. The phase-shift in the resonances (about 5%) is probably due to the differences in the model used for the simulations (implementation of the lumped elements, infinitely thin PEC sheet vs. 3D strips, etc.). Except for that systematic frequency shift , often encountered when comparing different methods with different meshings, the results are very similar. It can be seen that the presence of the PMC shielding is globally red-shifting the resonances. Adjusting the value of the capacitance, the second resonance can be blue-shifted

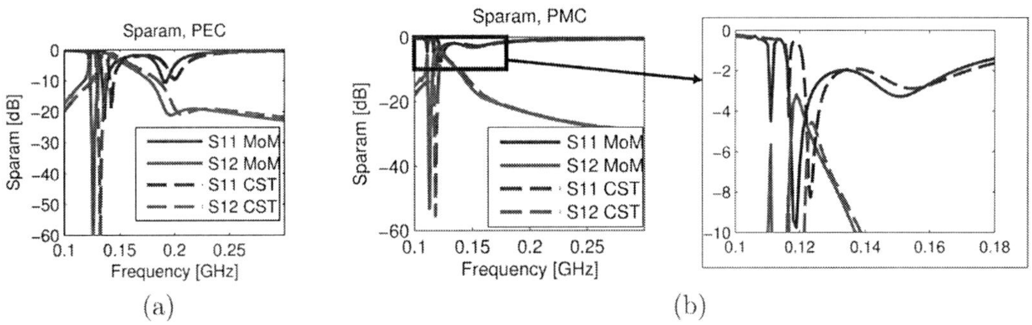

Figure 2. S-parameters of the birdcage antenna for (a) PEC and (b) PMC shielding. For the PMC case, a zoom on the results is provided for better readability.

back to 200 MHz. However, in this case, the homogeneity of the B_1^+ field remains better in the PEC case than in the PMC case. It is due to the fact that the whole geometry was originally designed for a PEC shielding. A complete redesign, with a possibly smaller radius for the shielding, should be conducted in order to provide a fair comparison, which is out of the scope the present paper.

The CST simulations took about 30 minutes per graph, while the MoM code took about 12 minutes for the first graph, and then 90 additional seconds for every successive graph. Note that the MoM code has not been parallelized so far, so that it is running on a single CPU, while the CST software is using 4 CPU by default.

Conclusion

We presented a numerical method that rapidly simulates a birdcage MRI antenna with a shielding whose surface impedance can be varied at will. Dealing with surface impedances rather than PEC boundary conditions only involves a small overhead in the computational time. Moreover, once the simulation has been carried out for a given surface impedance, many terms can be reused for subsequent simulations, so that rapid sweeping can be achieved. It proves to be very useful in design steps, where simulations have to be run for a wide range of surface impedances. The results obtained using this method were compared with results from a commercial software.

Acknowledgements

This project has received funding from the European Union Horizon 2020 research and innovation program under grant agreement No. 736937.

The authors would like to thank Marc Dubois for the design of the birdcage used as an example.

References

[1] Chen Z, Solbach K, Erni D and Rennings A 2015 *Microw. Conf. (GeMiC)* 111-4
[2] Saleh G, Solbach K and Rennings A 2012 *European Conf. on Antennas and Propagat. (EuCAP)* 1399-401
[3] Hurshkain A A *et al.* 2016 *J. Magn. Reson.* **269** 87-96
[4] Chen Z, Solbach K, Erni D and Rennings A 2016 *IEEE Trans. Antennas Propag.* **64** 972-81
[5] Bleszynski E, Bleszynski M and Jaroszewicz T 1993 *IEEE Trans. Antennas Propag. Mag.* **36** 14-25
[6] Yan S and Jin J-M 2013 *IEEE Trans. Antennas Propag.* **61** 5533-46
[7] Francavilla M A, Martini E, Maci S and Vecchi G 2015 *IEEE Trans. Antennas Propag.* **63** 2153-61
[8] Glisson A W 1992 *Radio Science* **27** 935-43
[9] Echeveste J I *et al.* 2018 *European Conf. on Antennas and Propagat. (EuCAP)*
[10] Poggio A and Miller E 1973 *Computer techniques for electromagnetics* (Oxford UK:Pergamon Press)
[11] (2018) CST *Microwave Studio* [Online]. Available: https:/www.cst.com

Self-organized aggregation of a triple of resonant nanoparticles into stable structures with various shapes controlled by a laser field

V S Kornienko[1,2], V A Tkachenko[1], A S Tsipotan[1], A S Aleksandrovsky[1,3] and V V Slabko[1]

[1]Siberian Federal University, 660041, 79 Svobodny pr., Krasnoyarsk, Russia
[2]Institute of Computational Modeling of SB RAS, 660036, 50/44Akademgorodok Krasnoyarsk, Russia
[3]L. V. Kirensky Institute of Physics Siberian Branch of the Russian Academy of Sciences, 660036, 50/43 Akademgorodok Krasnoyarsk, Russia

Abstract. The method of formation of nanostructures consisting of three particles in the field of quasiresonant laser radiation is considered. To obtain structures, two approaches were used: a third particle is added to a pre-formed particlespair at a certain angle; the necessary structure is formed from three initially isolated particles. Numerically shown that it is possibleto assemble structures of line and pyramid using both previously mentioned approaches.

1. Introduction

Inrecent decades, nanostructures with unique properties other than properties of a bulk sample and dependent on both composition and shape have been studying actively. Therefore, the problem of developing a universal method of nanostructures formation is a subject of interest of many scientists. The least expensive method of obtaining colloidal crystals, which doesn't require a local physical impact on the system, is based on the ability of nanoparticles to self-organize in the process of random Brownian collisions in real disperse systems [1]. However, in such case it is impossible to control the processes of formation of nanostructures with predetermined shapes. Therefore, one of the possible solutions of this problem is the physical impact on the ensembles of nanoparticles that allows the formation of complex nanostructures without additional surface modifications [2]. However, it isn't always possible to achieve the selectivity of the self-assembly process.

Earlier in [4-5], it was shown that the interaction of resonant nanoparticles in the field of laser radiation makes possible the formation of predetermined structures, since the energy of the induced dipole-dipole interaction is sufficient to overcome the barrier that prevents spontaneous aggregation (which makes possible the formation a stable structure), and their geometry depends on the wavelength and the polarization of the external field. In this case, the formation of structures, which scale is much smaller than the wavelength of the laser radiation caused this polarization, becomes possible.The presence of optical resonances in the particle leads to an increase of the interparticle interaction and is the basis for the selective formation of different structures with a predetermined position of the particles in the aggregate.It becomes possible because the energy of the interparticle interaction in the laser radiation field depends on its frequency, the resonance frequencies of the particles, and on the orientation of their group relatively to the plane of polarization.

IOP Conf. Series: Journal of Physics: Conf. Series **1092** (2018) 012153 doi:10.1088/1742-6596/1092/1/012153

The experiments [3-5] on the formation of pairs of colloidal quantum dots in the field of laser radiation have shown the possibility of realizing this method. The further formation of more complex structures (three or more particles) can be realized in the way of a step-by-step process, when a third particle is added to an already formed pair of particles, via selecting the wavelength of the external field.

2. Results

In this study, a dynamic self-assembly model for a triple of particles in a laser field using Brownian dynamics is proposed.The possibility of forming a three-particle structure with a predetermined geometry in two variants is studied: from a preformed pair of particles not fixed in a space (Fig. 1) and from triple of isolated particles (Fig. 2).

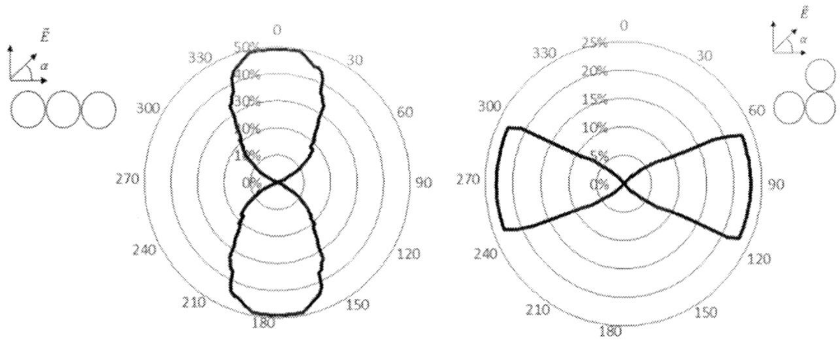

Figure 1. Probability of triple nanoparticles (from preformed pair of particles) aggregation as function of angel α at wavelength of external field λ_r=690 nm.

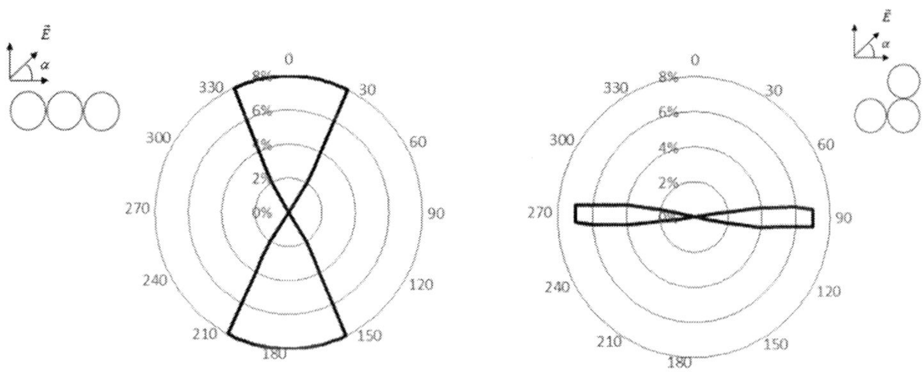

Figure 2. Probability of triple nanoparticles (from isolated particles) aggregation as function of angel α at wavelength of external field λ_r=675 nm.

The obtained results allows to make significant progress in the study of the method of structures self-assembly in the field of laser radiation, which can be use as universal method to form structures with specified properties that will find application as sensors, photodiode elements and solar cells.

3. References

[1] Karpov, S.V. General principles of the crystallization of nanostructured disperse systems / S.V. Karpov, I.L. Isaev, A.P. Gavrilyuk, et.al.// Colloid Journal.– 2009.– V.71.– I.3.– P. 313-328.

[2] Gong, J. Self-assembly of noble metal nanocrystals: Fabrication, optical property, and application/ J. Gong, G. Li, Z. Tang// Nano Today.–2012.–V.7.– Is. 6.– P. 564–585.
[3] Slabko V. V. Resonant light-controlled self-assembly of ordered nanostructures/ V. V. Slabko, A. S. Tsipotan, A. S. Aleksandrovsky// Photonics and nanostructures. Fundamental and Applications – 2012. – T.10. – P. 636-643
[4] Slabko V. V. Dynamics of self-organized aggregation of resonant nanoparticles in a laser field / V. V. Slabko, A. S. Tsipotan, A. S. Aleksandrovsky, E. A. Slyusareva// Springer-Verlag Berlin Heidelberg – 2014.
[5] Slabko V. V. Self-organised aggregation of a pair of practices with different resonant frequencies and electric dipole moments of transitions, controlled by an external quasi-resonant field / V. V. Slabko, A. S. Tsipotan, A. S. Aleksandrovsky // Quantum Electronic – 2013. – T. 43(5) – P. 458-462.

Acknowledgments
The reported study was funded by Russian Foundation for Basic Research, Government of Krasnoyarsk Territory, and Krasnoyarsk Regional Fund of Science to the research project: «Optical binding as a mechanism for self-assembly of stable complex nanostructures from resonant nanoparticles» (project № 18-42-243008). V.V.S. is grateful for the support from Ministry of Education and Science of the Russian Federation (Grant № 3.6341.2017/6.7).

IOP Conf. Series: Journal of Physics: Conf. Series **1092** (2018) 012154 doi:10.1088/1742-6596/1092/1/012154

Transient time in nonlinear interaction of counter-propagating waves with the effect of pump absorption and phase mismatching

V A Tkachenko[1], A K Popov[2], A S Tsipotan[1] and V V Slabko[1]

[1]Siberian Federal University, 660041, 79 Svobodny pr., Krasnoyarsk, Russia
[2]Birck Nanotechnology Center, Purdue University, 1205 W State St., West Lafayette, Indiana 47907, USA

Abstract. The parametric interaction of three waves, one of which travel in direction opposite to other, is associated with a number of unusual phenomena that find their applications. One of such phenomena is an extremely long transient process of amplitude changing of the output signal. In this paper, the impact of pump-wave absorption and the phase mismatch magnitude on the course of the process is investigated. It is shown that the absorption lead to significant changes in the process time, and the phase causes changes in dynamics but not in duration of the process.

1. Introduction

In recent years, in publications, much attention is paid to the possibilities of designing nanostructures with an unique electromagnetic response, which has prospects for use in various fields of both fundamental science and its various applications. Significant progress has been made in the field of nano engineering, and today it has become possible to create metamaterials with a negative refractive index (NIM) [1,2]. NIM, in turn, are comfort environments for the existence of backward waves, which are special waves whose phase vector direction and energy flux direction are opposite. These features open up profound entirely new possibilities of linear optics, unattainable with ordinary waves. The capabilities of NIM include: the creation of a sub-diffraction lens [3], sensitive sensors [4], optical cloaking [5], etc.

In the nonlinear optics, the backward waves in NIM are of interest as promising processes of parametric interaction of waves propagating counterwards each other - counterpropagating waves.The presence of a backward wave in this case, due to the inverse direction of the wave vector, makes it possible to reach phase-matching conditions in the medium [6]. The processes of counterpropagating waves interaction themselves are characterized by anomalous efficiency which can significantly exceed the conversion efficiency in the system of interaction with copropagating waves with the same values of nonlinearity and the medium length [7,8].

Another distinctive feature here is the presence of long transient processes[9]. As was shown in [10], the characteristic duration of a given process τ in the constant pump approximation $a_{30}KL < \pi/2$ also increases indefinitely with increasing gain parameters $\tau \sim 1/\cos(a_{30}KL)$ and can exceed, transient of copropagating case, by hundreds Δt .Such anomalous long-term processes should affect to the devices based on nonlinear NIMs. The influence of signal absorption and pump depletion on the process current was partially investigated earlier in [11, 12]. In particular, it was shown that outside the contant pump approximation threshold $a_{30}KL > \pi/2$, the process time decreases with increasing gain parameters, and the process get an abrupt form.

In this paper, some details of the medium absorption effect and phase mismatching on the transient process current are considered in more detail.

2. Model

Consider nonstationary problem three-wave interaction in one-dimensional quadratic nonlinear medium with length L. The magnitude of the electric field for each of the waves will be given by the expression $E_j(z,t) = A_j(z,t)\exp[i(\omega_j t - k_j z)]$ $j = 1,2,3$, where $A_j(z,t)$ are the amplitudes; ω_j - wave frequencies, bounded by the relation $\omega_3 = \omega_1 + \omega_2$; k_j - wave vectors. The system of equations corresponding to the problem, and taking into account phase mismatch $\Delta k = k_3 - k_2 - k_1$, can be written as follows [11]:

$$\begin{cases} \partial a_1/\partial z + (1/v_1)\partial a_1/\partial t = +[iKa_3a_2^*\exp(i\Delta kz) - (\alpha_1/2)a_1] \\ \partial a_2/\partial z + (1/v_2)\partial a_2/\partial t = -[iKa_3a_1^*\exp(i\Delta kz) - (\alpha_2/2)a_2] \\ \partial a_3/\partial z + (1/v_3)\partial a_3/\partial t = -[iKa_1a_2\exp(-i\Delta kz) - (\alpha_2/2)a_3] \end{cases} \quad (1)$$

Here $a_j = (\omega_j)^{-\frac{1}{2}}(\varepsilon_j)^{\frac{1}{4}}(\mu_j)^{-\frac{1}{4}}A_j$ is the normalized field amplitude; v_j - group velocity; $K = \chi^{(2)}8\pi c^{-1}(\mu_1\mu_2\mu_3)^{\frac{1}{4}}(\varepsilon_1\varepsilon_2\varepsilon_3)^{-\frac{1}{4}}(\omega_1\omega_2\omega_3)^{\frac{1}{2}}$ - coupling factor; $\chi^{(2)}$ - second-order nonlinear susceptibility of a medium; α_j is the absorption index of the medium at the frequency ω_j.

The wave with frequency ω_1 propagates in the negative direction of the axis, and the pump as well as the signal wave with frequencies ω_3 and ω_2 respectively propagate in the positive direction - this is determined by the signs in the equations.

To determine the temporal features of the transient process under absorption and phase mismatch conditions, the problem of the counterpropagating parametric amplifier was considered by equations (1) in the following configuration: The pump is continuously fed through the medium $a_3(z=0) = a_{30}$; the semi-infinite signal pulse in the form of a step $a_2(z=0) = a_{20}/2(1 - \tanh(-t/t_f))$ enters in the medium, where $a_{20} = 10^{-4}a_{30}$ - the input step amplitude, $t_f = 0{,}05L/v_3$ - the input step edge width, the counter wave is initially absent $a_1(z=L) = 0$. The problem was solved by numerical methods. The transient time τ was estimated as the time for which the output signal normalized amplitude was changed from a zero value to a $1 - e^{-1} \approx 0.63$ fraction of the steady state magnitude.

3. Absorption effect

Previous studies have shown that the problem has a characteristic threshold value of the gain parameters $a_{30}KL$ [10]. In the region below the threshold $a_{30}KL < Th$, the process time increases with the growth of the product of the parameters $a_{30}KL$, and it decreases in the region $a_{30}KL > Th$. In the absence of absorption $\alpha_1 = \alpha_2 = \alpha_3 = 0$, the threshold has a value $Th = \pi/2$ (Fig. 1 black circles). Also, in [11] it was shown that the stationary solution of system (1), obtained in [13], allows accurately predicting the position of the threshold with allowance for the amplified waves absorption. The stationary solution of the gain has the form:

$$\frac{a_{2L}}{a_{20}} = \frac{\exp((\alpha_1 - \alpha_2)L/2)}{\cos(RL) + \sin(RL)(\alpha_1 + \alpha_2)/4R} \qquad R = \sqrt{K^2 - (\alpha_1 + \alpha_2)^2/16} \quad (2)$$

The singularity in expression (2) arises when the denominator turns to zero, which corresponds exactly to the threshold value, which has a singular character. So for the case

$\alpha_1 = \alpha_2 = \alpha_3 = 0$, we get $Th = \pi/2$, and for the case $\alpha_1 = \alpha_2 = 2.3L^{-1}, \alpha_3 = 0$ - $Th \approx 1.5111\pi/2$, which corresponds to the data of numerical simulation (Fig. 1. Gray circles). The threshold shift as a function of absorption is easily explained by the fact that in the presence of absorption, the same result requires a greater system gain efficiency.

Expression (2) does not allow to determine the precise threshold position in the pump absorption case. However, it can be assumed that an increase in pumping absorption will lead to a similar threshold shift further into the direction of large values $a_{30}KL$. In Fig. 2 calculated τ dependence for $\alpha_1 = \alpha_2 = 0, \alpha_3 = 2.3L^{-1}$ is shown by black crosses and for the case $\alpha_1 = \alpha_2 = \alpha_3 = 2.3L^{-1}$ by gray crosses. These dependences have thresholds $Th \approx 1.7\pi/2$ and $Th \approx 2.5\pi/2$ approximately, that confirms the proposed assumption.

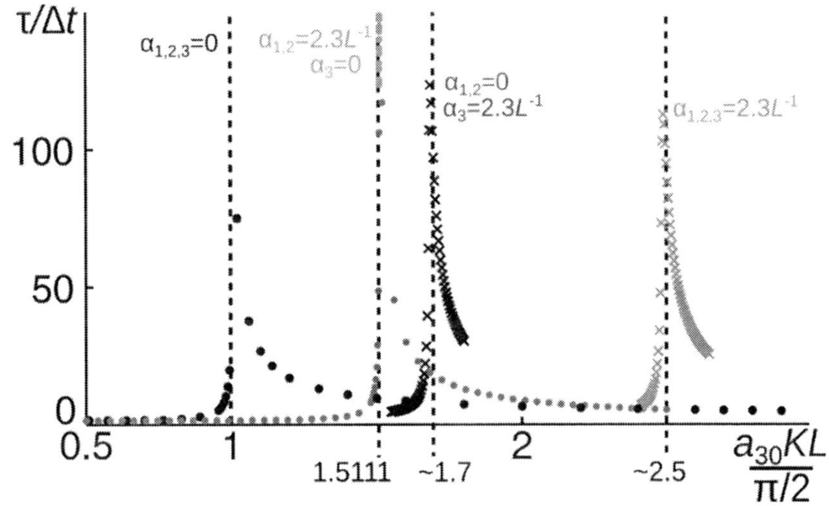

Figure 1. The dependence of the transient time τ on the gain parameters $a_{30}KL$: $\alpha_1 = \alpha_2 = \alpha_3 = 0$ - black circles (•••); $\alpha_1 = \alpha_2 = 2.3L^{-1}$, $\alpha_3 = 0$ - gray circles (•••); $\alpha_1 = \alpha_2 = 0$, $\alpha_3 L = 2.3$ - black crosses (**xxx**); $\alpha_1 = \alpha_2 = \alpha_3 = 2.3L^{-1}$ - gray crosses (xxx). Dashed lines show the corresponding threshold positions.

4. Phase mismatching effect

As a result of modeling, it was found that the change in the phase mismatch magnitude Δk does not affect the Th, but it affects the course of the transient process in the regions below $a_{30}KL < Th$ and above threshold $a_{30}KL > Th$.

In the region $a_{30}KL < Th$ the output signal decreses t with increasing the steady state magnitude (see Fig. 3a). At the same time, its growth rate does not change, and damped oscillations of the normalized amplitude appear, and their period is linearly related to the mismatch value $T_{os} \sim 1/\Delta k$. The transient process time here increases with growth of Δk, according to the previously chosen formal evaluation criterion. However, such a criterion is well suited only for describing monotonic transient processes. For phase mismatch case, the time of the steady-state mode establishing actually increases due to oscillations. As can be seen from the graph, despite the same amplitude growth rate, the stationary mode is reached approximately at the same time for different values Δk.

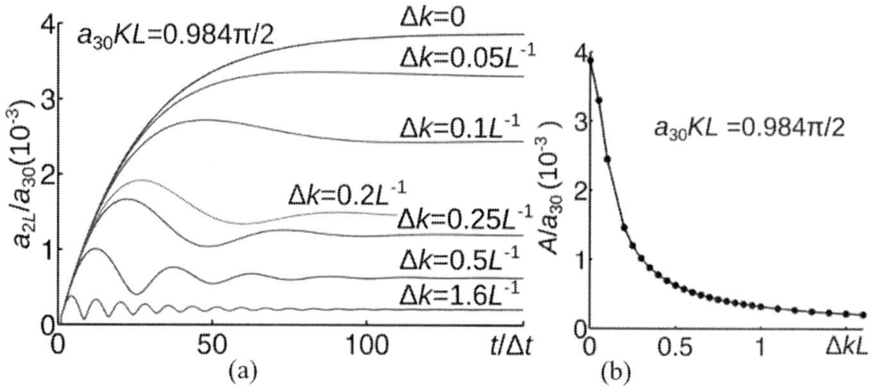

Figure 2. (a) Temporal behavior of the output signal normalized amplitude a_{2L}/a_{30} for different values of the phase mismatch magnitude Δk. (b) Dependence of the output signal normalized amplitude in the steady state mode A/a_{30} on the phase mismatch magnitude.

5. Conclusion

Thus, the absorption of the pump as well as the absorption of other waves leads to a shift in the threshold to a region of large values of $a_{30}KL$. This behavior is explained by the fact that achieving the same result requires a greater efficiency of amplification from the system.

However, the change in the phase mismatch magnitude Δk does not affect the position process threshold position. Moreover, in the region below the threshold, the process time does not decrease with growth of Δk due to the oscillations that arise, but this leads to a decrease in the signal level.

6. References

[1] Shalaev V, Cai W, Chettiar U, Yuan H, Sarychev A, Drachev V and Kildishev A 2005 *Opt. Lett.* **30**, pp 3356-58
[2] Drachev V, Cai W, Chettiar U, Yuan H, Sarychev A, Kildishev A, Klimeck G and Shalaev V 2006 *LaserPhys. Lett.* **3** pp 49-55
[3] Fang N, Lee H, Sun C and Zhang X 2005 *Science* **308** pp 534-537
[4] Tao C, Suyan L and Hui S 2012 *Sensors* **12** pp 2742-65
[5] Pendry J, Schurig D and Smith D 2006 *Science* **23** pp 1780-2
[6] Lan S, Kang L, Schoen D, Rodrigues S, Cui Y, Brongersma M and Cai W 2015 *Nat.Mater.* **14** pp 807-11
[7] Popov A and Shalaev V 2011 Proc. SPIE 8093
[8] Shadrivov I, Lapine M, and Kivshar Y Nonlinear, Tunable and Active Metamaterials 2015 (Springer) chap.10 pp 193–213.
[9] Volyak K and Gorshkov A 1973 *Radioteknika i Elektronika* **18**, 2075
[10] Slabko V, Popov A, Myslivets S, Rasskazova E, Tkachenko V and Moskelev A 2015 *Quantum Electronics* **45** 1151
[11] Slabko V, Popov A, Tkachenko V and Myslivets S 2016 *Opt. Lett.* **41** pp 3976-79.
[12] Tkachenko V, Popov A, Myslivets S and Slabko V 2016 Laser Optics International Conference 16251717
[13] Osborne M, and O'Key M 1992 *Optics Communications* **94** pp 346-352

Acknowledgments

The reported study was supported from Ministry of Education and Science of the Russian Federation (Grant № 3.6341.2017/6.7) , from RFBR according to the research (project № 18-32-00864) and by the U.S. Army Research Office (Grant Number W911NF-14-1-0619).

Silica microcarriers produced from a microfluidics-generated emulsion

A D Trofimov[1], A A Goncharenko[2], A R Muslimov[3], A S Timin[3], M V Zyuzin[1], and A A Evstrapov[1,4]

[1] Department of Nanophotonics and Metamaterials, ITMO University, St.Petersburg 197101, Russia

[2] Department of Biophysics, Peter the Great St. Petersburg Polytechnic University, St. Petersburg 195251, Russia

[3] Department of hematology, transfusiology and transplantation, First I. P. Pavlov State Medical University of St. Petersburg, St. Petersburg 197022, Russia

[4] Laboratory of information and measurement biosensor and chemosensor microsystems, Institute for Analytical Instrumentation RAS, Saint Petersburg 198095, Russia

E-mail: adtrofimov@corp.ifmo.ru

Abstract. Fabrication of micrometric and submicrometric capsules capable of carrying bioactive substances is important for many biological applications. Here we show the results of our work on creating water-in-oil emulsion and encasing the water droplets in silica shell with microfluidic techniques. The emulsion was generated in a polydimethylsiloxane (PDMS) flow-focusing microfluidic chip. TWEEN 80 and lecithin were used as surfactants to stabilize the emulsion. Tetraethylorthosilicate (TEOS) and 3-aminopropyl trimethoxysilane (APTMOS) were then added to the emulsion to form silica shell around water droplets via hydrolysis. We obtained silica microcapsules of about 2 μm in diameter and studied them using confocal fluorescent microscopy. This method allows encapsulating a wide range of sensitive substances including genetic material, enzymes, proteins for intracellular delivery and other biomedical applications.

1. Introduction

Droplet microfluidics is a field studying behavior of liquid droplets in small channels (from several micrometers to several hundred micrometers). Reynolds number in microchannels is typically very low which results in laminar flow, high precision and reproducibility. It allows high-performance generation of droplets with size ranging from hundreds of micrometers down to one micrometer and less. This process is suitable for generating emulsions out of wide range of substances, does not depend on unique chemical agents and happens under room conditions inside isolated channels of a microfluidic chip. These qualities has allowed microfluidics to be used in fabrication of many types of microparticles including microcapsules made of lipids, polylactide and other biopolymers. A huge variety of substances including commercial drugs have been successfully encapsulated into microcarriers, showing viability of microfluidics to be used in drug delivery applications [1].

Microparticles are promising carriers for delivering drugs inside living cells. Several types of microparticles have been employed for this: liposomes, polymeric particles, silica-covered and polymer-covered capsules and others [2]. Liposomes and polymeric particles are fragile, prone to leakage and exhibit low cell tranfection efficiency. Hard capsules, on the other hand, are more robust and biocompatible but harder to fabricate. A recent proof-of-concept study [3] has shown that silica-covered microcapsules are a viable and effective method of delivering genome-editing complex for cell gene therapy. Conventionally such microparticles can be fabricated via a complicated multistage chemical process based on calcium carbonate cores which have to be removed later by a solvent [4][5]. The chemical approach to microcapsule fabrication comprises many long stages leaving a lot of room for variabilities, errors and contamination. In the present study we show an alternative simpler approach to fabricating silica-covered microcapsules which is based on droplet microfluidics.

2. Materials and methods

The microchips were fabricated from polydimethylsiloxane (PDMS) using soft lithography method. PDMS (Sylgard 184, Dow Corning) was poured onto a structured mold, then cured at 70° C for 4 hours. Mold was fabricated using standard lithography from SU-8 negative photoresist on a silicon wafer. The microchip was then treated with O_2 plasma for 1.5 minutes and sealed with another PDMS slice. After that it was treated at 80° for 4 more hours to return hydrophobic properties to it. Liquids flowed into the microchip by a needle capillary system from two syringes operated by syringe pumps (Harvard apparatus PHD 2000 and Harvard apparatus 11 plus).

We utilized a flow-focusing asymmetric microchip design [6]. The dispersed liquid (water) is surrounded by the continuous liquid (oil) from both sides and squeezed through a 10 μm nozzle (Figure 1). The thin jet of liquid soon falls apart into separate droplets due to the capillary instability. The size of the resulting droplets depends on the jet diameter which in turn is determined by flow rates of liquids and nozzle dimensions. For aqueous phase of a water-in-oil emulsion we took a phosphate buffer solution with 1% (by mass) TWEEN 80 surfactant. For oil phase we used 4% (by mass) lecithin solution in mineral oil (M5310, Sigma-Aldrich) with BODIPY fluorescent dye.

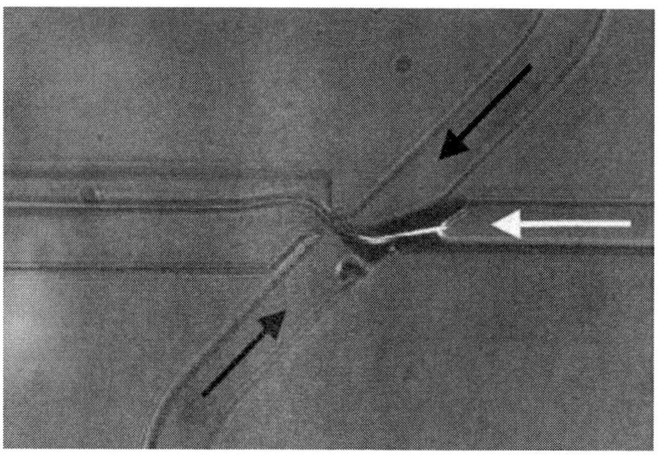

Figure 1. Micrograph of a flow-focusing microfluidic chip. White arrow indicates water flow, black arrows indicate oil flow. The nozzle diameter is 10 μm.

After the emulsion was obtained two organosilanes were added to it in order to encase droplets in silica shells [7]. Tetraethylorthosilicate (TEOS) was used as the main silica source, 3-aminopropyl trimethoxysilane (APTMOS) was chosen as an initiating agent for polycondensation reaction. After organosilanes diffused into water droplets TEOS and APTMOS became hydrolyzed and started silica shell polycondensation reaction. The reaction resulted in silica hollow particles with polar aqueous core. The particles were then examined with a Zeiss LSM 710 confocal laser scanning microscope.

3. Results

Normally BODIPY added to the oil phase remain evenly spread in it after formation of a water-in-oil emulsion (Figure 2).

Figure 2. Fluorescent micrograph of a water-in-oil emulsion with added BODIPY. Scale bar is 30 μm.

After the addition of organosilanes silica layer starts to grow around water droplets and microcapsules are formed. Micrographs of obtained silica capsules in bright field and BODIPY fluorescence channel are shown on Figure 3. Strong fluorescent signal from the microcapsules can now be seen despite BODIPY being initially dissolved in the oil surrounding the droplets and not inside them. This happens due to the dye with lecithin being absorbed onto the surface of silica microcapsules. High porous structure of silica allows it to absorb large amounts of the dye, resulting in strong fluorescence from the capsules. BODIPY in a water-in-oil emulsion does not exhibit such behaviour, proving that droplets in the emulsion were successfully covered in silica.

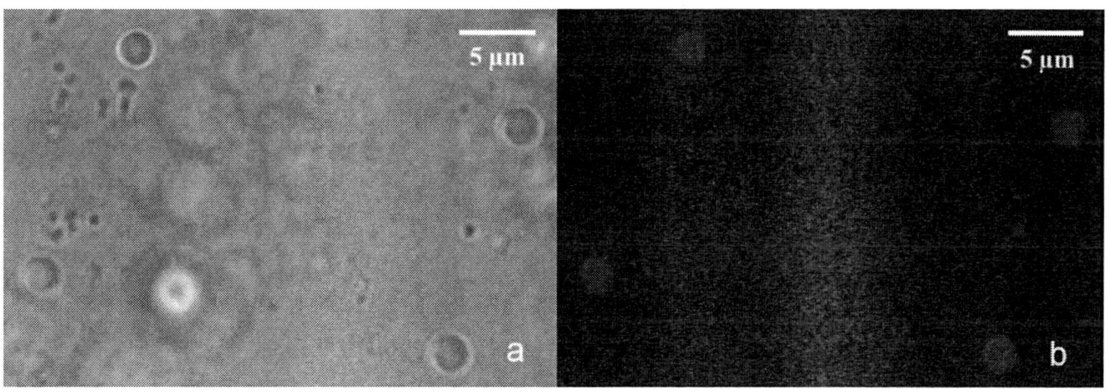

Figure 3. Silica microcapsules: a) bright field image; b) fluorescent image. Scale bar is 5 μm.

The capsules are shown to be about 2 μm in diameter which makes them small enough to be suitable for intracellular delivery of their cargo. Smaller capsules of less than 1 μm in diameter can also be seen which are likely formed out of satellite droplets that appeared during droplet formatting via jetting. Microfluidic method of initial droplets formation allows encapsulating wide range of water-soluble substances including genetic material without subjecting them to aggressive chemical environment.

Acknowledgements
MVZ thanks the Russian Science Foundation (grant No 17-19-01637).

References

[1] Zhao C 2013 Multiphase flow microfluidics for the production of single or multiple emulsions for drug delivery *Adv. Drug del. Rev.* **65** 1420–46
[2] Mintzer M A and Simanek E E 2009 Nonviral Vectors for Gene Delivery *Chem. Rev.* **109** 259–302
[3] Timin A S, Muslimov A R, Lepik K V, Epifanovskaya O S, Shakirova A I, Mock U, Riecken K, Okilova M V, Sergeev V S, Afanasyev B V, Fehse B and Sukhorukov G B 2018 Efficient gene editing via non-viral delivery of CRISPR–Cas9 system using polymeric and hybrid microcarriers *Nanomedicine: Nanotechnology, Biology and Medicine* **14** 97–108
[4] Zyuzin M V et al Comprehensive and Systematic Analysis of the Immunocompatibility of Polyelectrolyte Capsules *Bioconjugate Chem* **28** 556–64
[5] Sukhorukov G B, Volodkin D V, Günther A M, Petrov A I, Shenoy D B and Möhwald H 2004 Porous calcium carbonate microparticles as templates for encapsulation of bioactive compounds *J. Mater. Chem.* **14** 2073–81
[6] Filatov N A, Belousov K I, Kukhtevich I V, Bukatin A S and Evstrapov A A 2016 The study of mixing of reagents within a droplet in various designs of microfluidic chip *J. Phys. Conf. Ser.* **741** 012052
[7] Song L, Ge X, Wang M and Zhang Z 2006 Direct preparation of silica hollow spheres in a water in oil emulsion system: The effect of pH and viscosity *Journal of Non-Crystalline Solids* **352** 2230–5

Terahertz generation by periodic arrays of GaAs/AlGaAs core/shell nanowires

V N Trukhin[1,2,a], I A Mustafin[1,2], J P Kakko[3] and H Lipsanen[3]

[1]NRU ITMO, Saint-Petersburg 197101, Russia

[2]Ioffe Institute, St. Petersburg 194021, Russia

[3]Department of Micro- and Nanosciences, Micronova, Aalto University, P.O. Box 13500, FI-000a76 Aalto, Finland

[a]Corresponding author: valera.truchin@mail.ioffe

Abstract. This paper presents the experimental studies of the generation of terahertz radiation in periodic arrays of GaAs/AlGaAs core/shell nanowires via excitation by ultrashort optical pulses. It is demonstrated that terahertz emission occurs via excitation of photocurrent in the nanowires. The dynamics of photoinduced charge carriers is studied via the influence of an electron–hole plasma on terahertz radiation.

1. Introduction

Semiconductor nanostructures in the form of freestanding semiconductor nanowires are among the most promising 1D nano-objects for application in nanoelectronics, nanophotonics, and nanobioelectronics. In addition, the use of quasi-1D nanostructures, such as nanowires (NWs), as terahertz emitters is a promising way to improve existing emitters of terahertz radiation because of their unique electrical and optical properties [1-6]. First experimental studies of THz emission from InAs [4] and GaAs [5] NW arrays, excited by femtosecond laser pulses, demonstrated that the efficiency of terahertz (THz) generation can be increased by a factor of 15 when compared to bulk materials after accounting for the NW filling factor. Furthermore, since the efficiency of the THz emission is defined by the concentration and the mobility of charge carriers, studies of THz emission can be used to determine the transport, relaxation and recombination properties of electrons and holes in nanowires. In this work, we present the results of experimental studies of the generation of terahertz radiation in ordered arrays of GaAs/AlGaAs core/shell nanowires under excitation with ultrashort optical pulses.

2. Samples and experimental

The GaAs nanowires were grown on n-type GaAs (111) B substrates by metal-organic vapor-phase epitaxy (MOVPE) with the passivation of the surface of the nanocrystal with an AlGaAs layer. Fig. 1 shows 20^0 tilted scanning electron microscope (SEM) images of the fabricated GaAs NW arrays.

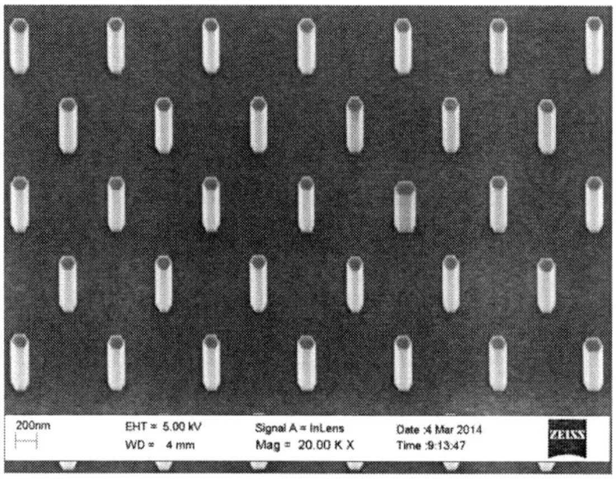

Fig.1. A typical SEM image of arrays with different distance (density\impact factor) between nanowires.

The main experiments were performed by using time-resolved spectroscopy: the amplitude of the electric field of a THz wave and the phase information of the THz pulse were recorded. A femtosecond Ti:Sa laser generating 15-fs pulses (central wavelength of the light pulse was 795 nm) served as the light source. The THz radiation was detected by electro-optical sampling.

3. Results

Investigations of THz generation processes in nanostructures based on GaAs/AlGaAs core/shell NWs were carried out. Waveforms of the THz pulse, obtained by exciting NWs with a diameter of ~ 150 nm and with different nanowire density, are shown in Fig.2. As already noted in [7], the field of terahertz radiation in periodic arrays of NWs with a distance between nanocrystals greater than the length of the exciting light (in this case ~ 800 nm) represents the sum of the fields of electromagnetic waves emitted by each nanocrystal. Correspondingly, the negative sign of the THz field for NWs' arrays with pitch of 1200 nm to 2100 nm indicates that the electron drift should be directed from the substrate towards the upper face of the nanocrystal. Since the contact field formed between the substrate and the nanocrystal is directed from the substrate to the upper face of the nanocrystal, the drift of the photoexcited electrons will be directed toward the substrate. Consequently, its contribution to THz generation should be negligible due to the experimental results (the amplitude of the terahertz pulse is negative). Thus, the obtained results confirm the band diagram shown in Fig. 2 on the right for the n-GaAs / n-AlGaAs heterojunction.

In this transition from the side of the GaAs nanocrystal there is an electric field directed toward the substrate. In this case, the photoexcited electrons will move in this field to the upper face of the NWs. The change in the sign of the THz pulse for NWs' arrays that have a period comparable to or shorter than the wavelength of the excited light indicates that for these samples the contribution of ambipolar diffusion that is associated with inhomogeneous excitation of the nanocrystal along its length (electrons move to the substrate) becomes substantial.

In this work, experiments were also conducted on the study of relaxation processes, recombination and transport of charge carriers in NWs by recording the waveform of the terahertz pulse with additional excitation of the NWs by a femtosecond pulse based on the optical-pump terahertz generation-probe time-domain spectroscopy method. To reduce the influence of THz pulse modulation generated by an additional femtosecond pulse (pump pulse), at the modulation frequency of the first femtosecond pulse (probe pulse), the pump pulse was directed toward the sample surface at an angle opposite to the dip angle of the probe pulse.

It was found that the efficiency of terahertz generation in NWs falls with the creation of an additional electron-hole plasma. The drop in the efficiency of generation of THz radiation in semiconductor nanostructures based on GaAs is mainly due to the screening of the internal field in the nanocrystal (the electric field of the n-GaAs / n-AlGaAs heterojunction near the upper face of the nanocrystal and the field of the n + -n junction), due to the separation of nonequilibrium electrons and holes, their drift in the contact fields. With a low level of excitation, the restoration of THz generation efficiency has a monoexponential character, whereas at a high excitation level the restoration of the efficiency of terahertz radiation generation becomes multi-exponential.

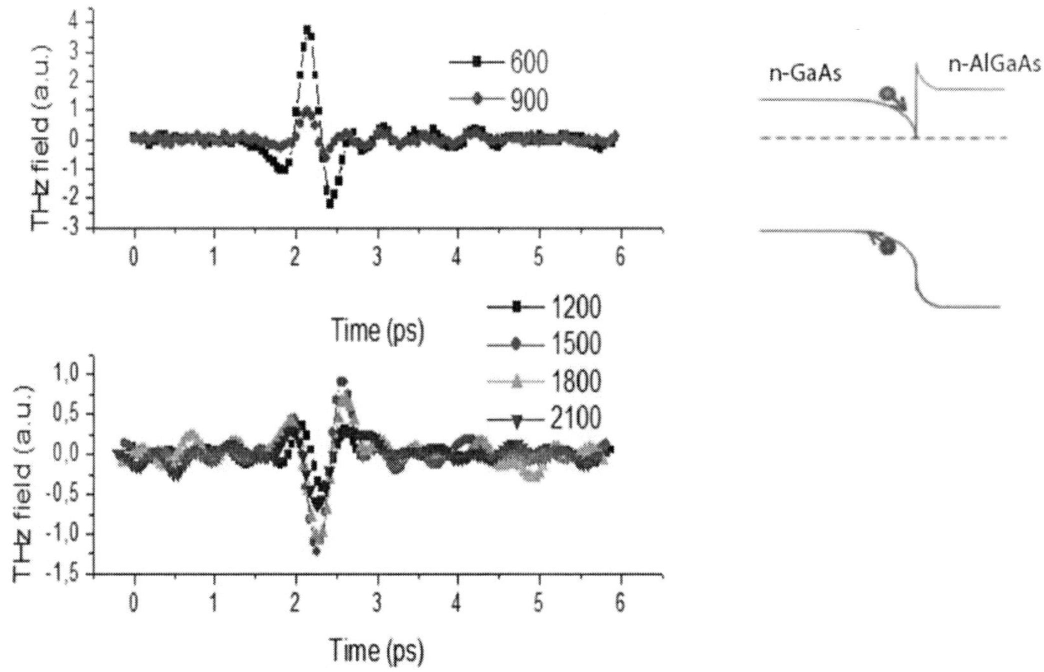

Fig.2. Waveforms of a THz pulse, obtained under excitation of the NW arrays with various densities and a NW diameter of 150 nm. The optical-excitation wavelength was 795 nm. On the right-the band diagram for the n-GaAs / n-AlGaAs heterojunction.

The observed phenomenon is due to the fact that at a high initial concentration of the electron-hole plasma, a gradient of the hole concentration is created due to their drift motion to the upper face of the NW, and then the charge of the depleted layer is recharged due to hole diffusion. The estimate of the hole diffusion time corresponds in order to the time of the fast recovery kinetics observed in the experiment, and subsequently, mainly, hole trapping at the surface levels occupied by electrons.

Thus, it was shown that the time dynamics of photoexcited charge carriers in semiconductor NWs is determined both by the rapid motion of electrons in a local electric field (charge of a depleted layer capacitance) and their rapid capture to surface centers, by diffusion of holes (charge-exchange of capacitance) and their slow capture to surface centers and nonradiative centers in NWs.

Acknowledgments

The work was partially supported by the Russian Foundation for Basic Research (Grant 18-51-16002) and the Program of the Presidium of the Russian Academy of Sciences.

References

[1] Reid M, Cravetchi I V, Fedosejevs R, Tiginyanu I M and Sirbu L 2005 *Appl. Phys. Lett.* **86** 021904
[2] He S, Chen X, Wu X. Wang G and Zhao F 2008 *J. Lightwave Technol.* **26** 1519
[3] Hoyer P, Theuer M, Beigang R and Kley E-B 2008 *Appl. Phys. Lett.* **93** 091106
[4] Seletskiy D V, Hasselbeck M P, Cederberg J G, Katzenmeyer A, Toimil-Molares M E, Leonard F, Talin A A and Sheik-Bahae M 2011 *Phys. Rev.* B **84**, 115421
[5] Trukhin V N, Buyskih A S, Buravlev A D, Cirlin G E, Horkov D P, Samoilov L L and Samsonenko Yu B 2012 *Proceedings of the 37th International Conference on Infrared, Millimeter, and Terahertz Waves IRMMW-THz 2012* doi: 10.1109/IRMMWTHz.2012.6380432
[6] Trukhin V N, Buyskikh A S, et al, 2013 *Appl.Phys. Lett.* **103** 072108
[7] Trukhin V N, Bouravleuv A D, Mustafin I A, et al 2015 *Appl.Phys.Lett.* **106** 252104

Direction of Arrival Estimation for Retrodirective Rotman Lens Antenna Array

P A Turalchuk

Microelectronics and Radio Engineering Department,
St. Petersburg Electrotechnical University "LETI", Saint Petersburg, Russia

pavel.turalchuk@mwlab.spb.ru

Abstract. A retrodirective array based on simplified direction-of-arrival estimation algorithm is presented. A beamforming network based on Rotman lens providing the direction of arrival estimation as well as the beam steering is proposed. Retrodirectivity is provided by analysing an interrogating pilot signal, which allows retrieving a data to estimate the source position and then steering a beam in response of a pilot signal. The proposed retrodirective array test system was fabricated and measured for operation at 5.85 GHz central frequency. Designed retrodirective Rotman lens-based antenna array forms eight beams with beamwidth of 10° to cover the scanning angular range from -40° to 40°. The evaluation of the direction of arrival by measuring the power level of the received signal depending on the port index for the different angle of incidence of the pilot signal source is experimentally demonstrated.

1. Introduction

As it is known, retrodirective antenna arrays reflect an incident wave towards the source without any prior information about its location. Such systems are widely used in radar, satellite, wireless communications, radio-frequency-identification (RFID), tracking systems, microwave power transmission applications etc. A retrodirective array is composed by analog circuitry that steer their beams automatically without any phase shifters or complex digital-signal-processing components in response to an interrogating signal [1].

However, well known direction of arrival (DoA) estimation algorithms could bring to the retrodirective arrays more flexibility. For example, it would allow implementing gain calibration, DoA correction techniques [2]. As a result, such algorithms are often used in wireless communications to increase the capacity and throughput of a network [3]. Typically, they are based on time-of-arrival, received signal strength, time-difference-of-arrival, or angle-of-arrival measurements determined from the mobile terminal signals received at the base stations [4].

This paper is focused on design of the retrodirective array based on simplified DoA estimation in order to steer the beam towards the location of electromagnetic source. The proposed architecture is based on a Rotman lens, which is used as a beamforming network of an antenna array to provide the DoA estimation as well as the beam steering.

2. Retrodirective Array Architecture

A general block diagram of the proposed retrodirective Rotman lens-based antenna array with DoA estimation is presented in Fig. 1. It consists of an antenna array, a Rotman lens, a RF switch, an amplitude power detector and a microcontroller, which is used to implement control of the system and perform a DoA estimation algorithm.

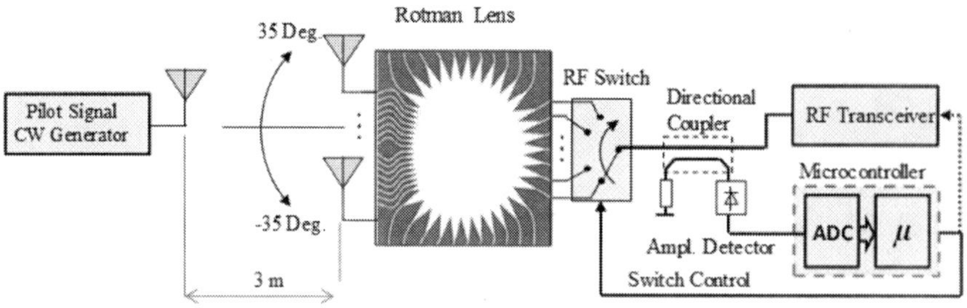

Figure 1. Block diagram of the retrodirective Rotman lens-based antenna array with DoA estimation.

The operational principle of the proposed architecture is based on receiving an interrogating pilot signal, which is used to retrieve the information about source location. It was decided to use a *Time-Division Duplexing* (TDD) mode because it has some advantages for this specific application. First of all, there is asymmetry of the uplink and downlink data rates, because for a DoA estimation algorithm a simple pilot signal is sufficient, while during the data transfer other signals are preferable for higher data rates. The Rotman lens (RL), which has M input (beam) ports and N output (array) ports, is using as a beamforming network (BFN) for a linear antenna array. In a transmitting mode, an individual beam port is fed that results in excitation of antenna elements with a certain phase distribution, defined by a position of the beam port relatively to the lens symmetry axis. The beam steering is realized by switching between different beam ports. An ideal Rotman lens is a reciprocal device. It means that in the receiving mode, an incident wave, which arrives with time delays across the linear array, is summarized in phase mainly at one of the beam ports. This fact can be efficiently used for a passive method of DoA estimation.

So, during the receiving time slot, a microcontroller performs a sequential switching between its M output ports for measuring the signal power levels at all beam ports using an amplitude power detector, which is connected to the switch through the coupled port of a 10-dB directional coupler, while the through port of the coupler is connected to a transmit chain. All measured values are analyzed by a microcontroller and the DoA of the received signal is estimated. In the simplest scenario, the DoA is defined by an index of the beam port, where the highest power level is detected. Based on the algorithm output, the microcontroller sets up the required state of the switch and during the transmitting time slot, the antenna array transmits the signal in the required direction.

3. Experimental Results

To demonstrate the DoA estimation operation of the proposed retrodirective array a test system was fabricated and measured for operation at 5.85 GHz central frequency. The antenna array consist of 8 series-fed 1×8 linear patch arrays, where a beam steering was realized in azimuth plane (Fig. 2a). Such aperture dimensions provide the main beam width of 10°. The antenna array was fabricated on a Rogers RO4003 substrate ($\varepsilon_r = 3.55$, $h = 1.52$ mm). The size of 8 × 8 patch sub-array was 220 × 250 mm^2. According to the measured reflection coefficient its bandwidth covers the range from 5.5 to 6 GHz.

The Rotman lens was designed for scanning in the range [-40°; 40°] with a step of 10° using Remcom Rotman Lens Designer (RLD) tool [5] (Fig. 2b). It includes 8 beam ports, 8 array ports and 16 dummy ports, which were introduced to reduce the amount of multipath interference within the lens. Dummy ports are loaded by 50 Ohms SMD chip resistors. The prototype Rotman lens was fabricated on the Arlon AD1000 substrate ($\varepsilon_r = 10.2$, $h = 1.27$ mm). The dimensions of the lens are 260 × 280 mm^2

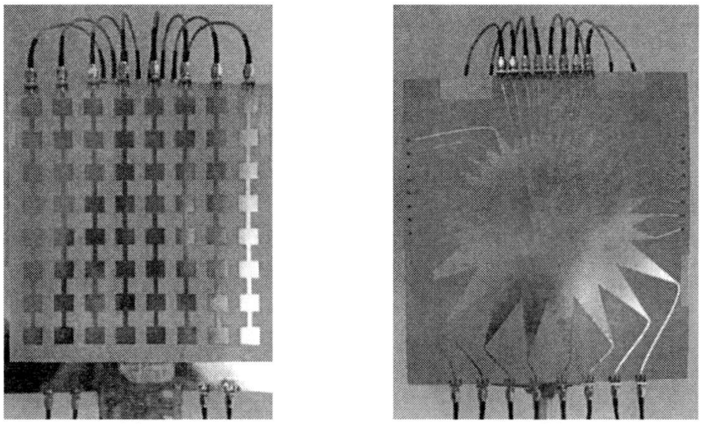

(a) (b)

Figure 2. Fabricated antenna array (a) and Rotman lens (b) composing the retrodirective Rotman lens-based antenna array.

The scattering parameters of the Rotman lens were measured and used for simulating radiation patterns. The normalized radiation pattern simulated at 5.85 GHz based on measured Rotman lens S-matrix is shown in Fig. 3a. As one can see, eight formed beams with beamwidth around 10° cover the angular sector from -40° to 40°. The gain of the antenna arrays drops by 2.5 dBi for the outermost direction of the beam.

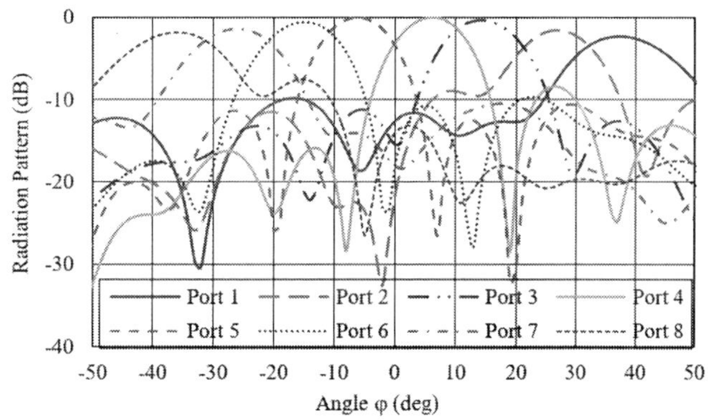

(a)

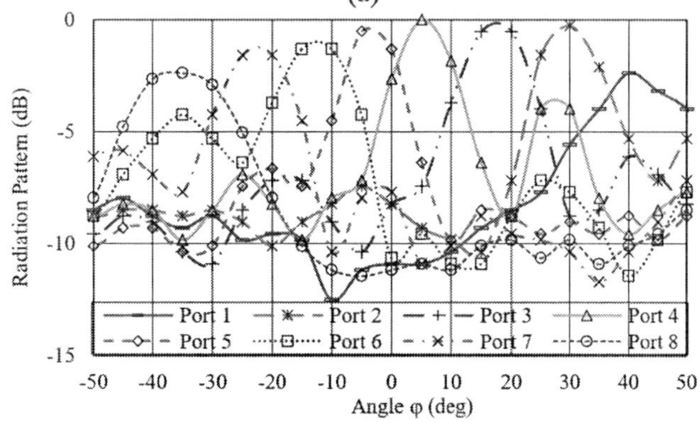

(b)

Figure 3. Normalized radiation pattern simulated based on the measured Rotman lens S-matrix data (a) and measured received signal power level (b).

The most critical part of the system is DoA estimation of an interrogating pilot signal because it has a direct impact on the direction of the transmitted signal afterwards. Thus, the retrodirective array test system presented in Fig. 1 was experimentally studied to evaluate DoA of an incident pilot signal. The continuous wave (CW) used as the pilot signal was radiated by a single 1×8 linear patch array. The distance between the retrodirective array and the pilot signal antenna is fixed at three meters. The azimuth angle of the Rotman lens-based antenna array was varied from -50° to 50°. For each position of an antenna array, the microcontroller changed the state of the *SP8T* HMC321LP4 switch [6] and recorded the voltage level at the amplitude detector ADL5906 [6] output every 150 msec. Thus, each receiving cycle was a little bit more than 1 sec. The absolute power is calculated using the measured dependence of the voltage level on the input power level for specified amplitude detector. As a result, the normalized values of the detected power at each beam port are presented in Fig. 3b as a function of the azimuth angle φ.

As can be observed, when the signal is coming from the specific direction the power level at a certain beam port is higher in comparison with power level at other ports by 3 - 6 dB. This dynamic range is sufficient to detect the received signal and estimate DoA by comparison the power levels. However, the dynamic range tends to decrease at the edges of the incident wave angular ranges. In order to achieve a high angular resolution between beams, a number of beam ports of the Rotman lens should be increased.

In general, Rotman lens array provides multiple beams toward different directions simultaneously. So, the functionality as well as operation rapidity of the system can be improved if M single pole single throw (*SPST*) switches will be used instead of one single pole M-throw switch. Therefore, several beam ports power level can be analyzed and fed simultaneously making the retrodirective array more flexible in the case two or more signals are coming from different directions.

4. Conclusion

In this work, the retrodirective array based on Rotman lens to steer beams towards the pilot signal sources is proposed. The simplified direction-of-arrival estimation circuitry allows extending a functionality of the proposed retrodirective array. The Rotman lens array antenna is designed to provide an 80-degree angular sector coverage. The evaluation of the direction of arrival by measuring the power level of the received signal depending on the port index for the different angle of incidence of the pilot signal source is experimentally demonstrated. The results of this study demonstrate that the Rotman lens can separate incident waves coming from different discrete directions providing a sufficient dynamic range to perform DoA evaluation. To achieve a high angular resolution between beams, beam ports number of the Rotman lens could be increased.

Acknowledgment

The work was supported by Russian Science Foundation (project 17-79-20374).

References

[1] Miyamoto, R.Y. and Itoh, T.: 'Retrodirective arrays for wireless communications', *IEEE Microwave Magazine*, 2002, **3** (1), pp.71-79.

[2] Chiang P. C., Liao W. J., Tu Y. T. and Liu H. C.: 'Implementation of direction-of-arrival estimation using Rotman lens array antenna', *International Symposium on Electromagnetic Theory*, Hiroshima, 2013, pp. 855-858.

[3] Zhao Y.: 'Mobile phone location determination and its impact on intelligent transport systems', *IEEE Trans. Intell. Transport. Syst.*, 2000, **1**, pp. 55–64.

[4] Cheung, K.W., So, H.C., Ma, W.K. and Chan, Y.T.: 'Least squares algorithms for time-of-arrival-based mobile location', *IEEE Trans. on Signal Process.*, 2004, **52**, (4), pp.1121-1130.

[5] Penney C. W.: 'Rotman lens design and simulation in software', *IEEE Microwave Mag.*, 2008, **9** (6), pp. 138- 149.

[6] Analog Devices: http://www.analog.com.

High index dielectric metasurfaces and colloidal solutions: from fabrication to application

Ruggero Verre,* Nils Odebo-Länk, Daniel Andrén, Hana Šípová, and Mikael Käll

Department of Physics, Chalmers University of Technology, 412 96 Göteborg, Sweden

E-mail: ruggero.verre@chalmers.se

Abstract. High index dielectric nanoparticles and meta-materials have been proposed for many different applications, including light harvesting, sensing and metalenses. However, widespread utilization in practice also requires large-scale fabrication methods able to produce homogeneous structures with engineered optical properties in a cost effective manner. Here, it is presented a facile fabrication method for silicon nanoparticles which is scalable to 4-inch wafers and can produce a wide range of nanoparticle shapes on demand. We also show that the fabricated nanoparticles can be detached from their support using a simple substrate removal technique and then transferred to colloidal suspension. We will finally discuss some uses of the fabricated systems. For the metasurfaces, we will demonstrate complete absorption due to far field interference effects. For the nanoparticles colloids we will show the possibility of realizing an intrinsically chiral structure composed of a low-loss dielectric resonator and we will study optical trapping phenomena for different particle sizes and shapes.

1. Introduction

Electromagnetic (EM) metasurfaces are structured material layers with subwavelength thickness that appears smooth to an impinging EM field because the feature spacing is small compared to the incident wavelength. In particular, the use of high-index dielectric (HID) nanoparticles as metasurface building blocks, or meta-atoms, allows for extremely flexible control of the optical properties by engineering their morphology. Similar to their metallic counterparts, HID nanoparticles support multiple resonances[1] and are able to efficiently convert propagating fields into sub-wavelength volumes. HID nanoparticles also have additional advantages, such as thermal stability, reduced Ohmic losses and CMOS fabrication compatibility. These features render HID metasurfaces highly interesting for applications like wavefront manipulation, light managment, directional antennas, light harvesting, ultra-fast optical switching and molecular sensing, to name a few.[2]

The diverse pool of potential applications highlights the need for flexible and cost-effective fabrication methods that would enable realization of large-scale nanostructured HID materials with well-defined physical and optical properties. Many of the available techniques have their limitation: lithographic methods cannot cover large areas and the features are generally planar. On the other hand, wet chemistry synthesis of HID nanoparticles is far from being as mature as for metal nanoparticles, several methods are capable of producing high quality spherical silicon particles with a range of sizes have been recently reported.

In this document, we first demonstrate a method for large-scale fabrication of different silicon nanostructures using a modified version of the hole-mask colloidal lithography (HCL) technique. Afterwards, we present a method for removing such Si particles from a supporting substrate and used it to realize colloidal HID solutions. We show that this methodology can be used to detach and dissolve nanoparticles of diverse shapes ranging from disks to chiral particles. In the final part of the document, we will discuss some applications of the different fabricated systems.

2. Metasurface fabrication.

Figure 1a shows the basic principle for the fabrication of HID metasurfaces. A sacrificial layer, for example a resist or a SiO2 film, is formed by spin coating or deposition, on top of a deposited Si film. Polystyrene beads are then dispersed on the substrate. After a thin metal film evaporation and tape stripping, a hole mask is formed. The sacrificial layer is then etched through the holes in the mask, followed by evaporation of hard mask (Ni or Cr) nanoparticles of a chosen shape. These hard mask particles can be transferred to the poly-Si film underneath by anisotropic etching.

The method described can be used to produce homogeneous metasurfaces over 4-inch wafers, as shown in Figure 1b. Most importantly, it is also highly versatile and capable of fabrication of nanoparticles of different shapes simply by changing the conditions for deposition of the Ni mask as shown in Figure 1c and d.

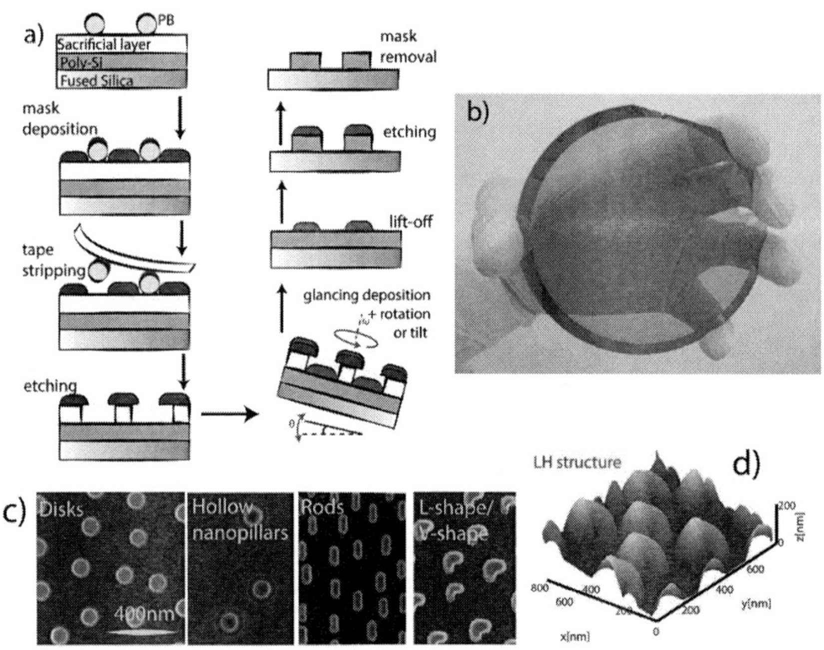

Figure 1. a) Schematic of HID metasurfaces fabrication. b) image of a 4-inch substrate covered with Si nanodisks. c) SEM images of different HID metasurfaces. d) Atomic force microscope image of a metasurface composed of 3D chiral Si nanocrescents.

3. Colloidal solutions

The fabricated nanoparticles can be removed from the substrate to realize colloidal. The as-fabricated metasurface is first dipped into a HF solution (5%) to selectively etch the substrate. Subsequently, diluted HF (0.1%) is drop-casted on the metasurface until the particles disperse into the solution. The HF solution is then left to evaporate. Finally, the particles are dispersed by sonication in an aqueous solution of cetrimonium bromide (CTAB, typically 1-5mM). The colloidal solution is stable for several months. Note that this method can in principle be adapted to substrates prepared by any large-scale fabrication method. The optical extinction measurements of the metasurface (b) and colloidal solution (d) composed of nanodisks, reveal the presence of tunable multiple resonances, typical of HID nanostructures.

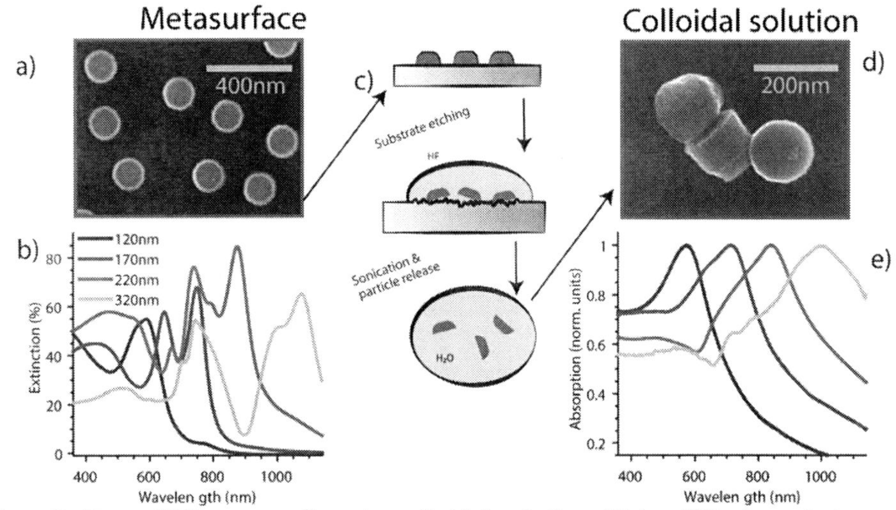

Figure2. From HID metasurface to colloidal solution. Using HF, the substrate supporting the Si nanoparticles can be etched. After HF dilution and evaporation, the particles can be released in solution by sonication, resulting in a colloidal suspension of HID nanoparticles. SEM image of a fabricated metasurface composed of Si nanodisks (a) and after redispersion onto a Si chip (d). The optical extinction spectra for nanodisks of different diameters on a substrate and in solution are shown in Figure b and e respectively.

4. Applications of the different systems

Figure 3 shows an overview of some of the applications that can be studied with the aforementioned fabrication methods. In figure 3a it is shown that a silicon metasurface excited in a total internal reflection configuration can absorb at least 97% of incident near-infrared light due to interferences between coherent electric and magnetic dipole scattering from the silicon nanopillars that build up the metasurface and the reflected wave from the supporting glass substrate. This "near perfect" absorption phenomenon loads more than 50 times more light energy into the semiconductor than what would be the case for a uniform silicon sheet of equal surface density, and it is irrespective of incident polarization.

Figure3b shows an optical trapping experiment using a solution composed of chiral colloidal HID nanoparticles. We found that the dispersed Si nanocrescents could be readily manipulated at the single particle level in solution using optical tweezers.[31] We demonstrate how a trapped nanoparticle can be moved across a surface using a power of the order of 50 mW at the sample. We also estimated the trapping stability based only on the probability density function of the average displacement.

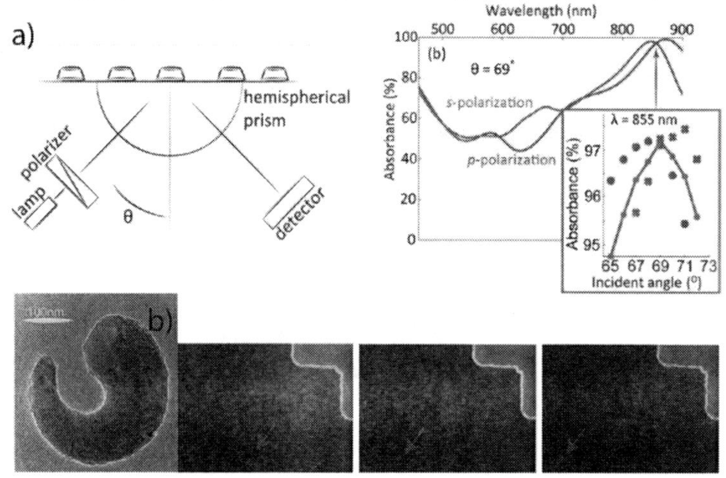

Figure3. Applications of HID nanostructures. a) Total internal absorption. A metasurface composed of nanodisks is illuminated in total internal reflection (left). By playing with the interference between the different modes, absorption can be boosted and it can reach up to 97% (right), irrespective to the polarization of the illuminated beam. b) Optical trapping of chiral nanocrescents. Using a 1064nm laser, chiral crescents (see left of a micrograph image) dispersed in solution can be trapped in 3D and easily manipulated. Selected images of a trapping experiment are shown on the right. The top right feature is a reference structure and the particle is indicated with an arrow in each movie frame.

5. Conclusions

We have demonstrated a method for efficient fabrication of large-scale metasurfaces composed of HID nanoresonators. Nanostructures of different shapes and sizes can be realized by modifying the process parameters. If desired, the nanostructures can be dispersed in a colloidal solution simply by selectively etching the substrate and then removing the particles by sonication. We present some applications for the two systems, namely perfect total internal absorption and the possibility of 3D optical trapping of such chiral nanostructures. Finally, the availability of tunable colloidal HID nanoparticles of non-spherical shapes opens wide avenues for further research applications. Similar to the case of colloidal plasmonic nanoparticles, we expect that many opportunities can be realized in the fields of bionanotechnology and nanomedicine.

These results are a collection of this previously published work:

- *Odebo Länk, et al. Nano Letter, 2017*
- *Verre et al., Advanced Materials, 2017*
- *Verre et al., Advanced Optical Materials, 2018*

References

[1] A. B. Evlyukhin, C. Reinhardt, B. N. Chichkov. *Phys. Rev. B* **2011,** *84*, 235429.

[2] A. I. Kuznetsov, A. E. Miroshnichenko, M. L. Brongersma, Y. S. Kivshar, B. Luk'yanchuk. *Science* **2016,** *354*.

Fast and unconditionally safe in vivo MR head protocol for home-made coil prototype assessment at 7T

Alexandre Vignaud[1], Franck Mauconduit[2], Vincent Gras[1], Nicolas Boulant[1], Olivier Girard[3], Alexander Raaijmakers[4], Frank Kober[3], Denis Le Bihan[1], Redha Abdeddaim[5]

[1]CEA, NeuroSpin, UNIRS & Université Paris-Saclay, Gif-Sur-Yvette, France ; [2]Siemens Healthineers, Saint-Denis, France; [3]CEMEREM, Université Aix-Marseille, Marseille, France; [4]University Medical Center, Department of Radiology, Utrecht, The Netherlands/Eindhoven University of Technology, Department of Biomedical Engineering, Eindhoven, The Netherlands; [5]Aix Marseille Univ, CNRS, Centrale Marseille, Institut Fresnel, Marseille, France

alexandre.vignaud@cea.fr

Abstract. Depending on the local IRB (Internal Regulation Board) regulations for safety reasons, homemade RF coil prototype assessment through an in vivo experiment can be painful and lengthy administrative process. It includes to document simulations and experimental validations on phantom before being able to proceed. The situation can be even worse, if for some reasons, once it has passed all the acceptance stages, the coil does not deliver as expected in vivo. The process to maybe then rebooted.

In this work, we introduce the concept of unconditional safe MR protocol allowing to safely use homemade coil in vivo, at any step of the RF coil development, securing valuable information for the developer, and making sure that neither local nor global SAR limits will be ever reached anywhere in the organ to image. The protocol includes in particular a fast B1+ mapping, which is essential to assess coil behaviour. The strategy can be easily extended to more contrasts, other organs and other magnetic field strengths.

1. Introduction

Ultra-High-Field (UHF) Magnetic Resonance Imaging (MRI) scanners hold great promises for clinical and neuroscientific researches. RF coil technologies, at the state of the art, have had hard time to deliver all the expectations put in UHF. The first reason, why they failed, is certainly the heterogeneous excitation of the nuclear spins obtained with a classic coil design, which typically leads to shadows across the human brain images, making the impacted regions non-exploitable. Parallel Transmission (pTx) is the dominant strategy to solve the issue, but such system holds also some weaknesses including long calibration time and a relative inefficiency because of strong coupling between coils [1,2]. Attempts to mitigate the B1+ field has also been done including in the RF coil high dielectric constant materials [3] or metamaterials [4,5]. Whatever the chosen strategy, RF coil delivers energy inside human body to excite the nuclear spins that will be used to form an image. This energy is also partially absorbed by the biological tissues and translates into a rise of body temperature. Therefore, it is necessary to be sure that globally and locally temperature stays within safe limitations. Because it is not possible to measure the temperature in real-time during an MR examination, regulation relies on Specific Absorption Ratio (SAR) limitations. The norm is based on IEC 60601-2-33 recommendations [6]. This question is specially scrutinized at UHF because local SAR can be reached faster than global SAR promoting insidious hotspots in the imaged organ [7]. Depending on the local IRB (Internal Regulation Board) regulations, homemade RF coil prototype assessment though an in vivo experiment can be painful and lengthy administrative process. It includes to document simulations and experimental validations on phantom. It relies on very well engineered processes which warrant that the SAR limitations are never exceeded [8,9] with any RF excitation scenarios for a given RF coil.

In this work, we propose unconditional MR protocol allowing to safely use home-made coil in vivo, at any step of the RF coil development, securing valuable information for the developer, and making sure that neither local nor global SAR limits will be ever reached. The idea is to warrant safety regardless of the coil to be used by adapting sequence parameters to deliver ultra-low averaged power. Within this mode if the whole RF power would focus in a single 10g of biological tissue, it would not exceed the regulatory global and local SAR limits even for this piece of tissue. Hence, the goal is to identify MR sequence parameters that could be played with such low input power and still deliver valuable MR images within a decent acquisition time (TA). In such conditions, it becomes possible to acquire MR image in vivo without any prior coil validations.

2. Materials and Methods

Demonstration has been carried on a Magnetom 7T MRI scanner (Siemens Healthineers, Erlangen, Germany) equipped with a SC72 whole body gradient (max gradient amplitude 100mT/m, and slewrate 200 T/m/s) with a standard transceiver birdcage (Invivo Corp, Gainesville, FL, USA). In vivo acquisitions have been done under local IRB rules on one healthy volunteer.

The calculation of the relevant sequence parameters has been done based on the fact that the calibration of the reference pulse (REF), a 500us boxcar 90° pulse, would require the full power available at the port of the coil (e.g. 8x1kW for configuration we used) which is again very restrictive because the coil would be excessively inefficient in this situation. Four sequences have been identified that can be played within 7min as described in the Table for single channel system. For pTx system, 1min per additional channel is required to monitor B1+. The so-called restricted SAR mode protocol relied on gradient recalled echo (GRE) for B0 shimming, XFL [10] for B1+ mapping and localizer and Echo-Planar (EPI) for Eddy current investigations.

3. Results

In Table, the sequences parameters are displayed with an estimation of the worst global and local SAR possible while using the worst possible REF. It is well below the limit even in the case where all the power would focus inside 10g of tissue. On Figure, the optimal setup of the XFL and the restricted mode version shows a very good agreement.

Table: *Parameters of the identified "restricted SAR mode" sequences. They have been validated on phantom and in vivo using a single channel volume coil to ensure to be sensitive enough to get reliable information on coil prototype behaviors.*

Goal	**localizer**	**B0 map**	**B1+ map**	**EPI**
Sequence type	**2D XFL (only ref scan)** [2]	**2D GRE**	**2D XFL**	**2D EPI**
Exc RF pulse type	Sinc	Sinc	Sinc	Sinc
Duration (us)	1000	1000	1000	1000
FA (°)	3	5	3	10
RF saturation	-	-	RECT	-
Sat duration (us)	-	-	1000	-
Sat FA (°)	-	-	60	-
REF Voltage (V)	470	470	470	470
TR (ms)	20000	400	20000	10000
Ny	128	64	64	64
Multislice mode	Sequential	Sequential	Sequential	Interleaved
Spatial resolution (mm^3)	2x2x4	4x4x4	4x4x4	4x4x4
Slice number	3	5	3	3
TA	**2min**	**2min**	**2min**	**10s**
SAR_{local} (W/kg) - 6min [10s]	**6 [12]**	**6.6 [6.6]**	**3.6 [8.3]**	**1.63 [1.63]**
SAR_{global} (W/kg) - 6min [10s]	**0.012 [0.024]**	**0.013 [0.013]**	**0.007 [0.016]**	**0.033 [0.033]**

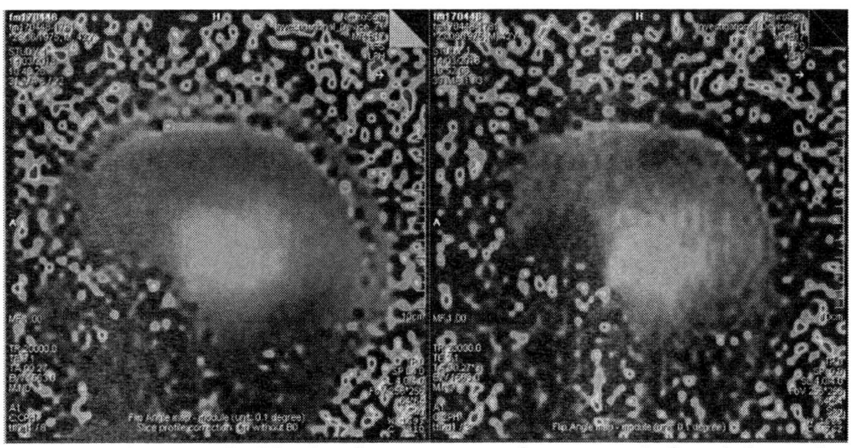

Figure : *In vivo sagittal views of a B1+ head maps acquired respectively from left to right with optimal XFL protocol and with "Restricted SAR mode" setting with InVivo Corp transceiver birdcage coil at the same resolution. The restricted SAR mode version brings SAR estimation toward zero. If all power focused into 10g of tissue, local SAR would only achieve 3.6W/kg.*

4. Discussion and conclusions

We successfully demonstrated the possibility to acquire reliable information using sequence powering very limited RF deposition with a short TA making it compatible with in vivo acquisition. Using this protocol, no prior simulation of any kind is needed but no change in the sequence parameters is allowed during the examination. The sequences should therefore be compiled into a "Restricted SAR mode" version to lock the sequence parameters.

More contrasts could be considered in this mode especially T1 using Turbo FLASH (TFL) like sequence and T2* weighted with GRE sequence. T2 might be more challenging in a decent time. The demonstration was done on purpose with a transceiver head birdcage, which is known to be weak in Signal to Noise Ratio. Thus, any tests on state of the art receiving coil array should return even better results.

References

[1] Katscher U., Börnert, P., Parallel RF transmission in MRI. NMR Biomed 2006; 19: 393–400

[2] Grissom W., Yip, C., Zhang, Z., Stenger, V.A., Fessler, J.A., Noll, D.C., Spatial domain method for the design of RF pulses in multicoil parallel excitation. Magn Reson Med 2006; 56: 620–629

[3] Webb A.G. Dielectric materials in magnetic resonance. Concept Magn Reson A 2011; 38: 148-184

[4] Slobozhanyuk A.P., Poddubny A.N., Raaijmakers A.J., van den Berg C.A., Kozachenko A.V., Dubrovina I.A., Melchakova I.V., Kivshar Y.S., Belov P.A. Enhancement of Magnetic Resonance Imaging with Metasurfaces. Adv Mater. 2016 Mar 2;28(9):1832-8.

[5] Marc Dubois, Lisa Leroi, Zo Raolison, Redha Abdeddaim, Tryfon Antonakakis, Julien de Rosny, Alexandre Vignaud, Pierre Sabouroux, Elodie Georget, Benoit Larrat, Gerard Tayeb, Nicolas Bonod, Alexis Amadon, Franck Mauconduit, Cyril Poupon, Denis Le Bihan, Stefan Enoch. Kerker Effect in Ultra High Field Magnetic Resonance Imaging. Nature Communication 2018 ***under review***

[6] International Electrotechnical Commission Medical electrical equipment-part 2–33: particular requirements for the basic safety and essential performance of magnetic resonance equipment for medical diagnosis - 3rd ed. Geneva: International Electrotechnical Commission: 2010- 2–33 :601

[7] Seifert, F., Wubbeler, G., Junge, S., Ittermann, B., Rinneberg, H., Patient safety concept for multichannel transmit coils. J. Magn. Reson. Imaging 2007; 26: 1315–1321

[8] N. Boulant, V. Gras, A. Amadon, M. Luong, G. Ferrand, A. Vignaud. Workflow proposal for defining SAR safety margins in parallel transmission. Proc Int Soc Magn Reson Med 2018;

[9] E.F. Meliado, A.J.E. Raaijmakers, P. Luijten, C.A.T. van den Berg, Magn Reson Med 2018 ***under review***

[10] Klose U. Mapping of the radio frequency magnetic field with a MR snapshot FLASH technique. Medical Physics 1992; 19: 1099–1104

Acknowledgements:

The project leading to this publication has received funding through the M-CUBE project by the European Unions Horizon 2020 Research and Innovation programme under Grant Agreement No 736937 and by France Life Imaging grant ANR-11-INBS-0006.
This work has also received funding support from the ERPT equipment program of the Leducq Foundation and DIM "Cerveau et pensée" from Région Ile-de-France.

IOP Conf. Series: Journal of Physics: Conf. Series **1092** (2018) 012160 doi:10.1088/1742-6596/1092/1/012160

Luminescent nanodiamonds as a material platform for production of single photon emitters

I I Vlasov[1], E A Ekimov[2], O S Kudryavtsev[1], O A Shenderova[3], M Metsch[4], and F Jelezko[4]

[1] Prokhorov General Physics Institute, Russian Academy of Sciences, Moscow, 119991, Russia
[2] Institute for High Pressure Physics, Russian Academy of Sciences, Troitsk, Moscow, 142190, Russia
[3] Adamas Nanotechnologies, Inc., Raleigh, North Carolina, 27617, United States
[4] Institute for Quantum Optics, University of Ulm, Ulm, 89081, Germany

E-mail: vlasov@nsc.gpi.ru

Abstract. The prospects of using individual nanodiamond particles as a source of single photons were studied for diamonds of various origin. It was shown that the single photon emitters could be produced on a base of "nitrogen-vacancy" centers in detonation nanodiamond synthesized from graphite-hexogen mixture. Studying meteoritic NDs we have found that single photon emitters based on "silicon-vacancy" centers could be realized even in molecular-sized diamond (less than 2 nm). Currently we are developing a new class of diamond materials, nanodiamonds synthesized from organic compounds at high pressure. In particular, we have demonstrated the possibility of synthesis of nanodiamond from its molecular analogue adamantane, controlling the size of crystallites in a wide range by changing synthesis temperature. A minimum size of obtained diamond particles about 3 nm was reached. Crystal structure of such diamonds is the most perfect in comparison with other classes of diamond materials, therefore they are the most promising material platform for production of single photon emitters.

1. Introduction

Single photon emitters (SPEs) are one of the key components in quantum communication technology. For today the most attractive SPEs, which operate at room temperature, are dye molecules, quantum dots, and color centers in wide-bandgap materials (diamond, silicon carbide and others). Owing to a longest photostability and a high quantum yield of luminescence, diamond with different color centers is the best candidate for practical implementation of single-photon devices.

This paper presents our results on studying luminescent nanodiamonds of various origin in terms of their applicability to be used as SPE, as well as on searching for smallest diamond crystallite capable to host active luminescent centers.

2. Results and discussion

2.1. Detonation Nanodiamonds (DNDs)

This material demonstrates a high potential for the production of SPE. It is produced at an industrial scale and contains large amount of N impurity, which is required for formation of luminescent "nitrogen-vacancy" (NV) centers. We have studied representative classes of DNDs produced from

IOP Conf. Series: Journal of Physics: Conf. Series **1092** (2018) 012160 doi:10.1088/1742-6596/1092/1/012160

different combinations of explosives and having various characteristic sizes of diamond nanoparticles. We have demonstrated, that (1) optically active NV centers are mainly detected in diamond particles larger than 30 nm in size; (2) DNDs synthesized from a mixture of graphite/explosive contain the largest amount of NV centers formed during the synthesis. Through autocorrelation function measurements the greatest number of crystallites containing single NV centers was found in detonation nanodiamonds synthesized from graphite-hexogen mixture [1]. Thus, this class of nanodiamond was concluded to be the best candidate among others to produce SPE based on DND.

2.2. Meteorite Nanodiamonds

Production of SPE was seen elusive for very small diamond nanoparticles because of their limited availability and a lack of fundamental understanding of impurity stability in such nanostructures. We have shown that isolated diamond nanoparticles with size below 2 nm, extracted from stone meteorites, are still capable of housing photoluminescent "silicon-vacancy" (SiV) colour centers [2]. Surprisingly, fluorescence from SiVs was stable (no bleaching) over time, and a few or only one colour centers were found per a diamond crystallite.

2.3. HPHT nanodiamonds

Currently we are developing a new class of diamond materials, nano- and micro-diamonds synthesized from organic compounds at high pressure and high temperature (HPHT technique). In particular, we have demonstrated the possibility of synthesis of nanodiamond from its molecular analogue adamantane, controlling the size of crystallites in a wide range by changing the synthesis temperature. A minimum size of synthesized diamond crystals about 3 nm was reached [3]. The possible presence of single NV centers in individual diamond particles, dispersed on a glass substrate was investigated by scanning confocal microscope and Hanbury-Brown and Twiss interferometer for two samples synthesized from adamantane at pressure of 8-9 GPa and temperatures of 1600 K and 1800 K. The

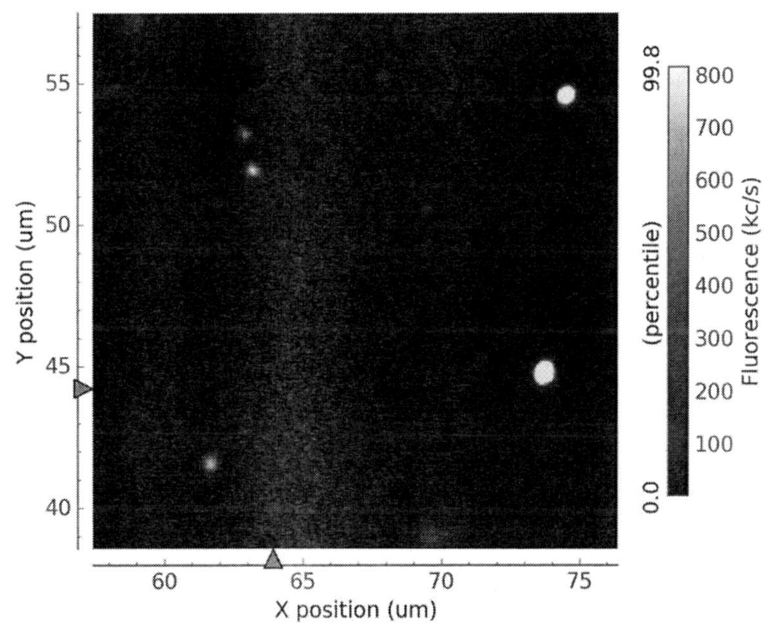

Fig. 1 Confocal luminescence scan image of "1800 K" sample dispersed on a glass substrate. Bright spots correspond to diamond crystallites containing NV centers.

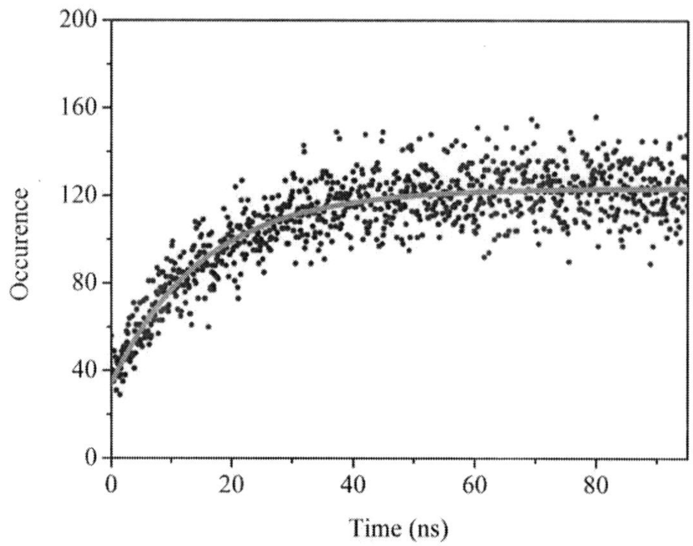

Fig.2 Autocorrelation function $g^{(2)}$ measured for one of the luminescent diamond particles. The fitting curve is shown in orange.

samples consisted of diamond particles with mean sizes of 30 nm and 500 nm, respectively. No NV centers were detected in “1600 K” sample. It is explained by small sizes of the diamond particles and low level of the sample contamination with nitrogen. From one to five centers was found in individual “1800 K” crystallites, whose volume was three orders of magnitude greater than the volume of “1600 K” crystallites. Figure 1 shows 20x20 μm^2 luminescence map of “1800 K” sample. Autocorrelation function $g^{(2)}$ measured for luminescent diamond particle containing one NV center is shown in Fig.2. Presently, controllable production of NV-containing nanodiamonds based on a mixture of adamantane with its derivatives, containing nitrogen, is under development.

3. Conclusion

It was established that nanodiamonds of various origins and sizes (down to ~ 1 nm) can be used for the production of single photon emitters. The most promising for this purpose are HPHT nanodiamonds synthesized from organics. They are produced in thermodynamically stable conditions, which are the most favorable to obtain the perfect diamond structure, and do not contain metal impurities inherent for HPHT synthesis in traditional metallic media.

Acknowledgments

This research was supported by Russian Science Foundation (grant No 14-12-01329).

References

[1] Shenderova O A, Vlasov I I, Turner S, Van Tendeloo G, Orlinskii S B, Shiryaev A A, Khomich A A, Sulyanov S N, Jelezko F, and Wrachtrup J 2011 *J. Phys. Chem. C* **115** 14014

[2] Vlasov I I, Shiryaev A A, Rendler T, Steinert S, Lee S.-Y., Antonov D, Vörös M, Jelezko F, Fisenko A V, Semjonova L F, Biskupek J, Kaiser U, Lebedev O I, Sildos I, Hemmer P R, Konov V I, Gali A, and Wrachtrup J 2014 *Nature Nanotechnology* **9** 54

[3] Ekimov E A, Kudryavtsev O S, Mordvinova N E, Lebedev O I, and Vlasov I I 2018 *ChemNanoMat* **4** 269

Hybrid graphene-nanometallic structures

V S Volkov[1, 2, 3], D I Yakubovsky[1], Y V Stebunov[1, 3], R V Kirtaev[1], K V Voronin[1] and A V Arsenin[1, 3]

[1]Moscow Institute of Physics and Technology, Institutskyi 9, 141700 Dolgoprudnyi, Russia
[2]SDU Nano Optics, Mads Clausen Institute, University of Southern Denmark, Campusvej 55, DK-5230, Odense, Denmark
[3]GrapheneTek, 7 Nobel Street, Skolkovo Innovation Center, 143026, Russia

vsv@mci.sdu.dk

Abstract. The development of functional hybrid nanomaterials based on graphene and metal nanostructures is currently the subject of intense research interest. In this study we discuss different aspects relevant to growth morphology, properties (optical, electrical, and structural), as well as the nanoscale architecture of hybrid graphene-nanometallic structures based on graphene as a multifunctional platform. The peculiarities of growth kinetics of thin gold and copper films on graphene are analysed (by use of the variety of experimental methods) and compared. It is shown that the presence of graphene affects strongly the optical, electrical, and structural properties of thin metal films.

1. Introduction

The discovery of graphene (and its extraordinary structural, chemical, and electronic properties) has excited high expectations for the development of a new paradigm for the design of nanostructures. However, these properties only emerge in the 2D planar direction of the graphene structure, limiting its scope and application. A promising approach to overcome this problem has arisen lately, by developing structures wherein graphene acts as a platform (or a 2D planar substrate) for supporting other nanomaterials [1-3]. Nowadays the research in hybrid nanostructures of 2D materials/metal covers different fields with the focus to understand (and apply) the interesting properties of these nanomaterials, and other parts have been directed to basic science with the target to modify structure and surface of thin metal films. This later has allowed the diversification of their properties and possible uses, opening new possibilities in chemistry, engineering, material science, and bio-sensing [4, 5].

2. Methods

The deposition of metals was made on SiO_2/Si substrates without and with graphene in order to systematically study and compare metal films properties. All depositions of films were performed using gold and copper pellets with the purity of 99.999% and the e-beam evaporation procedure. The nominal thicknesses of the deposited films (25 and 50 nm, correspondingly) and the deposition rate (0.5-15 A/s) were monitored by the quartz-crystal mass thickness sensor mounted in the vacuum chamber. The thickness of thin films deposition was independently estimated by step height atomic force microscopy measurements (and the root-mean-square surface roughness value was determined for each film). The properties of the polycrystalline structure of the fabricated thin metal films as a function of the film thickness were systematically investigated from X-ray diffraction measurements using conventional scanning methods. The average grain size was estimated by the Debye-Scherrer equation, which relates the size of metallic crystallites in a thin film to the broadening of the peak in the experimental X-ray diffraction pattern [6]. Finally, the dielectric function spectra of the studied thin metal films were evaluated from data measured using variable angle spectroscopic ellipsometer in the photon energy range from 4.13 to 0.62 eV.

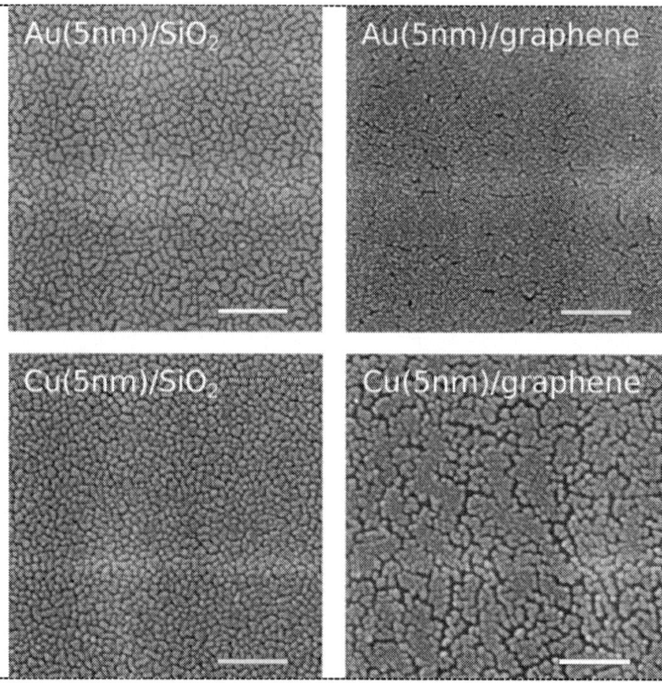

Figure 1. SEM micrographs (scale bar ≈ 200 nm) of thin gold and copper films (5-nm thick) deposited on SiO_2/Si substrates with/without graphene with a rate of 0.2-0.4 A/s at 20°C.

3. Results and discussion

Metal-graphene contact is a key interface in graphene-based device applications, and it has been recently studied in some detail [3], though, it is known that the growth of metals on the surface of graphene differs markedly from the growth of metals on ''typical'' substrates. For example, our results (see Figure 1) clearly demonstrate a significant difference in initial growth kinetics of ultrathin gold and copper films on SiO_2/Si substrates without and with graphene. It has also been shown, that this distinction in growth kinetics leads to a change of optical, electrical, and structural properties of thin metal films (20-200 nm) [7] that needs to be taken into account when developing and modelling systems/devices based on hybrid graphene-metal structures.

4. Conclusions

With the aim of preparing advanced functional materials, ultrathin gold and copper films have been integrated with graphene to study and fully exploit/optimize their properties. Thus, we studied optical, electrical and structural properties of continuous thin metal films on graphene transferred onto on SiO_2/Si substrates (grown by electron beam evaporation). The growth kinetics of thin gold and copper films on graphene has been investigated with the well-established microscopy techniques, including atomic force microscopy, scanning electron microscopy, and spectroscopic ellipsometry.

Acknowledgments
This work was supported by the Russian Science Foundation, grant no. 18-19-00684. The experiments were performed in part using the resources of the Moscow Institute of Physics and Technology Center of Shared Research Facilities, which is funded by the Ministry of Education and Science of the Russian Federation, grant no. RFMEFI59417X0014.

References

[1] Grigorenko A N, Polini M, Novoselov K S 2012 *Nature Photonics* **6** 749.
[2] Liu X, Zhu J, Wei B 2016 *Chem. Soc. Rev.* **45** 3145.
[3] Liu X, Han Y, Evans J W, Engstfeld A K, Behm R J, Tringides M C, Hupalo M, Lin H-Q, Li Huang, Ho K-M, Appy D, Thiel P A, Wang C-Z 2015 *Progress in Surface Science* **90** 397.
[4] Stebunov Y V, Aftenieva O A, Arsenin A V, Volkov V S 2015 *ACS Appl. Mater. Interfaces* **7** 21727.
[5] Stebunov Y V, Yakubovsky D I, Fedyanin D Yu, Arsenin A V, Volkov V S 2018 *Langmuir* (DOI: 10.1021/acs.langmuir.8b00276).
[6] Yakubovsky D I, Fedyanin D Yu, Arsenin A V, Volkov V S 2017 *AIP Conf. Proc.* **1874** 040057.
[7] Yakubovsky D I, Arsenin A V, Stebunov Y V, Fedyanin D Y, Volkov V S 2017 *Opt. Express* **25** 25574.

IOP Conf. Series: Journal of Physics: Conf. Series **1092** (2018) 012162 doi:10.1088/1742-6596/1092/1/012162

Integrated plasmonic biosensors based on microring resonators

K V Voronin[1], Y V Stebunov[1,2], A V Arsenin[1,2], V S Volkov[1,2,3]

[1]Moscow Institute of Physics and Technology, Institutskyi 9, 141700 Dolgoprudnyi, Russia
[2]GrapheneTek, 7 Nobel Street, Skolkovo Innovation Center, 143026, Russia
[3]SDU Nano Optics, Mads Clausen Institute, University of Southern Denmark, Campusvej 55, DK-5230, Odense, Denmark

kirill251097voronin@gmail.com

Abstract. We introduce a biosensor scheme comprising a plasmonic microring resonator coupled with a plasmonic bus waveguide. The detection principle is based on the violation of the surface plasmon polariton resonance condition for the microring (which is sensitive to refractive index changes). Analytical calculations and simulations in COMSOL Multiphysics are presented. As a result, we obtained main characteristics of the sensor, such as sensitivity (S=1200 nm) and Q-factor (Q≈100).

1. Introduction

In recent years, researchers have proposed various schemes of optical label-free biosensors, which simultaneously have high sensitivity and small footprint allowing their integration into electronic devices [1]. The operation principle of most optical biosensors is based on the detection of changes in optical properties induced by the adsorption of analyzed substances on the surface of optical structures. An efficient way to track changes in refractive index (RI) is by using photonic microring resonators due to the strong dependence of their optical resonance conditions on the optical properties of surrounding medium [2-6]. The development of plasmonics has significantly influenced the modern biosensing technology, and currently most commercial optical biosensors are based on the surface plasmon resonance excited using Kretschmann configuration [7-11]. In addition, various schemes with plasmon microresonators have also been proposed [12-14].

In this paper we present a compact biosensor based on an integrated plasmonic microring resonator. This scheme has some advantages over SPR biosensors that are commonly used today: it is much more compact and could be integrated into digital devices. The performance of the proposed biosensor is investigated using analytical calculations and numerical simulation in the COMSOL Multiphysics software package, which includes calculations of the biosensing sensitivity and quality factor of the resonator. The coupling between the eigenmodes of the bus waveguide and the resonator leads to the appearance of minima and maxima in transmission spectra. The change in the RI of an external medium causes a wavelength shift of resonant peaks. Despite the fact that plasmonic resonators have much lower quality factor comparing to photonic ones, a sensitivity of about 1.2 μm can be achieved for a biosensor based on a plasmonic microring resonator, whereas the typical sensitivity of biosensors based on photonic resonators does not exceed 0.5 μm. In the present study we use long-range plasmonic waveguides-based metal strips, which allows to increase the quality factor and slightly increase the sensitivity.

2. Optical biosensors design and simulation

The proposed scheme of a plasmonic biosensor comprises a plasmon waveguide based on a metal strip with a thickness of 20 nm and a width of 5 μm, and a microring resonator with a radius of 40 μm and a straight part with a length of 20 nm (figure 1). The plasmonic waveguide is embedded in a homogeneous dielectric medium, while the microring resonator is placed above it. Both the waveguide and microresonator are considered to be made of gold, which is the most widely used plasmonic material [15]. However, the use of thin gold films requires careful determination of their dielectric function [16-17]. In addition, gold can be replaced in plasmonic biosensors by copper, which demonstrates optical properties comparable to gold [15, 18]. A low-k polymer Cytop is chosen as a dielectric because its RI matches with the RI of aqueous solutions. This symmetry is necessary for the excitation of long-range plasmonic modes.

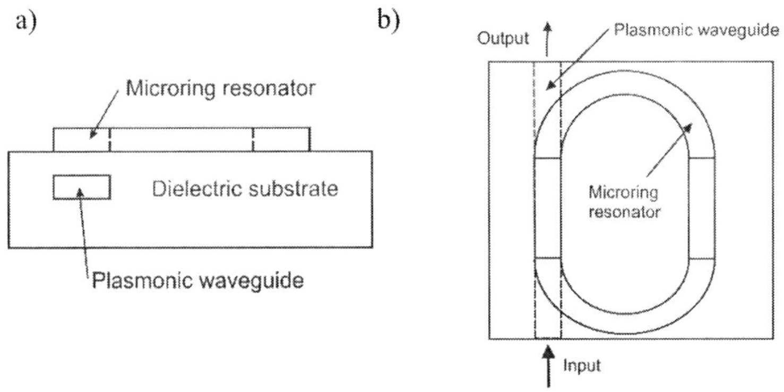

Figure 1. (a) Side and (b) top views of the proposed biosensor scheme comprising the plasmonic waveguide and the microring resonator. The waveguide is embedded in the dielectric substrate and optically coupled with the resonator located above.

A metal film embedded in a uniform dielectric can support two modes of a surface electromagnetic wave: with a symmetric and antisymmetric distributions of the normal component of the electric field. Their propagation constants are calculated from the following transcendental equation.

$$e^{-2k_1 t} = \frac{\dfrac{k_1}{\varepsilon_1} + \dfrac{k_2}{\varepsilon_2}}{\dfrac{k_1}{\varepsilon_1} - \dfrac{k_2}{\varepsilon_2}} \frac{\dfrac{k_1}{\varepsilon_1} + \dfrac{k_3}{\varepsilon_3}}{\dfrac{k_1}{\varepsilon_1} - \dfrac{k_3}{\varepsilon_3}} \qquad (1)$$

The field distributions for metal films with thicknesses of 10 and 100 nm are shown in Figure 2 (a). In a waveguide of a finite width, these modes are distorted, while demonstrating the similar optical properties. In the following, we consider only a symmetric mode of surface plasmons, since the antisymmetric mode strongly decays in the metal and its practical use is limited. The wave vectors of surface electromagnetic waves propagating in the considered waveguides were found using the simulations performed in the COMSOL Multiphysics software. Another parameter of the system is the coupling constant, which characterizes the transfer of electromagnetic energy between the waveguide and the microresonator. Instead of one long-range (symmetric) mode, the system of two parallel plasmonic waveguides supports two modes with close wave vectors, which have, respectively, symmetric and asymmetric distributions of the resulting electric field (figure 2 (b,c)). The difference between these wave vectors is a coupling constant, which characterizes energy transfer between plasmonic waveguides. Efficient coupling requires that the distance between the waveguides should be no more than the width of the waveguides. The coupling constant is determined by the following formula:

$$C = \frac{(\beta_1 - \beta_2)}{2} \tag{2}$$

Where β1 and β2 are the propagation constants for two eigenmodes existing in the system of two plasmonic waveguides. In the present study, the propagation constants were obtained using simulations in COMSOL.

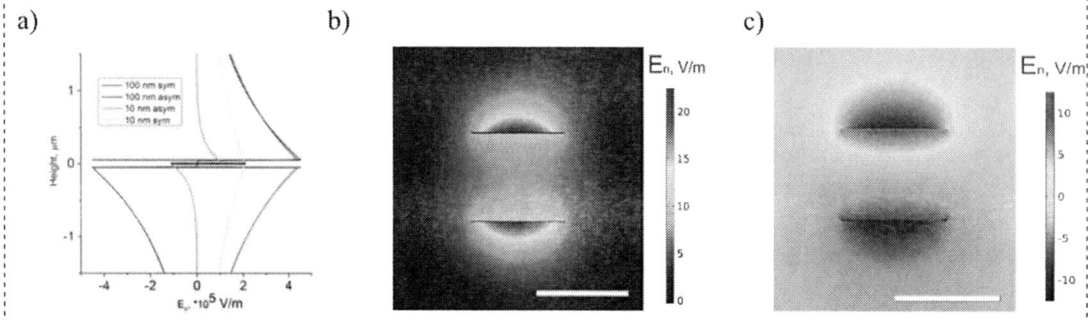

Figure 2. Transverse electric field distributions for symmetric (red line) and asymmetric (blue line) modes of a 100-nm waveguide and for symmetric (green line) and asymmetric (yellow line) modes of a 10-nm waveguide (a). Transverse electric field distributions for two eigenmodes of the system of two plasmonic waveguides (scale bar, 5 μm) (b, c).

3. Results and discussion

The change in the RI can be determined from the wavelength shift of the transmission spectrum. The efficiency of RI sensing is characterized by the sensitivity (S = δλ/δn) and the quality factor (Q) of the microresonator. From simulations, we obtain the wave vector of the plasmonic waveguide, the coupling constant and the attenuation coefficient. Thereafter, we calculate the transmission coefficient for the proposed system using the following equation as in [5]:

$$T = \left| \frac{\cos(CL) - e^{-\gamma l} e^{i\varphi}}{1 - e^{-\gamma l} e^{i\varphi} \cos(\mathrm{CL})} \right|^2 \tag{3}$$

The transmission spectra for different RIs of an analyzed medium are shown in figure 3a. The corresponding sensitivity equals 1.2 μm. The sensitivity to RI changes does not depend on the radius of the resonator and the coupling length but depends mainly on the optical properties of materials and on the width and thickness of the microring resonator (figure 3b).

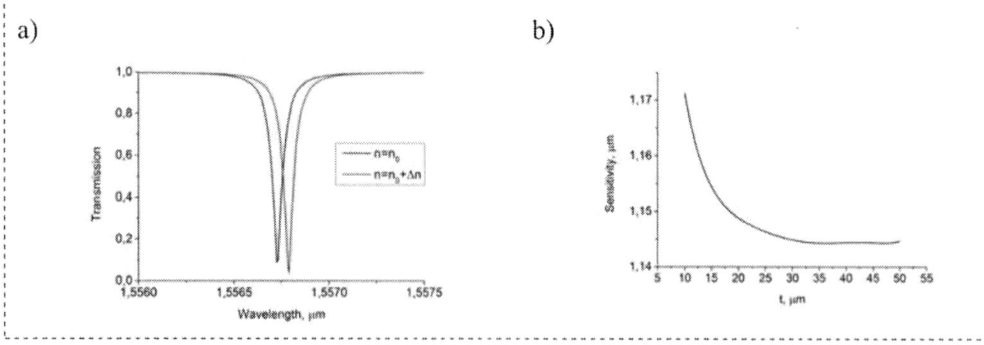

Figure 3. The transmission spectra of the proposed plasmonic biosensor obtained for different RIs of the analyzed solution (a). The sensitivity of the plasmonic biosensor depending on the waveguide thickness (b).

4. Conclusions

We have presented the integrated plasmonic biosensor based on the plasmonic bus waveguide and microring resonator. For considered long-range plasmonic modes, the microring resonator is characterized by the quality factor up to 100, which is higher than for most plasmonic resonators used in biosensing. This resonator is optically coupled to the plasmon waveguide, which is embedded into the dielectric substrate and, therefore, not influenced by the analyzed solution. The obtained sensitivity of 1.2 μm is significantly higher than the corresponding sensitivities of biosensors with photonic and plasmonic resonators, which usually do not exceed 0.3 and 0.6 μm, respectively [19-20]. This result was achieved because of using the metal strip waveguide which is stronger affected by RI changes than dielectric loaded waveguides.

Acknowledgments

This work was supported by the Russian Science Foundation, grant no. 18-19-00684.

References

[1] Roh S, Chung T and Lee B 2011 *Sensors* **11** 1565.
[2] Yang G, White I and Fan X 2008 *Sensors and Actuators* **133** 105.
[3] Xu D, Densmore A, Delâge A, Waldron P, McKinnon R 2008 *Optics express* **16** 15137.
[4] Martin M, Gleeson A, Spaugh B, Tybor F, Gunn W, Hochberg M, Baehr-Jones T, Bailey R and Gunn L 2010 *IEEE J* **16 654**.
[5] Boyd R and Heebner J 2001 *Applied optics* **40** 5742.
[6] Jin L, Li M and He J 2011 *Optics Communications* **284** 156.
[7] Iqbal M, Gleeson M, Spaugh B, Tybor F, Gunn W, Hochberg M, Baehr-Jones T, Bailey R and Gunn L 2010 *J Sel Top Quantum Electron* **16** 654.
[8] Mullett W, Lai E, Jupiter M. Yeung 1998 *Analytical biochemistry* **258** 161.
[9] Gupta G and Kondoh J 2007 *Sens. Actuat. B: Chem* **122** 381.
[10] Stebunov Y V, Aftenieva O A, Arsenin A V, Volkov V S 2015 *ACS Appl. Mater. Interfaces* **7** 21727.
[11] Volkov V S, Stebunov Y V, Yakubovsky D I, Fedyanin D Yu, Arsenin A V 2017 *AIP Conf. Proc.* **1874** 020008.
[12] Homola J, Yee S and Gauglitz G 1999 *Sensors and Actuators* **54** 3.
[13] Debackere P, Taillaert D, De Vos K, Scheerlinck S, Bienstman P and Baets R 2007 *Proc. of SPIE* **6477** 647719.
[14] Lee J, Song J, Sung G, Shin J 2014 *Nano Lett.* **14** 5533.
[15] McPeak K M, Jayanti S V, Kress S J P, Meyer S, Iotti S, Rossinelli A, Norris D J 2015 *ACS Photonics* **2** 326.
[16] Yakubovsky D I, Fedyanin D Yu, Arsenin A V, Volkov V S 2017 *AIP Conf. Proc.* **1874** 040057.
[17] Yakubovsky D I, Arsenin A V, Stebunov Y V, Fedyanin D Y, Volkov V S 2017 *Opt. Express* **25** 25574.
[18] Stebunov Yu V, Yakubovsky D I, Fedyanin D Yu, Arsenin A V, Volkov V S 2018 *Langmuir* (DOI: 10.1021/acs.langmuir.8b00276).
[19] Claes T, Molera J, De Vos K, Schacht E, Baets R and Bienstman P 2009 *IEEE Photonics Journal* **1** 196.
[20] Zhou L, Sun X, Li X and Chen J 2011 *Sensors* **11** 6856.

IOP Conf. Series: Journal of Physics: Conf. Series **1092** (2018) 012163 doi:10.1088/1742-6596/1092/1/012163

Bloch Surface Waves Fabry- Pérot Nanocavity

Babak Vosoughi Lahijani, Nicolas Descharmes, Gaël Osowiecki, Raphaël Barbey, and Hand Peter Herzig

Optics & Photonics Technology Laboratory (OPT), École Polytechnique Fédérale de Lausanne (EPFL), Rue de la Maladière 71b, CH-2002 Neuchâtel, Switzerland

Corresponding author: babak.vosoughilahijani@epfl.ch

Abstract. We demonstrate the first cavity confinement of Bloch surface waves in the fundamental mode of a Fabry–Pérot cavity. The cavity consists of two distributed Bragg mirrors separated by a spacer of dimensions smaller than the wavelength (633nm). The cavity mode exhibits quality factor of 400. This quality factor is the highest recorded thus far for a subwavelength optical surface waves cavity.

1. Introduction

The capacity to confine optical surface waves within a cavity has been the subject of intensive research over the last years. This research is motivated by the strong light matter interactions that such a configuration can enable. However most of the efforts have been focused on the confinement of surface plasmon polaritons (SPPs) [1-4]. Recently another type of optical surface waves named Bloch surface waves (BSWs) received a lot of interest [5-8]. Bloch surface waves are electromagnetic modes that can be excited on the surface of an all-dielectric multilayer stack. Because they can be made of dielectric materials, they do not suffer from metal absorption and can theoretically offer higher quality factors if confined. Cavity confinement of BSWs has been studied both theoretically [9] and experimentally [10,11]. However, the confinement of BSWs in the fundamental mode of a cavity has not been achieved yet.

In this work we report on the first subwavelength confinement of BSWs in a Fabry–Pérot type nanocavity. We demonstrate that the dielectric nature of BSWs leads to higher Q-factors compared to SPPs nanocavities.

2. BSW multilayer platform and cavity design

Figure 1 is a schematic representation of the Fabry–Pérot type nanocavity on the top of the BSW multilayer substrate. The cavity consists of high reflectivity distributed Bragg reflectors (DBR) separated by a subwavelength spacer. The multilayer substrate is designed to support transverse electric (TE) polarization of surface waves in the visible spectrum. It includes six layer pairs of silicon dioxide (SiO_2) and silicon nitride (SiN_x) with respective thicknesses of 159 and 118 nm. An additional 42 nm thin SiN_x defect layer terminates the multilayer periodicity and creates an allowed energy level inside the forbidden band of the multilayer stack. To realize the cavity mode, we introduce 100 nm ZEP on top of the multilayer structure. Figure 2 shows the dispersion diagrams correspond to the multilayer with and without 100 nm ZEP. The effective refractive index value is calculated to be 1.258 for multilayer and 1.389 for multilayer + 100 nm ZEP. The fundamental cavity mode design parameters (as shown in Fig. 1) are calculated using the effective refractive index values.

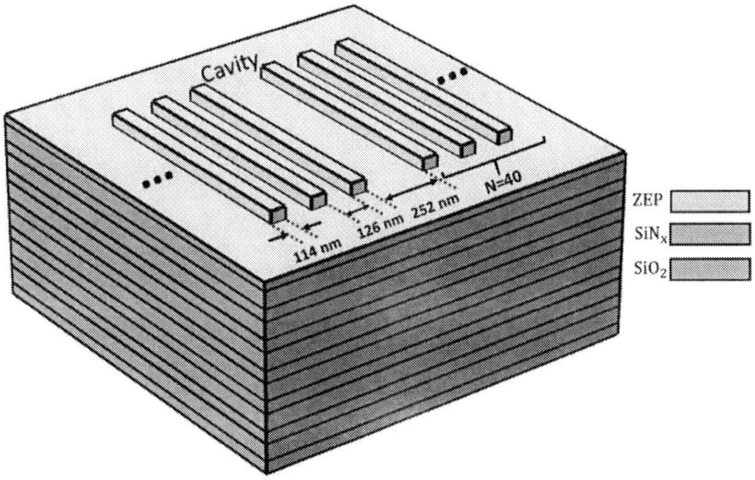

FIGURE 1. Schematic representation of the Fabry–Pérot cavity designed on a dielectric multilayer substrate. The 252 nm spacer is designed to support the fundamental resonance mode in the visible spectrum.

The multilayer stack is fabricated using standard plasma-enhanced chemical vapor deposition (PECVD). The electron beam lithography is used to transfer the Fabry–Pérot type cavity to the ZEP layer.

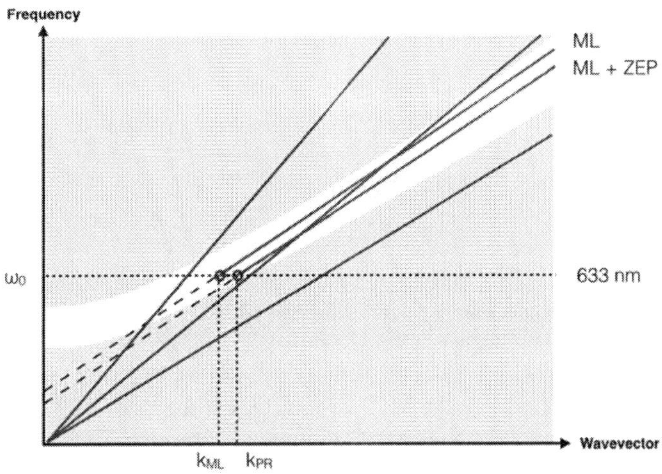

FIGURE 2. Dispersion diagram of the designed BSW substrate with/without 100 nm thin ZEP layer.

3. Results and discussions

A tunable light source in the visible spectrum with an appropriate coupling angle is illuminated to the left most edge of the entrance DBR using a Kretschmann coupling configuration. We use an optical spectrum analyzer (OSA) to measure the resonance spectrum and quality factor of the cavity. Figure 3 depicts the spectrum collected from the cavity. The constructive interference between Bloch surface waves being reflected back and forth by the DBRs lead to the formation of a standing waves, i.e. a resonance, at 627.8 nm wavelength. As can be seen from this figure, a Q-factor as high as 400 can be directly obtained by measuring the linewidth of the spectrum. To our knowledge, this is the highest Q reported for a fundamental surface wave cavity mode.

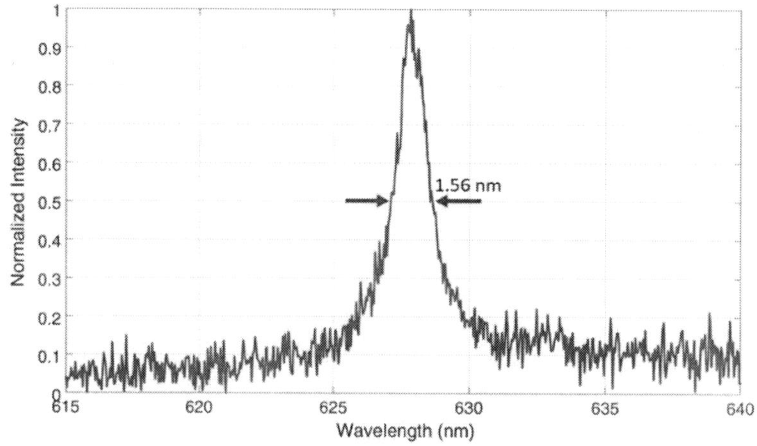

FIGURE 3. Spectrum of the BSW-based Fabry–Pérot nanocavity. A sharp resonance (1.56 nm linewidth) is observed at λ=627.8 nm and corresponding to a quality factor of ~ 400.

4. Conclusion

In summary, we experimentally demonstrate the first high quality factor fundamental mode cavity confinement of an optical surface wave. We show that the optical surface waves can be confined between two distributed Bragg reflectors separated by a subwavelength spacer. The surface waves are confined with a high quality factor of up to 400. This value is greater than reported quality factors for surface plasmon polaritons which is attributed to the dielectric nature of Bloch surface waves.

Acknowledgments

This research is supported by the Swiss National Science Foundation (SNSF FN200020-135455).

References

[1] B. Min, E. Ostby, V. Sorger, E. Ulin-Avila, L. Yang, X. Zhang, and K. Vahala, "High-Q surface-plasmon-polariton whispering-gallery microcavity," Nature **457**, 455-459 (2009).

[2] V. J. Sorger, R. F. Oulton, J. Yao, G. Bartal, and X. Zhang, "Plasmonic fabry-pérot nanocavity," Nano letters **9,** 3489-3493 (2009).

[3] R. Ameling, and H. Giessen, "Microcavity plasmonics: strong coupling of photonic cavities and plasmons," Laser & Photonics Reviews **7**, 141-169 (2013).

[4] N. P. De Leon, B. J. Shields, L. Y. Chun, D. E. Englund, A. V. Akimov, M. D. Lukin, and H. Park, "Tailoring light-matter interaction with a nanoscale plasmon resonator," Physical review letters **108**, 226803 (2012).

[5] R. Wang, H. Xia, D. Zhang, J. Chen, L. Zhu, Y. Wang and P. Wang, "Bloch surface waves confined in one dimension with a single polymeric nanofiber," Nature Communications **8**, 14330 (2017).

[6] G. M. Smolik, N. Descharmes, and H. P. Herzig, "Toward Bloch Surface Wave-Assisted Spectroscopy in the Mid-Infrared Region," ACS Photonics (2018).

[7] R. Dubey, E. Barakat, M. Häyrinen, M. Roussey, S. K. Honkanen, M. Kuittinen and H. P. Herzig, "Experimental investigation of the propagation properties of bloch surface waves on dielectric multilayer platform," J. Eur. Opt. Soc.**13**, (2017).

[8] R. Dubey, M. Marchena, B. Vosoughi Lahijani, M. S. Kim, V. Pruneri, and H. P. Herzig, "Bloch Surface Waves Using Graphene Layers: An Approach toward In-Plane Photodetectors," Applied Sciences **8**, 360 (2018).

[9] M. Menotti, and M. Liscidini, "Optical resonators based on Bloch surface waves." JOSA B **32,**

431-438 (2015).

[10] R. Dubey, B. Vosoughi Lahijani, E. Barakat, M. Häyrinen, M. Roussey, M. Kuittinen, and H. P. Herzig, "Near-field characterization of a Bloch-surface-wave-based 2D disk resonator," Optics letters **41**, 4867-4870 (2016).

[11] B. Vosoughi Lahijani, H. B. Ghavifekr, R. Dubey, M. S. Kim, I. Vartiainen, M. Roussey and H. P. Herzig, "Experimental demonstration of critical coupling of whispering gallery mode cavities on a Bloch surface wave platform," Optics letters **42**, 5137-5140 (2017).

IOP Conf. Series: Journal of Physics: Conf. Series **1092** (2018) 012164 doi:10.1088/1742-6596/1092/1/012164

Unusual penetration effect in ferromagnetics. Negative refraction under tangential wave incidence.

K. Vytovtov, E. Barabanova

Astrakhan State Technical University, Tatischeva 16, Astrakhan, Russia

vytovtov_konstan@mail.ru

Abstract. In this paper we study behaviour of a refracted wave within a ferrite medium and a reflected wave in vacuum if an incident wave propagates along the interface in an isotropic medium. It is shown analytically that a plane harmonic wave can transmit from an isotropic medium to anisotropic medium in this exotic case. It is shown also that wave behaviour in this case is analogous to wave bahavior in a metamaterial. The wavenumbers of the refracted waves and the Poyting vectors are found. The reflection and transmission coefficients are investigated also.

1. Introduction

Nowadays there are so many publications in the area of anisotropic material electromagnetic that one may think that all aspects of the electromagnetic theory for anisotropic structures and media have been extensively studied. However, we show in this paper, that some simple specific cases have not been considered deep enough in the literature despite the fact that in these cases interesting exotic phenomena can be observed.

The electromagnetic plane wave propagation in vacuum along an interface with an anisotropic medium has been studied in detail in [1-3]. It has been shown analytically for the first time in [1] that a bulk wave can be excited within a gyrotropic medium under tangential wave propagation. This phenomenon was identified as the "penetration effect".

Exotic behavior of electromagnetic waves in the vicinity of anisotropic media has been reported in [4-6]. In this case, authors have considered an oblique wave incidence on a metallic film. Extraordinary transmission has been described as being generated by surface plasmons, which are coherent oscillations of a metal's conducting electrons at the metal surface. Moreover, looking through the history of anisotropic media we find that unusual patterns of reflection and refraction by anisotropic structures have been studied in the middle of the last century [7-9]. For example, it has been shown in [7] that a reflected wave can be directed toward an interface but not from it. Unusual refraction from anisotropic media with negative dispersion have also been described in [8]. Other phenomena such as the decreasing of the refraction angle while increasing the incidence angle, the absence of reflected and refracted waves while an incident wave is presented have been discussed in [9].

In this paper we investigate the dependence of penetration effect parameters (an incident wave propagate parallel to an interface) on the anisotropy axis orientation when this axis is in the plane normal to the incidence plane.

2. Statement of the problem

It is assumed that a plane incident wave propagates along *y*-axis in an isotropic medium. The anisotropic medium is described as

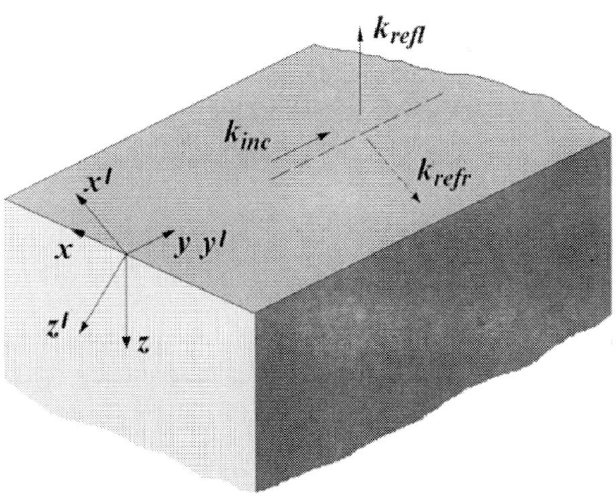

Fig.1. Statement of the problem

$$\mu = \begin{vmatrix} \mu_{xx} & j\mu_{xy} & 0 \\ -j\mu_{xy} & \mu_{xx} & 0 \\ 0 & 0 & \mu_{zz} \end{vmatrix} \qquad (1)$$

with $\mu_{xy} \to 0$. On other words we consider a nonresonance region of a ferromagnetic.

The geometry of the problem is shown in Fig.1. Here $x'y'z'$ is the coordinate system connected with the permeability dyadic (1), where $\mu_{xx}, \mu_{xy}, \mu_{zz}$ are the components of permittivity in the $x'y'z'$-system; xyz is the coordinate system connected with the normal to the interface, y is always coincide with y'. The x'- and z'-axes are orientated in the plane normal to the incidence (the *xOz*-plane) under an arbitrary angle β (Fig.1). Our main aim is investigation of behavior of reflected and refracted waves.

3. Analytical investigation

Taking into account the continuity conditions for the tangential components of the wavenumbers we can write the dispersion relation

$$a_4 k_z^4 + a_2 k_z^2 + a_0 = 0 \qquad (2)$$

were

$$\begin{gathered} a_4 = -\omega^2 \varepsilon_1 (\mu_{11} \sin^2\beta + \mu_{33} \cos^2\beta) \\ a_2 = \omega^4 \varepsilon_1^2 \mu_{11} (\mu_{33} + \mu_{33} \cos^2\beta + \mu_{11} \sin^2\beta) - \omega^2 \varepsilon_1 k_y^2 (\mu_{11} + \mu_{11} \sin^2\beta + \mu_{33} \cos^2\beta) \\ a_0 = -\omega^6 \varepsilon_1^3 \mu_{11}^2 \mu_{33} + \omega^4 \varepsilon_1^2 k_y^2 \mu_{11} (\mu_{11} + \mu_{33}) - \omega^2 \varepsilon_1 \mu_{11} k_y^4 \end{gathered} \qquad (3)$$

The expression (2),(3) gives us the normal components of the wavevector in an anisotropic media. In this case there are two independent waves: a right-polarized wave and a left-polarized wave. The electromagnetic field components of these waves can be obtained directly from Maxwell equations as

$$E_x = \frac{\omega \mu_{11} k_y}{k_z^2 - \omega^2 \varepsilon_1 \mu_{11}} H_z$$

$$E_y = \frac{k_z}{\omega \varepsilon_1 \tan\beta} + \frac{k_z \mu_{11} (\omega^2 \varepsilon_1 \mu_{11} - k_y^2 - k_z^2)}{\omega \varepsilon_1 \sin\beta \cos\beta \, (\mu_{33} - \mu_{11})(\omega^2 \varepsilon_1 \mu_{11} - k_z^4)} H_z$$

$$E_z = \frac{\omega^2 \varepsilon_1 \mu_{11} (k_z^2 - k_y^2) - k_z^4}{\omega \varepsilon_1 k_y (\omega^2 \varepsilon_1 \mu_{11} - k_z^2) \tan\beta} H_z +$$

$$\frac{\omega^2 \varepsilon_1 \mu_{11} \left[\mu_{33} (k_z^2 + k_y^2) + k_z^2 \mu_{11}\right] - k_z^2 \mu_{11} (k_z^2 + k_y^2) - \omega^4 \varepsilon_1^2 \mu_{11}^2 \mu_{33}}{\omega \varepsilon_1 k_y (\omega^2 \varepsilon_1 \mu_{11} - k_z^2) \sin\beta} H_z \qquad (4)$$

$$H_x = \left[\frac{\omega^2 \varepsilon_1 \mu_{11}^1 - \mu_{11}\left(k_z^2 + k_y^2\right)}{(\mu_{33} - \mu_{11})(\omega^2 \varepsilon_1 \mu_{11} - k_z^2) \sin\beta \cos\beta} + (\tan\beta)^{-1}\right] H_z$$

$$H_y = \frac{k_y k_z}{\omega^2 \varepsilon_1 \mu_{11} - k_z^2} H_z$$

The components of the Poyting vector are written by using the classical microwave theory

$$\begin{aligned} S_x &= \frac{c}{8\pi}(E_x H_z^* - E_z H_x^*) \\ S_y &= \frac{c}{8\pi}(E_z H_x^* - E_x H_z^*) \\ S_z &= \frac{c}{8\pi}\left(E_x H_y^* - E_y H_x^*\right) \end{aligned} \tag{5}$$

It is obviously that the z-components of the wavevectors and the Poyting vectors are nonzero (see (2) and (5)). Thus we can say about bulk wave propagation within an anisotropic material in the considered exotic case. Moreover the x-component of the Poyting vector is nonzero. Thus the Poyting vector is not lie in the incidence plane in contrast to the cases described in [3]. Therefore it is possible to obtain very exotic deflection of an optical beam. These properties can be used for example in switching elements of communication systems and photonic computers.

4. Numerical results

In Fig.2a the dependence of the wavenumber on the incidence angle is presented for $\mu_{xx} = 10.6$, $\mu_{xy} = 0$, $\mu_{zz} = 0.99$. The components of the wavenumber in Fig.2a and the Poyting vector are shown in Fig.2b. As we can see the positive z-component of the Poyting vector corresponds to the negative z-component of the wavevector. Analogously negative z-component of the Poyting vector corresponds to the positive z-component of the wavevector. Therefore the medium can be considered as a metamaterial.

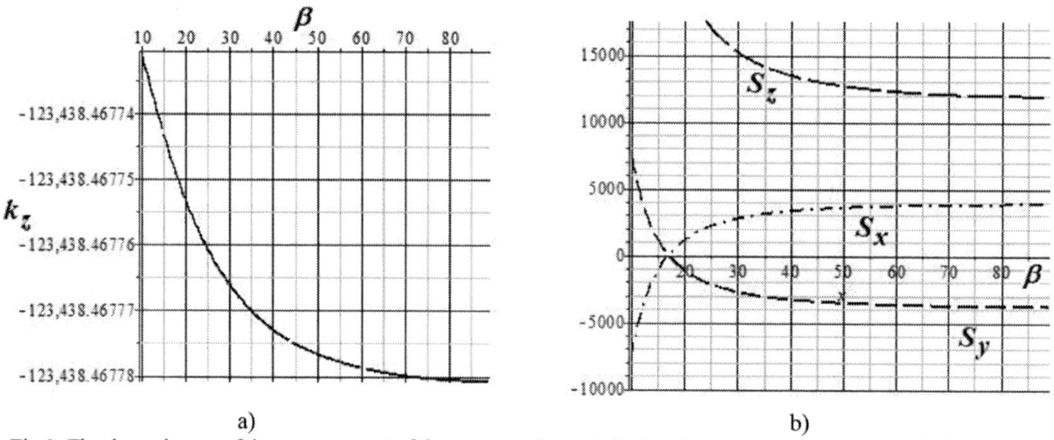

a) b)

Fig.2. The dependences of the z-component of the wavenumber and the Poyting vector components on the inclination angle β of the anisotropy axis

It is interesting that the resulting reflected wave is elliptically polarized and this wave can be represented as a superposition of the right- and left-polarized waves with the different reflection coefficients. The dependences of the reflection coefficients for the right- and left-polarization waves are presented in Fig.3.

5. Conclusions

The exotic penetration effect for a magnetic medium is studied analytically. It is shown that in this exotic case the anisotropic medium in this case is analogous to a metamaterial. The dependence of the wavenumbers, the field components, and the Poyting vector components on an inclination angle when the anisotropy axis is orientated in the plane normal to the incidence plane are investigated. It is shown that left- and right-polarized waves are existed within the anisotropic medium. Note that the Poyting vectors of these waves are not lying in the incidence plane in contrast to the case when the anisotropy axis is in the incidence plane. Moreover the reflected wave is elliptically polarized in contrast to the case described in [2,3]. Note that the presented results cannot be applied if $\beta = 0$.

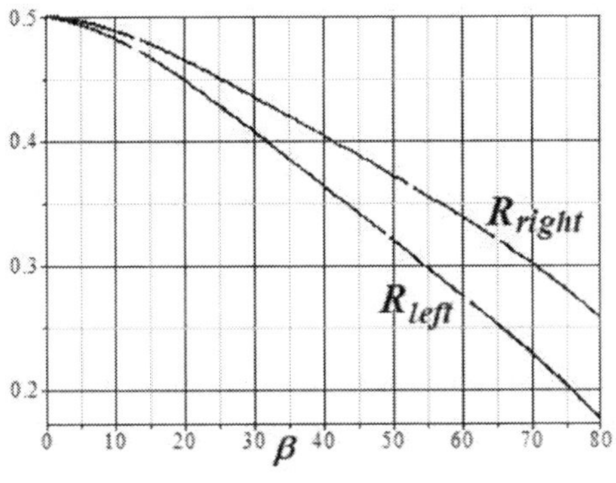

Fig.3 The dependence of the reflection coefficients on the inclination angle

The reported study was funded by RFBR according to the research project 18-37- 00059/18.

References

[1] *K. Vytovtov., L. Mospan,*"Penetration effect in gyrotropic slab: theory and applications",*J. Opt. Soc. Am. A***29**(6),877-882 (2012)

[2] *K. Vytovtov, S Zouhdi.,* "Exotic reflection of plane waves by anisotropic structures," in Proceedings of Metamaterials 2012: The Sixth International Congress on Advanced Electromagnetic Materials in Microwaves and Optics, (2012), pp. 648-650.

[3] K. *Vytovtov , O. Pischin O,* "Penetration Effect: Exotic Behavior of a Wave in Anisotropic Media,"*International Journal of Microwave Science and Technology*, Article ID 4082948, 7 pages (2017)

[4] *T.W. Ebbesen, H.J. Lezec, H.F. Ghaemi, T. Thio, and P.A. Wolff,*"Extraordinary optical transmission through sub-wavelength hole arrays,"*Nature***391**, 667 (1998).

[5] *R.Wannamacher,* "Plasmon-supported transmission of light through nanometric holes in metallic thin film,"*Opt. Commun.* **195,**107 (2001).

[6] *N. Yu, P. Genevet, M. A. Kats, F. Aieta, J.-Ph. Tetienne, F. Capasso, Z. Gaburro,*"Light Propagation with Phase Discontinuities: Generalized Laws of Reflection and Refraction,"*Science***21**,333-337 (2011).

[7] F. I. Fedorov, V. V. Filipov. *Reflection and refraction of light by transparent crystals* (Minsk 1976). 224p. (in Russian).

[8] L. I. Mandelshtam. *Compllete collection of works V.2. Group speed in a crystal latters* (Moskow 1950) pp. 334 – 338 (in Russian).

[9] R. A. Silin. *Unusual laws of refraction and reflection* (Moscow FAZIS 1999). 80p. (in Russian).

IOP Conf. Series: Journal of Physics: Conf. Series **1092** (2018) 012165 doi:10.1088/1742-6596/1092/1/012165

Direct near-field mapping of nano-sphere-excited leaky surface modes atanisotropic metasurface

F. Walla[1], F. Bürkle[1], I. Sinev[2], A. Bogdanov[2], F. Komissarenko[2],A. Soltani[1], R. Malureanu[3], A.V. Lavrinenko[3]and H. G. Roskos[1]

[1] Physikalisches Institut, Johann Wolfgang Goethe-Universität, Max-von-Laue-Strasse 1, 60438 Frankfurt am Main, Germany
[2] ITMO University, St. Petersburg 197101, Russia
[3]Technical University of Denmark, DK-2800 Kongens Lyngby, Denmark

Abstract. Leaky surface modes on an ultra-thin anisotropic gold metasurface are directly observed through scattering-type scanning near-field optical microscopy (s-SNOM). Surface modes are excited by focussing light onto a resonant silicon nano-sphere positioned on the surface by nano-manipulation under electron-beammonito-ring. The light scattered from the nano-sphere is ableto excitesurface modes with a broad range of wavevectors.The asymmetric air-metasurface-insulator structure supports a leaky surface mode which is confined at the top interface and leaks into the higher-index substrate.s-SNOM near-field mapping of the leaky surface wave is in good agreement with the results of the full-wave simulationsof the electro-magnetic modes on the structure.

1. Introduction

Metasurfaces are two-dimensional periodic arrays of subwavelength elements[1]. Due to strong interaction of the incident light and thesubwavelength scatterers forming the metasurface,abrupt changes of optical amplitude and phase can be introduced, which enablethe manipulation ofthe flow of light within ultra-smallvolumes. Metasurfacescan also be used to guide surface waves, such as surface plasmon polaritons (SPPs).The dispersion ofmodes on plasmonicmetasurfacescan be tailoredintothe isotropic, elliptical or hyperbolic regimes leading to exoticpropagation behaviourofsurface waves such as negative refraction, self-collimation and channelling of surface waves[2].Here, we report onthe excitation and direct near-field mapping of leaky surface modeson an ultra-thin anisotropic metasurface by using a resonant silicon nano-sphere for excitation and s-SNOMfor mode detection.

2. Methods

s-SNOM combines spectroscopic capabilities with subwavelength resolution down to the tens-of-nanometer scale. This technique is suited for direct mapping of optical surface modes on subwavelength structures. In our experiment, light ($\lambda = 850\,nm,\ P \approx 10\,mW$) is focused onto an oscillatingcantilever tipwhich is scanned withinthe surface near-field region along the sample-air interface. The probing tip thus acts as a nanoantenna, which scatters a part of theelectromagnetic surface wave from the near-fieldinto the far-field, where it is detected using pseudo-heterodyne detection.

The sample under study is a periodic array (with square unit-cell size of $(200\,nm)^2$) of gold nano-disks with elliptical base on a fused silica substrate. The lengths of the long and short axes are 175 and140 nm, respectively, and the thickness of the gold disks is 20 nm. Themetasurfaceis asymmetrically interfaced between air and its fused-silica substrate (n=1.45) [3]. Surface waves on the metasurfaceare excited by shining the focused TM-polarized light onto the resonant silicon nano-sphere which has a diameter of 210 nm and has been placed onto the sample using nano-manipulation techniques supervised by electron-beam microscopy.

3. Results

Figure 1(a)shows the s-SNOM near-field map of the metasurface. The silicon nano-sphere is visible near the centre of the figure (position marked by the dotted red line). The black arrow shows the direction of the incident light parallel relative to the long axis of the metastructure.We attribute the wave pattern, originating from the nano-antenna, to the interference of a surface wave,launched at the sphere, with the incident light [4],while the periodic fine-structure in the picture corresponds tothe near-field images of the gold nanodisks. In the picture of figure 1(b), this pattern is removed by Fourier filtering to leave only the wave interference signal. Figure 1(c) shows the two-dimensional Fourier transform of figure 1(b) centred at $k = 0\,\mu m^{-1}$. Interference fringes are visible as double ringpattern with an offset from the centreof the image which is equal to the projection of the incident light($k_{in}\theta_{in}$). For better visibility, the rings aremarked with green dashed lines. The radiusof the rings represents the wavevector of the surface wave which is found to be isotropic within the uncertainty of the measurement.

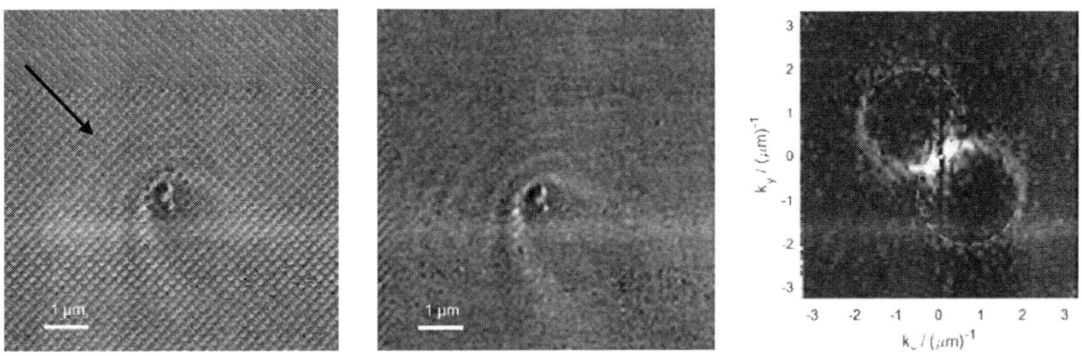

Figure 1.(a) Near-field image of interference fringes between incoming light (black arrow) and the leaky surface wave excited at the nano-sphere. **(b)** Near-field image after Fourier-filtering to remove the periodic pattern of the gold dots of the metasurface. **(c)** 2D Fourier-transform of (b), exposing the interference of incoming light with the leaky surface wave as ringswith an offset with respect tothe centre of the image.

Identification of the leaky mode issupported by full-wave simulationswhere the nano-sphereis modelled as a discrete electric dipole located slightly above the metasurface. For the given asymmetric interfacing, a leaky surface mode with $\lambda_{sw} \approx 820\,nm$ is found. The wave is confined at the metasurface-air interface, and leaks through the ultra-thin metasurface into the higher-index dielectric substrate.

We thus identify the observed surface mode as a leaky surface mode. Because of its small wavevector ($k_{sw} < k_{in}n$), this mode is not visible in total-internal-reflection measurements, as in [3],while bound modes studied there are not, or only weakly, excited by the nano-sphere under the illumination conditions of our experiment and are thus not identified in the s-SNOM measurements reported here.

References

[1] Glybovski S B, Tretyakov S A, Belov P A, Kivshar Y S and Simovski C R 2016*Physics Reports* **634** 1-72
[2] High A A, Devlin R C, Dibos A, Polking M, Wild D S, Perczel J, De Leon N P, Lukin M D and Park H 2015 *Nature***522** 192–6
[3] Samusev A, Mukhin I, Malureanu R, Takayama O, Permyakov D V., Sinev I S, Baranov D,

Yermakov O, Iorsh I V., Bogdanov A A and Lavrinenko A V. 2017 *Opt. Express***25** 32631
[4] Walla F, Wiecha M M, Mecklenbeck N, Beldi S, Keilmann F, Thomson M D and Roskos H G 2018 *Nanophotonics***7** 269–76
[5] Burke J J, Stegeman G I and Tamir T 1986 *Phys. Rev. B***33** 5186–201

Sn-nanothreads in GaAs matrix and their sub- and terahertz applications

D.S. Ponomarev[1,2], D.V. Lavrukhin[1,2], A.E. Yachmenev[1,2], R.A. Khabibullin[1], I.E. Semenikhin[3], V.V. Vyurkov[3], M. Ryzhii[4], T. Otsuji[5], V. Ryzhii[1,2,5]

[1]Institute of ultra high frequency semiconductor electronics of RAS, Moscow, Russia
[2]Center for Photonics and Infrared Technology, Bauman Moscow State Technical University, Moscow, Russia
[3]Institute of Physics and Technology RAS, Moscow, Russia
[4]Department of Computer Science, University of Aizu, Aizu-Wakamatsu, Japan
[5]Research Institute for Electrical Communication, Tohoku University, Sendai, Japan

E-mail: ponomarev_dmitr@mail.ru

Abstract. We report on theoretical and experimental studies of gated GaAs structures like the field-effect transistor with an array of parallel Sn nanothreads (Sn-NTs) embedded into GaAs matrix. Two device structures are proposed and considered: the pseudomorphic HEMT (PHEMT) with a doping profile of Sn-NTs and the terahertz hot-electron bolometer (THz HEB) both fabricated on vicinal GaAs substrates. PHEMT demonstrates the anisotropy of drain-source current $I_{\parallel}/I_{\perp} \sim 2.5$ at 300 K when current flows along and across Sn-NTs which we relate to formation of quasi-one-dimensional conductivity channels in InGaAs quantum well. Maximum power gain cut-off frequency $f_{max} \sim 0.15$ THz is shown. The operation of THz HEB is associated with an increase in the density of the delocalized electrons due to their heating by the incoming THz radiation. The quantum and the classical device models are developed. Since the fraction of the delocalized electrons strongly depends on the average electron energy (effective temperature), the proposed THz HEB can exhibit an elevated responsivity compared with the HEBs based on more standard heterostructures. Due to a substantial anisotropy of the device structure, the THz HEB may demonstrate a noticeable polarization selectivity of the response to the in-plane polarized THz radiation.

1. Introduction

Semiconductor devices based on quasi-one-dimensional electron systems might be attractive for sub-THz and THz regimes [1]. By utilizing vicinal GaAs substrates with the surface inclined with respect to the (100) direction by the certain angle one can fabricate GaAs structures with the surface that constitutes an array of the terraces with one monolayer height (0.28 nm) and the spacing between the terraces about of 50 nm. The δ-doping of this surface by Sn atoms results in the formation of the chains of these atoms (i.e., self-assembled Sn-nanothreads – Sn-NTs) at the terrace edges. On overgrowing the obtained structure, one acquires the array of the Sn-NTs inserted in the GaAs matrix. The free electrons accumulated around each Sn-NT form highly conducting channels along the NTs. The average electron density can be controlled by the gate voltage. The conductivity along these channels and perpendicular to them can be markedly different [2,3].

IOP Conf. Series: Journal of Physics: Conf. Series **1092** (2018) 012166 doi:10.1088/1742-6596/1092/1/012166

In this paper we propose and consider first pseudomorphic HEMT (PHEMT) and the lateral THz hot-electron bolometer (THz HEB) fabricated on vicinal GaAs substrates. The first utilizes InGaAs quantum well with doping profile of Sn-NTs while the second uses the gated GaAs structure like the field-effect transistor with Sn-NTs embedded into GaAs matrix as a sensing element of THz detector.

2. PHEMT fabrication and its frequency characteristics

The schematic view of the considered PHEMT is shown in Ref. [3]. The fabrication process in details is discussed in Ref. [4]. Figures 1,2 demonstrate measured I-V characteristics and frequency dependences of MSG/MAG (Maximum stable gain/Maximum available gain) and U_g (Mason's gain), respectively. As seen in figure 1 the anisotropy of drain-source current is obtained at 300 and 77 K; moreover the reduction of measurement temperature down to 77 K leads to the appearance of a negative differential resistivity, most likely attributed to the capture of the carriers by deep traps, warming up the electron gas and scatter carriers towards wide bandgap layers with smaller mobility [5]. The obtained experimental values (figure 2) are well agreed with PHEMT based on heterostructures with 20% content of InAs in InGaAs quantum well (QW) and exceeds its typical values (which are approximately 15 dB for MSG/MAG, for instance). We believe this is due to decrease of electron scattering during movement in QW along Sn-NWs, and as a result, due to higher drift electron velocity in the quasi-1D conductivity channels formed in QW. Besides, it should be pointed out that varying the width of the spacer layer, which spatially divides Sn-NWs and QW, it is possible to obtain a deeper modulation of the electron density in QW, which will result in a higher anisotropy of PHEMT characteristics.

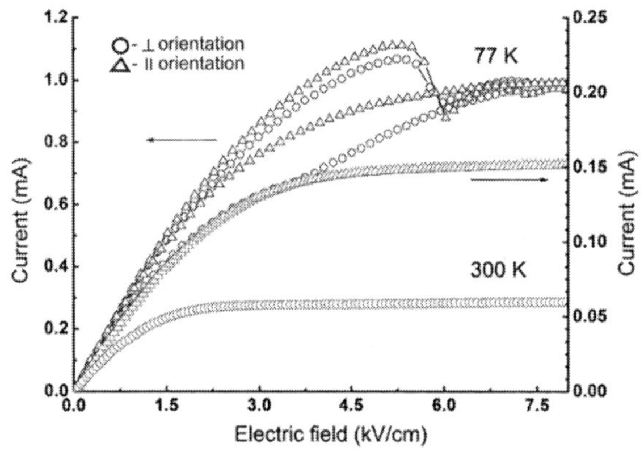

Figure 1. Measured I-V characteristics of in strong electric fields when current flow parallel (||) and perpendicular (⊥) to Sn-NTs for 300 K and 77 K.

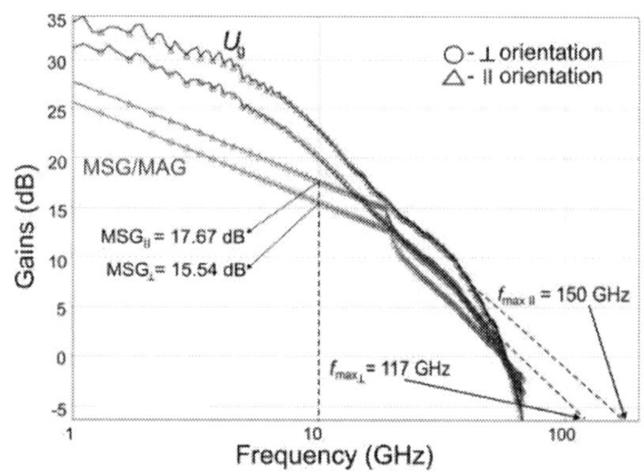

Figure 2. Power gain cut-off frequency f_{max} and MSG/MAG of PHEMT for parallel (||) and perpendicular (⊥) to Sn-NTs directions.

3. THz HEB device structure and operation principle

We consider THz HEB based on GaAs layer in which an array of Sn-NTs (which is a sensing element of detector) is inserted. There are the side contacts (the source and drain) and the gate electrode as schematically shown in figure 3. The operation of the proposed THz HEB is associated with the heating of electrons in the channels along the NTs resulting in an increase in the electron current perpendicular to these channels and, hence, between the source and the drain [6]. This current is created by the electrons with the energies exceeding the height of the potential barriers between the NTs. The average electron density can be controlled by the gate voltage (V_g) as seen in figure 3,c. Two device models (the classical and the quantum one) are developed: the quantum one was based on the self-consistent solution of the Poisson and Schrödinger equations, the classical model involved the Poisson equation and density of states omitting quantization. Some of simulation results (for the classical model) are depicted in figure 4.

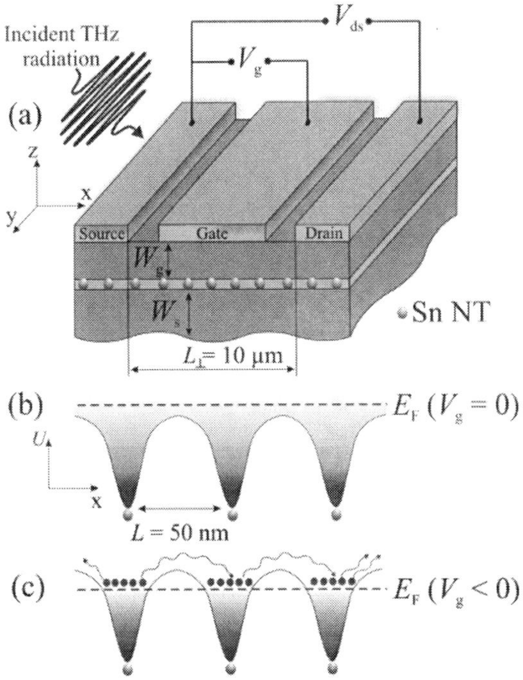

Figure 3. Schematic view of the proposed THz HEB (a). The potential profile (U) accumulated by Sn dopants for the ungated case is indicated in (b) and for the negative gate voltage (V_g) is depicted in (c) (see Ref. [2]). The vertical sizes of the calculation domain are indicated by W_g and W_s, the distance between the source and drain is shown by $L_\perp$ (W_g and W_s » L).

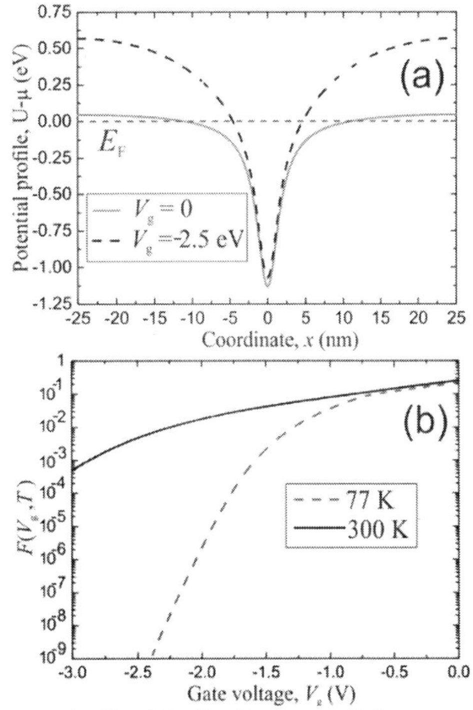

Figure 4. Spatial distributions of the potential energy in the source-drain direction for different gate voltages V_g at room temperature (a) and fraction of the electrons over barrier (the delocalized electrons) to all electrons $F(V_g,T)$ vs. gate voltage for two different temperatures (b). Both figures (a,b) are attached according to the classical simulation.

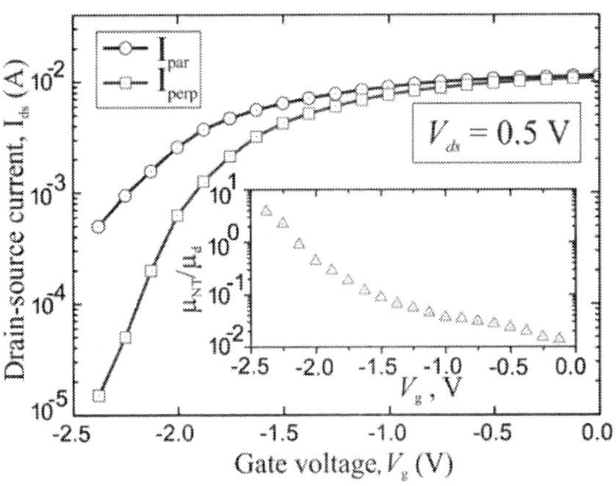

Figure 5. Extraction from series of the measured I-V curves: the drain-source current I_{ds} vs. gate voltage at V_{ds} = 0.5 V for current flow perpendicular to the Sn-NTs (I_{perp}) and parallel to them (I_{par}). The inset shows the ratio of mobilities vs. gate voltage.

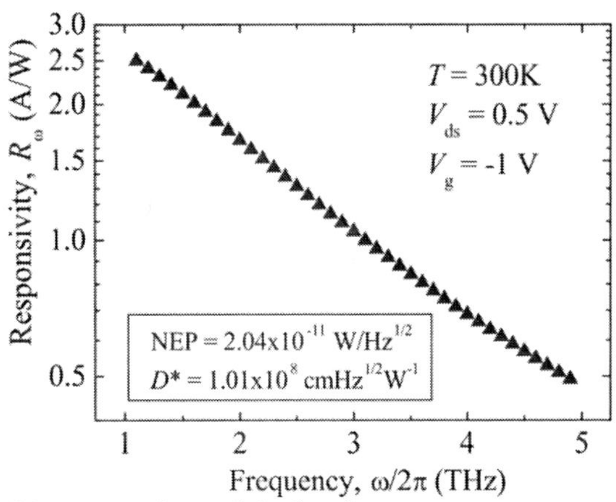

Figure 6. Calculated room-temperature responsivity R_w at the gate voltage V_g = -1 V and the drain-source voltage V_{ds} = 0.5 V. The noise-equivalent power (NEP) and the specific detectivity (D^*) of the THz HEB are presented.

It should be noted that according to the experimental data (see figure 5) the two device models qualitatively demonstrate similar behavior, nevertheless, the classical one is in the better quantitative agreement compared to the quantum one [7]. Figure 6 shows the calculated responsivity (R_ω) for room temperature at the gate voltage V_g = -1 V corresponding to the experimentally measured dark current (from I-V characteristics) with the drain-source voltage V_{ds} = 0.5 V. As seen for the frequency $\omega/2\pi$ = 1.0 THz the responsivity equals to R_ω = 2.51 A/W that exceeds the average values for HEBs on more standard heterostructures and some uncooled THz detectors [8].

The proposed HEB may possess the polarization selectivity (i.e. sensitivity) in the plane of the incident THz radiation which is useful for practical applications. The sensitivity of polarization dependence is associated with different absorption of radiation at which electric field is directed parallel or perpendicular to Sn-NTs. This is due to a distinction in the electron dynamics (virtually free motion in the former case and localization in the latter).Evidently, the HEB might possess higher responsivity when the THz radiation is incident parallel to the NTs.

4. Conclusion

In this paper we have proposed and considered first pseudomorphic HEMT and the lateral THz hot-electron bolometer fabricated on vicinal GaAs substrates. The first utilizes InGaAs quantum well with doping profile of Sn-nanothreads (Sn-NTs) while the second uses the gated GaAs structure like the field-effect transistor with an array of self-assembled Sn-NTs embedded into GaAs matrix which serves as a sensing element of THz detector. The theoretical and experimental studies of both devices were carried out and presented.

Acknowledgements

The work was supported by the Russian Scientific Foundation (Project No.14-29-00277) and the Russian Foundation for Basic Research, grants No. 16-29-03033 and 18-07-01145.

References

[1] Chuang S., Gao Q., Kapadia R., Ford A., Guo J., Javey A. 2013 *Nan. Lett.* **13+** 555.

[2] Ponomarev D.S., Lavrukhin D.V., Yachmenev A.E., Khabibullin R.A., Semenikhin I.E., Vyurkov V.V., Ryzhii M., Otsuji T., Ryzhii V. 2018 *J. Phys. D: Appl. Phys.* **51**(13) 135101.

[3] Khabibullin R.A., Yachmenev A.E., Bugayev A.S., Ponomarev D.S., Lavrukhin D.V., Maltsev P.P. 2015 *Semicond. Sci. Tech.* **30** 085009.

[4] Khabibullin R.A., Yachmenev A.E., Bugayev A.S., Ponomarev D.S., Lavrukhin D.V., Maltsev P.P. 2016 *Semicond.* **50**(2) 185.

[5] Houng M.P., Wang Y.H., Chen H.H., Pan C.C. 1992 *Sol. Stat. Electron.* **35**(1) 67.

[6] Ryzhii V., 1996 *Semicond. Sci. Tech.* **11** 759.

[7] Semenikhin I.E., Vyurkov V.V., Bugayev A.S., Khabibullin R.A., Ponomarev D.S., Yachmenev A.E., Maltsev P.P., Ryzhii M., Otsuji T., Ryzhii V. 2016 *Proc. SPIE* 102240R.

[8] Rogalski A. and Sizov F. 2011 *Opt. Electron. Rev.* **19** 346.

Morphology and effective dielectric functions of ultra-thin gold films

D I Yakubovsky[1], R V Kirtaev[1], Y S Stebunov[1,2], A V Arsenin[1,2] and V S Volkov[1,2,3]

[1]Moscow Institute of Physics and Technology, Institutskyi 9, 141700, Dolgoprudnyi, Russia
[2]GrapheneTek, 7 Nobel Street, Skolkovo Innovation Center, 143026, Russia
[3]SDU Nano Optics, Mads Clausen Institute, University of Southern Denmark, Campusvej 55, DK-5230, Odense, Denmark

dmitrii.yakubovskii@phystech.edu

Abstract. In the present study, a new method for analyzing the dielectric functions in dependence with the thickness of ultra-thin gold films is proposed. The reliability of the method is carefully verified through the correlation of optical properties and structural morphology of gold films at thicknesses around the percolation threshold (from 2 to 20 nm). The optical characterization of deposited films is performed in visible and NIR regions (300-1500 nm) by a spectroscopic ellipsometry along with transmission spectra measurements at normal incidence. Fitting of the ellipsometric data allows one to calculate the effective complex dielectric function of cluster-like and continuous films of different mass-equivalent thickness. Surface morphology of the films is analysed by scanning electron microscopy.

1. Introduction

Gold is one of the most frequently used plasmonic metals due to its low optical absorption, high conductivity and chemical stability. Thin films of gold are commonly used in various applications such as surface plasmon resonance (SPR) sensing [1,2], nanoantennas [3], surface enhanced Raman spectroscopy (SERS) [4], plasmonic waveguides [5] and many others. Efficiency of these devices is strongly dependent on dielectric function, thickness and structural features of films. Thus, knowledge of the refractive index and absorption coefficient is important for design and fabrication of plasmonic and nanophotonic devices. It is known that gold film prepared by conventional physical deposition methods onto glass and silicon substrate exhibits a weak surface adhesion, which results to different stages of film growth: nucleus formation, increase and merging of isolated islands and subsequent percolation up to a continuous film [6]. This indicates that ultrathin (1-10 nm) film has island-type structure (causing difficulties in depositing smooth and uniform ultrathin layers). However, manipulation of the growth process by changing deposition parameters, surface adhesion layers and activation of a substrate can help one achieve a favorable morphology of the film at low thicknesses [7]. Nevertheless, island films consisting of isolated particles found their wide application in SERS analysis due to the effect of the local electromagnetic field enhancement near the metal particle. The efficiency of this sensing method is also governed by the optical response of the film and its nanoparticle-network structure.

Recently, the comprehensive study examined the effect of the variation of film thickness (20-180 nm) on the optical, electrical and structural properties of thin gold films (and their relationship) [8,9]. However, for the lower thicknesses (near the percolation stage, i.e. <20 nm) the dependence between optical properties and structural morphology is worth to be considered as well in more detail. In order to find effective optical properties of discontinuous film an ellipsometry model fitting can be used with oscillator approach [10] or point by point fitting n and k which we have applied in this study.

2. Methods

Gold films were fabricated by conventional e-beam deposition procedure under vacuum conditions. The nominal thicknesses of the deposited films were monitored by the quartz-crystal mass-thickness sensor mounted in the vacuum chamber. The dielectric function spectra of the studied Au thin films were evaluated from data measured using variable angle spectroscopic ellipsometer in photon wavelengths range from 300 to 1500 nm. Ψ and Δ ellipsometry parameters were measured at two angles of incidence (70° and 75°) and the corresponding experimental dielectric function values were extracted by the numerical iteration fitting procedure. Optical transmission spectra were taken at normal incidence using the same spectroscopic ellipsometer setup. Finally, the surface morphology of the deposited films was analyzed using the scanning electron microscopy (SEM).

3. Results and discussion

Several series of measurements were conducted using the SEM equipment. These measurements were designed to study and illustrate the features of the films growth dynamics and thickness-dependent structural morphology changes. SEM images of some of these films are shown in figure 1. Thus, it is clearly seen that gold of mass thicknesses of 2 and 5 nm exhibit structures of isolated clusters, which lateral size is between the values of 10-20 nm and 25-50 nm, for corresponding island films. From thickness of 5 nm to 10 nm there is a beginning of the percolation stage which occurs up to the formation of a continuous film at 20 nm. Statistical image analysis in table 1 also indicates an increase of the metal filling fraction from 43% in 2 nm to 99% in 20 nm thick film.

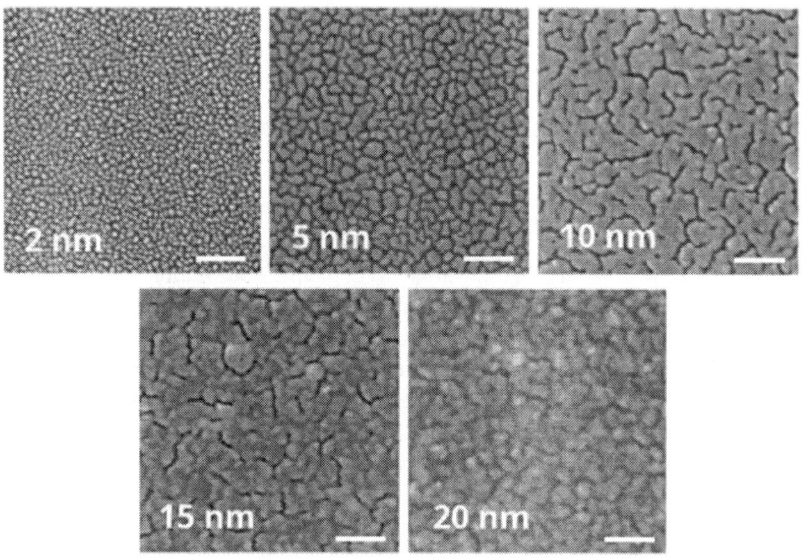

Figure 1. SEM images of deposited gold films (2-20 nm). Scale bar is 100 nm.

Table 1. Gold films surface coverage estimated by statistical analysis of SEM images using ImageJ software [11].

Thickness (nm)	2	5	10	15	20
Filling factor (%)	43.2	60.8	79.8	94.5	99.0

To investigate this phenomenon more thoroughly we used spectroscopic ellipsometry measurements so that optical constants, refractive index n and extinction coefficient k could be extracted accurately for all deposited Au films. Our results (shown in figure 2) demonstrate substantially different dependencies of measured ellipsometry parameters Ψ and Δ (a, b) and calculated dielectric curves (c, d) which should be related to the influence of film morphology. In particular, discontinuous films of low thickness (2-10 nm) demonstrate highly reduced metallic optical response which should be attributed to a high volume fraction of voids in films. Furthermore, samples with the thickness of 2 and 5 nm don't even indicate metallic dispersion in NIR range, as their real permittivity component is positive. For the thinnest film of 2 nm the imaginary component exhibits strong absorption peak at the wavelength around 650 nm, which is the result of the contribution of localized surface plasmon resonance (LSPR) excitation in gold nanoparticles. While the film thickness increases the isolated particles grow and coalesce, that leads to broadening and disappearing of the absorption peak shown in figure 2(d). In contrast to this, for 15- and 20-nm-thick gold films the obtained real and imaginary dielectric components demonstrate the typical Drude-Lorentz dielectric behavior, which can be seen by comparison with the data from Johnson and Christy for a guaranteed continuous gold film [12] (plotted in figure 2(c,d)). But due to discontinuous morphology, the presence of voids and isolated metal particles, optical properties of 2-10 nm films highly deviate from Johnson and Christy dependences.

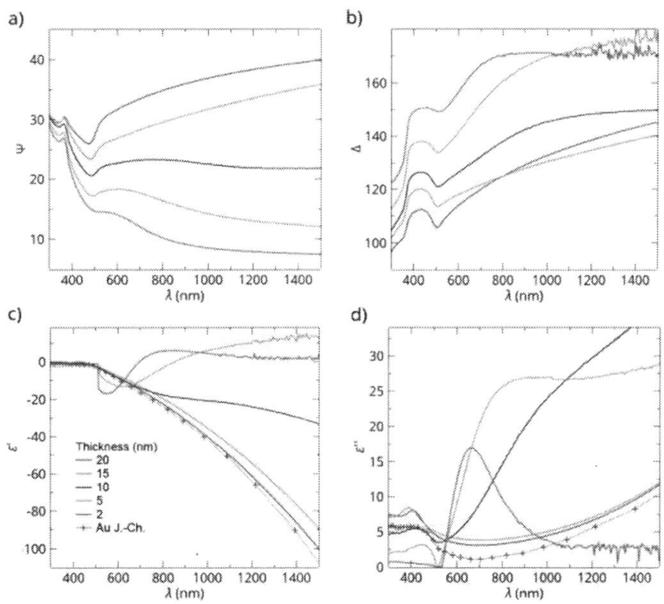

Figure 2. Results of measured ellipsometry parameters Ψ and Δ (a, b) and extracted effective dielectric functions (c, d) for all deposited films (2-20 nm).

In addition, the transmission spectra measurements were performed with films on transparent glass substrates (see figure 3). Minima in transmission spectra are observed only for island films of 2 and 5 nm. It shifts from 600 nm to 700 nm as the mass thickness increases from 2 nm to 5 nm, which is related to the particle size increase. The latter is also confirmed with the film surface morphology visualization by SEM presented in figure 1 and correlates well with the absorption maximum of measured ε'' curve in figure 2(d).

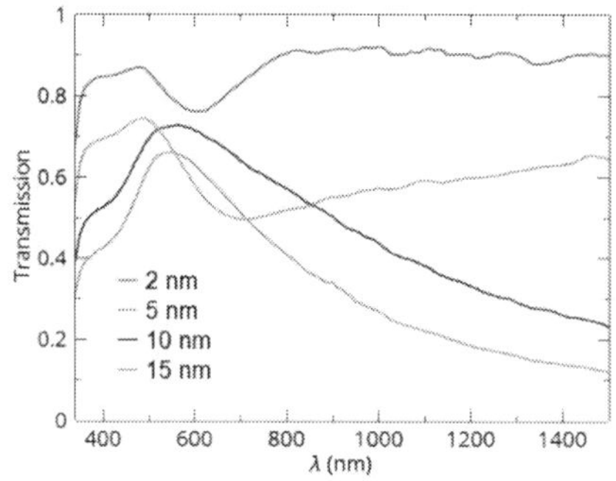

Figure 3. Transmission spectra of ultra thin gold films 2-15 nm.

4. Conclusions

To summarise, we have measured and analyzed the correlation between optical properties and morphology of ultra-thin gold films. Our study demonstrates that the results of ellipsometry measurements are significantly different for island, percolated and continuous films. The results obtained for ultra-thin (2-5 nm) films show that their optical response strongly deviates from percolated films with higher thicknesses, which is adequately explained by the excitation of LSPR. Gold polycrystalline films on SiO_2 substrates exhibit Drude-Lorentz dielectric response for thicknesses ranging from 15 to 20 nm. Thus, the influence of changes from island-like morphology to a continuous film on effective permittivity was studied.

Acknowledgments
This work was supported by the Russian Science Foundation, grant no. 17-79-20345. A portion of this work was performed at the Shared Facilities Center of the Moscow Institute of Physics and Technology, which is supported by the Ministry of Education and Science of the Russian Federation, grant no. RFMEFI59417X0014.

References

[1] Stebunov Y V, Aftenieva O A, Arsenin A V, Volkov V S 2015 *ACS Appl. Mater. Interfaces* **7** 21727
[2] Volkov V S, Stebunov Y V, Yakubovsky D I, Fedyanin D Yu, Arsenin A V 2017 *AIP Conf. Proc.* **1874** 020008
[3] Schumacher T, Kratzer K, Molnar D, Hentschel M, Giessen H, Lippitz M 2011 *Nat. Commun.* **2** 333
[4] Tian Z, Ren B, Wu D 2002 *J. Phys. Chem. B* **106** 9463
[5] Fang Y, Sun M 2015 *Light Sci. Appl.* **4** e294
[6] Kossoy A, Merk V, Simakov D, Leosson K, Kéna-Cohen S and Maier S A 2015 *Adv. Optical Mater.* **3** 71
[7] Schwartzkopf M, Hinz A, Polonskyi O, Strunskus T, Löhrer F C, Körstgens V, Müller-Buschbaum P, Faupel F and Roth S V 2017 *ACS Appl. Mater. Interfaces* **9** 5629
[8] Yakubovsky D I, Arsenin A V, Stebunov Y V, Fedyanin D Y and Volkov V S 2017 *Opt. Express* **25** 25574
[9] Yakubovsky D I, Fedyanin D Yu, Arsenin A V and Volkov V S 2017 *AIP Conf. Proc.* **1874** 040057
[10] Lončarić M, Sancho-Parramon J and Zorc H 2011 *Thin Solid Films* **519** 2946
[11] Schneider C A, Rasband W S, Eliceiri K W 2012 *Nat. Methods* **9** 671
[12] Johnson P B and Christy R W 1972 *Phys. Rev. B* **6** 4370

Circularly polarized antenna for coherent manipulation of NV-centers in diamond

Vitaly Yaroshenko[1], Anastasia Zalogina[1], Dmitriy Zuev[1], Polina Kapitanova[1], Ilya Shadrivov[1,2]

[1] Department of Nanophotonics and Metamaterials, ITMO University, 49, Kronverksky pr., 197101, Saint-Petersburg, Russia

[2] Nonlinear Physics Centre, Research School of Physics and Engineering, Australian National University, Canberra ACT 2601, Australia

E-mail: v.yaroshenko@metalab.ifmo.ru

Abstract. Dielectric resonator antenna with circular polarization of microwave magnetic field for efficient, coherent and uniform spin manipulation at nitrogen-vacancy centers in diamond is discussed. The results of numerical simulations of the microwave magnetic field generated by the antenna are reported and analyzed. The uniform magnetic field with circular polarization is obtained inside the antenna. Using the simulated amplitude of the magnetic field inside the antenna the Rabi frequency is estimated. The Rabi frequency of 41 MHz with inhomogeneity less then 1 % for the diamond volume of 3 mm^3 is demonstrated under 10 W of input microwave power.

Introduction

Nitrogen-vacancy (NV) center is one of numerous point defects in a diamond with unique properties, which could find a wide range of application in quantum information [1], sensing of magnetic fields [2] and temperature [3] etc. Electron spins in NV-centers, localized at atomic scales, can be manipulated at room temperature by applying a microwave (MW) magnetic field at the frequency of 2.87 GHz. For sensing applications it is highly important to coherently manipulate the spins of NV-centers in relatively high volume of a diamond. Thus the MW antenna producing large and uniform MW magnetic field in a large volume is in a high demand. Moreover, the circular polarization of the magnetic field is useful for addressing different orientations of NV-centers. Several designs of MW antennae and devices with linear and circular polarizations of MW magnetic field have been proposed recently [4–6]. However, they provide high intensity and uniform MW magnetic field over a large area and can be used to coherently manipulate the ensemble of NV-centers only in a plane.

In this paper we propose the design of circularly polarized MW antenna producing uniform MW magnetic field that will be able to coherently manipulate the ensemble of NV-centers in diamond in a volume at the room temperature. We study the antenna performance numerically using CST Suite Studio 2017 and demonstrate that high intensity microwave field can indeed be generated by the antenna. The obtained results of magnetic field simulations are used to estimate mean value and inhomogeneity of the Rabi frequency. We found that the Rabi frequency of 41 MHz with inhomogeneity of less than 1 % for the diamond volume of 3 mm^3 can be achieved under 10 W of input microwave power.

Dielectric Resonant Antenna Design

The dielectric resonator antennae (DRA) are well studied and became a promising candidate for replacing traditional metallic radiating elements at microwave frequencies [7]. This is mainly attributed to the fact that DRAs do not suffer from conduction losses and are characterized by high radiation efficiency. In this study, we consider the DRA which consists of a dielectric hollow cylindrical resonator with two metallic loops, placed in orthogonal planes as demonstrated in Fig. 1 (a). With this type of excitation we are able to excite the $HEM_{11\delta}$ mode of the dielectric resonator [7,8]. Providing the phase difference between the loops of 90° it is possible to obtain the circular polarization of the MW magnetic field inside the DRA [8,9]. Here we consider the microwave ceramic with permittivity of $\varepsilon = 520$ and loss tangent of $\tan(\delta) = 0.002$ measured at 3 GHz [10]. To estimate the resonance frequency of $HEM_{11\delta}$ mode [8] of the dielectric resonator as a function of its geometrical parameters we use the equation [8]:

$$f_{HEM_{11\delta}} = 2.735\frac{c\varepsilon^{-0.436}}{2\pi R}\left[0.543 + 0.589\frac{R}{T} - 0.05\left(\frac{R}{T}\right)^2\right], \quad (1)$$

where c is the speed of light, R is the resonator's outter radius, T is the resonator's height, ε is the permittivity of the dielectric material. Figure1 (b) demonstrates the calculated resonance frequency of $HEM_{11\delta}$ mode as a function of resonator's geometrical parameters for the fixed value of the permittivity. The $HEM_{11\delta}$ mode at the frequency of 2.87 GHz exists for the resonator of the height of T = 5 mm and the radius of R = 3 mm.

Results of Numerical Simulation

The numerical simulation of the DRA performances was done in CST Suite Studio 2017. Note, that now we consider the hollow dielectric resonator with the bore radius of 1.6 mm. The hole in the DRA is needed to place a diamond in the maximum of the magnetic field. It is important to mention that the presence of the bore affects the $HEM_{11\delta}$ mode frequency. Thus, a numerical optimization of the resonator's geometrical parameters to adjust the operational frequency to 2.87 GHz has been done. After this optimization the required dielectric resonator's height of T = 5.05 mm and outter radius of R = 2.76 mm have been found. To achieve a good matching of the DRA the position of feeding loops was also optimized. The feeding loops are identical in size, having the radius of 1.5 mm and wire size of 0.25 mm. The ports with 50 Ohm impedances

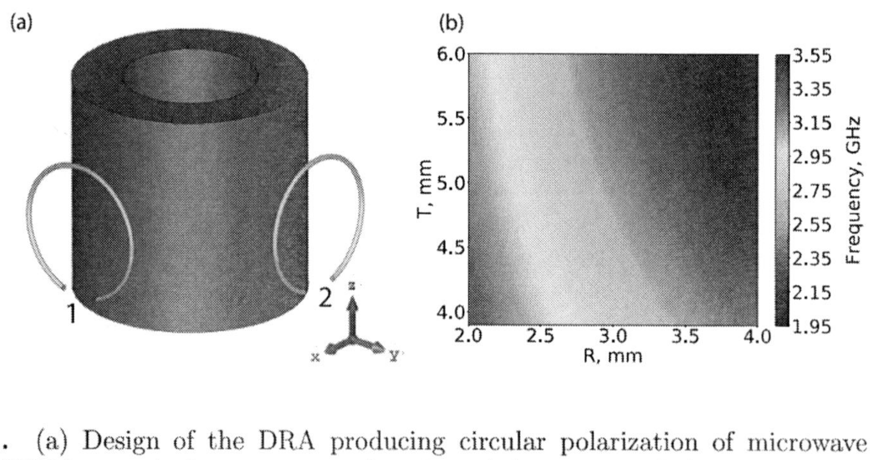

Figure 1. (a) Design of the DRA producing circular polarization of microwave magnetic field; (b) $HEM_{11\delta}$ mode frequency as a function of the resonator's height and radius for the permittivity of $\varepsilon = 520$.

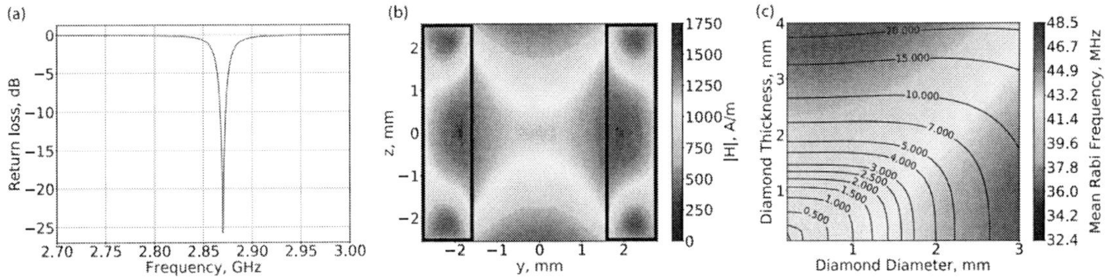

Figure 2. (a) Simulated return loss of the DRA; (b) Simulated distribution of the microwave magnetic field in $y-z$ plane for 10 W of input power; (c) Mean Rabi frequency as a function of the diamond volume (diameter and thickness) calculated by the Eq. (3) for 10 W of input power. The color represents the value of the mean Rabi frequency (MHz). Isolines represent the inhomogeneity (%) of the mean value of the Rabi frequency calculated by the the Eq. (3).

are placed in the 1 mm gap of the loops. The location of the loops at the center of the long axis of the resonator at a distance of 1.1 mm from its surface provides the return loss better than -25 dB at the frequency of 2.87 GHz (see Fig. 2 (a)).

The numerical simulation of the magnetic field of the DRA has been performed for the case when the feeding loops are excited with the 90° phase difference. The input power at each feeding loop was set at 5 W. The simulated magnetic field distribution inside the DRA is shown in Fig. 2(b). One can observe that the magnetic field is mostly concentrated in the dielectric walls of the DRA. The amplitude of the magnetic field reaches 1750 A/m. However, one can find the area in the middle of the bore where the magnetic field is uniform and its amplitude is about 1000 A/m.

To understand the efficiency of the coupling of the microwave magnetic field to the spins of NV-centers one needs to estimate the Rabi oscillation frequency. Normally, this value reaches tens of MHz at the relatively low power level of 5 W [4–6]. In the case of circular polarization of MW magnetic field the Rabi frequency can be calculated as [11]:

$$\Omega = \frac{\mu B}{h}, \tag{2}$$

where B is the amplitude of the magnetic field, h is the Planck constant, μ is the dipole moment. Here we consider the Rabi frequency not in a single point or a plane, but in volume. Thus we propose to calculate the mean Rabi frequency over the volume of a diamond which is placed in the middle of the entire bore of the DRA, where the homogeneous circularly polarized magnetic field is obtained. In this case we use the following equation, where the Rabi frequency is integrated over the diamond volume:

$$\langle\Omega\rangle = \frac{\int_V \Omega dV}{\int_V dV}, \tag{3}$$

where V is the volume of a diamond. To estimate the inhomogeneity of the mean Rabi frequency the standard deviation of its value is calculated as:

$$\sigma = \sqrt{\left(\frac{\int_V \Omega dV}{\int_V dV}\right)^2 - \left(\frac{\int_V \Omega^2 dV}{\int_V dV}\right)}. \tag{4}$$

For simplicity, we suppose that the diamond has a cylindrical shape and characterized by the diameter and thickness. Substituting the amplitude of the magnetic field numerically obtained inside the entire bore of the DRA to the Eq. (2), (3) and (4) and varying the diamond volume (diamond radius changes from 0.01 to 3 mm, diamond thickness changes from 0.01 to 4 mm) we obtain the mean value of the Rabi frequency and its inhomogeneity as demonstrated in Fig. 2 (c). The maximum Rabi frequency of 48.5 MHz with inhomogeneity less than 10% can be achieved for the diamond volume of 28 mm^2 x 1.5 mm. By reducing the diamond volume down to 3 mm^2 x 1 mm the Rabi frequency of 41 MHz with inhomogeneity less than 1% can be obtained.

Conclusion

The design of DRA with circular polarization of microwave magnetic field has been proposed. The antenna return loss and magnetic field distribution have been studied numerically. The obtained results of magnetic field have been used to estimate the mean value and inhomogeneity of the Rabi frequency. Over a diamond volume of 3 mm^3 the predicted value of Rabi frequency of 41 MHz with inhomogeneity of less than 1% can be achieved for 10 W of input power. The DRA prototype is under fabrication. We believe that after the optical examination it can be used for efficient, coherent and uniform spin manipulation of NV-centers in diamond.

Acknowledge

This work was supported by the Russian Science Foundation (Grant 16-19-10367).

References

[1] Guido Burkard, V. O. Shkolnikov, and D. D. Awschalom. Designing a cavity-mediated quantum cphase gate between nv spin qubits in diamond. *Phys. Rev. B*, 95:205420, May 2017.

[2] M. Ummal Nusran, N. M.and Momeen and M. V. Gurudev Dutt. High-dynamic-range magnetometry with a single electronic spin in diamond. *Nature Nanotechnology*, 7:109 EP –, Dec 2011.

[3] Yan-Kai Tzeng, Pei-Chang Tsai, Hsiou-Yuan Liu, Oliver Y. Chen, Hsiang Hsu, Fu-Goul Yee, Ming-Shien Chang, and Huan-Cheng Chang. Time-resolved luminescence nanothermometry with nitrogen-vacancy centers in nanodiamonds. *Nano Letters*, 15(6):3945–3952, 2015. PMID: 25951304.

[4] Khadijeh Bayat, Jennifer Choy, Mahdi Farrokh Baroughi, Srujan Meesala, and Marko Loncar. Efficient, uniform, and large area microwave magnetic coupling to nv centers in diamond using double split-ring resonators. *Nano Letters*, 14(3):1208–1213, 2014.

[5] Kento Sasaki, Yasuaki Monnai, Soya Saijo, Ryushiro Fujita, Hideyuki Watanabe, Junko Ishi-Hayase, Kohei M. Itoh, and Eisuke Abe. Broadband, large-area microwave antenna for optically detected magnetic resonance of nitrogen-vacancy centers in diamond. *Review of Scientific Instruments*, 87(5):053904, 2016.

[6] M. Mrzek, J. Mlynarczyk, D. S. Rudnicki, and W. Gawlik. Circularly polarized microwaves for magnetic resonance study in the ghz range: Application to nitrogen-vacancy in diamonds. *Applied Physics Letters*, 107(1):013505, 2015.

[7] K. M. Luk and K. W. Leung. *Dielectric Resonator Antennas*. Research Studies Pr Ltd, 1 edition, 2002.

[8] Sebastian A. Aussenhofer and Andrew G. Webb. Design and evaluation of a detunable waterbased quadrature hem11 mode dielectric resonator as a new type of volume coil for high field mri. *Magnetic Resonance in Medicine*, 68(4):1325–1331.

[9] C. A. Balanis. *Antenna Theory: Analysis and Design*. John Wiley and Sons, Inc., 4 edition, 2016.

[10] E. Nenasheva, N. Kartenko, I. Gaidamaka, O. Trubitsyna, S. Redozubov, A. Dedyke, and A. Kanareykin. *Journal of the European Ceramic Society*, 30:395–400, 2010.

[11] Johannes Herrmann, Marc A. Appleton, Kento Sasaki, Yasuaki Monnai, Tokuyuki Teraji, Kohei M. Itoh, and Eisuke Abe. Polarization- and frequency-tunable microwave circuit for selective excitation of nitrogen-vacancy spins in diamond. *Applied Physics Letters*, 109(18):183111, 2016.

Eigenmodes degeneracy with Huygens-like all-dielectric metasurface

Oleh Y. Yermakov, Andrey A. Bogdanov

Department of Nanophotonics and Metamaterials, ITMO University, St. Petersburg, Russia

E-mail: o.yermakov@metalab.ifmo.ru

Abstract. The spectrum of the free space is double degenerated. One of the possible eigenmodes bases is right- and left-handed circularly polarized plane waves. All classical bulk optics is based on this fact, e.g. polarizers and interferometers. Surface eigenmodes can be degenerated only for several fixed frequencies. This work focuses on the development of a metasurface design, which enables to support the broadband degeneracy of surface eigenmodes spectrum. The final results can become the significant milestone towards the switching to flat optics.

1. Introduction

Metasurfaces have recently gained significant attention due to their extraordinary properties and advantages over bulk metamaterials such as planar geometry, low losses, light weight, etc. They are widely used to manage the far-field electromagnetic fields in a number of applications such as reflection, transmission, refraction, absorption, polarization transformers, antennas, lenses, holograms [1]. Besides, metasurfaces provide the control over surface waves properties, namely dispersion, polarization, localization degree, etc [2, 3].

Another appealing direction of all-dielectric metasurfaces has been emerged due to their extremely low losses in the optical range [4] and magnetic dipole resonance [5]. Recently, it has been shown that a metasurface composed of silicon cylinders can act as a Huygens-like source, i.e. to have around unity transmission with full phase coverage [6].

In this work, we focus on the development of a Huygens-like metasurface design, which supports the eigenmodes degeneracy. Specifically, we investigate the near-field properties of a metasurface based on the periodical array of silicon cylinders placed into a homogeneous environment. It opens a route to the switching from bulk three-dimensional to flat two-dimensional optics.

2. Results

Metasurface design is shown in Fig. 1. Our metasurface consists of silicon nanodisks of 134 nm radius (R) and 308 nm height (H) with a subwavelength period (a) 420 nm. The metasurface is embedded into a homogeneous host medium with refractive index $n = 1.46$. It can be achieved by using a fused silica as a substrate and the poly-dimethyl-siloxane (PDMS) layer as a top layer. Similar experimental implementations have been done by using electron beam lithography on silicon-on-insulator wafer [7, 8]. The silicon permittivity was taken from [9].

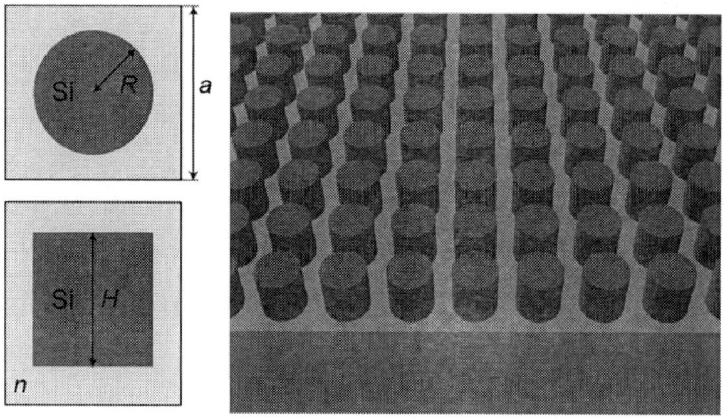

Figure 1. Design of a Huygens-like silicon metasurface.

The operational principle of high-transmissive Huygens metasurface is the full overlapping of collective electric (α_e) and magnetic (α_m) dipole resonances [6, 10]:

$$\eta_0 \alpha_e = \frac{1}{\eta_0} \alpha_m, \tag{1}$$

where η_0 is the free-space wave impedance.

However, this condition is not sufficient to achieve the overlapping of the eigenmodes. The reason is the strong spatial dispersion, i.e. the interaction between the nanoparticles. We have numerically optimized the related term via the design configuration with CST Microwave Studio in order to achieve the broadband degeneracy of the eigenmodes spectrum. The final results for two surface modes of different polarization are shown in Fig. 2.

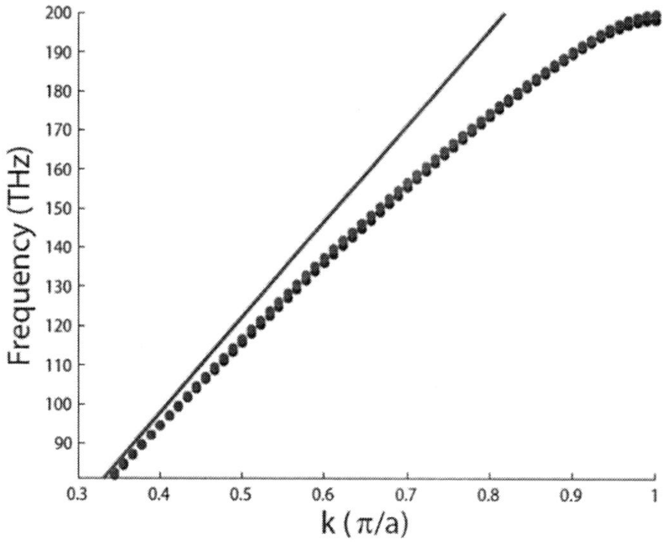

Figure 2. Degenerate spectrum of the eigenmodes dispersion. Blue dots correspond to the first TE-polarized mode, green dots correspond to the second TM-polarized mode. Red solid line is the light line in a host medium.

The coincidence of both modes is perfect. The obvious analogue is the double-degenerated free-space spectrum, where one of the possible basis is the right- and left-handed plane waves. Thus, the metasurface under consideration make possible to create the surface two-dimensional analogues of polarizers, interferometers and other bulk optical devices.

3. Conclusions

To conclude, we have proposed the platform for the flat optics based on the all-dielectric metasurface with degenerated eigenmodes spectrum. The numerical characterization has been performed. The experimental implementation and further perspectives have been offered. Practical significance of the results is highlighted by the intensive commercialization of the transition to flat optics and on-chip optical devices.

Acknowledgements

This work was supported by Russian Foundation for Basic Research (16-37-60064, 17-02-01234, 18-32-00739), the Ministry of Education and Science of Russian Federation (3.1668.2017/4.6), President of Russian Federation (MK-403.2018.2). O.Y. acknowledges the support of the Foundation for the Advancement of Theoretical Physics and Mathematics "BASIS" (17-15-604-1).

References

[1] Glybovski S B, Tretyakov S A, Belov P A, Kivshar Y S and Simovski C R 2016 *Phys. Rep.* **634** 1-72
[2] Yermakov O Y, Ovcharenko A I, Song M, Bogdanov A A, Iorsh I V and Kivshar Y S 2015 *Phys. Rev. B* **91(23)** 235423
[3] Yermakov O Y, Ovcharenko A I, Bogdanov A A, Iorsh I V, Bliokh K Y and Kivshar Y S 2016 *Phys. Rev. B* **94(7)** 075446
[4] Jahani S, Jacob Z 2016 *Nat. Nanotechnol.* **11(1)** 23
[5] Kuznetsov A I, Miroshnichenko A E, Fu Y H, Zhang J and Lukyanchuk B 2012 *Sci. Rep.* **2** 492
[6] Decker M, Staude I, Falkner M, Dominguez J, Neshev D N, Brener I, Pertsch T and Kivshar Y S 2015 *Adv. Opt. Mater.* **3(6)** 813-820
[7] Yu Y F, Zhu A Y, Paniagua-Dominguez R, Fu Y H, Luk'yanchuk B and Kuznetsov A I 2015 *Las. Photon. Rev.* **9(4)** 412-418
[8] Chong K E et al. (2016) *ACS Photon.* **3(4)** 514
[9] `www.refractiveindex.info`
[10] Asadchy V S, Faniayeu I A, RaDi Y, Khakhomov S A, Semchenko I V and Tretyakov S A 2015 *Phys. Rev. X* **5(3)** 031005

IOP Conf. Series: Journal of Physics: Conf. Series **1092** (2018) 012170 doi:10.1088/1742-6596/1092/1/012170

Formation of AlSi10Mg surfaces via selective laser melting: scanning electron microscopy and Raman spectroscopy study

S V Zabotnov[1,a], F V Kashaev[1], D V Shuleiko[1], A V Skobelkina[1], A A Vasyakov[1], A K Petrov[1], A V Chetvertukhin[1], P V Evdokimov[2], A V Garshev[2] and V I Putlayev[2]

[1] Faculty of Physics, Lomonosov Moscow State University, 1/2 Leninskie Gory, Moscow 119991, Russia

[2] Faculty of Material Sciencies, Lomonosov Moscow State University, 1/73 Leninskie Gory, Moscow 119991, Russia

[a] Corresponding author: zabotnov@physics.msu.ru

Abstract. Scanning electron microscopy analysis of surfaces produced via the selective laser melting technique by microsecond laser pulses of AlSi10Mg powders in vacuum and argon has been carried out. Because of such modification the structures are formed with micro- and nanopores with concentration depending on the geometric parameters of the particles in the starting powders, as well as on the atmosphere in which the treatment was performed. For the first time by means of the Raman spectroscopy a considerable change in the mechanical stresses and fraction of crystalline state silicon in the studied structures before and after selective laser melting was registered.

1. Introduction

Nowadays aluminum-based micro- and nanostructures are widely used as metasurfaces in optics and photovoltaics [1–3]. Aluminum is enough promising for a following implementation because of this material is sufficiently cheaper than gold and silver which are also used to produce the plasmonic structures. Usually, different lithographic techniques are involved in technological processes of metasurface formation from aluminum. We offer to discuss and study a possibility of surface fabrication via the selective laser melting (SLM) from aluminum powders. Such technology allows to efficiently fabricate not only 2D but also 3D structures. Also, this approach is almost wasteless technology because of the unsintered powder can be used repeatedly.

However, a set of problems exists to fabricate the appropriate plasmonic structures by means of the SLM: (i) such technological features of the process as the laser beam diameter and powder component size at the microscale does not allow to form the nanorelief and, therefore, the plasmon excitation is highly improbable in the visible spectral range; (ii) to avoid SLM-sample cracking it is preferably to use the aluminum powders doped by Si, while Mg and Cu are beneficial in sense of strengthening [4, 5], as a result the additional studies of optical and structural properties of the melted products are needed; and (iii) also, it is necessary to find ways to reduce the number of pores and roughness, which are inherent to the SLM, on the formed surfaces. The problem (i) can be solved by the large-area [6] metasurface fabrication for the infrared and THz ranges. Two other tasks demand the further technological and searching efforts.

In our work we have studied SLM-surfaces produced in vacuum and argon (Ar) from the widely used AlSi10Mg powders [5, 7] with different micromorphology. The main emphasis was done to the analysis of data obtained by the scanning electron microscopy (SEM) and Raman scattering spectroscopy techniques.

2. Materials and methods

Three types of the commercial AlSi10Mg powders named as A, B and C were used in the SLM. The difference between them is the distinction of particle size and form. Typical SEM images (Carl Zeiss LEO SUPRA 50VP) of the initial powders are shown in figure 1. The form of the particles in A and C powders is almost spheroidal. In contrast to, B powder contains the elongated elements also. The particle size is over the range of 2–35, 7–120 and 25–150 μm for powders A, B and C respectively.

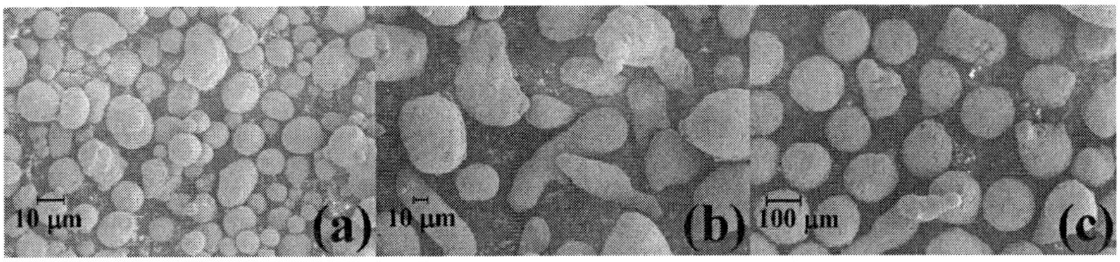

Figure 1. SEM images of the powders A (a), B (a) and C (c).

The SLM-process was implemented using a Solar LS LQ529 laser (1064 nm, 150 μs, 46 mJ, 10 Hz), vacuum chamber and two-coordinate automated positioner. The laser beam was focused into the screeded powder. The motion of the vacuum chamber by means of the positioner in a horizontal plane allowed the formation of monolayers 10x10 mm of the melted AlSi10Mg alloy from the initial powder. The vacuum chamber was evacuated until $2\cdot10^{-5}$ bar or filled with Ar under the atmospheric pressure.

To study Raman scattering a Raman spectrometer Horiba Jobin Yvon HR 800 was employed and pumped by a He-Ne laser at the wavelength 632 nm.

3. Results and discussion

Each of the SLM powder was melted at the same focusing and energy in vacuum and argon. The obtained plain structures were named as A-SLM-V, B-SLM-V, C-SLM-V, A-SLM-Ar, B-SLM-Ar and C-SLM-Ar. The first letter in such marking corresponds to the initial powder, SLM means the realized selective laser melting, V and Ar are the vacuum and argon atmosphere used at the processing respectively. Typical SEM images of C-SLM-V sample are shown in figure 2 as an example. It is worth to note the presence of pores with the typical size on the order of tens and hundreds of micrometers (figure 2(a)), as well as the ones with the size ~ 200 nm (figure 2(b)). Most likely, the first mode of pores is defined by the initial powder packing density, while the second one – by solidification features of the eutectic Al-Si melt [7].

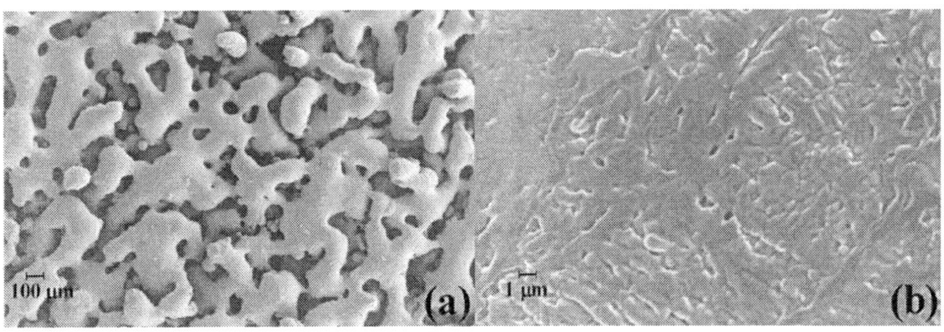

Figure 2. SEM image of the sample C-SLM-V at low (a) and high (b) magnifications.

All melted samples are the porous solid-state matrices with the different porosity (table 1) which was estimated by quantification of the SEM images with a photo editor. The samples based on A and C powders have smaller porosity in comparison with one for the B-type samples. The main reason is a difference between the packing densities. The sphere-like particles in A and C powders can be arranged closer in contrast to the elongated particles from B powder.

The porosities in the cases of SLM in argon and vacuum also differ. The presence of the buffer gas at a certain pressure causes collisions of their atoms with the melt and, therefore, defines the agglomeration and solidification efficiency under following cooling [8]. In our case this efficiency affects the pore formation. A and B powders are characterized by smaller particle sizes than powder C. Thus, A and B structures have lager surface area in comparison with the last powder. Lager surface induces more effective interactions with the Ar molecules and pore formation in A-SLM-Ar and B-SLM-Ar samples in contrast to A-SLM-V and B-SLM-V fabrication (table 1). On the contrary, there is no sufficient difference between the SLM-samples with large enough grains of C-type in Ar and vacuum.

Table 1. The porosity and fraction of amorphous silicon in the samples produced via SLM.

Sample	Porosity (%)	Amorphous silicon fraction (%)
A-SLM-V	28	1.2
B-SLM-V	36	0.9
C-SLM-V	33	5.3
A-SLM-Ar	37	16.1
B-SLM-Ar	51	8.5
C-SLM-Ar	24	10.1

It is necessary to mention that the obtained porosity values are typical for SLM experiments as reported in [4, 5]. From the point of view of metasurface fabrication it is desirable to reduce porosity as much as possible even if the plasmon wavelength exceeds a typical pore size.

Another question concerns possible phase modifications in AlSi10Mg alloy. The Si fraction in such alloy is about 10% [4, 5]. The different phase modifications of Si have an influence not only on such mechanical properties as the tensile strength and durability but also on electrical properties, for example, the conductivity. The last one affects the plasmon excitation efficiency. To study Si phase transformations under SLM treatment the Raman spectroscopy technique proved itself. Typical spectra for a starting and melted powder are shown in figure 3(a) and figure 3(b) respectively.

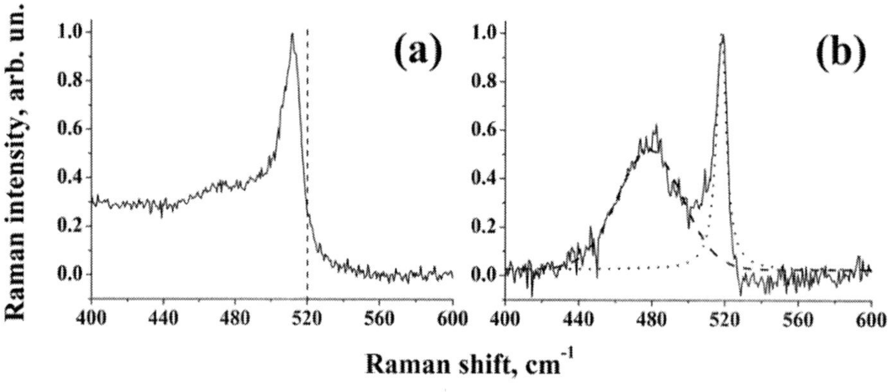

Figure 3. Raman spectra of A powder (a) and A-SLM-Ar sample (b). All spectra are normalized to the signal maximum. The dashed straight in part (a) corresponds to the line 520 cm^{-1} for monocrystalline Si. The dashed and dotted curves in part (b) are approximations of the Raman lines for amorphous and monocrystalline Si respectively.

All powders prior to SLM are characterized by a Raman line near 520 cm^{-1} which corresponds to monocrystalline Si. However, this line is shifted to a lower frequency (red shift) toward 512 cm^{-1} (figure 3(a)). Such behavior can be explained by a tensile stress and corresponding lattice deformation [9] which are consequences of the Al alloy powders fabrication by means of the "atomization of melt" technique.

The Raman spectra for the melted powders look quite different (figure 3(b)). The red shift for the line 520 cm^{-1} disappears. So, the stresses are negligible after the SLM treatment. Instead a broad line 480 cm^{-1} appears. The nature of this line may be interpreted both as a proper mode of the Si-solid solution in Al-matrix or as the mode of an amorphous silicon. In the both cases the Raman line must be situated at lower frequencies in comparison with the monocrystalline Si mode due to transformation of the long-range crystalline order to the short-range one during the SLM. In our work we suggested the amorphous silicon presence. The similar conclusion has been made in the work [10]. To estimate the amorphous phase fraction the ratio between the Raman line intensities at 480 cm^{-1} and 520 cm^{-1} was analyzed for all samples [11]. The results are presented in table 1. The amorphous phase fraction lies within the limits from 8% to 16% for the SLM in Ar, while for the modification in vacuum this value does not exceed several percent. The observed difference can be explained by noticeable gas pressure during the SLM when a sufficient number of Ar atoms prevent the uniform Si crystallization due to heat conductivity of the gas. At the same time, the similar mechanism in vacuum is impossible. Also, a correlation between the maximal porosities and minimal amorphous silicon fraction in samples B-SLM-V and B-SLM-Ar in comparison with other samples at the same atmosphere fabrication is observed. Coming from the non-spherical form of the particles in B powder we have supposed that the particle form in the studied powders affects not only the porosity but also the crystallinity of the samples after SLM. In any case, the amorphous silicon concentration in all samples is insufficient to reduce the conductivity of the surface significantly.

In conclusion, the studied AlSi10Mg alloy powders with the different particle sizes and forms treated by the SLM in Ar and vacuum revealed possibility to use them as a base for the surfaces with the porosity and phase content depending on the buffer gas atmospheres and properties of starting powder. The obtained results seem useful and promising for following investigations of the aluminum-based structures fabricated via SLM and their application in plasmonics.

Acknowledgments
This R&D work was done in the Prime contractor organization (Lomonosov Moscow State University) in the frameworks of the Contract #PMSH-218-2017/01 from 09.02.2017 at the financial support of the Ministry of Education and Science of the Russian Federation under the Contract #03.G25.31.0248 from 28.04.2017.

References
[1] Gao S, Yue W, Park C-S, Lee S-S, Kim E-S and Cho D-Y 2017 *ACS Photonics* **4** 322–8
[2] Højlund-Nielsen E, Zhu X, Carstensen M S, Sørensen M K, Vannahme C, Mortensen N A and Kristensen A 2015 *Opt. Express* **23** 28829–35
[3] Panchenko E, Cadusch J J, James T D and Roberts A 2016 *ACS Photonics* **3** 1833–9
[4] Olakanmi E O 2013 *J. Mater. Process. Tech.* **213** 1387–405
[5] Dadbakhsh S and Hao L 2012 *J. Alloys Compd.* **541** 328–34
[6] Matsuno Y and Sakurai A 2017 *Opt. Mater. Express* **7** 618–26
[7] Thijs L, Kempen K, Kruth J-P and Humbeeck J V 2013 *Acta Mater.* **61** 1809–19
[8] Pereira A, Cros A, Delaporte P, Georgiou S, Manousaki A, Marine W and Sentis M 2004 *Appl. Phys. A* **79** 1433–7
[9] Gao Y, Qin Y, Dong C and Li G 2014 *Appl. Surf. Sci.* **311** 413–21
[10] Hao Y, Gao B, Tu G F, Li S W, Dong C and Zhang Z G 2011 *Nucl. Instr. Meth. Phys. Res. B* **269** 1499–505
[11] Golubev V G, Davydov V Yu, Medvedev A V, Pevtsov A B and Feoktistov N A 1997 *Phys. Solid State* **39** 1197–201

IOP Conf. Series: Journal of Physics: Conf. Series **1092** (2018) 012171 doi:10.1088/1742-6596/1092/1/012171

Emission rate studying through nanodiamonds with embedded nitrogen vacancy centers

A.S. Zalogina[1*], F.A. Benemetskiy[1], D. Pidgayko[1], P. Kapitanova[1], V.V. Yaroshenko[1], S.M. Makarov[1,2], D.A. Zuev[1], I.V Shadrivov[1,2]

[1]Department of Physics and Engineering, ITMO University, St. Petersburg 197101, Russia
[2]Nonlinear Physics Centre, Research School of Physics and Engineering, Australian National University, Canberra ACT 2601, Australia

*a.zalogina@metalab.ifmo.ru

Abstract. High-refractive index nanoparticles, including diamond nanoparticles, can exhibit strong optical resonances, whose spectral position depends on particle size and shape. In this work we demonstrate that these resonances can be used to control the emission of active defects within the particles. We present the theoretical concept of an active dielectric nanoantenna, and study the dependence of the radiation properties on the position of the defect within the particle, as well as on particle size.

1. Introduction

Dielectric nanoparticles are actively studied within many research directions of nanophotonics since resonant behavior of small particles can substantially enhance light-mater interaction at the nanoscale. One of the potential applications areas of this enhancement is the emission control of single photon emitters through excitation of Mie resonances [1]. Efficient coupling of resonances of quantum sources with localized modes of high-index dielectric materials offers new opportunities for emission enhancement. In previous works, it was shown how to couple resonances of quantum emitters with resonances of dielectric and plasmonic nanoparticles [2,3], photonic crystals [4,5], and optical cavities [5,6]. By placing emitters inside the high-index dielectric nanoparticle we obtain the smallest possible structure with enhanced emission control, and we refer to it as an active nanoantenna [6,7].

The simplest structure that can serve as an active nanoantenna is a diamond nanosphere with an embedded quantum emitter – nitrogen vacancy center (NV-center). Diamonds have excellent optical properties, including high refractive index (~2.4), low losses in visible and high stability.

In this work we present the theoretical concept of active dielectric nanoantenna. We analyze the effect of high-order optical modes of relatively large diamond particles on the emission rate as compared to the small nanoparticles that do not support Mie resonances.

2. Concept of active dielectric nanoantenna

The optical properties of nanoantennas, and, as a result, the emission properties of the NV-Centres will depend on size, shape and material of nanoparticles. Larger refractive index leads to larger enhancement of the intensity of the optical field inside the particle. As the size of the particle increases, the number of supported resonances increases and high-order resonances appear.

To analyze multiple resonances and to reveal the relation between sizes and emission rate enhancement, it is common to use the Purcell factor, and here we calculate it for diamond nanoparticles with embedded NV-centers. Utilizing the theory developed in Refs. [8,9], we study the Purcell Factor dependence on dipole position inside the sphere (see Fig. 1). Purcell factor is calculated as a ratio of emission rate of the dipole inside a diamond sphere ($n = 2.4$) to that without the sphere. Due to the fact that the positions of NV centers in realistic nanodiamonds are not known, we assume that they can be randomly located inside a diamond. Moreover, larger nanodiamonds will contain multiple defects, and the observed emission enhancement will be proportional to the average Purcell factor:

$$\bar{F}_p = \frac{\int_{v_{min}}^{v_{max}} \iiint_V F_p(r,v) dV P_0(v) dv}{V \int_{v_{min}}^{v_{max}} P_0(v) dv}, \quad (1)$$

where V – volume of the diamond particle, v is frequency, and $P_0(v)$ – typical luminescence intensity spectrum of the NV-center.

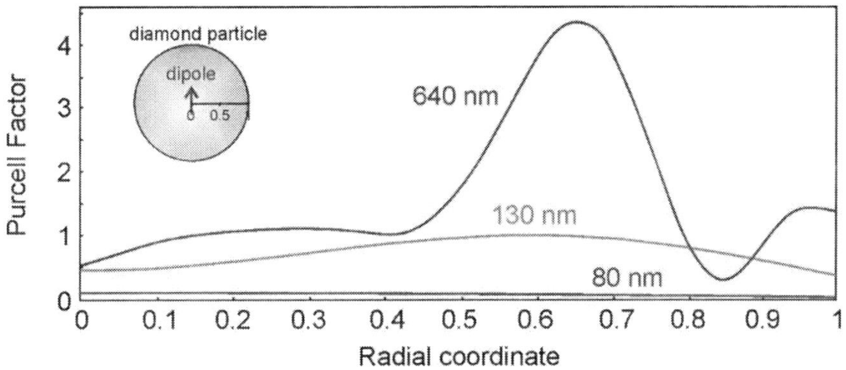

Figure 1. Purcell factor dependence on the normalized radial position of the particle within a spherical nanodiamond for threeThe diamond particle sizes and dipole radial coordinate dependence on Purcell factor particle sizes.

From the results presented in Fig. 1 it is clear that in the small nanoparticles (less than 100 nm), the emission rate is suppressed due to low local density of optical states [10]. Higher order resonances in large diamonds produce large Purcell factors that should lead to significant emission rate enhancement. We see that the Purcell factor in the diamonds of the diameter of 130 nm reaches the values of 1, while for the sphere of 640 nm it is 4.5. Importantly, as Fig. 1 shows, the values of Purcell factor change dramatically as the position of the defect varies along the radius. There is also an optimal position for the NV-center inside the nanodiamond that will produce the strongest emission enhancement. This position depends on the size of the nanodiamond, and in case of a spherical particle of the radius of 640 nm, the normalized optimal position is 0.66, while for the 130 nm sphere it is 0.6. In real experiments, however, various effects including roughness, irregular shape, and a large number of emitters of different orientations in different locations will reduce the observable emission rate. By further calculating the effects of randomness of the emitter orientation and location, and comparing the results to the deeply subwavelength nanodiamonds, we reveal that the Purcell factor for large resonant nanodiamonds is 4.5 times larger than in subwavelength nanoparticles.

While such technology is not available at the moment, we envisage that in future, by correctly choosing the size and shape of dielectric nanodiamonds and controllably placing the NV-center at a desired depth, we will be able to efficiently overlap the optical Mie-type resonances of particles with the photoluminescence spectrum and achieve manifold emission rate enhancement.

3. Conclusion

We have studied the Purcell effect in spherical nanodiamonds with multiple embedded NV-centers and found that it is possible to obtain the emission rate increase of 4.5. We predict that by controlled positioning of the NV-centers inside the nanoparticle one can achieve much larger values of the Purcell factor.

4. Acknowledgements

This work was supported by the Ministry of Education and Science of Russian Federation (Project 2.2267.2017/4.6) and Russian Foundation for Basic Research (Grant No. 18-37- 00384, 16-29- 05317).

References

[1] A. S. Zalogina, R. S. Savelev, E. V. Ushakova, G. P. Zograf, F. E. Komissarenko, V. A. Milichko, S. V. Makarov, D. A. Zuev, and I. V. Shadrivov, "Purcell effect in active diamond nanoantennas," Nanoscale, doi:10.1039/C7NR07953B (2018).

[2] Baranov, D.G., Savelev R.S., Li S.V., Krasnok A.E. and Al`u A., "Modifying magnetic dipole spontaneous emission with nanophotonic structures," Laser & Photonics Reviews, 2017.

[3] Schietinger, S., Aichele T., Wang H.Q., Nann T. and Benson O., "Plasmon-enhanced upconversion in single NaYF4: Yb3+/Er3+ codoped nanocrystals," Nano letters, Vol. 10, No. 1, 134–138, 2009.

[4] Noda, S., Fujita M. and Asano T., "Spontaneous-emission control by photonic crystals and nanocavities," Nature photonics, Vol. 1, No. 8, 449–458, 2007.

[5] Englund, D., Fattal D.,Waks E., Solomon G., Zhang B., Nakaoka T., Arakawa Y., Yamamoto Y. and Vuckovic J., "Controlling the spontaneous emission rate of single quantum dots in a two-dimensional photonic crystal," Physical review letters, Vol. 95, No. 1, 013904, 2005.

[6] Rutckaia, V., Heyroth, F., Novikov, A., Shaleev, M., Petrov, M., & Schilling, J. (2017). Quantum dot emission driven by Mie resonances in silicon nanostructures. Nano letters, 17(11), 6886-6892..

[7] Tiguntseva, E. Y., Zograf, G. P., Komissarenko, F. E., Zuev, D. A., Zakhidov, A. A., Makarov, S. V., & Kivshar, Y. S. (2018). Light-Emitting Halide Perovskite Nanoantennas. Nano letters, 18(2), 1185-1190..

[8] L. D. Landau and E. M. Lifshitz, Electrodynamics of Continuous Media, Course of Theoretical Physics Vol. 8, 1st ed., Pergamon Press, 1960.

[9] H. Chew, The Journal of Chemical Physics, 1987, 87, 1355– 1360

[10] H. Chew, Physical Review A, 1988, 38, 3410.

Dispersion relations and lattice dynamics of diphenylalanine nanotubes

Pavel S Zelenovskiy

School of Natural Sciences and Mathematics,
Ural Federal University, 620000 Ekaterinburg, Russia

Corresponding author e-mail: zelenovskiy@urfu.ru

Abstract. In this work we demonstrate the structural similarity between nanotubes of dipeptide diphenylalanine and one-dimensional locally resonant metamaterials. We developed dynamical model of nanotubes, derived corresponding dispersion relations, and analyzed the origin of the behavior of their optical branch leading to band gap formation. We demonstrate also that the width of the band gap can be varied by tuning the interaction between core molecules. Obtained results are general for many locally resonant metamaterials.

1. Introduction

Acoustic metamaterials are defined as artificial solids composed of meta-atoms with specific interaction with acoustic waves [1,2] endowing bulk material with unconventional effective properties such as acoustic band structure [2,3], negative refraction [2, 4], negative elastic modulus [5,6], and negative mass density [2, 7]. Such metamaterials can be used for transformation of acoustic waves and cloaking objects [1,8], for realizing unidirectional transmission in acoustic and thermal diodes [8], for acoustic imaging beyond the diffraction limit [1] etc.

Bioorganic materials acquiring multilevel hierarchical structure in the result of self-assembly processes are recognized nowadays as advanced materials for manufacturing various functional micro- and nanodevices [9]. The nanotubes of the simplest aromatic dipeptide diphenylalanine (D-Phe-D-Phe, FF) has recently attracted attention of researchers due to its pronounced piezoelectric [10,11], ferroelectric [12,13], pyroelectric [14] and elastic [15] properties. A lot of new biocompatible microelectronic and micromechanical devices based on FF nanotubes have been suggested [10,16,17].

In this work we consider lattice dynamics of diphenylalanine nanotubes. We developed simple dynamical model of such nanotubes, which is quite similar to that used for description of one-dimensional (1D) locally resonant metamaterials. We derived corresponding dispersion relations and revealed the origin of a non-typical behavior of their optical branch leading to band gap formation. We demonstrate also the way for tuning the width of the band gap.

2. Model description

A simple 1D dynamical model of FF nanotubes considers a helical FF nanotube as a linear chain of mass-in-mass units equivalent to the nanotube's unit cell [18]. The unit cell of the nanotube consists of two interacting subsystems: FF ring (6 monomers) and water core (24 molecules, 4 molecules per FF monomer [19]) inside the ring (Fig. 1(a)). Water molecules are held in the nanochannel core by radially oriented hydrogen bonds with FF monomers, whereas the latter interact among themselves via both hydrogen bonds and van der Waals forces [20,21]. Interaction between the FF ring and the water subsystem for transversal a and b directions is characterized by the same effective spring constant α. Longitudinal interaction between adjacent rings and water therein occurs with effective spring constants β and γ, respectively. Individual nanotubes bond together by aromatic rings and form microtubes, which can be considered as a hexagonal crystal [21,22].

This model was successively used recently for determination the origin of high rigidity of FF nanotubes [18] and for description of high temperature phase transitions [23]. It is worth noting that suggested model is similar to that used for description of acoustic metamaterials consisting of chains of local resonators [3,6]. Such structural similarity allows expecting dynamical and elastic properties of FF nanotubes similar to those inherent to such quite exotic metamaterials.

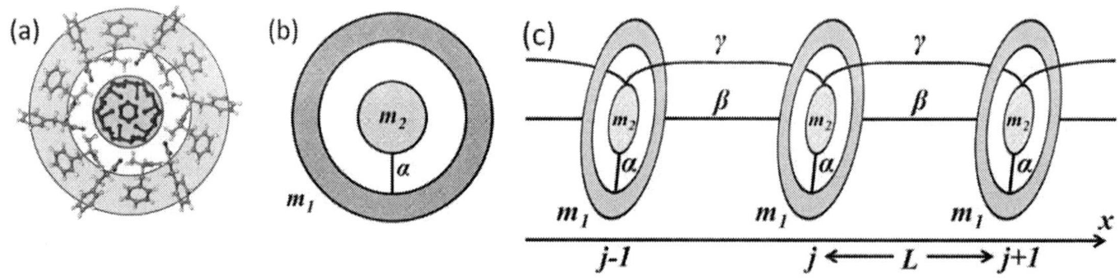

Figure 1. Representation of FF nanotube as 1D lattice of interacting mass-in-mass units. (a) Typical unit cell of FF nanotube; (b) its mass-in-mass representation; (c) 1D lattice of interacting mass-in-mass units. Solid lines show interactions between masses with spring constants denoted by α, β and γ.

3. Results and discussion

The equations of motion for suggested dynamical model can be written in the form:

$$\begin{cases} m_1 \dfrac{d^2 u_1^j}{dt^2} + \alpha\left(u_1^j - u_2^j\right) + \beta\left(2u_1^j - u_1^{j-1} - u_1^{j+1}\right) = 0 \\ m_2 \dfrac{d^2 u_2^j}{dt^2} + \alpha\left(u_2^j - u_1^j\right) + \gamma\left(2u_2^j - u_2^{j-1} - u_2^{j+1}\right) = 0 \end{cases} \tag{1}$$

Here u_1^j and u_2^j are displacements of the j-th FF ring and water core respectively. These equations are similar to those written for local resonance materials [6], but include additional interaction between cores. Solution of Eqs. (1) for ($j+n$)-th unit can be found in accordance with conventional calculation method [24] in the form of harmonic plane waves:

$$u_1^{j+n} = B_1 \exp[i(q(x+nL) - \omega t)] \quad \text{and} \quad u_2^{j+n} = B_2 \exp[i(q(x+nL) - \omega t)] \tag{2}$$

Here L is a distance between mass-in-mass units equivalent to cell parameter c of FF nanotube ($c = 5.46$ Å), q – wavenumber, ω – vibration frequency. Substitution of Eqs. (2) into equations of motion (1) and solving the corresponding eigenvalue problem provide the dispersion relations for the system under consideration [18]:

$$\omega^2 = \frac{\alpha}{2}\lambda(qL)\left\{1 \pm \sqrt{1 - 4\frac{f(qL)g(qL) - 1}{\mu M\,\lambda(qL)^2}}\right\}, \tag{3}$$

where $\mu^{-1} = m_1^{-1} + m_2^{-1}$ – reduced mass, $M = m_1 + m_2$ – total mass of the unit cell, $f(qL) = 1 + 2(\beta/\alpha)[1 - \cos(qL)]$, $g(qL) = 1 + 2(\gamma/\alpha)[1 - \cos(qL)]$ and $\lambda(qL) = (f(qL)m_2 + g(qL)m_1)/(m_1 m_2)$.

Numerical values of parameters used in Eq. (3) can be found from literature data. The masses of FF ring and water core are: $m_1 = 3.11 \times 10^{-24}$ kg and $m_2 = 0.72 \times 10^{-24}$ kg [18]. Spring constants α and γ can be found by classical formula $\omega^2 = \alpha/m$: $\alpha = 9.34$ N/m (from Raman spectra of FF nanotubes [18]) and $\gamma = 0.353$ N/m (from frequency of water dimer translation vibrations [25]). Spring constant $\beta = 438$ N/m can be found by Eq. (3) in Ref. 18 using elastic constant C_{33} obtained in Ref. 15.

The dispersion relations (3) provide two vibration branches presented in Fig. 2 in the interval $qL \in [-\pi, \pi]$. A non-typical curvature of the optical branch (the sign "+" in the Eq. (3)) should be noted. In case of classical crystal with two atoms per primitive cell the frequency of this dispersion branch gradually decreases from the center of the Brillouin zone to its edge [24]. In the present case an opposite behavior is observed. Such behavior was observed earlier for 1D locally resonance metamaterials [6,26], but its origin almost was not discussed.

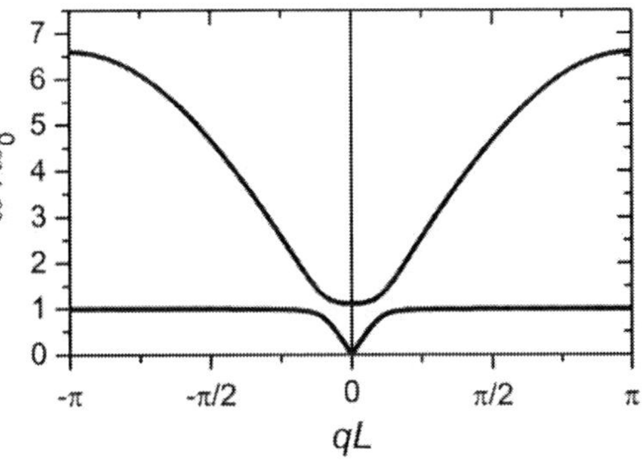

Figure 2. Dispersion curves for the mass-in-mass lattice representing FF nanotube. Vibration frequency ω is normalized over ω_0, the local resonance frequency of mass-in-mass unit: $\omega_0^2 = \alpha/m_2$.

We found that the reason of the observed behavior of the optical branch is interaction (with spring constant α) of acoustic vibrations of two chains: (1) a chain consisting of FF rings, and (2) a chain consisting of water cores (Fig. 1c). The optical branch in mass-in-mass model far from the center of Brillouin zone is determined by acoustic vibrations of the first chain of masses m_1 (red lines in Fig. 3). In case of $\gamma = 0$ (no interaction between water cores) inherent acoustic vibrations of the second chain are absent, and the bottom acoustic branch in mass-in-mass model is formed due to vibrations transmission from first chain to the second. In the vicinity of the Brillouin zone center the interaction between two chains distorts both branches and leads to formation of the phononic band gap.

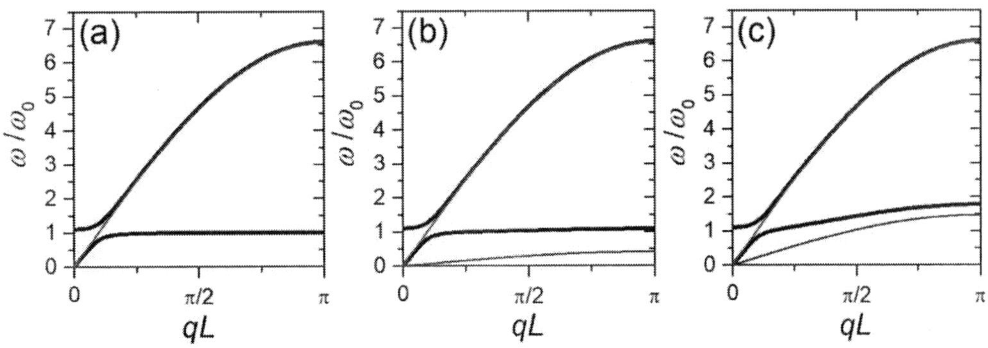

Figure 3. Dispersion curves for different γ: (a) $\gamma = 0$ N/m, (b) $\gamma = 0.353$ N/m and (c) $\gamma = 5$ N/m. Black lines – mass-in-mass model. Red lines – acoustic dispersion branch of a linear chain consisting of masses m_1 only. Blue lines – acoustic dispersion branch of a linear chain consisting of masses m_2 only.

Taking into account an interaction between water cores leads to appearance of another acoustic dispersion branch inherent to the second chain vibrations (blue lines in Figs. 3b,c). In the vicinity of the center of Brillouin zone these vibrations provide negligible effect on the mass-in-mass model, whereas away the center they lead to distortion of the resulted acoustic branch (bottom black curve in Figs. 3b,c) and band gap narrowing. Thus, tuning the interaction between cores (e.g., by the variation of the core molecules) allows control the width of the phononic band gap in FF nanotubes.

4. Conclusions

Finally, in this work we found that lattice dynamics of FF nanotube is quite close to that of locally resonant metamaterials. Derived dispersion relations demonstrate non-typical behavior of optical branch and formation of the band gap. Additional interaction between cores in local resonators, usually excluded from consideration in acoustic metamaterials, provides a tool for variation of the band gap width. Thus, FF nanotubes represent a convenient object for experimental study of various unconventional effects inherent to acoustic metamaterials at the nanoscale level.

5. References

[1] Cummer S A, Christensen J, Alu A, 2016 *Nat. Rev. Mater.* **1** 16001
[2] Lu M-H, Feng L, Chen Y-F 2009 *Materials Today* **12** 34-42
[3] Hussein M I, Leamy M J, Ruzzene M 2014 *Appl. Mech. Rev.* **66** 040802
[4] Lu M-H, Zhang C, Feng L, Zhao J, Chen Y-F, et al. 2007 *Nature Mater.* **6** 744-8
[5] Fang N, Xi D, Xu J, Ambati M, et al. 2006 *Nature Mater.* **5** 452–6
[6] Huang H H, Sun C T 2012 *J. Acoust. Soc. Am.* **132** 2887-95
[7] Yao S, Zhou X, Hu G 2008 *New J. Phys.* **10** 043020
[8] Maldovan M 2013 *Nature* **503** 209-17
[9] Lehn J-M 2002 *Science* **295** 2400-3
[10] Kholkin A, Amdursky N, Bdikin I, Gazit E and Rosenman G. 2010 *ACS Nano* **4** 610-4
[11] Vasilev S, Zelenovskiy P, Vasileva D, et al. 2016 *J. Phys. Chem. Solids* **93** 68-72
[12] Gan Z, Wu X, Zhu X and Shen J 2013 *Angew. Chem. Int. Ed. Engl.* **52** 2055-9
[13] Bdikin I, Bystrov V, Kopyl S, Lopes R P G, et al. 2012 *Appl. Phys. Lett.* **100** 043702
[14] Esin A, Baturin I, Nikitin T, Vasilev S, et al. 2016 *Appl. Phys. Lett.* **109** 142902
[15] Azuri I, Adler-Abramovich L, Gazit E, Hod O and Kronik L 2014 *J. Am. Chem. Soc.* **136** 963-9
[16] Nguyen V, Zhu R, Jenkins K and Yang R 2016 *Nat. Commun.* **7** 13566
[17] Yemini M, Reches M, Rishpon J and Gazit E 2005 *Nano Lett.* **5** 183-6
[18] Zelenovskiy P, Kornev I, Vasilev S and Kholkin A 2016 *Phys. Chem. Chem. Phys.* **18** 29681-5
[19] Wang M, Xiong S, Wu X and Chu P K 2011 *Small* **7** 2801–7
[20] Kim J, Han T H, Kim Y-I, Park J S, Choi J, et al. 2010 *Adv. Mater.* **22** 583–7
[21] Gorbitz C H 2006 *Chem. Commun.* 2332–4
[22] Wu X, Xiong S, Wang M, Shen J and Chu P K 2012 *J. Phys. Chem. C* **116** 9793–9
[23] Zelenovskiy P S, Davydov A O, et al. 2017 *J. Raman Spectrosc.* **48** 1401-5
[24] C. Kittel, Introduction to Solid State Physics, John Wiley & Sons, New York, 1953.
[25] Carey D M, Korenowski G M 1998 *J. Chem. Phys.* **108** 2669-75
[26] Huang H H, Sun C T, Huang G L 2009 *Int. J. Eng. Sci.* **47** 610-7

Acknowledgements

The research was carried out under financial support by the grant of the President of the Russian Federation for young scientists (Contract 14.Y30.17.2294-MK) and the Government of the Russian Federation (Resolution 211, Contract 02.A03.21.0006).

IOP Conf. Series: Journal of Physics: Conf. Series **1092** (2018) 012173 doi:10.1088/1742-6596/1092/1/012173

Experimental stochastic systems based on optical forces

Oto Brzobohaty[a], Stephen Simpson[a], Martin Siler[a], Petr Jakl[a], Jana Damkova[a], Vojtech Svak[a], Alejandro Arzola[c], Karen Volke-Sepulveda[c], Radim Filip[b] and Pavel Zemanek[*a]

[a]Czech Academy of Sciences, Institute of Scientific Instruments, Kralovopolska 147, 612 64 Brno, Czech Republic
[b]Department of Optics, Palacky University, 17. listopadu 1192/12, 771 46 Olomouc, Czech Republic
[c]Instituto de Física, Universidad Nacional Autónoma de México, Apdo. Postal 20-364, 01000 Cd. México,
*zemanek@isibrno.cz

Abstract. We will present our recent theoretical and experimental results related to the behaviour of micro- and nanoparticle placed into nonlinear optical potentials under overdamped or underdamped regime. The two-dimensional optical ratchet rectifies motion of Brownian particles in any direction in the plane, unstable cubic optical potential results in noise-induced particle motion and action of non-conservative optical spin-force leads to orbiting of a levitated particle.

1. Introduction

Recent progress in optical control of mechanical systems opens new ways to investigate a wide variety of mechanical effects in thermodynamics and statistical physics, stochastic dynamics and quantum mechanics. In near future, the most interesting in these fields are nonlinear stochastic effects which autonomously transform environmental noise to useful mechanical effects. A nonlinearity of potential brings mechanical systems out of thermal equilibrium with their environment, changing their statistics far away from thermal statistics of the environment. Such mechanical processes can be therefore principally used as a primary source of coherent mechanical displacement and mechanical oscillations. They might allow to do autonomously mechanical work at micrometer and nanometer distances. Yet their theoretical and experimental investigation is still far from being finished because of the strong nonlinearity and significantly complicated statistical analysis of the dynamics. Thinking forward, about a long term vision, such nonlinear processes might be extremely interesting in the underdamped regime that has been already reached for some cases.

2. Results and Discussion

2.1. Omnidirectional transport in 2D optical ratchets

A fully reconfigurable two-dimensional (2D) rocking ratchet system was created with holographic optical micromanipulation. Optical potentials with the geometry of any Bravais lattice in 2D can be formed and introduce a spatial asymmetry with arbitrary orientation. Nontrivial directed transport of Brownian particles along different directions is demonstrated numerically and experimentally, including on axial, perpendicular, and oblique angle with respect to an unbiased ac driving. The most important aspect to define the current direction is shown to be the asymmetry and not the driving orientation, and yet we show a system in which the asymmetry orientation of each potential well does not coincide with the transport direction, suggesting an additional symmetry breaking as a result of a coupling with the lattice configuration. Our experimental device, due to its versatility, opens up a new range of possibilities in the study of nonequilibrium dynamics at the microscopic level[1].

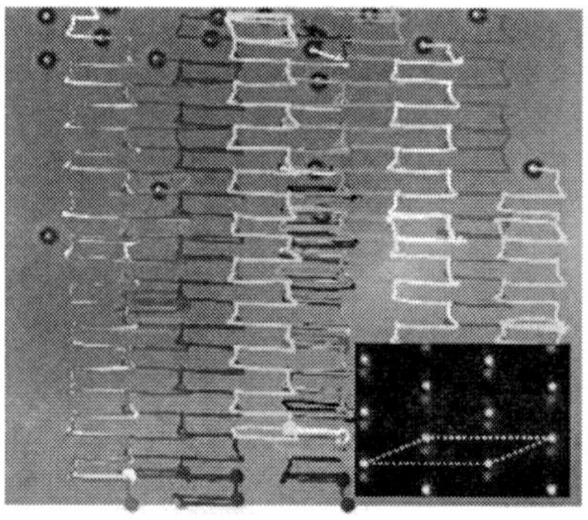

Figure 1. The bottom right part shows the regular lattice of laser beam foci (bright points) and its elementary unit is marked with yellow dashed lines. If microparticles dispersed in liquid are illuminated with such a system of asymmetric potential wells, periodic motion of the sample in horizontal direction is sufficient to induce rectified upwards motion of particles along highlighted colour trajectories.

2.2. *Cubic optical potential*

In the proof-of-principle theoretical analysis, we focus on the short-time approximation of the noise-induced position displacement of an overdamped Brownian particle in cubic optical potential[2,3]. Among the key results, confirmed experimentally[4], belong that the mean particle displacement and signal-to-noise ratio are linearly powered by the temperature of the environment and both monotonously increase in time. The long-term particle dynamics we addressed using the first-passage properties[5]. Such an approach eliminates diverging trajectories and could be reliably determined even from a small sample of trajectories.

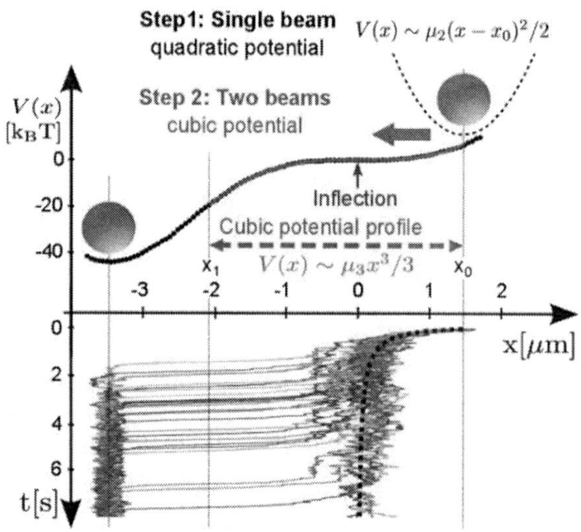

Figure 2. Principle of the experiment in cubic potential. Step 1: The particle (blue spot) is trapped in a single beam (quadratic potential) Step 2: The single beam is replaced with two co-propagating beams displaced in x-axis to form a non-linear single-well potential (dotted) with a cubic part (blue) along x-axis. The particle moves down the potential well (indicated by a blue arrow) passing through the cubic part and reaching the global minimum. Then the two beams are replaced with a single beam which confines the particle near the former global minimum and transports it to the initial position x_0. Steps 1 and 2 are repeated with the same particle to get a statistical ensemble of trajectories (e.g. color curves at the bottom part of the figure).

2.3. *Spin force induced particle dynamics*

We provide a vivid demonstration of the mechanical effect of Belifante spin momentum – a component of optical momentum previously thought to be virtual and unmeasurable[6]. Working in vacuum with a circularly polarized Gaussian optical trap, transverse spin momentum drives the underdamped motion of a probe particle far beyond thermodynamic equilibrium. Constrained by optical gradient forces, we first observe spin-driven Brownian motion and, subsequently, the formation of thermally excited orbits. Ultimately, centripetal forces overcome the trap, causing the expulsion of the particle. Our results complement and corroborate recent measurements of spin momentum in evanescent waves, and extend them to a new geometry. In doing so, we exhibit fundamental, generic features of the mechanical interaction of circularly polarized light beams with matter. Our work shows how observations of the under-damped motion of probe particles can illuminate our understanding of the nature and morphology of momentum flows in arbitrarily structured light fields as well as providing a test bed for elementary non-equilibrium statistical mechanics.

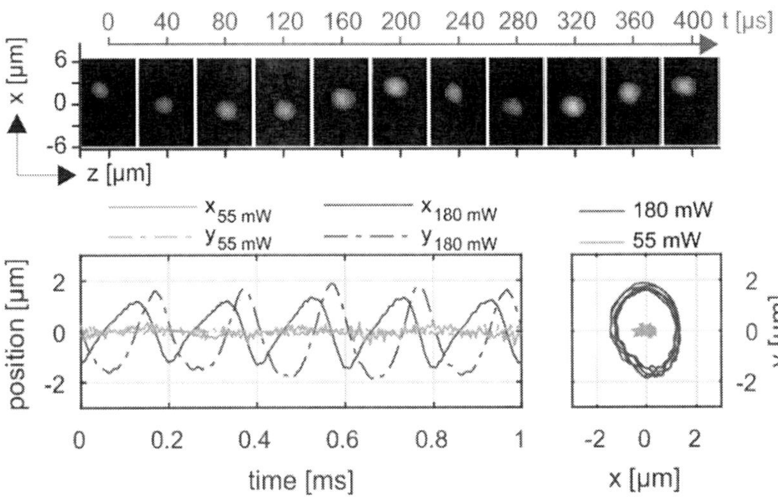

Figure 3.Particle orbiting due to optical spin force. Top: An example of orbiting particle taken by the CCD camera looking perpendicularly to the beam propagation. Bottom: Trajectories of the particle for low power regime (cyan), when the particle moves near the center of the Gaussian beam, and higher power regime, when the particle orbits around the beam..

Acknowledgement
The research was supported by the CSF (project GB14-36681G), TACR (TE01020233), MEYS CR (LO1212), CAS (RVO:68081731) and EC (CZ.1.05/2.1.00/01.0017).

References
[1] Arzola, A., Villasante-Barahona, M., Volke-Sepulveda, K., Jakl, P., Zemanek, P. Phys. Rev. Lett. 118, 138002 (2017).
[2] Filip, R., Zemanek, P. J. Opt. 18, 65401 (2016)
[3] Zemanek, P., Siler, M., Brzobohaty, O., Jakl, P., Filip, R., J. Opt. 18, 65402 (2016)
[4] Siler, M., Jakl, P., Brzobohaty, O., Ryabov, A., Filip, R., Zemanek, P. Sci. Rep. 7 1697 (2017)
[5] Ryabov, A., Zemanek, P., Filip, R. Phys. Rev. E 94, 42108 (2016)
[6] Antognozzi, M., Bermingham, C.R., Harniman, R.L., Simpson, S., Senior, J., Hayward, R., Hoerber, H., Dennis, M.R., Bekshaev, A.Y., Bliokh, K.Y., Nori, F. Nature Phys. 12, 731(2016)

Nonlinear ring-shaped waves in spinor polariton condenates

Dmitry A. Zezyulin
ITMO University, St. Petersburg 197101, Russia
E-mail: dzezyulin@corp.ifmo.ru

Abstract. Systems of interacting polaritons are known to support a rich variety of propagating states whose properties are essentially conditioned by the geometry of the potential and by the nonlinear nature of polariton-polariton interactions. We consider theoretically one-dimensional polariton ring accounting for both longitudinal-transverse (TE-TM) and Zeeman splitting of spinor polariton states and spin-dependent polariton-polariton interactions. We present the class of solutions which feature oscillations of the density in both spinor components and rotate with constant angular velocity. We show that the effects of the geometric phase arising from the interplay between external magnetic field and TE-TM splitting introduce chirality in the system and make solitons propagating in clockwise and anticlockwise directions non equivalent. This can be interpreted as solitonic analog of Aharonov-Bohm effect.

Cavity polaritons are hybrid light-matter quasiparticles emerging in the regime of the strong coupling between excitonic resonance and photonic mode of planar semiconductor microcavity. An important property of cavity polaritons is their spin (or pseudo-spin), inherited from the spins of quantum well excitons and cavity photons. Similar to photons, polaritons have two possible spin projections on the structure growth axis corresponding to two opposite circular polarizations. States with opposite spins can be mixed by effective magnetic fields of various origin. Magnetic field applied along the structure growth axis and acting on excitonic component splits energies of the spin positive and spin negative polariton states, while TE-TM splitting of the photonic modes hybridizes these states via the linear coupling. Another important feature of polaritons consists in the presence of polariton-polariton interactions which render the system nonlinear and enable propagation of self-sustained nonlinear entities (i.e., solitons) whose properties depend heavily on the underlying topology.

In Figure 1 we provide a schematic illustration of our system which consists of a polariton ring placed into external magnetic field **B** perpendicular to its interface. Interacting spinor polaritons trapped in such a quasi one-dimensional ring resonator can be described by the following system of dimensionless Gross-Pitaevskii equations [1]:

$$\begin{aligned} i\dot{\psi}_+ &= -\partial_\varphi^2\psi_+ + \left(|\psi_+|^2 + \alpha|\psi_-|^2\right)\psi_+ + \Omega\psi_+ + \kappa e^{-2i\varphi}\psi_-, \\ i\dot{\psi}_- &= -\partial_\varphi^2\psi_- + \left(|\psi_-|^2 + \alpha|\psi_+|^2\right)\psi_- - \Omega\psi_- + \kappa e^{+2i\varphi}\psi_+. \end{aligned} \quad (1)$$

Here, $\psi_\pm(t,\varphi)$ are the components of the exciton-polariton spinor wavefunction $\{\psi_+,\psi_-\}$ in the basis of circular polarizations satisfying $\psi_\pm(t,\varphi) = \psi_\pm(t,\varphi+2\pi)$, parameter $\alpha < 0$ characterizes attractive interaction of the cross-polarized polaritons, Ω is half of the Zeeman energy splitting

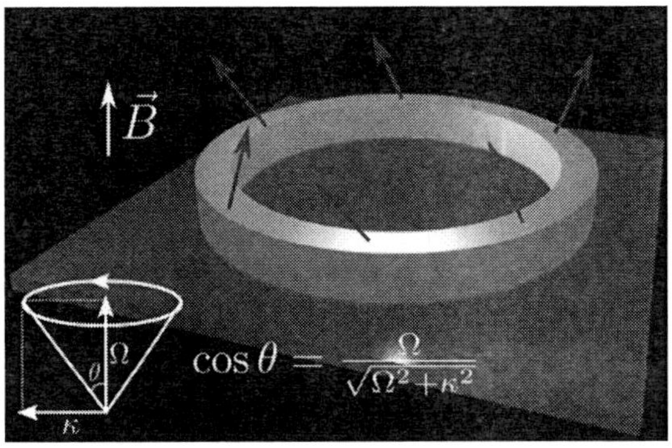

Figure 1. Schematic illustration of the geometry of the considered system. Polariton ring is placed into external magnetic field **B** perpendicular to its interface. The total effective magnetic field acting on polariton's spin is a combination of the real magnetic field and the field provided by TE-TM splitting. The direction of the total effective magnetic field changes along the ring as it is shown by the red arrows, if one moves along the ring it covers a cone characterized by the angle θ.

proportional to the amplitude of the applied magnetic field, and κ is half of the momentum independent TE-TM energy splitting. Parameters Ω and κ are dimensionless and scale in units of $\hbar^2/(2m^*R^2)$, where R is the ring radius and m^* is the exciton-polariton effective mass. We consider moving solutions which circulate with the angular velocity ω. To this end we introduce a rotating frame variable $x = \varphi - \omega t$ and seek for the wave functions in the form

$$\psi_\pm(t,\varphi) = u_\pm(x)e^{-i\mu t \mp i\varphi}, \tag{2}$$

where μ is the chemical potential, and 2π-periodic functions $u_\pm$ characterize spatial shapes of rotating solutions.

Our main results can be summarized as follows. First, we demonstrate that the system supports a very rich variety of rotating solutions which can be characterized by the angular velocity ω, chemical potential μ, and polariton density integral $\rho = \frac{1}{2\pi}\int_0^{2\pi}\left(|\psi_+|^2 + |\psi_-|^2\right)d\varphi$. Spatial profiles of rotating solutions $|u_\pm|$ feature dips and peaks of the density, and the number of the peak-dip pairs can be different for solutions of different complexity. Second, in the absence of the Zeeman splitting (i.e., for $\Omega = 0$) the solution propagating with angular velocity always has a counterpart propagating with the opposite angular velocity $-\omega$. These two solutions are equivalent: they are characterized by the wavefunctions of the same form and have the same stability properties. Third, nonzero Zeeman splitting $\Omega \neq 0$ makes the two counterpropagating solutions not equivalent: they have different shapes and, generically speaking, different stability, i.e., one of them can be stable while another can be unstable. In view of this non-equivalence, one can say that the solutions propagating in such a polariton ring are inherently *chiral* [2]. We additionally demonstrate that chirality can be explained using the concept of effective gauge field stemming from the combined effect of the TE-TM splitting and the external magnetic field.

This is a joint work with Dmitry Gulevich, Dmitry Skryabin and Ivan Shelykh.

Acknowledgments

This work was supported by megagrant 14.Y26.31.0015 of the Ministry of Education and Science of Russian Federation and by the Government of Russian Federation (Grant 08-08).

References

[1] D. R. Gulevich, D. V. Skryabin, A. P. Alodjants, and I. A. Shelykh, Phys. Rev. B **94**, 115407 (2016).
[2] D. A. Zezyulin, D. R. Gulevich, D. V. Skryabin, I. A. Shelykh, Phys. Rev. B **97**, 161302 (2018).

IOP Conf. Series: Journal of Physics: Conf. Series **1092** (2018) 012175 doi:10.1088/1742-6596/1092/1/012175

Near-IR resonant response of Ge microparticles fabricated by femtosecond laser printing

D M Zhigunov[1,*], A B Evlyukhin[2], U Zywietz[3], S M Novikov[4] and B N Chichkov[3,5]

[1]Faculty of Physics, M. V. Lomonosov Moscow State University, Leninskie Gory 1, 119991 Moscow, Russia
[2]ITMO University, 49 Kronversky Avenue, 197101, St. Petersburg, Russia
[3]Laser Zentrum Hannover e.V., Hollerithallee 8, D-30419 Hannover, Germany
[4]Centre for Nano Optics, University of Southern Denmark, Campusvej 55, DK-5230 Odense, Denmark
[5]Institut für Quantenoptik, Leibniz Universität Hannover, D-30167 Hannover, Germany

*E-mail: dmzhigunov@physics.msu.ru

Abstract. Recently introduced femtosecond laser printing approach was applied for the controlled fabrication of single germanium microparticles (MPs). The characterization of MPs was performed by scanning electron microscopy and near-IR optical spectroscopy. The size of obtained MPs is about 1-1.5 μm while their shape is close to spherical. Theoretical calculations of MP scattering cross section were performed on the basis of Mie theory. It is shown that extinction spectra of Ge MPs possess sharp resonances in near IR spectral region, which are identified by means of multipole decomposition analysis.

1. Introduction

Semiconductor and dielectric micro- and nanostructures attract increasing attention of researchers nowadays due to their high potential as a substitution for plasmonic metal-based metamaterials. The latter suffer from intrinsic Ohmic energy losses at optical frequencies, which limit considerably the device efficiency. In contrast, so-called all-dielectric metamaterials are characterized by much lower absorption losses, and, moreover, can exhibit strong Mie-type resonances in their optical response, which is favorable for the realization of the fundamental task of nanophotonics – light manipulation at the nanoscale [1]. Different types of metamaterials have been shown by employing silicon – most widely used material to date [2]. One of the progressive techniques for the fabrication of nearly ideally spherical metal and semiconductor micro- and nanoparticles is known as femtosecond (fs) laser printing [3, 4]. Si nanoparticles with sizes of a few hundred nanometers, produced by this method, demonstrated outstanding optical properties in visible spectral range due to excitation of electric and magnetic dipole resonances [5]. Going beyond silicon, in our most recent paper we further developed the fs laser printing technique and fabricated single spherical Ge and SiGe nanoparticles [6]. Such novel objects are found to possess a crystalline structure as revealed by Raman scattering measurements. Moreover, both Ge and SiGe nanoparticles also exhibit Mie-type resonances in the visible light scattering spectra which make them the promising candidates for the needs of nanophotonics [6].

In present paper we demonstrate the fabrication of single Ge microparticles (MPs) by the fs laser printing method. Optical characterization of obtained MPs in near-IR spectral region is also provided. Germanium is known as a material used in photodetectors for near-IR range due to its appropriate band gap. On the other hand, relatively low absorption losses of Ge in near-IR comparing to visible region allow one to consider Ge MPs as elementary building blocks of all-dielectric metamaterials for use in near-IR range.

2. Experimental details

Thin Ge film with the thickness of 100 nm was used as a donor for MP fabrication by means of fs laser printing approach [6]. The film was deposited on 150 μm-thick standard borosilicate wafer (Borofloat 33, Plan Optik AG) by the e-beam evaporation at room temperature using ultra high vacuum growth system [7]. A commercial femtosecond laser system (Spectra Physics) with 800 nm central wavelength, 50 fs pulse duration, 3 mJ maximum pulse energy and 1 kHz repetition rate was used for fs laser printing of MPs. Laser beam was focused on a transparent substrate with donor Ge thin film by means of 20 mm focal length lens. MPs were transferred on a receiver glass substrate located directly underneath the thin film. The distance between the donor and receiver substrates was equal to about 10 μm. Scattering spectra were collected in forward-scattering geometry using a microscopic setup equipped with CCD camera. Visualization of generated MPs was performed by means of dark-field microscopy. Theoretical calculations of scattering cross sections for ideally spherical Ge MPs were performed using Mie theory, while dispersion relations of real and imaginary parts of complex dielectric permittivity were taken from the respective database [8].

3. Results and discussion

Figure 1 shows dark-field microscopic image of Ge MP arrays printed by 800 nm single fs laser pulses. Bright spots correspond to 3x3 MP subarrays obtained at fixed laser pulse energy, while the latter gradually increased from left to right side of the image. The inset of Figure 1 demonstrates SEM image of a chosen Ge MP, which size is about 1.43 μm and shape is close to spherical. Overall MP size is varied from about 1 to 1.5 μm depending on the energy of employed laser pulses. The crystalline structure of MPs is evident from Raman spectroscopy measurements (not shown here), which revealed a sharp peak at 300 cm^{-1} characteristic for bulk crystalline Ge. The modification of the donor Ge film surface under fs laser treatment is shown in Figure 2. The initial formation and eventual ejection of Ge MP upon high enough pulse energies can be clearly seen. An increase of laser pulse energy was also accompanied by the rise of molten film area from about 3 to 4.5 μm in diameter. The reproducibility of fabricated MPs can be estimated to amount to 90 %. At the same time it is worth noting that the change of the laser wavelength to 400 nm (laser second harmonic) led to the formation of invariably multiple nanoparticles instead of single ones. The latter emphasize the necessity of proper adjustment of experimental parameters to achieve the control over the laser printing approach.

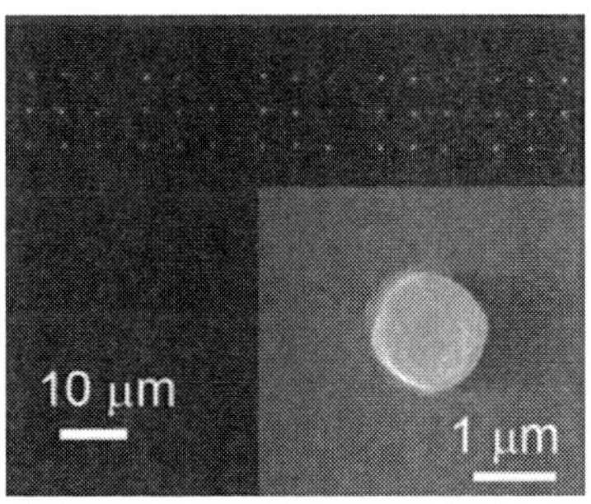

Figure 1. Dark-field microscopic image of Ge MP arrays printed by 800 nm single fs laser pulses. The inset shows SEM image of a chosen Ge MP.

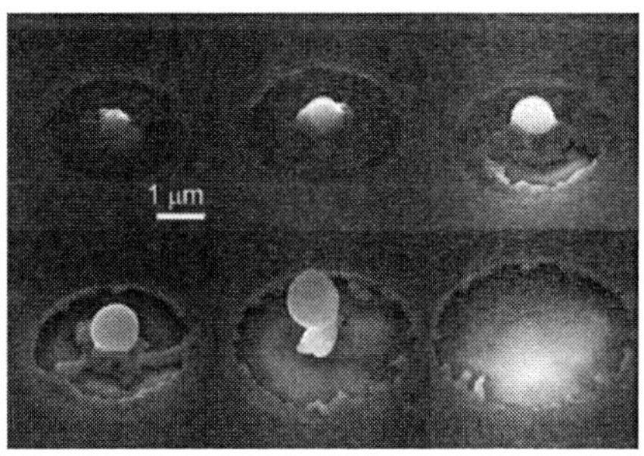

Figure 2. SEM images of the donor Ge film surface irradiated by single 800 nm fs laser pulses with gradually increasing energy. Scale bar is common for all images.

An optical response of fabricated MPs was studied in the size-related near-IR spectral range. Figure 3 demonstrates for comparison the experimental extinction spectrum of Ge MP, which is shown in the inset of Figure 1, and both theoretical scattering and extinction cross sections for equal-sized ideally spherical Ge MP (1430 nm in diameter). While the experimental spectrum is quite noisy due to the lack of the signal, it reproduces well in general the corresponding theoretical dependence. An absence of total agreement may be also explained by the deviation of MP shape from the ideal sphere. Multipole decomposition analysis of the scattering cross section revealed the dominance of magnetic-type contributions (dipole and quadrupole) responsible for the appearance of sharp peaks in the spectrum. The demonstrated resonant behavior of fabricated Ge MP optical response may be used both in photonics and IR sensorics.

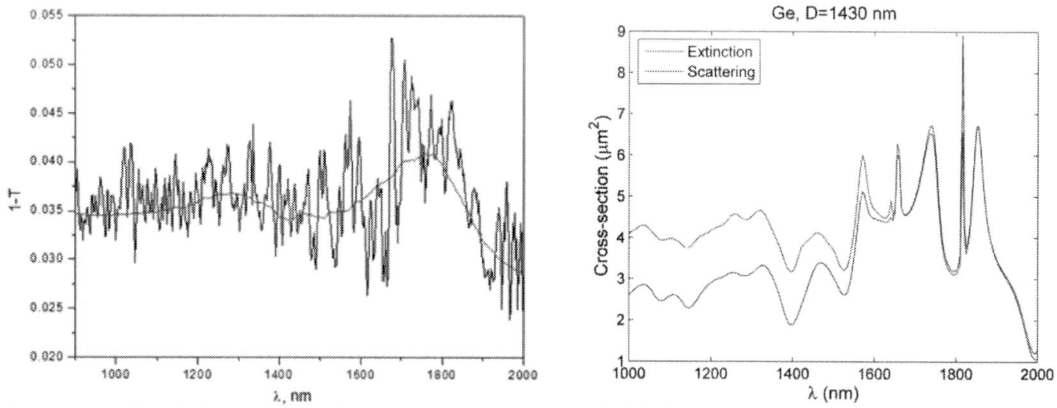

Figure 3. (Left) Experimental extinction spectrum of fabricated Ge MP (shown in the inset of Figure 1) and (right) theoretical scattering and extinction cross sections for equal-sized ideally spherical Ge MP.

4. Conclusions

In conclusion, we realized the reproducible fs laser printing of single Ge microparticles, which were characterized by SEM and visualized by dark-field microscopy. Near-IR optical response of fabricated Ge MPs was studied both experimentally and theoretically with a satisfactory agreement showing the dominance of magnetic-type multipole contributions.

Acknowledgments

The financial support from the Russian Science Foundation (RSF) of theoretical calculations (Grant No. 16-12-10287) is acknowledged.

References

[1] Zhao Q, Zhou J, Zhang F, Lippens D 2009 *Mater. Today* **12** 60–69
[2] Staude I, Schilling J 2017 *Nat. Photonics* **11** 274–284
[3] Zywietz U, Reinhardt C, Evlyukhin A B, Birr T, Chichkov B N 2014 *Appl. Phys. A* **114** 45–50
[4] Kuznetsov A I, Evlyukhin A B, Reinhardt C, Seidel A, Kiyan R, Cheng W, Ovsianikov A and Chichkov B N 2009 *J. Opt. Soc. Am. B* **26** B130
[5] Zywietz U, Evlyukhin A B, Reinhardt C, Chichkov B N 2014 *Nat. Comm.* **5** 3402
[6] Zhigunov D M, Evlyukhin A B, Shalin A S, Zywietz U and Chichkov B N 2018 *ACS Photonics* **5** 977–983
[7] Drozdov M N, Drozdov Yu N, Zakharov N D, Lobanov D N, Novikov A V, Yunin P A, Yurasov D V 2014 *Tech. Phys. Lett.* **40** 601–605
[8] Jellison G.E. Jr. Refractive index database. https://refractiveindex.info/

Experimental observation of spin-locked propagation of topological edge states in an open non-Hermitian metasurface

D. V. Zhirihin [1,a)], M. A. Gorlach[1], X. Ni[2,3], D. A. Smirnova[2], D. Korobkin[2], A. P. Slobozhanyuk[1], P. A. Belov[1], A. Alù[4] & A. B. Khanikaev[1,2,3]

[1]ITMO University, Saint Petersburg, 197101, Russia
[2]The City College of the City University of New York, NY, 10031, USA
[3]The Graduate Center of the City University of New York, NY, 10016, USA
[4]The University of Texas at Austin, Austin, TX, 78712, USA

[a)]Corresponding author: d.Zhirihin@metalab.ifmo.ru

Abstract. In this work we explore photonic topological edge states arising in dielectric metasurface based on a deformed honeycomb lattice. We demonstrate numerically the effect of pseudo-spin locking of the edge states and verify its existence experimentally in microwave frequency range. Our findings thus demonstrate that open non-Hermitian metasurface can support the topological edge modes despite the radiative losses.

1. Introduction

Inspired by the discovery of the topological phases in condensed matter systems, photonics analogues of topological insulators nowadays are under active investigation. Their fascinating properties such as disorder robustness and backscattering-immune propagation of the topological states have opened novel ways to control propagation of electromagnetic waves.

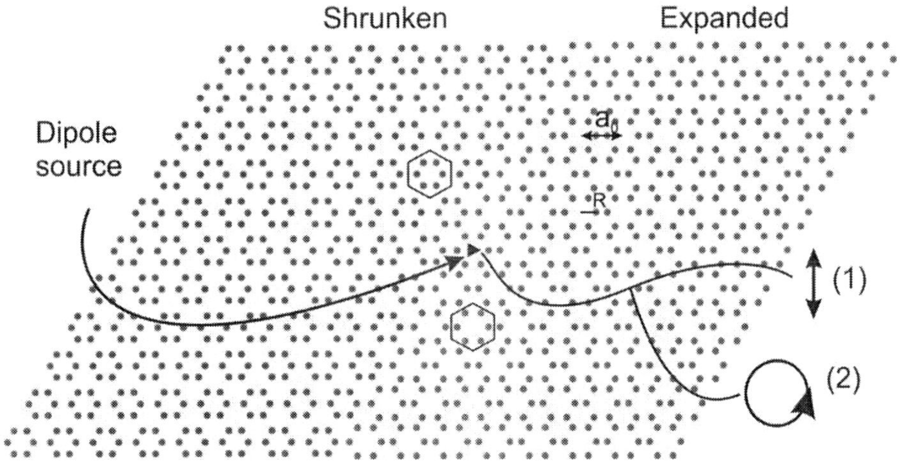

Figure 1. Sketch of the dielectric metasurface for observation of pseudo-spin locking in microwave frequency range

In the recent years, photonic topological systems have been studied theoretically [1,2] and investigated experimentally from microwave to optical frequency ranges [3-5]. But the number of recent theoretical proposals is still waiting for their experimental verification.

In this work, we consider dielectric metasurface with triangular honeycomb lattice and demonstrate numerically and experimentally the emergence of photonic topological edge states even if the system is open and consequently non-Hermitian.

2. Design and results

To prove our concept we utilize the recently proposed design based on a triangular honeycomb lattice [6]. The unit-cell of this lattice represents hexagon with specially arranged 6 parallel dielectric cylinders (Figure 1). The geometry of lattice possesses C_6 symmetry which gives in combination with time-reversal symmetry double-degenerate Dirac cone at the Γ-point of the first Brillouin zone in the case of $a_0/R = 3$. By deforming the honeycomb lattice topologically trivial (for shrunken structure – $a_0/R > 3$) and nontrivial (for expanded structure – $a_0/R < 3$) bandgaps are opened [6]. Thus, the interface between shrunken and expanded structures can host topological modes which was previously proved for pure Hermitian system [6]. For open non-Hermitian systems the situation can be different due to the leakage coupling of the wave to the free space. And it is needed to prove that in this case the topological edge states still persist.

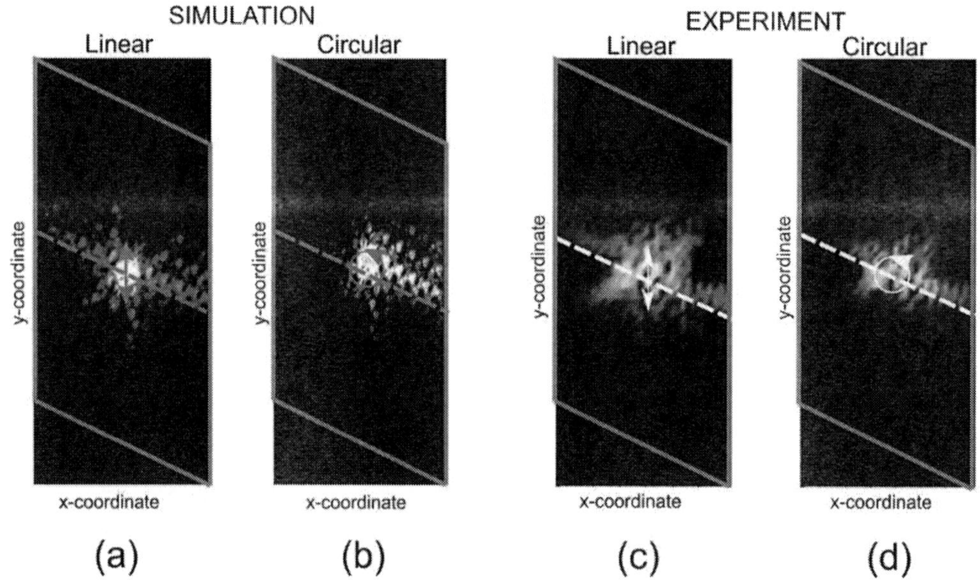

Figure 2. Observation of the pseudo-spin locking. Comparison between numerical and experimental results for two different sources: (a,c) linear dipole; (b,d) circular dipole. Orange parallelograms indicate the edges of metasurface, dashed lines – position of the domain wall.

To prove this idea, we create a finite numerical model in the license program package CST Microwave Studio (Figure 1). The model consists of dielectric metasurface with two neighbor domains (shrunken and expanded) and subwavelength dipole located on the center of a domain wall. Each domain comprises of 8x11 unit-cells with the same lattice constant a_0 = 37.5 mm, height and radius of the cylinders 10 mm and 1.5 mm, respectively. The permittivity of each cylinder is equal to 10. The sizes of the clusters R were chosen to be $a_0/R = 3.33$ and $a_0/R = 2.73$ for shrunken and expanded structures, respectively. We calculate electric field intensity map at the 2 mm distance above the metasurface and find the edge states excited by linear and circular dipoles at the frequency 6.84 GHz (Figure 2 a,b). The results show bi-directional excitation of edge states for linearly polarized dipole and one-way propagation for circularly polarized dipole that attests the locking of the polarization an propagation direction and thereby confirms the topological nature of the excited modes.

To verify our simulation results we fabricate metasurface and perform near-field measurements for linearly and circularly polarized excitations. Metasurface is fabricated on a machine-processed substrate of extruded polystyrene foam with drilled holes. Each hole is filled by commercial high-index dielectric powder (Eccostock HiK with the permittivity 10 and negligible losses). As a receiving antenna, we utilize subwavelength vertical electric dipole. The results of near-field measurements are in a perfect agreement with our simulations (Figure 2 c,d).

Aknowledgements
This work was supported by the Russian Science Foundation (Grant No. 16-19-10538).

References

[1] A. Khanikaev et al., "Photonic topological insulators," *Nature Materials* **12**, pp. 233–239 (2013).
[2] L.Lu et al., "Topological Photonics," *Nature Photonics* **8**, pp. 821–829 (2014).
[3] M. Hafezi et al., "Imaging topological edge states in silicon photonics," *Nature Photonics* **7**, pp. 1001-1005 (2013).
[4] M. C. Rechtsman et al., "Photonic Floquet topological insulators," *Nature* **496**, pp.196-200 (2013).
[5] A. P. Slobozhanyuk et al., "Experimental demonstration of topological effects in bianisotropic metamaterials," *Scientific Reports* **6**, 22270 (2016) .
[6] L.-H. Wu and X. Hu., "Scheme for Achieving a Topological Photonic Crystal by Using Dielectric Material," *Phys. Rev. Lett.* **114**, 223901 (2015).

IOP Conf. Series: Journal of Physics: Conf. Series **1092** (2018) 012177 doi:10.1088/1742-6596/1092/1/012177

Investigation of mechanical properties of transformed living cells by means of atomic-force microscopy with high aspect ratio probes

M V Zhukov[1,2], F E Komissarenko[1,4], V I Chubinskiy-Nadezhdin[3], M M Khalisov[1,5]

[1] ITMO University, 197101 Kronverkskiy pr., 49, Saint-Petersburg, Russian Federation
[2] Institute for Analytical Instrumentation RAS, 198095 Ivan Chernykh st., 31-33, Saint-Petersburg, Russian Federation
[3] Institute of Cytology RAS, 194064 Tikhoretsky avenue 4, Saint-Petersburg, Russian Federation
[4] St. Petersburg Academic University RAS, 194021 Khlopina st., 8/3, Saint-Petersburg, Russian Federation
[5] Institute of Physiology by I.P. Pavlov RAS, Makarova embankment, 6, Saint-Petersburg, Russian Federation

E-mail: cloudjyk@yandex.ru

Abstract. Fabrication of high aspect nanowhisker structures at the apex of standard probes by the focused electron beam was carried out. The protocols of cultivation, structural modification and transportation of the cell structures of the selected line (fibroblasts) were performed. The mechanical properties of living cells in normal and malignant transformation (cancer cells) in a liquid were investigated by high aspect nanowhisker probes and compared with standard probes results. Elevation of the adhesion force and reduction of the elasticity modulus on cells with pathology was detected.

1. Introduction

Atomic force microscopy (AFM) is a widely known tool for biological research. AFM allows to obtain images of the topography of biological objects with a resolution of up to several angstroms, investigate interactions between single molecules and measure the mechanical characteristics of various biological samples at the micro- and nanoscale level.

One of the directions of AFM development in the field of biological research is the creation of specialized probes. The quality of the AFM data and the resolution of images directly depends on the cantilever and its tip [1, 2]. Despite the fact that there are currently many different cantilevers on the market for studying biological objects in liquids, the issue of creating new types of probes remains the most important task since there are still disputes about the advantages and disadvantages of different types and shapes of the tips [3 ,4].

Thus, there is an obvious demand for specialized probes, which would allow high-resolution visualization of the surface of the studied objects. One of the most suitable materials in this case are high-aspect structures such as nanotubes [5], nanowhiskers [6], nanorods, etc.

Therefore, the aim of this work was to fabricate high-aspect probes to study mechanical properties of living cells in normal and malignant transformation by means of atomic-force microscopy and give a comparison with standard tip.

2. Experimental setup

The nanowhiskers (NW) were fabricated and measured at the top of the standard probes by electron beam induced deposition of precursor gases in a vacuum chamber of scanning electron microscope.

The investigation of cellular structures was carried out by scanning probe microscope combined with an inverted optical microscope. The investigation of native cells was carried out in PBS (phosphate-buffered saline) liquid by QNM PeakForce technique [7]. Heating platform in the main block of the microscope allowed to thermostating the cells during the measurements. The confocal laser scanning microscopy was used to study the location of the cells by using a TRITC fluorescent dye and a 100x/1.4 N.A lens.

Normal and transformed (cancer) mouse fibroblasts cells were cultured in a DMEM (Biolot, Russia), containing 10% of fetal bovine serum (Biowest, France) and 80 μg/ml gentamicin (Biolot, Russia) in an incubator at 37 °C and 5% CO_2. Cells were dissected in a multiplicity of 1:3 - 1:5, confluence of the cell line was 70-80%. The temperature maintenance during transportation of cells was carried out by Petri dish, covered (inside and outside) with several layers of aluminum foil.

3. Results and Discussions

Formation of a single NW was carried out by means of electron microscope with an integrated gas inlet system with $C_9H_{16}Pt$ precursor (figure 1, a). Optimal parameters for the formation of NW structures were: accelerating voltage 5 kV, exposure time 5-10 s, aperture of electron microscope 20 μm. The composition of NW was about 30% of platinum, the rest is carbon. The working pressure in the vacuum chamber of the microscope was about 10^{-6} mBar.

The typical geometric parameters of the Pt/C whisker were about 400 nm in length, about 50 nm in diameter, and about 20-25° of the growth angle, which allowed to achieve orthogonality orientation of the probe relative to the substrate and obtain stable AFM images of cells (figure 1, b). The angle was set to compensate the angle of the probe holder in AFM microscope.

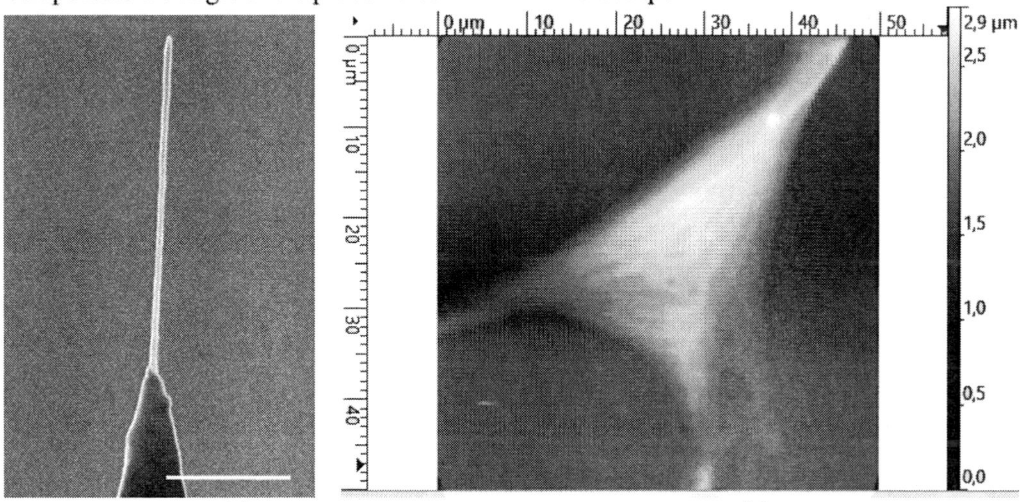

(a) (b)

Figure 1. SEM image of NW formed at the top of the standard probe (a), and AFM image of single fibroblast (b). The scale bar on SEM image is 30 μm.

Before studying biological objects a calibration grid with array of small rectangles of constant height 20nm±1,5 nm and period 3.0±0,05 μm (TGQ1, NT-MDT) was investigated (figure 2, a, b). The edges of rectangles are close to vertical orientation, therefore deviation from vertical at the edge of rectangles on AFM image shows lateral resolution of the probes. The measured value of deviation from vertical position at the edge of rectangles in the PBS buffer was about 28±9 nm for standard probe and 8±4 nm for NW probe (along X and the Y axis). Thus, improvement of lateral resolution by 3-5 times compared with standard probes was found. The roughness of the rectangles surface on the 1x1 μm region was 0.32 nm for the standard probe and 0.89 nm for NW probe, which shows an increase in the accuracy of the vertical measurement (along Z axis) and indicate an improvement of image contrast.

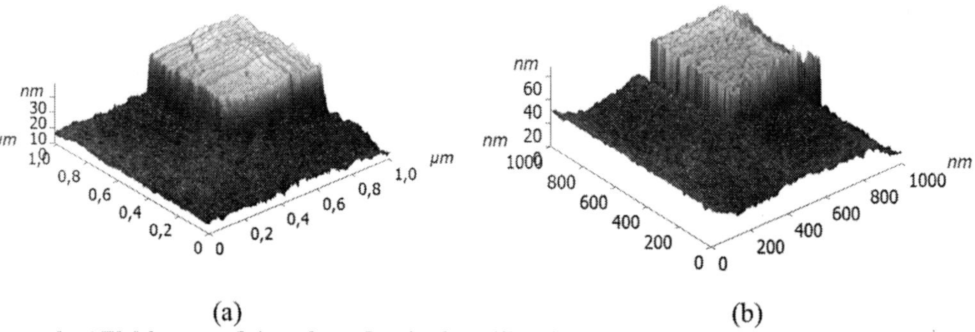

Figure 2. AFM image of the edge of a single calibration grid rectangle in the PBS buffer obtained by the standard probe (a) and the NW probe (b).

In the study of fibroblast membranes an enlargement of adhesion force on fibroblast membranes was found using NW probes (figure 3, b). The maximum value of the adhesion force was about 2.4 nN for NW probe and about 1.1 nN for standard probe. It should be noted that the native cell change during the study, therefore study of adhesion forces should take place at flat surface, where adhesion force was about (~ 0.9 nN) for NW probe, that 2 times bigger compared to standard probe (~ 0.45 nN) (figure 3, a).

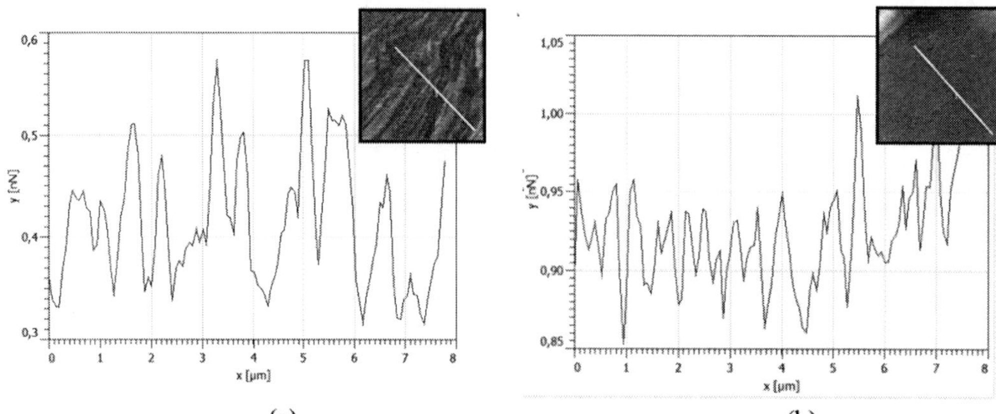

Figure 3. AFM images of the adhesion force distribution and the cross-section of the surface measured in the central region of fibroblast using the standard probe (a) and the NW probe (b) in PBS buffer.

The adhesion forces in both cases increase with the transition from normal fibroblasts to fibroblasts with pathology, while the adhesion for the NW probe are large (table 1). This can be an additional factor along with a high aspect ratio explaining the better penetrating ability and hydrophilic properties of NW probes, which corresponds to past data on the adhesion strength of HB to dry erythrocyte membranes [8].

Table 1. Average values of cell parameters in normal and with pathology, obtained by different types of probes.

Cell type	Probe Type	Adhesion force, nN	Young's modulus, kPa	Deformation, nm
Normal fibroblast	Standard Probe	0,42±0,06	43,17±8,19	269,0±19,8
	Probe with NW	0,95±0,02	160,60±21,08	158,3±36,7
Fibroblast with pathology	Standard Probe	0,76±0,48	36,19±3,21	250,0±47,4
	Probe with NW	1,07±0,33	177,60±36,59	102,2±31,4

The Young's modulus for the standard probe decreases with the transition to cells with pathology, whereas for the NW probe, on the contrary, it slightly increases. That can be explained by the bending of the whisker during the approach to normal fibroblast and sticking of the whisker, having better hydrophilic properties, to the object surface.

It should be noted that increasing the adhesion force and reducing the Young's modulus is usually characteristic of cells with pathologies that acquire great softness and elasticity (cancer cells have greater elasticity for incorporation into normal cells). Wherein the deformation of the cell has lower values using NW probes, what can be explained by the elasticity of NW and the comparable sizes of the whisker tip and the tip of pyramid of the standard probe (~10 nm), which contact with the investigated surface.

4. Conclusions

Thus, the optimal parameters for the nanowhisker formation at the axis of the standard probes was found. It was shown that specialized nanowhisker probes give better results when measured objects in liquid medium. The protocols of cultivation, structural modification and transportation of the fibroblast cells in normal and malignant transformation were fulfilled. An increase of the adhesion strength and a decrease of the elastic modulus at cells with pathology was revealed. The results obtained in this study can be used for a more detailed and accurate study of native objects in liquid by means of atomic force microscopy.

Acknowledgments

The work was carried out with the support of Russian Foundation for Basic Research (16-32-00806), the Federal Agency for Scientific Organizations (AAAA-A16-116041110123-5), the leading universities of the Russian Federation (grant 08-08).

References

[1] Payam A F, Fathipour M 2015 *Micron* **70** 50–54
[2] Dörig P, Ossola D, Truong A M, Graf M, Stauffer F, Vörös J, Zambelli T **2013** *Biophysical journal* **105** 463–472
[3] Gavara N. A 2017. *Microscopy research and technique* **80** 75–84
[4] Kreplak L. 2016 *Current protocols in protein science* **85** 17.7. 1–17.7. 21
[5] Jiang A N, Gao S, Wei X L, Liang X L, Chen Q 2008 *J. Phys. Chem.* **112** 15631–15636
[6] Zhukov M V, Belousov K I, Mozharov A M, Mukhin I S, Golubok A O 2015 *J. Phys.: Conf. Ser.* **643** 012095 (5pp)
[7] Pittenger B., Slade A. 2013 *Microscopy today* **21** 12–17
[8] Zhukov M V, Mukhin I S, Levichev V V, Golubok A O 2015 *Technical Physics Letters* **41** 149–152

IOP Conf. Series: Journal of Physics: Conf. Series **1092** (2018) 012178 doi:10.1088/1742-6596/1092/1/012178

Charge transport mechanisms in macro-scale CNT films

E S Zhukova[1,*], B P Gorshunov[1], A P Tsapenko[2], A K Grebenko[1,2], A V Bubis[2], S S Zhukov[1], E A Simchuk[1], V I Tsebro[3], A A Tonkikh[4], D V Rybkovskiy[4], E I Kauppinen[5], A G Nasibulin[2,5] and E D Obraztsova[4]

[1] Moscow Institute of Physics and Technology, Dolgoprudny, Moscow Region, 141700 Russia
[2] Skolkovo Institute of Science and Technology, Nobel str. 3, 143026, Moscow, Russia
[3] P.N. Lebedev Physical Institute, RAS, 53 Leninsky Prospect, 119991 Moscow, Russia
[4] A.M. Prokhorov General Physics Institute, RAS, Moscow, 119991 Russia
[5] Department of Applied Physics, Aalto University, School of Science, P.O. Box 15100, FI-00076 Espoo, Finland

*zhukova.es@mipt.ru

Abstract. Carbon nanotubes (CNT) attract considerable attention due to their unique physical properties and potential application in optoelectronics. Despite of intensive studies there is still a lack of agreement in experimental data on electrical properties of the material. Here we report on extremely broad-band conductivity and dielectric permittivity spectra of macro-scale thin films composed of large number of randomly distributed pristine and p-doped CNTs of different length, measured in the frequency range 5 – 24 000 cm^{-1} and at temperatures from 5 to 300 K. We show that terahertz-infrared spectra of the films are determined by response of delocalized charge carriers. Controversially to the existing experimental results we did not clearly observe the so-called terahertz conductivity peak. Yet, a weak bump-like feature in conductivity spectra around 30 cm^{-1} showed no signs of tube length dependence. We associate its origin with plasmonic excitation due to reflections of charge carrier plasma at the CNT intersections. Applying the Drude-model to describe the low frequency conductivity and dielectric permittivity spectra of CNT films we obtained effective values of carries parameters. Our results can shed light on electromagnetic waves absorption mechanisms and will be useful while designing new CNT-based devices.

1. Introduction

The most important factors determining the progress in the development of optoelectronics are the increase in speed, capacity, compactness, chemical stability, radiation and mechanical resistance of elements and devices. The convenience of using various devices in practice, for example, associated with sufficient transparency of conductive coatings, the flexibility of sensor-sensitive coatings, etc., also plays an important role. Along with the improvement of existing devices and technologies, intensive research is being carried out in the field of implementation of novel materials. Of course, these include graphene and carbon nanotubes (CNT). On the basis of CNT, two-dimensional layers (films) of macroscopic dimensions (centimeters) can be created with thicknesses of only a few tens of nanometers.

The possibility of regulating the electronic characteristics of CNT layers by doping, variation in morphology, thickness, and the creation of complex (e.g., multilayered, hybrid) systems and heterostructures opens up new prospects for creating qualitatively new optoelectronic devices.

Given that the operating frequencies of such devices are already close to the lower limit of the terahertz range (0.1-10 THz), the study of the electrodynamic properties of the corresponding components and assemblies becomes particularly relevant. It is important to note, however, that the data available today in the literature on the terahertz characteristics of individual CNTs, as well as structures based on them, are very fragmented and have an unsystematic, often contradictory character; it is also difficult to find reliable enough quantitative data on the characteristics of such objects.

Here we have performed a comprehensive study of the terahertz electrodynamic properties of macro-dimensional films composed of high quality single wall carbon nanotubes, while their electronic properties and nanomorphology varied in a controlled manner. We measured temperature dependence of terahertz-infrared spectra of conductivity and dielectric permittivity of the films in a wide frequency range from 5 cm^{-1} up to 24000cm^{-1}. The data obtained allowed us to uncover the fundamental mechanisms that determine the low energy electrodynamic response of macro-scale films and define effective values of charge carrier parameters of the CNT network: mobility, concentration, mean-free path, collision time, plasma frequency. We show that the terahertz-infrared spectroscopy is an effective contactless technique that allows to study microscopic transport mechanisms in carbon nanotube layers.

2. Experimental details

The films were synthesized by an aerosol CVD (floating catalyst) method which allows the formation of extended single-layer films of different thicknesses and then deposited on a nitrocellulose filter in the form of individual and small diameter bundles. Such films can easily be transferred to various substrates and apertures allowing a "free-standing" geometry, optimal for optical studies. Some films were subjected to a special treatment described in [1] to obtain p-doped (with $AuCl_3$, CuCl, Iodine) samples. Quality of the films was checked via electron microscopy and Kelvin-probe methods [2, 3]. Table 1 presents parameters of studied samples in the form of free-standing pristine and doped films with different transparencies. Additionally, we measured several films composed of nanotubes with different length and deposited on n nitrocellulose.

Table 1. Parameters of pristine and doped free-standing CNT films.

Transparency (at λ = 555 nm)	55%	60%	65%	70%	75%	80%	85%	90%
Film thickness (nm)	83.2	75.32	75.55	61.18	49.30	35.26	32.44	21.3

Terahertz-infrared measurements were performed in transmission geometry with the use of two commercially available spectrometers: pulsed terahertz time-domain spectrometer TPS spectra 3000 and infrared Fourier-transform spectrometer Vertex80v. Such combination of the setups allowed us to obtain spectra of electrodynamic parameters of the films in extremely wide spectral range from 5 cm^{-1} to 24000 cm^{-1}. The home-made optical cryostats were used to examine the materials in the temperature interval from 5 to 300 K.

3. Experimental results and discussion

Figure 1 presents measured spectra of transmission coefficient (a) and real and imaginary parts (b, c) of conductance ($Y=\sigma_1*d+i\sigma_2*d$, where d is a layer thickness) of a 65% transparency pristine film at two different temperatures 300 K and 5 K. The observed temperature-frequency behavior of the terahertz-infrared spectra is typical for response of delocalized charge carriers: real conductance shows an increase towards low frequencies and imaginary conductance displays distinct signs of a characteristic broad peak. Applying Drude conductivity model, we determined temperature and doping dependences of effective parameters of charges. In figure 2 the CNT network charge carriers scattering rate, mobility, mean-free path and collision time are plotted for films with different transparencies (a - d) together with the temperature behavior of the parameters (e-g).

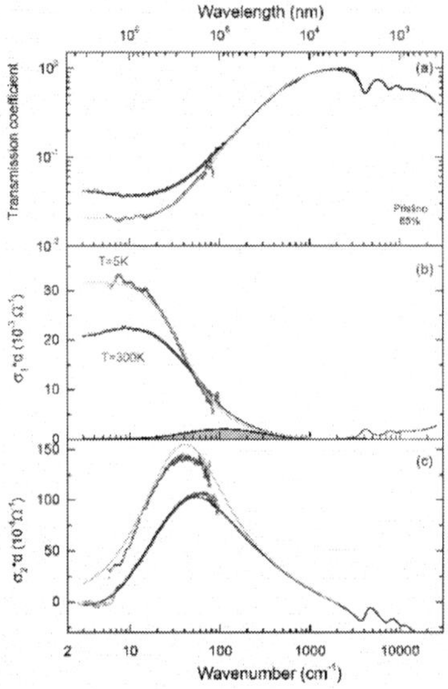

Figure 1. Room temperature (red lines) and 5 K (blue lines) spectra of (a) transmission coefficients and (b) effective real and (c) imaginary parts of conductance of a pristine CNT films with 65% transparency. Pink-shaded area in (b) shows absorption peak due to plasmon resonance.

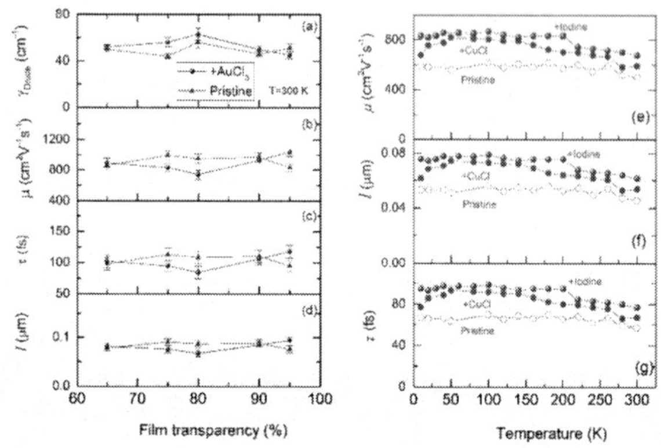

Figure 2. Effective parameters of charge carriers in pristine and doped CNT films as dependent on the films transparency (graphs on the left) and temperature (graphs on the right): (a) the charge carriers scattering rate, (b), (e) the mobility, (c), (f) the collision time and (d), (g) the mean-free path.

At the same time, we did not observe any clear signs of a so-called terahertz conductivity peak which is supposed to govern the low frequency response of CNT networks. Its origin is usually associated with a plasmon resonance due to scattering of charge carriers at the ends of individual CNTs and thus its frequency position strongly depends on the length of nanotubes [4]. Still, we detected a weak resonance-like excitation in the 30 – 100 cm^{-1} region in the conductivity spectra (a shaded area in figure 1 (b)). To analyze nature of this absorption band we performed a systematic study of series of CNT films with an average length of individual tubes of 1, 2, 4 and 9 μm. As seen from conductivity spectra in figure 3 the position of the absorption resonance does not change drastically with the length of the tubes and stays almost at the same value for all films. Furthermore, the band intensity is larger for the films with doped nanotubes and decreases with growth of transparency of the films. We associate these absorption peaks with plasmonic oscillations of charge carriers that are partly localized by defects, impurities and intersections of tubes. We can estimate the distance L between such reflecting centers as $L=V_p(\pi\nu_p)^{-1}$, where ν_p is the peak frequency, V_p is the plasmon velocity that is several times larger than the Fermi velocity. With $\nu_p\approx100$ cm^{-1}, $V_p\approx4V_F$ and the Fermi velocity $V_F\approx10^8$ cm s^{-1} we obtain $L\approx0.4$ μm. With the mean-free path in our films $l\approx0.1$ μm (figure 2) we have $l/L\approx0.25$ meaning that our films are rather 'clean'. The value $L\approx0.4$ μm correlates well with the average distance between intersections of the CNTs. We thus assume that the plasmonic excitations occur due to reflections of the charge carrier plasma at the CNT intersections.

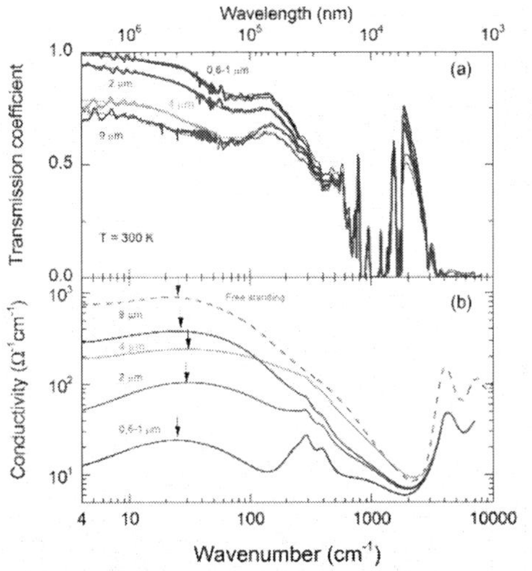

Figure 3. Room temperature spectra of (a) transmission coefficient and (b) conductivity of macro-scale films composed of CNTs of different lengths and deposited on nitrocellulose substrate. Pink dotted line on panel (b) shows a conductivity spectrum of a 85% transparency free-standing pristine film. Arrows on panel (b) indicate positions of the plasmon resonance.

4. Conclusion

We performed broad-band measurements of electrodynamic response of macro-scale CNT films. We show that at terahertz-far infrared frequencies the response is fully governed by dynamics of delocalized charge carriers whose scattering occurs mainly at tubes intersections. Applying the Drude model, we determined temperature and doping dependences of effective parameters of the charge carriers in the films. We also show that the terahertz-infrared spectroscopy is an effective contactless technique that allows to study microscopic transport mechanisms in carbon nanotube layers.

Acknowledgments

The work was supported by the Russian Ministry of Education and Science (Program '5top100'), RFBR project №18-32-002406, RFBR project № 16-02-00979 and RSF project No 17-19-01787 (the synthesis, doping and characterisation of CNTs).

References

[1] Tonkikh A A, Tsebro V I, Obraztsova E A, Suenaga K, Kataura H, Nasibulin A G, Kauppinen E I and Obraztsova E D 2015 *Carbon* **94** 768

[2] Zhukova E S, Grebenko A K, Bubis A V, Prokhorov A S, Belyanchikov M A, Tsapenko A P, Gilshteyn E P, Kopylova D S, Gladush Yu G, Anisimov A S, Anzin V B, Nasibulin A G and Gorshunov B P 2017 *Nanotechnology* **28** 445204

[3] Gorshunov B P, Zhukova E S, Starovatykh Ju S, Belyanchikov M A, Grebenko A K, Bubis A V, Tsebro V I, Tonkikh A A, Rybkovskiy D V,Nasibulin A G, Kauppinen E I and Obraztsova E D 2018 *Carbon* **126** 544

[4] Shuba M V, Paddubskaya A G, Plyushch A O, Kuzhir P P, Slepyan G Ya, Maksimenko S A, Ksenevich V K, Buka P, Seliuta D, Kasalynas I, Macutkevic J, Valusis G, Thomsen C and Lakhtakia A 2012 *Phys. Rev. B* **85** 165435

IOP Conf. Series: Journal of Physics: Conf. Series **1092** (2018) 012179 doi:10.1088/1742-6596/1092/1/012179

Photoluminescence spectral position shift governed by optical heating of perovskite resonant nanoparticles

Pavel Tonkaev, George Zograf, and Sergey Makarov
ITMO University, Kronverksky pr. 49 197101 St. Petersburg, Russia
E-mail: g.zograf@metalab.ifmo.ru

Abstract. Recent studies demonstrated high interest in halide perovskite materials for different applications including solar cells, light emitting diodes and many others. Moreover, it was revealed that resonant nanoantennas made of such materials can serve as efficient nanoscale light sources with tunable spectra in broad range by changing the composition of the initial perovskite material. However, resonant dielectric and semiconductor structure undergo considerable optical heating which affects the spectral position of the photoluminescence. In this paper we demonstrate theoretically photoluminescence shift for single resonant $MAPbI_3$ nanoparticle governed by temperature increase. Thus, our results can make resonant perovskite nanoparticles a promising platform for temperature-feedback optical heating and precise tuning of photoluminescence emission.

1. Introduction

Nanoscale light sources are basic elements for advanced nanophotonics application [1]. Among the most frequently used active light sources are nanoparticles with active defects [2] and quantum dot emitters [3]. But still, it remains a challenging problem to effectively steer and control the emission radiation pattern, even though various designs were introduced. Conventional methods of routing the emission often required multistage fabrication steps which resulted in sufficient increase of cost for practical implementations or placing nanoantennas near the active emitter, thus making the whole system non-robust against chemical pollution and mechanical damages.

On the other hand, it was revealed that nanoparticles made of organic-inorganic (hybrid) $MAPb(I,Br)_3$ perovskite materials can support Mie-resonance in the visible range, thus enhancing and guiding the photoluminescence (PL) emission from the nanoantennas [4]. Moreover, by selecting the initial perovskite compound, one can tune the PL spectra throughout whole visible range. However, even under moderate laser pump powers dielectric and semiconductor structures suffer from optical heating [5, 6], which affects spectral position of the PL peak [7].

In this work, we theoretically study PL spectral shift governed by optical heating of resonant $MAPbI_3$ nanoparticles. We believe that such approach can be promising for various applications, which require optical heating with fine tuning of emission spectrum, e.g. nanothermometry.

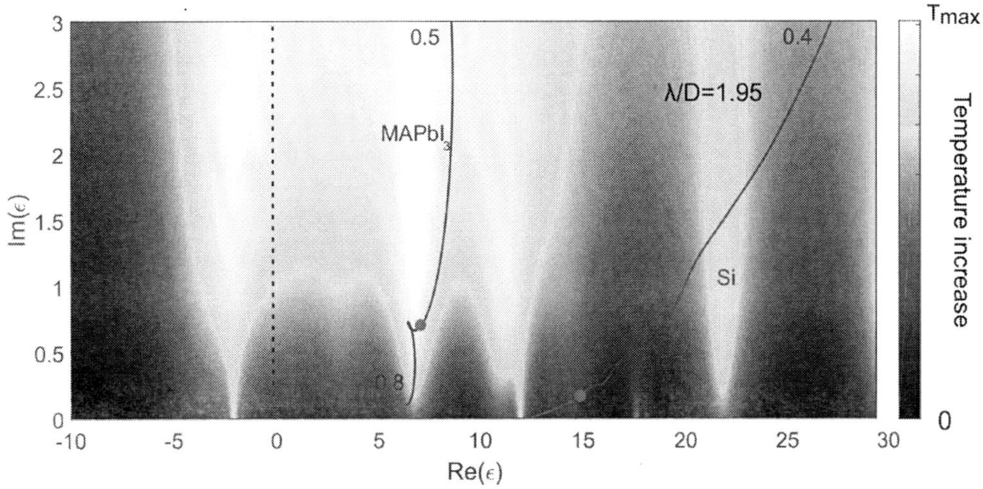

Figure 1. Optical heating of a spherical nanoparticle. Theoretically calculated (Mie theory coupled with with thermal equation [5]) heating maps for spherical nanoparticles with fixed ratio of the wavelength to the radius of nanoparticle λ/D=1.95 for different real and imaginary parts of permittivity in homogeneous medium (air). Green lines depict values of Re(ϵ) and Im(ϵ) for different materials (dispersions). The numbers correspond to the values of the wavelengths in microns. Blue dots correspond to permittivities at 633 nm for given materials.

2. Results

In order to obtain deep understanding in optical heating of perovskite spherical nanoparticles, theoretical calculations should be applied. It was shown that the absorbed power P_{abs} by the nanoparticle is proportional not only to conductivity *sigma* of the material, but also to the field enhancement factor F inside the nanostructure [5]. Thus:

$$P_{abs} = \int J^* E dV \sim \sigma E^2 F \ , \tag{1}$$

where J is optically induced current in the nanoparticle, E electric field. It means that efficiency of conversion of incident light to thermal heat energy is strongly dependent on resonant behaviour of the nanostructure. Indeed, temperature increase of the nanoparticle under CW illumination in homogeneous media can be found by following expression [8]:

$$\Delta T = \frac{I C_{abs}}{4\pi\kappa R} \ , \tag{2}$$

where, I intensity of the incident light, C_{abs} absorption cross section, κ thermal conductivity of the surrounding media and R radius of the spherical nanoparticles.

Figure 1 demonstrates optical heating of a single spherical nanoparticle in homogeneous air media for different real and imaginary parts of permittivity for fixed wavelength over diameter ration $\lambda/D = 1.95$. Blue dots represent dispersion of the materials at incident wavelength of 633 nm, thus making perovskite nanoparticles of $MAPbI_3$ compound a better structure for efficient optical heating than crystalline silicon for given conditions. However, heating of the perovskite affects the optical properties of its PL emission. Indeed, Figure 2(a) shows experimental results on shifting of the bulk $MAPbI_3$ PL peak spectral postion with increase of the temperature [7]. With higher temperatures PL spectrum undergoes blue-shift. Figure 2(b) shows theoretical calculations on PL peak spectral position governed by optical heating of the

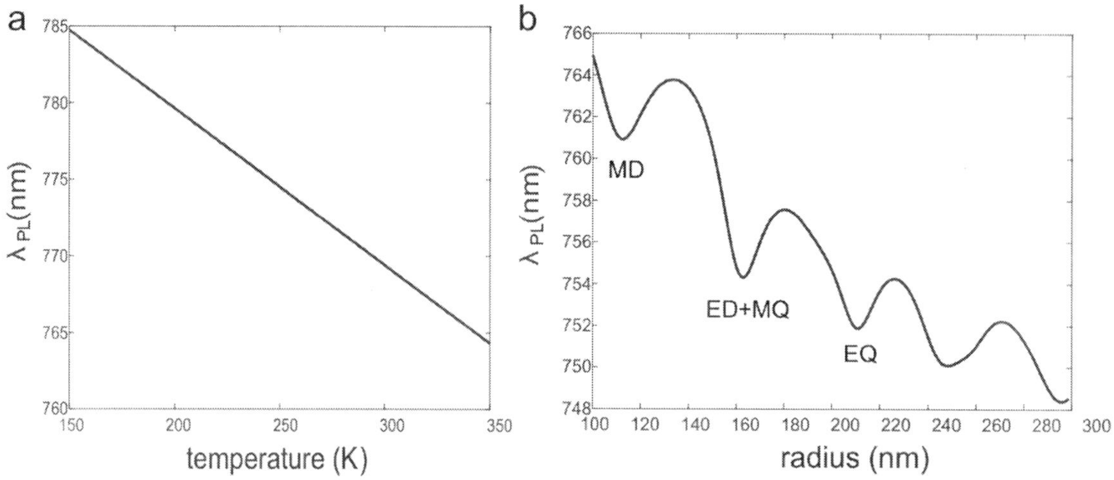

Figure 2. (a) Experimental dependence of photoluminescence spectral position on temperature for $MAPbI_3$ [7]. (b) Theoretical photoluminescence spectral position dependence of a single spherical nanoparticle on radius, which undergoes optical heating by HeNe laser (λ=633 nm) with intensity of 0.05 mW/μm^2.

resonant $MAPbI_3$ nanoparticle. The general tendency of stronger blue shift with the increase of the size is reasonable due to Eq. 2, where absorption cross-section increases with the second power of the size parameter (radius), whereas in the denominator radius appears in the first power, thus for non-resonant spherical structures heating grows linearly with size. However, if we study resonant nanoparticle one can see resonant behaviour in the PL peak shift, which occurs due to excited optical electric and magnetic modes, thus enhancing absorption of light . Mode decomposition for different sizes of $MAPbI_3$ nanoparticles reveals magnetic dipole (MD) resonance with approximately 110 nm radius, mixture of electric dipole (ED) and magnetic quadrupole (MQ) with 160 nm and electric quadrupole (EQ) with 210 nm radius at a wavelength λ of 633 nm.

3. Conclusions

In summary, we have theoretically demonstrated the ability for optical heating of resonant $MAPbI_3$ nanoparticles with possibility of precise tuning of their PL emission spectra, thus making such structures a promising platform for optical nanothermometry and thermo-optical detectors.

Acknowledgments

The work was supported by Russian Foundation for Basic Researches (Grants 16-02-00684 and 17-03-00621)

References

[1] Tame M S, McEnery K, Özdemir Ş, Lee J, Maier S and Kim M 2013 *Nature Physics* **9** 329–340
[2] Zalogina A, Saveliev R, Ushakova E V, Komissarenko P, Milichko V, Makarov S, Zuev D, Shadrivov I *et al.* 2018 *Nanoscale*
[3] Curto A G, Volpe G, Taminiau T H, Kreuzer M P, Quidant R and van Hulst N F 2010 *Science* **329** 930–933
[4] Tiguntseva E, Zograf G P, Komissarenko F E, Zuev D A, Zakhidov A A, Makarov S V and Kivshar Y S 2018 *Nano letters* **18** 1185–1190
[5] Zograf G P, Petrov M I, Zuev D A, Dmitriev P A, Milichko V A, Makarov S V and Belov P A 2017 *Nano letters* **17** 2945–2952

[6] Aouassa M, Mitsai E, Syubaev S, Pavlov D, Zhizhchenko A, Jadli I, Hassayoun L, Zograf G, Makarov S and Kuchmizhak A 2017 *Applied Physics Letters* **111** 243103
[7] Ha S T, Shen C, Zhang J and Xiong Q 2016 *Nature Photonics* **10** 115
[8] Baffou G and Quidant R 2013 *Laser & Photonics Reviews* **7** 171–187

Small animal large field of view magnetic resonance imaging with metamaterial-inspired resonator

M Zubkov[1], A A Hurshkainen[1], E A Brui[1], S B Glybovski[1], M V Gulyaev[2], N V Anisimov[2], D V Volkov[3], Y A Pirogov[4], I V Melchakova[1]
[1]Department of Nanophotonics and Metamaterials, ITMO University, Birjevaja line V.O., 14, Saint-Petersburg, 199034, Russia
[2] Laboratory of Magnetic Resonance and Spectroscopy, Faculty of Fundamental Medicine, Lomonosov Moscow State University, 31-5 Lomonosovsky Prospekt Moscow, 117192, Russia
[3]Department of Physics of Accelerators and Radiation Medicine, Faculty of Physics, Lomonosov Moscow State University, 1-2 Leninskiye Gory, Moscow, 119991, Russia
[4]Department of Photonics and Microwave Physics, Faculty of Physics, Lomonosov Moscow State University, 1-2 Leninskiye Gory, Moscow, 119991, Russia

Abstract. Large field of view magnetic resonance imaging is often required during small animal imaging to cover the entire body of an animal. Large field of view imaging should nevertheless be performed with sufficient signal to noise ratio and resolution. A combination of large signal to noise ratio and high resolution is usually achieved via using small loop coils with small field of view, which is inapplicable in case whole animal or a large portion of its body needs to be imaged. Here, a metamaterial-inspired coil based on an array of parallel wires is employed to perform large field of view imaging. A number of mice are imaged in a 7 T preclinical magnetic resonance scanner. Properties of the resulting images are compared to the ones acquired with commercial coil designated for whole-body imaging as well as a with a commercial small loop antenna designated to provide high signal to noise ratio images in a limited area. The signal to noise ratio is compared between the coils in different tissues and on different distances from the metamaterial-inspired coil plane. The coil is found to provide signal to noise ratio a few times higher than the commercial coil in the optimal reception region and comparable signal to noise ratio off the optimal observation region. The metamaterial-inspired coil is also shown to provide rivaling signal to noise ratio in small field-of-view application, when compared to a small-region loop coil.

1. Introduction

Some application in preclinical magnetic resonance imaging (MRI) require acquisition of large field of view (FoV) images of small animals (e.g., angiography [1], fat quantification [2], contrast agent or drug delivery [3-5]). In order to obtain a large FoV image an animal should either be placed in a receiver or transceiver coil covering the required area of interest (usually, a birdcage coil) or sequentially scanned with small surface loop coils. The first approach allows large FoV imaging at the cost of low signal to noise ratio (SNR), the latter leading to low image quality, longer acquisition and increased chance of image inaccuracies induced by physiological motion. The second approach provides higher SNR, but requires multiple scans to cover the entire animal body due to smaller FoV of a single scan, thus leading to multifold experiment time increase.

This work aims to show the possibility of combining both the large receptive field and high SNR in single-channel preclinical imaging by using a metamaterial-inspired coil. Such medium-sized flat coil should have both high radiofrequency field homogeneity and high sensitivity, allowing it to compete with large volume coils and small coil arrays.

2. Methods

The metamaterial-inspired coil was fabricated for a 300 MHz MR-scanner employing an array of six parallel telescopic tubes connected at both ends to copper patches on a dielectric substrate with a common ground plane at the opposite side of the substrate. A 40-mm diameter feeding loop was used to excite the lowest order eigenmode of the manufactured resonator, having the most homogeneous magnetic field distribution at the scanner transmitter frequency. The structures were held together by a 3D-printed holder. A more detailed description of the coil structure and its operation principles is presented elsewhere [6].

In-vivo large FoV imaging was performed on a 7 T Bruker BioSpec 70/30 USR (Bruker BioSpin GmbH, Germany) horizontal 200 mm bore scanner. These experiments were carried out in the Centre for Collective Usage "Biospectrotomography" supported by the Faculty of Fundamental Medicine of Lomonosov Moscow State University, Moscow, Russia. Images of 2 outbred mice BALB/c of age 5-6 months and weight 35-40g were acquired. Experimental procedures were conducted in accordance with the European Community Council directives 2010/63/EU and were approved by the local institutional animal ethics committee. Animal body motion was reduced by providing isoflurane via face mask during the whole scanning procedure. Three type of coils have been used: first mouse was scanned with a RF RES 1H T6594 birdcage coil and metamaterial-inspired coil. Second mouse was scanned with the latter coil as well as with a rat-brain receive-only surface coil. Excitation in the latter case was provided by the RF RES 300 1H/13C T10334 72 mm birdcage coil.

The imaging sequence for large FoV imaging was 2D SE with the following scan parameters: FoV = 40×100 mm^2, matrix = 200×200, slice thickness = 1.5 mm, flip angle= 90°, refocusing flip angle = 180°, TR = 800 ms and TE = 14 ms. Additionally, small FoV kidney images were acquired with all the aforementioned coils using a similar pulse sequence with scan parameters FoV = 40×30 mm^2, matrix = 266×200, slice thickness = 1 mm, flip angle= 90°, refocusing flip angle = 180°, TR = 800 ms and TE = 13.64 ms. In an experiment, where small FoV image was acquired with a small surface coil TE was changed to14.34 ms due to longer refocusing pulses required to produce a 180° rotation.

3. Results

The mouse *in-vivo* large FoV images obtained with all coils allowing large FoV imaging are presented in Figure 1. In order to compare the coils, SNR was measured in three sets of images presenting the whole mouse body as well as in the focused small FoV mouse kidney images. The SNR values were calculated as the ratio of the mean intensity values in muscle tissue (or kidney parenchyma in the case of targeted kidney imaging) to the standard deviation of the intensity values in the tissue-free region. Intensities from 4 muscle/kidney regions per image were averaged for mean signal calculation and variances from 4 tissue-free regions were averaged for mean noise calculation. The results for large FoV and targeted kidney imaging are presented in Table 1 and Table 2 correspondingly.

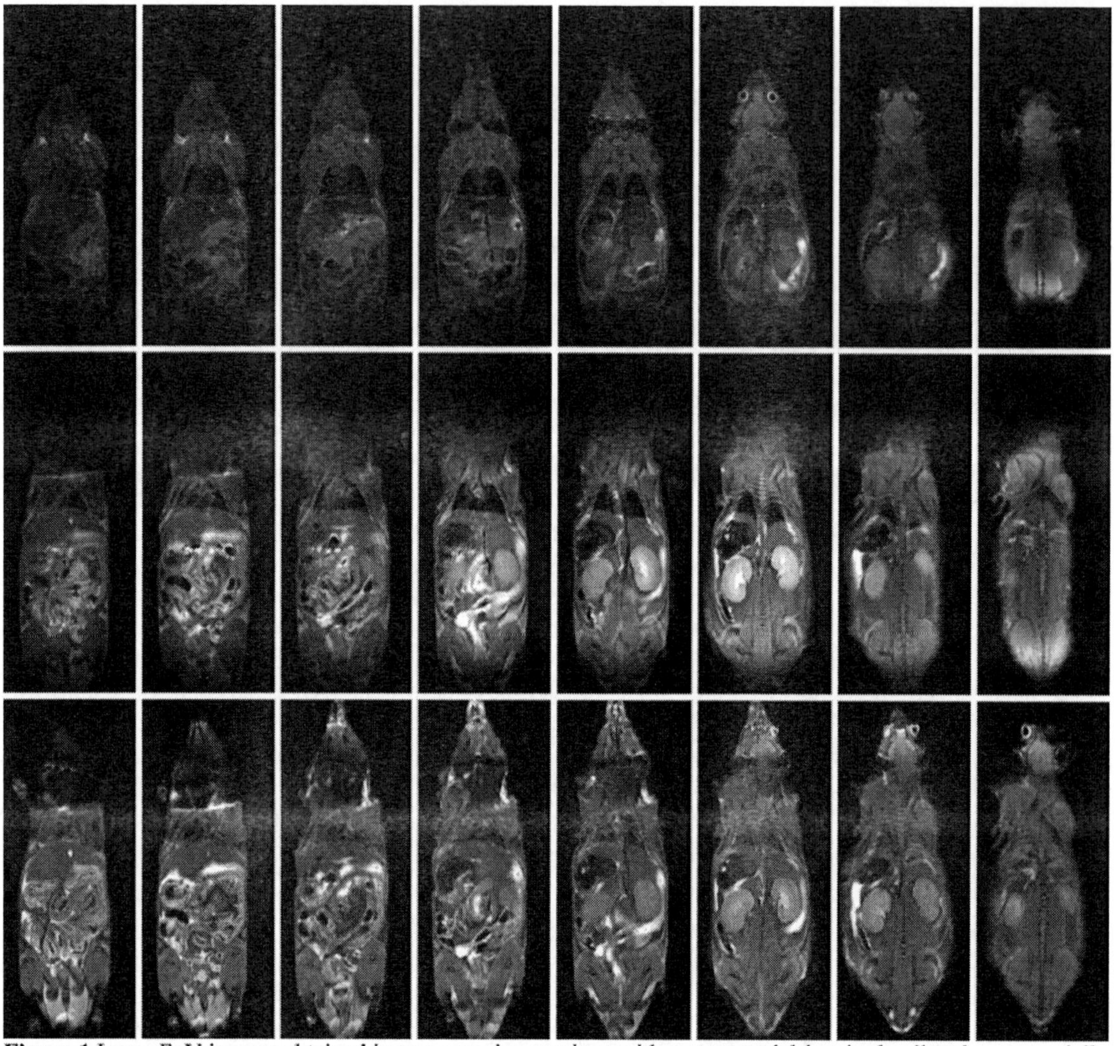

Figure 1 Large FoV images obtained in two scanning sessions with metamaterial-inspired coil and commercially available birdcage coil. Top row: mouse one, metamaterial-inspired coil. Middle row: mouse two, metamaterial-inspired coil. Bottom row: mouse two, birdcage coil.

Table 1 SNR measurements summary in muscle tissue for large FoV imaging with metamaterial-inspired coil and birdcage coil.

Slices (ventral to dorsal)	Metamaterial-inspired coil		154 mm birdcage
	Mouse 1	Mouse 2	
3	39 ± 3	55 ± 5	36 ± 3
4	40 ± 5	54 ± 9	33 ± 4
5	67 ± 10	63 ± 7	35 ± 3
6	65 ± 6	85 ± 9	33 ± 2
7	99 ± 11	84 ±9	33 ± 2
8	96 ± 9	101 ± 9	24 ± 8
9	118 ± 8	112 ± 11	30 ± 3
10	97 ± 9	75 ± 19	30 ± 2
11	88 ± 11	73 ± 16	32 ± 2

Table 2 SNR measurements in mouse kidney tissue with different coils.

	Mouse 1	Mouse 2
Metamaterial-inspired coil	19.7 ± 1.6	15.5 ± 2.1
154 mm birdcage	6.8 ± 1.6	-
Surface coil	-	25.0 ± 4.1

As has been shown in two scanning sessions, field distribution of the metamaterial-inspired coil allows imaging the whole mouse torso with addition of either the head or the caudal portions of the mouse. The SNR measurements in the images from two sessions obtained with the metamaterial-inspired coil tend to fall within error limits of each other with some deviations resulting from different mouse positioning, physiological movement and varying slice location.

4. Summary

The SNR measurements presented in Table 1 show similar values for metamaterial-inspired and birdcage coils on the slices most distant from the metamaterial-inspired coil surface with over 300% SNR enhancement in the area of best reception. The small FoV SNR values show similar (about 290%) benefit of using the metamaterial-inspired coil, when compared to the birdcage coil. An expected superiority of the small loop antenna in small FoV imaging is also visible in SNR values in Table 2. Nevertheless, the metamaterial-inspired coil shows only 38% less SNR than a dedicated loop coil.

The measured SNR values show the metamaterial-inspired coil to be suitable for whole-body small animal imaging with SNR surpassing one of the birdcage coils. Additional small FoV experiments show the coil can also be used in multiresolution experiments, where images with different FoVs and spatial resolutions need to be acquired without mouse repositioning.

Acknowledgements

This work was supported by the Ministry of Education and Science of the Russian Federation (project No. 14.587.21.0041 with the unique identifier RFMEFI58717X0041). This project has received funding from the European Union's Horizon 2020 research and innovation programme under grant agreement No 736937

References

[1] Lefrançois W, Thiaudière E, Hassen WB, Sanchez S, Franconi JM and Miraux S 2011 Fast whole-body magnetic resonance angiography in mice. *Magn Reson Med* **66(1)** 32-9.

[2] Schmidt MV, Schmidt MV, Czisch M, Schmidt MV, Czisch M, Sterlemann V, et al. 2009 Chronic social stress during adolescence in mice alters fat distribution in late life: Prevention by antidepressant treatment. *Stress* **12(1)** 89-94.

[3] Odintsov B, Chun JL, Mulligan JA and Berry SE 2011 14.1 T whole body MRI for detection of mesoangioblast stem cells in a murine model of duchenne muscular dystrophy. *Magn Reson Med* **66(6)** 1704-1714.

[4] Markides H, Kehoe O, Morris RH and El Haj AJ 2013 Whole body tracking of superparamagnetic iron oxide nanoparticle-labelled cells – a rheumatoid arthritis mouse model. *Stem Cell Research & Therapy* **4(5)** 126.

[5] Odintsov B, Chun JL and Berry SE 2013 Whole body MRI and fluorescent microscopy for detection of stem cells labeled with superparamagnetic iron oxide (SPIO) nanoparticles and DiI following intramuscular and systemic delivery. *Imaging and Tracking Stem Cells: Methods and Protocols. Totowa, NJ: Humana Press* 177-193.

[6] Hurshkainen A, Nikulin A, Glybovski S, Abdeddaim R, Vilmen C, Enoch S, et al. 2017 A metamaterial-inspired MR antenna independently tunable at two frequencies. *11th International Congress on Engineered Materials Platforms for Novel Wave Phenomena (Metamaterials)*

Intensity-dependent reflectance modulation of femtosecond laser pulses in GaAs nanocylinders with magnetic resonances

Varvara V Zubyuk[1], Polina P Vabishchevich[1,2,3], Maxim R Shcherbakov[1,4], Alexander S Shorokhov[1], Anna A Fedotova[1,5], Sheng Liu[2,3], Gordon A Keeler[2], Tatyana V Dolgova[1], Isabelle Staude[5], Igal Brener[2,3] and Andrey A Fedyanin[1]

[1] Quantum Electronics Department and Quantum Technology Center, Faculty of Physics, Lomonosov Moscow State University, Moscow 119991, Russia
[2] Sandia National Laboratories, Albuquerque, New Mexico 87185, United States
[3] Center for Integrated Nanotechnologies, Sandia National Laboratories, Albuquerque, New Mexico 87185, USA
[4] School of Applied and Engineering Physics, Cornell University, Ithaca, NY 14853, USA
[5] Institute of Applied Physics, Abbe Center of Photonics, Friedrich-Schiller-University Jena, 07743 Jena, Germany

E-mail: zubyuk@nanolab.phys.msu.ru, fedyanin@nanolab.phys.msu.ru

Abstract. We experimentally demonstrate modulation of reflectance in periodic arrays of subwavelength gallium arsenide nanocylinders with Mie-type resonances due to absorption saturation and changes in the refractive index of the semiconductor material of metasurface. The intensity-dependent reflectance modulation of up to 30% in the vicinity of the magnetic dipole resonance at a low laser fluence below 200 $\mu J/cm^2$ is shown by I-scan measurements.

1. Introduction

Phenomenon of saturable absorption of semiconductors is widely used in optical devices such as semiconductor saturable absorber mirror (SESAM) [1]. These devices have become a key component of ultrafast passively mode-locked and Q-switched laser sources [2, 3]. The saturation of optical absorption was demonstrated in different materials and systems such as plasmonic nanocomposites [4, 5]. Recently, semiconductor nanoparticles and metasurfaces with Mie-type resonance have proven themselves as potentially useful for nanophotonic applications [6, 7]. This happened since they possess completely new properties in comparison with bulk materials [8, 9] and reveal enhanced interaction with electromagnetic radiation [10, 11]. The dielectric metasurfaces also demonstrate low absorption losses as compared to plasmonic structures. In this work, we experimentally demonstrate the intensity-dependent reflectance modulation of femtosecond laser pulses in GaAs nanocylinders samples that exhibit the Mie-type resonances.

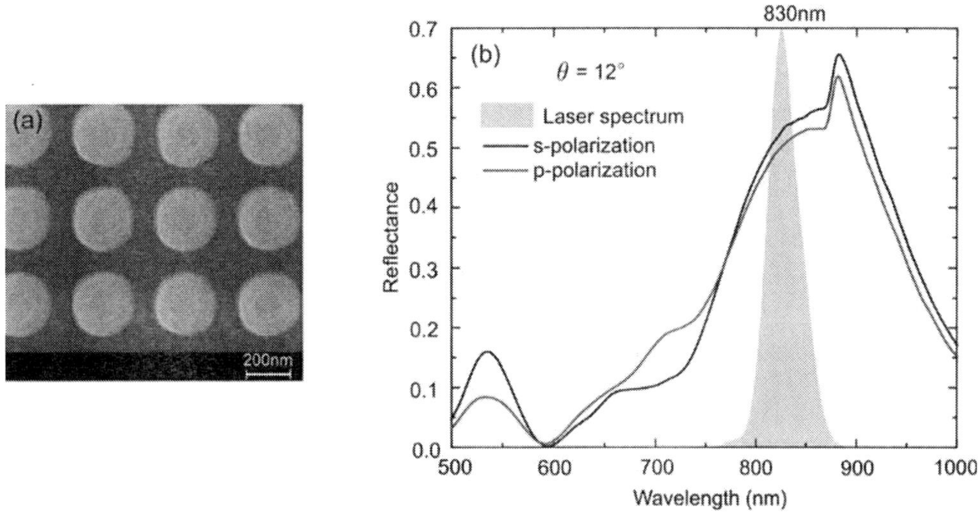

Figure 1. (a) A scanning electron microscope image of the GaAs metasurface. (b) Reflection spectra of the GaAs nanocylinders sample with a magnetic dipole (MD) resonant peak at $\lambda \sim 880$ nm for p- and s-polarization of the incident laser radiation at $\theta = 12°$.

2. Results and discussion

2.1. Metasurface sample

The samples are periodic arrays of GaAs nanocylinders with silica caps on top, nanocylinders situated on AlGaO nanocylinders that are situated on a bulk GaAs substrate (Fig.1(a)). The metasurface was fabricated using a previously published procedure [11, 12]. The diameter and height of the nanocylinders are 200 nm and 300 nm, respectively, the diameter-to-pitch ratio is 0.5, and the height of the AlGaO nanocylinders is also approximately 300 nm. Measured reflectance spectra of GaAs nanocylinders arrays at $\theta = 12°$ are close to the normal angle of incidence and exhibit peaks both for p- and s-polarized light that correspond to the magnetic dipole (MD) resonance at the same wavelength in the near infrared (Fig.1(b)). The MD resonance for p-polarized light shifts to the long-wavelength range as the angle of incidence is increased.

2.2. I-scan measurements

We measure the intensity of femtosecond laser pulses that are reflected from the GaAs metasurface using I-scan measurements in reflection geometry. The wavelength $\lambda = 830$ nm is set not far from the bandgap of GaAs ($\lambda = 870$ nm) to maximize the effect of absorption saturation. This laser radiation generates electron-hole pairs due to the single photon absorption processes. The free carrier plasma affects the complex dielectric permittivity; the intensity-dependent reflectance is detected at the normal angle of incidence. The GaAs sample shows the enhanced modulation order of magnitude larger in comparison with the GaAs substrate (Fig.2(a)). The reflectance coefficient as a function of the incident laser intensity for the GaAs nanocylinders sample is indicated by black dots and for the substrate by blue dots. We demonstrate intensity-dependent laser pulse modulation of up to 30% in the vicinity of the magnetic dipolar resonance at a low pump fluence below 200 $\mu J/cm^2$. We found excellent agreement of the experimental results with fullwave simulations that use a model with changes in the complex dielectric permittivity induced by free carriers.

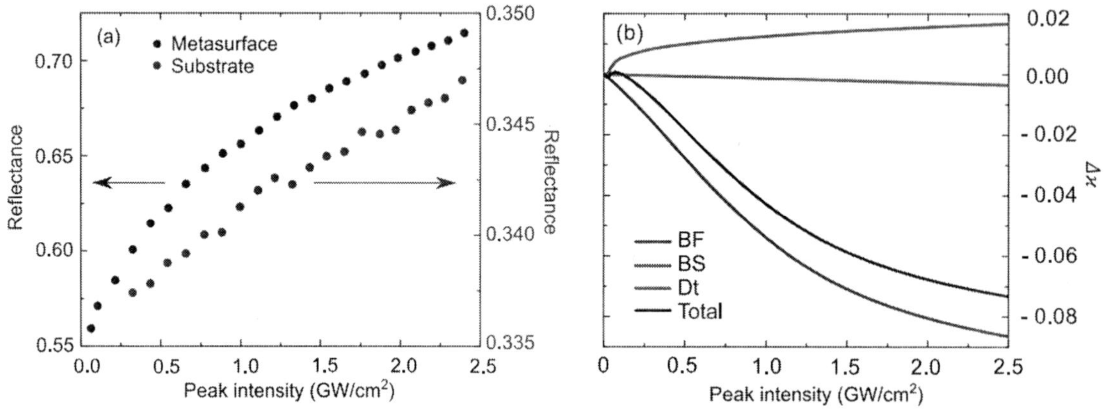

Figure 2. (a) Reflectance coefficient as a function of the incident laser intensity for the GaAs nanocylinders sample (black dots) and for substrate (blue dots). (b) Different contribution to the changes in extinction coefficient as a function of the incident laser intensity for the GaAs material induced by free carriers: BF – bandfilling contribution, BS – bandgap shrinkage contribution, Dt – Drude term and Total contribution (Total=BF+BS+Dt).

2.3. Reflectance modulation mechanism

The laser radiation generates electron-hole pairs due to the single photon absorption processes and this leads to the changes in the complex dielectric permittivity. We provided the full-wave simulations with different changes in complex refractive index and find out that reflectance modulation occurs mainly due to changes in extinction coefficient. These changes are determined by three contributions [13] and total change become negative:

$$\Delta\varkappa = \Delta\varkappa_{BF} + \Delta\varkappa_{BS} + \Delta\varkappa_{Dt} < 0, \tag{1}$$

where the indexes of $\Delta\varkappa_{BF,BS,Dt}$ correspond to BF – bandfilling contribution, BS – bandgap shrinkage contribution, Dt – Drude term and $\Delta\varkappa$ is the total changes: Total=BF+BS+Dt. The main contribution to the total change is made by the bandfilling. The reflectance spectrum of the GaAs metasurface is strongly modified under intense laser radiation due to the high sensitivity of Mie-type resonance to material parameters.

3. Conclusions

In conclusion, we have experimentally demonstrated intensity-dependent reflectance modulation of the laser pulse in subwavelength GaAs nanocylinders exhibiting localized Mie-type resonances. The modulation of reflectance coefficient is up to 30% and occurs under free carriers generation at a low laser fluence below 200 $\mu J/cm^2$. The reflectance modulation mechanism is complex and is determined primarily by absorption changes due to bandfilling effect.

Acknowledgments

This work was partly supported by financial support from Ministry of Education and Science of the Russian Federation (grant N14.W03.31.0008, calculation of optical spectra). Experimental part was performed with the support of the grant of the President of the Russian Federation for leading scientific schools (grant No. NSh-6862.2018.2 — optical spectra), supported in part by Russian National Technology Initiative, Russian Foundation for Basic Research (grant No. 16-29-11811 — I-scan measurements), and by the German Research Foundation through the Emmy Noether Programme (STA 1426/2-1). Parts of this work were supported by the

U.S. Department of Energy, Office of Basic Energy Sciences, Division of Materials Sciences and Engineering and performed, in part, at the Center for Integrated Nanotechnologies, an Office of Science User Facility operated for the U.S. Department of Energy (DOE) Office of Science. Sandia National Laboratories is a multi-mission laboratory managed and operated by National Technology and Engineering Solutions of Sandia, LLC, a wholly owned subsidiary of Honeywell International, Inc., for the U.S. Department of Energys National Nuclear Security Administration under contract DE-NA0003525.

References

[1] Keller U *et al.* 1996 *IEEE J. Sel. Top. Quantum Electron.* **2** 435–451
[2] Keller U 2003 *Nature* **424** 831
[3] Saraceno C *et al.* 2012 *IEEE J. Sel. Top. Quantum Electron.* **18** 29–41
[4] Plaksin O *et al.* 2008 *J. Appl. Phys.* **103** 114302
[5] Kumar M *et al.* 2014 *Plasmonics* **9** 129–136
[6] Shcherbakov M *et al.* 2015 *Nano Lett.* **15** 6985–6990
[7] Shcherbakov M *et al.* 2017 *Nat. Commun.* **8** 17
[8] Evlyukhin A B *et al.* 2012 *Nano Lett.* **12** 3749–3755
[9] Decker M *et al.* 2015 *Adv. Opt. Mat.* **3** 813–820
[10] Shorokhov A *et al.* 2016 *Nano Lett.* **16** 4857–4861
[11] Liu S *et al.* 2016 *Nano Lett.* **16** 5426–5432
[12] Liu S *et al.* 2016 *Adv. Opt. Mat.* **10** 1457–1462
[13] Bennett B *et al.* 1990 *IEEE J. Quantum Electron.* **26** 113–122

IOP Conf. Series: Journal of Physics: Conf. Series **1092** (2018) 012182 doi:10.1088/1742-6596/1092/1/012182

Hybrid capsules as efficient non-viral vector for gene delivery into cells

M V Zyuzin[1], A S Timin[2,3]

[1] Department of Nanophotonics and Metamaterials, ITMO University, St.Petersburg 197101, Russia

[2] First I. P. Pavlov State Medical University of St. Petersburg, St. Petersburg 197022, Russia

[3] National Research Tomsk Polytechnic University, Lenin Avenue, 30, 634050, Tomsk, Russian Federation

E-mail: mikhail.zyuzin@metalab.ifmo.ru

Abstract. Advances in the field of nanotechnology make it possible to employ nano- and microparticles as delivery carriers of genetic materials. In this study, hollow capsules of different composition were used as non-viral vectors to deliver plasmids into human cervical carcinoma cells (HeLa). Plasmids were encoded to express green fluorescent protein (eGFP) inside cells. Hybrid capsules were produced from degradable polyelectrolytes with an outer silica layer. Capsules were characterized with scanning electron and bright field microscopy. Successful cell transfection was verified with the confocal laser scanning microscope (CLSM).

1. Introduction

The choice of the optimal delivery carrier depends on the used cells and on the size of the delivered cargo. There are viral and non-viral gene delivery systems (vectors). Non-viral vectors include, for example, nano- and microparticles, which can be loaded with DNA. Nowadays, viral vectors are used as delivery vehicles in a higher rate because of high transfection efficiency and the feature of selectivity (targeted delivery of cargo into cells) [1]. However, the viral DNA carriers can be immunogenic; they can be destroyed with the immune system before they realize their function. Moreover, these viral systems can induce strong immune responses after introduction of viruses. The first unsuccessful clinical trials of viruses as gene delivery systems led to an inflammatory response to the viral vector [2]. Thus, the non-viral delivery carriers can be a real alternative to the viral vectors because of their biocompatibility, low pathogenicity, stable synthesis and low costs. The use of the non-viral vectors in the clinical trials has significantly increased since the past years, while the tests of the viral delivery systems have, on the contrary, decreased. Unfortunately, there are some biological barriers, which prevent the successful non-viral drug/gene delivery into the cells. These biological barriers include

1. Degradation of cargo, for example, polynucleotides, in the extracellular environment,
2. Internalization pathway of drug/gene carrier with cells,
3. Endosomic release of cargo/carrier into the cytosol of cells,

4. Release of cargo from the carriers inside cells,
5. Penetration of polynucleotides into the nucleus of the cells, where the mechanism of the protein expression is triggered.

These problems can be solved by encapsulation of DNA or other organic substances into the hollow particles (capsules). One of the main advantages of capsules is the stability of their synthesis, which takes place under neutral conditions and allows loading of sensitive substances into their cavity. Moreover, the capsules walls can perform a protective function, thus the cargo does not degrade in the extra- and intracellular environment, for example, in the cell culture medium. The use of silica capsules as carriers makes it possible to deliver substances to the cytosol of cells, and the non-toxicity and biocompatibility of these delivery systems allow their employment as non-viral vectors [3].

In the present study, hybrid silica capsules were used as non-viral vectors to deliver plasmids into HeLa cells.

2. Materials and methods

Hollow hybrid capsules were prepared as published elsewhere [4,5]. Sacrificial $CaCO_3$ core-particles were synthesized by precipitation of $CaCl_2$ and Na_2CO_3 water solutions in presence of plasmids. Afterwards, polyelectrolyte-based capsules were prepared with Layer-by-Layer method, where oppositely charged polyelectrolytes were deposited onto the top of each other forming polymer-shell around the $CaCO_3$ core-particles. Shell of hybrid capsules constituted of 6 monolayers of polyelectrolytes (3 bilayers). To obtain hybrid capsules, additional layer of silica was deposited onto the last positively charged layer of polyelectrolytes using solution of tetraethyl orthosilicate as a silica source. Finally, $CaCO_3$ cores were dissolved under mild conditions with ethylenediaminetetraacetic acid.

Synthesized capsules as well as transfected cells were visualized with confocal laser scanning microscopy (CLSM, Carl Zeiss LSM 710), Objective EC Plan-Neofluar 40x/1.30 Oil DIC, Objective Plan-Apochromat 63x/1.40 Oil DIC. Scanning electron microscopy was carried out on a Quanta 200, FEI, Netherlands with an acceleration voltage of 100 kV.

Human cervical carcinoma cells (HeLa) were obtained from the American Type Culture Collection (ATCC) and were grown in Dublecco's Modified Eagle's Medium supplemented with 10% fetal serum bovine (FBS) and 1% penicillin/ streptomycin (P/S) at 37 °C in 5% CO_2.

3. Results

Bright field microscopy images and scanning electron microscopy images of hybrid silica capsules are shown in the Figure 1. As it can be seen, hybrid capsules were fully covered by a dense layer of inorganic silica shell, which make capsules stiffer (Fig. 1B). As evidenced with bright field microscopy (Fig. 1A), capsules were not aggregated and the average size of capsules was around 2 μm.

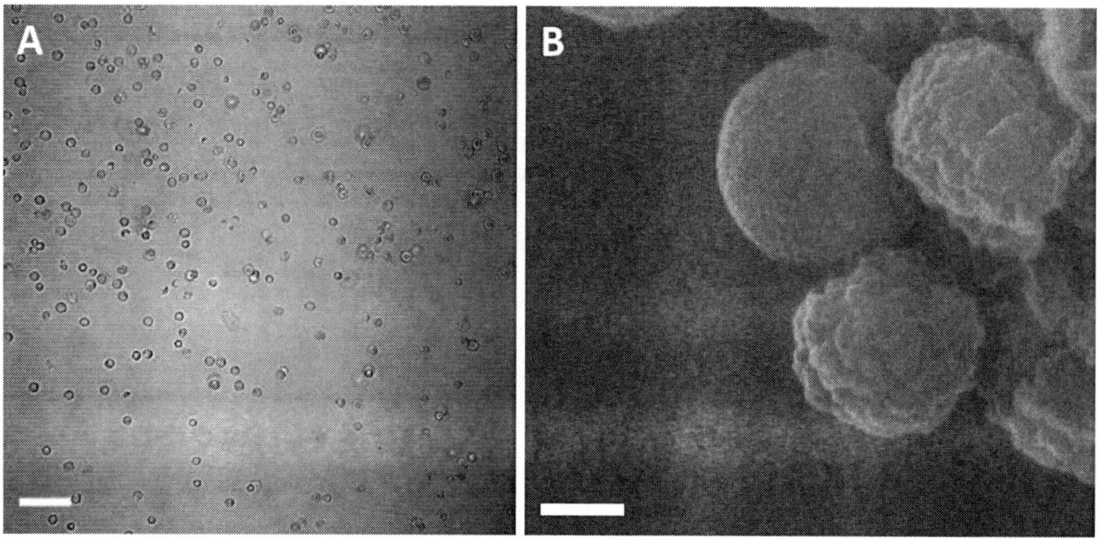

Figure 1. A. Representative bright field microscopy image of hybrid silica capsules loaded with plasmids. Scale bar corresponds to 20 μm. B. Scanning electron microscopy image of hybrid silica capsules loaded with plasmids. Scale bar corresponds to 1 μm.

Capsules were added to the HeLa cells at amount 5 capsules per cell and incubated for 24 h. Next day, cells were observed under confocal microscope (Fig. 2). Green fluorescence signal coming from HeLa cells was an indication of the expression of eGFP inside cells and, thus, successful cell transfection.

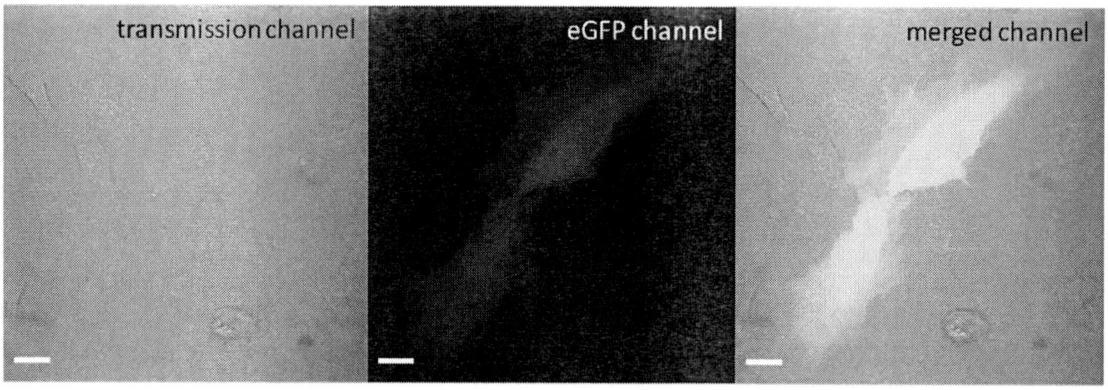

Figure 2. Representative confocal laser scanning microscopy image of transfected HeLa cells with hybrid capsules loaded with plasmids. Scale bar corresponds to 20 μm.

According to the obtained results, hybrid capsules based on silica can overcome biological barriers and deliver genetic material into cells. Thus, these capsules can be employed as non-viral vectors to deliver bioactive compounds into cells. This confirms their great potential in the field of gene therapy.

Acknowledgements
Synthesis of capsules was supported by the Russian Science Foundation (grant No 17-19-01637), cell experiments were supported by the Russian Foundation for Basic Research (grant No 18-015-00100). MVZ thanks the President's Scholarship SP-1576.2018.4.

References

[1] Yin H, Kanasty R L, Eltoukhy A A, Vegas A J, Dorkin J R and Anderson D G 2014 Non-viral vectors for gene-based therapy *Nat. Rev. Gen.* **15** 541–555

[2] Gloves D J, Lipps H J 2005 Towards safe non-viral therapeutic gene expression in humans *Nat. Rev. Gen.* **6** 299–310

[3] Ott A, Yu X, Hartmann R, Rejman J, Schütz A, Ochs M, Parak W J and Carregal-Romero S 2015 Light-addressable and degradable silica capsules for delivery of molecular cargo to the cytosol of cells *Chem. Mater.* **27** 1929–1942

[4] Zyuzin M V, Díez P, Goldsmith M, Carregal-Romero S, Teodosio C, Rejman J, Feliu N, Escudero A, Almendral M J, Linne U, Peer D, Fuentes M and Parak W J 2016 Comprehensive and systematic analysis of the immunocompatibility of polyelectrolyte capsules *Bioconjugate Chem.* **28** 556–564

[5] Kantner K, Rejman J, Kraft K V L, Soliman M G, Zyuzin M V, Escudero A, del Pino P and Parak W J 2018 Laterally and temporally controlled intracellular staining by light-triggered release of encapsulated fluorescent markers *Chem. Eur. J.* **24**, 2098-2102